Peptides

Chemistry, Structure and Biology

Peptides

Chemistry, Structure and Biology

Proceedings of the Thirteenth American Peptide Symposium
June 20-25, 1993, Edmonton, Alberta, Canada

Edited by

Robert S. Hodges
Department of Biochemistry
University of Alberta
Edmonton, Alberta, T6G 2H7, Canada

and

John A. Smith
Department of Pathology
University of Alabama at Birmingham
Birmingham, Alabama 35233-7331, U.S.A.

ESCOM ▪ Leiden ▪ 1994

CIP-Data Koninklijke Bibliotheek, Den Haag

Peptides

Peptides: Chemistry, Structure and Biology: Proceedings of the Thirteenth American Peptide Symposium, June 20-25, 1993, Edmonton, Alberta, Canada/ed. by Robert S. Hodges and John A. Smith. - Leiden : ESCOM. - Ill.
With index, ref.
Subject headings: Peptides/Proteins.

ISBN 90-72199-19-7 (hardbound)

Published by:

ESCOM Science Publishers B.V.
P.O. Box 214
2300 AE Leiden
The Netherlands

Preface

The Thirteenth American Peptide Symposium was held in the Edmonton Convention Centre, Edmonton, Alberta on June 20-25, 1993. This Symposium was held under the auspices of the American Peptide Society and the University of Alberta. Compared to previous symposia, this meeting was the largest in terms of attendance (1447), and truly an international Symposium with participants from 32 countries and more than 30% of participants from outside North America. It was a pleasure to see the large number of first-time participants at the Symposium. The scientific program consisted of plenary lectures, poster presentations, workshops and exhibits. Of the 762 presentations (58 plenary lectures involving 14 sessions; 3 workshops involving 13 additional speakers; and 691 poster displays) the Program Committee selected 372 articles for publication in the Proceedings Volume. These manuscripts were selected on the basis of originality and scientific significance.

As with past symposia, the Thirteenth Symposium continued to show the exponential growth of the use of peptides in the diverse fields of medical science.

We began the Symposium with an exciting lecture on "Recollections of the Past 40 Years in Peptide Chemistry" by Dr. Bruce Merrifield, Nobel laureate, which focused on past scientists and their contributions that have paved the way to what is now a mushrooming field of peptide chemistry in all aspects of biology. Sessions I and II and 189 posters discussed Synthetic and Analytical Methods, the cornerstone of our field. It is interesting to observe the continuing improvements to solution and solid-phase peptide chemistry such as new supports, coupling agents and optimization procedures, fragment condensation methods and enzymatic synthesis. The development of new instrumentation (*e.g.*, capillary electrophoresis, on-line HPLC with electrospray mass spectrometry, micro-peptide sequencing and ultra-fast HPLC) is aiding, and will continue to aid greatly, the peptide chemist in purification and characterization of peptides.

Session III dealt with the exciting developments in designing peptide mimetics, cyclic and constrained peptides and secondary structure mimetics. Session IV discussed the synthesis of glycopeptides and lipopeptides, the role glycosylation plays on the biological properties of synthetic peptides, and the role of lipids as delivery systems for peptides.

Sessions V, VI and VII and 230 posters discussed the area of Biologically Active Peptides/Neuropeptides/Peptide Hormones/Peptide Inhibitors/Peptide Receptor Interactions. This area continues to be of fundamental importance, not only for understanding mechanism of action, but also in the development of new potent agonists and antagonists in the pharmaceutical sector as well.

Antibiotics are facing a losing battle with the resurgence of antibiotic resistant pathogens, which is predicted to be "nothing short of a medical disaster" in the 1990s and beyond. Session VIII titled Peptide Vaccines and Immunology again emphasized the important contributions that peptides are making in the field of immunology and the increasing importance of peptides to immunologists, microbiologists and virologists in the development of synthetic peptide vaccines against bacterial and viral pathogens.

Session X emphasized the important breakthroughs that are occurring in peptide delivery; strategies to enhance intestinal permeability of peptides; protease-resistant peptide analogs for oral delivery; and delivery of peptides into the central nervous system by sequential metabolism. In addition, the use of peptides as diagnostics, a field in its infancy, was included.

Two sessions, IX and XI, discussed the area of Conformational Analysis and the use of Computational Biochemistry to study peptide structure. Session XII, Peptide Macromolecular Interactions, went to the next step in complexity with the study of the interactions of peptides with biological macromolecules, the important first steps in signal transduction and understanding ligand-receptor interactions.

Session XIII focused our attention on the rapidly developing field of using Peptide Libraries as screening tools in research and drug discovery. Novel methods of synthesis and strategies of using these libraries are being developed. Libraries of peptides, free in solution, bound to the solid support or being generated by phage can contain cyclic peptides, unnatural amino acids, and colored and fluorescent markers. We look forward to the exciting new developments that this technology will bring in the coming years in terms of unique peptide chemistry and application.

Session XIV on *de novo* design represented the inroads peptide chemistry is bringing to understanding protein folding and stability. The future will bring small chemically synthesized protein molecules with novel enzymatic and binding properties to the forefront of the biotechnology industry.

The Symposium introduced three workshops, which provided participants with an opportunity to familiarize themselves at the basic level of an area outside their own expertise: Workshop I, Approaches and Advances in Peptide Synthesis, Purification and Analysis; Workshop II, An Introduction to NMR Spectroscopy of Peptides; and Workshop III, An Introduction to Energy Minimization, Molecular Dynamics, Molecular Modelling and Conformational Analysis of Peptides. These workshops ran simultaneously and the workshop concept was an overwhelming success and will continue at future meetings.

This year's Alan E. Pierce Award recipient, Dr. Victor J. Hruby, was recognized for his outstanding contributions to the chemistry and biology of peptides, especially the design of peptides, pseudopeptides and peptidomimetics to understand the relationship of structure and biological activity. His fascinating lecture illustrated the wide-ranging contributions Victor has made to the field of peptide chemistry and, combined with his outstanding scientific and personal qualities, clearly demonstrated his deserving position among the previous eight awardees.

The use of the Edmonton Convention Centre and the success of this venue indicates the need for this type of facility to provide an atmosphere which can deliver the high quality of scientific lectures, poster presentations, workshops and exhibits in an effective and efficient manner to such a large group of participants. The large lounge area in the center of the exhibit/poster hall provided an excellent atmosphere for discussion and exchange of information.

This meeting was organized solely with the use of volunteers and the success of the meeting was directly related to their dedication and hard work. As Chairman, I am forever grateful to the Department of Biochemistry personnel including technologists, graduate students and postdoctoral fellows from my laboratory. A special thanks goes to Dr. Colin Mant, who as Conference Manager contributed to all aspects of the meeting from fund raising to managing the exhibition and without his help I could not have brought the Symposium to Edmonton; Colleen Iwanicka, who as Financial Administrator diligently handled and maintained all financial transactions; Janice MacDonald, who as Social Director made this aspect of the meeting a success and for many who were involved in the post conference tour of the Canadian Rockies an unforgettable experience; and lastly Janet Wright who as head of Office Administration and Dawn Lockwood together organized the daily correspondence, created the Symposium database required to correspond with all conference attendees, kept track of adjudicated abstracts, typed all printed material and lastly, literally corrected, typed and reformatted the Proceedings Volume as camera ready, altogether an enormous task.

To members of the Program Committee who assisted in organizing the scientific program, who volunteered to organize the workshops and who evaluated all abstracts and selected papers for publication in the Proceedings Volume, your hard work and commitment is greatly appreciated. To the Chairmen who stimulated discussion and kept the meeting on schedule, we are indebted.

We are especially grateful to the generous support by the Benefactors, Sponsors, Donors and Contributors (see the following pages). Such support is critical to the quality and success of the American Peptide Symposium. We wish to express our special thanks to Dr. Alan E. Pierce and Pierce Chemical Company for providing funds for the Alan E. Pierce Award and partial support for the Awards Banquet and symposium photographer, and to the sponsors of the American Peptide Society Travel Awards (Applied Biosystems Inc., Millipore and Star Biochemicals) which enabled young scientists from around the world to attend the Symposium.

We congratulate the Student Affairs Committee of the American Peptide Society who again organized the Job Fair and the Student Poster Competition which facilitate career development of young scientists and reward their individual collective achievements.

As Chairman, I would like to acknowledge the Department of Biochemistry, Faculty of Medicine, and the University of Alberta for their support of my efforts to hold the Thirteenth American Peptide Symposium and The Alberta Heritage Foundation for Medical Research for providing travel funds to speakers at the Symposium.

Lastly, the Chairman would like to acknowledge his wife, Phyllis, who understands the challenges and workload involved in this endeavour.

Robert S. Hodges
John A. Smith

Thirteenth American Peptide Symposium

Edmonton Convention Centre, Edmonton, Alberta, Canada
June 20-25, 1993

THE AMERICAN PEPTIDE SOCIETY

OFFICERS

Charles M. Deber	President	University of Toronto
Jean E. Rivier	President-Elect	The Salk Institute
Arthur M. Felix	Secretary	Hoffmann-La Roche Inc.
John A. Smith	Treasurer	University of Alabama at Birmingham

COUNCILLORS

Irvin M. Chaiken	SmithKline Beecham
David H. Coy	Tulane University
Bruce W. Erickson	University of North Carolina
Ralph F. Hirschmann	University of Pennsylvania
Maurice Manning	Medical College of Ohio
Garland R. Marshall	Washington University School of Medicine
Arno E. Spatola	University of Louisville
James P. Tam	The Vanderbilt University
Daniel F. Veber	Merck Sharp and Dohme Research Laboratories
Janis D. Young	Skyline Peptides

CHAIRMAN - THIRTEENTH AMERICAN PEPTIDE SYMPOSIUM

Robert S. Hodges	University of Alberta

PROGRAM COMMITTEE

Murray Goodman	University of California at San Diego
Robert S. Hodges	University of Alberta
Randy T. Irvin	University of Alberta
Pravin T. Kaumaya	Ohio State University College of Medicine
Cyril M. Kay	University of Alberta
Rachel E. Klevit	University of Washington
Gilles Lajoie	University of Waterloo
Garland R. Marshall	Washington University School of Medicine
Ruth F. Nutt	Merck Sharp and Dohme Research Laboratories
Jean E. Rivier	The Salk Institute
Peter W. Schiller	Clinical Research Institute of Montreal
Bhagirath Singh	University of Alberta and University of Western Ontario
John A. Smith	University of Alabama at Birmingham
Brian D. Sykes	University of Alberta
James P. Tam	The Vanderbilt University

ORGANIZING COMMITTEE - UNIVERSITY OF ALBERTA

Robert S. Hodges   Symposium Chairman
Colin T. Mant   Conference Manager
Colleen Iwanicka   Financial Administrator
Janice E. MacDonald   Social Director
Janet Wright   Administrative Assistant
Morris R. Aarbo
T.W. Lorne Burke
Cyril M. Kay
Dawn Lockwood
Gail Redmond
Brian D. Sykes

LOCAL VOLUNTEERS - UNIVERSITY OF ALBERTA

Christi Andrin   Jack Moore
Sherron Becker   Sai Ming Ngai
Jim Black   Stacy Oare
Lori Burke   Bob Parker
Rose Caday   Paul Semchuk
Michael Carpenter   Terry Sereda
Irene Church   Cindy Shaughnessy
Len Daniels   Sue Smith
Beth-Anne Exham   Natalie Strynadka
Rod Gagne   Kathy van Denderen
Devon Husband   Gary Van Domselaar
Shari Kasinec   Jennifer Van Eyk
Terri Keown   Iain Wilson
Xin Liu   Wah Wong
Vicki Luxton   Shirley Woywitka
Joyce MacDonald   Peter Wright
Jane Miller   Mae Wylie
Oscar D. Monera

The Thirteenth American Peptide Symposium greatly appreciates the support and generous financial assistance of the following organizations:

BENEFACTORS

Alberta Economic Development and Tourism
Alberta Heritage Foundation for Medical Research
Applied Biosystems
Bachem California
Mallinckrodt Specialty Chemicals Company
Millipore Corporation
Novabiochem/Calbiochem
Pierce Chemical Company
Protein Engineering Network of Centres of Excellence, Canada
Sigma Chemical Company
U.S. Army Medical Research Acquisition Activity

SPONSORS

Abbott Laboratories
Advanced ChemTech
Amgen, Inc.
Bachem Bioscience, Inc.
Bachem Feinchemikalien AG
Biogen, Inc.
Hoffmann - La Roche, Inc.
Propeptide
Vydac/The SEP/A/RA/TIONS Group, Inc.

DONORS

BioChem Pharma, Inc.
Biomeasure Incorporated
Boehringer Mannheim GmbH
Bristol-Myers Squibb Company
ESCOM Science Publishers B.V.
Glaxo
Hewlett-Packard
Hoechst-Roussel Pharmaceuticals, Inc.
Merck Research Laboratories
Orpegen
Peninsula Laboratories, Inc.
Peptide Institute / Protein Research Foundation
Peptides International
Peptisyntha and Cie, SNC
Pfizer

Senn Chemicals AG
Sterling Winthrop, Inc.
Syntex Research
Synthetech, Inc.
The DuPont Merck Pharmaceutical Company
The Upjohn Company
UCB-Bioproducts S.A.
Wyerst-Ayerst Research

CONTRIBUTORS

Alberta Tourism, Parks and Recreation
Astra Pharma, Inc.
Bio-Mega, Inc.
Ciba-Geigy Canada, Inc.
CSPS
Cytogen
Edmonton Convention and Tourist Authority
Immunobiology Research Institute
Lilly Research Laboratories
Sandoz Canada, Inc.
Sandoz Pharmaceuticals Corporation
Tanabe Research Laboratories, USA, Inc.
The Peptide Laboratory, Inc.
Warner-Lambert
Zeneca Pharmaceuticals

Alan E. Pierce Award

(Sponsored by Pierce Chemical Company)

The recipient is an individual who has made outstanding contributions to techniques and methodology in the chemistry of amino acids, peptides and proteins.

1993	Victor J. Hruby	University of Arizona Tucson, AZ, U.S.A.
1991	Daniel Veber	Merck Sharp and Dohme Research Laboratories, West Point, PA, U.S.A.
1989	Murray Goodman	University of California - San Diego San Diego, CA, U.S.A.
1987	Choh Hao Li	University of California - San Francisco San Francisco, CA, U.S.A.
1985	Robert Schwyzer	Swiss Federal Institute of Technology Zürich, Switzerland
1983	Ralph Hirschmann	Merck Sharp and Dohme Rahway, NJ, U.S.A.
1981	Klaus Hofmann	University of Pittsburgh School of Medicine, Pittsburgh, PA, U.S.A.
1979	Bruce Merrifield	The Rockefeller University New York, NY, U.S.A.
1977	Miklos Bodanszky	Case Western Reserve University Cleveland, OH, U.S.A.

American Peptide Symposia

Symposium		Chair(s)	Location
First	1968	Saul Lande Yale University, New Haven Boris Weinstein University of Washington, Seattle	Yale University New Haven, CT, U.S.A.
Second	1970	F. Merlin Bumpus Cleveland Clinic, Cleveland	Cleveland Clinic Cleveland, OH, U.S.A.
Third	1972	Johannes Meienhofer Harvard Medical School, Boston	Children's Cancer Research Foundation, Boston, MA, U.S.A.
Fourth	1975	Roderich Walter University of Illinois Medical Center Chicago	The Rockefeller University and Barbizon Plaza Hotel New York, NY, U.S.A.
Fifth	1977	Murray Goodman University of California-San Diego	University of California- San Diego San Diego, CA, U.S.A.
Sixth	1979	Erhard Gross National Institutes of Health Bethesda	Georgetown University Washington, DC, U.S.A.
Seventh	1981	Daniel H. Rich University of Wisconsin-Madison	University of Wisconsin-Madison Madison, WI, U.S.A.
Eighth	1983	Victor J. Hruby University of Arizona, Tucson	University of Arizona Tucson, AZ, U.S.A.
Ninth	1985	Kenneth D. Kopple Illinois Institute of Technology Chicago Charles M. Deber University of Toronto, Toronto	University of Toronto Toronto, Ontario, Canada
Tenth	1987	Garland R. Marshall Washington University School of Medicine, St. Louis	Washington University St. Louis, MO, U.S.A.

Symposium		Chair(s)	Location
Eleventh	1989	Jean E. Rivier The Salk Institute for Biological Studies, La Jolla	University of California- San Diego San Diego, CA, U.S.A.
Twelfth	1991	John A. Smith Massachusetts General Hospital Boston	Massachusetts Institute of Technology, Cambridge, MA, U.S.A.
Thirteenth	1993	Robert S. Hodges University of Alberta, Edmonton	Edmonton Convention Centre Edmonton, Alberta, Canada

Abbreviations

Abbreviations used in the proceedings volume are defined below:

α-AE	α-amidating enzyme	ADE	atrial peptide degrading enzyme
A$_2$bu	2,4-diaminobutyric acid	ADR	adriamycin
A$_2$pr	2,3-diaminopropionic acid; *see* Dpr	AEC	3-amino-9-ethylcarbazol
AA, aa	amino acids	Agm	4-guanidino-butyl-amine (descarboxy arginine)
AAA	amino acid analysis		
Aab	3-aminomethyl-4-aminobutanoic acid	AFP	antifreeze protein
αAT	α-antitrypsin	AGSP	atrial granule serine proteinase
Ab	antibody		
Aba	2-aminobutyric acid	Aha	7-aminoheptanoic acid
ABTS	2,2'-azido-bis(3-ethyl-benzthiazoline sulfonic acid)	AHPBA	3-amino-2-hydroxy-4-phenylbutanoic acid; phenyl-norstatine
ABZ	aminobenzoic acid		
AC	adenylate cyclase	AHPPA	4-amino-3-hydroxy-5-phenylpentanoic acid
AC$_2$O	acetic anhydride		
ACE	angiotensin-converting enzyme	Ahx	aminohexyl
ACh	acetylcholine	Aïb	α-aminoisobutyric acid
ACHPA	4-amino-3-hydroxy-5-cyclohexylpentanoic acid	Aic	2-aminoindan-2-carboxylic acid
AChR	acetylcholine receptor	AIDS	acquired immune deficiency syndrome
Acm	acetamidomethyl		
ACN; Acn	acetonitrile		
AcOH	acetic acid	Alloc, Aloc	allyloxycarbonyl
ACP	acyl carrier protein	Allyl	allyl ester
ACSA	adenylate cyclase-stimulating activity	AMD	actinomycin D
		Amf	aminophenylalanine; *see* Aph
Acsc	aminocyclopentane-carboxylic acid	AMP	aminomethyl-piperidine
ACT	Advanced ChemTech, Inc.	AMPA	aminomethylphenyl-acetic acid
AcT	N^{α}-acetyltransferase	ANF	atrial natriuretic factor
ACTH	corticotropin	Ang II, AII	angiotensin II
AD	Alzheimer's disease	Ang, ANG	angiotensin; angiotensinogen
Ada	adamantyl		

ANP	atrial natriuretic peptide	BN, Bn	bombesin
Anq	anthraquinone	BnPeOH	2,2-[bis(4-nitro-phenyl)]-ethanol
AO	antiovulatory	Boc	tert-butyloxycarbonyl
Aoc	1-azabicyclo[3.3.0]-2-carboxylic acid	Boc-ON	2-tert-butyloxy-carbonylamino-2-phenylacetonitrile
AOGO	5-amino-4-oxo-8-guanidinooctanoic acid	BOI	2-(benzotriazol-1-yl)-oxy-1,3- dimethyl-imidazolinium
AP	aminopeptidase		
APC	antigen presenting cell	Bom	benzyloxymethyl
APG	azidophenyl glyoxal	BOP	benzotriazolyloxy tris-(dimethylamino) phosphonium hexafluoro-phosphate
Aph	aminophenylalanine		
APM	aminopeptidase M		
Apo	apolipoprotein		
APP	avian pancreatic polypeptide		
APY	anglerfish peptide YG	BOP-Cl, BopCl	bis(2-oxo-3-oxazo-lidinyl) phosphinic chloride
AR	adrenergic receptor		
ARC	AIDS related complex	BPA	benzylphenoxyacet-amidomethyl
Asa	azidosalicylic acid		
ASF	African swine fever	Bpa	benzoylphenylalanine
Asu	aminosuberic acid	Bpo	D-α-benzoyl-penicilloyl
AT	antithrombin		
Atc	2-aminotetralin-2-carboxylic acid	Bpoc	biphenylpropyloxy-carbonyl
ATIII	antithrombin III	BPTI	bovine trypsin inhibitor
ATR	attenuated total internal reflection	bR	bacterial rhodopsin
ATZ	anilinothiazolinone	Br$_2$Dmb	3,5-bis(bromo-methyl)benzoate
AVP	arginine-8-vasopressin		
AZT	3'-azido-3'-deoxy-thymidine; zidovudine	BroP	bromo tris(dimethyl-amino)phosphonium hexafluorophosphate
		BSA	bovine serum albumin
b	bovine	Bt	biotinoyl
Bab	3,5-bis(2-aminoethyl) benzoic acid	BTD	bicyclic β-turn dipeptide
Bal	β-alanine	BTU	O-benzotriazolyl-N,N,N',N'-tetra-methyluronium hexafluorophos-phate
BAP, βAP	beta amyloid protein		
BBB	blood brain barrier		
BGG	bovine gamma globulin		
BHAR	benzhydrylamine resin	Bum	tert-butyloxymethyl
BHI	biosynthetic human insulin	BU^t, ^{t}BU	tert-butyl
		Butaz	1,2-diphenyl-pyrazolidine-3,4-dione
Biot	biotin		
BK	bradykinin		
BME	β-mercaptoethanol	BW	body weight

Bzl	benzyl	ClZ	2-chlorobenzyloxy-carbonyl
c-3-PP	cis-3-propyl-L-prolyl	Cle	cycloleucine (1-amino-1-carboxyl-cyclopentane)
CA	chemical acetylation		
cAMP	cyclic adenosyl mono-phosphate	CM	chloromethyl; carboxymethyl
CAP	core amyloid peptide		
CAT	chloramphenicol acyl transferase	CNS	central nervous system
Cbz, Z	carbobenzoxy; benzyloxycarbonyl	COSY	correlated NMR spectroscopy
CCD	countercurrent distribution	Cpa	4-chlorophenylalanine
		CPD, CP	carboxypeptidase
CCK	cholecystokinin	CPF	caerulin precursor fragment
CD	circular dichroism		
CD	complement domain	CPMAS	cross-polarization/ magic angle spinning
cDNA	complementary DNA		
CE	carbetocin; capillary zone electro-phoresis; *see* CZE	CR	chain recombination
		CRF	corticotropin releasing factor
CEC	cation exchange chromatography	CRP	C-reactive protein
		CsA	cyclosporin A
CFA	complete Freunds adjuvant	CT	carboxy terminus; calcitonin; chymotrypsin; cholera toxin
CGRP	calcitonin gene-related peptide		
CgTx	conotoxin		
CHA	cyclohexylamine	CTAB	cetyl trimethyl ammonium bromide
Cha	cyclohexylalanine		
CHAPS	3-[(3-cholamido-propyl)-dimethyl-ammonio]-1-pro-pane-sulfonate	CTL	cytotoxic T-lympho-cytes
		CTMS	chlorotrimethylsilane
cHex, cHx	cyclohexyl	Ctp	chloroacetyl-trypto-phan
CHF	congestive heart failure	CVAP	cerebrovascular amyloid peptide
CHO	Chinese hamster ovary; aldehyde	CVS	cardiovascular system
		Cyp	cyclophilin
ChTX	charybdotoxin	CZE	capillary zone electrophoresis
CID	chemically ionized desorption; collision-induced dissociation		
		DA	D/Ala substitution factor
CINC	cytokine-induced neutrophil chemoattractant	Dab	diaminobutyric acid
		DABCYL	4-dimethyl-amino-azobenzene-4'-sulfonyl chloride
CLA	cyclolinopeptide A		

DABITC	4-dimethylamino-phenyl-4'-isothio-cyanate
DADLE	[D-Ala^2,D-Leu^5] enkephalin
DAGO	[D-Ala^2,N-MePhe^4,Gly^5-ol] enkephalin
Dah	1,6-diaminohexane
Dap	diaminopimelic acid
Dat	desamino tyrosine
DBU	1,8-diazabicyclo [5.4.0]-undec-7-ene
Db$_z$g	dibenzylglycine
DCCI, DCC	dicyclohexyl-carbodiimide
DCHA	dicyclohexylamine
DCI	3,4-dichloro-isocoumarin
DCM	dichloromethane
Dcp	dichlorophenyl
DCU	dicyclohexylurea
DDDA	2,9-diamino-4,7-dioxadecanedioic acid
Dde	N-(1-(4,4-dimethyl-2,6-dioxocyclo-hexylidene)ethyl)
DDQ	dichlorodicyano-quinone
DEAE	diethylaminoethanol
DEAM	diethylacetamido-malonate
Deg	diethylglycine
DEPC	diethylphosphoro-cyanidate
Dφg	diphenylglycine
DG/SA	distance geometry/ simulated annealing
Dha	dehydroalanine
Dhc	S-(2,3-dihydroxy-propyl)cysteine
DHO	dihydroorotic acid
DHP	dihydroxypropyl
DIBAL	diisobutyl aluminium hydride
DIC	N,N'-diisopropyl-carbodiimide

DIEA	diisopropylethylamine
DIP	4,7-diphenyl phenanthroline
DIPCDI	diisopropylcarbo-diimide; *see* DIC
DKP	[Asp^{B10},Lys^{B28},-Pro^{B29}]-insulin; diketopiperazine
DLPS	dilauroylphosphatidyl-serine
DMA	dimethylacetamide
DMAP	dimethylamino-pyridine
DMBA	9,10-dimethyl-1,2-benzathracene
DMBHA	2',4'-dimethoxybenz-hydryl amine
DMF	dimethylformamide
Dmp, DMP	dimethylphosphinyl
DMPC	dimyristoylphospha-tidylcholine
DMPG	dimyristoylphospha-tidylglycerol
DMPSE	dimethylphenylsilyl ethane
DMPTU	1-dimethyl-3-phenyl-2-thioures
DMS	dimethyl sulfide
DMSO	dimethyl sulfoxide
Dmt-OH	2,2-dimethyl-L-thiazolidine-4-carboxylic acid
Dncp	2,4-dinitro-6-carboxy-phenyl
DNP	dinitrophenyl
Dns	dansyl
DOACl	dimethyl-dioctadecyl ammonium chloride
DOC	deoxycholate
DOPC	dioleoyl-*sn*-glycero-phosphocholine
Dpa	β,β-diphenylalanine
Dpa	diphenylalanine
DPBT	diphenylphosphoryl-benzoxazolthione
DPCDI	diisopropyl-carbodiimide;

	see DIC
DPDPE	cyclo[DPen²-DPen⁵]enkephalin
Dpg	dipropylglycine
DPI	despentapeptide insulin
DPP	dipeptidyl peptidase
DPPA	diphenylphos-phorylazide
DPPC	dipalmitoyl phosphatidyl-choline
DPPG	dipalmitoyl phosphatidyl-glycerol
DPP-IV	dipeptidylpeptidase-IV
Dpr	2,3-diaminopropionic acid
DPTU	*N,N'*-diphenylthioures
DQF	double quantum focused
DSB	4-(2,5-dimethyl-4-methylsulfinyl-phenyl)-4-hydroxy-butanoic acid
DSP	dimethylsulfonium methyl sulfate
DTC	dimeric tripeptide chemoattractants
Dtc	*S,S*-dimethyl-thiazolidine-4-carboxylic acid
DTH	delayed type hyper-sensitivity
DTNB	dithiobis(2-nitro-benzoic acid)
DTPA	diethylenetriamine pentaacetic acid
Dts	dithiasuccinoyl
DTT	dithiothreitol
Dyn	dynorphin
e	eel; equine
EA	ergotamine
EAE	experimental autoimmune encephalomyelitis

EDAC, EDC	1-(3-dimethyl-amino-propyl)-3-ethyl-carbodiimide hydrochloride
EDRF	endothelium-derived relaxing factor
EDTA	ethylenediamine-tetraacetic acid
EGF; EGFR	epidermial growth factor; epidermial growth factor receptor
EI	epidermal cell inhibitor
EIAV	equine infectious anemia virus
ELAB	enantiomer labeling
ELISA	enzyme-linked immunosorbent assay
Enk	enkephalin
EP	endorphin
EPNP	1,2-epoxy-3-(*p*-nitro-phenoxy)propane
ESI-MS	electrospray ionisation mass spectrometry
ESR	electron spin resonance
ET	endothelin
Et₃N	triethylamine; *see* TEA
EtA	α-ethylalanine
Etm	ethyloxymethyl
FAA	fatty amino acid
FAB-MS	fast atom bombardment mass spectrometry
Farn	farnesyl
FeLV	feline leukemia virus
Fg	fibrinogen
FGF	fibroblast growth factor
FID	flame ionization detector
FITC	fluorescien isothiocyanate
Flg	fluorenylglycine

Fm, fm	fluorenylmethyl	GnRH	gonadotropin-releasing hormone
FMDV	foot-and-mouth disease virus	GPI	guinea pig ileum
FMOC, Fmoc	9-fluorenylmethoxy-carbonyl	GRF, GHRH	growth hormone-releasing factor
Fpa	4-fluorophenylalanine	GRP	gastrin releasing peptide
FPLC	fast protein liquid chromatography	GS	gramicidin S
FRET	fluorescence resonance energy transfer	GSH	reduced glutathione
		GSSG	oxidized glutathione
FSH	follicle stimulating hormone; follitropin	GTP	guanosine triphosphate
FTIR	Fourier transform infrared	GvH	graft versus host
		GVIA	conotoxin G VIA
		h	human
GA	gramicidin A	Hat	6-hydroxy-2-aminotetralin-2-carboxylic acid
GABA	gamma aminobutyric acid		
GAL	galanin	hBP	human serum binding protein
GAP	growth associated protein; gonadotropin-releasing hormone associated protein	HBTU	*O*-benzotriazolyl-*N,N,N',N'*-tetramethyl-uronium hexa-fluorophosphate
GC	gas chromatography		
gCSF	granulocyte-colony stimulating factor	HBV	hepatitis B virus
		HC	heparin cofactor II
GDA	glutaraldehyde	hCG	human chorionic gonadotropin
GEMSA	guanidino ethyl-mercaptosuccinic acid	hCGRP	human calcitonin gene-related peptide
GH	growth hormone	hCys, Hcys	homocysteine
GHRH, GRF	growth hormone releasing hormone	HEL	hen egg lysozyme
		HEL	human erythroleukemia
GHRP	growth hormone releasing peptide	Hepes	*N*-[2-hydroxy-ethyl]piperazine-*N'*2-ethanesulfonic acid]
GITC	2,3,4,6-tetra-*O*-Ac-β-D-glucopyranosyl isothiocyanate		
Gla	D-galactopyranosyl; gamma carboxy-glutamic acid	HF	hydrogen fluoride
		Hfa	homophenylalanine
		HFBA	heptafluorobutyric acid
GLP	glucagonlike peptide		
GM	gramicidin M; [Phe^{9,11,13,15}] gramicidin A	hGH	human growth hormone
		HHM	humoral hypercalcemia of malignancy
Gn, Gu	guanidine		

HI	hemoregulatory cell inhibitor	HR	histamine release
HIC	hydrophobic interaction chromatography	HR, hr	human recombinant
		HSA	human serum albumin
		Hse	homoserine
HILIC	hydrophilic interaction chromatography	HSPS	high speed peptide synthesis
HIV	human immuno-deficiency virus	HSV	herpes simplex virus
		Htc	7-hydroxytetrahydro-isoquinoline-3-carboxylic acid
HIV-PR	human immuno-deficiency virus protease		
		HTE	hamster trachea epithelial cell
HLE	human leukocyte elastase	HTLV	human T-cell leukemia virus
HMP	hydroxymethyl-phenoxyacetic acid; hydroxymercapto-propionic acid	HUB	hamster urinary bladder
		HUVEC	human umbilical vein endothelial cell
HMP	hydroxymethyl-phenoxymethyl	Hyp	hydroxyproline
		Hz	hertz
HMPA, HMPT	hexamethylphosphoric triamide	ICE	interleukin-1 beta converting enzyme
HNE	human neutrophil elastase	i.c.v.	intra-cerebro-ventricular
hNP	human neutrophil peptide	i.m.	intramuscular
		i.v.	intravenous
HOBt	N-hydroxybenzo-triazole	IC	inhibitory concentration
HODhbt	hydroxyoxodihydro-benzotriazine	ICAM	intracellular adhesion molecule
HONp	nitrophenol	ICE	interleukin convertase
HOOBt	hydroxyoxodihydro-benzotriazine	IEC	ion-exchange chromatography
HOSu	N-hydroxysuccinimide	IEF	isoelectric focusing
HOTic	7-hydroxy-1,2,3,4-tetrahydroquinoleic-3-carboxylic acid	IFNα	interferon α
		Ig	immunoglobulin
HPA	hypothalamic-pituitary-adrenal axis	IGF	insulin-like growth factor
		IL	interleukin
HPI	human proinsulin	ILys	lysine(N^ϵ-isopropyl)
HPLC	high performance liquid chromatography	im	imidazole
		in	indole
		INEPT	insensitive nuclei enhancement by pulse transfer
Hpp	3-(4-hydroxyphenyl)-propionyl		
HPSEC	high performance size exclusion chromatography	Ing	indenylglycine
		IP	inositol phosphate

IR	infrared; insulin receptor	MBP	myelin basic protein
IRMA	immunoradiometric assay	MBS	*m*-maleimidobenzoyl-*N*-hydroxy-succinimide ester
IU	international units	MCH	melanin concentrating hormone
KLH	keyhole limpet hemocyanin	MCPBA	*m*-chloroperbenzoic acid
KP	[Lys^{B28},Pro^{B29}]-insulin	MCPS	multiple constrained peptide synthesis
		MD	molecular dynamics
LDH	lactate dehydrogenase	Me	methyl
LDTOF	laser desorption time-of-flight	Mea	2-mercaptoethylamine
		MeA	α-methylalanine
LEC	ligand-exchange chromatography	MeCN	acetonitrile
LFA	lymphocyte associated antigen	MeNTI	*N*-methyl noroxy-morphindole
LH	luteinizing hormone; lutropin	MeOH	methanol
		MHC	major histocompati-bility complex
LHRH	*see* GnRH	MIC	minimal inhibitory concentration
LPC	lauroylphosphoryl-choline	MIR	main immunogenic region
LPH	lipotropin		
LPS	lipopolysaccharide	Mls	minor lymphocyte stimulating gene
LSIMS	liquid secondary ion mass spectrometry	MMC	migrating motor complex
LVP	lysine-8-vasopressin	Mom	methyloxymethyl
		MoMuLV	Moloney murine leukemia virus
m	murine; messenger		
MAb	monoclonal antibody	Mot	motilin
Man	2-mercaptoanaline	Mpg	3-methoxypropyl-glycine
MAP, MAp	membrane-anchored protein; multiple antigen peptide; mean arterial pressure	MPM	p-methoxyphenyl-methyl ester
		Mpr	3-mercaptopropionyl
MAPs	macromolecule-associated proteins	Mpr	mercaptopropionic acid
MARCKS	myristolated alanine-rich C kinase substrate	MS	mass spectrometry
		MSH	melanocyte stimulating hormone; melanotropin
MAS	magic angle spinning		
Mba	2-mercaptobenzoic acid	Msob	methylsulfinylbenzyl
		Msz	methylsulfinylbenzyl-oxycarbonyl
Mbh	methoxybenzhydryl		
MBHA	methylbenzhydryl-amine		

Mtr	4-methoxy-2,3,6-trimethylbenzene-sulfonyl	NVOC	nitroveratryloxy-carbonyl
Mts	mesitylene sulfonyl	o	ovine
MuLV	murine leukemia virus	OEt	ethyl ester
MxAn	mixed anhydride	Oic	2,3,4,5,6,7,8-octa-hydroindole-2-carboxylic acid
Nal	2-naphthylalanine		
Nbb	nitrobenzamidobenzyl	OMe	methyl ester
Nbb	o-nitrobenzamido-benzyl	OMP	outer membrane protein
NBD	7-nitrobenz-2-oxa-1,3-diazole	ONb	o-nitrobenzyl
		OPA	o-phthaldialdehyde
NBS	N-bromosuccinimide	OSu	O-succinimide ester
NCp	nucleocapsid protein	OT	oxytocin
NCS	neocarcinostatin	OTf	O-triflate
NEM	N-ethyl maleimide	OVA	ovalbumin
NFT	neurofibrallary tangles	OVLT	organum vasculosum laminate terminalis
Nic	nicotinoyl		
NIDD	non-insulin dependent diabetes	Ox	oxazolidines
		OXT	oxytocin
NIS	N-iodosuccinimide		
NK	neurokinin	P_3CSS	macrophage activator tripalmitoyl-S-glycerylcysteinyl-serylserine
NM	neuromedin		
NMB	neuromedin B		
NMDA	N-methyl-D-aspartate		
NMM	N-methylmorpholine	PA	parent antagonist
NMP	N-methyl-pyrrolidinone	PAB	p-alkoxybenzyl
		PAC, Pac	phenacyl
NMR	nuclear magnetic resonance	PAF, Paf	p-aminophenylalanine
		PAGE	polyacrylamide gel electrophoresis
NMT	N-myristoyl transferase	PAL	photoaffinity labeling
NOE	nuclear Overhauser effect	Pal	3-pyridylalanine
		Paloc	3-(3-pyridyl)allyloxy-carbonyl
NOESY	nuclear Overhauser enhanced spectroscopy	PAM	phenylacetamido-methyl
NP	neutrophil peptide; neurophysin	Pas	6,6-pentamethylene-2-aminosuberic acid
Npp	nitrophenyl-pyrazolinone	Pbf	2,2,4,6,7-pentamethyl-dihydrobenzofuran-5-sulfonyl
NPY	neuropeptide Y		
Npys	3-nitro-2-pyridyl-sulfenyl	PBM	peripheral blood monocytes
NT	N-terminus; amino terminus; neurotensin	PBS	phosphate-buffered saline
		PCOR	peptide cyclization on oxime resin

PCP	phencyclidine	Pmc	2,2,5,7,8-penta-methylchroman-6-sulfonyl
PDB	phorbol 12,13-dibutyrate		
PD-MS	plasma desorption mass spectrometry	Pmp	3,3-pentamethylene-3-mercaptopropionic acid
PEG	polyethylene glycol		
Pen	penicillamine	Pmp	phosphonomethyl-phenylalanine
PEPS	polystyrene-grafted polyethylene film	pNA	*p*-nitroanaline
PFC	plaque forming cell	PND	principal neutralizing determinant
Pfp	pentafluorophenyl ester	PON	periodically oscillating neuron
PG	proteoglycan		
PGF	proteoglycan growth factor	POPS	palmitoyl-oleoyl-phosphatidylserine
Pgl	n-pentylglycine	PP	pancreatic polypeptide
PGL[a]	peptide glycine leucine amide	PPA	n-propylphosphoric anhydride
Phaa	phenylacetic acid	PPE	porcine pancreatic elastase
PHBT	polymeric hydroxy-benzotriazole	PPIase	peptidyl-prolyl isomerase
PHF	paired helical filaments	PPL	porcine pancreatic lipase
Phi	4-iodophenylalanine		
Phpa	3-phenylpropanoic acid	Ppt	diphenylphosphino-thionyl
pI	isoelectric point	PQ	paraquat
Pic	picolinoyl	Pqt	3-(1'-methyl-4,4'-bipyridinium-l-yl)propyl
Pilot	peptide identification and lead optimization	Pra	propargylglycine
Pin	β-pineyl	PRL	prolactine
Pip	pipecolinyl; piperidine	PRP	platelet-rich plasma
Pipes	piperazine-*N,N'*-bis[2-ethanesulfonic acid]	PT	pertussis toxin
		PTFCE	polytrifluorochloro-ethylene
PITC	phenylisothiocyanate	PTH	phenylthiohydantoin; parathyroid hormone
Piv	pivaloyl		
Piz	piperazic acid	PTK	protein tyrosine kinase
PK	protein kinase		
PLA$_2$	phospholipase A$_2$	Ptm	phenyloxymethyl
PLP	poly-L-proline	PTPase	protein-tyrosine phosphatase
PMA	phorbol myristate acetate	PTZ	phenothiazine
PMB	polymyxin B	Ptz	3-(10-phenothiazinyl)-propanol
pMBHA	4-methylbenz-hydrylamine	PVA	polyvinyl alcohol
		PVDF	polyvinylidene fluoride

PVN	paraventricular nuclei	Sar	sarcosyl, sarcosine
PyBOP	(benzotriazolyl)-*N*-oxy-pyrrolidinium phosphonium hexafluoro-phosphate	SCAL	safety catch amide linkage
		SCC	short circuit current
		SCLC	small cell lung carcinoma
PYL[a]	peptide tyrosine leucine amide	SEC	size exclusion chromatography
PYY	peptide tyrosine-tyrosine	SEM	standard error of the mean
		SH	sulfhydryl
QUIS	quisqualate	SHMT	serine hydroxymethyl transferase
recDNA	recombinant DNA	SLE	systemic lupus erythematoius
REDOR	rotational echo double resonance	SMPS	simultaneous multiple peptide synthesis
RGD	Arg-Gly-Asp fibrinogen binding sequence	SP	substance P
		SPCL	synthetic peptide combination libraries
RIA	radioimmunoassay		
RMSD, rmsd	root mean square deviation	SPPS, SPS	solid phase peptide synthesis
RNase	ribonuclease	SRIF	somatostatin
ROE	rotating frame nuclear Overhauser effect	SRP	signal recognition peptide
ROESY	rotating frame nuclear Overhauser enhanced spectroscopy	ssDNA	single stranded DNA
		ST	heat stable enterotoxin
		Sta	statin
ROS	rat osteosarcoma cells	Su	succinimide
RP	reversed phase	Suc	succinoyl
RP-HPLC (RPC)	reversed-phase high performance liquid chromatography	SWM	sperm whale myoglobin
		t-3-PP	*trans*-3-propyl-L-prolyl
s	salmon		
s.c.	subcutaneous	Tacm	*S*-trimethyl-acetamidomethyl
SA	symmetrical anhydrides	TAP	tick anticoagulant peptide
SAH	*S*-adenosylhomo-methionine	TASP	template-assembled synthetic protein
SAM	*S*-adenosylmethionine	TBDMSCl	tert-butyldimethylsilyl chloride
SAMBHA	(4-succinylamido-2',2',4'-trimethoxy)benz-hydryl amine	TBS	tert-butyldimethylsilyl
		TBTA	tert-butyl-2,2,2-tri-chloroacetamide
SAP	serum amyloid protein		
SAR	structure-activity relations	TBTU	*O*-(benzotriazol-l-yl)-*N,N,N',N'*-tetra-

	methyluronium tetrafluoroborate
ᵗBu	tert-butyl
Tca	trichloroacetamide
TCEP	tris(2-carboxyethyl) phosphine
TCR	T lymphocyte antigen receptor
TCS	trypsin-catalyzed semi-synthesis
TCT	tracheal cytotoxin
TEA	triethylamine
TEAP	triethylamine phosphate
TEDOR	transferred echo double resonance
Teoc	trichloroethyloxy-carbonyl
TEP	triethylphosphite
TFA	trifluoroacetic acid
TFE	trifluoroethanol
TFM	trifluoromethyl
TFMSA	trifluoromethane-sulfonic acid
TGF	transforming growth factor
Th	thiazolidines
THF	tetrahydrofuran
Thi	*see* Dtc
THTP	tetrahydrothiophene
Thz	thiazolidine carboxylic acid
Tic, Tiq	1,2,3,4-tetrahydro-quinoleic-3-carboxylic acid
TicOH	7-hydroxy-1,2,3,4-tetrahydro-quinoleic-3-carboxylic acid
TLC	thin layer chromatography
TM	transmembrane
Tm	melting temperature
TMBD	tetramethylbenzadine
Tmob	trimethoxybenzyl
TMP	3,4,7,8-tetramethyl-phenanthroline
TMS	trimethylsilyl
TMSCN	trimethylsilyl nitrile

TMSE	β-trimethylsilyl ethane
TMSOTf	trimethylsilyl trifluoromethane-sulfonate
Tn	troponin
TNF	tumor necrosis factor
TPA	12-*O*-tetradecanoy-phorbol-13-acetate; tissue plasminogen activator
TPI	triose phosphate isomerase
TPK	tyrosine-specific phosphate kinase
TPTU	1,1,3,3-tetramethyl-2-(2-oxo-1(2H)-pyridyl)uronium tetrafluoroborate
TPyClU	1,1,3,3-bis(tetra-methylene)-chlorouronium tetrafluoroborate
TR	time resolved
TR-COSY	transferred rotational correlated NMR spectroscopy
TRH	thyrotropin releasing hormone
Tris	tris(hydroxymethyl)-aminomethane
TRNOE	transferred nuclear Overhauser effect
Trt	trityl
TSH	thyroid stimulating hormone
TT	tetanus toxoid
TxA$_2$	thromboxane A$_2$
UK	urokinase
UNCA	urethane protected α-amino acid N-carboxy anhydride
UNCA's	urethane N-carboxy-anhydrides
UV	ultraviolet
VCD	vibrational circular dichroism

Vly	valeryl	XAL	5-(9-aminoxanthen-2-oxy)valeric acid
VMH	ventro medial hypothalamus		
Vn	vitronectin	Z, Cbz	carbobenzoxy; benzyloxycarbonyl
VNA	virus neutralizing antibody		
VSMC	vascular smooth muscle cells	Ψ	pseudo
WSCI	water soluble carbodiimide		

Contents

Ninth Alan E. Pierce Award Lecture

Session I: Synthetic Methods

Contents

Contents

Session II: Synthetic and Analytical Methods

Contents

Session III: Peptide Mimetics

Session IV: Glycopeptides/Lipopeptides

Contents

Session V: Biologically Active Peptides

Contents

Contents

Session VI: Peptide Hormones/Neuropeptides

Contents

Session VII: Peptide Inhibitors/Peptide-Receptor Interactions

Contents

Session VIII: Peptide Vaccines and Immunology

Session IX: Conformational Analysis

Session X: Peptide Pharmaceuticals/Diagnostics and Peptide Delivery

Session XI: Computational Biochemistry

Session XII: Peptide Macromolecular Interactions

Contents

Session XIII: Peptide Libraries

Contents

Session XIV: *De Novo* Design of Peptides and Proteins

Contents

Contents

Indexes

Ninth Alan E. Pierce Award Lecture

Dr. Victor J. Hruby

Introduced by: Robert S. Hodges
University of Alberta
Edmonton, Alberta, Canada

and

Clark W. Smith
The UpJohn Company
Kalamazoo, Michigan, U.S.A.

Dr. Victor J. Hruby

Recipient of the Ninth Alan E. Pierce Award

Peptide chemistry: Designing peptides pseudopeptides and peptidomimetics for biological receptors

V.J. Hruby

*Department of Chemistry, University of Arizona
Tucson, Arizona 85721, U.S.A.*

Introduction

I am greatly honored to have been chosen the recipient of the Ninth Alan E. Pierce Award. I am especially grateful to those who nominated me and the Awards Committee who selected me. Most of all I am grateful to the over 90 graduate students and postdoctoral associates, numerous undergraduates, and 50 professors throughout the world, past and present, who have collaborated with me. Their creativity, dedication and friendship have been an inspiration and source of great joy.

I owe an enormous debt of gratitude to my mentors. A. William Johnson, my M.S. Professor, introduced me to scientific research and the pleasure of working in a new area of chemistry he discovered, the conversion of sulfur ylids to epoxides. A. T. Blomquist, my Ph.D. Professor, allowed me to develop my own creative scientific ideas, and instilled in me the importance of all areas of chemistry. Vincent du Vigneaud, my postdoctoral mentor, convinced me to work in peptide chemistry when I had other plans. He greatly encouraged me to pursue my interests in structural and physical chemistry in the peptide field. I also want to thank Carl S. Marvel, my scientific grandfather twice, who strongly supported my collaborations with biologists, at a time when it was considered decidedly gauche for an academic chemist to do so.

My introduction to the field of peptide chemistry in du Vigneaud's laboratory was the determination of the acetone-oxytocin structure, a long standing problem of 20 years. The solution to this problem, which greatly pleased du Vigneaud, was accomplished [1] by the relatively new method in peptide chemistry of high resolution nuclear magnetic resonance spectroscopy. When du Vigneaud moved from the Cornell Medical School in New York to the Cornell Chemistry Department in Ithaca, the lab was dismantled for a few months. During this time and as a result of my math background and interest in the chemical basis for bioactivity, du Vigneaud allowed me to present a minicourse on the kinetic and thermodynamic aspects of bioassays that greatly influenced my scientific development.

In the space allotted, it is not possible to review our published work. Rather I will offer a perspective on the status of Peptide Chemistry related to some of its critical scientific issues. I will use selected examples from our laboratory, to illustrate how we came to our current state, and hopefully to point to a few future directions.

Synthetic Chemistry In Peptide Research

It goes without saying that chemical synthesis is of central importance in peptide research. The remarkable developments in peptide chemistry during the past 50 years, especially the development of solid-phase synthesis pioneered by Merrifield [2] and its optimization, makes peptide chemistry the most quantitative synthetic methodology available to organic chemistry. For me this method was a godsend, and as I began my career at Arizona, we were among the first outside of Merrifield's laboratory to adopt this methodology for incorporation of specifically dueterated and carbon-13 labeled amino acids (many of which we synthesized ourselves [e.g., 3]) into oxytocin (OT) and vasopressin derivatives, using optimized solid phase methods [e.g., 4]. These labeled hormone derivatives were critical to our correctly assigning all of the proton and carbon-13 [5-7] spectra of these hormones long before the advent of 2-D NMR, and allowed us to examine the conformational and dynamic properties of oxytocin and several analogues in solution and on binding to its carrier protein in neurosecretory granules, neurophysin [7, 8-10]. These studies demonstrated that different parts of the oxytocin molecule had different dynamic properties when bound to neurophysin, and established for the first time the microscopic on-off rates for binding of a hormone (OT) to a macromolecular protein receptor.

During this time, we were preparing an increasing number of constrained cyclic peptides, of increasing synthetic difficulty which led us to speculate that the solid phase support might provide a suitable environment for cyclization. This was tested by M. Lebl in our laboratory with some success [11] and subsequently optimized by F. Al-Obeidi for the synthesis of cyclic superagonist lactam analogues of α-MSH [12]. Recently we have quantitatively converted reduced [D-Pen^2,D-Pen^5] enkephalin to its cyclic disulfide (Misicka & Hruby, unpublished). Other types of macrocyclic peptides are needed and will require the development of new synthetic methodologies (Table 1). It is exceptionally satisfying to note that in synthesis of very large (> 10^6) peptide libraries on solid phase supports, preview analysis at the single bead level has demonstrated a > 99.5% per step fidelity of synthesis using the Selectide Method.

The critical need for further developments in asymmetric synthesis cannot be overemphasized. Asymmetric synthesis of dipeptide mimetics, generally used for amide bond replacements, is well developed for enzyme inhibitor design. There is a need for the design and synthesis of complex amino acids, amino acid analogues, and di- and tri-peptide mimics that constrain in chi (χ) space [13, 14]. We have emphasized this theme since we proposed that differences in χ space can distinguish oxytocin agonists and antagonists [15]. More recently, we have synthesized a variety of specialized amino acids with constraints in χ_1 and χ_2 space [e.g., 16-19]. There is much to be done in this area that will be critical for the future development of rational design of peptides, pseudopeptides and peptidemimetics. We look forward to participating in this exciting, under-developed area of synthetic chemistry (Table 1).

Table 1 *Some future goals in peptide synthesis*

I.   Asymmetric Synthesis
    A.   Specialized Amino Acids - Constrained, Heterocyclic, etc.
    B.   Specialized Pseudopeptides
    C.   Cyclic (Linear) Templates
    D.   With Amide Bond Replacements
II. Macrocyclic Synthesis
    A.   Lactams, Lactones, Thioethers, Disulfides, Ethers, etc.
    B.   Templated Ring and Turns
    C.   Orthogonal Methods
    D.   Bicyclic, Tricyclic, etc. - Highly Constrained
III. "Clean" Fragment Solid Phase Synthesis
    A.   Large Peptide Synthesis
    B.   Protein Synthesis
IV. "Clean" Pseudopeptide and Peptoid Libraries
V.   Peptide and Peptidemimetic Conjugates
    A.   Nucleic Acid;
    B.   Lipids & Steroids
    C.   Sugars
    D.   Designed Heterocyclics; Suitable Linkers

Peptide Conformation and Topography and Their Relationships to Biological Activity

A central goal of our research has been to obtain an understanding of the physical/chemical basis for biological activity and information transfer in biological systems. Our major premise has been that peptide hormones, neurotransmitters, growth factors, cytokines, antibodies, immunogenes, etc. are chemical messengers and that their information is dependent on conformational and dynamic properties: conformation because biological systems exist in 3D chiral space; dynamics because conformational changes in time are essential to living systems. Thus we have used spectroscopy (NMR, CD, IR, Raman), X-ray crystallography, model building, and computational methods to examine peptide conformation and dynamics, as well as binding and bioassay data (thermodynamics and kinetics) to provide insights into the conformational, topographical, and dynamic properties that are critical to molecular recognition and biological activity.

Determining the bioactive conformations of oxytocin in its many biologically relevant actions has been a central problem, and we have been fortunate to participate in determining four different bioactive conformations: 1) oxytocin as an agonist at the uterine receptor; 2) oxytocin as an antagonist at the uterine receptor; 3) oxytocin binding to its carrier protein neurophysin; and 4) oxytocin in the crystalline state. We were among the first to recognize the significance of *cis-trans* isomerization about a Pro bond [20]. Subsequent studies of OT and deaminooxytocin by Urry and Walter [21], and of these and other analogues in DMSO [7] and water [6] demonstrated the conformational flexibility of the molecule, but that a β-turn could occur at the Tyr^2-Ile^3 residue. These studies led Walter to propose his "cooperative model" of OT activity at the uterine receptor [21]. At the same time, we examined the conformation of the oxytocin antagonist [Pen[1]]oxytocin [15] and determined that the conformation, topography and dynamic properties of oxytocin agonists and antagonists were different. We postulated a "dynamic model" [15, 22] of oxytocin agonist and competitive antagonist actions at uterine receptors. Subsequent CD and Raman studies with carefully chosen analogues [23] led to the realization that agonists and antagonist favored different chiralities of the disulfide bridge. The determination of the X-ray structure of deamino oxytocin [24], in which two different conformations of OT were found, supported the proposed models for oxytocin agonist and antagonist activity. This led to the design of highly constrained bicyclic OT analogues in which both disulfide and lactam bridges were used together for the first time in design [25]. Recently, 2D NMR, molecular modeling, and computational studies reported at this meeting (Wilke, Kövér, Chow, et al.), have demonstrated that oxytocin agonists and antagonists present different side chain topographical motifs for interaction with the oxytocin receptor. Two valuable lessons for the future are provided: 1) peptide agonists and competitive antagonists interact differently with their receptor proteins and have different structure-activity relationships; and 2) the availability of multiple conformational space is critical for bioactive peptides in the course of their biological lifetimes. Selective peptides for opioid receptors represent another example as to how agonists and antagonists have different structure activity relationships. The cyclic pentapeptide [D-Pen[2],D-Pen[5]] enkephalin (DPDPE) [27, 28] a highly δ_1 opioid receptor selective ligand has a completely different conformation and structure-activity relationships than the linear δ antagonists Allyl$_2$-Tyr-Aib-Aib-Phe-Thr-OH and H-Tyr-Tic-Phe-Phe-OH [29]. Similarly, the highly constrained cyclic μ receptor antagonist D-Tic-Cys-Tyr-D-Trp-Arg-Thr-Pen-Thr-NH$_2$ [30, 31] has no conformational structure relationships to μ receptor agonists such as H-Tyr-D-Ala-Gly-MePhe-Gly-ol.

Conformational constraint by stabilization of secondary structure has been a recurrent theme in our research. The development of superpotent, biostable, superagonists of α-MSH are an excellent illustration of this rational design approach. The recognition that the high potency of the designed analogue [Nle[4],D-Phe[7]]α-MSH [32] might be due to the stabilization of a reverse turn about the D-Phe[7] residue led to the design of the superpotent cyclic analogue [Cys[4],Cys[10]]α-MSH using the concept of pseudoisosteric cyclization [33, 34]. Further conformational analyses and molecular dynamics investigations led to the design of the superagonist, prolonged acting (days to weeks) cyclic constrained MSH agonists such as Ac-[Nle[4],Asp[5],D-Phe[7],Lys[10]]α-MSH$_{4-10}$-NH$_2$ [12, 35]. These molecules are found to have prolonged *in vitro* and *in vivo* bioactivity, exceptional stability against biodegradation, and long times in circulation *in vivo*. These studies led us to propose a pharmacophore which has a reverse turn template involving the D-Phe[7] and Arg[8] residues and in which the His[5], D-Phe[7] and Trp[9] residues are presented on the same surface, with the amino terminal, carboxyl terminal, lactam bridge (or disulfide bridge) on the other (ancillary) surface. We now have designed superpotent, super-prolonged acting bridged analogues using the ancillary surface [36] to place a dipeptide basic motif for interaction with the membrane in the receptor compartment. We have demonstrated that these peptides partition into the plasma membrane of cells in preference to the aqueous phase, quite unlike the native hormone [Sharma, Hadley and Hruby, unpublished). Since many biologically active peptides appear to segregate their molecular recognition sites onto a specific surface, we believe that the other (ancillary) surfaces can be used for a variety of secondary, but critical purposes (Table 2).

Table 2   *Proposed uses of ancillary sites on biologically active peptides for rational design of specific properties*

I.   Design of Sites for Specific Interaction With the Membrane Compartment of Receptors

II.   Design of Sites for Specific Interaction With Ancillary Sites on Receptors (and Acceptors Including Nucleic Acids and Polysaccharides).

III. Design of Specific Chemical Functionality to Enhance Biodistribution, Penetration Through Membrane Barriers, etc.

IV.   Design of Enhanced Stability Against Biodegradation, or to Provide a Site Specific for Biodegradation (Prodrug Approach or Specific Degradation Site).

V.   Utilize for Attaching Reporter Groups to be Used In Chemical, Physical and Clinical Research, and for Diagnostics.

Our work on the design, synthesis, and pharmacophore examination of peptide ligands that are highly potent and extraordinarily selective for μ, δ, and κ opioid receptor types and subtypes has been intellectually challenging and scientifically satisfying. Space does not permit discussion of even a small fraction of the science that has been done, but a few highlights will illustrate a few lessons we have learned. In the process, we have obtained a number of highly potent and highly selective ligands for the μ, κ_1, δ_1, δ_2, μ-δ, and CCK-δ (see Table 3). For several of these receptors, we were the first to discover them and/or the selective ligands for them.

[D-Pen^2,D-Pen^5]enkephalin (DPDPE) was the first highly selective (>1000 fold) ligand for the δ opioid receptor [27] and utilized two design concepts; 1) pseudo-isosteric cyclization; and 2) topographical transannular constraint in medium size rings. Subsequent NMR studies, molecular mechanics and molecular dynamics calculations, molecular modeling and X-ray analysis, and further topographical constraints determined that DPDPE and several more potent and δ opioid receptor ligands [28, 37-40 for selected references] had a bioactive conformation with an amphipathic topography. The X-ray structure of DPDPE revealed three different, related conformations that closely resembled the one predicted to be the "bioactive conformation" [Flippen-Anderson, Collins, et al., to be published; 41]. Frank Porreca has found that δ ligands might reverse the addictive effect of μ opioids and perhaps of other addictive drugs such as cocaine. A structural highlight of this research was the development of topographical insights by the use of β-MePhe^4, β-MeTyr^1 and Hat^1 enantiomers and diastereoisomers [38, 39, 41]. The recent discovery of the first naturally occurring δ-opioid receptor ligands [42, 43], the Deltorphins, led us to investigate how these and other structurally and conformationally diverse ligands as DPDPE could be so selective for the δ opioid receptor. Our basic hypothesis was that they must share common topochemical properties for molecular recognition of the δ receptor. Using computational chemistry in conjunction with NMR studies, topographically constrained analogues containing β-MePhe$^{3(4)}$ and β-MeTyr^1, and evaluation of their conformations by NMR in conjunction with conformational calculations [40, 44-46] led to the suggestion of a common topographical motif. Based on these studies, we have designed cyclic deltorphins and dermenkephalins with high δ receptor potency and selectivity [47].

Table 3 *High potent and selective ligands for opioid receptor types and subtypes*

Structure	Receptor Class	Agonist/ Antagonist	Selectivity	Receptor Subtypes
[D-Pen^2,p-ClPhe^4,D-Pen^5]Enk	δ	Agon.	>10,000	δ$_1$
[D-Pen^2,L-Cys^5,p-FPhe^6]Enk	δ	Super Agon.	>30,000	δ$_1$(P)
[D-Pen^2,Ala^3,D-Pen^5]Enk	δ	Agon.	>10,000	δ$_1$
Tyr-D-Pen-Phe-Asp-Pen-Val-Gly-NH$_2$	δ	Agon.	>10,000	δ$_2$
Tyr-D-Ala-Phe-Cys-Val-Val-Gly-NH$_2$	δ	Antag.	>1,000	δ$_2$
D-Tic-Cys-Tyr-D-Trp-Arg-Thr-Pen-Thr-NH$_2$	μ	Antag.	>10,000	μ
D-Tca-Cys-Tyr-D-Trp-Arg-Thr-Pen-Thr-NH$_2$	μ,δ	μ Antag./ δ Agon.	N.K.	μδ$_{cx}$
(Tyr-D-Ala-Gly-Phe-NH-)$_2$	μ,δ	μ Agon./ δ Agon.	None	μδ$_{cx}$
Tyr-Gly-Gly-Phe-Cys-Arg-Arg-Ile-Arg-Pro-Cys-OH	κ	Agon.	>10,000	κ$_1$
Tyr-Gly-Gly-Phe-Nle-Arg-Arg-N-MeNle-Arg-Pro-Asn-NH$_2$	κ	Agon.	>20,000	κ$_1$

A particularly interesting and instructive effort was our conversion of somatostatin from a somatostatin receptor ligand to a highly potent, and exceptionally μ-receptor selective opioid receptor antagonist with little or no somatostatin-like activity. Structure-activity studies in several laboratories led Veber and co-workers at Merck to conclude that essentially all bioactivity of the tetradecapeptide somatostatin could be accounted for by the tetrapeptide segment -Phe-Trp-Lys-Thr- [48], which existed as a β-turn, and than this β-turn was essentially all that was required for receptor recognition. With these ideas they were able to design cyclic hexapeptides that were as potent or more potent that somatostatin in inhibiting the release of growth hormone, glucagon and insulin. It had been suggested that somatostatin had opioid analgesic activities albeit at concentrations 10^3 to 10^4 greater than those that elicit somatostatin-like activity. These and other results led us to ask whether we could convert somatostatin into a receptor selective opioid. Two basic assumptions went into our original design: 1) the β-turn tetrapeptide would serve primarily as a conformational template; and 2) the N- and C-terminal residues would serve as the major sites to "build on" an opioid receptor recognition site. With the total synthesis of about ten or so analogues of a somatostatin cyclic octapeptide we were able to find that appropriate substitution of the disulfide bridge (Cys in **2** position, Pen in **7** position), of Tyr for Phe in the **3** position and of a carboxamide terminal, provided an analogue D-Phe-Cys-Tyr-D-Trp-Lys-Thr-Pen-Thr-NH$_2$ (CTP) [49] that was about 200 fold selective for the μ vs. somatostatin receptor and ~1000 fold selective for the μ vs. $\delta(\kappa)$ receptor. Hence, we had converted somatostatin from a somatostatin receptor ligand to a highly μ opioid receptor ligand. Substitution of the D-Phe^1 with the topographically constrained D-Tic^1 residue and Lys^5 with Orn^5 or Arg^5 led to analogues that were nanomolar binders to the μ receptor, 10^3-10^4 selective for μ vs. somatostatin receptors and 10^3-10^5 selective for the μ vs. δ (κ) receptors [30, 31]. Conformation analysis and molecular dynamic simulations using state-of-the-art 2D NMR and computational chemistry demonstrated [31, Kazmierski et al., unpublished] that a stable reverse β-turn was present in the -Tyr-D-Trp-Orn-Thr- sequence, that only a specific disulfide chirality was compatible with strong μ receptor preference, and that correct topographical presentation of the aromatic group in positions 1 and 3 and of Thr^8 were essential. A topographical model for the pharmacophore for μ receptor recognition of an inhibitor could be suggested [31]. Comprehensive *in vitro* and *in vivo* bioassays demonstrated the exceptional inhibitory potency [pA$_2$ values of 11-12] and prolonged action [t$_{1/2}$ = >250 min] of these compounds [e.g., 50-51].

Recently we have designed constrained analogues of dynorphin with high potency and extraordinary selectivity (>10,000 fold) for the κ_1 (vs. κ_2, etc.) opioid receptors [52-53, Porreca et al., unpublished] using concepts of cyclic conformational constraints and relationships of address regions of opioid receptors. Similar concepts were used for developing the structure-biological activity relationships of melanocyte concentrating hormone agonists and antagonists [e.g. 54], also using second messenger effects and a solution conformation [55]. Using similar concepts, we also have developed highly selective cholecystokinin analogues for the CCK-B and CCK-A receptors [56; Fang, Hruby, et al., manuscripts in preparation].

Conformational Constraint in Ramachandran (ϕ, ψ) Space and in Chi (χ) Space

ϕ, ψ **Space.** From the beginning of our independent studies in 1968, we assumed that peptide conformation would be a critical aspect of biological activity. A basic assumption we made for design purposes, since we did not have the 3D structure of the receptors, was that we could nonetheless determine the 3D stereostructural requirements of the receptor by design of a suitably constrained ligand whose conformation would be complementary to the receptor. Thus, both binding studies and bioassays (in conjunction with biophysical measurements) would be needed to provide the thermodynamic, kinetic, dynamic and structural insights needed for rational design. Thus, a good binding study or bioassay is a chemical experiment. We further assumed that the peptide would likely prefer some common secondary structures for maximal interaction at the receptor. Thus, our design efforts always included conformational considerations. In the beginning this was done intuitively, and by model building, but once secondary structure prediction methods and statistical and force field methods of conformational evaluation became available, we used calculations to supplement and supplant our intuitive conformational evaluations. We further assumed that the primary sites of interaction involved the side chain moieties properly arranged in 3D space, and that the backbone served primarily as a template for proper presentation of the side chains. Perhaps too influenced by the putative structures of the gramicidins and cyclic peptide antibiotics, we saw H-bonding as mainly of use for intramolecular stabilization and not intermolecular recognition. These assumptions have greatly aided our design, and in almost every problem, we have quickly found a constrained, secondary structure that could serve as a template for the purpose of design and evaluation of conformational structure-biological activity relationships. Indeed, the use of conformational constraints as the basic design mode has been the central theme of our research for the past 20 years. Some of the methods used in our design of conformational constraints were listed in ref. [34]. The specific application of these methods to particular problems is outside the scope of this presentation, but several overviews have appeared in the literature [e.g., 14, 34].

     Conformational Constraint in Chi (χ) Space. In the past decade, we have increasingly turned our attention to constraints in Chi space. In my view, this approach will become increasingly the major focus of peptide and peptide mimetic design. This is because molecular recognition both in terms of potency and selectivity is highly effected by both entropic and enthalpic factors. The side chains of most amino acids tend to be quite flexible and, even in highly ordered secondary structural motifs, have considerable flexibility. Nevertheless, a true understanding of a pharmacophore requires careful attention to the side chain conformation. As already discussed, studies of the bioactive conformation of oxytocin agonists and antagonists strongly supported the notion of a transannular Tyr[2] to Asp[5] side chain interaction as important to agonist activity, whereas prevention of that interaction led to antagonist activity. Furthermore, conformation-activity studies with oxytocin clearly demonstrated that topography not conformation was the major determinant of molecular recognition for agonist vs. antagonist activity [26]. We, therefore, began in the early 1980s to ask how we might address the problem of χ space more directly. Since such constraints inevitably could require additional substituents on the β-carbon, aromatic carbons, or other carbons where additional chiral centers would be created, it was clear that asymmetric synthetic methods would be required. Fate again intervened. We were getting started and had come to the conclusion that use of a chiral auxiliary in conjunction with a nucleophilic or electrophilic nitrogen were needed, when I went off to Harvard for a sabbatical with Martin Karplus and ran into David Evans who had just moved to Harvard and already had an auxiliary. With David's help, we adopted his chiral auxiliary for asymmetric synthesis of β-methyl aromatic amino acids [16, 18, 19]. In our latest efforts, using a new chiral auxiliary, we can fix both α and β stereochemistry in a one pot reaction with very high chiral purity. Compounds such as β-Me-2',6'-Me$_2$Tyr that have severely hindered rotation about the χ_2 angle (ΔG^+ = 15-22 kcal (mole) [57], and hence can form a third chiral surface, have been prepared. Their application to topographical conformation-activity relationships should be most exciting. Space does not allow a discussion of the use of conformation constraints in χ space, but I outline in Table 4 some of the approaches in our laboratory that have led to peptides and peptidomimetics with extraordinary potency and selectivity including β-MePhe[4] [39] and β-MeTyr[1] [40] analogues of DPDPE and β-MePhe[3] analogues of CCK-8 (56) with extraordinary selectivity and potency for δ opioid and CCK-B receptors, respectively.

Table 4   *Conformational constraints in chi (χ) space (designed topographical structures) - some approaches*

I.   Penicillamine and Other β-Substituted Cysteine Analogs
    A.   S-S chirality can be determined
    B. χ_1 and χ_2 about disulfide bridge hindered
    C.   Transannular gem-dimethyl effects in medium size rings
II.   Phenylglycines, t-Butylglycine, etc.
    A.   Only χ_1 - constrained by adjacent amino acid residues
III. β-Methyl(Alkyl)-Aromatic Amino Acids
    A.   Generally χ_1 constrained to 2 values
    B.   Steric and stereoelectronic effects on recognition surfaces
IV. β,β-Dimethyl-Aromatic Amino Acid
    A.   Generally χ_1 constrained to 1 value; χ_2?
V. β-Methyl-2'6'-Dimethyl-Aromatic Amino Acids
    A. χ_1 Generally constrained to 2 values
    B. χ_2 Generally constrained to 1 value
VI.   Tic and Related "Cyclic" Amino Acids
    A. χ_1 constrained to 2 values - preferred
    B. χ_2 constrained to 1 value
VII. α-Methyl-β-Methyl-Tic and Related Cyclic Amino Acids
    A.   Gauche(-) and Gauche(+) allowed for internal residues
VIII.   Amino Acids Substituted in the β-Position with -SH, -OH, -CH$_2$X, NH$_2$ etc.
    A.   New stereoelectronic effects
    B.   Generally χ_1 and χ_2 constraints;
    C.   To construct new classes of cyclic peptides and peptidemimetics
IX.   Amino Acids Substituted in the β-Position With Heterocyclic, Pseudoaromatic, Aromatic and Other Functional Groups
    A.   Used to design specific aromatic and other stereoelectronic properties
    B.   Generally specific χ_1 and χ_2 constraints
X. α,β-Dehydro- and Other Unsaturated Amino Acids
    A. χ_1 can be constrained to "unnatural" side chain topography
    B.   Stereochemistry can be determined precisely
    C. α,β-Unsaturation can influence backbone conformation

Some Lessons Learned From Studies of Conformational-Structure-Biological Activities Relationships

Peptide agonists and antagonists for the same receptor have different structural, conformational and topographical requirements for the same receptor site. We have learned this and a number of other lessons that we hope will be useful in the future, and a few will be discussed here. Development of antagonists of bioactive compounds is very important, but as we have discussed [58], though there is no direct route to go from agonists to antagonists, there are several useful approaches that have led to past success and can serve as inspiration for the future. The use of conformational constraint, topographical constraint, and amide bond replacements have all been used successfully in our laboratory to obtain peptide hormone and neurotransmitters antagonists. A potentially rational approach is the use of partial agonist activity as a road to a pure antagonist. Partial agonists are interesting in that they can bind to the receptor, but at no concentration will they fully activate the receptor transduction system. Thus, the ligands have available topochemical properties that are compatible for molecular recognition of both states of the receptor. The trick then is to eliminate those aspects of structure necessary for agonist activity. This was our major approach for developing glucagon antagonists for the adenylate cyclase coupled receptor [59-62], some of which could lower blood glucose levels in diabetic animals [63]. Unfortunately, the glucagon receptor (and this appears to be true for many receptors) has multiple transduction states [64] and apparently these can lead to glucose release *in vivo* though the cAMP pathway is completely blocked. From these and many other studies, it is now clear that proper development of potent, receptor specific peptide antagonists, enzyme inhibitors, antigens, etc. requires the use of multiple assay systems that provide chemically relevant information including binding constants, on and off rates, thermodynamic data, chemical transduction mechanisms, and so forth. Peptide chemists need to more fully recognize the value and importance of collaboration with biologists so that the full impact of our work can be realized.

Peptide Stability and Biodistribution

It is often said, even by peptide chemists who should know better, that peptides are unstable and thus not suitable for use as drugs. This is nonsense. We humans could not exist unless many of the peptides and proteins that constitute our cells, blood constituents, membrane constituents, etc. were stable. Nonetheless, many endogenous peptides are unstable. However, our experience is that conformational constraint alone or in conjunction with other chemical "tricks" (D-amino acids, amide bond replacements, non-standard amino acids) can stabilize peptides against proteolytic breakdown in the serum, the brain, the kidney or the liver. Space does not permit an extensive discussion, but we have successfully stabilized or increased by orders of magnitude the stability of peptide ligands to proteolytic degradation. Peptides such as Ac-[Nle^4,Asp^5,D-Phe^7,Lys^{10}]α-MSH$_{4\text{-}10}$-NH$_2$, H-Tyr-D-Pen-Gly-Phe-D-Pen-OH (DPDPE), and D-Phe-Cys-Tyr-D-Trp-Arg-Thr-Pen-Thr-NH$_2$ (CTAP) are completely stable to proteolytic digestion, and analogues such as [Nle^4,D-Phe^7]α-MSH, [Cys^4,Cys^{10}]α-MSH, [β-Mpa^1,p-MePhe^2,Thr^4,Orn^8]OT, and R-CO-[D-Phe^7]α-MSH have greatly enhanced stability. We have published extensively with our biological colleagues about this [e.g., see 65-70]. We believe the problem has been basically solved and requires a commitment to further science to make it a totally rational process. Biodistribution of peptides is still a major problem. However peptides and proteins are everywhere in the body and get everywhere in the body. In terms of science, this means we need to determine how Nature does it so that we can do Nature one better. I believe a variety of approaches can be successful (Table 5).

Table 5 *Approaches to improve biostability and biodistribution*

I.   Stabilize Peptide Against Biodegradation
      Conformational Constraint; Amide Bond Replacements; D- and Non-Standard
      Amino Acids
II.   Enhance Time in Circulation/Receptor Compartment
      Conformational Constraint; Lipophilicity; Conjugates
III. Design For Receptor Compartment Using Ancillary Sites of Interaction
      Lipophilicity; Electrostatic Interaction
IV.   Physicochemical Approaches to Membrane Barriers
      Lipophilicity; Transport Systems; Hydrophilicity
V.   Use Carrier Mediated Mechanisms for Design
VI.   Use Pseudopeptide and Non-Peptide Templates
VII. Design Dynamic Lipophiles/Hydrophiles

In addition to our efforts to stabilize peptides against biodegradation, we have sought new methods to improve their time in circulation and/or in the receptor compartment. Some useful, and we hope universal, strategies have been developed, such as conformational constraint. In addition, various increases in lipophilicity can have the same affect [e.g., see 71]. Such increased lipophilicity may provide increased interaction of these compounds with certain circulating blood components. Another area of interest is the possibility of increasing the lifetime of the peptide in the receptor compartment of the membrane by using ancillary sites that can interact with the membrane or the receptor (Table 5). We have developed prolonged-acting α-melanotropin analogues that can maintain activity after several cell divisions even though all of the peptide has been removed after short exposures of a few hours or less [72]. This is due to maintenance of these peptides, e.g. [Nle^4,D-Phe^7]α-MSH, in the receptor compartment. More recently, we have designed bridged analogues of α-MSH, specifically containing dibasic acids on the ancillary surface for interaction with the receptor and have demonstrated that these molecules partition into the membrane compartment of melanophores much more so than the native hormone [36]. We also have a large effort to find methods to get peptide hormones and neuro-transmitter across the blood brain barrier, and have had some success with several of our opioid analogues [73]. A particularly dramatic effect was recently obtained in collaboration with Robin Polt where we were able to show that potent centrally mediated analgesia could be obtained following peripheral administration of specific glycopeptide conjugates [R. Polt et al., submitted for publication]. This is an important area for further research (Table 5).

We have many other interests that we have actively pursued over the years including new analytical methods development in the areas of chromatography and mass spectrometry, the development of new NMR methods [e.g., 74], and in recent years, the synthesis and use of very large peptide libraries [e.g., 75, 76]. I believe there is in place a logical approach to peptide design that works. Most of these steps need to be refined, but fortunately we have or can develop the tools for continued progress. It is an exciting and creative time for peptide science.

Summary Conclusions

I have enjoyed tremendously this research trail in peptide chemistry. I have especially enjoyed the opportunity to interact with so many bright and dedicated students, postdoctoral fellows, and colleagues throughout the world. I firmly believe that scientists bring to human endeavor a special quality of highly critical thinking, but a collective spirit that, given further development and maintenance of high ethical standards (currently under severe stress), could educate the world to friendship and cooperation rather than the enmity and destructive competition that has marked so much of human history. I hope we will develop our field with this spirit.

Acknowledgements

I want to thank the National Institutes of Health and National Institute of Drug Abuse that have supported most of our current research, and the National Science Foundation for significant past support. Also, small short term support from Abbott, Smith Kline & French, Merck, Selectide and Research Corporation have been helpful.

References

1. (a) Hruby, V.J., Yamashiro, D. and du Vigneaud, V., J. Am. Chem. Soc., 90 (1968) 7106; (b) Hruby, V.J. and du Vigneaud, V., J. Am. Chem. Soc., 91 (1969) 3624.
2. Merrifield, R.B., J. Am. Chem. Soc., 85 (1963) 2149
3. (a) Spatola, A.F., Cornelius, D.A., Hruby, V.J. and Blomquist, A.T., J. Org. Chem., 39 (1974) 2207; (b) Viswanatha, V. and Hruby, V.J., J. Org. Chem., 44 (1978) 2892 and references therein.
4. Hruby, V.J., Upson, D.A. and Agarwal, N.S., J. Org. Chem., 42 (1977) 3552.
5. Brewster, A.I., Hruby, V.J., Spatola, A.F. and Bovey, F.A., Biochemistry, 12 (1973) 1643.
6. Brewster, A.I. and Hruby, V.J., Proc. Natl. Acad. Sci. U.S.A., 70 (1973) 3806.
7. Brewster, A.I., Hruby, V.J., Glasel, J. and Tonelli, A.E., Biochemistry, 12 (1973) 5294.
8. Glasel, J.A., Hruby, V.J., McKelvey and Spatola, A.F., J. Mol. Biol., 79 (1973) 555.
9. Blumenstein, M. and Hruby, V.J., Biochemistry, 16 (1977) 5169.
10. Blumenstein, M., Hruby, V.J. and Viswanatha, V., Biochem. Biophys. Res. Commun., 94 (1980) 431, and references therein.
11. Lebl, M. and Hruby, V.J., Tetrahedron Letts., 25 (1984) 2067.
12. Al-Obeidi, F., Castrucci, A.M. de L., Hadley, M.E. and Hruby, V.J., J. Med. Chem., 32 (1989) 2555.
13. Kazmierski, W. and Hruby, V.J., Tetrahedron, 44 (1988) 697.
14. Hruby, V.J., Kazmierski, W. and Al-Obeidi, F., Biochemical J., 268 (1990) 249.
15. Meraldi, J.-P., Hruby, V.J. and Brewster, A.I.R., Proc. Natl. Acad. Sci. U.S.A., 74 (1977) 1373, and references therein.
16. (a) Dharanipragada, R., Nicolas, E., Toth, G. and Hruby, V.J. Tetrahedron Letts., 30 (1989) 6841;(b) Nicolas, E., Dharanipragada, R., Toth, G. and Hruby, V.J., Tetrahedron Letts. 30 (1989) 6845.
17. Kazmierski, W.M. and Hruby, V.J., Tetrahedron Letts., 32 (1991) 5769.
18. Boteju, L.W., Wegner, K. and Hruby, V.J., Tetrahedron Letts., 33 (1992) 7491.
19. Nicolás, E., Russell, K.C. and Hruby, V.J., J. Org. Chem., 58 (1993) 766.

20. Hruby, V.J., Brewster, A.I. and Glasel, J.A., Proc. Nat. Acad. Sci. U.S.A., 68 (1971) 450.

21. Urry, D.W. and Walter, R., Proc. Natl. Acad. Sci. U.S.A., 68 (1971) 956; Walter, R. Fed. Proc., 36 (1977) 1872.

22. Hruby, V.J. and Mosberg, H.I., In D.H. Schlesinger, (Ed.), Neurohypophyseal Peptide Hormones and Other Biologically Active Peptides, Elsevier, N.Y., 1981, p.227.

23. Hruby, V.J., Deb, K.K., Fox, J., Bjarnason, J. and Tu, A.T., J. Biol. Chem., 253 (1978) 6060, and subsequent references.

24. Wood, S.P., Tickle, I.J., Treharne, A.M., Pitts, J.E., Mascarenhas, Y., Li, J.Y., Husain, J., Cooper, S., Blundell, T.L., Hruby, V.J., Wyssbrod, H.R., Buku, A. and Fishman, A.J., Science, 232 (1986) 633.

25. Hill, P.S., Smith, D.D., Slaninova, J. and Hruby, V.J., J. Am. Chem. Soc., 112 (1990) 3110.

26. (a) Hruby, V.J., In A.S.V. Burgen and G.C.K. Roberts, (Eds.), Topics in Molecular Pharmacology, Vol. 1, Elsevier/North-Holland, Amsterdam, Holland, 1981, p.99; (b) Hruby, V.J., In A. Eberle, R. Geiger and T. Wieland, (Eds.), Perspectives in Peptide Chemistry, S. Karger, Basel, Switzerland, 1981, p.207.

27. Mosberg, H.I., Hurst, R., Hruby, V.J., Gee, K., Yamamura, H.I., Galligan, J.J. and Burks, T.F., Proc. Natl. Acad. Sci. U.S.A., 80 (1983) 5871.

28. Hruby, V.J., Kao, L.-F., Pettitt, B.M. and Karplus, M., J. Am. Chem. Soc., 110 (1988) 3351.

29. Schiller, P.W., Nguyen, T.M.-D., Weltrowska, G., Wilkes, B.C., Mardsen, B.J., Lemieux, C.L. and Chung, N.N., Proc. Natl. Acad. Sci. U.S.A., 89 (1992) 11871.

30. Kazmierski, W. and Hruby, V.J., Tetrahedron, 44 (1988) 697.

31. Kazmierski, W.M., Yamamura, H.I. and Hruby, V.J., J. Am. Chem. Soc., 113 (1991) 2275.

32. Sawyer, T.K., Sanfilippo, P.J., Hruby, V.J., Engel, M.H., Heward, C.B., Burnett, J.B. and Hadley, M.E., Proc. Natl. Acad. Sci. U.S.A., 77 (1980) 5754.

33. Sawyer, T.K., Hruby, V.J., Darman, P.S. and Hadley, M.E., ibid., 79 (1982) 1751.

34. Hruby, V.J., Life Sciences, 31 (1982) 189.

35. Al-Obeidi, F., Hadley, M.E., Pettitt, B.M. and Hruby, V.J., J. Amer. Chem. Soc., 111 (1989) 3413.

36. Sharma, S.D., Nikiforovich, G., Jiang, J., de L. Castrucci, A.M., Hadley, M.E. and Hruby, V.J., In C.H. Schneider and A.N. Eberle, (Eds.), Peptides 1992, Escom, Leiden, The Netherlands, 1993, p.95.

37. Hruby, V.J., Kao, L.-F., Hirning, L.D. and Burks, T.F., In C.M. Deber, V.J. Hruby, and K.D. Kopple, (Eds.), Peptides: Structure and Function, Pierce Chemical Co., Rockford, IL, 1985, p.487.

38. Toth, G., Kramer, T.H., Knapp, R., Lui, G., Davis, P., Burks, T.F., Yamamura, H.I. and Hruby, V.J., J. Med. Chem., 33 (1990) 249.

39. Hruby, V.J., Toth, G., Gehrig, C.A., Kao, L.-F., Knapp, R., Lui, G.K., Yamamura, H.I., Kramer, T.H., Davis, P. and Burks, T.F., J. Med. Chem., 34 (1991) 1823.

40. Toth, G., Russell, K.C., Landis, G., Kramer, T.H., Fang, L., Knapp, R., Davis, P., Burks, T.F., Yamamura, H.I. and Hruby, V.J., J. Med. Chem., 35 (1992) 2384.

41. Nikiforovich, G.V., Hruby, V.J., Prakash, O. and Gehrig, C.A., Biopolymers, 31 (1991) 941.

42. Erspamer, V., Melchiorri, P., Falconieri-Erspamer, G., Negri, L., Corsi, R., Saverine, C., Barra, D., Simmaco, M. and Kreil, G., Proc. Natl. Acad. Sci. U.S.A., 86 (1989) 5188.

43. Amiche, M., Sagan, S., Mor, A., Delfour, A., and Nicolas, P. Mol. Pharmacol., 35 (1989) 774.

44. Misicka, A., Lipkowski, A.W., Horvath, R., Davis, P., Kramer, T.H., Yamamura, H.I. and Hruby, V.J., Life Sciences, 51 (1992) 1025.

45. Nikiforovich, G., Prakash, O., Gehrig, C.A. and Hruby, V.J., Int. J. Peptide Protein Res., 41 (1993) 347.
46. Nikiforovich, G.V., Prakash, O., Gehrig, C. and Hruby, V.J., J. Am. Chem. Soc., 115 (1993) 3399.
47. Misicka, A., Nikiforovich, G.V., Lipkowski, A.W., Horvath, R., Davis, P., Kramer, T.H., Yamamura, H.I. and Hruby, V.J., Bioorg. & Med. Chem. Letts., 2 (1992) 547.
48. Veber, D., Freidinger, R.M., Perlow, D.S., Palaveda Jr., W.J., Holly, F.W., Strachan, R.G., Nutt, R.F., Arison, B.H., Homnick, C., Randall, W.C., Glitzer, M.S., Saperstein, R. and Hirschmann, R., Nature, 292 (1981) 55.
49. Pelton, J.T., Gulya, K., Hruby, V.J., Duckles, S.P. and Yamamura, H.I., Proc. Natl. Acad. Sci. U.S.A., 82 (1985) 236.
50. Shook, J.E., Pelton, J.T., Hruby, V.J. and Burks, T.F., J. Pharmacol. Exp. Therap., 243 (1987) 492 and references therein.
51. Shook, J.E., Lemcke, P.K., Gehrig, C.A., Hruby, V.J. and Burks, T.F., J. Pharmacol. Exp. Therap., 249 (1989) 83 and references therein.
52. Kawasaki, A.M., Knapp, R.J., Kramer, T.H., Wire, W.S., Vasquez, O.S., Yamamura, H.I., Burks, T.F. and Hruby, V.J., J. Med. Chem., 33 (1990) 1874.
53. Kawasaki, A.M., Knapp, R.J., Kramer, T.H., Walton, A., Wire, W.S., Hashimoto, S., Yamamura, H.I., Porreca, F., Burks, T.H. and Hruby, V.J., J. Med. Chem., 36 (1993) 750.
54. Lebl, M., Hruby, V.J., de L. Castrucci, A.M., Visconti, M.A. and Hadley, M.E., J. Med. Chem., 31 (1988) 949; Matsunaga, T.O., de L. Castrucci, A.M., Hadley, M.E. and Hruby, V.J., Peptides, 10 (1989) 349.
55. Matsunaga, T., Gehrig, C.A. and Hruby, V.J., Biopolymers, 30 (1990) 1291.
56. Hruby, V.J., Fang, S., Knapp, R., Kazmierski, W., Lui, G.K. and Yamamura, H.I., Int. J. Peptide Protein Res., 35 (1990) 566; Fang, S., Nikiforovich, G.V., Knapp, R.J., Jiao, D., Yamamura, H.I. and Hruby, V.J., In J.A. Smith and J.E. Rivier (Eds)., Peptides: Chemistry and Biology, 1992, p.42.
57. Jiao, D., Russell, K.C. and Hruby, V.J., Tetrahedron, 49 (1993) 3511.
58. Hruby, V.J., In J. Joosse, R. M. Buijs, F.J.H. Tilders, (Eds.), Progress in Brain Research, Vol. 92, Elsevier Sci. Publ., Chapter 18, 1992, p.215.
59. Bregman, M.D., Trivedi, D., and Hruby, V.J., J. Biol. Chem., 255 (1980) 11725.
60. Hruby, V.J., Krstenansky, J., Gysin, B., Pelton, J.T., Trivedi, D. and McKee, R., Biopolymers, 25 (1986) S135.
61. Gysin, B., Trivedi, D., Johnson, D.G. and Hruby, V.J., Biochemistry, 25 (1986) 8278; Gysin, B., Johnson, D.G., Trivedi, D. and Hruby, V.J., J. Med. Chem., 30 (1987) 1409.
62. Zechel, C., Trivedi, D., and Hruby, V.J., Int. J. Peptide Protein Res., 38 (1991) 131, and references therein.
63. Johnson, D.G., Goebel, C.U., Hruby, V.J., Bregman, M.D. and Trivedi, D., Science, 215 (1982) 1115.
64. Wakelam, M.J.O., Murphy, G.J., Hruby, V.J. and Houslay, M.D., Nature, 323 (1986) 68.
65. Akiyama, K., Yamamura, H.I., Wilkes, B.C., Cody, W.L., Hruby, V.J., de L. Castrucci, A.M. and Hadley, M.E., Peptides, 5 (1984) 1191.
66. de L. Castrucci, A.M., Hadley, M.E., Sawyer, T.K. and Hruby, V.J., Comp. Biochem. Physiol., 78B (1984) 519.
67. Chan, W.Y., Rockway, T.W. and Hruby, V.J., Proc. Soc. Exp. Biol. Med., 185 (1987) 187.
68. de L. Castrucci, A.M., Hadley, M.E. and Hruby, V.J., In M.E. Hadley, (Ed.), The Melanotropic Peptides. CRC Press, Boca Raton, FL, U.S.A., 1988, p.171.
69. Dawson, B.V., Hadley, M.E., Kreutzfeld, K., Dorr, R.T., Hruby, V.J., Al-Obeidi, F. and Don, S., Life Sciences, 43 (1988) 1111.

70. Weber, S.J., Greene, D.L., Hruby, V.J., Yamamura, H.I., Porreca, F. and Davis, T.P., J. Pharm. Exp. Therap., 263 (1992) 1308.
71. Al-Obeidi, F., Hruby, V.J., Yaghoubi, N., Marwan, M.M. and Hadley, M.E., J. Med. Chem., 35 (1992) 118.
72. Abdel Malek, Z.A., Kreutzfeld, K.L., Marwan, M.M., Hadley, M.E., Hruby, V.J. and Wilkes, B.C., Cancer Research, 45 (1985) 4735.
73. Weber, S.J., Abbruscato, T.J., Brounsen, E.A., Lipkowski, A.W., Polt, R., Misicka, A., Haaseth, R.C., Bartosz, H., Hruby, V.J. and Davis T.P., J. Pharm. Exp. Therap., 226 (1993) 1649.
74. Kövér, K.E., Prakash, O. and Hruby, V.J., J. Magn. Reson., 99 (1992) 426.
75. Lam, K.S., Salmon, S.E., Hersh, E.M., Hruby, V.J., Kazmierski, W.M. and Knapp, R.J., Nature, 354 (1991) 82.
76. Hruby, V.J., Lam, K.S., Lebl, M., Kazmierski, W., Hersh, E.M. and Salmon, S.E., In R. S. Rapaka, (Ed.), Medications Development: Drug Discovery, Databases and Computer-Aided Drug Design, Washington, D.C., NIDA Res. Monogr., in press, 1993.

Session I
Synthetic Methods

Chairs: Arthur M. Felix
Hoffmann-La Roche Inc.
Nutley, New Jersey, U.S.A.

and

Stephen B.H. Kent
The Scripps Research Institute
La Jolla, California, U.S.A.

Fig. 1. Scientists: Emil Fischer (top left); Max Bergmann (top right); Leonidas Zervas (bottom left) and Joseph Fruton (bottom right).

Remarks on peptide chemistry

R.B. Merrifield

The Rockefeller University, New York, NY 19921, U.S.A.

A count of the attendance at this meeting and a look at the range and quality of papers to be presented tells me that peptide chemistry and biology remains an active, and important field.

I have been asked to comment on the growth and development of peptide research as I have observed it over the years and to point out the events and achievements that I consider to be most important. My theme is that we all owe a debt of gratitude to those who have gone before us and have laid the foundations of our field. We should not forget the struggles the pioneers went through to build a viable field.

One cannot begin such a talk without going to the very beginning, when Theodor Curtius made the first amide bond between two α amino acids and Emil Fischer (Figure 1) made the first free dipeptide. Fischer coined the name **peptide** and went on to establish peptide chemistry as a true field of research with a great potential. He saw it as a subdiscipline of chemistry, but could imagine it as encompassing a much broader area of interest. I have been much impressed by his published remarks of 1907, which I take from a translation by Greenstein and Winitz. Fischer wrote the following:

"Whereas cautious professional colleagues fear that a rational study of this class of compounds [proteins], because of their complicated structure and their highly inconvenient physical characteristics, would still uncover insurmountable difficulties, other optimistically endowed observers, among which I will count myself, are inclined to the view that an attempt should at least be made to besiege this virgin fortress with all the expedients of the present; because only through this hazardous affair can the limitations of the ability of our methods be ascertained."

Fischer was, of course, far ahead of his time. No structure of a protein was known and, in fact, there was a school of thought stating that they were not even true compounds. It was many years before Sumner crystallized urease and Sanger deduced the sequence of insulin and the idea that proteins each had a unique structure was accepted. So Fischer could not hope to synthesize a real protein. But he could begin by making large polypeptides that might simulate their general properties. The important point was that he took a very optimistic point of view and he had a vision that the chemist could eventually achieve such a goal and would learn something important while doing so. Basically, Fischer was right and 60 years later a protein was synthesized. Now they can be made so cleanly that crystals can be obtained and X-ray structures determined.

Before that could be done, much groundwork had to be laid. As we all recognize, the first major breakthrough was by Fischer's student Max Bergmann (Figure 1) and his colleague Leonidas Zervas (Figure 1) when they finally devised a reversible amino protecting group - the carbobenzoxy, or benzyloxycarbonyl group. That was done in Dresden in 1932. A short time later Bergmann came to the United States and was given a position at the Rockefeller Institute in New York City. Zervas soon followed, and Bergmann began to assemble a most remarkable group of young men who succeeded in making this the most outstanding group of peptide chemists in this hemisphere. He first added Joseph Fruton (Figure 1), then Klaus Hofmann (Figure 2), William Stein, Stanford Moore, Emil Smith, Mark Stahman, Hans Frankel-Conrat, Paul Zamecnik and Sidney Fox.

Their ability to synthesize peptides allowed them to do the pioneering studies on specificity of proteolytic enzymes. This was an application for which peptide synthesis was uniquely suited, and it was the forerunner of the vast amount of structure-function studies that have followed. But we should not forget how difficult this work was. Nearly all starting materials and reagents had to be made in house, and the chemistry was still very limited.

I came to the Institute in 1949 after Bergmann had died and his group had disbanded, but I have come to know all of the members of the group except Bergmann himself. One of the things I am very proud of is the fact that I have occupied his old office and laboratories for the past 27 years - and that has been a source of great satisfaction and inspiration.

Following the Bergmann period we began to see the introduction of new synthetic chemistry; mainly new protecting groups and new methods of activation and coupling. In addition, concerns over optical purity of starting materials and racemization during the coupling reaction came under investigation, together with ways to assess and eliminate this side reaction. Now-a-days we hear little about this subject and tend to take it for granted that it is not a real problem - without any thought about the years of work that went into its study. Recall the Young test, the Anderson test, and later the Kemp isotope dilation test or the carboxyanhydride diastereomer test and Benoiton's work on hydantoins. The life saver, of course, was the finding that urethane protection prevents racemization during activation and coupling of amino acids.

The keys to much of the great advances in the 1950's were the discoveries of activation of amino acids by phenyl and thiophenyl esters by Wieland (Figure 2), cyanomethyl esters by Schwyzer (Figure 2), nitrophenyl esters by Bodanszky (Figure 2) and the chloro- and fluorophenyl esters by Kovacs, Kisfaludy and others. These were supplemented by further contributions from Wieland's laboratory: the mixed anhydrides between the amino acid and aliphatic carboxylic acids, carbonic acid derivatives, or inorganic acids, and the symmetric anhydrides, crystalline or prepared in situ, which became important activation methods. These were followed by the introduction of carbodiimides into peptide chemistry by Sheehan.

These and many other new reagents opened the way to an avalanche of new synthetic chemistry on biologically active peptides. DuVigneaud (Figure 3) and his group prepared the nonapeptides oxytocin and vasopressin, the first synthetic hormones. This was soon followed by Hofmann's α-MSH and S-peptide, Li's (Figure 3) β-MSH and Schwyzer's ACTH (1-39) and so many others, among whom can be mentioned George Kenner, Helmut Zahn, Erich Wünsch, Bernd Gutte and Ralph Hirschmann. Each of these pioneers spawned a school of peptide chemists and many of you can no doubt trace your beginnings to one of them.

Fig. 2. Scientists: Klaus Hofmann (top left); Theordor Wieland (top right); Robert Schwyzer (bottom left) and Miklos Bodanszky (bottom right).

Fig. 3. Scientists: **Vincient** *du Vigneaud (top left);* Choh Hao **Li (top** *right); George Kenner (bottom left) and Miklos Bodanszyky and Helmut Zahn (bottom right).*

In my view the next major improvement in peptide synthesis was the use of physical-organic chemical principles by Carpino to propose the tert-butyloxycarbonyl group as an acid-labile nitrogen protecting group that should be applicable to peptide chemistry. Very soon, Anderson and McGregor, and Albertson and MacKay demonstrated that this was the case and that the standard workhorse Cbz group could be replaced by a group that was removeable under mild non-destructive acidic conditions, without the need for catalytic hydrogenolysis or sodium in liquid ammonia reduction. This, of course, was followed by other very acid-labile groups that have their own special advantages.

In 1958 an historic event occurred - the 1st European Peptide symposium was held. Josef Rudinger (Figure 4), with the encouragement of F. Sörm (Figure 4), organized it in Prague, and invited about a dozen leaders of the new discipline to meet and discuss peptide chemistry. Among them were Wieland, Wünsch (Figure 4), Rothe, Goldschmidt, Brenner, Liberek, Taschner, Medzihradszky, Schemyakin, Russell and Young (Figure 4) and all the members of the Prague group.

It was just at this period (1959-1962) that I began to develop the idea of solid phase peptide synthesis. It was a new principle for the preparation of peptides, and a range of other oligomeric organic compounds. It depended on the assembly of a peptide chain while it was covalently anchored at one end to an insoluble support, and this provided several advantages that have made it a useful technique. John Stewart and I (Figure 5) built a machine for that purpose, which further simplified and accelerated the process. It, of course, was eventually replaced by more sophisticated commercial instruments.

It is important to point out that most of the chemical reactions that were used for the solid phase method were taken from methods already worked out for classical synthesis by the people I have been discussing. This is a good example of how current work is usually dependent on work that has gone before.

By 1968 peptide chemistry had expanded in America, and a group of us believed it was time for a symposium here, especially because it was not possible for most of us to attend the series of continuing European meetings. Boris Weinstein and Saul Lande (Figure 6) took the initiative and organized the 1st American Peptide Symposium at Yale University, with about 80 in attendance. It was a success, and attendance has grown to over 1400 for the current meeting.

The Japanese and Chinese have also organized their peptide symposia under the leadership of Sakakibara (Figure 6), Yajima, Shiba and Yanaihara (Figure 6) and Wang, Du, Kung and many others (Figure 7).

Shortly after our first symposium a new postdoc agreed to come to my laboratory - on the condition that he could represent his country in the 1972 winter Olympics in Sapporo, Japan. Some of you may recognize him in this photo, but others may need a more formal picture to help identify Bob Hodges, the chairman of this 13th American Peptide Symposium (Figure 7).

In the time remaining I would like to describe briefly what I consider to be some of the major developments in peptide chemistry during the past few years and the major applications of synthesis to physical or biological questions. Nearly all of the key people are in attendance this week.

Protecting groups are a key component in synthesis and they have been improved in several ways. The Boc group is still standard, but in many cases it has been replaced by more acid labile N^{α}-protecting groups such as Bpoc, MeOZ, Nps, etc., or by thiol cleavable groups like Dts. Carpino's base-labile Fmoc group has

Fig. 4. Scientists: Josef Rudinger (top left); F. Sörm (top right); Erich Wünsch (bottom left) and Geoffrey Young (bottom right).

Fig. 5. Scientists: Bruce Merrifield and John Stewart (top left and right, respectively); Bernd Gutte and Ralph Hirschmann (bottom left and right, respectively).

Fig. 6. Scientists: Boris Weinstein and Saul Lande (top left and right, respectively);
Noboru Yanaihara and Shumpei Sakakibara (bottom left and right, respectively).

Fig. 7. Scientists of The Shanghai Peptide Group. In the front row are Libby Merrifield, Drs. Kung and Du from left to right respectively, top panel; Bob Hodges, left and right of bottom panel.

now become the most popular temporary N^{α}-blocking group and has added much flexibility to synthetic design. For C^{α}-protection or for anchoring to resin supports a large number of acid-cleavable groups have been devised, including benzhydryl amines, which produce peptide amides. Cleavage conditions cover the range from anhydrous HF to acetic acid. One of the most promising new anchoring handle utilizes the allyl function, introduced by Kunz, which is removed by Pd(O) under very mild and selective conditions. Multidetachable resins and orthogonal combinations of protecting groups have provided much flexibility to synthetic design, leading from a single synthesis to; free peptides, protected peptides or protected peptides containing a C-terminal handle for reattachment to resin supports for further synthesis.

For coupling reactions the introduction of N-hydroxysuccinimide and N-hydroxybenztriazole (HOBt) as additives to accelerate the condensation and minimize racemization was an important step forward and much developmental work has come in the last few years based on these components. Castro made the big breakthrough in 1975 with his BOP reagent, a phosphonium salt derived from HOBt. It rapidly forms the Fmoc-aminoacyl-OBt ester, which is aminolysed by the amino component to give the peptide bond. The reagent has been widely used, with much enthusiasm. Many variations of this phosphonium class have now been studied. The newest of this general group of reagents incorporate uronium salts and are considered by some to be even more reactive. The amino acid fluorides and urethane protected carboxyanhydrides also contribute to the armamentarium of new coupling reagents.

Fischer's original goal of synthesizing a protein continues as an objective of many at this meeting. The target may be a natural protein or one designed de novo. The approaches can be by stepwise or by segment synthesis and with or without a solid support.

The newest trends are for mild methods of segment ligation that can be conducted in mixed solvents or in water alone. Blake found this to be facilitated by silver ion-catalysed condensation of a thioacid and the amine component, and Kent used thioacids to condense with a bromoacyl peptide. The very extensive work of Dan Kemp has led to an elegant and promising thiol capture method. The carboxyl and amine components are thus brought close together for an intramolecular reaction at high effective concentrations, in the molar rather than millimolar range. Perhaps the most promising of the new methods is an aldehyde capture technique of Tam. A 3,3,0 bicyclo transition state structure brings the nitrogen and carbonyl derivatives very close together and allows an efficient O→N acyl transfer to give the final peptide bond. The technique occurs in aqueous solution and requires no protecting groups. All of these methods have utilized solid phase synthesis and all have been demonstrated on large peptides or small proteins.

Other very important areas that we will hear about at this symposium include synthetic vaccines, de novo design of active proteins, protein-nucleic acid hybrids and peptide libraries. I believe we can look forward to an exciting week.

Synthesis and disulfide structure determination of ω-agatoxin IVA

H. Nishio, K.Y. Kumagaye, Y-N. Chen, T. Kimura and S. Sakakibara

Peptide Institute Inc., Protein Research Foundation, Minoh-shi, Osaka 562, Japan

Introduction

Many types of calcium channels exist in mammalian brains, but high-affinity blockers are available only for L- and N-types. The venom of the funnel web spider *Agelenopsis aperta* contains various toxins that have been shown to inhibit calcium channels. Among them is the recently isolated ω-agatoxin IVA, a new channel blocker that inhibits both calcium entry into rat brain synaptosomes and P-type calcium channels in rat Purkinje neurons [1]. This peptidyl toxin has also been shown to inhibit glutamate release from rat synaptosomes, suggesting that glutamate release is coupled to P-type calcium channels [2]. The primary structure of ω-agatoxin IVA was determined to be a 48-amino acid peptide containing eight cysteine residues.

Lys-Lys-Lys-Cys-Ile-Ala-Lys-Asp-Tyr-Gly-Arg-Cys-Lys-Trp-Gly-Gly-Thr-Pro-Cys-Cys-Arg-Gly-Arg-Gly-Cys-Ile-Cys-Ser-Ile-Met-Gly-Thr-Asn-Cys-Glu-Cys-Lys-Pro-Arg-Leu-Ile-Met-Glu-Gly-Leu-Gly-Leu-Ala

In the present study, we report the synthesis of ω-agatoxin IVA and the assignment of four intramolecularly linked disulfide bonds by a combination of Edman degradation and mass spectra measurement of enzymatic digests.

Results and Discussion

The synthesis was carried out by the solution procedure applying our maximum protection strategy [3]. The fully protected peptide was assembled from six segments as shown in Figure 1. All of the side-chain functional groups were protected by benzyl-type protecting groups except for Cys and Trp residues which were protected by Acm and For groups, respectively. Every segment was synthesized in the form of Boc-peptide-OPac except for the C-terminal segment which was protected by benzyl ester instead of Pac ester. Removal of Pac ester was carried out by treatment with zinc powder in AcOH. However, due to a solubility problem, in the cases of segment I, IV and V, a mixture of dichloromethane and trifluoroethanol (TFE) in a ratio of 3:1 (v/v) was used as the solvent [4]. The peptide dissolved easily in this solvent system, and removal of the Pac ester by zinc powder proceeded smoothly at 35°C within 1 hr after diluting with AcOH. Each segment thus obtained was coupled from the C-terminal segment using water-soluble carbodiimide (WSCI) in the presence of

HOBt or HOOBt. The final two coupling reactions were carried out by the WSCI/HOOBt method using a mixture of CHCl₃ and TFE (3:1, v/v) as solvents [4].

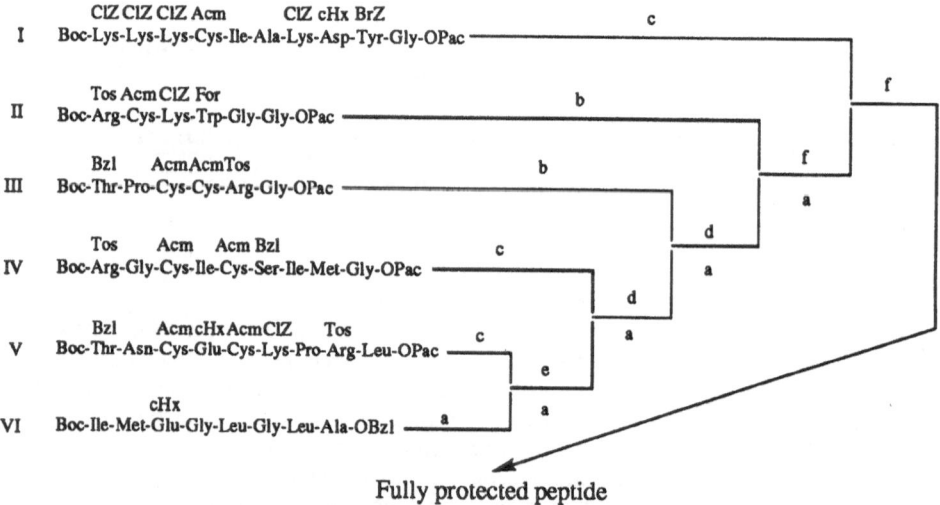

Fig. 1. Synthesis of the protected peptide. a: TFA, b: Zn/AcOH, c: Zn/DCM-TFE-AcOH, d: WSCI/HOBt, e: WSCI/HOOBt, f: WSCI/HOOBt in CHCl₃-TFE (3:1).

The fully protected peptide thus obtained was treated with trifluoroacetic acid to remove the amino terminal Boc group and then treated with HF in the presence of anisole (9:1) at -5°C for 1 hr to remove all of the protecting groups except the Acm and For groups. After evaporating HF and anisole, the peptide was treated again with HF in the presence of butanedithiol (HF:butanedithiol = 7:3) at -5°C for 30 min to remove the remaining For group [5]. The crude product thus obtained was purified by CM-cellulose chromatography followed by RP-HPLC. The purified hexa-Acm peptide was treated with Hg(OAc)₂ (1.1 eq for 1 Acm group) in 5% AcOH for 2 hr to remove all of the Acm groups. Hg ions were removed completely by adding β-mercaptoethanol followed by gel filtration on Sephadex G-25 and RP-HPLC, successively. During the cleavage reaction of the Acm groups by Hg(OAc)₂, we found side products in which the Trp residue had been alkylated by mercaptoethanol when a large amount of Hg(OAc)₂ (5 eq for 1 Acm group) was used. Details of this side reaction will be reported elsewhere. To fold and oxidize the peptide, the octa-sulfhydryl peptide was stirred in 0.2 M NH₄OAc (pH 7.8) at a peptide concentration of 1×10^{-5} M in the presence of reduced and oxidized glutathione (peptide:GSH:GSSG = 1:100:10) at 4°C for 1 day. The principal product was isolated to homogeneity by RP-HPLC followed by ion-exchange HPLC. The homogeneity of the final product was confirmed by RP- and IEX-HPLC as well as by capillary zone electrophoresis (CZE) (Figure 2). The results of amino acid analysis after acid hydrolysis agreed well with the expected values. The molecular weight of the product, [M+H] = 5202.7, obtained by plasma desorption mass spectrometry, agreed with the calculated value (5203.3). The inhibitory activity by the final product of the high-threshold current in rat Purkinje neuron was measured and found to have the same activity as that reported for the natural product (1).

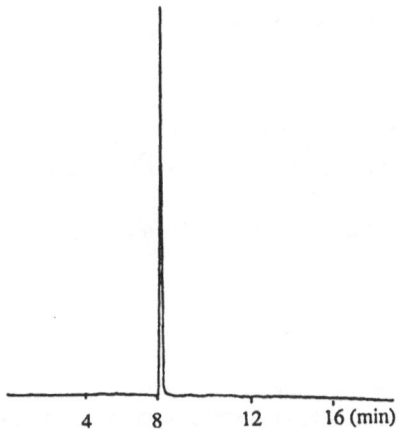

Fig. 2. CZE of the final product. Column: Uncoated Fused Silica (50 μm x 72 cm); 30°C; 22 kv Buffer: 20 mM Citrate (pH 2.5); Absorbance: 200 nm.

In order to determine the disulfide structure of synthetic ω-agatoxin IVA, the product was treated successively with thermolysin, V8 protease and proline specific endopeptidase. Peaks containing cystine peptides obtained in each enzymatic hydrolysis were isolated, and their structures were detemined by amino acid analysis, mass spectrometry measurement and gas-phase protein sequence analysis. From these analytical data, we determined the disulfide structure as shown below.

KKKCIAKDYGRCKWGGTPCCRGRGCICSIMGTNCECKPRLIMEGLGLA
      4   12   1920   25 27   34 36

Acknowledgements

The authors are grateful to Professor Tomoyuki Takahashi of Kyoto University for measuring the biological activity of synthetic ω-agatoxin IVA.

References

1. Mintz, I.M., Venema, V.J., Swiderek, K.M., Lee, T.D., Bean, B.P. and Adams, M.E., Nature, 355 (1992) 827.
2. Turner, T.J., Adams, M.E. and Dunlap, K., Science, 258 (1992) 310.
3. Kimura, T., Takai, M., Masui, Y., Morikawa, T. and Sakakibara, S., Biopolymers, 20 (1981) 1823.
4. Kuroda, H., Chen, Y-N., Kimura, T. and Sakakibara, S., Int. J. Peptide Protein Res., 40 (1992) 294.
5. Nishio, H., Kumagaye, S. Kuroda, H., Chino, N., Emura, J., Kimura, T. and Sakakibara, S., Peptide Research, 5 (1992) 227.

Synthesis of beta amyloid protein [1-40] scope and limitations of convergent solid phase synthesis

J.-P. Durieux, F. Dick, M. Schwaller, G. Haas, U. Wixmerten, S. Mundwiler and R. Nyfeler

Bachem Feinchemikalien AG, CH-4416 Bubendorf, Switzerland

Introduction

Convergent solid phase synthesis [1] has been defined as the approach in which a peptide fragment is synthesized on a solid support, cleaved fully protected from the support and, after purification and characterization, is coupled to another fragment anchored to a polymeric support. The use of this strategy for the preparation of a long peptide chain should give a product of higher quality as compared to stepwise synthesis. One would indeed avoid the accumulation of deletion and truncated sequences which are the origin of the "broad foot" very often observed in the HPLC profiles of crude products prepared by a stepwise approach.

Beta amyloid protein (1-40) **I** is the main component of the extracellular deposit found in the cerebral cortex of patients suffering from Alzheimer's disease [2].

DAEFRHDSGYEVHHQKLVFFAEDVGSNKGAIIGLMVGGVV   **I**

A convergent solid phase synthesis of the elongated beta amyloid (1-42) has been described [3]. We have chosen to compare different approaches in order to design the most efficient way to prepare large quantities of high quality beta amyloid (1-40).

Results and Discussion

A first stepwise synthesis (Fmoc/tBu strategy, TBTU/DIPEA coupling) gave unsatisfactory results. Nearly 20 out of the 40 amino acids required a double coupling; in addition, the crude product cleaved from the resin contained a high proportion of peptide oxidized at the Met residue. This led us to undertake the preparation of the title compound following the above mentioned principles of convergent solid phase synthesis. Fragments **II** to **IV** were chosen with C-terminal Gly to exclude any risk of racemization during coupling; the C-terminal fragment **V** was assembled on p-alkoxybenzylalcohol resin. For the preparation of the fragments **II** to **IV** we used the super acid labile Sasrin resin which allows the cleavage of fully protected fragments by 1% TFA in methylene chloride [4].

Met was introduced as its sulfoxide to avoid any complication due to oxidation. The results of these syntheses are given in Table 1.

Table 1 *Synthesis of the protected fragments[a]*

Fmoc-(1-9)-OH **II** Yield: 80% - Purity (HPLC): 85%
Fmoc-Asp(OtBu)-Ala-Glu(OtBu)-Phe-Arg(Pmc)-His(Trt)-Asp(OtBu)-Ser(tBu)-Gly-OH

Fmoc-(10-25)-OH **III** Yield: 88% - Purity (HPLC): 75%
Fmoc-Tyr(tBu)-Glu(OtBu)-Val-His(Trt)-His(Trt)-Gln(Mtt)-Lys(Boc)-Leu-Val-Phe-Phe-Ala-Glu(OtBu)-Asp(OtBu)-Val-Gly-OH

Fmoc-(26-33)-OH **IV** Yield: 90% - Purity (HPLC): 90%
Fmoc-Ser(tBu)-Asn(Mtt)-Lys(Boc)-Gly-Ala-Ile-Ile-Gly-OH

Fmoc-(34-40)-Wang resin **V** Purity (HPLC): 90%
Fmoc-Leu-Met(O)-Val-Gly-Gly-Val-Val-Wang resin

---

[a]Purity has been determined after TFA cleavage. FAB-MS was used to confirm the structure of the fragments.

The coupling of fragment **IV** to peptide resin **V**, after the cleavage of the Fmoc group, was mediated by TBTU/DIPEA and afforded H-[26-40](O)-Wang **VI** after piperidine treatment. The poor solubility of fragment **III** was at the origin of the problems encountered in its coupling to **VI**. Different attempts were made using various combinations of solvents, temperature and prolonged reaction time; it was nearly impossible to bring the reaction to completion and the best incorporation (still less than 50%) used a 5 - 7 fold excess of fragment **III**. Finally Fmoc-[10-40] (O)-Wang was assembled by stepwise elongation (TBTU/DIPEA as coupling agent) starting from **VI**.

The coupling of **II** to H-[10-40](O)-Wang **VII** was mediated by TBTU/DIPEA and gave the expected Fmoc-[1-40](O)-Wang **VIII**. When this coupling was performed with a Wang resin having a high level of substitution (0.6 mmoles/g), it was necessary to make a double coupling of **II** and one had to use a 2-3 fold excess of the fragment whereas a low substitution of the resin (0.15 mmoles/g) allowed complete acylation with a single coupling and the use of only a 1.2 - 1.5 fold excess of the acylating component.

Finally, the Fmoc group was removed from **VIII** and the peptide was cleaved from the resin by TFA:H$_2$O:EDT (85:5:10). The peptide was then purified by preparative HPLC (RP C18-triethylammonium phosphate buffer at pH 7.5 with an acetonitrile gradient; desalting with 0.1% TFA-acetonitrile gradient). The fractions were pooled to provide a product of a purity of approximately 80-85%. Met(O) was finally reduced to Met using the conditions described by Tam [5] (TFMSA:TFA:DMS (1:6:3)) and the obtained product, namely the title compound **I**, was further purified to provide a product of a purity greater than 95% in 2 different HPLC solvent systems. Electrospray mass spectrometry and amino acid analysis confirmed the structure. However, the yield of the last two steps (reduction-purification) was much too low (10-15%) to make the synthesis efficient and we therefore repeated it with a 0.2 mmoles/g Wang resin, and Met instead of Met(O), without encountering special problems. This pathway produced a crude **I** showing an acceptable level of purity (45-50%), and the purification gave a product with a high degree of purity (greater than 97%) in amounts of several hundred milligrams.

We have also explored the possibility offered by the combination synthesis [6], i.e. coupling in solution of the fragments prepared by solid phase. The C-terminal fragment H-(34-40)-OtBu was synthesized in solution and the peptide was assembled as in the convergent approach. It must be noted that the coupling of the fragment Fmoc-(10-25)-OH proceeded under normal conditions, although the reaction medium was more a gel than a solution, whereas coupling of the same fragment onto resin proved unsuccessful. The results of the combination synthesis were comparable to the ones obtained from the convergent synthesis both in terms of yield and quality. However, due to the poor solubility of the intermediates, we could not make use of the advantages of the combination approach, i.e. to purify the intermediates.

Convergent solid phase synthesis is a valuable approach for the preparation of long and complex peptides. We could demonstrate that in the case of fragment coupling onto solid support, the level of substitution of the resin is an important parameter influencing the quality of the reaction product, and also the economy of the synthesis. We have also shown that coupling of identical fragments may proceed differently in solution and on solid support. For the preparation of the title compound, a combination of stepwise synthesis and of fragment couplings provided the best approach. Similar results were obtained from a more laborious combination approach.

References

1. Albericio, F., Lloyd-Williams, P., Gaini, M., Jou, G., Eritja, R., Giralt, E., In Smith, J.A. and Rivier J.E. (Eds.), Peptides: Chemistry and Biology (Proceedings of the 12th American Peptide Symposium), Escom, Leiden, The Netherlands, 1992, p.607.
2. Mori, H., Takio, K., Ogawara, M., Selkoe, D., J., Biol. Chem., 267 (1992) 17082.
3. Hendrix, J.C., Halverson, K.J., Lansbury, Jr., P.T., J. Am. Chem. Soc., 114 (1992) 7930.
4. Mergler, M., Nyfeler, R., Tanner, R., Gosteli, J., Grogg, P., Tetrahedron Lett., 29 (1988) 4004.
5. Tam, J.P., In Shiba T. and Sakakibara S. (Eds.), Peptide Chemistry 1987, Protein Research Foundation, Osaka, Japan, 1988, p.199.
6. Nyfeler, R., Wixmerten, U., Seidel, C., Mergler, M., In Smith J.A. and Rivier, J.E. (Eds.), Peptides: Chemistry and Biology (Proceedings of the 12th American Peptide Symposium), Escom, Leiden, The Netherlands, 1992, p.661.

Oxazolidines and thiazolidines as secondary structure disrupting, solubilizing protection techniques for serine and cysteine

A. Nefzi, T. Haack, B. Dhanapal, R. Flögel,
J. Kapron and M. Mutter

*Institute of Organic Chemistry, University of Lausanne,
Rue de la Barre 2, CH-1005 Lausanne, Switzerland.*

Introduction

The finding, that conformational effects of the growing peptide chain play a crucial role in peptide synthesis [1] has stimulated extensive research on this subject over the past years. The formation of secondary structures, in particular β-sheet conformations, has been shown to reduce the solvation of the peptide chain resulting in a decrease of the reaction kinetics and in the solubility [2]. In order to overcome some of these limitations, the reversible modification of the backbone for disrupting ordered conformations has recently been suggested [3]. To this end, we have introduced oxazolidines (Ox) and thiazolidines (Th) as protection techniques for Ser, Thr and Cys [3a]. Due to their structural similarity to proline, these readily accessible derivatives ('pseudo-prolines') may act as solubilizing, secondary structure disrupting building blocks in peptide synthesis (Figure 1).

Results and Discussion

Pseudo-prolines [Ser(Ox)]; [Cys(Th)] are prepared via side chain cyclization of Ser or Cys according to established procedures [4]. For their use in SPPS, the corresponding dipeptide derivatives are prepared as depicted in Fig 1. The chemical stability of the ring systems strongly depends on the substituent in position C^2 of the heterocycle, allowing for a variable combination with commonly applied protection schemes (Table 1). For example, Ser(Ox) and Cys(Th) are stable towards base and nucleophiles, but can be cleaved by 10-100% TFA, depending on the substituent at C^2. Conformational energy calculations and NOE studies of model dipeptides of the type Ac-Xaa-Ser(Ox)-NHCH$_3$ reveal the preference of a *cis* amide bond in the presence of bulky substituents at C^2.

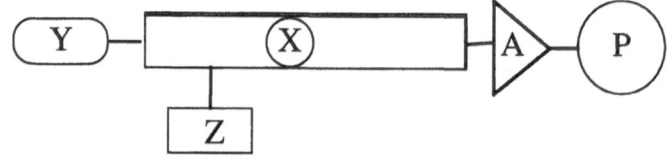

Fig. 1. Synthesis and incorporation of a Cys(Th) into a peptide backbone and effect upon the conformational properties of the host peptide before and after ring opening.

When incorporated in switch-peptides [5] which show medium-dependent transitions from α-helical (pH 11) to β-sheet (pH 4) conformations, the peptide (X= Cys(Th)) no longer adopts an ordered conformation as shown by the CD spectra (Figure 2). Most notably, the 'enforcement' of a flexible random-coil conformation in pseudo-proline containing peptides is paralleled by a significant increase in solvation, solubility and reaction kinetics. The protected Cys(Th) containing peptide could be condensed to a topological template [6] resulting in a template-assembled synthetic protein (TASP). This opens new ways for the construction of large peptides and in protein de novo design [6].

Table 1 *Synthetic strategies for pseudo - proline containing peptides*

Strategy	Ⓧ	A	Y	Z	cleaved product
I	Ser(Ox) Cys(Th)	Sasrin Rink	Fmoc	Boc/ᵗBu	protected Ⓧ - peptide
II	Ser(Ox^R) Cys(Th)	Hycram	Fmoc	Alloc/Allyl	deprotected Ⓧ-peptide
III	Ser(Ox^R) Cys(Th)	Wang Pam	Boc	Cbo/Bzl	deprotected peptide

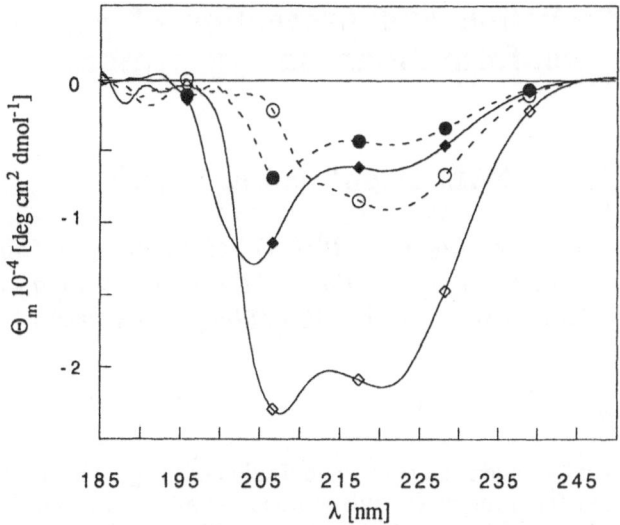

Fig. 2. CD spectra of switch - peptide (5) Ac-EAALKAXLELAAKLAA- NH_2 at pH 4 (X = Cys (Th) –•–; X = Cys – o –) and pH 11 (X = Cys(Th) – ♦ –; X = Cys –◊–).

Acknowledgements

This work was supported by the Swiss National Science Foundation.

References

1.   Mutter, M., Pillai, V.N.R., Anzinger, H., Bayer, E and Toniolo, C., In Peptides 1980, Brunfeldt, K., (Ed.), Scriptor, Copenhagen, 1981, p.660.
2.   Anzinger, H. and Mutter, M., Biopolymers, 19 (1980) 173.
3.   (a) Haack, T. and Mutter, M., Tetrahedron Lett., 33 (1992) 1589; (b) Johnson, T., Quibell, M., Owen, D. and Sheppard, R.C., J. Chem. Soc., Chem. Comm. (1993) 369.
4.   (a) Lewis, N.J., Inloes, R. L., Hes, J., Matthews, R.H. and Milo, G., J. Med. Chem., 21 (1978) 1070; (b) Wolfe, S., Militello, G., Ferrari, C., Hasan, S.K. and Lee, S.L., Tetrahedron Lett., 41 (1979) 3913.
5.   Mutter, M., Gassman, R. and Buttkus, U., Angew. Chem. Int. Ed. Engl., 30 (1991) 1514.
6.   Mutter, M. and Vuilleumier, S., Angew. Chem. Int. Ed. Engl., 28 (1989) 535.

Prediction and prevention of peptide conformations during synthesis

W.E. Rapp[1] and E. Bayer[2]

[1]Rapp Polymere, Eugenstr. 38/1, D 7400 Tübingen, Germany,
[2]University of Tübingen, Institute of Organic Chemistry Auf der
Morgenstelle 18, D 7400 Tübingen, Germany

Introduction

Peptide conformation and poor solubilization of the growing peptide chain are the main reason for incomplete acylation and deprotection reactions. There are several methods available to predict such difficulties [1, 2], however, unexpected sequence dependent difficulties still appear during synthesis. Several approaches are described to avoid peptide conformations which may hinder synthesis, such as chaotropic salts, solvent mixtures or new amino acid derivatives which modify the peptide backbone [3, 4]. Based on monosized grafted microspheres we have developed a method where the peptide is screened by a preview synthesis within a rather short time to recognize such problematic regions. To overcome the conformation dependent kinetic hindrance we investigated the influence of solvent systems, activation methods and structure destroying reagents. It turned out that surfactants in combination with solvent mixtures are the most potent systems and in general useful for preventing peptide conformations during synthesis.

Results and Discussion

Beside resin parameters like polarity, particle size and solvation, mass transport and diffusion are of prime importance in all polymer supported reactions [5]. The driving force for mass transport is diffusion and diffusion is dependent on path length. Normally solid supports show a particle size distribution and therefore for each bead exists an individual diffusion time whereas the largest one controls the overall reaction rate. To overcome these problems we have developed monosized 25 μm beads of our polystyrene polyethyleneglycol graft copolymer [6, 7].

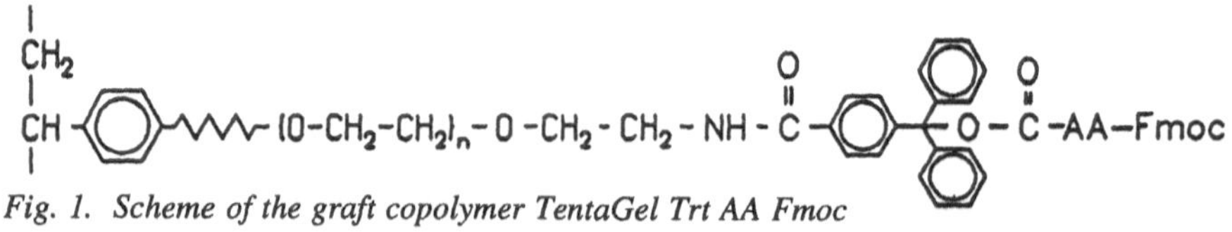

Fig. 1. Scheme of the graft copolymer TentaGel Trt AA Fmoc

To a low crosslinked PS-matrix, PEG is grafted by an anionic graft copolymerization. The reactive sites which are located at the end of the PEG-spacers behave kinetically like in solution. The monosized nature of the beads divides the total reaction space (represented by all beads) into exactly the same small reaction compartments. The monosized nature and the uniform architecture of the beads allow optimization of the reaction conditions to a great extent because of identical reaction conditions for each individual bead. In contrast to 90 μm particles where the release of the Fmoc group and the wash out was finished within 2.5 min, the time could be reduced to 80 sec by using 25 μm particles at 45°C. This time corresponds to the kinetic rate of the Fmoc deprotection. There is no longer any influence of the mass transport to the overall reaction rate.

For peptide synthesis the following conditions were used: ABI 431 A synthesizer (200 mg resin, 180 sec acylation, 2x2.5 min Fmoc deprotection) and MilliGen 9050/synthesizer (200 mg resin, 180 sec acylation, 120 sec. Fmoc deprotection, 60 sec washout, 45°C, total cycle time: 7 min 37 sec). The starting Fmoc amino acid is attached to the tentacle polymer via an new trityl linkage (Figure 1). Cleavage of the peptide could be performed by either 50% acetic acid in CH_2Cl_2 to get the completely protected peptide or with TFA to get the free peptide. Applying said conditions the C-terminal sequence KAKAAPKKAPKSPAKAKAVKPKAA of the histone H1b (174 - 198) was synthesized within 3h 10 min in 90% purity using Fmoc protected N - carboxy-anhydride amino acids (Fmoc NCA). This indicates the success of this method.

We have then screened the complete C-terminus of the histone H1c (176-220, **KVKTPQP**KKAAKSPAKAKAPKPKAAKPKSG**KPKV**TKAKKAAPKKK). The preview synthesis of this 45 mer was completed within 7h 15 min. We recognized breakdowns for the coupling efficiency and Fmoc deprotection in the regions 209-205 and 176-183. This phenomenon is caused by bad solvation or any kind of peptide conformations on the resin. This finding was very unexpected because up to now the C-terminal sequences of the histones are known to be random. For further investigations and synthesis optimizations the H1c sequence 176-220 was cut into two overlapping sequences:

176-200: **KVKTPQP**KKAAKSPAKAKAPKPKAA
196-220: KPKAAKPKSG**KPKV**TKAKKAAPKKK

The most dramatic decrease in the Fmoc deprotection yield was recognized for K 206: 54% by TBTU activation and 57% with Fmoc NCA's (Figure 2). IR investigations of the resin bound peptide (196-220) show shifts for the amide I and II bands which indicates β-sheet and helical conformations. Therefore it is very understandable that a change in the activation method results in a negligible change in yield because the reactive sites are not accessible. To overcome this problem we changed the solvent system from DMF to a mixture of DMF/NMP/CH_2Cl_2 containing 1% Triton, an nonionic surfactant. The yield increased to 69% and with prolonged coupling in the region 206-209 to 86%. Due to the surfactant properties of Triton the peptide conformations on the resin are destroyed. Bayer *et al.* reported that ethylene carbonate achieves in some cases complete coupling [8]. We created a "magic mixture": DMF/NMP/CH_2Cl_2, 1% Triton and 1,5-2,5 molar ethylene carbonate. The total coupling and deprotection yield for K 206 rose to 93% and we obtained the crude peptide with 90% purity. This effect of the "magic mixture" was confirmed by the very successful syntheses of several other difficult sequences.

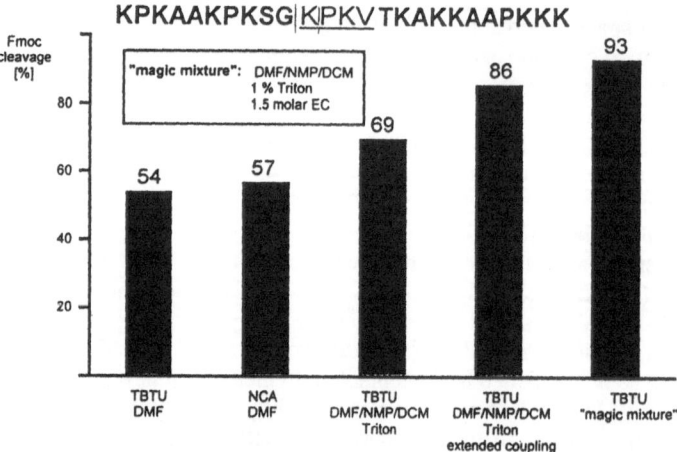

Fig. 2. Influence of solvents, activation method and conformation avoiding reagents to the coupling yield.

CD investigations of both, the free peptide and the completely protected peptide confirmed the presence of suspected peptide conformations: each peptide contained helical structures. These results encouraged us to synthesize the complete H1c sequence 176-220 again applying such optimized conditions. Figure 3 shows the HPLC and ion spray ms of the crude completely protected peptide. Due to the still incomplete coupling of K 206 we find the deletion peptide -Lys(Boc) at 1711.5 [M + 4H]$^{4+}$ and a second failure sequence where the dipeptide Lys(Boc) Ala at 1693 [M + 4H^{4+} is deleted.

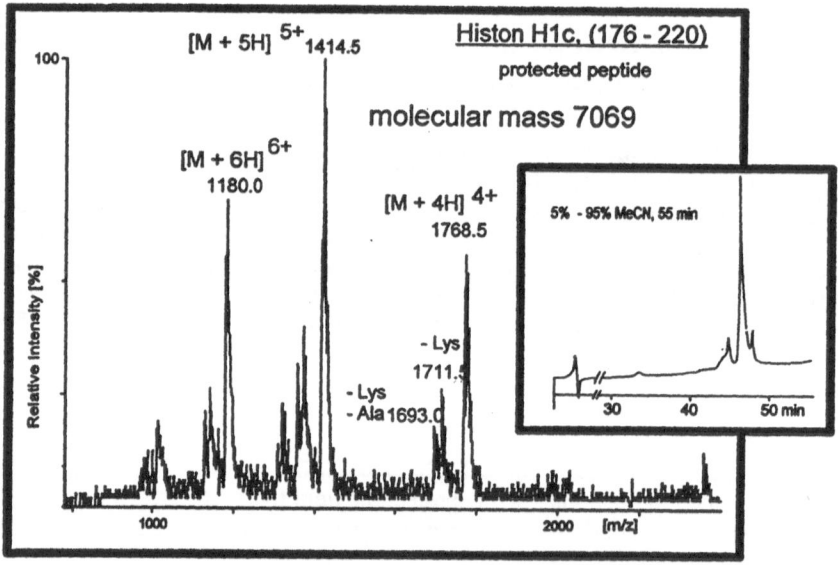

Fig. 3. HPLC and Ion Spray MS of the crude completely protected H1c (176-220).

Preview synthesis on grafted microspheres allow very effective screening of a peptide synthesis within shortest time. Peptide conformations on the resin are recognized and confirmed by solid state IR of the resin attached peptide. CD investigations of the free peptides (completely protected and unprotected) show helical structures and confirm the IR data. The use of a "magic mixture" could avoid peptide conformations during synthesis and even long peptides with difficult regions in the sequence can be synthesized very successfully.

References

1.   Sarin, V.K., Kent, S.B.H., Tam, J.P. and Merrifield, R.B., Anal. Bio. Chem., 237 (1981) 927.
2.   Pillai, V.N. and Mutter, M., Ex. Acc. Chem. Res., 14 (1981) 122.
3.   Eckert, H. and Seidel, C., Angew. Chem., 98 (1986) 168.
4.   Bartl, R., Kloppel, K.D. and Frank, R., in Smith, J.A. and Rivier, J.E., (Eds.), Peptides, Chemistry and Biology (Proceedings of the Twelfth American Peptide Symposium), Escom, Leiden, The Netherlands, 1992, p.505.
5.   Rapp, W. and Bayer, E., In R. Epton (Ed.), Innovations and Perspectives in Solid Phase Synthesis, Peptides, Polypeptides and Oligonucleotides, Intercept Ltd., Andover, 1992, p.259.
6.   Bayer, E. and Rapp, W., German patent DOS 3714258 (1988), US Patent No. 4,908,405, March 13, 1990.
7.   Monosized polystyrene polyethylene glycol graft copolymers are the improvement of TentaGel resins, purchased by Rapp Polymere, Tübingen.
8.   Bayer, E. and Goldammer, C., In Smith, J.A. and Rivier, J.E., (Eds.), Peptides, Chemistry and Biology (Proceedings of the Twelfth American Peptide Symposium), Escom, Leiden, The Netherlands, 1992, p.589.

Guanidine hydrochloride assists intramolecular disulfide bond formation in cysteine peptides

J.G. Adamson and G.A. Lajoie

*Department of Chemistry, University of Waterloo,
Waterloo, Ontario, N2L 3G1, Canada*

Introduction

Intramolecular disulfide bonds are an attractive means of imposing conformational constraints in structure/activity studies of peptides. Air oxidation in aqueous solution (pH ~ 8) is a straightforward protocol, but suffers in that steric factors and/or hydrophobic aggregation may hinder thiol interaction, increase reaction times and promote intermolecular disulfides. Also, peptides with pI values close to that of the oxidation medium may precipitate before oxidation occurs. Oxidation in guanidine hydrochloride (GuHCl) solution was found to overcome these problems and to accelerate intramolecular disulfide formation in all peptides examined. Although urea is frequently used in protein denaturation studies, GuHCl is a more potent denaturant at equal concentrations [1]. As well, aqueous urea generates cyanate ion which is capable of irreversibly modifying thiol and amino groups [2].

Results and Discussion

Thiol forms of the peptides (Table 1) were initially soluble in AcOH or DMSO, but quickly precipitated upon dilution with water. Other peptides with pI values of 8.3 precipitated once the aqueous AcOH or DMSO solution was adjusted to pH 8.5. HPLC of the redissolved precipitates revealed a number of covalent products. Presumably, peptide aggregation resulted in intermolecular disulfide formation giving polymeric products. Yield and purity were dramatically improved by air oxidation in a GuHCl solution. Typically, peptide was dissolved to a final concentration of 1 mM in 2 to 8 M GuHCl, 0.2 M TrisHCl, pH 8.5 aqueous buffer and stirred vigorously with air. Progress was monitored by RP HPLC. Large volume solutions were desalted by dialysis or by ultrafiltration prior to lyophilization and preparative scale HPLC purification. Smaller oxidation solutions (<10 mL) were not desalted prior to HPLC purification. Peptides remained in solution throughout the oxidation which proceeded more quickly than in other conditions (Table 2), and very little dimerization or polymerization was observed.

DMSO has been advocated as a mild oxidant for the formation of disulfide bonds over a wide pH range [3, 4]. Two test peptides (Table 2) were oxidized approximately twice as quickly in buffered GuHCl solution as in 20 or 80% DMSO solutions. Oxidation rates were not proportional to DMSO concentration, suggesting that its ability to catalyze disulfide bond formation is not solely governed by its oxidizing property but also by its solvating or denaturing properties [4, 5].

Table 1 *Intramolecular disulfide formation by air oxidation in 8M GuHCl*

Peptide	Calc'd pI	Oxidation yield (%)	Purified yield (%)
Ac-CASPVSIEVKFNKPRSFC	8.3	95	79
Ac-CAIPVSIEVKFNKPRSFC	8.3	97	67
Ac-CGSPVTLDLRYNRTRSFC	8.3	95	80
Ac-CAIPVSIEVKFNKPFSFC	6.2	90	45
Ac-CAIPVSIPPEVKFNKPFSFC	6.2	97	64

Table 2 *Half-times of intramolecular disulfide formation in various media[a]*

Air Oxidation Media	$t_{1/2}$ (hr.) Peptide A	$t_{1/2}$ (hr.) Peptide B
0.2M TrisHCl, pH 8.5	9.3	insoluble
8M GuHCl, 0.2M TrisHCl, pH 8.5	3.0	0.8
20% DMSO	5.2	5.3
80% DMSO	20	1.7

[a] Peptide A = Ac-CAAIPVSIC, peptide B = Ac-CASPVSIEVKFNKPRSFC

GuHCl was shown to assist in the mild oxidation of cysteine-containing peptides by overcoming limitations in solubility and conformational flexibility to cleanly yield intramolecular disulfide bonds. This protocol is especially efficient for hydrophobic or neutrally charged peptides.

References

1. Lapanje, S., Physicochemical Aspects of Protein Denaturation, J. Wiley and Sons, New York, NY, U.S.A., 1978, p.8.
2. (a) Stark, G.R., Stein, W.H. and Moore, S., J. Biol. Chem., 235 (1960) 3177. (b) Volkin, D.B. and Klibanov, A.M., in Creighton, T.E. (Ed.) Protein Function: A Practical Approach, IRL Press, New York, NY, U.S.A., 1989, p.8.
3. (a) Fujii, N., Otaka, A., Okamachi, A., Watanabe, T., Arai, H., Tamamura, H., Funakoshi, S. and Yajima, H. in Jung, G. and Beyer, E. (Eds.) Peptides 1988, Walter de Gruyter, New York, NY, U.S.A., 1989, p.58. (b) Otaka, A., Koide, T., Shide, A. and Fujii, N. Tetrahedron Lett., 32 (1991) 1223.
4. Tam, J.P., Wu, C-R., Liu, W. and Zhang, J-W., In Smith, J.A. and Rivier, J. (Eds.) Proc. 12th Am. Peptide Symp., Escom, Leiden, 1992, 499.
5. Wallace, T.J. and Mahon, J.J., J. Am. Chem. Soc., 86 (1964) 4099.

Condensation of minimally protected peptide segments catalyzed by the Glu/Asp-specific endopeptidase: Convergent solid-phase/enzymatic synthesis of GRF

J. Bongers

Hoffmann-La Roche Inc., Nutley, NJ 07110, U.S.A.

Introduction

Proteolytic enzymes can catalyze racemization-free condensations of peptide segments with unprotected side chains [1]. Larger polypeptides invariably contain multiple proteolytic sites and thus would appear to be practically inaccessible by this approach. This obstacle can, however, be kinetically circumvented. By use of certain acyl-esters, enzymatic synthesis is accelerated many-fold over competing proteolyses [2, 3]. This principle is illustrated by the recent semisynthesis of a superpotent analog of the human growth hormone-releasing factor [desNH$_2$Tyr^1,D-Ala^2,Ala^{15}]-GRF(1-29)-NH$_2$ by a V8 protease catalyzed 3+26 segment condensation [3]. The use of a "highly activated" benzyl ester of the acyl-donor segment permitted enzymatic synthesis of Asp^3-Ala^4 to proceed 100-fold faster than competing proteolyses at Asp^3-Ala^4 and Asp^{25}-Ile^{26}.

This report describes a solid-phase/enzymatic synthesis of the parent hormone GRF(1-44)-NH$_2$ employing a 33+11 segment condensation catalyzed by the Glu/Asp-specific endopeptidase (GSE) from *B. licheniformis* [4]. This synthesis involves competing proteolyses at Glu^{33}-Ser^{34}, Glu^{37}-Arg^{38}, Asp^3-Ala^4, and Asp^{25}-Ile^{26}, and is thus a simple test of the general synthetic scheme (Figure 1).

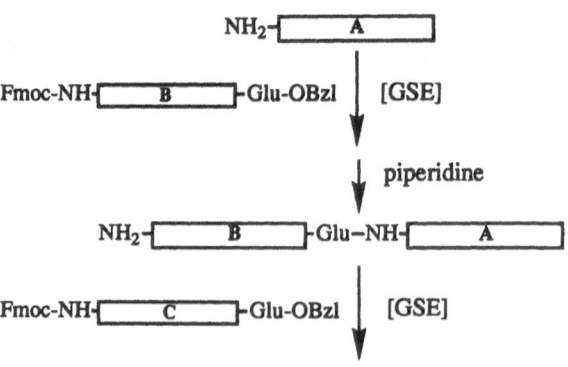

Fig. 1.   *Scheme for successive GSE catalyzed condensations of peptide segments.*

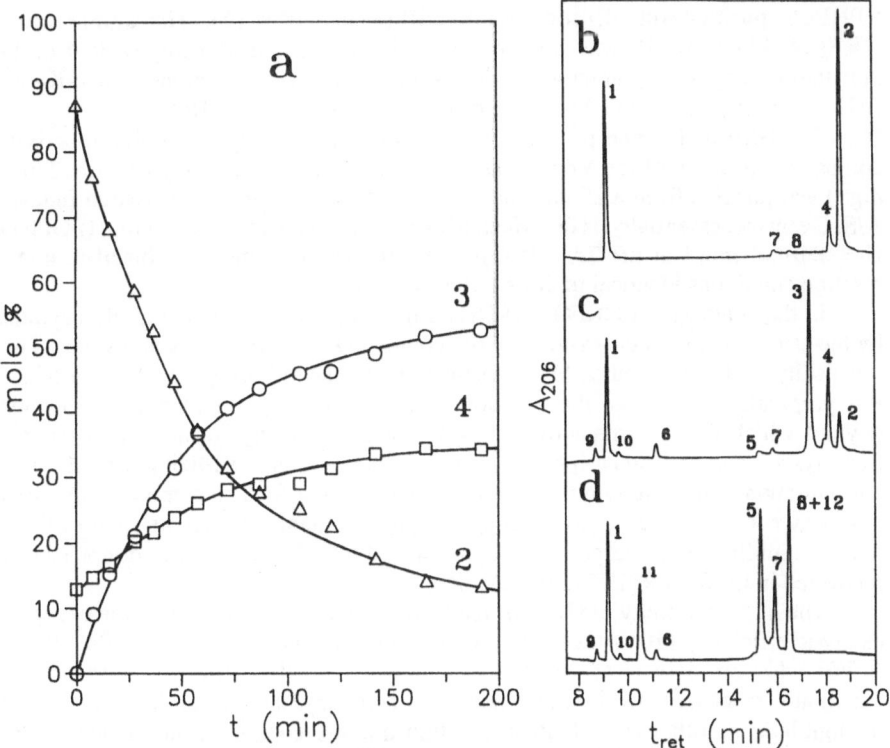

Fig. 2.   *GSE catalyzed condensation of GRF(34-44)-NH₂ (1) (317 mg mL⁻¹, 185 mM) and Fmoc-GRF(1-33)-OBzl (2) (382 mg mL⁻¹, 67.6 mM) to produce Fmoc-GRF(1-44)-NH₂ (3) in 60% v/v DMF, pH 8.8, 22°C (37.5 μg mL⁻¹ enzyme): (a) time-course of the aminolysis/hydrolysis of 2; RPHPLC of the reaction mixture (b) immediately before adding enzyme, (c) at 192 min and (d) after deprotection. Other labels are Fmoc-GRF(1-33)-OH (4), GRF(1-44)-NH₂ (5), BzlOH (6), GRF(1-33)-OH (7), GRF(1-33)-OBzl (8), GRF(38-44)-NH₂ (9), GRF(34-37)-(34-44)-NH₂ (10), piperidine (11), and Fmoc-piperidine (12). RPHPLC: Waters μBondapak C₁₈, 0 to 54% CH₃CN/H₂O (0.025% TFA) in 15 min. Reaction: Dissolved 1·4TFA (11.4 mg, 6.7 μmol) and 2·5TFA (13.8 mg, 2.9 μmol) in 21.7 μL DMF, added 3.3 μL of 3.0 M N(CH₂CH₂OH)₃, titrated to pH 8.8 with 5 M NaOH (5.0 μL), and added 1.0 μL of 1.35 mg mL⁻¹ GSE. Aliquots (4.0 μL total) were removed for RPHPLC monitoring and the reaction was stopped (192 min) by adding acetic acid and lyophilizing. The residue was deprotected in 30% v/v piperidine/DMF at 22°C for 60 min and lyophilized. After RPHPLC purification, 5·9TFA (5.3 mg, 33%) was characterized by RPHPLC, tryptic mapping, AAA, FABMS (MH⁺ calc 5040.7; found 5039.5), and showed full potency in an in vitro rat pituitary cell bioassay.*

Results and Discussion

Suitable N-protected polypeptide benzyl esters can be readily prepared by existing Fmoc/tBu solid-phase methods. Fmoc-Glu(OH)-OBzl was attached to 4-alkoxybenzyl alcohol (Wang) resin and automated synthesis performed on an ABI 431A (FastMoc™ protocol [5]). Final deprotection/cleavage with TFA and

RPHPLC purification afforded Fmoc-GRF(1-33)-OBzl (**2**). The amino-donor GRF(34-44)-NH$_2$ (**1**) was also prepared by this route starting with 4-(2,4'-dimethoxyphenyl-Fmoc-aminomethyl)phenoxyacetyl-benzhydrylamine (Fmoc-Knorr-BHA) resin. Peptides **1** and **2** were characterized by AAA and FABMS.

The GSE is a serine protease with substrate specificity and specific activity comparable to those of the V8 protease. The GSE used in this study (71 mg, 270 U mg^{-1}) was purified from a 25 mL portion of the detergent Alcalase^{TM} (Novo Industri A/S, Denmark) essentially as described [4] and assayed against Boc-Glu(OH)-OPh (1 U = 1 µmol min^{-1} at pH 7.8, 20°C). V8 protease (Sigma) was inactive under reaction conditions identical to those used for GSE (Figure 2).

In the synthesis of GRF(1-44)-NH$_2$ (Figure 2), the enzyme is rapidly acylated by the ester **2** and then deacylated by the competing nucleophiles, **1** and water. This kinetically controlled method [1] relies on a rapid nonequilibrium "burst" of product. For preparative purposes, the reaction is halted at the point of maximum yield beyond which the enzyme will merely degrade the newly formed bond ("back-hydrolysis") and any other specific cleavage sites present. Utilizing a high initial concentration of the amino-donor **1** is important in order for **1** to compete effectively with water for the binding cavity of the enzyme. Organic cosolvents such as DMF aid in solubilizing the larger polypeptide reactants. Excess amino-donor can be recovered in the final RPHPLC purification.

The high specificity of GSE for the benzyl ester **2** is evidenced by the excellent catalytic efficiency (10^4 wt/wt ratio **2**/GSE) and minimal proteolysis at Glu^{37}-Arg^{38} (<5%). No proteolysis at the Asp-X sites was detected. An 18% loss of **2** to enzymatic ester-hydrolysis was incurred. Nonenzymatic ester hydrolysis was negligible in an otherwise identical reaction mixture containing no enzyme. The 12% ester hydrolysis of **2** prior to adding GSE and slight loss of Fmoc (<1%) probably occurred while adjusting the pH with NaOH.

References

1. Schellenberger, V. and Jakubke, H.-D., Angew. Chem. Int. Ed. Engl., 30 (1991) 1437.
2. Schellenberger, V., Görner, A., Könnecke, A. and Jakubke, H.-D., Peptide Res., 4 (1991) 265.
3. Bongers, J., Lambros, T., Liu, W., Ahmad, M., Campbell, R.M., Felix, A.M. and Heimer, E.P., J. Med. Chem., 35 (1992) 3924.
4. Svendsen, I. and Breddam, K., Eur. J. Biochem., 204 (1992) 165.
5. Applied Biosystems Inc. Bulletin 33 (1990).

Synthesis of an anti-freeze protein

S. Brandtner, J. Schleucher and C. Griesinger

*Institut für Organische Chemie, Universität Frankfurt,
Marie-Curie-Str. 11, D-60439 Frankfurt/Main, Germany*

Introduction

Animals exposed to temperatures below 0°C have developed strategies to avoid lethal freezing of their body fluids. Among these is the stabilization of body fluids in a supercooled state by inhibiting the growth of ice crystals. Antarctic fish living in ice-ladden environments achieve this by the synthesis of antifreeze proteins (AFPs) [1]. These proteins lower the *freezing* point of their body fluids, whilst the melting point is not affected [2]. However, the exact mechanism of this interaction has not been established to date [2, 3, 4].

Of the groups of AFPs known to occur in fish [1], a group of proteins of about 60 amino acids is of special interest, since their sequence cannot be correlated with the models proposed so far for the interaction of AFPs with water. Recently, the structure of one such protein has been determined by NMR [5].

Proteins of this group have so far only been accessible by isolation or in low yield by heterologous expression [6]. Here, we report the first synthesis of one of the proteins by solid-phase peptide synthesis. The strategy of fragment condensation allows the easy modification of the sequences to be synthesized and is therefore well suited for the study of structure-activity relationships.

Results and Discussion

The complete AFP is synthesized using 8 protected fragments (Figure 1). The fully protected peptides were synthesized on an automatic peptide synthesizer (ACT 200) from ACT on the 2-chlorotrityl resin [7] using the Fmoc/t-Butyl strategy [8]. The coupling with TBTU/DIEA and the Fmoc-cleavage were monitored by TLC. The yields of the crude fragments were around 95%, only fragment AFP34-30 had to be purified by RP-HPLC using ACN/H$_2$O gradient with 0.1%TFA. The intactness of the protection groups was proven by 1D and 2D ^{1}H-NMR spectroscopy. The correct masses were established by FAB-MS.

To obtain the complete protein, we coupled the protected fragments sequentially on the same resin. The coupling, which was done in DMSO with HOBt/DIC, and the Fmoc-cleavage were monitored by the Ninhydrin-reaction and by TLC.

After deprotection the protein was characterized by ESI-MS.

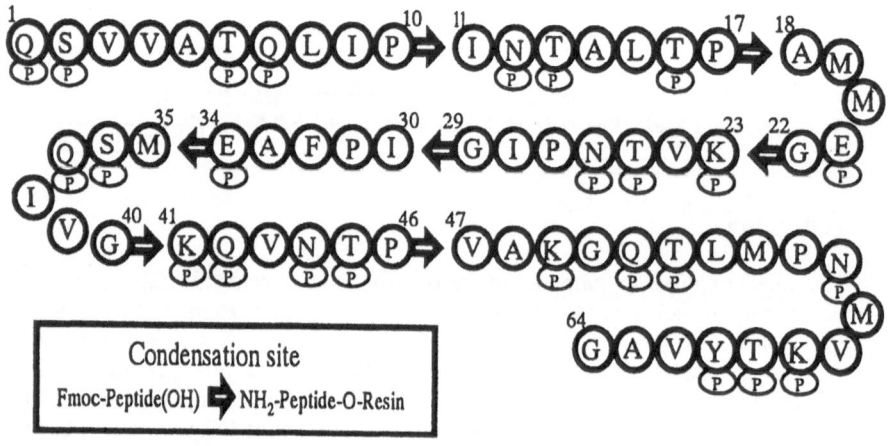

Fig. 1. *The sequence of the AFP-Type III HPLC-6 is shown. We used tBu as protection group at S;T;E;Y and Trt as protection group at Q;N and K they are marked as* **P**.

Acknowledgements

This work is supported by DFG under grant Gr 1211/4-1. We thank Dr. Zechel, BASF, Ludwigshafen, and Dr. J. W. Metzger, Universität Tübingen, for the recording of mass spectra.

References

1. Davies, P.L. and Hew, C.L., FASEB J., 4 (1990) 2460.
2. Knight, C.A., Cheng, C.C. and DeVries, A.L., Biophys. J., 59 (1991) 409.
3   Hew, C.L. and Yang, D.S.C., Eur. J. Biochem., 203 (1992) 33.
4. Chou, K.-C., J. Mol. Biol., 223 (1992) 509.
5. Sönnichsen, F.D., Sykes, B.D., Chao, H. and Davies, P.L., Science, 259 (1993) 1154.
6. Li, X., Trinh, K.Y. and Hew, C.L., Protein Eng., 4 (1991) 995.
7. Barlos, K., Chatzi, O., Gatos, D. and Stavropoulos, G., Int. J. Peptide Protein Res., 37 (1991) 5132.
8. Fields, G.B. and Noble, R.L., Int. J. Peptide Protein Res., 35 (1990) 161.

A convenient, straightforward solid phase synthesis of γ-endorphin on polystyrene matrix, grafted with polyoxyethylene spacer arms, using Fmoc amino acid N-carboxy anhydrides

G. Grübler[1], S. Stoeva[2], H. Echner[2] and W. Voelter[2]

[1]Eppendorf/Biotronik, Edmund-Seng-Straße 4, D-6457 Maintal, Germany, and [2]Abteilung für Physikalische Biochemie des Physiologisch-chemischen Instituts der Universität Tübingen, Hoppe-Seyler-Straße 4, D-7400 Tübingen, Germany

Introduction

Urethane-protected α-amino acid N-carboxy anhydrides (UNCAs) have been demonstrated to form peptide bonds rapidly in high yields, with practically no racemization [1-3]. These unique properties prompted us, to apply Fmoc UNCAs for the solid phase synthesis of the heptadecapeptide γ-endorphin (H-Tyr-Gly-Gly-Phe-Met-Thr-Ser-Glu-Lys-Ser-Gln-Thr-Pro-Leu-Val-Thr-Leu-OH).

Results and Discussion

To demonstrate the flexibility of the fully automatic batch-synthesizer ECOSYN P (Eppendorf/Biotronik, Maintal, Germany) concerning its applicability for any kind of peptide synthesis strategy (e.g. resins, coupling methods, protecting groups), the heptadecapeptide γ-endorphin was taken as an example, using TentaGel[R] [4] as the solid support and Fmoc amino acid-N-carboxy-anhydrides as coupling agents. For the synthesis, the following Fmoc-protected NCAs were applied: Fmoc-Leu-NCA, Fmoc-Thr(tBu)-NCA, Fmoc-Val-NCA, Fmoc-Gln(Trityl)-NCA, Fmoc-Ser(tBu)-NCA, Fmoc-Lys(Boc)-NCA, Fmoc-Glu(OtBu)-NCA, Fmoc-Met-NCA, Fmoc-Phe-NCA, Fmoc-Gly-NCA and Fmoc-Tyr(OtBu)-NCA [5]. All UNCAs were reacted in a one-coupling step in DMF in threefold excess. After completion of the synthesis, the peptide resin bond and the side chain protecting groups were cleaved, using a mixture of trifluoroacetic acid/anisole/ethanedithiol (15:2:0.5 ml, 2 hr, room temperature). The crude synthetic product was purified on a Nucleosil $7C_{18}$ column (Macherey-Nagel, Düren, Germany), using a semipreparative HPLC system (Eppendorf/Biotronik, Maintal, Germany; Figure 1). Furthermore, the purity was controlled by capillary electrophoresis [6] (Figure 2), applying a Bio-Rad HP 100 apparatus (München, Germany). The HPLC-purified peptide showed the expected molecular mass (m/z:1859), determined by matrix-assisted laser desorption mass spectrometry (Lasermat, Finnigan, MAT Bremen, Germany, Figure 3; [7]). Prior to amino acid analysis, the peptide was hydrolyzed with 6N HCl (24 h, 110°C). The AAs were separated by cation exchange chromatography and determined after post-column reaction with ninhydrin (Amino Acid Analyser LC 3000, Eppendorf/Biotronik, Maintal, Germany), and the amino acid ratios are in good agreement with the calculated values.

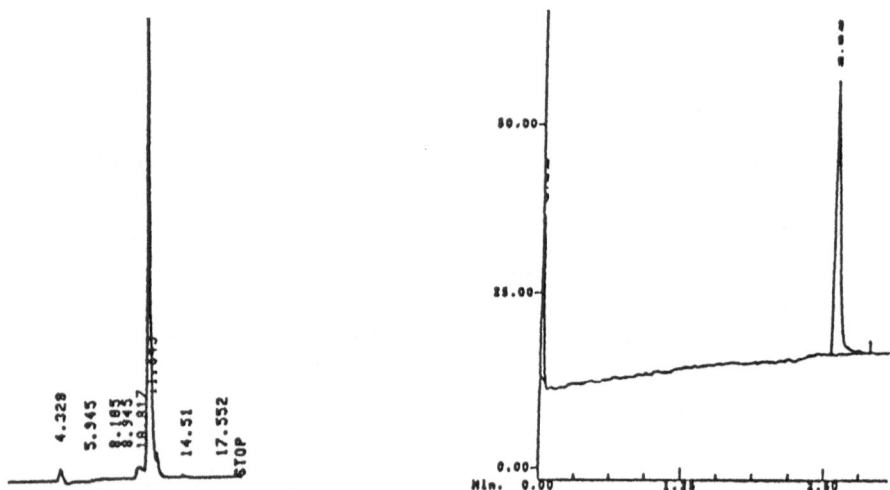

Fig. 1. HPLC of the synthetic γ-endorphin. Column: Nucleosil $7C_{18}$ (250x100mm); solvent A: 0.05% TFA/H₂0, B: 60% acetonitrile /0.1% TFA/H₂0; gradient elution: 20-60% in 30 min; flow rate: 2,5 ml/min; detection: 220 nm; injection: 25 µg (left side).

Fig. 2. HPCE of the synthetic γ-endorphin. Capillary: coated, 25µmx20cm; voltage: 8KV, positive polarity; buffer: 0.1 M phosphate, pH 2.5; detection: UV, 220 nm, 0.02 AUFS (right side).

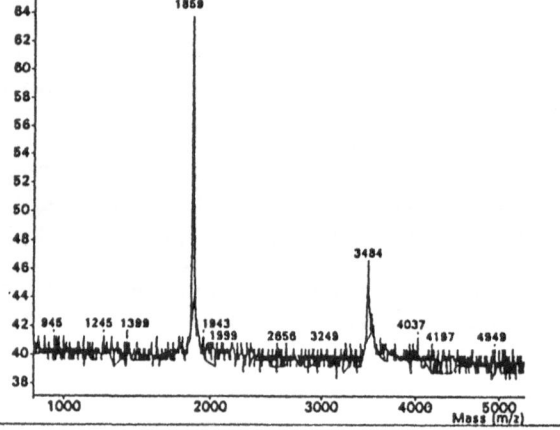

Fig. 3 Matrix-assisted laser desorption mass spectrum of the synthetic γ-endorphin. Number of pulses: 6; wavelength: 337 nm; matrix: α-cyano-hydroxycinnamic acid; internal standard: glucagon (m/z: 3484 Da).

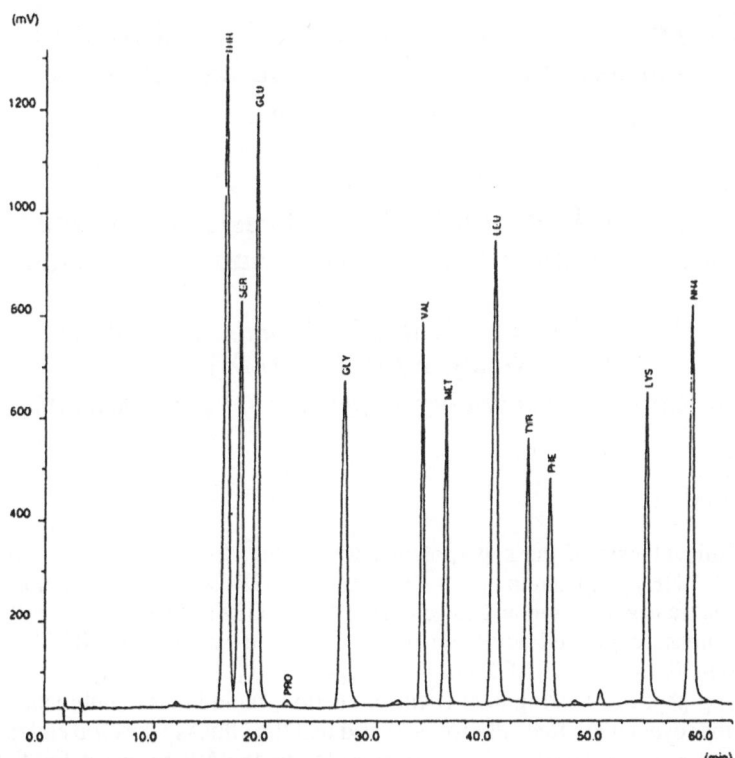

Fig. 4. Amino acid analysis hydrolyzed of γ-endorphin. Chromatographic conditions: column: 125x4mm; resin: BTC 2410; buffer: commercial from Eppendorf/Biotronik, Maintal, Germany; flow rate: 0.2ml/min; detector: VIS 570 nm; method: ninhydrin post-column reaction.

References

1. Akiyama, M., Hasegawa, M., Takeuchi, H. and Shimizu, K., Tetrahedron Lett., 28 (1979) 2599.
2. Fuller, W.D., Cohen, M.P., Shabankareh, M. and Blair, R.K., J. Am Chem. Soc., 112 (1990) 7414.
3. Kolbeck, W., Fuller, W.D., Cohen, M.P., Shabankareh, M., Naider, F.R., Loffet, A. and Goodman, M., In Brandenburg, D., Ivanov, V. and Voelter, W. (Eds.), Chemistry of Peptides and Proteins, vol. 5/6, part A, Verlag Mainz, Wissenschaftsverlag, Aachen, Germany, 1993, p.41.
4. Rapp Polymere, Eugenstraße 38/1, D-7400 Tübingen, Germany.
5. Propeptide, BP 12, F-91710, Vert-Le-Petit, France.
6. Yildiz, E., Grübler, G., Hörger, S., Zimmermann, H., Echner, H., Stoeva, S. and Voelter, W., Electrophoresis, 13 (1992) 683.
7. Karas, M. and Hillenkamp, F., Anal. Chem., 60 (1988) 2299.

Enzymatic semisynthesis: Efficient fragment condensation using a glu/asp-specific endopeptidase

J. Bongers[1], W. Liu[1], T. Lambros[1], K. Breddam[2], R.M. Campbell[1], A.M. Felix[1] and E.P. Heimer[1]

[1]Roche Research Center, Hoffmann-La Roche Inc., Nutley, NJ 07110, U.S.A.
[2]Carlsberg Laboratory, Copenhagen, Valby, Denmark

Introduction

Semisynthesis of the superpotent GRF analog, [desNH$_2$Tyr^1,D-Ala^2,Ala^{15}]-GRF(1-29)-NH$_2$ (**4**), from the precursor, [Ala^{15,29}]-GRF(4-29)-OH (**1**), employed an enzymatic transpeptidation of **1** to generate [Ala^{15}]-GRF(4-29)-NH$_2$ (**2**), followed by a V8 protease catalyzed acylation of **2** with the tripeptide, desNH$_2$Tyr-D-Ala-Asp(OH)-OR (**3**) (R = CH$_3$CH$_2$ or 4-NO$_2$C$_6$H$_4$CH$_2$) [1]. The utility of enzymes for modifying peptides lacking side-chain protection, including recombinant peptides, was further evaluated. Recently we observed that the Glu/Asp-specific endopeptidase (GSE) from *B. licheniformis* [2] is superior to the V8 protease for the above acylation. We also investigated the influence of the ester leaving-group (R) of the acyl component on the acylation rate and yield for the series, desNH$_2$Tyr-D-Ala-Asp(OH)-OR [R = CH$_3$CH$_2$-(**3a**), CH$_3$- (**3b**), ClCH$_2$CH$_2$- (**3c**), C$_6$H$_5$CH$_2$- (**3d**), 4-NO$_2$C$_6$H$_4$CH$_2$- (**3e**)].

Results and Discussion

The GSE proved to be a more stable and economical alternative to the V8 protease for the condensation of fragments **2** and **3**. As shown in Table 1, the ester leaving-group (R) of the acyl component has a major influence on both catalytic efficiency and yield. The yield is determined by the relative rates of enzymatic synthesis of Asp^3-Ala^4, which is influenced strongly by R, and proteolyses at Asp^3-Ala^4 and Asp^{25}-Ile^{26}, which are independent of R. The rate and yield of synthesis appears to increase with increasing electrophilicity of R. However, there is slight break in this trend in going from the benzyl ester **3d** to the 4-nitrobenzyl ester **3e**. GSE does not show a preference for more hydrophobic esters as has been noted for α-chymotrypsin [3]. These studies showed that the most favorable substrate was the benzyl ester **3d** which is processed by the enzyme nearly 100-fold more rapidly than the ethyl ester **3a**, the least favorable substrate (Figure 1).

Table 1 *Influence of the Ester Leaving-Group (R) on GSE Catalyzed Acylations of [Ala^{15}]-GRF(4-29)-NH$_2$ (2) with desNH$_2$Tyr-D-Ala-Asp(OH)-OR (3)*

cmpd	R	%Yield$_{max}$	t$_{max}$ (min)
3a	CH$_3$CH$_2$-	60	700
3b	CH$_3$-	70	300
3c	ClCH$_2$CH$_2$-	99	70
3d	C$_6$H$_5$CH$_2$-	99	10
3e	4-NO$_2$C$_6$H$_4$CH$_2$-	93	35

Initial conditions: $[2]_0$ = 13.5 mM (50 mg mL^{-1}), $[3]_0$ = 54 mM (45 mg mL^{-1}), [GSE] = 2.5 mg mL^{-1}, 20% v/v DMF, 150 mM triethanolamine, pH 8.2, 37°C.

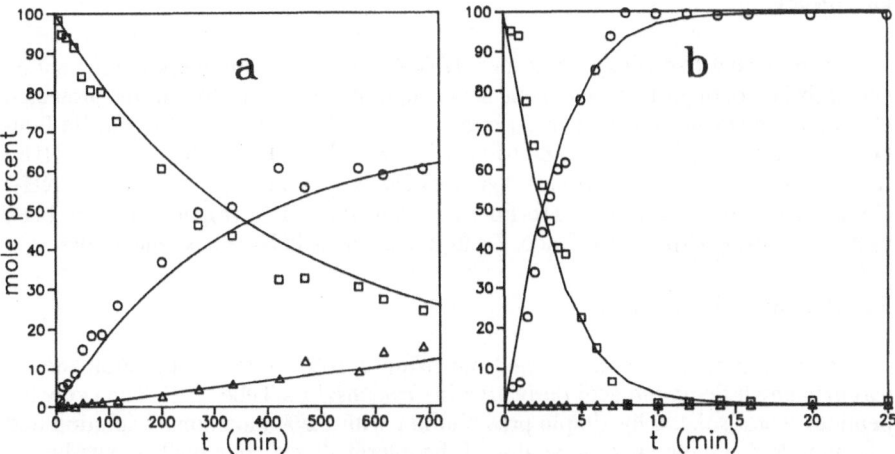

Fig. 1. *Time courses of the GSE catalyzed syntheses of **4** (O) by acylation of **2** (□) with (a) the ethyl ester **3a** and with (b) benzyl ester **3d**. Side-products (Δ) represent the combined loss of **2** and **4** to proteolysis at Asp^{25}-Ile^{26}; i.e. [side-products] = $[2]_0$ - [2] - [4]. See Table 1 for initial reaction conditions.*

References

1. Bongers, J., Lambros, T., Liu, W., Ahmad, M., Campbell, R.M., Felix, A.M. and Heimer, E.P., J. Med. Chem., 35 (1992) 3924.
2. Svendsen, I. and Breddam, K., Eur. J. Biochem., 204 (1992) 165.
3. Schellenberger, V., Görner, A., Könnecke, A. and Jakubke, H.-D., Peptide Res., 4 (1991) 265.

Evaluation of urethane N-carboxyanhydrides (UNCA's) for the synthesis of C-terminal fragments of bombesin, cholecystokinin, gastrin and neurotensin

M. Llinares[1], E. Bourdel[1], J.C. Califano[1], M. Rodriguez[1], A. Loffet[2] and J. Martinez[1]

[1]EP CNRS 51, Faculté de Pharmacie, 15 Av. C. Flahault, 34060 Montpellier and [2]Propeptide, 91710 Vert le Petit, France

Introduction

N-protected-N-carboxyanhydrides (UNCA's) were recently proposed in peptide synthesis [1]. In order to evaluate this new coupling method, analogs of the protected C-terminal fragments of neurotensin [Fmoc-Lys(Boc)-Lys(Boc)-Pro-Tyr(tBu)-Ile-Leu-OtBu] (Figure 1), bombesin [Z-D-Phe-Gln-Trp-Ala-Val-Gly-His-Leu-Leu-NH$_2$] (Figure 2), cholecystokinin [Fmoc-Tyr(tBu)-Met-Gly-Trp(For)-Met-Asp-Phe-NH$_2$] (Figure 3), and gastrin [Boc-Trp(For)-Leu-Asp-Phe-NH2] (Figure 4), have been synthesized using N-protected Fmoc, Z and Boc amino acid N-carboxyanhydrides.

Results and Discussion

Coupling reaction were carried out using a 10% excess of the urethane N-carboxyanhydride amino acid derivative in "Bodanszky's Tube" [2]. The expected peptides were isolated by simple precipitation with 0.5% bicarbonate solution and filtration, and without characterization of the intermediates. In a typical experiment, the C-terminal protected amino acid or peptide derivative with free α-amino group (4 mmol for the first coupling) is dissolved in 10 ml of DMF and treated with the respective Fmoc, Boc or Z urethane N-carboxyanhydride (Fmoc-, Boc- or Z-NCA) derivative in 10% excess for 30 minutes at room temperature; DIEA (0.5 equivalent) was added in the reaction mixture. Precipitation of the protected peptide was obtained by pouring the reaction mixture in 200 ml of 0.5% NaHCO3; the precipitate is then collected, washed with water, 0.1 N KHSO4, water and finally dried. In some cases, washing with ethyl ether was also performed. The purity of the crude peptides obtained in these syntheses were evaluated by HPLC. Boc-Trp(For)-Leu-Asp-Phe-NH$_2$ was obtained with an overall yield of 85%, Fmoc-Tyr(tBu)-Met-Gly-Trp-Met-Asp-Phe-NH$_2$ with an overall yield of 57%, Fmoc-Lys(Boc)-Lys(Boc)-Pro-Tyr(tBu)-Ile-Leu-OtBu with an overall yield of 85% and Z-D-Phe-Gln-Trp-Ala-Val-Gly-His-Leu-Leu-NH$_2$ with an overall yield of 66%. All crude peptides were obtained in a high degree of purity (85 to 97%).

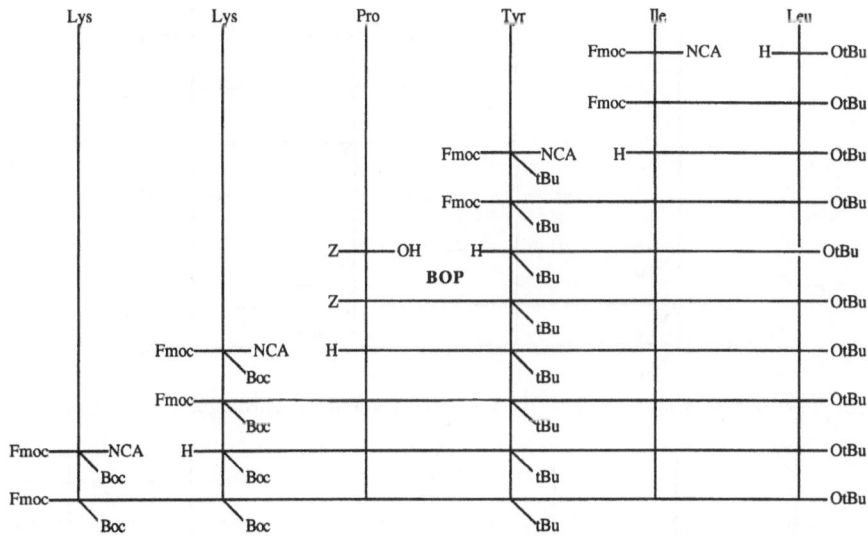

Fig. 1. Synthesis of the analog of the protected C-terminal hexapeptide of neurotensin.

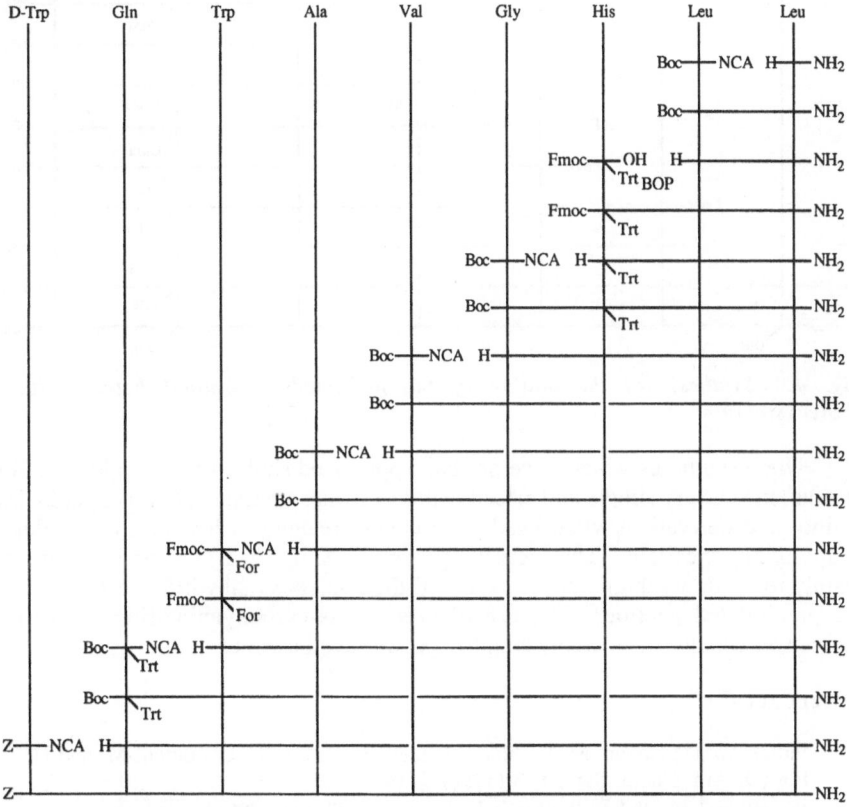

Fig. 2. Synthesis of the analog of the protected C-terminal nonapeptide of bombesin.

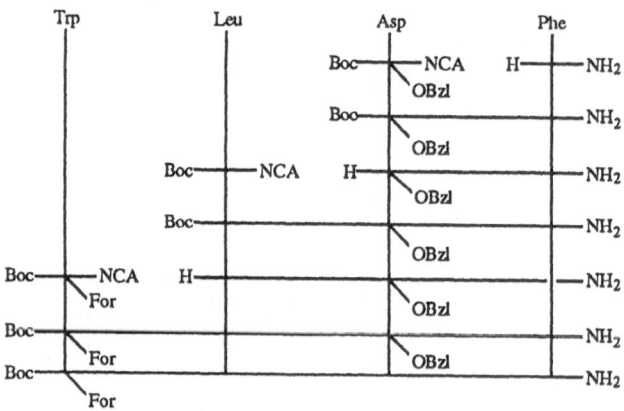

Fig. 3. Synthesis of the analog of the protected C-terminal tetrapeptide of gastrin.

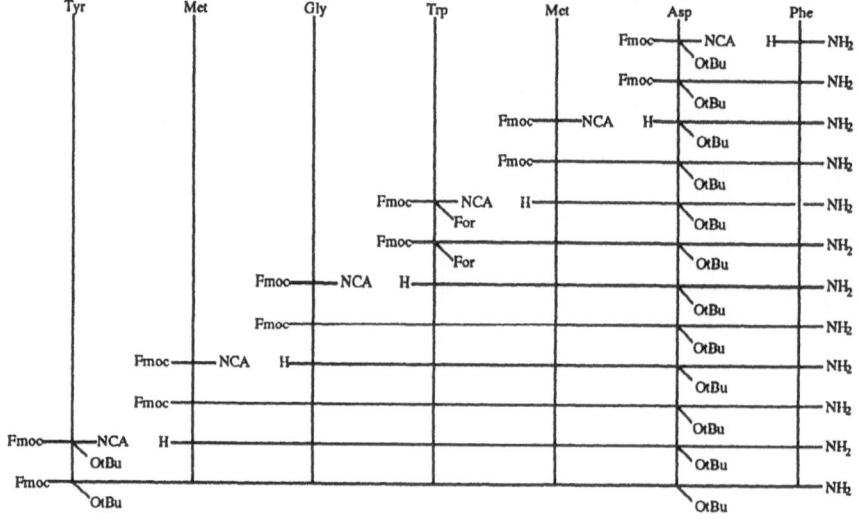

Fig. 4. Synthesis of the analog of the protected C-terminal heptapeptide of cholecystokinin.

These syntheses which have not been optimized showed that peptide coupling in solution is a very simple and effective process when urethane-N-carboxyanhydride amino acid derivatives were used. The excess reagent is low and the yields of coupling are excellent. The work up process is simplified to a great extent as it mainly relies on washing out the excess of the N-carboxyanhydride derivative from the precipitated peptide. The use of urethane-N-carboxyanhydride amino acid derivatives should considerably simplify peptide synthesis in solution.

References

1. Fuller, W. D., Cohen, M. P., Shabankarech, M., Blair, R., Goodman, M. and Naider, F.R., J. Am. Chem. Soc., 112 (1990) 7414.
2. Bodanszky, M. and Williams, N.J., J. Am. Chem. Soc., 89 (1967) 685.

Synthesis of human CCK-58-related analogs

M.T.M. Miranda[1], C. Miller[2], A.G. Craig[2], J. Dykert[2], R.A. Liddle[3] and J.E. Rivier[2]

[1]Department of Biochemistry, Institute of Chemistry, University of São Paulo, C. P. 20780, 01498, São Paulo, SP, Brazil
[2]The Salk Institute for Biological Studies, 10010 North Torrey Pines Rd., La Jolla, CA 92037, U.S.A.
[3]Duke University Medical Center, Sands Building, Research Drive, Durham, NC 27710, U.S.A.

Introduction

Several different molecular forms of cholecystokinin have been detected *in vivo* (CCK-58, CCK-39, CCK-33, CCK-8) [1, 2, 3]. The understanding of the biological function of each of these forms is essential for the elucidation of CCK's physiology in mammals. While descriptions of the synthesis of CCK-8 and analogs are numerous, that of CCK-33 was carried out by three groups only [4, 5, 6]. The total synthesis of larger entities such as that of CCK_{20-58} and CCK-58 or analogues has never been reported most likely because of synthetic difficulties. The cumulative effect of peptide length and sensitivity to acids may be the cause of undisclosed earlier failures.

We report here the synthesis, purification, characterization and bioactivity of analogues of CCK_{20-58} and CCK-58 in which $Tyr^{52}(SO_3H)$ was replaced by Phe(p-CH_2SO_3Na), methionines by norleucines (Nle) and tryptophan by 2-naphtylalanine (Nal). Those modifications were introduced in order to increase the chemical stability of those peptides during the synthetic process.

Results and Discussion

Fmoc-L-Phe(p-CH_2SO_3Na)-OH [7] was prepared in an alternative and shorter way than that described in the literature [8].

Peptide syntheses were carried out manually on a 2,4-dimethoxy-benzhydrylamine (2,4-DMBHA) resin using the Fmoc strategy and DIC or BOP as coupling reagents. Acetylation was avoided as much as possible. The chain elongation was interrupted several times for cleavage and analysis of the peptides as they were built on the resins. This constant monitoring of the synthetic process was important in optimizing the syntheses. The most effective method for the cleavage/full deprotection of the peptide-resins was reagent K (TFA/thioanisole/water/phenol/ ethanedithiol, 83:5:5:5:2.5) at room temperature. Reaction times for cleavages had to be increased with increasing chain length (from 3 to 24 hr for CCK-8 to CCK-58). The crude peptide preparations were extremely complex when analyzed by RP-HPLC, capillary zone electrophoresis (CZE) and ion-exchange chromatography (IEC). CCK-39 and CCK-58 analogues were purified by CEC (cation-exchange chromatography) followed by RP-HPLC using a 0.1% TFA system.

The purified peptides (purity greater than 90%) were characterized by RP-HPLC using highly resolutive systems different from those used during the purification, CZE, mass spectrometry (LSIMS) and amino acid analysis. The results showed that the products corresponded to the desired peptides. In the case of [Tyr^{52},Nle^{32,53,56}, Nal^{55}]-CCK-58 and [Phe(p-CH$_2$SO$_3$Na)^{52},Nle^{32,53,56},Nal^{55}]-CCK-58, size exclusion HPLC (SEC) and IEC were also used. The importance of doing these orthogonal analyses is that the longer and more difficult the synthesis, the greater the chances that the desired material be seen as homogeneous by RP-HPLC, CZE and mass spectrometry only. [Tyr^{52},Nle^{32,53,56},Nal^{55}]-CCK-58 was also submitted to Edman sequence analysis which confirmed not only its identity, but also its high degree of purity.

Besides the fact that it describes the successful total synthesis of some difficult CCK-58-related analogs, this report offers complementary strategies for the purification and characterization of peptides/proteins prepared on a solid support.

Those peptides were tested for their ability to stimulate amylase release from isolated rat pancreatic acini [9]. We found that [Phe(p-CH$_2$SO$_3$Na)^{52},Nle^{32,53,56}, Nal^{55}]-CCK$_{20-58}$ had marginal activity and [Phe(p-CH$_2$SO$_3$Na)^{52},Nle^{32,53,56}, Nal^{55}]-CCK-58 was inactive at the highest doses tested (10^{-9} M) [10]. Considering the fact that CCK-58 is considered to be only five times less potent than CCK-8 (unpublished data), we were not expecting this significant difference between the relative potencies of the standard used and our analogs. Our explanation is that, while each of the modifications was shown to increase the chemical stability of those molecules, the cumulative effects of each substitution might be incompatible with the receptor recognition or activation in rat pancreatic acini. On the other hand, we did not expect [Tyr^{52},Nle^{32,53,56},Nal^{55}]-CCK-58 to be active for the lack of a sulfate group in position 52.

Acknowledgements

Research was supported in part by NIH grants DK26741 and DK38626, the Hearst Foundation, the Department of Veterans Affairs and by FAPESP (90/2689-2 - Brazil). We thank Duane Pantoja for amino acid analysis, G.C. Jiang for suggestions during the synthesis of Phe(p-CH$_2$SO$_3$Na).

References

1.   Dockray, G.J., Gregory, R.A., Hutchison, J.B., Harns, J.I. and Runswick, M.J., Nature, 274 (1978) 711.
2.   Liddle, R.A., Goldfine, I., Rosen, M.S., Taplitz, R.A. and Villiams, J.A., J. Clin. Invest., 75 (1985) 1144.
3.   Reeve, J.R., Jr., Eysselein, V., Walsh, J.H., Ben-Avram, C.M. and Shively, J.E., J. Biol. Chem., 261 (1986) 16392.
4.   Fujii, N., Futaki, S., Morimoto, H., Ionue, K., Doi, R., Tobe, T. and Yajima, H., J. Chem. Soc., Chem. Commun., 4 (1988) 324.
5.   Penke, B. and Nierges, L., Peptide Research, 4 (1991) 289.

6.   Kurano, Y., Kimura, T. and Sakakibara S., J. Chem. Soc., Chem. Commun. (1987) 323.

7.   Gonzalez-Muniz, R., Cornille, F., Bergeron, F., Ficheux, D., Pothier, J., Durieux, C. and Roques, B.P., Int. J. Peptide Protein Res., 37, (1991) 331.

8.   Miranda, M.T.M., Liddle, R. A. and Rivier, J., J. Med. Chem. (1993) (in press).

9.   Liddle, R.A., Elashoff, J.E., Reeve, J.R., Jr., Peptides, 7 (1986) 723.

10. Miranda, M.T.M., Craig, A. G., Miller, C., Liddle, R. A. and Rivier, J., J. Protein Chem., (1993) (in press).

A comparative study on the synthesis of cysteinyl-peptides by conventional methods in solution and on solid supports

H.-J. Musiol, D. Quarzago, R. Scharf and L. Moroder

Max-Planck-Institut für Biochemie, Martinsried, Germany

Introduction

Bis-cysteinyl-octapeptides related to the active site sequences 31-38 of E. coli thioredoxin and 134-141 of thioredoxin reductase, i.e. Ac-Trp-Cys-Gly-Pro-Cys-Lys-His-Ile-NH$_2$ (1) and Ac-Ala-Cys-Ala-Thr-Cys-Asp-Gly-Phe-NH$_2$ (2), were synthesized as N-acetyl and C-amide derivatives in order to define the role of interactions in the folded state of these thiol-protein oxidoreductases in determining the redox properties of the functional domains [1].

Results and Discussion

The sequence-dependent difficulties expected for the synthesis of the two octapeptides derive from the presence of two Cys residues in combination with His and Trp (peptide 1) and with Asp-Gly (peptide 2). Besides the known facile alkylation of Trp in acidolytic deprotection steps, this side reaction seems to be significantly enhanced by the proximity of a Cys residue [2]. Aspartimide formation with concomitant α→β transpeptidation is known to occur in acidic deprotection steps, but could also result from repetitive piperidine treatments for Fmoc removal [3] in Fmoc/tBu synthetic strategies. The octapeptides were synthesized in solution and on solid supports.

Synthesis in solution: Acid-labile side chain protecting groups on tert-butanol and trityl basis were combined with Cys(StBu) and N$^\alpha$-Z and N$^\alpha$-Nps derivatives in chain-elongation steps. The protected octapeptides were obtained by fragment assembly with DCC/HOSu in good overall yields without encountering particular difficulties. Upon treatment with aqueous TFA (in presence of 2-methylindole for peptide 1) crude products of a high degree of purity were obtained as shown by their hplc (Figure 1). Subsequent reduction with (C$_4$H$_9$)$_3$P in aqueous TFE [4] and oxidation of bis-cysteinyl-peptides with azodicarboxylic acid derivatives led to the desired monomeric cyclic octapeptides as well characterized compounds in high yields [1].

Solid phase synthesis: Fmoc-amino acids were coupled in DMF to the modified Rink-linker on the resins Rapp S RAM (0.2mmol/g), Nova Syn KR (0.1mmol/g) and Nova Syn PR 500 (0.47mmol/g). Couplings were performed in high and low pressure continuous flow reactors via NCA's [5] for Fmoc-Ile, Fmoc-His(Trt), Fmoc-Lys(Boc), Fmoc-Gly, Fmoc-Phe, Fmoc-Thr(tBu) and Fmoc-Ala and via TBTU/HOBt/DIEA (1:1:2) for Fmoc-Cys(StBu)-OH (or Fmoc-Cys(Acm)-OH),

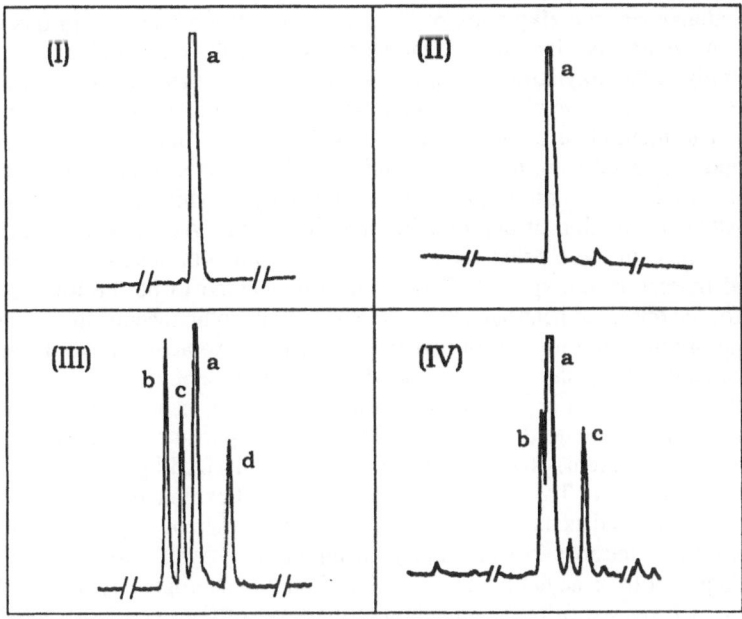

Fig. 1. HPLC of crude deprotection products as Cys(StBu) derivatives from synthesis in solution of peptide 1 (I) and 2 (II), and from synthesis on solid support of peptide 1 (III) and 2 (IV). Ion spray MS: molecular ion of 1160.8 for (a) of panel I and for (a), (b), (c) and (d) of panel III; molecular ion of 1004.6 for (a) of panel II and for (a), (b) and (c) of panel IV.

Fmoc-Asp(OtBu)-OH and Fmoc-Trp-OH (or Fmoc-Trp(Boc)-OH), whilst the fluoride/DIEA (1:1) method was used for Fmoc-Pro [6]. Excesses of acylating species over the growing chain were chosen to keep their concentrations in the loops $\geq$ 0.07M and reaction times (1-3h) were optimized for each acylating step upon spectrophotometric monitoring of Fmoc-removal with piperidine/DMF (1:4, 8 min). Acetylation was performed with $(CH_3CO)_2O$/DIEA (1:1, 20 eq.). Cleavage of the peptides from the modified Rink-linker with aqueous TFA in the presence of various scavengers (or of their mixtures) led to surprisingly low recoveries. Percentages of uncleaved or reattached peptide was determined by quantitative amino acid analysis upon acid hydrolyses of the resins prior and after the cleavage step. Recoveries of peptide **1** as 2xStBu derivative ranged from 62% (from Nova Syn KR) to 54% (from Nova Syn PR 500) and 20% (from Rapp S RAM) with TFA/H_2O/EDT, 90:5:5, 2h. Best cleavage of this peptide from Rapp S RAM resin was obtained with TFA/H_2O/triisopropylsilane, 93:5:2, 2h (54%) whereas exposure to TFA/H_2O/phenol/thioanisole/ethanedithiol, 82.5:5:5:5:2.5, 2h, led to a relatively good recovery (56%), but concomitantly to removal of the StBu groups [7]. Recoveries of peptide **2** as 2xStBu derivative from Rapp S RAM varied from 63% with TFA/H_2O/triethylsilane, 93:5:2, 2h to 85% with TFA/H_2O, 95:5. Thereby minimal effects were detected by exchanging StBu with Acm as thiol protecting group. Uncleaved Ile-resin bond was found to contribute to above values in the range of 8-12%, reattachment via Cys(StBu) or Cys(Acm) between 10 and 40%, the rest being reattachment via alkylation of Trp, where N^{in}-Boc protection in this context proved to be of marginal benefit at least when EDT was used as scavenger. Surprisingly, even alkylation of His is contributing to the reattachment as deduced from model experiments with C-terminal fragments of peptide **1**. Independently of the cleavage methods used and of the synthetic batches, the crude thiol-protected octapeptides **1** and **2** were found to exhibit in hplc a cluster of mainly 4 peaks (see Figure 1) besides some minor impurities the intensity of which was found to vary to some extent in function of deprotection methods used. As the 4 peaks related to peptide **1** and 3 of the 4 peaks related to peptide **2** (the 4th peak n.d.) revealed the identical masses in ion spray MS, racemization seemed to be the most probable source for the observed heterogeneity. Gas chromatographic racemization analysis of the crude products did not reveal percentages of D-amino acids above the standard values of acid hydrolysates; thereby the 10% D-Ala (in peptide **2**) fully agrees with the known hydrolysis-dependent racemization X in X-Cys sequences. Since this method does not allow to determine the enantiomeric purity of Cys, the C-terminal fragment Ac-Cys(Acm)-Asp-Gly-Phe-NH$_2$ was synthesized by TBTU coupling of Fmoc-Cys(Acm)-OH. The resin-cleaved crude product exhibited 2 peaks in the hplc which were identified as the L-Cys and D-Cys related diastereoisomers by comparative analysis with the tetrapeptides prepared by using Fmoc-L-Cys(Acm)-OPfp and Fmoc-D-Cys(Acm)-OPfp as acylating agents. Therefore the heterogeneity observed in the crude **1** and **2** derives from racemization at the level of the two Cys coupled to the resin-bound peptides via the TBTU/HOBt/DIEA procedure. Thereby the thiol-protecting group does not affect the extent of racemization, which is reduced to undetectable extents by using Pfp-esters.

Besides the loss of product via reattachment to the resin, under optimized conditions the octapeptides were obtained in SPPS in high quality, too.

References

1. Siedler, F., Rudolph-Böhner, S., Doi, M., Musiol, H.-J. and Moroder, L., Biochemistry, Biochemistry, 32 (1993) 7488.
2. Ponsati, B., Giralt, E. and Andreu, D., In Rivier, J.E. and Marshall, G.R. (Eds.), Peptides: Chemistry, Structure and Biology, Escom, Leiden, The Netherlands, 1990, p.960.
3. Nicolás, E. Pedroso, E. and Giralt, E., Tetrahedron Lett., 30 (1989) 497.
4. Moroder, L., Gemeiner, M., Göhring, W., Jaeger, E., Thamm, P. and Wünsch, E., Biopolymers, 20 (1981) 17.
5. Fuller, W.D., Cohen, M.P., Shabankareh, M., Blair, R.K., Goodman, M. and Naider, F.R., J. Am. Chem. Soc., 112 (1990) 7414.
6. Carpino, L.A., Sadat-Aalaee, D., Chao, H.G. and DeSelms, R.H., J. Am. Chem. Soc., 112 (1990) 9651.
7. Atherton, E., Sheppard, R.C. and Ward, P.J., Chem. Soc. Perkin Trans., I (1985) 2065.

A convergent solid phase synthesis of growth inhibitory factor

Y. Nishiyama and Y. Okada

Faculty of Pharmaceutical Sciences, Kobe-Gakuin University, Nishi-ku, Kobe 651-21, Japan

Introduction

Metallothionein-like growth inhibitory factor (GIF) [1], which inhibits survival and neurite formation of cortical neurons, is abundant in the normal human brain, but remarkably reduced in the Alzheimer's disease brain. It was suggested that the loss of GIF would cause the elevated neurotrophic activity, followed by unusual sprouting, structural alterations and neural death. In order to study the relationship between the structure and growth inhibitory activity, our studies have been directed to the synthesis of GIF and related peptides.

To overcome the difficulties in the synthesis of large peptides or proteins, the convergent solid phase strategy, which includes 1) stepwise solid phase synthesis of fully protected peptide-resins, 2) their detachment from the resin to release the α-carboxy free fragments, 3) purification of the protected peptide fragments, and 4) fragment assembly on solid support, is one of the most advantageous methods [2]. This report deals with the synthesis of GIF, which consists of 68 amino acid residues including 20 cysteines (Figure 1), by slightly modified convergent solid phase method.

```
        1                    10                  20                  30
Ac-M-D-P-E-T-C-P-C-P-S-G-G-S-C-T-C-A-D-S-C-K-C-E-G-C-K-C-T-S-C-
                         40                  50                  60
K-K-S-C-C-S-C-C-P-A-E-C-E-K-C-A-K-D-C-V-C-K-G-G-E-A-A-E-A-E-A-E-
        68
K-C-S-C-C-Q
```

Fig. 1. Primary structure of GIF.

Results and Discussion

At the beginning of this investigation, we employed the following synthetic scheme; 1) synthesis of the fully protected fragments by SPPS in combination with N$^\alpha$-Fmoc, TFA-stable side-chain protecting groups and TFA cleavable Wang resin, 2) their purification, and 3) fragment condensation on the solid support. However, the purification of Fmoc-bearing peptide fragments was problematic. We evaluated some purification procedures i.e. reprecipitation (DMF-MeOH or TFA-ether), HPLC

(H$_2$O/MeCN/DMF) and silica-gel column chromatography (CHCl$_3$:MeOH:H$_2$O =

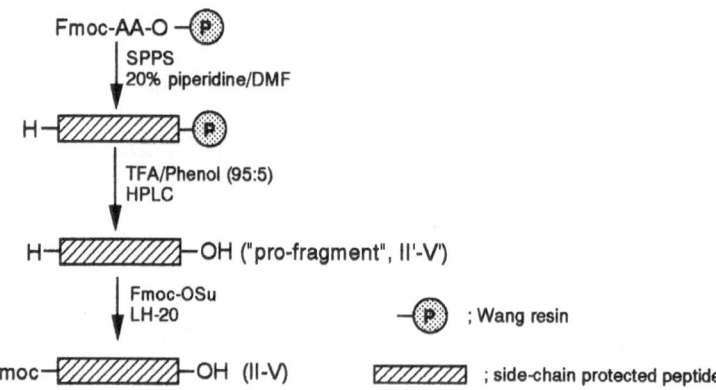

Fig. 2. SPPS of the protected peptide via "pro-fragment".

8:3:1, lower phase), however, all methods gave unsatisfactory results. Therefore, a
slightly modified synthetic scheme was employed; 1) synthesis of α-amino and α-
carboxy free fragments ("pro-fragments") by Fmoc-SPPS (Figure 2), 2) their
purification by reversed-phase HPLC, 3) introduction of N$^\alpha$-Fmoc group to the
purified "pro-fragments", and 4) fragment-condensation on the resin. For side-chain
protections of Asp and Lys residues, 2-adamantyl (2-Ada) [3] and 2-
adamantyloxycarbonyl (2-Adoc) [4] groups were employed to increase solubility of
the "pro-fragments" and fragments. All thiol groups of Cys residues were protected
by Acm to prevent the disulfide formation. Other side-chain protecting groups
employed were as follows: Bzl for Ser and Thr, cycloheptyl (Chp) for Glu, Mbh for
Gln and tBu for Glu and Ser only in the fragment (VI). All amino acid derivatives
were incorporated by Bop-mediated coupling procedure, and N$^\alpha$-Fmoc group was
removed by 20% piperidine/DMF. The resultant protected peptide resins having the
sequences of (I), (II'), (III'), (IV') and (V') were treated with TFA in the presence of

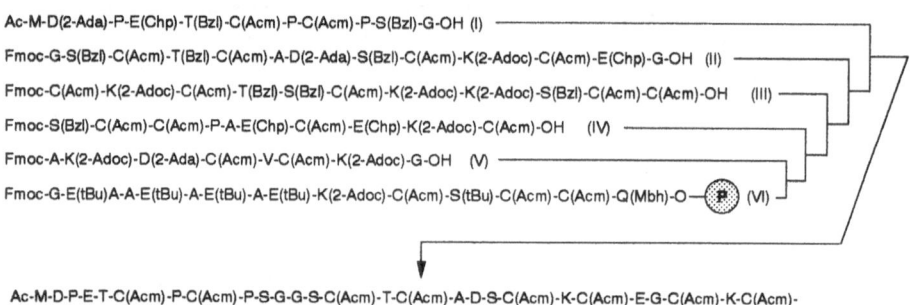

Fig. 3. Synthetic Scheme for GIF.

phenol to release the fragment (I) and "pro-fragments" (II'), (III'), (IV') and (V'). The fragments containing no or a few Bzl group(s), (I), (IV') and (V'), were highly soluble in the mixture of H_2O and MeCN, or H_2O and THF, and could be purified easily by HPLC with H_2O/MeCN solvent system, while the "pro-fragments" containing many Bzl groups, (II') and (III'), were soluble only in the mixture of DMF and H_2O, and their purification by reversed-phase HPLC was somewhat difficult. Subsequently, N^α-Fmoc group was introduced to "pro-fragments", (II')-(V'), to give the fragments, (II)-(V). Fragments (I)-(V) were coupled with the fragment (VI) built on the solid support by Bop-HOBt or DIPC-HOBt method as shown in Figure 3, followed by the capping with 1-acetylimidazole, successively, to give the fully protected peptide resin corresponding to the entire amino acid sequence of GIF.

The protected peptide resin was treated with 1 M TMSBr-thioanisole/TFA and HF, successively, to give Acm-GIF. After the purification with HPLC, Acm groups were removed by the treatment with $Hg(AcO)_2$ in 50% AcOH containing 8 M guanidine•HCl, followed by the incubation with DTT and gel-filtration to give apo-GIF molecule.

Side-chain protected peptides, "pro-fragments", containing 2-Ada or 2-Adoc group were efficient intermediates for the convergent SPPS in terms of the ease of their purification by reversed-phase HPLC. This synthesis suggests that 2-adamantyl-type protecting groups are an alternative to benzyl-type, and additional 2-adamantyl-type protecting groups for other side-chain functions which may increase solubility of the fragments, are now required to enhance the convergent solid phase peptide synthetic approach.

Acknowledgements

This work was supported in part by a grant from The Science Research Promotion Fund of the Japan Private School Promotion Foundation.

References

1. Uchida, Y., Takio, K., Titani, K., Ihara, Y. and Tomonaga, M., Neuron, 7 (1991) 337.
2. Albericio, F., Williams, P.-L., Gairi, M., Jou, G., Celma, C., Cordonier, N.-K., Grandas, A., Ertja, R., Pedroso, E., van Reitschoten, J., Barany G. and Giralt, E., In Epton, R. (Ed.) Innovation and Perspectives in Solid Phase Synthesis; Peptides, Polypeptides and Oligonucleotides, Intercept Ltd., Andover, 1992, p.39, and references cited therein.
3. Okada, Y. and Iguchi, S., J. Chem. Soc. Perkin Trans., I (1988) 2129.
4. Nishiyama, Y. and Okada, Y., J. Chem. Soc. Chem. Commun., 1083.

Instrumentation for semi-large scale continuous-flow peptide synthesis with solvent and acyl component recovery systems

K. Nokihara and S. Nakamura

Biotechnology Instruments Department, Shimadzu Corporation, Nishinokyo-Kuwabaracho 1, Nakagyoku, Kyoto, 604, Japan

Introduction

The need for high-quality synthetic peptides has greatly increased with the expansion of life-science research. The cost of starting materials and purification in down stream processing are the most important factors for production of peptide pharmaceuticals. In addition waste is a serious environmental problem in the chemical industries. Regeneration and re-use of acyl components, which usually account for the largest production costs in SPPS, have already been mentioned [1]. We have constructed a prototype semi-large scale continuous-flow solid-phase peptide synthesizer, which is equipped with recovery lines for acyl components as well as solvents.

Results and Discussion

The present synthesizer consists of a unit containing the acyl components (amino acid reservoir with the same number of ports for activation reagents and an activation flask with a temperature controller), a pump and a reaction column. After the reaction column a UV detector was inserted in the line to monitor Fmoc groups. Using this small plant several peptides were synthesized by the Fmoc strategy using DMF as a solvent, after evaluation of coupling protocols as well as excess amounts of the acyl component by the use of a simultaneous multiple peptide synthesizer, PSSM-8 [2]. The N^{α} Fmoc group was removed by 20% piperidine in DMF. Cleavage was carried out in the same reaction column, *in situ*, (before cleavage the peptide bound resin was dried in a stream of nitrogen after washing with methanol followed by t-butylmethyl ether) or in a flask with a TFA-scavenger cocktail.

Recovered excess amino acid derivatives dissolved in DMF were concentrated, quenched with saturated $NaHCO_3$ solution, acidified with HCl with cooling and solidified or extracted with ethyl acetate. Regenerated Fmoc amino acids were purified in the conventional manner to yield ca 70 % of theoretical value and analyzed by TLC, RP-HPLC, FAB-MS and DL amino acid analysis [3]. DMF in the washing steps was recovered, collected and distilled in the conventional manner. The quality of purified DMF was checked by GC-MS for re-use in SPPS.

Leucine enkephalin and human angiotensin I were prepared by the stepwise coupling in the presence of PyBOP and HOBt with NMM (**Table 1**). **Figure 1** shows RP-HPLC profiles of these peptides after cleavage. The resulting material was characterized by LSIMS, D/L amino acid analysis and was found to be identical with authentic peptides prepared by the PSSM-8.

Table 1 *Synthesis of [Leu]^5-enkephalin (a) and angiotensin I (b)*

(a) Tyr–Gly–Gly–Phe–Leu–OH
(b) Asp–Arg–Val–Tyr–Ile–His–Pro–Phe–His–Leu–OH

		a	b
Starting Resin	NαFmoc–Leu–TentaGel S AC (0.22meq/g)	50 g	40 g
Acylcomponent	4 fold excess, concentration ca. 0.25 mol/L		
Activation	PyBOP, HOBt, NMM (5 min)		
Flow rate		30 mL/min	50 mL/min
Coupling time		25 min	15 min
Cleavage (Cocktail)	a = in column, 300 mL (95% TFA, 5% EDT, 5% anisol) 1 hr b = in flask, 250 ml (82% TFA, 5% water, 5% EDT, 2.5% EMS, 5% thioanisole, 2.5% thiophenol) 6 hrs		
Yield (cleaved peptide)		88 %	85 %

In the present study several problems were solved to enable construction of a fully automated synthesizer. The system is useful for peptide pharmaceutical research and semi-large scale production of commercial peptides with high efficiency and low running costs.

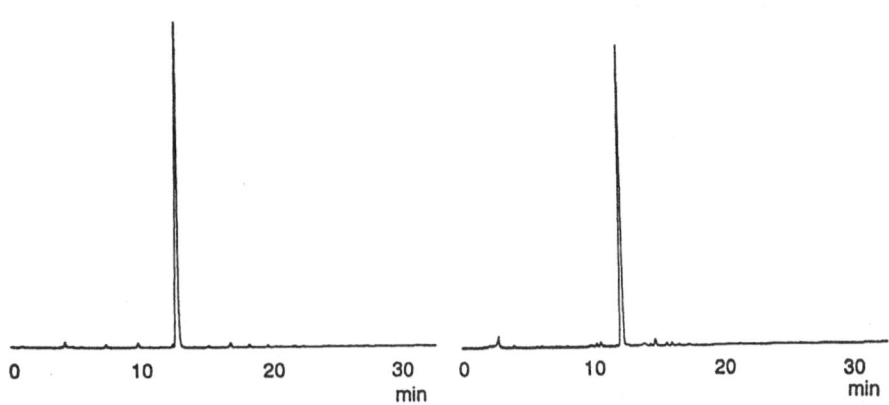

Fig. 1. *HPLC-profiles of synthetic [Leu]^5-enkephalin and angiotensin I after cleavage. Column: SynProPep RPC18 (4.6 x 150 mm); Eluent: 0.01N HCl/CH$_3$CN=85/15-55/45 (30 min); Flow rate: 1.0 ml/min; Absorbance 210 nm.*

References

1. Nokihara, K., Hellstern, H. and Hoefle, G., In Rivier, J.E., and Marshall, G.R. (Eds.)
    Peptides: Chemistry, Structure and Biology, ESCOM, Leiden, 1990, p.1046.
2. Nokihara, K., Yamamoto, R., Hazama, M., Wakizawa, O. and Nakamura, S., In
    Epton, R. (Ed.), Innovation and Perspectives in Solid-Phase Synthesis 1992,
    Intercept Limited, Andover, 1992, p.445.
3. Nokihara, K., Gerhardt, J., Yamamoto, R. and Nishine, T., In Takai., K. (Ed.),
    Frontiers and Horizons in Amino Acid Research, Elsevier Science Publishers B.V.
    Amsterdam, The Netherlands, 1992, p.391.

SPPS using high temperature

A.K. Rabinovich and J.E. Rivier

The Salk Institute for Biological Studies, 10010 N. Torrey Pines Rd., La Jolla, CA 92037, U.S.A.

Introduction

Elevated temperature has been previously used in SPPS, but to date, only during the coupling step [1]. It has also been reported that coupling at elevated temperatures (30-80 °C) using 1-benzotriazolyl esters of amino acids did not induce racemization even in the presence of excess triethylamine [2]. In order to shorten cycle time (<30 min/cycle) without having to reduce resin substitution (>0.4 meq/g) or increase excesses of reagent (2-3 fold), we have studied the effects of elevated temperatures on all steps of several syntheses [rat CRF (a 41 peptide amide), porcine NPY (a 36 peptide amide), and α-MSH (a 13 peptide amide)] with protocols that were otherwise the standard used in our laboratories. In addition, we synthesized two model peptides, YLRPDWHQA-amide and AQFFYNNTTQLFNNA-amide encompassing most trifunctional amino acids or rich in Asp, Glu, Asn and Gln reported to give problematic side reactions [1].

Results and Discussion

Syntheses were carried out manually using a MBHA resin in a water (75°C) jacketed vessel, maintaining an elevated temperature during the coupling, wash and deblocking steps. The side-chain protecting groups were His(Tos), Tyr(2BrZ), Ser(Bzl), Glu(cHex), Asp(cHex), Arg(Tos), Cys(Mob), Cys(Acm), Lys(2ClZ), Thr(Bzl) when using the Boc-strategy and His(Trt), Asp(tBu), Arg(Mtr), Tyr(tBu) when using the Fmoc strategy. In our search for an ideal solvent system that would solubilize all amino acid derivatives, be environment friendly and obtainable at a reasonable cost, we tested the following solvents and mixtures thereof: ethylene carbonate/toluene, DMF, DMF/toluene and DMSO/toluene. We found that 25% DMSO in toluene was the overall best solvent system for coupling. Toluene, as shown below could also be used in combination with TFA, piperidine and Et₃N for deblocking the α-amino function and neutralization in either the Boc or Fmoc strategies. As an example, the following protocols were used (volume/g resin, reaction time): **Boc-strategy** 1) deblocking, TFA/toluene/1,2-ethanedithiol = 10/10/1 (15 mL, 3 min); 2- Toluene wash (15 mL, 15 sec) ; 3) neutralization, 10% Et₃N/toluene (15 mL, 15 sec); 4) MeOH wash (15 mL, 15 sec); 5) repeat steps 3 and 4; 6) coupling in 25% DMSO/toluene (15 mL): Boc-amino acid/HOBT = 1/1 (2 eq), DIC (2 eq, 10 min) [exception: use BOP with Boc-His(Tos)·DCHA and Boc-amino acid/HOBT = 1/2 for Gln and Asn]; 7) 25% DMSO/toluene wash (15 mL, 15 sec); 8) MeOH wash (15 mL, 15 sec); 9) repeat steps 7 and 8; 10) toluene wash (15 mL, 15 sec). **Fmoc-strategy** (for model peptide YLRPDWHQA-amide) 1) deblocking, 20% Piperidine/Toluene (20 mL, 5 min); 2) toluene wash (15 mL, 15 sec); 3) MeOH wash (15 mL, 15 sec); 4) repeat steps 2 and 3; 5) coupling in 25% DMSO/toluene (15 mL) : Fmoc-amino acid/HOBT = 1/1 (2 eq), DIC (2 eq, 10 min) [exception : use BOP with Fmoc-His(Trt)·DCHA and Fmoc-amino acid/HOBT = 1/2 for Gln and Asn], 6) 25% DMSO/toluene wash (15 mL, 15 sec); 7) MeOH wash (15 mL, 15 sec); 8) repeat steps 6 and 7; 9) toluene wash (15 mL, 15 sec). In these studies, couplings were monitored by the ninhydrin test of Kaiser and driven to completion by recoupling when necessary. Acetylation at high temperature was not performed in these studies but was used with other peptides and found to give results similar to those obtained at room temperature.

Peptide resins were cleaved (HF/anisole/ethylmethylsulfide: 10/1/1; 15 mL/g peptide resin) and analyzed by RP-HPLC. The crude peptides appeared to be of equivalent or greater purity than the preparations obtained using similar conditions (but different solvents and reaction time) at room temperature. Figure 1 and 2 show the elution profile of crude rCRF and of pNPY. YLRPDWHQA-amide was synthesized to evaluate the stability of most trifunctional AA under high temperature conditions while AQFFYNNTTQLFNNA-amide rich in Asn and Gln and an analog of the previously studied QFFYNNTTQLF-NN (1) was synthesized for comparative purposes with no apparent problem. In order to minimize problems associated with Asn as the C-terminus benzyl ester and Gln as the N-terminus (the former providing a substrate for as yet unresolved side reactions and the latter having a propensity to cyclize to give the pyroglutamy N-terminus), we added Ala residues at the N- and C-termini. Peptides were purified by HPLC and characterized by MS, AAA and coelution with authentic samples. rCRF and NPY were tested in specific bioassays and found to be equipotent to standard preparations. Under these conditions, Gln was the only AA with which difficulties were encountered. It is well known that peptides with NH_2-terminal deblocked Gln can be directly cyclized to pyroglutamyl peptides when heated in TFA or when exposed to dilute carboxylic acids.

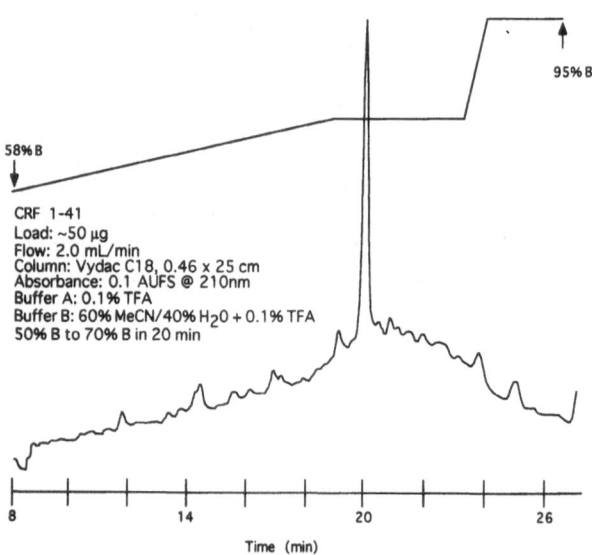

Fig. 1. Rat CRF sequence: SEEPPISLDLTFHLLREVLEMARAEQLAQQAHSMRKLMEII-amide

72

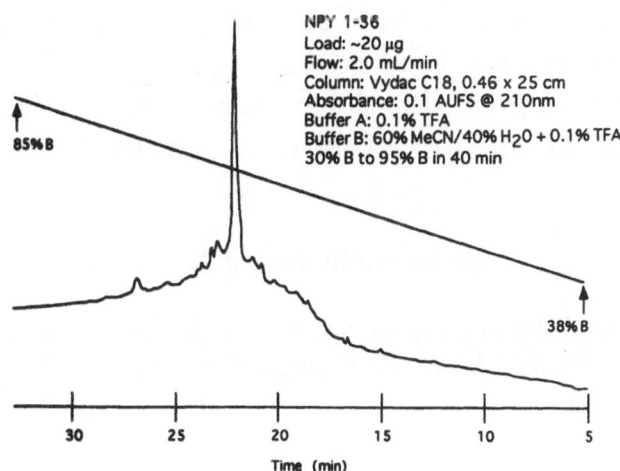

Fig. 2. NPY sequence: YPSKPDNPGEDAPAEDMARYYSALRHYINLITRQRT-amide

In the synthesis of our model peptide YLRPDWHQA-amide, Pyr-Ala-amide was also found. When this synthesis was repeated, this side reaction was prevented by deblocking the Boc-Gln peptides at room temperature and by adding the preformed Boc-AA·HOBT esters during the following coupling step.

In separate studies, we also used the Merrifield resin and the Boc strategy for the synthesis of a number of peptides (including endothelin) with a free carboxyl. Weight gain was as expected, suggesting that loss of peptide from the resin during deblocking was not a major problem.

With few modifications to usual protocols (such as those presented here), this method will increase efficiency by reducing cycle times to less than 30 min. Our choice of solvents is also compatible with SPPS at room temperature and should solve some aspects of disposal associated with chlorinated compounds. Finally, it is obvious that these protocols could be easily adapted to automation.

Acknowledgements

This work was supported by NIH grant DK26741 and HD13527 and the Hearst Foundation. We would like to thank Dr. Anthony Craig for mass spectral analysis and Duane Pantoja for technical assistance.

References

1. Lloyd, D., Petrie, G., Noble, R. and Tam, J., In J. Rivier and G. Marshall (Eds.), Peptides, Chemistry, Structure and Biology (Proc. 11th Amer. Peptide Symp.), Escom Scientific Publ., Leiden, The Netherlands, 1989, p.909.
2. Barlos K., Papaioannou D., Patrianakou S. and Tsegenidis T., Liebigs Ann. Chem., (1986) 1950.

A new approach to "carba"-peptide isosteres: Synthesis of Boc-(S)-PheΨ[CH$_2$CH$_2$]-(R/S)Val and Boc-(S)-TrpΨ[CH$_2$CH$_2$]-(R/S)-Val via Michael addition of sulfone-stabilized anions to acrylates

A. Spaltenstein and J.J. Leban

Division of Organic Chemistry, Burroughs Wellcome Co, 3030 Cornwallis Rd., Research Triangle Park, NC 27709, U.S.A.

Introduction

Backbone modifications by isosteric amide bond replacement have attracted widespread interest in peptide chemistry as a means to alter and/or improve chemical and biological properties of peptides [1]. Although increased stability toward enzymatic degradation and thus prolonged duration of action of potential peptide drugs is usually the major objective when utilizing peptide isosteres, other pharmacological parameters may be affected as well. For example, it has recently been shown that in some cases agonistic peptides can be converted into antagonists by applying such principles [2]. While the synthesis of some more commonly used amide replacements, such as Ψ(NH-CO), Ψ(NH-CH$_2$), or Ψ(COCH$_2$) has received considerable attention and has become routine, the "carba" replacement Ψ(CH$_2$CH$_2$) has only recently been dealt with in a more general fashion [3]. Unfortunately, the reported procedure requires the homologation of an amino acid using diazomethane and silver oxide, which can be problematic, particularly for larger scale operations. For this reason, we investigated new approaches to the synthesis of "carba", peptide isosteres and we report here a short and convergent procedure starting from readily accessible materials. Our approach which utilizes a 1,4-addition of a sulfone-stabilized anion to an appropriate acrylate is exemplified by the synthesis of Boc-PheΨ(CH$_2$CH$_2$)Val and Boc-TrpΨ(CH$_2$CH$_2$)Val below.

Results and Discussion

The required protected amino sulfones were accessible [4], from commercially available Boc-amino alcohols [5] **1**, (Scheme 1) usually in two steps [6] by treatment with tributylphosphine/diphenyldisulfide to form the corresponding phenylsulfides, followed by mCPBA oxidation to the desired sulfones **2**. The key coupling step was carried out by addition of the acrylate **3** [7] to the dianion of sulfone **2**, generated with n-butyllithium in the presence of 2 equivalents of HMPA at -30°. The resulting set of four diastereomeric phenylsulfones **4** was not normally characterized further, but was instead subjected directly to reductive desulfonylation with sodium amalgam in phosphate buffered methanol [8] to give esters **5**. In the case of tryptophan analogs, these reaction conditions also conveniently removed the indolyl (Boc) protecting group. Saponification with methanolic potassium hydroxide afforded the desired Boc-protected isosteres **6**. Based on NMR analysis after conversion to the cyclic lactam derivatives **7** [9], no stereochemical induction occurred during the Michael addition and a 1:1 mixture of diastereomers at the α-carbon was obtained. Although a method for the separation of such diastereomers via cyclic amides **7** has been reported [9], we generally prefer to delay separation until after incorporation of the isosteres into peptides.

Scheme 1

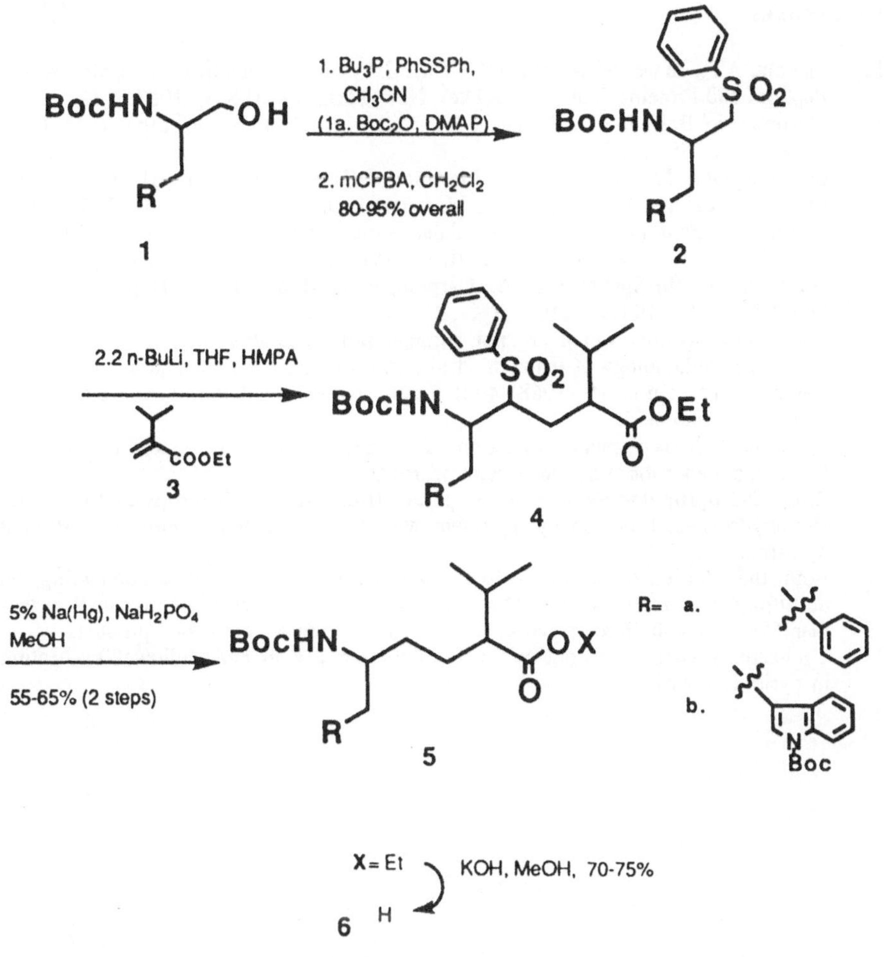

In conclusion, we have developed a short and efficient method for the synthesis of Boc protected "carba" peptide isosteres. Incorporation of these and other isosteres into longer peptides and biological evaluation are in progress.

7   8

References

1. Spatola, A.F., In B. Weinstein (Ed.), Chemistry and Biochemistry of Amino Acids, Peptides and Proteins, Vol 7, M.Dekker, New York, NY, U.S.A., 1983, p.1.
2. Martinez, J., Bali, J.-P., Rodriguez, M., Castro, B., Laur, J. and Lignon, M.-F., J. Med. Chem., 28 (1985) 1874.
3. (a) Rodriguez, M., Heitz, A. and Martinez, J.,Tetrahedron Lett., 31 (1990) 7319. (b) Rodriguez, M., Aumelas, A. and Martinez, J., Tetrahedron Lett., 31 (1990) 5153. A general route to the related *trans*- alkene isosteres has also been reported [4].
4. (a) Lehman de Gaeta, L.S.,Czarniecki, M. and Spaltenstein, A., J. Org. Chem., 54 (1989) 4004. (b) Spaltenstein, A., Carpino, P.A., Mijake, F. and Hopkins, P.B., J. Org. Chem., 52 (1987) 3759.
5. For amino alcohols which are not commercially available, we prefer the mixed anhydride methodology of Ishizumi et al (Ishizumi, K., Koga, K. and Yamada, S.-I., Chem. Pharm. Bull., 16 (1968) 492) for the reduction of the corresponding Boc protected amino acids.
6. In the case of tryptophan, protection of the oxidatively labile indole moiety with Boc is required prior to treatment with mCPBA.
7. Ethyl 2-isopropylacrylate was prepared from diethyl isopropylmalonate by monohydrolysis, followed by treatment with formaldehyde/pipiridine in refluxing pyridine.
8. Both, the addition of phosphate to the reaction mixture as well as conducting the desulfonylation at 0° are necessary to suppress the formation of alkene 8 to less than 5%. The unbuffered reaction at 25° leads to 40-60% elimination product.
9. Cyclizations were carried out by treating the esters 5 with TFA, followed by heating in pyridine [3a,b].

In vitro amidation for preparation of recombinant peptides: Enzymatic coupling with specific endopeptidases

H. Togame, T. Inaoka and T. Kokubo

*International Research Laboratories, Ciba-Geigy Japan Ltd.,
Takarazuka, Hyogo 665, Japan*

Introduction

Many biologically active peptides have the α-amide structure at their *C*-termini, and in most cases the amide structure is essential for their biological activities. These amidated peptides cannot be directly produced with microorganisms by means of recombinant DNA, since microbes lack the function of the post-translational mechanism to form the *C*-terminus amide. The free acid form of an expressed peptide must therefore be converted to the amide form by an *in vitro* amidation process. Enzymatic coupling of an amino acid amide (AA-NH$_2$) to the *C*-terminus of a precursor peptide has been known as one of such *in vitro* amidation methods. However, its application was limited only for short oligopeptides, because non-selective transpeptidation reactions, which are due to the broad specificities of the peptidases in common use, were associated with the coupling reaction to give various by-products. To overcome this drawback of the conventional enzymatic coupling, we have examined the use of specific endopeptidases rather than the peptidases with broad specificities. We report here prolyl endopeptidase (PEP) [1] and arginyl endopeptidase (AEP) were successfully applied to the enzymatic coupling for the *in vitro* amidation of acid form precursors of clinically useful peptides.

Results and Discussion

Since both LH-RH and oxytocin have a proline residue near the *C*-terminus, they are subjected to specific cleavage with PEP to give a large *N*-terminus fragment terminated with the proline residue and the rest containing the *C*-terminus amide. We prepared these α-amidated peptides by the enzymatic coupling in the reverse manner of the hydrolytic cleavage. In the case of LH-RH, G-NH$_2$ was coupled to the *N*-terminal nonapeptide <EHWSYGLRP (1 mM) with recombinant PEP, which was cloned [2] and expressed in *E. coli* by us. The reaction proceeded rapidly even with a

$$\text{<EHWSYGLRP-OH} + \text{H-G-NH}_2 \underset{}{\overset{r\text{-PEP}}{\rightleftharpoons}} \text{<EHWSYGLRPG-NH}_2$$
$$LH\text{-}RH$$

$$\text{H-CYIQNCP-OH} + \text{H-LG-NH}_2 \underset{}{\overset{r\text{-PEP}}{\rightleftharpoons}} \text{H-CYIQNCPLG-NH}_2$$
$$Oxytocin$$

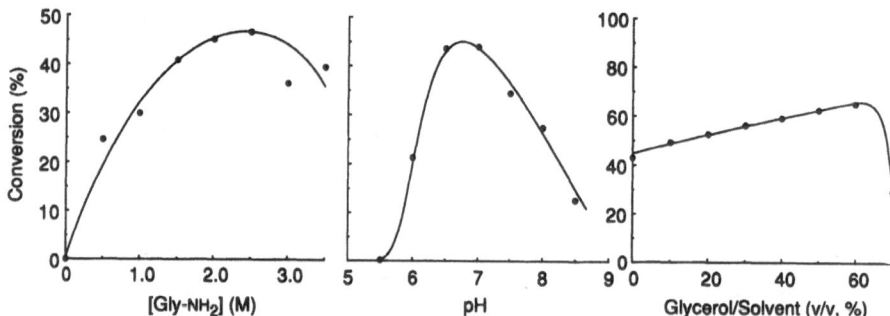

Fig. 1. Effects of concentration of Gly-NH₂, pH and addition of glycerol on the coupling reaction of LH-RH(1-9) with Gly-NH₂.

catalytic amount of PEP (0.08 μM) at 30°C to give LH-RH in the equilibrium with the precursor. The concentration of G-NH₂ and pH significantly affected conversion of the precursor to LH-RH (Figure 1) and the maximum conversion of 45% was attained with 2.5 M of G-NH₂ at pH 6.5-7.0. To improve the conversion water-miscible organic solvent was added to the reaction mixture. Among the examined solvents glycerol was found to be most effective with PEP; the conversion reached 67% in 60% (v/v) glycerol (Figure 1). PEP catalyzed the coupling reaction so selectively that no side product was detected in the HPLC analysis of the reaction mixture (Figure 2). Since the unreacted substrate was easily recovered in the purification of the produced LH-RH, its chemical yield was quantitative (>95%) on the basis of the consumed substrate.

In the similar manner to LH-RH, oxytocin was obtained selectively by the coupling of LG-NH₂ to the precursor fragment, CYIQNCP. With 0.13 μM (0.5 unit/ml) of r-PEP the precursor (1 mM) was converted to oxytocin at 30°C. The maximum conversion was observed with 0.8 M LG-NH₂ at pH 6.0-7.0, but it was rather unsatisfactory (27%) in the absence of an organic solvent. The addition of glycerol was found to be effective also for the improvement of the conversion to oxytocin and the best conversion of 55% was attained with 60% glycerol.

AEP was employed to prepare vasopressin by the coupling of G-NH₂ to its precursor fragment terminated with an arginine residue, CYFQNCPR. AEP, as well as PEP, catalyzed the coupling reaction very selectively to give vasopressin without

$$\text{H-CYFQNCPR-OH} + \text{H-G-NH}_2 \underset{}{\overset{AEP}{\rightleftharpoons}} \text{H-CYFQNCPRG-NH}_2$$
$$\textit{Vasopressin}$$

forming any by-product. From 1 mM of the precursor, vasopressin was obtained in a 51% conversion with 4.0 M G-NH₂ at pH 7.0 by use of 2.1 μM AEP. Although PEP was found to lose activity in the presence of a low concentration of a chaotropic solvent such as DMF and DMSO, AEP was quite tolerant for the chaotropic solvent and allowed its addition to the reaction mixture at a high concentration. DMSO was especially effective for the AEP-catalyzed coupling reaction, and the maximum conversion of 79% was attained in 45% DMSO with 2.0 M G-NH₂ at pH 7.0.

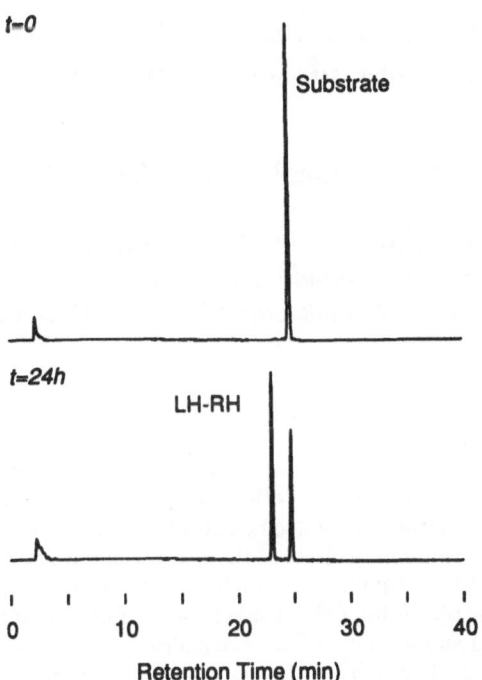

Fig. 2. HPLC analysis of the coupling reaction of LH-RH(1-9) with Gly-NH₂. Conditions for HPLC: Column: YMC-Pack ODS-AM AM-302 (4.6 x 150 mm). Mobile phase: A; 0.1% TFA in 5% (v/v) acetonitrile; B: 0.1% TFA in 55% (v/v) acetonitrile. Gradient: 10-90%B in 40 min. Flow rate: 1 ml/min. Detection: 280 nm.

For *in vitro* amidation of the acid form of a precursor peptide, enzymatic coupling with an amino acid-amide (or a dipeptide-amide) was examined by use of a specific endopeptidase. Without any by-product LH-RH and oxytocin were prepared with r-PEP and vasopressin with AEP, demonstrating an advantage over conventional methods by use of non-specific peptidases. Since endopeptidases specific for a single amino acid residue, e.g., PEP and AEP, have been intensively studied, the enzymatic coupling would increase its usefulness as an *in vitro* amidation method of recombinant precursor peptides.

References

1. Yoshimoto, T., Walter, R. and Tsuru, D., J. Biol. Chem., 255 (1980) 4786.
2. Yoshimoto, T., Kanatani, A., Shimoda, T., Inaoka, T., Kokubo, T. and Tsuru, D., J. Biochem., 110 (1991) 873.

Semisynthesis of mutants at residues 70 and 71 in a sharp bend of the cytochrome c fold

C.J.A. Wallace[1] and I. Clark-Lewis[2]

[1]Departments of Biochemistry, Dalhousie University, Halifax, Nova Scotia, B3H 4H7, Canada and [2]University of British Columbia, Vancouver, British Columbia, V6T 1W5, Canada

Introduction

Cytochrome c is one of the most intensively studied proteins and has been a subject of structural engineering by both semisynthetic and genetic methods [1]. It functions as a mobile surface-associated electron carrier between major multienzyme proton pumps in the respiratory chain of the mitochondrial inner membrane and thus possesses the essential attributes of a high and stable redox potential, interaction surfaces for many physiological partners, and rapid electron transfer through electrostatic steering and special conductance pathways; all conferred by a polypeptide chain of 103-112 residues. Of these some 21 are absolutely conserved and thus particular targets for protein engineers. Two conserved residues, Asn^{70} and Pro^{71}, are not associated with any of the above-mentioned functions and thus may be crucial to the maintenence of the active conformation of the protein: the characteristic 'cytochrome fold' is evident even in bacterial cytochromes with different physiological roles. Possible structural roles are the rigidification of the bend between the 60s and 70s helices (Asn-Pro are the optimal residues for an Asx-turn [2]) or stabilization of the helices themselves by capping [3], or both. We set out to examine these possibilities by making synthetic peptides containing the non-coded amino acids homoserine at position 70, and norvaline at position 71, for incorporation into semisynthetic analogs of the horse sequence using our previously published methods [4].

Results and Discussion

The two protein analogs, $[Hse^{70}]$ cytochrome c and $[Nva^{71}]$ cytochrome c were obtained in 45% yield from 1:1 mixtures of synthetic peptides corresponding to residues 66-104 and the CNBr fragment 1-65 (H) derived from the native protein under the standard conditions of autocatalytic fragment religation [4]. This simple and efficient means of fragment condensation depends on the propensity of large peptides to complex and assume the native conformation, thus bringing the breakpoint termini into proximity and catalysing the aminolysis of a carbonyl activated, in this case, by the lactonization of homoserine generated in the CNBr cleavage [5]. Crude protein was purified by cation-exchange chromatography and checked by high-performance liquid chromatographic methods. Purified products were subject to a wide range of physicochemical and biological tests. These included

Cytochrome	pH	E'm (mv)	pK695 III→IV	pK695 III→II	$t_{1/2}$ redn. (min)	$t_{1/2}$ oxdn. (min)	Succinate oxidase activity @ pH 7.0	Cytochrome oxidase activity @ pH 7.5
Native Horse	7.0	272-276	9.2	2.7	0.8	>1000	100%	100%
	6.0	274			n.d.	n.d.		
	5.0	274						
[Hse[70]]	7.0	272	8.4	3.3	1.5	300	110%	90%
[Nva[71]]	7.0	136	5.3	4.7	23	<1	>15%	>0%
	6.5	165 (214)			n.d.			
	6.0	194 (230)			n.d.			
	5.5	203 (227)			5			
	5.0	212 (225)			n.d.			

Table 1:   *A compilation of the physical and biological properties of horse cytochrome c and the two analogs. The redox potential of the protein is normally stable in the range pH 5-7. The changing value of E'm for [Nva[71]] can be corrected for the proportion of the active state III conformer determined from the titration of the 695nm absorbance (values in parentheses). Titrations were performed of both the alkaline (from state III to state IV) and acid (III to II) transitions to alternative protein-heme iron ligation patterns. $T_{1/2}$ for ascorbate reduction was determined by the increase of the 550nm band for 35µm cytochrome c in the presence of 280µm ascorbic acid in 50mM phosphate buffer. $T_{1/2}$ for oxidation was calculated from the drop in 550 asborbance on buffer exchanging reduced protein into O_2-saturated 50mM phosphate buffer.*

comparison of the visible-light spectra with those of the native horse protein, pH titrations of certain spectral bands and redox titrations with the ferro/ferricyanide couple to determine E'_m. Non-physiological oxidation and reduction rates, by O_2 and ascorbic acid, respectively, were checked, and biological assays of oxidation by cytochrome oxidase and reduction, in the succinate oxidase system, of the analogs were compared to those of the native protein. All tests were performed by established methods [4, 5, 6] and the results are summarized in Table 1.

The in vivo assays of the [Hse^{70}] analog show biological activity is little affected by the change, but some structural destabilization is evident from decreased resistance to H^+ and OH^- induced conformational state change, and the increased access of environmental O_2 to the heme crevice. The effect of the Pro$^{71} \rightarrow$Nva change is dramatic. The cytochrome is only slowly reduced by ascorbate and instantly autoxidises. It has near zero biological activity. The visible spectrum is changed in significant ways (that of [Hse^{70}] is not) at pH 7, including the absence of the 695 nm absorbance characteristic of normal heme iron ligation. Most of the changes can, in fact, be rationalized in terms of a shift in pK for the alkaline transition from 9.2 to 5.3, a change exceeded by no other analog thus far studied. Thus, at pH 7, essentially all the protein is in the inert conformational state IV.

Our data confirm that both Asn^{70} and Pro^{71} contribute to the stability of cytochrome c and its heme crevice, though the effects of the Asn$^{70} \rightarrow$Hse change are not so severe as to offer an immediate explanation of Asn^{70}'s effective invariance. However, since homoserine was deliberately chosen as a replacement to introduce only minimal change (it is nearly isosteric and isopolar), possibly any natural substitution would be more destabilizing. The contribution of Pro^{71} is very significant, particularly in terms of the preservation of the normal ligation in favor of the strongly competing alternative that characterizes state IV, and implies its conservation is for the rigidity imposed at the right-angle bend between helices (to which Asn^{70} probably also contributes) that establishes the direction of the 70s loop. This in turn will dictate the relative affinity of the heme iron for the normal methionine ligand or that which replaces it, thought by many to be a lysine residue in the 70s loop. The availability of a mutant with type IV conformation at pH 7 provides the opportunity for a structure determination to clarify this still perplexing question.

Acknowledgements

Thanks are due to Anne Rich and Angela Brigley for technical assistance, and NSERC and PENCE of Canada for financial support.

References

1.   Wallace, C.J.A., FASEB J., 7 (1993) 505.
2.   Baker, E.N. and Hubbard, R.E., Progr. Biophys. Mol. Biol., 44 (1984) 97.
3.   Richardson, J.S. and Richardson, D.C., Science, 240 (1988) 1648.
4.   Wallace, C.J.A. and Clark-Lewis, I., J. Biol. Chem., 267 (1992) 3852.
5.   Proudfoot, A.E.I., Rose, K. and Wallace, C.J.A., J. Biol. Chem., 264 (1989) 8764.
6.   Craig, D.B. and Wallace, C.J.A., Protein Sci., 2 (1993) 966.

Two-step selective formation of three disulfide bridges in the C-terminal epidermal growth factor-like module of human blood coagulation factor IX

Y. Yang[1], W.V. Sweeney[1] and J.P. Tam[2]

[1]Department of Chemistry, Hunter College of CUNY, 695 Park Ave., New York, NY 10021, U.S.A.
[2]Department of Microbiology and Immunology, Vanderbilt University, A5321 M.C.N., Nashville, TN 37232, U.S.A.

Introduction

Formation of multiple-disulfide bridges is a common problem in the synthesis of cysteine-rich modules. Human blood coagulation factor IX contains two epidermal growth factor-like modules, each with three disulfide-bridges (1-3, 2-4, 5-6) [1]. There are fifteen possible combinations of disulfide bridges. Unlike the N-terminal EGF-like module, the C-terminal EGF-like module (FIX_{EGF-C}) exhibited significant disulfide bond scrambling under a wide range of conventional folding conditions. In this work, a two-step selective formation of disulfide-bridge strategy under FastMoc SPPS has been developed to resolve the problem. The possible disulfide-pairing patterns can be reduced from 15 to 3 with this approach.

In the synthesis, six cysteines are selectively grouped, protected with two different blocking-groups, Acm and trityl, and deblocked in two steps. This allows the differently blocked cysteines to complete the folding process separately. In the initial step, TFA cleavage of the peptide from the resin simultaneously deblocks all the side chain protecting groups except Acm, which remains on two cysteines. Thus 4 cysteines with free sulfhydryls may form two S-S bridges, yielding 3 possible isomers. Then in the next step, removal of Acm with iodine for the last two cysteines, allows formation of the final disulfide bridge [2]. Since the formation of the disulfide-bridges in the first-step has 3 possibilities while in the second-step it is exclusive, identification of the folded products from the first-step is crucial. Proteolytic digestion combined with matrix-assisted laser desorption mass spectrometry and HPLC is used for the assignment of the disulfide-bridge location.

Results and Discussion

1. Synthesis of the peptides
    The proper position where the cysteines should be blocked with Acm was investigated. Three peptides, A, B, C, each with same sequence but a different pair of cysteines protected by Acm, were synthesized. The sequence of the FIX_{EGF-C} is:

```
    I       II  III        IV V                    VI
H-LDVTCNIKNGRCEQFCKNSADNKVVCSCTEGYRLAENQKSCEPAV-OH
  1(84)        10        20        30        40    45(128)
```

2. Folding in the first-step

In the first-step air oxidation was used. Peptide solution was slowly stirred at room temperature about 24-30 hr and folding was monitored by HPLC. The location of the disulfide bridges was determined by proteolytic digestion (data not shown).

Table 1 *Studies on the location of the disulfide-bridges in the first-step folding*

Peptide	Cys-Acm	Expected Two Disulfides	Disulfides Found
A	5,6	1-3, 2-4	1-2, 3-4
B	2,4	1-3, 5-6	1-3, 5-6
C	1,3	2-4, 5-6	several isomers

Peptide A generated an unexpected folding isomer. Peptide B gave a properly folded product. In peptide C, folding yielded a mixture of several isomers. This disulfide bond scrambling may have been caused by the close proximity of cysteines 4 and 5.

3. Folding in the second step

The second-step folding was started by removal of the Acm with iodine. Under inert gas and avoiding light, I_2 in MeOH was added to the peptide solution containing 10% HAc. After cooling the solution in ice bath, $Na_2S_2O_3$ was added to react the excess I_2. The folding was monitored by HPLC.

An EGF-like module containing three disulfide-bridges has been synthesized with two-step selective formation of disulfide bridges, under FastMoc SPPS. This approach reduced the possible disulfide-pairing patterns from 15 to 3 in this peptide. For FIX$_{EGF-C}$, only the peptide with a middle pair of cysteines blocked with Acm yielded a properly folded product.

Acknowledgements

The authors gratefully acknowledge Dr. R.B. Merrifield for helpful discussion, Dr. K. Schneider and Dr. B.T. Chait for excellent work on matrix-assisted laser desorption mass spectrometry. This work was supported in part by a grant from US PHS HL.

References

1. Yoshitake, S., Schach, B.G., Foster, D.C., Davie, E.W. and Kurachi, K., Biochem., 24 (1985) 3736.
2. Hiskey, R.G., In Gross, E. and Meienhofer, J. (Eds.), The Peptides: Analysis, Synthesis, Biology, Volume 3, Academic Press, 1981, p.137.

Highly efficient peptide-amide preparation using a simultaneous multiple-peptide synthesizer with two novel acid labile linkers

K. Nokihara[1,2,3], E. Ando[2], M. Yamaguchi[2], K. Takeda[3] and M. Noda[3]

[1]Tokyo University of Agriculture and Technology, Koganei, Tokyo 184, Japan
[2]Bio-application Laboratory of Biotechnology Instruments Department and [3]Central Research Institute of Shimadzu Corp., Nishinokyo-Kuwabaracho 1, Nakagyoku, Kyoto, 604, Japan

Introduction

Many neuropeptides and hormones have an amide group at their C-terminus, which cannot be easily obtained genetically, hence their chemical synthesis is very important. Two linkers [1, 2] for the Fmoc SPPS are commercially available, although they require fairly high concentrations of TFA for complete cleavage. We have developed two novel acid-labile linkers, which contain the 10, 11-dihydro-5H-dibenzo[a,d]cycloheptene-5-yl group (5-dibenzosuberyl group: CHA), and the 5-H-dibenzo[a,d]cycloheptene-5-yl group (5-dibenzosuberenyl group: CHE), respectively [3]. We have further evaluated these linkers and established a commercial synthesis route and applied this to the synthesis of peptides related to the VIP-secretin family.

Results and Discussion

The optimized synthetic routes for CHA and CHE are illustrated in Figure 1. CHA was obtained in 90% and CHE in 30% over-all yield calculated from 2-metoxydibenzosuberone (MDBS), respectively. Both linkers (CHA and CHE) were incorporated onto the polyoxyethylene-polystyrene support and successfully used for synthesis of several peptides using a newly developed simultaneous multiple-peptide synthesizer [4], Shimadzu Model PSSM-8, equipped with eight independent channels. The linkers were compared with conventional linkers, RAM, introduced by Rink [1] and PAL [2].

Neurokinin A fragment, positions 6-10, which shows biological activity [5], was simultaneously prepared by the same protocol using the linkers listed in Table 1. After completion of assembly the peptidyl resin was treated with a TFA-scavenger cocktail. Aliquots of the cleavage mixture were analysis by RP-HPLC and compared with each other. Table 1 shows the half life time and complete time for cleavage. The results show that CHA-and CHE-peptide can be easily cleaved in a lower concentration of TFA. Disadvantage of PAL-linker are undesirable isomer formation in its preparation, low over-all yield, and the side reactions during storage of PAL-resin reported by A. Lobbia and W. Schafer [personal communication].

Table 1 *Cleavability of C-terminal amide linkers*

	90% TFA		25% TFA
	Half Life Time	Complete Time	Complete Time
CHA	~ 1 min	5 min	15 min
CHE	~ 1 min	< 5 min	7 min
RAM	22 min	50 min	< 60 min
PAL	17 min	45 min	< 60 min

PHI, Secretin, PACAP-38 and GRF were successfully synthesized using these novel linkers and easily purified by a single step of RP-HPLC. The peptides were analyzed by LSIMS, HPLC, sequencing and amino acid analysis, and found to be

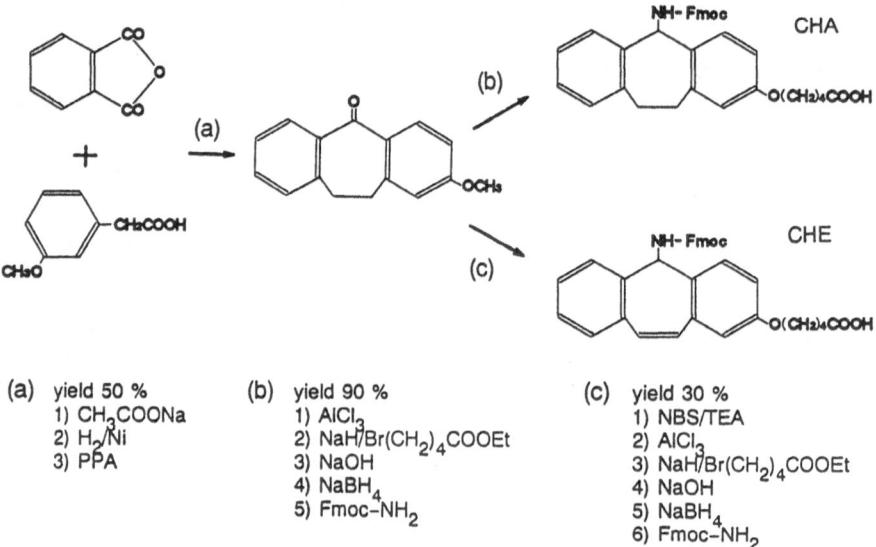

(a) yield 50 %	(b) yield 90 %	(c) yield 30 %
1) CH$_3$COONa	1) AlCl$_3$	1) NBS/TEA
2) H$_2$/Ni	2) NaH/Br(CH$_2$)$_4$COOEt	2) AlCl$_3$
3) PPA	3) NaOH	3) NaH/Br(CH$_2$)$_4$COOEt
	4) NaBH$_4$	4) NaOH
	5) Fmoc–NH$_2$	5) NaBH$_4$
		6) Fmoc–NH$_2$

Fig. 1. *Synthetic route for CHA- and CHE-linkers.*

86

highly homogeneous. As an example HPLC profiles of GRF using CHA and CHE are shown in Figure 2.

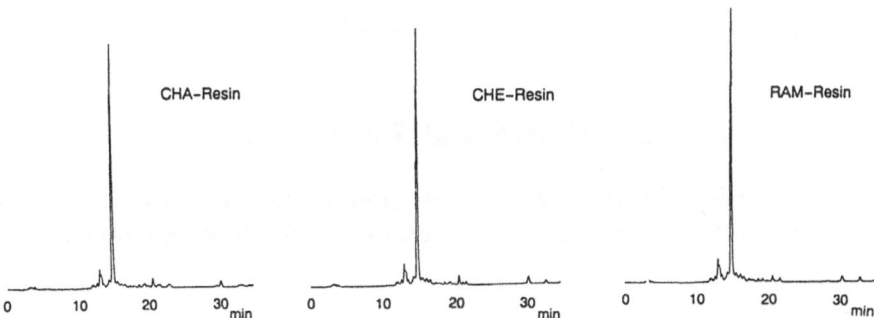

Fig. 2. *RP-HPLC of crude GRF (44 AA), Column: SynProPep RPC18 (4.6 x 150 mm); Eluent: 0.01N HCl/CH₃CN=80/20-50/50 (30 min); Flow rate: 1.0 ml/min; Absorbance 210 nm.*

As CHE was not particularly stable during storage at room temperature and over-all yield was much lower, we have concluded that CHA is a superior linker for preparation of peptide amides which contains acid labile residues such as Trp and/or Tyr(SO_3H).

References

1. Rink, H., Tetrahedron Lett., 28 (1987) 3787.
2. Albericio, F., Kneib-Cordonier, N., Biancalana, S., Gera, L., Masada, R.I., Hudson, D. and Barany, G., J. Org. Chem., 55 (1990) 3730.
3. Noda, M., Takeda, K., Ando, E., Yamaguchi, M. and Nokihara, K., In Yanaihara, N. (Ed.), Peptide Chemistry 1992, Escom Science Publishers BV, Leiden, The Netherlands, 1993, p.110.
4. Nokihara, K., Yamamoto, R., Hazama, M., Wakizawa, O. and Nakamura, S., In Epton, R. (Ed.), Innovation and Perspectives in Solid-Phase Synthesis 1992, Intercept Limited, Andover, 1992, p.445.
5. Nokihara, K., Yamaguchi, M., Ohmori, T. and Kuwahara, A., In Schneider, C.H. and Eberle, A. (Eds), Peptides 1992, Escom Science Publishers BV, Leiden, The Netherlands, 1993, p.667.

Development of a fully automated multichannel peptide synthesizer with an integrated TFA cleavage capability

J. Neimark and J.P. Briand

*UPR 9021 Immunochimie des Peptides et des Virus,
IBMC du CNRS, 15 rue Descartes, 67084 Strasbourg, France*

Introduction

A fully automated multichannel peptide synthesizer has been constructed which performs the simultaneous and rapid assembly of peptides on a 20 to 200 μmole scale. In situ activation of amino acids using BOP or PyBOP was chosen to give an optimized coupling chemistry. Specially designed blocks of valves with a zero dead volume, combined with an original circuitry, permit the distribution of amino acids derivatives and reagents pre-dissolved in DMF. Either Boc or Fmoc chemistry can be adapted on the synthesizer. In Boc synthesis a very rapid protocol involving Boc group deprotection in neat TFA followed by the concomitant steps of neutralization and coupling allow the addition of three amino acids per hour on each channel. In Fmoc chemistry we have integrated into the synthesizer an automatic TFA cleavage system which allows the peptides to be cleaved from the resin directly in the reactors used for synthesis. The stability of the Fmoc amino acid derivatives in solution in DMF was investigated and decomposition was found to be insignificant during the time - span of a synthesis. The satisfactory performance of the instrument was demonstrated by routine synthesis of 10 - 40 mer peptides.

Results and Discussion

The core of the multichannel synthesizer are new and highly reliable blocks of valves with a zero dead volume (Patented). Each block is a monolithic construction, consisting of several electrovalves. It is provided with a PTFCE - manifold body, in which a number of cells have been processed. A schematic diagram of the body of a zero dead volume 3 valve block is illustrated in Figure 1. Every cell is equipped with a chamber providing a central inlet (R1) with a truncated cone section, and with an eccentric bore-hole providing an outlet (R2) at the chamber bottom. A distribution channel connects the R2 cell outlets. Each cell is covered by a diaphragm which seals the R1 outlet by means of the driving power of type 127 Bürkert actuator. Each valve cell is free of dead volume, the cell being completely closed by the diaphragm when the valve is de-energized. The standardized zero dead volume valve block is available with 2 to 10 independent inlets and outlets.

Combined with an original circuitry, a series of valve blocks permit the amino acid derivatives and reagents to be transferred to different reaction vessels (4 on the prototype and 8 on the industrial machine) without cross contamination. The automatic module for delivery of amino acids is illustrated in Figure 2.

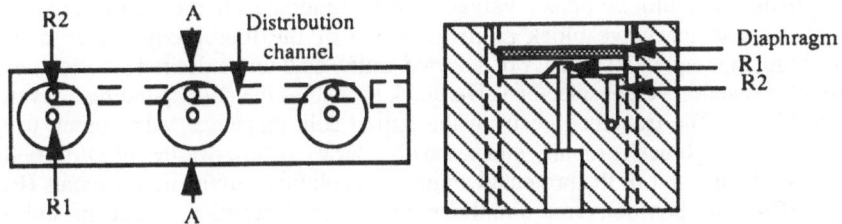

Fig. 1. Schematic diagram of the body of a zero dead volume 3 valve block.

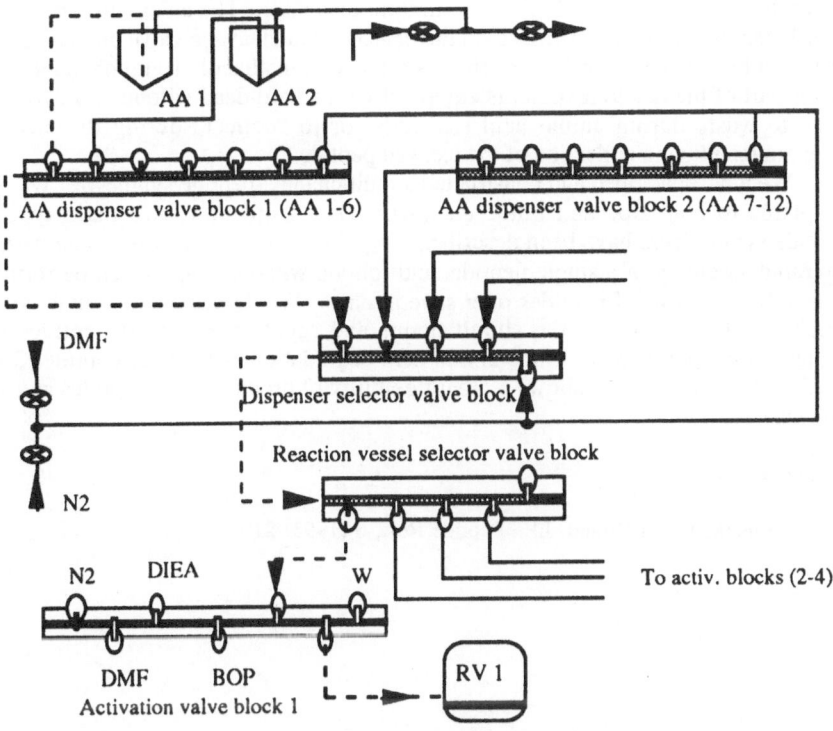

Fig. 2. Diagrammatic representation of the module for delivery of amino acids.

Protected amino acids are pre-dissolved in DMF at a concentration of 0.25 M in 40 ml conical glass tubes. Solutions of 0.5 M BOP and 1.0 M or 1.5 M DIEA in DMF are also pre-prepared. All these solutions are then maintained and delivered under a nitrogen pressure of 250 ±0.1 mbar. This highly stabilized nitrogen pressure is obtained by virtue of a medium pressure delivery unit and allows accurate time-controlled delivery of reagents. Amino acids are then delivered to the reaction vessels through a series of blocks of valves with zero dead volume as illustrated for the loading of amino acid AA1 shown as a dotted line in Figure 2. The amino acid is selected from the bank of stock solutions by a series of valves (one for each AA) comprising four blocks of six valves to accommodate all the amino acids. A dispenser selector valve block controls which of the four groups is used, thus permitting the sequential transfer of selected amino acid to the desired reaction vessel through a reaction vessel selector valve block as well as the corresponding activation valve block. The transfer line from the amino acid dispenser valve block to the activation valve block is rapidly cleaned with a series of flow-washes of DMF before and after the transfer of the protected amino acid solution. Activation reagents (BOP and DIEA) are then delivered simultaneously to each reaction vessel through the activation valve blocks.

For Fmoc chemistry an integrated TFA cleavage and deprotection system permits the automatic collection of the crude peptides. The cleavage reagent is placed separately in a conical glass tube (40 ml) under nitrogen pressure (300 mbar) and connected to a multichannel cleavage mixture dispenser. The acidic mixture is then distributed to each reaction vessel via the deprotection/cleavage valve blocks. This is followed by extensive washing of the circuitry with methanol (10 ml/channel). The lower part of the reaction vessel is equipped with a zero dead volume 2 valve block open to waste during amino acid assembly, or to "collect" during the cleavage program resulting in collection of the cleaved peptide.

We have developed and constructed a multichannel peptide synthesizer which is adaptable to both Boc and Fmoc chemistry [1]. Until now, two main types of peptide synthesizers have been described, i.e., monosynthesizers which can perform the rapid assembly of a single peptide, and robotic workstations which perform the assembly of dozens of peptides over several days. We describe here a new type of machine which performs the simultaneous and rapid assembly of peptides (not subject to the problems of Fmoc amino acid degradation at least for peptides 30-40 residues long) and which allows a great flexibility of production of peptides at a very low cost.

References

1.   Neimark, J. and Briand, J.P., Peptide Res., 6 (1993) 219.

A facile, efficient method for the simultaneous synthesis of high-quality individual peptides: Application of ACT Model 350 for rapid multiple synthesis of peptides by solid phase method

J. Zsigo and H. Saneii

Advanced ChemTech, Inc., 5609 Fern Valley Road
Louisville, KY 40228, U.S.A.

Introduction

There is an urgent, ever-increasing need for peptides in drug discovery and research programs, for developing monoclonal antibodies and vaccines as well as in diagnostics and therapeutics. More and more peptides with unusual amino acids, modified peptide bonds, linker or spacer molecules, and peptide mimetics are being prepared. Great progress has also been made in the development of antagonists and inhibitors of biologically active peptides. The demand exists for new strategies, faster syntheses, better coupling reagents and protecting groups and especially for methods for simultaneous preparation and analysis of very large numbers of peptides in a short time.

Automated peptide synthesizers were recently developed [1] which utilize known valving and agitation systems and are limited to the synthesis of three peptides without reprogramming. Here we report the use of an improved prototype of a fully automatized computer controlled peptide synthesizer with a robotic device for liquid transfer and with a motorized orbital shaker for resin agitation. After cleaving the peptide from the resin and removing the side protecting groups, a simple Sep-Pak purification step usually gave the desired peptides with a 95%+ quality, satisfactory for most biological investigations.

Results and Discussion

The ACT Model 350 peptide synthesizer first described by Groginsky [2] is equipped with a computer controlled orbital shaker and a high density polyethylene work table resistant to solvents and reagents. We applied FMOC-based chemistry and utilized both DIC/HOBt and TBTU activation for the simultaneous synthesis of up to 96 peptides.

The FMOC-protected amino acids were stored in stock solutions and were transferred to small reactor tubes. Only DMF was used as system fluid for washings. Table 1 below shows the protocol followed during the synthesis:

Table 1 *Synthesis protocol*

Step	Reagent	Time	Repetition
1 Deprotection	25% Piperidine/DMF	1min	1
		10 min	1
2 Wash	DMF	4 min	8
3 AA Addition	Stock solutions	3 min	1
	8-10 X excess		
4 Activation	Equimolar DIC/HOBt		1
5 Coupling		20 min	2
6 Wash	DMF	4 min	7
7 Acetylation	30% AC$_2$O in DMF	10 min	1
(Optional)			
8 Wash	DMF	4 min	7

Twenty-four hour per day operation is possible. After the initial calculations and set-up, only replenishing of solvent solution (DMF) and amino acids is required. Technician time amounts to under 10% of the total operating time. Despite the large excess of FMOC-amino acids used in each coupling reaction, the cost per cycle is very low due to the small quantities of solvents and reagents consumed. After the final FMOC deprotection, resins are washed with MeOH and ether and dried in high vacuo. Since TFA cleavage (scavengers are usually added depending on the peptide sequence) is done in the same tube as the synthesis, automation is possible at this step, too. The precipitated and lyophilized peptides are resolubilized and applied onto Sep-Pak cartridges. Depending on the component of interest, various cartridges can be used.

The chemically filtering C18 Sep-Pak Plus Cartridges with long body proved to be the most useful for peptides. The luer fittings for connection to pumps and vacuum manifolds can easily serve for further automatization. Changing the eluting MeCN concentration in 3 steps (10%, 25%, 40%), purity of 85-92% crude peptides can easily be increased up to 97-98%, enough for most *in vitro* and *in vivo* studies.

References

1. Jung, G. and Beck-Sickinger, A.G., Angewandte Chemie, 31 (1992) 367.
2. Groginsky, C.M., Amer. Biotech. Lab., 8 (1990) 13.

Orthogonal solid-phase synthesis of bicyclic analogues of α-conotoxin SI

N.A. Solé[1], S.A. Kates[2], F. Albericio[2] and G. Barany[1]

[1]Department of Chemistry, University of Minnesota, Minneapolis, MN 55455, U.S.A., and [2]Millipore Corporation, 75A Wiggins Avenue, Bedford, MA 01730, U.S.A.

Introduction

Conotoxins are a family of small, multiple disulfide-containing peptides isolated from the venom of marine snails, and are of interest because they are competitive antagonists of nicotinic acetylcholine receptors [1]. Recent work from one of our laboratories [2] evaluated several regioselective routes to the synthesis of α-conotoxin SI, which has the sequence H-Ile-Cys-Cys-Asn-Pro-Ala-Cys-Gly-Pro-Lys-Tyr-Ser-Cys-NH_2 and disulfide bridges connecting Cys^2 with Cys^7 and Cys^3 with Cys^{13}. In other ongoing work, we have optimized orthogonal solid-phase procedures for the construction of side-chain lactams by an Fmoc/tBu/allyl protection strategy [3]. The present contribution demonstrates and contrasts several strategies for the preparation of analogues in *each* of which one of the disulfide bridges of α-conotoxin SI is replaced by a side-chain lactam between Glu and Lys: Analogue 1: Glu^2 with Lys^7; analogue 2: Glu^3 with Lys^{13}; analogue 3: Lys^2 with Glu^7; and analogue 4: Lys^3 with Glu^{13}.

Results and Discussion

Linear chain assembly proceeded smoothly by Fmoc chemistry on tris(alkoxy)benzylamide (PAL) polyethylene glycol-polystyrene (PEG-PS) graft supports [4]. There followed resin-bound selective removal of the allyl ester and urethane respectively protecting Glu and Lys, using Pd(PPh$_3$)$_4$ (4 equiv) in NMM–HOAc–DMF (1:2:10), 3 h at 25°C under Ar (followed by washings with a solution of DIEA (0.5% v/v) and sodium diethyldithiocarbamate (0.5% w/v) in DMF). Next, side-chain cyclization was mediated by a number of activating reagents in DMF. The best cyclization yields (75 to 95%) were achieved upon 1 h activation at 25°C with BOP/HOBt/NMM (1:1:2), PyBOP/HOBt/NMM (1:1:2), HBTU/NMM (1:2), and particularly HATU/NMM (1:2). (HATU is a new activating agent; see ref. 5.) Monocyclic monomers were the predominant products, but some cyclodimerization also occurred under these conditions, as judged by gel filtration on Sephadex G-25 and FABMS. The solid-phase cyclization to form lactams proved to be sequence-dependent, with the small loop (e.g., analogues **1** and **3**) forming relatively quickly, whereas closing of the large loop (e.g., analogues **2** and **4**) was slower with possible side reactions. The carboxyl component of the lactam can be interchanged without affecting rates.

Syntheses continued by removal of the *N*-terminal Fmoc group, and then a TFA/scavenger cocktail [Reagent K, TFA–phenol–water–thioanisole–1,2-ethanedithiol (82.5:5:5:5:2.5), or Reagent B, TFA–phenol–water–tri(*iso*-propyl)silane (88:5:5:2)], for 2 to 3 h at 25°C, released peptide chains from the support concomitant with deblocking of Cys(Trt). Finally, the single disulfide bridge in these analogues was closed in dilute pH 8 solution (1 µM peptide) by air oxidation, 5 h to 3 d at 25°C. As with lactamization, disulfide bicyclization was also found to be sequence dependent, i.e., oxidation to form the small loop was 3-fold more rapid with respect to the large loop.

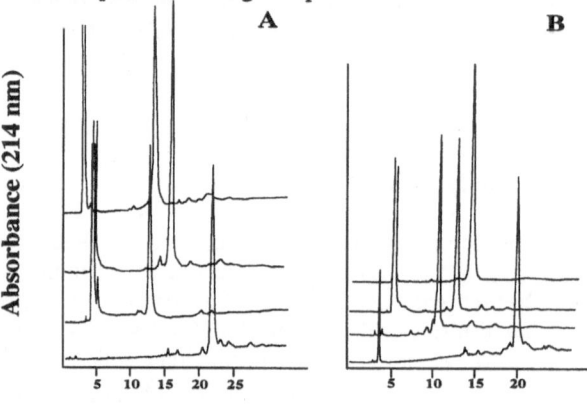

Retention time (min)

Fig. 1. HPLC chromatograms of peptides directly after cleavage. HPLC was carried out on a C18 reversed-phase column (0.46 x 25 cm), eluted with a linear gradient of 0.012 N aq. HCl and CH₃CN (9:1 to 3:2, over 30 min), flow 1 mL/min, detection 214 nm. Panels A and B are respectively analogues 1 and 2. Starting from bottom are shown protected intermediate (Al, Aloc, two SH), free peptide (SH form), monocyclic lactam (SH form), and bicyclic compound (lactam and S-S).

All desired intermediates and products from these syntheses were obtained with excellent initial purities, as judged by reversed-phase HPLC (Figure 1) and gel filtration, and by analytical characteristics including FABMS. Isolated yields after preparative chromatography were in the range of 40 to 60%.

Acknowledgements

We are grateful to NIH (GM 28934 and GM 43552) for financial support.

References

1. Zafaralla, G.C., Ramilo, C., Gray, W.R., Karlstrom, R., Olivera, B.M. and Cruz, L.J., Biochemistry, 27 (1988) 7102.
2. Munson, M.C. and Barany, G., J. Am. Chem. Soc., 115 (1993) 10203.
3. Kates, S.A., Daniels, S.B., Solé, N.A., Barany, G. and Albericio, F., In Hodges, R.S. and Smith, J.A. (Eds.), Peptides: Chemistry, Structure and Biology, Escom, The Netherlands, this volume.
4. Barany, G., Albericio, F., Solé, N.A., Griffin, G.W., Kates, S.,A. and Hudson, D., In Schneider, C.H. and Eberle, A.N. (Eds.), Peptides 1992 (Proceedings of the 22nd European Peptide Symposium), Escom, Leiden, The Netherlands, 1993, p.267.
5. Carpino, L.A., J. Am. Chem. Soc., 115 (1993) 4397.

Synthesis of cyclic peptides via efficient new coupling reagents

A. Ehrlich[1], S. Rothemund[1], M. Brudel[1], M. Beyermann[1], L.A. Carpino[2] and M. Bienert[1]

[1]Research Institute of Molecular Pharmacology, Alfred-Kowalke-Str. 4, D-10315 Berlin, Germany; [2]Department of Chemistry, University of Massachusetts, Amherst, MA 01003, U.S.A.

Introduction

Due to their restricted conformational flexibility, cyclic peptides are of great interest in connection with structure-activity relationships, especially the elucidation of bioactive conformations.

The aim of the present study was to compare the utility of various coupling reagents for peptide cyclization, including some newly developed reagents based on 1-hydroxy-7-azabenzotriazole (HOAt) [1, 2].

Results and Discussion

Using the linear GnRH-derived decapeptide **3** as a model, it was found that the newly developed uronium salts **1** (HAPyU) and **2** (TAPipU) [2], were highly effective for the cyclization (Figure 1A). These reagents led to complete cyclization within less than 30 min at a peptide concentration of 1.5 mM, whereas TBTU, TOPPipU [3] and DPPA gave only 60%, 10% and 12% cyclization, respectively. A 10%-excess of **1** or **2** was found to be sufficient for quantitative ring closure. The pyrrolidine salt 1 proved to be more active than the piperidine analog 2.

The cyclization was considerably accelerated by increasing the peptide concentration. Thus, the linear decapeptide was cyclized within 2 min at a peptide concentration of 0.1 M when reagents **1** and **2** were used. Surprisingly, even at high peptide concentrations (0.1 - 0.2 M), no intermolecular reactions were observed, indicating that at least for the head-to-tail and side-chain cyclizations studied, application of the principle of dilution is not required.

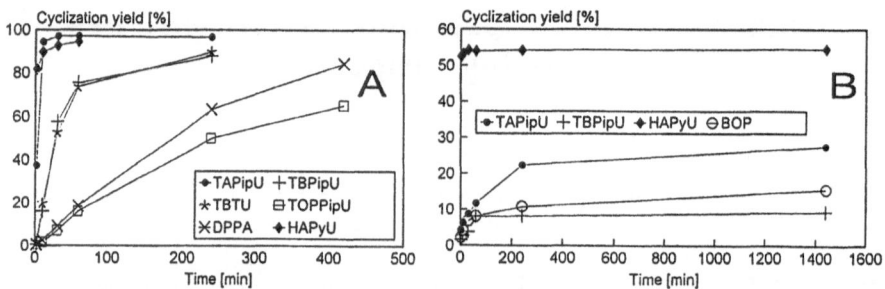

Fig. 1. Comparison of different coupling reagents for peptide cyclization step.
A: Cyclization of 3 c=0.0015 M); B: Cyclization of 4 (c=0.01 M); 3: H-Nal-D-Cpa-D-
Pal-Gln-Tyr-D-Arg-Leu-Arg-Pro-Lys(Ac)-OH; 4:H-Val-Arg-Lys(Ac)-Ala-Val-Tyr-OH.

Linear hexapeptides, constructed exclusively from L-amino acids and not
containing glycine or proline, are known to cause problems during attempted ring
closure reactions [4]. Indeed, our attempts to cyclize the hexapeptide **4**, $(10^{-2}M)$
failed, when BOP and TBPipU [5] were used (Figure 1B). A moderate yield (25%)
was obtained by use of compound **2**. The cyclization is however accompanied by
extensive racemization (BOP 24%, TBPipU 7%, TAPipU 8%) of the C-terminal
tyrosine residue. In remarkable contrast to these results, with HAPyU, **1**, the all L-
cyclohexapeptide, was formed in 55% yield within 30 min and less than 0.5% of the
D-Tyr-isomer was detected in the reaction mixture.

These results demonstrate the superiority of coupling reagents derived from 1-
hydroxy-7-azabenzotriazole, especially HAPyU, **1**, for promoting peptide cyclization
quickly and with a minimum of racemization.

Naturally, even these new activating agents cannot overcome difficulties arising
from unfavourable conformations of linear pentapeptides devoid of glycine and
proline and containing only L-amino acids [6]. Thus, our initial attempts to cyclize
H-Arg-Lys(Ac)-Ala-Val-Tyr-OH using HAPyU **1** resulted predominantly in the
formation of the corresponding dimer and cyclodimer.

References

1.   Sacher, R.M., Alt, G.H. and Darlington, W.A., J. Agr. Food Chem., 21 (1973) 132.
2.   Carpino, L.A., J. Amer. Chem. Soc., in press.
3.   Henklein, P., Beyermann, M., Bienert, M. and Knorr, R., In Giralt, E. and Andreu,
      D. (Eds.), Peptides 1990, Escom, Leiden, 1991, p.67.
4.   Kessler H. and Haase, B., Int. J. Peptide Protein Res., 39 (1992) 36.
5.   Knorr, R., Trzeciak, A., Bannwarth, W. and Gillessen, D., Peptides 1990, Escom,
      Leiden, The Netherlands, 1991, p.62.
6.   Kessler. H. and Kutscher, B., Liebigs Ann. Chem., 869-892, 1986, p.885.

Side chain to ψ(CH$_2$NH) backbone cyclizations in antiplatelet RGD peptides

S.L. Harbeson*, A.J. Bitonti, P.S. Wright and T.J. Owen

Marion Merrell Dow Research Institute, 2110 E. Galbraith Rd., Cincinnati, OH 45215, U.S.A.

Introduction

GPIIbIIIa is an integrin found on the surface of platelets [1] which has been shown to bind fibrinogen [2] and other adhesion proteins. These proteins contain the Arg-Gly-Asp (RGD) sequence, which is a critical recognition site for GPIIbIIIa [3]. Since cyclic peptides containing the RGD sequence have been shown to retain significant affinity for GPIIbIIIa [4, 5] we sought to develop a strategy for incorporating localized conformational constraints into a peptide chain by cyclization of a side chain onto the peptide backbone. The ψ(CH$_2$NH) reduced amide bond provides a convenient and reactive moiety in the peptide backbone which can be covalently linked to a side chain such as an Asp or Glu. Depending upon the synthetic strategy, the 2^0 amine of the ψ(CH$_2$NH) moiety (see compound **9** in Scheme 1) can be extended by either acylation or reductive alkylation. Since the conformational constraint has been introduced using a side chain functionality, the C-terminus could also be extended by replacing the benzyl amide with an amino acid or peptide. Cyclic peptides containing the RGD sequence were chosen as the test system for this cyclization strategy.

Results and Discussion

The cyclic peptides were synthesized by the general methodology shown in Scheme 1. The first two residues were incorporated as a dipeptide which was coupled through the side chain carboxylic acid to Kaiser's oxime resin [6, 7]. This dipeptide was prepared by standard solution phase methods. The N-terminus was then extended using the solid phase technique with iterative deprotection and coupling of N-t-Boc protected amino acid symmetrical anhydrides. Removal of the amino terminal Boc group allowed the simultaneous cyclization of the peptide and the cleavage of the peptide-resin bond [8]. This methodology was used to prepare the compounds shown in Tables 1 and 2. The data shown in Table 1 were used to optimize the structure of the cyclic peptide prior to initiating the synthesis of compounds containing the ψ(CH$_2$NH) moiety shown in Table 2. These peptides were prepared analogously except that the last residue was incorporated as an N-t-Boc-amino aldehyde by reductive alkylation of the amino terminal arginine using sodium cyanoborohydride

---

*Present address: Alkermes, Inc., 64 Sidney St., Cambridge, MA 02139, U.S.A.

[9]. The cyclic peptide with intact side chain protecting groups was further derivatized at the $\psi(CH_2NH)$ moiety by either reductive alkylation or by acylation. In all cases, the cyclic peptides were globally deprotected using anhydrous HF and purified by RP-HPLC. *In vitro* activity is reported as the ability to block GPIIbIIIa binding to fibrinogen in an ELISA and to inhibit ^{125}I-fibrinogen binding to GPIIbIIIa [10].

Scheme 1 *General synthetic route to conformationally constrained peptides*

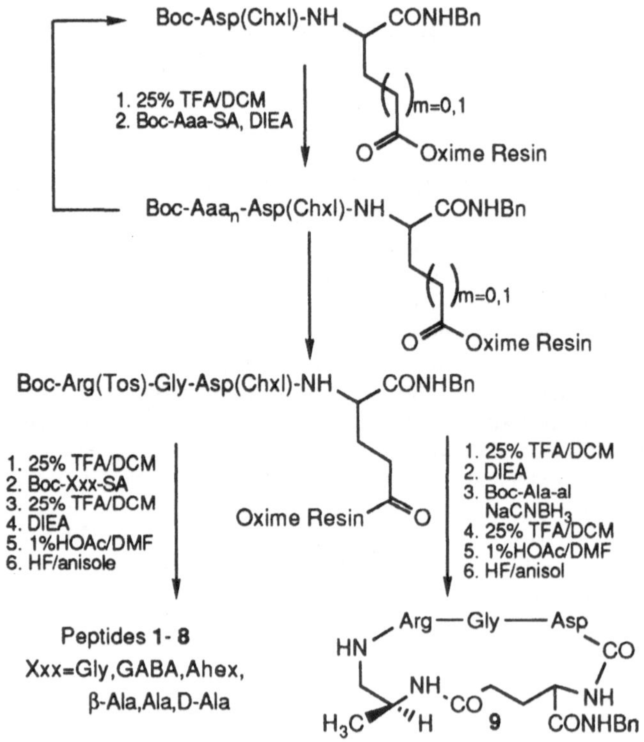

The results of varying the ring size of the cyclic RGD peptides are shown in Table 1. The side chain of either Asp or Glu (m= 1 and 2, respectively) was used in the cyclization to the α-amino of Arg via a second amino acid residue where n was varied from 1 to 5. Although the differences in the *in vitro* assays did not vary dramatically, these results showed that the best activity was obtained when m=2 and n=1 (**3**). The next structural modification was the use of a chiral amino acid residue to link the side chain of Glu to the α-amino group of Arg. A D-Ala residue (**7**) does result in greater affinity for the GPIIbIIIa receptor than an L-Ala (**8**). Although peptide **7** is more active, the compounds containing the $\psi(CH_2NH)$ moiety were synthesized using L-Ala for convenience.

Table 1 *Effects of ring size and chirality of linker*

Peptides	m	n	Linker	Size(atoms)	IC$_{50}$ (μM)	
					^{125}I-Fn	ELISA
1	1	1	β-Asp-Gly	16	4.0	1.0
2	1	2	β-Asp-β-Ala	17	4.0	1.0
3	2	1	γ-Glu-Gly	17	1.0	1.0
4	2	2	γ-Glu-β-Ala	18	1.5	1.0
5	2	3	γ-Glu-GABA	19	9.0	7.0
6	2	5	γ-Glu-Ahex	21	n.d.	11.8
7	2	1	γ-Glu-D-Ala	17	1.0	0.6
8	2	1	γ-Glu-Ala	17	4.5	0.8

Table 2 *Effect of cyclization via* $\psi(CH_2NH)$

Peptides	R	IC$_{50}$ (μM)	
		^{125}I-Fn	ELISA
9	-H	2.4	0.2
10	-(CH$_2$)$_3$CO$_2$H	1.2	0.54
11	-(CH$_2$)$_3$-phenyl	0.27	0.3
12	-CO-(CH$_2$)$_2$-NH$_2$	2.4	3.0

In order to extend the peptide chain from the α-amino group of Arg, the Ala-Arg amide bond must be converted into the reduced amide bond. This modification was accomplished to yield peptide **9** (Table 2) which did show greater affinity for the

GPIIbIIIa receptor than the parent peptide **8**. The observation that the ψ(CH$_2$NH) moiety is compatible with receptor binding suggests that R can be either acyl or alkyl. Reductive alkylation of this nitrogen with either succinic semialdehyde or dihydrocinnamaldehyde provided **10** and **11**, respectively, which still retained good affinity for the GPIIbIIIa receptor. Likewise, acylation of this nitrogen with β-Ala yielded **12** which appears to have similar activity to the parent compound **9**. These results show that cyclization of a side chain to the peptide backbone via a ψ(CH$_2$NH) moiety can be used to introduce a localized conformational constraint into a peptide.

References

1. Jennings, L.K. and Phillips, D.R., J. Biol. Chem., 257 (1982) 10458.
2. Bennett, J.S., Hoxie, J.A. and Leitman, S.F., Proc. Natl. Acad. Sci. U.S.A., 80 (1983) 2417.
3. Ruoslahti, E. and Pierschbacher, M.D., Cell, 44 (1986) 517.
4. Barker, P.L., Bullens, S., Bunting, S., Burdick, D.J., Chan, K.S., Deisher, T., Eigenbrot, C., Gadek, T.R., Gantzos, R., Lipari, M.T., Muir, C.D., Napier, M.A., Pitti, R.M., Padua, A., Quan, C., Stanley, M.S., Struble, M., Tom, J.Y.K. and Burnier, J.P., J. Med. Chem., 35 (1992) 2040.
5. Samanen, J., Ali, F., Romoff, T., Calvo, R., Sorenson, E., Vasko, J., Storer, B., Berry, D., Bennett, D., Strohsacker, M., Powers, D., Stadel, J. and Nichols, A., J. Med. Chem., 34 (1991) 3114.
6. Kaiser, E.T., Mihara, H., Laforet, G.A., Kelley, J.W., Walters, L., Findeis, M.A. and Sasaki, T., Science, 243 (1989) 187.
7. DeGrado, W.F. and Kaiser, E.T., J. Org. Chem., 45 (1980) 1295.
8. Ösapay, G. and Taylor, J.W., J. Am. Chem. Soc., 112 (1990) 6046.
9. Sasaki, Y. and Coy, D.H., Peptides, 8 (1986) 119.
10. Wright, P.S., Saudek, V., Owen, T.J., Harbeson, S.L. and Bitonti, A.J., Biochem. J., 293 (1993) 263.

Peptide cyclization on an oxime resin (the PCOR method): Reaction conditions and applications

G. Ösapay[1], H. Shao[1], M. Goodman[1] and J.W. Taylor[2]

1Department of Chemistry, University of California at San Diego, La Jolla, CA 92093, U.S.A.; 2Department of Chemistry, Rutgers University, Piscataway, NJ 08855, U.S.A.

Introduction

Cyclic peptides are usually synthesized to investigate the relationship between chiroptical properties or biological activities and the conformation of peptides. Consequently, the development of reliable and convenient methods to prepare cyclic peptides remains an important objective in peptide chemistry.

The strategy of the synthesis of cyclic peptides involves preparation of an open-chain peptide intermediate followed by cyclization. Side reactions such as dimerization and even cyclodimerization represent problems in the synthesis of cyclic peptides. Our recently developed approach of Peptide Cyclization on an Oxime Resin (the PCOR method) [1-2] overcomes many of the difficulties in the syntheses of cyclic peptides. The peptide chain is assembled on Kaiser-oxime resin followed by cyclization: the cyclic product is released in high yield and purity from the solid support during a carboxyl-catalyzed [3] intramolecular aminolysis of the peptidyl oxime ester bond.

Many factors influence the reaction yields: resin type; substitution level; amino acid sequence; character and chirality of N- and C-terminal amino acids; protecting groups, etc. Several of these factors are surveyed in this paper.

Results and Discussion

Peptide cyclizations can be accomplished both in solution and by solid phase reactions. In order to avoid intermolecular side reactions, high dilution is necessary in both cases. A low peptide substitution level on an oxime resin corresponds to a high dilution reaction in solution. The characteristics of an oxime resin strongly influence its applicability for cyclization reactions as will be discussed in this brief paper.

Oxime resins with 1 and 2% crosslinking

The Kaiser-oxime resin [4-5] is a 1% crosslinked, chemically modified polystyrene resin, which is usually used for batchwise peptide synthesis. In an attempt to use the oxime resin in a continuous flow peptide synthesizer, we prepared an oxime resin with 2% crosslinks to provide a more rigid resin support. The resin possessed a 0.38 mmol/g content of oxime groups and was usable in the normal method for the synthesis of the Merrifield's test peptide (LAGV). However, the ammonolysis of the peptide from the resin gave a very low (21%) yield in comparison with the results (85%) achieved by using of 1% crosslinked resin (0.54 mmol/g). The major portion of the peptide remained on the resin and could be titrated with picric acid. By using of a manually operated synthesizer, no better results were obtained. Thus we believe that the 2% crosslinked resin is not suitable for peptide synthesis in the PCOR method.

Substitution levels

A peptide segment of Cbz-Ala$_L$(BocAla$_L$OMe)-Gly-Phe-O-N=C<(oxime resin), where Ala$_L$ denotes each of the lanthionine amino acid ends constitutes an intermediate in the preparation of a lanthionine enkephalin [6]. This peptide has been assembled on a Kaiser-oxime resin (substitution level of oxime groups: 0.65 mmol/g) with two different peptide substitution levels, 0.22 and 0.38 mmol/g, respectively (Figure 1). The cyclization reaction resulted in 70-72 % yield of cyclic product in both cases. But at the higher peptide substitution level, the product was a mixture of cyclic monomer and cyclic dimer in the ratio of 2:1. Our experiments made it clear that higher cyclization yield of the monomer peptide is achieved at the lower substitution level. Also, at a very low substitution level (< 0.1 mmol/g), when acetyl capping of excess oxime groups was used, a transacetylation reaction was observed when the reactive acetyl oxime ester bonds react with the free amino groups of the peptide [2, 7].

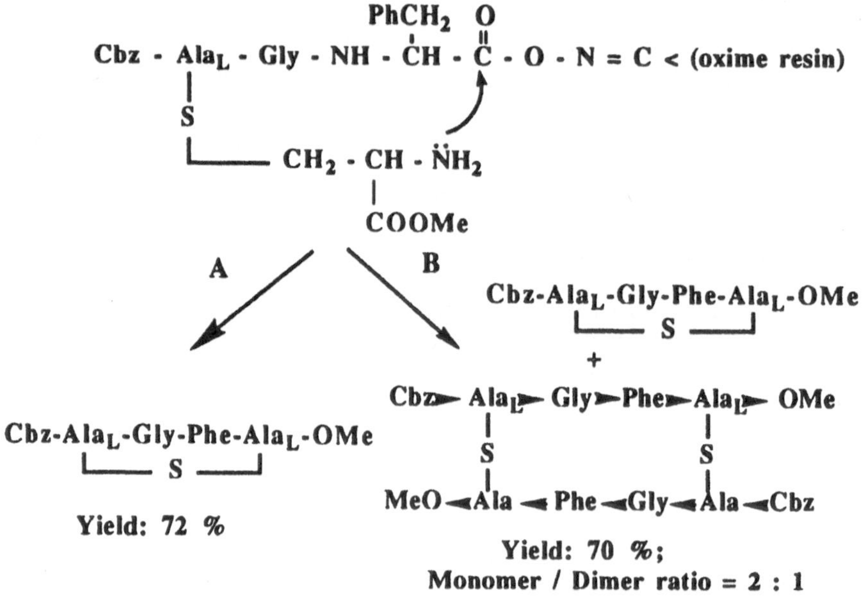

Fig. 1. *Effect of high substitution level for product formation. (A) Optimal peptide substitution level (0.22 mmol/g); (B) High peptide substitution level (0.38 mmol/g). This scheme shows a novel application of the PCOR method for preparation of lanthionine peptides.*

An optimal substitution level (0.2-0.3 mmol/g) can be achieved by allowing 1 g oxime resin to react with 0.4-0.45 mmol Boc-amino acid + DIC in DCM for 3 hrs. For loading of a cysteine derivative on an oxime resin, 2 molar equiv. of ethyl 2-(hydroximino)-2-cyanoacetate + DIC reagents resulted in a good substitution level.

Peptide couplings

For peptide couplings on an oxime resin, the BOP and UNCA activation methods, and the use of symmetrical anhydrides are commonly applied. We tried to carry out a segment condensation reaction on oxime resin through azide coupling: BocLys(2Cl-Cbz)-LeuN$_3$ + Lys(2Cl-Cbz)Glu(O-resin)OPac. During the long reaction time a large portion of the peptide chain cleaved from the resin, and a very low yield (16%) was observed. In conclusion, the azide coupling method cannot be applied in oxime resin syntheses.

Sequence dependence

The rate of oxime resin cyclizations (usually 24-72 hr) strongly depends on the nature of the amino acids at the C- and N-termini of a peptide chain [8]. For the synthesis of the monomeric lanthionine peptide in Figure 1, we attempted to cyclize a modified sequence, H-Gly-Phe-Ala$_L$[CbzAla$_L$(O-resin)]-OMe. No product formation could be detected. This result confirmed the observaton that to improve the yield in the PCOR method, the sequences of residues in the peptide chain play a key role in successful cyclization reactions.

Acknowledgements

This work was supported by grants NIH DA-05539, NIH DK-15410 (M.G.) and NIH GM-38811 (J.W.T.).

References

1.   Ösapay, G. and Taylor, J.W., J. Am. Chem. Soc., 112 (1990) 6043.
2.   Ösapay, G., Bouvier, M. and Taylor, J.W., In Villafranca, J.J. (Ed.), Techniques in Protein Chemistry II, Academic Press, Inc., San Diego, CA, U.S.A., 1991, p.221.
3.   Gisin, B.F. and Merrifield, R.B., J. Am. Chem. Soc., 94 (1972) 3102.
4.   DeGrado, W.F. and Kaiser, E.T., J. Org. Chem., 45 (1980) 1295.
5.   Nakagawa, S.H. and Kaiser, E.T., J. Org. Chem., 48 (1983) 678.
6.   Ösapay, G., Wang, S., Shao, H. and Goodman, M., In Yanaihara, N. (Ed.), Peptide Chemistry 1992 (Proceedings of the 2nd Japan Symposium on Peptide Chemistry), Escom, Leiden, The Netherlands, 1993, p.152.
7.   Bouvier, M. and Taylor, J.W., In Smith, J.A. and Rivier J.E. (Eds.), Peptides: Chemistry and Biology (Proceedings of the 12th American Peptide Symposium), Escom, Leiden, The Netherlands, 1992, p.535.
8.   Nishino, N., Xu, M., Fujimoto, T., Ueno, Y. and Kumagai, H., Tetrahedron Lett., 33 (1992) 1479.

Cyclic peptides and the MAP strategy

I.L. Picard, D.L. Wiegandt and A.F. Spatola

Department of Chemistry, University of Louisville, Louisville, KY 40292, U.S.A.

Introduction

The MAP (multiple antigen peptide) strategy has been used for preparing synthetic vaccine candidates but also has proven useful as a template for more complex immunological probes [1, 2] as well as for designing protein mimetics [3, 4]. We have been interested in using the MAP strategy to help answer the following questions: 1. Can cyclic peptides provide more readily identifiable and effective epitopes in preparing polyclonal and monoclonal antibodies? 2. To what extent can dimers and even higher order peptide analogs contribute to greater potency, selectivity, and biostability? 3. Can the combination of cyclization and the MAP amplification, together with appropriate ligand extensions, lead to binding to multiple receptors and thus help us to understand better the nature of the initial steps of hormone-receptor interactions?

Results and Discussion

Boc-Tyr(Bzl)-D-Lys(Z)-Gly-Phe-Asp(OBzl)-Xxx-OMe (Xxx=Ala and Leu) were prepared using solution phase methods with Boc protection and DCC/HOBt as coupling reagent. Ammonium formate-catalytic transfer hydrogenation (AF-CTH) afforded the product Box-Tyr-D-Lys-Gly-Phe-Asp-Xxx-OMe where Xxx=Ala in 94% yield and where Xxx=Leu in 97% yield. Side-chain to side-chain cyclization was achieved using 1 eq BOP and 5 eq $NaHCO_3$ in DMF in 51% yield with Xxx=Ala and in 49% where Xxx=Leu. Removal of the Boc group in HCl/dioxane resulted in the desired cyclic peptides H-Tyr-c[D-Lys-Gly-Phe-Asp]-Xxx-OMe [Xxx=Ala (IA) and Leu (IIA)]. The results of the GPI and MVD in vitro opioid assays are listed in Table 1, and indicate that IA shows high potency on both assays.

Table 1 *In vitro Opioid Assay of Cyclic Peptides*

	IC_{50}	
	GPI (nM)	MVD (nM)
H-Tyr-c[D-Lys-Gly-Phe-Asp]-Ala-OMe (IA)	3.14	1.54
H-Tyr-c[D-Lys-Gly-Phe-Asp]-Leu-OMe (IIA)	14.3	1.5

GPI = guinea pig ileum; MVD = mouse vas deferens

The dimeric and tetrameric lysine cores were synthesized by solid phase techniques on Merrifield resin using the Boc protecting group strategy and BOP/DIEA as the coupling reagent in DMF. After saponification of cyclic peptide IA, the free carboxylic acid was obtained. Using the solid phase synthesis strategy the cyclic peptide was coupled to the dimeric lysine core and also to the tetrameric lysine core. HF cleavage afforded the dimeric cyclic product (III) and the tetrametric cyclic product (IV) (Figure 1), whose structures were confirmed by FAB-MS.

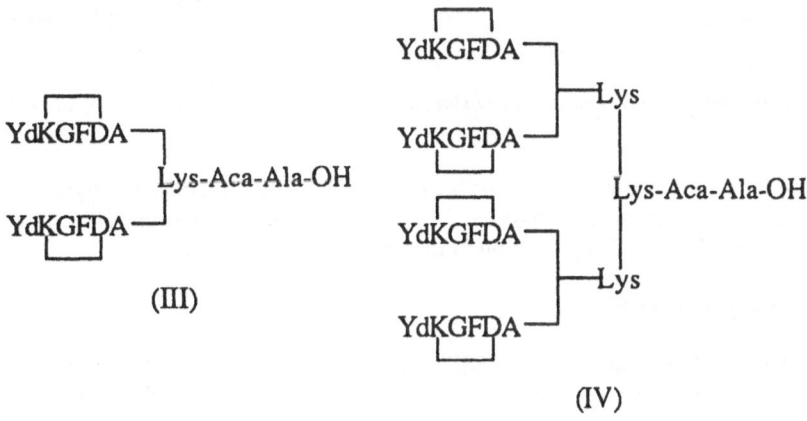

Fig. 1. *Structures of dimeric (III) and tetrameric (IV) MAPS-cyclic peptides.*

Various groups have reported the synthesis of dimeric peptides. In some cases these reports have suggested that the analogs were able to simultaneously bind to two receptor molecules while other workers report that only single binding occurred [5, 6]. Conn and coworkers were apparently able to convert a dimeric GnRH antagonist into an agonist by linkage through the corresponding dimeric GnRH antibody [7]. It seems reasonable to assume that, while the exact dimensions should be receptor specific, the distance spanned by two ligands must be reasonably large. The traditional MAP concept does not generally use variable linkers between the peptide and the lysine core. But this variable (position Z in Figure 2) would be the most appropriate method of altering this critical distance.

As a first step for extending the MAP concept to cyclic peptides, we required an appropriately configured biologically active sequence. This was accomplished by the design and synthesis of two new cyclic opioids (shown in Table I). The biological activities obtained for the methyl esters IA and IIa were quite good, especially for the alanine variant. On the other hand, neither of the two compounds proved very selective, so binding to either opioid receptor class is appropriate for assay.

Since the use of cyclic peptides in the MAP system is, to our knowledge, unprecedented, our first goal will be to test the immunogenicity of cyclic peptides compared to their closest linear counterparts. Such testing is currently in progress. In the meantime we have been able to achieve acceptable synthetic results (AAA, FAB-Ms) using both dimeric and tetrameric synthetic variants. An octameric version (Figure 2) is also being prepared.

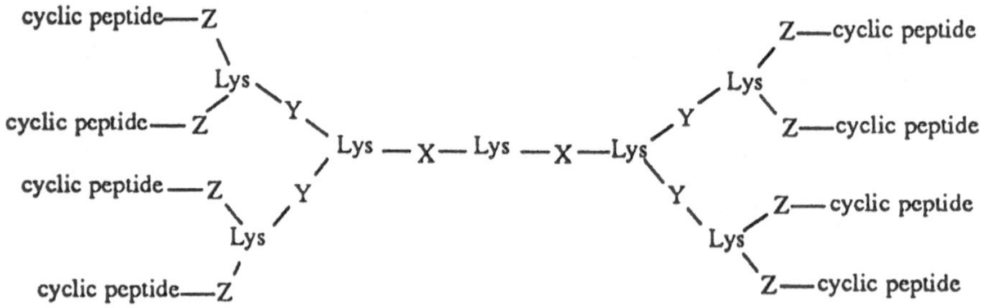

Fig. 2. Schematic representation of variables (X, Y, and Z) in the synthesis of MAP-based cyclic peptides.

Future studies will include verification of synthetic protocols, biological assays of various monomeric vs. oligomeric analogs, and determination of the effect of variables X, Y, and Z (Figure 2) on both synthesis and biology.

Acknowledgements

This work was supported by NIH-GM 33376, NIDA DA 04504 and fellowship support from the GAANN Program from the Dept. of Education. We thank Dr. Tom Burks for the bioassay results.

References

1. Tam, J.P., In Tam, J.P. and Kaiser, E.T. (Eds.), Synthetic Peptides: Approaches to Biological Problems, Alan R. Liss, Inc., New York, NY, U.S.A., 1989, p.3.
2. Kaumaya, T.P., Gerdau, A. and Kobs-Conrad, S.F., In Smith, J.A. and Rivier, J.E. (Eds.), Peptides: Chemistry and Biology (Proceedings of the Twelfth American Peptide Symposium), Escom, Leiden, The Netherlands, 1992, p.886.
3. Mutter, M., Carey, R.I., Dorner, B., Ernest, I., Flogel, R., Giezendanner, U., Rivier, J.E., Servis, C., Sigel, C., Steiner, V., Tuchscherer, G., Vuilleumier, S. and Wyss, D., In Smith, J.A. and Rivier, J.E. (Eds.), Peptides: Chemistry and Biology (Proceedings of the Twelfth American Peptide Symposium), Escom, Leiden, The Netherlands, 1992, p.326.
4. Stewart, J.M., Hahn, K.W. and Klis, W.A., Science, 248 (1990) 1544.
5. Schiller, P.W., In Ellis, G.P. and West, G.B. (Eds.), Progress in Medicinal Chemistry, 28, Elsevier, 1991, p.301.
6. Yagi, K., Shimohigash, Y., Ogasawara, T., Koshizaka, T., Waki, M., Kato, T., Izumiya, N. and Kurono, M., Biochem. Biophys. Res. Commun., 146 (1987) 1109.
7. Conn, P.M., Rogers, D.C., Stewart, J.M., Neidel, J. and Sheffield, T., Nature, 296 (1982) 654.

A novel Fmoc-protected ornithine derivative useful for the synthesis of arginine peptides by a guanylation approach

M.S. Bernatowicz and G.R. Matsueda

Bristol-Myers Squibb Pharmaceutical Research Institute, Princeton, NJ 08543, U.S.A.

Introduction

A method of solid phase synthesis of arginine-containing peptides by an approach which employs guanylation of ornithine-containing precursors and the "Boc/benzyl" protecting group combination has been recently reported [1]. The development and characterization of 1-guanylpyrazole hydrochloride 1 (Figure 1) as a guanylating reagent was key to the demonstrated success of the guanylation approach to Arg peptides by the "Boc/benzyl" strategy.

To apply this approach to synthesis by the "Fmoc/t-butyl" protection strategy, which may be advantageous for some peptides, it was necessary to develop a suitable protecting group for the δ-amino group of ornithine. This protecting group must fulfill the requirements of stability to peptide synthesis conditions used for chain elongation and must be specifically cleaved without affecting t-butyl based protection used for the other amino acids with nucleophilic sidechains. It was found that the novel 1-(4-methoxyphenyl)ethyloxycarbonyl (Mpeoc) group 2 (Figure 1), cleavable under mildly acidic conditions fulfills those requirements.

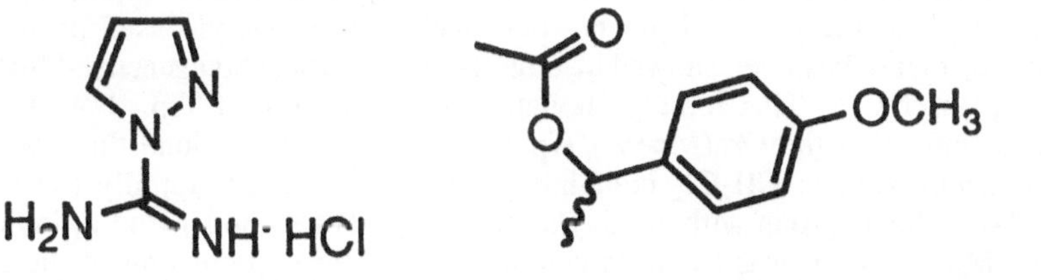

Fig. 1. Structures of 1-guanylpyrazole hydrochloride 1 and the Mpeoc group 2.

Results and Discussion

The synthesis of Fmoc-Orn(Mpeoc)-OPfp 3 is outlined in Figure 2. The copper(II) complex of ornithine [2] was allowed to react with Mpeoc-4-nitrophenylate to yield Cu{Orn(N$^\delta$-Mpeoc)}$_2$. The Fmoc group was introduced with Fmoc-Nsuccinimidate in the presence of EDTA and NaHCO$_3$ and the isolated Fmoc-Orn(Mpeoc)-OH product was converted to its pentafluorophenyl ester (OPfp) using pentafluorophenyltrifluoroacetate. The overall yields of Fmoc-Orn(Mpeoc)-OPfp **3** starting from free ornithine ranged from 21-35% but may improve with optimization. The authenticity of the Fmoc-Orn(Mpeoc)-OH and Fmoc-Orn(Mpeoc)-OPfp derivatives thus obtained was confirmed by ^{1}H NMR, FABMS and elemental analysis.

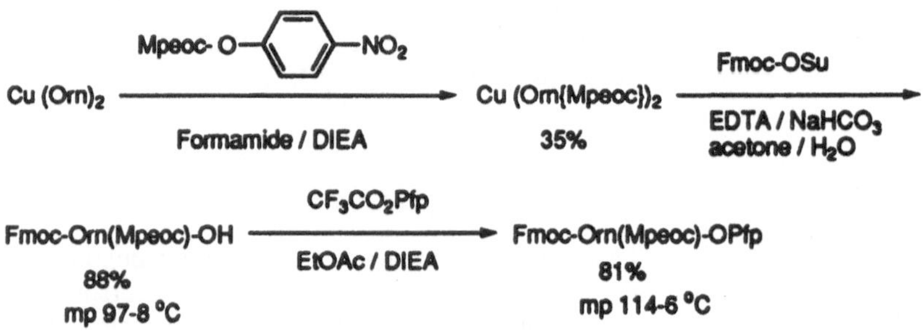

Fig. 2. Synthesis of Fmoc-Orn(Mpeoc) derivatives.

Because the Mpeoc-4-nitrophenylate reagent used to introduce the Mpeoc group slowly decomposes at room temperature, it (prepared by reaction of 1-(4-methoxyphenyl)ethanol with 4-nitrophenylchloroformate in pyridine/p-dioxane at 4°C followed by filtration to remove pyridine hydrochloride) was directly used without isolation. Because it was thought that the Fmoc-Orn(Mpeoc)-OH derivative might be unstable due to the acidity of its carboxyl group, it was initially promptly converted to its Pfp ester for long term storage. It was later found that the Fmoc-Orn(Mpeoc)-OH derivative showed no detectable decomposition after one year at room temperature.

Solution studies showed that the Mpeoc group was completely cleaved from Fmoc-Orn(Mpeoc)-OH with 1% TFA in CH$_2$Cl$_2$ in less than 10 min at room temperature, conditions which do not detectably affect t-butyl based protection. Additional studies, however, showed that the stabilized carbocation generated from the Mpeoc group by 1% TFA rapidly alkylated the indole ring of Trp. For use with peptides containing the Orn(Mpeoc)/Trp combination, it was found that 3% (v/v) dichloroacetic acid in CH$_2$Cl$_2$ containing 20% (wt/v) skatole rapidly (<10 min) cleaved the Mpeoc group without alkylation of Trp. Other model studies indicated that the Mpeoc group was stable to conditions used for deprotection of the Fmoc group as well as those used for coupling Fmoc-amino acids.

Table 1 lists arginine peptides prepared using **3** with guanylation by **1**. These peptides were prepared manually starting with Fmoc-Linker AM on Merrifield polystyrene using Fmoc-amino acid-Pfp esters and standard coupling and deprotection protocols. The N-terminal amino acid was Boc-protected (peptides **4** and **5**) since the Fmoc group is cleaved under the conditions employed for guanylation of Orn [1]. All the crude peptides obtained were >70% pure as ascertained by HPLC and were purified to homogeneity. Ornithine could not be detected by amino acid analysis of the crude products indicating complete guanylation reactions. Authenticity of the purified products was established by FAB MS and amino acid analysis.

Table 1 *Arginine peptides prepared using 3 with guanylation of Orn by 1*

Gly-His-Arg-Pro-Leu-Asp-Lys-NH$_2$ (**4**)
Gln-Gly-Val-Val-Pro-Arg-Gly-Val-Asn-Leu-Gln-Glu-NH$_2$ (**5**)
Ac-Trp-Trp-Pro-Trp-Arg-Arg-NH$_2$ (**6**), (Skatolicidin)

Peptide **6** (acetyl-Indolicidin 8-13 [3]) was chosen for synthesis to challenge stringently the method. Crude peptide **6** obtained using the Mpeoc approach and Reagent K [4] final cleavage for 1 h was 77% pure (Figure 3), an improvement over the 49% purity obtained for the same peptide synthesized by optimized conventional Fmoc chemistry (Reagent K, 2 h) utilizing the Pmc group for protection of Arg. In this case substantial peptidic impurities retaining the Pmc group were evidenced by FABMS and UV spectral data. Interestingly, when TFA/anisole/H$_2$O (80:15:5) was employed in place of Reagent K for final cleavage of **6** prepared by the Mpeoc approach, a crude product of similar quality (70% purity) was obtained.

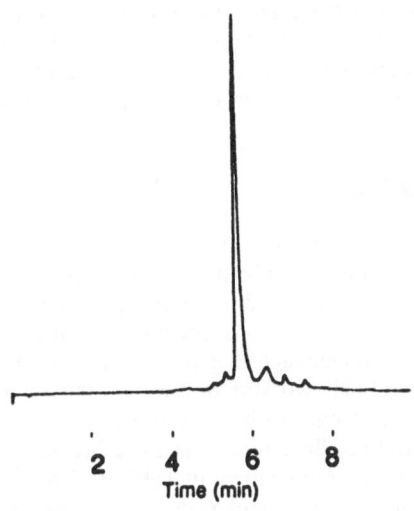

2   4   6   8
Time (min)

*Fig. 3. RP (C18, YMC, 4 mm X 5 cm column) HPLC of crude **6**. Linear gradient (6-64% ACN/H$_2$O, 0.1% TFA) at 2.0 ml/min.*

References

1. Bernatowicz, M.S., Wu, Y. and Matsueda, G.R., J. Org. Chem., 57 (1992) 2497.
2. Kurtz, A.C., J. Biol. Chem., 180 (1949) 1253.
3. Selsted, M.E., Novotny, M.J., Morris, W.L., Tang, Y.-Q., Smith, W. and Cullor, J.S., J. Biol. Chem., 267 (1992) 4292.
4. King, D.S., Fields, C.G. and Fields, G.B., Int. J. Pept. Protein Res., 36 (1990) 255.

Development of a new photo-removable protecting group for the amino and carboxyl groups of amino acids

C.P. Holmes and B. Kiangsoontra

Department of Chemistry, Affymax Research Institute, 4001 Miranda Ave., Palo Alto, CA 94304, U.S.A.

Introduction

We have reported recently a technique capable of generating arrays of immobilized peptides suitable for ligand binding assays through a combination of photochemistry and solid phase peptide synthesis [1]. Critical to this method is the use of a photo-removable protecting group for the amino terminus of the growing peptide chain. The now classic photo-protecting group is the nitroveratryloxycarbonyl (Nvoc) group first reported by Patchnornik [2]. As part of a program in novel photochemistry, we have begun to explore other nitrobenzyl compounds as alternatives to the Nvoc group.

Results and Discussion

A key element necessary for the productive photolysis of nitrobenzyl compounds is the presence of a benzylic hydrogen. Photolysis of carbamate protected amines proceeds through a complicated mechanism involving removal of the benzylic hydrogen, addition of the oxygen atom of the nitro group to the benzylic carbon, and ultimately through an acetal cleavage to generate a nitrosoaldehyde, carbon dioxide, and the liberated amine. Early workers in the field have noted the requirement for either an acid or other scavenger to be present in the reaction mixture in order to achieve a quantitative yield of the amine. We have investigated the photokinetics of several analogs of the Nvoc group shown in Table 1. The novel photo-groups were evaluated as both their acetates (acid protection) and their carbamates (amine protection). Aliquots of each compound (100 μM concentration) were photolyzed in 5 mM H_2SO_4/dioxane for specific times and the rate of the reaction was determined via HPLC analysis of the remaining starting material. The reactions were first order in light intensity. The UV light source was a Hg(Xe) ARC lamp with the 365 nm output selected with a monochrometer and adjusted to 10 mW/cm^2.

The series of nitrobenzyl compounds are equally effective as both amine and carboxyl protecting agents. We observed a dramatic rate enhancement upon substitution of the phenyl ring with two alkoxy substituents which parallels the increase in absorbance at 365 nm. A further increase in rate was observed by the introduction of a methyl group at the benzylic carbon. Presumably this stabilizes the intermediate carbon radical and promotes the cleavage reaction. The α-methyl-nitropiperonyloxycarbonyl (Menpoc, **5**) and the α-methyl-nitroveratryloxycarbonyl

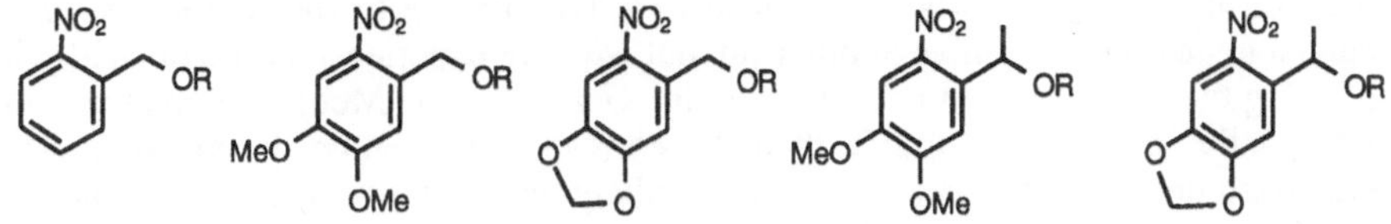

Fig. 1. Structures of nitrobenzyl derivatives.

Table 1 Photolysis half lives for the disappearance of starting material for various
Nvoc derivatives under 365 nm irradiation

Compound	R = COCH₃ (sec)	R = CONHCH₂CO₂Et (sec)
1	1941	--
2	129	97
3	65	59
4	42	38
5	33	40

(Menvoc, **4**) groups offer both faster photokinetics and higher quantum yields than the known nitrobenzyloxycarbonyl (Nboc **1**) and Nvoc (**2**) groups. The photolysis of the nitropiperonyloxycarbonyl (Npoc **3**) group is particularly enhanced in acidic dioxane, as its photokinetics more closely match that of Nvoc when performed in other solvents.

We have begun to further explore the use of the Menpoc group for the amine protection of amino acids. The crystalline chloroformate can be prepared in large scale and in high yield in three steps from commercially available 3,4-methylenedioxyacetophenone. This material serves as a convenient derivatization reagent and we have observed no discernible decomposition of the chloroformate after several months storage at 4°C. The intermediate α-methylnitrobenzyl alcohol can be coupled to a carboxylic acid directly, or alternatively converted to its chloride for derivatization of acids and other functional groups.

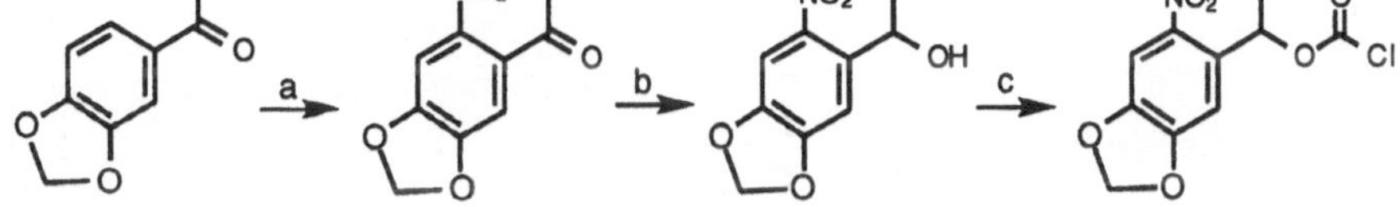

Fig. 2. Synthesis of Menpoc-Cl: (a) HNO3; (b) NaBH4; (c) phosgene.

Several common amino acids were chosen to examine in more detail. Experiments have shown that treatment of a free amino acid with Menpoc-Cl under water-dioxane conditions provides good yields of the desired products and material of high purity can be obtained after simple chromatography on silica gel. The methyl group on the benzylic carbon does lead to a 1:1 mixture of diastereomers when coupled to an optically active amino acid, but we have experienced little difficulty in obtaining the Menpoc amino acids as solids. Generally, the Menpoc derivatives are more easily induced to solidify than the corresponding Fmoc derivatives. The derivatized amino acids can be stored indefinitely when kept cool, dry and protected from light. The comparison of the Nvoc and Menpoc groups as photo-removable protecting groups for the amine of an amino acid is shown in Table 2.

Table 2 *Comparison of the photolysis half lives for carbamate protected Nvoc and Menpoc amino acids under 365 nm irradiation*

Amino acid[a]	Nvoc derivative (sec)	Menpoc derivative (sec)
PG-L-Ala-OH	92	23
PG-Gly-OH	110	29
PG-L-Phe-OH	96	27
PG-L-Leu-OH	115	24
PG-L-Tyr(OtBu)-OH	109	28

[a]PG denotes protecting group.

We have described a new photoprotecting group for protecting either the amino or carboxyl groups of amino acids. The α-methylpiperonyl group offers superior photokinetics and a higher quantum yield as compared to the nitroveratryl group. This new group should also prove useful for caging purposes and the protection of other functional groups in addition to amines and carboxyls where higher quantum yields are important.

References

1. Fodor, S. P. A., Read, J. L., Pirrung, M. C., Stryer, A. T., Lu, A. and Solas, D., Science, 249 (1991) 404.
2. Patchnornik, C. A., Amit, B. and Woodward, R. B., J. Am. Chem. Soc., 92 (1970) 6333.

Automated allyl chemistry for solid-phase peptide synthesis: Applications to cyclic and branched peptides

S.A. Kates[1], S.B. Daniels[2], N.A. Solé[3], G. Barany[3] and F. Albericio[1]

[1]Millipore Corporation, 75A Wiggins Avenue, Bedford, MA 01730, U.S.A., [2]Waters Chromatography Division, Millipore Corporation, 34 Maple Street, Milford, MA 01757, U.S.A., and [3]Department of Chemistry, University of Minnesota, Minneapolis, MN 55455, U.S.A.

Introduction

Allyl-based protecting groups have been used extensively in organic synthesis and have been applied recently to DNA, carbohydrate and peptide synthesis. The mild conditions to remove allyl functions are compatible with classical Fmoc/tBu based methods for solid-phase peptide synthesis [1, 2]. The combination of allyl/Fmoc/tBu provides a flexible orthogonal strategy that permits construction of more complex peptides including peptides which are cyclic, have branching, contain post-translational modifications or are conjugated to sugars and oligonucleotides [2-4].

Allyl removal is generally achieved with a suspended palladium catalyst under heterogeneous conditions and inert atmosphere. After completion of the reaction, the catalyst must be separated. This overall approach is not feasible on batch or continuous-flow peptide synthesizers due to difficulties in initial delivery and later removal of the insoluble palladium catalyst. We have now examined and optimized solvents to solubilize the catalyst, prevent undesired N^α-Fmoc deprotection, and be compatible with glyco- and sulfo-peptides. The continuous-flow Millipore 9050 Plus PepSynthesizer was modified using new conditions and software to carry out allyl deprotection schemes for the facile preparation of cyclic and branched peptides [5].

Results and Discussion

Model peptides Y-Ala-X-Gln-Lys-Thr-Asp(OAl)-Thr-Pro-NH$_2$ where X = Met, Trp and Y = Fmoc, H were prepared. We found that the previously described conditions for allyl removal, i.e., Pd(PPh$_3$)$_4$ (3 equiv) and 0.3 M morpholine in DMSO-THF-0.5 N HCl (2:2:1) [6], were compatible with Trp, but led to partial oxidation of Met and partial loss of N^α-Fmoc. Furthermore, the earlier cocktail was biphasic and a poor solvent for the catalyst. To avoid a biphasic system, HCl was replaced with HOAc, a mild organic acid that is compatible with sensitive functions. CHCl$_3$ was chosen as the solvent since it dissolves completely the palladium catalyst and is compatible with the PEG-PS resins which are used for all of the work described. Lastly, a weak, tertiary base, N-methylmorpholine [which does not remove N^α-Fmoc functions] was added. Hence, treatment of the model peptides with

Pd(PPh$_3$)$_4$ (3 equiv) in CHCl$_3$-HOAc-NMM (37:2:1), followed by washings with a solution of DIEA (0.5%, v/v) and sodium diethyldithiocarbamate (0.5%, w/v) in DMF, provided the allyl-deblocked peptide while preserving N^α-Fmoc.

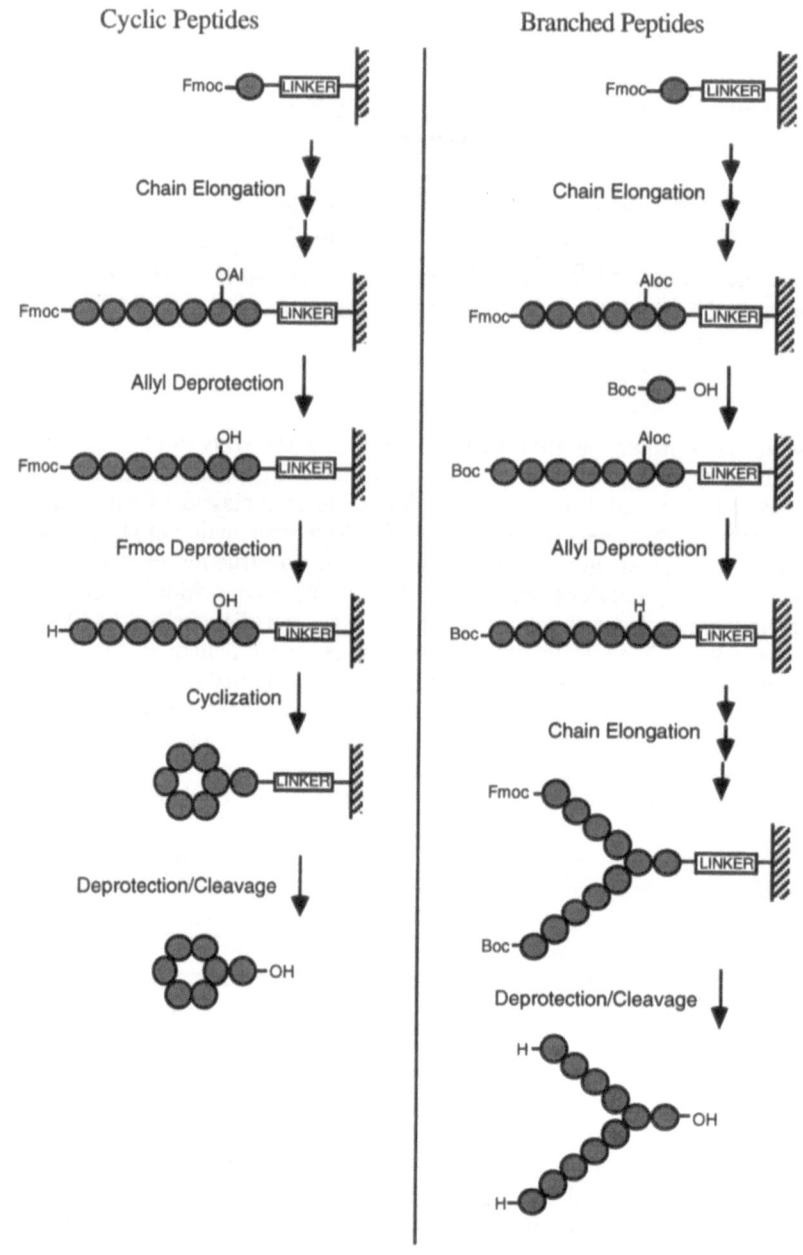

Scheme 1. General strategy for synthesis of cyclic and branched peptides using Fmoc/tBu/allyl chemistry.

Cleavage of allyl groups on continuous-flow or batchwise peptide synthesizers was automated, as follows: the catalyst was added to standard amino acid vials, purged with Ar, and sealed in order to maintain an inert atmosphere. The $CHCl_3$-HOAc-NMM cocktail described above was delivered to the vial containing $Pd(PPh_3)_4$, and bubbling of the mixture with Ar for 5 min ensured that the catalyst was dissolved completely before delivery to the reaction column. In the continuous-flow mode, the deprotection solution was delivered to the column containing the resin and recycled for 2h. An additional resin wash cycle with the DIEA and sodium diethyldithiocarbamate in DMF solution was used to remove side-products that can produce back-pressure. For the preparation of cyclic peptides, phosphonium or uronium salts were added to an amino acid vial, and a solution of 0.6 M DIEA in DMF was delivered to dissolve the solids. This activation solution was then delivered to the peptide-resin with free end groups in order for the cyclization to occur; recycling continued until the peptide-resin was ninhydrin negative.

The effectiveness of our new conditions was demonstrated first by the automated synthesis of a cyclic peptide, tachykinin antagonist peptide MEN 10355, *cyclo*(Asp-Tyr-DTrp-Val-DTrp-DTrp-Arg)(Scheme 1, left side). The linear sequence was assembled using the Fmoc/*t*Bu strategy, and selective side-chain deblocking of allyl groups followed. Next, N^α-Fmoc removal, PyBOP-HOBt-DIEA-mediated cyclization and acidolytic cleavage of the peptide-resin and of *t*Bu side-chain protecting groups gave the cyclic peptide. For the synthesis of branched peptides (Scheme 1, right side), a Fmoc-Lys(Aloc)-PAL-PEG-PS resin was the starting point for a linear synthesis of prothrombin(1-9) by the Fmoc/*t*Bu strategy [terminating with N^α-Boc]. Side-chain deprotection of the Aloc group of Lys was followed by a linear synthesis of Leu-enkephalin off the ε-amino side-chain of lysine.

Multiple antigenic peptides (MAP) are of interest since they allow production of immunogens without the use of a protein carrier. A small lysine core index is used to amplify the peptide antigen to multiple copies. The current strategy for obtaining MAP containing two epitopes starts with Fmoc-Lys(Boc)-OH and requires Boc chemistry. With our current method, the preparation of MAP containing two epitopes can be achieved and automated with Fmoc chemistry. Starting with [Fmoc-Lys(Aloc)]$_4$-βAla-PAL-PEG-PS resin, a bis-epitope of the HIV protease ([HIV(16-29)]$_4$[HIV(70-80)]$_4$)-(Lys)$_7$-βAla-NH$_2$ was synthesized.

References

1. Lyttle, M.H. and Hudson, D., In Smith, J.A. and Rivier, J.E. (Eds.), Peptides: Chemistry and Biology (Proceedings of the Twelfth American Peptide Symposium), 1992, p.583.
2. Albericio, F., Barany, G., Fields, G.B., Hudson, D., Kates, S.A., Lyttle, M.H. and Solé, N.A., In Schneider, C.H. and Eberle, A.N. (Eds.), Peptides 1992 (Proceedings of the Twenty-Second European Peptide Symposium), Escom, Leiden, The Nethlerlands, 1993, p.191.
3. Kates, S.A., Solé, N.A., Johnson, C.R., Hudson, D., Barany, G. and Albericio, F., Tetrahedron Lett., 34 (1993) 1549.
4. Solé, N.A., Kates, S.A., Albericio, F. and Barany, G., this volume.
5. Kates, S.A., Daniels, S.B. and Albericio, F. Analytical Biochem., 203 (1993) 245.
6. Lloyd-Williams, P., Jou, G., Albericio, F., and Giralt, E., Tetrahedron Lett., 32 (1991) 4207.

S-Phenylacetamidomethyl (Phacm). A versatile cysteine protecting group for Boc and Fmoc solid-phase synthesis strategies

M. Royo[1], J. Alsina[1], E. Giralt[1], U. Slomczynska[1,2] and F. Albericio[1,3]

[1]Department of Organic Chemistry, University of Barcelona, E-08028 Barcelona, Spain
[2]Institute of Organic Chemistry, Technical University, 90-924 Lódz, Poland, and
[3]Millipore Corporation, 75A Wiggins Avenue, Bedford, MA 01730, U.S.A.

Introduction

Side-chain cysteine protection is essential during the coupling steps of solid-phase peptide synthesis [1]. The synthesis of peptides having all cysteine residues as the free thiol or with one or more disulfide bridges formed in a controlled fashion is dependent upon to the availability of appropriate orthogonal protecting groups for the thiol function. In addition to the common protecting groups, an alternative group removable by an independent method such as enzymatic cleveage, would be desirable.

During the last few years, enzymatic methods have been more frequently used to remove blocking groups from α-amino and α-carboxy groups rather than from side-chain functionalities in peptide synthesis. Enzymes often operate at neutral, weakly acidic or weakly basic pH media and in many cases combine a high selectivity for the reactions they catalyze and the structures they recognize with a broad substrate specifity. Thiol protecting groups of the acylamidomethyl type can be split in a two-step process [2] by first eliminating the acyl component and then generating the thiol function by spontaneous hydrolysis of the intermediate aminomethyl mercapto compound. Best results were obtained with the phenylacetamidomethyl group (Phacm) [3]; this group being deprotected by penicillin G acylase from *E. coli* (EC 3.5.1.11) with a P1 specificity for the phenylacetyl residue.

The selective enzymatic deprotection of the S-phenylacetamidomethyl is orthogonal to the chemical deprotection of several groups used for the protection of thiol side-chain of cysteine in peptide synthesis. This characteristic encouraged us to synthezise the Boc and Fmoc derivatives of S-phenylacetamidomethyl-L-cysteine (Figure 1) for use in the most important strategies in solid-phase synthesis (Boc/Bzl and Fmoc/tBu) to obtain cystine-peptides.

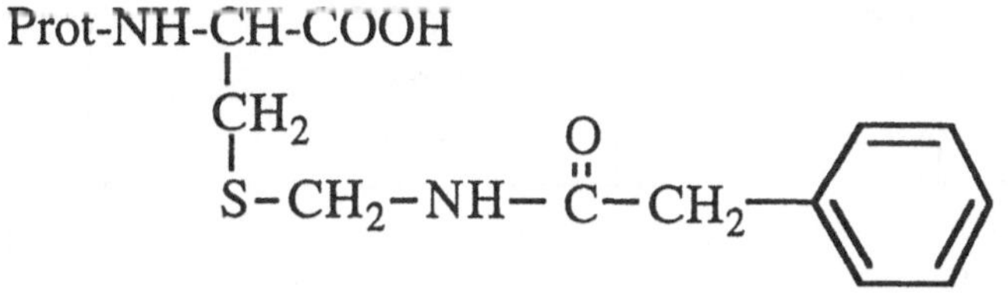

Prot-NH-CH-COOH
 |
CH$_2$
 |
S-CH$_2$-NH- C-CH$_2$-〈benzene ring〉
           ‖
           O

Prot = Boc or Fmoc

Fig. 1. N$^{\alpha}$-Protected S-phenylacetamidomethyl-L-cysteine.

Results and Discussion

The S-phenylacetamidomethyl (Phacm) group can be introduced using the same protocol as that used for the S-acetamidomethyl (Acm) group. L-Cysteine was reacted with N-hydroxymethylphenylacetamide in the presence of trifluoromethanesulfonic acid, and subsequent treatment with Fmoc-succinimide or di-*tert*-butyl dicarbonate gave the appropriate fully protected derivatives in 67% and 56% yields, respectively. The Phacm group has the same stability/lability as Acm, but Phacm is additionally cleaved by the action of penicillin amidohydrolase (PHA). The best conditions involve use of amidohydrolase immobilized on acrylic beads (Eupergit C, Röhm Pharma, Germany) in sodium phosphate, ammonium acetate or triethylammonium acetate (pH 7.8) at 38°C. The latter two buffers have the advantage that the salt can be removed by lyophilization. If the enzymatic hydrolysis is carried out under argon in the presence of 2% β-mercaptoethanol, the free thiol is isolated; in their absence direct oxidation to the cystine derivative takes place. Thus, the Phacm group is orthogonal with respect to both base and acid-labile cysteine protecting groups, and is compatible with Acm (initial enzymatic deprotection of Phacm followed by deprotection of Acm with iodine or thallium salts).

Several cystine-containing peptides were synthesized to demonstrate the scope of this protecting group in solid-phase synthesis.

Dimer β-turn model peptide was prepared to show the compatibility of Phacm with Acm (Figure 2). This parallel dimer was prepared from a single peptide [(Ac-Cys(Phacm)-Pro-D-Val-Cys(Acm)-NH$_2$)], which was synthesized manually using a Boc strategy, starting with an MBHA-resin. Treatment of the peptide-resin with anhydrous HF-anisole (19:1), 1 h at 0°C, provided the crude protected peptide I. The protected peptide I was then incubated with immobilized penicillin amidohydrolase in 0.05 M ammonium acetate buffer pH 7.9 at 38°C and after 24 h the open dimer II was obtained (90% purity). The second disulfide bond was formed upon treatment of the bis-S-Acm peptide with I$_2$ (10 equiv.) in HOAc-H$_2$0 (4:1), 2 h at 25°C.

The synthesis of somatostatin-14-amide was accomplished automatically on a continuous-flow synthesizer (Millipore 9050), according to the standard Fmoc/^{t}Bu strategy using the PAL-handle [4] on a PEG-PS-resin. Treatment of the peptide-resin with TFA-H$_2$O (19:1), 1 h at 25°C, provided the crude bis(Phacm) protected peptide, which subsequently was incubated with inmobilized PHA in 0.05 M triethylammonium acetate buffer, pH 7.8, at 38°C for 24 h. The reaction was followed by HPLC. After 20 h, the desired final product was obtained (85% purity).

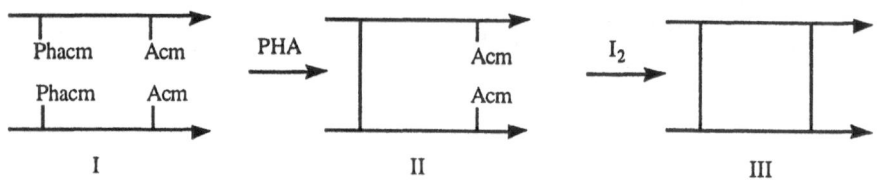

Fig. 2. Scheme of synthesis of dimer β-turn model peptide.

Acknowledgements

This work was partially supported by CICYT (PB91-283) and Commission of the European Communities (SC1-CT91-0748). The immobilized amidohydrolase was generously provided by Dr. Annette Brenmer (Röhm Pharma, Germany) and Antibióticos S.A. (Spain).

References

1. Andreu, D., Albericio, F., Solé, N.A., Munson, M.C., Ferrer, M. and Barany, G., In Pennington, M.W. and Dunn, B.M. (Eds), Peptide Synthesis and Purification Protocols, Humana Press, Clifton, NJ, U.S.A., 1993, in press.
2. Hermann, P., Hoffmann, G., In Loffet, A., (Ed.), Peptides 1976, Editt. de l'Univ. Bruxelles, Belgium, 1977, p.121.
3. Greiner, G. and Hermann, P., In Giralt, E. and Andreu, D. (Eds), Peptides 1990. Proceedings of the Twenty-First European Peptide Symposium, Escom, Leiden, The Netherlands, 1991, p.277.
4. Albericio, F., Kneib-Cordonier, N., Biancalana, S., Gera, L., Masada, R.I., Hudson, D. and Barany, G., J. Org. Chem., 55 (1990) 3730.

The carba-modification of cystine-containing peptides: Synthesis of selectively protected cystathionines and their incorporation into the oxytocin molecule

P. Šafář[1], J. Slaninová[2] and M. Lebl[1]

[1]Selectide Corp., 1580 E. Hanley Blvd., Tucson, AZ 85737, U.S.A.
[2]Institute of Organic Chemistry and Biochemistry AVČR,
Flemingovo nam. 2, 166 10 Prague 6, Czech Republic

Introduction

The reduction of disulfide bridge in cystine containing peptidic hormones results in many cases in loss of biologically active conformation. The replacement of the chemically labile disulfide group by an isostere structure is of interest for new stable structures of hormone analogs. The methylene-thio-group ($-CH_2-S-$) was introduced as a suitable replacement of the disulfide bond [1, 2]. Several types of cystine-substituting amino acids were prepared [1-9], but only several examples maintaining all functional groups and the length of the bridge are known [10, 11]. In this study, the selectively protected monocarba-isosteres of cystathionines were prepared as building blocks and applied in a combined stepwise solid phase/solution synthesis for a series of analogs of the neurohypophyseal hormone oxytocin.

Results and Discussion

Selectively protected cystathionines **1-5** (Figure 1) were synthesized using S-alkylation [12] of cysteine with fully protected derivatives of 2-amino-4-bromobutyric acid [10] and subsequent introduction of Fmoc-group (compound **3-5**). The derivatives **2** and **3** were used for the synthesis of carba-1-oxytocin (C^1-OXT) chosen as a model peptide. The procedure using derivative **2**, pMBHA resin and Boc/Bzl strategy gave in 12 synthetic and purification steps 6.8% of C^1-OXT. In contrast, our new approach using Fmoc-protected derivative **3**, acid labile handle and Fmoc/Bzl strategy (Figure 2) yielded 63.8% of C^1-OXT after optimization. The cyclization was carried out by DPPA/DIEA (10eq) in DMF with 78% preparative yield. For the final deprotection, a mixture TFMSA/TFA/thioanisole/m-cresole/DMS (5/85/5/2.5/2.5) was used with the yield of 81%. As an alternative method, derivative **3** and SCAL-handle [13] with on-the-resin cyclization was tested with the yield 34% of C^1-OXT. The synthetic procedure using cystathionine derivative **3** was further applied for a series of analogs of oxytocin modified in position 4 and/or 7 (agonists) and positions 2,4,8 (inhibitors). Biological evaluation of these analogs allowed to differentiate the influence of deamination in position 1 and carba modification of the bridge on their agonistic and/or antagonistic properties.

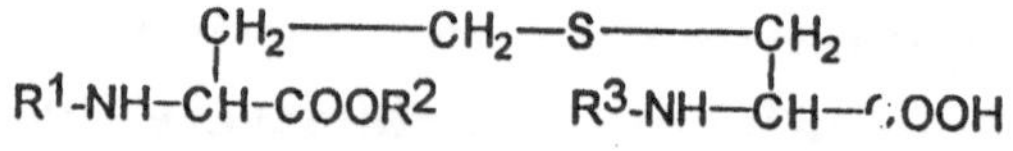

$$CH_2\text{------}CH_2\text{--}S\text{-------}CH_2$$
$$R^1\text{-NH--CH--COOR}^2 \qquad R^3\text{-NH--CH--COOH}$$

1. R^1=ClZ, R^2=Me, R^3=Boc   **3.** R^1=ClZ, R^2=Bu^t, R^3=Fmoc
2. R^1=Z, R^2=Me, R^3=Boc   **4.** R^1=Z, R^2=Bu^t, R^3=Fmoc
                                    5. R^1=Boc, R^2=Bu^t, R^3=Fmoc

Fig. 1. Selectively protected cystathionines.

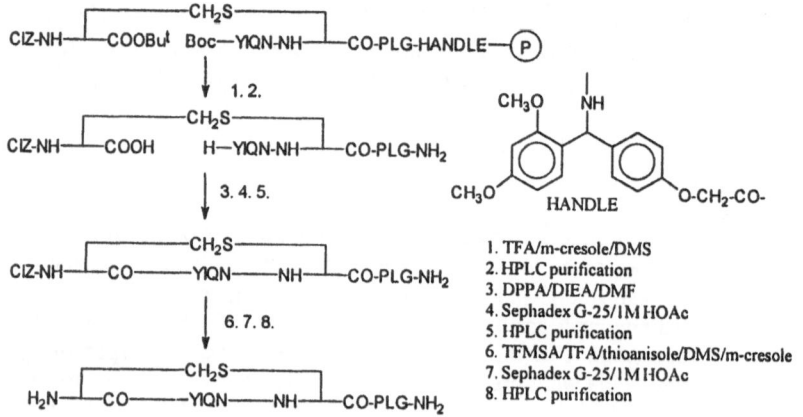

Fig. 2. Optimized strategy for the synthesis of C^1-OXT.

References

1. Rudinger, J. and Jošt, K., Experientia, 20 (1964) 570.
2. Jošt, K., Collect. Czech. Chem. Commun., 36 (1971) 218.
3. Lebl, M., Hruby, V.J., Slaninová, J. and Barth, T., Collect. Czech. Chem. Commun., 50 (1985) 418.
4. Brtník, F., Barth, T. and Jošt, K., Collect. Czech. Chem. Commun. 46 (1981) 278.
5. Jošt, K. and Šorm, F., Collect. Czech. Chem. Commun., 36 (1971) 234.
6. Lebl, M., Hrbas, P., Škopková, J., Slaninová, J., Machová, A., Barth, T. and Jošt, K., Collect. Czech. Chem. Commun., 47 (1982) 2540.
7. DiMayo, J., Jaramillo, J., Wernic, D., Grenier, L., Welchner, E. and Adams, J., J. Med. Chem., 33 (1990) 661.
8. Jošt, K. and Šorm, F., Collect. Czech. Chem. Commun., 36 (1971) 2795.
9. Procházka, Z., Jošt, K. and Šorm, F., Collect. Czech. Chem. Commun., 37 (1972) 289.
10. Jošt, K. and Rudinger, J., Collect. Czech. Chem. Commun., 32 (1967) 2485.
11. Procházka, Z. and Jošt, K., Collect. Czech. Chem. Commun., 45 (1980) 1982.
12. Logush, E.W., Tetrahedron Lett., 29 (1988) 6055.
13. Pátek, M. and Lebl, M., Tetrahedron Lett., 32 (1991) 3891.

The 2,2,4,6,7-pentamethyldihydrobenzofuran-5-sulfonyl group (Pbf): A new acid-labile sulfonamide type side-chain protecting group for arginine

H.N. Shroff[1,2], L.A. Carpino[2], H. Wenschuh[2], E.M.E. Mansour[2], S.A. Triolo[1,2], G.W. Griffin[1] and F. Albericio[1]

[1]Millipore Corporation, 75A Wiggins Avenue, Bedford, MA 01730, U.S.A.; [2]Department of Chemistry, University of Massachusetts, Amherst, MA 01003, U.S.A.

Introduction

Although the guanidino group of arginine is relatively inert because it remains protonated except under extrememly basic conditions it can undergo a variety of side reactions during peptide coupling processes such as acylation, both internal and external, at the guanidino nitrogen atoms. Arylsulfonyl protection decreases the tendency of many arginine derivatives to be converted to the corresponding ornithine derivatives. The most widely used guanidino protecting groups are the Mtr [1] and Pmc [2] groups. Indeed the Pmc group was developed out of the Mtr function on the basis of mechanistic considerations of the sensitivity toward acids of various arylsulfonyl residues. Based on these studies and x-ray crystallographic data the dihydropyran ring moiety present in the Pmc unit was selected as being optimal[3]. On the other hand, experimental studies by Baddeley [4, 5] suggested that the dihydrobenzofuran analogs of the Pmc residue might be more sensitive to acidic deblocking agents. In the event this proved to be the case.

Results and Discussion

The synthesis of 2,2,4,6,7-pentamethyldihydrobenzofuran was achieved in a single step from 2,3,5-trimethylphenol 1 and isobutyraldehyde using H_2SO_4 as catalyst in toluene. Chlorosulfonation gave the key intermediate Pbf-Cl 2 which by reaction with Z-Arg-OH under the usual conditions gave Z-Arg(Pbf)-OH. Hydrogenolysis by means of 10% Pd/C followed by reaction with Fmoc-OSu gave Fmoc-Arg(Pbf)-OH 3 which could be readily purified by column chromatography.

Kinetic studies of the deblocking of Z-Arg(X)-OH via 80% TFA showed that with X=Pbf conversion to Z-Arg-OH was 1.2-1.4 times as fast as with X=Pmc.

In order to evaluate the applicability of Fmoc-Arg(Pbf)-OH in solid phase peptide synthesis a number of arginine-rich model peptides were synthesized and results compared with parallel syntheses using the corresponding Pmc analog. All peptides were synthesized via Fmoc/tBu chemistry on a Millipore 9050+ automated peptide synthesizer. Acylation was carrried out with 5 equiv. of the Fmoc-amino acid, preactivated with PyBOP/HOBt/DIEA or with Pfp/Dhbt esters/ HOBt using 60-min coupling times. Fmoc deprotection was carried out with 2% DBU/2 % piperidine in DMF for 6 minutes. Treatment with reagent R (TFA-thioanisole-EDT-anisole, 90:5:3:2) for 2 hr was used to deblock and remove the peptide from the resin. Comparison of peptides synthesized via Fmoc-Arg(Pbf)-OH and Fmoc-Arg(Pmc)-OH showed that the yields and quality of the final products in the two cases were similar. Upon treatment of the protected form of the multi-arginine peptide **vi** (tryptophan Boc-protected) with reagent R for 20 min., significantly more of the undeblocked and/or partially deblocked peptides ($t_R \cong 26$ min.) remain unaffected in the case of the Pmc as opposed to the Pbf derivatives (Figure 1).

1   **2**   **3**

Table 1 *Examples of peptides made via Fmoc-Arg(X)-OH   X=Pbf or Pmc*

i.	H-Tyr-Gly-Lys-Arg-NH$_2$[6]
ii.	H-Tyr-Arg-Gln-Arg-Tyr-NH$_2$[6]
iii.	H-Arg-Lys-Asp-Val-Tyr-NH$_2$[7]
iv.	H-Gly-Arg-Ala-Asp-Ser-Pro-Lys-NH$_2$
v.	H-Tyr-Pro-Ser-Lys-Aca-Arg-His-Tyr-Ile-Asn-Leu-Ile-Thr-Arg-Gln-Arg-Tyr-NH$_2$[6]
vi.	Ac-Trp-Arg-Arg-Arg-Arg-Val-OH[8]

In view of the ease of synthesis of key intermediates and the somewhat faster deblocking process, the Pbf residue is likely to prove at least as useful for arginine side-chain protection as the Pmc residue, and could be especially advantageous for acid-sensitive systems.

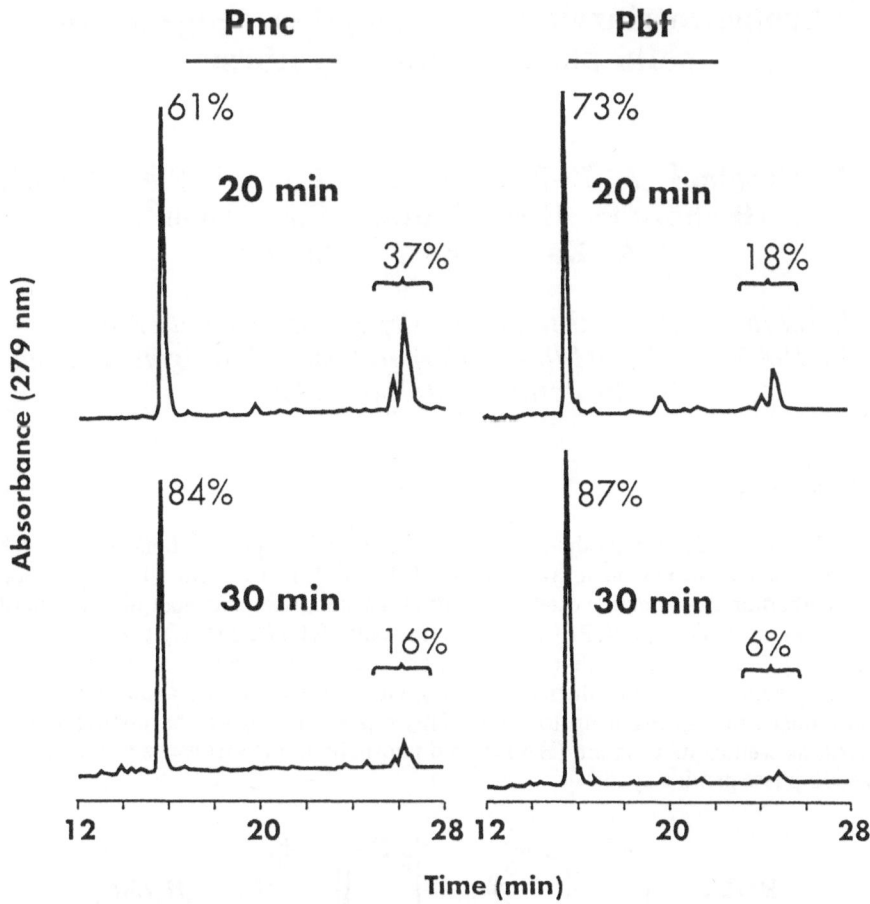

Fig. 1. *Comparison of the deblocking of Pbf vs. Pmc in the case of peptide vi after 20 and 30 min with reagent R.*

References

1.   Fujino, M., Wakimasu, M. and Kitada, C., Chem. Pharm. Bull., 29 (1981) 2825.
2.   Ramage, R. and Green, J., Tetrahedron Lett., 28 (1987) 2287
3.   Ramage, R., Green, J. and Blake, A., Tetrahedron, 47 (1991) 6353.
4.   Baddeley, G., Smith, N.H.P. and Vickars, M.A., J. Chem. Soc., (1956) 2455.
5.   Baddeley, G. and Cooke, J.R., J. Chem. Soc., (1958) 2797.
6.   Beck-Sickinger, A.G., Schnorrenberg, G., Metzger, J. and Jung, G., Int. J. Peptide Protein Res., 38 (1991) 25.
7.   Heinzel, W. and Voelter, W., Chem.-Zeitung, 105 (1981) 291.
8.   White, P., In Smith, J.A., Rivier, J.E. (Eds.) Peptides: Chemistry and Biology. (Proc. of the 12th Am. Pept. Symp.), Escom, Leiden, The Netherlands, 1992, p.537

Azabenzotriazole-based coupling reagents in solid-phase peptide synthesis

L.A. Carpino[1], A. El-Faham[1], G.A. Truran[1], S.A. Triolo[2], H. Shroff[2], G.W. Griffin[2], C.A. Minor[2], S.A. Kates[2] and F. Albericio[2]

[1]Department of Chemistry, University of Massachusetts, Amherst, MA 01003, U.S.A., [2]Millipore Corporation, 75A Wiggins Avenue, Bedford, MA 01730, U.S.A.

Introduction

The most common methods for the formation of the peptide bond involve the presence of 1-hydroxybenzotriazole (HOBt)[1]. HOBt is used either in conjunction with carbodiimides or active esters, or built into a stand-alone reagent in the form of phosphonium (BOP, PyBOP) [2] or uronium salts (HBTU, TBTU) [3]. Recently, 1-hydroxy-7-azabenzotriazole (HOAt) [4] has been described as a superior peptide coupling additive, which enhances coupling yields in solution by about 6-32 times and reduces racemization up to 50%. This report focuses on the application of HOAt, as well as its uronium (HATU) and phosphonium (PyAOP) salt derivatives, to solid-phase peptide synthesis.

HOBt

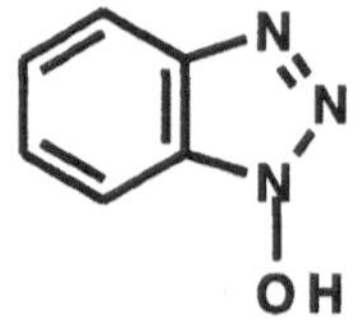

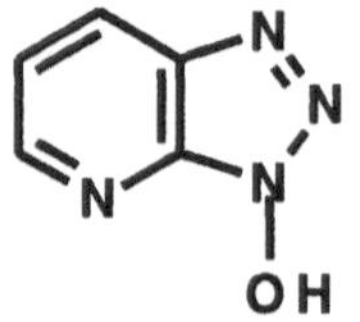

HOAt

Results and Discussion

HOAt was prepared by methylation of 2-nitro-3-pyridinol followed by treatment with excess hydrazine [5, 6]. The corresponding phosphonium and uronium salts were obtained by methods previously described [2, 3].

In order to demonstrate the suitability of HOAt derivatives, several syntheses of the fragment 65-74 (H-Val-Gln-Ala-Ala-Ile-Asp-Tyr-Ile-Asn-Gly-NH$_2$) of the acyl carrier protein were carried out by a Fmoc/tBu protection scheme (tBu for Tyr and Asp, and Trt for Asn and Gln) on Fmoc-PAL-PEG-PS using a continuous-flow Millipore 9050 Plus synthesizer. Coupling times were shortened and excesses of reagents were reduced in order to emphasize the differences between HOAt- and HOBt-mediated couplings. Under these conditions, incomplete incorporations were detected for Asn onto Gly, Ile^{72} onto Asn, Ile^{69} onto Asp, and Val onto Gln. The purity of the peptides was judged, after cleavage of the peptides from the resin with TFA-H$_2$O (9:1) for 2h, by reversed-phase HPLC. The results shown in Table 1 indicated clearly that HOAt, HATU, and PyAOP are superior to HOBt, HBTU, PyBOP, and PyBroP. Furthermore, uronium and phosphonium salts were preferred to carbodiimide/active ester methods. Finally, addition of HOXt in HXTU couplings did not significantly improve the coupling yields, with the exception of Asn.

Table 1 *Synthesis of ACP(65-74) with HOAt- and HOBt-based coupling reagents*

Coupling Method	equiv	Time (min)	ACP	-2Ile	-Ile^{72}	-Ile^{69}	-Val	-Asn
DIPCDI-HOAt	4	3	65	2	7	9	1	2
DIPCDI-HOBt	4	3	31	13	15	18	3	1
DIPCDI	4	3	14	4	7	22	2	32
Pfp-HOAt	4	3	82	-	-	15	1	1
Pfp-HOBt	4	3	53	1	1	33	5	1
Pfp	4	3	0	-	-	-	-	-
HATU	1.5	1.5	53	3	6	12	3	16
HBTU	1.5	1.5	18	16	11	19	2	7
PyAOP	1.5	1.5	45	5	7	16	1	12
PyBOP	1.5	1.5	12	19	12	16	1	2
PyBroP	1.5	1.5	0	-	-	-	-	-
HATU-HOAt	1.5	1.5	50	5	9	12	2	8
HBTU-HOBt	1.5	1.5	17	18	12	18	3	3

Similar results were obtained when HATU was applied to the solid-phase synthesis of the pentapeptide H-Tyr-Aib-Aib-Phe-Leu-NH$_2$ using the same strategy described above. This peptide contains the sequence Aib-Aib, which is described as being particularly difficult [7]. After 2 h of coupling using 4 equiv of both Fmoc-Aib-OH, Fmoc-Tyr(*t*Bu)-OH, and HXTU plus 8 equiv of DIEA (all remaining couplings were carried out for 30 min), the pentapeptide was obtained with a purity of 94% for the HATU synthesis, and only 43% for HBTU.

Finally, the synthesis of peptide H-D-Ala-MeLeu-MeLeu-MeVal-Phe-Val-OH, on a KB-PEG-PS resin was attempted. The *N*-terminal tetrapeptide section represents a fragment of cyclosporin. The incorporation of the last three amino acids required a double coupling protocol for 2 h. Using these conditions and after cleavage of the peptide with TFA-CH$_2$Cl$_2$ (2:98) for 1 h, the hexapeptide was obtained in a purity of 85% in the case of HATU, and only 8% for HBTU. In this latter synthesis, the major peak (48%) corresponded to the tripeptide H-MeVal-Phe-Val-OH.

In conclusion, the efficiency of HATU and HOAt has been demonstrated. These derivatives enhance reactivity, reduce racemization, and are suitable for the preparation of peptides containing hindered amino acids.

Acknowledgements

This work was supported in part by NIH (GM-09706) and NSF (CHE-9003192).

References

1. König, W. and Geiger, R., Chem. Ber., 103 (1970) 788.
2. Castro, B., Dormoy, J.R., Evin, G. and Selve, C., Tetrahedron Lett., (1975) 1219.
3. Knorr, R., Trzeciak, A., Bannwarth, W. and Gillessen, D., Tetrahedron Lett., 30 (1989) 1927.
4. Carpino, L.A., J. Am. Chem. Soc., 115 (1993) 4397.
5. Yutilov, Y.M., Ignatenko, A.G. and Khim, Prom-st., Ser. Reakt. Osobo Chist. Veshchestva, (1981) 27 (Chem. Abstr., 96 (1982) 68883s).
6. Mokrushina, G.A., Azev, Y. and Postovoskii, I.Ya., Chem. Heterocyclic. Cmpds., (1975) 880.
7. Belton, P., Cotton, R., Giles, M.B., Atherton, E., Horton, J. and Richards, J.D., In Jung, G. and Bayer, E. (Eds.), Proceedings to the 20th European Peptide Symposium, Walter de Gruyter, Berlin, 1989, p.619.

A novel method of solid phase peptide synthesis

R.P. Sharma, D.A. Jones, D.L. Corina and M. Akhtar

*Department of Biochemistry, School of Biological Sciences,
University of Southampton, Bassett Crescent East,
Southampton, SO9 3TU, U.K.*

Introduction

Solid phase peptide synthesis was introduced by Merrifield in 1963 [1] and has been widely used ever since. However, little has been mentioned of advances towards reversing the conventional C-to-N direction of synthesis. This would create a new approach to synthesising peptides with modifications at the C-terminal and it may facilitate fragment coupling using solid phase methodology. Earlier attempts at solid-phase peptide synthesis in the N-to-C direction [2, 3] were hindered by the use of amino acid esters that were effectively too stable. As a consequence the conditions required for the removal of the ester protection before commencing the next addition cycle were very harsh. In order to improve this situation a more suitable amino acid ester building block had to be employed. Silyl-based protecting groups have found ever-increasing use in organic synthesis. Amino acid trialkyl and trialkoxy silyl esters have been investigated in this research as possible derivatives suitable for use in solid-phase peptide synthesis in the N-to-C direction. Three types of amino acid silyl esters have been prepared, trimethylsilyl, *tert*-butyldimethylsilyl and tri-*tert*-butoxysilyl esters. Of these three it has been the amino acid tri-*tert*-butoxysilyl esters, first described by Gruszecki and co-workers [4] that have exhibited the greatest potential. These derivatives are easily prepared, are stable throughout the coupling reaction and are conveniently and rapidly removed before commencing the next cycle. A novel method of solid phase peptide synthesis from the N → C direction has been developed and a number of peptides have been synthesised, characterised and evaluated using this methodology. The protocol involves the incorporation of alkoxysilyl group into the carboxyl function of amino acids for use in peptide synthesis.

Results and Discussion

Amino acid tri-*tert*-butoxysilyl esters were prepared by the procedure of Gruszecki and co-workers [3] and their identity was confirmed by infra-red, ^{1}H nuclear magnetic resonance (NMR) spectra and accurate mass measurements. Attachment of the N-terminal residue to the solid support via its amino function had previously been achieved through a benzyloxycarbonyl linkage [2]. This approach provides a peptide-resin linkage that is stable to all reagents used for ester hydrolysis and is cleavable at the end of the synthesis by treatment with strong acid such as hydrogen fluoride or trifluoromethanesulphonic acid.

Employing the benzyloxycarbonyl resin-amino acid linkage, amino acid tri-*tert*-butoxysilyl esters and 1,3-diisopropylcarbodiimide/1-hydroxybenzotriazole mediated couplings peptides have been successfully synthesised on the solid-phase in the N-to-C direction. The test tetrapeptide, Leu-Ala-Gly-Val, and leucine-enkephalin were both prepared in good yield. The quality of the product in both cases was comparable to the product obtained when the peptides were synthesised in the conventional C-to-N direction using Fmoc chemistry. In the case of the leucine-enkephalin synthesis a somewhat longer coupling time was required for the addition of the C-terminal leucine than was expected. The problem of the difficult leucine addition was easily overcome by simply extending the duration and coupling.

The unpredictable nature of the coupling rates encountered during solid-phase peptide synthesis means that the strategy benefits from a monitoring system that can detect slow couplings as they occur. A great deal of effort has been expended in the development of such a system, but without definitive success. The detection of residual carboxyl groups on the resin would first require the decomposition of any active ester present. Decomposition of active esters followed by the detection of the resultant carboxyl group could still be a one-step process in practice provided the assay is performed in aqueous solution after the correct choice of activation. The correct choice of activation is governed by the fact that any active ester present would need to be sufficiently unstable in an aqueous medium to be hydrolysed to the acid. Detection of the carboxylic acid by its reaction with sodium carbonate in the presence of phenolphthalein shows great promise in this type of monitoring system. The levels of acid that can be detected with phenolphthalein were as low as 1.0 µmole which would provide adequate sensitivity in most cases. A drawback of performing the assay in aqueous medium is that the polystyrene based resin used in this research does not show extensive swelling in such a solution. This is currently under investigation.

The successful synthesis of leucine-enkephalin in the N-to-C direction using 1,3-diisopropylcarbodiimide/1-hydroxbenzotriazole mediated couplings gave the product in good yield (77%) requiring minimal purification. The synthesis was repeated employing BOP as the activating agent, but using dichloromethane as the solvent rather than N,N-dimethylacetamide. The coupling times were thus reduced to 2 hours duration increasing the overall synthetic procedure. On the other hand, the quality of the product was not as good when this approach was used giving a product that was only around 80% pure. Furthermore, the hydroxyl functions of tyrosine, threonine and serine may not require protection during the synthesis.

The possibility of racemisation during the synthesis was always considered and the chiral purity of the synthesised leucine-enkephalin was extensively scrutinised. No racemisation was detected when analysing the product by RP-HPLC, enzymatic digestion with Carboxypeptidase A or specific optical rotation. The coupling procedures employed thus far, i.e. 1,3-diisopropylcarbodiimide/1-hydroxbenzotriazole and BOP in dichloromethane are suitable for the solid-phase synthesis of peptides in the N-to-C direction.

The solid-phase synthesis was finally extended to the preparation of peptides possessing C-terminal modifications. The important advantage of this methodology is that it allows the *in situ* modification of carboxyl function (aldehyde, alcohol, etc.) which has important applications in the synthesis of therapeutically active peptides. The peptide analogues synthesised were based on leucine-enkephalin with the addition of a modified leucine residue at the C-terminus. The leucine-enkephalin-diol was synthesised using 1,3-diisopropylcarbodiimide/1-hydroxybenzotriazole mediated couplings and afforded the desired product in excellent yield (73%) and purity. The synthesis of a number of enkephalin analogues which are otherwise inaccessible has been accomplished. Further work in this area is in progress.

The method developed offers an alternative approach to the preparation of peptides possessing C-terminal modifications. The implementation of solid-phase strategies for the production of such peptide analogues will greatly enhance the efficiency of their synthesis over the currently employed solution-phase techniques. The solid-phase approach to the synthesis of C-terminally modified peptides provides a procedure that is simple to perform and possesses the potential to become an automated process due to its repetitive nature. The advantages of this method should have a beneficial effect in both preparation of peptide analogues for research and in the efficient production of some peptide analogues currently used for diagnostic and therapeutic purposes.

Acknowledgements

We thank Porton Developments Limited, U.K. for financial support.

References

1. Merrifield, R.B., J. Amer. Chem. Soc., 85 (1963) 2149.
2. Letsinger, R.L. and Kornet, M.J., J. Amer. Chem. Soc., 85 (1963) 3045.
3. Felix, A.M. and Merrifield, R.B., J. Amer. Chem. Soc., 92 (1970) 1385.
4. Gruszecki, W., Gruszecka, M. and Bradaczek, H., In Giralt, E. and Andreu, D. (Eds.), Peptides 1990: Proceedings of the Twenty-First European Peptide Symposium, Escom, Leiden, The Netherlands, 1991, p.27.

Comparative studies on the coupling of sterically hindered amino acid residues

H. Wenschuh[1], M. Beyermann[1], E. Krause[1], M. Brudel[1], L.A. Carpino[2] and M. Bienert[1]

[1]Research Institute of Molecular Pharmacology, Alfred-Kowalke Str. 4, D-10315 Berlin, Germany
[2]Department of Chemistry, University of Massachusetts, Amherst, MA 01003, U.S.A.

Introduction

The chemical incorporation of Aib (α-aminoisobutyric acid) into peptides is of growing interest due to the increasing number of isolated naturally occurring peptides bearing such units. The conventional peptide-chain assembly for the coupling of one or more adjacent Aib-residues either fails completely or requires long reaction times, as reported for Fmoc-Aib symmetrical anhydride [1] and recently for the use of Fmoc-Aib-NCA and PyBroP activation [2]. In this paper a comparative examination of Fmoc-Aib-F and of the above mentioned techniques is described.

Results and Discussion

In order to compare different coupling procedures for the incorporation of sterically hindered amino acid residues in SPPS, four adjacent Aib-units were coupled to a model hexapeptide [h-CRF(36-41): H-Lys-Leu-Met-Glu-Ile-Ile-NH$_2$] by means of Fmoc-Aib-F, Fmoc-Aib-NCA, Fmoc-Aib symmetrical anhydride and Fmoc-Aib-OH/PyBroP. Coupling of the Aib-residues was carried out for 15 min in a 0.2 M solution of DMF with 3 equivalents of the Fmoc amino acid derivative (Fmoc-Aib-F and preformed (Fmoc Aib)$_2$O: 1 eq. DIEA; PyBroP: 1.8 eq. DIEA / 0.2 eq. DMAP, 3 min preactivation; Fmoc-Aib-NCA: no base added). The extend of reaction was monitored by U.V. analysis of the Fmoc deprotection step (20% piperidine/DMF, 15 min) (Figure 1).

The results show that only the fluoride derivative gave the desired decapeptide with all four Aib residues incorporated (Figure 2b), whereas all three of the other techniques led to formation of the heptapeptide bearing only a single Aib unit as the major product (Figures 2a, c, d). All of the major components were analyzed by ES-MS.

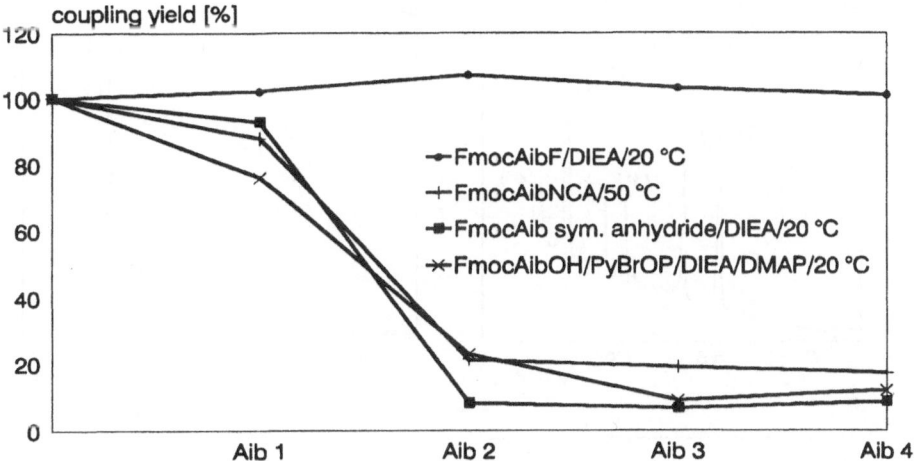

Fig. 1. Aib coupling efficiency, stepwise elongation of h-CRF(36-41).

Since we found similar acylating yields of the used coupling techniques for the coupling of Fmoc-Aib onto an amino-resin (TG-SRAM), the obtained differences for the coupling of adjacent Aib residues are most likely due to the small size of the fluoride as leaving group.

In the case of chain-assembly by Fmoc-Aib-NCA the formation of side products was detected, which could, according to ES-MS (M+44) and ^{13}C-NMR (additional carbon: DMSO-d$_6$, δ (ppm): 157.13) be explained by a "wrong" opening of the UNCA, as described by Kopple [3] for unprotected Aib-NCA and by Kaminski et. al. [4] for sterically hindered mixed anhydrides. Interestingly, the reproducible side product formation did not occur by means of BOC-Aib-NCA and more surprisingly by substitution of Lys by Gly in the used model peptide [h-CRF(36-41)] either.

In view of the expectation of prolonged reaction times by the incorporation of sterically hindered amino acids into peptides, the stability of the Fmoc-group becomes important to avoid side reactions. In order to examine possible differences regarding the Fmoc-cleavage, Fmoc-Aib-F, Fmoc-Aib-OH, Fmoc-Aib-NCA and Fmoc-Aib symmetrical anhydride have been checked towards bases routinely used in SPPS at ambient temperature under synthesis relevant conditions (amino acid derivative: 0.2M in DMF, 1 eq. DIEA or 0.1 eq. NMM). The results obtained indicate a significant loss of the Fmoc-group for the Fmoc-Aib-NCA (1 eq. DIEA: 35% after 90 min, 0.1 eq. NMM: 12% after 90 min), whereas the other species show Fmoc-deprotection less than 5% towards both bases.

These examinations suggest an avoidance of long coupling times and exclusion of bases for the Fmoc-protected NCA's where the use of 1 eq. DIEA is recommended to obtain products with higher purity [5].

Conclusively, the Fmoc-amino acid fluorides are because of their small leaving group and their sufficient stability well suited for the incorporation of highly hindered amino acids.

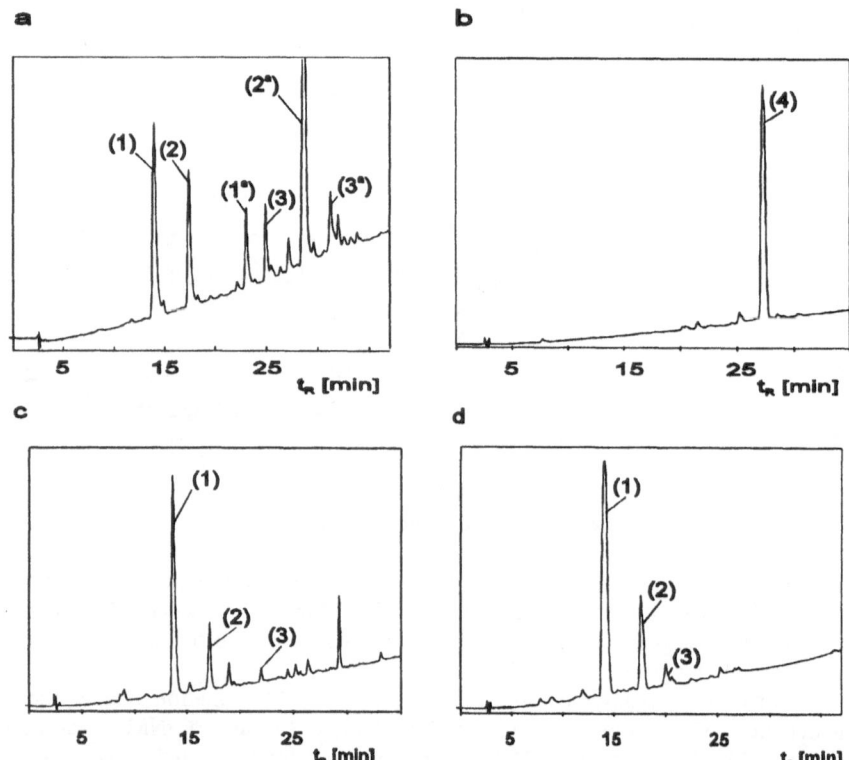

Fig. 2. HPLC profiles of crude products from the stepwise elongation of h-CRF(36-41) by four Aib-residues using: (a) Fmoc-Aib-NCA, (b) Fmoc-Aib-F, (c) Fmoc-Aib-symmetrical anhydride, (d) Fmoc-Aib-OH/PyBroP. Peaks 1-4 show the products with 1, 2, 3 and 4 Aib-residues incorporated. Peaks 1^a-3^a show side products obtained in the case of Fmoc-Aib-NCA with masses of +44 relative to the corresponding peptide with 1, 2, and 3 Aib-units incorporated. Additionally smaller peaks with detected masses of +56 relative to the corresponding peptides with 1, 2, and 3 Aib-units incorporated were detected by using the UNCA.

References

1. Belton, P., Cotton, R., Ciles, M.B., Atherton, E., Horton, J. and Richards, J.D., In Jung, G. and Bayer, E. (Eds.), Peptides 1988, Walter de Gruyter, Berlin, New York, NY, U.S.A., 1989, p.619.
2. Auvin-Guette, C., Frerot, E., Coste, J., Rebuffat, S., Jouin, P. and Bodo, B., Tetrahedron Lett., 34 (1993) 2481.
3. Kopple, K.D., J. Am. Chem. Soc., 79 (1957) 6442.
4. Kaminski, Z., Leplawy, M.T., Olma, A. and Redlinski, A., In Brunfeldt, K. (Ed.), Peptides 1980, Scriptor, Copenhagen, Denmark, 1981, p.201.
5. Xing, D.S. and Naider, F., J. Org. Chem., 58 (1993) 350.

Peptide synthesis by insertion reactions

J.C.H.M. Wijkmans, J.H. van Boom and W. Bloemhoff

*Gorlaeus Laboratories, University of Leiden, P.O. Box 9502,
2300 RA Leiden, The Netherlands*

Introduction

In an attempt to reduce racemization accompanying several peptide bond forming reactions, Brenner [1] proposed already in the sixties the 'low power approach'. Typically, the individual amine and carboxyl components are linked to a template and peptide bond formation occurs *via* intramolecular aminolysis. The approach is exemplified by the acid-catalyzed rearrangement of N-acyl-N'-α-aminoacylhydrazine **A** into dipeptide hydrazide **B** [2]:

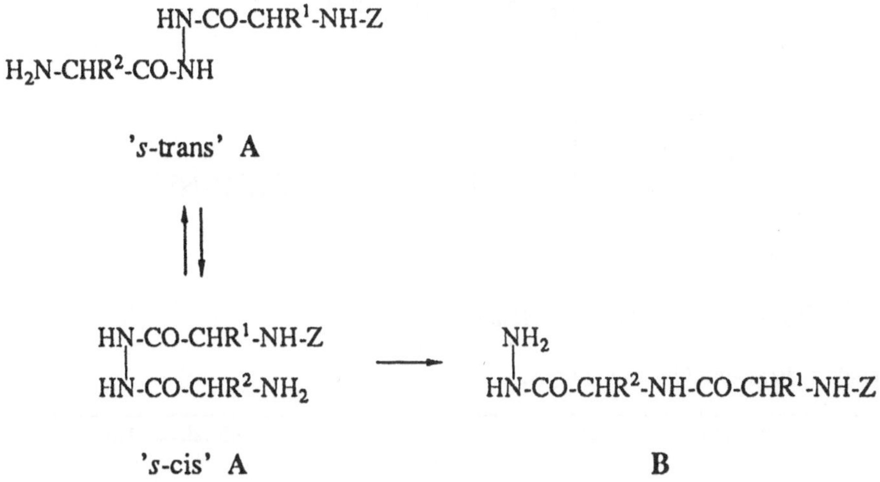

HN-CO-CHR^1-NH-Z

H$_2$N-CHR^2-CO-NH

's-trans' **A**

HN-CO-CHR^1-NH-Z

HN-CO-CHR^2-NH$_2$

NH$_2$

HN-CO-CHR^2-NH-CO-CHR^1-NH-Z

's-cis' **A**   **B**

Fig. 1. Rearrangement of di-acyl hydrazine A [2].

A serious drawback of this amino acid insertion method using hydrazine as a template is that di-acyl compound **A** tends to adopt the 's-trans conformation' while only the 's-cis conformation' rearranges. However, as Bodanszky [3] already recognized, "... *the insertion principle is a thought-provoking and challenging concept.*" In this respect, we reasoned that the use of a cyclic hydrazine, *i.e.* tetrahydrophthalazine (THPhth), leading to the optimal 'cis' orientation of hydrazide **A** would favour insertion. The advantageous application of THPhth as a template for intramolecular peptide synthesis will be illustrated by the synthesis of the protected tripeptide Z-Val-Gly-Ala-OMe [5].

Results and Discussion

The requisite di-acylated THPhth derivatives **1a-g** (Figure 2, Table 1), were prepared by mono-acylation of the HCl salt of THPhth [4] with BOP-activated Z-amino acids followed by acylation with Fmoc-amino acid chlorides and subsequent Fmoc-deprotection. Isomerization of **1a-g** into **2a-g** was achieved by the catalytic action of acetic acid. In most cases, the reaction was complete in less than 1 h at r.t. as gauged by TLC analysis. In contrast, the comparable hydrazine derivatives **A** rearrange much slower (reaction times ranging from a few hours to several weeks) or even do not react at all.

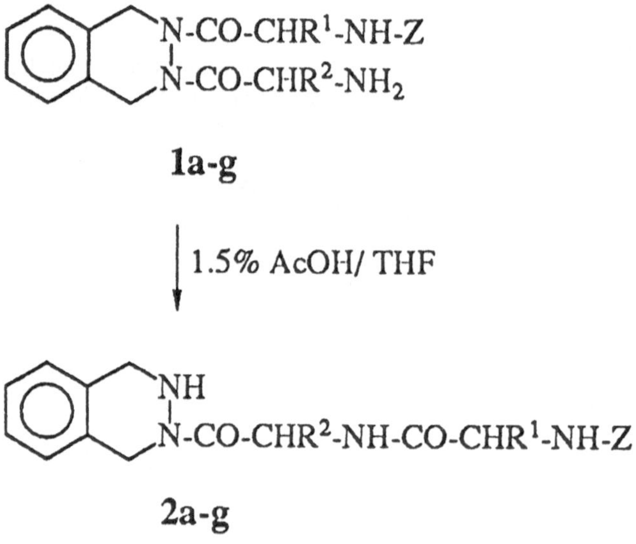

$$\text{N-CO-CHR}^1\text{-NH-Z}$$
$$\text{N-CO-CHR}^2\text{-NH}_2$$

1a-g

1.5% AcOH/ THF

$$\text{NH}$$
$$\text{N-CO-CHR}^2\text{-NH-CO-CHR}^1\text{-NH-Z}$$

2a-g

Fig. 2. Rearrangement of di-acyl THPhth 1.

Table 1 *Relevant data on the rearrangement of 1*

Entry	Substrate 1	R^1		R^2		Product 2	Time (min)	Yield (%)
1	1a	H	(Gly)	H	(Gly)	2a	25	81
2	1b	H	"	Me	(Ala)	2b	20	85
3	1c	H	"	i-Pr	(Val)	2c	10	70
4	1d	Me	(Ala)	H	(Gly)	2d	45	79
5	1e	Me	"	i-Bu	(Leu)	2e	20	90
6	1f	i-Pr	(Val)	s-Bu	(Ile)	2f	3.5 h	69
7	1g	i-Pr	"	H	(Gly)	2g	9 h	73

The feasibility of the described amino acid insertion for peptide synthesis is demonstrated by the synthesis of the protected tripeptide **5** (Figure 3). Thus, **2g** was effectively coupled with Fmoc-Ala-Cl to afford the di-acylated THPhth derivative **3**.

Fmoc-deprotection of **3** followed by acid-catalyzed rearrangement yielded the mono-acylated THPhth **4**. Release of the tripeptide from the template by oxidative treatment of **4** with *N*-bromosuccinimide [6] in the presence of MeOH resulted in the isolation of Z-Val-Gly-Ala-OMe [5] in 34% overall yield from THPhth•HCl.

Peptide synthesis by amino acid insertion *via* intramolecular rearrangement can effectively be realized by application of THPhth as a template.

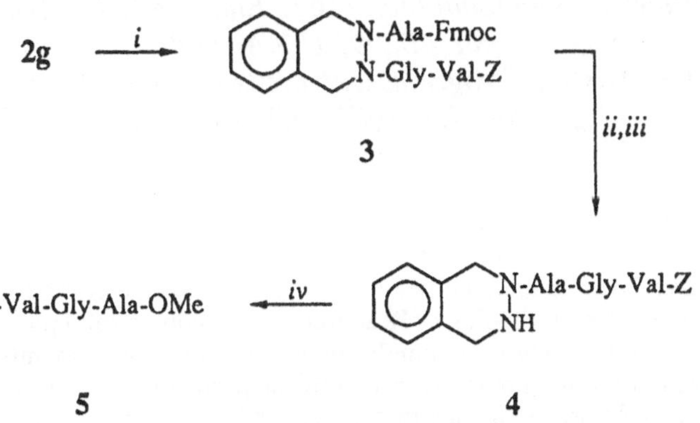

Fig. 3. The synthesis of a tripeptide by the insertion principle. **Conditions:** (i) Fmoc-Ala-Cl, DIEA, CH_2Cl_2, 5 min, 94%; (ii) 40% aq. $HN(CH_3)_2$/THF, 1 h; (iii) 1.5% AcOH/THF, 35 min, 91% (from **3**); (iv) NBS, MeOH/CH_2Cl_2, 2 h, 67%.

References

1. Brenner, M., In Beyerman, H.C., Linde, A. van de, and Maassen van den Brink, W. (Eds.), Peptides (Proceedings of the 8th European Peptide Symposium), North-Holland Publishing Co., Amsterdam, Holland, 1966, p.1.
2. Brenner, M. and Hofer, W., Helv. Chim. Acta, 44 (1961) 1794.
3. Bodanszky, M., Principles of Peptide Synthesis, Springer-Verlag, Berlin, 1984, p.14.
4. Carpino, L.A., J. Am. Chem. Soc., 85 (1963) 2144.
5. Hofer, W. and Brenner, M., Helv. Chim. Acta, 47 (1964) 1625.
6. Wieland, T., Lewalter, J. and Birr, C., Liebigs Ann. Chem., 740 (1970) 31.

Membrane based peptide synthesis (MBPS)

D.E. Benovitz[1], K. Darlak[1], W.A. Klis[1], E. Klein[2], and A.F. Spatola[1]

[1]Peptides International,Inc., 11621 Electron Drive, Louisville, KY 40299, U.S.A. and
[2]Kidney Disease Program, University of Louisville, School of Medicine, Louisville, KY 40292, U.S.A.

Introduction

We have investigated membrane based peptide synthesis (MBPS) as a viable alternative to existing technology. Reaction rates in porous solid phases are limited by diffusion and have correspondingly longer reaction times. Membranes in flow through configuration should permit uniform perfusion [1]. When converting advantageous membrane kinetics to peptide synthesis, proper membrane choice is critical. One example in peptide chemistry is the polypropylene membrane coated with polyhydroxypropylacrylate introduced by Milligen/Biosearch [2].

Results and Discussion

Various synthetic membranes were monitored for stability to reagents and physical processing. The most stable membranes had high cellulose content. Four membranes were then chemically modified. Functionalization of the membrane with diaminohexane or diaminobutane was first achieved by carbonyldiimidazole treatment to form covalent bonding via a urethane group. Two additional carbohydrate modification chemistries were used including cyanuric chloride (trichlorotriazene) [3]. The most stable linkage chemistry involved formation of an ether bond to cellulose using ethylene glycol diglycidyl ether (EGDGE) [4] or a diepoxide followed by treatment with a diamine (Table 1).

Amplification of active sites on the membrane with lysine branching ("MAPS" approach) [5] yielded substitution levels (0.5-1.0 mmol/g) acceptable for peptide synthesis. Following amplification, membranes were submitted to various linker chemistries. 4-hydroxymethyl benzoic acid, linked to the amine by way of an aminocaproic acid spacer, was chosen for stability during peptide synthesis and compatibility with mild methods of peptide resin cleavage.

Rapid flow rates were observed using NPC-100 membranes and various common solvents. Residence times measured under pressurized flow (200 mm Hg) were 46 and 44 sec, respectively, in water and DMF. Peptides synthesized using the NPC-100 membrane support and Fmoc/HBTU chemistry included leucine-enkephalin and des-Arg[8]-bradykinin.

Table 1 *Selection of Membranes with Functionalization*

Solid Support	Function (Chemistry)	Substitution (μmol/gram)
Mac[a]	1,4-diaminobutane (CDI)	35
Amino-membrane (CUNO)	None	48
Cellulose fiber (rayon)	1,4-diaminobutane (CNCl)	31
NPC-100	1,6-diaminohexane (CDI)	139*
		171*
NPC-100	1,4-diaminobutane(CDI)	27
NPC-100	1,6-diaminohexane(CNCl)	
	and Aca spacer	80
NPC-100	1,6-diaminohexane(EGDGE)	140

*unstable to peptide chemistry
[a]Abbreviations: Mac (microcrystalline amino-cellulose); NPC-100 (microporous cellulose membrane); Aca (aminocaproic acid); CDI (carbonyldiimidazole); CNCL (cyanuric chloride); EGDGE (ethylene glycol diglycidyl ether).

Membrane Based Peptide Synthesis (MBPS) has unique qualities that should make it a useful candidate for rapid large scale peptide synthesis. Its utility is increased by lysine amplification and by attachment of benzyl-based linkers suitable for mild cleavage techniques.

References

1. Klein, E., Affinity Membranes: Their Chemistry and Performance in Adsorptive Separation Processes, John Wiley, NY, U.S.A., 1991.
2. Daniels, S.B., Bernatowicz, M.S., Coull, J.M. and Koster, Tetrahedron Lett., 30 (1989) 4345.
3. Smith, N.A. and Lenhoff, H.M., Anal. Biochem., 61 (1974) 392.
4. Sundberg, L. and Porath, J., J. Chrom., 90 (1974) 87.
5. Tam, J.P., In Tam, J.P. and Kaiser, E.T. (Eds.), Synthetic Peptides: Approaches to Biological Problems, Alan R. Liss, Inc., NY, U.S.A., 1989, p.3.

SAL: A silicon-based linkage agent for Fmoc solid phase synthesis: Improved yields of peptide amides containing tryptophan

H.-G. Chao, M.S. Bernatowicz and G.R. Matsueda

Bristol-Myers Squibb Pharmaceutical Research Institute, Princeton, NJ 08543, U.S.A.

Introduction

Recently, SPPS using the Fmoc/t-butyl strategy has gained great popularity. Consequently, a variety of TFA cleavable linkage agents have been developed for the production of C-terminal peptide amides using Fmoc/t-butyl strategy. Many of the linkage agents share common properties and all of them have the tendency to produce stable carbonium ions under the cleavage conditions [1]. The side reactions caused by such stable carbonium ions have been well documented [1, 2]. In order to suppress such problems, the use of scavengers or scavenger combinations during TFA cleavage is common but not always completely satisfactory [1, 3].

Results and Discussion

In this paper we describe the use of the novel silicon-based linkage agent, 4-{1-(amino)-2-(trimethylsilyl)ethyl} phenoxyacetic acid (SAL), for the preparation of C-terminal peptide amides using the Fmoc/t-butyl based solid phase peptide synthesis strategy.

SAL linker

The SAL linker was specifically designed to minimize the problems associated with the carbonium ions generated under TFA cleavage conditions. The design of SAL linker takes advantage of the β-elimination propensity of organosilicon compounds which facilitates the quenching of the carbonium ions formed during cleavage with TFA [1]. The SAL linker was prepared in six steps in 17-21% overall yield [1]. The coupling of SAL onto aminomethylpolystyrene support was performed using standard coupling chemistry (e.g., BOP reagent, HBTU, DCC/HOBt etc.) and satisfactory results were obtained in all cases. In comparative studies, Fmoc-Val-linker-resin and Fmoc-Met-linker-resin derivatives were treated with TFA/phenol (95:5) and 97% cleavage was obtained within 15 min for the SAL linker, in contrast to 75% and 60% obtained for commercially available PAL linker and Linker AM, respectively. A more stringent test was performed using Fmoc-Val-Trp-linker-resin derivatives as models. In these studies, maximum cleavage (90%) of SAL linker was obtained after 45 min with TFA/EDT/phenol/thioanisole (90:5:3:2) (Reagent TEPT). Under the same conditions, Fmoc-Val-Trp-PAL linker and Linker AM derivatives gave maximum cleavage yields of about 20-30% (Figure 1).

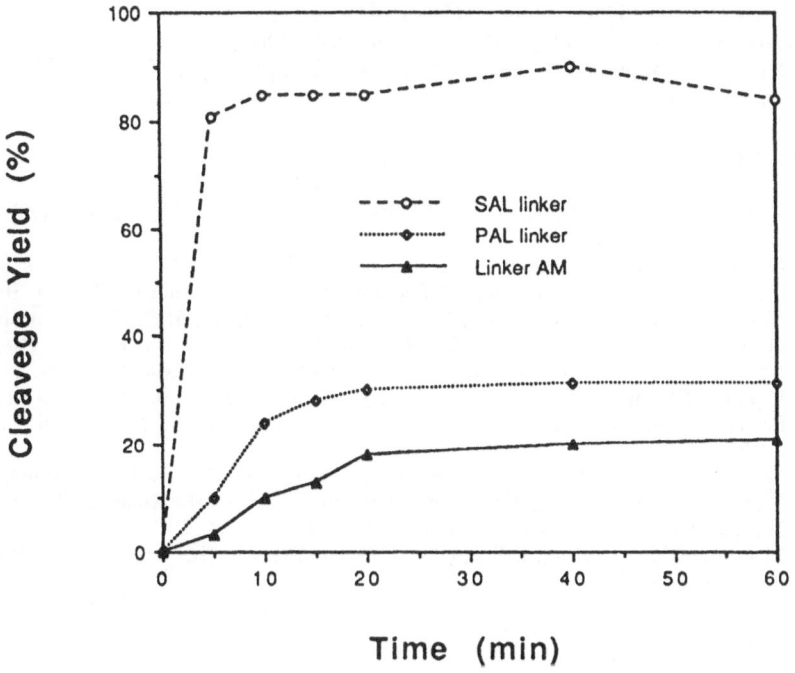

Fig. 1. Time course for the cleavage of Fmoc-Val-Trp-linker resin derivatives using reagent TEPT. An aliquot of the resin was removed at various times, washed, dried and subjected to UV analysis.

These results clearly demonstrate that the use of SAL linker has advantages compared to PAL linker and Linker AM in terms of both peptide yields and speed of cleavage. In addition, several peptide amides including two C-terminal tryptophan peptide amides were prepared using SAL linker. All were obtained in high yields and high purity (Table 1). Authenticity of the purified peptides was established by FABMS and amino acid analysis.

Table 1 *Peptide amides prepared using SAL linker*

Sequence	Yields(%)	
	crude	purified
(1) GDFEEIPEEYLQ	87	56
(2) KKLESHNDC	92	51
(3) SSIPDAPW	91	43
(4) SSIPDAPWW	87	40

A novel silicon-based linkage agent that we call "SAL", has been synthesized and characterized. The benefits of using the SAL linker compared to conventional linkers have been clearly demonstrated, particularly with respect to tryptophan-containing peptides. Development of silicon-based linkage agents for the synthesis of C-terminal peptide carboxylic acids has been completed will be the subject of a future publication.

References

1. Chao, H.-G., Bernatowicz, M.S. and Matsueda, G.R., J. Org Chem., 58 (1993) 2640 and references therein.
2. (a) Riniker, B. and Kamber, B., In Jung G., Bayer E. (Eds.), Peptides 1988, (Proceedings for the 20th European Peptide Symposium), Walter de Gruyter, Berlin, New York, 1989, p.115. (b) Atherton, E. and Sheppard, R.C., Tetrahedron, 44 (1988) 843. (c) Gesellchen, P.D., Rothenberger, R.B., Dorman, D.E., Paschal. J.W., Elzey, T.K. and Campell, C.S., In Rivier, J. E. and Marshall G.R. (Eds.), Peptides; Chemistry, Structure and Biology, (Proceedings for the Eleventh American Peptide Symposium), Escom Science Publisher, Leiden, The Netherlands, 1990, p.957.
3. (a) Barany, G. and Merrifield, R.B., In Gross, E. and Meienhofer, J. (Eds.), The Peptides, Academic Press, New York, 1979, Vol. 2, p.72. (b) Kemp, D.S., Fotouhi, N., Boyd, J.G., Carey, R.I., and Ashton, C., Hoare, J. Int. J. Peptide Protein Res., 31 (1988) 359. (c) King, D.S., Fields, C.G. and Fields, G.B., Int. J. Peptide Protein Res., 36 (1990) 255.

Synthesis of protected peptide fragments following the Fmoc/^{t}Bu-strategy and fragment condensations on a solid support

S. Ihringer, E. Lichte and C. Griesinger

*Institute of Organic Chemistry, University of Frankfurt,
Marie Curie Str.11, D-60439 Frankfurt/Main, Germany*

Introduction

Recently the sequence of a gene was published which is conserved and Y-specific among a wide range of mammals. It could be established that the transcript of this gene is testis-determining (at least in mice) [1] and that it binds to DNA selectively [2]. This gene was called SRY (Sry for mice) - Sex-determining Region on the chromosome **Y**.

An 78 aa peptide fragment out of the protein encoded by SRY shares high homology with a conserved DNA-binding motif present in the nuclear high-mobility group proteins HMG1 and HMG2. This peptide fragment seems to have an important structural role in the protein and investigations concerning its structure and interactions with DNA may provide insight into the genetic control of developmental decisions in mammals.

In order to make the SRY-peptide available for NMR structure elucidation we started its chemical synthesis by the assembly of fully protected peptide fragments on the 2-chlorotrityl resin [3].

SRY: DRVKRPMNA^9 FIVWSR ^{15} DQRRKMA^{22} LENP^{26} RMRNSEISKQKG^{38}
YQWKMLTEA^{47}EKWPFFQEA^{56} QKLQAMHREKYP^{68}NYKYRPRRKA^{78}

Results and Discussion

The protected peptide fragments were synthesized via solid-phase methods on the 2-chlorotrityl resin (0.5-0.8mmol/g Fmoc-aa) using Fmoc/^{t}Bu-strategy and in situ DIC/HOBT coupling. The side chain protecting groups used throughout the syntheses were: ^{t}Bu for S, T, D and E; Trt for H, N and Q; Boc for K; Mtr or Pmc for R. Couplings were performed employing a 2 fold molar excess of Fmoc-aa/HOBT/DIC (1:1.5:1) in DMF for one to four hours and repeated if necessary. Completeness of the couplings was verified by the reaction with ninhydrin. Fmoc-deprotection was carried out using 40% piperidine in DMF for 40 minutes at RT and monitored by TLC (SiO$_2$60-plates, toluol/methanol/acetic acid 7:1.5:1.5) [5]. This monitoring often enhanced the purity of the crude products. The protected fragments were cleaved from the resin by treatment with DCM/TFE/AcOH 7:2:1 and 3:1:1, respectively, for 2 hours at RT. They all were obtained in very good yields and purity, except for SRY(57-68). This peptide contains a His(Trt) which is partially deprotected even under the mildest acidic conditions of peptide-resin cleavage. RP-HPLC with ACN/H_2O+0.1%TFA gradients could be applied for the purification. In some cases, however, there was an enormous loss in yield due to the poor solubility of the peptides. The fragments were characterized by RP-HPLC, FAB-MS and standard sequential 2D-NMR assignment (DQF-COSY, NOESY).

For the condensations the purified SRY(69-78) was attached to 2-chlorotrityl resin (0.06 mmol/g) and the following fragments were coupled with an 6-8 fold molar excess (protected fragment/HOBT/DIC 1:2:1) in DMSO at RT for one to three days. After each condensation small amounts of the protected peptide were cleaved from the resin, treated with reagent K (82.5% TFA, 5% phenol, 5% water, 5% thioanisole, 2.5% ethanedithiol [4]) for 3 hours at RT and characterized by RP-HPLC and ESI-MS. We found that the condensations of the fragments could easily be carried out up to the 40 aa residue peptide SRY(39-78) with good yield (>65%) and high homogeneity of the crude product. Because of the inhomogeneity of the sample, the mass of the target peptide SRY(1-78) could only be obtained by ESI-MS with maximum-entropy deconvolution. The purification of SRY(1-78) has still to be elaborated.

SRY(39-78) was investigated by NMR (3.86 mM in water, pH 4.0, 283.1 K) and showed no secondary structure. Following the Chou-Fasman predictions this fragment should contain two α-helical segments. And indeed, the CD-spectrum of SRY(39-78) shows two negative bands at 206 and 222 nm and one positive band near 192 nm when 50% TFE is added which are characteristic for α-helices.

Acknowledgements

We are indebted to Dr. Fehlhaber (Höchst AG, Frankfurt, Germany), to Dr. Beck (research group of Prof. Jung, Tubingen, Germany) and to Dr. Langeland Johansen (Novo, Denmark) for the mass spectra. We thank Dr. Grell (MPI for Biophysics, Frankfurt, Germany) for the possibility to do the CD measurements.

References

1. Sinclair, A.H., Lovell-Badge, R. and Goodfellow, P.N., Nature, 346 (1990) 240 and 351 (1991) 117.
2. Nasrin, N., Buggs, C., Fu Kong, X., Carnazza, J., Goebl, M. and Alexander-Bridges, M., Nature, 354 (1991) 317.
3. Barlos, K., Gatos, D. and Schäfer, W., Angew. Chem., 103 (1991) 572.
4. King, D.S., Fields, C.G. and Fields, Int. J. Pep. Prot. Res., 36 (1990) 255.
5. Barlos, K., Gatos, D., Papaphotiou, G. and Schäfer, W., Liebigs Ann. Chem., (1993) 215.

A new resin for the synthesis of peptide aminoalkylamides

K. Kaljuste and A. Undén

Department of Neurochemistry and Neurotoxicology, Stockholm University, S-10691 Stockholm, Sweden

Introduction

Solid phase synthesis of peptides has traditionally been carried out by linking an amino acid to a polymer by an ester or amide bond. For certain applications, it is desirable to be able to link an amino group to the resin. The synthesis of the peptides with C-terminal alkylamino terminus has been carried out in homogeneous solution, combining the solid phase and solution synthesis as well as on solid phase employing a trifluoroacetic acid labile linker and Fmoc chemistry [1, 2].

Below we describe a method that allows anchoring of a primary amino group to the solid support for a further solid phase synthesis. Peptides with terminal alkylamino group can be synthesised using both t-Boc and Fmoc strategy, and the cleavage from the resin is carried out under very mild reaction conditions.

Results and Discussion

The total route for the synthesis of the spider venom component pseudoargiopinine III [3] on solid phase is shown on the Scheme 1. The DCHA salt of the allyl handle Trt-O-CH$_2$-CH=CH-CH$_2$-O-CH$_2$-COOH was directly coupled to the aminomethylresin by a BOP reagent, resulting in the trityl protected allylresin (III). This handle was described by Guibe et al. [4] and is now commercially available. After the removal of the Trt group with TFA the hydroxyl group was converted to p-nitrophenylcarbonate by 12 hour treatment with 4-nitrophenyl chloroformate at the presence of diisopropylethylamine (IV). Diamines carrying a single protective group (like N-trityl-1,5-diaminopentane) can then be coupled to the resin (V). After the removal of the Trt group by 1 hour treatment by TFA/CH$_2$Cl$_2$ and coupling of a Boc-Asn and indolacetic acid with conventional DCC - HOBt activation resulted in a resin linked pseudoargiopinine III (VII). The product was cleaved from the resin by (Ph$_3$P)$_2$PdCl$_2$ and Bu$_3$SnH in CH$_2$Cl$_2$/(CH$_3$)$_2$SO. The HPLC elution profile of the crude pseudoargiopinine III (VIII) is shown on the Figure 1. The molecular mass of the product was determined by plasma desorption mass spectrometry and was within the error limit of the instrument (±0.1%).

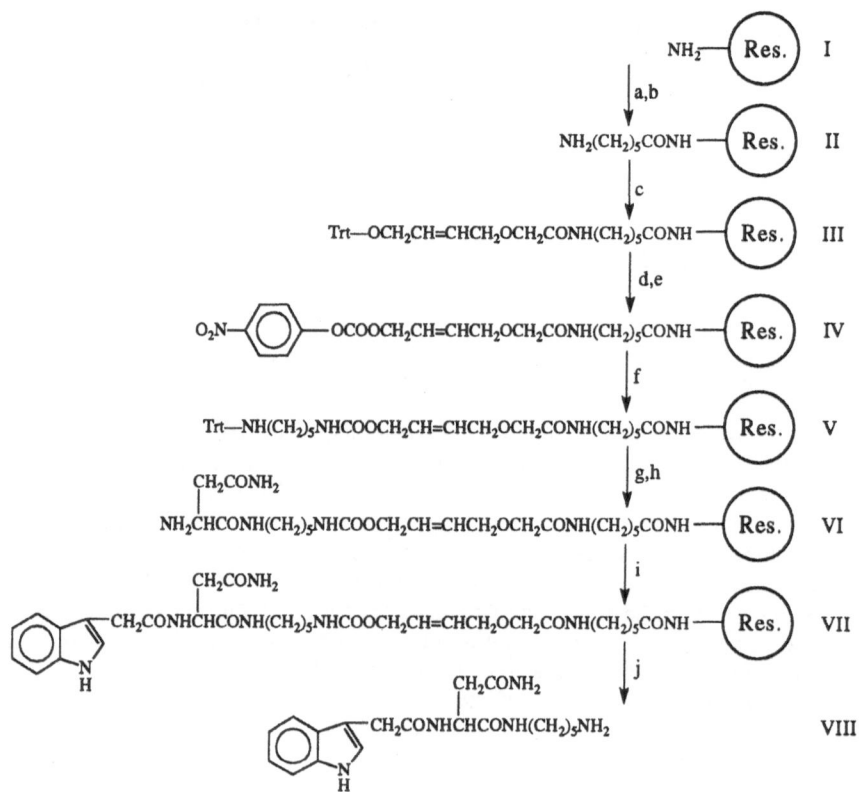

Scheme 1. The total route of the synthesis of pseudoargiopinine III.
(a) Boc-Aca, DCC/HOBt, DMF; (b) TFA/CH₂Cl₂ 50/50, 20 min; (c) Trt-O–CH₂–CH=CH–CH₂–O–COOH • DCHA, Bop, DIEA, DMF; (d) TFA/CH₂Cl₂ 50/50, 1 hour; (e) 4-Nitrophenyl chloroformate, DIEA, CH₂Cl₂ 12 hours; (f) N-trityl-1,5-diaminopentane, CH₂Cl₂, 4 hours; (g) Boc-Asn, DCC/HOBt, DMF; (h) TFA/CH₂Cl₂ 50/50, 20 min; (i) indole-3-acetic acid, DCC/HOBt, DMF; (j) Bu₃SnH, (Ph₃P)₂PdCl₂, CH₂Cl₂/(CH₃)₂SO 50/50, 1 hour.

The main advantage with this resin is that the allyloxycarbonyl link between resin and peptide is both base and acid stabile. Only about 10% of the peptide was cleaved by hydrogen fluoride during one hour at 0°C (data not shown). This stability allows a wide range of synthetic procedures to be carried out without cleaving the peptide from the resin. Furthermore, mild selective reaction conditions employed during the cleavage of the peptide from the resin result in peptides fully protected on all functional groups except the terminal amino group. This can be used for selective modification of the free terminal amino group. This approach can prove to be of value in the solid phase synthesis of naturally occurring polyamines that often contain terminal guanidino groups.

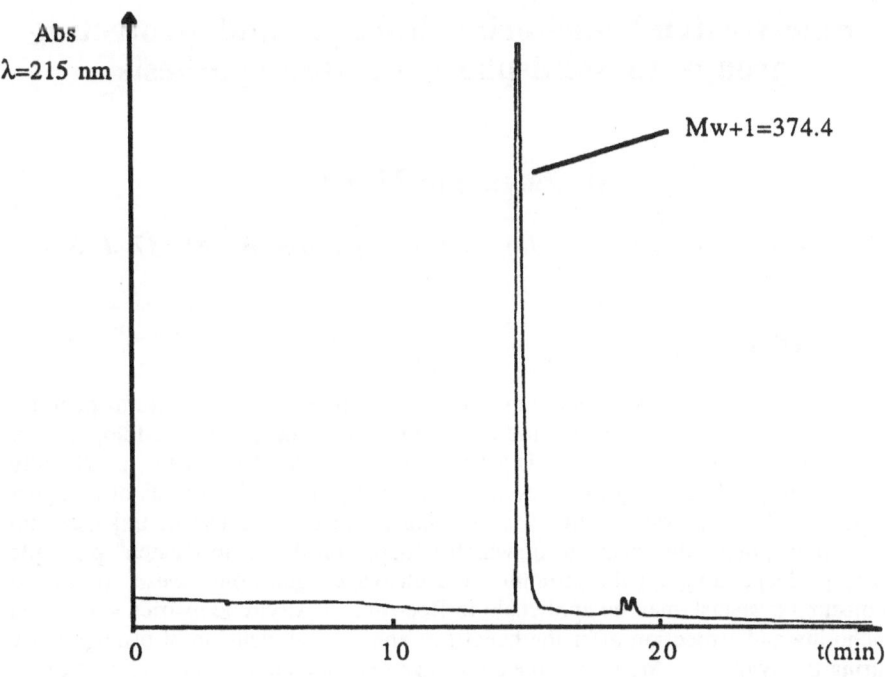

Fig. 1. The HPLC elution profile of crude Pseudoargiopinine III synthesised on solid phase.

References

1.   Geiger, R., Liebigs Ann. Chem., 750 (1971) 165.
2.   Breipohl, G., Knolle, J. and Geiger, R., Tetrahedron Lett., 28 (1987) 5647.
3.   Grishin, E.V., Volkova, T.M. and Arseniev, A., S. Tetrahedron, 27 (1989) 541.
4.   Guibé, F., Dangles, D., Balavoine, G. and Loffet, A., Tetrahedron Lett., 30 (1989) 2641.
5.   Lloyd-Williams, P., Jou, G., Albericio, F. and Giralt, E., Tetrahedron, 32 (1991) 4207.

"Safety-catch" anchoring linkages and protecting groups in solid-phase peptide synthesis

M. Patek and M. Lebl

Selectide Corp., 1580 E. Hanley Blvd., Tucson, AZ 85737, U.S.A.

Introduction

The C-terminal amide functionality which is found in many natural peptides often serves as a focal point for the development of synthetic methodology. The recent advances in peptide synthesis have opened vast new ideas for the intelligently designed use of anchoring linkages and protecting groups in synthesis of complex peptides [1]. In particular, there is growing interest in the use of linkages and protecting groups, the cleavage of which is based on the "safety-catch" principle [2,3,4]. Depending on the structure and cleavage conditions, these "protected protecting systems" may be applicable to Boc-/Fmoc-/Allyl-chemistries since such anchoring and protection offer the benefit of enhanced dimension of orthogonality during the synthesis. Moreover, the advantage that side-chain deprotection may be accomplished prior to cleavage from the support makes this methodology even more attractive. In our preliminary communication we have reported on the "safety-catch" anchoring linkage (SCAL) for synthesis of peptide amides [5]. The present report summarizes the results obtained with the SCAL anchoring and introduces a new "safety-catch" 2-methoxy-4,4'-bis(methylsulfinyl)benzhydryl (Msbh) (Figure 1) and 4,4'-bis(methylsulfinyl)trityl (Strt) protecting groups which have been evaluated as potential side-chain amide and SH protecting groups, respectively (Figure 2).

Results and Discussion

All three protecting blocks studied incorporate an interconvertible sulfoxide/sulfide system attached in the para position of the aromatic ring. In the first

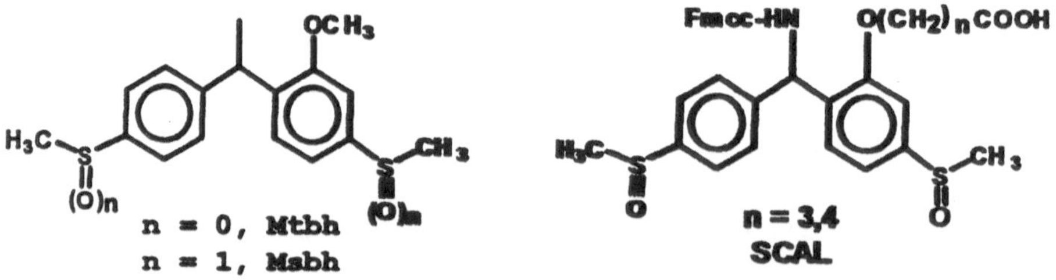

Fig. 1. Structures of Msbh/Mtbh protecting groups and SCAL linkage.

146

series of experiments we evaluated the stability/lability properties of Msbh and SCAL protecting groups under typical conditions for peptide synthesis. It was found that both the SCAL anchoring and the Msbh group are stable to anhydrous TFA (24h), 1M thioanisole/TFA (8h), TFA/thioanisole/EDT/phenol/water-82.5:5:2.5:5:5 (mixture K, 8h), anhydrous HF (0°C, 1h), 0.5M Ac_2O/NEt$_3$/ DCM (1h), and $Pd(PPh_3)_4$/AcOH/THF (1h). The 3.5M HCl/dioxane, 3M HBr/AcOH, and 1M Me_3SiBr/TFA, on the other hand, were found to cleave the benzhydrylamine C-N bond in SCAL skeleton within a few minutes by reductive acidolysis (one-step procedure). Alternatively, a two-step deprotection procedure using 1M Me_3SiCl/PPh$_3$/DCM (2h), 1M PhSeSiMe$_3$/DCM (1h), 20% $(EtO)_2P(S)SH$/DMPU (2h), 1M Me_3SiCl/PBu$_3$/DCM (2h) for reduction of the sulfoxide moieties, followed by the acidolytic cleavage with TFA/DCM/^{i}Bu$_3$SiH (90:8:2, 1h) offers the advantage of prior side-chain deprotection while the peptide chain remains attached to the carrier. A model peptide H-Tyr-Arg-Gln-Gly-NH$_2$ was synthesized on TentaGel (TG) resin using Fmoc-N^a-protected amino acids. The required N^a-fluorenylmethyloxycarbonyl-N$^\omega$-2-methoxy-4,4'-bis(methylsulfinyl)benzhydryl-L-glutamine [Fmoc-Gln(Msbh)-OH] was prepared in three steps from Fmoc-Glu-O^{t}Bu and 2-methoxy-4,4'-bis(methylthio)benzhydrylamine. Oxidation of sulfur atoms to sulfoxides with NaIO$_4$ followed by the DCC/HOBt-mediated coupling afforded fully protected derivative Fmoc-Gln(Msbh)-O^{t}Bu. Subsequent treatment with 95% TFA/H$_2$O gave the Fmoc-Gln(Msbh)-OH in 37% overall yield. The sequence of reactions for the final cleavage of peptide from the support provides a basis for evaluation of acid stability/lability properties of both SCAL anchoring and Msbh protecting group. Firstly, after Fmoc group deprotection, the acid-labile ^{t}Bu and Pmc groups were removed from Tyr and Arg, respectively, by treatment with mixture K at 20°C for 2h. Secondly, the remaining Msbh group on the side chain of glutamine as well as the SCAL linker were reduced with 1M Me_3SiCl/PPh$_3$/DCM at 20°C for 2h to give the corresponding sulfide forms. Finally, cleavage of the peptide from the resin using a mixture of TFA/DCM/^{i}Bu$_3$SiH (90:8:2) afforded the pure peptide (95% by RPHPLC) in 84% yield. Further utility of the SCAL anchoring is illustrated in the preparation of the more complex peptide, Human Gastrin I (pGlu-Gly-Pro-Trp-Leu-(Glu)$_5$-Ala-Tyr-Gly-Trp-Met-Asp-Phe-NH$_2$). The synthesis was performed manually on TentaGel (PEG-PS) resin (Nle as internal standard) using Fmoc/^{t}Bu chemistry, DMF as a solvent, and DIC/HOBt for carboxyl activation. After completion of the synthesis, samples of peptidyl resin were treated with TFA/EDT/anisole (80:10:10, 1h) to remove all acid-labile side-chain protecting groups. Final treatment of peptidyl resin with Me_3SiBr/TFA/^{i}Bu$_3$SiH/DCM (12:78:5:3, 2h, 0°C) afforded crude material which after purification by preparative RPHPLC furnished the title peptide in overall isolated yields of 10-12%. Purity and identity of product were assessed by RPHPLC (purified sample co-eluted with authentic Gastrin I), AAA, and FABMS.

The success of the synthesis of cysteine-containing peptides depends mostly on the protection chosen for the reactive thiol groups. Several thiol-protecting groups have been reported during the last decades, most of which are used in synthesis of simple monocyclic or more complex polycyclic peptides [6]. In the context of the "safety-catch" principle touched on above, a new "safety-catch" trityl protecting group was developed and evaluated as a potential S-protecting group (Figure 2). The synthesis of key intermediate, 4,4'-bis(methylsulfinyl)triphenylmethanol, involved initial treatment of 4,4'-bis(methylthio)-benzophenone with 1M PhMgBr/THF followed by the oxidation of sulfide moieties to sulfoxides with NaIO$_4$ (overall yield

91%). As a model system, N^α-fluorenylmethyloxycarbonyl-S-4,4'-bis(methyl-sulfinyl)trityl-L-cysteine [Fmoc-Cys(Strt)-OH] was prepared in 55% overall yield by the TFA-catalyzed reaction of L-cysteine with the derivative of trityl alcohol followed by the treatment with Fmoc-succinimide. In its oxidized form, the Strt group was found to be stable to 50% TFA/DCM (30min), 50% piperidine/DMF, iodine, Tl(Tfa)$_3$, and β-mercaptoethanol. On the other hand, the reduced form of this group can be easily removed by treatment with the mixture of 1% CCl$_3$COOH/DCM/scavenger. The applicability of Strt group was demonstrated by the synthesis of oxytocin (Fmoc/^{t}Bu chemistry) using SCAL anchoring to the TentaGel resin. After removal of Strt group (reduction with 1M Me$_3$SiCl/PPh$_3$/DCM (2h) followed by cleavage with CCl$_3$COOH/EDT/thio-anisole/^{i}Bu$_3$SiH/DCM (1:3:3:5:88) for 1h), the disulfide bridge was formed on the resin under pseudo-dilution conditions using CCl$_4$/NEt$_3$/NMP (10 equiv, 5h, 35°C) [7]. Final cleavage of the cyclized peptide from the resin with TFA/DCM/^{i}Bu$_3$SiH (90:8:2) followed by the preparative RPHPLC afforded pure oxytocin in 25% overall yield.

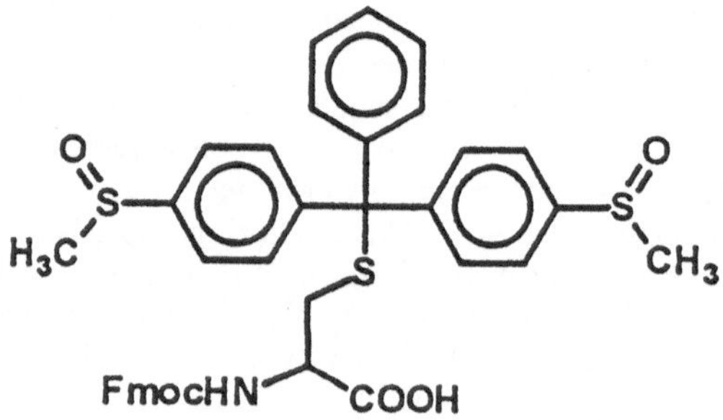

Fig. 2. Structure of Fmoc-Cys-(Strt)-OH.

References

1. Fields, G.B. and Noble, R.L., Int. J. Peptide Protein Res., 35 (1990) 161.
2. Osborn, N.J. and Robinson, J.A., Tetrahedron, 49 (1993) 2873.
3. Kiso, Y., Kimura, T., Itoh, H., Tanaka, S. and Akaji, K., In Smith, J.A. and Rivier, J.E. (Eds.) Peptides: Chemistry and Biology (Proceedings of the Twelfth American Peptide Symposium), Escom, Leiden, The Netherlands, 1992, p.533.
4. Samanen, J.M. and Brandeis, E., J. Org. Chem., 53 (1988) 561.
5. Patek, M. and Lebl, M., Tetrahedron Lett., 32 (1991) 3891.
6. Munson, M.C., Garcia-Echeverria, C., Albericio, F. and Barany, G., J. Org. Chem., 57 (1992) 3013 and references cited therein.
7. Munson, M.C., Garcia-Echeverria, C., Albericio, F. and Barany, G., In Smith, J.A. and Rivier, J.E. (Eds.) Peptides: Chemistry and Biology (Proceedings of the Twelfth American Peptide Symposium), Escom, Leiden, The Netherlands, 1992, p.605.

A dilute TFA-labile resin linkage for the Fmoc solid phase synthesis of protected peptide amides

J. Shao and W. Voelter

Institute of Biochemistry, University of Tübingen,
Hoppe-Seyler-Str. 4, D-7400 Tübingen, Germany

Introduction

The preparation of protected peptide segments is central to the peptide synthesis by the segment condensation method and also highly desirable for peptide post-synthesis modification. Fmoc chemistry, with its mild, orthogonal protection combination, provides a useful synthetic method for preparing protected peptide segments. Among the existing cleavage methods, the acidolytic cleavage with dilute TFA has the advantages of its simplicity in use and the high purity of the products. For synthesis of protected peptide acids employing Fmoc chemistry, several highly acid-labile resin linkages have been developed and their application has been well established [1]. For the synthesis of peptide amides, although several resin linkages based on multiple alkoxyl substituted benzyl amine or benzhydryl amine have been developed [1, 2], the concentration of TFA required in final cleavage, removing the side chain protecting groups simultaneously, makes them unsuitable for preparation of protected peptide amides [1, 3]. Here we describe the development of a novel trityl-based amide resin linkage. Peptide amides were cleaved from this resin with dilute TFA and side chain protected peptide amide was obtained after cleavage.

Results and Discussion

The 4-benzyloxy-4',4"-dimethoxytrityl amine resin (BDMTA-resin, Structure **VI**) was prepared starting from Merrifield resin (**I**) in four steps (Scheme 1): i) methyl-4-benzyloxy benzoate resin (**II**) was prepared from the chloromethyl resin (**I**), ii) then it was converted to trisalkoxytrityl alcohol resin (**IV**) via the subsequent Grignard reaction, iii) the alcohol resin was then transformed to chloride resin (**V**) with acetyl chloride, iv) finally the amine resin (**VI**) was obtained after the reaction of the resultant chloride resin with ammonia. Fmoc amino acids were attached onto the amine resin (**VI**) with DCC. The cleavage condition was investigated by reacting the Fmoc-Ala-(**VI**) resin with dilute TFA. 80.0% and 88.0% of Fmoc-Ala-NH_2 were released in 1 and 2 h respectively if 1% TFA was used for cleavage, while 89.5% and 91.2% were released if treated with 3% TFA. Several peptide amides including TRH (pGlu-His-Pro-NH_2), a partial sequence of substance P (pGlu-Phe-Phe-Gly-Leu-Met-NH_2 = SP[6-11]) and Leucopyrokinin (Phe-Thr-Pro-Arg-Leu-NH_2 = LPK[4-8]) were synthesized using this linkage under dilute TFA cleavage. The cleavage of TRH was performed with 3% TFA to remove His[im]-trityl simultaneously, 1% TFA was used for other cleavages. In the case of LPK[4-8] the peptide was obtained in the protected form. All peptide amides were obtained in high purity (>90% according to HPLC) and satisfactory yields (>80% based on resin loading). The products are confirmed by analytical methods such as FAB-MS, amino acid analysis and cochromatography with authentic samples.

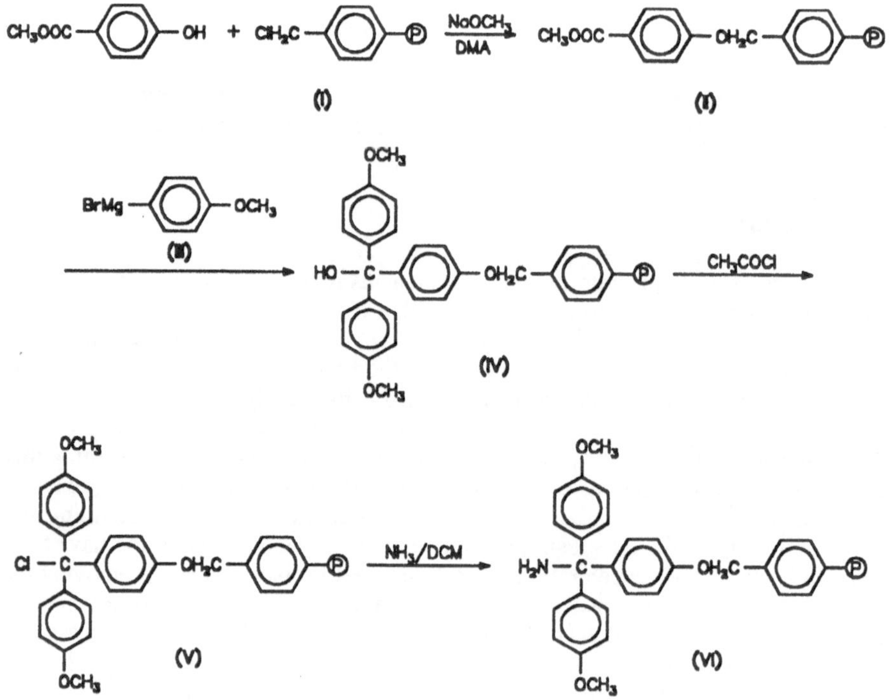

Scheme 1. Synthesis of the 4-benzyloxy-4',4"-dimethoxytrityl amine resin (VI)

The use of highly acid-labile 4-benzyloxy-4',4"-dimethoxytrityl amine resin linkage provides a new approach to prepare peptide amides with Fmoc strategy under dilute TFA cleavage. This is particularly useful if peptide amides are sensitive to acid or side chain protected peptide amides are needed for post-synthesis modification.

References

1.   Fields, G.B. and Noble, R., Int. J. Peptide Protein Res., 35 (1990) 161.
2.   Shao, J., Li, Y.-H., and Voelter, W., Int. J. Peptide Protein Res., 36 (1990) 182.
3.   Story, S.C. and Aldrich, J.V., Int. J. Peptide Protein Res., 35 (1992) 87.

Protected peptide intermediates using a trityl linker on a solid support II

A. van Vliet, B.H. Rietman, S.C.F. Karkdijk, P.J.H.M. Adams and G.I. Tesser

Catholic University of Nijmegen, Department of Organic Chemistry, Toernooiveld, 6525 ED Nijmegen, The Netherlands

Introduction

In order to develop a new method for synthesizing small protected peptides, a new acid labile linker has been designed [1, 2]. The prepared resin is composed of aminomethyl polystyrene and a trityl function; both are connected by an alkylspacer (Figure 1). Trityl was chosen for its suitability in the temporary protection of nitrogen (in hydrazides, amines and amides) and sulfur (in thiols). Immunogenic determinants which are to be coupled onto carriers by means of a C-terminally located amino or thiol function, have been synthesized using the new resin as protected peptide hydrazides and (ω-aminoalkyl)amides [2]. Another application is the synthesis of protected peptide disulfides, involving iodine-controlled cleavage from the resin and concomitant regiospecific forming of the disulfide bridge.

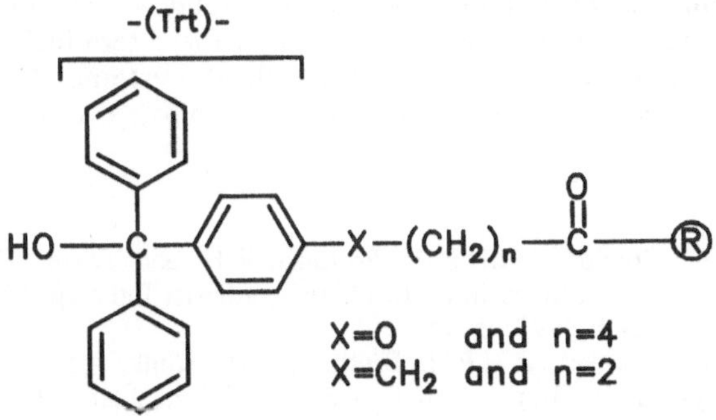

Fig. 1. Trityl resin.

Results and Discussion

An application of the new resin has been found in the synthesis of the following peptide: DNP-Pro-Gln-Gly-Ile-Ala-Gly-Lys(Cum)-Arg-Dah-Acr. It is composed of DNP-Pro-Gln(Trt)-Gly-Ile-Ala-Gly-N_2H_3 and H-Lys(Cum)-Arg-Dah-Acr, these are built up stepwise starting from Fmoc-Gly-N_2H_2-(Trt)-O-Resin and Fmoc-Dah-(Trt)-O-Resin (DNP = dinitrophenyl; Cum = 7-methoxycoumarin-4-acetic acid; Acr = acryloyl chloride). The solid phase synthesis was performed as follows: initial deprotection with 20% piperidine in DMF, introduction of Fmoc amino acids using DIPCDI and HOBt. The eventual cleavage of the peptides from the resin can be accomplished by 1 to 3 treatments with 0.5% TFA in CH_2Cl_2 in the presence of 4% MeOH or H_2O. A second application has been found in the synthesis of cyclic peptide disulfides. The general way for the synthesis of peptide disulfides is the following: Fmoc-cysteamine is immobilized by its thiol function. The peptide is synthesized stepwise. For cyclic disulfides, a Cys(Acm) residue should be introduced later in the chain. Detachment will be performed in one operation by oxidative cyclization with iodine in DMF only and yields the fully protected cyclic peptide disulfide in accordance with the observations of Kamber [3]. The use of DMF as the solvent ensures the full profit of the pseudo dilution: nearly pure cyclic disulfides were obtained in one operation. An example of the synthesis of a real cystine-peptide is the synthesis of oxytocin. Fmoc-Cys-OH was immobilized by its thiol function, H-Pro-Leu-Gly-NH$_2$ was attached C-terminally using TBTU. The peptide was further built up normally. Detachment and cyclization took place with iodine in DMF. Detachment with CHCl$_3$/MeOH 1:1 gave, besides the cyclized product also substantial amounts of the linear dimer [2]. Another application could be the synthesis of peptide amides. Attempts to attach amides to the linker in the way Sieber prepared Fmoc-Gln(Trt)-OH [4a] failed. Some other promising model experiments were performed. They imply the reactivity of the nitrile function. Starting from trityl formate and acetonitrile in acidic medium or from trityl chloride and silver nitrate in acetonitrile N-tritylacetamide was obtained [4b,c]. Aminonitriles arising from Strecker syntheses would thus be converted to immobilized amino acid amides. The diphenyl-4-pyridyl methyl group [4d] was also introduced in Z-Gly-NH$_2$ and Fmoc-Gln-OH.

Using the new linkers protected peptide hydrazides are obtained for fragment condensation in solution. Subsequent deprotection can then be performed in dilute solution preventing unwanted alkylations (Tyr). Peptide-(ω-aminoalkyl)amides or (ω-sulfhydryl)amines can be prepared as epitopes for C-terminal immobilization. Oxidation including cyclization of immobilized cysteine derivatives gives surprisingly pure cyclic disulfides. Finally the resin can be recycled.

References

1. van Vliet, A., Smulders, R.H.P.H., Rietman, B.H. and Tesser, G.I. (Epton, R.), Innovation and Perspectives in Solid Phase Synthesis, Intercept Limited, Andover, Hampshire, England, 1992, p.475.
2. van Vliet, A., Smulders, R.H.P.H., Rietman, B.H., Klink, A.M.E., Rijkers, D.T.S., Eggen, I.F., van de Werken, G. and Tesser, G.I., In Schneider, C.H. and Eberle, A.N. (Eds.), Peptides 1992 (Proceedings of the 22nd European Peptide Symposium), Escom, Leiden, The Netherlands, 1993, p.279.
3. Kamber, B., Hartmann, A., Riniker, B., Rink, H., Sieber, P. and Rittel, W., Helv. Chim. Acta 63 (1980) 899.
4. a) Sieber, P. and Riniker, B., Tetrahedron Lett., 32 (1991) 739; b) Bacon, R.G.R. and Köchling, J., J. Chem. Soc., (1964) 5609; c) Cheeseman, G.W.H., J. Chem. Soc., (1959) 452; d) Coyle S., and Young, G.T., J. Chem. Soc. Chem. Commun., (1976) 980.

Evaluation of a new membrane support for peptide synthesis and affinity adsorption purification

J.J. Wen, E. Klein and A.F. Spatola

*Department of Chemistry, University of Louisville,
Louisville, KY 40292, U.S.A.*

Introduction

Membrane-based peptide synthesis and affinity adsorption chromatography have attracted increased attention because of the characteristic high convective flow rate of membranes [1]. We were interested in developing a new type of membrane support for peptide synthesis and for peptide or peptide analogue-based immuno-affinity chromatography. Mild peptide cleavage methods which we have developed including phase transfer catalysis [2] (PTC) and ammonium formate-catalytic transfer hydrogenation [3] (AF-CTH), have been applied for the cleavage of peptides from the membrane.

Results and Discussion

We have established that ethylene-vinyl alcohol copolymer (EVAL) can be used in a flat membrane form for Fmoc-based solid phase peptide synthesis (Fmoc-SPPS) when it is appropriately modified. The modification procedures are shown below:

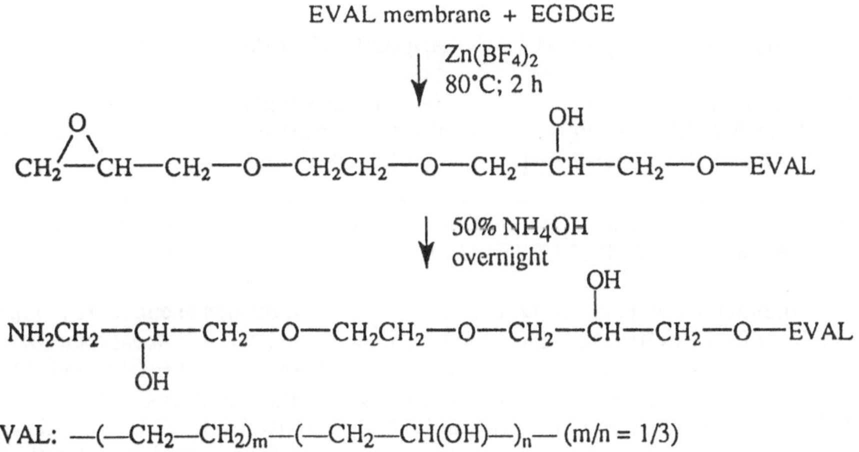

EVAL membrane + EGDGE

$$\downarrow \quad \begin{array}{l} Zn(BF_4)_2 \\ 80°C; \; 2 \; h \end{array}$$

$$CH_2\overset{O}{\diagup}CH-CH_2-O-CH_2CH_2-O-CH_2-\overset{\overset{\displaystyle OH}{|}}{CH}-CH_2-O-EVAL$$

$$\downarrow \quad \begin{array}{l} 50\% \; NH_4OH \\ overnight \end{array}$$

$$NH_2CH_2-\underset{\underset{\displaystyle OH}{|}}{CH}-CH_2-O-CH_2CH_2-O-CH_2-\overset{\overset{\displaystyle OH}{|}}{CH}-CH_2-O-EVAL$$

EVAL: $-(-CH_2-CH_2)_m-(-CH_2-CH(OH)-)_n-$ (m/n = 1/3)
EGDGE: ethylene glycol diglycidyl ether

The modifications applied to EVAL have several purposes: 1) a new amine functionality is introduced; 2) many of the reactive hydroxyls are replaced; and

3) cross-linking induced by the bifunctional EGDGE significantly increases structural stability while retaining the microporous membrane characteristics. The stability of modified EVAL membrane toward various solvents has been determined. The modified EVAL membrane proved stable to most organic solvents used in Fmoc-SPPS, such as DMF, and piperidine, but was very labile to trifluoroacetic acid (TFA). The free amine content of the membrane was determined to be 0.1 mmol/g by ninhydrin assay. This membrane can be directly used for Fmoc-SPPS or further reacted with various linkers, such as 4-hydroxymethylphenoxyacetic acid to be ready for Fmoc-SPPS.

The first amino acid was attached directly to the EVAL membrane by BOP coupling reagent (in the case of affinity purification) or by symmetrical Fmoc amino acid anhydride in the presence of DMAP catalyst coupled to the phenoxy-acetic acid handle (when it is necessary to cleave the final peptide from the membrane either for synthesis or for analytical purposes). The peptide derived from I (Table 1) was cleaved from the EVAL by AF-CTH, and peptide-EVAL II was cleaved by PTC. The peptides were checked by analytical HPLC and FABMS. The results are summarized in Table 1. Yields of both peptides could not be quantitatively determined but were judged in the acceptable range.

Table 1

Peptide supports	Amino acid analysis	Retention time[a] (min)	FABMS
I. H-AVGFG-linker-EVAL	Ala 1.00, Val 1.08 Gly 1.97, Phe 0.95	13.60	calc: 449 found: 450
II. H-[^{14}C]AFVGL-linker-EVAL	Ala 1.00, Val 1.01 Gly 1.00, Phe 1.04 Leu 1.10	16.45	calc: 505 found: 506
III. H-Arg(Mtr)-Gly-Asp(Bu^t)-EVAL	Arg 0.91, Gly 1.00 Asp 0.98		
IV. α-endorphin fragment H-Ser(Bu^t)-Gln-Thr(Bu^t)- Pro-Leu-Val-Thr(Bu^t)-EVAL	Ser 1.10, Glx 0.91 Thr 1.87, Pro 0.96 Leu 1.14, Val 1.00		

[a]HPLC conditions: Vydac C_{18}4.6x250 mm. B from 5-90% in 25 min, 230 nm. Buffer A: 0.1% TFA/water; Buffer B: 0.1% TFA/CH_3CN

In order to compare various cleavage procedures, the pentapeptide, H-[^{14}C]Ala-Phe-Val-Gly-Leu-OH, which incorporates a radioactive [^{14}C] Ala, was synthesized on four different solid supports: Merrifield resin, PAM resin, PEG-PS and EVAL membrane. The success of peptide synthesis was confirmed by amino acid analysis. These four peptide-supports were individually subjected to PTC and AF-CTH cleavage and the cleavage process was monitored by removal of aliquots and assessment of relative radioactive counts (1900TR-Scintillation counter). The results are summarized in Figure 1.

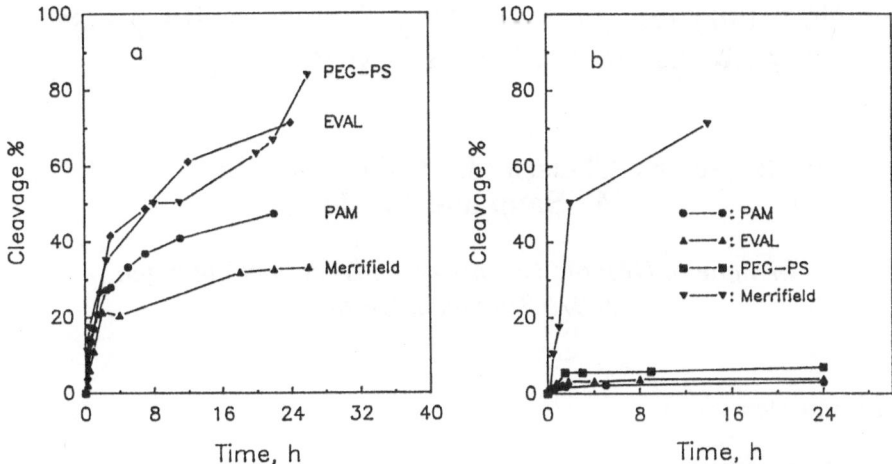

Fig. 1. *Kinetic studies of peptide cleavage from various solid supports by using (a) PTC and (b) AF-CTH. Peptide sequence is H-[^{14}C]Ala-Phe-Val-Gly-Leu-OH.*

Figure 1(a) shows that in the case of the EVAL membrane, the PTC cleavage method was comparable in its efficiency with the polyethyleneglycol/polystyrene (PEG-PS) support. The PAM and Merrifield resins gave somewhat less cleavage per unit time using PTC. Initial rates for all procedures were similar. In terms of the AF-CTH cleavage procedure, the results shown in Figure 1(b) were more surprising. The AF-CTH hydrogenation method is known to be affected by hydrophobic peptide sequences (such as those used here). Thus the poor results with three of the four supports might be expected. But at this time, we have no explanation as to why the standard Merrifield support gave relatively acceptable cleavage results.

The major drawback of the EVAL membrane is its lability toward TFA treatment. Furthermore, deprotection of acid-labile side chain protecting groups is still a requirement for use of the EVAL membrane in peptide or peptide analogue based affinity chromatography. A recent successful trial synthesis of a dipeptide, H-Gly-Leu-EVAL (AAA: Gly, 1.00; Leu, 0.95), by using BF$_3$/Et$_2$O as the Boc-protecting group cleavage reagent, provides a more optimistic analysis regarding the compatibility of EVAL membranes for both Fmoc-SPPS and Boc-SPPS methodologies.

Acknowledgements

Supported by NSF EPSCOR EHR-9108764.

References

1. Daniels, S.B., Bernatowicz, M.S., Coull, J.M. and Koster, H., Tetrahedron Lett., 30 (1989) 4345.
2. Anwer, M.K. and Spatola, A.F, Tetrahedron Lett., 33 (1992) 3121.
3. Anwer, M.K., Spatola, A.F., Bossinger, C.D., Flanigan, E., Liu, R.C., Olsen, D.B. and Stevenson, D., J. Org. Chem., 48 (1993) 3503.

New polymer and strategy for the solid-phase synthesis of protected peptide fragments

E. Bayer, N. Clausen, C. Goldammer, B. Henkel, W. Rapp and L. Zhang

Institute of Organic Chemistry, University of Tübingen, 72076 Tübingen, Germany

Introduction

A restriction of solid-phase peptide synthesis is still the sequence-dependent difficulties during assembly. Therefore, the combination of solution-phase and solid-phase peptide synthesis appears to be an attractive approach towards the goal of avoiding such difficulties. This requires a strategy for obtaining protected peptides by SPPS after cleavage from the polymer. We investigated available linkers and report upon a new strategy for the rapid preparation of side chain protected fragments via trityl-linker using polystyrene-polyethyleneglycol graft copolymers in conjunction with the continuous flow technique.

Results and Discussion

During synthesis of HIV-1 protease with Fmoc-strategy difficulties are experienced in the coupling and deprotection steps 12-18 and 28-38 of the sequence, as is revealed by kinetic investigations [1]. Despite improvement in the synthesis of this sequence by the use of a "magic mixture" containing nonionic detergents as an additive to the solvents, the final product is still contaminated with failure sequences, rendering isolation by RP-HPLC difficult. We have therefore attempted to prepare protected fragments corresponding to HIV-1 protease by the continuous flow technique and combine them using solution or convergent solid-phase method. HIV-1 protease was divided into 11 fragments with 6-12 residues, five of which contain Gly, three contain Pro and two contain Leu at the C-terminus. The fragments as well as the protecting groups are shown in Figure 1.

PQ(Pmc)IT(tBu)LWQ(Trt)R(Pmc)P ✂ LVT(tBu)IR(Pmc)IG ✂ GQ(Trt)LK(Boc)E(OtBu)ALL-
       1-9   10-16   17-27
D(OtBu)T(tBu)G ✂AD(OtBu)D(OtBu)T(tBu)VL ✂E(OtBu)E(OtBu)MN(Trt)LPG ✂K(Boc)W
       28-33   34-40
K(Boc)PK(Boc)MIGGIG ✂GFIK(Boc)VR(Pmc)Q(Tmob)Y(tBu)D(OtBu)Q(Tmob)IP ✂V
       41-51   52-63
E(OtBu)IC(Trt)GH(Trt)K(Boc)AIG ✂T(tBu)VLVGPT(tBu)P ✂VN(Trt)IIGR(Pmc)N(Trt)L ✂
       64-73   74-81   82-89
LT(tBu)Q(Tmob)IGC(Trt)T(tBu)LN(Tmob)F
       90-99

Fig. 1. HIV-1 protease sequence and its fragments to be prepared.

Several strategies for solid-phase synthesis of protected fragments have been employed, including peptide resin linkers which are labile to dilute acid and to light. The 1% TFA or acetic acid labile linkers (known as Riniker- and Rink-linkers) seem to be useful for the preparation of protected fragments bearing Fmoc as N-terminal protection [2, 3]. The Riniker-linker was first checked for our purpose. 4-(4'-hydroxymethyl-3-methoxyphenoxy)-butyric acid was coupled to amino-functionalized PS-PEG (amino-PEG-PS) and applied to synthesize the most sensitive fragment 64-73 since it contains Boc and Trt protecting groups for Lys and His respectively. We found that Boc and especially Trt protecting groups do not survive under the cleavage conditions with 1% TFA in DCM even for a very short time. We decided then to use the Rink approach because Fmoc-His(Trt) is stable to 50% acetic acid in DCM up to 4 hours or more according to HPLC assays. The cesium salt of 2,4-dimethoxy-4'-hydroxybenzophenone was added to brominated PS-PEG in DMF and shaken for 18 hours at 50¤C. The reduction of ketone resin was carried out similarly to that described by Rink [3]. With this new resin, the purity of the fragment 64-73 after cleavage from the resin could be dramatically improved. Unfortunately, the yield on a one gram resin scale synthesis was too low because of the significant loss of the hydroxyfunction during the ketone reduction.

Barlos et al. introduced the 2-chlorotrityl chloride resin for synthesis of protected peptide fragments [4]. In this case, protected fragments can be cleaved from the resin by dilute acetic acid. However, its application is restricted to polystyrene (1% DVB) as polymer support used only for batchwise synthesis. This limitation led us to search for a new trityl-linker which may be employed in CFSPPS on PS-PEG resin. *p*-Carboxy-trityl-alcohol is readily obtained and therefore chosen for this purpose. We found that electron withdrawing substituents in the para position ensure the stability of the ester bond between the trityl group and the carboxyl group of the C-terminal amino acid during synthesis. In order to test the usefulness of the designed linker, p-carboxytritylalcohol was easily coupled to amino-PEG-PS via DICI/HOBt activation and the hydroxyl substituted polymer converted into its chlorosubstituted analog with acetyl chloride in DCM. The first amino acid was incorporated onto the resin using the method recommended by Barlos et al. [4]. With this new system, the fragment 64-73 was first synthesized in 80% yield based on Gly-resin. Comparison of analytical HPLC of both crude products obtained from the Riniker-linker and new trityl-linker shows that the purity of the protected fragment can thus be significantly improved. The undesired side reactions during cleavage of the fragment were not observed according to LC-MS investigation. Boc- and Trt-protections for Lys and His remained unchanged. In order to further explore the efficacy of this new method, fragment 41-51 was assembled on three resins derivatised with Riniker-, Rink- and trityl-linkers. After cleavage from the resins, protected fragments from different syntheses were directly analysed by the HPLC-ISMS technique. In contrast to the product obtained from the Riniker-linker, Rink- and trityl-linkers provided the expected fragment in sufficient purity according to analytical HPLC.

The established approach proved suitable for synthesis of protected fragments. The 11 fragments illustrated in Figure 1 were successfully synthesized on a MilliGen 9050 peptide synthesizer with TBTU activation using this new method. All protected fragments were well characterized by analytical HPLC, ion spray MS (IS-MS) or LC-MS with the exception of fragment 10-16. HPLC analysis of this fragment performed on RP columns showed a broad peak probably due to aggregation. However, sufficient purity was shown by TLC and IS-MS. Results obtained from the syntheses along with the solubilities of individual protected sequences are compiled in Table 1.

Table 1 *Results of analytical data and syntheses of the fragments*

Fragment	MS(calc.)	HPLC(%)	Yield(%)	Synthesis Duration	Solubility
1-9	2042	98	78	4 h 37 min	DCM
10-16	1317	90(TLC)	76	3 h 30 min	DMF, CHCl$_3$
17-27	1878	97	75	4 h 43 min	DCM/DMF
28-33	1024	98	90	2 h 18 min	DCM
34-40	1366	98	91	2 h 53 min	DCM/DMF
41-51	1737	96	82	5 h 32 min	DCM
52-63	2525	91	72	6 h 54 min	DCM
64-73	1889	92	80	4 h 40 min	DCM
74-81	1117	95	77	3 h 50 min	DCM
82-89	1871	97	82	5 h 03 min	DCM/DMF
90-99	2046	95	75	5 h 07 min	DCM

It is worth mentioning that a fragment to be prepared must have sufficient solubility in common solvents used for the fragment condensation. At the beginning we selected the sequences 87-94 as the first fragment because it contains Gly as C-terminus. Surprisingly, synthesis difficulties occurred at the same position as observed during assembly of the total sequence and could not be overcome by simply reducing the C-terminus by 5 residues. The aggregation seems to be sequence-dependent. In fact, the fragment 82-94 is sparingly soluble in DMF and insoluble in DCM. The same is true for the fragment 28-40. During the preview synthesis of HIV-1 protease, synthesis difficulties were also encountered in this region. This part of the sequence had to be divided into two pieces, 28-33 and 34-40. The fragments given in Table 1 are optimized regarding to their solubilities. Most of them are readily soluble in DCM, some of them in DCM/DMF. The heptapeptide fragment 10-16 could not be dissolved in DCM, but in CHCl$_3$ and DMF. This peptide was difficult to synthesize owing to the extremely hydrophobic nature of the sequence, LVTIRI. For almost every step a double coupling is required. It is evident that sequence-dependent synthesis difficulties are correlated with the solubility of the corresponding fragment. Therefore, a preview synthesis of a total sequence with continuous monitoring can be helpful for predicting the optimum fragments.

References

1. Bayer, E., Zhang, L., Rapp, W. and Waidelich, D., In Giralt, E. and Andrea, D. (Eds.), Peptides 1990 (Proceedings of the 21st European Peptide Symposium), Escom, Leiden, The Netherlands, 1991, p.352.
2. Flörsheimer, A. and Riniker, B., In Giralt, E. and Andrea, D. (Eds.), Peptides 1990 (Proceedings of the 21st European Peptide Symposium), Escom, Leiden, The Netherlands, 1991, p.131.
3. Rink, H., Tetrahedron Lett., 28 (1987) 3787.
4. Barlos, K., Gatos, D., Kallitsis, J., Papaphotiu, G., Sotiriu, P., Yao, W. and Schäfer, W., Tetrahedron Lett., 30 (1989) 3943.

Rapid dimerization of an N-terminal cysteine containing peptide and dimer inhibition through N-terminal capping

M.R. Routhe, G.J. Bowman, J.M. Gangitano, J.L. Lambing, S.M. Pratt and T.P. Hopp

Multiple Peptide Systems, Inc., San Diego, CA 92121, U.S.A.

Introduction

N-terminal cysteine containing peptides are commonly used by researchers in various disciplines, especially in immunological studies. The cysteine thiol moiety is required for the conjugation of the peptide to a carrier proteins amino group through the use of cross-linkers [1]. Peptides synthesized with amino terminal cysteine residues have been observed to have an unusually high level of dimerization, even in acidic media. It is proposed that the N-terminal free amine lowers the activation energy of the thiol moiety towards oxidation, resulting in increased dimerization. This lowers monomer yields and decreases the amount of peptide that can be cross-linked to the carrier protein. A representative example of this phenomenon is presented here using the test sequence, CLSVKAMG-amide.

Results and Discussion

Analysis of the N-terminal cysteine analog by RP-HPLC demonstrates that the dominant form of the peptide is the dimer (Figure 1-left). The mass spectral analysis (Figure 1-right) confirms the existence of dimer. Similar RP-HPLC analysis of the acetylated analog (Figure 2-left) shows a significant reduction in the amount of dimer formation. The mass spectral analysis (Figure 2-right) confirms that small amounts of dimer still exist.

These results suggest that N-terminal capping prevents a significant amount of dimerization resulting in higher peptide yields. It is hypothesized that when the N-terminal amino group is capped in the acetylated peptide, it cannot interact with the thiol group on the cysteine, therefore little dimerization can occur. These results suggest that unless the peptide must mimic the N-terminus of a protein, an acetylated-amino terminus may be desirable [2]. We conclude that when a N-terminal cysteine peptide is synthesized for the purpose of conjugating a carrier protein to the thiol moiety, capping the N-terminal by acetylation may prove to be a useful tool.

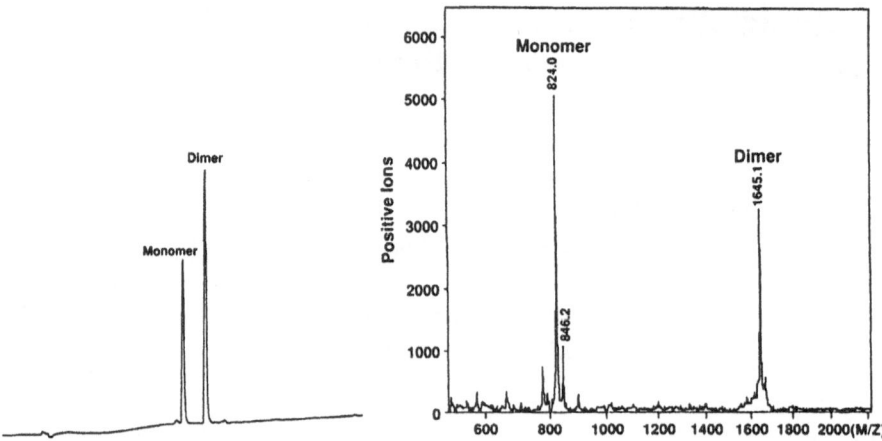

Fig. 1. RP-HPLC analysis of the test sequence H₂N-CLSVKAMG-amide showed the dimer comprised >50% of the final material (left). Qualitative mass spectral analysis for this sequence indicates the presence of both monomer and dimer (right).

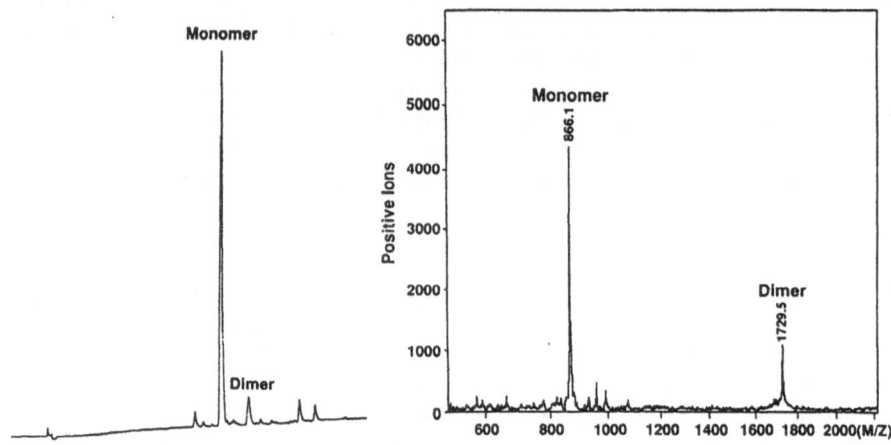

Fig. 2. RP-HPLC analysis of the test sequence Acetyl-CLSVKAMG-amide showed the dimer comprised <8% of the final material (left). Qualitative mass spectral analysis for this sequence indicates the presence of both monomer and dimer (right).

References

1.   Lee, A.C.J., Powell, J.E., Tregear, G.W., Niall, H.D. and Stevens, V.C., Mol. Immunol., 17 (1980) 749.
2.   Grant, G.A., Synthetic Peptides, A User's Guide, 1992, p.50.

Synthesis of α-fluoroglycine derivatives and their incorporation into peptides

P.D. Bailey[1], J. Clayson[1], S.R. Baker[2] and A.N. Boa[3]

*[1]Department of Chemistry, Heriot-Watt University, Riccarton,
Edinburgh, EH14 4AS, U.K.*
*[2]Lilly Research Centre Limited, Erl Wood Manor, Windlesham,
Surrey, GU20 6PH, U.K.*
*[3]Department of Chemistry, University of York, Heslington,
York, YO1 5DD, U.K.*

Introduction

The potent biological activity of fluorinated peptides has stimulated great interest in the development of asymmetric synthetic routes towards fluorinated amino acid derivatives [1].

Since fluorine is isosteric with hydrogen it may be introduced into molecules such as peptides without causing extensive conformational change. Its introduction into amino acid substrates can confer several new properties upon them such as: (i) causing them to act as suicide substrates, (ii) introducing additional centres of hydrogen bonding, (iii) altering the hydogen bonding ability of adjacent amide groups and (iv) altering the stability of surrounding peptide bonds towards hydrolysis.

Free α-fluoroamino acids readily decompose with concommitant loss of fluoride (Figure 1): because of this intrinsic instability work involving α-fluorinated amino acids has been severely limited to date.

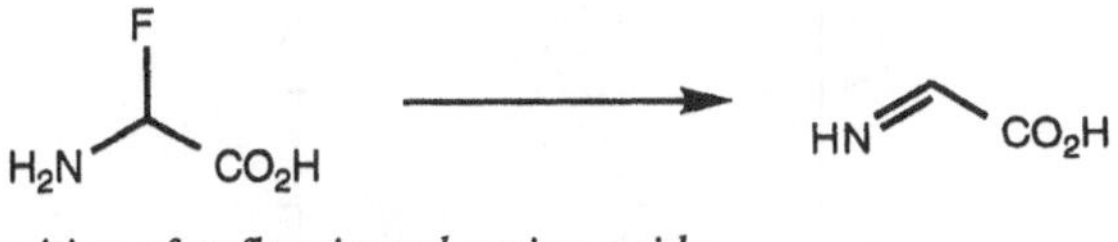

Fig. 1. Decomposition of α-fluorinated amino acids.

We have sought to overcome the inherent instability of α-fluoroamino residues by the development of new synthetic methodology.

The standard peptide disconnection cannot be applied to the synthesis of target peptides containin the α-fluoroglycyl residue (Figure 3). However, we envisaged that a nitrogen atom could be introduced by the nucleophilic displacement of a suitable leaving group from the fluorine bearing carbon.

The initial fluorine containing precursors were chosen to allow both amino and carboxy terminal extension. Literature precedent suggested that compounds such as (1) and (2) should be amenable to nucleophilic substitution, and that extension of the carboxy terminous *via* the corresponding acid chloride should be possible (Figure 2).

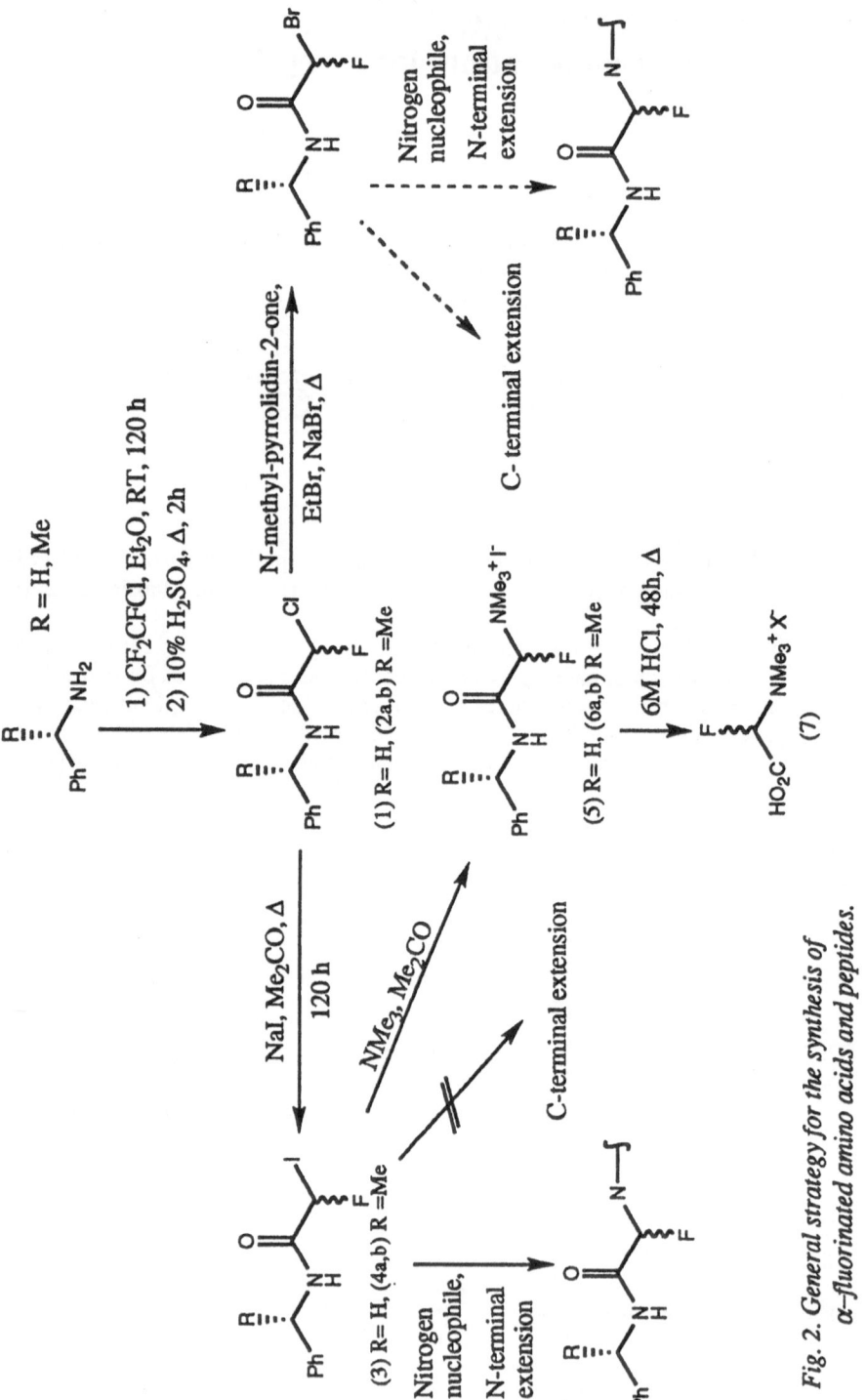

Fig. 2. *General strategy for the synthesis of α–fluorinated amino acids and peptides.*

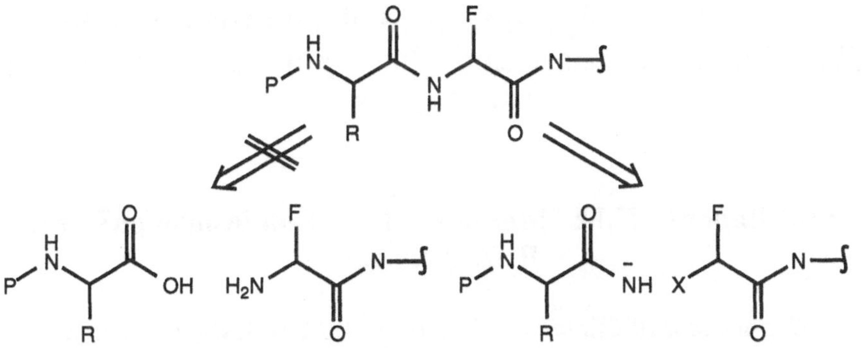

Fig. 3. Proposed disconnections of α-fluorinated amino acids.

Results and Discussion

Starting materials (1) and (2a,b) were prepared using established literature methodology [2]. In the case of the latter, no asymmetric induction was observed, but the diastereoisomers were separable by flash chromatography. A range of nitrogen nucleophiles failed to displace chloride from these compounds. Consequently, the analogous fluoroiodoethanamides (3) and (4a,b) were prepared (Figure 2). These compounds underwent the desired reaction with nitrogen nucleophiles to yield α-fluoroglycine derivatives.

Crucially, reaction of (3), (4a) and (4b) with trimethylamine followed by acidic hydrolysis yielded the first 'free' α-fluorinated amino acid, α-fluorobetaine (7). Work is currently underway to incorporate this residue into peptides using standard coupling techniques.

Evidence has been obtained which suggests that it is possible to react (3), (4a) and (4b) with protected amino acids e.g. piperazine triones (Figure 4) to give protected Phe-Gly(F) dipeptides. We are attempting to find conditions under which such products may be isolated.

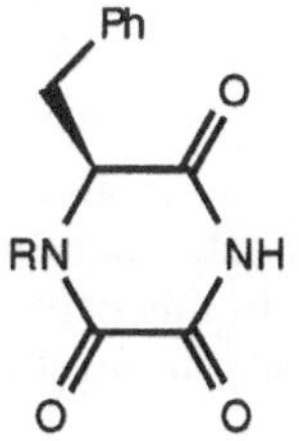

R = H, PhCH$_2$

Fig. 4. Protected amino acid nitrogen nucleophile: piperazine trione.

Acknowledgements

We wish to thank analytical staff both at York and Heriot-Watt University, SERC for studentships to JC and ANB, and Lilly Research for a CASE award to JC.

References

1.   Welch, J.T., Tetrahedron, 43 (1987) 3123.
2.   Molines, H. and Wakselman, C., Synthesis, (1984) 838.

Synthesis of pipecolic acid derivatives via the Diels-Alder reaction and their potential as proline analogues

P.D. Bailey[1], K.M. Morgan[1], D.J. Londesbrough[2] and R.D. Wilson[2]

[1]Department of Chemistry, Heriot-Watt University, Riccarton, Edinburgh, EH14 4AS, U.K., and [2]Department of Chemistry, University of York, Heslington, York, YO1 5DD, U.K.

Introduction

Substituted piperidine ring systems are ubiquitous in nature, and short stereo-controlled routes to functionalised piperidines are of great synthetic value. We recently developed an asymmetric aza-Diels-Alder reaction that does indeed give direct access to optically active piperidine derivatives [1] - see Figure 1 and Table.

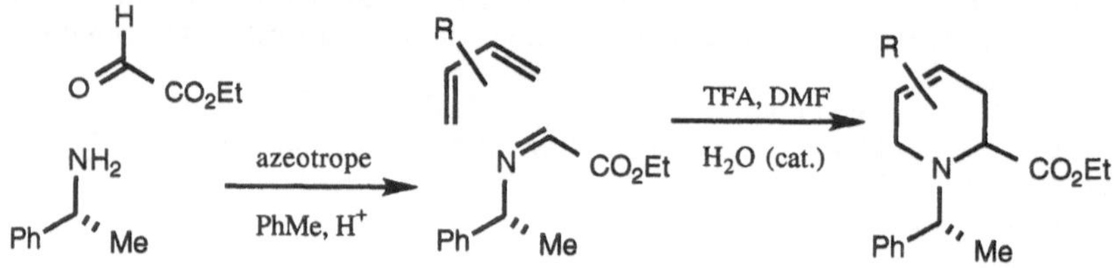

Fig. 1. General strategy.

This procedure is especially valuable as it allows the rapid synthesis of pipecolic acid derivatives that can be incorporated into biologically active peptides such as the MQPA thrombin inhibitors [2]. In this paper, we describe how we are utilising the asymmetric aza-Diels-Alder reaction towards a synthesis of carpamic acid.

Results and Discussion

Carpamic acid, the monomer of carpaine has generated much synthetic interest due to the pharmacological properties of the parent molecule which exhibits a range of anti-hypertensive and anti-bacterial effects. Our basic synthetic route is outlined in Figure 2. In order to simplify the early synthetic steps, we have used the achiral benzyl auxiliary on nitrogen for the aza-Diels-Alder reaction. The chiral (R)-α-methylbenzyl auxiliary will be used in later studies, in order to generate optically active piperidines. The initial aza-Diels-Alder reaction displays remarkable regio- and diastereo-control, exclusively generating the *cis*-2,6-disubstituted adduct.

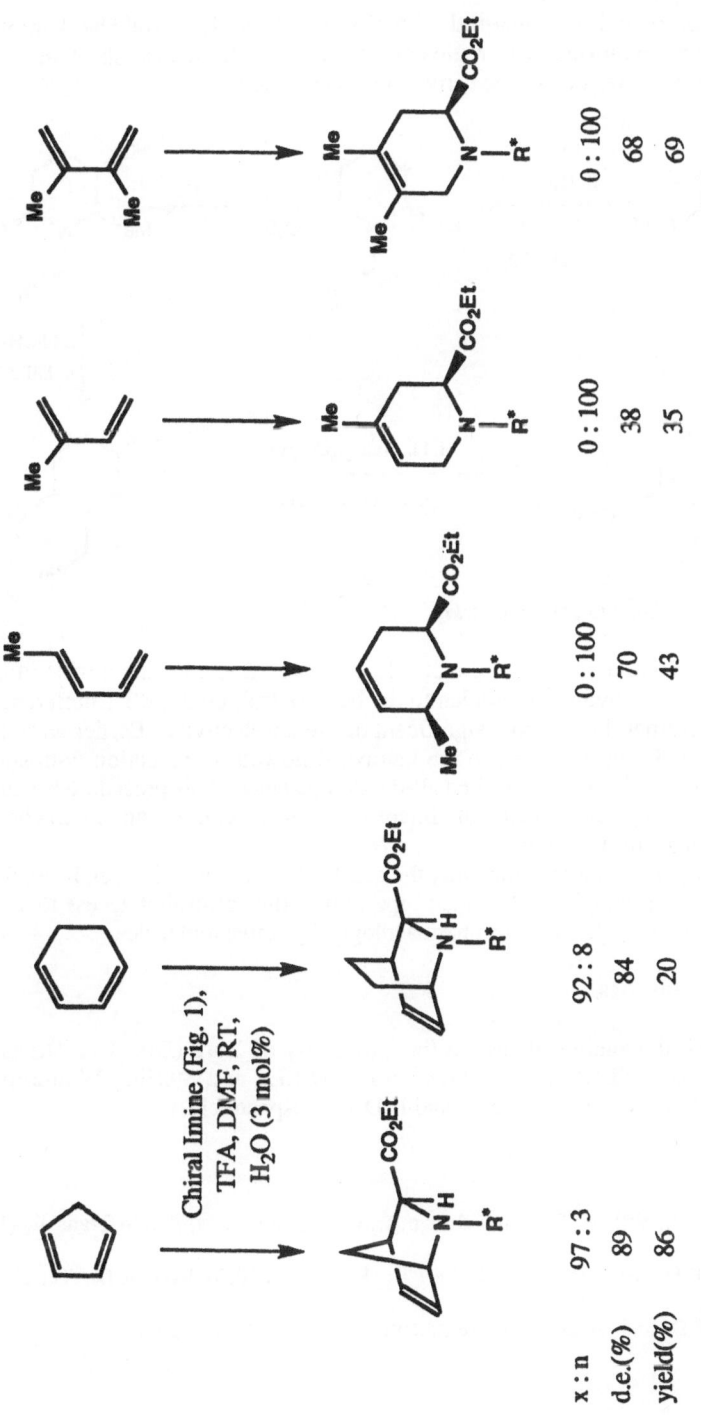

Table. *Some examples of aza-Diels-Alder reactions of the type outlined in Fig. 1.*

This compound is apparently ideally functionalised for the subsequent transformations, requiring only regio- and stereo-specific hydration of the double bond, and simple extension of the carboxylic ester moiety.

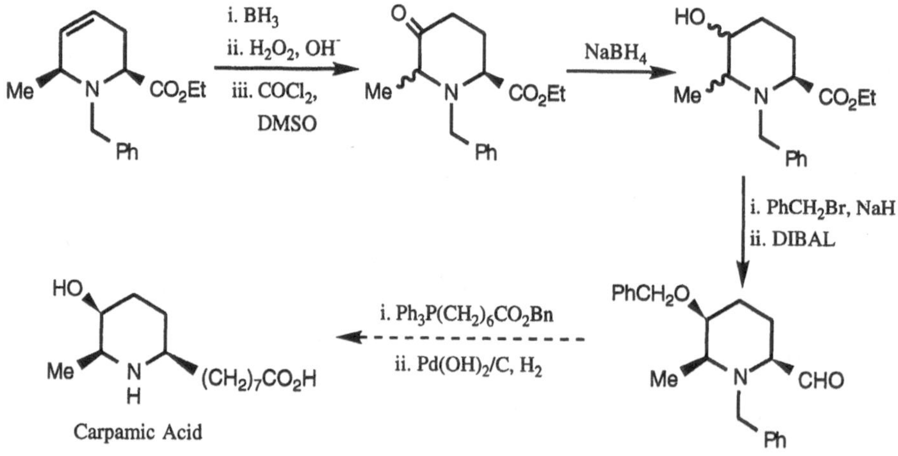

Fig. 2. *Synthetic route to carpamic acid.*

Regioselective hydration of the double bond was studied using a wide range of reagents, but simple hydroboration led to the best control, giving a 3:1 preference for the desired 5-isomer, but without significant diastereoselectivity. Earlier work by us [3] had indicated that oxidation to the ketone, followed by reduction with sodium borohydride, would lead to the desired all-*cis* stereoisomer. This procedure has indeed afforded some of the desired material. Interestingly, some epimerisation was observed in the early stages of this route.

In this paper, we demonstrate that the aza-Diels-Alder reaction can be exploited in the synthesis of complex substituted piperidines, thus affording access to a range of advanced intermediates towards pharmacologically active molecules.

Acknowledgements

We thank the analytical staff at the University of York (Drs. T.A. Dransfield and G.K. Barlow) SERC for a studentship to D.J.L., and Sterling-Winthrop and Zeneca for CASE awards to K.M.M. and R.D.W., respectively.

References

1. Bailey, P.D., Wilson, R.D. and Brown, G.R., J. Chem. Soc., Perkin Trans 1., (1991) 1337.
2. Bailey, P.D., Brown, G.R., Korber, F., Reid, A. and Wilson, R.D., Tetrahedron Asymmetry, 2 (1991) 1263.
3. Bailey, P.D. and Bryans, J.S., Tetrahedron Lett., 29 (1988) 2231.

Stereoselective synthesis of β–substituted-β-hydroxy α-amino acids and β,γ-unsaturated α-amino acids from serine and threonine

M.A. Blaskovich and G.A. Lajoie

Guelph-Waterloo Centre for Graduate Work in Chemistry, University of Waterloo, Waterloo, Ontario, N2L 3G1, Canada

Introduction

Non-proteinaceous α-amino acids are of considerable interest as they often possess biological activity, both as individual compounds, or when incorporated in peptides. They are also useful for structure-activity relationship studies and as intermediates in the synthesis of other structures [1]. Many synthetic routes to these challenging targets have been devised, but they often require harsh hydrolysis or oxidation steps which limit the types of derivatives that can be prepared [2]. We have developed a novel synthetic route based on elaboration of the side chain of a protected serine aldehyde or threonine ketone derivative. This strategy has not been successfully employed in the past due to the tendency of these amino acids to racemize or undergo elimination when oxidation is attempted. However, we have discovered that by masking the carboxyl group as a cyclic ortho ester, the acidity of the α-proton is sufficiently reduced to allow for oxidation of the side chain with no racemization. Details of the synthetic strategy and its application to the synthesis of chiral β-hydroxy α-amino acids have recently been published [3, 4]. We report here further use of this versatile synthon, with entry to the β-substituted-β-hydroxy and β,γ-unsaturated (vinylglycine) classes of α-amino acids.

Results and Discussion

The cyclic ortho esters derivatives (OBO ester, 2,6,7-trioxabicyclo[2.2.2]octane ortho ester) of urethane protected serine and threonine are readily synthesized in high yields via $BF_3.Et_2O$ catalyzed rearrangement [5] of an oxetane ester (Scheme 1). Oxidation of the hydroxyl side chain proceeds in quantitative yields under Swern conditions, with the resulting aldehyde or ketone possessing >97% ee as determined by chiral shift ^1H NMR studies or by deprotection, derivatization, and HPLC analysis. Grignard additions to the aldehyde produce protected β-hydroxy amino acids with good diastereoselectivity in favor of the L-*threo* (2S,3R) isomer (generally >4:1, as high as 20:1 for R = Me at -78°C). Oxidation to the ketone followed by reduction with $LiBH_4$ gives the opposite L-*erythro* (2S,3S) diastereomers, with even better diastereoselectivity (6:1 to >50:1). The free amino acids are obtained in high yields via a mild three step deprotection procedure.

Scheme I

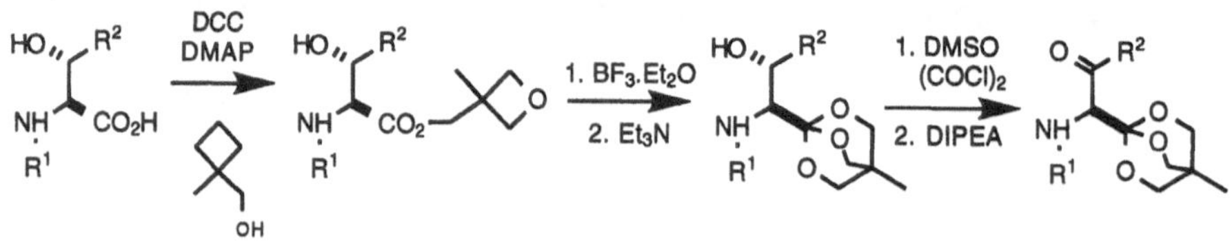

R^1 = Fmoc, Cbz, Boc R^2 = H, CH$_3$

Grignard additions to the threonine derived ketone give access to β-methyl-β-hydroxy amino acids, such as β-hydroxyvaline or β-hydroxyisoleucine (Table 1, Scheme 2). The L-*threo* (2S,3R) isomer is produced with very high diastereoselectivity (>20:1) and good yields (>60%). The erythro diastereomers can also be selectively synthesized, by Grignard addition to the appropriate ketone derived from an earlier Grignard addition to the protected serine aldehyde. Diastereoselectivity and yields are slightly reduced. Thus, depending on the order of organometallic addition and the chirality of the serine aldeyde synthon, all four diastereomers of any combination of β-substituted-β-hydroxy α-amino acids can be synthesized. Isotopically labelled side chains are readily introduced.

Table 1 *Diastereoselectivity and yields of Grignard additions to the protected ketones, and of the subsequent deprotection*

Ketone side chain R^1	Added alkyl group R^2.	protected adduct diastereomeric ratio[a] 2S,3R:2S,3S	yield(%)	enantiomeric purity of deprotected amino acid[b]	overall yield[c](%)
CH$_3$	CH$_3$	-	70	99.8% ee	37
CH$_3$	CD$_3$	95:5	73	98.6%ee	52
CH$_3$	Et	98:2	77	99.2%ee	31
Et	CH$_3$	3:97	33	83.8%ee	13
CH$_3$	4-MeOPh	97:3	58	n.d.[d]	25
4-MeOPh	CH$_3$	8:92	36	n.d.[d]	15

[a]determined by ^{1}H NMR, or by deprotection, derivatization, and HPLC analysis.
[b]determined by derivatization and HPLC analysis.
[c]measured for recrystallized free a.a. from Fmoc-Thr-OBO ester or Fmoc-Ser-OBO ester.
[d]not determined as enantiomers not resolved by HPLC assay procedure.

Scheme 2

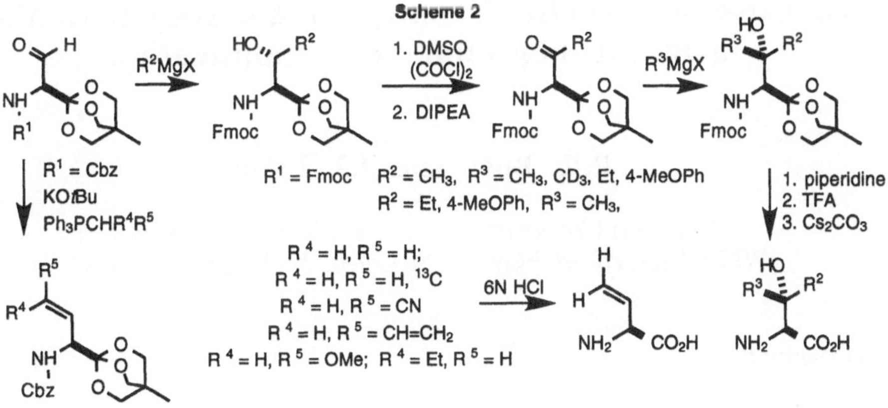

Wittig additions to the Cbz protected serine aldehyde proceed in good yields (35-71%) with stereoselectivity as expected for the type of ylide employed (Scheme 2). Protected vinylglycine was prepared in 70% yield, as was the γ-^{13}C labelled analog. Deprotection gave the free labelled and unlabelled vinylglycines in 72% recrystallized yield, though some racemization had occurred (71% ee).

Further investigations of the recently developed OBO ester protected serine aldehyde and threonine ketone synthons have led to the first diastereoselective synthesis of all four isomers of β-substituted-β-hydroxy α-amino acids. The same synthons can be employed to prepare β,γ-unsaturated amino acids, though some racemization was observed in the synthesis of vinylglycine. Isotopically labelled derivatives are readily prepared.

Acknowledgements

We would like to thank NSERC (Canada) for grants to G.L. and for a graduate scholarship to M.B.

References

1. Coppola, G.M. and Schuster, H.F., Asymmetric Synthesis. Construction of Chiral Molecules Using Amino Acids, Wiley-Interscience, Toronto, Canada, 1987.
2. Williams, R.M., Synthesis of Optically Active α-Amino Acids, Pergamon Press, Toronto, Canada, 1989.
3. Blaskovich, M.A. and Lajoie, G.A., J. Am. Chem. Soc., 115 (1993) 5021.
4. Blaskovich, M.A. and Lajoie, G.A., Tetrahedron Lett., 34 (1993) 3837.
5. Corey, E.J. and Raju, N., Tetrahedron Lett., 24 (1983) 5571.
6. Bruckner, H., Wittner, R. and Godel, H., J. Chromatogr., 476 (1989) 73.

An expedient synthetic entry to heterocyclic-amino acid for peptidomimetic application

P.R. Bovy and J.J. Likos

Monsanto Corporate Research, Chemical Sciences,
700 Chesterfield Pkw N., St Louis, MO 63198, U.S.A.

Introduction

β-amino acids are increasingly used in peptidomimetic as surrogates for α-amino acids. Renin inhibitors [1], HIV protease inhibitors and RGD mimetics [2] incorporate β-amino acids as part of their structure. These compounds are usually accessible through a variety of reactions which include reductive amination of precursor ketones, modified Knoevenagel reaction or Michael addition of ammonia or a synthetic equivalent to acrylic derivatives [3]. We describe an expedient synthesis of a novel β-amino acid, β-amino-5-pyrimidinepropanoic acid by Michael addition of ammonia to an acrylic derivative. It has been argued that β-amino acids are much more flexible than α-amino acids [4]. We have examined the compounds described by ^{1}H NMR and observed, in some situations, a strong conformational preference.

Results and Discussion

The palladium-catalyzed Heck reaction between 5-bromopyrimidine and methyl acrylate was found to give the corresponding β-pyrimidine acrylates of Econfiguration [5]. When the t-butyl acrylate **2b** is used, the reaction produces t-butyl 3-(5-pyrimi-

a. $Pd(OAc)_2$, 48 hr, 80°C. b. $MeOH/NH_3$, 60°C, 72 hr. c. CbzCl, NMM; Trifluoroacetic acid, CH_2Cl_2, 25°C, 3 hr.

dinyl)-2-propenoate, **3b**, which can be used as Michael acceptor for addition of ammonia in dry methanol.

   The formation of the amino ester **4** competes with the formation of the unsaturated amide and addition of methanol which produces t-butyl-3-methoxy-3-(5pyrimidinyl)propanoate. The amino ester can be purified by precipitation as a salt or by chromatography. The free amino acid is easily obtained by treatment with trifluoroacetic acid. The ^{1}H NMR spectrum (Figure) of **4** reveals, even in DMSO solutions, a very characteristic ABX pattern for the superficially identical protons of the α-methylene.

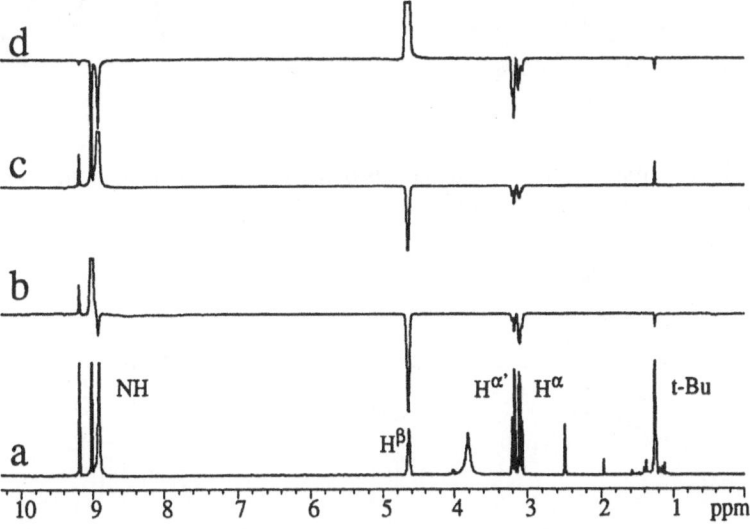

Fig. 1. a is the 1D-500 Mhz proton spectrum of the trifluoroacetate salt of 4 (0.05M in d6-DMSO); b, c and d are slices of the ROESY spectrum through the diagonal peaks of, respectively, the 4,6-pyrimidine protons (9.03 ppm), the NH proton (8.93 ppm) and the β proton (4.6 ppm). The intensities of the negative peaks (NOE) are inversely proportional to the sixth power of the distance between the proton under consideration and the proton on the diagonal. The protons of the α-methylene, Hα and Hα', are magnetically non-equivalent (3.2 and 3.1 ppm, referenced to DMSO) and have a geminal coupling constant of 15.4 Hz. The vicinal coupling constants, Jαβ and Jα'β, are respectively 9.2 Hz and 5.3 Hz.

   The ^{1}H NMR data suggest that, although **4** could exist as an average of the available stable conformations I, II and III (see below), a preferred conformation is III, possibly stabilized by a strong hydrogen bond between the amine NH and the carbonyl carboxyl. The ROESY experiment supports the conformation preference hypothesis since a the NOE between Hβ and Hα' than Hβ and Hα. Further, the NOE between the 4,6-pyrimidine aromatic protons and Hα is stronger than that of Hα'.

   The aminoester **4** can also be reacted with benzylchloroformate to produce the Cbz protected species **5** isolated as an oil. Deprotection with TFA gave the crystalline NCbz-β-amino acid **6** which is ready to be used in further peptide coupling reactions. Thus, **6** was coupled to methyl glycinate using the mixed anhydride method. ^{1}H NMR pattern of compounds **5-7** were examined for indication of conformational preference. While the acid **6** showed the same ABX pattern for the

α-methylene, neither of the two compounds **5** or **7** showed conformational preference by NMR, even in chloroform solutions.

      6   **7**

     The observation of a conformational preference in the N-protected acid **6** may indicate that the β-amino acid conformation plays a role in the biological activity of biologically active peptide or peptidomimetic having β-amino acids at the C-terminus. Studies to further characterize these observations are underway.

References

1. Bursztyn, M., Gavras, I., Tifft, C.P., Luthur, R., Boger, R. and Gavras, H.J., Cardiovasc. Pharmacol., 15 (1990) 493.
2. Bovy, P.R., Rico, J.G., Rogers, T.E., Tjoeng, F.S. and Zablocki, J.A., PCT Patent WO 93/08164.
3. For a review: Barrett, G.C. (Ed.), Chemistry and Biochemistry of the Amino Acids, Chapman and Hall, London, New York, 1985, p.25.
4. Schulz, G.E. and Schirmer, R.H., Principle of Protein Structure, Springer Verlag, New York, Heidelberg, 1979, p.6.
5. Nishikawa, Y., Shindo, T., Ishii, K., Nakamura, H., Kon, T. and Uno, H., Chem. Pharm. Bull., 37 (1989) 684.

Facile synthesis of β-substituted glutamic acids

W.M. Bryan and M.L. Moore

Department of Medicinal Chemistry, SmithKline Beecham Pharmaceuticals, 709 Swedeland Road, P.O. Box 1539, King of Prussia, PA 19406, U.S.A.

Introduction

The O'Donnell synthesis [1] of α-amino acids generally consists of the phase-transfer mediated alkylation of the benzophenone Schiff base of a glycine ester with an alkyl halide. This synthesis represents an attractive route to non-proteinogenic α-amino acids since it employs readily available reagents and starting materials [2] and proceeds under mild reaction conditions. Several unusual amino acids have been prepared employing this procedure, such as 1-aminocyclopropane-1-carboxylic acid [3], 3-fluorophenylalanine [4], and α-amino-(2,2'-bipyridine)-6-propanoic acid [5]. However, while this method usually works well with primary or activated halides, it sometimes proceeds with difficulty or not at all with secondary halides.

Results and Discussion

As shown in Figure 1, we have found that the α-anion of the benzophenone imine of glycine *tert*-butyl ester generated in a heterogeneous mixture of aqueous

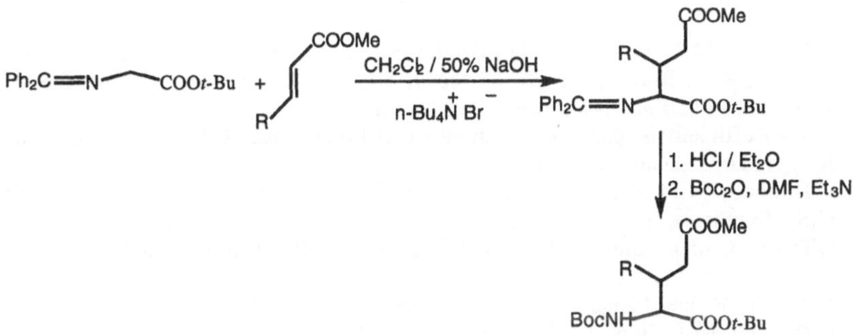

Fig. 1. Synthesis of β-substituted glutamic acids.

sodium hydroxide and methylene chloride in the presence of a catalytic amount of tetrabutylammonium bromide can react with an α,β-unsaturated ester (R=methyl or phenyl) to yield the corresponding β-substituted glutamic acid ester. Because these acid labile imines tended to decompose during flash chromatography on silica gel, they were hydrolyzed by stirring overnight in a two phase mixture of dilute hydrochloric acid and ether. The resulting amine hydrochlorides were then converted to the corresponding N-*tert*-butyloxycarbonyl α-*tert*-butyl β-methyl glutamates with di-*tert*-butyl dicarbonate in dimethylformamide with triethylamine to yield after flash

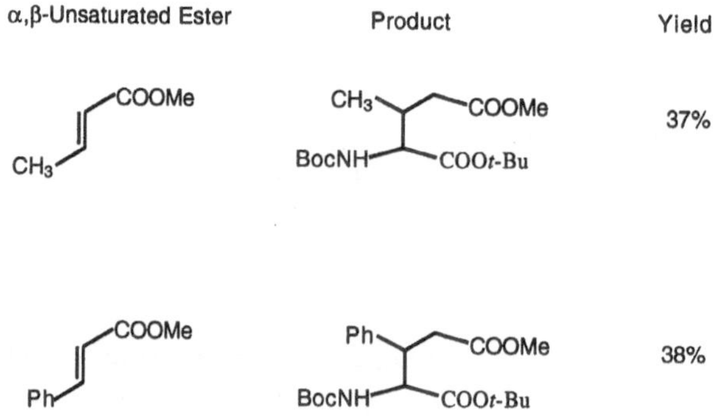

α,β-Unsaturated Ester	Product	Yield

chromatography on silica gel the N-*tert*-butyloxycarbonyl protected diesters in 37% and 38% overall yield respectively as shown above. The ^{1}H NMR showed these products to consist of a mixture of diastereomers (7:1 and 15:1 respectively). It should be noted that the free amines were not stable, readily cyclizing to the corresponding pyrrolidones on standing at room temperature.

Although further work is needed to define the scope of this Michael addition, we feel it represents an exceptionally mild and convenient synthesis of these substituted glutamic acids. We are currently investigating the use of various chiral catalysts [6] in this conjugate addition as well as exploring its use for the synthesis of other amino acids.

References

1. O'Donnell, M.J., Bennett, W.D., Bruder, W.A., Jacobsen, W.N., Knuth, K., LeClef, B., Polt, R.L., Bordwell, F.G., Mrozack, S.R. and Cripe, T.A., J. Am. Chem. Soc., 110 (1988) 8520 and references cited therein.
2. For an efficient preparation of amino acid imines, see: O'Donnell, M.J. and Polt, R.L., J. Org. Chem., 47 (1982) 2663.
3. O'Donnell, M.J., Bruder, W.A., Eckrich, T.M., Schullenberger, D.F. and Staten, G.S., Synthesis, 4 (1984) 313.
4. O'Donnell, M.J., Barney, C.L. and McCarthy, J.R., Tetrahedron Lett., 26 (1985) 3067.
5. Imperiali, B. and Fisher, S.L., J. Org. Chem., 57 (1992) 757.
6. O'Donnell, M.J., Bennett, W.D. and Wu, S., J. Am. Chem. Soc., 111 (1989) 2353.

Solid-phase synthesis of endothiopeptides by the use of amino monothioacids

T. Høeg-Jensen, C.E. Olsen and A. Holm

Research Center for Medical Biotechnology, Chemistry Department, The Royal Veterinary and Agricultural University, Thorvaldsensvej 40, DK-1871 Copenhagen F, Denmark

Introduction

Among the growing family of modified peptides, endothiopeptides (peptide thioamides) possess perhaps the most gentle alteration. Endothio-analogues of biologically active peptides have shown several interesting properties, such as partial protease resistance, enhanced activity or receptor selectivity [1]. Preparations of endothiopeptides have most often been achieved by thionation of dipeptides by the use of Lawessons or similar reagents. Extensions of endothiopeptides from the N-terminal have been done by several groups, but attempts to extend from the C-terminal have most often failed due to formation of thiazolones, which are not easily reopened, and are easily epimerised [2]. An alternative strategy involves thioacylation, but known reagents such as dithio methyl esters of N-protected amino acids are often not sufficiently reactive [3]. Additionally, such compounds, which are prepared by use of Lawessons Reagent etc., do not allow for the presence of amides in the amino acid side chain. We have recently demonstrated that Fmoc amino monothioacids, which are potentially available from all amino acids, can be activated by use of the phosphorus-containing PyBOP, to generate new thioacylating species [4], probably thiono HOBt esters (Figure 1, route a). Thioamides made by this method are formed in mixtures with their corresponding oxoamides, but they can be separated chromatographically. We now wish to report the use of this strategy for the solid-phase synthesis of the following endothiopeptides:

pGlu-ψ[CSNH]-His-Pro-NH$_2$ (Thyrotropin Releasing Hormone)
pGlu-ψ[CSNH]-Ala-Lys-Ser-OH (Serum Thymic Factor 1-4)
H-Leu-Gln-ψ[CSNH]-Leu-Lys-OH (α-Mating Factor 1-4)
H-Pro-Gln-ψ[CSNH]-Ala-Lys-Ser-OH (Pro-STF 1-4)

Results and Discussion

Fmoc-Gln(Trt)-SH was prepared from the HOSu ester and H$_2$S/TEA [5], yield 96%. The hydrophilic pGlu-SH, however, could not be prepared by the usual methods. Extraction from water was difficult, and resulted in a seriously hydrolysed product (>20% pGlu-OH). Therefore pGlu-SH was isolated as its DCHA salt upon reaction of pGlu-OPfp with H$_2$S/DCHA in THF. PfpOH was removed from the product by washing with ether, to leave analytically pure pGlu-SH•DCHA, yield

92%. It has been previously described, that DCHA salts can be used directly in BOP-couplings [6].

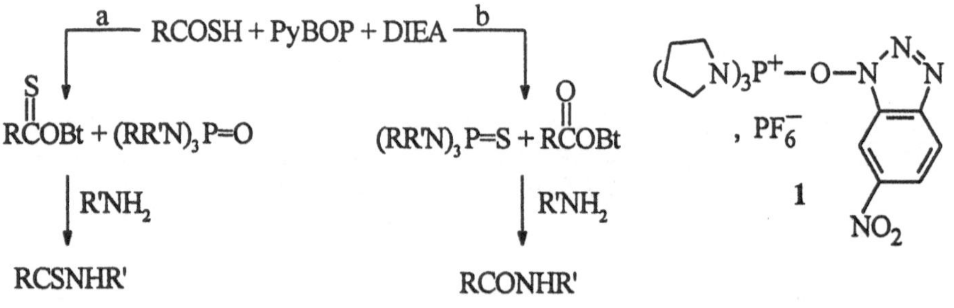

Fig. 1. *Formation of the thiono HOBt ester and phosphine oxide (route a), or of the HOBt ester and phosphine sulphide (route b). Structure of NO_2-PyBOP (1).*

As demonstrated by the use of ^{31}P-NMR, monothioacids are activated by PyBOP significantly slower than their oxygen analogues [4]. This observation led us to test PyBrOP [7], as well as several new, substituted BOP's and PyBOP's [8] for monothioacid activation. It turned out that 6-nitro-benzotriazol-1-yl-oxy-tris(pyrrolidino)phosphonium hexafluorophosphate (1), used in lipophilic solvents, gave the best results in terms of reaction speed and yield of thioamide (Table 1). While PyBrOP also gave fast activation, its O/S-selectivity (Figure 1, route b) resulted in low yield of thioamide. For pGlu-SH.DCHA, treatment with PyBOP gave practically no reaction at all, whereas satisfactory activation was obtained with 1. Though the solubility of 1 in lipophilic solvents is less than that of PyBOP, it quickly dissolves as activation proceeds.

Table 1 *Activation of CH_3COSH by use of PyBOP and analogues*

Reagent	$t_{1/2}$ [a]	$C=S$ [b]	$P=S$ [c]
NO_2-PyBOP	< 2 min	79 %	8 %
CF_3-PyBOP	2 min	77 %	6 %
PyBOP	41 min	71 %	8 %
BOP	91 min	12 %	5 %
PyBrOP	4 min	4 %	73 %

[a] $t_{1/2}$ was measured by the disappearance of the phosphonium signal in ^{31}P-NMR.
[b] $C=S$ was measured as CH_3CSNHR relative to CH_3CONHR, by the use of ^{1}H-NMR.
[c] $P=S$ was measured relative to P=O, by the use of ^{31}P-NMR.

Since endothiopeptides are often sensitive towards acid [9], we chose the Fmoc-strategy, where only one acid treatment is used, for the solid-phase synthesis. Though thioamides are sometimes reactive towards nucleophilic amines, this study suggests that Fmoc-deprotection with piperidine leaves the endothio function intact, at least for short peptides. An alternative deprotection method could involve non-nucleophilic DBU [10]. Peptide synthesis was done in 25 μmol scale on Wang or Rapp resins, which both have good swelling properties in lipophilic solvents. Normal couplings were done in DMF by use of Pfp/Dhbt esters (3 eq., 2 hours), whereas deprotections were done with 20% piperidine in DMF (3+7 min). Before coupling of the monothioacid, the peptide-resin was washed thoroughly with dried $CHCl_3$, and purged with N_2. The monothioacid (2 eq.), **1** (2 eq.) and DIEA (3 eq. for Fmoc-Gln(Trt)-SH, 1 eq. for pGlu-SH•DCHA) were mixed in dried, N_2-purged $CHCl_3$ containing a few drops of CH_3CN, added to enhance solubility. This mixture was added to the resin immediately. After leaving overnight, the peptide was extended by use of standard methods. Following treatment with $TFA/CH_2Cl_2/EDT$, 48:48:4, for 30 min, peptides were precipitated in ether, lyophilised and subjected to preparative RPHPLC. Use of longer periods or higher concentration of TFA, or use of 4M HCl/dioxane, gave less or no yield of endothiopeptides. The chromatograms demonstrated the presence of several products (oxo-analogues, residual protecting groups etc.), but the endothiopeptides were easily identified on basis of their strong UV-absorption in the thioamide area (≈260 nm). Yields of endothiopeptides were 8-14%, calculated from the resin substitutions. The isolated peptides showed the correct molecular ion in FABMS, and gave the expected AAA. The relatively low yields of endothiopeptides can possible be attributed to the single acid treatment. An alternative to the use of acid labile protection groups may be desirable.

Acknowledgements

THJ would like to thank the American Peptide Society and Acta Chemica Scandinavica for travel awards.

References

1. Lankiewicz, L., Bowers, C.Y., Reynolds, G.A., Labroo, V., Cohen, L.A., Vonhof, S., Sirén, A.L. and Spatola, A.F., Biochem. Biophys. Res. Commun., 184 (1992) 359.
2. Unversagt, C., Geyer, A. and Kessler, H., Angew. Chem., 104 (1992) 1231.
3. Jensen, O.E. and Senning, A., Tetrahedron, 42 (1986) 6555.
4. Høeg-Jensen, T., Jacobsen, M.H., Olsen, C.E. and Holm. A., Tetrahedron Lett., 32 (1991) 7617.
5. Yamashiro, D. and Li, C.H., Int. J. Peptide Protein Res., 31 (1988) 322.
6. Le-Nguyen, D., Heitz, A. and Castro, B., J. Chem. Soc., Perkin Trans I (1987) 1915.
7. Frérot, E., Coste, J., Pantaloni, A., Dufour, M.N. and Jouin, P., Tetrahedron, 47 (1991) 259.
8. Høeg-Jensen, T., Olsen, C.E. and Holm, A., J. Org. Chem., accepted for publication.
9. Brown, D.W., Campbell, M.M., Chambers, M.S. and Walker, C.V., Tetrahedron Lett., 28 (1987) 2171.
10. Wade, J.D., Bedford, J., Sheppard, R.C. and Tregear, G.W., Peptide Res., 4 (1991) 194.

Synthesis and biological activity of cionin and its mono-tyrosine-sulfate-containing derivatives

K. Kitagawa[1], S. Futaki[1], T. Yagami[2], S. Sumi[3] and K. Inoue[3]

[1]Faculty of Pharmaceutical Sciences, The University of Tokushima, Tokushima 770, Japan
[2]National Institute of Hygienic Sciences, Setagaya-ku, Tokyo 158, Japan
[3]First Department of Surgery, Faculty of Medicine, Kyoto University, Sakyo-ku, Kyoto 606, Japan

Introduction

Cionin, a newly discovered octapeptide amide from the neural ganglion of the protochordate *Ciona intestinalis*, has a remarkable structural homology with the C-terminal amino acid sequences of gastrin and CCK [1]. In addition, cionin contains two consecutive tyrosine-sulfate [Tyr(SO$_3$H)] residues at the "gastrin-position" and the "CCK-position" (6th and 7th position from the C-terminus, respectively). In order to give insight into the role of the respective Tyr(SO$_3$H) residue for the biological activity of cionin, two derivatives in which one of the two Tyr(SO$_3$H) residues was replaced by Tyr were prepared (Figure 1). Here we describe the solid-phase synthesis of cionin and the regioisomerically mono-sulfated derivatives using two different approaches. The effects on exocrine pancreas of these synthetic peptides were examined in dog.

cionin	H-Asn-**Tyr**-**Tyr**-Gly-Trp-Met-Asp-Phe-NH$_2$
cionin B-peptide	H-Asn-**Tyr**–Tyr-Gly-Trp-Met-Asp-Phe-NH$_2$
cionin C-peptide	H-Asn-Tyr-**Tyr**-Gly-Trp-Met-Asp-Phe-NH$_2$
cionin N.S.	H-Asn–Tyr–Tyr-Gly-Trp-Met-Asp-Phe-NH$_2$
CCK-8	H-Asp-**Tyr**–Met-Gly-Trp-Met-Asp-Phe-NH$_2$

Tyr =Tyr(SO$_3$H)

Fig. 1. Structure of cionin and related peptides.

Results and Discussion

In the first set of the synthesis, cionin and two regioisomerically mono-sulfated derivatives (cionin B- and C-peptide in Figure 1) were prepared directly with the Fmoc-based solid-phase methodology (**Approach 1**). This approach is based on the following strategies: (i) an acid-labile PAL-linked resin is used and peptide-chain assembly is conducted with the Fmoc-chemistry; (ii) Tyr(SO$_3$Na) residue(s) is incorporated into the peptide-chain by the use of Fmoc-Tyr(SO$_3$Na)-OH [2] as a building block; (iii) final deprotection of the protecting groups and cleavage of peptide from the resin support are concurrently carried out with 90% aqueous TFA in the presence of m-cresol and 2-methylindole, under which condition the deterioration of Tyr(SO$_3$Na) moiety can be effectively minimized [2]. Each peptide was obtained in a reasonable purity after acid treatment (4°C, 5 h), however, yield was not satisfactory in each case because of the insufficient cleavage of the peptide from the resin.

Alternatively, two regioisomerically mono-sulfated cionin derivatives were synthesized with an aid of orthogonally cleavable protecting groups (**Approach 2**) [3]. In Figure 2, the synthetic scheme of cionin B-peptide is shown. Tyr(Msib) [Msib=p-(methylsulfinyl)benzyl] derivative was used for the peptide-chain assembly to achieve the sulfation on the selective Tyr residue; the Tyr residue not to be sulfated was incorporated with Fmoc-Tyr(Msib)-OH and the Tyr residue to be sulfated was incorporated with Fmoc-Tyr(^{t}Bu)-OH, respectively. The Msz protecting group [Msz=p-(methylsulfinyl) benzyloxycarbonyl] was also used for an amino protection of the N-terminus to avoid a possible sulfamic acid formation during the sulfation. The fully protected peptide-resin was treated with TFA/EDT/m-cresol (85:10:5, 25°C, 2h). The partially protected peptide was then subjected to the sulfation on the free Tyr residue and the reduction of the Msib/Msz groups to the acid-labile Mtb/Mtz groups [Mtb=p-(methylthio)benzyl, Mtz=p-(methylthio)benzyloxycarbonyl] with the action of DMF-SO$_3$ complex in the presence of EDT, concurrently [4, 5]. A final deprotection of the Mtb/Mtz groups was conducted with 90% aqueous TFA/m-cresol/2-methylindole (4°C, 1 h). HPLC purification afforded a highly purified mono-sulfated peptide. Overall yields were around 20% for both cionin B- and C-peptide. Thus, our new approach was successfully applied to the selective sulfation of the multiple-Tyr-containing peptide.

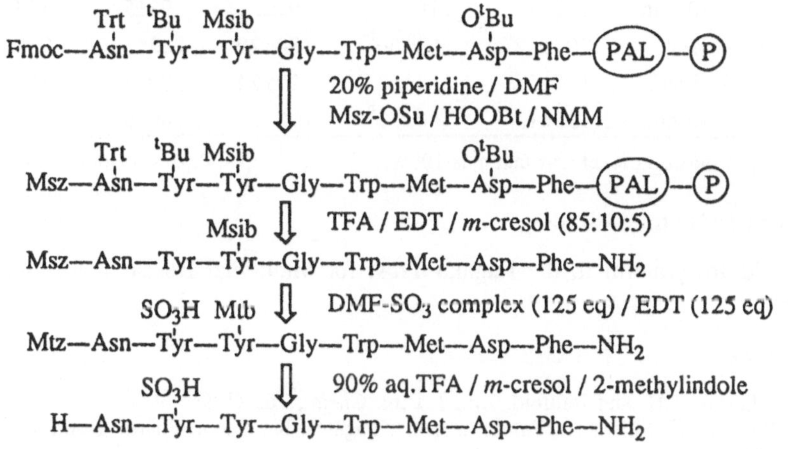

Fig. 2.   Synthetic scheme of cionin B-peptide with **Approach 2**.

The effects of the synthetic cionin, two mono-sulfated derivatives, and cionin non-sulfate (N.S.) on exocrine pancreas were examined in dogs. In both experiments, CCK-8 was used as a reference. In Table 1, the increment of pancreatic protein output (mg/kg/10 min) in response to the synthetic peptides is shown. Both cionin and cionin B-peptide increased pancreatic protein output almost in the equal potency with CCK-8, while the potency of cionin C-peptide and cionin N.S. were substantially reduced. In Table 2, the increase of pancreatic tissue blood flow (PTBF) after injection is shown. In each case, the peak was observed about 3 min after injection. From these experiments, the CCK-like effects of the peptides appeared to be an order of cionin = cionin B-peptide = CCK-8 > cionin C-peptide >>cionin N.S. Thus it will be conceived that the $Tyr(SO_3H)$ residue at position 7 (from the C-terminus) is an important structural feature of cionin and controlling the activities on the pancreas. For the further evaluation on the role of the sulfate ester at position 6 (gastrin-position), gastrin-like activity of cionin and related derivatives is now under investigation.

Table 1 *Increment of pancreatic protein output (mg/kg/10 min)*

Dose (pmol/kg)	cionin	cionin-B	cionin C	cionin N.S.	CCK-8
20	4.8±0.9 [a]	4.8±1.0 [a]	0.5±0.2 [b]	0.2±0.2	5.2±0.9 [a]
200	8.2±1.0 [a]	9.1±1.0 [a]	2.8±0.6 [b]	0.7±0.2	8.5±1.0 [a]

[a] $p < 0.05$ vs. cionin C and cionin N.S.   [b] $p < 0.05$ vs. cionin N.S.

Table 2 *Change of pancreatic tissue blood flow (PTBF) (%)*[a]

Dose (pmol/kg)	Time	cionin	cionin-B	cionin-C	cionin N.S.	CCK-8
20	peak	178±28	194±31[b]	114±6	107±4	188±26[b]
	5 min	124±10	115±12	101±2	100±3	120±15
	10 min	120±11	116±5	95±5	98±5	114±5
200	peak	207±22[b]	263±60[b]	184±29	123±5	210±13[b]
	5 min	185±24[b]	212±63[b]	129±24	90±7	187±19[b]
	10 min	136±15[b]	146±33[b]	105±6	89±6	137±16[b]

[a] The pre-injection level was taken as 100%.   [b] $p < 0.05$ vs. cionin N.S.

Acknowledgements

We are grateful to the Fugaku Trust for Medicinal Research for financial support.

References

1. Johnsen, A.H. and Rehfeld, J.F., J. Biol. Chem., 265 (1990) 3054.
2. Yagami, T., Shiwa, S., Futaki, S. and Kitagawa, K., Chem. Pharm. Bull., 41 (1993) 376.
3. Futaki, S., Taike, T., Akita, T. and Kitagawa, K., Tetrahedron, 48 (1992) 8899.
4. Futaki, S., Taike, T., Yagami, T., Ogawa, T., Akita, T. and Kitagawa, K., J. Chem. Soc. Perkin Trans., 1 (1990) 1739.
5. Futaki, T., Taike, T., Yagami, T., Akita, T. and Kitagawa, K., J. Chem. Soc. Perkin Trans., 1 (1990) 653.

A new method for the synthesis of the optically pure precursors to β-methyl derivatives of tryptophan, histidine, tyrosine and phenylalanine

G. Li, L.W. Boteju, D. Patel and V.J. Hruby

Dept. of Chemistry, University of Arizona, Tucson, AZ 85721, U.S.A.

Introduction

A major goal in peptide and protein research is to develop a rational approach to highly selective peptide and protein ligands with specific conformational and topographical features [1]. For this purpose, it is necessary to develop novel and efficient methods for the enantioselective synthesis of unusual amino acids and their key intermediates. We have proposed a one-pot asymmetric synthesis of key intermediates of β-methyl-α-amino acids [2a] and completed the total synthesis of all four individual isomers of β-methyl-Trp and Tyr derivatives [3b & c]. We have also synthesised a systematic series of (2S, 3S) and the (2R, 3R) precursors via the one-pot asymmetric 1,4-conjugate addition and electrophilic bromination using 3(2E)-(1-oxobut-2-eneyl)-4-phenyl-2-oxazolidinone as the reaction substrate with a variety of different aromatic organocuprates as the addition reagents [2b]. Here we report the asymmetric synthesis of the (2S, 3R) and (2R, 3S) precursors, by a new method using a series of aromatic α, β-unsaturated analogs as the reaction substrates (**1a-1i**) and the same cuperate ligand as the addition reagent.

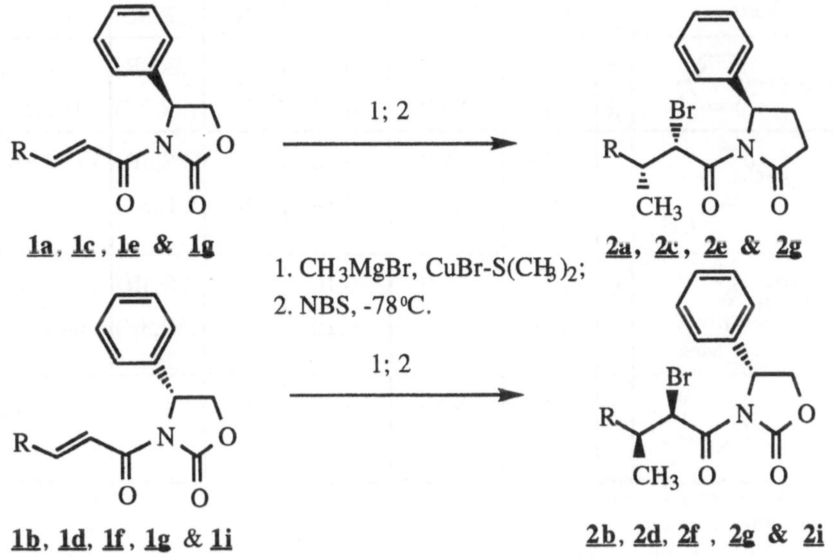

Scheme 1

Results and Discussion

The overall syn-addition of organocuprates and the tandem NBS bromination for directing the two new chiral centers is controlled by the improved Evans auxiliary, resulting in higher stereoselectivities than those obtained from our former system in most cases [2b]. A representative procedure is described, for example **2i**. To a copper(I)bromide-dimethyl sulfide complex (0.23g, 1.10mmol, 1,1 eq) was added dimethyl sulfide (1.31 ml) and dry THF (2.2 ml). The resultant solution was cooled to -78°C to form an opaque mixture. Methyl magnesium bromide (0.49 ml of 3 M solution in ethyl ether, 1.47 mmol, 1.47 eq) was added to the solution under a nitrogen atmosphere to yield a yellow slurry which was stirred at -78°C for 10 min, 0°C for another 10 min, and re-cooled to -78°C before being transferred via a Teflon cannula to a pre-cooled (-78°C) slurry of **1i** (0.53 g, 1.0 mmol) in THF (4 ml) and dicholoromethene (2.4 ml). The resulting mixture was stirred at -78°C for 30 min, -10°C for 15 min (45 min for reactions **a-h**), (the color of the reaction changing from brown/yellow to green during this period). Following being re-cooled to -78°C, the solution was transferred to a -78°C solution of NBS (2.3 g) in THF (65 ml).

The resultant mixture was stirred at -78°C for 45 min (90 min for reactions **a-h**), quenched by sodium sulfite (1.3 M), washed by water (100 ml) and brine (100 ml). Organic extracts were dried over magnesium sulfate and concentrated *in vacuo*. The crude residue was evaluated by ^{1}H-NMR prior to column chromatography to yield a glassy solid (0.50g, 80%) (Table 1, below).

Group **R** (scheme 1)		Crude[a] d.e. %	Purified[b] yield %	$\delta\alpha$-H of bromides (ppm)	Chiral aux.	Config.	$[\alpha]D^{25}$ (CHCl₃)
(phenyl)	**a**	99.0	81.4	6.06	S	(2S,3R)	+67.7 (c=2.8)
	b	99.0	77.5	6.06	R	(2R,3S)	-68.0 (c=2.4)
H₃CO-(phenyl)	**c**	78.0	71.9	6.00	S	(2S,3R)	+149.0 (c=1.5)
	d	80.0	89.9	6.00	R	(2R,3S)	-145.0 (c=1.4)
H₃CO-(phenyl) CH₃	**e**	90.3	74.0	6.18	S	(2S,3R)	+119.5 (c=2.4)
	f	99.0	76.0	6.18	R	(2R,3S)	-120.0 (c=2.6)
(indole) N Mes	**g**	70.2	70.0	6.10	S	(2S,3R)	+69.0 (c=0.60)
	h	82.0	69.0	6.10	R	(2R,3S)	-64.9 (c=0.99)
(Ph)₃C-N N=N	**i**	99.0	80.3	5.92	R	(2S,3R)	-67.5 (c=1.5)

[a]Only two isomers were observed in the case of **c, e, g** and **h**.
[b]After purification only one isomer was obtained in all of the cases; 99.0% indicates that only one isomer was observed.

TLC and ^{1}H NMR (250 MHz) proved a convenient method to monitor the progress and determine the stereoselectivities for both the asymmetric addition and bromination processes. TLC conditions employed EtOAc:hexane:CH_3CN (2.7:6.3:1) for reaction **i**, and EtOAc:hexane (3:7) for reactions **a-h** respectively. The absolute stereochemistry was confirmed by the X-ray structure analysis of **2e** [3b] and by converting **2a** to (2R,3R) β-methyl-phenylalanine which was compared to an authentic sample obtained by alternative procedures [3d]. The downfield chemical shifts (5.9-6.47 ppm) of the α-protons of the bromides and several ^{1}H signals from the chiral auxiliary can be used to evaluate the stereoselectivities following the bromination reactions. Only one isomer was observed for the reactions **a**, **b**, **f** and **i** (Table 1). The solid products obtained of **e** and **f**, could be easily crystallized, all other products remained as oils or glassy solids and failed most attempts at recrystallization.

The established asymmetric one-pot Micheal-like addition and electrophilic bromination reaction is highly efficient in conjunction with aromatic α, β-unsaturated systems for the synthesis of (2R, 3S) and (2S, 3R) precursors to b-methyl aromatic amino acids in high stereoselectivities and high yields. The amino acids derived from several of these precursors have demonstrated their importance in our efforts to develop a rational approach to the design of highly selective peptide and protein ligands, with specific conformational and topographical features [4].

Acknowledgements

Support has been provided by U.S. Public Service Grants NS 19972 and CA 57723, and NIDA Grant DA 06284.

References

1. (a) Hruby, V.J., Biopolymers, in press (1993), (b) Hruby, V.J., Progress in Brain Research, 92 (1992) 215, (c) Hruby, V.J., Life Sciences, 31 (1982) 189, (d) Kazmierski, W.M., Yamamura, H.I. and Hruby, V.J., J. Am. Chem. Soc., 113 (1991) 2275, (e) Hruby, V.J., Al-Obeidi, F. and Kazmierski, W.M., Biochem. J., 268 (1990) 249.
2. (a) Li, G. and Hruby, V.J., Personal Communication, Dec. 1991, and (b) Li, G., Jarosinski, M.A. and Hruby, V.J., Tetrahedron Lett., 34 (1993) 2561.
3. (a) Li, G. and Hruby, V.J., Personal Communication, Dec. 1991, and (b) Li, G., Russell, K.C., Jarosinski, M.A. and Hruby, V.J., Tetrahedron Lett., 34 (1993) 2565, (c) Boteju, L.W., Wegner, K. and Hruby, V.J., Tetrahedron Lett., 33 (1992) 7491, (d) Dharanipragada, R., Van Hulle, K., Bannister, A., Bear, S., Kennedy, L. and Hruby, V.J., Tetrahedron, 48 (1992) 4733, (e) Nicolas, E., Dharaniprada, R., Toth, G. and Hruby, V.J., Tetrahedron Lett., 30 (1989) 6845.
4. (a) Toth, G., Russell, K.C., Landis, G., Kramer, T.H., Fang, L., Knapp, R., Davis, P., Burks, T.F., Yamamura, H. and Hruby, V.J., J. Med. Chem., 35 (1992) 2383, (b) Hruby, V.J., Fang, S., Toth, G., Jiao, D., Matsunaga, T., Collins, N., Knapp, R. and Yamamura, H., In Geralt, E. and Andreu, D. (Eds.), Peptides 1990, (Proc. 21st Eur. Peptide Symp.), Escom Sci. Publ., Leiden, The Netherlands, 1991, p.707.

Synthesis of peptide-oligonucleotide hybrids containing a KDEL signal sequence

K. Arar, M. Monsigny and R. Mayer

Laboratoire de Biochimie des Glycoconjugués, Centre de Biophysique Moléculaire, CNRS et Université, F-45071 Orléans Cedex 02, France

Introduction

Synthetic antisense oligonucleotides, able to control gene expression in various systems including virus infected cells are putative therapeutic agents [1]. It is known that: i) fluoresceinylated oligonucleotides are localized into vesicular compartments, ii) no or a minute amount of fluorescent material is found in the cytosol (or in the nucleus) where the antisense oligonucleotide targets are located, iii) newly synthetized proteins with a Lys-Asp-Glu-Leu (KDEL) sequence at their C-terminal end are selectively retained in the endoplasmic reticulum [2], and iv) reticulum endoplasmic is involved in the presentation of cytosolic peptide by the MHC I system.

On these bases, we prepared peptide-oligonucleotide hybrids made of an antisense oligodeoxynucleotide linked to a peptide with a KDEL sequence at its C-terminal end, with the aim of helping these conjugates to reach the endoplasmic reticulum and from there to cross the membrane to enter the cytosol or/and the nucleus where the antisense oligonucleotide targets are located.

The synthesis of peptide-oligonucleotide conjugates, directly on solid-phase peptide synthetizers or on DNA synthetizers, has been reported, but this method is not satisfactory due to the lack of "universal" protecting groups for both strategies and to side reactions during coupling cycles or cleavage from the solid support. A postsynthesis conjugation appears to be a better alternative.

The conjugation through disulfide linkage of an oligonucleotide to a separately synthetized peptide has been reported [3], but the S-S bridge being reduced in endosomes [4], is therfore not appropriate for the assigned purpose. For these reasons, we prepared oligonucleotides and peptides separately, each one being adequately functionalized to obtain a conjugate through a stable thioether bond.

In this paper, we describe an improved method which allows to synthetize oligonucleotide-peptide hybrids by linking a 3'-thiol oligonucleotide to a Nα-maleimidocaproyl-peptide. The oligonucleotide used is a 12-mer with a sequence specific for Ha-*ras* around the point mutation in the 12th codon [5].

Results and Discussion

Similar conjugations using the reaction of a maleimide group with a thiol group have been previously reported [6], but the maleimide group was appended at pH 8.0. Gregory [7] showed that the maleimide group is stable between pH 5.5 to 7.0, but is readily hydrolyzed at pH 8.0 during the coupling step to give maleamic acid. In our procedure the maleimide group is appended to the α-NH$_2$ of the peptide, at neutral pH, to preserve the ε-NH$_2$ of lysine side chain which is essential for the peptide biological activity. The synthesis strategy is outlined in Figure 1.

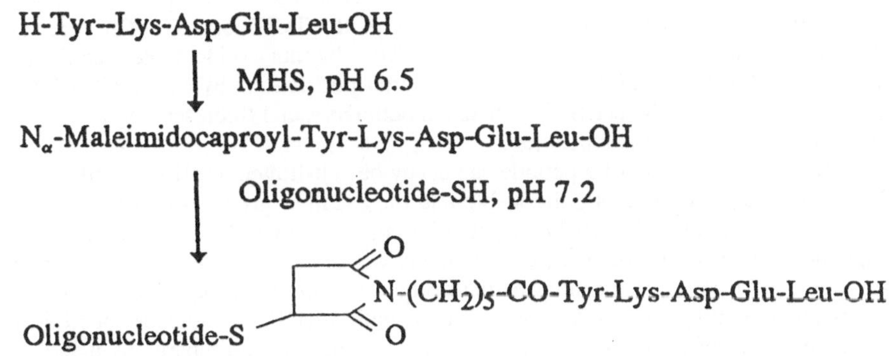

H-Tyr--Lys-Asp-Glu-Leu-OH

     ↓ **MHS, pH 6.5**

N$_\alpha$-Maleimidocaproyl-Tyr-Lys-Asp-Glu-Leu-OH

     ↓ **Oligonucleotide-SH, pH 7.2**

  ↓

Oligonucleotide-S ⎯⎯ **N-(CH$_2$)$_5$-CO-Tyr-Lys-Asp-Glu-Leu-OH**

Fig. 1. Synthesis strategy of an oligonucleotide-peptide hybrid. MHS: ε-maleimidocaproic acid N-hydroxysuccinimide ester.

The peptide NH$_2$-Tyr-Lys-Asp-Glu-Leu-OH (YKDEL), bearing a tyrosine residue for subsequent radiolabeling of the conjugate, was synthesized in solution using commercially available protected amino acids : Boc-Tyr(Bzl)-OH, Boc-Lys(Z)-OH, Boc-Asp(OBzl)-OH, Boc-Glu(OBzl)-OH and HCl,H-Leu-OBzl. Peptide coupling was performed by BOP mediated reaction in dichloromethane (DCM), 40% trifluoroacetic acid in DCM was used in the Boc deprotecting step. After synthesis, the protecting groups were removed by catalytic hydrogenolysis. The peptide was purified by reversed-phase chromatography using a 100 RP 18 Merck column (250 x 10 mm) eluted with a linear A-B gradient (5-90% B), where A is 0.1% ammonium acetate in water and B is 0.1% ammonium acetate in acetonitrile, for 25 min. The peptide is eluted at 14.0 min.

The oligonucleotide bearing a 3'-disulfide bridge and a 5'-NH$_2$ group was synthetized by solid phase synthesis on a disulfide derivatized solid support [8] (10 μmole scale) using the phosphoroamidite method and the fast oligonucleotide deprotection phosphoroamidite base protecting groups (FODs^{TM}). The oligomer 3'-disulfide was substituted on its 5'-end by introducing an amino group which was further fluoresceinylated in order to study the intracellular traffic of the conjugate. The 5'fluoresceinyl-oligonucleotide was purified by ion-exchange chromatography on a DEAE 8HR Waters column (100 x 10mm), using two buffers: A [25 mM Tris HCl, 1mM EDTA, pH 8.0 (90%)/acetonitrile (10%)] and B (buffer A +1.0 M NaCl). A linear gradient was run from 10 to 80% B over 40 min. The disulfide bridge of the fluorescent oligonucleotide was reduced with [tris(2-carboxyethyl)phosphine] TCEP [9] (3 eq) to give the 12 mer 3'-thiol oligonucleotide

ε-Maleimidocaproic acid N-hydroxysuccinimide ester (4.5 μmoles) in 50μL DMF and peptide (5 μmoles) were incubated in 1mL 0.1M phosphate buffer, pH 6.5 in the presence of 0.3M NaCl for 1 hour at room temperature [10]. Under these conditions we obtained 46% of α-monosubstituted peptide. The α-monosubstituted peptide, isolated by RP 18 HPLC had a UV spectrum with the characteristics of both tyrosine and maleimide groups. Oligonucleotide 0.1 μmole bearing a disulfide bridge at its 3' end and a fluoresceinyl moiety at its 5' end was reduced with TCEP (0.3 μmole) in 500 μL 2M NaCl, 0.1M phosphate buffer, pH 7.2, for 3 h at room temperature under nitrogen.

One equivalent of the oligonucleotide-SH was added to one equivalent of Nα-maleimidocaproyl-peptide without elimination of the TCEP excess. The reaction mixture was stirred for 5h at room temprature. The oligonucleotide-peptide conjugate was collected in 82% yield). The conjugate, after purification by RP 18 HPLC, has the expected spectral characteristics related to both DNA and fluorescein moieties and the expected amino acid composition.

The α-amino group of a peptide can easily be substituted at pH 6.5, under such conditions the maleimide is quite stable and the ε-amino group is poorly reactive. Furthermore, the use of TCEP as a reducing agent, allows the preparation of the oligonucleotide peptide conjugate even in the presence of a TCEP excess and without any purification step of the intermediate oligonucleotide mercaptan avoiding the oligonucleotide dimerization.Various oligonucleotide-KDEL conjugates specific for human immunodeficiency virus genomic sequences have been prepared by using the above described method. The intracellular traffic and biological efficiency of the covalent peptide-oligonucleotide conjugates synthetized are currently under investigation.

Acknowledgements

This work was financially supported by the Agence Nationale de Recherches sur le SIDA.

References

1   Hélène, C. and Toulmé, J.J., Biochim. Biophys. Acta, 1049 (1990) 99.
2.   Munro, S. and Pelham, H.R.B., Cell, 48 (1987) 899.
3.   Eritja, R., Pons, A., Escarceller, M., Giralt, E. and Albericio, F., Tetrahedron, 47 (1991) 4113.
4.   Feener, E.P., Shen, W.C. and Ryser, H.J., J. Biol. Chem., 265 (1990) 18780.
5.   Reddy, E.P., Reynalds, R.K., Santos, E. and Barbacid, M., Nature, 300 (1982) 149.
6.   Tung, C-H., Rudolph, M.J. and Stein, S., Bioconjugate Chem., 2 (1991) 464.
7.   Gregory, J. D., J. Am. Chem. Soc., 77 (1955) 3922.
8.   Bonfils, E. and Thuong, N.T., Tetrahedron Lett., 35 (1991) 3053.
9.   Burns, J.A., Bulter, J.C., Moran, J. and Whitesides, G.M., J. Org. Chem., 56 (1991) 2648.
10.   Shechter, Y., Schlessinger, J., Jacobs, S., Chang, K.J. and Cuatrecasas, P., Proc. Natl. Acad. Sci. U.S.A., 75 (1978) 2135.

Methylthiomethyl (MTM) methodology for the synthesis of phosphoserine, phosphothreonine and thiophosphoserine synthons

N. Mora, J.M. Lacombe and A.A. Pavia

Laboratoire de Chimie Bioorganique, Université d'Avignon, 33, rue Louis Pasteur, 84000 Avignon, France

Introduction

Reversible protein phosphorylation mediated by protein kinases/phosphatases is widely recognized as one of the major mechanisms by which eukaryotic cells regulate various cellular processes. Many biological important proteins such as enzymes, growth factors receptors, cytoskeletal, contractile and oncogenic proteins are known to exist as phosphoproteins [1-9]. Ready access to phosphopeptides related to biological phosphoproteins is essential for the study of such fundamental processes. For this reason, the development of efficient methods for the synthesis of phospho-peptides and analogs has became a goal of utmost importance.

There are currently two approaches for the synthesis of phosphopeptides. The first, known as *"global strategy"* involves post-synthetic phosphorylation of unprotected side-chain hydroxyl group either in solution or in the solid phase. The second, called *"sequential or stepwise strategy"* involves incorporation of a phosphoaminoacid derivative in the growing peptide. It requires the preparation of adequately protected phosphoaminoacid bearing a free carboxyl group. Although virtually suitable for both liquid and solid-phase methodologies, the synthesis of phosphopeptides has been so far hampered by the extreme lability of phosphoserine and phosphothreonine derivatives which are readily converted by β-elimination into the corresponding α,β-dehydro derivative [10].

Results and Discussion

We now propose a simple convenient and versatile method which proceeds through the intermediacy of a methylthiomethyl (MTM) ester. As shown in the reaction sequence below, the amino group can be protected with either benzyloxyl (Z), *tert*-butyloxycarbonyl (Boc) or allyloxycarbonyl (Alloc) [11-12]. Moreover, MTM ester which is compatible with both phenyl, 2,2,2-trichloroethyl and benzyl phosphate protecting groups is smoothly removed without production of β-elimination by-products.

Phosphoaminoacids **3** (X=O) were obtained by treatment of **1** with either diphenyl or bis(2,2,2-trichloroethyl) phosphochloridate, two commercially available reagents. Thiophospho analog **3** (X=S) was prepared by the phosphoramidite methodology using bis-(2,2,2-trichloroethyl)-*N,N*-diethyl phosphoroamidite. The intermediate phosphite was sulfurized by sulfur leading to the fully protected thiophosphoaminoacid derivatives. No β-elimination product was detected during this step.

$$R^1\text{-NH-CH-COOH} \xrightarrow{\text{i)}} R^1\text{-NH-CH-COOCH}_2\text{SCH}_3 \xrightarrow{\text{ii)}} R^1\text{-NH-CH-COOCH}_2\text{SCH}_3$$

CH-R²	CH-R²	CH-R²
OH	OH	O-P(OR³)₂
		‖
1	**2**	**3** X

$$\xrightarrow{\text{iii)}} R_1\text{-NH-CH-CO}_2\text{-H}$$

CH-R₂

O-P(OR³)₂
‖
4-16   X

R¹ = Z, Boc or Alloc
R² = H (Serine) or CH₃ (threonine)
R³ = Ph or CH₂CCl₃
X = O or S

i) To a solution of **1** (1 eq.) and NaHCO₃ (10 eq.) in DMSO (5 ml) was added t-butylbromide (10 eq.) in DMSO (5 ml). The mixture was stirred for 12 h at room temperature: yield 70-80% [13].

ii) **2** ((X=O), 1 eq.) was dissolved in distilled pyridine (5 ml) under nitrogen atmosphere and cooled to -30°C. To this stirred solution, the phosphochloridate (3 eq.) was added. The mixture was stirred under nitrogen atmosphere at -20°C (R=H) or at room temperature (R=CH₃) overnight: yield 80-90%.

ii) **2** ((X=S), 1 eq.) 1-H tetrazole (3 eq.) was added in one portion to a solution of **2** (1 eq.) in dry THF (2.5 ml) containing bis-2,2,2-trichloroethyl N,N-diisopropyl phosphoramidite (1.05 eq.) at +20°C. The reaction was monitored by TLC. Then, a saturated solution of S₈ (1.66 eq.) in toluene/2,6-lutidine (19:1; v/v) was added at room temperature. The reaction was monitored by ³¹P-NMR: yield 60% [15].

iii) To a solution of **3** (1 eq.) in diethylether (50 ml) was added MgBr₂ Et₂O (4 eq.) at room temperature. The reaction was monitored by TLC: yield 70-90%.

Table 1 *Phospho and thiophosphoserine and threonine synthons*

R²	H	H	H	CH₃	CH₃	CH₃
R¹	Boc	Z	Alloc	Boc	Z	Alloc
X=O ; R³=Ph	4	5	6	7	8	9
X=O ; R³=CH₂CCl₃	10	11	12	13	14	15
X=S ; R³= CH₂CCl₃	16					

Finally, the MTM group was readily removed by treatment with magnesium bromide in ether [14] to afford the expected phosphoaminoacids **4-16** (Table 1) in excellent yields: 75 to 95% for compound **4-15** (X=O), 50% for compound **16** (X=S).

All compounds reported in Table 1 were obtained in a pure crystalline form or as DCHA salts and fully characterized by ¹H-, ¹³C- and ³¹P-NMR spectroscopy.

Two of them (**4** and **16**) were used for the preparation of a phosphopeptide sequence related to the rat liver pyruvate kinase (Ala-Ser(P)-Val-Ala) in both the phosphorylated (**17**) and thiophosphorylated (**18**) form. The synthesis of **17** and **18** was performed by liquid-phase methodology with BOC as the amino protecting group and DCC and/or BOP as coupling reagent.

The approach reported herein based on methylthiomethyl ester intermediates allowed us to prepare phosphoserine, phosphothreonine and thiophosphoserine synthons in good yield without β-elimination by-products. Such building blocks were used to prepare phospho- and thiophospho derivative of Ala-Ser(P)-Val-Ala, an epitope of the rat liver pyruvate kinase.

References

1. Weller, M., Protein Phosphorylation, Pion Ltd., London, 1979.
2. Kemp, B.E. (Ed.), Peptides and Protein Phosphorylation, CRC Press, 1990, p.289.
3. Cohen, P., Nature, 296 (1982) 613.
4. Ingebritten, T.S. and Cohen, P., Science, 221 (1983) 331.
5. Cohen, P., Eur. J. Biochem., 151 (1985) 439.
6. Shenolikar, S.J., Cyclic Nucl. Pro. Phys. Res., 11 (1987) 531.
7. Edelman, A.M., Bumenthal, D.K. and Krebs, E.G., Ann. Res. Biochem., 56 (1987) 567.
8. Huganir, R.L. and Greengard, P., Trends Pharmacol. Sci., 8 (1987) 472.
9. Shenolikar, S., FASEB J., 2 (1988) 2753.
10. Paquet, A., Tetrahedron Lett., 31 (1990) 5269.
11. Ross, C.E., Barnabé, P., Hiemstra, H. and Speckamp, W.H., Tetrahedron Lett., 32 (1991) 6633.
12. Lacombe, J.M., Andriamanampisoa, F. and Pavia, A.A., Int. J. Pept. Protein Res., 36 (1990) 275: the alloc protecting group can also be removed by the hydrostannolytic method.
13. Dossena, A., Palla, G., Marchelli, R. and Lodi, T., Int. J. Peptide Protein Res., 23 (1984) 198.
14. Kim, S., Park, Y.H. and Kee, I.S., Tetrahedron Lett., 32 (1991) 3099.
15. Yau, E.K., Ma, Y.-X. and Caruthers, M.H., Tetrahedron Lett., 31 (1990) 1953.

Mechanistic studies on the hydrodechlorination reaction using sodium formate-catalytic transfer hydrogenation

S. Rajagopal and A.F. Spatola

Department of Chemistry, University of Louisville, Louisville, KY 40292, U.S.A.

Introduction

Catalytic transfer hydrogenation (CTH) is an important synthetic technique in organic chemistry. This method is applicable to numerous functional groups including halo, nitro, cyano, allyl, and benzyl groups as well as aldehydes and ketones. Though many organic compounds have been employed as hydrogen donors for transfer hydrogenation [1], formic acid and its salts occupy a special place because of their efficiency. Recent studies have demonstrated that formate salts such as ammonium formate are superior to formic acid as hydrogen donors [2, 3]. A detailed mechanistic study of transfer hydrogenation is of great significance since this technique is gaining considerable importance. Unfortunately, such investigations on these reactions are very scarce [4, 5]. The present study was undertaken to evaluate the efficiencies of various formate salts as hydrogen donors and elucidate the mechanism of hydrodechlorination, a reaction useful for peptide applications as well as for detoxifying environmental chlorocarbons [6].

Results and Discussion

Formic acid and some of its salts such as HCOOLi, HCOONa, HCOOK, $HCOONH_4$ and $HCOONHEt_3$ were selected to assess their efficiency as hydrogen donors. The well proven and efficient 10% Pd/C was chosen as a catalyst. A mixed solvent system ($EtOH:H_2O$ - 88:12) was utilized to bring both substrate (2-chlorotoluene) and the hydrogen donor into solution. The kinetic profiles for the chosen formates are presented in Figure 1. Initial rates show that the hydrogen-donating abilities of the Group 1A formates follow the order: $K^+ > Na^+ > Li^+ > H^+$. The other formates used in this study are grouped separately and ranked as follows: $NH_4^+ > NHEt_3^+ > H^+$. The following arguments may explain the observed activities of Group 1A formates. The separation of the formate ion from its cation, which is necessary for the ionization of alkali metal formates, depends on both the polarity of the bond and the internuclear distance between the oxygen of the formate and the alkali metal ion. Assuming equal polarities of the M-O bond, where M is H, Li, Na, K, the ease of separation of ions is greatly influenced by the initial distance between their centers of charge. Thus, a plot of initial rates with ionic radii has demonstrated that the relation is indeed quite striking.

From the base strength of NH$_3$ (pK_a = 9.25) and Et$_3$N (pK_a = 10.72), one might expect HCOONHEt$_3$ to show greater activity. The size effect also dictates HCOONHEt$_3$ to display higher activity. But the observed results display a reverse trend. Apparently, the cation size comparison is not valid in these two instances (NH$_4^+$ and NHEt$_3^+$) where steric hindrance predominates in the latter case. Therefore, the results can be explained on the premise that the adsorption of formate ion on palladium is hindered when the counter ion is bulkier (NHEt$_3^+$).

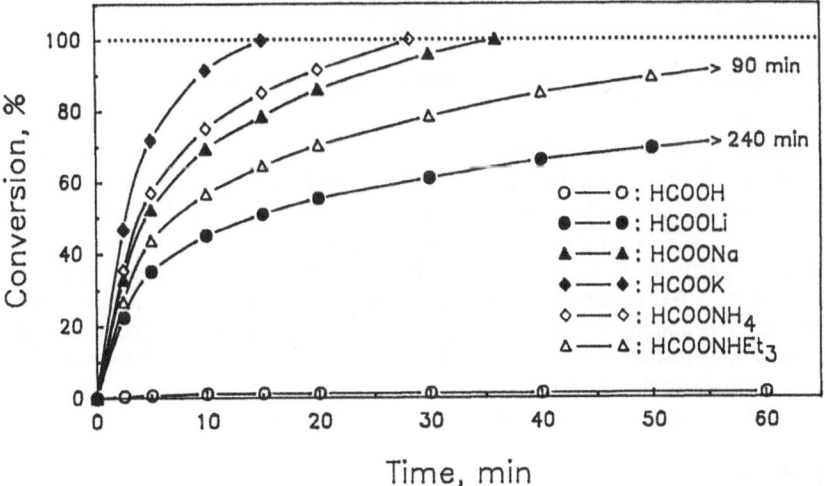

Fig. 1. *Time - conversion profiles for hydrodechlorination of 2-chlorotoluene. Reaction conditions: 2-chlorotoluene (0.2 M), formic acid or salt (0.24 M), 10% Pd/C (200 mg), volume (25 mL) and temperature (20°C).*

As shown in Figure 1, formic acid is a very poor hydrogen donor for the dechlorination of 2-chlorotoluene. The time-conversion plot appears as if HCl has poisoned the catalyst irreversibly. Further experiments have helped to conclude that HCOO$^-$ ion is the precursor for the reaction and HCl deactivates the catalyst but does not poison it completely. The kinetic data on hydrodechlorination of 2-chlorotoluene were obtained by changing variables such as the concentration of substrate, hydrogen donor and the catalyst. The plots of initial rate vs. initial concentration revealed that the order of the reaction is zero with substrate (the rate is independent of the substrate concentration), first order with hydrogen donor (up to 0.08 M) and first order with respect to catalyst. In the case of HCOONa, the order becomes fractional beyond 0.08 M. The isotope effect was calculated to be 1.87 when the HCOONa/EtOH/H$_2$O is replaced with DCOONa/EtOD/D$_2$O. Though HCOONa is considered as a strong electrolyte, complete dissociation into HCOO$^-$ and Na$^+$ ions in 88% ethanol may be limited especially at higher concentrations. Therefore, the dissociation of the sodium formate is considered the first step of this transfer hydrogenolysis mechanism. The second step in the process may be the adsorption of HCOO$^-$ onto the Pd surface. This adsorption leads to dissociative chemisorption resulting in PdH$^-$ and CO$_2$. The observed kinetic isotopic effect helps us to conclude that the dissociation of formate is the *rate-determining step* in the overall process. The zero order dependence of the substrate reveals that adsorption as well as the surface reaction of 2-chlorotoluene occurs very quickly. From the kinetic results, the following mechanism has been proposed for the dechlorination of 2-chlorotoluene by HCOONa under the catalytic influence of 10% Pd/C.

$$HCOONa \overset{K_1}{\leftrightarrow} HCOO^- + Na^+ \tag{1}$$

$$Pd^0 + HCOO^- \overset{K_2}{\leftrightarrow} Pd(HCOO^-)_{ad} \tag{2}$$

$$Pd(HCOO^-)_{ad} \overset{slow}{\underset{k}{\rightarrow}} PdH^- + CO_2 \tag{3}$$

$$PdH^- + Cl\text{-}C_6H_4\text{-}CH_3 \overset{fast}{\rightarrow} Pd^0 + C_6H_5\text{-}CH_3 + Cl^- \tag{4}$$

The rate expressions derived from the above scheme explain all the observations and hence the proposed mechanism is considered more appropriate.

The hydrogen-donating abilities of formate salts have been shown to depend on the counter ion. The poor hydrogen-donating efficiency of formic acid to aryl chlorides is partly attributed to the deactivating effect of one of the products (HCl) on the catalyst. All evidence shows that $HCOO^-$ is the active species during hydrogenolysis and not HCOOH. The kinetic data and the isotopic effect demonstrate that scission of formyl C-H is the rate-determining step. The hydride or hydride like species formed on the Pd surface reacts with the substrate in a fast step resulting in the desired hydrogenolysis reaction.

References

1. Johnstone, R.A.W., Wilby, A.H. and Entwistle, I.D., Chem. Rev., 85 (1985) 129.
2. Anwer, M.K. and Spatola, A.F., Synthesis, (1980) 929.
3. Wiener, H., Blum, J. and Sasson, Y., J. Org. Chem., 56 (1991) 4481.
4. Wiener, H., Blum, J. and Sasson, Y., J. Org. Chem., 56 (1991) 6145.
5. Anwer, M.K., Sherman, D.B., Roney, J.G. and Spatola, A.F., J. Org. Chem., 54 (1989) 1284.
6. Anwer, M.K. and Spatola, A.F., Tetrahedron Lett., 26 (1985) 1381.

Enantiospecific synthesis of 2-hetero aralkyl substituted statine analogues, novel highly potent HIV-1 protease inhibitors

D. Scholz, H. Retscher, B. Charpiot, H. Gstach, P. Lehr, B. Rosenwirth and A. Billich

Sandoz Forschungsinstitut, Brunner Strasse 59, A 1235 Vienna, Austria

Introduction

The human immunodeficiency virus (HIV) encodes an aspartyl protease necessary for viral replication. Following the nucleoside analogue inhibitors of reverse transcriptase (AZT and congeners) protease inhibitors are envisioned to create a "second front" against AIDS, because inactivation of the HIV-protease (HIVP) leads to the formation of non-infectious virions due to their immaturity. Significant efforts to identify potent inhibitors of HIVP have been made, mostly based on incorporation of transition state mimics into substrate analogues [1]. We introduce here a novel scissile bond replacement, namely the 2-aminobenzyl statine moiety as a non hydrolyzable dipeptide isostere. The cornerstone of the synthesis is a two-step sequence, namely the diastereospecific epoxidation of 2,3-unsaturated carboxylic esters, obtained by Wittig olefination of appropriately protected α-aminoaldehydes, and the enantio- and regiospecific opening of these epoxides with N-, S- and O-nucleophiles.

Results and Discussion

The inhibitors were synthesized as shown in Figure 1. The key reaction is the epoxidation of the 2,3-unsaturated carboxylic esters, easily obtained in a two-step sequence starting from BOC-protected aminoalcohols or the appropriate dipeptides [2]. This reaction is unusually slow (4-5 days, r.t., CH_2Cl_2, m-chloroperbenzoic acid), but yields exclusively the 2S,3R epoxide (when starting from the natural L-amino acids) in 50-75% yield. The structure has been verified by X-ray crystallography. These epoxides react enantio- and regiospecific with amines, mercaptans and phenols exclusively in position 2, leading to 2(R)-hetero aryl, alkyl or aralkyl substituted statine derivatives. Conventional saponification and coupling gives the target compounds. Using this new, highly flexible synthetic approach to the hitherto unknown 2-amino-benzyl statine compounds, a whole range of derivatives has been produced [5], leading to SDZ 282-870.

This compound, synthesized congruently in 6 + 2 steps starting from Z-Val-phenylalaninol in an overall yield of 30%, was, due to its promising biological activity (see Table 1) selected for further evaluation. In addition, synthetic efforts to provide even stronger inhibitors of HIV replication are ongoing.

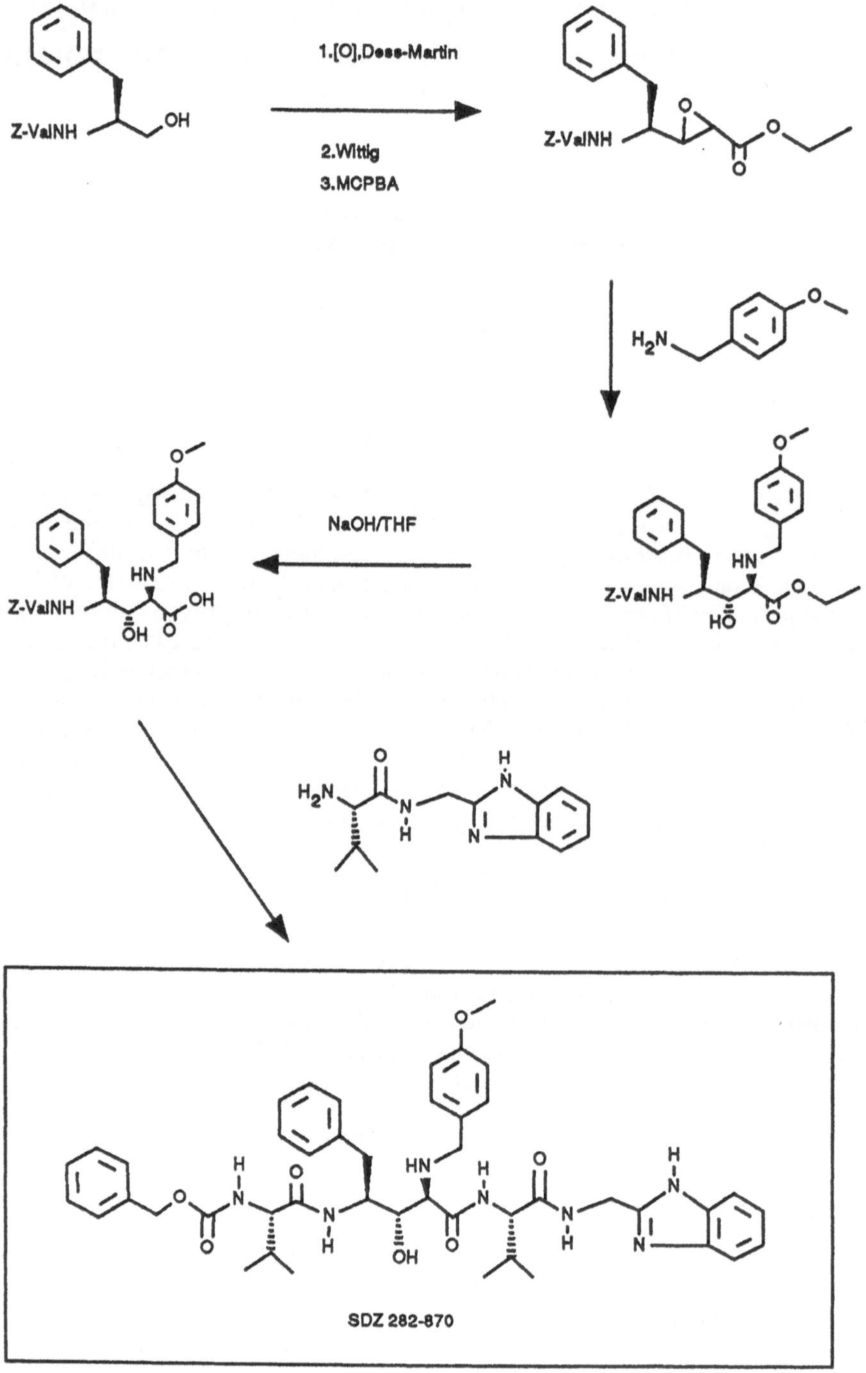

Fig. 1. Inhibitor synthesis.

Table 1

Compound	K_i [nM]		ED_{50} [nM]	
	HIV-1P [3]	HIV-2P [3]	HIV-1 [4]	IIIB/MT4
SDZ 282-870	3.4	120	29[a]	
Ro 31-8959	1.4	1.4	12[b]	

No inhibition of hu-renin, hu-trypsin, hu-elastase and porcine pepsin at 10 μM.
[a] Mean of 18 determinations.
[b] Mean of 16 determinations.

References

1. (a) Debouck, C. and Metcalf, B.W. (Review), Drug Develop. Res., 21 (1990) 1; (b) Huff, J.R., J. Med. Chem., 34 (1991) 2305; (c) Rich, D.H., Vara Prasad, J.V.N., Sun, C.H., Green, J., Mueller, R., Hauseman, K., MacKenzie, D. and Malkovsky, M., J. Med. Chem., 35 (1992) 3803, and references therein.
2. During our work a similar sequence, starting from dibenzylamino-aldehydes has been published by Reetz, M.T. and Lauterbach, E.H. Epoxidation with t.BuOOH/KOt.Bu/THF/NH3 leads to the same 2S,3R epoxide in a de>96%., Tet. Let., 32 (1991) 4477.
3. Billich, A., Hammerschmid, F. and Winkler, G., Biol. Chem. Hoppe-Seyler, 371 (1990) 265.
4. Pauwels, R., Balzarini, J., Baba, M., Snoeck, R., Schols, D., Herdewijn, P., Desmyter, J. and DeClerq, E., J. Virol. Meth., 20 (1988) 309.
5. Billich, A., Charpiot, B., Gstach, H., Lehr, P. and Scholz, D., WO 9301166, (93-01-21).

Molecular imprinting: Synthesis of 3-helix bundle proteins on modified silica gel

D.C. Tahmassebi and T. Sasaki

*Department of Chemistry, University of Washington,
Seattle, WA 98195, U.S.A.*

Introduction

Organic templates can assist peptide strands to assemble into a protein-like molecule [1]. In this paper, silica gel is "imprinted" with the geometry of a tridentate template so that 3-helix bundle proteins can be built on the silica surface. We describe the synthesis of a tridentate aldehyde "TRIPOD", modification of silica surface with the tridendate template, and preliminary experiments for the assembly of surface-bound helix bundle proteins.

Results and Discussion

We synthesized a trialdehyde, "TRIPOD", which when reacted with 4-methoxydimethylsilylaniline (MODSA) forms a tridentate Schiff's base [2].

Scheme 1

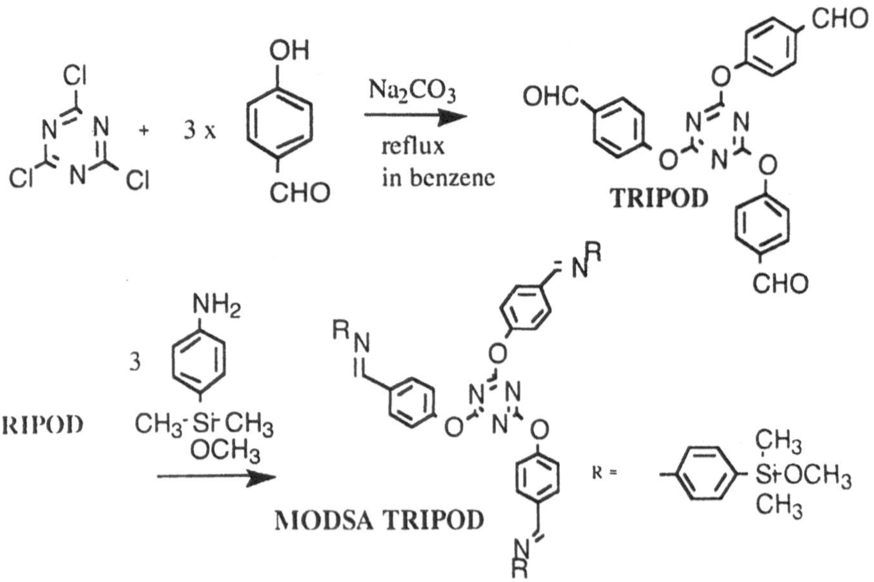

This molecule, "MODSA tripod" was refluxed with silica gel in dry toluene to covalently attach the template molecule to the surface [3]. The modified silica gel was subjected to mild acidic hydrolysis to remove the tripod portion, and to leave the silica surface "imprinted" with amine groups in the geometry of the tripod molecule. To confirm the structure of the aminophenyl groups on the surface, the silica gel was treated with aqueous HF [4] to release 4-fluorodimethylsilylaniline, which was analyzed by MS.

Scheme 2

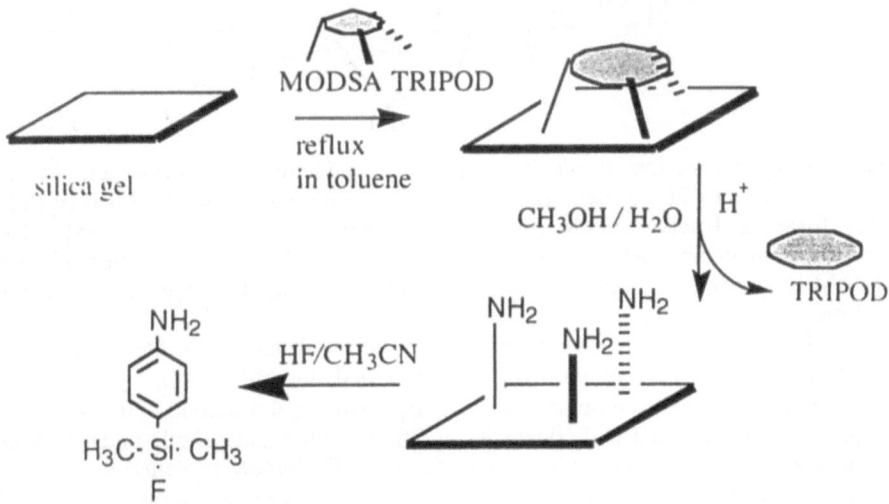

The distance between the amine groups on the surface should be appropriate for the synthesis of a 3-helix bundle protein on the modified silica gel. A 15-residue peptide, NH_2-AEQLLQEAEQLLQEL-$CONH_2$ was modified at the N-terminus with 4-carboxybenzaldehyde for attachment to the modified silica gel by reductive amination. This peptide has a potential to form an amphiphilic α-helix, and the same sequence has been used to organize a stable 3-helix bundle on a metal template [5]. The formyl group was protected as a cyclic acetal to avoid the competitive formation of a Schiff's base during the coupling reaction. Benzaldehyde-modified Ala-amide was also synthesized as a model compound using the same reaction.

R = Ala-amide

A-E-Q-L-L-Q-E-A-E-Q-L-L-Q-E-L-amide

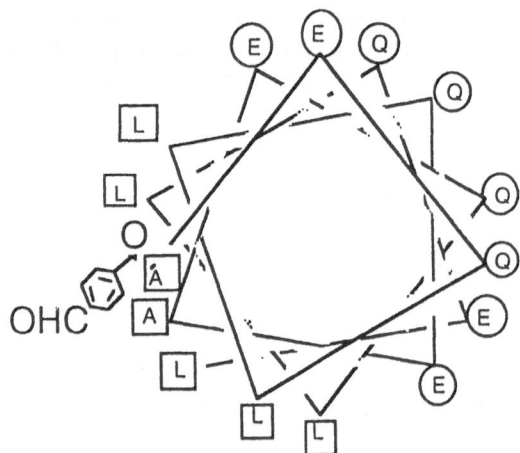

CD analysis showed that the aldehyde-modified peptide has modest α-helicity (44%) in pH 4.8 buffer. The α–helicity is comparable to the bipyridine-modified peptide that forms a 3-helix bundle protein in the presence of Fe(II) ion [5]. No concentration dependence of α-helicity was observed below a concentration of 2.6 x 10^{-5} M suggesting a monomeric state of the aldehyde-modified peptide in aqueous solution.

The aldehyde-modified alanine was reacted with aniline in the presence of $NaBH_3CN$. The reductive amination [6] proceeded smoothly, and the product was characterized by NMR and mass spectrometry. Silica gel modified with "MODSA tripod" is being reacted with aldehyde-modified alanine and peptide under similar conditions. To characterize the surface-bound helix bundle proteins, we plan to modify quartz slides so that the protein structures can be studied using CD spectroscopy [7].

Acknowledgements

This work was supported by the Office of Naval Research (N00014-92-J-1861). D.T. thanks the American Peptide Society for a travel award.

References

1.   (a) Sasaki, T. and Kaiser, E.T., J. Am. Chem. Soc., 111 (1989) 380   (b) Sasaki, T. and Kaiser, E.T., Biopolymers, 29 (1990) 79   (c) Mutter, M. and Vuilleumier, S., Angew. Chem., Int. Ed. Engl., 28 (1989) 535.
2.   Weisenfeld, R.B., J. Org. Chem., 51 (1986) 2434.
3.   (a) Wulff, G, Heide, B. and Helfmeier, G., Reactive Polymers, 6 (1987) 299   (b) Wulff, G., Heide, B. and Helfmeier, G., J. Am. Chem. Soc., 108 (1986) 1089   (c) Shea, K.J. and Dougherty, T.K., J. Am. Chem. Soc., 108 (1986) 1091   (d) Tao, Y-T. and Ho, Y-H., Chem Soc., Chem. Commun., 417 (1988) 417.
4.   Newton, R. and Reynolds D., Tetrahedron Lett, 41 (1979) 3981.
5.   (a) Lieberman, M. and Sasaki T., J. Am. Chem. Soc., 113 (1991) 1470   (b) Ghadiri, M.R., Soares, C. and Choi, C.J., Am. Chem. Soc., 114 (1992) 825   (c) Ghadiri, M.R., Soares, C. and Choi, C., J. Am. Chem. Soc., 114 (1992) 4000.
6.   Borch, R., J. Am. Chem. Soc., 93 (1971) 2897.
7.   Taylor, J., Acc. Chem. Res., 23 (1990) 338.

Chemical synthesis of a phosphorylated and thiophosphorylated EGFR-related peptide

W. Tegge

Gesellschaft für Biotechnologische Forschung (GBF),
38124 Braunschweig, Germany

Introduction

Most cellular processes are regulated by the phosphorylation of specific serine, threonine and tyrosine residues in target proteins, mediated by protein kinases. Especially tyrosine phosphorylation is the focus of many investigations, because it is involved in the regulation of cell growth and differentiation [1]. For several studies directed at protein phosphatases one would like to use phosphopeptide analogues that bind to the enzymes but are resistant to hydrolytic cleavage. Phosphonate and thiophosphate analogues have proven useful for this purpose [2, 3]. The use of phosphonate analogues necessitates first the elaborate synthesis of special suitably protected amino acid building blocks and second the repeated assembly of the same sequence if the unphosphorylated and normally phosphorylated peptides are also needed. Alternatively, the well established procedure of global phosphorylation of an unprotected hydroxyl amino acid on the resin offers the advantage, that the unphosphorylated and phosphorylated sequence can be obtained from one batch [4]. Especially the use of the very reactive phosphoamidites has been reported many times. These reagents also offer the possibility to synthesize phosphatase-resistant thiophosphate-analogues by oxidizing the intermediate phosphite with elemental sulfur or sulfur transferring agents [3]. Here this approach is demonstrated with the synthesis of a 13 amino acid tyrosine-phosphorylated and -thiophosphorylated sequence of the EGF receptor, containing the autophosphorylation site Tyr 1173. The quality of the peptides after phosphorylation with and without N-terminal protection was compared. Deprotection and workup conditions were investigated especially for the thiophosphate analogue with respect to side reactions and problems. The tyrosine thiophosphate group turned out to be hydrolytically labile under acidic conditions. The hydrolysis rate was determined with the thiophosphorylated model dipeptide Ac-Tyr(TP)-Gly.

Results and Discussion

The synthesis of the unphosphorylated EGFR sequence 1168-1180 was carried out with Fmoc amino acids on a commercial peptide synthesizer, employing TBTU and diisopropylethylamine for activation. Since the strategy of global phosphorylation on the resin was used, tyrosine was incorporated with an unprotected hydroxyl function. The sequence was synthesized in two batches, batch A with an N-terminal Boc-alanine (**1**, Figure 1) and batch B with an unprotected N-terminus (**2**). HPLC and FAB-MS analyses of small samples of both peptides after cleavage and workup gave no significant differences between the two batches. During the synthesis of this and other peptides we never observed any O-acylation of unprotected hydroxyl functions with this coupling chemistry. Global phosphorylation of both peptide batches was carried out with a large excess (50 fold) of di-t-butyl N,N-diethylphosphoramidite and 1*H*-tetrazole in DMF [5]. Oxidation of the phosphites was achieved in situ either by the addition of t-butyl hydroperoxide to give the phosphate **3** or by the addition of a solution of dibenzoyltetrasulfide in THF [6] for the preparation of the thiophosphate **4**. The quality of crude **3** obtained from the N-terminal Boc-protected sequence was slightly better than that from the unprotected sequence. In both cases the product was obtained almost quantitatively. A N-terminal phosphoramidate, if it is being formed from **2**, is cleaved by 95% TFA in the timespan of the deprotection (3 hours).

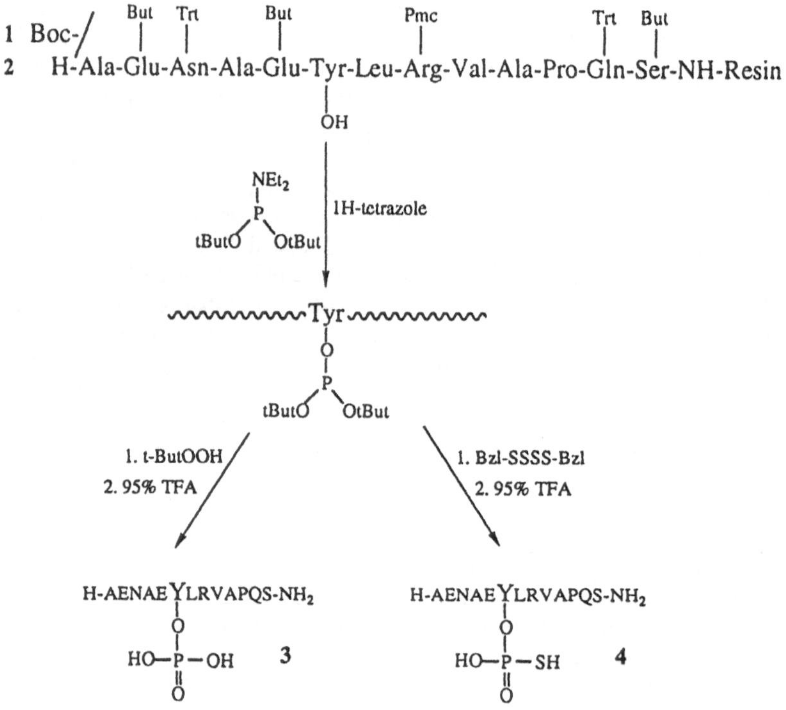

Fig. 1. Reaction scheme for the phosphorylation and thiophosphorylation of the Na-Boc-protected (1) and unprotected (2) EGFR-related sequence by global phosphorylation on the resin.

In contrast, the HPLC profiles of the thiophosphorylated products from both batches showed peaks for the desired peptide, the unphosphorylated sequence and the sequence containing a P-S-t-butyl group, that is probably formed by a reattachment of 2-methylpropene to P-SH or P=S. Formation of this reattachment product amounted to about 30% in neat TFA and could be lowered by the addition of scavengers like triisobutylsilane or DTT, although it could never be completely suppressed. Treatment of the compound with 95% TFA released the t-butyl group very slowly, only to a degree of about 50% in 24 hours. The unprotected thiophosphate group of **4** turned out to be hydrolytically labile under acidic conditions. The hydrolysis between pH 0.1 and 8 was determined with the thiophosphorylated model dipeptide Ac-Tyr(TP)-Gly (Table 1). Hydrolysis was fastest at pH 3 with a half-time of 12.5 hours. The compound was completely stable at pH 8, although for most practical purposes the stability under neutral conditions (pH 7) is probably sufficient. The thiophosphorylated EGFR peptide could be purified by omitting TFA in the solvent, accompanied by only a slight peak-broadening. Analysis of the phosphorylated and thiophosphorylated peptides by FAB-MS, NMR and sequencing gave the expected results. Tyrosine thiophosphate and tyrosine phosphate both gave blanks during the gas-phase as well as the pulsed liquid sequencing.

Table 1 *Half-time of the hydrolysis of the thiophosphate group between pH 0.1 and 8.0. *extrapolated*

pH	$t_{1/2}$ (h)
8.0	no hydrolysis
7.0	760*
6.0	170*
4.7	21.5
4.0	14.0
3.0	12.5
2.0	16.0
1.1	21.2
0.1	34.0

Acknowledgements

I thank R. Frank for stimulation of this work and valuable discussions, D. Marewski for excellent technical assistance, R. Christ for obtaining the FAB mass spectra, V. Wray for NMR measurements and M. Kieß for gas phase sequencing.

References

1.   Ullrich, A. and Schlessinger, J., Cell, 61 (1990) 203.
2.   Shoelson, S.E., Chatterjee, S., Chandhuri, M., and Burke Jr., T.R., Tet. Lett., 32 (1991) 6061.
3.   Kitas, E.A. and Bannwarth, W., Proceedings of the 22nd European Peptide Symposium (1992), Interlaken, Switzerland, p.345.
4.   Bannwarth, W. and Trzeciak, A., Helv. Chim. Acta, 70 (1987) 175.
5.   Perich, J.W. and Johns, R.B., Tet. Lett., 29 (1988) 2369.
6.   Rao, M.V., Reese, C.B., and Zhengyun, Z., Tet. Lett., 33 (1992) 4839.

Synthesis of acetylated and phosphorylated peptide fragments of histone H2A

W.O. Wachs[1], R. Hoffmann[1], M. Zeppezauer[1], K. Schmeer[2] and E. Bayer[2]

[1]*Department of Biochemistry, University of the Saarland, POB 1150, D-66041 Saarbrücken, Germany*
[2]*Institute of Organic Chemistry, University of Tübingen, Auf der Morgenstelle 18, D-72076 Tübingen, Germany*

Introduction

During the cell cycle histones are subject to various posttranslational modifications, e.g. ubiquitination, acetylation and phosphorylation [1, 2]. For structure-function studies it is desirable to have access to defined histone fragments with and without corresponding chemical modifications. Here we describe the synthesis of the N-terminal 14mer peptide of calf histone H2A including acetylated and phosphorylated derivatives.

Results and Discussion

The peptide S G R G K Q G G K A R A K A (H2A 1-14 from calf thymus [1]) was synthesised by the Fmoc/HBTU-strategy using a modified FastMoc-cycle of Applied Biosystems (Foster City, CA, USA) monitoring the peptide synthesis by measuring the absorption of the fulvene-piperidine adduct after cleavage of the Fmoc-groups at 301 nm. Serine was incorporated as FmocSerOH. The unprotected hydroxyl group showed no side reaction [5]. The following modifications were introduced:

$$\text{S G R G K Q G G K A R A K A}$$
$$1 \quad\quad 5 \quad\quad\quad 10 \quad\quad 14$$

S1 = (P), (Ac); K5 = (Ac) where (P) denotes O-phosphorylation of serine and (Ac) acetylation of the ε-amino group of lysine or the α-amino group of serine. All seven possible derivatives of the peptide were obtained.

Acetylation at pos 1 was achieved on resin utilising a novel cycle on the 430 A Applied Biosystems Peptide Synthesizer coupling 10% acetic anhydride in DCM twice to the unprotected N-terminus of the resin bound peptide. Peptides acetylated at pos 5 were synthesised using the FmocLys(Ac)OH-derivative in the course of the solid phase synthesis. Subsequent phosphorylation of the respective peptides was achieved via di-t-butyl-N,N-diethyl-phosphoramidite with subsequent oxidation [3, 5] and alternatively via dibenzophosphochloridate [4, 5]. In the case of the phosphoramidite procedure the tetrazole-catalysed phosphitilation was carried out in acetonitrile followed by oxidation with m-chloroperbenzoic acid. In the case of dibenzophosphochloridate the resins with the peptides were exposed to the phosphorylating reagent twice for two hours and a third time over night, the first addition taking place at -25°C.

All peptides were cleaved from the resin by exposure to 95% TFA/H2O for 1h at room temperature. After precipitation with ether and lyophilisation the peptides were deprotected completely by a mixture containing thioanisole (111μl), ethane dithiol (56μl), m-cresol (19μl), trimethylbromosilane (125μl) and TFA (691μl) for another hour at 0°C. The often employed scavenger mixture including phenol, ethane dithiol, thioanisole, water and TFA resulted in several incompletely deprotected peptides, even after prolonged cleavage time (up to 5h). All peptides were characterised and purified on a Nucleosil 100 C18 5 μm column by a linear gradient of acetonitrile in 0.1% TFA/water.

Capillary electrophoresis was performed using the Model 270HT of Applied Biosystems, the samples being injected for one second at a pressure of 5 inch on a capillary of 72 cm (50 μm I.D.). In 20mM sodium citrate buffer, pH 2.3, all phosphopeptides were separated from their unphosphorylated parent species at 30 kV within 15 min, the differences of the respective migration times lying between 0.3 and 1.9 min.

A TAGA 6000E triple-quadrupole mass spectrometer equipped with a standard atmospheric pressure ionisation source (Sciex, Toronto, Canada) was used to sample ions from the ion spray source [6]. The samples (about 50 μg freeze dried phosphopeptide in 100 μl 50% methanol and 50% water) were continuously infused into the spray probe by means of a syringe pump (Harvard Apparatus, Southnatick, Mass., U.S.A., Model 22) at a flow rate of 3 μl/min. All peptides and phosphopeptides gave the expected masses.

Although both phosphorylation procedures yield the desired phosphopeptides, the phosphoramidite procedure results in higher yields and fewer by-products, which is evident from the RP-HPLC profiles and the ion spray-mass spectra.

References

1. Matthews, H.R. and Waterborg, J.H. (Eds.), The Enzymology of Posttranslational Modifications of Proteins, Academic Press, London, 1985, p.125.
2. Böhm, L. and Crane-Robinson, C., Bioscience Reports, 4 (1984) 365.
3. Bannwarth, W. and Trzeciak, A., Helv. Chim. Acta., 70 (1987) 175.
4. Otvos, L. Jr., Elekes, I., and Lee, V.M.-Y., Int. J. Pept. Prot. Res., 34 (1989) 129.
5. Hoffmann, R., Wachs, W.O., Zeppezauer, M., Kalbitzer, H.-R., Waidelich, D. and Bayer, E., 1994, submitted for publication.
6. Bruins, A.P., Weidolf, L.O.G., Henion, J.D., and Budde, W.L., Anal. Chem., 59 (1987) 2642.

Assessment of long and difficult sequences in FMOC solid-phase peptide synthesis: The protease from HIV-1

P.M. Fischer[1], J.C. Golding[2], K.V. Retson-Yip[2] and M.I. Tyler[2]

[1]Nycomed Bioreg AS, Gaustadalleen 21, N-0371 Oslo, Norway,
[2]Deakin Research Ltd., CSIRO Div. Food Processing, Delhi Rd.,
North Ryde, NSW 2113, Australia

Introduction

The chemistry of peptide bond formation and protection strategies, as well as the practical methodologies of SPPS have been refined to such an extent that it is now possible to synthesise small proteins by this method. The resources required for such syntheses are considerable and the development of methods to optimise synthetic protocols is important. Quantitative solid-phase Edman degradation for the evaluation of extended SPPS, often referred to as "preview sequencing", using the Boc/polystyrene synthesis approach has been used successfully for some time [1]. This method is particularly valuable in assessing syntheses because it is capable of revealing accurately incomplete aminoacylation reactions. It has been pointed out [2] that sequencing of protected resin-bound peptides is not applicable to the products of standard Fmoc-chemistry SPPS since both amino acid side chain protecting groups as well as resin linkage agents are labile to the acid conditions of the Edman degradation. The fact that polystyrene peptidyl resins with acid-labile linkers such as p-hydroxymethyl-phenoxyacetic acid can be sequenced, either using external or *in situ* resin detachment prior to analysis, has been demonstrated [3]. We find that direct quantitative preview sequence analysis of peptides attached to a variety of synthesis resins by acid-labile anchors can be achieved. The utility of preview sequence analysis to extended Fmoc continuous flow SPPS is exemplified by the synthesis of the 99-residue protease from HIV-1.

Whereas examination of the results of pilot syntheses by preview sequence analysis of peptidyl resins and analysis of the cleaved materials (e.g., using mass spectrometry) are useful in the design of scale-up syntheses, it would be of great advantage to be able to accurately *predict* difficulties likely to be encountered in a synthesis. Much is known about the side-reactions associated with various aspects of SPPS, particularly acidolytic deprotection and resin detachment, and problems emanating from this source are more or less predictable. The chief remaining problem is accurate prediction of so-called "difficult couplings". Much progress has been made in this area by attempting to identify those combinations of carboxyl and amino components likely to cause difficulties in SPPS acylation reactions. The results from such studies (e.g., [4]) are derived from databases of peptide synthesis coupling tests. It has been realised for some time that the occurrence of "difficult sequences" in SPPS is not so much a consequence of the nature of the solid support, amino acid protecting groups or coupling chemistry applied but is inherent to particular sequences. Indeed it would appear that the same phenomena which cause troublesome solubility problems of protected peptide segments in convergent syntheses, i.e. , the inherent propensity of some sequences to adopt β-type secondary structures, are responsible for acylation difficulties with resin-bound protected peptides and are predictable [5]. We have tried to contribute to the reliability of such prediction methods by assessment of the reactivities not of the solid-phase bound amino components, but the relative efficiency of the activated carboxyl components in peptide bond formations.

Results and Discussion

Preview sequence analysis of peptides bound to Wang resin (*i.e.*, the same polystyrene-divinylbenzene support as in Boc-type syntheses but with the TFA-labile *p*-alkoxybenzylalcohol linker) presents no problems. We have found that flow synthesis supports such as macroporous kieselguhr/polydimethylacrylamide [6] and Polyhipe™/polydimethylacrylamide [7] composite resins, suitable for automatic flow synthesis, are compatible with gas-phase sequence analysis. Peptides assembled on such resins using either *p*-alkoxybenzylalcohol handles or acid-labile linkers yielding peptide amides give equally satisfactory results. Although we find that in most cases peptides can be sequenced up to and including the *C*-terminal residue, repetitive sequencing yields gradually decrease at a similar rate as when sequencing free peptides. In several instances we have sequenced both the complete Fmoc-deprotected peptidyl resin as well as the crude cleaved peptide in order to demonstrate that the presence of the solid support and side-chain protecting groups does not adversely affect sequencing results. We have found that wash-out can be minimised if the peptidyl resins to be sequenced are placed into the sequencer cartridge (ABI model 470 Sequenator) sandwiched between Polybreen™-soaked filter paper discs. The side-chain protecting groups of the *t*-butyl and trityl type are sufficiently acid labile to the conditions of phenylthiocarbamoyl peptide cleavage so that only the corresponding free side chain PTH-amino acids are observed after conversion. The Pmc protecting group of the Arg guanidino function is less acid labile [8] and, unlike all the other amino acids, assessment of Arg preview is somewhat unreliable. Low yields of PTH-Arg are observed if Arg occurs close to the *N*-terminus of a sequence, if Arg residues are located further into the sequence, the repetitive exposure to acid at the various cleavage reactions of the Edman chemistry results in improved PTH-Arg yields.

Several syntheses of HIV-1 protease have been reported. Generally in these syntheses Cys[67] and Cys[95] residues were replaced by isosteric Abu residues in order to simplify synthetic manipulations. We have achieved the synthesis of HIV-1 protease with the native sequence using the Fmoc/But-polyamide flow resin synthesis strategy. The Cys residues were protected in the side chain with the *S*-acetamidomethyl group to permit purification prior to liberation of the thiol groups. The successful progress of the chain assembly was followed using preview sequence analysis (see Table 1). Folding of the synthetic material into the biologically active form was achieved by dilution of the protease from a solution containing a high concentration of urea. The activity of the crude folded synthetic material was sufficient to permit routine screening of potential protease inhibitors in RP-HPLC assays or using continuous spectro-photometric methods. Further purification by RP-HPLC and affinity chromatography on immobilised pepstatin yielded highly pure enzyme by comparison with authentic samples.

Table 1 *Quantitative sequence analysis of synthetic HIV-1 protease peptidyl resin*

Sampled at Residue	Preview at Residue	Cycles (no.)	Preview (%)	Yield (%)	
				Correct Chains	Average per Step
78	96	19	26.1	73.9	98.3
51	75	24	33.5	66.5	98.2
40	47	7	14.6	85.4	97.8
27	36	9	15.1	84.9	98.2
16	22	6	21.4	78.6	96.1
1	10	9	25.3	74.7	96.8

The steric and electronic availability of an *N*-terminal amino group on a peptidyl resin, in combination with the chemical reactivity of the incoming activated carboxyl component, determines the quantitative outcome of any coupling reaction in SPPS. It should be noted that compilations of acylation tests from various peptide syntheses represent a combination of these two factors; it is obviously difficult to determine to what extent the results are biased towards one or the other. We sought to shed some light on this question by determining the relative reactivities of Fmoc-amino acid derivatives with invariant amino components. This was accomplished by analysis of 20 different dipeptidyl resins which had been reacted with an equimolar mixture of Fmoc-amino acids. We chose PyBOP/HOBt/NMM coupling chemistry and those side-chain protected Fmoc-amino acids currently favoured in our laboratory (*t*-butyl type for Asp, Glu, Ser, Thr, Tyr and Trp, trityl for Asn, Gln, His and Cys, Pmc for Arg). The derivatives of Gly, Pro, Ser and Tyr were among the better "couplers" whereas those of His, Arg, Thr, Val, Ile, Leu and Lys were comparatively poorly reactive.

A combination of preview sequence analysis, use of predictive methods such as that of Milton *et al.* [5], as well as assessment of the relative reactivities of amino acid derivatives should facilitate long and difficult solid-phase peptide syntheses.

References

1. Matsueda, G.R., Haber, E. and Margolies, M.N. , Biochemistry, 20 (1981) 2571.
2. Kent, S.B.H., Ann. Rev. Biochem., 57 (1988) 957.
3. Kochersperger, M.L., Blacher, R., Kelly, P., Pierce, L. and Hawke, D.H., Am. Biotech. Lab., 7 (1989) 26.
4. van Woerkom, W.J. and van Nispen, J.W., Int. J. Peptide Protein Res., 38 (1991) 103.
5. Milton, R.C. de L., Milton, S.C.F. and Adams, P.A. , J. Am. Chem. Soc., 112 (1990) 6039.
6. Atherton, E., Brown, E. and Sheppard, R.C., J. Chem. Soc., Chem. Commun., (1981) 1151.
7. Small, P.W. and Sherrington, D.C., J. Chem. Soc., Chem. Commun., (1989) 1589.
8. Fischer, P.M., Retson, K.V., Tyler, M.I. and Howden, M.E.H., Int. J. Peptide Protein Res., 40 (1992) 19.

Optimization of real time coupling efficiency determination using counterion distribution monitoring

C. Van Wandelen

Millipore Corporation, 34 Maple Street, Milford, MA 01757, U.S.A

Introduction

The use of quinoline yellow to monitor coupling reactions provides real time qualitative kinetic information and quantitation of the percent coupling. This type of analysis described by Young [1] is commonly known as counterion distribution monitoring (CDM). As the coupling goes to completion, the concentration of dye in solution increases. When the dye concentration reaches a steady state, the percent coupling can be calculated. The 9050Plus Pepsynthesizer is uniquely suited to use this type of monitoring due to its continuous flow design, and integrated UV/Vis detectors. This method was originally described using a kieselguhr type of synthesis support. The following paper describes the use of PEG-PS, polystyrene & glass, and Polyhipe with CDM. A method for evaluating the performance is also discussed.

Results and Discussion

A study was performed to determine the linearity of detector response with respect to dye concentration, and to ascertain whether any non-specific dye interaction with the support may affect data interpretation. The results (not shown) indicate that there is a linear response vs dye concentration for all of the supports. This linearity indicates that any non-specific dye interaction is not concentration related.

The shape of the 9050Plus detector readouts are a direct indication of coupling kinetics (Figure 1). Study of these shapes provides the peptide chemist with a unique perspective on identifying troublesome sequence motifs. In order to determine if the various supports were capable of exhibiting good dye release responses, they were subjected to a known slow coupling. The results (Figure 1) indicate that all of the supports respond similarly and may be used with CDM. Furthermore, to improve quantitiation accuracy, a two point calibration protocol was designed to replace the existing one point calibration. The one point calibration would add the quinoline solution to the support when all the N-termini were blocked. This value was used as the 100% coupling value against which all of the other coupling reactions were measured. In addition to the 100% value, a new step which added the dye solution to the support when completely deblocked, was added. This two point calibration yields a formula which no longer assumes the detector response of 0.0 AU for zero % coupling (Figure 2). This calibration also improves the sensitivity of the quantitation (% coupling/AU).

In order to evaluate the performance of the system, the sensitivity (S) of the calibration must be determined. Since the non-linear nature of the calibration allows

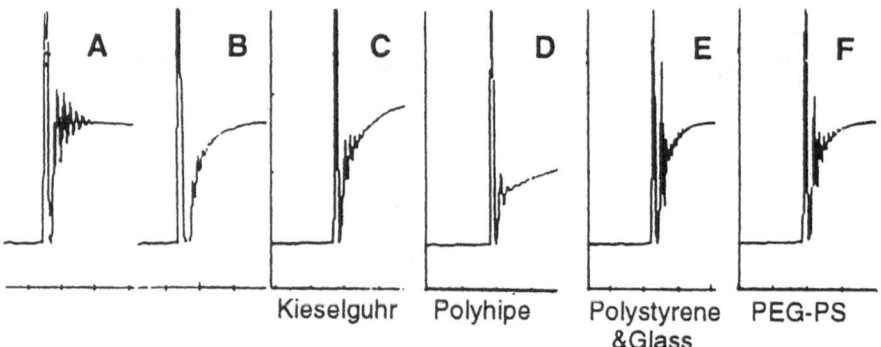

Fig. 1. Traces A & B are representative of fast and slow coupling rates, respectively, as
normally seen on the 9050Plus Pepsynthesizer. Traces C–F are from the Ile to Asn
coupling of the acyl carrier protein decamer using the various supports indicated.

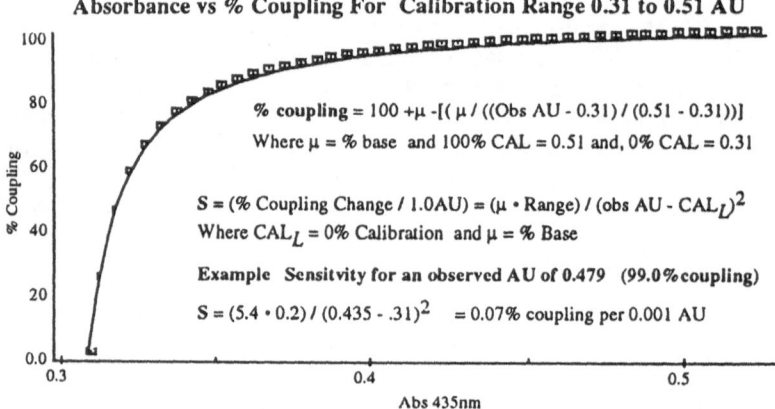

Fig. 2. Graph of the % coupling using the two point calibration equation is shown. Also
given is the formula to determine the sensitivity (S) of a calibration at a given %
coupling. The example given is for the 99.0% coupling where accuracy is critical.

sensitivity to decline very rapidly below 90%, these parameters must be examined
carefully. To determine the sensitivity of a calibration at any given absorbance (%
coupling) the formula shown in Figure 2 is used. This formula can be used to
evaluate the sensitivity of a one point calibration by using a value of 0.0 AU for the
CAL_L. By determining the sensitivity of a given calibration, a more informed choice
can be made about where to set the minimum acceptable % coupling. All of the
supports tested show a linear response to dye concentration, the expected slow
coupling shape during actual synthesis, and appear to be suitable for use with CDM.
Once calibration is complete, it is then possible to evaluate the sensitivity of the
system, which is a function of various parameters, including type of support,
concentration of dye, type of calibration, and maintenance of detector optics.

References

1.   Young, S.C., White P.D., Davies, J.W., Owen, D.E.I.A., Salisbury, S.A. and
      Tremeer, E.J., Biochem. Soc. Trans., 18 (1990) 1131.

Session II
Synthetic and Analytical Methods

Chairs: Arne Holm
The Royal Veterinary & Agricultural University
Frederiksberg, Denmark

and

Gilles A. Lajoie
University of Waterloo
Waterloo, Ontario, Canada

Facile peptide bond cleavage reactions

M. Goodman, J.R. Spencer, A. Toy-Palmer, J.X. Jiang, H.T. Li, T.A. Tran and T.T. Romoff

Department of Chemistry, University of California San Diego, La Jolla, CA, 92093-0343, U.S.A.

Introduction

The synthesis of highly hindered peptides [1] is part of the program in our laboratories to investigate conformationally constrained molecules. Hence, a cyclic hexapeptide related to somatostatin was designed with the sequence c[(NMe)Aib-Phe[1]-D-Trp-Lys-Thr-Phe[2]]. The synthesis was carried out by standard methods in solution. However, during the final stage acidolytic deprotection reaction, an unusual peptide bond cleavage was observed [2]. Analysis of this reaction has been carried out and indicates susceptibility of the (NMe)Aib-Phe[1] amide bond to nucleophilic attack.

Results and Discussion

Simultaneous acidolysis of *tert*-butyl ether and N^ε-Boc protections from the compound c[(NMe)Aib-Phe[1]-D-Trp-Lys(Boc)-Thr(tBu)-Phe[2]] was accomplished with TFA, in the presence of ethanedithiol and anisole as scavengers, over 1 h at 0°C and 2 h at 25°C [2]. Two linear products were isolated from the reaction mixture. The proportion of the two products was 42:58 as determined by integration from RP-HPLC.

The high resolution FAB MS of one product from the deprotection reaction reveals a molecular weight of 827.4441. This suggests the compound is a linear product of a peptide hydrolysis reaction. From 1-D proton NMR spectral analysis, the broad singlet at 11.93 ppm attributed to a carboxyl group supports this identification. Analysis by 2-D NMR spectroscopy indicated that Phe[1] is at the N-terminus since there is no correlation between the (NMe)Aib resonances and the protons of the Phe[1] residue. This was confirmed by N-terminal amino acid sequencing which clearly demonstrated the presence of a Phe residue followed by a Trp residue.

The high-resolution FAB MS for the other product reveals a molecular weight of 903.4254. We determined that the theoretical molecular weight of the linear hexapeptide thioester, in which the scavenger ethanedithiol acts as a nucleophile to cleave a peptide bond, is 903.4266. This value is in excellent agreement with the experimentally determined molecular weight. From 1-D and 2-D NMR experiments, we established that the cleavage occurs between the (NMe)Aib and Phe[1] residues as was observed for the structure arising from hydrolysis.

In the 1-D ^{1}H-NMR spectrum, the Phe[1] NH resonance appears broad and integrates for more than a single proton (Figure 1). NMR titration studies showed that under increasingly basic conditions, the Phe[1] NH peak broadened to the point where it could no longer be observed. This behavior is similar to the behavior of the Lys εNH amine resonance under the same conditions. These results also support a structure where the Phe[1] amine is N-terminal.

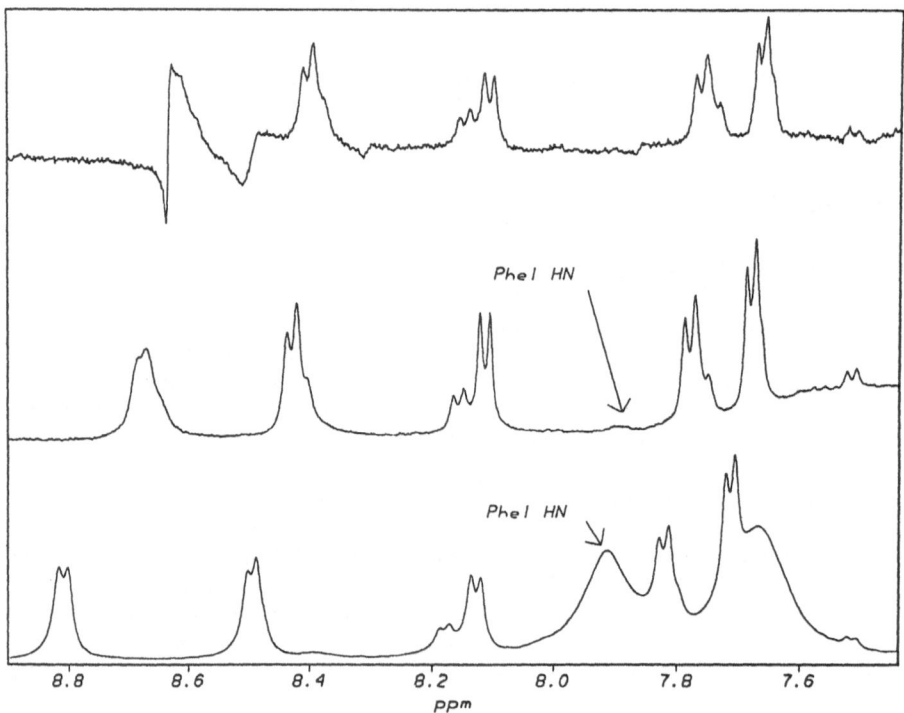

Fig. 1. The ^1H NMR spectrum of H-Phe^1-D-Trp-Lys-Thr-Phe^2-(NMe)Aib-S-CH$_2$-CH$_2$-SH without addition of N-methylmorpholine (bottom), with approximately 2 equivalents of N-methylmorpholine (middle) and with an excess of N-methylmorpholine (top).

In the 2-D ROESY spectra, no NOEs were observed between the (NMe)Aib and Phe[1] residues which would be expected for a cyclic compound. In addition, resonances that could be attributed to the methylenes of ethanedithiol were resolved in the β-region of the 2-D TOCSY spectra. These resonances were not linked to any α-protons. The identification of the linear hexapeptide thioester was thus confirmed by ^{1}H-NMR spectroscopy.

Finally, we characterized the two linear products and the related cyclic analog c[Aib-Phe[1]-D-Trp-Lys-Thr-Phe[2]] by ^{13}C 1-D and heteronuclear multiple bond correlation (HMBC) [3] NMR experiments. Assignment of the amide carbonyls reveals a downfield shift to 200.4 ppm for the (NMe)Aib thioester containing analog as compared with shifts of 174.6 ppm for the linear (NMe)Aib compound and 174.0 ppm for the cyclic Aib molecular (Figure 2). The downfield ^{13}C carbonyl chemical shift of the thioester was compared to the ^{13}C 1-D spectrum of CH$_3$-CH$_2$-C(O)-S-CH$_3$ [4], which also shows a carbonyl shift of 200.5 ppm. In the HMBC spectrum, no correlation was observed between the α-proton of Phe¹ and the carbonyl of (NMe)Aib. The crosspeaks to the carbonyl of (NMe)Aib arose only from the β-methyls and the N-methyl of the (NMe)Aib residue. These experimental observations support the linear structure for this hexapeptide derivative. Therefore, one of the products from the TFA deprotection reaction resulted from an unexpected nucleophilic attack by the scavenger ethanedithiol on the (NMe)Aib-Phe¹ amide bond.

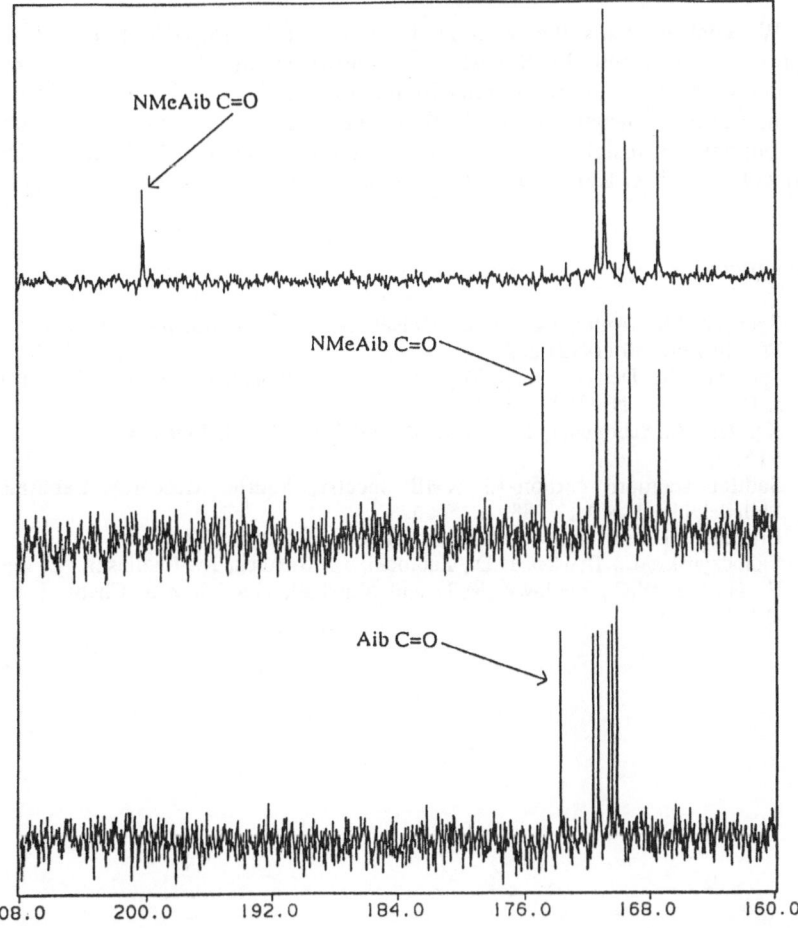

Fig. 2. Carbonyl region of (top) H-Phe¹-D-Trp-Lys-Thr-Phe²-(NMe)Aib-S-CH₂-CH₂-SH, (middle) H-Phe¹-D-Trp-Lys-Thr-Phe²-(NMe)Aib-OH, and (bottom) c[Aib-Phe¹-D-Trp-Lys-Thr-Phe²]. The ¹³C-NMR spectra illustrate the downfield shift of the (NMe)Aib carbonyl carbon involved in a thioester.

Studies by Jung [5] and Marshall [6] have shown that peptides which are rich in Aib sequences are sensitive to selective trifluoroacetolysis of Aib-Xaa bonds, especially when Xaa is proline. However, the reactions are observed only after treatment of the peptides with anhydrous TFA at 37-60°C over 10-20 h. Our observations indicate that the presence of strain caused by steric bulk and cyclization increases the lability of the (NMe)Aib-Phe[1] peptide bond, such that cleavage by sulfhydryl nucleophilic attack or hydrolysis can occur under mild acidic conditions. This lability may arise from the twisting or tilting of the specific (NMe)Aib-Phe[1] amide bond.

Acknowledgements

We wish to thank the National Institutes of Health (DK 15410) for their support of this research. J.R.S. wishes to acknowledge the NIH for support from the Cell and Molecular Biology Training Grant, GM 07313. We also thank Dr. Richard Kondrat and his colleagues at the UC Riverside Mass Spectrometry Facility for their contributions and insights. We are very grateful to Nancy G.J. Delaet, Valery V. Antonenko, Michael Kurz and Victor L. Hsu for their valuable contributions to this paper.

References

1. Spencer, J.R., Antonenko, V.V., Delaet, N.G.J. and Goodman, M., Int. J. Pept. Protein Res., 40 (1992) 282.
2. Spencer, J.R., Delaet, N.G.J., Toy-Palmer, A., Antonenko, V.V. and Goodman, M., J. Org. Chem., 58 (1993) 1635.
3. Kessler, H., Schmieder, P., Köck, M. and Kurz, M., J. Magn. Reson., 88 (1990) 615.
4. Sadtler standard carbon-13 NMR spectra, Sadtler Research Laboratories, Philadelphia, U.S.A., 1988, p.25066.
5. Bruckner, H. and Jung, G., Chromatographia, 13 (1980) 170.
6. Slomczynska, U., Beusen, D.D., Zabrocki, J., Kociolck, K., Redlinski, A., Reusser, F., Hutton, W.C., Leplawy, M.T. and Marshall, G.R., J. Am. Chem. Soc., 114 (1992) 4095.

Synthesis of some biologically active peptides of marine origin

T. Shioiri, T. Imaeda and Y. Hamada

Faculty of Pharmaceutical Sciences, Nagoya City University
Tanabe-dori, Mizuho-ku, Nagoya 467, Japan

Introduction

We have been interested in the synthesis of structurally unique and biologically intriguing peptides of marine origin [1]. Since many of these marine peptides are usually produced in minute amounts in nature, the method suitable for the large scale production through an efficient and convenient way should be developed to investigate their biological profiles in detail. We recently succeeded in the synthesis of nazumamide A (1), a potent thrombin-inhibiting linear tetrapeptide, isolated from a marine sponge *Theonella* sp. [2]. We now wish to report an efficient synthesis of geodiamolide A (2) which was isolated from a Caribbean sponge *Geodia* sp., and a sponge *Pseudaxinyssa* sp. collected in Papua New Guinea.

Nazumamide A (1)   Geodiamolide A (2)

Geodiamolide A (2) is a cyclic depsipeptide containing the peptide and polypropionate units. The other geodiamolides (B-F) which are different from each other in the peptide portion are also known in addition to jaspamide, a similar cyclic depsipeptide containing the same polypropionate portion as that of geodiamolides. The combination of interesting cytotoxic and antifungal activities with their structural aspects has made these cyclic depsipeptides an attractive target for total synthesis. To date, several syntheses of these peptides have appeared [3] but their efficiencies are not excellent. Now, we wish to report an efficient synthesis of geodiamolide A (2), which will be suitable for the large scale production and applicable to the synthesis of the other geodiamolides and jaspamide.

Results and Discussion

The key steps of our synthesis are (1) the asymmetric alkylation using the Evans chiral oxazolidinone, (2) the coupling of the tripeptide unit with the polypropionate unit by the Mitsunobu reaction, and (3) the intramolecular amide bond formation using diphenylphosphorazidate (DPPA, $(C_6H_5O)_2P(O)N_3$).

The preparation of the tripeptide unit (**6**) was achieved from the C-terminal alanine trichloroethyl ester. Condensation of the ester with O-tert-butyldimethylsilyl-N-tert-butoxycarbonyl-N-methyl-(R)-tyrosine (**3**) was carried out by use of diethyl

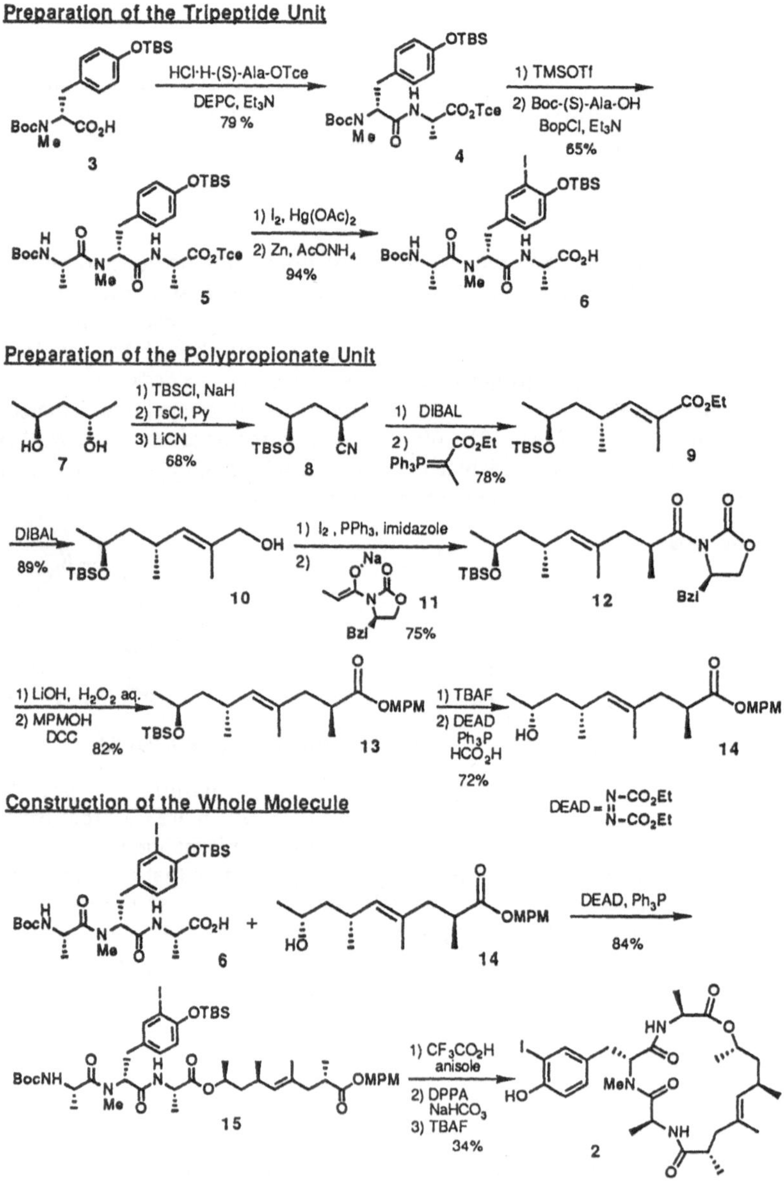

Fig. 1. *Total synthesis of geodiamolide A.*

216

phosphorocyanidate (DEPC, $(C_2H_5O)_2P(O)CN$). Selective removal of the Boc group from the dipeptide (4) by use of trimethylsilyl trifluoromethanesulfonate (TMSOTf), followed by the coupling with Boc-(S)-alanine using bis(2-oxo-3-oxazolidinyl)-phosphinic chloride (BopCl) afforded the tripeptide (5), which was iodinated with iodine and mercuric acetate to give the fully protected tripeptide unit from which reductive removal of the trichloroethyl ester afforded the required tripeptide carboxylic acid (6). Each step efficiently proceeded in more than 80% yield.

The synthesis of the polypropionate unit (14) started from commercially available (2S,4S)-2,4-pentanediol (7). After protection of one of the hydroxyl functions with the tert-butyldimethylsilyl (TBS) group, the resulting mono-protected alcohol was converted to the cyanide (8) through the tosylate. Reduction of the cyanide (8) with diisobutylaluminum hydride (DIBAL), followed by the Wittig homologation preferentially afforded the (E)-ester (9), which was reduced with DIBAL to give the alcohol (10). Although the bromination of the alcohol (10) failed, treatment with iodine afforded the corresponding iodide. The sodium enolate (11) of the oxazolidinone derived from (R)-phenylalaninol was alkylated with the iodide to give the polypropionate (12) as the major product (94% diastereomeric excess). Removal of the chiral auxiliary from the polypropionate (12) with alkaline hydrogen peroxide afforded the carboxylic acid, which was converted to the p-methoxyphenylmethyl (MPM) ester (13). After removal of the TBS group with tetra-n-butylammonium fluoride (TBAF), the hydroxyl group was inverted by use of the Mitsunobu reaction to give the required polypropionate unit (14). The overall yield for the preparation of the polypropionate unit was 21% in 12 steps.

Construction of the full carbon skeleton of geodiamolide A (2) was efficiently achieved by the coupling of the tripeptide unit (6) with the polypropionate one (14) using the Mitsunobu reaction. Deprotection of the N- and C-terminal protective groups from the MPM ester (15) was easily achieved with trifluoroacetic acid. Cyclization of the linear precursor smoothly proceeded to give the depsipeptide by use of DPPA, in the presence of sodium hydrogen carbonate. Final treatment with TBAF afforded geodiamolide A (2) in an overall yield of 34% in 3 steps from the MPM ester (15).

Thus, we could complete the efficient synthesis of geodiamolide A (2). The overall strategy adopted here can be applied to the synthesis of jaspamide as well as the other geodiamolides.

References

1. Shioiri, T. and Hamada, Y., In Atta-ur-Rahman (Ed.), Studies in Natural Products Chemistry, Vol. 4, Stereoselective Synthesis (Part C), Elsevier, Amsterdam, The Netherlands, 1989, p.83.
2. Hayashi, K., Hamada, Y. and Shioiri, T., Tetrahedron Lett., 33 (1992) 5075.
3. (a) Grieco, P.A., Hon, Y.S. and Perez-Medrano, A., J. Am. Chem. Soc., 110 (1988) 1630; (b) Grieco, P.A. and Perez-Medrano, A., Tetrahedron Lett., 29 (1988) 4225; (c) White, J.D. and Amedio, J.C., Jr., J. Org. Chem., 54 (1989) 736; (d) Hirai, Y., Yokota, K., Sakai, H., Yamazaki, T. and Momose, T., Heterocycles, 29 (1989) 1865; (e) Chu, K.S., Negrete, G.R. and Konopelski, J.P., J. Org. Chem., 56 (1991) 5196.

A chemical ligation strategy to form a peptide bond between two unprotected peptides

C.F. Liu, J. Shao, C. Rao and J.P. Tam

*Department of Microbiology and Immunology,
A-5119 Medical Center North, Vanderbilt University,
Nashville, TN 37232-2363, U.S.A.*

Introduction

The synthesis of peptides or proteins has become highly efficient with the advances of the solid-phase peptide synthesis [1] and the recombinant technology. Solid-phase peptide synthesis can produce, in short duration, a peptide of greater than 100 amino acids. The recombinant technology with an optimal expression system can produce proteins accurately and in large quantity. Ideally, a chemical ligation method to form proteins utilizing the efficiency of the solid-phase or recombinant method to generate specific segments would be desirable. In such a way, proteins can be engineered to contain unusual structures or nongenetically coded amino acids by a specific ligation method. A strong impediment in this approach is a lack of an efficient method for their synthesis. Recently, we have developed a "domain ligation strategy" to couple selectively two unprotected peptide segments to form an amide bond [2]. The present report describes the basic idea behind this new method and the verification of its feasibility using simple model compounds and peptides.

Results and Discussion

The general concept underlying the new method is outlined in Figure 1. It consists of three steps: (1) aldehyde initiation, (2) ring formation, and (3) O,N-acyl rearrangement. The key to the carboxyl component is an amino acid esterified to an α-hydroxyl aldehyde to form an ester α-aldehyde. This ester aldehyde masked as an acetal is coupled to the carboxyl terminal of an unprotected peptide segment by reverse proteolysis. Liberation of the acetal under acidic conditions to the aldehyde would allow its reaction with the N-terminal but β-functionalized amino acid to form either thiazolidine (e.g. Cys) or oxazolidine (e.g. Thr). The net result is that carboxyl and amino components are brought together by a ring formation leading to a well positioned and facile intermolecular O,N-acyl rearrangement to form the desired amide bond.

Aldehydes condense with amines in proteins to form imines that are unstable and reversible in aqueous solution unless a ring is formed. We made use of the reversibility of the reaction of aldehydes with amines and the stability of ring formation with β-functionalized amines in designing our strategy. Reactions of amino acids containing the ester-aldehyde with cysteine derivatives showed that the formation of the cyclic thiazolidine derivatives occurred in a wide pH range. To avoid hydrolysis of the ester and the unwanted side reaction of aldehyde with other amino

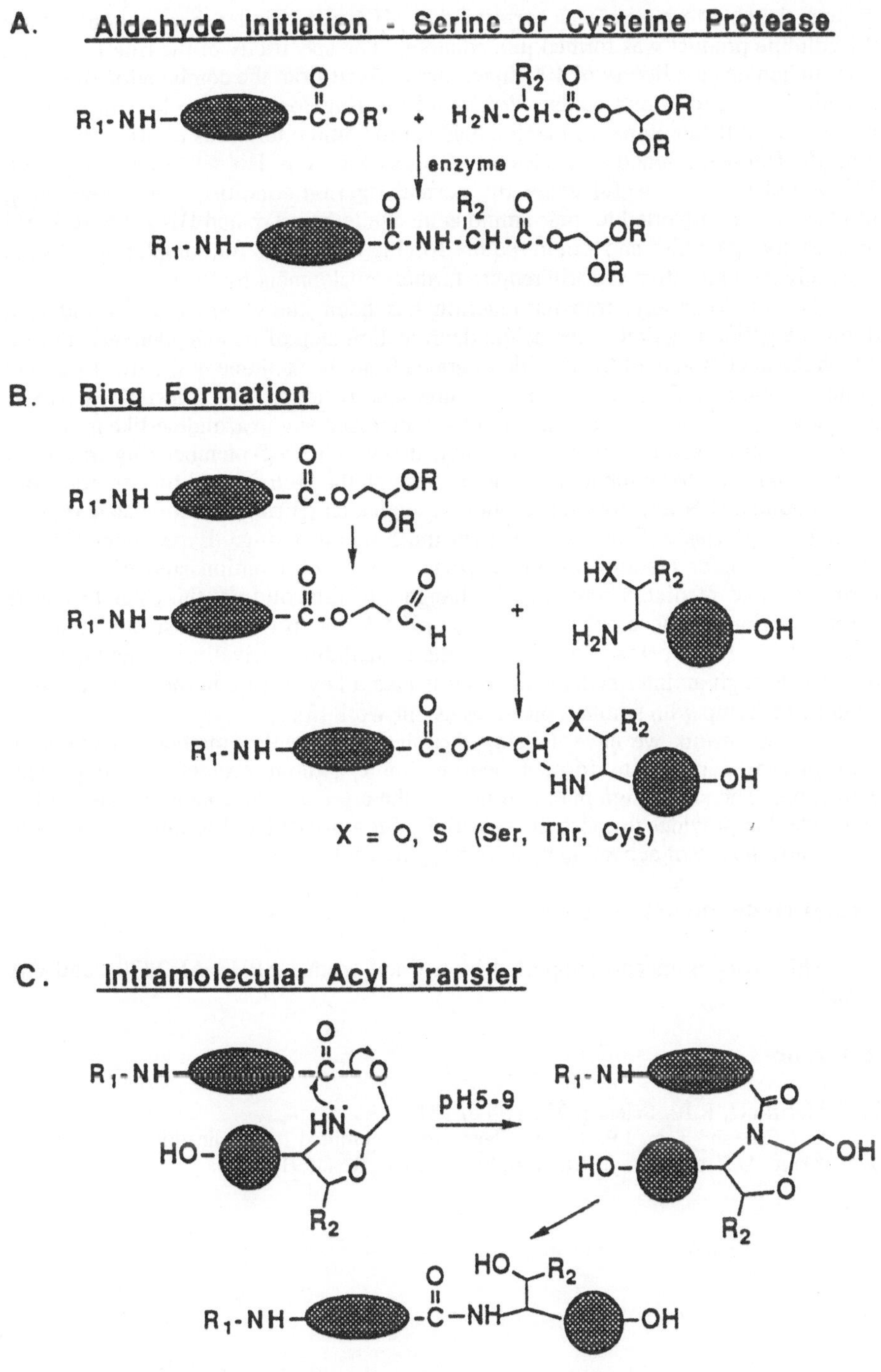
A. Aldehyde Initiation - Serine or Cysteine Protease
enzyme
B. Ring Formation
X = O, S (Ser, Thr, Cys)
C. Intramolecular Acyl Transfer
pH5-9

Fig. 1. Principle of domain ligation strategy.

groups, the ring formation was performed at pH 4 to 5. Under these conditions the thiazolidine product was formed immediately. The specificity of the ring formation was studied using a library of 400 dipeptides derived from the combination of the 20 genetically coded amino acids. We found the ring formation to be specific and limited to six β-functionalized amino acids at the amino terminus of 400 dipeptides with the following reactivity: Cys >>> Thr >> Ser, Trp, His >>>> Asn. Cys and Thr would be most useful under the normal ligation conditions and whose ring products can be reverted to their amino acids while Ser, Trp and His are > 100 fold less reactive than Cys and would require forcing conditions for completion. Asn is generally too unreactive and will require further development for its use.

The O- to N-acyl transfer reaction has been known since 1923 and is a dominating side reaction in the acidic deprotection step of peptide synthesis during which the acyl moiety of the peptide migrates from the α-amine to the free hydroxyl group on the side chain of a serine or threonine residue. The transfer reaction is reversible upon base treatment, involving a 5-member ring oxazolidine-like transition state. We proposed a similar rearrangement involving a 5-member ring transition state for our ester to amide transformation through the O- to N-acyl transfer reaction. We found that O,N-acyl transfer reaction through a larger ring transition state (e.g. 6-member ring) was >30 fold slower than the 5-member ring. Experimentally, the O,N-acyl transfer reaction in our strategy was a continuation step of the ring-formation and mediated through pH change. In thiazolidine, the ester to amide transformation was accomplished between pH 5-9 due to weakly basic amine of the thiazolidine. Basic pH was required for the oxazolidine derivatives. The O,N-acyl transfer through an intermediate size ring is also a key feature in the Thiol Capture strategy of Kemp with similar objectives as our work [3].

In conclusion, we have developed a chemical ligation method for coupling segments of peptides to form a peptide bond without protecting groups and activation. The key design in our strategy is the ester linkage containing an α-alkyl aldehyde that provides the absolute selectivity for a particular N-terminal amino acid and avoids the use of activation by a coupling reagent.

Acknowledgements

This work is in part supported by a grant from USPHS AI 28701 and CA 36544.

References

1.   Merrifield, R.B., Science, 232 (1986) 341.
2.   Liu, C.F. and Tam, J.P., J. Am. Chem. Soc., submitted for publication.
3.   Kemp, D.S. and Carey, R.I., Tetrahedron Lett., 32 (1991) 2845.

Preparation of synthetic polymers selective for amino acid derivatives and peptides by molecular imprinting

M. Kempe and K. Mosbach

*Pure and Applied Biochemistry, University of Lund,
P.O. Box 124, S-221 00 Lund, Sweden*

Introduction

Molecular imprinting, sometimes referred to as template polymerization, is a technique for preparing recognition sites of predetermined specificity in synthetic polymers [1-3]. The print molecule, also called template molecule, is either non-covalently prearranged to functional monomers or covalently coupled to them. After polymerization, the polymer is ground and sieved to particles of appropriate size. The print molecule is removed by either extraction or chemical cleavage. This results in a polymer with specific recognition sites, complementary to the print molecule in positioning of the functional groups and in shape. Due to the cognitive properties of the polymer, it is able to rebind selectively the print molecule (Figure 1).

This technique has proven to be useful in preparation of chiral stationary phases for HPLC, since the polymers are selective for the enantiomer used as print molecule. The non-covalent approach of molecular imprinting has been used for preparation of polymers selective for amino acid derivatives [4-6] and pharmaceuticals [7].

Results and Discussion

Molecularly imprinted polymers, selective for various amino acid derivatives and peptides, were prepared by using functional monomers such as methacrylic acid [8] and 4-vinylpyridine [9]. The polymers were used as stationary phases and evaluated by HPLC.

Dissociation constants and number of binding sites were determined by frontal chromatography of a Boc-L-Phe-OH-imprinted polymer. The dissociation constants for Boc-L-Phe-OH and Boc-D-Phe-OH were determined to be 6.3 mM and 8.1 mM, respectively, which means that the polymer has higher affinity for the L-enantiomer than for the D-enantiomer. The polymer was shown to contain an equal number of sites for each enantiomer (28 μmol/g dry polymer) [8].

The ligand selectivity of this type of polymer was shown to be high. Despite the small difference between Cbz-aspartic acid and Cbz-glutamic acid, the polymer is able to discriminate between these molecules. A polymer imprinted against Cbz-L-Asp-OH resolved Cbz-D,L-Asp-OH but not Cbz-D,L-Glu-OH, and vice versa when Cbz-L-Glu-OH was used as print molecule [9].

221

To obtain a stereospecific recognition, a three-point interaction is necessary [10]. In the case of a Cbz-L-phenylalanine-imprinted polymer prepared with methacrylic acid as functional monomer, it is believed that the carboxy-functions of the polymer interact via hydrogen bonds with the carboxy- and carbamate-functions of the print molecule.

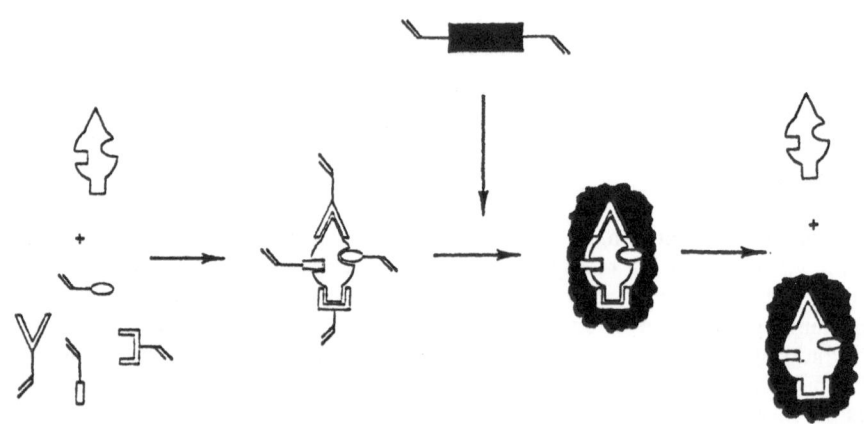

Fig. 1. Schematic representation of the concept of molecular imprinting. Functional monomers are mixed with the print molecule. The print molecule and the monomers prearrange and form a solution adduct. Crosslinker is added and the polymerization is initiated. A rigid bulk polymer is formed. The functional groups of the polymer interact with the functional groups of the print molecule. The print molecule is extracted from the polymer, leaving specific recognition sites, complementary to the print molecule in positioning of the functional groups and in shape.

However, this gives only two interaction points. Separation of Boc-D,L-Phe-OH, Fmoc-D,L-Phe-OH and Cbz-D,L-Ala-OH on this polymer resulted in lower separation factors than for the separation of Cbz-D,L-Phe-OH. This implies that both the N-protecting group and the amino acid side chain are recognized by the polymer. Separation of Boc-D,L-Phe-p-nitroanilide and Boc-D,L-Phe-Gly-OEt on a polymer imprinted against Boc-L-Phe-anilide resulted in lower separation factors than that obtained for the separation of the racemate of the print molecule. This implies that also the C-protecting group is recognized by the polymer.

Recently, successful attempts to improve the separation and resolution factors of molecularly imprinted polymers have been performed. For example, Cbz-L-Ala-L-Ala-OMe and Cbz-D-Ala-D-Ala-OMe were resolved on a Cbz-L-Ala-L-Ala-OMe-imprinted polymer with a separation factor of 3.19 and a resolution factor of 4.50 (Figure 2).

We believe that these cognitive polymers are promising alternatives to conventional chiral stationary phases, since they are inexpensive, easy to prepare and possess excellent chemical and physical stability.

Molecular imprinting offers a way of preparing recognition sites of predetermined specificity in synthetic polymers. Functional monomers are polymerized in presence of a print molecule, which is subsequently extracted from the polymer, leaving recognition sites complementary to the print molecule. The polymers possess high ligand- and enantioselectivity.

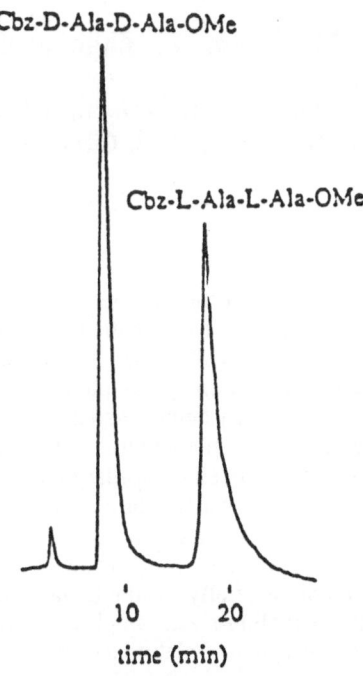

Fig. 2. *Separation of a mixture of 0.1 mg Cbz-L-Ala-L-Ala-OMe and Cbz-D-Ala-D-Ala-OMe on a Cbz-L-Ala-L-Ala-OMe-imprinted polymer. Column size: 250 mm x 5 mm. Eluent A: Chloroform/acetic acid 99.75/0.25. Eluent B: Chloroform/acetic acid 8/2. Gradient: 0-10 min, 0% B; 10-18 min, 0-5% B; 18-22 min, 5% B; 22-24 min, 5-0% B.*

Acknowledgements

Travel awards from Kungliga Fysiografiska Sällskapet in Lund and the American Peptide Society are gratefully acknowledged.

References

1. Arshady, R. and Mosbach, K., Makromol. Chem., 182 (1981) 687.
2. Ekberg, B. and Mosbach, K., Tibtech, 7 (1989) 92.
3. Wulff, G., Am. Chem. Soc. Symp. Ser., 308 (1986) 186.
4. Sellergren, B., Lepistö, M. and Mosbach, K., J. Am. Chem. Soc., 110 (1988) 5853.
5. O'Shannessy, D.J., Ekberg, B. and Mosbach, K., Anal. Biochem., 177 (1989) 144.
6. Andersson, L.I. and Mosbach, K., J. Chromatography, 516 (1990) 313.
7. Fischer, L., Müller, R., Ekberg, B. and Mosbach, K., J. Am. Chem. Soc., 113 (1991) 9358.
8. Kempe, M. and Mosbach, K., Anal. Lett., 24 (1991) 1137.
9. Kempe, M., Fischer, L. and Mosbach, K., J. Mol. Recogn., 6 (1993) 25.
10. Dalgliesh, C.E., J. Chem. Soc., 137 (1952) 3940.

Development of a highly miniaturized peptide sequencer

K.C. Waldron, M. Chen, Y. Zhao and N.J. Dovichi

Department of Chemistry, University of Alberta, Edmonton, Alberta, T6G 2G2, Canada

Introduction

Many fundamental techniques in the biosciences are limited by the minimum amount of product that can be detected. In conventional peptide sequencing by Edman degradation, for example, detection limits for phenylthiohydantoin (PTH) amino acids are in the low picomole range; at least one picomole of peptide is normally required for sequence analysis. We have recently demonstrated sub-femtomole detection limits for PTH amino acids using laser-based photothermal absorbance detection [1]. Fast, efficient separation is achieved by micellar capillary electrophoresis (CE); analysis times are less than fifteen minutes for a standard sample containing nineteen PTH amino acids. Compared with traditional HPLC methods, CE/laser-based PTH analysis provides a potential 1000-fold improvement in sequencing sensitivity. Direct interface of CE with commercially available gas-liquid phase sequencers is, however, not feasible because nanolitre-sized samples are required for CE/laser-based analysis. We have designed, instead, a miniature gas-phase peptide sequencer similar in operation to the ABI 470A Sequencer [2]. The volume requirements for phenylisothiocyanate (PITC) degradation in the miniaturized sequencer match those for the CE system. This instrumentation will prove useful for sequencing peptides and proteins that can only be isolated and purified in trace quantities.

Results and Discussion

The highly miniaturized peptide sequencer consists of three parts: a small-volume reaction chamber, in which PITC degradation chemistry is performed, a small-volume conversion chamber, in which anilinothiazolinone (ATZ) amino acids are converted to the more stable phenylthiohydantoin (PTH) form, and a CE separation column. Figure 1 is a schematic of the miniaturized sequencer. Reagents and solvents are delivered via capillary tubes to an electrically actuated multiposition distribution valve (V1). The reaction chamber volume (0.2 mL) is defined by the area of Polybrene-treated glass fibre filter onto which the peptide is immobilized, and by the length of tubing heated by thermoelectric modules. Coupling of PITC to the peptide and cleavage of the N-terminal residue are performed in the reaction chamber. Cleaved ATZ amino acid is extracted to the conversion chamber which is defined by the tubing volume between the thermoelectric modules (20 nL) and is also part of the separation column. After conversion, the entire amount of PTH amino acid is directly analysed from the conversion chamber. Electrophoresis proceeds from pneumatically actuated valve V2 to valve V3 with on-column laser-based detection of the cleaved residue just before V3. Details of the detection scheme have been previously described [1].

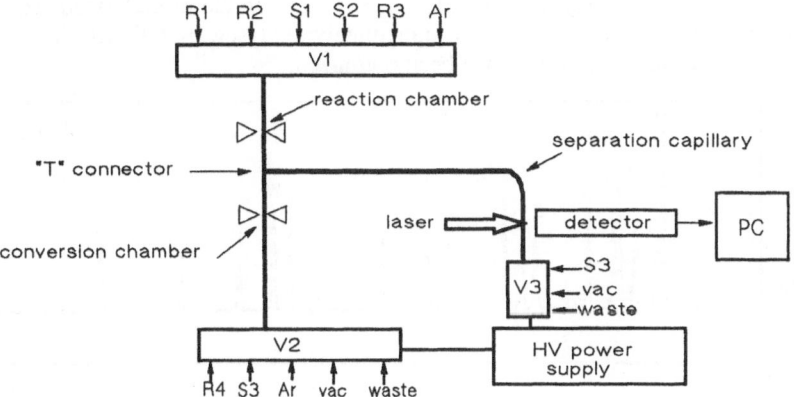

Fig. 1. Miniaturized peptide sequencer, R1: 3% PITC in heptane, R2: 12.5% trimethylamine, R3: anhydrous TFA, R4: 25% TFA, S1: ethyl acetate, S2: benzene, S3: separation buffer, V1-V3: multiposition distribution valves, ><: thermoelectric modules for rapid heating/cooling. Reaction chamber (40 x 0.4 mm ID), conversion chamber (120 x 0.075 mm ID) and separation column (500 x 0.050 mm ID) are fused silica capillaries. Vacuum (vac) is used to aid in solvent removal. The HV (high voltage) power supply is operated at 10 to 15 kV to drive electrophoresis. Data are recorded directly on a personal computer.

Development of the miniature peptide sequencer has proceeded in five stages:

1. The electrophoretic separation was optimized to provide baseline resolution of 19 PTH amino acids plus two degradation by-products, DPTU (*N,N'*-Diphenylthiourea) and DMPTU (1-Dimethyl-3-phenyl-2-thiourea) (Figure 2A). A simple program was written to normalize analyte retention times relative to DPTU resulting in less than 1% relative standard deviation in retention time for peak identification.

2. The CE/laser-based detection system was tested with residues obtained from both manually and commercially sequenced peptides [3]. Some residues from the ABI 470A and ABI 477A instruments were difficult to identify because concentrations were near the detection limit. Only 1 to 2 nL of 1000 nL sample (e.g., 0.1%) could be injected into the CE system. This is in contrast to HPLC analysis where 50% of the sample is injected for analysis.

3. Electrophoresis through pneumatically actuated multiposition valves and through a "T" connector were optimized. Several types of zero-dead-volume fittings produced band broadening in the separation so we decided to design our own miniature "T" fitting.

4. The coupling and cleavage steps of PITC degradation were performed in the miniaturized sequencer reaction chamber with ATZ residues extracted into micro-tubes and manually converted to the PTH form. Sample was then manually injected into the CE system. Figure 2B shows the results of the first cycle of PITC degradation of 50 pmol of bovine insulin chain B. Compared to traditional HPLC analysis of

degradation products, the electropherogram in Figure 2B is very clean and represents only 0.1% of the total residue cleaved from the peptide. We are currently optimizing the instrument to extend the number of degradation cycles.

    5. The conversion of ATZ to PTH is performed in the miniaturized sequencer conversion chamber and product is immediately analysed by CE with laser-based detection. This last step has not yet been optimized.

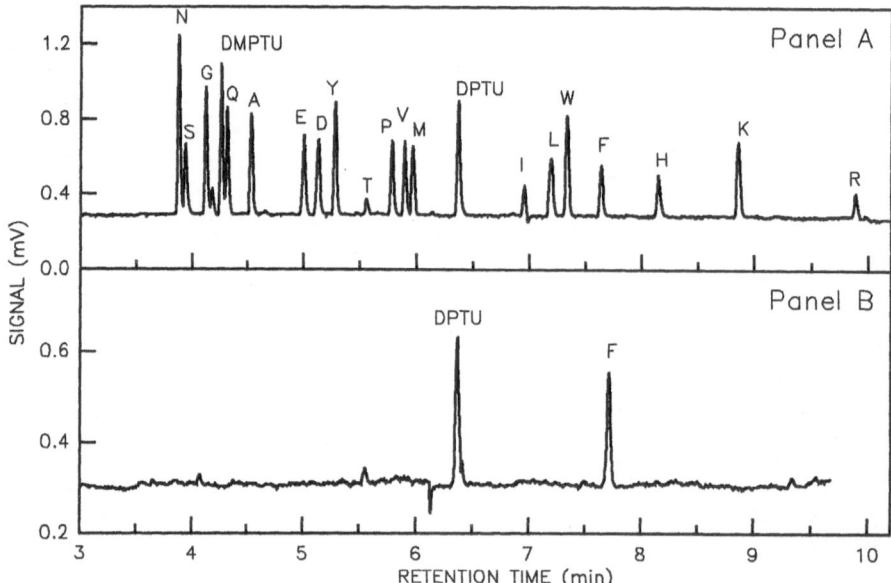

Fig. 2. Panel A shows the CE separation at 250 V/cm of 19 PTH amino acids (labeled by their single letter codes), DMPTU and DPTU. Separation buffer: 25 mM SDS, 10.7 mM sodium phosphate and 1.8 mM sodium tetraborate, pH 6.7. Panel B shows the result of one degradation cycle of 50 pmol insulin chain B (see text). Only 1 nL of solution was injected--the analyte peak represents 50 fmol of PTH-Phe (F).

    Careful adjustment of the delivery times for reagents/solvents, the temperatures for coupling/cleavage/conversion, and the conditions for electrophoresis from the conversion chamber will provide equivalent or better cycle performance to commercial sequencing instruments, but on femtomole-sized peptide samples. Given the low volume requirements of the miniaturized sequencer, we predict cycle times of 20-30 minutes, including residue identification. We have recently separated seventeen fluorescein thiohydantoin (FTH) amino acids by CE with laser-induced fluorescence (LIF) detection [4]. Detection limits are on the order of 10^{-21} mol. Incorporation of double coupling with FITC/PITC and analysis by CE/LIF detection will provide sequence information on attomole peptide samples.

References

1. Waldron, K.C. and Dovichi, N.J., Anal. Chem., 64 (1992) 1396.
2. Smillie, L.B. and Carpenter, M.R., In Mant, C.T. and Hodges, R.S., (Eds.), HPLC of Peptides and Proteins: Separation, Analysis and Conformation, CRC Press Inc., Boca Raton, FL, U.S.A., 1991, p.875.
3. Waldron, K.C., Chen, M., Zhao, Y., and Dovichi, N.J., University of Alberta, Edmonton, AB, Canada, unpublished data.
4. Wu, S. and Dovichi, N.J., Talanta, 39 (1992) 173.

Separation and determination of molecular masses of histone H1 subtypes from calf thymus

R.G. Berger[1], R. Hoffmann[1], D. Waidelich[2], E. Bayer[2], A. Ingendoh[3], F. Hillenkamp[3] and M. Zeppezauer[1]

[1]Fachrichtung Biochemie, Universität des Saarlandes,
Postfach 1105, D-66041 Saarbrücken, Germany
[2]Institut für Organische Chemie, Universität Tübingen,
Auf der Morgenstelle 18, D-72076 Tübingen, Germany
[3]Institut für Medizinische Physik, Universität Muenster,
Robert-Koch-Str. 31, D-48149 Münster, Germany

Introduction

There have been many attempts to separate the H1-histones from calf thymus into their subtypes using reversed phase chromatography (RPC) [1]. So far none of the reported homogeneous fractions obtained by chromatography or capillary electrophoresis (CE) has been unambiguously identified in terms of molecular mass and amino acid sequence. Here we introduce a combination of RPC and hydrophilic interaction chromatography (HILIC) to separate the subtypes of histone H1. Starting with material prepared according to Pehrson and Cole [2] different acetonitrile/water (0.1% trifluoroacetic acid (TFA)) gradients and flow rates were optimized for the repetitive RP-columns: Nucleosil 100Å C18 5μm, Eurosil Bioselect 200Å C18 10μm, Compax 300Å C4 5μm, Nucleosil 120Å C18 3μm, all 250 mm x 4 mm ID and a polymer column PRP Eurogel 250Å 5μm 120 mm x 4 mm, to get the best resolution of the H1 subtypes from calf thymus. HILIC is a mixed-mode hydrophilic and ionic interaction chromatography developed by Alpert [3]. We used a sodium perchlorate gradient with 80% acetonitrile/0.025% TFA on a strong cation-exchange column (TSK SP 5 PW). The collected fractions from HILIC were then used as samples for the RPC, where they were characterised by their retention times, desalted and separated further in one step. The purified subtypes were characterised by SDS-PAGE, ion-spray mass spectrometry (ion-spray MS) and matrix-assisted laser desorption/ionization mass spectrometry (MALDI-MS).

Results and Discussion

All tested C18 columns and the polymer column yielded two fractions with the following pattern in SDS PAGE: the first fraction resulted in the upper band and the second fraction in double bands, typically for histone H1 from most species. The best results were obtained with a gradient of 0.2% acetonitrile/min and 1ml/min. With shallower gradients or lower flow rates the fractions were broader without better resolution. In agreement with theory the tailing for the very basic histones was lower using the polymer column. Dividing the second peak in two fractions, namely the ascending and the descending part, the ascending yields the upper and the descending part the lower band. With the C4 phase we obtained two not baseline-separated peaks with similar electrophoretic behavior as the material from other RP-columns. But in contrast to the C18-columns the high mobility group (HMG) proteins and the core histones were not well separated from the H1 histones. Therefore we chose Nucleosil 100Å C18 5μm for the subtype separation as second step after HILIC. The RP-fraction between the two H1 fractions is ubiquitin characterised by SDS-PAGE, ion-spray MS, MALDI MS and protein sequencing. With HILIC we obtained three histone H1 fractions and all of them showed either the upper or lower single band in SDS-PAGE. One of these HILIC fractions could be separated into two peaks by RPC. As a result we obtained at last four histone H1 fractions by this chromatographic combination. The first histone fraction obtained by RPC showed at least four subtypes with ion-spray MS and MALDI MS with molecular weights between 22500 and 22950 D. The signals for the second fraction from RPC were so complex in ion-spray MS that they could not be evaluated. The evaluation from the MALDI MS revealed at least five subtypes with molecular weights between 21150 and 22300 D. After RPC fraction 3 from HILIC resulted in one pure subtype with the molecular weight of 21365 ± 2.4 D as determined by ion-spray MS.

The above described two-dimensional chromatography makes it possible to obtain four fractions from histone H1 from calf thymus including one pure subtype. With an additional chromatographic step (IEC) it should be possible to achieve further separation. Mass spectrometry has proven to be an indispensable tool for assessing the homogeneity of obtained chromatographic fractions and determining the molar masses of their constituents.

References

1.   Lindner, H., Helliger, W. and Puschendorf, B., Anal. Biochem., 158 (1986) 424.
2.   Pehrson, J.R. and Cole, R.D., Biochemistry, 20 (1981) 2298.
3.   Alpert, A.J., J. Chromatogr., 499 (1990) 177.

Amino-terminal serine to glycine post-translational modification observed in nerve growth factor biosynthesized in Chinese hamster ovary cells

E. Canova-Davis, V.T. Ling, M.L. Eng and S.M. Skieresz

Genentech, Inc., Medicinal and Analytical Chemistry, 460 Point San Bruno Boulevard, South San Francisco, CA 94080, U.S.A.

Introduction

Recombinant proteins are often biosynthesized in mammalian cells to effect necessary post-translational modifications. The most commonly required reactions include disulfide linkages to aid in proper folding and glycosylation at asparagine and/or serine/threonine residues. However, vigilance must still be exerted to insure that the secreted protein contains the proper amino-terminal amino acid confirming that no nonspecific processing by proteases has occurred. In recent studies, an unexpected post-translational amino-terminal serine to glycine conversion has been observed in human recombinant nerve growth factor isolated from Chinese hamster ovary (CHO) cells. Utilization of the technique of electrospray mass spectrometry was crucial to the detection and identification of this variant.

Results and Discussion

Recombinant human nerve growth factor is a dimer secreted from CHO cells with its six cysteine residues correctly linked to form three disulfide bridges. Care was taken to closely examine the integrity of both the amino and carboxyl termini since it has been reported that the 120-residue monomer can be reduced by limited enzymatic digests to 118 or 117 residues [1]. Electrospray mass spectrometry was instrumental in corroborating these cleavages as proposed from amino acid composition data. Hence, a careful analysis of an electrospray mass spectrum was conducted. A species with a mass of 30 units less than expected was observed (Figure 1). This difference is consistent with a serine to glycine conversion. Sequence analysis suggested that this modification occurred at the amino-terminal serine. A mass spectral analysis of the amino-terminal peptides released by trypsin digestion and isolated using reversed-phase chromatography revealed the presence of two species of masses 1068.1 and 1038.1 corresponding to SSSHPIFHR and GSSHPIFHR, respectively, in approximately a 4:1 ratio. A tandem mass spectral analysis of the two parent ions located the amino-terminal residue as the site of the conversion.

A chemically synthesized amino-terminal peptide SSSHPIFHR was covalently attached to polystyrene beads and incubated for 7 days at 37°C with the supernatant fluid from a CHO cell culture. An increase in glycine and a decrease in serine in the first cycle of a sequence analysis was observed in comparison to an unincubated sample indicating a probable post-translational mechanism. This observation is consistent with the involvement of the enzyme serine hydroxymethyltransferase [2].

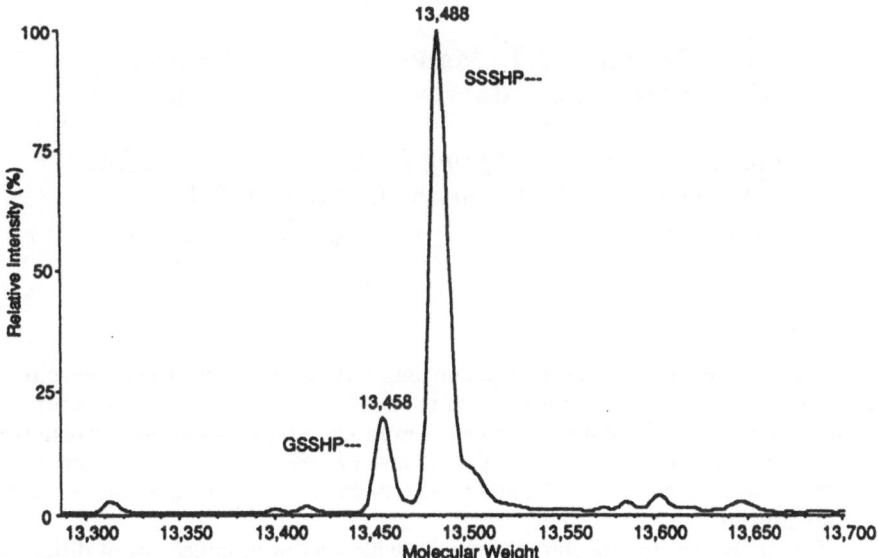

Fig. 1. Electrospray mass spectrum of a recombinant nerve growth factor preparation.

Acknowledgements

The authors wish to thank Dr. Michael Mulkerrin for conducting the CHO cell culture experiment.

References

1. Schmelzer, C.H., Burton, L.E., Chan, W.-P., Martin, E., Gorman, C., Canova-Davis, E., Ling, V.T., Sliwkowski, M.B., McCray, G., Briggs, J.A., Nguyen, T.H. and Polastri, G., J. Neurochem., 59 (1992) 1675.
2. Schirch, L., Adv. Enzymol. Relat. Areas Mol. Biol., 53 (1982) 83.

Optimization of preparative RP purification of peptides by correctly matching media pore size and capacity to the needs of the purification

P.G. Cartier[1], J.J. Maikner[1], K.C. Deissler[1], J.R. Fisher[2], L.E. Barstow[3] and G.E. Fuentes[3]

[1]Rohm and Haas Co., Spring House, PA 19477, U.S.A.
[2]TOSOHAAS, Montgomeryville, PA 18936, U.S.A.
[3]Protein Technologies, Inc., Tucson, AZ 85719, U.S.A.

Introduction

An understanding of the key parameters - resolution, speed and capacity - affecting reversed-phase (RP) purification is essential in order to design an optimum purification process. Such understanding is even more important to the preparative purifier as the preparative purifier might not have the resources to use high pressure, high theoretical plate count media and must rely on choosing the optimum medium pressure packing.

This study uses frontal chromatography with a series of molecules of different molecular weight to assist optimization. The frontal loading runs provide data directly useful for choosing sample loading conditions and since frontal performance is related to "effective" pore size, allows the selection of the optimum pore size resin for purification. The experimental results are summarized in a grid (Figure 1) that facilitates the choice of the optimum resin for the purification of a given sized molecule. Values of key purification parameters - flow rate, maximum capacity and pressure drop - are given for each resin in the grid.

Results and Discussion

For a small molecule like cephalosporin C (415 D) kinetic capacity to 1% leakage is independent of flow rate and average pore diameter (APD) and proportional to surface area. For the larger glycopeptide antibiotic vancomycin (1449 D), the 150 Å APD resin is seen to have pore diffusion limitations. With molecules the size of insulin (5700 D) and BSA (66,000 D) fast chromatography is possible with the 1000 Å resin, for high capacity the 300 Å resin must be operated below about 150 cm/hr. For BSA and insulin sized molecules the 150 Å resin offers the potential for the removal of small molecular impurities. Above 200 cm/hr large molecules will pass through the column unbound while considerable capacity remains for low molecular weight impurities (Figure 2).

	CG-161md				CG-300md				CG-1000sd			
	Ceph C	Vanco-mycin	INS	BSA	Ceph C	Vanco-mycin	INS	BSA	Ceph C	Vanco-mycin	INS	BSA
Average Pore Diameter	150Å				300Å				1000Å			
Capacity												
Dynamic Saturation Capacity (1)	102	87	91	43	72	50	97	74	22	25	38	34
Max. Loading Flow Rate (2)	769	153	38	38	769	769	153	38	769	769	769	769
Capacity to 1% Leakage at Max. Loading Flow	88	67	26	7	60	40	70	57	20	22	36	25
Purification Flow Rate Range (2)	38-150				38-150				38-769			
Pressure Drop												
at 150 cm/hr water	16 psig / Meter				6 psig / Meter				6 psig / 15 cm bed			
50/50 MeOH/Water	40 psig / Meter				16 psig / Meter				12 psig / 15 cm bed			

(1) All capacities are in mg/ml of resin   (2) All flows are empty column linear velocity/cm/hr

Fig. 1. Resin performance grid: The grid lists the resin physical properties of the resins studied along with column capacities and recommended values of key purification parameters.

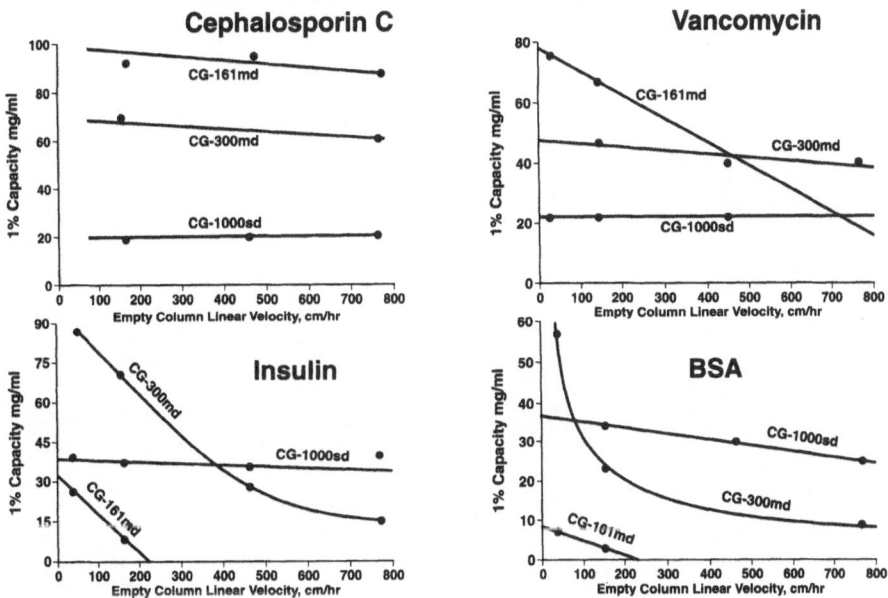

Fig. 2. Capacity of 1% leakage as a function of linear velocity: The effect of loading velocity on the adsorption capacity at 1% breakthrough of a series of molecules of different sizes is shown. The capacity at 100% breakthrough (dynamic saturation capacity, Figure 1) is the same at all velocities. The influent concentration was 5 mg/cc (10 mg/cc for cephalosporin C;); influent pH was 2-3.

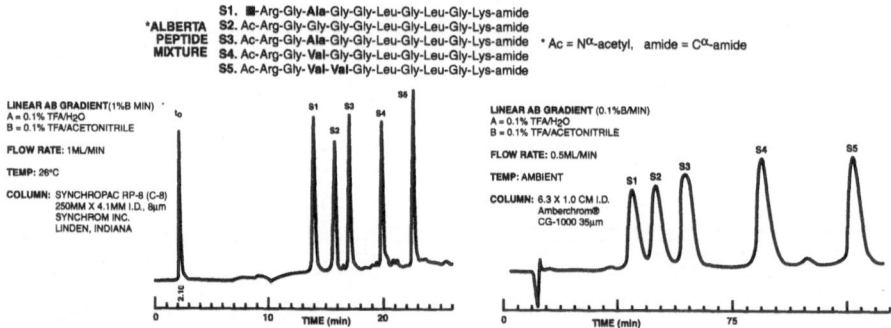

Fig. 3. Effect of particle size on the resolution of the Alberta peptide mixture: The resolution of an 8 µm silica base and a 35 µm styrenic resin.

The Alberta peptide mixture is well resolved on a 378 plate column of 35 µm resin particles (Figure 3). Although a shallower gradient is used than with the 8 µm column, this shows that well packed columns of medium pressure media can resolve chemically similar molecules.

The purification of peptides and proteins by preparative and production scale liquid chromatography

G.B. Cox and H. Colin

Prochrom R & D, 54250 Champigneulles, France

Introduction

The past decade has seen a revolution in purification techniques for many peptides and proteins, both in the laboratory and in an industrial setting. High Performance Liquid Chromatography (HPLC), for example, has moved from being regarded as an expensive, primarily analytical tool to a position where it is being used more and more as a purification methodology. The reasons for this include the greatly improved understanding of the fundamentals of preparative HPLC as well as the increased need for purification techniques capable of meeting the stringent regulatory requirements for recombinant products. The situation today is that HPLC is recognized as having a major part to play in purification; it is highly cost-effective in that it is able to replace multiple steps in conventional purification processes, thus eliminating handling, labour costs etc., and it is also able to perform purifications which cannot be carried out by any other process. One of the major problems in designing preparative HPLC methodologies is that of optimization. It is not always clear exactly how best to optimize in any given situation, nor are there good guidelines as to the influence of the operating parameters on the optimum separation. This paper addresses some of the issues relevant to these questions.

Separations on a large scale must be carried out under optimized conditions in order to be most effective. The question of optimization has been addressed in a number of publications, most of which result in the optimization of production rate - i.e. to maximize the quantity of product produced in unit time from a given piece of equipment. In production liquid chromatography, the most important parameter is not the production rate, but is the cost of the process. The optimal conditions for cost minimization are not necessarily those which give the maximum production rate. In order to begin to assess how the optimization of peptide separations may be approached, computer simulations based upon the separation of bovine and porcine insulins were carried out.

Experimental. The calculations were performed assuming a total annual production of 10 metric tonnes of a peptide starting at a 95% pure product, arriving at a material of 99.5% purity. The development of competitive adsorption isotherms is the most difficult part of simulation of separations. This is especially true for peptide and protein solutes since the measurement of even the single solute isotherms is fraught with difficulty. This problem was met by finding parameters for isotherms based upon the Ideal Adsorbed Solution theory which allowed computer simulations that resulted in predictions which matched the observed behaviour of the insulins studied. The mobile phase was a gradient of 27 to 31% acetonitrile in 10 column volumes. Log k' and S values were taken to be those previously determined for the insulins (5.31 and 14.9 for "peptide 1" and 5.48 and 15.2 for "peptide 2"). Saturation capacities of 30 and 45 respectively were employed in the calculations. The diffusion coefficient for the peptides was taken to be 2.10^{-11} and the Knox equation coefficients A, B and C were 1,2 and 0.1 respectively. Operation at 20 bar pressure was assumed. Calculations were made using proprietary Prochrom software using a 66 MHz 80486-based personal computer.

The costs of solvent were taken to be $5/1 for acetonitrile and $0.1/1 for the cost of recovered acetonitrile. Packing materials were estimated to be $6000/kg for 10μm packings and $3500/kg for 15μm material. Labour cost was set at $30/hour and a total of 7000 working hours were assumed for the production. Equipment costs were for Prochrom Industrial Scale HPLC systems of appropriate size (from 20 to 100 cm id columns) and were amortized over a 10 year period. The cost of crude product was either $100, $1000 or $5000/kg.

Results and Discussion

Figure 1 shows the results of cost calculations for a column 33 cm in length packed with 10μm particles for three different values of crude cost. The cost of the process in terms of cost per kilo of purified product is plotted against the recovery of the product. Since the required annual production is fixed, the recovery is changed by adjusting the loading in the column by changing the diameter. This in turn changes the quantity of solvent and packing material used in the process. The cost of the loss of the crude material is balanced against the cost required to achieve a given recovery. The optimum for the low cost products occurs at recoveries considerably below 90%.

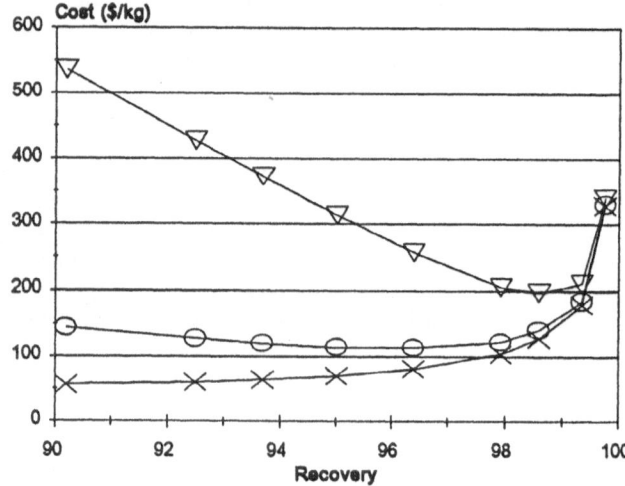

Fig. 1. Plot of product cost vs recovery at crude costs of $100/kg (crosses), $1000/kg (circles) and $5000/kg (triangles).

Optimum production rates are found at recoveries of 60 to 70%, and the minimum cost in this instance is likely to be somewhere between this value and around 80%.

Since the curve is very flat, the optimum is broad and the actual value chosen does not matter greatly. This means that the separation can be fitted into existing equipment and custom units are not required. As the crude cost is increased (the open circles correspond to $1000/kg, the triangles to $5000/kg), the optimum cost is found at higher values of recovery. Very expensive crudes force the separation conditions to very high recoveries. This results in expensive processes since large, lightly loaded columns are used. The efficiency required for the columns is a function of the cost of the crude material.

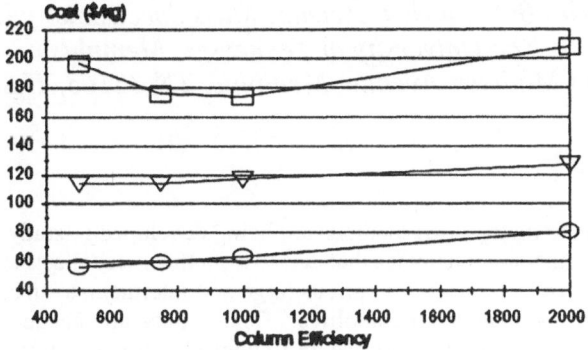

Fig. 2. Plot of product cost vs column efficiency. Legend as Fig. 1.

Figure 2 shows a plot of product cost against column efficiency for the three crude cost levels. At low crude cost, the column efficiency is very low, at around 400 plates. (Note that the efficiencies are low despite the use of 10μm particles because the high intrinsic efficiency of the column is traded for the speed of separation). Higher crude costs, which result in the need for higher recoveries, also demand higher column efficiencies. Again, the optima are relatively broad. The simulations were carried out for packing materials of both 10 and 15 μm. In all cases, for this separation, the cost of the product resulting from the 10 μm packing was less than that from the 15 μm material. This is not a general conclusion for preparative chromatography, but may be specially relevant to peptide and protein separations due to the extremely low diffusion coefficients observed.

The computer optimization of preparative peptide separations results in a number of observations. The cost of the separation process is strongly dependent upon the cost of the crude material. Low cost crudes result in low recovery of material and operation of the column at high load. Higher cost crudes require increasingly high values of recovery. This forces the loading of the column down; thus larger column sizes must be employed. This necessity increases the cost of the production. For high cost crudes, typical of biotechnological production or multiple reaction synthetic procedures, the recovery of product has to be very high and the optimum cost is a strong function of operating conditions. There is an optimum column efficiency which depends both upon the selectivity found in the separation, the displacements between the solutes and the cost of the crude material. Finally, the particle diameter of the packing material can have unexpected influence on the optimum process cost.

Electrospray ionization mass spectrometry analysis of opioid peptide precursors

D.M. Desiderio[1,2,3], L. Yan[3], G. Fridland[3] and J. Tseng[3]

*[1]Departments of Neurology and [2]Biochemistry and
[3]The Charles B. Stout Neuroscience Mass Spectrometry Laboratory
The University of Tennessee, Memphis,
800 Madison Avenue, Memphis, TN 38163, U.S.A.*

Introduction

Mass spectrometry (MS) methods are being developed to study the production of several important neuropeptides that derive from the metabolic cascade that occurs in a neuron: DNA $\rightarrow$ mRNA $\rightarrow$ various large and intermediate-sized precursors $\rightarrow$ neuropeptides $\rightarrow$ inactive metabolites. Those data are needed to clarify the metabolism of neuropeptides that occurs in control and pathophysiological tissues and fluids. Electrospray ionization (ESI) MS has been used to characterize intermediate-sized neuropeptide precursor molecules that are produced from their respective precursors, preproenkephalin A_{1-263}, which produces methionine enkephalin (ME) and proopiomelanocortin$_{1-265}$ (POMC), which produces β-endorphin (BE). "Intermediate-sized" means fewer than the *ca.* 260 amino acids in a precursor, but more than the five or 31 amino acids in ME or BE, respectively.

Results and Discussion

Bovine pituitaries (n=5) were homogenized at neutral pH with enzyme inhibitors. Gel permeation (G75) and RP-HPLC (linear gradient over 90 min of 15~60% acetonitrile in TEAF buffer pH 3.15) were used to isolate intermediate-sized precursors from bovine pituitaries. Commercial ME RIA kits were used to monitor the peaks containing ME-like immunoreactivity (ME-li) in GP and RP-HPLC fractions after treatment with CNBr and trypsin. All ME-li containing fractions were analyzed by ESI-MS on a VG Auto Spec Q (EBEqQ) mass spectrometer. The source was operated at 4 kV. The samples were introduced into the source at a flow-rate of 3 ml/min in 1:1 (v:v) $CH_3OH:H_2O$ containing 1% CH_3COOH. ESIMS analysis of one immunoreactive fraction showed three major components, with M.W. of *ca.* 6550, 8440 and 9855, respectively (Figure 1). Trypsin treatment of this mixture, when analyzed by LSIMS, showed the presence of ions at m/z 558 and 1134 (among others) (Figure 2), which correspond to the tryptic peptides BE_{20-24} (=$POMC_{254-258}$) and BE_{10-19} (=$POMC_{244-253}$). The peaks at m/z 558 and 1134 were also present in the FAB spectra from trypsin treatment of endogenous BE (M.W. 3465) in our previous studies [1]. Therefore, that peak may well be an indication that this fraction contains intermediate-sized precursors derived from POMC. At this stage, we feel that the component with a mass of 9,855 is the intermediate-sized precursor of BE.

Therefore, this HPLC fraction will be subjected to further HPLC and/or CZE separation, and the putative precursor ion at 9,855 will be studied by trypsin treatment, ESIMS, LSIMS, and MS/MS sequencing of each tryptic fragment. In that manner, we will locate the intermediate-sized precursor of 9,855 mass units within the larger POMC precursor molecule itself. We are also searching for the intermediate-sized precursors of ME by the same method. For example, ESIMS analysis of another active fraction showed the presence of a major peak with a mass of *ca.* 15,034 Da, as shown by the series of $(M+nH)^{n+}$ ions 12^+-23^+ at m/z 1254-655, and a smaller peak (*ca.* 5%) at 14,879 (Figure 3). Trypsin treatment of this fraction yielded 12 tryptic fragments, which were analyzed by LSIMS and ESIMS and which showed very small peaks at m/z 701 and 729, corresponding to the $(M+H)^+$ ions of ME-K and ME-R.

Finally, this analytical system will be used to analyze human pituitaries (post-mortem controls and post-surgical tumors).

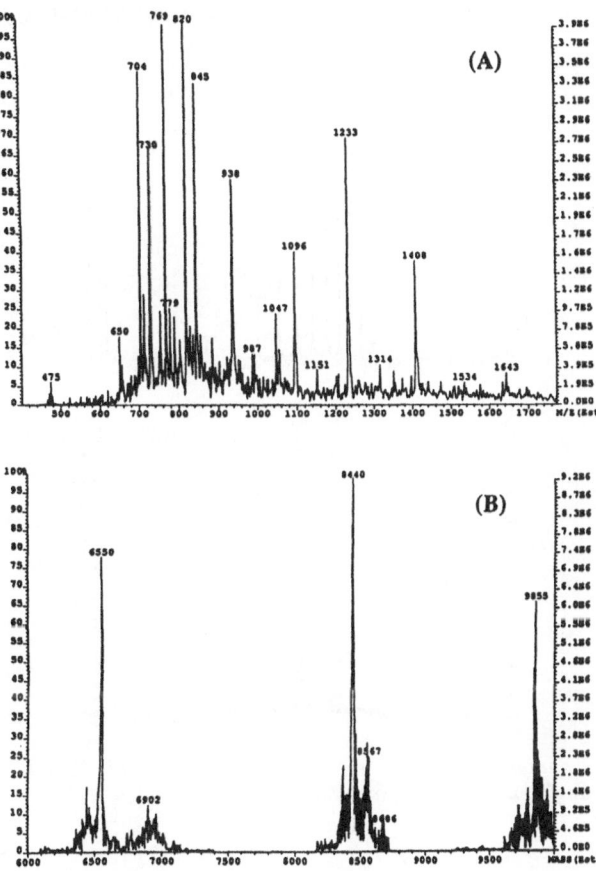

Fig. 1. ESIMS analysis of one immunoreactive fraction extracted from bovine pituitaries. (A) ESI mass spectrum; (B) transformed ESI mass spectrum.

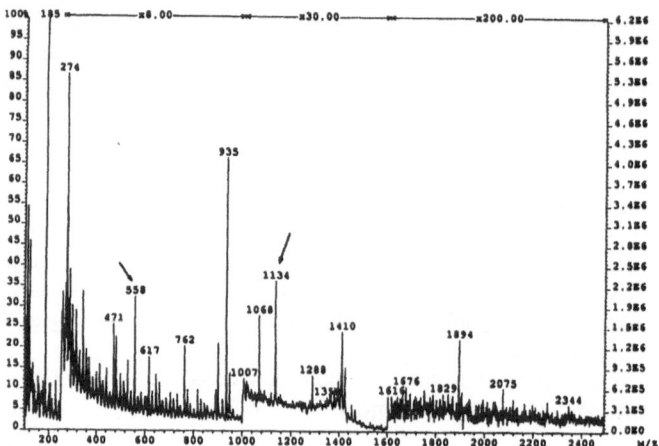

Fig. 2. LSIMS mass spectrum of an unfractionated tryptic digest that contained ME-li activity. Note the ions at m/z 558 and 1134, which correspond to BE_{20-24} and BE_{10-19}, respectively.

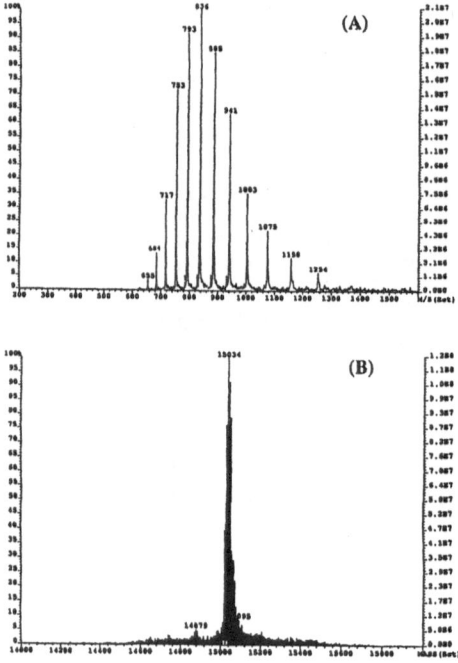

Fig. 3. ESIMS analysis of peptide extracted from bovine pituitaries. (A) ESI mass spectrum; (B) transformed ESI mass spectrum.

References

1. Dass, C., Kusmierz, J.J. and Desiderio, D.M., Biol. Mass Spectrom., 20 (1991) 130.

Unambiguous determination of the optical purity of peptides via GC-MS

J. Gerhardt[1] and G.J. Nicholson[2]

*[1]C.A.T. GmbH & Co. Chromatographie und Analysentechnik KG,
Heerweg 10-12, D-72070 Tübingen, Germany
[2]Universität Tübingen, Institut für Organische Chemie, Auf der
Morgenstelle 18, D-72076 Tübingen, Germany*

Introduction

One of the most accurate and sensitive methods for the analysis of the optical purity of amino acid residues in peptides involves hydrolysis in 6N HCl/H_2O, derivatization of the free amino acids and gas chromatographic separation of the enantiomers on a chiral stationary phase (e.g., Chirasil Val [1]). This analysis has been automated by the novel instrument DLAA1 consisting of the automatic chemical derivatizer Autoderivat 100/2 (C.A.T.), gas chromatograph GC-14/1 and autoinjector AOC-14 (both Shimadzu) [4]. The method however suffers from one weakness: under the conditions of hydrolysis, racemization occurs and the amount of racemate determined represents the sum of the amount originally present in the peptide plus that generated during acidolysis.

A technique for the unambiguous quantitation of the racemate content of peptides and proteins of synthetic or natural origin was presented by Frank et al. [2]. The protein or peptide is hydrolyzed in 6N DCl/D_2O, whereby racemization is accompanied by deuterium exchange in the α-C position. The standard detector (FID or NSD) is replaced by a mass spectrometer operating in the scan mode and from the relative amounts of deuterium-labeled to non-labeled species of a characteristic ion necessarily containing the α-H/D, the proportion of racemate arising from the hydrolysis could be calculated. The contribution of the natural isotopes ^{13}C, ^{15}N, ^{17}O etc. to the intensity of the $(I+1)^+$ ion must be considered. Using multiple ion detection, Liardon et al. [3] improved the methodology, developing a series of equations with which the interference from neighboring ions $(I+1)^+$ and $(I-1)^+$ could be taken into account.

According to our experience, a simplification of the method described by Liardon but employing single-ion detection is viable. With this method, mass spectrometers not equipped with MID hardware may also be used. Using the instrumental setup as described below, it is possible to quantitate down to approximately 0.1% of D-enantiomers in peptides.

Results and Discussion

Sample preparation: With the AHST-1 station (C.A.T.), peptides and proteins are typically hydrolyzed under vacuum in 6 N deuterium chloride in deuterium oxide (isotopic purity >99.5 atom% D, Fluka, Buchs, Switzerland) for 24 hours at 110° C. If tryptophan or cystein are to be determined, 1% thioglycolic acid is added as an antioxidant. After hydrolysis, excess reagent is evaporated at 110° C using a gentle N_2 stream. 400 μl of 4N HCl in 1-propanol is added to the dry hydrolysate, the vial flushed with nitrogen before sealing, and heated to 110° C for 30 min after which excess reagent is evaporated under a nitrogen stream at 110° C. 200 μl trifluoroacetic anhydride/ethyl trifluoroacetate (1:1) are added to the dry amino acid ester hydrochlorides and the mixture is heated to 150° C for 10 min. After cooling to room temperature, the solution is carefully brought to dryness and the residue dissolved in 150 μl toluene. If histidine is to be determined, 50 μl isopropyl chloroformate are added to the sample and the vial heated to 40° C for 20 min.

Gas chromatography: The N(O,S)-trifluoroacetyl amino acid esters are separated on a deactivated glass or fused silica capillary (20 m x 0.28 mm) coated with Chirasil-Val ($d_f = 0.13$ μ) with hydrogen as carrier gas. The column is programmed from 67° C to 200° C at 4° C/min and the peaks identified by way of their retention times. Standard detection is done with FID or NSD. The derivatization procedure, combined with the GC separation may be automated with the Autoanalyzer DLAA1 (Shimadzu/C.A.T.).

Mass spectrometry: An MAT112 (magnetic field) mass spectrometer with electron impact ionization (80eV) was operated in the SIM mode. Adjustment of the magnetic field was facilitated by the accurate mass display of the data system (AMD Intectra). The SIM signal was monitored with a Trivector integrator.

Racemization during acid hydrolysis involves protonation and deprotonation of the carboxyl group with tautomeric enol rearrangement at the α-C position. If the hydrolysis is carried out in a fully deuterated environment, the racemate formed is deuterium-labeled at the α-position with a consequent shift of one mass unit of all fragments containing this moiety. The proportion of D-amino acids originally present in the peptide is thus represented by the relative amounts of the unlabeled form:

$$\text{Original racemate content} = \frac{A_{D(I)} \cdot 100}{A_{D(I)} + A_{L(I)}} \quad (\%)$$

where $A_{D(I)}$ is the peak area of the D-enantiomer measured on the ion (I).

Two prerequisites must be fulfilled :
1. It is imperative that the ion selected for monitoring $(I)^+$ includes the α-H.
2. The ion $(I-1)^+$ should be of low intensity (less than 5%) relative to the monitored ion $(I)^+$. Incorporation of deuterium would lead to its being detected together with the unlabeled ion $(I)^+$ and would result in a positive error with a value of the product of the relative intensity of $(I-1)^+$ and the degree of racemization during hydrolysis.

In Table 1, the recommended ions for the determination of each amino acid together with their relative intensity and the intensity of the ion $(I-1)^+$ relative to $(I)^+$ are listed.

Although usually only the original racemate content is of importance, the degree of racemization during hydrolysis may also be determined by subtraction of the racemate content originally present from the total racemate content determined with non-specific detection (FID, NSD or MS-TIC).

Despite its unquestionable advantages over the standard gas chromatographic method with non-specific detection, the described method has its own sources of error:

Table 1 *Ions for determination of amino acids*

Amino acid	$(I)^+$	$(I)^+$ rel. abundance	$(I-1)^+$ rel. abundance to $(I)^+$
Ala	140	100	1.5
Val	168	100	3
Ile	182	60	2.5
Pro	253	10	<4
Leu	182	45	0.7
Ser	252	3	<6
Hyp	365	1.5	<10
Asp	228	40	2.5
Met	287	26	<1.5
Phe	303	1.5	<10
Tyr	302	48	<1
Lys	180	100	3.6
Arg	435	47	<1
Trp	438	4.3	<6
Hyp	365	1.5	<10
Cl-Phe	337	7.9	<5
Nal	353	9.6	<4

The small dynamic range of 10^4 of the quadrupol and magnetic sector mass spectrometer and only 10^3 of the ion trap (cf. 10^7 for FID) is one of the main drawbacks of this analysis. In consequence, the sample has to be injected within a small concentration range to allow determination of 0.1% of the D-enantiomer. In addition the relative abundance of the various ions measured may differ by a factor of 30.

The error arising from the deuterium-labeling of the $(I-1)^+$ ion has been mentioned above. Its maximum magnitude may be estimated from the intensity of the $(I-1)^+$ ion in the mass spectrum of the non-labeled amino acid and the degree of racemization during hydrolysis. If the ion $(I-1)^+$ however results from the abstraction of the α-H, it will not contribute to the intensity of the ion $(I)^+$, and will thus not falsify the results.

The lower limit of determination will probably be governed by racemization during the acid catalysed esterification in ^{1}HCl/1-propanol. We find it possible to determine down to approximately 0.1% for most amino acids under the conditions described. Liardon noted a non-linearity in the racemate content of spiked samples at this level, which was presumably a result of racemization with concomitant de-labeling during derivatization.

References

1. Frank, H., Nicholson, G.J. and Bayer, E., J. Chrom. Sci., 15 (1977) 175.
2. Frank, H., Woiwode, W., Nicholson, G.J. and Bayer, E., Stable Isotopes, Proceedings of 3rd Int. Conference, Academic Press, 1979, p.165.
3. Liardon, R., Ledermann, S. and Ott, U., J. of Chromatogr., 203 (1981) 385.
4. Gerhardt, J., Nokihara, K. and Yamamoto, R., Peptides, Chemistry and Biology (Proceedings of the 12th American Peptide Symposium), Escom, Leiden, The Netherlands, 1992, p.531.

Purity control and sample identification of synthetic peptides by capillary electrophoresis (CE)

T. Kaiser[1], S. Stoeva[1], G. Grübler[1], H. Echner[1],
A. Haritos[2] and W. Voelter[1]

[1]*Abteilung für Physikalische Biochemie des Physiologisch-chemischen Instituts der Universität Tübingen, Hoppe-Seyler-Straße 4, D-7400 Tübingen, Germany*
[2]*Zoological Laboratory, Faculty of Science, Departement of Biologie, University of Athens, Greece*

Introduction

Due to rapid separation (up to 10-15 min), high resolution (several million theoretical plates) and high sensitivity (attomole range), CE will become an unreplaceable tool in the laboratories of peptide chemists [1 and references cited therein]. Especially, the most recent commercially available CE systems, such as the BioFocus 3000 CE (Bio-Rad, Munich, Germany) [2], offering automatic constant volume sample injection, efficient capillary temperature control, cooling or heating of the sample and fraction collection compartment and coated or uncoated capillaries in cartridges of different capillary lengths, allow rapid, reproducible and sensitive separations.

For the development of specific antibodies and epitope mapping experiments, solid phase peptide synthesis of parathymosin α fragments (26-30), (20-30), (16-30), (11-30), (7-30), (1-30) was performed, and the crude and purified products were investigated by HPLC and CE.

Ac-SEKSVEAAAELSAKDLKEKKEKVEEKASRK-OH

Fig. 1. Primary structure of parathymosin α (1-30). Underlined amino acid residues were coupled twice during peptide synthesis.

Results and Discussion

The peptide synthesis of the parathymosin α fragments (sequence 1-30, shown in Figure 1) was performed batchwise by the Fmoc/Bu^t solid phase technique, using an automatic synthesizer (Ecosin P, Eppendorf-Biotronic, Maintal, Germany) and p-benzyloxybenzyl alcohol/polystyrene/1%-divinylbenzene resin (Wang resin; 0.5g), loaded with 0.3 mM Fmoc-Lys(Boc)-OH.

A)   B)

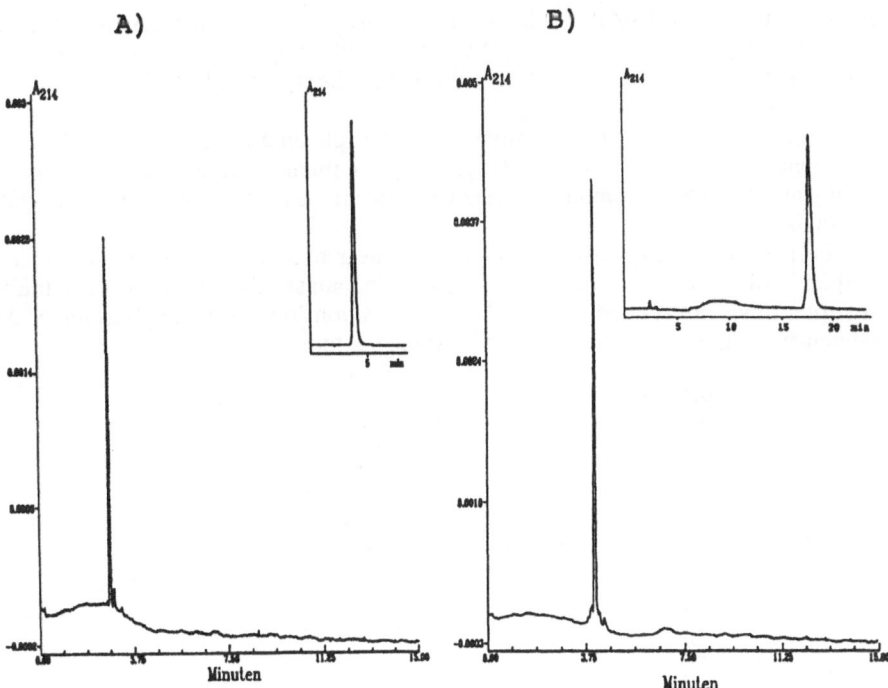

Minuten   Minuten

Fig. 2. Electropherograms and Chromatograms (RP-HPLC) of purified synthetic parathymosin α (26-30), panel A and parathymosin α (1-30), panel B.
Electropherograms: BioRad, BioFocus 3000; capillary: coated (25 μm x 24 cm); loading: 6 sec/6KV; running conditions: 10 KV; buffer: 0.1M phosphate, pH 2.5; concentration: 141 μg/100 μl(A), 129 μg/100 μl(B); detection: UV(λ = 214 nm).
RP-HPLC: high precision pump model 480 (Gynnkotek, Munich, Germany), UVIDEC-100 III UV spectrometer (Jasco, Groβ-Zimmern, Germany), detection: UV (λ = 214 nm).
Conditions A: Column: C$_2$ + C$_{18}$-PEP-S (5 μm, 4 x 250 nm; LKB-Pharmacia, Bromma, Sweden), flow rate: 1 ml/min; eluent (isocratic):" H$_2$O/TFA (1000:0,17).
Conditions B: H$_2$O/TFA (1000:0,17), B: CH$_3$CN/H$_2$O/TFA (600:400:0.15); flow rate: 1ml/min; gradient: 5-95% B in 30 min.

The Fmoc amino acids (Calbiochem-Novabiochem, Bad Soden, Germany) are coupled in 2.5 fold excess with benzotriazol-1-yl-oxy-tris-(dimethylamino)-phosphonium hexafluorophosphate (BOP; 2.5 equivalents) and diisopropylethylamine (DIEA; 3.5 equivalents). The following side chain protections were applied: Arg(Pmc), Asp(OtBu), Glu(OtBu), Ser(tBu) and Lys(Boc). The syntheses were performed in dimethylformamide (DMF), applying single and double coupling procedures (underlined in Figure 1). The Fmoc amino acids, BOP and DIEA, dissolved in DMF, were preactivated and transferred to the reaction vessel. Deprotection of the Fmoc groups were carried out with 25% piperidine in DMF (1 x 3; 1 x 15 min). The peptides (Fmoc-deprotected) were cleaved from the resin with simultaneous removal of the side chain-protecting groups in trifluoroacetic acid (TFA)/thioanisole/H_2O (4:1:0.5 v/v/v) within 3h. The peptides were precipitated in 10 fold excess ice cooled diethyl ether, collected by filtration and lyophilized from 1% acetic acid in water.

Figure 2 shows, for comparison, the chromatograms (HPLC) and electropherograms of the purified synthetic fragments (26-30) and (1-30) of parathymosin α, which demonstrate that CE separations proceed in shorter time and with higher resolution.

Furthermore, using coated capillaries, a linear relationship was found for the synthetic fragments, if their $M^{2/3}/Z$ values (M=mass, Z=charge) versus their migration times are plotted (Figure 3). This relationship can be applied for peak assignments in the electropherograms of peptide mixtures.

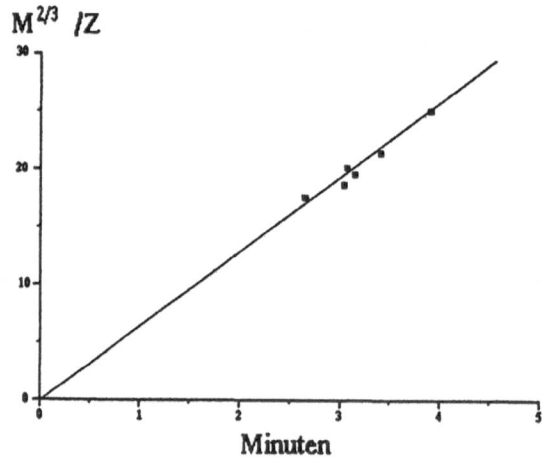

Fig. 3. *Standard curve for the synthetic peptide fragments of parathymosin α.*

References

1. Yildiz, E., Grübler, G., Hörger, S., Zimmermann, H., Echner, H., Stoeva, S. and Voelter, W., Electrophoresis, 13 (1992) 683.
2. Bio-Radiations, Bio-Rad Laboratories, Richmond, CA, U.S.A., 81 (1992) 6.

Correlation of α-helical secondary peptide structure and retention behaviour in reversed-phase HPLC

E. Krause, M. Beyermann, M. Dathe, H. Wenschuh, S. Rothemund and M. Bienert

Institute of Molecular Pharmacology, Alfred-Kowalke Str. 4, D-10315 Berlin, Germany

Introduction

Beside the overall hydrophobicity, also the secondary structure induced by reversed-phase interaction may affect the HPLC retention behaviour of peptides. The formation of an amphipathic α-helical structure induces a preferred hydrophobic binding domain and leads to a considerably stronger retention on reversed-phase columns [1, 2]. In order to evaluate the relation between the presence of amphipathic α-helical secondary structure and the RP-HPLC retention time we synthesized D-amino acid replacement sets of KLALKLALKALKAALKLA-amide and NPY 1-36, since incorporation of D-amino acids is known to destroy helical structures.

Results and Discussion

Inversion of the amino acid configuration at position Leu[4] to Ala[18] of the amphipathic α-helical model peptide KLALKLALKALKAALKLA-NH$_2$ leads to a significant drop of both, CD determined helicity as well as reversed-phase HPLC retention time (Figure 1). The effect is stronger in the middle of the molecule and is pronounced by substitution of two adjacent amino acids (DD-replacement). Replacement of an amino acid in the N-terminal region of the helical peptide with its D-enantiomer enhances helicity and leads to an increase of the hydrophobic interaction in reversed-phase chromatography.

The investigation of a DD-replacement set of NPY 1-36 has shown that substitutions in the N-terminal sequence which is known to have random coil structure have rather no influence on helicity and retention behaviour (Figure 2). In contrary, double D-amino acid substitutions in positions, which are described to be involved in the helical structure of NPY (residues 19-34 [3], or residues 15-26 and 28-35, linked by a hinge inducing a 100° angle [4]) cause a significant decrease of CD determined helicity as well as of retention time of the analogs, indicating the influence of D-amino acid substitution on the helical propensity.

It is obvious that the influence of the D-amino acid substitution position on helicity correlates with change of peptide retention data. Conclusively, HPLC retention behaviour of a D-amino acid replacement set of peptides provides an indication of the presence and location of amphipathic α-helical secondary structure.

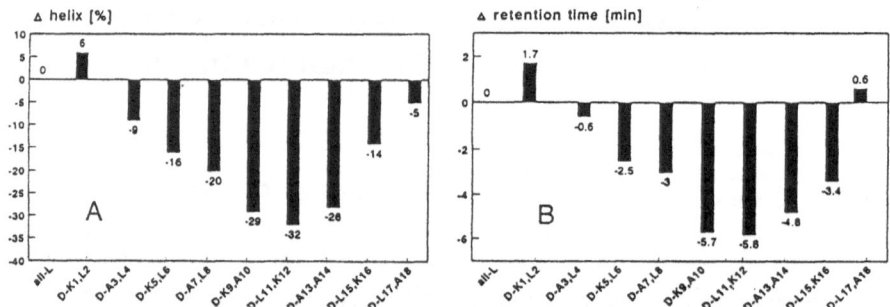

*Fig. 1. Influence of the position of D-amino acid substitution on the CD determined helicity [5] **A** and RP-HPLC retention time [6] **B** of KLALKLALKALKALKAALKLA-NH$_2$.*

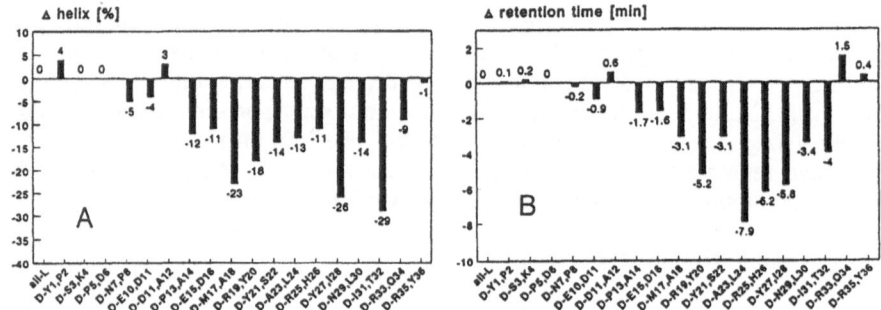

*Fig. 2. Influence of the position of D-amino acid substitution on the CD determined helicity [5] **A** and RP-HPLC retention time [6] **B** of NPY 1-36.*

References

1. Zhou, N.E., Mant, C.T. and Hodges, R.S., Pept. Res., 3 (1990) 8.
2. Steiner, V., Schär, M., Börnsen, K.O. and Mutter, M., J. Chromatogr., 586 (1991) 43.
3. Mierke, D.F., Dürr, H., Kessler, H. and Jung, G., Eur. J. Biochem., 206 (1992) 39.
4. Darbon, H., Bernassau, J.M., Deleuze, C., Chenu, J., Roussel, A. and Cambillau, C., Eur. J. Biochem., 209 (1992) 765.
5. CD measurements were carried out in 50% TFE/50% water (v/v), pH 3 over 185-260 nm (Jasco 720). The α-helicity was estimated from Θ_{222}.
6. HPLC retention times were determined on a PolyEncap A300 column (250x4.6 mm, 5 μm, Bischoff Analysentechnik GmbH). Mobile Phase A: 0.1% TFA in water, B: 0.1%TFA in 50% acetonitrile/50% water (v/v), linear gradient 5-95% B in 40 min.

Antibody binding constants by capillary electrophoresis

L.M. Martin, K.S. Rotondi and R.B. Merrifield

The Rockefeller University, New York, NY 10021, U.S.A.

Introduction

In the course of our studies of McPC603, its F_{ab} fragments, and its semi-synthetic recombinants, we have developed a capillary electrophoretic technique to measure antibody binding constants quickly and with a minimal amount of sample. Existing methods include but are not limited to equilibrium dialysis, NMR studies of the active site micro environment, sedimentation equilibrium and fluorescence quenching. Although all of these methods allow for binding constant determination, they suffer from requirements of either large amounts of time or large amounts of material or both. Previously researchers have obtained the binding constants for the McPC603 F_{ab} fragments utilizing equilibrium dialysis, the calculated values are 1.7 x 10^5 M^{-1} [1], 1.21 ± 0.06 x 10^5 M^{-1} [2] and 6.59 x 10^4 M^{-1} [3]. The only two requirements for the implementation of the capillary electrophoretic technique are conditions that suitably resolve the immunoglobulin and an antigen that possesses a charge at the pH at which the determination is made. A charged antigen affects an electrophoretic mobility shift on the binding peak (Figure 1). The experiment is done under physiological conditions, and sample volumes consumed do not exceed 1 µl of a concentrated sample. The detection is performed by UV spectroscopy, and therefore no special chemistry is required.

Results and Discussion

A common problem encountered by researchers attempting CE analysis on large proteins is adherence to the bare silica walls. A Supelco C1 coated capillary was used to overcome this problem. There are a number of advantages to using such a system, among them are decreased endoosmotic flow, decreased electrostatic attraction between positively charged proteins and the capillary wall, and the avoidance of buffer additives which might interfere with the antibody, antigen complex. Using the aforementioned capillary the buffer system which gave the best resolution was 20 mM borate pH 9.5. Our hapten, phosphorylcholine (PC) is negatively charged above pH 5 and does not absorb significantly at 210 nm. The buffers used in the binding constant determination were made up from dilutions of stock solutions containing 20 mM borate, 1 x 10^{-3} M PC, 50 mM NaCl, and the borate buffer with NaCl added to give buffer the same conductivity as the PC containing buffer. We have noted that reduced currents produced by lower ionic strengths yield better results.

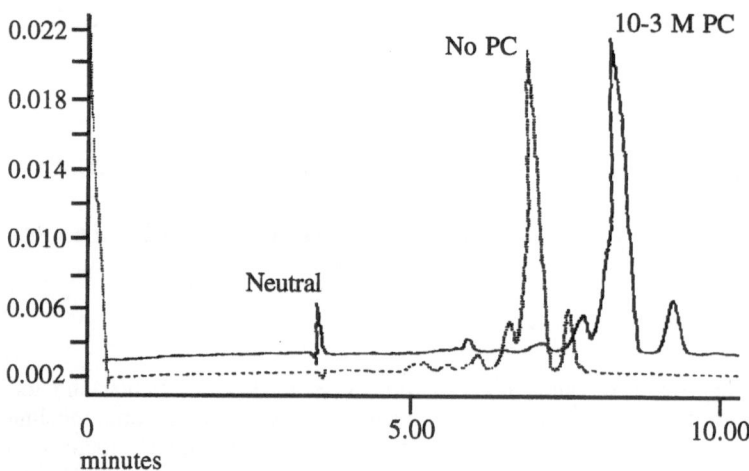

Fig. 1. *Electrophoretic mobility shift upon addition of PC to CE buffer.*

The electrophoretic mobility (μ) of a species is determined by the elution time of the species in question compared to the elution time of a sample of known electrophoretic mobility. In our case the reference molecule is mesityl oxide which is uncharged and thus has a μ of 0. Additionally μ is related to the total capillary length, the length to the detector, and the applied voltage [4]. Upon binding the PC the charge of the IgA or F_{ab} is changed by -1 and its electrophoretic mobility (μ) will thus be altered. By plotting μ vs. $\mu/[PC]$ and extrapolating to large [PC] the μ_{max} and v (μ/μ_{max}) can be calculated for a series of runs (Table 1).

$$\mu = \frac{L_T L_D}{V}\left(\frac{1}{t_{nm}} - \frac{1}{t_s}\right)$$

Table 1 *Recorded and calculated data for the binding of PC by the F_{ab} fragment*

[PC]x10⁴ M⁻¹	μx10³ (cm²/Vsec)	μ/[PC] (cm²/VsecM)	v=μ/μmax	v/[PC] M⁻¹
1.0	-8.93	-89.34	0.94	9400
2.5	-9.22	-36.88	0.97	3900
5.0	-9.44	-18.87	0.99	2000
10.0	-9.70		1	1000

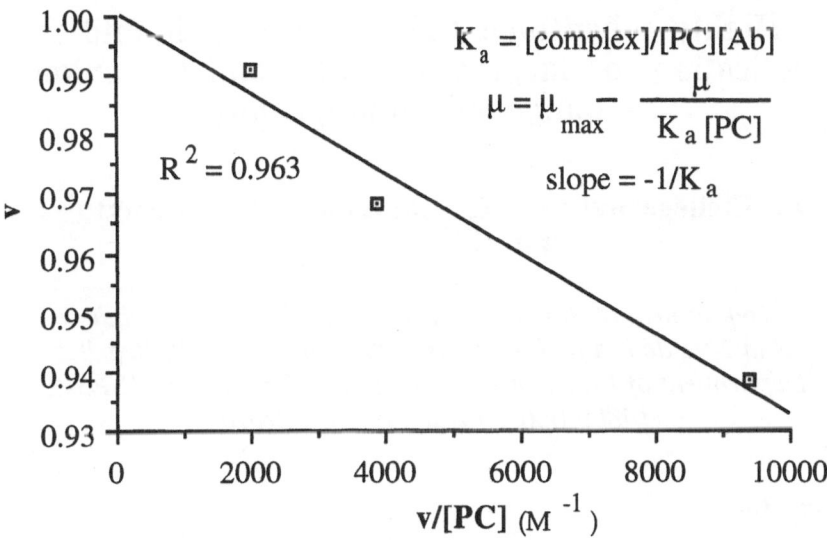

Fig. 2. Binding plot of McPC603 Fab fragment.

Plotting v vs. v/[PC] produces a graph with a slope that is the negative inverse of the binding constant (Figure 2). The Ka which has been determined for the Fab fragment is 1.5×10^5 at 30°C which is in good agreement with the published values.

The binding constant determination of the intact McPC603 by CE has proven elusive thus far, due primarily to the large net charge (-17) present on the immunoglobulin at the pH at which the experiment must be performed. Pepsin cleavage of McPC603 produced a F_{ab} fragment with a MW of 62,000 by SDS-PAGE. This fragment was calculated to have a charge of -8 at the experimental pH and thus showed a more dramatic shift upon binding the hapten. The speed of sample analysis coupled with minimal sample consumption makes this method of binding constant determination highly promising for those samples which can be adequately separated by CE.

References

1. Rudikoff, S., Potter, M., Segal, D.M., Padlan, E.A. and Davies, D.R., Proc. Natl. Acad. Sci. U.S.A., 69 (1972) 3689.
2. Skerra, A. and Plückthun, A., Science, 240 (1988) 1038.
3. Goetze, A.M. and Richards, J.H., Biochemistry, 17 (1978) 1733.
4. Moring, S.E., Colburn, J.C., Grossman, P.D. and Lauer, H.H., LC•GC, 8 (1990) 34.

Highly substituted BHAR: Synthesis and application for large scale SPPS and for anion exchange chromatography

A. Etchegaray[1], R.S.H. Carvalho[1], R. Marchetto[2] and C.R. Nakaie[1]

[1]Department of Biophysics, Escola Paulista de Medicina,
Rua Tres de Maio 100, 04044-020, São Paulo, SP, Brazil
[2]Department of Biochemistry, Institute of Chemistry, UNESP,
14800-060, Araraquara, SP, Brazil

Introduction

Carefully controlled kinetic studies [1] had allowed the synthesis of BHAR batches containing higher amounts of amino groups than previously reported and, important for large scale SPPS, these resins have been useful for syntheses of α-carboxamide peptides [2]. Besides this use, there is indeed a possibility of employing highly substituted BHARs also as novel anion exchange resins in aqueous solution. This is due to our earlier findings which demonstrated that, the higher the amount of positively charged ammonium groups in resins, the better the swelling in more polar solvents including water [1]. Thus, the purpose of this work was to test, by purifying negatively charged model peptides, the true potentiality of the anion exchanger properties of BHAR-type resins.

Results and Discussion

Forcing more the benzoylation reaction step of our previously reported BHAR synthesis method [1], it was possible to increase the maximum substitution degree of BHAR from 2.2 mmol/g to 3.3 mmmol/g. The hydration capacity of the two BHAR batches was estimated by the microscopic measurement of resin beads [3] and, as expected, the average volume of solvent (0.02 M NH_4Ac aqueous buffer, pH 5.0) absorbed by each bead (calculated by subtracting the volume of dry from that of the swollen bead) increased from 2.6×10^5 μm^3 to 5.9×10^5 μm^3 as the degree of substitution ranged from 2.2 mmol/g to 3.3 mmol/g. These swelling values of amino protonated BHARs in aqueous solution are rather similar to those of amino deprotonated BHAR in DCM or DMF during the coupling reaction in the peptide synthesis cycle [1] thus suggesting that, in principle, highly substituted BHARs have sufficient hydration capacity to be tested as anion exchange resins in aqueous media.

The two acidic peptides, DEVYIHPF and DEVYIEPF with net charges (at pH 5.0) near (-1) and (-3), respectively, were submitted to chromatographic purification with 2.2 mmol/g and in 3.3 mmol/g BHARs. The (+2) DRVYIHP-amide was also tested as a control and among commercial resins, the strong anion exchange AG1-X2 resins (BioRad) was also chosen for comparison. Figure 1 shows

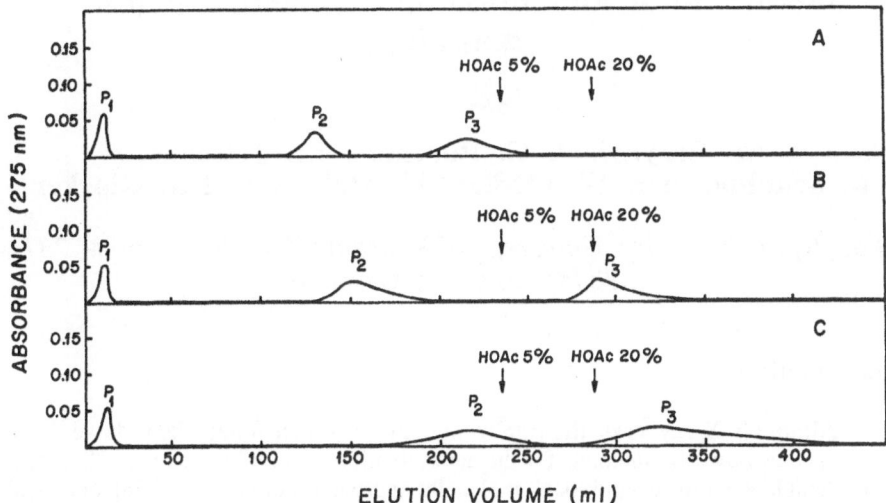

Fig. 1. Anion exchange chromatography of DRVYIHP-amide (P₁), DEVYIHPF (P₂) and
DEVYIEPF (P₃) on AG1-X2 resin (A), 3.3 mmol/g BHAR (B) and 2.2 mmol/g BHAR (C).
Linear pH-gradient from 0.02 M ammonium acetate, pH 5.5 to 10% HOAc, pH 2.3, 120 ml
each) and 3.5 mmol of ammonium groups per column.

the chromatographic profiles obtained in a linear pH-gradient (from pH 5.5 to 2.3).
The comparison of the anion exchange properties of the two BHAR batches indicated
improved elution and resolution of peaks as the amount of ammonium groups in the
resin increased (better swelling). The 3.3 mmol/g BHAR presented an elution profile,
rather similar to that seen with commercial AG1-X2 resin. As expected, the
positively charged control peptide (P₁) eluted in the void volume of the three
columns. These findings confirmed the feasibility of anion exchange purification
with highly substituted BHAR-type polymer, so far employed as the solid support
for peptide synthesis. Furthermore, differently from commercial resins which contain
quaternary, tertiary or secondary amino functions, the presence of a more reactive
primary amino group bound to the BHAR matrix might be used advantageously to
obtain other resin-derivatives for various purposes.

References

1.   Marchetto, R., Etchegaray, A. and Nakaie, C.R., J. Braz. Chem. Soc., 3 1-2 (1992)
     30.
2.   Cilli, E.M., Oliveira, E., Marchetto, R. and Nakaie, C.R., In Schneider, C.H. and
     Eberle, A.N. (Eds.), Peptides 1992, Escom, Leiden, The Netherlands, 1993, p.425.
3.   Sarin, V.K., Kent, S.B.H. and Merrifield, R.B., J. Am. Chem. Soc., 102 (1980)
     5463.

Molecular approaches for the characterization of protein tertiary structures by selective chemical modification and mass spectrometric peptide mapping

M. Przybylski, C. Borchers, M. Jetschke, R. Schuhmacher, W. Fiedler, M. Mák and M.O. Glocker

Faculty of Chemistry, University of Konstanz, P.O. Box 5560 M 731, 78434 Konstanz, Germany

Introduction

Although X-ray crystallography and multi-dimensional NMR have been developed as powerful methods for the determination of protein structures, there is a considerable need for unequivocal, molecular approaches of characterizing chemical properties of specific residues in proteins. Protein-chemical modification has long been employed for evaluation of chemical reactivities in structure-function studies [1]; however, conventional analytical methods frequently do not provide unambiguous identification of modification sites, and require radioactive labelling. The development of desorption-ionization mass spectrometric methods, such as fast atom bombardment (FABMS), 252-Cf plasma desorption (PDMS) and electrospray (ESMS) has enabled accurate molecular weight determinations of polypeptides and proteins up to >100 kDa, as well as the direct analysis of multicomponent peptide mixtures. The combination of specific chemical modification with direct mass spectrometric peptide mapping of proteolytic mixtures has recently been developed as a new approach to obtain chemical reactivity data in proteins [2] (Figure 1). This method has already been successfully shown with several specific reactions such as amino acylation, Tyr-iodination and -nitration and carboxylate-amidation, to yield direct information on surface topology and microenvironment effects in protein structures.

Results and Discussion

Principal analytical steps of the mass spectrometric approach are: (1) limited protein-chemical modification at different reaction conditions (time, reagent excess); (2) direct molecular weight determinations revealing the number and distribution of modified residues of protein derivatives; (3) proteolytic digestion and mass spectral peptide mapping analysis to identify the modification sites; and (4) estimation of relative reactivities from the increase and decrease in molecular ion abundances of modified vs unmodified peptide fragments. In addition, the ESMS analysis of protein derivatives in step (2) provides direct information on structural changes with increasing modification by comparison of the charge structure (envelope) of multiple protonated molecular ions, and analysis by deuterium exchange [3, 4].

For the amino-acetylation and -succinylation with a range of molar excess of acylanhydrides at pH 6.5, the application of this approach for surface topology probing has been systematically studied with proteins of different globular structures, such as hen egg white lysozyme (HEL), bovine RNAse and horse heart myoglobin. A gradual increase of modification (mol. weights) was found, yielding selective acetylation of the total number of 7 (HEL), 11 (RNAse) and 21 (myoglobin) lysine and N-terminal amino groups at 100-300-fold excess of anhydride/NH_2-group. PD mass spectrometric peptide mapping analysis after trypsin digestion provided the unequivocal identification of the acetylation sites by the shifts (Lys → Arg) of tryptic cleavage sites, and the corresponding molecular ions of acetylated peptides [3]. Relative reactivities were derived from the corresponding molecular ion abundances by PDMS; in addition a direct quantification of acetylated peptides was obtained by FABMS peptide mapping after differential acylation of proteins with acetyl- and trideuteroacetyl-anhydride [5].

The relative reactivities for the acetylation of amino groups determined from the peptide mapping data were compared with the relative surface accessibilities (SA) for a 1.4 Å van der Waals sphere [6] from the X-ray structures. A clear linear correlation of reactivities with the SA values of Lys residues was found as shown in Figure 1 for HEL and RNAse A, while $pK_{1/2}$ values had no significant effect. Besides the expected large deviation in reactivity of the N-terminal amino groups, the only significant exception was found for the Lys^{41} residue of high reactivity in RNAse A (SA = 0.15) which together with the His^{19} and His^{112} residues is part of the active hydrolytic centre of the enzyme (Figure 2) [7].

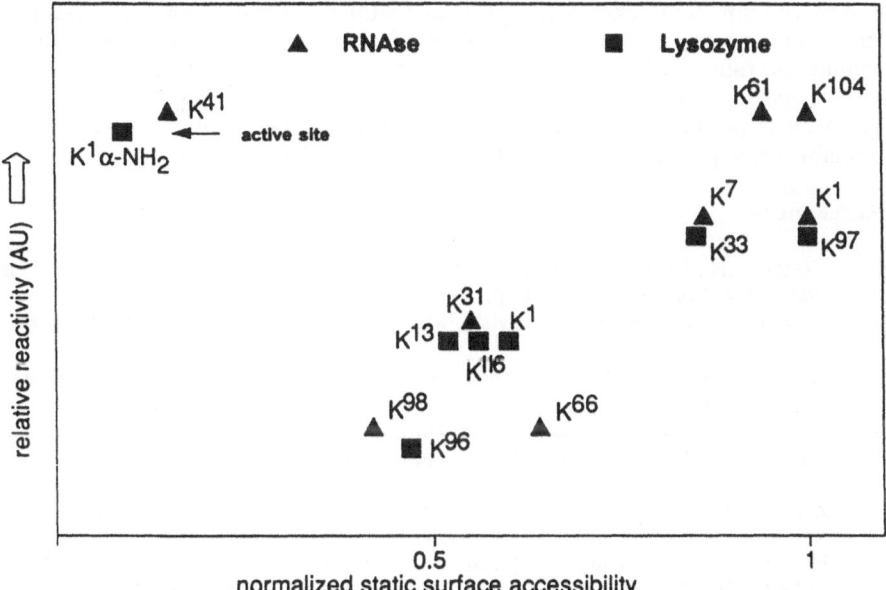

Fig. 1. *Correlation of relative reactivities of amino-acetylation in HEL (■) and bovine RNAse A (▲) with surface accessibilities. SA values are referred to the tripeptide Ala-Lys-Ala [6].*

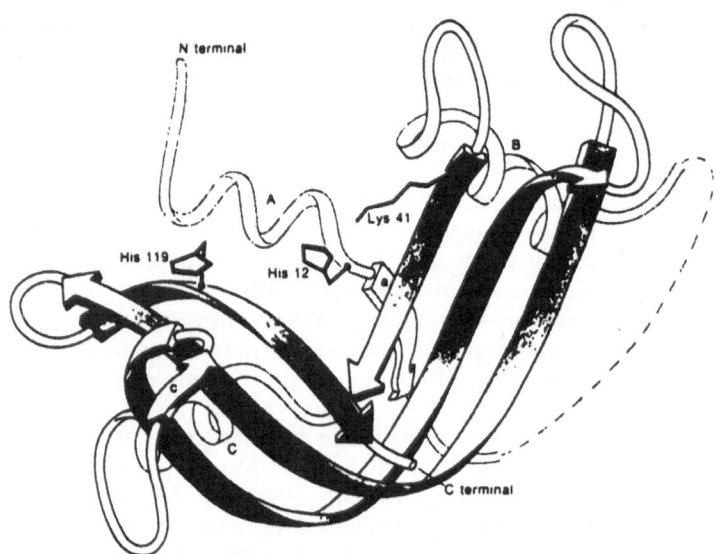

Fig. 2. Tertiary structure scheme of RNAse A highlighting the amino acid residues involved in the active hydrolytic centre.

The results obtained so far with model proteins of known tertiary structures are promising, and first applications are studied at present of mass spectrometric peptide mapping to surface topology probing, such as in helical polypeptides and leucine zipper protein complexes with oligonucleotides. Furthermore, an exciting application potential is the selective modification of protein antigen-antibody complexes, followed by proteolytic digestion and mass spectrometric analysis of dissociated antigen peptides (mass spectrometric epitope mapping [8]). This approach is presently being applied to the characterization of conformationally assembled epitopes of myoglobin with monoclonal antibodies.

References

1. Glazer, A.N., In Neurath, H. and Hill, R.L. (Eds.), The Proteins, Vol. 2, Academic Press, New York, U.S.A., 1976, p.1.
2. Suckau, D., Mák, M. and Przybylski, M., Proc. Natl. Acad. Sci. U.S.A., 89 (1992) 5630.
3. Glocker, M.O., Borchers, C., Fiedler, W., Suckau, D. and Przybylski, M., Bioconj. Chem., in press.
4. Przybylski, M., Borchers, C., Suckau, D., Mák, M. and Jetschke, M., In Schneider, C.H. and Eberle, A.N. (Eds.), Peptides 1992, Escom, Leiden, The Netherlands, 1993, p.83.
5. Jetschke, M., Schuhmacher, R., Borchers, C., Glocker, M.O. and Przybylski, M., Rap. Commun. Mass Spectrom., in press.
6. Matthew, J.B. and Richard, F.M., Biochemistry, 21 (1982) 4989.
7. Roberts, G.C.K., Dennis, E.A., Meadows, D.H., Cohen, J.S. and Jardetsky, O., Proc. Natl. Acad. Sci. U.S.A., 62 (1969) 1151.
8. Suckau, D., Köhl, J., Karwarth, G., Schneider, K., Casaretto, Bitter-Suermann, D. and Przybylski, M., Proc. Natl. Acad. Sci. U.S.A., 87 (1990) 9848.

Identification of the active site of CoA transferase by peptide sequencing

J.-C. Rochet and W.A. Bridger

*Department of Biochemistry, University of Alberta,
Edmonton, Alberta, T6G 2H7, Canada*

Introduction

The recently sequenced pig heart CoA transferase [succinyl-CoA:3-ketoacid CoA transferase, EC 2.8.3.5.] consists of two identical subunits of 481 residues each [1]. A hydrophilicity plot of the monomer suggests two interacting domains linked by a solvent-accessible hinge region. SDS-PAGE analysis indicates that this linker region is nicked in a preparation of active enzyme purified in the absence of PMSF.

Previously, it was demonstrated that the catalytically active moiety of this enzyme, to which CoA is thioesterified to form a covalent intermediate, is a glutamate sidechain [2]. In order to identify this glutamate residue within the sequence, we have made use of an autolytic fragmentation shown to occur at the active site when the enzyme-CoA intermediate is heated [3]. This fragmentation, which can be carried out on intact or proteolytically nicked CoA transferase, occurs approximately 2/3 of the distance from the enzyme's N-terminus, and generates an N-terminally blocked C-terminal fragment. To locate the active site glutamate more precisely, we have attempted to apply the methods of HPLC and mass spectrometry.

Results and Discussion

The nicked form of CoA transferase, both prior and subsequent to autolysis, is represented schematically in Figure 1. Proteolysis alone is shown to generate segments A and B, while the autolytic fragmentation releases peptides C and D. Significantly, peptide D is bordered at its N-terminus by the proteolytic cleavage site, and at its C-terminus by the site of autolysis (*i.e.* the active site). Because the site of proteolysis has been identified in the sequence of CoA transferase, accurate determination of the mass of peptide D allows for precise identification of the active site glutamate residue. In order to isolate this peptide, the two forms of nicked CoA transferase were analyzed by reversed-phase (RP) HPLC; their corresponding elution profiles are compared in Figure 1. Given that peaks labeled 1-5 appear in both plots, it is clear that three peptides, denoted X, Y, and Z, are specifically associated with the autolytically fragmented sample. The identity of these three peptides will very shortly be addressed by their analysis on a benchtop electrospray mass spectrometer.

The design of these experiments illustrates the value of peptide sequencing by mass spectrometry to the study of a protein's primary structure.

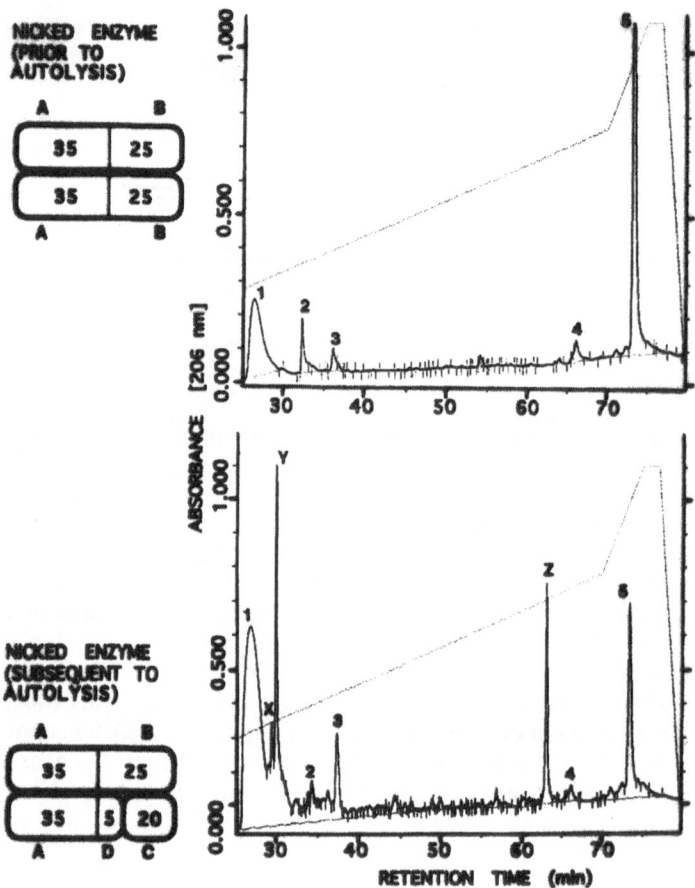

Fig. 1. Elution profiles for RP-HPLC of CoA transferase, either nicked (top) or nicked/autolytically fragmented (bottom). Separations were performed at 0.2 ml/min on a Vydac 218TP C18 column (250 mm x 2.1 mm ID, 5-μm particle size, 300-Å pore size). Gradient curve indicates 25%B -> 70%B -> 100%B -> 100%A, where A is 0.1% TFA/water, B is 0.1% TFA/acetonitrile. Peaks appearing in both plots are labeled 1-5, whereas those unique to the lower plot are labeled X, Y, and Z. Also shown to the left of each plot is a schematic diagram of the appropriate form of the enzyme. The four peptides are labeled A-D and are assigned approximate molecular weight values. Peptides C and D are shown to be released from only one of the two subunits to reflect the observed 50% efficiency of the autolytic fragmentation.

Acknowledgements

We are indebted to Dr. Charles Holmes for the use of his HPLC facility.

References

1. Lin, T. and Bridger, W.A., J. Biol. Chem., 267 (1992) 975.
2. Solomon, F. and Jencks, W.P., J. Biol. Chem., 244 (1969) 1079.
3. Howard, J.B., Zieske, L., Clarkson, J. and Rathe, L., J. Biol. Chem., 261 (1986) 60.

Hydrogen-deuterium exchange as a method facilitating unambiguous assignment of fragment ions in CID spectra of peptides

N.F. Sepetov[1], O.L. Issakova[1], M. Lebl[1], K. Swiderek[2], D.C. Stahl[2] and T.D. Lee[2]

[1]The Selectide Corporation, Tucson, AZ 85737, U.S.A.
[2]Division of Immunology, Beckman Research Institute of the City of Hope, Duarte, CA 91010, U.S.A.

Introduction

During the last years mass spectrometry has been proven to be a powerful and sensitive tool for peptide sequencing and is becoming a more and more useful alternative and complementary approach to automated Edman degradation [1, 2]. But sequence determination of an unknown peptide is still a difficult task due to the initially huge number of possible sequences consistent with the molecular weight of peptide among which the correct one must be chosen by using spectral information about fragment ions and (if any) additional data about peptide. Different derivatization procedures are known to be used to facilitate interpretation of mass spectra [3, 4, 5]. Recently we reported about a facile method of derivatizing peptides for sequence analysis of peptides based on hydrogen/deuterium exchange. When a peptide is dissolved in a deuterated solvent containing labile protons (D_2O deuterated methanol etc.) there is exchange of all hydrogens in the peptide attached to oxygen, nitrogen and sulfur. The number of labile protons in the peptide can be easily determined from measurement of the mass difference between the intact peptide and the peptide after hydrogen/deuterium exchange. This information places significant restraints on the number of different amino acid compositions corresponding to the determined molecular weight of the peptide and provides an efficient filter for a list of candidate sequences provided by a computer interpretation program [6]. In an extension of this work, the present paper reports results of collisionally induced dissociation (CID) experiments of peptides before and after hydrogen-deuterium exchange to demonstrate efficiency of our approach to increase accuracy of fragment ions assignment.

Results and Discussion

Twenty different peptides have been analysed before and after hydrogen/deuterium exchange of labile protons of a peptide using a TSQ-700 triple sector quadrupole mass spectrometer (Finnigan MAT, San Jose, CA, U.S.A.) equipped with a standard electrospray ion source; low energy CID spectra were obtained for all peptides studied using Argon collision gas introduced into the octapole collision cell. Deuterium exchanged samples were prepared by dissolving the peptide samples in 50% D$_2$O, methanol-D$_4$. Analysis of data of CID spectra of peptides before and after hydrogen/deuterium exchange has shown that pathways for fragmentation are not significantly affected by deuterium and shifts of masses of fragments after deuterium exchange coincide with calculated values. Thus the CID spectrum of a peptide after hydrogen/deuterium exchange can be used to confirm fragment ion assignments and is especially valuable when there is ambiguity in assignment. For example, one of the main difficulties which arises in sequencing of an unknown peptide is the mass redundancy associated with certain combinations of amino acid residues: for example -Ala, Gly- and -Gln/Lys-; -Gly,Gly- and -Asn-; -Gly,Val- and -Arg-; -Ala,Asp- and -Trp-; -Ser,Val- and -Trp-; -Gly,Leu- and -Ala,Val- etc. Failure to observe an ion resulting from cleavage between amino acid pairs can result in misassignment of a single amino acid residue for what is actually an amino acid pair. For example, the only peptide which was not sequenced correctly in [7] contained -Ala-Gly-. The computer interpretation of the CID spectrum of this peptide gave -Gln-. Similar problems in sequencing an unknown peptide are described in [8] when -Trp- was misassigned in place of -Val-Ser-. The authors of [9] emphasized that this kind of problem is one of the limitations of their approach. In most such instances the number of exchangeable protons in fragment ions which is available from comparison of CID spectra of intact peptide and deuterated one, enables one to resolve ambiguities in assignment of these fragment ions.

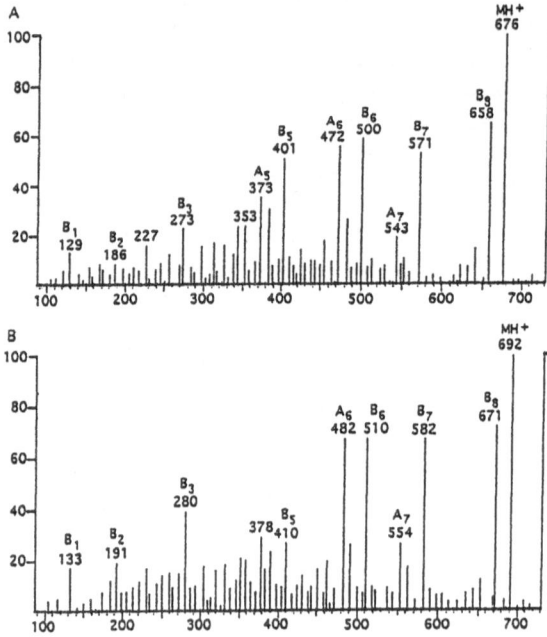

Fig. 1. Collision-induced dissociation fragment-ion spectra of the single protonated peptide KGSGAVAS-amide. Spectrum A is for the ion at m/z 842 formed by electrospray from 20 pm/μl solution in 50% aqueous methanol. Spectrum B is for the hydrogen/deuterium exchanged ion at m/z 864 obtained by electrospray using deuterated water and deuterated methanol.

Figure 1 presents CID spectra of peptide KGSGAVAS-amide with and without deuterium exchange. It is clear from those spectra that no bond cleavage between glycine and alanine observed. However, the shift of B_5 ion from 401 to 410 after deuterium exchange provides the basis to make correct assignment of this fragment to KGSGA (9 labile protons) rather than to KGSQ (10 labile protons).

Comparison of CID spectra of peptides before and after deuterium exchange can be especially valuable when -A,G-; -G,G-; -G,V-; -A,D- etc. are the first two residues of a peptide since very often there is no cleavage between the first and the second residues in CID experiments.

Results presented show that hydrogen/deuterium exchange is a facile technique which has the potential to increase the efficiency of peptides sequencing by mass spectrometry. CID spectra of peptides after hydrogen/deuterium exchange provide important additional information about peptide which can be especially valuable when incorporated in computer interpretation programs.

References

1. Biemann, K., Annu. Rev. Biochem., 61 (1992) 977.
2. Hunt, D.F., Henderson, R.A, Shabanovitz, J., Sakaguchi, K., Michel, H., Sevilir, N., Cox, A.L., Apella, E. and Engelhard, V.H., Science, 255 (1992) 1261.
3. Vath, J.E. and Biemann, K., Int. J. Mass Spectrom. Ion Processes, 100 (1990) 287.
4. Hunt, D.F., Yates, J.R., Shabanowitz, J. and Winston, S., Proc. Natl. Acad. Sci. U.S.A., 83 (1986) 6233.
5. Biemann, K., In McCloskey (Ed.), Methods in Enzymology, Vol. 193, Academic Press, San Diego, CA, U.S.A., 1990, p.445.
6. Sepetov, N.F., Issakova, O.L., Lebl, M., Swiderek, K., Stahl, D.C. and Lee, T.D., Rapid. Comm. Mass Spectrom., 7 (1993) 58.
7. Johnson, R.S. and Biemann, K., Biomed. Environ. Mass Spectrom., 18 (1988) 945.
8. Yates, J.R., Griffin, P.R., Hood, L.E. and Zhou, J.X., Techniques Protein Chem., 2 (1991) 447.
9. Zidarov, D., Thibault, P., Evans, M.J. and Bertrand, M.J., Biomed. Environ. Mass Spectrom., 19 (1990) 13.

Matrix-assisted laser desorption mass spectrometry of enzymatic digests of peptides and proteins

S. Stoeva, G. Grübler, S. Hörger, A. Keen and W. Voelter

Abteilung für Physikalische Biochemie des Physiologisch-chemischen Instituts der Universität Tübingen, Hoppe-Seyler-Straße 4, D-7400 Tübingen, Germany

Introduction

The formation of single-charged molecular ions via the matrix-assisted laser desorption mass spectrometer system (Lasermat, Finnigan MAT), enables the characterization of biomolecules of different chemical structures [1, 2]. Various research groups [3-12] have reported measurements of peptide and protein ions by the matrix-assisted laser desorption ionization method. Because of the mild ionization conditions used, LDMS provides an excellent method of fingerprinting the peptides released by chemical or proteolytic cleavage of proteins and glycoproteins. Unlike many other ionization techniques in mass spectrometry, LDMS is extremely tolerant to some buffer salts and detergents. Analysis of the peptides produced is therefore possible directly from buffer solutions. The advantage of LDMS over chromatography is that the determined molecular weights provide a basis for the suggested identification of the peptide fragments from the known primary structure of the protein.

Results and Discussion

Each spectrum exhibits mainly peaks corresponding to the protonated molecular ion for each fragment in the digest mixture. The molecular weights, determined for thymosin β_4 and β_9 at 0 min incubation time, are 4965 Da and 4718 Da, respectively, corresponding to the theoretical values from the sequence. The spectra indicate that the starting material was homogeneous. Figures 1 and 2 show the primary structures of thymosin β_4 and β_9 and the expected Asp-N cleavage sites.

Ac-Ser-Asp-Lys-Pro/-**Asp**-Met-Ala-Glu-Ile-Glu-Lys-Phe/-**Asp**-Lys-Ser-Lys-Leu-Lys-Lys-Thr-Glu-Thr-Gln/-**Glu**-Lys-Asn-Pro-Leu-Pro-Ser-Lys-Glu-Thr-Ile-Glu-Gln-Glu-Lys-Gln-Ala-Gly-Glu-Ser-OH

Fig. 1. Primary structure and cleavage sites of thymosin β_4 caused by the endoproteinase Asp-N.

Ac-Ala-Asp-Lys-Pro/-**Asp**-Leu-Gly-Glu-Ile-Asn-Ser-Phe/-**Asp**-Lys-Ala-Lys-Leu-Lys-Lys-Thr-Glu-Thr-Gln/-**Glu**-Lys-Asn-Thr-Leu-Pro-Thr-Lys-Glu-Thr-Ile-Glu-Gln-Glu-Lys-Gln-Ala-Lys-OH

Fig. 2. Primary structure and cleavage sites of thymosin β_9 caused by the endoproteinase Asp-N.

Tables 1 and 2 list the masses of the thymosin β_4 and β_9 Asp-N fragments obtained after LDMS measurements. The data are averages from 6 spectra.

Table 1 *Molecular masses of the Asp-N fragments of thymosin β_4*

Fragment	Calc. (MH+)	Determined	Error	% Dev.
1-4	488.50	488.57	0.07	0.014
5-12	983.10	983.02	0.08	0.009
13-23	1306.48	1307.21	0.73	0.056
13-34	3532.12	3533.58	1.46	0.041
24-43	2243.40	2244.93	1.53	0.068
1-43	4964.48	4965.02	0.54	0.011

Table 2 *Molecular masses of the Asp-N fragments of thymosin β_9*

Fragment	Calc. (MH+)	Determined	Error	% Dev.
1-4	472.50	471.93	0.57	0.120
5-12	894.93	895.96	1.03	0.115
13-23	1290.48	1291.04	0.56	0.043
13-41	3388.17	3388.23	0.06	0.002
24-41	2116.35	2116.11	0.24	0.011
1-41	4717.26	4718.76	1.50	0.032

These values are compared with the theoretical ones, calculated on the basis of the sequence. Four of the mass values are between 1 and 2 Da in error. In the case of the 5-12 β_9 fragment (894,93 Da) this is due to the peak of low intensity. The errors for the peaks 3532 Da (β_4 fragment 13-43), 2243 Da (β_4 fragment 24-43) and 4717 Da (β_9) are caused by the relatively high laser power applied, necessary to detect the peaks of minor intensity. In the course of digestion a shifting of the peaks from higher to lower masses was observed. It should be mentioned that fragments with masses below 500 Da (1-5 amino acid residues) are difficult to detect due to the interference with matrix peaks. For this reason, spectra of the tryptic fragmentation of thymosin β_4, β_9 and ubiquitin are not presented. After an incubation time of 2 and 4 hours of thymosin β_4 and β_9 with the Asp-N proteinase, still a considerable amount of starting material was detected. Final digestion of both substrates was observed after 8 hours. No peaks due to thymosin β_4 (4965 Da) or β_9 (4717 Da) were found in the corresponding reaction mixtures.

These two examples clearly show how the kinetics of the digestion can rapidly be monitored using the Lasermat mass spectrometer. This method allows to optimize the digest conditions, particularly for fragmenting a new protein. It requires minimal time and picomole quantities for sample preparation and is also tolerant to buffer salts.

References

1. Karas, M., Bachmann, D., Bahr, U. and Hillenkamp, F., Int. J. Mass Spectrom. Ion Proc., 78 (1987) 35.
2. Karas, M., Bachmann, D., Bahr, U. and Hillenkamp, F., Int. J. Mass Spectrom. Ion Proc., 92 (1989) 231.
3. Spengler, B. and Cotter, R.J., Anal. Chem., 62 (1990) 793.
4. Karas, M. and Hillenkamp, F., Anal. Chem., 60 (1988) 2299.
5. Karas, M., Ingendoh, A., Bahr, U. and Hillenkamp, F., Biomed. Environ. Mass Spectrom., 18 (1989) 841.
6. Beavis, R.C. and Chait, B.T., Rapid Commun. Mass Spectrom., 3 (1989) 432.
7. Beavis, R.C. and Chait, B.T., Proc. Natl. Acad. Sci. USA, 87 (1990) 6873.
8. Zhao, S., Somayajula, K.V., Sharkey, A.G., Hercules, D.M., Hillenkamp, F., Karas, M. and Ingendoh, A., Anal. Chem., 63 (1991) 450.
9. Overberg, A., Karas, M. and Hillenkamp, F., Rapid Commun. Mass Spectrom., 5 (1991) 128.
10. Hillenkamp, F. and Karas, M., Methods Enzymol., 193 (1991) 280.
11. Yildiz, E., Grübler, G., Hörger, S., Zimmermann, H., Echner, H., Stoeva, S. and Voelter, W., Electrophoresis, 13 (1992) 683.
12. Karas, M., Bahr, U., Ingendoh, A. and Hillenkamp, F., Angew., Chem., 101 (1989) 805.

Matrix-assisted laser desorption mass spectrometry to monitor synthesis and folding of *Manduca sexta* eclosion hormone and its analogs

Y. Wang[1], L. Yurttas[1], B.E. Dale[1], D.H. Russell[2], G.R. Kinsel[2], L.M. Preston[2], M.S. Wright[3] and T.K. Hayes[3]

[1]Department of Chemical Engineering; [2]Department of Chemistry; [3]Institute of Biosciences and Technology; [3]Department of Entomology, Texas A&M University, College Station, TX 77843, U.S.A.

Introduction

Eclosion hormone (EH) is an excellent small protein model system for studying protein folding, stability, neuropharmacology and developmental neurobiology. In insects, EH triggers eclosion behavior and during metamorphosis, rewires the nervous system to serve new functions [1]. Thus EH is a specific neurodevelopmental stimulator and neurotoxin. *M. sexta* EH is 62 amino acids long and its native conformation is stabilized by 3 disulfide bonds [2]. This sequence is long enough to contain information for the study of how a small protein neurohormone folds to form a surface that interacts with a specific biological receptor. The small size of EH makes successful synthesis possible. Through chemical synthesis unnatural amino acids can be incorporated as probes to assess different factors governing the folding, stability and biological activity of proteins. Analogs of EH are being developed in our laboratories to determine the important structural components for stability, receptor interaction and biological action of the hormone. Native EH (EHQ), [Asn^{20}]-EH analog (EHN), and a chemically stable [Nle^{11}, Asn^{20}, Nle^{24}, NpA^{28}]-EH analog (EHS) were synthesized with Fmoc solid-phase peptide synthesis (NpA: L-3-(2-Naphthyl)-Alanine). By using matrix-assisted laser desorption mass spectrometry (MALDI-MS), a purification scheme including gel filtration chromatography, anion exchange chromatography, and reverse-phase HPLC was developed. The purified peptides have been characterized via MALDI-MS, amino acid analysis, sequencing, and bioassay. MALDI-MS was then used to monitor the protein folding process of EHS and bovine pancreatic trypsin inhibitor (BPTI).

Results and Discussion

The peptides were synthesized using Fmoc chemistry on a MilliGen 9050 synthesizer with the general methods similar to those described for Tat protein [3] except for the cleavage. Upon assembling on the synthesizer, the peptide-resins were cleaved twice, each for two hours at 25°C with 93% trifluoroacetic acid (TFA)+7% triisopropylsilane (TIPS) for EHQ and EHN, or +7% triethylsilane for EHS. The crude peptides were precipitated in ice cold anhydrous ethyl ether containing 1%

β-mercaptoethanol and washed thoroughly with the same. Crude EHQ was analyzed with MALDI-MS (Figure 1A). The components in crude mixture were identified

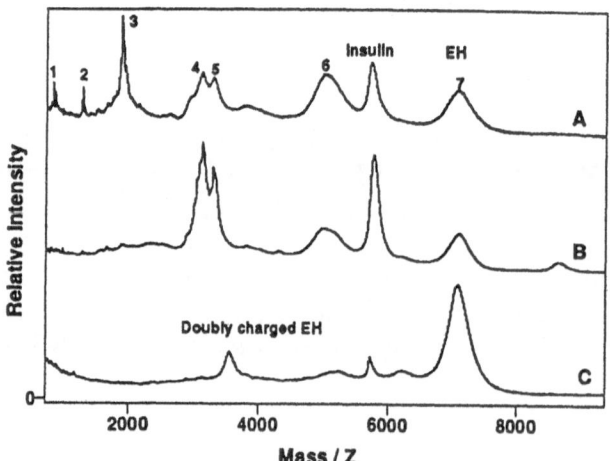

Fig. 1. *MALDI mass spectra of EHQ during each purification step (panel A: crude; panel B: after gel filtration; panel C: after ion exchange and RP-HPLC).*

using their molecular masses (Table 1); assumed peptide sequences are based on the assumption that impurities are due to the N-terminal truncated peptides in the synthesis. With this information, a purification scheme was designed as follows: (1) Sephadex G-25 gel filtration chromatography to remove peaks 1-3 of lower mass (Figure 1B); (2) anion exchange chromatography and reverse phase HPLC to remove peaks 4-6 of higher pI (Figure 1C). In the final spectrum, only EHQ and its doubly charged signal were detected. The molecular weight of purified peptides was found to be: EHQ 6840 (calc. 6813); EHN 6840 (calc. 6799); EHS 6777 (calc. 6774). The first 22 amino acid residues from N-terminus for all three peptides were correct by sequence analysis. The amino acid analysis of EHS confirmed the expected composition. EH activity in partially folded protein was detected in all three peptides by using pharate adult of *Heliothis virescens in vivo* assay [2].

Table 1 *Population analysis of the crude EHQ by MALDI-MS*

Peak No.	Mass	Assumed Peptide Sequence	pI
1	814	APFLNKL	9.67
2	1272	FASIAPFLNKL	9.67
3	1879	IPECEDFASIAPFLNKL	4.00
4	3104	AE...LIPECEDFASIAPFLNKL	6.35
5	3270	CAE...LIPECEDFASIAPFLNKL	6.35
6	5005	ANC...LCAE...LIPECEDFASIAPFLNKL	7.74
7	7071	NPAIA...NCANC...LCAE...LIPECEDFASIAPFLNKL	4.68

In protein folding studies, disulfide bond interactions are used as probes to detect and to trap folding intermediates in a stable form by alkylation of free thiols with iodoacetate to form carboxymethyl Cys (CM-Cys). Usually, the trapped intermediates are analyzed with electrophoresis or ion exchange chromatography. However, these techniques are time consuming and less sensitive. With MALDI-MS, the mass difference between the reduced species and the partially or fully folded then alkylated species can be detected with only a fraction of the amount required for other methods. Based on this difference, one can calculate the number of carboxymethyl moieties, thus the free thiols at the stage of quenching. Figure 2 is the mass spectra of EHS refolding intermediates. Protein was denatured and reduced initially with 6 M guanidine hydrochloride (Gdn•HCl), then dialyzed against redox buffer with decreased Gdn•HCl concentration in a Pierce System 100 microdialyzer at a continuous flowrate of 0.2 ml / min. The preliminary study showed that in 2 M Gdn•HCl, BPTI started to fold up, and after dialysis in 0 M Gdn•HCl, it folded to the conformation with three disulfides (data not shown). For EHS, the folding did not start until 1.5M Gdn•HCl and went to completion within 1 concentration unit of Gdn•HCl, suggesting that EHS folding was more cooperative than BPTI.

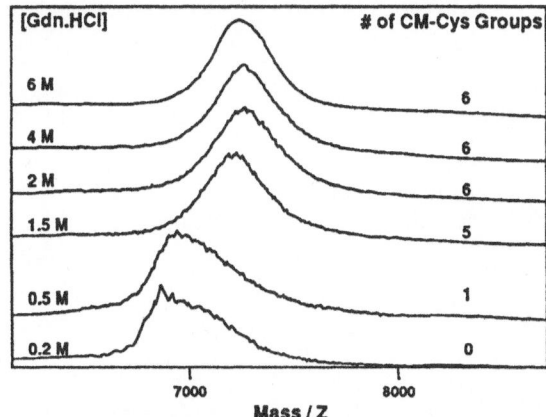

Fig. 2. MALDI mass spectra of EHS folding intermediates at different [Gdn•HCl].

MALDI-MS is a powerful tool for accurate mass analysis of molecules up to hundreds kilodaltons with an accuracy of ±0.1% and picomole sensitivity. The analysis takes about 5 minutes. Most importantly, it doesn't require sample purification. By virtue of these properties, it has broad application in biomolecular analysis. It provides valuable information for synthetic peptide purifications and characterizations. Preliminary studies indicate that it has a potential as a better alternative to electrophoresis for monitoring protein folding and identifying folding intermediates.

References

1. Hewes, R.S. and Truman, J.W., J. Comp. Physiol., A168 (1991) 697.
2. Kataoka, H., Troetschler, R.G., Kramer, S.J., Cesarin, B.J. and Schooley, D.A., Biochem. Biophys. Res. Commun., 146 (1987) 746.
3. Frankel, A.D., Biancalana, S., and Hudson, D., Proc. Natl. Acad. Sci. U.S.A., 86 (1989) 7397.

Session III
Peptide Mimetics

Chairs: Miklos Bodanszky
Case Western Reserve University
Cleveland, Ohio, U.S.A.

and

Tomi Sawyer
Parke-Davis/Warner Lambert
Ann Arbor, Michigan, U.S.A.

Peptide secondary structure mimetics: Recent advances and future challenges

M. Kahn[1,2], M.S. Lee[2], H. Nakanishi[2], J. Urban[2] and B. Gardner[2]

[1]*Department of Pathobiology, University of Washington, Seattle, WA 98195, U.S.A. and* [2]*The Molecumetics Institute, 2023 120th Avenue N.E., Suite 400, Bellevue, WA 98005, U.S.A.*

Introduction

Molecular mimicry of peptides depends on two factors: the ensemble of pharmacophoric information present in the primary sequence, and the three-dimensional presentation of this information. A successful peptidomimetic must meet these two criteria in order to retain significant bioactivity. One approach toward the discovery of peptidomimetics, which has met with some success, involves the screening of chemical libraries or natural products [1]. However, a rational strategy for the discovery of nonpeptide ligands remains elusive. Our approach to the design of peptidomimetics has been guided by the simple elegance which nature has utilized in the molecular architecture of proteinaceous species. Three basic building blocks (α–helices, β–sheets and reverse turns) are utilized for the construction of all proteins. We are involved in the design and synthesis of peptidomimetic prosthetic units to replace these three architectural motifs. This is affording us the opportunity to dissect and investigate complex structure-function relationships in proteins through the use of small synthetic conformationally restricted components. This is a critical step toward the rational design of low molecular weight nonpeptide pharmaceutical agents, devoid of the shortcomings of conventional peptides.

Results and Discussion

The belief that turns play critical roles in a myriad of molecular recognition events led us to focus our initial efforts on the design of reverse turn mimetics [2]. Beta–turns comprise a rather diverse group of structures; therefore, no one rigid framework will suffice. Synthetic expediency is a major concern, particularly at an early stage when the delineation of structure–activity relationships is critical, and requires the synthesis and evaluation of a series of related structures. With these criteria in mind, we have developed system 1, whose retrosynthetic strategy is outlined in Scheme 1.

Scheme 1

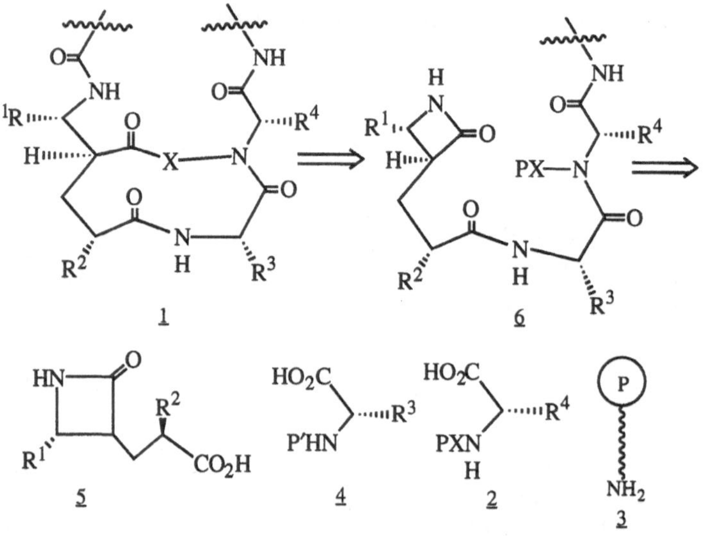

The synthesis of the reverse turn mimetic can be performed in solution; however, it is designed to be and is fully compatible with SPPS protocols [3]. In essence, it involves the coupling of the first modular component piece 2 to the amino terminus of a growing peptide chain 3. Coupling of the second component 4, removal of the protecting group P' and subsequent coupling of the third modular component 5 provides the nascent β–turn 6. The critical step in this sequence involves the use of an azetidinone as an activated acyl species to effect the macrocyclization reaction [4, 5]. Upon deprotection of P, nucleophilic opening of the azetidinone by the X moiety closes the ring and generates a new amino terminus for continuation of the synthesis. An important feature of this scheme is the ability to alter the X-group linker, both in regard to length and degree of rigidity/flexibility. The requisite stereogenic centers are readily derived, principally from the "chiral pool." The synthesis allows for the introduction of natural or unnatural amino acid side chain functionality in either L or D configuration. Additionally, deletion of the second modular component 4 provides access to γ-turn mimetics [6].

Recently, we have been addressing the problem of the topological relationship between morphine and enkephalin. The inherent mobility of the enkephalin framework, its rapid degradation *in vivo* and the existence of multiple receptor subtypes have hampered the assessment of its bioactive conformations [7, 8]. Conformationally constrained peptides or peptidomimetics should facilitate this task [9, 10]. Several turn conformations have been proposed based upon computational models, X-ray crystallography and spectroscopic studies [10, 11]. In an effort to probe the receptor bound conformation of leucine enkephalin, we have synthesized a family of nonpeptide mimetic structures which incorporate a 4→1 β–turn prosthetic unit (Figure 1).

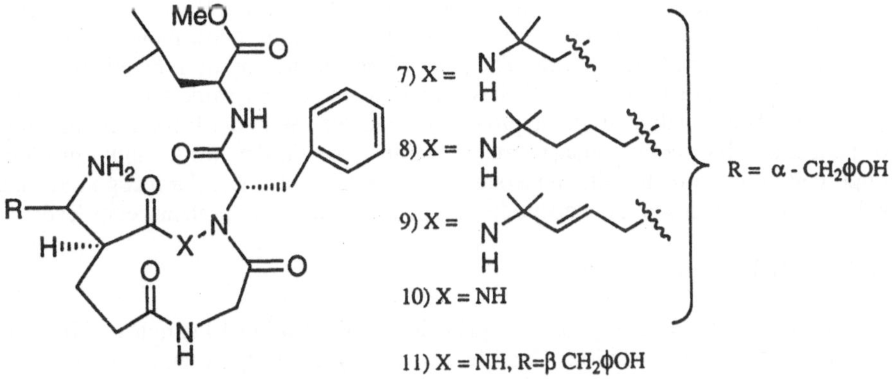

Fig. 1. *Nonpeptide mimetic structures.*

The lowest energy conformer of the 10 membered ring system 7 is an excellent mimic of an idealized type I' β–turn (6 atom rms deviation 0.22 Å) and overlays well with the critical Phe[4] aromatic ring and tyramine moieties of PET. Yet it is essentially devoid of biological activity. Only the 14-membered ring analog 9, which has a rather expanded loop structure demonstrates any, albeit minimal, binding activity at the μ or δ receptor. The results of this investigation can be interpreted as casting significant doubt on the biological relevance of a 4→1 β–turn conformation for enkephalin. During the molecular graphics investigation, we noted that by merely inverting the asymmetric center containing the tyramine moiety, we could achieve a superior fit between the 10 membered ring analog 11 and PET (Figure 2). The synthesis of 11 has been completed and its biological evaluation is in progress.

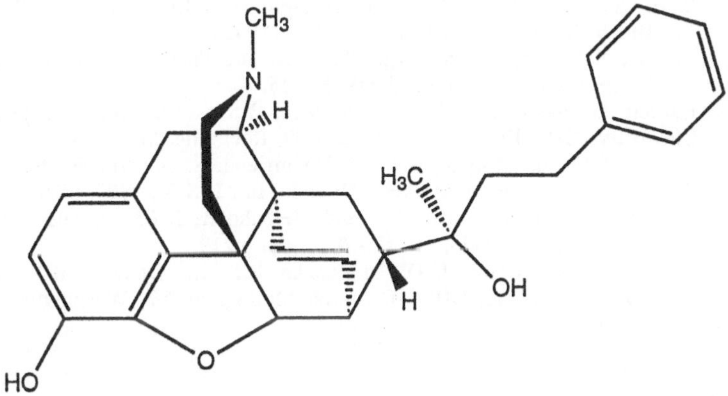

PET: 7-[1-phenyl-3-hydroxybutyl-3-]endoehenotetrahydrothebaine

Fig. 2. *The tyramine and phenyl ring moieties of enkephalin mimetic 11 (solid line) can be perfectly aligned with those of the morphine analog PET, 7-[1-phenyl-3-hydroxybutyl-3-] endoethenotetrahydrothebaine (dashed line) by a flexible fitting procedure without any steric conflict.*

Peptidomimetics are powerful tools for the study of molecular recognition. As such, they represent a critical step toward the rational design of low molecular weight nonpeptide pharmaceutical agents. The possibilities opened up by synthetic mimetic technology are in their infancy. We have focused here on our work with reverse turn mimetics. Additionally, similar approaches are underway in our laboratory utilizing α–helical and β–sheet secondary structure mimetics. Further work should provide important information to better understand the processes of molecular recognition and aid in the rational design of nonpeptide therapeutic agents from proteinaceous leads.

Acknowledgements

Generous financial support was provided by the National Institutes of Health, National Science Foundation, Camille and Henry Dreyfus Foundation, and the American Heart Association, Established Investigator Award.

References

1. Goetz, M.A., Lopez, M., Monaghan, R.L., Chang, R.S.L., Lotti, V.J. and Chen, T.B., J. Antibiot., 38 (1985) 177.
2. For a recent review see: Kahn, M., Synlett (1993) 821.
3. Reverse turn mimetics have been synthesized on both PAM and MBHA resins. Couplings are typically conducted using BOP/HOBT/DIEA in a mixture of CH_2Cl_2 and DMF at room temperature. The Boc(P) and Fmoc(P') groups are used for X-group linker and second modular component protection, respectively.
4. Saragovi, H.U., Fitzpatrick, D., Raktabutr, A., Nakanishi, H., Kahn, M. and Greene, M.I., Science, 253 (1991) 792.
5. Chen, S., Chrusciel, R.A., Nakanishi, H., Raktabutr, A., Johnson, M.E., Sato, A., Weiner, D., Hoxie, J., Saragovi, H.U., Greene, M.I. and Kahn, M., Proc. Natl. Acad. Sci. U.S.A., 89 (1992) 5872.
6. Sato, M., Lee, J.Y.H., Nakanishi, H., Johnson, M.E., Chrusciel, R.A. and Kahn, M., Bioch. Biophys. Res. Commun., 187 (1992) 999.
7. Su., T., Nakanishi, H., Xue, L., Chen, B., Tuladhar, S., Johnson, M.E. and Kahn, M., Biorg. Med. Chem. Lett., 3 (1993) 835.
8. Barnett, G. and Hawks, R.L., In Rapaka, R.S. (Ed.), Opioid Peptides: Medicinal Chemistry NIDA Research Monograph 69, Rockville, MD, U.S.A., 1986.
9. Hansen, P.E. and Morgan, B.A., In Udenfriend, S. and Meienhofer, J. (Eds.), The Peptides, Vol. 6, Academic Press, Orlando, FL, U.S.A., 1990, p.269.
10. Schiller, P.W., In Udenfriend, S. and Meienhofer, J. (Eds.), The Peptides, Vol. 6, Academic Press, Orlando, FL, U.S.A., 1990, p.219.
11. Griffin, J.F. and Smith, G.D., In Rapaka, R.S. and Dhawan, B.N. (Eds.), Opioid Peptides: An Update, NIDA Research Monograph 87, Washington, DC, U.S.A., 1988, p.41.

Orally active, non-peptide oxytocin antagonists

P.D. Williams[1], R.G. Ball[2], M.G. Bock[1], L.A. Carroll[1],
S.-H.L. Chiu[2], B.V. Clineschmidt[1], M.J. Cook[3],
M.A. Cukierski[1], C. Culberson[1], J.M. Erb[1], B.E. Evans[1],
R.M. Freidinger[1], G.J. Haluska[3], M.J. Kaufman[1],
J.L. Leighton[1], G.F. Lundell[1], M.J. Novy[3],
J.M. Pawluczyk[1], D.S. Perlow[1], D.J. Pettibone[1]
and D.F. Veber[1]

[1]*Merck Research Laboratories, West Point, PA 19486, U.S.A. and*
[2]*Rahway, NJ 07065, U.S.A. and the* [3]*Oregon Regional Primate
Reseach Center, Beaverton, OR 97006, U.S.A.*

Introduction

The nonapeptide hormone oxytocin (OT) plays a key role in the initiation and maintenance of uterine contractions of labor. The potential utility of an antagonist of OT at the uterine receptor has been demonstrated in recent clinical studies in which intravenous (i.v.) atosiban (ORF 22164; Figure 1), an antagonist analogue of OT, was shown to be efficacious in inhibiting uterine contractions in women with threatened and established preterm labor [1]. Historically, OT antagonists have been derived from structural modifications to OT and the related hormone, arginine vasopressin (AVP). We recently reported the development of a new structural class of cyclic hexapeptide OT antagonists, e.g., L-366,948 (Figure 1), based on a fermentation product identified through receptor-based screening [2]. The poor oral bioavailability of peptidyl OT antagonists such as ORF 22164 and L-366,948 places certain limitations on their potential utility as therapeutic agents. An ideal tocolytic agent would have suitable aqueous solubility to allow i.v. formulation for acute administration as well as good oral bioavailability for subsequent outpatient use until the end of gestation. Advances toward this goal were recently reported by Evans [3] and Pettibone [4] with the development of the non-peptide OT antagonist, L-366,509 (Table 1), starting with a lead identified from receptor-based screening of the Merck chemical sample collection.

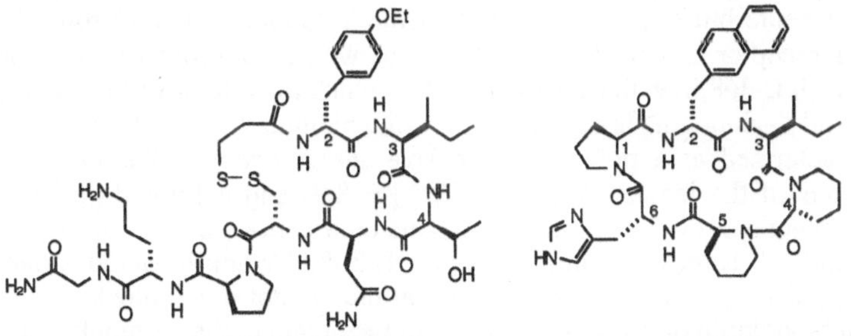

Atosiban (ORF 22164)   L-366,948

Fig. 1. *Examples of peptidyl oxytocin antagonists from the oxytocin analogue class and cyclic hexapeptide class.*

Herein we report potency-enhacing modifications to the L-366,509 structure that have led to L-368,899, a potent, soluble, and orally bioavailable OT antagonist with properties suitable for clincial testing as an i.v. and oral tocolytic agent.

Results and Discussion

A number of modifications to the spiroindenylpiperidine (SI) portion of L-366,509 were investigated with the finding that the SI group could be replaced with an *ortho*-tolylpiperazine (TP) group without loss of OT receptor affinity (cf. **1** and **2**, Table 1). The SI and TP groups are topologically similar as evidenced by the X-ray crystal structures of the oxime of **1** and ketone **2**. The phenyl and piperidine rings of the SI group are orthogonal due to the covalent constraints of the spiro linkage, and the *ortho*-methyl substituent of the TP group provides a non-covalent means of favoring a twisted orientation of the phenyl and piperazine rings. This out-of-plane orientation is preferred at the receptor as inferred from the low affinity of the analogue of **2** lacking an *ortho*-substituent (IC_{50} >30,000 nM), the X-ray crystal structure of which shows a nearly coplanar relationship of the phenyl and piperazine rings.

Molecular modeling studies using the X-ray crystal structure of **2** and an NMR-consistent, low energy conformer of the cyclic hexapeptide OT antagonist L-366,948 reveals a good alignment of **2** with the critical D-Nal[2]-Ile[3] portion of the peptide, suggesting that the TP- or SI-camphorsulfonamide group functions as a dipeptide mimetic. This alignment also indicates that substitutions on the *endo* face of the camphor ring would be oriented toward other regions of the peptide structure, consistent with SAR studies in which certain *endo* but not *exo* substitutions on the camphor ring were found to enhance OT receptor affinity. A prominent example of such a potency-enhancing substitution is the analogue which contains an *endo*-L-glutaminylamino group (**3**, Table 1). SAR studies with analogues of **3** demonstrated the important role of a properly placed hydrogen bond accepting group in the *endo* substituent side chain for obtaining high OT receptor affinity (e.g., Gln = $Met(O)_2$ > Asn > D-Gln > Leu). Molecular modeling studies using the dipeptide mimetic paradigm indicate that the important hydrogen bond accepting group in **3** can overlap with the Pip[5] or D-His[6] amide oxo groups of L-366,948. Modeling studies are currently underway to establish a structural relationship of compounds such as **3** with antagonist analogues of OT, e.g. as a mimetic of the D-Tyr(Et)[2]-Ile[3]-Thr[4] tripeptide of ORF 22164.

Compound **3** antagonized OT-stimulated uterine contractions *in vivo* in the rat by the i.v. route but only very weakly by the intraduodenal (i.d.) route (Table 1). Good OT receptor affinity and selectivity, as well as improved *in vivo* properties resulted with L-368,899, the analogue of **3** in which the dihydro-SI group is replaced with TP and L-glutamine is replaced with L-methionine sulfone (Table 1). L-368,899 was characterized as a potent, competitive antagonist of OT-stimulated uterine contractions of the isolated rat uterus (pA_2 = 8.9) and exhibited potent and long lasting antagonism of OT-stimulated uterine contractions *in vivo* in the rat by both the i.v. and i.d. routes of administration. L-368,899 administered alone did not exhibit agonist-like effects on the *in situ* rat uterus, nor did it block the contractile effects of bradykinin or prostaglandin $F_{2\alpha}$. In the pregnant rhesus monkey, L-368,899 given i.v. was a potent antagonist of both OT-induced and spontaneous nocturnal uterine contractions. L-368,899 also exhibits good affinity for the human uterine receptor (IC_{50} = 26 nM), has good aqueous solubility, is orally bioavailable in the rat, dog, and chimpanzee, and shows no adverse effects in acute toxicological studies [Pettibone, in preparation]. Because of these favorable characteristics, L-368,899 has recently entered clinical studies for testing as an i.v. and oral tocolytic agent.

Table 1 *In vitro receptor affinities and in vivo antagonist potencies for inhibition of oxytocin-induced uterine contractions*

compound	X	R_{exo}	R_{endo}
L-366,509	SI	OH	CH_2CO_2H
1	SI	=O	
2	TP	=O	
3	SI[a]	H	$NHCOCH(NH_2)CH_2CH_2CONH_2$
L-368,899	TP	H	$NHCOCH(NH_2)CH_2CH_2SO_2CH_3$

	IC_{50} (nM)			AD_{50} (mg/kg) *in situ* rat uterus	
compound	$[^3H]$-OT rat uterus	$[^3H]$-AVP rat liver V_{1a}	$[^3H]$-AVP rat kidney V_2	i.v.	i.d.
L-366,509	780	89,000	83,000		
1	1,800	>100,000	>100,000		
2	770	29,000	>100,000		
3	11	2,300	5,100	ca. 2	>30
L-368,899	8.9	370	570	0.35	7.0

[a]the indene double bond is reduced to the corresponding indane analogue

References

1. Goodwin, T.M., Paul, R.H., Silver, H., Parsons, M., Chez, R., Spellacy, R.W., Hayashi, R., North, L. and Merriman, R., Am. J. Obstet. Gynecol., 166 (1992) 359.
2. Williams, P.D., Bock, M.G., Tung, R.D., Garsky, V.M., Perlow, D.S., Erb, J.M., Lundell, G.F., Gould, N.P., Whitter, W.L., Hoffman, J.B., Kaufman, M.J., Clineschmidt, B.V., Pettibone, D.J., Freidinger, R.M. and Veber, D.F., J. Med. Chem., 35 (1992) 3905.
3. Evans, B.E., Leighton, J.L., Rittle, K.E., Gilbert, K.F., Lundell, G.F., Gould, N.P., Hobbs, D.W., DiPardo, R.M., Veber, D.F., Pettibone, D.J., Clineschmidt, B.V., Anderson, P.S. and Freidinger, R.M., J. Med. Chem., 35 (1992) 3919.
4. Pettibone, D.J., Clineschmidt, B.V., Kishel, M.T., Lis, E.V., Reiss, D.R., Woyden, C.J., Evans, B.E., Freidinger, R.M., Veber, D.F., Cook, M.J., Haluska, G.J., Novy, M.J. and Lowensohn, R.I., J. Pharm. Exp. Ther., 264 (1993) 308.

The synthesis of peptide secondary structure mimetics with covalent hydrogen bond mimics on the solid support

L.-C. Chiang, E. Cabezas, J.C. Calvo and A.C. Satterthwait

Department of Molecular Biology, The Scripps Research Institute, La Jolla, CA 92037, U.S.A.

Introduction

We have proposed a general approach for constraining peptides with covalent hydrogen bond mimics [1]. It is based on observations that different structures in proteins share a common feature, the amide-amide hydrogen bond (NH•••O=CRNH). On average, one-half of the amino acids in globular proteins form main chain amide-amide hydrogen bonds. Furthermore, secondary and irregular structures can be defined by distinct main chain hydrogen bonding patterns. We have developed synthetic techniques for replacing this weak hydrogen bond with two covalent mimics, an amidinium link (N-CR=N(H+)CH$_2$CH$_2$) and a hydrazone link (N-N=CHCH$_2$CH$_2$). The amidinium link has been shown to fold tetrapeptides into the Type 1 reverse turn [2, 4, 5] while the hydrazone link folds peptides into the alpha helix [2-6].

In order to facilitate and extend the use of hydrogen bond mimics, we have developed a protocol for the insertion of the hydrazone link into peptides during solid phase synthesis. This protocol has been adapted in our laboratory for multiple constrained peptide syntheses (MCPS) of alpha helices and loops which are finding novel uses in structural studies [6] and in vaccine research [5, 7, 8, 10, 11].

Results and Discussion

A protocol for the insertion of the hydrazone link into peptides on the solid support is outlined in Fig. 1. These syntheses require the preparation of three modified amino acids and otherwise employ standard Fmoc synthesis on the TFA labile Rink's amide resin. [(1-propylidene-2-Fmoc)hydrazino] acetic acid is employed for inserting Z in the peptide chain. Fmoc amino acid chlorides are used to extend Z by one amino acid and the peptide is end-capped with J, 5,5-dimethoxy-1-oxopentanoic acid. Cyclization is carried out on the solid support with two equivalents of HCl in 20% trifluoroethanol/DCM. The complete synthesis can be carried out manually or in part with machine assistance. We currently synthesize 20-30 cyclic peptides simultaneously using "T-bags" [9] and machine assistance from an ACT 350 multiple peptide synthesizer (Advanced Chem. Tech., Louisville, KY). Cyclized peptides are cleaved from the resin and protecting groups removed with 5% water: trifluoroacetic acid. Crude cyclic peptides are precipitated from the cleavage solution with diethyl ether and purified by HPLC on reverse phase C-18 columns using water-acetonitrile (0.1% TFA) gradients. Loops enclosing up to eighteen amino acids have been synthesized. Yields of purified products range from about 5-10% for loops to about 20% for helical peptides. However, yields are determined by the size and sequence of the peptide and can be lower. Purified constrained peptides are routinely identified by FAB MS. Selected loops and helices have been analyzed using 2D NMR spectroscopy to confirm the composition and position of the hydrazone link. Further studies are directed to conformational analyses [5-7, 10].

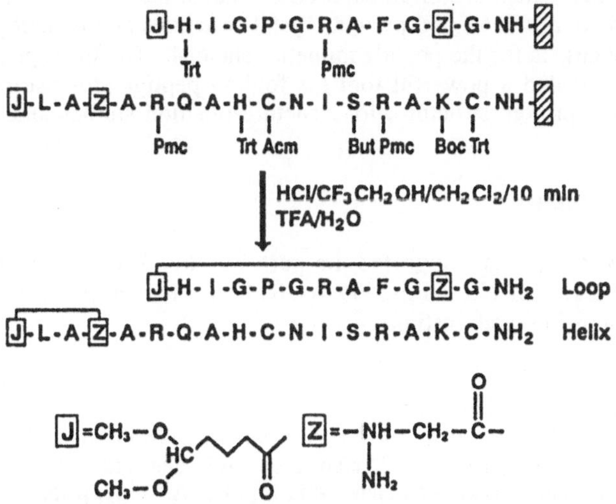

$$\boxed{J}\text{-H-I-G-P-G-R-A-F-G-}\boxed{Z}\text{-G-NH}$$
                Trt   Pmc

$$\boxed{J}\text{-L-A-}\boxed{Z}\text{-A-R-Q-A-H-C-N-I-S-R-A-K-C-NH}$$
       Pmc   Trt Acm   But Pmc   Boc Trt

HCl/CF$_3$CH$_2$OH/CH$_2$Cl$_2$/10 min
TFA/H$_2$O

$$\boxed{J}\text{-H-I-G-P-G-R-A-F-G-}\boxed{Z}\text{-G-NH}_2 \quad \text{Loop}$$

$$\boxed{J}\text{-L-A-}\boxed{Z}\text{-A-R-Q-A-H-C-N-I-S-R-A-K-C-NH}_2 \quad \text{Helix}$$

$$\boxed{J} = CH_3-O\diagdown \atop CH_3-O\diagup HC \cdots \cdots$$

$$\boxed{Z} = -NH-CH_2-\overset{O}{\overset{\|}{C}}- \atop NH_2$$

Fig. 1. *Modular synthesis of constrained peptides on the solid support with a covalent hydrogen bond mimic.*

More than a hundred peptides have been constrained to alpha helices and loops on the solid support by substituting structure defining hydrogen bonds with the hydrazone covalent hydrogen bond mimic. These peptides are chemically stable in water at neutral pH. The substitution of an (i, i + 4) hydrogen bond in the first turn of an alpha helix yields a nucleation site (NucSite) which stabilizes peptides as full length alpha helices in water [4-6] while the substitution of predicted hydrogen bonds more remote in sequence yields midsized loops with enhanced biological properties [5, 7, 8, 10, 11].

Paradoxically, an important property of the hydrazone linker may be its flexibility. Amino acids linked by hydrogen bonds in proteins assume different spatial orientations determined by the amino acids and their local and global contexts. While a covalent hydrogen bond mimic is designed to replace the hydrogen bond, its wider use is dependent on its adaptability to different spatial and steric requirements. The hydrazone link appears to meet these requirements. As amino acids are substituted in NucSite, a 13-membered ring, its ability to nucleate alpha helix formation has been found to be exquisitely sensitive to the particular substitution [6]. This demonstrates both the flexibility and steric compatibility of the linker and shows how different amino acids in the context of the link can be brought to bear on structure.

The same phenomenon may be important for loops. The hydrazone link has been used to constrain malaria peptides to loops that bind monoclonal and polyclonal antibodies with conformational requirements [7, 8]. Remarkable improvements in antigenicity are observed, providing a method for identifying antibodies to conformational epitopes [7, 8]. These antigenic loops have been used to co-crystallize neutralizing Fabs for the malaria parasite [7] and HIV-1[10] and for isolating a neutralizing human Fab to HIV-1 from a combinatorial Fab-phage library derived from an infected individual [11].

Approximately 50% of the amino acids in globular proteins fold into alpha helices and loops which are often exposed on the surface of proteins. Secondary structure predictions, NMR and protein crystal structures and homologous proteins provide many targets for the peptide mimetic chemist. The hydrogen bond mimic approach has yielded a powerful tool for folding peptides for use in identifying determinants of protein structure, for structure-function studies and for drug and vaccine design.

Acknowledgements

This research was supported in part by the Agency for International Development Contract No. DPE-5979-A-1035-00. This is publication 8126-MB from The Scripps Research Institute.

References

1. Arrhenius, T., Lerner, R.A. and Satterthwait, A.C., In Oxender, D. (Ed.), Protein Structure and Design (UCLA Symposia on Molecular and Cellular Biology, New Series 69), Alan R. Liss, Inc., New York, NY, U.S.A., 1987, p.453.
2. Arrhenius, T., Chiang, L.-C., Lerner, R.A. and Satterthwait, A.C. In Lerner, R.A., Ginsberg, H., Chanock, R.M., Brown, F. (Eds.), Vaccines 89: Modern Approaches to Vaccines Including Prevention of AIDS, Cold Spring Harbor Laboratory, New York, NY, U.S.A., 1989, p.17.
3. Arrhenius, T. and Satterthwait, A.C. In Rivier, J.E. and Marshall, G.R. (Eds.), Peptides: Chemistry, Structure and Biology (Proceedings of the Eleventh American Peptide Symposium), ESCOM, Leiden, The Netherlands, 1990, p.870.
4. Chiang, L., Cabezas, E., Noar, B., Arrhenius, T., Lerner, R.A. and Satterthwait, A.C. In Giralt, D. and Andreu, A. (Eds.), Peptides 1990 (Proceedings of the 21st European Peptide Symposium), Escom, Leiden, The Netherlands, 1991, p.465.
5. Chiang, L.-C., Cabezas, E., Calvo, J. and Satterthwait, A.C. In Du, Y. (Ed.), Peptides: Biology and Chemistry (Proceedings of the Chinese Peptide Symposium 1992), ESCOM, Leiden, The Netherlands, 1993, p.204.
6. Cabezas, E., Chiang, L.-C. and Satterthwait, A.C., this volume.
7. Stura, E., Kaslow, D. and Satterthwait, A.C., this volume.
8. Calvo, J., Perkins, M. and Satterthwait, A.C., this volume.
9. Houghten, R.A., Proc. Natl. Acad. Sci. U.S.A., 82 (1985) 5131.
10. Wilson, I., Abst. LW1, Thirteenth American Peptide Symposium.
11. Barbas, C.F. III, Collett, T.A., Amberg, W., Roben, P., Binley, J.M., Hoekstra, D., Cababa, D., Jones, T.M., Williamson, R.A., Pilkington, G.R., Haigwood, N.L., Cabezas, E., Satterthwait, A.C., Sanz, I. and Burton, D.R., J. Mol. Biol., 230 (1993) 812.

Potent renin inhibition activity of tetrapeptide mimetics with a 1,2-hydroxyazidoethylene group connecting the P_1 and P_1' residues

R.G. Almquist[1], A. Nakazato[2], K. Kameo[2],
H. Fukushima[2] and W.-R. Chao[3]

[1]*Biogen Inc., Cambridge, MA 02142, U.S.A.*
[2]*Taisho Pharmaceutical Co., Ltd., Omiya-Shi, Saitama, 330 Japan*
[3]*SRI International, Menlo Park, CA 94025, U.S.A.*

Introduction

In an effort to produce compounds with enhanced binding affinity for renin, we prepared a series of tetrapeptide mimetics with a 1,2-hydroxyaminoethylene group connecting the P_1 and P_1' residues. We hoped to obtain increased activity compared to the 1,2-dihydroxyethylene class of renin inhibitors [1] by adding a potential ionic bond between the amino group of the 1,2-hydroxyaminoethylene linkage and the carboxylate group of Asp 215 at the cleavage site of renin. During the course of this work, a publication by Rosenberg et al., [2] called our attention to the possibility that 1,2-hydroxyazidoethylene derivatives obtained as intermediates in our syntheses could be potent renin inhibitors themselves. The synthesis and renin inhibition activities of these derivatives is the subject of this report.

Results and Discussion

Scheme 1 outlines the synthetic approach that was used to prepare both the 1,2-hydroxyamino- and hydroxyazidoethylene containing renin inhibitors. The stereochemistry and location of the hydroxy and azido centers in compounds 4a-d and 14 were determined by NMR analyses of epoxides 3a-d, 13 and of oxazolidinones resulting from treatment of 4a-d and 14 with NaH.

The results of renin inhibition testing of various 1,2-hydroxyamino- and hydroxyazidoethylene derivatives are shown in Table 1. These results show that the hydroxyamino derivatives are poor inhibitors of renin with IC_{50}s at best in the micromolar range. Primary amines as in 6b, 10 and 18 are known to form solvated ammonium groups in aqueous solution at the pH (6.0) of the renin inhibition assay. Perhaps the energy required to desolvate the ammonium group is not compensated for by an ionic interaction with Asp 215 in the renin active site. Alternatively, maybe the amino group is not positioned properly within the active site to undergo an ionic interaction.

The 1,2-hydroxyazidoethylene derivatives have much greater renin inhibition activities than their corresponding hydroxyamino analogs. The stereochemistry of the most active isomer in the hydroazido series is the same as that reported for the most active isomer in the dihydroxy series [1]. In most cases the renin inhibition activities of these isomers are also similar to their dihydroxyethylene counterparts. However, in the case of compound 20, the 1,2-hydroxyazidoethylene compound has exceptional renin inhibition potency. Renin binding of the amino terminal t-butylsulfonyl group of 20 must position the azido group so that it can participate in exceptionally tight binding with renin. A Lineweaver-Burk plot of the interaction of 20 with human renin shows that it is a competitive inhibitor with a K_i of 7.2 pM. This makes it one of the most potent renin inhibitors known.

Scheme 1

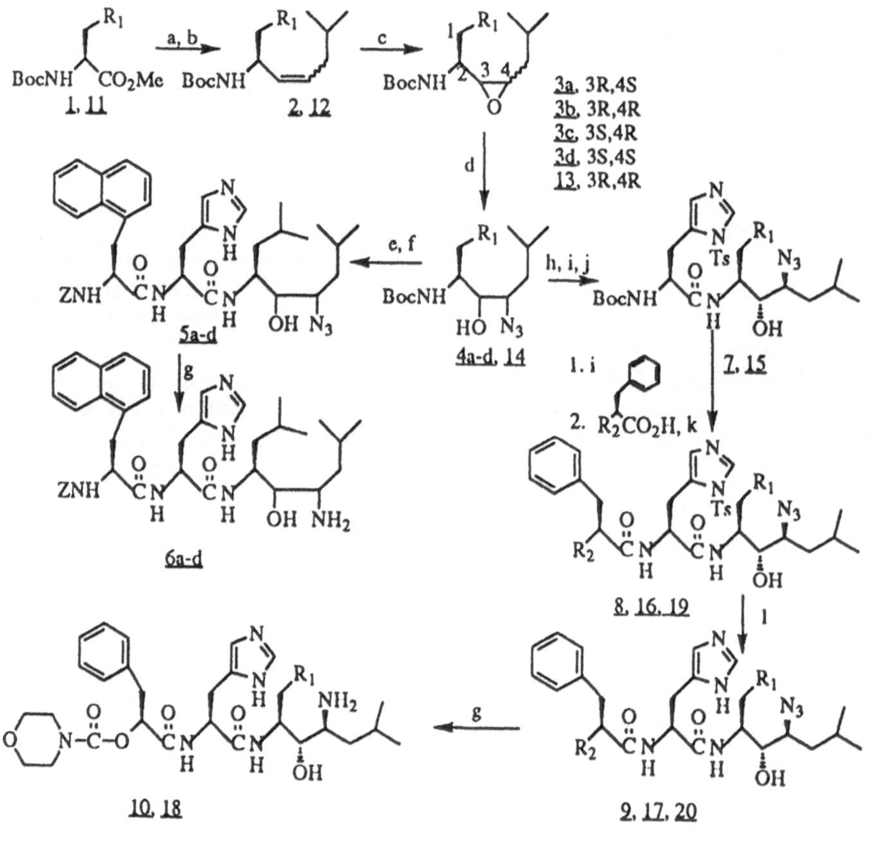

Compounds *1-4* and *7-10*, R_1 = $CH(CH_3)_2$, R_2 = Morpholinyl-CO_2; Compounds *11-18*, R_1 = cyclohexyl, R_2 = morpholinyl-CO_2; Compounds *19, 20*, R_1 = Cyclohexyl, R_2 = $tBuSO_2CH_2$. (a) DIBAL-H. (b) Ph_3P = $CHCH_2CH(CH_3)_2$. (c) mCPBA. (d) NaN_3, NH_4Cl. (e) HCl/Dioxane. (f) Z-L-Nal(1)-His-OH, EDC.HCl, HOBt, NMM. (g) $HS(CH_2)_3SH$, Et_3N. (h) separate *4b* (3R,4S isomer). (i) TFA-CH_2Cl_2. (j) Boc-L-His(Ts)OH, Et_3N, DEPC. (k) DEPC, Et_3N. (l) HOBt, MeOH.

Table 1 *Human renin inhibition activities*

No.	R_1	R_2	R_3	Human Renin IC_{50} (μM)
5b	Z-L-Nal(1)	$CH(CH_3)_2$	N_3	0.090
6b	Z-L-Nal(1)	$CH(CH_3)_2$	NH_2	>10
9	Morpho-COOCH(Bzl)CO	$CH(CH_3)_2$	N_3	0.110
10	Morpho-COOCH(Bzl)CO	$CH(CH_3)_2$	NH_2	>10
17	Morpho-COOCH(Bzl)CO	C_6H_{11}	N_3	0.0016
18	Morpho-COOCH(Bzl)CO	C_6H_{11}	NH_2	2.8
20	*t*-BuSO$_2$CH$_2$CH(Bzl)CO	C_6H_{11}	N_3	0.000007

Morpho = Morpholine

The azide group with its partial negative charge on the nitrogen attached to carbon may be able to form a strong hydrogen bond with a proton donor such as Ser 76 in human renin, as has been proposed for the carboxyterminal hydroxyl group of the dihydroxyethylene inhibitor class. The idea of substituting an azido group for a hydroxyl group may prove useful in designing other enzyme inhibitors or peptide mimetics.

Acknowledgements

This research was supported by Taisho Pharmaceutical Co. Ltd..

References

1. Luly, J.F., BaMaung, N., Soderquist, J., Fung, A., Stein, H., Kleinert, H., Marcotte, P., Egan, D., Bopp, B., Merits, I., Bolis, G., Greer, J., Perun, T. and Plattner, J., J. Med. Chem., 31 (1988) 2264.
2. Rosenberg, S.H., Woods, K.W., Plattner, J.J., Stein, H.H., Kleinert, H.D., Cohen, J., In Marshall, G.R. (Ed.), Peptides: Chemistry and Biology, (Proceedings of the Tenth Amer. Pept. Symp.), Escom, Leiden, The Netherlands, 1988, p.500.

Gramicidin S: A general model for β-turn mimics?

A.C. Bach, II, R.S. Pottorf*[1a], J.M. Blaney*[1b],
G.V. De Lucca and W.C. Ripka*[1c]

*The Dupont Merck Pharmaceutical Company, Experimental Station
Box 80336, Wilmington, DE 19880-0336, U.S.A.*

Introduction

As part of our study of β-turn mimics [1, 2, 3], particularly benzodiazepine-based (BZD) mimics (1), we examined gramicidin S (GS) as a model system. NMR studies of BZD-GS 2 showed no preferred conformation. This was surprising since native GS has a very stable solution conformation, with two type II' β-turns connected by a short anti-parallel sheet [4], and is considered to be a good physical model for studying potential β-turn mimics [3]. Altering the GS native sequence of Val-Pro-D-Phe-Leu to the BZD mimic changes four residues at once, alters both the symmetry and turn type, and removes the Leu and Val sidechains. Molecular modeling suggested that the two most likely reasons for the observed instability are the missing Leu and Val sidechains and the turn asymmetry that 2 may have. We synthesized and analyzed several peptide analogs of GS (Fig. 1) to determine which of these possibilities cause the conformational instability. The Leu and Val residues were replaced with Gly in 3. γ-lactams [5, 6] were used to constrain one turn in GS to type II', (4), or type II (5), to assess the effects of turn asymmetry.

Results and Discussion

NMR studies of 3 in methanol-d_3 (Table 1), indicate it adopts a solution conformation which is very similar to that of native GS in methanol-d_3. The same pattern of intra-molecular hydrogen bonding is found. Both the amide $\Delta\delta/\Delta T$ values and $J_{N-\alpha}$ coupling constants support this observation. Molecular dynamics (MD) studies also indicate that 3 can adopt a low energy conformation which is very similar to that of native GS. The RMS difference between the 200 ps MD average conformation of 3 and a low energy conformation of GS [4] is 0.52 Å. These data indicate that 3 does adopt a GS-like solution conformation, and the removal of the Leu and Val sidechains is not a critical factor in the conformational instability of 2.

Similar NMR results were obtained for 4 in methanol-d_3 (Table 1). In 4 the same intra-molecular hydrogen bonding pattern is found with approximately the same $\Delta\delta/\Delta T$ and $J_{N-\alpha}$ values. MD studies show that the 200 ps average conformation of 4

* Present addresses: [1a]Marion Merrell Dow Research Institute, 2110 E. Galbraith Road, Cincinnatti, OH 45215, U.S.A.; [1b]Chiron Corporation, 4560 Horton Street, Emeryville, CA 94608, U.S.A.; [1c]Corvas International, Inc., 3030 Science Park Road, San Diego, CA 92121-1102, U.S.A.

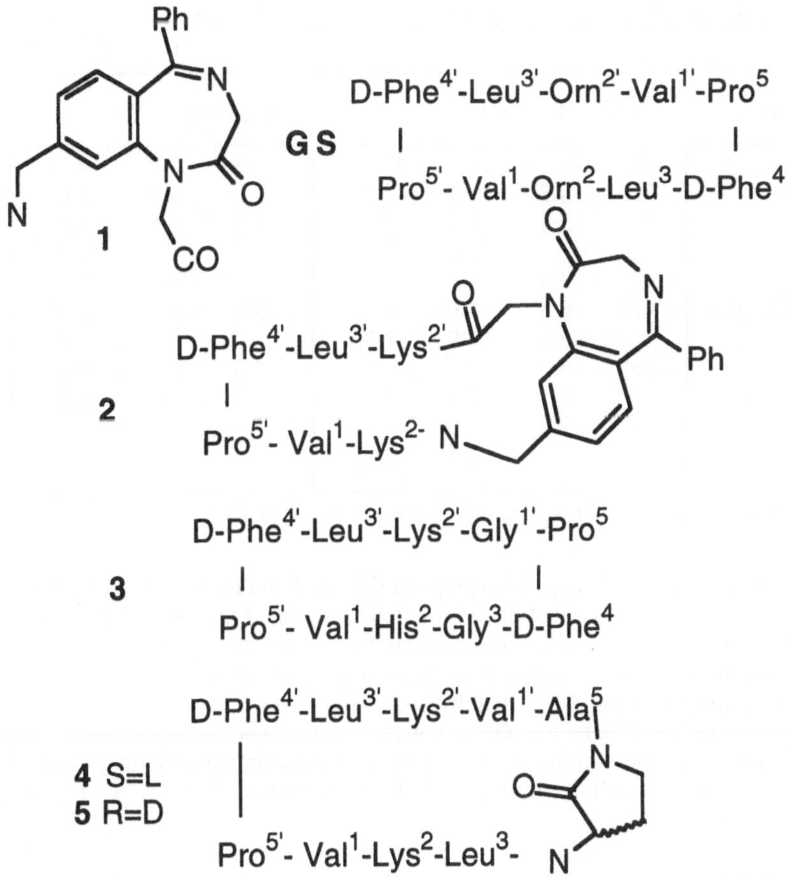

Fig. 1. **1**, benzodiazepine β-turn mimic; GS, gramicidin S; **2**, c(Val-Lys-BZD-Lys-Leu-D-Phe-Pro); **3**, [Gly$^{3,1'}$, His^2, Lys$^{2'}$]GS; **4**, [S-Lct^4, Ala^5, Lys$^{2,2'}$]GS; **5**, [R-Lct^4, Ala^5, Lys$^{2,2'}$]GS.

has a RMS difference of 0.48 Å from GS. The results for **4** indicate that an adequate type II' β-turn mimic is compatible with the solution conformation of GS.

5 had roughly similar NMR parameters as **4** but also showed some striking differences (Table 1); most notable were differences in $\Delta\delta/\Delta T$ for the Leu^3. Also, the D-Phe and lactam $J_{N-\alpha}$'s are not resolved and the Leu^3 coupling changes with temperature. In general, the $J_{N-\alpha}$'s are not as large in **5**, indicating the solution structure is not as conformationally stable as GS. MD studies show that the 200 ps average conformation of **5** has a RMS difference of 0.86 Å from GS. These results indicate **5** is less conformationally stable than **4** or GS. Since the D-lactam mimics a type II turn, and the resulting mixed turn type compound is less conformationally stable, then **2** may also be unstable because the BZD mimic is not particularly good at mimicking a type II' turn.

Table 1 *NMR data compared to GS in methanol-d$_3$*

Residue	$\Delta\delta/\Delta T$ (ppb/°C)				$J_{N-\alpha}$ (Hz)			
	GS	3	4	5	GS	3	4	5
D-Phe	-7.1	-8.2[a]	-7.1	-7.2	3.0	4.0	3.0	-[b]
Leu	-2.1	-2.8	-2.0	+0.2	9.5	9.1	8.7	8.4[c]
			-2.6	-1.8			9.1	7.6
Gly		-2.6				Σ=9.1		
Orn(Lys,His)	-6.0	-6.5	-5.0	-5.1	9.5	8.1	9.0	8.5
		-6.5	-5.9	-4.3		8.6	9.0	8.5
Val	-2.4	-3.5	-2.0	-1.9	9.0	9.1	9.2	8.0
			-1.6	-3.1			9.2	8.0
Gly		-2.9				Σ=9.4		
Lct			-5.1	-4.9			6.6	-[b]

[a]resonances degenerate. [b]not resolved. [c]coupling disappears with ΔT.

Using asymmetric peptide analogs of GS we demonstrated that: the loss of the Leu and Val sidechains is not the reason **2** is unstable; **4** adopts a conformation very similar to that of native GS, which appears to be even more stable than GS by MD; **5** also adopts a conformation very similar to that of native GS, but is not as conformationally stable as GS by NMR and MD.

This study indicates that GS is sensitive to the type of β-turn or mimic which is incorporated into the molecule. Since **1** was designed to be a general β-turn mimic, it may not mimic the type II' turn well enough for GS to adopt a stable conformation.

References

1. Bach, II, A.C., Markwalder, J.A. and Ripka, W.C., Int. J. Pept. Prot. Res., 38 (1991) 314.
2. Ripka, W.C., De Lucca, G.V., Bach, II, A.C., Pottorf, R.S. and Blaney, J.M., Tetrahedron: Symposium in Print, 49 (1993) 3593.
3. Ripka, W.C., De Lucca, G.V., Bach, II, A.C., Pottorf, R.S. and Blaney, J.M., Tetrahedron: Symposium in Print, 49 (1993) 3609.
4. Dygert, M., Go, N. and Scheraga, H., Macromolecules, 8 (1975) 750.
5. Freidinger, R.M., Perlow, D.S. and Veber, D.F., J. Org. Chem., 47 (1982) 104.
6. Freidinger, R.M., In Rich, D.H. and Gross, E. (Eds.), Peptides Synthesis Structure Function (Proceedings of the Seventh American Peptide Symposium), Pierce Chemical Co., Rockford, IL, U.S.A., 1981, p.673.

A role for amino acids in the cooperative interactions between an alpha helix nucleation site and appended peptides

E. Cabezas, L.-C. Chiang and A.C. Satterthwait

Department of Molecular Biology, The Scripps Research Institute,
La Jolla, CA 92037, U.S.A.

Introduction

The substitution of the structure defining (i, i + 4) amide backbone hydrogen bond (NH··O=RNH) in one turn of the alpha helix with a hydrazone covalent hydrogen bond mimic (N-N=CH-CH$_2$CH$_2$) yields a nucleation site (NucSite). When NucSite is covalently linked to the N-terminus of test peptides, AE(OEt)E(OEt)-E(OEt)E(OEt)$_2$ or AEAAKA-NH$_2$, it induces full length alpha helix formation in trifluoroethanol [1] and water [2] at ambient temperatures.

The hydrogen bond mimic is part of a flexible linker which allows peptides to adapt different conformations depending on the amino acid context. Here we explore the role that different amino acids in NucSite can play in alpha helix induction. By varying amino acids in NucSite, alpha helical content in AEAAKA-NH$_2$ can be controlled over a wide range. In addition, the effect of Aib on conformation and helix stabilization is examined. A series of peptides differing only in the extent of alpha helix formation could find important uses in structure-reactivity studies.

Results and Discussion

The peptides used in this study are found in Fig. 1. They include a series of NucSite-AEAAKA-NH$_2$ peptides with amino acid substitutions in NucSite, a corresponding linear peptide and a parallel series of NucSite-AEAAibKAib-NH$_2$ peptides. The syntheses of these peptides including the insertion of a hydrogen bond mimic were carried out on the solid support [3]. The yields of purified products (20% in most cases) were sufficient for NMR studies.

Conformational analysis were carried out using NMR spectroscopy which provides critical distance information for distinguishing helix types [4]. All NMR studies were carried out with peptides in 10% D$_2$O/H$_2$O at 22°C. Proton chemical shift assignments for the linear and NucSite peptides were made on the basis of 1D-NMR (Fig. 2), DQCOSY, TOCSY and ROESY 2D-NMR spectra. NOE profiles were consistent with alpha helix formation. In particular, sequential αN(i, i + 3) and αβ(i, i + 3) NOE's spanning the full length of the molecules were observed for the [LA]NucSite and [LP]NucSite peptides providing strong evidence for full length alpha helix formation. Ratios of αN(i, i + 3)/αβ(i, i + 3) for each pair of protons spanning LAZ (L-leucine-L-alanine-N-aminoglycine) in NucSite and AEA in the

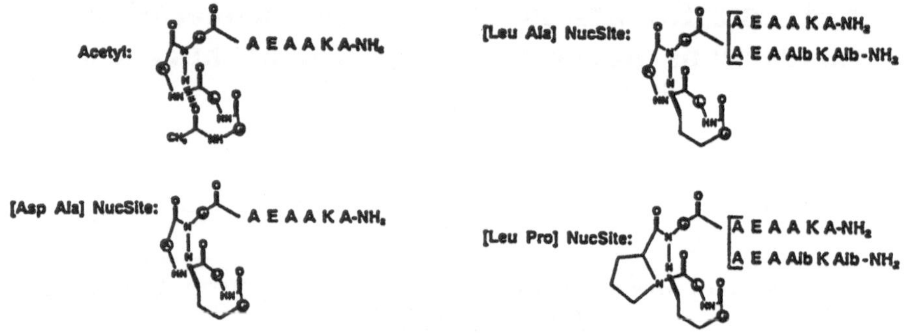

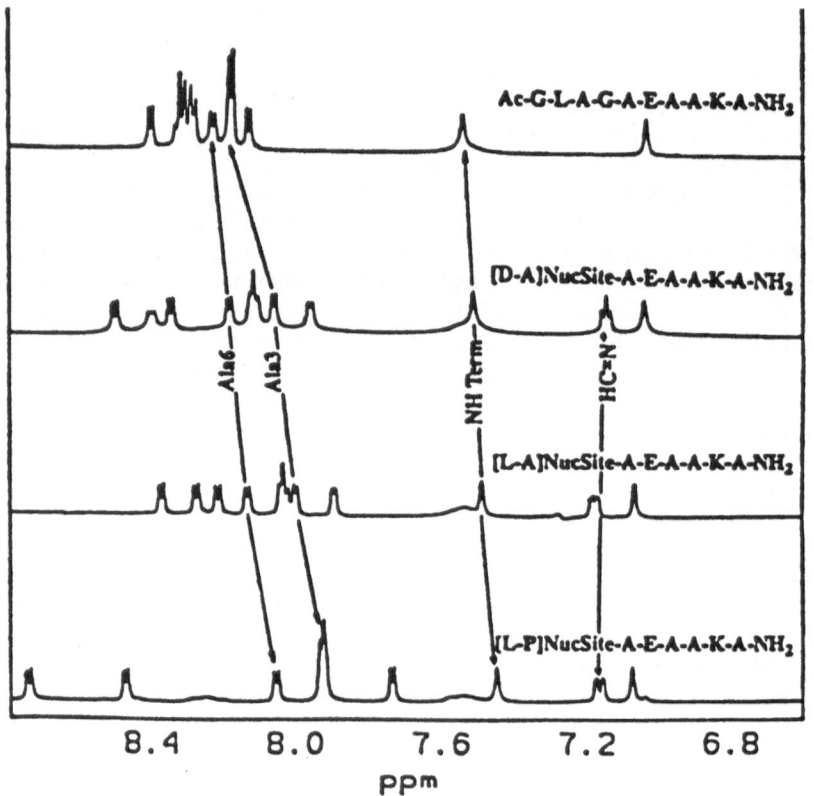

Fig. 1. *Peptide series. The one letter amino acid code for the amino acids in NucSite are placed on the alpha carbons to indicate the side chain.*

Fig. 2. *Amide NH region of the 1D NMR spectra for a series of a linear and NucSite peptides which differ in the composition of the nucleation site (10% D_2O/H_2O, 22°C). Signals for the hydrazone proton N=CH showing changes in coupling constants and upfield shifts for the A^3, A^6 and terminal NH protons are indicated with the solid lines. The changes in coupling constants are correlated with an upfield shift in the amide NH resonances and increases in alpha helix formation.*

first turn of the nucleated alpha helix were 0.1-0.4 providing strong support for alpha helix nucleation as opposed to 3_{10} helix nucleation [4]. The substitution of Aib for Ala in the C-terminal fragment did not significantly alter these ratios.

Signals for the N=CH proton show large changes in coupling constants for N=CHCH$_2$ indicating a change in the conformation of NucSite (Figure 2). This is accompanied by upfield shifts in the signals for amide NH protons for the C-terminal fragment, AAKA-NH (Figure 2). These changes are consistent with NucSite conformation governing an increase in alpha helical content of the added peptide with helical content for [LP]NucSite peptide > [LA]NucSite peptide > [DA]NucSite peptide > acetyl-GLAGAEAAKA-NH$_2$. Ratios of αN(i, i + 1)/NN(i, i + 1) have been used to estimate the extent of alpha helix formation [5]. By these measurements, [LA]NucSite nucleates alpha helix formation in about 40% of the added peptide while [LP]NucSite nucleates about 65%. Aib substitution at the C-terminus may increase the extent of nucleation slightly. The best evidence for an increase in alpha helicity is provided by an increase in intensities for the critical αN(i, i + 3) and $\alpha\beta$(i, i + 3) NOE's. Intensities for [LP]NucSite peptides are about double those for [LA]NucSite peptides. The [DA]NucSite AEAAKA-NH$_2$ shows only one weak $\alpha\beta$(i, i + 3) NOE.

A consistent picture emerges from these studies. Amino acid substitution in NucSite can be used to govern the extent of alpha helix formation in an appended peptide over a wide range. Thus the helix potential of individual amino acids in the context of the flexible hydrazone covalent hydrogen bond mimic in NucSite are expressed. Hydrogen bond mimics were designed to take advantage of position and sequence to fold peptides into conformations governed by these parameters as is evident here.

Acknowledgements

This research was supported in part by the Agency for International Development Contract No. DPE-5979-A-1035-00. This is publication 8127-MB from The Scripps Research Institute.

References

1. Arrhenius, T. and Satterthwait, A.C. In Rivier, J.E., Marshall, G.R. (Eds.), Peptides: Chemistry, Structure and Biology (Proceedings of the Eleventh American Peptide Symposium), Escom, Leiden, The Netherlands, 1990, p.870.
2. Chiang, L.-C., Cabezas, E., Calvo, J. and Satterthwait, A.C. In Du, Y. (Ed.), Peptides: Biology and Chemistry (Proceedings of the Chinese Peptide Symposium 1992), Escom, Leiden, The Netherlands, 1993, p.204.
3. Chiang, L.-C., Cabezas, E., Calvo, J. and Satterthwait, A.C., this volume.
4. Wüthrich, K., Billeter, M. and Werner, B., J. Mol. Biol., 180 (1984) 715.
5. Bradley, E.K., Thomason, J.F., Cohen, F.E., Kosen, P.A. and Kuntz, I.D., J. Mol. Biol., 215 (1990) 607.

Design and synthesis of a conformational mimetic of β-turn secondary structure

J.F. Callahan[1], P.W. Schiller[2], C. Lemieux[2] and W.F. Huffman[1]

[1]Department of Medicinal Chemistry, SmithKline Beecham Pharmaceuticals, 709 Swedeland Rd., King of Prussia, PA 19406, U.S.A. and [2]Laboratory of Chemical Biology and Peptide Research, Clinical Research Institute of Montreal, Montreal, Quebec, Canada

Introduction

Mimetics of peptide secondary structure, that constrain and thereby limit the various conformations available to a particular peptide, have become valuable tools in the medicinal chemist's repertoire for the design of novel agonists or antagonists of peptide action [1]. These mimetics can define various aspects of the biologically active conformation of a peptide and may serve as a template for the assembly of pharmacophore elements, allowing for the ultimate design of non-peptide agonists and antagonists. The most common form of reverse turn found in small peptides is the β-turn (**I**), which is defined by a hydrogen bond between the carbonyl of the first (i) residue and the amide nitrogen of the fourth (i+3) residue of the reverse turn [2]. The β-turn mimetic (**II**) was designed (Figure 1) so that the amide bond between the

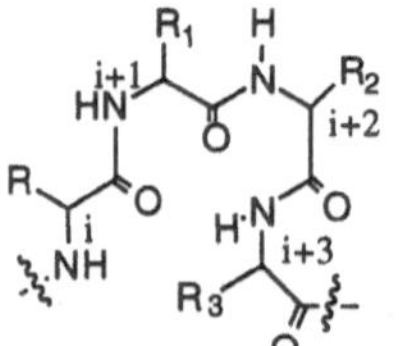

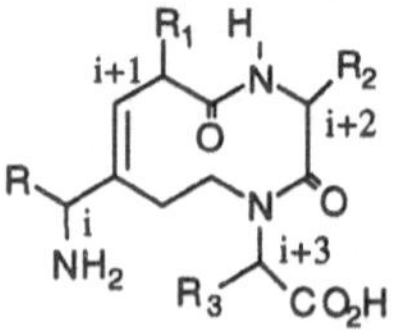

Fig. 1. Design of β-turn mimetic (II).

first (i) and second (i+1) residue of the turn was replaced with a *trans* olefin which was then bridged via two methylene units to the nitrogen of the fourth residue (i+3). The result is a β-turn mimetic (**II**) that constrains the backbone conformation of the turn containing residues while allowing for the incorporation of the requisite side chains. The synthesis of **6**, an example of the mimetic (**II**), and its incorporation into the model enkephalin peptide **8** is described below. Enkephalin was chosen as a model due to its multiple receptor pharmacophores, some of which have been proposed to contain a β-turn, its relative structural simplicity and its use as a model for several other conformational mimetics [3].

Results and Discussion

The synthesis of the mimetic-containing peptide **8** is outlined in Fig. 2. Reductive amination of the aldehyde **1** [4] with protected leucinol gave the amine **2** which is then coupled with (Boc-Phe)$_2$O to give **3**. Hydrolysis of **3** followed by treatment with anhydrous 4N HCl in dioxane and cyclization gave the 10-membered ring mimetic **5**. The primary alcohol in **5** was converted, via the Mitsunobu reaction [5], to the phthalimide **6** which was then deprotected and coupled to Boc-Tyr-OH to give **7**. Final deprotection with anhydrous HF gave the mimetic-containing peptide **8**.

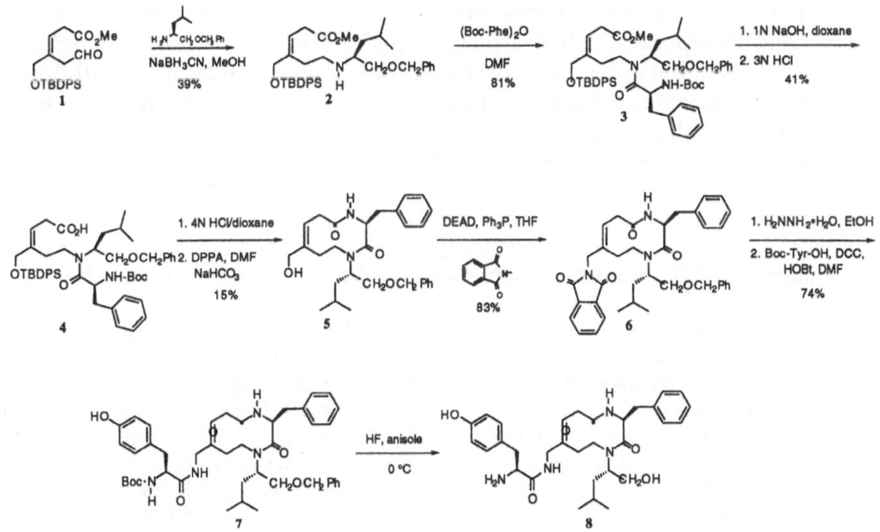

Fig. 2. Synthesis of β-turn mimetic-containing peptide 8.

The affinity and biological activity of **8** are presented in Table 1. The mimetic containing peptide **8** retains about 1% of the affinity of [Leu^5]-enkephalin for the μ receptor while showing the same moderate preference for μ receptors over δ receptors as its parent peptide, [Leu-ol^5]-enkephalin **9**. In the guinea pig ileum (GPI) assay the mimetic containing peptide **8** behaved as a partial agonist, producing inhibition of the electrically evoked contractions to a maximal extent of 30% at a concentration of 1x10^{-4} M. These results indicate that the introduction of the mimetic not only reduces receptor affinity but also affects signal transduction. In the mouse vas deferens (MVD) assay **8** was virtually inactive.

Preliminary molecular modeling studies [6] have shown that the 10-membered ring of mimetic (**II**) has several accessible low energy conformations, the lowest of which most closely resembles a Type III β-turn. Further work is needed to correlate these low energy conformations to the solution conformation of **8**.

Table 1 *Receptor affinity and biological activity* [7]

Compound	[³H]DAGO Displacement K_i (nM)	[³H]DSLET Displacement K_i (nM)	GPI IC_{50} (nM)	MVD IC_{50} (nM)
8	801±39	5,640±340	>24,600	>114,000
9	1.94±0.21	31.1±0.4	124±23	324±14
[Leu⁵]-Enk	9.43±2.07	2.53±0.35	246±39	11.4±1.1

In conclusion, we have successfully designed and synthesized a novel β-turn mimetic and incorporated it into a linear peptide. The low affinity and reduced intrinsic activity of **8** may be due to the fact that the mimetic favors a Type III β-turn conformation instead of a Type I or Type II β-turn conformations that have been proposed for the enkephalin pharmacophores [3]. Alternatively, the loss of affinity may also be caused by some structural feature inherent in the mimetic itself.

References

1. Holtzemann, G., Kontakte (Darmstadt), (1991) 3; Holtzemann, G., Kontakte (Darmstadt), (1991) 55.
2. Smith, J.A. and Pease, L.G., In Fasman, G.D. (Ed.), CRC Crit. Rev. Biochem., Vol. 8, CRC Press, Inc., Boca Raton, FL, U.S.A., 1980, p.315.
3. Gardner, B., Nakanishi, H. and Kahn, M., Tetrahedron, 49 (1993) 3433; Currie, B.L., Krstenansky, J.L., Lin, Z.-L., Ungwitayatorn, J., Lee, Y.-H., del Rosario-Chow, M., Sheu, W.-S. and Johnson, M.E., Tetrahedron, 49 (1993) 3489.
4. Huffman, W.F., Callahan, J.F., Eggleston, D.S., Newlander, K.A.,Takata, D.T., Codd, E.E., Walker, R.F., Schiller, P.W., Lemieux, C., Wire, W.S. and Burks, T.F., In Marshall, G.R. (Ed), Peptides: Chemistry and Biology, Proceedings of the Tenth American Peptide Symposium, Escom, Leiden, The Netherlands, 1988, p.105.
5. Mitsunobu, O., Synthesis, (1981) 1.
6. The lowest energy conformation was calculated using the multiconformer and Batchmin utility found in MacroModel Version 3.1x, W. Clark Still, Department of Chemistry, Columbia University, New York, NY 10027, U.S.A.
7. For details regarding opiate binding and smooth muscle contraction assays see: Bryan, W.M., Callahan, J.F., Codd, E.E., Lemieux, C., Moore, M.L., Schiller, P.W., Walker, R.F. and Huffman, W.F., J. Med. Chem., 32 (1989) 302 and references cited therein.

"Shifty peptides" a novel topochemical modification; model peptides and substance P related analogs

R. Turgeman, G. Bar-Akiva, Z. Selinger, A. Goldblum and M. Chorev

Departments of Pharmaceutical and Biological Chemistry, The Hebrew University of Jerusalem, Jerusalem 91120, Israel

Introduction

Interaction between ligands and their respective macromolecular biological targets relies very much on topochemical complementarity. Peptides play different roles in many regulatory biological processes. They act as hormones, neurotransmitters, neuroregulators and neurotoxins. Nevertheless, due to their peptidic nature very few peptides or peptide analogs are used as therapeutic agents. One of the major efforts in modern drug design is the search for non-peptidic structural modifications which will reproduce, as close as possible, the putative bioactive topochemistry and therefore mimic the biological activity of the original peptide. To this end, we have been working on the development of a structural isomerization of amino acids in which the side-chain on the $C\alpha$ is shifted to the adjacent nitrogen. The isomeric peptide obtained as a result of this operation is designated as "Shifty Peptide" (e.g. see Scheme 1). To compensate for the loss of chirality at the $C\alpha$, that this modification entails, we decided to study also the D- and L-alanyl residues substituted on the N by the natural side-chain. These residues were termed "L-Ala- and D-Ala-Shifty Analogs" (e.g. see Scheme 1).

Results and Discussion

We chose substance P-derived hexapeptide, pGlu-Phe-Phe-Gly-Leu-Met-NH$_2$, as our model structure for testing the effect of a single "Shifty Modification" on biological activity. The following "Shifty Analogs" were prepared by SPPS purified by RP-HPLC and characterized by amino acid analysis and FAB-MS:

Phe[7]-position "Shifty Analogs"
I. [pGlu[6],N-BzlGly[7]]SP(6-11), II. [pGlu[6],N-BzlAla[7]]SP(6-11), III. [pGlu[6],N-Bzl-D-Ala[7]]SP(6-11)
Phe[8]-position "Shifty Analogs"
IV. [pGlu[6],N-BzlGly[8]]SP(6-11), V. [pGlu[6],N-BzlAla[8]]SP(6-11), VI. [pGlu[6],N-Bzl-D-Ala[8]]SP(6-11)
Leu[10]-position "Shifty Analogs"
VII. [pGlu[6],N-iBuGly[10]]SP(6-11), VIII. [pGlu[6],N-iBuAla[10]]SP(6-11), IX. [pGlu[6],-N-iBu-D-Ala[10]]SP(6-11)

Met[11]-position "Shifty Analogs"

X. [pGlu[6],N-(CH₂)₂SCH₃Gly[11]]SP(6-11)

Scheme 1

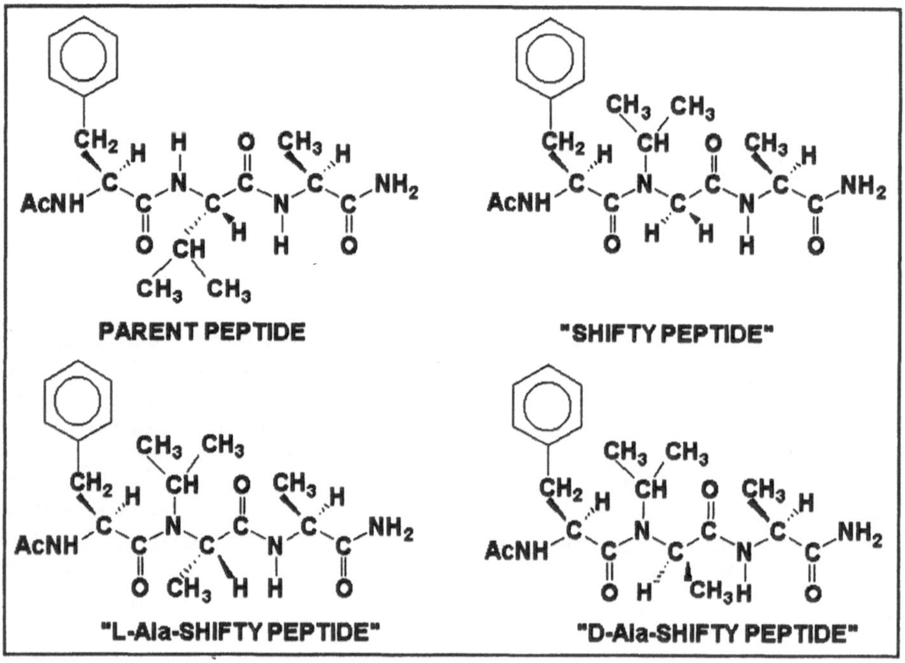

"Shifty peptides" **I-VI** and **VII-X** were synthesized on hydroxymethyl- and *p*MBHA-resin•HCl, respectively, employing DCC or DCC/HOBt as the coupling agents. Amino acid derivatives were protected by the Boc group which was removed by TFA-DCM (1:1) in the presence of 1% thioanisole. Cleavage of peptides from resins was achieved by either NH₃/MeOH (Shifty Peptides **I-VI**) or liquid HF in presence of 10% anisole (Shifty Peptides **VII-X**). In all cases Boc groups were used for α-amino protection.

Different approaches were undertaken to synthesize the N-alkylated amino acids required for the preparation of the above mentioned substance P-derived "Shifty Analogs". Boc-NBzlGly-OH and Boc-NBzlAla-OH were prepared by the method of Benoiton employing Bzl-Br and NaH in THF [1]. N-iBuGly and N-iBuAla residues were prepared by reductive alkylation of the corresponding resin-bound amino acid residue following the procedure of Sasaki and Coy [2]. The free α-amino function on the resin-bound peptide was alkylated by isobutyraldehyde and NaCNBH₃ in the presence of 1% AcOH - DMF. The "Shifty modification" of Met was achieved by the following reaction:

$$CH_3SO(CH_2)_2\text{-}NH_2 + BrCH_2CO\text{-Resin} \longrightarrow CH_3SO(CH_2)_2\text{-}NHCH_2CO\text{-Resin}$$

The substance P-derived "Shifty Analogs" were tested in the guinea pig ileum smooth muscle contraction assay following procedure described by Wormser et al. [3]. Analogs **II**, **V** and **VII** were the only ones with significant potency (ED_{50} = 7, 1.5 and 10 μM, respectively). All the rest were less potent by at least one order of magnitude.

Energy calculation of model "Shifty dipeptide" (Ac-N(Bzl)-CH_2-$CONH_2$) and "Ala Shifty dipeptide" (Ac-N(Bzl)-CH(CH_3)-$CONH_2$) and a related dipeptide model Ac-Phe-NH_2 used Discover® program to minimize the energy by employing the CVFF force field. Energies were obtained by rotating the torsion angles Φ and Ψ through 360° in 30° increments. The energy minima of the dipeptide model as well as of the two "Shifty Peptides" are located in the C_7-conformational domain.

Recently, Simon and coworkers published a structural modification similar to the one described in this report [4]. Their non-chiral "Peptoids" were prepared by completely different synthetic approaches looking for an entire different outcome. Simon et al. are interested in structural diversity and perceive the excessive conformational flexibility as a major advantage of their design. Our approach searches for topochemical resemblance and therefore we appreciate a certain degree of structural rigidity. For this reason, we have introduced chirality into the "Shifty Peptides" in the form of "L- and D-Ala-based Shifty Peptides".

The low potency of this series of substance P-related "Shifty Analogs" may suggest that "simple" shift of side-chain from Cα to N is not sufficient to reproduce closely the original molecular topology. Future work will aim for the optimization of the N-alkylating side-chain to allow better topological fit. Interestingly, the higher potency of the L-Ala-based substance P-related "Shifty Peptides" may identify a requirement for a certain amount of conformational rigidity. The similarity of the conformational profiles as emerging from energy calculations of the model "Shifty Dipeptide" may not represent fully the steric hindrance caused in a more extended structure. For example, in the reported substance P-related "Shifty Analogs" the N-substituting side-chain is closer to the adjacent down-stream side-chain than in the peptidic structure.

Acknowledgements

The authors acknowledge support from grant US-Israel BSF (M.C.).

References

1. Cheung, S.T. and Benoiton, N.L., Can. J. Chem., 55 (1977) 906.
2. Sasaki, Y. and Coy, D.H., Peptides, 8 (1987) 119.
3. Wormser, U., Laufer, R., Hart, Y., Chorev, M., Gilon, C. and Selinger, Z., EMBO J., 5 (1986) 2805.
4. Simon, R.J., Kania, R.S., Zuckermann, R.N., Huebner, V.D., Jewell, D.A., Banville, S., Ng, S., Wang, L., Rosenberg, S., Marlowe, C.K., Spellmeyer, D.C., Tan, R., Frenkel, A.D., Santi, D.V., Cohen, F.E. and Bartlett, P.A., Proc. Natl. Acad. Sci. U.S.A., 89 (1992) 9367.

Nonpeptide angiotensin II (AII) receptor antagonists: N-substituted indole, dihydroindole, phenylaminophenylacetic acid and acylsulfonamide-based AII receptor antagonists

D.S. Dhanoa[1]*, S.W. Bagley[1], R.S.L. Chang[2],
V.J. Lotti[2], T. Chen[2], S.D. Kivlighn[2], G. Zingaro[2],
P.K.S. Siegl[2] and W.J. Greenlee[1]

[1]Merck Research Laboratories, Rahway, NJ 07065, U.S.A.
[2]Merck Research Laboratories, West Point, PA 19486, U.S.A.

Introduction

The renin-angiotensin system (RAS) is a major regulator of blood pressure, and electrolyte and fluid homeostasis [1, 2]. Inhibition of the RAS by angiotensin II (AII) receptor antagonists continues to be the most active area of drug discovery for the treatment of hypertension and congestive heart failure [3]. Peptidic AII antagonists such as saralasin, [Sar[1]-Ala[8]]-AII, have been known for sometime, but their use as therapeutic agents is limited by their lack of oral absorption, rapid clearance and partial agonist activity. The development of nonpeptide AII receptor antagonists has attracted much attention recently. In our continuing efforts to discover novel classes of nonpeptide AII antagonists, we herein report the design and biological activity of new series of AII receptor antagonists derived from N-substituted indoles, dihydroindoles, phenylaminophenylacetic acids, and their bioisosteric acyl-sulfonamides and tetrazoles.

Results and Discussion

Structure-activity relationship (SAR) of the indole- and dihydroindole-based AII antagonists for AT_1 (rabbit aorta) receptor subsite is shown in Table 1. The SAR data demonstrates that 3,6-dichloro substitution of the parent indole 1 resulted in a 25-fold increase in AT_1 activity of the subnanomolar antagonist 2. Replacement of the carboxyl with the bioisostere tetrazole gave 4 with increased AT_1 binding affinity. Indole series of AT_1 antagonists was found to be more potent than the corresponding dihydroindole series (1, 2, 3 and 4 vs 8, 9, 10 and 11 respectively). The N-alkylated indoles (X = H, H) were synthesized to alleviate the amide bond of N-acylated indoles (X = O, Table 1). Although a decrease in potency was observed for the N-alkylated indole 5, the tetrazole analog 6 was found to be nearly equipotent to its corresponding N-acylated counterpart 4. Incorporation of 6-Cl substitution enhances the AT_1 potency of the antagonist 7. Evaluation of the *in vivo* activity of the potent compounds 2 (AT_1 IC_{50} = 0.8 nM) and 7 (AT_1 IC_{50} = 1 nM) at 1 mg/kg

---

* Current Address: Synaptic Pharmaceutical Corp., 215 College Road, Paramus, NJ 07652, U.S.A.

i.v. dose showed that, while **2** blocked the AII induced pressor response in conscious normotensive rats for only 0.5 h, **7** inhibited the AII pressor response for 5.5 h [4].

Table 1 *AII antagonist activity of N-substituted indoles and dihydroindoles*

Compd (Racemic)	C	R	X	Z	AT_1 IC_{50} nM
1	C = C	H	O	COOH	20
2	C = C	3,6-di-Cl	O	COOH	0.8
3	C = C	4,5-di-Cl	O	COOH	15
4	C = C	H	O	Tetrazole	5
5	C = C	H	H, H	COOH	230
6	C = C	H	H, H	Tetrazole	8.7
7	C = C	2-Cl	H, H	Tetrazole	1
8	C - C	H	O	COOH	48
9	C - C	3,6-di-Cl	O	COOH	18
10	C - C	4,5-di-Cl	O	COOH	170
11	C - C	H	O	Tetrazole	78
12	C - C	H	H, COOH	H	82
13	C - C	H	H, Ttrzl	H	18

A structurally distinct class of AII antagonists derived from phenylaminophenylacetic acids, which is a ring-opened form of the N-alkylated dihydroindoles **12** and **13** (Table 1), was explored for AT_1 activity. The SAR data shown in Table 2 demonstrates that the N-alkylation of the phenylaminophenylacetic acids is important for high binding affinity for the AT_1 receptor. N-Et, N-n-Pr and N-allyl side chains were found to be the optimal chain lengths for AT_1 binding. Substitution at the central as well as the bottom phenyl ring decreases the AT_1 activity of these phenylaminophenylacetic acids. Replacement of the carboxyl with the bioisostere acylsulfonamide enhances the AT_1 potency (**23-25** vs **16-18**), while the replacement with the tetrazole resulted in a loss of AT_1 activity (**26** vs **15**). The acylsulfonamides (**23-25**) showed excellent *in vivo* activity in conscious rats, for >6 hours duration of action, after oral administration at a dose of 3 mg/kg [4].

Table 2   *Phenylaminophenylacetic acids, acylsulfonamides and a tetrazole*

Compd (Racemic)	R_1	R_2	R_3	Z	AT_1 IC_{50} nM
14	H	H	H	COOH	200
15	Me	H	H	COOH	8.2
16	Et	H	H	COOH	4
17	Allyl	H	H	COOH	5.3
18	n-Pr	H	H	COOH	5.3
19	n-Bu	H	H	COOH	94
20	Et	H	Me	COOH	10
21	Allyl	2-Me	Me	COOH	82
22	Et	3-Me	H	COOH	66
23	Et	H	H	CONH-SO$_2$Ph	0.9
24	Allyl	H	H	CONH-SO$_2$Ph	1
25	n-Pr	H	H	CONH-SO$_2$Ph	0.8
26	Me	H	H	Tetrazole	15

In summary, the development of our potent new classes of nonpeptide AT_1-selective AII antagonists derived from indoles/dihydroindoles and phenylamino-phenylacetic acids/acylsulfonamides is disclosed.

References

1. Johnson, C.I., Drugs, 39 (1990) 21.
2. Vallotton, M. B., Trends Pharmacol. Sci., 8 (1987) 69.
3. Greenlee, W.J. and Siegl, P.K.S., Ann. Rep. Med. Chem., 27 (1992) 59.
4. Dhanoa, D.S., Bagley, S., Chang, R., Lotti, V., Chen, T., Kivlighn, S., Zingaro, G., Siegl, P., Patchett, A. and Greenlee, W., J. Med. Chem., 36 (1993) 4230, 4239.

Synthesis and properties of the first all-aza-amino acid analogue of a biologically active peptide

J. Gante, M. Krug, G. Lauterbach and R. Weitzel

E. Merck, Pharmaceutical Research Department,
Frankfurter Straße 250, D-64271 Darmstadt, Germany

Introduction

Some years ago analogues of the potent peptidic renin inhibitor Boc-Phe-Gly-ACHPA-Ile-3-pyridylmethylamide **1** [1] with single amino acid residues being replaced with the corresponding α-aza-amino acid residues were synthesized [2]. These analogues of **1** had lower biological activity than the native compound.

Our present aim was to investigate the influence of multiple aza-replacement in **1** on the biological activity. Therefore, we synthesized the peptide analogue **2** with all three positions of natural amino acids being occupied by their aza-counterparts. To our knowledge, this is the first synthesis of an all-aza-analogue of a biologically active peptide [3].

Compound **2** was synthesized in the following reaction sequence:

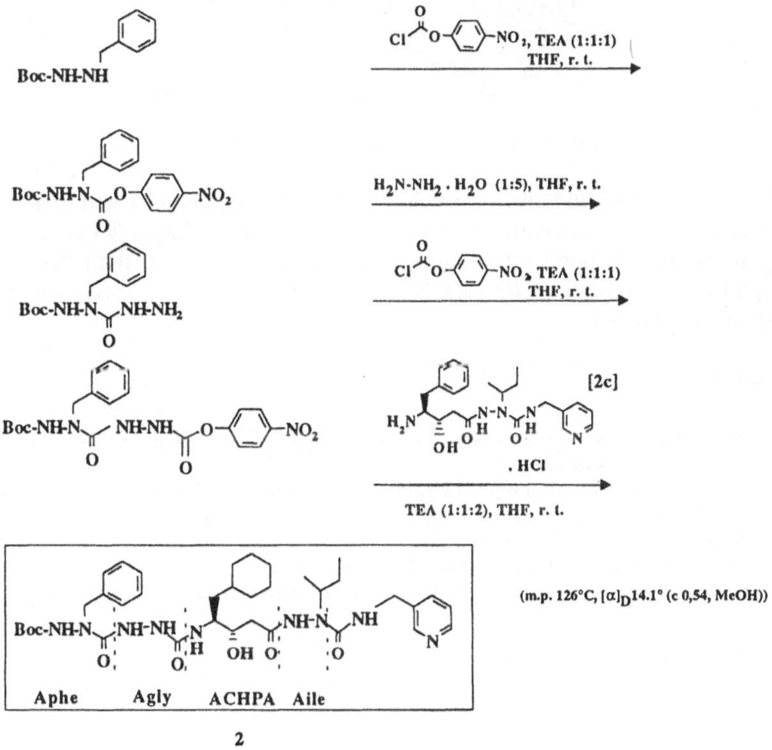

2

299

Results and Discussion

The all-azapeptide **2** could be shown to have no renin inhibiting activity.

As **2** could be crystallized from acetone, an X-ray structure analysis was performed. As can be seen from Fig. 1, **2** shows a bend conformation in the crystal with an internal H-bond between the carbonyl oxygen of Agly and the amide hydrogen of the 3-pyridylmethyl group. Additionally, the turn is held by π interactions between the phenyl ring of Aphe and the pyridine ring of the 3-pyridylmethyl group. Moreover, from the X-ray structure enhanced planarity at the position analogous to C^{α} can be derived.

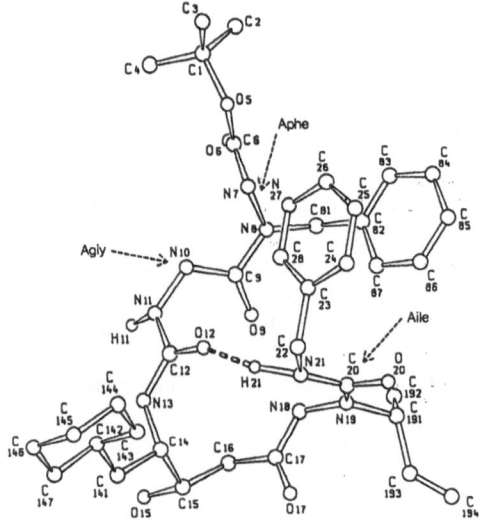

Fig. 1. Structure from X-ray analysis of 2 (space group $P2_12_12_1$; No.19).

From computer modelling investigations with compounds **1** and **2** fitted into the active site of human renin [4] it was observed that analogue **2** fits worse and clearly forms less H-bond interactions to the renin active site than the original compound **1**. These findings may be the main reason for the lack of renin-inhibiting activity in compound **2**.

References

1. Raddatz, P., Jonczyk, A., Minck, K.-O., Schmitges, C.J. and Sombroek, J., J. Med. Chem., 34 (1991) 3267.
2. a) Gante, J. and Kahlenberg, H., Liebigs Ann. Chem., (1989) 1085; b) Gante, J. and Weitzel, R., Liebigs Ann. Chem., (1990) 349; c) Gante, J. and Kahlenberg, H. Chemiker-Zeitung, 115 (1991) 215.
3. For a review see: Gante, J., Synthesis, (1989) 405.
4. Rahuel, J., Priestle, J.P. and Gruetter, M.G., J. Struct. Biol., 107 (1991) 227.

Substituted methyleneamino pseudopeptides: Studies on the transformation of the Ψ[CH(CN)NH] peptide bond surrogate

R. Herranz, M.L. Suárez-Gea and M.T. García-López

*Instituto de Química Médica, C.S.I.C., Juan de la Cierva 3,
E-28006 Madrid, Spain*

Introduction

Isosteric peptide bond replacements in biologically active peptides have been widely used to increase stability towards proteolytic enzymes, to achieve receptor selectivity and/or to obtain peptide antagonists [1]. Some of these amide bond surrogates have also been incorporated into peptidase inhibitors to mimic the enzyme-bound tetrahedral transition state of the scissile amide bond [2]. Among these backbone modifications, one of the simplest is the reduced peptide bond Ψ[CH₂NH], which has been successfully used for the design of metabolically stable agonists [3]/antagonists [4] of natural peptides and enzyme inhibitors [5].

Taking into account that semiempirical quantum mechanic calculations for the (cyano)methyleneamino [CH(CN)NH] and methyleneamino [CH₂NH] groups indicate that the first one could be a better mimic of the amide peptide bond and of the tetrahedral transition state involved in the peptide hydrolysis than the second one [6], we have reported the synthesis of Ψ[CH(CN)NH] pseudopeptides [6]. Now, taking advantage of the chemical versatility of the cyano group, the conversion of (cyano)methyleneamino pseudopeptides into other substituted methyleneamino pseudopeptides, such as Ψ[CH(CONH₂)NH] and Ψ[CH(CH₂NH₂)NH] has been studied.

Synthesis of Ψ[CH(CONH₂)NH] pseudopeptides

The replacement of one hydrogen atom of the methyleneamino Ψ[CH₂NH] peptide bond surrogate with a carbamoyl group could counteract, in part, the increase in flexibility and the decrease in *H*-bonding properties that this peptide bond surrogate causes on the peptide backbone. With this aim, (carbamoyl)methyleneamino Ψ[CH(CONH₂)NH] pseudodi- and -tripeptides **2** were obtained by oxidative hydratation of the (cyano)methylene pseudopeptides **1** with hydrogen peroxide, in phase transfer conditions (Scheme I). This transformation was carried out maintaining the chirality of compounds **1**, and, in general, with good yields (50-80%). In the basic medium of this reaction the pseudodipeptides methyl esters **2a-c** (R³ = OMe) cyclized to the dipeptide analogues 2,6-dioxopiperazines **3a-c**. Removal of the Boc-protecting group under standard conditions (TFA or HCl) gave the deprotected pseudopeptides **4**.

It is interesting to note that the pseudodipeptide ValΨ[CH(CONH$_2$)NH]His has been used as Val-His replacement in angiotensin II analogues [7]. This pseudodipeptide was synthesized by hydratation of the corresponding Boc-protected Ψ[CH(CN)NH] pseudodipeptide with concentrated H$_2$SO$_4$. However, in this procedure the yield was low (36%), and the removal of the Boc-protecting group took place simultaneously with the cyano hydratation. Therefore, in this method, it was necessary to reintroduce this protecting group to continue the peptide synthesis.

Scheme I

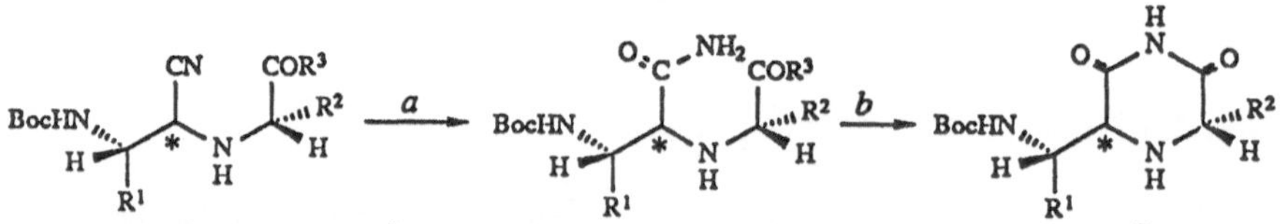

Boc-XaaΨ[CH(CN)NH]Yaa-R³   R⁴-XaaΨ[CH(CONH$_2$)NH]Yaa-R³   3a-c

1a-i

2a-i: R⁴ = Boc
4a-i: R⁴ = H

a : Xaa = Phe; Yaa = Leu; R³ = OMe   f : Xaa = Phe; Yaa = Leu; R³ = Pro-OMe
b : Xaa = Phe; Yaa = Ala; R³ = OMe   g : Xaa = Phe; Yaa = Leu; R³ = Ala-OH
c : Xaa = Phe; Yaa = Phe; R³ = OMe   h : Xaa = Phe; Yaa = Leu; R³ = Ala-NH$_2$
d : Xaa = Phe; Yaa = Leu; R³ = NH$_2$ i : Xaa = Trp; Yaa = Leu; R³ = NH$_2$
e : Xaa = Phe; Yaa = Ala; R³ = Pro-OMe

Reagents: a) H$_2$O$_2$, NaOH, (n-Bu)$_4$NHSO$_4$; b) NaOH

Synthesis of Ψ[CH(CH$_2$NH$_2$)NH] pseudopeptides

The (aminomethylene)methyleneamino Ψ[CH(CH$_2$NH$_2$)NH] pseudodi- and -tripeptides **5a,d,e**, chiral building blocks for the preparation of branched pseudopeptides and conformational restricted analogues, were obtained by catalytic hydrogenation of the appropriate (cyano)methyleneamino pseudopeptides **1a,d,e** in MeOH/AcOH, at r.t., and in the presence of 10% Pd/C (Scheme II). Pseudodipeptides **5a,d** (R³ = OMe, NH$_2$) partially cyclized to the 2-oxopiperazines **6a,d** during their isolation from the reaction medium. This lactamization was prevented when the hydrogenation was carried out in the presence of di-*tert*-butyl-dicarbonate, to give the respective Boc-protected Ψ[CH(CH$_2$NHBoc)NH] pseudopeptides **7a,d,e** as the only reaction products. On the other hand, when the epimeric mixture of pseudodipeptides Boc-PheΨ[(*R,S*)CH(CN)NH]Leu-OMe (**1a**) was hydrogenated in the presence of Ala-OMe.HCl, the reductive amination of these pseudodipeptides took place, obtaining the branched pseudotripeptides **8a**. These compounds in refluxing xylene cyclized to the 2-oxopiperazines **9a**, which could be considered as conformational constrained tripeptide analogues.

Scheme II

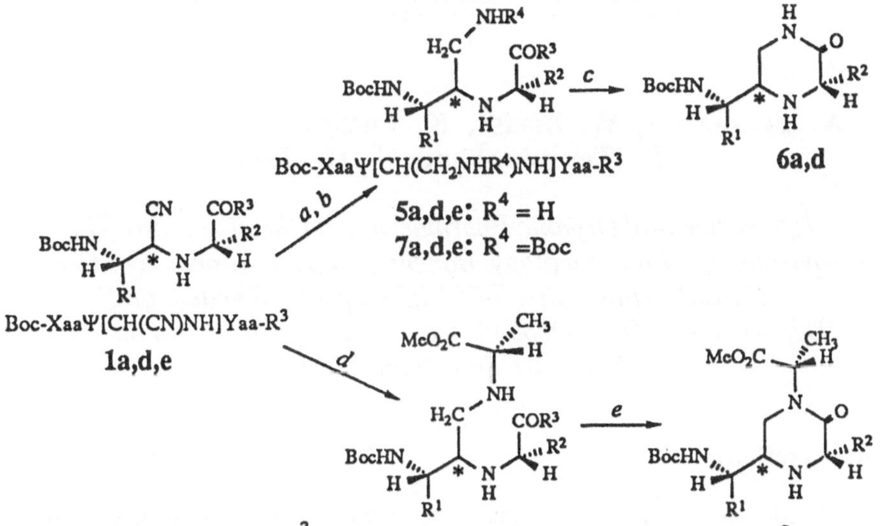

Boc-XaaΨ[CH(CH$_2$NHR^4)NH]Yaa-R^3

5a,d,e: R^4 = H

7a,d,e: R^4 = Boc

Boc-XaaΨ[CH(CN)NH]Yaa-R^3

1a,d,e

8a

9a

a : Xaa = Phe; Yaa = Leu; R^3 = OMe

d : Xaa = Phe; Yaa = Leu; R^3 = NH$_2$

e : Xaa = Phe; Yaa = Ala; R^3 = Pro-OMe

Reagents: a) H$_2$, Pd/C; b) O[CO$_2$C(CH$_3$)$_3$]$_2$; c) MeOH; d) H$_2$, Pd/C,
H$_2$N-Ala-OMe.HCl; e) Δ

In summary it can be concluded: a) The Ψ[CH(CN)NH] pseudopeptides are good intermediates for the synthesis of Ψ[CH(CONH$_2$)NH] and Ψ[CH(CH$_2$NH$_2$)NH] analogues; b) These substituted methyleneamino pseudo-peptides are easily transformed into conformational restricted building blocks, bearing different side chains and having different conformational parameters, which could be useful for the construction of peptide mimetics.

References

1. Weinstein, B. (Ed.), Chemistry and Biochemistry of Amino Acids, Peptides and Proteins, Vol 7, Marcel Dekker, New York, U.S.A., 1983, p.267.
2. Hansch, C., Sames, P.G. and Taylor, J.B. (Eds.), Comprehensive Medicinal Chemistry, Vol 2, Pergamon Press, New York, USA, 1990, p.391 and p.61.
3. Sasaki, Y., Murphy, W.A., Lance, V.A. and Coy, D.H., J. Med. Chem., 28 (1985) 1162.
4. Jensen, R.T. and Coy, D.H., Trends Pharmacol. Sci., 12 (1991) 13.
5. Greenlee, W.J., Med. Res. Rev., 10 (1990) 173.
6. Herranz, R., Suárez-Gea, M.L., Vinuesa, S., García-López, M.T. and Martínez, A., Tetrahedron Lett., 32 (1991) 7579.
7. Mohan, R., Chou, Y.L., Bihovsky, R., Lumma, W.C., Erhardt, P.W. and Shaw, K.J., J. Med. Chem., 34 (1991) 2402.

Design and synthesis of novel *bis*-phenylalanine mimetics - analogues of JA48

A. Jenmalm[1], W. Berts[1], K. Luthman[1,*], F. Nyberg[2], L. Terenius[3] and U. Hacksell[1]

[1]Department of Organic Pharmaceutical Chemistry, Box 574,
[2]Department of Pharmacology, Box 591, Uppsala Biomedical Centre,
Uppsala University, S-751 23 Uppsala, Sweden, and
[3]Department of Drug Dependence Research, Karolinska Institute,
S-104 01 Stockholm, Sweden

Introduction

Peptide-derived drugs have a number of shortcomings which are associated with their metabolic instability and the hydrophilic character of the peptide bond. Consequently there is a need for peptide mimicking agents - peptidomimetics - which exhibit adequate pharmacodynamic and pharmacokinetic properties. As part of a current program aimed at the synthesis of dipeptidomimetics which can be used as building blocks in biologically interesting peptides we have synthesized some novel *bis*-phenylalanine mimetics. The synthesized compounds have been used as replacements of Phe-Phe in the neuropeptide substance P (SP).

SP is known to be involved in several important physiologial processes such as neurogenic inflammation, pain transmission and asthma [1]. SP exerts its action by activation of the NK$_1$-receptor. The *bis*-phenylalanine moiety of SP is believed to be of importance in the recognition and binding of SP to the NK$_1$-receptor. This structural element is also the primary cleavage site of a SP cleaving enzyme, Substance P endopeptidase (SPE), which is present in human cerebrospinal fluid [2]. SPE cleaves SP also between Phe-Gly.

Results and Discussion

The peptidomimetics presented in this study were designed to fulfill one or several of the following criteria for structural mimicry: i) geometrical similarity, ii) the same conformational flexibility, iii) electrostatic complementarity and iv) the same hydrogen bonding properties. All these factors have to be taken into account in the design of peptidomimetic structures based on the amino acid sequence of a peptide.

The vinyl isostere was considered interesting since the *trans* double bond has about the same geometry as an amide bond and also shows about the same conformational flexibility. However, the electrostatic and hydrogen bonding properties are considerably different. The vinyl isostere is not only interesting in itself but also serves as an important starting material in the synthesis of other peptidomimetics.

The key reaction in the synthesis of the vinyl isostere is the Julia olefination [3] starting from the Boc-protected sulfone derivative of L-Phe and an aldehyde (Scheme 1). The Julia reaction has been reported to give exclusively the *trans*-olefin but we isolated also significant amounts (15-20%) of the *cis*-isomers. Desilylation and oxidation completed the synthesis of the vinyl isostere as a mixture of epimers, which were separated by chromatography.

Scheme 1

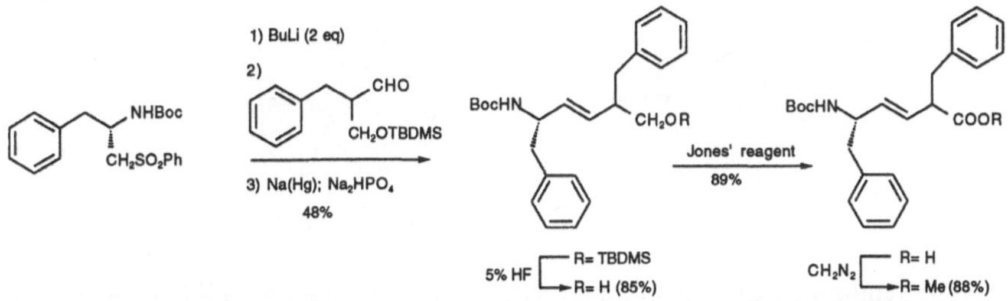

The vinyl isosteres were functionalized into epoxides and allylic alcohols by efficient and stereoselective reactions. Treatment of the ester derivatives with *m*-CPBA gave the corresponding epoxides in good yield (Scheme 2). When the Phe-Gly vinyl isostere was used as starting material we could only isolate one epoxide [4], however, epoxidation of the Phe-Phe analogue produced two isomers. The stereoselectivity in the epoxidation reaction is believed to result from a coordination effect which outweighs the steric effect; the peracid is hydrogen bonded to the carbamate and ester functionalities, and thereby preferentially directed towards one face of the alkene [4].

Scheme 2

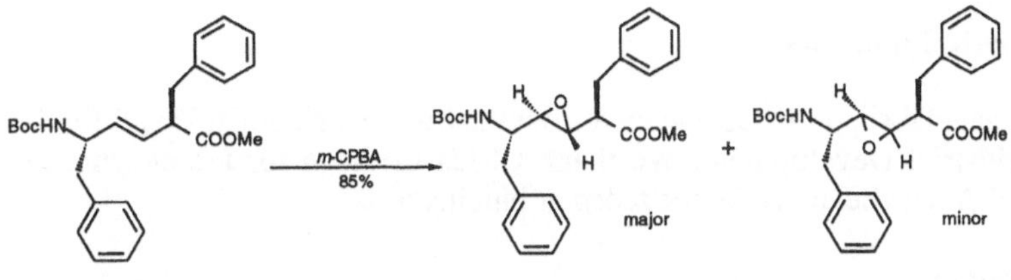

The allylic alcohols were synthesized by a stereoselective ring opening of the epoxides using TBAF (Scheme 3) [4].

Scheme 3

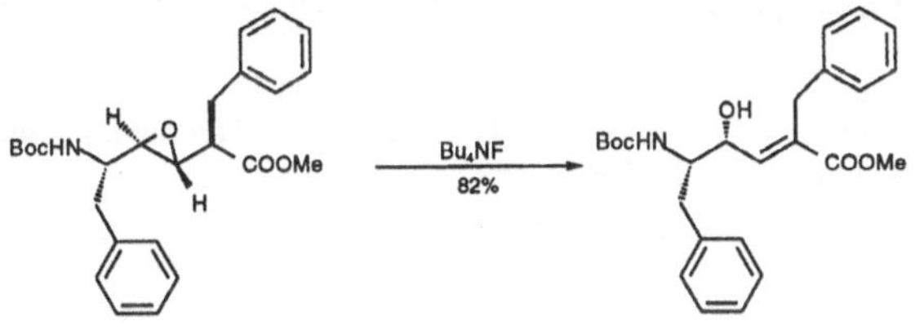

Further modifications of the alcohol derivative have been described in the Phe-Gly series, *e.g.* the fluoro-containing Phe-Gly mimetic was synthesized from the corresponding alcohol by treatment with DAST [4].

The fully deprotected dibenzylated derivatives **JA48** and **JA62**, synthesized via the Julia reaction, were found to be highly potent inhibitors of SPE, however the compounds did not show any appreciable affinity for rat brain NK_1-receptors [5]. Other derivatives are currently being studied as potential SPE inhibitors; so far **JA48** is the most potent nonpeptidergic inhibitor known for this enzyme.

The synthesized dipeptidomimetics described were used as building blocks in the preparation of modified, "mimo-mutated", SP analogues. These artificial peptides were synthesized by the solid phase method using Boc chemistry. The peptides were purified using reversed phase HPLC and were characterized by mass spectrometry and NMR spectroscopy.

The modified SP analogues also showed activity as SPE inhibitors with the dibenzylated derivative being the most potent. However, the change in peptide structure prevented recognition and binding to the NK_1-receptor. This result indicates that a double bond is a poor mimic of an amide bond despite the geometrical similarities. The introduction of an alcohol functionality did not improve the affinity although the electrostatic and hydrogen bonding properties are in better agreement with the amide bond.

Acknowledgements

Financial support was obtained from the Swedish National Board for Technical and Industrial Development. We thank Ulrica Ericsson for the enzyme inhibition tests and Anette Selander for the receptor binding tests.

References

1.   Pernow, B., Pharmacol. Rev., 35 (1983) 85.
2.   Nyberg, F., LeGreves, P., Sundqvist, C. and Terenius, L., Biochem. Biophys. Res. Commun., 125 (1984) 244.
3.   Spaltenstein, A., Carpino, P.A., Miyake, F. and Hopkins, P.B., J. Org. Chem., 52 (1987) 3759.
4.   Li, Y.-L., Luthman, K. and Hacksell, U., Tetrahedron Lett., 33 (1992) 4487.
5.   Jenmalm, A., Luthman, K., Lindeberg, G., Nyberg, F., Terenius, L. and Hacksell, U., Bioorg. Med. Chem. Lett., 2 (1992) 1693.

Synthesis of 5-alkylprolines and their use in the design of X-proline *cis*-amide surrogates

H.H. Ibrahim, E. Beausoleil, M. Atfani and W.D. Lubell

Département de chimie, Université de Montréal,
C.P. 6128, Succursale A, Montréal, Québec, H3C 3J7, Canada

Introduction

Secondary amides exist primarily as the energetically favored *trans*-rotamer which avoids steric interactions between adjacent α-carbon substituents. In contrast, the *cis*- and *trans*-rotamers of tertiary amides *N*-terminal to proline have more similar energies [1, 2]. Significant populations of X-Pro *cis*-rotamer have been observed by NMR in many biologically active peptides under physiological conditions [3-6]. Furthermore, increasing evidence suggests that X-Pro rotamer geometry plays an important role in the recognition and reactivity of bioactive peptides [1]. Hydrolysis of X-L-Pro peptide bonds by proline-specific peptidases requires the *trans*-rotamer [7], and isomerization of X-Pro amide bonds has been proposed as a rate-limiting step in protein folding [8]. The X-Pro *cis-trans* isomerization energy barrier of 13 kcal/mol can be overcome by peptidyl prolyl isomerase enzymes, which can facilitate refolding of denatured proteins [1, 8]. The rotamer equilibrium of acylprolyl residues can also be influenced by local peptide sequence patterns [6, 9], protonation of *N*-terminal residue side-chains [5], solvent conditions [10], and artificial hydrogen bonding receptors [11].

Interest in understanding the relation of X-Pro *cis*-amide rotamers to peptide bioactivity has led to the synthesis of prolines with ring alkyl substituents designed to constrain rotamer and torsion angle geometries [12-14]. Methyl substituents on the 5-position of proline can exert a steric effect on the *N*-acyl residue that alters the rotamer isomerization barrier and augments the population of *cis*-amide in *N*-acetylproline *N'*-methylamide [13]. Similarly, the use of 5,5-dimethyl proline was recently demonstrated to produce 90% of the *cis*-peptide bond isomer in a dipeptide [14]. Addition of bulkier alkyl substituents to the 5-position of proline should thus be expected to produce even more pronounced effects on rotamer and torsion angle geometry.

Results and Discussion

We have developed an efficient synthesis of 5-alkylprolines based on the reductive amination of δ-oxo α-amino esters derived from glutamic acid [15]. This method provides unnatural prolines substituted at the 5-position with primary, secondary and tertiary alkyl groups, as well as with aromatic substituents (Scheme 1).

Scheme 1

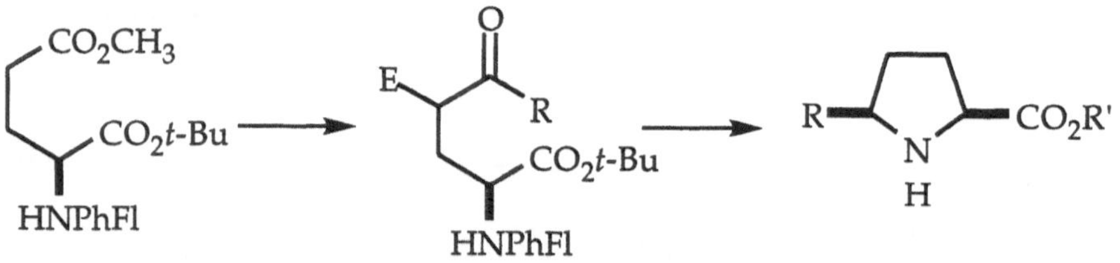

1

2 : E = CO_2CH_3

3 : E = H

4 : R' = *t*-Bu

5 : R' = H

R = Et, *n*-Bu, PhCH$_2$, *i*-Pr, *t*-Bu

Synthesis of 5-substituted prolines begins with α-*tert*-butyl γ-methyl *N*-(9-(9-phenylfluorenyl))glutamate (**1**) [16]. The 9-(9-phenylfluorenyl), (PhFl) group for nitrogen protection allowed selective enolization of the γ-ester without racemization of the chiral α-amino center; in addition, it shielded the amine from acylation [15-17]. Treatment of **1** with lithium bis(trimethylsilyl)amide provided the γ-lithium enolate, which reacted with acid chlorides to give good yields of β-keto esters **2**. Selective hydrolysis of the γ-methyl ester and decarboxylation was achieved on exposure of β-keto esters **2** to lithium hydroxide in a THF : water solution to provide δ-oxo α-amino esters **3**. Hydrogenation of δ-oxo α-amino esters **3** with palladium-on-carbon in a 9:1 methanol : acetic acid solution proceeds by cleavage of the phenylfluorenyl group, intramolecular imine formation, protonation, and hydrogen addition to the less hindered face of the iminium ion intermediate to furnish *cis*-5-alkylprolines **4** with the 5-substituent and carboxylate on the same side of the proline ring. Although imine products from incomplete hydrogenation were sometimes observed by proton NMR analysis of the crude reaction mixtures, *trans*-diastereomer **4** was not detected. The *tert*-butyl ester was cleaved by treatment with trifluoroacetic acid to give prolines **5** as their TFA salts after evaporation of the volatiles. 5-*n*-Butylproline **4** was determined to be of >99% enantiomeric purity as ascertained by preparation of diastereomeric ureas on reaction with α-methylbenzylisocyanate and observation of the *tert*-butyl ester singlets in the proton NMR spectrum during incremental additions of the diastereomer prepared from D-glutamate. Initial modeling studies of *N*-acetyl-5-*tert*-butylproline *N'*-methylamides suggest that the bulkier 5-*tert*-butyl substituent exerts effects on both the *N*-acyl and α-carboxylate groups to favor the *cis*-rotamer as well as to constrain the proline Ψ torsion angle.

Acknowledgements

This research was supported in part by the Natural Sciences and Engineering Research Council of Canada, the Ministère de l'Éducation du Québec, and Novo Nordisk A/S.

References

1.   MacArthur, M.W. and Thornton, J.M., J. Mol. Biol., 218 (1991) 397.
2.   Stewart, D.E., Sarkar, A. and Wampler, J.E., J. Mol. Biol., 214 (1990) 253.
3.   Yamazaki, T., Ro, S., Goodman, M., Chung, N.N. and Schiller, P.W., J. Med. Chem., 36 (1993) 708.
4.   Larive, C.K., Guerra, L. and Rabenstein, D.L., J. Am. Chem. Soc., 114 (1992) 7331.
5.   Mapelli, C., Van Halbeek, H. and Stammer, C.H., Biopolymers, 29 (1990) 407.
6.   Dyson, H.J., Rance, M., Houghten, R.A., Lerner, R.A. and Wright, P.E., J. Mol. Biol., 201 (1988) 161.
7.   Lin, L.-N. and Brandts, J.F., Biochemistry, 18 (1979) 5037.
8.   Schönbrunner, E.R. and Schmid, F.X., Proc. Natl. Acad. Sci. U.S.A., 89 (1992) 4510.
9.   Grathwohl, C. and Wüthrich, K., Biopolymers, 15 (1976) 2025.
10. Kofron, J.L., Kuzmic, P., Kishore, V., Colón-Bonilla, E. and Rich, D.H., Biochemistry, 30 (1991) 6127.
11. Vicent, C., Hirst, S.C., Garcia-Tellado, F. and Hamilton, A.D., J. Am. Chem. Soc., 113 (1991) 5466.
12. Montelione, G.T., Hughes, P., Clardy, J. and Scheraga, H.A., J. Am. Chem. Soc., 108 (1986) 6765.
13   Delaney, N.G. and Madison, V., Int. J. Peptide Protein Res., 19 (1982) 543.
14. Magaard, V.W., Sanchez, R.M., Bean, J.W. and Moore, M.L., Tetrahedron Lett., 34 (1993) 381.
15. Ibrahim, H.H. and Lubell, W.D., J. Org. Chem., 58 (1993) 6438.
16. Koskinen, A.M.P. and Rapoport, H., J. Org. Chem., 54 (1989) 1859.
17. Lubell, W.D. and Rapoport, H., J. Am. Chem. Soc., 110 (1988) 7447.

The design and synthesis of protein mimics

M.L. Smythe and M. von Itzstein

Department of Pharmaceutical Chemistry, Victorian College of Pharmacy, Monash University, 381 Royal Parade, Parkville, Victoria 3052, Australia

Introduction

Our interests in sialidases, in particular the design and synthesis of novel sialidase inhibitors [1], has led us to the study of an N9 sialidase-NC41 antibody crystal complex [2, 3]. When this antibody binds to the influenza viral enzyme sialidase N9 it inhibits the enzyme, by apparently binding to an active site loop and sterically interfering with the approach of the substrate [4]. We have undertaken an extensive investigation of this complex in order to try and develop: (a) antibody mimics that may inhibit enzyme activity by binding to the active site loop and sterically interfering with the approach of substrate and (b) antigen mimics which may be useful for chemical vaccination strategies. In this paper we report preliminary results on the design and synthesis of antigen mimetics, as well as our successes in the development of a biologically active antibody mimic from the protein antigen-antibody crystal structure.

Results and Discussion

The determination of the components of recognition between the energy refined N9 sialidase-NC41 antibody complex involved a geometrical non-bonded analysis (using an in-house program CONTACT [5]) and an enthalpic calculation (using the program AMBER 3.0) [6]. From these data it was concluded that one loop on the antibody surface (H96-H99, CDR H3) contributes a significant portion of the interaction energy, and three loops (714-720, 746-750 and 779-784) of the antigenic surface.

Based on the receptor bound conformation of these loops (from the protein crystal structure) a design strategy towards suitable cyclic constraints that would hold the peptide sequence in the preferred receptor bound conformation was required. Organic scaffolds were designed to bridge the *C* and *N* termini of the protein loops of interest. These organic scaffolds were rigid aromatic compounds that contained a carboxyl and amine functional group that could energetically adopt the required initiation and termination conformation of the loops of interest. They were designed by a combination of restrained minimisations, systematic search, template forcing and high temperature molecular dynamics calculations [5].

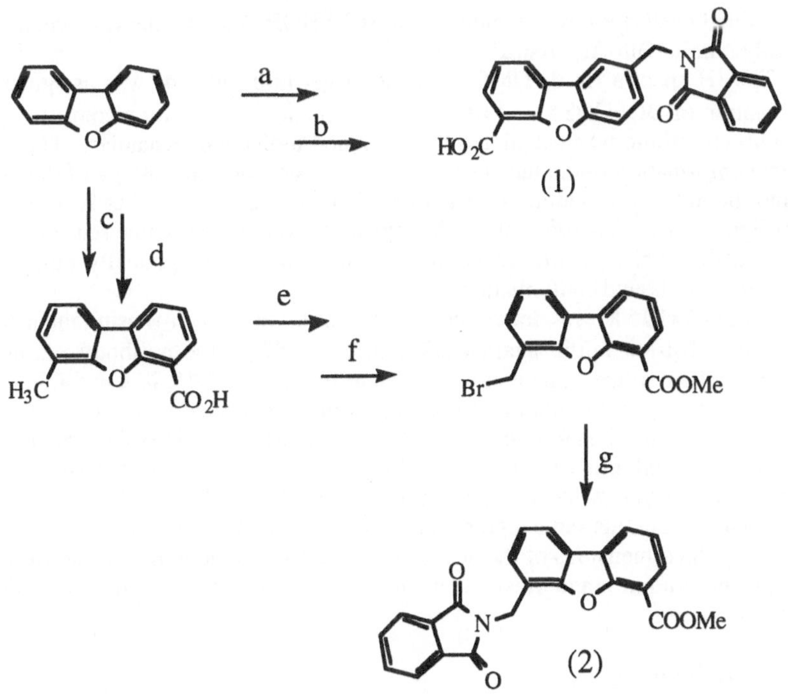

Fig. 1. *Synthesis of the organic scaffolds for the antigen mimetics. (a) nBuLi, -78°C, CO_2; (b) N-hydroxymethylphthalimide, H_2SO_4, 0°C, 30 min; (c) nBuLi, -78°C, dimethylsulfate; (d) nBuLi, -78°C, CO_2; (e) MeOH/H+; (f) N-bromosuccinimide, CCl_4, reflux, 4 h; (g) potassium phthalimide, DMF.*

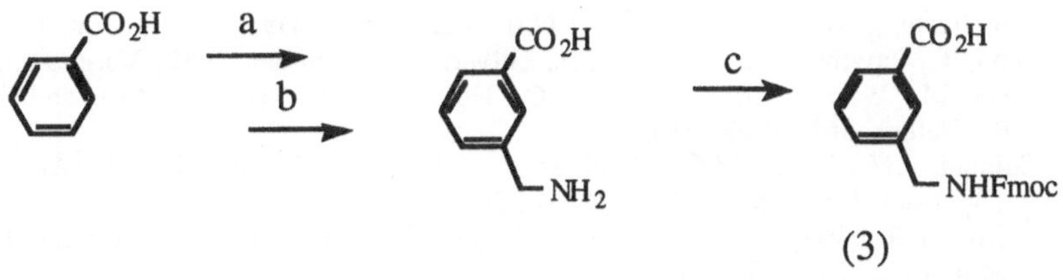

Fig. 2. *Synthesis of the antibody mimic organic scaffold. (Amb) (a) N-hydroxy-methylphthalimide, H_2SO_4, 0°C, 24 h; (b) $NaBH_4$, 2-propanol/H_2O, (c) 9-fluorenyl-methylsuccinimidyl carbonate, H_2O, pH 8, 3 h at room temperature.*

The synthesis of the designed organic scaffolds for the antigen [1, 2] and antibody mimics [3] are shown in Figures 1 and 2. The next stage was to synthesise the required constrained cyclic peptides that utilise these scaffolds as constraints. Currently, only the constrained cyclic peptide of the antibody mimic has been synthesised. The synthesis involved the addition of the β-carboxyl of Asp onto pMBHA resin, followed by the addition of Asp Glu, Constraint [3] (Amb) and Phe.

Resin bound cyclisation was achieved with BOP/HOBt and HF cleavage yielding the antibody mimic [Glu-Asp-Asn-Phe-Amb] (4).

The ^{1}H spectra (in d_6-DMSO) of the target molecule (4) was assigned by a combination of DQFCOSY and ROESY experiments. The solution conformation of the antibody mimic was calculated by restrained molecular dynamics. The lowest energy conformation found had an energy of 90 kcal/mol. The energy of the desired receptor bound conformation of the antibody mimic was calculated as 95 kcal/mol in the design process. Therefore it would appear as though the organic scaffold (3) has stabilised the cyclic peptide to conformations that are energetically close to the desired receptor bound conformation.

A K_i of 1×10^{-4} M was determined for the antibody mimic (4) using the naturally occurring substrate fetuin against N9 sialidase. Since the antibody mimic was designed to bind at the same epitope region as antibody NC41, it is possible that an inhibition similar to the inhibition observed with NC41 occurs. Such an argument is supported from the sialidase-sialic acid crystal structure [4] which suggests that antibody NC41 inhibits enzymic action by binding to active site loop 368-370 and sterically interfering with the approach of the substrate. The antibody mimic was designed to bind at this same region, and is considered unlikely to bind directly in the active site (the dimensions of the active site appear to preclude the mimic from both entering and binding), and thus may competitively inhibit the enzyme by binding to this loop.

Acknowledgements

We would like to thank Drs. P.R. Andrews, P. Colman, W. Tulip and J. Varghese for their assistance and the National Health and Medical Research Council for financial support.

References

1.   Von Itzstein, M., Wu, W.Y., Kok, G.B., Pegg, M.S., Dyason, J.C., Jin, B., Van Phan, T., Smythe, M.L., White, H.F., Oliver, S.W., Colman, P.M., Varghese, J.N., Ryan, M., Woods, J.M., Bethell, R.C., Hotham, V.J., Cameron, J.M. and Penn, C.R., Nature, 363 (1993) 418.
2.   Colman, P.M., Laver, W.G., Varghese, J.N., Baker, A.T., Tulloch, P.A., Air, G.M. and Webster, R.G., Nature, 326 (1987) 358.
3.   Tulip, W.R., Varghese, J.N., Laver, W.G., Webster, R.G. and Colman, P.M., J. Mol. Biol., 227 (1992) 122.
4.   Varghese, J.N., McKimm-Breschkin, J.L., Caldwell, J.B., Kortt, A.A. and Colman, P.M., Proteins: Struct., Func. and Genetics, 14 (1992) 327.
5.   Smythe, M.L., Ph.D. Thesis, The University of Melbourne, 1992.
6.   Weiner, S.J., Kollman, P.A., Case, D.A., Singh, U.C., Ghio, C., Alagona, G., Profeta, S. and Weiner, P.J., J. Am. Chem. Soc., 106 (1984) 765.

Structure-activity relationship studies of RGD analogs containing β-amino acids

F.S. Tjoeng[1], P.R. Bovy[1], R.B. Garland[2],
R.J. Lindmark[1], D.E. McMackins[1], J.G. Rico[1],
T.E. Rogers[1], M.V. Tóth[1], J.A. Zablocki[2],
M.E. Zupec[1], S.G. Panzer-Knodle[2], N.S. Nicholson[2],
A. Salyers[2], B.B. Taite[2], M. Miyano[2],
L.P. Feigen[2] and S.P. Adams[1]

[1]Monsanto Company, St. Louis, MO 63198, U.S.A.
[2]G.D. Searle Co., Skokie, IL 60077, U.S.A.

Introduction

Fibrinogen (Fg) binding to activated platelets is an obligatory step for platelet aggregation and arterial thrombosis. This binding is mediated by the platelet fibrinogen receptor, gp IIb/IIIa, which recognizes the RGD sequence in fibrinogen. It is well known that peptides containing the RGD sequence block fibrinogen binding to gp IIb/IIIa and prevent platelet aggregation in the face of a variety of activating agents [1]. The concept of inhibiting platelet mediated thrombosis in human disease using antagonists of gp IIb/IIIa is therefore extremely attractive. Peptide mimetics of the RGD sequence have also been shown to inhibit platelet aggregation *in vitro* and *in vivo* and prevent platelet mediated thrombosis [2]. Recently, a potent gp IIb/IIIa antagonist, 4-amidinophenylpentanoyl-Asp-Phe-OH (**1**, IC_{50} = 53 nM; dog platelet rich plasma aggregation assay), was identified [3] and is now being developed for acute indications. These drugs are designed to potentiate thrombolysis, prevent reocclusion, and provide symptomatic relief in patients with unstable angina. Our goal was to prepare orally active analogs of **1**, which could be useful in chronic treatment of thrombosis. To achieve this goal, we have focused our effort on the peptide portion of **1** and substituted it with various aromatic and heterocylic β-amino acids.

Results and Discussion

Several analogs of **1** have been synthesized in which the peptide portion of the molecule was replaced with a variety of substituted β-amino acids. These β-amino acids were prepared by Knoevenagel condensation from corresponding aldehydes and malonic acid in the presence of ammonium acetate, and subsequently coupled with 4-amidinophenylpentanoic acid. Initial screening for inhibition of platelet aggregation was accomplished *via* an *in vitro* dog platelet-rich plasma assay (PRP) on the free acid form of the compounds. Those which exhibited significant biological activity in this primary screen were then tested in ester form for oral activity in an *ex vivo* dog

PRP assay (Table 1). Monohalogen substitution at any position in the ring has resulted in analogs (**3-5**) with activity similar to the unsubstituted compound (**2**). Significant increase in potency was observed in analogs substituted with hydroxy, ethoxy, dimethoxy and methylenedioxy groups (**6-9**). All these compounds showed submicromolar activity in the *in vitro* PRP assay. In addition, ethyl esters of 3,5-dimethoxyphenyl and 1,3-benzodioxole analogs were also active orally in the *ex vivo* PRP assay. The 1,3-benzodioxole analog (**9**) completely inhibits collagen-induced platelet aggregation at 10 mpk IG with long duration of activity. The effects of aromaticity of the β-phenyl ring were also evaluated. While the cyclohexenyl analog (**10**) was equally potent, the saturated cyclohexyl analog (**11**) was found to be less active than the parent phenyl analog (**2**). However, the most active compound in this series is the 3-pyridyl-β-alanine containing analog (**12**) with an IC_{50} of 120 nM. The *ex vivo* activity of the ethyl ester of this compound was still detectable after 12 h. Interestingly, even the acid form (**13**) was found to be orally active and bioavailable at 8.4 mpk. It is remarkable because, within this series, for these compounds to be orally active and bioavailable, they normally have to be administered in the "prodrug" ester form. In conclusion, substitution of the dipeptide portion of **1** with β-amino acids and systematic modification of its side chain have resulted in long acting orally active analogs.

Table 1 *Structure-activity relationships of β-amino acid containing RGD mimetics*

Compound	R	PRP [Acid] IC_{50}, [nM]	*ex vivo* IG dose [mpk]	[ethyl ester] duration [h] (%max. Inh)
2	phenyl	5,300		
3	4-fluorophenyl	3,000		
4	3-fluorophenyl	1,300		
5	2-fluorophenyl	5,200		
6	4-hydroxyphenyl	800		
7	4-ethoxyphenyl	400		
8	3,5-dimethoxyphenyl	330	20	~6 (68)
9	1,3-benzodioxole	280	10	~6 (100)
10	3,4-cyclohexenyl	4,000		
11	cyclohexyl	10,000		
12	3-pyridyl	120	10 [ester]	>6<24 (100)
13	3-pyridyl	120	8.4 [acid]	>6<24 (100)

References

1. Ruoslahti, E. and Pierschbacher, M.D., Science, 238 (1987) 491.
2. Tjoeng, F.S. and Adams, S.P., U.S. Patent 4,879,313, November 7, 1989.
3. Garland, R.B., Miyano, M. and Zablocki, J., EP 502536A, September 9, 1992.

Preferred conformation of (αMe)Leu-rich peptides

C. Toniolo[1], M. Pantano[1], F. Formaggio[1], M. Crisma[1],
A. Aubry[2], D. Bayeul[2], G. Précigoux[3], A. Dautant[3],
W.H.J. Boesten[4], H.E. Schoemaker[4] and J. Kamphuis[4]

[1]*Biopolymer Research Center, CNR, Department of Organic
Chemistry, University of Padova, I-35131 Padova, Italy*
[2]*Laboratory of Mineralogy and Crystallography, URA-CNRS 89,
University of Nancy I, F-54506 Vandoeuvre, France*
[3]*Laboratory of Crystallography and Crystalline Physics, URA 144,
University of Bordeaux I, F-33405 Talence, France*
[4]*DSM Research, Bioorganic Chemistry Section, 6160 MD Geleen,
The Netherlands*

Introduction

The rational design of conformationally restricted peptides which mimic or block the effects of their physiologically important parent compounds represents a major goal of pharmaceutical companies. These conformational analogues should not only retain the properties of the original peptide, but they should also exhibit improved bioavailability and pharmacokinetics. In this connection in the last decade medicinal chemists have become interested in bioactive peptides containing chiral $C^{\alpha,\alpha}$-disubstituted glycines since these stereochemically constrained, backbone modified, non-coded residues tend to freeze specific conformations and dramatically slow down enzymatic processes. In our continuing investigation of the preferred conformation of peptides containing C^{α}-methylated α-amino acids we report here a crystal-state structural study by X-ray diffraction of a number of peptides rich in (αMe)Leu, a C^{α}-methylated amino acid with a γ-branched, aliphatic side chain.

Results and Discussion

An intriguing observation can be extracted from the analysis of the Z-D-(αMe)Leu-(Aib)$_2$-OtBu tripeptide. Both independent molecules are folded in a β-bend conformation, but despite the presence of a chiral residue [D-(αMe)Leu] in the sequence, no preference is seen in the β-bend screw sense, *i.e.* the two diastereomeric screw senses (right- and left- handed) simultaneously occur in the asymmetric unit. An analogous observation (right- and left-handed helices concomitantly present in the asymmetric unit) has been made for the 3_{10}-helix forming pBrBz-(Aib)$_2$-D-(αMe)Leu-(Aib)$_2$-OtBu. Conversely, only small differences are noted in the backbone torsion angles of the two *right*-handed 3_{10}-helical molecules of the octapeptide Ac-(Aib)$_5$-D-(αMe)Leu-(Aib)$_2$-OtBu monohydrate.

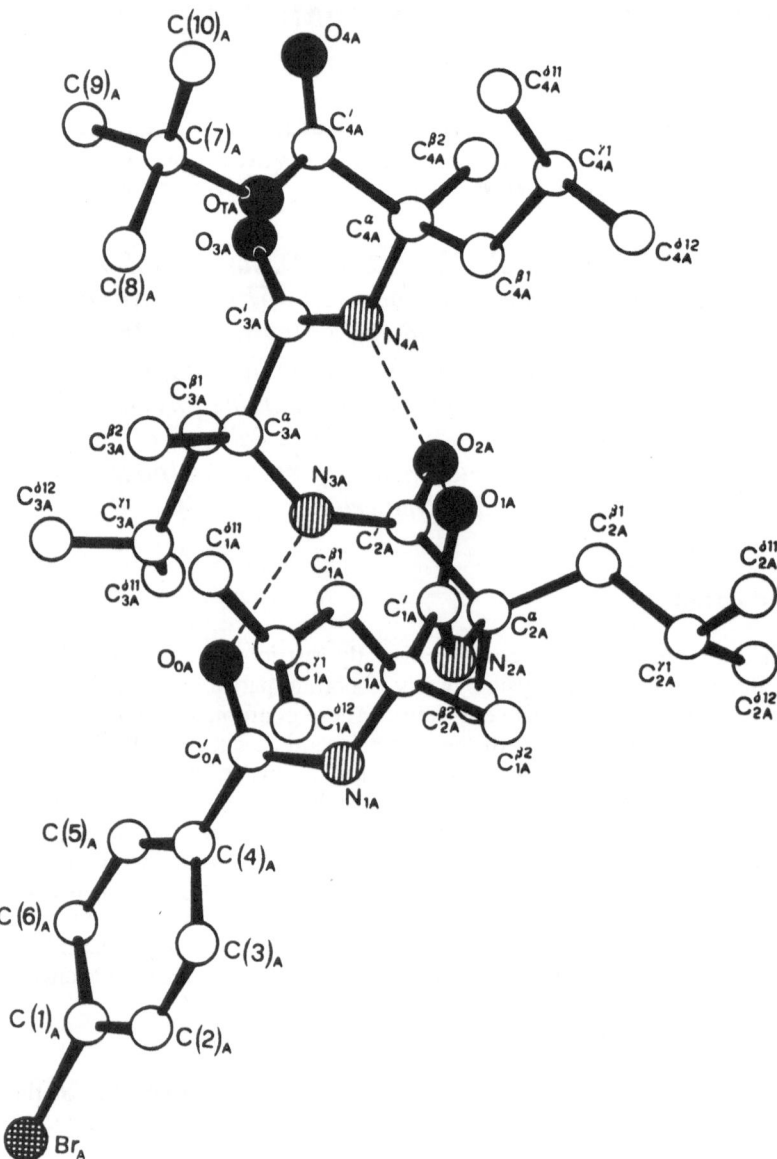

Fig. 1. X-ray structure of molecule A of pBrBz-[D-(αMe)Leu]₄-OtBu.

The homochiral homo-trimer *p*BrBz-[D-(αMe)Leu]₃-OH monohydrate is folded in a type-III (*right*-handed) β-bend. Interestingly, this structure is followed by an unusual "oxy-analogue" of a type-III β-bend conformation (where the -OH group of the carboxylic acid moiety is the H-bonding donor). No significant difference is seen in the backbone torsion angles of the three *right*-handed 3_{10}-helical molecules in the asymmetric unit of the homo-tetramer *p*BrBz-[D-(αMe)Leu]$_4$-O*t*Bu (Figure 1).

All 21 (aMe)Leu residues are found in the helical region of the conformational map. The average absolute values for the ϕ, ψ torsion angles are 57, 29°, closer to those expected for a 3_{10}-helix (57, 30°) than for a α-helix (63, 42°) [1]. It is worth noting that in eight out of the ten molecules the signs of the ϕ, ψ values for the N-terminal and central *D*-(αMe)Leu residues are *negative*. The distribution of the most relevant torsion angle for the characterization of the side-chain conformation (χ^1) for the 21 D-(αMe)Leu residues is 18 g^+ and 3*t*.

The results obtained support the view that (αMe)Leu, like the (αMe)Val and (αMe)Phe residues already investigated [2], is an extremely efficient β-bend and 3_{10}/α-helix promoter, much stronger than its unmethylated counterpart (Leu). Interestingly, in the peptides where (αMe)Leu is the only chiral residue, the relationship between C^α configuration and helix screw sense is predominantly opposite to that exhibited by protein amino acids (including Leu) and (αMe)Val. This behavior is also typical of (αMe)Phe. In other words, this *inverse* relationship appears to be dictated by the position of side-chain branching (at the γ-carbon), rather than by the electronic nature (whether aliphatic or aromatic) of the side chain.

References

1. Toniolo, C. and Benedetti, E., TIBS, 16 (1991) 350.
2. Toniolo, C., Crisma, M., Formaggio, F., Valle, G., Cavicchioni, G., Précigoux, G., Aubry, A. and Kamphuis, J., Biopolymers, 33 (1993) 1061.

Session IV
Glycopeptides/Lipopeptides

Chairs: Richard M. Epand
McMaster University
Hamilton, Ontario, Canada

and

James T. Sparrow
Baylor College of Medicine
Houston, Texas, U.S.A.

Synthesis of glycopeptides: A challenge for chemical selectivity

H. Kunz, K. von dem Bruch, C. Unverzagt and W. Kosch

Institut für Organische Chemie, Universität Mainz,
D - 55099 Mainz, Germany

Introduction

During the past decades it has been recognized that most of the natural proteins of multicellular organisms in reality are glycoproteins. The saccharide side chains of glycoproteins influence the physical properties as well as the conformations, and can also protect these protein conjugates against proteolytic attack. Moreover, the carbohydrate portions of the glycoproteins have been found to be important recognition labels, e.g., in infectious processes, in cell differentiation and in the regulation of cell growth [1]. For investigations of these challenging recognition processes, glycoproteins of exactly specified structure are of particular interest. As glycoproteins are accessible in only small amounts from biological sources and these isolates, furthermore, suffer from the biological microheterogeneity of natural glycoproteins, synthetic glycopeptides of confirmed structure and purity are required. The natural glycoproteins exclusively contain glycosidic linkages between the peptide and the carbohydrate portions. These bonds as well as the intersaccharidic linkages, being acetal-like bonds, are potentially acid-sensitive. In addition O-glycosyl serine and threonine derivatives are prone to an easy base-catalyzed β-elimination of the carbohydrate. Therefore, special emphasis has to be given to mild reaction conditions throughout the glycopeptide synthesis [2].

Results and Discussion

One of the early solutions we found for the synthesis of glycopeptides via controlled and selective deblocking and chain extension consisted in the application of the Fmoc group and its selective removal with morpholine [3]. The weak base morpholine does not attack the base-sensitive O-glycosyl serine linkage, e.g., in the xylosyl serine derivative 1. But, its basicity is sufficient for a complete and selective removal of the Fmoc group to furnish the selectively deblocked derivative 2 susceptible to further chain extension.

This method was used in the synthesis of tumor-associated antigen structures, e.g., in the synthesis of the T antigen glycopeptide **3** [4]. Coupling of **3** to bovine serum albumin gave a synthetic glycoprotein antigen **4**. According to its carbohydrate analysis it contained on average 38 T antigen glycopeptide units per macromolecule.

The monoclonal antibody obtained with **4** reacted with all types of epithelial tumor tissues tested. But it also showed reactivity against normal cells of the same tissues. The antibody obtained from **4** exhibited interesting selectivity towards asialoglycophorins. As the structure of **4** corresponds to the N-terminus of glycophorin of M blood group specificity, the antibody was distinctly more reactive to M- than to N-type asialoglycophorin. This suggests that not only the T antigen disaccharide but also the amino acid sequence contributes to the recognized epitope [5].

Particularly efficient protecting groups in glycopeptide synthesis are the allylic protecting groups. Both, the allyl ester [6] and the allyloxycarbonyl (Aloc) group [7] are stable to acids and bases commonly applied in peptide chemistry. However, they can selectively be removed under neutral conditions via palladium(0)-catalyzed allyl transfer to weakly basic or neutral nucleophiles. This methodology was successfully extended to a new allylic anchoring principle in the solid phase synthesis of peptides and glycopeptides [8]. Utilizing the efficiency of the allylic protecting groups, glycopeptides with Tn and T antigen structure [9, 10], with fucosylated chitobiose asparagine structures [11] and with Lewis^a antigen side chains [12] were successfully synthesized. During these studies, an effect of indirect protection of the O-glycosidic bonds exhibited by the O-acyl groups of the saccharide part was found. Of particular interest are glycopeptides with Lewis^x antigen side chains, which have been described as tumor-associated antigens. For example, the Lewis^x asparagine conjugate **5** was selectively and alternatively deblocked either by Pd(0)- or Rh(I)-catalyzed cleavage of the allyl ester to give **6** or by acidolytic removal of the Boc group to form **7**. Chain extension and subsequent N-terminal deblocking furnished **8**, which was condensed with **6** to form the glycopeptide **9** containing two Lewis^x antigen side chains. Further chain extension, deprotection and coupling with BSA gave a synthetic Lewis^x antigen glycoprotein. According to its clustered Lewis^x antigen structure, this synthetic glycoprotein should exhibit enhanced activity towards both, receptors and the immune system.

The examples given illustrate that glycopeptides of biologically interesting structure can be synthesized in preparative amounts and high purity using the developed selective methods.

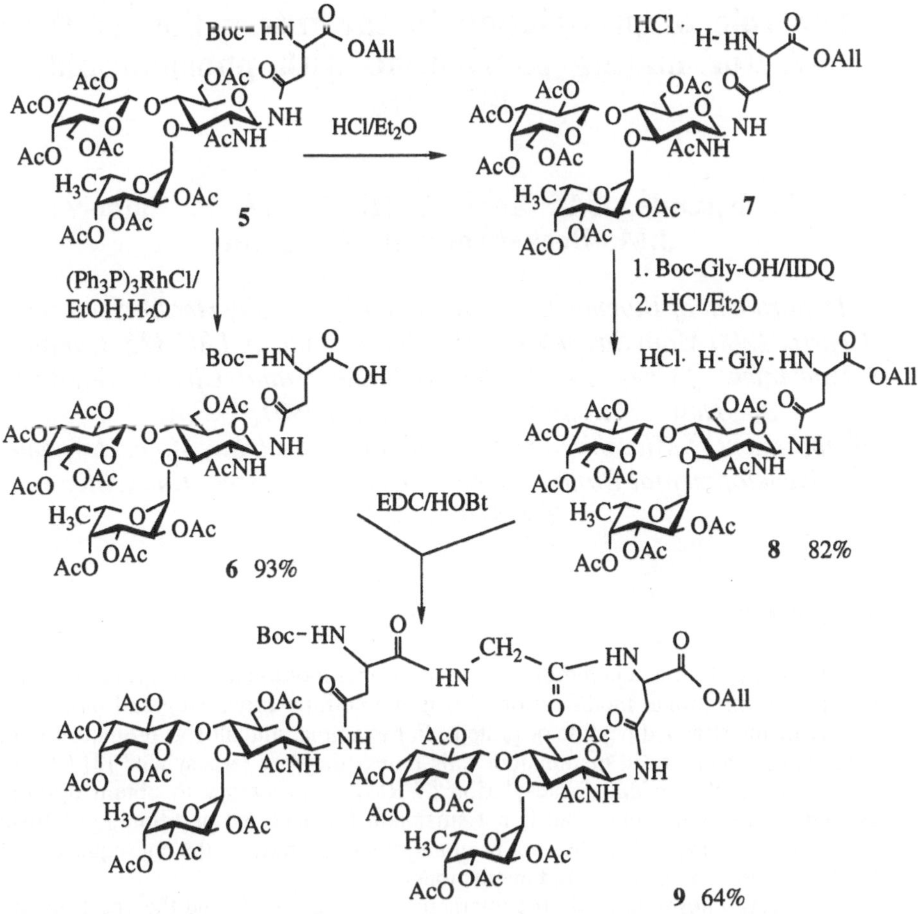

Boc–HN O OAll

AcO AcO OAc OAc O NH AcNH H₃C OAc AcO OAc **5**

HCl/Et₂O →

HCl· H–HN O OAll

AcO AcO OAc O NH AcNH H₃C OAc AcO OAc **7**

(Ph₃P)₃RhCl/ EtOH,H₂O

1. Boc-Gly-OH/IIDQ
2. HCl/Et₂O

Boc–HN O OH

AcO AcO OAc O NH AcNH H₃C OAc AcO OAc **6 93%**

EDC/HOBt →

HCl· H·Gly·HN O OAll

AcO AcO OAc O NH AcNH H₃C OAc AcO OAc **8 82%**

Boc–HN O HN CH₂ C HN O OAll

AcO AcO OAc O NH AcNH AcO OAc O NH AcNH H₃C OAc AcO OAc H₃C OAc AcO OAc **9 64%**

References

1. Ivatt, R.J. (Ed), The Biology of Glycoproteins, Plenum Press, New York, U.S.A., 1984.
2. Kunz, H., Angew. Chem. Int. Ed. Engl., 26 (1987) 294.
3. Schultheiss-Reimann, P. and Kunz, H., Angew. Chem. Int. Ed. Engl., 22 (1983) 62.
4. Kunz, H., Birnbach, S. and Wernig, P., Carbohydr. Res., 202 (1990) 207.
5. Dippold, W., Steinborn, A., Meyer zum Büschenfelde, K.-H., Birnbach, S. and Kunz, H., unpublished results.
6. Kunz, H. and Waldmann, H., Angew. Chem. Int. Ed. Engl., 23 (1984) 71.
7. Kunz, H. and Unverzagt, C., Angew. Chem. Int. Ed. Engl., 23 (1984) 436.
8. Kunz, H. and Dombo, B., Angew. Chem. Int. Ed. Engl., 27 (1988) 711.
9. Friedrich-Bochnitschek, S., Waldmann, H. and Kunz, H., J. Org. Chem., 54 (1989) 751.
10. Ciommer, M. and Kunz, H., Synlett, (1991) 593.
11. Kunz, H. and Unverzagt, C., Angew. Chem. Int. Ed. Engl., 27 (1988) 1697.
12. Kunz, H. and März, J., Synlett, (1992) 591.

The role of prenylation in the interaction of the a-factor mating pheromone with phospholipid bilayers

R.F. Epand[1], C.B. Xue[2], S.-H. Wang[2], F. Naider[2], J.M. Becker[3] and R.M. Epand[1]

[1]Department of Biochemistry, McMaster University Health Sciences Centre, 1200 Main Street West, Hamilton, Ontario, L8N 3Z5, Canada
[2]Department of Chemistry, College of Staten Island, City University of New York, Staten Island, NY 10301, U.S.A.
[3]Department of Microbiology and Program in Cellular, Molecular, and Developmental Biology, University of Tennessee, Knoxville, TN 37996, U.S.A.

Introduction

Isoprenylation is a common post-translational modification of proteins [1, 2]. This post-translational modification affects the cellular processing and biological activity of the attached peptide or protein. For example, mutant Ras proteins do not manifest oncogenicity in mammalian cells unless they are farnesylated [3]. It has recently been shown that farnesylation of Ras2 is necessary to obtain optimal binding to adenylyl cyclase but is not sufficient for maximal partitioning of Ras2 into membranes [4]. There has not been a systematic study of the consequences of this modification on peptide-lipid interactions.

We have chosen to study the interaction between lipids and the yeast mating pheromone, a-factor, and its analogs. This lipopeptide with the sequence YIIKGVFWDPAC is farnesylated on the side-chain thiol group and the terminal carboxyl group is in the form of a methyl ester. It is particularly attractive to study a-factor because its relatively small size allows for the synthesis of analogs.

Results and Discussion

We assessed the penetration of a-factor and its analogs into a phospholipid bilayer using the fluorescence emission spectrum of the Trp residue as a criterion. We first demonstrated that both the farnesylated lipopeptide, a-factor, as well as the S-methyl-a-factor, which has only a methyl group replacing the farnesyl group on the Cys residue, have the same fluorescence emission properties when dissolved in methanol with a maximum at 343 nm. This indicates that farnesylation alone does not induce a change in the polarity of the Trp environment when the lipopeptide is dissolved in methanol. In addition, both the a-factor and the S-methyl-a-factor have similar fluorescence emission spectra in aqueous solution in the presence of lysolecithin micelles. The polarity of the Trp environment in lysolecithin is similar to that in methanol for both peptides. The **a**-factor was not sufficiently soluble in water to measure its fluorescence emission under these conditions, however, the S-methyl-**a**-factor had a fluorescence emission maximum in 10 mM sodium phosphate buffer, pH 7.4, at 352 nm, compared with 341 nm in the presence of lysolecithin. This indicates that even without farnesylation the S-methyl-**a**-factor will spontaneously partition from the aqueous to the membrane phase. In the presence of sonicated vesicles of dimyristoylphosphatidylcholine, however, there is a marked difference, with the **a**-factor appearing to insert more deeply into the lipid bilayer than the S-methyl-**a**-factor.

The shift of the bilayer to hexagonal phase transition temperature (T_H) of dielaidoylphosphatidylethanolamine (DEPE), with the addition of a membrane additive is a measure of how this additive affects the relative stability of the bilayer vs nonlamellar inverted phases. This information has proven to be of value in predicting the effect of these additives on certain membrane functional properties [5]. In order to compare various **a**-factor derivatives, we measured the slope of a plot of T_H versus the mole fraction of additive (Table 1). Positive slopes indicate bilayer stabilizers (relative to the H_{II} phase) whereas negative slopes indicate compounds that cause a relative stabilization of the hexagonal phase.

Table 1 *Effect of **a**-factor and related compounds on the T_H of DEPE*

Substance	Slope (Degrees/mol fraction additive)
a-factor	217 ± 14
trans,trans-farnesol	-188 ± 54
trans,trans-farnesol acetate	-133 ± 27
hexadecanol	-145 ± 8
S-hexadecanyl-a-factor	106 ± 19
benzyl alcohol	49 ± 32
S-benzyl-a-factor	112 ± 50
S-methyl-a-factor	31 ± 16
S-methyl-a-factor + farnesol[a]	-194 ± 10

[a]Equimolar amounts of these materials were used.

We have examined the influence of the nature of the hydrocarbon group on the Cys^{12} sulfur on the bilayer stabilizing tendency of the peptide. The natural form of **a**-factor is farnesylated at this position. The farnesyl group in isolation is a hexagonal phase promoter (Table 1). Despite this, when the farnesyl group is attached to the peptide, the resulting **a**-factor becomes a good bilayer stabilizer. Indeed, when equimolar amounts of S-methyl-**a**-factor and farnesol were added to DEPE, the T_H was lowered by an amount similar to that found with farnesol alone (Table 1). Thus, the types of membrane interactions exhibited by **a**-factor are a consequence of the covalent linkage of the peptide and lipid moieties and cannot be simulated by a simple mixture of these two components. The change as a result of farnesylation is not simply a consequence of increased hydrophobicity and/or membrane partitioning of the peptide. This is clearly demonstrated by comparing the effects of S-hexadecanyl-a-factor with a-factor on lipid polymorphism. Farnesylation leads to a lipopeptide (**a**-factor) which is a much better bilayer stabilizer than is S-hexadecanyl-**a**-factor (Table 1). This is not likely to be a consequence of different membrane partitioning of the two lipopeptides since both the hexadecanol and farnesol would be expected to have similar hydrophobicities [6]. If anything, the hexadecanyl peptide may be somewhat more hydrophobic as indicated by its greater retention on reversed-phase HPLC (data not shown). The greater bilayer stabilizing effect of **a**-factor is also not a consequence of an intrinsic difference in the effects of the corresponding hydrocarbons on lipid polymorphism since both farnesol and hexadecanol have similar effects on T_H (Table 1). Even a benzyl group, which as illustrated by benzyl alcohol is a bilayer stabilizer by itself, is not able to shift the T_H of the **a**-factor peptide as much as farnesylation does. It thus appears that farnesylation results in a particular conformation and/or orientation of the peptide moiety of **a**-factor at the membrane surface.

Modification of the **a**-factor peptide by farnesylation alters the way in which it interacts with membranes. This is likely to have consequences not only on membrane properties, but also on which groups or protein sites in a biological membrane will have access to the lipoprotein. We suggest, therefore, that the interaction of the farnesyl side chain with the bilayer results in a presentation of the pheromone to the receptor which is an important determinant of activity.

Acknowledgements

This work was supported by grants from the Medical Research Council of Canada (MA-7654) and the National Institutes of Health, General Medical Sciences (GM-46520).

References

1. Clarke, S., Ann. Rev. Biochem., 61 (1992) 355.
2. Schafer, W.R. and Rine, J., Ann. Rev. Genet., 30 (1992) 209.
3. Kato, K., Cox, A.D., Hisaka, M.M., Graham, S.M., Buss, J.F. and Der, C.J., Proc. Natl. Acad. Sci., U.S.A., 89 (1992) 6403.
4. Kuroda, Y., Suzuki, N. and Kataoka, T., Science, 259 (1993) 683.
5. Epand, R.M., In Ohki, S. (Ed.), Cell and Model Membrane Interactions, Plenum Press, New York, U.S.A., 1991, p.135.
6. Black, S.D., Biochem. Biophys. Res. Commun., 186 (1992) 1437.

Delivery studies using peptide-fatty acid conjugates

V.J. Bender, F.H. Cameron, P. Hendry, T.J. Lockett and R.G. Whittaker

CSIRO Division of Biomolecular Engineering, Sydney Laboratory, North Ryde, NSW 2113, Australia

Introduction

Delivery of peptides and other pharmaceutical compounds to an appropriate location in an animal cell is still a major problem in biology and medicine and the large effort to develop safe and efficient delivery systems indicates this. The attachment of lipid moieties has been one of the more popular approaches employed for the delivery of peptide antigens (or other molecules of interest) [1] and for the potentiation of the immune response to these compounds [1, 2].

A novel method of conjugating amino acids and peptides to fatty acids was developed in these laboratories [3, 4, 5] using Tris (2-amino-2-hydroxymethyl-1,3-propanediol) as a multifunctional linker. This method enables the conjugation of 1, 2 or 3 fatty acids to the peptide resulting in a molecule with a structure that mimics those of triglycerides. We have synthesised a number of such compounds where fatty acids of various lengths are attached to different peptides, and/or to other molecules of interest. Investigation of the biological properties of a number of variants showed that fatty acid number influenced their rate of uptake into cells but did not affect their specificity of sub-cellular localisation.

$$^+NH_3 - (aa)_n - CO - NH - \overset{\displaystyle CH_2O - R_1}{\underset{\displaystyle CH_2O - R_3}{C}} - CH_2O - R_2$$

Fig. 1. General structure of lipopeptide conjugates where R_1 R_2 and R_3 are fatty acids.

Results and Discussion

Amino acid/peptide-Tris conjugates were prepared by incubation of their corresponding N-protected esters with Tris [5]. Products were purified by preparative HPLC and then acylated with fatty acids in the presence of DCC and dimethylaminopyridine (DMAP). Palmitic, myristic and lauric acids were coupled in this fashion. The reaction products were purified either by chromatography on silica gel or by preparative HPLC.

After the removal of the N-protecting moiety the free amino acid could be reacted with fluorescein isothiocyanate (FITC) and the labelled products purified by HPLC.

We have found that these peptide-fatty acid conjugates interact very rapidly with mammalian cells, readily crossing the lipid bilayer of the plasma membrane into the cytoplasm under physiological conditions and localising in specific intracellular membranes and organelles. The presence of a different number of fatty acid chains (from 1 to 3 molecules) has a profound effect on both the rate of entry of the conjugates into the cell and their persistence over time.

Preliminary experiments on whole live *Drosophila* embryos have shown that the membrane of a dechorionated embryo is readily traversed by the fluorescein-labelled tris-palmitate conjugate. Over a period of two hours some embryos appeared to concentrate the conjugate in the pole cells, which subsequently become the animal's germ cells.

The clearest conclusion from these experiments was that the monopalmitate (alanine-Tris-palmitate, ATP_1) has a greater capacity to rapidly interact with the cells. A number of separate points were noted: (a) ATP_1 crosses into the cell more rapidly; (b) it crosses the nuclear membrane more rapidly than conjugates containing two or three fatty acids; (c) ATP_1 was also able to cross membranes of paraformaldehyde fixed cells whereas the other conjugates could not, indicating that it did not require cellular activity for efficient entry; (d) ATP_1 persisted in the cell to a greater degree than the di-and tri-palmitates, (ATP_2 and ATP_3).

References

1. Toth, I., Hughes, R.A., Munday, M.R., Mascagni, P. and Gibbons, W.A., In Rivier, J.E. and Marshall, G.R. (Eds.), Peptides: Chemistry, Structure and Biology, (Proc. 11th Amer. Peptide Symp., U.S.A.), Escom, The Netherlands, 1989, p.1078.
2. Wiesmuller, K.H., Bessler, W. and Jung, G., Hoppe-Seyler's Z. Physiol. Chem., 364 (1983) 593.
3. Whittaker, R.G., Int. Pat. Appl. No. PCT/AU90/00599 (1990).
4. Whittaker, R.G. and Bender, V.J., In Epton, R. (Ed.), Innovations and Perspectives in Solid Phase Synthesis, (Collected papers Second International Symposium Intercept Limited), U.K., 1992, p.495.
5. Whittaker, R.G., Hayes, P.J. and Bender, V.J., Peptide Res., 6 (1993) 125.

Glycopeptide mapping and structural analysis by electrospray mass spectrometry

T. Hutton[1], B.N. Green[2] and M.J. Geisow[2]

[1]Fisons Instruments/VG BioTech, Altrincham, WA14 5RZ, U.K.
[2]Biodigm, Bingham, Nottingham, NG13 8SS, U.K.

Introduction

Electrospray mass spectrometry (ESMS) is a fast evolving analytical technique which can be used to determine the molecular weights of intact proteins and glycoproteins. The mass accuracy of the technique is in many instances better than 0.01% and masses differing by as little as 0.1% can be resolved. This resolution can be increased by the application of maximum entropy (MaxEnt) to the data. Electrospray is therefore ideal for detecting and assigning post translational modification of intact proteins. Furthermore, the coupling of liquid chromatography and electrospray mass spectrometry provides a fast and highly sensitive method for the detection of glycopeptides produced from an enzyme digestion of a glycoprotein.

Glycoprotein glycoforms affect the biological activity and pharmacokinetic behaviour of many glycoconjugates. Their detection and analysis has become increasingly important in structure function studies as well as in the manufacture of recombinant glycoproteins. Although full characterisation of glycoprotein glycan requires their release from the protein, it would be of considerable advantage to be able to detect particular glycoforms in the intact protein.

In this study we demonstrate the resolving power of electrospray mass spectrometry using MaxEnt applied to the intact plant glycoprotein horseradish peroxidase showing that the individual glycoforms can be resolved. This can be compared with data obtained from a recombinant horseradish peroxidase and two variant strains. The glycopeptides produced by enzyme digestion of HRP were mapped and analysed by LC/MS and compared to those peptides produced from the recombinant non glycosylated protein produced in E. coli.

Results and Discussion

The data were obtained on Fisons Instruments/VG BioTech Quattro triple quadrupole and Platform single quadrupole mass spectrometers using electrospray ionisation at atmospheric pressure. The horse radish peroxidase was obtained from Sigma Chemical Co., St Louis, MO. The recombinant HRP and the two variants were supplied as part of a collaboration with Dr Andrew Smith, University of Sussex, Brighton, U.K.

HRP and recombinant HRP were digested with trypsin and the resulting peptides and glycopeptides were analysed by on-line LC/MS.

Using the MaxEnt algorithm on the multiply charged ion electrospray mass spectrum of natural horseradish peroxidase, identifiable peaks corresponding to both full length and C terminally truncated HRP molecules with either 8, 7 or 6 N-linked glycans could be seen in the MaxEnt spectrum. Variants of these differing by one hexose unit (e.g., mannose) and 13 other peaks were also resolved. These included both hexose, deoxyhexose and N-acetylhexosamine variants.

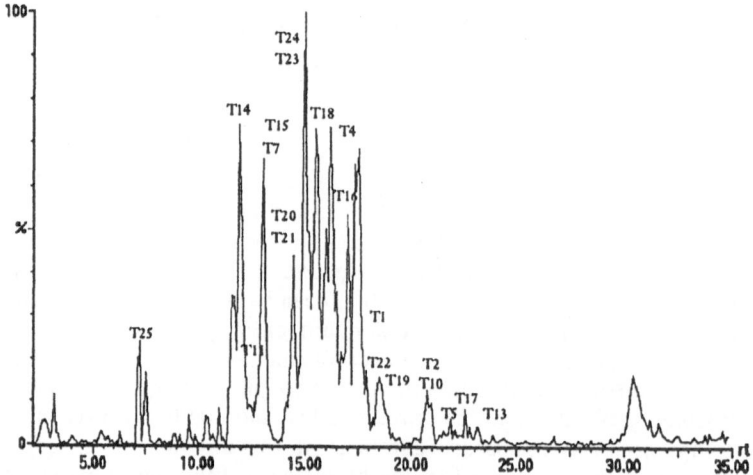

Fig. 1. Base peak intensity (BPI) chromatogram of separated peptides from recombinant horseradish peroxidase protein. The peptides are labelled with their tryptic fragment number.

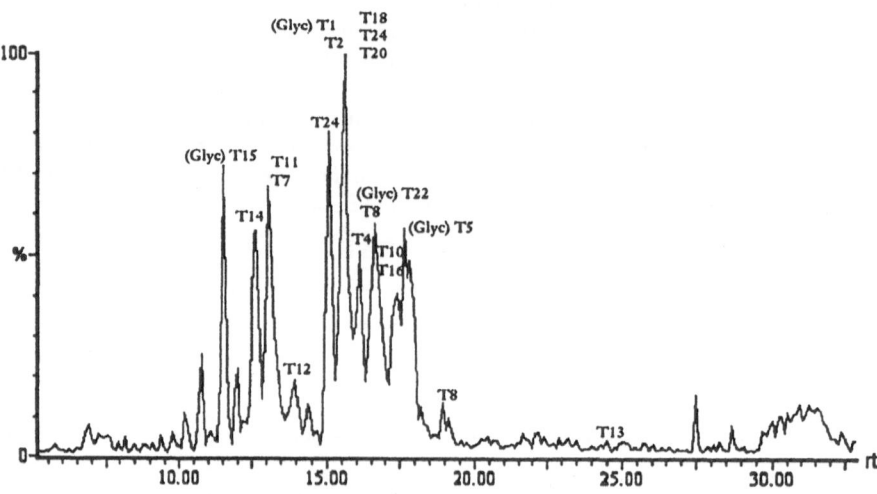

Fig. 2. BPI chromatogram of separated peptides and glycopeptides from natural horseradish peroxidase.

The mean mass error between observed and calculated glycoforms was 1.5Da. The molecular masses determined for the recombinant protein and variant horseradish peroxidase proteins F142A and R38K were 34041.8, 33965.5 and 33979.3, respectively. The differences between the expected and calculated molecular weights for the recombinant protein and the F142A variant were 0.3 and 0.1Da respectively. The variant R38K is outside this range and may in fact be a double mutant. Other possible mutations accounting for the mass difference could be either His-Cys, Met-Pro, Phe-Ile/Leu or Tyr-Gly.

On line separation of the tryptic peptides and glycopeptides obtained from both recombinant HRP and the natural glycoprotein, analysed in positive ion electrospray, produced the chromatograms shown in Figures 1 and 2. The digested natural HRP was also analysed by on-line LCMS in the negative ion mode). The peaks in the chromatogram are labelled according to the corresponding tryptic peptide or glycopeptide fragment number. Several of the chromatographic peaks contain more than one component. These unseparated components can be easily identified by mass analysis.

The glycoforms inferred in the intact proteins are made on the basis of the known structure of the principle glycan component present and the mass differences between this and other components in the spectrum. These components correspond in many but not all cases, with the glycans released from horseradish peroxidase by hydrazinolysis [1]. The recombinant form of horseradish peroxidase corresponds to that of the known amino acid sequence plus an additional methionine residue at the N-terminus. The F142A variant also has the correct molecular weight. The second mutant R38K does not have the correct molecular mass calculated for this variant, but other experiments have also suggested that this may in fact be a double mutation. On-line LC/MS analysis of the peptides produced from the natural and recombinant proteins allowed identification of the peptides and glycopeptides by mass measurement.

References

1.   Harthill, J., Ph.D. Thesis, University of Oxford, 1991.

Synthesis of glycopeptides containing 6'-*O*-phosphorylated disaccharides as ligands for the mannose 6-phosphate receptors

M.K. Christensen, M. Meldal and K. Bock

Department of Chemistry, Carlsberg Laboratory, Gamle Carlsberg Vej 10, DK-2500 Valby, Copenhagen, Denmark

Introduction

Mannose 6-phosphate (Man-6-P) has been shown to be an inhibitor of inflammation in the central nervous system [1]. The anti-inflammatory effect of Man-6-P may be due to an inhibition of the interaction between the Man-6-P receptors (MPR's) and lysosomal enzymes, as these interactions are partly responsible for the intracellular transport and binding of lysosomal enzymes [2, 3, 4].

Man-6-P residues are present at the terminal positions of asparagine linked oligosaccharides of the high mannose type in lysosomal enzymes [5]. Binding inhibition studies have shown that mannobiosides phosphorylated at the terminal position and linked $\alpha(1-2)$ are better inhibitors of the binding to the MPR's than Man-6-P itself, and also than the corresponding $\alpha(1-6)$ and $\alpha(1-3)$ linked isomers. Branched divalent ligands with two terminal Man-6-P units gave much stronger inhibitory effects (6-20 times higher) than the corresponding monophosphorylated compounds [5, 6].

We therefore found it interesting to synthesize glycopeptides containing two 6'-*O*-phosphorylated mannobiosides as bidental ligands for the MPR's, where the peptide backbone mimics the rest of the oligosaccharide. A force field calculation using the GEGOP program [7] was performed on the core structure of the *N*-linked oligosaccharide and this showed the distance between the phosphorylated disaccharides to be 9-13 Å, which corresponds to three to five amino acids.

We here report a convenient multiple column solid-phase peptide synthesis (MCPS) [8] of glycopeptides containing two mannobiosides phosphorylated at the terminal position and linked $\alpha(1-2)$ and/or $\alpha(1-6)$. The glycopeptides also contain a fluorescence probe, which allows easy monitoring in biological assays. Eleven glycopeptides were synthesized containing either 3, 4 or 5 amino acids, as presented below.

Results and Discussion

The most convenient method for the synthesis of the glycopeptides containing 6'-*O*-phosphorylated $\alpha(1-2)$ and/or $\alpha(1-6)$ linked disaccharides utilizes phosphorylated, glycosylated threonine (or serine) building blocks in MCPS. Man-6-P is the nonreducing carbohydrate unit in both the $\alpha(1-2)$ and $\alpha(1-6)$ linked disaccharides, and the optimal strategy was to synthesize a protected, phosphorylated monosaccharide unit, which could be incorporated into both disaccharides by glycoside synthesis [9]. Therefore phenyl 2,3,4-tri-O-benzoyl-1-thio-α-D-mannopyranoside was phosphorylated using bis(2,2,2-trichloroethyl)-phosphochloridate and pyridine. The resulting compound was converted into the corresponding glycosyl bromide by addition of bromine and used in silver triflate promoted glycosylations of phenyl 2,3,4-tri-O-benzoyl-1-thio-α-D-mannopyranoside and 1,3,4,6-tetra-O-acetyl-β-D-mannopyranose, respectively. The resulting disaccharides were converted into glycosyl bromides with bromine and hydrogen bromide, respectively, and used in silver triflate promoted glycosylations of Fmoc-Thr-O-Pfp to give the two phosphorylated building blocks in high yield.

To find optimal cleavage conditions for the 2,2,2-trichloroethyl groups the peptide Ac-Thr(Bz$_3$-Man-6-O-P(O)(OCH$_2$CCl$_3$)$_2$-α(1-6)-Bz$_3$-Man)-NH$_2$ was synthesized. Various conditions were studied, but only deprotection in pyridine containing 10% acetic acid using Zn and Ag$_2$CO$_3$ gave quantitative deprotection. The phosphate moiety has to be deprotected before the hydroxyl groups to avoid cyclic phosphate formation and migration.

The building blocks were used in MCPS, using the Fmoc strategy on a PEGA 1900/130 resin with an amide linker [10]. Eleven glycopeptides were synthesized, each containing two 6'-O-phosphorylated mannobiosides. All the peptides contain the fluorescence probe anthranilic acid linked to the N$^\epsilon$ of Lys:

Ac-Thr(α-D-Man-6P-(1-2)-α-D-Man)-Lys(ABz)-Thr(α-D-Man-6P-(1-2)-α-D-Man)-NH$_2$
Ac-Thr(α-D-Man-6P-(1-2)-α-D-Man)-Lys(ABz)-Thr(α-D-Man-6P-(1-6)-α-D-Man)-NH$_2$
Ac-Thr(α-D-Man-6P-(1-6)-α-D-Man)-Lys(ABz)-Thr(α-D-Man-6P-(1-2)-α-D-Man)-NH$_2$
Ac-Thr(α-D-Man-6P-(1-6)-α-D-Man)-Lys(ABz)-Thr(α-D-Man-6P-(1-6)-α-D-Man)-NH$_2$
Ac-Thr(α-D-Man-6P-(1-2)-α-D-Man)-Lys(ABz)-Gly-Thr(α-D-Man-6P-(1-2)-α-D-Man)-NH$_2$
Ac-Thr(α-D-Man-6P-(1-2)-α-D-Man)-Lys(ABz)-Gly-Thr(α-D-Man-6P-(1-6)-α-D-Man)-NH$_2$
Ac-Thr(α-D-Man-6P-(1-6)-α-D-Man)-Lys(ABz)-Gly-Thr(α-D-Man-6P-(1-2)-α-D-Man)-NH$_2$
Ac-Thr(α-D-Man-6P-(1-2)-α-D-Man)-Gly-Lys(ABz)-Gly-Thr(α-D-Man-6P-(1-2)-α-D-Man)-NH$_2$
Ac-Thr(α-D-Man-6P-(1-2)-α-D-Man)-Gly-Lys(ABz)-Gly-Thr(α-D-Man-6P-(1-6)-α-D-Man)-NH$_2$
Ac-Thr(α-D-Man-6P-(1-6)-α-D-Man)-Gly-Lys(ABz)-Gly-Thr(α-D-Man-6P-(1-2)-α-D-Man)-NH$_2$
Ac-Thr(α-D-Man-6P-(1-6)-α-D-Man)-Gly-Lys(ABz)-Gly-Thr(α-D-Man-6P-(1-6)-α-D-Man)-NH$_2$

The phosphate moieties were deprotected as described above, and the acyl groups were removed with hydrazine in chloroform:methanol, 1:4. The ^{1}H NMR spectrum of one of the fully deprotected glycopeptides is shown below.

The glycopeptides will be used in inhibition studies of binding to the MPR's. In conclusion, a facile method for the synthesis of glycopeptides containing phosphorylated carbohydrate moieties has been developed.

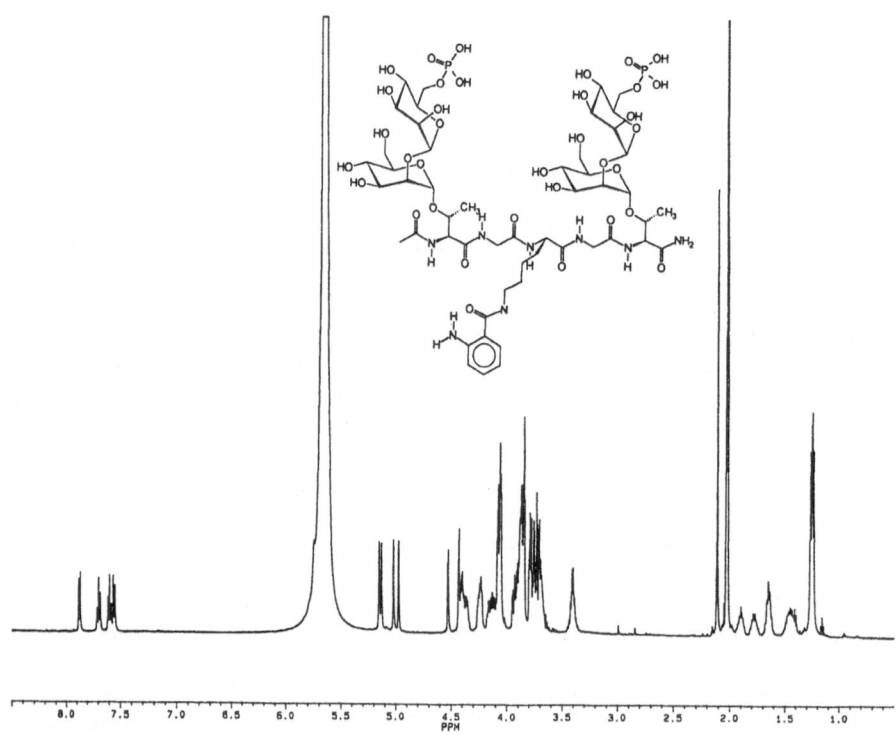

Fig. 1. 500 MHz ^{1}H NMR spectrum of a fully deprotected phosphorylated glycopeptide in D$_2$O:CD$_3$COOD, 1:1.

References

1. Willenborg, D.O., Parish, C.R., Cowden, W.B. and Curtin, J., FASEB J., 3 (1989) 1968.
2. von Figura, K. and Hasilik, A., Ann. Rev. Biochem., 55 (1986) 167.
3. Kornfeld, S., FASEB J., 1 (1987) 462.
4. Kornfeld, S., Ann. Rev. Biochem., 61 (1992) 307.
5. Distler, J.-J., Guo, J., Juordian, G.W., Srivastava, O.P. and Hindsgaul, O., J. Biol. Chem., 266 (1991) 21687.
6. Tomoda, H., Ohsumi, Y., Ichikawa, Y., Srivastava, O.P., Kishimoto, Y. and Lee, Y.C., Carbohydr. Res., 213 (1991) 37.
7. Stuike-Prill, R. and Meyer, B., Eur. J. Biochem., 194 (1990) 903.
8. Meldal, M., Holm, C.B., Bojesen, B., Jakobsen, M.H. and Holm, A., Int. J. Peptide Protein Res., 41 (1993) 250.
9. Christensen, M.K., Meldal, M. and Bock, K., J. Chem. Soc., Perkin Trans., 1 (1993) 1453.
10. Meldal, M., Tetrahedron Lett., 33 (1992) 3077.

The effect of glycosylation on various biological properties of synthetic peptides

L. Otvos Jr.[1], D.C. Jackson[2], L.E. Brown[2], H.E. Drummer[2], T. Stewart[3], M.F. Powell[3] and L. Urge[1]

[1]The Wistar Institute of Anatomy and Biology, 3601 Spruce Street,
Philadelphia, PA 19104, U.S.A.,
[2]Department of Microbiology, The University of Melbourne, Parkville,
Victoria 3052, Australia, and
[3]Genentech, Inc., 460 Point San Bruno Boulevard, South San
Francisco, CA 94080, U.S.A.

Introduction

Synthetic glycopeptides may be useful targets in many areas of chemistry, biology, and medicine. Recently, three observations prompted us to search for biotechnological uses of glycosylation of peptides. First, we demonstrated that glycosylating synthetic medium-sized peptides results in the break of dominant α-helical structures and stabilizes β-turns [1], and showed that existing turns can be further stabilized or the turn geometry modified, by glycosylation [2, 3]. Second, glycosylation may protect major histocompatibility complex (MHC)-binding peptides from proteolytic attack [4]. Third, post-translationally modified versions of T cell epitopic peptides seem to bind to MHC, but fail to bind to the T cell receptor, or be appropriately processed [5]. Since many of the peptide-drug leads assume different reverse-turn structures, these features make the synthetic glycopeptides attractive compounds for rational drug design.

Results and Discussion

The stability of glycosylated T peptides [3] in 25% human serum was investigated. The octapeptide (ASTTTNYT), its C-terminal pentapeptide analog (TTNYT), and N-terminally asparagine-containing derivatives (NASTTTNYT and NTTNYT) exhibited half-lives of less than 300 minutes. No detectable degradation was found in a 4000 minute study, however, when the extra N-terminal asparagine was glycosylated with N-acetyl-glucosamine (GlcNAc). This is understandable since T peptide degradation has been shown to proceed via aminopeptidase activity [6]. An approximate 4-fold increase of the serum half-life was detected after incorporation of GlcNAc to the internal asparagine of the pentapeptide (Figure 1), but not into the octapeptide. The stability of these peptides is correlated with the markedly changed conformation of the previous peptide, but not of the latter one, after glycosylation [4].

The biologically active conformation of the RGD peptide family has been shown to be a β-turn. We synthesized glycopeptide variants of peptide GRGDSPK in which GlcNAc was walked through the molecule in order to determine whether stabilization of the secondary structure of the peptide results in increased inhibition of binding of fibrinogen to the integrin receptor GPIIbIIIa on a solid-phase ELISA assay. Although only minor conformational changes were detected by CD after glycosylation, the *in vitro* biological activity was altered, ranging from a 2-fold decrease to a 3-fold increase (Table 1).

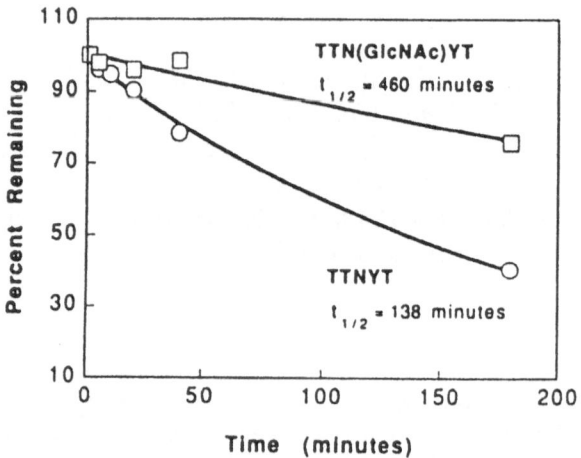

Fig. 1.   *Inhibition of serum degradation of peptide H-Thr-Thr-Asn-Tyr-Thr-NH$_2$ by incorporation of GlcNAc into the asparagine residue. Peptide analysis was carried out by reversed-phase high performance liquid chromatography after digestion of the samples with 25% human serum.*

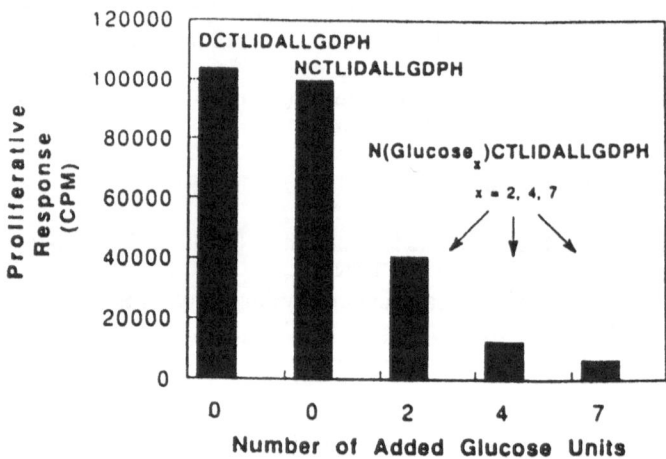

Fig. 2.   *Stimulation of T cell clone F1-36 by synthetic peptides DCTLIDALLGDPH, NCTLIDALLGDPH and corresponding glycopeptides. The oligosaccharides consisted of oligomers of α-glucose units attached to the asparagine residue via a β-glycosylic bond. The T cell clone was cultured with 250 mmolar peptide and antigen presenting cells for 4 days and ^{3}H-thymidine incorporation during the last 18 hours determined.*

Table 1 *Inhibition of GPIIbIIIa/fibrinogen binding by glycosylated RGD peptides*

Peptide sequence	Relative potency by ELISA[a]
Ac-Gly-Arg-Gly-Asp-Ser-Pro-Lys-NH$_2$	1.0
Ac-Asn(GlcNAc)-Gly-Arg-Gly-Asp-Ser-Pro-Lys-NH$_2$	1.1
Ac-Gly-Arg-Gly-Asp-Ser-Pro-Lys-Asn(GlcNAc)-NH$_2$	0.5
Ac-Asn(GlcNAc)-Gly-Arg-Gly-Asp-Ser-Pro-Lys-Asn(GlcNAc)-NH$_2$	0.7
Ac-Gly-Arg-Gly-Asp-Ser(GlcNAc)-Pro-Lys-NH$_2$	3.4

[a]A relative potency of 1.0 corresponds to an $IC_{50} = 5.1 \times 10^{-5}$ nM.

The influence of the length of the sugar on the stimulation of proposed glycosylation-dependent [7] and glycosylation-independent T cell clones (derived from the hemagglutinin of the influenza virus) was investigated by testing the *in vitro* activity of glycopeptides containing 2-7 glucose moieties (oligomers of maltose). Peptides DCTLIDALLGDPH and NCTLIDALLGDPH stimulated clone F1-36 well, but this stimulation was gradually eliminated by glycosylation of the asparagine with sugars of increasing length (Figure 2). Glycosylation of the N-terminal asparagine residue of a different influenza T cell determinant analog, NKYVKQNTLKLA with di- or hexasaccharides did not affect the T cell stimulation of clone 4.51.

Our results indicate that glycosylated synthetic peptides may be more resistant to serum proteases. Glycosylation offers a viable alternative to the peptide mimetics approach in rational drug design, possibly including the prevention of unwanted immune responses.

Acknowledgements

The authors would like to thank Drs. Wai-Lee Wong, Randy Yen, and Thomas R. Gadek (Genentech) for the ELISA of the glycosylated RGD peptides, and Jessica Burdman for technical editing comments. This work was supported by NIH grant GM 45011.

References

1. Otvos Jr., L., Thurin, J., Kollat, E., Urge, L., Mantsch, H.H. and Hollosi, M., Int. J. Peptide Protein Res., 38 (1991) 476.
2. Laczko, I., Hollosi, M., Urge, L., Ugen, K.E., Weiner, D.B., Mantsch, H.H., Thurin, J. and Otvos Jr., L., Biochemistry, 31 (1992) 4282.
3. Urge, L., Gorbics, L. and Otvos Jr., L., Biochem. Biophys. Res. Commun., 184 (1992) 1125.
4. Gaeta, F.C.A., Sette, A., Arrhenius, T., Thomson, D., Soda, K. and Colon, S.M., Pharm. Res., 10 (1993) 1268.
5. Larson, J.K., Otvos Jr., L. and Ertl, H.C.J., J. Virol., 66 (1992) 3996.
6. Marastoni, M., Salvadori, S., Balboni, G., Spisani, S., Gavioli, L., Traniello, S. and Tomatis, R., Int. J. Peptide Protein Res., 35 (1990) 81.
7. Drummer, H.E., Jackson, D.C. and Brown, L.E., Virology, 192 (1993) 282.

Primary structure elucidation, surfactant function and specific formation of supramolecular dimer structures of lung surfactant associated SP-C proteins

M. Przybylski[1], C. Maier[1], K. Hägele[1], E. Bauer[2],
E. Hannappel[2], R. Nave[3], K. Melchers[3],
U. Krüger[3] and K.P. Schäfer[3]

*[1] Faculty of Chemistry, University of Konstanz, P.O. Box 5560 M731,
78434 Konstanz, Germany; [2] Institute of Biochemistry, University of
Erlangen; [3] Byk Gulden GmbH, Konstanz, Germany*

Introduction

Pulmonary surfactant is a complex mixture of phospholipids, and ca. 5% proteins which are secreted by alveolar pneumocytes and exert an important role, predominantly by reducing the surface tension at the air-water interface of the lung epithel [1]. Among the surfactant proteins known thus far, the hydrophobic SP-B (ca. 16 kD) and SP-C (ca. 5 kD) proteins have found high interest in the development of surfactant preparations for the therapy of respiratory distress syndrome [2]. Although amino acid sequences of SP-B and SP-C have been derived from the gene structures, structural studies of the proteins proved difficult or unfeasible due to their extreme hydrophobicity. By contrast, desorption-ionization mass spectrometry (252-Cf plasma desorption, PDMS; electrospray, ESMS) were particularly successful. Combined with selective chemical modification and cleavage, PDMS and ESMS enabled the complete primary structure determination of natural and recombinant SP-C proteins. Of particular interest has been the identification by ESMS of specific dimeric structures which in analogy to leucine zipper proteins are formed by the hydrophobic C-terminal part of the protein (see Figure 2). This dimerization and the modified, palmitoylated structure are suggested to constitute essential elements for the biological function of SP-C.

Results and Discussion

Natural SP-C proteins from several species were isolated by lung lavage using organic solvent extraction, followed by Sephadex LH-60 chromatography and final C4-RP-HPLC purification [3]. For primary structure determination, PDMS and ESMS were used as predominant methods in combination with selective fatty acid-deacylation by DTT, S-alkylation, and N-terminal Edman/PDMS sequence analysis [4], and yielded abundant protonated ($[M + nH]^{n+}$) or $[M + Na]^+$ molecular ions and partially sequence-specific fragment ions [3]. Complete primary structures are compared in Table 1 including a recombinant human SP-C expressed in *E. coli*. All natural SP-C proteins contain a strictly conserved N-terminal -PCC/XP- sequence (X=Phe) with one or two palmitoylated cysteine residues, Lys-Arg residues in positions 10/11, followed by an extremely hydrophobic C-terminal domain of 24/25 residues. Mass spectrometric analyses revealed that SP-C proteins from proteinosis patients, and from amniotic fluid had unchanged sequences, but consistently showed partial or complete lack of palmitoylation. Attempts to synthesize the entire SP-C protein by standard SPPS procedures failed due to its extreme hydrophobicity, while an N-terminal palmitoylated model peptide was amenable by chemical synthesis. However, human-identical SP-C (rh-SP-C) has been prepared by selective Cys-palmitoylation of recombinant SP-C (see Figure 1).

Table 1   *Primary structures of SP-C proteins from different species.*

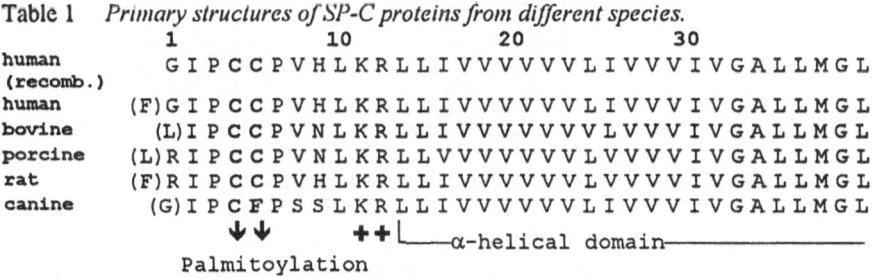

Fig. 1. *ES mass spectra of natural human SP-C (left) and palmitoylated rh-SP-C (right). Molecular ions A,B in natural SP-C are due to heterogeneity by one N-terminal Phe-residue.*

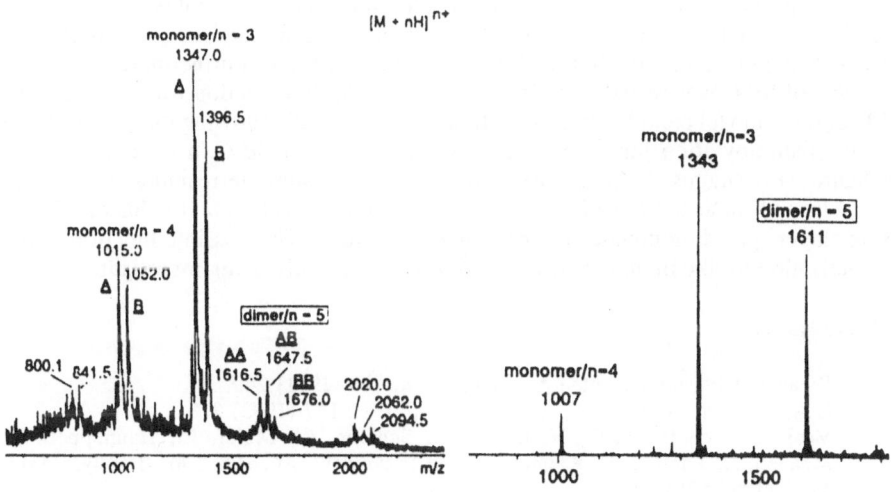

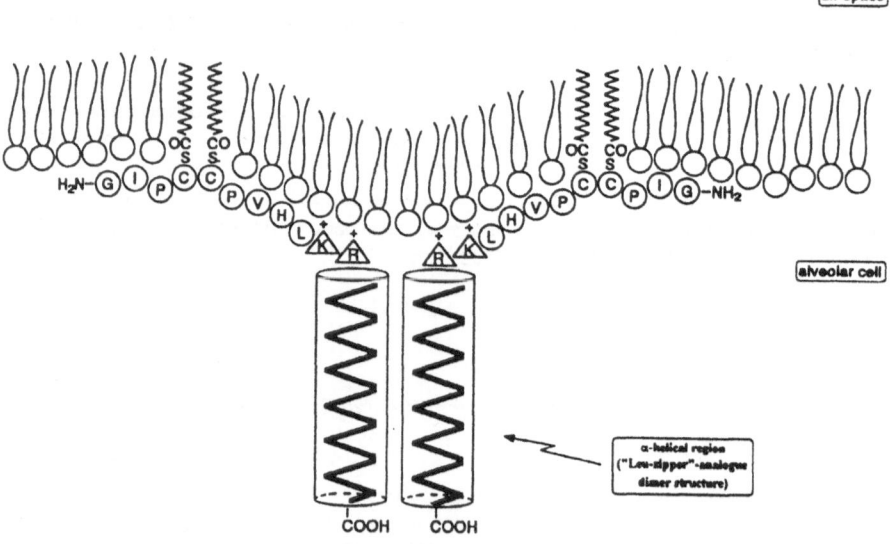

Fig. 2. Structural model and possible lipid monolayer insertion of SP-C dimer.

Evidence for the formation of specific, "leucine zipper-analogue" dimeric structures was obtained from selective protein-chemical reactions, spectroscopic and, particularly, ESMS data. ESMS provided the direct identification of dimers by specific *odd charge* molecular ions ($[M + 5H]^{5+}$), showing homodimer formation for rh-SP-C and heterodimers for the two forms (A,B.) of natural human SP-C (Figure 1). Protein-chemical, secondary structure prediction and molecular modelling data are consistent with a stable, α-helical dimerization motif (12-35) flanked by two basic (R^{11}, K^{10}) residues, which are resistant to trypsin digestion; these residues in the SP-C(1-14) peptide and Arg^2 in rat SP-C are readily digested. Furthermore, the Met^{32} residue of SP-C was resistant to BrCN cleavage. No dimerization was found for the SP-C(1-14) model peptide. Based on these data and results showing the essential role of the palmitoylation for the *in vitro* surface activity of SP-C, a structural model (Figure 2) is suggested for the insertion of the cys-palmitate residues in the lipid monolayer film at the alveolar surface, and their stabilization and enhancement of monolayer spreading kinetics by the dimeric structure. The possible function of the dimerization for the biological activity of SP-C is currently being investigated.

References

1.   Weaver, T.E. and Witsett, J.A., Biochem. J., 273 (1991) 249.
2.   Lewis, J.F. and Jobe, A.H., Am. Rev. Resp. Dis., 147 (1993) 218.
3.   Voss, T., Schäfer, K.P., Nielsen, P.F., Schäfer, A., Maier, C., Hannappel, E., Maaßen, J., Landis, B., Klemm, K. and Przybylski, M., Biochim. Biophys. Acta, 1138 (1992) 261.
4.   Nielsen, P.F., Svoboda, M., Schneider, K., Landis, B. and Przybylski, M., Anal. Biochem., 191 (1990) 302.

Solid phase synthesis of muramyl dipeptides and analogues

J. Tschakert and W. Voelter

Abteilung für Physikalische Biochemie des Physiologisch-chemischen Instituts der Universität Tübingen, Hoppe-Seyler-Straße 4, D-7400 Tübingen, Germany

Introduction

Muramyl dipeptide (N-acetylmuramyl-L-alanyl-D-isoglutamine, MDP) is the smallest subunit of bacterial cell walls possessing immunoadjuvant activity with the ability to replace micobacteria in Freund's Complete Adjuvant [1]. Due to their prophylactic and therapeutical uses, a large number of analogues has been synthesized in the past and their immunostimulant activities were tested [2, 3, 4]. In this communication we describe a solid phase approach which allows to synthesize MDP and its derivatives via a convenient route with high purity using our recently developed 4-benzoxy-dimethoxy-benzylamine-(styrene-1%-divinylbenzene) resins with aminomethylene functions in *o*-, *p*-, in *o,o*- or *o,p*-positions. These resins are coupled, applying the conventional Fmoc/Bu^t strategy, with the D- and L-isomers of Fmoc-Glu(OBu^t)-OH and Fmoc-Ala-OH, to yield dipeptide resins to which the carbohydrate derivatives and polycyclocarbonic acids are attached.

Results and Discussion

Figure 1 shows the structures of the resins **1a-1d** and the subsequent coupled carbohydrates **2**, **4** and **6** and polycyclocarbonic acids **3** and **5**.

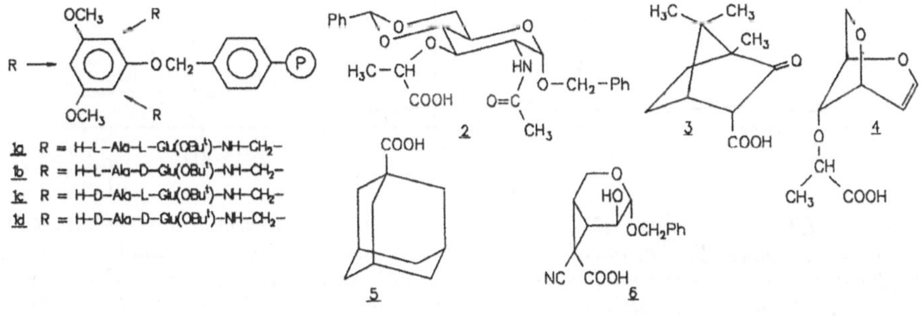

*Fig. 1. Structures of the synthesized resins **1a-1d**, carbohydrate derivatives **2**, **4**, **6** and polycyclocarbonic acids **3** and **5**.*

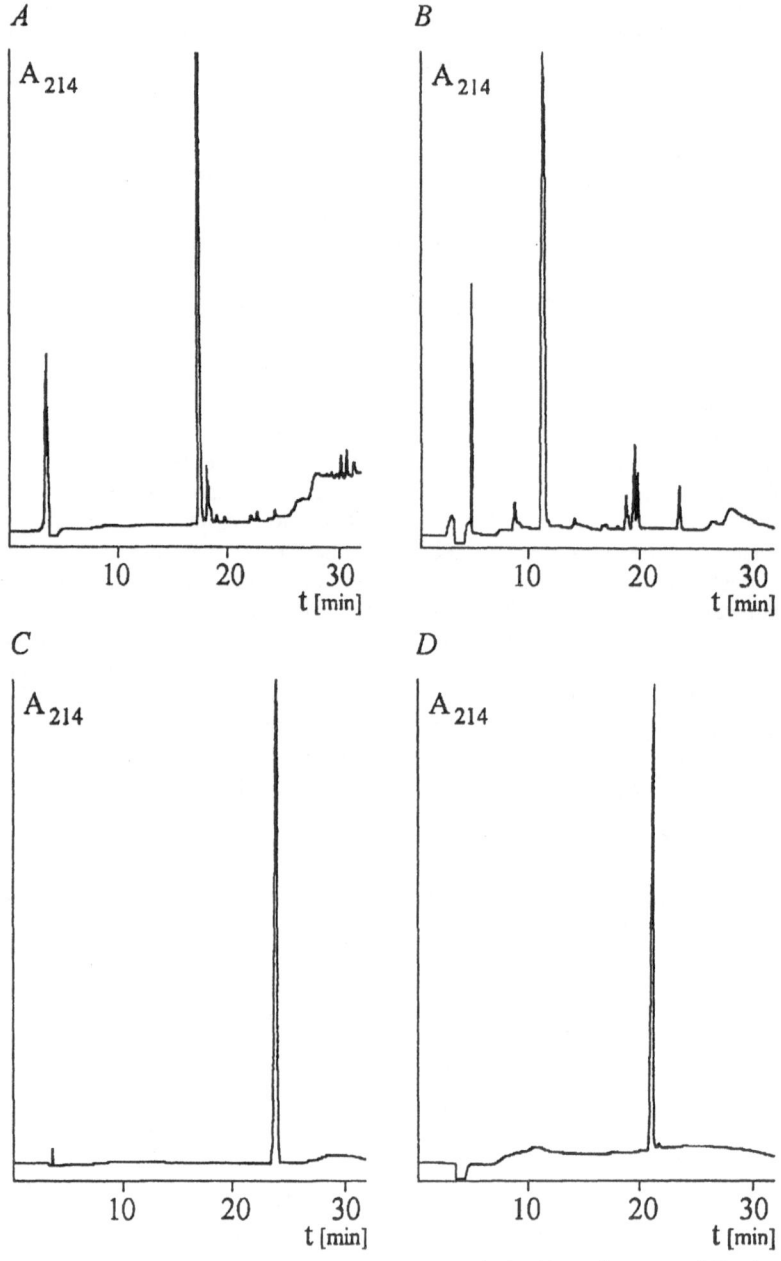

Fig. 2.   HPLC of crude MDP analogues; panel A: Benzyl-muramyl-D-alanyl-D-
isoglutamine, panel B: 3,6-Anhydro-4-O-[D-1-(carbonyl-L-alanyl-D-isoglutamine)ethyl]-
D-glucal, panel C: Adamantyl-1-carbonyl-L-alanyl-D-isoglutamine, panel D: Camphor-±-
3-carbonyl-L-alanyl-L-isoglutamine; conditions: Column: LiChrospher® RP-18 5μm;
eluents: A: 0.05% TFA in H_2O (v/v), B: TFA/H_2O/CH_3CN 0.05:40:60 (v/v); gradient: 0-
1min 95% A, 1-31min linear 95-5% A; flow rate: 1ml/min; detection: λ=214nm.

The protected muramic acid is synthesized from 2-acetamido-2-deoxy D glucopyranoside via benzyl 2-acetamido-2-deoxy-4,6-benzylidene-α-D-glucopyrano-side. To a dioxane solution of the latter sodium hydride is added and the mixture stirred at 95°C for 1 h. After cooling to 65°C, D,L-α-chloropropionic acid and after 1 h again sodium hydride are added. The mixture is stirred over night at 65°C and after workup and recrystallization **2** is obtained with 90% optical purity (D-lactyl residue). This racemic mixture is used for coupling to the dipeptide resins. After removal of the benzyl muramyl-D-alanyl-D-isoglutamine from the resin, the HPLC of the crude product (Figure 2, panel *A*) shows besides the major peak a neighbouring one at somewhat higher retention time, caused by benzyl isomuramyl-D-alanyl-D-isoglutamine resulting from the starting material. While the synthesis of 3,6-anhydro-4-O-[D-1-(carbonyl-L-alanyl-D-isoglutamine)ethyl]-D-glucal suffers from minor side reactions, adamantyl-1-carbonyl-L-alanyl-D-isoglutamine and camphor-±-3-carbonyl-L-alanyl-L-isoglutamine are received in almost 100% purity (Figure 2, panels *B*, *C*, *D*). The structures of the synthesized products are confirmed by FAB, FD, and LD MS (Finnigan MAT, Bremen, Germany) and ^{1}H NMR spectroscopy demonstrating that the new anchor resins give easy access to muramyl dipeptide, respectively its derivatives. The four resins **1a-1d** are all equally suitable for their syntheses. Furthermore the biological activities of the synthetic muramyl dipeptide derivatives are tested according to reference [5], using the luminol-dependent chemiluminescence associated with the phagocytosis of opsonized zymosan by granulocytes. The data show that only adamantyl-carbonyl-L-alanyl-D-isoglutamine has an activity similar to MDP, but contrary to MDP, the activity increases at higher concentrations.

References

1. Ellouz, F., Adam, A., Ciorbaru, R. and Lederer, E., Biochem. Biophys. Res. Commun., 59 (1974) 1317.
2. Adam, A. and Lederer, E., Med. Res. Rev., 4 (1984) 111.
3. Lefrancier, P., Derrien, M., Jamet, X., Choay, J. and Lederer, E., J. Med. Chem., 25 (1982) 87.
4. Zaoral, M., Jezek, J., Krchnák, V. and Straka, R., Coll. Czech. Chem. Commun., 45 (1980) 1424.
5. Bruchelt, G. and Schmidt, K.H., J. Clin. Chem. Biochem., 22 (1984) 1.

Synthesis, analysis, and circular dichroism studies of glycopeptides and their building blocks containing extended sugar systems

L. Urge[1], D.C. Jackson[2], G. Graczyk[3], L. Gorbics[1] and L. Otvos, Jr.[1]

*[1]The Wistar Institute of Anatomy and Biology, 3601 Spruce Street, Philadelphia, PA 19104, U.S.A.,
[2]Department of Microbiology, The University of Melbourne, Parkville 3052, Australia, and
[3]Department of Biochemistry and Biophysics, The University of Pennsylvania, Philadelphia, PA 19104, U.S.A.*

Introduction

The automated synthesis of glycopeptides is a focus of research in many laboratories. Recently we showed that Fmoc-Asn(sugar)-OH synthons couple to resin-bound amino functions with acceptable efficiency, provided that the hydroxyl groups of the carbohydrates remain unprotected [1, 2]. We did not detect significant acylation of the sugar-hydroxyls during syntheses or decomposition of the O-glycosidic bonds during the acidic cleavage in a great many di- and trisaccharide-containing N-glycopeptides [3, 4]. Although some characteristic conformational and immunological properties of the fragments of glycoproteins are present in the synthetic glycopeptides after incorporation of the first or the first two sugar moieties [5], others require longer sugar chains to be added [6]. Here we report the synthesis and secondary structural analysis of T-cell epitopic peptides carrying extended carbohydrate systems.

Results and Discussion

Maltobiose (Glc_2), maltotetrose (Glc_4), maltohexose (Glc_6), and maltoheptose (Glc_7) were converted to their 1-amino derivatives with the conventional NH_4HCO_3 method [7], and the resulting amino-sugars were coupled to Fmoc-Asp-^tBu. The free acids were obtained after cleavage with trifluoroacetic acid (TFA), and purification by reversed-phase high-performance liquid chromatography (RP-HPLC). In spite of the long carbohydrate chain, the glycosylated asparagine derivatives are still soluble in DMF. After activation, their pentafluorophenyl esters couple to the resin with a decreased efficiency as the size of the sugar increases (Table 1). The ^tBu esters, and the free acids, were characterized by fast-atom bombardment mass spectroscopy (FAB-MS) and NMR. Asparagine-bound Glc_2, Glc_4, or Glc_7 was incorporated into the N-terminus of CTLIDALLGDPH (peptide I), and Asn-bound Glc_2 or Glc_6 was coupled to the N-terminus of KYVKQNTLKLA (peptide II). Peptide I was found to respond to T-cell clones directed against the hemagglutinin of influenza virus in a sugar-dependent manner, and peptide II to respond in a sugar-independent manner. Many T-cell epitopes are believed to be amphipathic α-helices, and indeed both non-glycosylated peptides assume strong α-helical structure. N-terminal glycosylation does not interfere with the dominant conformations. In fact, the identical circular dichroism (CD) spectra of peptide I and N(Glc$_7$)-peptide I, and peptide II and N(Glc$_6$)-peptide II, in TFE after subtraction of the molar contributions of the appropriate H-Asn(sugar)-OH derivatives, indicate that such mathematical processing of the CD spectra of the glycopeptides is justifiable (Figure 1). The heptasaccharide itself may form one turn of a sugar-helix, and it is comparable in size with the natural carbohydrate antennae (Figure 2).

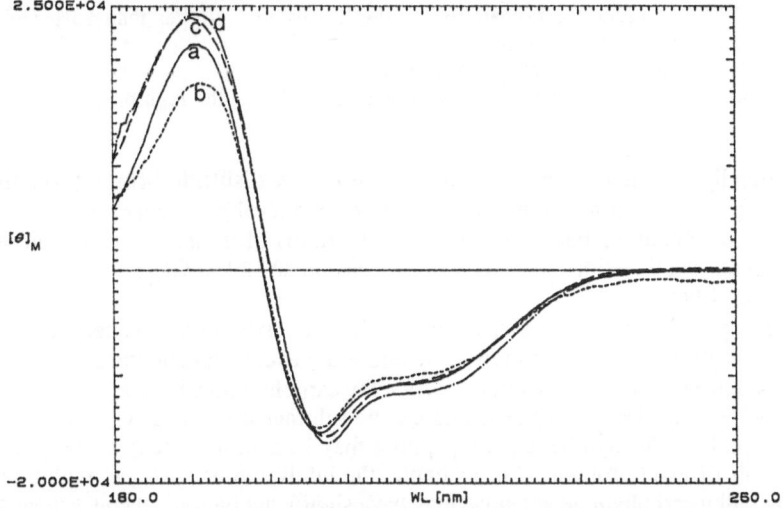

Fig. 1.   *CD spectra of non-glycosylated and glycosylated influenza hemagglutinin peptides in 90-100% trifluoroethanol (TFE). Curve a (solid line) NCTLIDALLGDPH, curve b (dotted line) N(Glc$_7$)CTLIDALLGDPH, curve c (dashes) NKYVKQNTLKLA, curve d (dots and dashes) N(Glc$_6$)KYVKQNTLKLA. Spectra were taken on a Jasco J720 CD instrument at room temperature. Peptide concentration was about 0.5 mg/ml.*

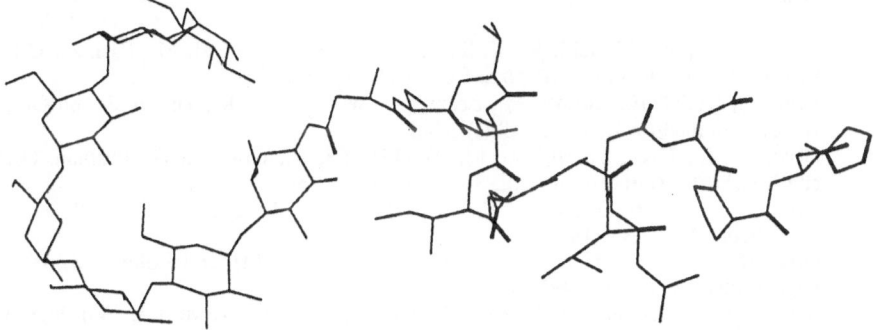

Fig. 2.   *Structure of N(Glc$_7$)CTLIDALLGDPH as represented by the molecular modeling program MM2.*

Table 1 *Characterization of Fmoc-Asn(sugar)-OH derivatives*

Sugar	RP-HPLC retention time (min)	FAB-MS (M+H)	Yield[a]	Coupling efficiency to resin
Glc_2	29.9	679	35%	78%[b]
Glc_4	28.7	1003	23%	53%[b]
Glc_6	27.8	1327	14%	24%[c]
		1349 (M+Na)		
Glc_7	27.5	1512 (M+Na)	15%	30%[c]

[a]These yields represent the preparation of the ^{t}Bu esters. Higher than 95% yield was observed after the TFA treatment.
[b]A 4 molar excess of the acylating glycoamino acid.
[c]A 0.9 molar excess of the acylating glycoamino acid (due to the high price of the starting sugars).

Finally, we investigated the rate of the *in situ* disulfide bridge formation of peptide I as a function of the length of the proximal oligosaccharide chain. After standing for about 22 hours in 80% TFE, the ratios of dimerized glycopeptides to monomeric glycopeptides were as follows: No sugar: 39%; Glc_2: 27%; Glc_4: 10% and Glc_7: 13%.

The preparation of Fmoc-Asn(sugar)-OH derivatives with extended sugar chains proceeds without difficulty and these synthons are excellent candidates for the solid-phase synthesis of N-glycopeptides. When the carbohydrates lack acetamido groups, the spectral contribution of the sugars can be subtracted from the CD spectra of the glycopeptides. The synthetic glycopeptides may be useful to study surface physical properties of glycoproteins. For example, the inhibition of cysteine oxidation by a nearby bulky carbohydrate antennae may have significant biological consequences.

Acknowledgements

This work was supported by NIH grant GM 45011.

References

1. Otvos Jr., L., Wroblewski, K., Kollat, E., Perczel, A., Hollosi, M., Fasman, G.D., Ertl, H.C.J. and Thurin, J., Peptide Res., 2 (1989) 362.
2. Urge, L., Kollat, E., Hollosi, M., Laczko, I., Wroblewski, K., Thurin, J. and Otvos Jr., L., Tetrahedron Lett., 32 (1991) 3445.
3. Otvos Jr., L., Urge, L., Hollosi, M., Wroblewski, K., Graczyk, G., Fasman, G.D. and Thurin, J., Tetrahedron Lett., 31 (1990) 5889.
4. Urge, L., Otvos Jr., L., Lang, E., Wroblewski, K., Laczko, I. and Hollosi, M., Carbohydr. Res., 235 (1992) 83.
5. Otvos Jr., L., Thurin, J., Kollat, E., Urge, L., Mantsch, H.H. and Hollosi, M., Int. J. Pept. Protein Res., 38 (1991) 476.
6. Jackson, D.C., Drummer, H.E., Urge, L., Otvos Jr., L. and Brown, L.E., Virology, in press.
7. Likhosherstov, L.M., Novikova, O.S., Derevitskaja, V.A. and Kotchetkov, N.K., Carbohydr. Res., 146 (1986) c1.

Session V
Biologically Active Peptides

Chairs: Daniel F. Veber
Merck Sharp & Dohme Research Laboratories
West Point, Pennsylvania, U.S.A.

and

Richard D. DiMarchi
Lilly Research Laboratories
Indianapolis, Indiana, U.S.A.

Research and clinical potentials of bradykinin antagonists

J.M. Stewart

Department of Biochemistry, University of Colorado Medical School, Denver, CO 80262, U.S.A.

Introduction

Applications of kinin antagonists in physiological research have brought about a veritable revolution in improved understanding of biological regulation and control, whereas applications of bradykinin antagonists to clinical medicine are likely to bring about a revolution in drugs for antiinflammatory therapy. Although kinins evidently participate in regulation of every major physiological system, the most striking actions of kinins are in pathology, where such conditions as shock, asthma, pain, and many forms of inflammation involve kinin mediation [1]. Kinins can evoke the "cardinal signs" of inflammation: pain, redness, edema and hyperthermia.

The mammalian kinins are bradykinin (BK), a nonapeptide having the structure:

$$Arg^1\text{-}Pro^2\text{-}Pro^3\text{-}Gly^4\text{-}Phe^5\text{-}Ser^6\text{-}Pro^7\text{-}Phe^8\text{-}Arg^9$$

and its homologs kallidin (Lys-BK) and methionyl-lysyl-BK. These peptides are released from precursor kininogens by the action of kallikrein enzymes and exert their biological effects through specific receptors to trigger many physiological and pathological processes [2, 3]. The two kinin-forming enzymes, plasma kallikrein and tissue kallikrein, are trypsin-like proteases that cleave kinins from two protein precursors, "high molecular weight kininogen" (HMWK) and "low molecular weight kininogen" (LMWK). Plasma kallikrein cleaves HMWK to yield BK, whereas tissue kallikrein cleaves principally LMWK to yield kallidin. Met-Lys-BK is produced in inflammation by the concerted action of tissue kallikrein and neutrophil elastase.

Physiological functions of kinins include relaxation or contraction of smooth muscles, regulation of ion transport, stimulation of secretion and neurotransmitter-neuromodulator action. Relaxation of vascular smooth muscle, mediated by endothelial nitric oxide [4], causes lowering of blood pressure, especially important in opposing the pressor action of angiotensin (ANG). On the other hand, bronchial, intestinal and uterine smooth muscles are contracted by kinins. Kinins play an important role in overall ion regulation by controlling kidney function. Here, also, BK opposes physiologically the action of ANG; BK stimulates sodium excretion, whereas ANG stimulates sodium retention. The renal action of atrial natriuretic peptides appears to be mediated by kinins. Kinins also regulate ion transport in intestine and in sweat glands. While kinins for most actions are produced non-neuronally, certain brain neurons do produce kinins and use them as neurotransmitters or neuromodulators, and actions of kinins on peripheral sensory neurons are very important.

In pathology, bradykinin is produced whenever vascular injury exposes negatively-charged sulfated polysaccharides of underlying tissues [5]. HMWK, inactive pre-plasma kallikrein (PK) and the clotting Factors XII and XI adsorb to the negatively charged surface, with concomitant activation of Factor XII and PK to initiate blood clotting and liberate BK. BK is also liberated whenever tissue perfusion and oxygenation are compromised; in this case the fall in tissue pH caused by accumulation of acidic metabolites is important, and may be adequate to evoke full kinin release and concomitant production of severe pain [6]. In tissue injury and infection, accumulation, attachment and activation of neutrophils add other mediators to the inflammatory site; these may potentiate the actions of kinins. In many tissues, BK stimulates phospholipase A2 to liberate arachidonic acid; the prostaglandins and leukotrienes produced from the arachidonate are especially important in pain and inflammation.

There is clear evidence for several classes of receptors for kinins. Most kinin functions are mediated by "B2" receptors that require the full kinin chain for activation [7]. Recently several laboratories have reported cloning and sequencing of B2 receptors; these are standard G-protein linked structures having seven trans-membrane helical segments [8, 9, 10]. Differences in tissue selectivity of kinin antagonists has led investigators to propose several sub-classes of B2 receptors, but as yet no molecular basis for such classes has been reported. In tissue injury and in chronic inflammation another class of receptors ("B1") is also expressed [11]. The B1 receptors are activated primarily by BK (1-8) or kallidin (1-9); these peptides lacking the C-terminal arginine residues of the native kinins are formed by action of carboxypeptidase N ("kininase I"), the principal kinin-destroying enzyme of plasma. The other major kininase is angiotensin I converting enzyme (ACE, "kininase II"), a membrane-bound enzyme localized primarily to the pulmonary vasculature. ACE normally destroys 98-99% of kinins in blood on a single passage through the pulmonary circulation, but in serious infection and inflammation the glycophosphatidyl link that attaches ACE to the vascular wall is cleaved by phospholipases, and the very important protective action of this enzyme is lost. Loss of ACE and appearance of B1 receptors are important causative factors of the serious hypotension seen in shock.

Bradykinin Antagonists

Antagonists for kinin B1 receptors were reported in 1976 by Regoli [12], but these were primarily of academic interest at the time, since they did not block normal kinin functions. Extensive attempts to develop competitive antagonists for B2 kinin systems were unsuccessful until 1984 when the first antagonists, [D-Phe^7]-BK and [Thi^{5,8}-D-Phe^7]-BK (Thi=β-(2thienyl)-alanine) were reported by Vavrek and Stewart [13]. The key modification was replacement of the proline at position seven with a D-aromatic amino acid residue. One antagonist, DArg-[Hyp^3, Thi^{5,8}, DPhe^7]-BK (known as NPC 349; Hyp=*trans*-4-hydroxyproline), was used by many biologists and helped demonstrate the importance of kinins in physiology and pathology [14]. Many BK antagonists of this type have been synthesized [14]. The "first generation" antagonists are degraded *in vivo* and have short durations of action. Although the D-amino acid residue at position seven of the antagonists blocks action of ACE, action of kininase I is not inhibited. Indeed, this cleavage can yield B1 antagonists. Structural modifications of BK antagonists by several groups, including the authors' laboratory, led to development of "second generation" BK antagonists by incorporating unusual amino acids, conformationally constrained and stable to enzymatic degradation, in the C-terminal part of the peptides. Perhaps most remarkable among these is the Hoechst antagonist HOE-140 [15], that has the structure of NPC-349 but with D-tetrahydroisoquinoline-3-carboxylic acid (DTic) at position 7 and octahydroindole-2-carboxylic acid (Oic) at position 8. The Oic residue at position 8 confers resistance to kininase I, and HOE-140 has a long duration of action on blood pressure *in vivo*. In receptor studies, HOE-140 has been found to be very "sticky:" it has a very slow "off" time. Oic is a proline homolog, and we had earlier reported potent BK antagonists having proline at position 8. Then we and the Nova group reported BK antagonists having D-aliphatic residues in position seven in combination with L-aliphatic residues at position 8 [16, 17]. Thus, contrary to the original belief, effective BK antagonists do not require a D-aromatic residue at position seven. The Nova modification with D-4-propoxyproline [16] and our D-α-cyclopentylglycine (Cpg) [17] at position seven yielded potent antagonists. We then found that when the residues at positions seven and eight are aliphatic, the residue at position five can also be aliphatic [18]. Bulky branched or cyclic side chains at these positions are most effective. Some of these new antagonists show quite prolonged activity *in vivo* [Gera, et al., this volume].

The approach of the Cortech group [19; see Blodgett et al., this volume] to BK antagonists is very different. They have described a series of dimers in which a [Cys[6]]BK antagonist is crosslinked with bis-maleimidohexane. Most recently they have reported a series of heterodimers in which B2-B1 antagonists are combined, and heterodimers in which a BK antagonist is linked to a tachykinin antagonist, an opiate agonist, or an elastase inhibitor. Such hybrids offer the possibility of developing drugs directed simultaneously against more than one mediator of inflammation, and may be particularly useful in such conditions as asthma [20], shock [21] and arthritis [22], where several mediators are involved. The field of kinins and their antagonists is thus at a very high level of interest and activity. Important new discoveries in the kinin field and clinical applications can be anticipated.

Acknowledgements

This work was supported by grants from the US NIH (grant HL-26284) and from Nova Pharmaceutical Corp. We thank Floyd W. Dunn and Charles H. Stammer for providing unusual amino acids, Robin Reed and Robert Binard for assistance with chemistry, and Frances Shepperdson for bioassays.

References

1.   Stewart, J.M., In Henson, P.M. and Murphy, R.C., (Eds.), Mediators of the Inflammatory Process, vol. 6, Elsevier, Amsterdam, 1989, p.189.
2.   Erdos, E.G. (Ed.), Handbook of Experimental Pharmacology, vol 25 (suppl.), Springer, Heidelberg, Germany, 1979.
3.   Bhoola, K.D., Figueroa, C.D. and Worthy, K., Pharmacol. Rev., 44 (1992) 1.
4.   Moncada, S., Palmer, R.M.J. and Higgs, E.A., Pharmacol. Rev., 43 (1991) 109.
5.   Kozin, F. and Cochrane, C.G., In Gallin, J.I, Goldstein, I.M. and Snyderman, R. (Eds), Inflammation: Basic Principles and Clinical Correlates, Raven Press, New York, U.S.A., 1988, p.101.
6.   Dray, A. and Perkins, M., Trends Neurosci., 16 (1993) 99.

7. Stewart, J.M., In Cheronis, J.C. and Repine, J.E. (Eds), Proteases, Protease Inhibitors and Protease-Derived Peptides, Birkhaeuser, Basel, Switzerland, 1993, p.145.
8. McEachern, A.E., Shelton, E.R., Bhakta, S., Obernolte, R., Bach, C., Zuppan, P., Fujisaki, J., Aldrich, R.W. and Jarnagin, K., Proc. Nat. Acad. Sci. U.S.A., 88 (1991) 7724.
9. Hess, J.F., Borkowski, J.A., Young, G.S., Strader, C.D. and Ransom, R.W., Biochem. Biophys. Res. Comm., 184 (1992) 260.
10. Eggerickx, D., Raspe, E., Bertrand, D., Vassart, G. and Parmentier, M., Biochem. Biophys. Res. Comm., 187 (1992) 1306.
11. Bouthillier, J., Deblois, D. and Marceau, F., Brit. J. Pharmacol., 92 (1987) 257.
12. Regoli, D. and Barabe, J., Pharmacol. Rev., 32 (1980) 1.
13. Vavrek, R.J. and Stewart, J.M., Peptides, 6 (1985) 161.
14. Stewart, J.M. and Vavrek, R.J., In Burch, R.M. (Ed), Bradykinin Antagonists, Dekker, New York, U.S.A., 1991, p.51.
15. Lembeck, F., Griesbacher, T., Eckhardt, M., Henke, S., Breipohl, G. and Knolle, J., Brit. J. Pharmacol., 102 (1991) 297.
16. Kyle, D.J., Martin, J.A., Burch, R.M., Carter, J.P., Lu, S., Meeker, S., Prosser, J.C., Sullivan, J.P., Togo, J., Noronha-Blob, L., Sinsko, J.A., Walters, R.G., Whaley, L.W. and Hiner, R.N., J. Med. Chem., 34 (1991) 2649.
17. Vavrek, R.J., Gera., L. and Stewart, J.M., In Bonner, G., Fritz, H., Schoelkens, B., Dietze, G. and Lupetz, K. (Eds), Recent Progress on Kinins, Agents Actions Suppl. 38 (I), 1992, p.572.
18. Stewart, J.M., Gera, L. and Srivastava, V., In Schneider, C.H. and Eberle, E. (Eds), Peptides 1992, Escom, Leiden, The Netherlands, 1993, p.691.
19. Cheronis, J.C., Whalley, E.T., Nguyen, K.T., Eubanks, S.R., Allen, L.G., Duggan, M.J., Loy, S.D., Bonham, K.A. and Blodgett, J.K., J. Med. Chem., 35 (1992) 1563.
20. Page, C.P. and Barnes, P.J., (Eds.), Pharmacology of Asthma (Handbk. Exp. Pharmacol. vol. 98) Springer, Heidelberg, Germany, 1991.
21. Jochum, M. and Fritz, H., In Faist, E., Ninnemann, J.L. and Green, D.R. (Eds.), Immune Consequences of Trauma, Shock and Sepsis, Springer, Heidelberg, Germany, 1989, p.165.
22. Kelley, W.H., Harris, E.D., Ruddy, S. and Sledge, C.B., Textbook of Rheumatology, Saunders, Philadelphia, 1993.

Development of dual action heterodimeric compounds containing bradykinin antagonist activity

J.K. Blodgett, L.G. Allen, V.S. Goodfellow, S.D. Loy, M.V. Marathe, D.A. McLeod, L.W. Spruce, E.T. Whalley and J.C. Cheronis

Cortech, Inc., 6850 N. Broadway, Denver, CO 80221, U.S.A.

Introduction

The pathological effects of bradykinin and related kinins and their key role in the inflammatory process have resulted in bradykinin antagonists being a therapeutic target for the pharmaceutical industry for well over a decade. This type of compound could be useful in the treatment of such diverse disorders as septic shock, asthma and rhinitis [1, 2]. Work in our laboratories related to the discovery of new bradykinin antagonists resulted in a series of peptide-based BK_2 antagonist homodimers which had significantly improved activities both *in vitro* and *in vivo* relative to their corresponding monomeric peptide precursors [3]. This work culminated in the discovery of CP-0127; alternatively Bradycor^{TM} (Table 1), a bis-succinimidohexane cross-linked peptide homodimer based on the monomeric BK_2 antagonist (D)Arg-Arg-Pro-Hyp-Gly-Phe-Cys-(D)Phe-Leu-Arg [3]. Bradycor^{TM} is currently in Phase II human clinical trials for sepsis.

While investigating the SAR of Bradycor^{TM} and related compounds, it became apparent that a significant component of the activities of these compounds at the BK_2 receptor could be obtained by a portion of the dimer, specifically by the monomeric BK_2 antagonist and an appropriately chosen linker moiety. Furthermore, we ascertained that additional therapeutic activity could then be incorporated into a single molecule by employing a heterodimeric structure (Table 1).

Results and Discussion

There is increasing evidence in the literature indicating that, while BK_2 receptors appear to be constitutive in both normal and pathophysiologic conditions, a dynamic up-regulation of BK_1 receptors occurs whenever there is a chronic or more prolonged inflammatory condition present [4]. In an effort to address both of the major bradykinin receptor populations simultaneously, we developed CP-0364 (Table 1), a bis-succinimidohexane cross-linked peptide-based heterodimer with potent combined BK_1 and BK_2 antagonist activities. This compound is currently being evaluated in a number of animal models of acute and chronic inflammatory conditions in which BK_1 and BK_2 kinin activities have been implicated.

Table 1   *In vitro activities of dual action heterodimeric compounds containing bradykinin (BK_2) antagonist (ant) activity*

(D)Arg-Arg-Pro-Hyp-Gly-Phe-Cys-(D)Phe-Leu-Arg
                 | (=Monomeric BK_2 antagonist, $pA_2 = 7.2 \pm 0.2$)
           Linker
            |

Additional Therapeutic Monomer

Heterodimer	Linker[a]	Additional Therapeutic Monomer[b]	In Vitro Activities[c]
CP-0127 (Bradycor)[TMd]	BSH	(D)R-R-P-J-G-F-C-(D)F-L-R (BK_2 ant; pA_2=7.2±0.2)[e]	$pA_2(BK_2)$=8.5±0.3
CP-0364	BSH	(D)R-C-P-J-G-F-S-P-L (BK_1 ant; $-logIC_{50}$=6.9±0.1)[e]	$pA_2(BK_2)$=8.3±0.2 $-logIC_{50}(BK_1)$=7.5±0.1
CP-0394	ESC	(D)R-(D)P-K-P-K-N-(D)F-F-(D)W-L-Nle-$CONH_2$ (NK_1 ant; pA_2=5.6 ± 0.2)[e]	$pA_2(BK_2)$=7.9±0.1 $pA_2(NK_1)$=5.8±0.3
CP-0416	BSH	D-Y-(D)W-V-(D)W-C-R-$CONH_2$ (NK_2 ant; pA_2=6.9±0.2)[e]	$pA_2(BK_2)$=8.7±0.2 $pA_2(NK_2)$=5.6±0.2
CP-0494	BSH	Mercaptoethyloxymorphamine (opioid agonist; $-logIC_{50}$=7.7)[e]	$pA_2(BK_2)$=8.4±0.2 $-logIC_{50}$(opioid)=7.6
CP-0502	NHS	4-(3'-carboxy-propylsulfonyl)phenyl-4-tert-butylphenylisobutyrate (NEI; Ki=10.5 nM)[e]	$pA_2(BK_2)$=7.5 Ki(NEI)=6.6nM
CP-0460	NHS	Indomethacin (free carboxyl) (COI)	$pA_2(BK_2)$=7.8

[a]BSH=bis succinimidohexane; ESC=epsilon succinimido n-caproyl; NHS=N-hexyloxysuccinimido; BK_2 antagonist monomer always attached to the linker via a thioether bond between the Cys sulfhydryl group and the succinimido moiety.
[b]Attachment to indicated linker made via underlined residue/moiety.
[c]*In vitro* activities are those of intact homo- or heterodimers evaluated in the following tissue/enzyme preparations: BK_2, rat uterus; BK_1, rabbit aorta; NK_1, guinea pig ileum; NK_2, rat pulmonary artery; opioid, guinea pig ileum; elastase, human neutrophil elastase.
[d]BK_2 homodimer; included for comparison purposes.
[e]Indicated monomer activity is that of the unmodified parent monomer.

As important as bradykinin is, it is not considered to be solely responsible for the ultimate clinical picture observed in inflammatory conditions. Rather, bradykinin interacts with a variety of additional mediators, including the neuropeptides substance P and neurokinin A, and actually promotes the release of these neuropeptides from C-fiber afferent nerve terminals. Release of these and other neuropeptides is also under the local negative control of endorphins acting at opioid receptors located on the peripheral terminals themselves. Given this dynamic relationship of positive and negative control elements, we proposed to control both the kinin-related as well as the neurogenic aspects of inflammation by developing heterodimers that could either block the effects of both bradykinin and the neuropeptides (substance P and neurokinin A) or block the effects of bradykinin and stimulate peripheral opioid receptors. Heterodimers found to be capable of exerting this type of "combined influence" included CP-0394, a combined BK_2/NK_1 antagonist, CP-0416, a combined BK_2/NK_2 antagonist, and CP-0494, a combined BK_2 antagonist/opioid agonist (Table 1).

In addition to bradykinin and the neuropeptides, the activities of the enzymes human neutrophil elastase (NE) and cyclooxygenase (CO) are critically important to the overall inflammatory process. Inhibitors of these enzymes (NEI, COI) were modified for incorporation into heterodimers in order that inhibition of elastase-mediated tissue degradation or cyclooxygenase-mediated prostaglandin biosynthesis, respectively, might be accomplished in addition to bradykinin antagonism. The resulting heterodimers, CP-0502 (combined BK_2 antagonist/NEI) and CP-0460 (combined BK_2 antagonist/COI), were also found to be capable of interacting with their respective biological targets *in vitro* in an unencumbered manner (Table 1).

"Dual action" heterodimeric compounds possessing bradykinin antagonist activity are capable of interacting with two different receptor populations or with a receptor and an enzyme and offer the opportunity of intervening at multiple points in the inflammatory process with a single compound.

References

1. Farmer, S.G. and Burch, R.M., In Burch, R.M. (Ed.), Bradykinin Antagonists: Basic and Clinical Research, Marcel Dekker, Inc., New York, NY, U.S.A., 1991, p.1.
2. Burch, R.M., Farmer, S.G. and Steranka, L.R., Med. Res. Rev., 33 (1990) 237.
3. Cheronis, J.C., Whalley, E.T., Nguyen, K.T., Eubanks, S.R., Allen, L.G., Duggan, M.J., Loy, S.D., Bonham, K.A. and Blodgett, J.K., J. Med. Chem., 35 (1992) 1563.
4. Dray, A. and Perkins, M., TINS, 16 (1993) 101.

Total synthesis of aureobasidin A, an antifungal cyclic depsipeptide

T. Kurome[2], K. Inami[1], T. Inoue[2], K. Ikai[2], K. Takesako[2], I. Kato[2] and T. Shiba[1]

[1]Peptide Institute, Protein Research Foundation, Minoh, Osaka 562, and [2]Biotechnology Research Laboratories, Takara Shuzo Co., Ltd., Otsu, Shiga 520-21, Japan

Introduction

A new cyclic depsipeptide aureobasidin A (**1**), isolated as a major component from the culture medium of the black yeast *Aureobasidium pullulans* R 106, exhibits a strong antifungal activity against pathogenic fungi with a low toxicity [1]. A whole structure of the peptide was determined mainly by the spectroscopic and chemical techniques [2]. The structures of more than twenty congeners of aureobasidin were determined in comparison with aureobasidin A [3]. We attempted a total synthesis of aureobasidin A aiming an establishment of a synthetic technique of the cyclic depsipeptide containing *N*-methyl amino acids for purpose of investigation of a structure-activity relationship of the aureobasidin family antibiotic.

Aureobasidin A (**1**)

Results and Discussion

As a site of the final cyclization, the linkage between *a*Ile[1] and Pro[9] was chosen to avoid the coupling at an *N*-methyl amino acid as an amine component, which otherwise may reduce a yield of the cyclization reaction due to its steric characters. A linear nonapeptide(1-9) was synthesized according to the Boc strategy mainly using PyBroP [4] as a coupling reagent. The fragment condensation between Leu[3]-HOMeVal[4]-Hmp[5] (Segment B) and MeVal[6]-Phe[7]-MePhe[8]-Pro[9] (Segment C) was first attempted, followed by coupling between the coupling product (3-9) and *a*Ile[1]-

MeVal^2 (Segment A) as shown below.

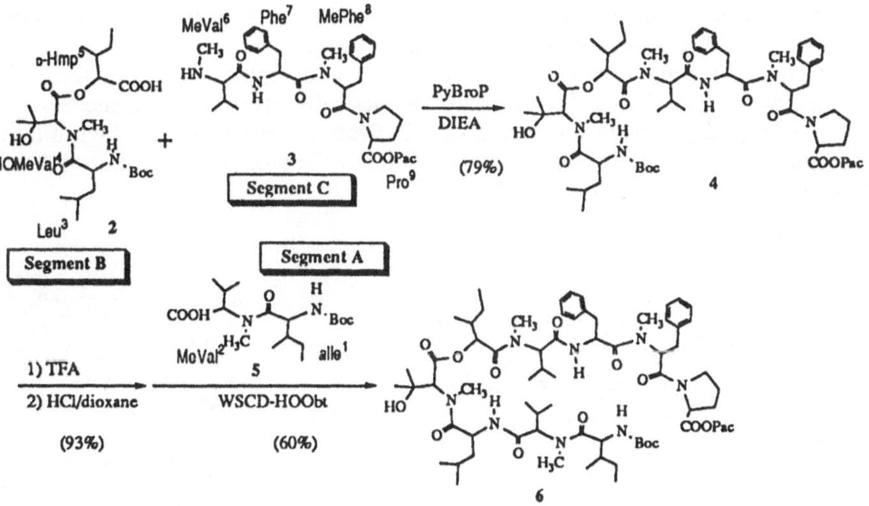

DL-β-Hydroxymethylvaline (HOMeVal) in Segment B was prepared according to the literature [5], and its amino group was then protected with Boc group after activation and solubilization by trimethylsilylation with BSTFA. The protection of its carboxyl function with the benzyl group and the removal of the Boc group were carried out, followed by coupling with Boc-L-Leu-OH using PyBroP without protection of β-OH group in HOMeVal afforded the desired dipeptide with no *O*-acyl derivative. After deprotection of Bzl group, a mixture of the diastereomers of the product was coupled with H-D-Hmp-OPac with DCC and 4-pyrrolidinopyridine as an acylation catalyst. In this condensation reaction, DCC is superior to PyBroP in the yield, and both coupling reagents without the base catalyst did not afford the desired product due to the formation of the oxazolinium compound of Boc-Leu-HOMeVal as a main intermediate [6]. This oxazolinium-forming phenomenon was recently reported in case of *N*-MeVal and *N*-methylamino acids by B. Castro et al. independently [7]. Then the diastereomers Boc-L-Leu-DL-HOMeVal-D-Hmp-OPac were separated by silica gel column chromatography, and the L-L-D compound was deprotected with zinc/acetic acid to give **2** (Segment B). These diastereomeric peptides (L-L-D or L-D-D) after separation were hydrolyzed with 6N HCl for 19 hours to give respectively L-HOMeVal and D-HOMeVal, which were assigned in comparison with the authentic amino acids on Daicel Chiralpak WH.

Segment C was prepared by a stepwise elongation method starting from Boc-L-Pro-OPac as a carboxyl terminus by the successive condensation using PyBroP and DIEA in dichloromethane. The coupling reaction of Boc-L-MePhe-OH with HCl•H-L-Pro-OPac by the WSCD-HOBt method gave the product in a yield of 87%, accompanying L-D isomer (A ratio of L-L and L-D isomers was 10:3). This racemization was due to the intramolecular formation of the charged Schiff base in the Pro residue between the imino group and the carbonyl function of the Pac ester. The reaction with PyBroP afforded the desired product without racemization.

The coupling of **2**, whose *C*-terminus is a hydroxy acid, Hmp, with **3** was carried out using PyBroP. After removal of the Boc group in the peptide **4** with

TFA, a fragment condensation with **5** (Segment A) by the WSCD-HOOBt method was carried out to give the protected linear nonapeptide **6** with a slight racemization as shown in the scheme, although the same condensation by the PyBroP method gave a 1:1 mixture of the completely racemized diastereomers.

The protecting groups of both carboxyl and amino groups in the linear nonapeptide **6**, which was purified diastereomerically, were removed successively. The free peptide **7** was then cyclized with PyBroP in CH_2Cl_2 under a high-dilution condition (10^{-3} M) to afford predominantly the cyclic monomeric peptide (PD-MS: Found, M^+ 1100; Calcd 1100).

In contrast, the cyclization of *N*-hydroxysuccinimide ester of the nonapeptide at the same concentration in CH_2Cl_2 as in that of PyBroP reaction gave a cyclic dimer as the major product. In case of PyBroP, the reaction seemed to proceed promptly compared that in the active ester. Therefore, a condensation in a high-dilution condition was actually achieved preventing the dimerization in that case. On the other hand, the cyclization of the active ester in DMF proceeded very slowly even though the cyclic monomer was obtained as a major product besides the dimer. Such difference depending on the solvent seemed to arise from the solvation feature of the peptide molecule in DMF which may be advantageous to the monomeric cyclization.

The monomeric cyclic peptide thus obtained is completely identical with the natural antibiotic in all respects (TLC, HPLC, ^{1}H-NMR, and antifungal activity).

References

1. Takesako, K., Ikai, K., Haruna, F., Endo, M., Shimanaka, K., Sono, E., Nakamura, T., Kato, I. and Yamaguchi, H., J. Antibiot., 44 (1991) 919.
2. Ikai, K., Takesako, K., Shiomi, K., Moriguchi, M., Umeda, Y., Yamamoto, J., Kato, I. and Naganawa, H., J. Antibiot., 44 (1991) 925.
3. (a) Ikai, K., Shiomi, K., Takesako, K., Mizutani, S., Yamamoto, J., Ogawa, Y. and Kato, I., J. Antibiot., 44 (1991) 1187; (b) Ikai, K., Shiomi, K., Takesako, K. and Kato, I., J. Antibiot., 44 (1991) 1199.
4. Coste, J., Frérot, E., Jouin P. and Castro, B., Tetrahedron Lett., 32 (1991) 1967.
5. Izumiya, N. and Nagamatsu, A., J. Chem. Soc. Jpn., 72 (1951) 336.
6. Benoiton, N.L. and Chen, F.M.F., Can. J. Chem., 59 (1981) 384.
7. Frérot, E., Jouin, P., Coste, J. and Castro, B., Tetrahedron Lett., 33 (1992) 2815.

Chemical synthesis, purification and characterisation of the murine pro-inflammatory protein CP-10

P.F. Alewood[1], D. Alewood[1], A. Jones[1], M. Lackmann[2], C. Cornish[2] and C. Geczy[2]

[1]*Centre for Drug Design and Development, The University of Queensland, Queensland 4072, Australia*
[2]*Heart Research Institute, 145 Missenden Rd, Camperdown, NSW 2050, Australia*

Introduction

A central event in cell-mediated immune reactions is cytokine induced accumulation of immunocompetent cells at sites of antigen challenge. A variety of chemoattractants may determine the nature of the inflammatory filtrate during lesion formation. We recently described the purification and structural analysis of a protein termed chemotactic protein 10 (CP-10) according to its relative molecular mass of approximately 10kD, derived from supernatants of activated murine spleen cells [1]. The homogeneous cytokine had maximal chemotactic activity for murine neutrophils at 10^{-12} to 10^{-13}M, making it one of the most potent chemotactic factors known to date. Comparison of the 88 amino acid sequence with known proteins revealed no structural homologies to previously described cytokines or growth factors, but indicated extensive sequence similarities to S-100 Ca^{2+} binding proteins [2]. A peptide synthesised according to a unique domain within CP-10 (42-55) is chemotactic within the same concentration range and elicits a transient inflammatory response, thus indicating the approximate position of the active region of the protein [3]. Here we report the synthesis of full length CP-10 (Figure 1).

Results and Discussion

CP-10 was chemically synthesised on an Applied Biosystems 430A peptide synthesiser using highly optimised Boc chemistry protocols [4]. Each residue was double coupled and coupling efficiencies monitored by the quantitative ninhydrin reaction [5]. The average yield of chain assembly was 99.68% per residue. Chain assembly took 3.5 days.

Partial deprotection of the protein was carried out on the resin prior to HF cleavage. DNP groups were removed by thiolysis (20% mercaptoethanol : 10% DIEA : DMF). The N-terminal Boc group was removed with 100% TFA. The resin was washed with DMF followed by DCM and air dried. The polypeptide chain was deprotected and cleaved from the resin with standard high HF treatment using p-cresol and p-thiocresol as scavengers. Following HF removal and trituration with diethyl

**PSELEKALSNLIDVYHNYSNIQGNHHALYKNDFKKMVTTECPQ
FVQNINIENLFRELDINSDNAINFEEFLAMVIKVGVASHKDSHKE**

Fig. 1. The sequence for chemotactic protein 10 (CP-10).

ether, the crude protein was extracted with 50% acetic acid and following aqueous dilution, lyophilized.

The crude protein was dissolved in 0.15% TFA and loaded directly onto a Vydac C4 RP-preparative chromatography column (2.2cm x 25cm) then eluted using a linear gradient of 0-80% B over 80 min at a flow rate of 8ml/min (A = 0.1% TFA in water, B = 90% acetonitrile in water + 0.09% TFA). Fractions containing the desired product in pure form were identified using analytical RP-HPLC and Ion Spray MS (Figure 2). These fractions were combined and lyophilized (21% yield).

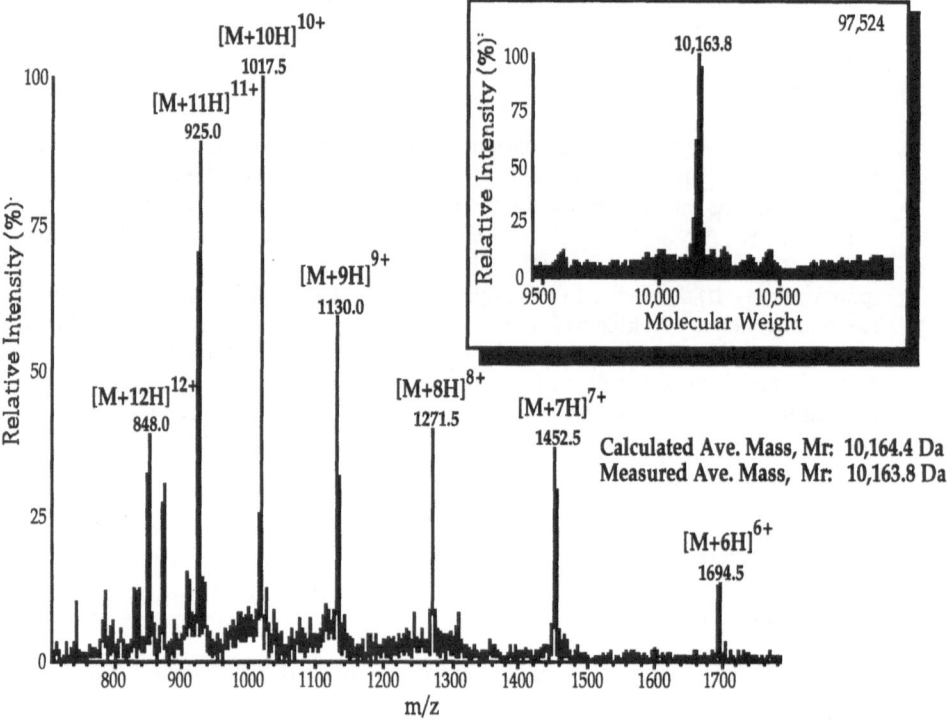

Fig. 2. Ion spray mass spectrum of synthetic CP-10.

Mass spectometric measurements were performed on a PE-Sciex Ion Spray mass spectrometer at a mass accuracy of ±0.2 amu. The peptide (1mg/ml) was dissolved in 45% acetonitrile in water plus 0.1% TFA and introduced to the machine via a glass capillary using an infusion of the peptide solution at a flow rate of 5µl/min.

Ion Spray mass spectrometry also allowed some preliminary experiments into the nature of the interaction of CP-10 with metal ions. As a member of the S-100 family of proteins two calcium ion binding sites were likely; a putative N-terminal and 'typical' C-terminal Ca^{2+}-binding domain at positions 20-32 and 58-68 respectively. Indeed, addition of CP-10 to 10mM $CaCl_2$ followed by immediate MS analysis indicated that two calcium ion adducts were observable. Similar experiments with $ZnCl_2$ indicated the presence of at least one zinc-binding site.

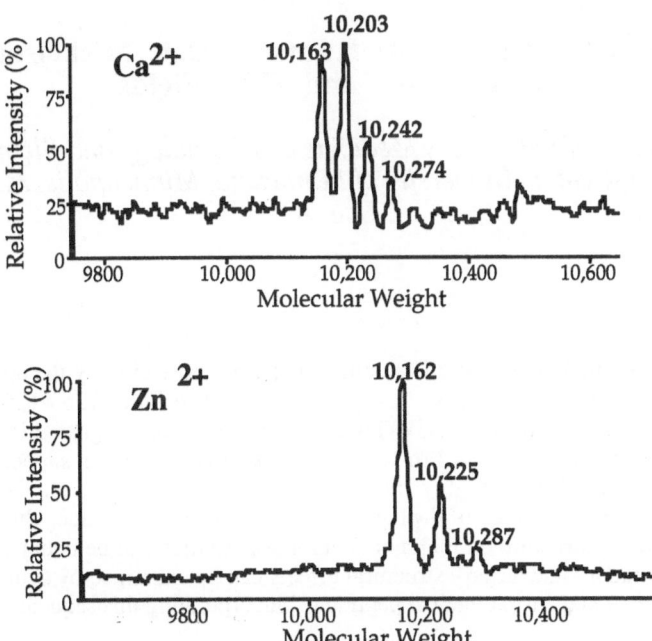

Fig. 3. *Ion spray mass spectra of CP-10 and metal ions.*

References

1.   Lackmann, M., Cornish, C.J., Simpson, R.J., Moritz, R.L. and Geczy, C.L., J. Biol. Chem., 267 (1992) 7499.
2.   Kligman, D. and Hilt, D.C., TIBS, 13 (1988) 437.
3.   Lackmann, M., Rajasekariah, P., Iismaa, S.E., Cornish, C.J., Simpson, R.J., Moritz, R.L. and Geczy, C., J. Immunol., 150 (1993) 2981.
4.   Kent, S.B.H., Alewood, D., Alewood, P.F., Baca, M., Jones, A. and Schnollzer, M., In Epton, R. (Ed.), Innovation and Perspectives in Solid Phase Synthesis, Intercept limited, Andover, 1992, p.1.
5.   Sarin, V.K., Kent, S.B.H., Tam, J.P. and Merrifield, R.B., Anal. Biochem., 117 (1981) 147.

The identification of essential structural components for melanoma cell adhesion, spreading and motility on basement membrane collagen

J.B. McCarthy, J.R. Knutson, D.J. Mickelson, C.G. Fields and G.B. Fields

Department of Laboratory Medicine and Pathology and Biomedical Engineering Center, University of Minnesota, Minneapolis, MN 55455, U.S.A.

Introduction

Basement membrane (type IV) collagen can promote directly the adhesion and migration of various tumor cells. A peptide model of residues 1263-1277 from the gene-derived sequence of the α 1(IV) human collagen chain, designated IV-H1 (Gly-Val-Lys-Gly-Asp-Lys-Gly-Asn-Pro-Gly-Trp-Pro-Gly-Ala-Pro), can support adhesion, spreading and motility of highly metastatic melanoma cells [1, 2]. By applying chemical mutagenesis via solid-phase synthesis to this sequence, the effects of collagen primary structure on melanoma cellular activities can be better defined. In addition, secondary and tertiary structural effects can be examined by synthesizing an all-D enantiomer and a triple-helical peptide [3] incorporating this sequence.

Results and Discussion

Analogs of IV-H1 containing single and multiple substitutions of Lys_{1265} and Lys_{1268} by amino acids of similar, opposite, or no charge were synthesized in order to evaluate the importance of basic residues on cell adhesion. Replacement of both Lys residues by amino acids of either no charge [norleucine (Nle)] or opposite charge (Glu) resulted in less than half of the melanoma cell adhesion activity promoted by native IV-H1. Decreased melanoma cell adhesion due to scrambling of basic residues in fibronectin-model peptides has been correlated to cell surface chondroitin sulfate proteoglycan-mediated activity [4]. However, some integrin-mediated cellular activities have also been correlated to substrate basic domains [5]. Since the $\alpha 1\beta 1$ and $\alpha 2\beta 1$ integrins have been implicated in melanoma cell adhesion to fragments of type IV collagen [6], the loss of activity may be due to a decreased integrin-mediated adhesion. Individual substitution of each Lys residue by Nle showed equal loss of activity, while substitution by Arg showed loss of activity from Lys_{1268} only. Activity at Lys_{1268} is thus dependent on both charge and steric effects. Circular dichroism (CD) spectroscopy of IV-H1 and the single-substitution peptides indicated that only the Lys_{1268} to Nle substitution gave a secondary structural change, reflected by a decrease in the molar ellipticity from 220-250 nm (Figure 1). It is not clear if this change reflects the peptide backbone or indole dichroism.

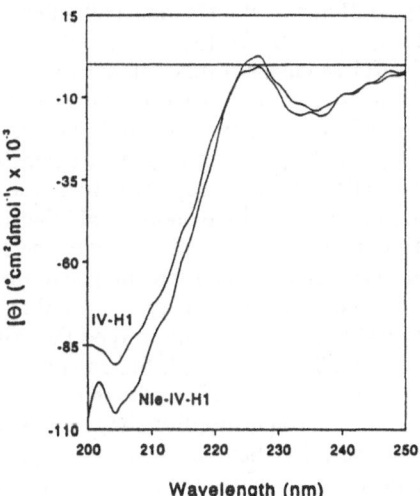

Fig. 1. CD spectra of IV-H1 and IV-H1 with Nle substituted for Lys$_{1268}$ (Nle-IV-H1). Peptide concentrations were 0.1 mM in phosphate-buffered saline.

An analog of IV-H1 containing a Phe substitution of Trp$_{1273}$ was synthesized to evaluate the functional significance of Trp, which is a relatively rare residue in collagen. This analog had less than 10% of the cell adhesion activity of native IV-H1, demonstrating definitively that melanoma cell adhesion to IV-H1 is not mediated solely by charged residues.

A D-enantiomer of IV-H1 was synthesized to identify chiral and non-chiral cellular interactions. Approximately 20% of input melanoma cells adhered to substrata coated with as little as 0.63 μM of L-IV-H1, with higher concentrations of peptide supporting the adhesion of essentially all of the input cells. In contrast, while melanoma cells also adhered to D-IV-H1, the actual amount of adherent cells (maximum 30-35% of input) was far less than for L-IV-H1. Similarly, while L-IV-H1 can completely inhibit melanoma cell adhesion to substrata coated with L-IV-H1 [1], D-IV-H1 can only partially inhibit cell adhesion to L-IV-H1. Thus, melanoma cell recognition of IV-H1 is partially dependent on chirality of the substrate. Interestingly, although the D-IV-H1 could partially support melanoma cell adhesion, it was completely ineffective at supporting melanoma cell cytoskeletal reorganization or haptotactic cell motility. Similarly, while soluble L-IV-H1 inhibited melanoma cell invasion of reconstituted basement membrane *in vitro*, D-IV-H1 had no effect on invasion in a 18 h invasion assay. Cell spreading, motility, and invasion, unlike cell adhesion, are completely dependent upon substrate chirality. These results suggest that the molecular basis for melanoma cell recognition of IV-H1 includes chiral-dependent and chiral-independent cell surface receptors. As different cell surface biomolecules (chondroitin sulfate proteoglycans and integrins) are now implicated for melanoma cellular activities [see also 7], the recently introduced "receptor cluster" model for cell adhesion to fibronectin [4] should be considered for type IV collagen.

A branched triple-helical peptide (THP) of 124 residues incorporating the IV-H1 sequence was synthesized to examine substrate secondary and tertiary structural effects on cellular behavior [3]. Half maximal melanoma cell adhesion occurred at [THP] = 1.12 μM and [linear IV-H1] = 170 μM [3]. Thus, triple-helical conformation in combination with the IV-H1 sequence resulted in a 100-fold increase in melanoma cell adhesion activity compared with the IV-H1 sequence alone. This result is the first direct demonstration of the significance of triple-helicity for cell adhesion to a specific collagen sequence. Cell spreading was more extensive on the THP, as cell areas averaged 0.048 and 0.012 mm^2/μM peptide in response to THP and linear IV-H1, respectively [3]. As in the case of melanoma cell adhesion, cell spreading was most efficient when triple-helicity was combined with the IV-H1 sequence. The combined melanoma cell THP adhesion and spreading activities supports the concept that tumor cell adhesion and spreading on type IV collagen involves multiple, distinct domains, as at least two domains within type IV collagen *in triple-helical conformation* are tumor cell adhesion sites [3, 6]. In addition, the enhancement of cellular activities due to triple-helicity confirms the α1(IV) 1263-1277 sequence as a specific melanoma cell adhesion and motility site, as this sequence in its native conformation has greater activity than the isolated sequence, and implies that basement membrane type IV collagen is a site for tumor cell invasion based on collagen primary, secondary, and tertiary structures.

Acknowledgements

This work is supported by the NIH (KD 44494, CA 43924, and CA 54263), McKnight-Land Grant Professorship, and Leukemia Task Force.

References

1. Chelberg, M.K., McCarthy, J.B., Skubitz, A.P.N., Furcht, L.T. and Tsilibary, E.C., J. Cell Biol., 111 (1990) 261.
2. Mayo, K.H., Parra-Diaz, D., McCarthy, J.B. and Chelberg, M., Biochemistry, 30 (1991) 8251.
3. Fields, C.G., Mickelson, D.J., Drake, S.L., McCarthy, J.B. and Fields, G.B., J. Biol. Chem., 268 (1993) 14153.
4. Iida, J., Skubitz, A.P.N., Furcht, L.T., Wayner, E.A. and McCarthy, J.B., J. Cell Biol., 118 (1992) 431.
5. Vogel, B.E., Lee, S.-J., Hildebrand, A., Craig, W., Pierschbacher, M.D., Wong-Staal, F. and Ruoslahti, E., J. Cell Biol., 121 (1993) 461.
6. Vandenberg, P., Kern, A., Ries, A., Luckenbill-Edds, L., Mann, K. and Kühn, K., J. Cell Biol., 113 (1991) 1475.
7. Mickelson, D.J., Faasen, A.E. and McCarthy, J.B., J. Cell Biol., 115 (1991) 287a.

Synthesis of prothymosin α deduced from nucleotide sequence of the murine cDNA and its effect on the impaired T-lymphocytes of uremic patients

T. Abiko and H. Sekino

Kidney Research Laboratory, Kojinkai, 1-6 Tsutsujigaoka 2-chome, Miyagino-ku, Sendai 908, Japan

Introduction

Thymosin α 1, a peptide containing 28 amino acid residues, was isolated by Goldstein et al. [1] from thymosin fraction 5, a mixture of peptides from calf thymus. Thymosin α 1 was found to be active in some of the in vitro tests used for thymosin fraction 5 [2], and it was considered to be one of the factors that modulated steps in the maturation of T cells [1, 2]. In our preceding papers [3-6], we have demonstrated that not only our synthetic thymosin α 1 but also our synthetic C-terminal fragments (positions 14-28 and 19-28) exhibit restoring activity of low E-rosette-forming lymphocytes from patients suffering from impaired immunological function due to minimal change nephrotic syndrome or chronic renal failure. Recently, Schmidt and Werner [7] detected a full length cDNA for the murine prothymosin α mRNA which was sequenced at the DNA level. The amino acid sequence deduced from the nucleotide sequence shows a high degree of positional identities with prothymosin α from man and rat. The sequence of 28 amino acids at the N-terminal except for the N-terminal methionine is identical to that of calf thymosin α 1. We reported earlier [8] that our synthetic rat prothymosin fragment 29-111 could increase the peripheral E-rosette-forming lymphocytes when incubated in vitro with blood from uremic patients. In addition to this result, the restoring effect of the synthetic prothymosin α fragment 29-111 was greater than that of our

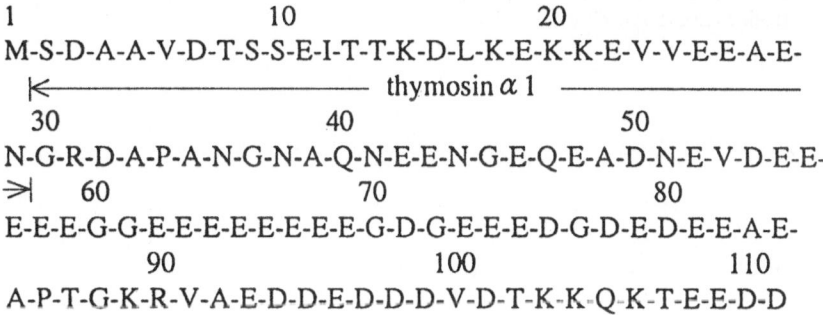

Fig. 1. Amino acid sequence of prothymosin α deduced from murine cDNA.

365

synthetic thymosin α 1. These results prompted us to synthesize the deduced complete amino acid sequence of murine prothymosin α from the nucleotide sequence of cDNA encoding precursor. We describe here the solid-phase synthesis of cDNA clone encoding murine prothymosin α (110 residue-peptide) in order to examine whether our synthetic peptide has an enhancing effect on the marked reduction of E-rosette-forming lymphocytes of uremic patients.

Results and Discussion

Peptide synthesis was performed manually by the stepwise solid-phase method using the base-labile Fmoc group for protecting the α-amino groups. At the end of the synthesis and after the last deprotection step, the deprotected peptide was purified by gelfiltration, ion-exchange chromatography and HPLC, and we finally obtained highly pure peptide by analytical HPLC. Incubation of peripheral venous blood from uremic patients in the presence of various amounts of the synthetic peptide from 0.01 μg/ml to 10 μg/ml resulted in recovery of E-rosette formation. The restoring effect of the synthetic murine prothymosin α was far greater than that of the synthetic thymosin α 1 [5]. Harritos et al. [9] reported that in the mouse protection test and in assays for lymphokine release, rat thymus prothymosin α was found to be 10-20 times as potent as thymosin α 1. As we also mentioned in our previous paper on rat prothymosin α fragment 29-111 [8], these results seem to suggest that the C-terminal region of murine prothymosin α (positions 30-111) also contains important molecular signal functions for immunological activity.

References

1. Goldstein, A.L., Low, T.L.K., McAdoo, M., McClure, J., Thurman, G.B., Rossio, J., Lai, C.Y., Chang, D., Wang., S.S., Harvey, C., Ramel, A.H. and Meinhofer, J., Proc. Natl. Acad. Sci. U.S.A., 74 (1977) 725.
2. Low, T.L.K. and Goldstein, A.L., J. Biol. Chem., 254 (1979) 891.
3. Abiko, T., Onodera, I. and Sekino, H., Chem. Pharm. Bull., 27 (1979) 3171.
4. Abiko, T., Sekino H., and Higuchi, H., Chem. Pharm. Bull., 28 (1980) 3511.
5. Abiko, T., Onodera, I. and Sekino, H., Chem. Pharm. Bull., 28 (1980) 3542.
6. Abiko, T. and Sekino, H., Chem. Pharm. Bull., 30 (1982) 1776.
7. Schmidt, G. and Werner, D., Biochim. Biophys. Acta, 1088 (1991) 442.
8. Abiko, T. and Sekino, H., Biotech. Appl. Biochem., 13 (1991) 406.
9. Harritos, A.A., Blacher, R., Stein, S., Caldarella, J. and Horecker, B.L., Proc. Natl. Acad. Sci. U.S.A., 82 (1985) 343.

Synthesis of a novel hematopoietic peptide, SK&F 107647

D.P. Alberts[1], E. Agner[3], J.S. Silvestri[1], C. Kwon[1],
K. Newlander[1], A. King[2], L.M. Pelus[2], P. DeMarsh[2],
C. Frey[2], S. Petteway[2], W.F. Huffman[1]
and P.K. Bhatnagar[1]

[1]Department of Medicinal Chemistry and
[2]Department of Molecular Virology and Host Defense, SmithKline
Beecham Pharmaceuticals, King of Prussia, PA, U.S.A., and
[3]Hafslund Nycomed-Bioreg, Gaustadalleen 21, Oslo, Norway

Introduction

Stem cells, the blood-forming cells in bone marrow, are capable of self-renewal and differentiation to produce lymphoid and myeloid blood cell lineages. This process of differentiation and maturation, known as hematopoiesis, is affected by a number of hematopoietic growth factors (cytokines) including M-CSF, G-CSF, Kit ligand, GM-CSF, Il-1, and Il-3 [1, 2]. In 1982, a synthetic hematoregulatory peptide, pGlu-Glu-Asp-Cys-Lys, was reported to have selective inhibitory effect on myelopoeitic cells [3]. In 1987, Laerum et al. [4] reported that when this linear peptide was oxidized, the resulting dimer (HP-5), had the opposite effect of the monomer. The dimer stimulates colony formation of both human and murine CFU-GM *in vitro* and up-regulates murine myelopoeitic cells *in vivo*. We report the synthesis of a potent hematopoietic peptide (Figure 1).

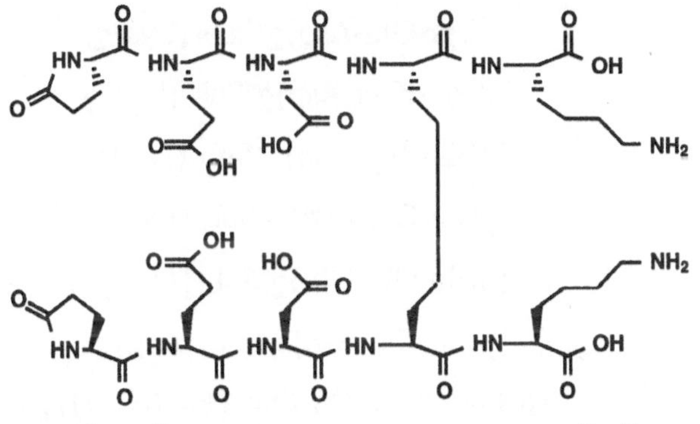

Fig. 1. Structure of N^2-[[N^8-((S)-5-amino-1-carboxypentyl)-N^2,N^7-bis[[N-(N-L-pyro-glutamyl)-L-α-glutamyl]-L-α-aspartyl]-8-oxo]-L-(S-(L-threo))-2,7,8-triamino-octanoyl)-L-lysine (1).

Results and Discussion

The ability of the HP-5 monomer to form a dimer and the possibility that the dimer can be reduced back to the monomer *in vivo*, made the study of these peptides difficult. Figure 2 showing the effects of SK&F 107647 and HP-5 on immuno-suppressed mice infected with *Candida albicans* is shown on the next page [5]. SK&F 107647 is a symmetrical dimer in which the unnatural amino acid, (S,S)-2,7-diaminosuberic acid, acts as an isosteric replacement for cystine. Di-Boc suberic acid (Boc$_2$-Sub) was synthesized via a Kolbe reaction [6]. SK&F 107647 was prepared using solid phase peptide synthesis. One equivalent of Boc$_2$-Sub was coupled to one equivalent of lysine (2Cl-Z)-resin. When the Kaiser test gave a negative result, usually after 4 hours of coupling, 1.5 equivalents of lysine (Cbz)-OBzl was added. The peptide yield from this procedure was low and this could be explained by the formation of the dipeptide Boc-Sub-[Lys(Cbz)-OBzl]$_2$ which was subsequently lost in the CH$_2$Cl$_2$ wash. The synthesis was then modified by using one equivalent of Boc$_2$-Sub and 2 equivalents of lysine (2Cl-Z)-resin. The coupling was slow (up to 4 days) but the yield was improved. In addition, less Boc$_2$-Sub was needed. The remaining amino acids were coupled as their HOBt active esters. The resin peptide was cleaved from the resin using anhydrous HF. The crude peptide was purified by RP HPLC using a Vydac C18 column. The analogues listed in Table 1 were made using the synthetic scheme described above.

Table 1 *The activity of SK&F 107647 and its analogues*

Compound	Peptide Sequence	EC$_{50}$ (pg/ml)
SK&F107647 (1)	{pGlu-Glu-Asp}$_2$-Sub-{Lys}$_2$	5
HP-5 dimer	pGlu-Glu-Asp-Cys-Lys	100
3	{Pro-Glu-Asp}$_2$-Sub-{Lys}$_2$	20,000
4	{Tyr-Glu-Asp}$_2$-Sub-{Lys}$_2$	NA[a]
5	{pGlu-dGlu-Asp}$_2$-Sub-{Lys}$_2$	1,000,000
6	{pGlu-Asp-Asp}$_2$-Sub-{Lys}$_2$	5
7	{pGlu-Glu-Asn}$_2$-Sub-{Lys}$_2$	NA[a]
8	{pGlu-Glu-Glu}$_2$-Sub-{Lys}$_2$	5
9	{pGlu-Glu-Asp}$_2$-Sub-{Arg}$_2$	NA[a]
10	{pGlu-Glu-Asp}$_2$-Sub-{Phe(NH$_2$)}$_2$	NA[a]

SK&F 107647 stimulates marrow fibroblasts to produce colony stimulating activity (CSA)[7]. This activity was measured using murine bone marrow CFU-GM as target cells (Table 1).

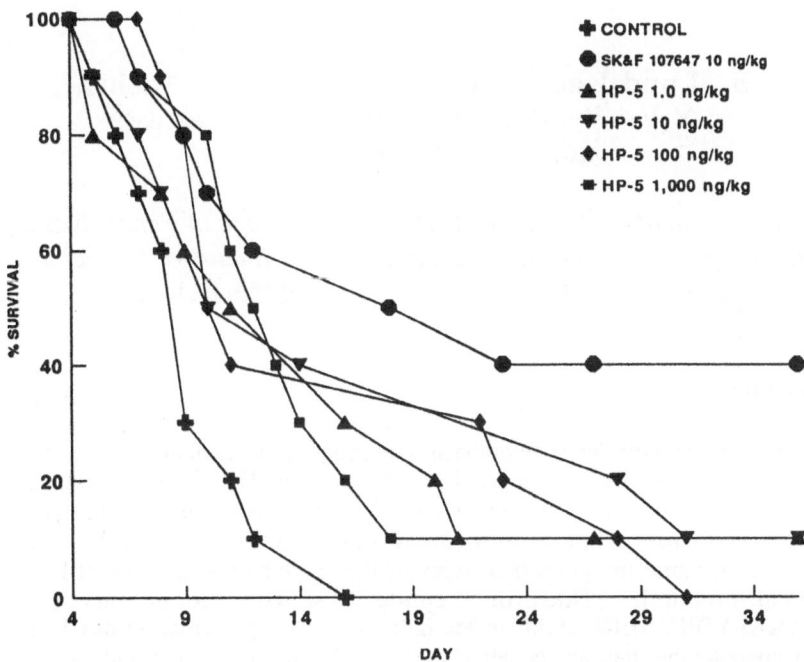

Fig. 2. The effect of SK&F 107647/HP-5 on immunosuppressed mice infected with Candida albicans.

Replacement of the disulfide bridge in HP-5 dimer with the dimethylene carbon bridge in SK&F 107647 resulted in a more potent hematopoietic peptide. Screening more analogues of this peptide should provide information that may help us understand the structure-activity relationships of SK&F 107647.

References

1. Moore, M.A.S., Blood, 78 (1991) 1.
2. Moore, M.A.S., Cancer, (1991) 2718.
3. Paukovits, W.R. and Naturforsch, Z., 37C (1982) 1297.
4. Laerum, O.D. and Paukovits,W.R., Differentiation, 27 (1984) 106.
5. DeMarsh, P., Sucoloski, S., Frey, C., Bhatnagar, P., Koltin, Y., Actor, P. and Petteway, S., Journal of Antimicrobial Agents and Chemotheraphy, 1993, submitted for publication.
6. Nutt, R., Strachan, R., Veber, D. and Holly, F., J. Org. Chem., 45 (1980) 3078.
7. King, A.G., Talmadge, J.E., Badger, A.M. and Pelus, L.M., Exp. Hematol., 20 (1992) 223.

Interfacial basic amino acids increase the lipid affinity of class A amphipathic helix: Evidence for snorkel hypothesis by ^{13}C-NMR spectroscopy

S. Lund-Katz[1], M.C. Phillips[1], V.K. Mishra[2],
Y.V. Venkatachalapathi[2], J.P. Segrest[2]
and G.M. Anantharamaiah[2]

[1]Department of Biochemistry, Medical College of Pennsylvania,
Philadelphia, PA 12129, U.S.A. and [2]Departments of Medicine and
Biochemistry, UAB Medical Center, Birmingham, AL 35294, U.S.A.

Introduction

The amphipathic helix hypothesis was proposed to explain the mechanism of lipid association of the exchangeable apolipoproteins [1]. We have addressed the question of whether or not the position of charged residues on the polar face of the α-helix plays a role in lipid affinity [2]. A peptide with 18 amino acid residues was designed to mimic the general features of the amphipathic helix described in the original proposal. Thus, the peptide 18A with the primary sequence DWLKAFYDKVAEKLKEAF in the helical wheel representation has Lys at the polar-nonpolar interface and negatively charged Asp and Glu at the center of the polar face [2]. To address the question of the contribution of interfacial Lys residues to the lipid affinity of 18A, a peptide with the same amino acids but with positions of charged residues reversed was synthesized with the following sequence: KWLDAFYKDVAKELEKAF (18R). A helical wheel representation of this sequence shows negatively charged amino acid residues Asp and Glu at the polar-nonpolar interface and positively charged Lys residues at the center of the polar face. Studies with these two peptides and many other analogs of these two peptides showed that 18A analogs had higher lipid affinity compared to the corresponding 18R analogs [5].

To explain the relatively high lipid affinity of 18A and the corresponding peptide analogs a snorkel model was proposed [3]. In this model, the bulk of the van der Waals surface which is hydrophobic, interacts with lipid alkyl chain when the peptide is bound to lipid. The charged Lys amino groups "snorkel" into the aqueous environment for hydration. In the snorkel orientation, the entire 18A peptide is thought to be buried within the interior of the phospholipid bilayer. In contrast, due to the short alkyl chains of interfacial Asp and Glu, this orientation is not possible with 18R. This effect may explain the high lipid binding ability of 18A and apolipoprotein (Class A) amphipathic helices [5].

^{13}C-NMR has been used by our laboratories to study the lipid-bound conformations of apolipoproteins and amphipathic helical peptide analogs [4]. The present paper describes the use of ^{13}C-NMR to determine the differences in the

microenvironments of Lys residues in lipid-bound amphipathic helical peptides 18A and 18R. If the snorkel hypothesis is correct, all four Lys residues in 18R are expected to be in the same microenvironment due to their equal exposure to the aqueous phase whereas the interfacial Lys residues of 18A are expected to be exposed to different extents and in microenvironments of differing polarities.

Results and Discussion

Peptides 18A and 18R were synthesized by the solid phase procedure as described previously [2]. The peptides were ^{13}C-dimethylated using ^{13}C-labelled formaldehyde and sodium cyanoborohydride. The synthesis of selectively labelled 18A analogs involves a selective protection and preferential deprotection scheme (Figure 1). Peptides were synthesized using the solid phase peptide synthesis method on a Wang resin. The C-terminal FMOC-Phe was coupled to the resin using DCC in the presence of catalytic amounts of 4-dimethylaminopyridine. The ε-NH_2 groups of Lys were protected with BOC-groups (cleavable with TFA) except for the Lys residue to be labeled. This was protected with the Z-group that is stable under the conditions of Boc cleavage. The other TFA-cleavable protecting groups used were t-butyl ester for Asp and Glu, and t-butyl ether for Tyr. Thus, at the end of the synthesis, cleavage of the peptide along with the cleavage of side chain protecting groups except for the ε-NH_2 of the Lys to be labeled, was accomplished with TFA. The peptide was then treated with HCHO and $NaBH_3CN$ to convert the free ε-NH_2 to dimethyl-amino groups. The only ε-Z group present on the sidechain of Lys to be labeled was then cleaved with HF. Dimethylation of this with $H^{13}CHO$-$NaBH_3CN$ gave the selectively ^{13}C-dimethylated peptide. Using this procedure, four 18A peptides were synthesized in which different Lys side chains were selectively labeled. The intermediates in the synthesis have been characterized by proton NMR as ascertained by the presence or absence of characteristic benzyl proton signals at around 5 to 5.5ppm.

Discoidal complexes with dimyristoyl phosphatidylcholine (DMPC) were prepared using ^{13}C-dimethylated peptides 18A and 18R and were studied using ^{13}C-NMR (Figure 2). Since the α-amino group was also labeled during ^{13}C-labeling, both peptides gave a signal due to α-^{13}C dimethyl groups at 40.8 ppm. The ^{13}C-dimethyl Lys groups in the 18R analog gave a single signal at 42.7 ppm from the four ε-NH_2 groups. This indicates that the four Lys residues in this peptide when bound to lipid are in the same microenvironment with the pKa being 10.2. In contrast, the [^{13}C-dimethylLys]18A:DMPC complex gave five signals with chemical shifts in the range of 42.1 to 43.5 ppm; the pKa values were in the range 8.9 to 11. To identify the pKa of individual Lys residues in 18A, NMR spectra of peptide:DMPC complexes containing selectively ^{13}C-dimethyl-labeled 18A analogs were obtained. These results are consistent with the snorkel model since in the lipid-bound conformation, interfacial basic residues in the 18A amphipathic helix are present in microenvironments of different polarities; the different accessibilities of positively charged Lys residues to the aqueous environment leads to different pK_a values.

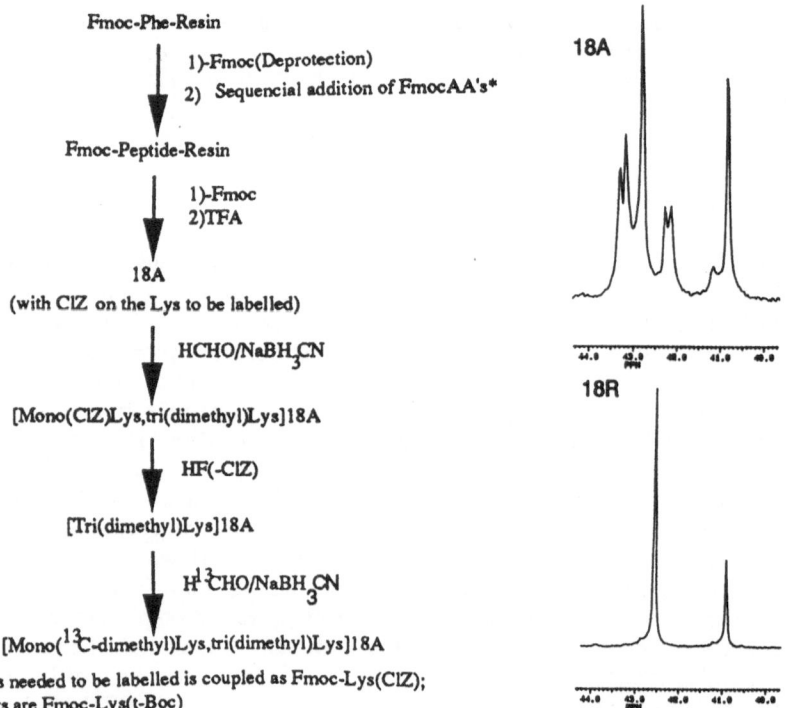

Fmoc-Phe-Resin

    1)-Fmoc(Deprotection)
    2) Sequencial addition of FmocAA's*

Fmoc-Peptide-Resin

    1)-Fmoc
    2)TFA

18A
(with ClZ on the Lys to be labelled)

    HCHO/NaBH$_3$CN

[Mono(ClZ)Lys,tri(dimethyl)Lys]18A

    HF(-ClZ)

[Tri(dimethyl)Lys]18A

    H^{13}CHO/NaBH$_3$CN

[Mono(^{13}C-dimethyl)Lys,tri(dimethyl)Lys]18A

* Lys needed to be labelled is coupled as Fmoc-Lys(ClZ);
others are Fmoc-Lys(t-Boc)

18A

18R

Fig. 1. Selective labeling of lysine residues in 18A.

Fig. 2. ^{13}C-NMR spectra of peptide: DMPC complexes.

References

1. Segrest, J.P., Jackson, J.L., Morrisett, J.D. and Gotto Jr., A.M., FEBS Lett., 38 (1974) 247.
2. Anantharamaiah, G.M., Jones, J.L., Brouillette, C.G., Schmidt, C.F., Chung, B.H., Hughes, T.A., Bhown, A.S. and Segrest, J.P., J. Biol. Chem., 260 (1985) 10248.
3. Segrest, J.P., DeLoof, H., Dohlman, J.G., Brouillette, C.G. and Anantharamaiah, G.M, Proteins, 8 (1990) 117.
4. Sparks, D.L., Phillips, M.C. and Lund-Katz, S., J. Biol. Chem., 267 (1992) 25830.
5. Segrest, J.P., Jones, M.K., Loof, H.D., Brouillette, C.G., Venkatachalapathi, Y.V. and Anantharamaiah, G.M., J. Lipid Res., 33 (1992) 141.

Helical character and aggregation effects on the activity of short cecropin-melittin hybrid peptides

D. Andreu[1], I. Fernández[1], J. Ubach[1], F. Reig[2], R.B. Merrifield[3] and M. Pons[1]

[1]Department of Organic Chemistry, University of Barcelona, E-08028 Barcelona, Spain
[2]CID-CSIC, Jordi Girona 18, E-08034 Barcelona, Spain
[3]Rockefeller University, New York, NY 10021, U.S.A.

Introduction

In the course of studies aimed at finding antibacterial peptides of improved activity, several short cecropin A-melittin hybrids have been shown to have promising antibiotic properties [1, 2]. These peptides retain the ability of the parent molecules to adopt an amphipathic α-helical conformation, but their small size (12-15 residues) prevents membrane spanning by a single helical molecule. In order to investigate the role of factors such as helical character, amphipathicity and charge on biological activity, we have compared the conformational properties and the effect on liposome permeability of two peptides, KWKLFKKIGAVLKVL-amide [cecropin A(1-7)melittin(2-9), I], and its hexasuccinyl derivative (II), with the same sequence but reverse polarity. Negative charges near the N-terminus of II are expected to stabilize its helical conformation. On the other hand, since both charged (but inverted) and non-polar residues retain their positions, the amphipathic character of the helix will be preserved. Since the anionic character of II would prevent its interaction with the negatively charged bacterial cell surface, the biological activity of both I and II has been compared by the changes in permeability induced by both peptides in model membranes with no net charge.

Results and Discussion

Peptide I was converted quantitatively into II by treatment with succinic anhydride (20 eq) at pH 9.5. The completeness of the reaction was verified by HPLC and FAB mass spectrometry. I and II contain a maximum of six positive or negative charges respectively, resulting in characteristically different HPCE behaviour.

I and II were tested for their effect on the permeability of neutral membranes formed by zwitterionic and neutral lipids and encapsulating carboxyfluorescein. A sharp increase in liposome permeability (80-90% dye release within 15 min) was found for peptide I. In contrast, peptide II did not show significant activity even at concentrations more than 10-fold higher and longer incubation times.

In aqueous solution I and II showed little or no structure by CD but readily acquired an α-helical conformation in the presence of HFIP or TFE. Minimal

373

ellipticities consistent with essentially 100% helical contents for peptides of this length were observed above 12% and 16% HFIP for **I** and **II**, respectively. While intrinsically less prone to become an α-helix, **I** had a more cooperative behaviour than **II** and reached maximal helicity at lower HFIP concentrations, suggesting that its α-helical conformation was related to aggregation. Indeed, concentration dependence studies showed clear evidence of aggregating behaviour for both **I** and **II**.

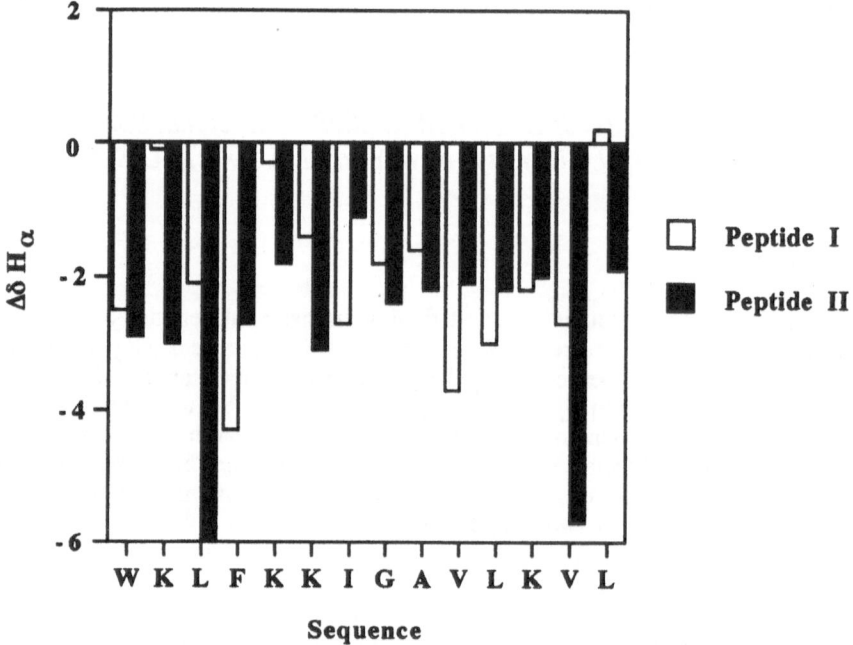

Fig. 1. Chemical shift differences of CH$_\alpha$ protons of peptides I (empty bars) and II (filled bars) in 12% HFIP relative to I in aqueous solution.

^{1}H-NMR studies (TOCSY, DQF-COSY, NOESY experiments) were done in 12% HFIP, where maximal helix formation for both peptides is predicted by CD. Comparison of CH$_\alpha$ signals of peptides **I** and **II** with those of **I** in aqueous solution (i.e., random-coil) showed high-field shifts in consecutive CH$_\alpha$ protons (Figure 1), characteristic of helical segments [3, 4]. By this criterium both **I** and **II** were found to form α-helices at the C-terminal region whereas **II** showed higher helical content than **I** at the N-terminus. Sequential NH-NH NOEs provided confirmatory evidence of the helical character of the C-terminal moieties of **I** and **II**. The presence of [i,i+3] NOE in the N-terminus of **I** supported at least some helical conformation in equilibrium with more disordered states. Results from D/H exchange experiments were also significant. While all NH protons of **II** exchanged within a few minutes, as expected for a small peptide, some NH protons in the C-terminus of **I** exchanged very slowly. Similar behaviour recently reported for a related peptide [5] was attributed to hydrogen bonding in the α–helix. An alternative explanation would be that slow rates result from tight aggregates, consistent with the CD data.

    Although it is not possible at this time to present a detailed model for the aggregation of **I** in solution, one could envisage several possible juxtaposition patterns (e.g., tail-to-tail, hydrophobic faces, melittin-like) producing helical aggregates large enough to span a lipid bilayer. This could provide a plausible explanation for the antibiotic activity of these short peptide antibiotics.

Acknowledgements

Work supported by CICYT (PTR93-0032) and NATO (CRG.920295).

References

1. Wade, D., Andreu, D., Mitchell, S.A., Silveira, A.V., Boman, I.A., Boman, H.G. and Merrifield, R.B., Int. J. Peptide Protein Res., 40 (1992) 429.
2. Andreu, D., Ubach, J., Boman, I.A., Wåhlin, B., Wade, D. and Merrifield, R.B., FEBS Lett., 296 (1992) 190.
3. Jiménez, M.A., Nieto, J.L., Herranz, J., Rico, M., and Santoro, J., FEBS Lett., 221 (1987) 320.
4. Wishart, D.S., Sykes, B.D. and Richards, F.M., Biochemistry, 31 (1992) 1647.
5. Sipos, D., Chandrasekhar, K., Arvidsson, K., Engström, Å. and Ehrenberg, A., Eur. J. Biochem., 199 (1991) 285.

Conformational and enzymatic phosphorylation studies on synthetic peptides corresponding to the main autophosphorylation site of pp$^{60\text{c-src}}$

P. Ruzza[1], A. Calderan[1], B. Biondi[1], A. Donella Deana[2], L.A. Pinna[2] and G. Borin[1]

[1]Biopolymer Research Center C.N.R., Department of Organic Chemistry, Via Marzolo, 1 and [2]Department of Biological Chemistry, Via Trieste, 75, University of Padova, 35131 Padova, Italy

Introduction

All tyrosine protein kinases (TPKs) encoded by cellular genes of the *src* family contain two major phosphoacceptor sites which are homologous to Tyr-416 and Tyr-527 of pp$^{60\text{c-src}}$. The former represents the main autophosphorylation site *in vitro* and its phosphorylation correlates with increased kinase activity. By using synthetic peptides with a modified sequence around the Tyr-416 (EDNEYTA), the influence of the charged side chains and the propensity to assume a structural conformation on the activity of different splenic TPKs has been reported [1]. To better examine the effect of these two structural factors we have synthesized by the classical methods in solution the N-terminal acetylated and C-terminal amidated analog of the parent heptapeptide and of its dimeric form. Since cyclization imposes constraints that enhance conformational homogeneity reducing the number of available conformations, compared to that corresponding to the linear peptides, we have also prepared the cyclic monomer and dimer analogs. These synthetic peptides were tested as phosphoacceptor substrates for *fgr*-TPK, an enzyme belonging to the *src*-family, and for TPK-IIB, a spleen TPK lacking autophosphorylation, active *in vitro* on the products of *src* protooncogenes [1].

Results and Discussion

The synthetic peptides have been assayed for their ability to serve as phosphoacceptor substrates for the above-mentioned *fgr*-TPK and TPK-IIB. As shown in Table 1, while the N^{α}, C^{α}-blocked and cyclic monomer (II and III) are poor substrates for *fgr*-TPK, the detrimental effect due to either cyclization or N^{α}, C^{α}- modification of the EDNEYTA is overcome by dimerization. All the three dimers (IV, V and VI) are in fact highly phosphorylated. This dramatic activity increase cannot simply be accounted for by the presence of two tyrosyl residues in the sequence. The cyclization of the monomer also causes inhibition of the phosphorylation efficiency of the TPK-IIB. The reduced activity of this enzyme for the N^{α}, C^{α}-modified heptapeptide (II) is clearly due to an affinity decrease, as denoted by higher Km value. The free and cyclic dimers (IV and VI) are better substrates, displaying however only 2-fold higher phosphorylation efficiency. This might simply be due to the presence of two phosphorylatable tyrosyl residues conformationally suitable for the enzyme.

A possible correlation between the secondary structure of the synthetic linear and cyclic peptides and their substrate recognition by the mentioned TPKs was studied using CD and fluorescence spectroscopy in three different media: 5 mM Tris/HCl pH 6.8, 98% TFE and 30 mM SDS. While the transfer of all peptides from water to the hydrophobic media induces spectral variations, the cyclic heptapeptide (III) shows the same CD spectra in the different solvents used. This result is indicative of the presence of an ordinate structure, characterized by a high degree of rigidity. This lack of flexibility due to the heptapeptide cyclization might explain the negligible phosphorylation degree of cyclic EDNEYTA by the enzymes. Also the cyclization of the dimer (VI) has a negative effect on the phosphoacceptor activity of fgr-TPK in comparison with the corresponding linear peptide (IV). However this effect is much less pronounced than in the case of the monomer, suggesting that the larger cycle consents a much greater flexibility of the sequence than the smaller one.

Table 1 *Phosphorylation rates and kinetic constants of fgr-TPK and TPK-IIB for synthetic peptides substrates. (Efficiency is expressed as 10^{-2})*

	PEPTIDES	PHOSPHORYLATION RATES (pmol·min^{-1})		fgr-TPK			TPK-IIB		
		fgr-TPK	TPK-IIB	Vmax pmol min^{-1}	Km (μM)	Efficiency Vmax Km^{-1}	Vmax pmol min^{-1}	Km (μM)	Efficiency Vmax Km^{-1}
I	H-EDNEYTA-OH	1.7	7.6	19.6	500	3.9	42.0	100	42
II	Ac-EDNEYTA-NH$_2$	0.2	2.5	---	---	---	33.4	400	8
III	c(EDNEYTA)	0.2	0.8	---	---	---	---	---	---
IV	H-(EDNEYTA)$_2$-OH	14.3	14.4	50.0	45	111.0	51.2	62	83
V	Ac-(EDNEYTA)$_2$-NH$_2$	8.0	3.8	29.4	47	62.5	---	---	---
VI	c(EDNEYTA)$_2$	10.6	9.8	38.0	91	41.8	26.8	30	89

These data confirm our previous suggestion that more than a defined secondary structure characterized by a fixed geometry, a flexible chain seems the major requirement for a suitable interaction between the substrates and the enzymes and their subsequent phosphorylation.

Acknowledgements

Funding by Consiglio Nasionale delle Ricerche (P.F. Chimica Fine).

References

1.   Marin, O., Donella Deana, A., Brunati, A.M., Fischer, S. and Pinna, L.A., J. Biol. Chem., 266 (1991) 17798.

Hybrid rodent-human growth hormone-releasing factor (GRF) analogs with enhanced stability and biological activity

R.M. Campbell, P. Stricker, R. Miller, J. Bongers, E.P. Heimer and A.M. Felix

Roche Research Center, Hoffmann-La Roche Inc., Nutley, NJ 07110, U.S.A.

Introduction

Human GRF(1-44)-NH_2 is subject to biological inactivation by both enzymatic and chemical routes. In plasma, hGRF(1-44)-NH_2 is rapidly degraded via dipeptidylpeptidase IV (DPP-IV) enzyme cleavage between residues Ala^2 and Asp^3 [1]. hGRF(1-44)-NH_2 may also undergo spontaneous rearrangement (Asp^8 to Iso-Asp^8) [2] and oxidation (Met^{27} to $Met(O)^{27}$) [3] in aqueous environments, greatly reducing its bioactivity [4]. It was postulated that longer acting GRF analogs could be developed if these degradative pathways could be inhibited, i.e., using specific amino acid replacements at the amino terminus (to prevent recognition by DPP-IV), residue 8 (to reduce isomerization) and residue 27 (Leu^{27} to prevent oxidation). Inclusion of Ala^{15}-substitution (for Gly^{15}), previously demonstrated to enhance receptor binding affinity [4], was expected to improve GRF analog potency [5]. To examine this hypothesis, a series of [Ala^{15}, Leu^{27}]-hGRF(1-32)-OH analogs was synthesized with His^1 and/or Val^2-substitutions (from the mouse GRF sequence) [6] for Tyr^1, Ala^2- (human sequence) [7, 8]; and replacement of Thr^8 (mouse GRF sequence) [6], Ser^8 (rat GRF sequence) [9] or Gln^8 (not naturally occurring) for Asn^8 (human GRF sequence) [7, 8]. His^1 and Val^2-substitutions were selected as they were previously observed to inhibit DPP-IV cleavage of GRF analogs *in vitro* [10]. Thr^8, Ser^8 or Gln^8 substitutions were utilized to suppress the isomerization associated with Asn^8 [2]. The analogs produced were bioassayed for GH-releasing activity (rat pituitary cell culture) and plasma stability (incubation in porcine plasma) *in vitro*. The most potent analogs were also examined for *in vivo* GH-releasing activity (s.c. injection into pigs).

Results and Discussion

1. *In Vitro Potency:* With the exception of [Val^2, Ala^{15}, Leu^{27}]-hGRF(1-32)-OH, all hybrid rodent-human hGRF(1-32)-OH analogs were highly active (potencies ≈2-3 fold that of hGRF(1-44)-NH_2). Replacement of Asn^8 with either Ser^8, Thr^8 or Gln^8 did not alter potency.
2. *Aqueous Stability:* The aqueous half-lives of hGRF(1-32)-OH analogs were greatly enhanced by substitution of either Ser^8, Thr^8 or Gln^8 (≈ 4-7 fold over [His^1, Val^2, (Asn^8), Ala^{15}, Leu^{27}]-hGRF(1-32)-OH).

Table 1 *Assays of rodent-human GRF(1-32)-OH (hGRF$_{32}$) analogs*

hGRF Analog	Relative Potency[a] *In vitro*	Relative Potency[b] *In vivo*	Aqueous Stability[c] T1/2 (hr)	Plasma Stability[d] T1/2 (min)
hGRF(1-44)-NH$_2$	1.0	1.0	76	7.4
hGRF(1-29)-NH$_2$	0.7	0.9	—	8.4
[A^{15}]-hGRF$_{32}$	2.1	—	—	8.0
[A^{15}, L^{27}]-hGRF$_{32}$	1.9	—	—	15
[H^1, A^{15}, L^{27}]-hGRF$_{32}$	3.0	1.2	—	25
[V^2, A^{15}, L^{27}]-hGRF$_{32}$	0.3	1.7	—	35
[H^1, V^2, A^{15}, L^{27}]-hGRF$_{32}$	2.8	5.4	208	45
[H^1, V^2, S^8, A^{15}, L^{27}]-hGRF$_{32}$	3.1	12.4	1470	42
[H^1, V^2, T^8, A^{15}, L^{27}]-hGRF$_{32}$	2.5	10.9	1046	49
[H^1, V^2, Q^8, A^{15}, L^{27}]-hGRF$_{32}$	2.8	12.5	826	43

[a]GH-releasing potency relative to hGRF(1-44)-NH$_2$ using cultured rat pituitary cells.
[b]GH-releasing potency relative to hGRF(1-44)-NH$_2$ using area under the curve data from pigs injected s.c. (1-30 µg/kg) and blood sampled at various times over 6 hrs.
[c]0.5 mg/ml analog incubated in 0.25M Na$_2$HPO$_4$/H$_3$PO$_4$ at 37°C (pH 7.4), peptide content of aliquots determined by analytical HPLC.
[d]0.1 mg/ml analog incubated in porcine plasma at 37°C and aliquots extracted using C18 SEP-PAK columns; peptide content determined by analytical HPLC.

3.   *Plasma Stability:* The half-lives of hGRF(1-32)-OH analogs in plasma were improved by all substitutions employed. The results of the selected position 1, 2, 8 and 27 replacements were approximately additive; with the ultimate penta-substituted analogs being 6-7 fold more resistant to degradation compared to hGRF(1-44)-NH$_2$.
4.   *In Vivo Potency:* As anticipated from *in vitro* potency and stability data (Table 1), the combined substitution of His^1, Val^2 dramatically increased the potency of [Ala^{15}, Leu^{27}]-hGRF(1-32)-OH analogs (≈5 fold that of hGRF(1-44)-NH$_2$). This potency was further increased by additional Ser^8, Thr^8 or Gln^8 substitution (≈11-13 fold that of hGRF(1-44)-NH$_2$).

In summary, greatly enhanced bioactivity was observed with three penta-substituted hGRF(1-32)-OH analogs. *In vitro*, these analogs were ≈3 fold more potent than hGRF(1-44)-NH$_2$, whereas *in vivo* they were 11-13 fold more potent. As the *in vitro* results reflect only receptor affinity and signal transduction, the increment in potency observed *in vivo* is likely due to the increased biological half-life of these analogs (i.e., the result of decreased enzymatic and chemical decomposition such that more bioactive peptide is available per unit time).

References

1.   Frohman, L.A., Downs, T.R., Williams, T.C., Heimer, E.P., Pan, Y.-C. and Felix, A.M., J. Clin. Invest., 78 (1986) 906.
2.   Bongers, J., Heimer, E.P., Lambros, T.L., Pan, Y.-C., Campbell, R.M. and Felix, A.M., Int. J. Pept. Prot. Res., 39 (1992) 364.

3. Rivier, J., Spiess, J. and Vale, W., In Munekata, E. (Ed.), Peptide Chemistry 1983, Prot. Res. Foundation, Osaka, 1984, p.11.
4. Campbell, R.M., Lee, Y., Rivier, J., Heimer, E.P., Felix, A.M. and Mowles, T.F., Peptides, 12 (1991) 569.
5. Felix, A.M., Wang, C.-T., Heimer, E.P., Fournier, A., Bolin, D., Ahmad, M., Lambros, T.L., Mowles, T.F. and Miller, L., In Marshall, G.R. (Ed.), Peptides: Chemistry and Biology, Escom, Leiden, The Netherlands, 1988, p.465.
6. Frohman, M.A., Downs, T.R., Chomczynski, P. and Frohman, L.A., Mol. Endocrinol., 3 (1989) 1529.
7. Rivier, J., Spiess, J., Thorner, M. and Vale, W., Nature, 300 (1982) 276.
8. Guillemin, R., Brazeau, P., Böhlen, P., Esch, F., Ling, N. and Wehrenberg, W., Science, 218 (1982) 585.
9. Spiess, J., Rivier, J. and Vale, W., Nature, 303 (1983) 532.
10. Heimer, E.P., Bongers, J., Ahmad, M., Lambros, T., Campbell, R.M. and Felix, A.M., In Smith, J.A. and Rivier, J.E. (Eds.), Peptides: Chemistry and Biology, Escom, Leiden, The Netherlands, 1992, p.80.
11. Barany, G. and Merrifield, R.B., In Gross, E. and Meienhofer, J. (Eds.), The Peptides: Analysis, Synthesis, Biology, Vol 2, Academic Press, New York, 1980, p.1.
12. Brazeau, P, Ling, N., Böhlen, P., Esch, F., Ying, S.-Y. and Guillemin, R., Proc. Natl. Acad. Sci. U.S.A., 79 (1982) 7909.

A systematic study of the SAR in second generation bradykinin antagonists leads to the design of the first high affinity cyclic peptide antagonists

S. Chakravarty, B.J. Mavunkel, S. Lu, D.E. Wilkins and D.J. Kyle

Scios Nova Inc., Baltimore, MD 21224, U.S.A.

Introduction

Over the past several years, at least two generations of peptide bradykinin antagonists have been reported. We have explored the N-terminal structure-activity-relationship associated with the highly potent, second generation antagonists extensively, and systemically. The results led directly to the synthesis of two, conformationally constrained, cyclic peptides.

Results and Discussion

Using NPC 18545 (D-Arg^0-Arg^1-Pro^2-Hyp^3-Gly^4-Phe^5-Ser^6-D-Tic^7-Oic^8-Arg^9) as a potent reference peptide (K$_i$ = 0.08 nM), systematic glycine replacements (i.e. side chain removals) were made for residues 1 - 5. Subsequent binding results suggested that the side chains of Arg^1 and Phe^5 contribute about the same toward receptor affinity, and a peptide containing them together shows an additive enhancement in potency (Table 1). Potencies in the range of the parent peptide can only be obtained by the further addition of either the Pro^2 or Hyp^3 side chains.

Following up on the discovery that the 2 chiral centers and the side chains of Arg^1 and Phe^5 are minimally required to maintain high receptor affinity, a second series of peptides was prepared to study the relative importance of the individual amide bonds between residues 1 - 5. Based on receptor affinities, there appears to be a gradient of importance for the amide bonds between residues 1 - 5. This importance could be structural, electronic, or both. Since (14) had a K$_i$ of 13.7 nM and (17) had a K$_i$ of 1 nM, which differs from the former only in that it lacks the amide between residues 2 - 3, it is clear that this amide contributes little to receptor affinity. For the other two amides studied, that linking residues 3 - 4 appears to be of lesser importance (K$_i$ = 813 nM) than that linking residues 4 - 5 (K$_i$ > 2000nM).

C$^\alpha$-Me and N-Me backbone constraints were introduced into either Gly^4, Phe^5, or both as a means of gaining insight into the bio-active conformation. The C$^\alpha$-methyl Phe^5 substitution results in no significant loss in receptor affinity, which suggests that the ϕ, ψ backbone dihedral angles about Phe^5 are in the vicinity of -50°, -50° in the biologically active conformation. This would represent a helical twist or "kink" in the mid-section of the peptide.

As a test of this hypothesis, two Cys-containing cyclic peptides were prepared which were expected to induce a similar "kink" via covalent side chain cyclization. These cyclic peptides, both found to be potent competitive antagonists, are shown in Table 1.

Table 1 *Peptides used to study conformation preferences and the relative importances of amide bonds and side chains in bradykinin receptor antagonists*

Compound	X_2	X_3	X_4	X_5	X_6	K_i (nM) GPI
1	R	P	HP	G	F	.08
2	G	G	G	G	G	inactive
3	R	G	G	G	G	285 ± 111.19
4	G	P	G	G	G	1620
5	G	G	HP	G	G	714
6	G	G	G	G	F	483
7	R	G	G	G	F	13.7
8	R	P	G	G	F	0.42 ± .09
9	R	G	HP	G	F	0.28 ± .03
10	G	P	HP	G	F	10
11	R	G	HP	G	F	0.28 ± .03
12	R	P	G	G	F	0.42 ± .09
13	R	P	HP	G	G	
14	R	G	G	G	F	13.7
15	R	G	G	(cyclic structure spanning X_5–X_6)		
16	R	G	(cyclic structure spanning X_4–X_5)		F	813
17	R	(cyclic structure spanning X_3–X_4)		G	F	15.1
18	R	P	HP	G	F	< 0.1
19	R	P	HP	NMG	F	30.77 ± 15.49
20	R	P	HP	G	NMF	26.45 ± 10.05
21	R	P	HP	NMG	NMF	1693.33 ± 26.03
22	R	P	HP	AMG	F	60.6
23	R	P	HP	G	AMF	0.1
24	R	P	HP	AMG	AMF	2520
25	R	C	P	G	C	1.5 ± .1
26	R	C	P	G	F	14.83 ± 3.42

Table 1 *Continued*

1   All peptides were of the general sequence H-DR-X_2-X_3-X_4-X_5-X_6-Ser-(D)Tic-Oic-Arg-Oic, except 26, which is:

H-DR-R-C-P-G-F-C-DTic-Oic-R-OH

2   All potent peptides were shown to be antagonists by functional assays in GPI.

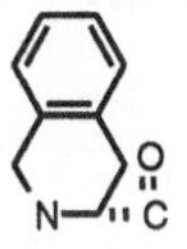

DTic

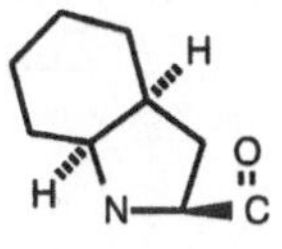

Oic

At the time of this report, these peptides represent the first example of potent, cyclic bradykinin antagonists. We anticipate that the elucidation of the solution structures, when combined with SAR data as we presented here could ultimately lead to the design of non-peptide bradykinin antagonists.

Design & synthesis of novel cyclic RGD peptides as highly potent & selective GPIIb/IIIa antagonists

S. Cheng, W.S. Craig, D. Mullen, J.F. Tschopp, D. Dixon and M.D. Pierschbacher

Telios Pharmaceuticals, Inc., 4757 Nexus Centre Drive San Diego, CA 92121, U.S.A.

Introduction

Adhesion of platelets to the endothelial surface of injured blood vessels as well as platelet aggregation are critical events for the development of coronary artery thrombosis. Adhesion and aggregation of platelets is mediated by fibrinogen that interacts with a platelet membrane glycoprotein complex, GPIIb/IIIa, at the platelet surface. The GPIIb/IIIa complex is a member of the family of cell adhesion receptors called integrins [1]. It has been shown that some of these receptors bind to a common recognition structure, an Arg-Gly-Asp (RGD) tripeptide sequence [2]. Inhibition of the binding of fibrinogen to GPIIb/IIIa via molecules based on the RGD tripeptide sequence is a promising approach for the inhibition of platelet aggregation and subsequent thrombus formation.

We have reported previously that we have identified a series of cyclic RGD-containing peptides that are highly potent and selective for GPIIb/IIIa with negligible affinity for vitronectin and fibronectin receptors [3]. One of these analogues, TP9201, has been shown to inhibit platelet-mediated thrombus formation without associated prolongation of template bleeding time [4]. Our current studies have focused on the design and synthesis of smaller ring analogues with the aim of optimizing binding affinity while reducing ring size.

Results and Discussion

The arginine, glycine and aspartic acid residues in the RGD tripeptide sequence are essential for activity and the arginine side chain and the aspartic side chain are critical for binding. From the structure-activity relationship studies [3], it has been shown that the hydrophobic residues at position 3 [5] and a positive charge residue at position 4 [5] play very important roles in binding of these analogues to GPIIb/IIIa. It is critical to retain these two features for our design of small ring analogues. First, an effort was made to explore the possibility of obtaining active antagonist analogues of analogue 1, but with a reduced ring size from 29-membered to 26-membered and 23-membered. Analogue 3 and analogue 4 were prepared and evaluated. Analogue 4, which contains a 23-membered ring, inhibited platelet aggregation with an IC_{50} of 0.27 μM. Deletion of the asparagine residue at position -2 [5] and the proline residue at position -1 [5] from analogue 1 resulted in a compound that retained approximately 80% of inhibitory activity. These two residues (N and P) served primarily to attain the bioactive conformations and both can be deleted provided that they are replaced by proper cyclic constraints.

Decreasing the ring size of analogue **4** from 23 to 20 by deleting the arginine residue at position 4 [5] induced a drastic decrease of potency. In addition, the 17-membered ring analogue **6** was made by deleting the O-methyl-tyrosine residue at position 3 [5] of analogue **5**. Analogue **6** also showed a dramatic decrease in potency.

A series of small cyclic disulfide heptapeptides containing the minimum active sequence, R-G-D-(O-Me-Y)-R, were designed and synthesized based on the sequence of analogue **4**. The substitution of Cys with an *N*-terminal Mpr residue resulted in a 2-3 fold increase in potency compared with analogue **4**. Analogue **8** (Mpr)RGD(O-Me-Y)RC-NH$_2$ inhibited platelet aggregation with an IC$_{50}$ of 0.12 μM which was approximately 2-fold more reactive than analogue **1** Ac-CNPRGD(O-Me-Y)RC-NH$_2$. Analogue **8** also exhibited very low affinities for FnR and VnR.

Next, we examined the effect of the substitution of the O-Me-Y residue at position 3 with Ala. The substitution of Ala for O-Me-Y preserves the overall peptide backbone and more carefully assesses the importance of the O-Me-Y side chain. This modification caused a drastic decrease of potency and reaffirmed the importance of the O-Me-Y side chain.

Table 1 *Systematic reduction in ring size*

Peptide[a]	Platelet Aggr. IC$_{50}$(μM)[b]	IIb/IIIa ELISA IC$_{50}$(μM)	FNR ELISA IC$_{50}$(μM)	VNR ELISA IC$_{50}$(μM)
1	0.22	0.029	8.20	4.70
2	0.17	0.001	6.8	6.7
3	0.46	0.0046	10.0	10.0
4	0.27	0.0094	5.2	1.7
5	19.0	nt[c]	6.6	nt[b]
6	53.0	0.18	nt[b]	1.7

Peptide[a]
1 Ac-CNPRGD(O-Me-Y)RC-NH$_2$(**TP9201**), 2 Ac-(D-Pen)NPRGD(O-Me-Y)RC -NH$_2$
3 Ac-(D-Pen)PRGD(O-Me-Y)RC-NH$_2$, 4 Ac-CRGD(O-Me-Y)RC-OH
5 Ac-CRGD(O-Me-Y)C-NH$_2$, 6 Ac-CRGDC-NH$_2$

[a]The unnatural amino acids used in this study have been given the following abbreviations: Mpr, β-mercaptopropionic acid; Pen, penicillamine; D-Pen, D-penicillamine; Pmp, β,β-pentamethylene-ß-mercaptopropionic acid; O-Me-Y, O-methyl-tyrosine; O-*n*-butyl-Y, O-*n*-butyl-tyrosine.
[b]The platelet aggregation in human platelet rich plasma (PRP) was induced by ADP.
[c]Not tested

Table 2 *Cys and Tyr(Me) replacements (modifications of Ac-CRGD(O-Me-Y)RC-OH)*

Peptide[a]	Platelet Aggr. IC$_{50}$(μM)[b]	IIb/IIIa ELISA IC$_{50}$(μM)	FNR ELISA IC$_{50}$(μM)	VNR ELISA IC$_{50}$(μM)
4	0.27	0.0094	5.20	1.70
7	0.34	0.016	4.90	7.04
8	0.12	0.016	3.50	4.30
9	0.20	0.038	7.70	10.00
10	0.46	0.05	1.70	0.35
11	1.70	0.63	9.30	7.30
12	4.0	4.6	nt[c]	2.40

Peptide[a]
7 Ac-CRGD(O-Me-Y)RC-NH$_2$, 8 (Mpr)RGD(O-Me-Y)RC-NH$_2$,
9 (Pmp)RGD(O-Me-Y)RC-NH$_2$, 10 Ac-CRGD(O-Me-Y)R(Pen)-NH$_2$
11 Ac-(D-Pen)RGD(O-Me-Y)RC-NH$_2$, 12 Ac-CRGDARC-OH
[a,b,c]See Table 1

References

1. Ruoslahti, E., J. Clin. Invest., 87 (1991) 1.
2. Ruoslahti, E. and Pierschbacher, M.D., Cell, 44 (1986) 517.
3. Cheng, S., Craig, W.S., Mullen, D., Tschopp, J.F., Dixon, D. and Pierschbacher, M.D., J. Med. Chem., 37 (1994) 1.
4. Tschopp, J.F., Bell, D.J., Bunting, S., Burnier, J.P., Cheng, S., Craig, W.S., Dixon, D., Gadeck, T., Mazur, C., McDowell, R.S., Mullen, D.G., Napier, M.A. and Pierschbacher, M.D., Blood, 80 (Suppl. 1) (1992) 320a.
5. The position assignments are as follows:
    X$_1$ -X$_2$ -X$_3$ -X$_4$ -R -G -D -X$_5$ -X$_6$ -X$_7$ -X$_8$
    -4 -3 -2 -1 0 1 2 3 4 5 6

Novel thrombin receptor analogs with potent agonist effects

D.-M. Feng[1], D.F. Veber[1], T.M. Connolly[2] and R.F. Nutt[1]

Departments of [1]Medicinal Chemistry, and [2]Biological Chemistry, Merck Research Laboratories, West Point, PA 19486, U.S.A.

Introduction

Thrombin-induced platelet activation plays a pivotal role in arterial thrombosis. According to a novel mechanism of receptor activation, thrombin cleaves its receptor to create a new amino terminus, which can then act as a tethered ligand to activate the receptor [1]. A 14-amino acid peptide, Ser-Phe-Leu-Leu-Arg-Asn-Pro-Asn-Asp-Lys-Tyr-Glu-Pro-Phe-OH, derived from the new amino terminus of the receptor, is also able to fully activate human platelets and cause aggregation in the absence of thrombin with an EC_{50} of 10,000 nM. Carboxy-truncated analogs of this 14-residue peptide have also been found to have full receptor activating potencies [2-4]. In order to develop a high affinity ligand, position modifications for agonist potency enhancement were carried out. SAR studies resulted in the discovery of potent thrombin receptor activating peptides. Potent peptides suitable for radiolabelling were also found.

Results and Discussion

Peptide analogs of Ser-Phe-Leu-Leu-Arg-Asn-Pro-Asn-Asp-Lys-Tyr-Glu-Pro-Phe-OH were synthesized incorporating modifications which would help elucidate structural features critical for thrombin receptor activation. Structure modifications involved: (a) peptides of shorter chainlength and (b) side-chain and backbone modifications. Chemical syntheses were carried out by the solid phase method using an ABI 430 A (t-Boc based chemistry) instrument, HF deprotection, and purification by HPLC. Products were characterized by HPLC (>99% purity), amino acid analysis and FABMS. Biological potencies of analogs were measured as peptide effects on the extent of human blood platelet aggregation.

As shown in Table 1, the N-terminal pentapeptide amide (analog 5) retains high potency and was used as reference structure for further SAR studies. C-terminal amides were found to be more potent than acids (see for example 3 versus 2). An alanine scan confirmed literature reports [2, 3] in that the aromatic sidechain of Phe-2 is critical for receptor activation (analog 6), and that Ser-1 and Leu-3 can be replaced by Ala (analog 7 and 8) with complete retention of potency. Leu-4 can be replaced with the more hydrophobic cyclohexylalanine (ChA) (analog 10) with a two fold increase in potency. The replacement of Phe-2 with Phe(p-F) (analog 9) has a potency increasing effect of four fold. The potency enhancing modifications of p-F-Phe-2 and ChA-4 were combined and resulted in Ala- Phe(p-F)-Leu-ChA-Arg-NH_2

387

(analog **11**) with EC_{50} = 140 nM. An arginine in position 3 (analog **12**) gives added solubility, while homo-Arg in position 5 increases potency slightly (analog **13**). The C-terminally extended hexapeptide (analog **4**) was shown to be more potent than the pentapeptide (analog **5**). Thus, in order to further increase potency and make available analogs for potential radiolabelling, the most potent pentapeptide (analog **14**) was elongated with a tyrosine amide. This modification resulted in Ala-Phe(p-F)-Arg-ChA-hArg-Tyr-NH_2 (analog **16**) with an EC_{50} of 10 nM, a 27-fold potency enhancement over the corresponding pentapeptide (analog **14**), the most potent agonist reported to date. The replacement of Tyr with Tyr(p-I) resulted in Ala-Phe(p-F)-Arg-ChA-hArg-Tyr(p-I)-NH_2 (analog **17**), which shows an EC_{50} of 30 nM. This level of potency is suitable for possible use as a radioligand in receptor binding assays.

Table 1 *Activity of novel thrombin receptor agonist peptides*

No.	Structure	Platelet Aggregation[a] (EC_{50} nM)
1	Ser-Phe-Leu-Leu-Arg-Asn-Pro-Asn-Asp-Lys-Tyr-Glu-Pro-Phe-OH	10,000
2	Ser-Phe-Leu-Leu-Arg-Asn-Pro-Asn-Asp-Lys-OH	8,000
3	Ser-Phe-Leu-Leu-Arg-Asn-Pro-Asn-Asp-Lys-NH_2	2,000
4	Ser-Phe-Leu-Leu-Arg-Asn-NH_2	180
5	Ser^1-Phe^2-Leu^3-Leu^4-Arg^5-NH_2	1,000
6	Ser-Ala-Leu-Leu-Arg-NH_2	>8000
7	Ala-Phe-Leu-Leu-Arg-NH_2	800
8	Ser-Phe-Ala-Leu-Arg-NH_2	1,000
9	Ala-Phe(p-F)-Leu-Leu-Arg-NH_2	200
10	Ala-Phe-Leu-ChA-Arg-NH_2[b]	400
11	Ala-Phe(p-F)-Leu-ChA-Arg-NH_2	140
12	Ala-Phe(p-F)-Arg-ChA-Arg-NH_2	130
13	Ala-Phe(p-F)-Leu-ChA-hArg-NH_2	120
14	Ala-Phe(p-F)-Arg-ChA-hArg-NH_2	100
15	Ala-Phe(p-F)-Arg-ChA-hArg-NHEt	160
16	Ala-Phe(p-F)-Arg-ChA-hArg-Tyr-NH_2	10
17	Ala-Phe(p-F)-Arg-ChA-hArg-Tyr(p-I)-NH_2	30

[a]EC_{50} for stimulation of aggregation of human platelets;
[b]ChA, cyclohexyl-Ala.

Structural features responsible for activation of the thrombin receptor have been elucidated. The potency of shortened peptides has been enhanced up to 1000-fold. These potent thrombin receptor activating peptides provide powerful tools for the development of a receptor radioligand and the potential establishment of a binding assay.

References

1. Vu, T.-K.H., Hung, D.T., Wheaton, V.I. and Coughlin, S.R., Cell, 64 (1991) 1057.
2. Scarborough, R.M., Naughton, M.A., Teng, W., Hung, D.T., Rose, J., Vu, T.-K.H., Wheaton, V.I., Turck, C.W. and Coughlin, S.R., J. Biol. Chem., 267 (1992) 13146.
3. Feng, D-M., Veber, D.F., Connolly, T. and Nutt, R.F., In Du, Y.-C., Tam, J.P. and Zhang, Y.-S. (Eds.), Receptor and Regulation, Proceedings of the Chinese Peptide Symposium 1992, Escom, Leiden, The Netherlands, 1993, p.141.
4. Vassallo, R.R. Jr., Kieber-Emmons, T., Cichowski, K. and Brass, L.F., J. Biol. Chem., 267 (1992) 6081.

Growth hormone releasing factor analogs with modifications at position 2 and position 19: Effects on growth hormone releasing activity

**A.R. Friedman[1], W.M. Moseley[1], A.K. Ichhpurani[1],
R.A. Martin[1], G.R. Alaniz[1], W.H. Claflin[1], D.R. Reeves[1],
D.L. Cleary[1], L.A. Frohman[2], T.R. Downs[2], J.F. Caputo[1]
and T.M. Kubiak[1]**

[1]*The Upjohn Company, Kalamazoo, MI 49001, U.S.A.*
[2]*University of Cincinnati, Cincinnati, OH 45267, U.S.A.*

Introduction

We have previously shown that within the format $[Thr^2,Ala^{15},Leu^{27}]bGRF(1-29)NH_2$, substitution of the native Ala^{19} residue with the hydrophobic residues Leu, Ile or Val, effected changes in the growth hormone (GH) releasing activity of the analogs [1]. Additional studies *in vitro* and *in vivo* [2] have shown that at position 2, Ile and Val were superior to Thr in the $[X^2,Ala^{15},Leu^{27}]bGRF(1-29)NH_2$ format. We have now prepared GRF analogs modified both at position 2 (Thr, Ile, Val) and at position 19 (Leu, Ile, Val). These compounds were examined for GH release *in vitro* (cultured rat anterior pituitary cells) and *in vivo* (steers). The contribution of each of these substitutions to GH release was determined.

Results and Discussion

The results of the GH release *in vitro* and *in vivo* are shown in Table 1. *In vitro*, the Ile^2,Val^{19} analog was most active, while the Thr^2,Leu^{19} analog was least active. Compounds with Leu^{19} substitution were less active than their Ile or Val counterparts. Free and Wilson analysis [3, 4] (which assumes substituent additivity) of the *in vitro* data showed that relative to the Thr^2,Ala^{19} analog the specific contributions (expressed as ng GH/10^5 cells/4 hrs) were as follows: Val^2 +895, Ile^2 +1704, Val^{19} +29181, Ile^{19} +1181, and Leu^{19} −4052. All of these contributions were significant (p=0.05) except Val^2. The correlation for the regression, r=0.988.

In vivo, differences in activity were more difficult to detect. Free and Wilson modelling the *in vivo* data did not yield a significant regression. Compounds with a Leu^{19} substituent tended to be least active, however, due to the variability of the assay and perhaps pharmacokinetic factors not accounted for, it was difficult to demonstrate effects of the other substitutions as could be observed *in vitro*.

Table 1 *Effects of GRF analogs on growth hormone release in vitro and in vivo*

Treatment	In vitro GH release[#]	In vivo GH release[*]
Vehicle	1963[a]	1.00[a]
[Leu^{27}]bGRF(1-29)NH$_2$	7561[d]	ND
Ile^2 Ala^{15}Val^{19}	12082[g]	4.25[c]
Ile^{2}Ala^{15}Ile^{19}	11006[f,g]	2.91[b,c]
Val^{2}Ala^{15}Val^{19}	11832[f,g]	3.77[b,c]
Val^{2}Ala^{15}Ile^{19}	10686[f]	4.47[c]
Thr^{2}Ala^{15}Val^{19}	9252[c]	4.10[c]
Thr^{2}Ala^{15}Ile^{19}	6263[c,d]	3.23[c]
Thr^{2}Ala^{15}Leu^{27}	5831[c]	3.60[c]
Ile^{2}Ala^{15}Leu^{19}	5835[c]	2.38[b]
Val^{2}Ala^{15}Leu^{19}	3976[b]	3.07[b,c]
Thr^{2}Ala^{15}Leu^{19}	2443[a]	3.20[b,c]
bGRF(1-44)NH$_2$	ND	2.29[b]

[#]The amount of rGH release in the pituitary cell cultures from treatments (0.1nM) is expressed as ng per 2×10^5 cells/4hrs (n=4). [*]Two hour bGH integrated area under the curve from steers treated (i.v.) at 30pmol/kg. Data was analyzed by ANOVA. Methods are described in reference 1. [a,b,c,d,e,f,g]Within a given column, means lacking a common superscript differ at $p < 0.05$. ND, not determined.

References

1. Friedman, A.R., Ichhpurani, A.K., Moseley, W.M., Alaniz, G.R., Claflin, W.H., Cleary, D.L., Prairie, M.D., Krueger, W.C., Frohman, L.A., Downs, T.R. and Epand, R.M., J. Med. Chem., 35 (1992) 3928.
2. Kubiak, T.M., Friedman, A.R., Martin, R.A., Ichhpurani, A.K., Alaniz, G.A., Claflin, W.H., Goodwin, M.C., Cleary, D.L., Kelly, C.R., Hillman, R.M., Downs, T.R., Frohman, L.A. and Moseley, W.M., J. Med. Chem., 36 (1993) 888.
3. Free, S.M. and Wilson, J.W., J. Med. Chem., 7 (1964) 395.
4. Ban, T. and Fujita, T., J. Pharm. Pharmacol., 15 (1963) 285.

Bradykinin antagonists: In vitro activity of BK(cysteine⁶) modified analogs containing extended hydrophobic side chains

V.S. Goodfellow, M.V. Marathe, D.A. Zummach, F. Wincott, L. Allen, E.T. Whalley, S. Loy, K.T. Nguyen and D. Cuadrado

Cortech Inc., 6850 N. Broadway, Denver, CO 80221, U.S.A.

Introduction

CP-0127 **1** is a potent, selective, antagonist of bradykinin type 2 receptors and is currently undergoing testing in phase II clinical trials for the treatment of SIRS/sepsis (Table 1). Compound **1** is formed by dimerization of a decapeptide **2**, which exhibits less potent BK_2 antagonist activity. The dimerized structure results in a compound with approximately 10 to 100 times the potency of the monomeric precursor in functional tissue assays. However, a monomer **3** containing an (N-hexyl)-succinimide moiety exhibited identical potency. Analogs were constructed to test the role of the succinimide carbonyls and hydrophobic alkyl side chains in receptor binding. Potent (in vitro) analogs were discovered which possess neither a dimeric peptide structure nor an (N-alkyl)-succinimide side-chain. We present here a preliminary study to optimize four parameters: potency in functional assays (rat uterus, pA_2), antagonist irreversibility as evidenced by low recovery percentages after wash-off experiments, compound stability, and minimization of partial agonism at very high doses (10^{-5} M).

Results and Discussion

Rat uterus in vitro pA_2 measurements were done as previously described [1]. Antagonist potency was calculated according to the method of Arunlakshana and Schild [2]. Compound **3** was synthesized as a mixture of diastereomers which were separated by HPLC. The data given here are for the mixture. The separated diastereomers gave essentially equivalent biological data. Compound **4**, which represents formal deletion of CH_2CO from the succinimide ring results in a significant loss in antagonist potency, an increase in wash-off, and a decrease in partial agonism at high doses. Compound **5**, a "retro" analog of **4** shows approximately equivalent activity, suggesting the lack of importance for the remaining carbonyl for antagonist potency. Formal deletion of the putative "non-essential" carbonyl results in compound **6**, with increased intrinsic potency (pA_2 = 8.7). However, antagonist wash-off is increased, and minor partial agonism at high doses is maintained. The BK_2 receptor is membrane-bound. It may be possible that the antagonists with hydrophobic side chains first interact with the cell membrane and then diffuse on, or in the membrane, to the receptor binding site. A farnesyl analog **7** was constructed to test this hypothesis. The olefinic carbons provide a

Table 1 *Bradykinin antagonists of the structure, D-Arg-Arg-Pro-Hyp-Gly-Phe-X-D-Phe-Leu*

X

1 pA$_2$ =8.5±0.3 (n > 3)
50% Recovery
No partial agonism at 10 uM.

2 pA$_2$ =7.1±0.10 (n > 3)
100% Recovery
No partial agonism at 10 uM.

3 pA$_2$ =8.4±0.14 (n=9)
43% Recovery
~15% Partial agonism at 10 uM.

4 pA$_2$ = 7.9 ±0.12 (n=6)
74% Recovery
5% Partial agonism at 10 uM.

5 pA$_2$ = 7.8 ±0.12 (n = 3)
100 % Recovery
No partial agonism at 10 uM.

6 pA$_2$ =8.7±0.14 (n=7)
75 % Recovery
~15 % Partial agonism at 10 uM.

7 pA$_2$ =7.7±0.10 (n=6)
0% Recovery
No partial agonism at 10 uM.

mimic of the succinimide scaffolding. Formal deletion of the "essential" carbonyl decreases intrinsic potency, but results in an essentially irreversible antagonist. The loss of potency of the n-dodecyl straight-chained analog **8** illustrates the necessity of a rigid scaffolding structure, as well as a hydrophobic chain. Although the "essential" carbonyl of the succinimide or lactam analogs is desirable for high intrinsic potency, it is also associated with poorer wash-off characteristics and partial agonism at very high doses. The farnesyl analog provided a template for exploring scaffolding geometry which may mimic the requisite geometry for potency but eliminate partial agonism at high doses, as demonstrated by compound **9** in which a benzyl group scaffold provides good intrinsic potency, no partial agonism, and an essentially irreversible antagonist. The necessity of a rigid scaffold of a well defined geometry for intrinsic potency is illustrated by the lack of potency of **10** containing a "para" phenyl scaffold.

$pA_2 = 7.3\pm0.10$ (n=9)
0% Recovery
No partial agonism at 10 uM.

8

$pA_2 = 8.1 \pm0.12$ (n=5)
0% Recovery
No partial agonism at 10 uM.

9

$pA_2 = 6.9\pm0.10$ (n=4)
21% Recovery
No partial agonism at 10 uM.

10

The linker structure of CP-0127 is responsible for a significant enhancement in potency of the dimer as compared to the monomeric structure. Using fragments of the bis-succinimidohexane linker it is possible to obtain compounds with equal potency, such as **3**, or even enhanced potency such as **6** compared to compound **1**. A specific carbonyl in compounds **3** or **6** appears to be desirable for intrinsic potency; however, this carbonyl also appears to be related to partial agonism at high doses and poor wash-off characteristics. Using a rigid scaffolding structure, lacking carbonyl units, it is possible to design potent and essentially irreversible antagonists of the BK_2 receptor.

References

1. Cheronis, J.C., Whalley, E.T., Nguyen, K.T., Eubanks, S.R., Allen, L.G., Duggan, M.J., Loy, S.D., Bonham, K.A. and Blodgett, J.K., J. Med. Chem., 35 (1992) 1563.
2. Arunlakshana, O. and Schild, H.O., Br. J. Pharmacol., 14 (1958) 48.

The total synthesis of biologically active toxin ω-Aga IVa from the spider *Agelenopsis aperta*

M.K. Ahlijanian[1], G.C. Andrews[1], L.G. Contillo[1],
B.C. Guarino[1], L.D. Hirning[2], A.L. Mueller[2],
D. Phillips[1], N.A. Saccomano[1], D.H. Singleton[1]
and R.A. Volkmann[1]

*[1]Pfizer Central Research, Eastern Point Rd, Groton, CT 06340,
U.S.A., [2]NPS Pharmaceuticals, 420 Chipeta Way,
Salt Lake City, UT, U.S.A.*

Introduction

Voltage dependent calcium channels are a diverse group of proteins that play a major role in regulation of calcium entry into neurons and thereby control diverse cellular and physiological responses like membrane excitability and neurotransmitter release [1]. The most recently discovered calcium channel subtype is termed the P-type channel [2]. The only selective antagonist of the P-channel is a peptide toxin isolated from the spider *Agelenopsis aperta* termed ω-Aga IVa. [3]. The native peptide is 48 amino acids in length and contains 4 disulfide bonds. This toxin represents an essential new tool to study the structure and function of P-channels per se and to understand the role of P-channels in central neuron function. Herein, we report the successful construction and biological activity of synthetic ω-Aga IVa.

Results and Discussion

Solid phase synthesis was performed on an Applied Biosystems model 430A using standard t-BOC protocol Version 1.4 for NMP/HOBt chemistry. Mass spectra were obtained on a Finnegan TSQ-700 operating in the electrospray ionization mode. Amino acid side chain protection used was: p-chlorocarbobenzoxy for Lys, p-toluenesulfonyl for Arg, 2-bromocarbobenzoxy for Tyr, N-benzyloxymethyl for His, O-benzyl for Ser and Thr, O-Cyclohexyl for Asp and Glu, formyl for Trp, S-4-methylbenzyl for Cys, and Met was protected as the oxide. The C-terminal acid peptide chain was assembled on t-BOC Ala derivatised PAM resin (0.66 g, 0.753 mmol/g). Total yield of dried resin was 1.9 grams. The peptide was cleaved using the "Low-High" HF method of Tam [4] providing 91 mg of crude product from 250 mg of resin. RPHPLC and ESMS analysis showed the major component to have the correct M^+ parent ion for the persulfhydryl peptide (5202.8). Crude HF product (179 mg) was dissolved in 3 ml of 6 M guanidine-HCl, 200 mM dithiothreitol, 50 mM Tris at pH 6.1 and incubated at 37°C for 30 min. The reduced peptide solution was diluted into 2 liters of folding buffer containing 2 mM glutathione, 2 mM glutathiol, 10 mM methionine, 50 mM Hepes at pH 8.0. This solution was stirred at 4°C for

144 hours (Figure 1). The folding mixture was purified on a Polysulfoethyl Aspartamide cation exchange HPLC column (KH_2PO_4 pH 8.2 mobile phase system) and desalted by C18 RPHPLC, affording 946 mg product, consistent by ESMS, AAA, DTNB, standard Edman degradation and co-injection with native authentic standard.

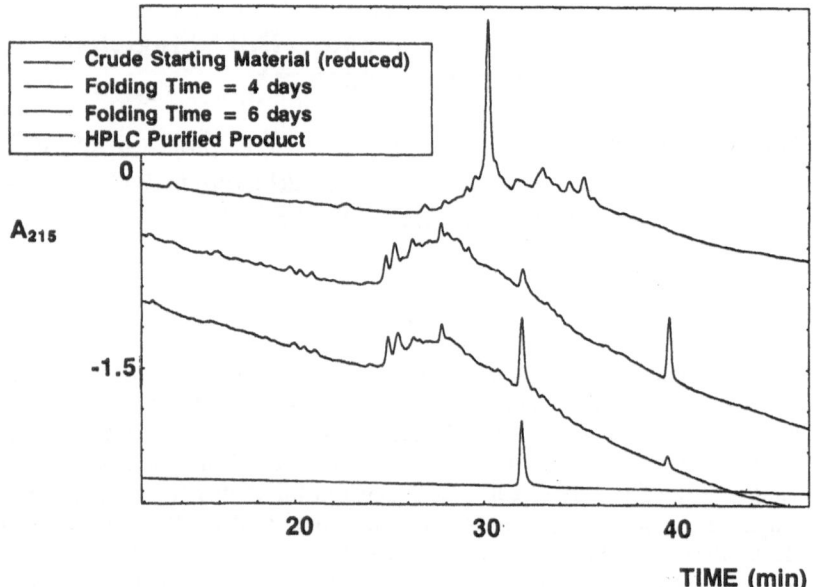

Fig. 1. HPLC time course for folding of crude HF product. (Waters μBondapak C-18, 3.9 x 300 mm, column eluting 22 - 44 % B over 45 min; flow rate = 1 ml/min; A = 5% CH3CN/water + 0.1% TFA, B = CH3CN).

The P-channel mediated currents are high threshold currents requiring large depolarizations from negative holding potentials to activate and are maximally available at -20 to -10 mV. High concentrations of ω-CgTx (3 μM) produced only a modest blockade of the current (21 ± 5%; n=11) and nifedipine (3 μM) did not block the Ca^{2+} current elicited from a holding potential of -90 mV (n=2). The P-channel current was also sensitive to Ca^{2+}, 30 μM of which blocked 87% (± 8; n=4) of the current elicited from a holding potential of -90mV. ω-Aga-IVA (native or synthetic) produced a rapid and near total blockade of the calcium current at a concentration of 100 nM. The IC_{50} values for native and synthetic ω-Aga-IVA in our system were 33.1 nM (21.2-51.8; 95% confidence interval) and 35.1 nM (20.1 - 61.2), respectively, as determined by Hill plot analysis.

The synthesis and refolding of ω-Aga IVa provided toxin which was chemically indistinguishable from native material. The IC_{50}'s obtained from native and synthetic toxin for blockade of P-channels in acutely dissociated Purkinje cell were equivalent within experimental error (Figure 2). Synthetic material was also shown to be biologically identical to native toxin in other models of P-channel activation (e.g., synaptic transmission, neurotransmitter release and $^{45}Ca^{2+}$ synaptosomal uptake.

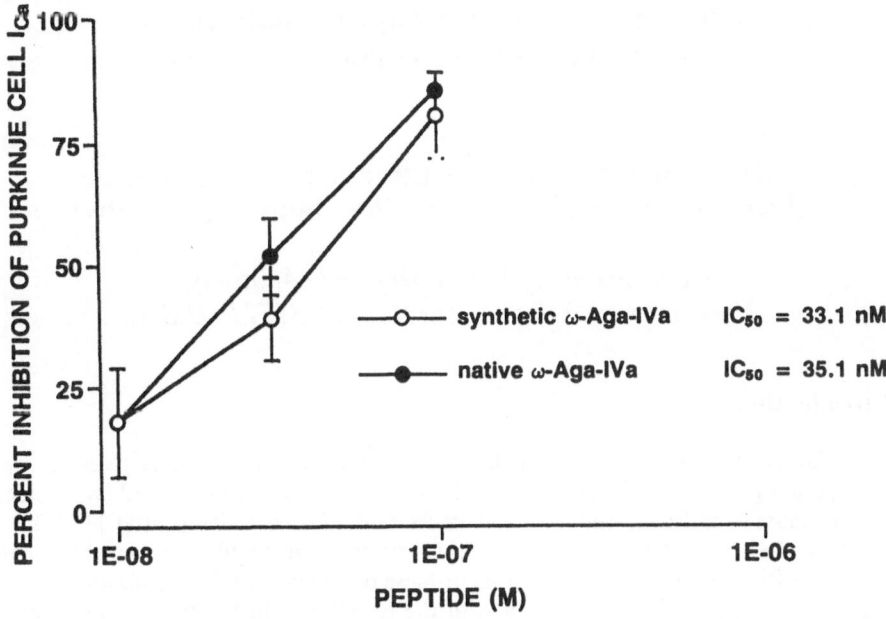

Fig. 2. Inhibition of Purkinje cell I_{ca}.

References

1. Bean, B.P., Annu. Rev. Physiol., 51 (1989) 367; Hess, P., Annu. Rev. Neurosci., 13 (1990) 337.
2. Llinas, R., Sugimori, M., Lin, J.W. and Cherksey, B., Proc. Natl. Acad. Sci. U.S.A., 86, (1989) 1689; Regan, L.J., Sah, D.W.Y. and Bean, B.P., Neuron, 6 (1991) 269; Mintz, I.M., Adams, M.E. and Bean, B.P., Neuron, 9 (1992) 85.
3. Phillips, D., Saccomano, N.A. and Volkmann, R.A., U.S. Patent #5, 122, 596. 16 June 1992, filed 29 September 1989; Mintz, I.M., Venema, V.J., Swiderek, K.M., Lee, T.D., Bean, B.P. and Adams, M.E., Nature, 355 (1992) 827.
4. Tam, J.P., Heath, W.F. and Merrifield, R.B., J. Am. Chem. Soc., 105 (1983) 6442.

Cationized melanotropin analogues: Structure-function relationships

S.D. Sharma[1], G.V. Nikiforovich[1], J. Jiang[2], A.M.L. Castrucci[2], M.E. Hadley[2] and V.J. Hruby[1]

Departments of [1]Chemistry and [2]Anatomy, University of Arizona, Tucson, AZ 85721, U.S.A.

Introduction

Cationization of a potent α-melanotropin analogue [Nle^4, D-Phe^7]α-MSH by incorporating a pair of dibasic amino acid residues that bridge Glu^5 and Lys^{11} side chains cause significant enhancement in its biological activity profile [1]. The electrostatic accumulation of these peptides on the cell membrane was shown to cause an efficient compartmentalization of these derivatives in the membrane phase, thereby facilitating their interaction with the receptors. In the present study this designing concept was further tested on another potent yet smaller melanotropin analogue Ac-Nle-Asp-His-D-Phe-Arg-Trp-Lys-NH$_2$ [2].

Table 1 *Comparative biological activities of cyclic bridged analogues*

Analogue	Potency relative to α-MSH	
	Frog skin	Lizard skin
1. Ac-[Nle^4, Asp^5, D-Phe^7, Lys^{10}]-α-MSH$_{4\text{-}10}$-NH$_2$	0.1 (-)^a	10 (+)
2. Ac-[Nle^4, Asp^5, D-Phe^7, Lys^{10}]-α-MSH$_{4\text{-}10}$-NH$_2$ (Arg–Arg)	0.14 (-)	6.7 (+)
3. Ac-[Nle^4, Asp^5, D-Phe^7, Lys^{10}]-α-MSH$_{4\text{-}10}$-NH$_2$ (Arg–Lys)	0.25 (-)	13.4 (+)
4. Ac-[Nle^4, Asp^5, D-Phe^7, Lys^{10}]-α-MSH$_{4\text{-}10}$-NH$_2$ (Lys–Arg)	0.05 (-)	10 (+)
5. Ac-[Nle^4, Asp^5, D-Phe^7, Lys^{10}]-α-MSH$_{4\text{-}10}$-NH$_2$ (Lys–Lys)	0.03 (-)	12.5 (+)
6. Ac-[Nle^4, Asp^5, D-Phe^7, Lys^{10}]-α-MSH$_{4\text{-}10}$-NH$_2$ (Arg)	0.1 (-)	1.0 (+)
7. Ac-[Nle^4, Asp^5, D-Phe^7, Lys^{10}]-α-MSH$_{4\text{-}10}$-NH$_2$ (Lys)	0.1 (-)	NA^b

^{a}Response is prolonged (+), or not prolonged (-); ^{b}NA means not assayed.

Results and Discussion

Macrocyclic analogues of the parent cyclic lactam peptide (**1**) containing dibasic amino acid residue(s) bridging Asp[5] and Lys[10] side chain functionalities were synthesized and evaluated for their biological activity on frog and lizard skins (Table 1). In general, all the analogues 2-7, being more potent on lizard skin than on frog skin, behaved like parent analogue **1**. The activity profile on lizard skin was either similar or slightly higher than **1**. Therefore, all these analogues are capable of presenting their message sequence (-His-$\underline{\text{D}}$-Phe-Arg-Trp-) favorably to the receptor. This has also been supported by their conformational analyses. As shown in Figure 1, geometric overlap of a low energy conformer of **5** with that of **1** exhibited a good overlap of C^α and C^β carbon atoms of His, $\underline{\text{D}}$-Phe, and Trp residues. Further, it was interesting to observe that **2** and **6** were potent stimulators of tyrosinase in mouse melanoma cells (minimal effective dose 10^{12}M) and exhibited activity up to 4 days after the removal of analogues from the incubation medium on the first day. These results confirm our earlier report [1] that the introduction of basic residues in a bio-compatible fashion can further modulate and enhance its biological activity profile. The results suggest general applicability of this peptide design concept for various ligands that act at membrane bound receptors.

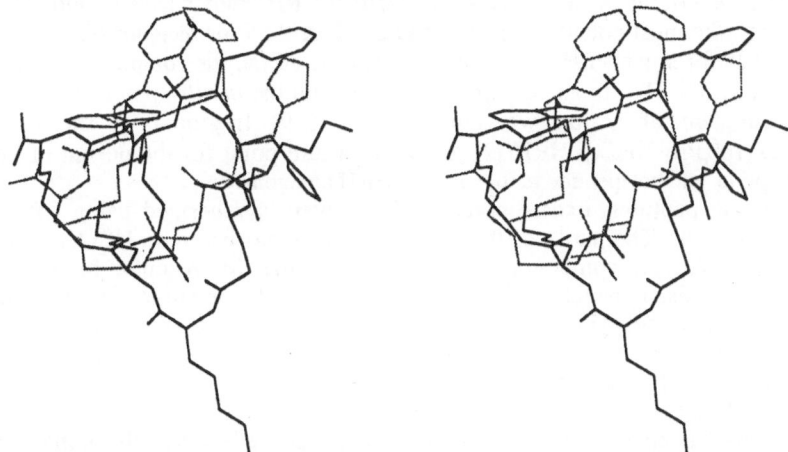

Fig. 1. Geometrical shape comparison of low energy conformers of 5 (bold line) and 1 (thin line). All hydrogens are omitted for clarity.

Acknowledgements

This research was supported by a grant from U.S. P.H.S.

References

1.   Sharma, S.D., Nikiforovich, G.V., Jiang, J., de L. Castrucci, A.M., Hadley, M.E. and Hruby, V.J., In Schneider, C.H. and Eberle, A. (Eds.), Peptides 1992, Escom Sci. Publ., Leiden, The Netherlands, 1993, p.95.
2.   Al-Obeidi, F., de L. Castrucci, A.M., Hadley, M.E. and Hruby, V.J., J. Med. Chem., 32 (1989) 2555.

Active reduced size octapeptide analogues of luteinizing hormone-releasing hormone

F. Haviv[1], T.D. Fitzpatrick[1], C.J. Nichols[1], E.N. Bush[1], G. Diaz[1], G. Bammert[1], A.T. Nguyen[1], N.S. Rhutasel[1], E.S. Johnson[1], J. Knittle[2] and J. Greer[1]

[1]Pharmaceutical Products Division, Abbott Laboratories and [2]TAP Pharmaceuticals Inc., Abbott Park, IL 60064, U.S.A.

Introduction

We previously reported a series of reduced size hexapeptide analogues of luteinizing hormone-releasing hormone (LHRH), which were designed based on the (3-9) fragment of the agonist [DLeu[6],Pro[9]NHEt]LHRH, known as leuprolide [1]. In that series, the most potent antagonist had a pK_I of 9.55 for receptor binding and a pA_2 of 9.28 for *in vitro* LH inhibition; in the castrated rat, the compound was active only by infusion [1]. In an attempt to increase both the *in vitro* potency and the *in vivo* duration of action we selected the (2-9) fragment of the agonist [Phe[2],DTrp[6],Pro[9]NHEt]LHRH [2] as the departure point for the design of novel heptapeptide and octapeptide reduced size LHRH antagonists.

All the peptides, except compound **15**, were synthesized using SPPS on Merrifield resin. The purity of the final compound was based on HPLC, FABMS and AAA. All the compounds were tested *in vitro* for receptor binding, LH inhibition and histamine release [3]. Selected compounds were tested in the castrated rat for LH suppression [3].

Results and Discussion

In the LH release from rat pituitary cultured cells assay the nonapeptide [Phe[2],Trp[6],Pro[9]NHEt]LHRH, (**1**), was found to be a potent agonist with a pD_2 of 10.81 [2]. Elimination of residue pGlu[1], (**2**), caused over 1000-fold loss in activity, but still maintained the agonist response (Table 1). Acylation of fragment **2** increased the binding affinity by 500-fold to 9.78 and converted the compound to antagonist with a pA_2 of 8.02. We have previously observed a similar switch from agonist to antagonist response upon N-methylation of position 2 [2]. Substitutions of compound **3** with phenylacetyl[2] or 4-Cl-phenylacetyl[2] slightly reduced the *in vitro* activity, whereas substitutions with 4-Cl-phenylpropionyl[2] and 4-F-phenylpropionyl[2] increased both receptor binding and LH inhibition by over 10-fold. Both compounds **6** and **7** were found inactive when tested for LH suppression in the castrated rat following sc administration of 30 µg/kg. Subsequently, in an attempt to increase metabolic stability we substituted DTrp at position 3 of compound **7**. This structural modification yielded antagonists **8** and **9** which had pK_I greater than 10.0. Peptide **9** showed for the first time a modest suppression of LH *in vivo*. Encouraged by these results we continued to further stabilize compound **9** by substituting NMeTyr[5] [2]. This modification raised the pA_2 value to 11.15 and significantly improved *in vivo* LH suppression (data not shown). Further substitutions at positions 3 and 6 yielded compounds **11-13**, which were active *in vitro* and *in vivo*, however when tested for histamine release (HR) from rat peritoneal mast cells showed ED_{50} values <1 µg/ml, which were considered inadequate. To improve the safety profile we substituted Lys(Isp)[8] and DAla[10] in compound**13** yielding antagonist A-76154, which *in vitro* and *in vivo* was active in the range of antide and A-75998 (Table 1; Figure 1). The ED_{50} for HR of A-76154 was 10.0 µg/ml, identical to that of A-75998 [3]. These peptides represent the first series of hepta- and octapeptide LHRH antagonists which had activities in the range of standard decapeptide antagonists in both *in vitro* and *in vivo* tests.

Table 1 *In vitro rat receptor binding and LH release or inhibition of LHRH analogs*

H[1]-Phe[2]-Trp[3]-Ser[4]-Tyr[5]-DTrp[6]-Leu[7]-Arg[8]-Pro[9]NHEt

Cmpd	Substitution	$pK_I{}^a$	$pD_2{}^a$	$pA_2{}^a$
1	pGlu[1]	10.61	10.81	
2	as above	7.08	7.64	
3	NAcPhe[2]	9.78		8.02
4	Phenylacetyl[2]	9.08		8.28
5	4Cl-Phenylacetyl[2]	9.24		8.24
6	4Cl-Phenylpropionyl[2]	10.28		9.90
7	4FPP[2]	10.48		9.23
8	4Cl-Phenylpropionyl[2],DTrp[3]	10.12		9.78
9	4FPP[2],DTrp[3]	10.54		10.41
10	4FPP[2],DTrp[3],NMeTyr[5]	10.48		11.15
11	4FPP[2],D1Nal[3]	10.59		11.15
12	4FPP[2],D1Nal[3],NMeTyr[5]	10.32		11.25
13	4FPP[2],DTrp[3],NMeTyr[5],DLys(Nic)[6]	10.78		10.46
14	4FPP[2],D1Nal[3],NMeTyr[5],DLys(Nic)[6],Lys(Isp)[8]	11.62		11.43
15	4FPP[2],D1Nal[3],NMeTyr[5],DLys(Nic)[6],Lys(Isp)[8],DAla[10]NH$_2$	10.69		11.13
	A-76154			
	A-75998[b]		10.50	11.23
	antide[b]		10.21	10.63

[a]pK_I = the negative logarithm of the equilibrium dissociation constant in the rat pituitary receptor binding assay. pD_2 = the negative logarithm of the concentration of agonist that produces 50% of the maximum release of LH from cultured rat pituitary cells in response to the test compound. pA_2 = the negative logarithm of the concentration of antagonist that requires 2-fold higher concentration of agonist to release LH from pituitary cells. 4FPP = 4F-Phenylpropionyl.
[b]Ref. 3.

We were able to reduce the size of standard decapeptide LHRH antagonists by two D-amino acids and still maintain all the *in vitro* and *in vivo* activities while achieving an acceptable ED_{50} value for histamine release. A-76154 represents the first octapeptide LHRH antagonist which, in the castrated rat, was as effective in suppressing LH as the standard decapeptide antagonists. In the intact rat A-76154, following infusion for two weeks of 259.2 µg/kg/day dose, suppressed testosterone to castrate level causing atrophy of reproductive organs.

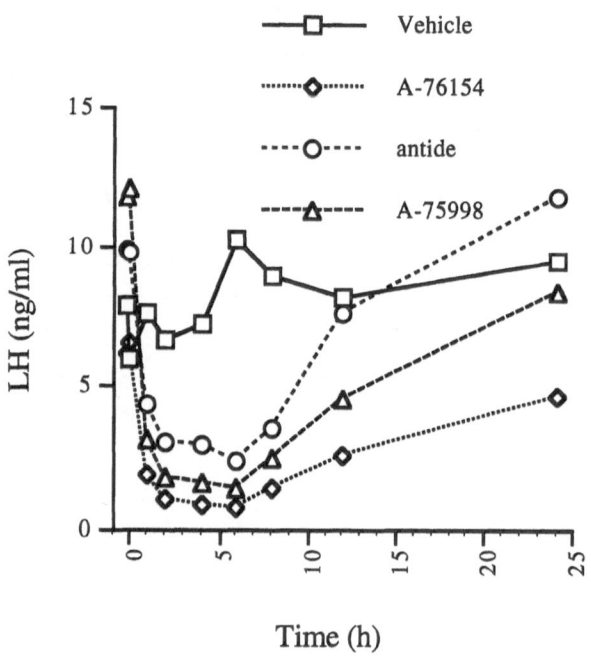

Fig. 1. *Suppression of LH in the castrated rat following sc injection of 30 µg/kg.*

References

1. Haviv, F., Palabrica, C.A., Bush, E.N., Diaz, G., Johnson, E.S., Love, S. and Greer, J., J. Med. Chem., 32 (1989) 2340.
2. Haviv, F., Fitzpatrick, T.D., Swenson, R.E., Nichols, C.J., Mort, N.A., Bush, E.N., Diaz, G., Bammert, G., Nguyen, A., Rhutasel, N.S., Nellans, H.N., Hoffman, D.J., Johnson, E.S. and Greer, J., J. Med. Chem., 36 (1993) 363.
3. Haviv, F., Fitzpatrick, T.D., Nichols, C.J., Swenson, R.E., Mort, N.A., Bush, E.N., Diaz, G., Nguyen, A., Holst, M.R., Cybulski, V.A., Leal, J.A., Bammert, G., Rhutasel, N.S., Dodge, P.W., Johnson, E.S., Cannon, J.B., Knittle, J. and Greer, J., J. Med. Chem., 36 (1993) 928.

Gonadotropin-releasing hormone antagonists containing novel amino acids

G.-C. Jiang, J. Porter, C. Rivier, A. Corrigan, W. Vale and J.E. Rivier

*Clayton Foundation Laboratories for Peptide Biology,
The Salk Institute for Biological Studies, 10010 N. Torrey Pines Road,
La Jolla, CA 92037, U.S.A.*

Introduction

We recently reported that azaline B [Ac-D2Nal[1],D4Cpa[2],D3Pal[3],4Aph[5] (Atz),D4Aph[6](Atz),ILys[8],DAla[10]]GnRH was among the most potent and long acting antagonists of GnRH with adequate solubility in aqueous solutions at neutral pH [1]. In order to further improve its properties, several GnRH antagonists containing novel amino acids at positions 5 (L-isomers), 6 (D-isomers) or 8 (L-isomers) have been synthesized, characterized and tested in an *in vitro* pituitary cell culture assay and in an antiovulatory assay. The synthetic amino acids (Figure 1) are D- and L-3-aminophenylalanine (3Aph), D- and L-4-thiomorpholinophenylalanine (Tmf), D and L-4-Aminomethyl- phenylalanine (4Amf), L-4-isopropyl-aminomethylphenylalanine (4IAmf), L-4-Isopropyl-amino-phenyl-alanine (4IAph) and N^{α}-methyl-4-amino-phenylalanine (NMe4Aph). Their additional distal amino groups were protected either by Fmoc or by Z except for D- and L-Tmf which have distal tertiary amino groups.

Results and Discussion

The desired phenylalanine derivatives were synthesized following five synthetic routes and their structures shown in Figure 1. D- and L-N^{α}-Boc-N^{ω}-Fmoc-3Aph [2] were prepared via condensation of 3-nitrobenzyl chloride with diethyl acetamidomalonate, followed by resolution with α-chymotrypsin, hydrogenation and Fmoc-protection of the ω-amino groups. D- and L-N^{α}-Boc-Tmf [3] were prepared via chloromethylation and bromomethylation of D- and L-N-acetyl-phenylalanine ethyl ester respectively, followed by amination (thiomorpholine), thorough hydrolysis and Boc-protection of α-amino groups. D- and L-N^{α}-Boc-N^4-Fmoc-4Amf were prepared with an improved method [4]. The synthesis was started from trichloro- or trifluoro-acetamidomethylation of D- and L-phenylalanine, followed by N^{α}-Boc-protection, selective deprotection of 4-amino groups by 20% sodium hydroxide in a mixture of methanol and water (v/v, 1:1) for 0.5 hr with subsequent Fmoc-protection of the exposed 4-NH$_2$. L-N^{α}-Boc-N^4-Z-4IAmf and L-N^{α}-Boc-N^4-Z-4IAph [5] were prepared via reductive isopropylation of L-N^{α}-Boc-4Amf and L-N^{α}-Boc-4Aph respectively and Z-protection of the resulting secondary amino groups. L-N^{α}-Boc-N^4-Fmoc-NMe4Aph was prepared [6] via N^{α}-methylation of N-Boc-phenylalanine, nitration, hydrogenation and Fmoc-protection of the 4-amino group.

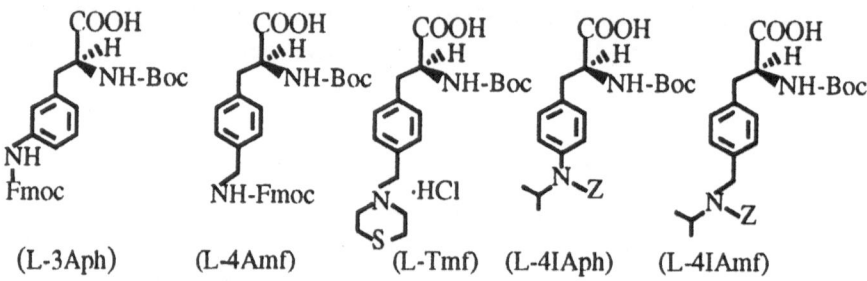

(L-3Aph)   (L-4Amf)   (L-Tmf) (L-4IAph)   (L-4IAmf)

Fig. 1. Partial structure of novel phenylalanine derivatives.

Table 1 Biological characterization of GnRH antagonists with novel amino acids at position 5, 6 and/or 8

Ac-D2Nal[1]-D4Cpa[2]-D3Pal[3]-Ser[4]-Aaa[5]-Aaa[6]-Leu[7]-Aaa[8]-Pro[9]-DAla[10]-NH$_2$	in vitro relative potencies[a]	AOA[b]	in vitro histamine release[c]
1. [Lys[5](Atz),DLys[6](Atz),ILys[8]] **Azaline A**	0.23 (0.15-0.36)	2.0(1/10), 2.0(8/20)*	139 ±8.7
2. [Orn[5](Atz),DOrn[6](Atz),ILys[8]]	0.2 (0.1-0.3)	2.0(1/10)	158 ±10
3. [4Aph[5](Atz),D4Aph[6](Atz),ILys[8]] **Azaline B**	1.3 (0.8-2.1)	0.5(7/9), 1.0(0/7)	224 ±23
4. [4Amf[5](Atz),D4Amf[6](Atz),ILys[8]]		2.5(0/7)	
5. [3Aph[5](Atz),D3Aph[6](Atz),ILys[8]]	1.5 (0.9-2.8)	2.5(0/5)	17
6. [4Aph[5](Atz),D4Aph[6](Atz),4IAph[8]]		10(2/2)	
7. [4Aph[5](Atz),D4Aph[6](Atz),4IAmf[8]]	1.6 (1.0-2.6)	2.5(3/7)	
8. [NMe4Aph[5](Atz),D4Aph[6](Atz),ILys[8]] **Azaline C**	1.9 (0.85-2.2)	1.0(2/9), 2.5(0/8)	72
9. [NMe4Aph[5](Atz),D4Aph[6](Atz),4IAph[8]]		10(2/2)	
10. [NMe4Aph[5](Atz),D4Aph[6](Atz),4IAmf[8]]	1.5 (1.0-2.3)	5.0(0/6)	
11. [Tmf[5],DTmf[6],ILys[8]]		2.5(0/5)	
12. [Tmf[5],DTmf[6],Tmf[8]]		50(7/8)	
13. [Tyr[5],DPal[6],Tmf[8]]		10(0/3)	
14. [4Aph[5](Ser),D4Aph[6](Ser),ILys[8]]		1.0(4/5)	
15. [4Aph[5](DSer),D4Aph[6](DSer),ILys[8]]		2.5(2/5)	
16. [4Aph[5](Ac-Ser),D4Aph[6](Ac-Ser),ILys[8]]		1.0(4/7)	
17. [4Aph[5](Ac-DSer),D4Aph[6](Ac-DSer),ILys[8]]	0.5 (0.4-0.7)	1.0(2/7)	

[a]Relative to [Ac-Δ^3Pro[1],DFpa[2],DTrp[3,6]]GnRH = 1.0. [b]AOA = antiovulatory assay: dosage in micrograms/rats (rat ovulating/total), peptides were dissolved in ca.1% DMSO/saline or (*) in corn oil. [c]ED$_{50}$ ± SEM, μg/mL. ED$_{50}$ for [Ac-DNal[1],DFpa[2], DTrp[3],DArg[6]]GnRH (internal standard) was 0.17 ± 0.01 μg/mL. Peptides fully active at 10 μg, 2.5 μg and 1.0 μg were only partially active at 5.0 μg, 1.0 μg or 0.5 μg, respectively.

The difference between peptide **1** and **2** is the shortening of the side chains at positions 5 and 6 with no significant effect on biological potencies. Peptide **3** on the other hand is significantly more potent in the AOA and in the castrated male rat assay where it was found to be considerably longer acting than peptide **1** [1]. We suggested that this difference resulted from the presence of the aromatic ring which shielded access to the backbone against enzymatic hydrolysis. Introduction of a methylene group on the para position of the phenyl ring (peptide **4**) and introduction of the triazolyl function on meta-aminophenylalanines (peptide **5**) at positions 5 and 6 resulted in a significant lowering of potency in the AOA. Because ILys at position 8 was recognized to maintain AOA potency while contributing to a major reduction of the histamine releasing activity, we investigated the possibility to substitute it by an aromatic containing amino acid such as 4IAph or 4IAmf (peptides **6** and **7**). In both cases, considerable loss of potency resulted from these substitutions. We also synthesized three analogs (**8-10**) containing NMe4Aph in position 5 based on the observation of Haviv et al. [7] that N-methylation at position 5 conferred increased solubility. Analog (**8**) was more soluble, however, significantly less potent in the AOA and furthermore released histamine at a significantly lower concentration than (**3**). Introduction of IAph and IAmf at position 8 of the NMe substituted Azaline B (**9, 10**) also resulted in further loss of potency. Another approach to increasing water solubility was to increase basicity at positions 5, 6 and 8 (peptides **11-13**) by the introduction of one or several Tmf residue(s) which encompass a thiomorpholino moiety. Analogs **12** and **13** were significantly less potent while **11** was fully potent at 2.5 µg in the AOA. Finally, we investigated the possibility of increasing solubility by the introduction of a D- or L-serine (or an Acetyl-D- or L-serine) on the 4Aph side chain at positions 5 and 6 of Azaline B (peptides **14-17**). These peptides were also less potent than Azaline B.

Acknowledgements

This work was supported in part by NIH under Contract N01-HD-9-2903, the Hearst Foundation and the World Health Organization (WHO) Research Training Grant funded by the United Nations Fund for Population Activities (UNFPA) under Project CPR/90/P.25. We acknowledge the outstanding technical contributions of Ron Kaiser, Laura Cervini, Charleen Miller and Duane Pantoja. We thank Dr. A.G. Craig for the mass spectra.

References

1.   Rivier, J.E., Porter, J., Hoeger, C., Theobald, P., Craig, A.G., Dykert, J., Corrigan, A., Perrin, M., Hook, W.A., Siraganian, R.P., Vale, W. and Rivier, C., J. Med. Chem., 35 (1992) 4270.
2.   Porter, J., Dykert, J. and Rivier, J., Int. J. Peptide Protein Res., 30 (1987) 13.
3.   He, B., Liu, K.-L. and Shaobo, X., Chinese Sci. Bull., 22 (1988) 1712.
4.   Stokker, G.E., Hoffman, W.F. and Homnick, C.F., J. Org. Chem., 58 (1993) 5015.
5.   Augustine, R.B., Catalytic Hydrog., M. Dekker, Inc., New York, NY, U.S.A., 1965, p.102.
6.   Cheung, S.T. and Benoiton, N.L., Can. J. Chem., 55 (1977) 906.
7.   Haviv, F., Fitzpatrick, T.D., Nichols, C.J., Swenson, R.E., Mort, N.A., Bush, E.N., Diaz, G., Nguyen, A.T., Holst, M.R., Cybulski, V.A., Leal, J.A., Bammert, G., Rhutasel, N.S., Dodge, P.W., Johnson, E.S., Cannon, J.B., Knittle, J. and Greer, J., J. Med. Chem., 36 (1993) 928.

Role of Ile^8 on the antibacterial and channel forming properties of shortened cecropin A - melittin hybrids

P. Juvvadi[1], E.L. Merrifield[1], D. Andreu[2],
H.G. Boman[3] and R.B. Merrifield[1]

*[1]The Rockefeller University, New York, NY 10021, U.S.A.
[2]The University of Barcelona, Spain, and [3]Arhennius Laboratories,
Stockholm University, S-10691 Stockholm, Sweden*

Introduction

The cecropins are a class of broad spectrum antibacterial peptides produced by *Hyalophora cecropia* [1, 2], while the bee venom toxin, melittin resembles cecropins with a reversed polarity but with the ability to lyse the red cells [3, 4]. Our interest in the design and synthesis of chimeric peptides that are cecropin-melittin hybrids is mainly due to the similarities in the structure of these two peptides but differences in their lytic properties. Our earlier results on 15-residue analogs with essentially comparable potency to the larger size peptides has shown that CA(1-7)M(2-9) was the most active [5]. We presumed that the presence of Ile^8 has an important effect on the antibacterial activity and studied the role of Ile^8 in the 15-residue analogs towards five test bacteria. The antibacterial activity is also correlated to the formation of voltage dependent ion-channels in planar lipid membranes. The mechanism of action of these peptides appears to be similar on bacteria, red blood cells and artificial lipid bilayers.

Results and Discussion

The primary sequence of the most active peptide CA(1-7)M(2-9) is considered and seven hybrid peptides were synthesized by a combination of manual and automated (Applied Biosystems 320A) solid phase techniques. Deletion of Leu^{15}, **2**, decreased activity, but addition of Thr^{16}, **3**, maintained good activity. The presence of Ile^8 appears to have dramatic effect on the antibacterial activity, which is shown by the reduction of activity of 14-residue analog **4**, obtained by the deletion of Ile^8 and was not restored by lengthening to 15-residues by addition of Thr at the C-terminus. The presence of Ile^8 makes a significant contribution to the bacterial activity and the length of the analogs need not be restored at 15-residues.

In addition replacement of Ile^8 by hydrophobic Leu residue, **6**, maintained good activity and with Ala^8, analog **7**, was equally active for four organisms but less active against *Staphylococcus aureus*. A hydrophobic residue seems to be essential at position 8 to maintain the antibacterial activity, which is also confirmed by the reduced potency against all five organisms, when hydrophilic residue Ser replaced Ile^8 in **8**. The requirement for the interaction of the peptide with the bacterial membrane appears to be an appropriate hydrophobic environment.

The conductivity measurements on planar lipid bilayers were performed, as a function of voltage gradient and composition of added peptide antibiotic [6], which demonstrated the correlation of antibacterial activity with the voltage dependent channel formation. The 14-residue analog **2** without any significant antibacterial activity does not form channels even at a higher concentration, while **3** possessing Ile[8] interacts with the membrane and forms pores, which span the membrane. The analog **6** with the hydrophobic Leu residue forms channels, while the presence of hydrophilic Ser in analog **8** does not show the channel formation by increasing the concentration ninefold as compared to **6**. These results explain the involvement of Ile[8] in the interaction with the hydrophobic lipid core of the membrane which eventually produces a hollow conducting pore and spans the membrane.

Table 1 *Lethal and lysis concentrations (μM) for cecropin-melittin hybrids*

Peptide	Size	D21	OT 97	Bs 11	Sp 1	Sac 1	SRC Lysis
1) CA(1-7)M(2-9)	15	1.0	4.0	0.5	0.3	4.0	>300
2) CA(1-7)M(2-8)	14	8.6	7.5	8.0	0.5	14.8	>280
3) CA(1-7)M(2-10)	16	1.4	3.7	2.7	1.1	0.87	>400
4) CA(1-7)M(3-9)	14	3.9	14	10	0.88	2.9	>320
5) CA(1-7)M(3-10)	15	9.1	14	6.3	1.2	>200	>600
6) CA(1-7) L^2 M(2-9)	15	1.6	4.5	7.5	0.76	0.88	>360
7) CA(1-7) A^2 M(2-9)	15	1.5	3.5	5.2	0.47	13.5	>200
8) CA(1-7) S^2 M(2-9)	15	9.7	16.2	28.6	17.8	89.2	>300

[a]Lethal concentrations calculated from inhibition zones on agarose plates seeded with the respective organisms: D21: *E. coli*; OT 97: *P. aeruginosa*; Bs 11: *B. subtilis*; Sp 1: *S. pyogenes*; Sac 1: *S. aureus*; SRC: Sheep red cells lysis.

References

1. Steiner, H., Hultmark, D., Engstrom, A., Bennich, H. and Boman, H.G., Nature, 292 (1981) 246.
2. Boman, H.G. and Hultmark, D., Annu. Rev. Microbiol., 41 (1987) 103.
3. Habermann, E. and Jentsch, J., Z. Physiol. Chem., 348 (1967) 37.
4. Terwilliger, T.C. and Eisenberg, D., J. Biol. Chem., 257 (1982) 6016.
5. Andreu, D., Ubach, J., Boman, A., Wahlin, B., Wade, D., Merrifield, R.B. and Boman, H.G., FEBS Lett., 296 (1992) 190.
6. Christensen, B., Fink, J., Merrifield, R.B. and Mauzerall, D., Proc. Natl. Acad. Sci. U.S.A., 85 (1988) 5072.

Synthetic apolipoprotein C-I peptide inhibition of cholesteryl ester transfer protein activity in baboon plasma

P. Kanda[1], R.G. Dunham[1], S.Q. Hasan[2],
S.T. Weintraub[3] and R.S. Kushwaha[2]

*Departments of [1]Virology/Immunology
[2]Physiology and Medicine, Southwest Foundation for Biomedical
Research, San Antonio, TX 78227, U.S.A.
[3]Department of Pathology, University of Texas Health Sciences Center,
San Antonio, TX 78284, U.S.A.*

Introduction

The baboon, due to its similarities with humans in lipoprotein metabolism, represents a useful primate model for the investigation of dyslipoproteinemias resulting in atherosclerotic syndromes. Certain baboon families, when challenged with a high cholesterol-high saturated fat (HCHF) diet, display a plasma lipoprotein profile featuring an elevated level of a high density lipoprotein (HDL) species, designated HDL_1, which is rich in cholesteryl ester (CE) [1]. These baboons also appear to be protected against atherosclerotic processes. We have determined that impaired net CE transfer between this HDL_1 species and LDL+VLDL in high HDL_1 baboons is due to inhibition of the cholesteryl ester transfer protein (CETP) by the amino terminal 4.2 kD fragment of baboon apolipoprotein C-I (apo C-I) [2]. A 38 amino acid synthetic peptide corresponding to this region is also inhibitory toward CE transfer in an *in vitro* assay [2]. In this study, we have examined the effects of apo C-I peptide length and amino acid substitution on this inhibition.

Results and Discussion

The N terminal 38 amino acid baboon apo C-I sequence and its variants were synthesized by SPPS on a polyamide support using tBoc chemistry (2). All peptides gave satisfactory values when analyzed by electrospray ionization MS. The sequence of the 38-mer is APDVSSALDKLKEFGNTLEDKAWEVINRIKQSEFPAKT and is predicted to have a high degree of amphipathic helical content. When added to an *in vitro* assay measuring the CETP-catalyzed transfer of [^{3}H] cholesteryl esters from HDL to LDL+VLDL, CETP activity was maximally inhibited (40% of total transfer) at a peptide concentration of 10 nmol/ml. A truncated version of this sequence, amino acids 18-38, was only weakly inhibitory (<5%), as was the plasma-isolated full length 57 AA apo C-I. A 16 amino acid control peptide of sequence YEALEKALKEALAKLG had no effect. The 38-mer peptide, when added to lipoproteins from low HDL_1 baboons (which have normal CETP activity), was also able to associate with apo A-I in the HDL to yield a dimer of apo A-I - apo C-I peptide of 31 kD (gel electrophoresis). This "modified" apo A-I, when isolated from baboon plasma, was able to inhibit CETP activity at the same concentrations as the apo C-I 38-mer, suggesting that the modified apo A-I was the actual inhibitory species *in vivo*. The modification of apo A-I by the apo C-I 38-mer was confirmed by immunoblotting with antibodies specific for the apo C-I peptide.

Table 1 *Inhibition of CETP by human-baboon hybrid apo C-I peptides*[a]

Peptide		avg MH		% inhibition of
		Predicted	Found	CE transfer [a]
None				0
Baboon	Apo C-I 1-38	4276.9	4277.3	42
Human	Apo C-I 1-38	4231.8	4231.8	27
	Apo C-I R	4245.9	4245.3	36
	Apo C-I N	4258.9	4259.0	0
	Apo C-I W	4275.9	4276.2	20
Control				0

[a]Assays were also performed at 4° (with added CETP) to measure uncatalyzed transfer as described [2]. We have consistently found that the maximal CE transfer inhibition achievable with excess baboon apo C-I peptide is about 40-50% of total transfer in 1 hr.

We were interested as to whether the analogous region of human apo C-1 could inhibit CETP activity, and thus potentially provide a way to raise HDL levels in plasma. The human apo C-I 1-38 peptide and 3 "hybrid" human-baboon peptides, as shown below, were synthesized and tested in the baboon CETP transfer assay.

```
Baboon C-I    APDVSSALDKLKEFGNTLEDKAWEVINRIKQSEFPAKT
Human C-I     T------------------------------------ R-- L-S -----------LS--- Nle
      C-I N   ------------------------------------------R ----------------------- Nle
      C-I R   ------------------------------------------ R---L-S ------------------Nle
      C-I W   ------------------------------------------L-S ------------------Nle
```

As seen in Table 1, the human apo C-I peptide and 2 of the variants were able to significantly inhibit baboon CETP activity, while the C-I N sequence bearing the W→R and T→Nle substitutions was inactive. This suggests that the W→R non-conserved substitution alone may abolish inhibitory properties, but it cannot explain the partial inhibitory activities of the other sequences. Experiments are underway to evaluate these apo C-I peptides in CETP assays using human plasma lipoproteins and to determine the mechanism of CETP inhibition.

References

1.   Kushwaha, R.S., Rainwater, D.L., Williams, M.C., Getz, G.S. and McGill, H.C., Jr., J. Lipid Res., 31 (1990) 965.
2.   Kushwaha, R.S., Hasan, S.Q., McGill, H.C., Jr., Getz, G.S., Dunham, R.G. and Kanda, P., J. Lipid Res., 34 (1993) 1285.

Novel antitumor peptide hormones with selective antitumor activity

Gy. Kéri[1], I. Mezõ[1], A. Horváth[1], Zs. Vadász[1], T. Bajor[1],
M. Idei[1], T. Vántus[1], I. Teplán[1], J. Horváth[3], O. Csuka[2],
R.I. Nicholson[4] and B. Szende[1]

*1st Inst. of Biochem., [1]Exp. Cancer Res. Inst., Semmelweis Univ.
Med. Sch. 1444 Budapest 8. POB. 260.
[2]National Inst. of Oncology, [3]Med. Univ. P,cs, Hungary,
[4]Tenovus Cancer Research Institute, Cardiff, U.K.*

Introduction

The concept that carcinogenesis can be conceived as a distortion in signal transduction suggests a variety of new strategies for the development of novel antitumor drugs. These strategies include the use of peptide factors and mimetic drugs. We have developed a series of novel GnRH and Somatostatin analogs aimed for selective antitumor activity.

It is generally assumed that both in the case of GnRH and Somatostatin, different structural elements of the peptide molecule are responsible for the wide diversity of biological functions, for example, via the determination of certain preferred conformations. On the other hand, in these relatively small peptides there must be substantial overlapping between the structural elements or residues being responsible for the various biological functions. According to this concept, significant effort has been directed for developing GnRH and Somatostatin analogs with selective biological activity, mostly from an endocrine point of view since the antitumor activity of these peptide hormones was considered to be mediated mostly by their endocrine effect. The new concept of signal transduction modulation based antitumor drug development raised the possibility that the antitumor activity of these peptides is due to a direct effect through the activation or inhibition of certain signal transduction pathways and enhancement of the so called programmed cell death (apoptosis), respectively.

Results and Discussion

In our earlier work we have developed a Gonadotropin releasing hormone analog, (Folligen: D-Phe[6], Gln[8], desGly[10]-GnRH-ethylamide) which has antitumor activity on DMBA induced mammary carcinoma in rats without blocking ovarian functions. Folligen exerts antitumor activity via a different mechanism of action than the known superactive mammalian GnRH analogs, which act through desensitization and "hormonal castration" [1]. In *in vitro* experiments on the MDA-MB-231 human breast cancer cell line Folligen was found to be slightly more active in inhibiting cell proliferation than Buserelin (/D-Ser-(tBu)/[6], desGly[10]-GnRH-ethylamide; Hoechst), and significant differences were found in the signal transduction pathways activated by these analogs. Both analogs inhibited tyrosine kinases, Buserelin being more active, while Folligen stimulated phospholipid turnover and caused a larger shift in the protein kinase C activity from the cytosol to the membrane.

A further variant of Folligen, /L-β-Asp-(α-DEA)/^{6}, Gln^8-GnRH; Folligen II, has only a very small effect on LH release (less than native GnRH) in superfused rat pituitary cells, while it has a strong antitumor activity and induces apoptosis in human breast and prostate tumor cells. In PC3 human prostate tumor cells 10 μg/ml peptide during a 24 hr incubation strongly inhibited proliferation, increased apoptosis and decreased mitosis. The number of apoptotic cells increased more than 5 times in a 24 hr incubation and it remained high for 72 hr.

There has been considerable interest recently in the antitumor activity of another peptide hormone, somatostatin, because it should also be considered as a general growth inhibitory peptide and regulator of signal transduction.

The wide diversity of biological effects of somatostatin and their physiological significance led to defining an octapeptide structure as the main feature responsible for the various biological functions. Analogs of this octapeptide usually proved to be longer-acting than the native hormone in *in vivo* assays but only some improvement was achieved in the selectivity of their biological effects. We have developed a series of somatostatin analogs which strongly inhibited tyrosine kinase activity and had antitumor activity both *in vitro* and *in vivo* [2]. Some of these analogs had no effect on GH release, neither *in vitro* nor *in vivo*. One of the analogs containing a five residue ring structure (D-Phe-Cys-Tyr-D-Trp-Lys-Cys-Thr-NH$_2$, *TT-232*) was found to be unique, having no GH release inhibitory activity while it had strong tyrosine kinase inhibitory and antiproliferative effect. We also have demonstrated that these analogs exert their direct antitumor activity via signal transduction modulation and the induction of apoptosis. The induction of apoptosis with this analog was very much dose dependent, and the effect became dramatic at 20 μg/ml dose, during a 24 hr incubation.

Acknowledgements

This work was supported by grants of OTKA 2618, 2994 and by BioSignal Ltd.

References

1. Nicholson, R.I. and Kéri, Gy., Tumor Biology, 13 (1992) 44.
2. Kéri, Gy., Mezõ, I., Horváth, A., Vadász, Zs., Balogh, Á., Idei, M., Vántus, T., Teplán, I., Mák, M., Horváth, J., Pál, K. and Csuka, O., Biochem. Biophys. Res. Commun., 191 (1993) 681.

Human serum amyloid P component (hSAP): Binding to and degradation by human polymorphonuclear leukocytes

P. Landsmann[1], O. Rosen[1], M. Pontet[2],
E.G. Shephard[3] and M. Fridkin[1]

[1]Department of Organic Chemistry, the Weizmann Institute of Science,
Rehovot, Israel 76100; [2]UER Biomedicale des Saints-Peres,
Laboratoire de Chimie Biologie, Rue des Saints-Peres, 75270, Paris
Cedex 06, France; [3]MRC Liver Research Center, University of Cape
Town Medical School, Cape Town, Republic of South Africa

Introduction

Recent studies concerning immunorelevant properties of human pentraxins, namely C-reactive protein (hCRP) and serum amyloid P component (hSAP), strongly indicate that both hCRP and hSAP appear to interact with a variety of ligands both *in vitro* and *in vivo* and thus might be involved in a number of defensive mechanisms of the body [1, 2, 3]. Particularly, there are numerous reports concerning hCRP receptor-mediated binding to and interaction with phagocytic cells [4, 5]. Aggregated hSAP (AP) was identified as a constituent in different pathological amyloid deposits [3]. The plasma circulating form of hSAP is still obscure. The possibility that free hSAP molecules exist in plasma has been questioned because of the tendency of this protein to aggregate at calcium concentrations exceeding 1.1 mM [3]. Studies on interactions of aggregated or soluble forms of hSAP with phagocytic cells have not been, as yet, reported.

Results and Discussion

Displacement experiments with ^{125}I-labeled hSAP (Figure 1, panel A) and hCRP (Figure 1, panel B) show that both proteins have specific high-affinity binding sites on normal human polymorphonuclear leukocytes and each can compete efficiently with the binding of the other. Scatchard analysis of hSAP homologous displacement curves reveals two populations of hSAP binding sites existing on the PMN cells (Figure 1, panel C). Further, AP was found to be degraded by enzymes from human neutrophilic origin to yield a mixture of low molecular weight peptides (LMWP), similarly to the case of CRP reported previously [5]. The binding of hSAP can be inhibited by this LMWP mixture (Figure 1, panel D). The sequence of some of the peptides present in this mixture was identified. The peptide VIIKPL inhibited superoxide production, in Cytochrome C assays, by fMLP-stimulated human neutrophils at a concentration of 50 nM (13.8 ± 10.4%). This peptide showed as well concentration-dependent inhibitory effect in fMLP-induced lucigenin enhanced chemiluminescence of leukocytes. At the concentrations 400 and 800 nM maximal values of inhibition (52 and 60%, respectively) were attained. The results indicate that both hCRP and hSAP, together with related peptides, may participate *in vivo* in certain mechanisms of regulation of human neutrophils.

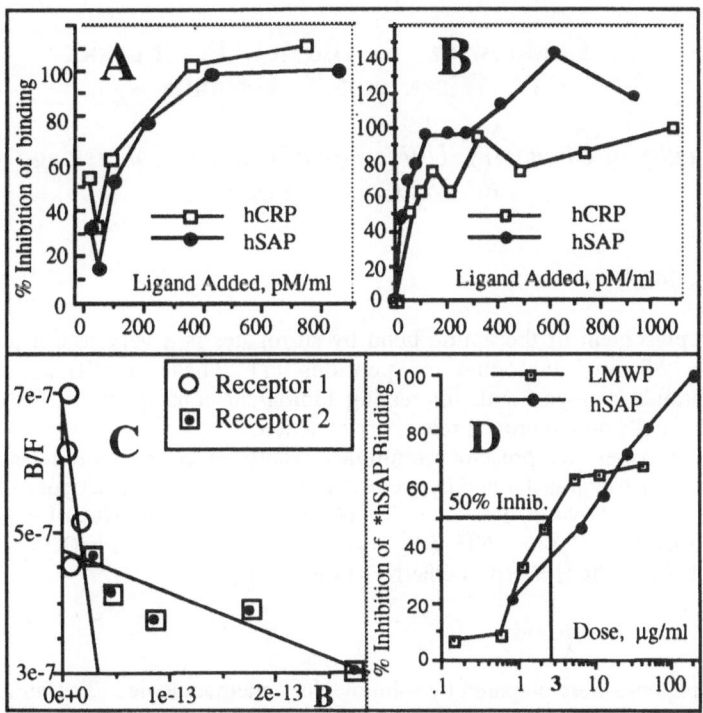

Fig. 1. Displacement experiments with ^{125}I-labeled human pentraxins. Panel A, B - Displacement of ^{125}I-hSAP (Panel A) and ^{125}I-hCRP (Panel B) by unlabeled hSAP and hCRP, in PBS containing 1 mM calcium and 0.2 w.% BSA. In both cases, maximal inhibition of protein total binding by the same unlabeled protein is taken as 100% inhibition. Panel C - Scatchard plot of ^{125}I-hSAP displacement by unlabeled hSAP. Receptor 1: $K_d = 4.0 \times 10^{-8}$, 30,000 sites per cell; receptor 2: $K_d = 7.5 \times 10^{-7}$, 270,000 sites per cell (mean of 5 experiments). Panel D - Displacement of ^{125}I-hSAP by the proteolytic mixture generated from hSAP by neutrophil membrane protease at pH 7.4.

References

1. Kottgen, E., Hell, B., Kage, A. and Tauber, R., J. Immunol., 149 (1992) 445.
2. Loveless, R.W., Floid-O'Sullivan, G., Raynols, J.G., Yuen, C.-T. and Feizi, T., EMBO J., 11 (1992) 813.
3. Schwalbe, R.A., Dahlback, B., Coe, J.A. and Nelsestuen, G.L., Biochemistry, 31 (1992) 4907.
4. Tebo, J.M. and Mortensen, R.F., J. Immunol., 144 (1990) 231.
5. Rosen, O., Landsmann, P., Pras, M., Levartovsky, D., Pontet, M., Shephard, E.G. and Fridkin, M. In Smith, J.A. and Rivier, J.E. (Eds.), Peptides: Chemistry and Biology, (Proceedings of the Twelfth American Peptide Symposium), Escom, Leiden, The Netherlands, 1992, p.879.

Conformational study of pseudopeptide analogues of enkephalin

L. Lankiewicz, S. Oldziej, P. Skurski, W. Wiczk and Z. Grzonka

Faculty of Chemistry, University of Gdansk, Sobieskiego 18, 80-952 Gdansk, Poland

Introduction

A replacement of the amide bond by surrogates is a very useful tool in the designing of new analogues of peptides [1]. These modifications cause conformational changes with interesting biological consequences, for example, increased activity and improved selectivity of action.

In this paper, we present preliminary results of conformational studies by fluorescence energy transfer and by molecular mechanics on the analogues of leucine-enkephalin: Tyr-Gly-Gly-Phe-Leu-NH$_2$ (I), Tyr-Gly-Gly-Phe(NO$_2$)-Leu-NH$_2$ (II), Tyr-Gly-Sar-Phe(NO$_2$)-Leu-NH$_2$ (III), Tyr-GlyΨ[CH$_2$CH$_2$]Gly-Phe(NO$_2$)-Leu-NH$_2$ (IV) and Tyr-GlyΨ[CH=CH]Gly-Phe(NO$_2$)-Leu-NH$_2$ (V).

Results and Discussion

All peptides were prepared by solution phase methodologies. The final products were purified and examined on RP-HPLC and characterized by amino acid analysis and FD-MS. The fluorescence spectra were obtained using a Perkin Elmer LS-50 spectrofluorimeter. The theoretical calculations were carried out using MMX force-field as implemented within commercial PCModel program [2].

Table 1 *Fluorescence and theoretical parameters for the series of enkephalin analogues*

	II	III	IV	V
Q.Y. x10^3	3.38	4.91	2.95	2.23
E$_T$	0.913	0.874	0.924	0.948
r$_{DA}$ [Å]^a	10.4±0.2	11.15±0.2	10.14±0.2	9.65±0.3
r$_{DA}$ [Å]^b	12.07	13.02	10.92	10.16

Q.Y. - fluorescence quantum yield
E$_T$ - energy transfer efficiency
r$_{DA}$ - average distance between donor and acceptor
[a] - calculated from fluorescence measurements
[b] - calculated from molecular mechanics data

The quantum yield of tyrosine in leucine-enkephalin analogue (I) Q.Y. = 0.039, is in a good agreement with the results obtained previously [3]. The replacement of Phe[4] by Phe(NO$_2$) causes an additional strong tyrosine fluorescence quenching in the analogues (II)-(V); quantum yield of tyrosine is about 10 times lower for these analogues in comparison with the analogue (I) (Table 1 and Figure 1).

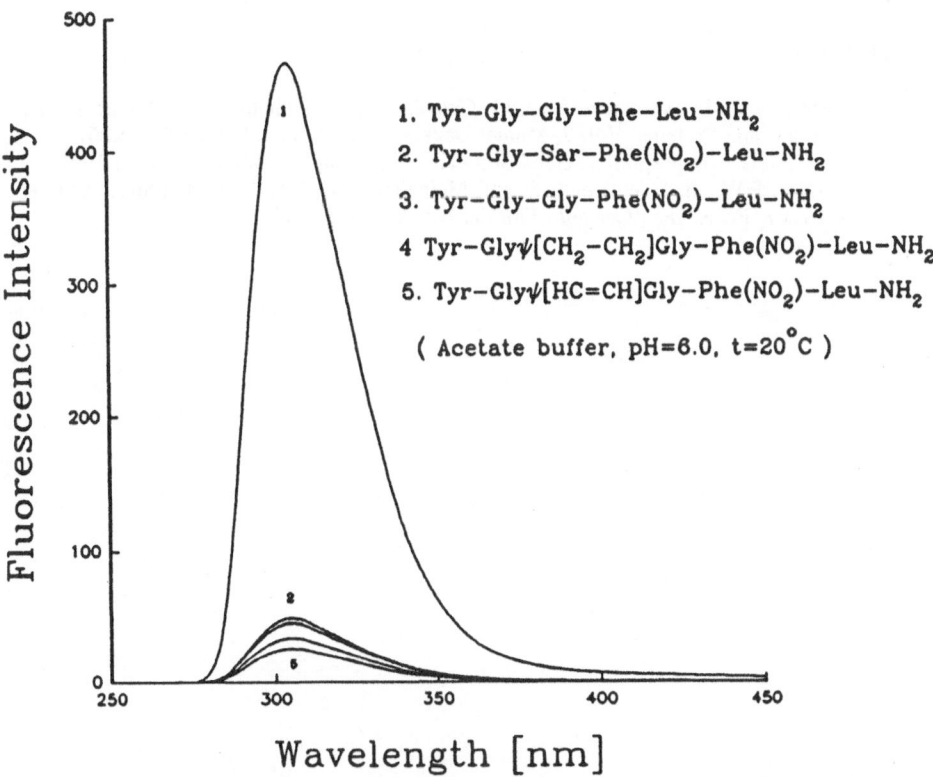

Fig. 1. Emission spectra of the enkephalin analogues.

We calculated also an average distance between chromophores (tyrosine and modified phenylalanine) in the analogues (II)-(V). For the analogue (II), without peptide bond modification, the average distance (10.4±0.2Å) is very similar to the values known from literature [3]. We did not observe very significant differences in an average distance between chromophores in the analogues containing the surrogates (Table 1). However, we found an increase of the average distance for the analogue (III). This suggests a more extended conformation for the analogue (III) than for the parent compound (II). On the other hand, we observed a decrease of the average distance between the chromophores for the analogues (IV) and (V) (Table 1). The same order of changes of the average distances were obtained from molecular mechanics calculations (Table 1). These results do not demonstrate very significant influence of surrogate between Gly[2] and Gly[3] residues for an average conformation of enkephalin. Comparison of these conformational changes with biological activity are in process.

Acknowledgements

Supported by Polish State Committee for Scientific Research (KBN) under grant: 1981/4/91. Participation of L. Lankiewicz supported by American Peptide Society.

References

1. Spatola, A.F., In Weinstein, B. (Ed.), Chemistry and Biochemistry of Amino Acids, Peptides, and Proteins, Vol. 7, Marcel Dekker, New York, U.S.A., 1983, p.267.
2. PCMODEL v.04 program, Serena Software, Bloomington, Indiana, U.S.A., 1991.
3. Schiller, P.W., In Udenfriend, S. and Meienhofer, J. (Eds.), The Peptides, Vol. 7, Academic Press, Inc., Orlando, Florida, U.S.A., 1985, p.115.

Structure-activity relationship study on GRP antagonists having a C-terminal $\underline{D}$ProΨ(CH$_2$NH)Phe-NH$_2$

J.J. Leban[1], A. Landavazo[1], J.D. McDermed[1], E.J. Diliberto, Jr.[2], M. Jansen[2], B. Stockstill[3] and F.C. Kull, Jr.[3]

[1]*Organic Chemistry, *[2]*Pharmacology, and *[3]*Cell Biology Divisions, Burroughs Wellcome Co., 3030 Cornwallis Road, Research Triangle Park, NC 27709, U.S.A.*

Introduction

We developed GRP antagonists using their ability to inhibit ^{3}H-Thymidine incorporation into quiescent Swiss 3T3 cells. Our analog design started with a reduced bond antagonist derived from the sequence of BN as developed by Coy et al. [1]. An octapeptide GRP analog having a C-terminal $\underline{D}$ProΨ(CH$_2$NH)Phe-NH$_2$ was a potent GRP antagonist and used as our lead structure (**1**).

Results and Discussion

In the original report by Coy et al. on GRP antagonists, both Ψ^{9,10}-BN and Ψ^{13,14}-BN were reported to have some antagonistic properties [1]. Therefore, we made **2** with Ψ^{9,10} and Ψ^{13,14} together, but we lose activity. The Statine analog **3** is almost as potent as our lead, **1**. When we introduced a "$\underline{D}$Pro-statine" (**4**), the activity drops somewhat. It is interesting to note that the stereochemistry of the hydroxyl group in the "Pro-statine" series (**4,5**) does seem to make a difference in binding activity. Replacement of Trp and His (**6,7**) with sulfur isosteres does retain some activity. Replacing Trp with BzthAla leads to a slightly active analog in the mitogenic assay (**6**); there is, however, an unexplained 1000-fold increase in the 3T3 membrane assay. Replacing histidine with ThiAla leads to a slightly less active compound (**7**); here the values of both assays match, as expected. In compounds **8** through **11**, we intended to investigate the influence of optically active aromatic acids at the N-terminal. Although there is some difference in the mitogenic assay, the 3T3 membrane assay shows no clear preference between the R and S isomers. We observed very little, if any, change in activity among the ortho-, meta-, and para-substituted aromatic rings of **12-17**, however, there is a clear preference for an aromatic propionic acid residue versus an aromatic acetic acid residue.

Compounds **22** and **23**, utilizing different aliphatic rings to fuse the position of the aromatic ring of **1**, exhibit a 10-fold increase in 3T3 membrane binding affinity. There was a small change in mitogenic activity for these analogs.

Compound **24** is the first of a series to explore variations at the C-terminal aromatic amino acid.

Table 1 GRP Antagonists

	A^a	B^b	C^c

1 (phenyl)-CH$_2$CH$_2$(CO)HisTrpAlaValDAlaHisDProΨPhe-NH$_2$
$\qquad$ 1.0×10^{-9} $\quad$ 1.0×10^{-12} $\quad$ 2.2×10^{-10}

2 (phenyl)-CH$_2$CH$_2$(CO)HisTrpAlaYValDAlaHisDProΨPhe-NH$_2$
$\qquad$ 4.7×10^{-8} $\qquad$ nt $\qquad$ $> 1.0\times10^{-6}$

3 (phenyl)-CH$_2$CH$_2$(CO)HisTrpAlaValDAlaHisStaPhe-NH$_2$
$\qquad$ 6.0×10^{-9} $\quad$ 1.0×10^{-10} $\quad$ 7.3×10^{-9}

4 (phenyl)-CH$_2$CH$_2$(CO)HisTrpAlaValDAlaHis[D—N(pyrrolidine)—CH(OH)CH$_2$(CO)]Phe-NH$_2$
$\qquad$ na $\qquad$ 6.0×10^{-8} $\quad$ 1.4×10^{-7}

5 (phenyl)-CH$_2$CH$_2$(CO)HisTrpAlaValDAlaHis—N(pyrrolidine)—CH(OH)CH$_2$(CO)Phe-NH$_2$
$\qquad$ 2.6×10^{-8} $\quad$ 5.0×10^{-8} $\quad$ 3.1×10^{-8}

6 (phenyl)-CH$_2$CH$_2$(CO)His(BzthAla)AlaValDAlaHisDProΨPhe-NH$_2$
$\qquad$ 4.3×10^{-7} $\qquad$ nt $\qquad$ 7.5×10^{-10}

7 (phenyl)-CH$_2$CH$_2$(CO)(ThiAla)TrpAlaValDAlaHisDProΨPhe-NH$_2$
$\qquad$ 5.0×10^{-8} $\quad$ 1.0×10^{-9} $\quad$ 5.6×10^{-9}

8 (R)-(phenyl)-CH(CH$_3$)CH$_2$(CO)HisTrpAlaValDAlaHisDProYPhe-NH$_2$
$\qquad$ 1.7×10^{-8} $\qquad$ nt $\qquad$ 4.9×10^{-10}

9 (S)-(phenyl)-CH(CH$_3$)CH$_2$(CO)HisTrpAlaValDAlaHisDProΨPhe-NH$_2$
$\qquad$ 1.6×10^{-10} $\quad$ 1.0×10^{-9} $\quad$ 2.1×10^{-10}

10 (R,-)-(phenyl)-CH(CH$_3$)(CO)HisTrpAlaValDAlaHisDProΨPhe-NH$_2$
$\qquad$ 9.0×10^{-10} $\quad$ 3.0×10^{-9} $\quad$ 9.5×10^{-10}

11 (S,+)-(phenyl)-CH(CH$_3$)(CO)HisTrpAlaValDAlaHisDProΨPhe-NH$_2$
$\qquad$ 7.0×10^{-9} $\quad$ 1.0×10^{-9} $\quad$ 1.3×10^{-9}

12 (H$_3$CO-phenyl)-CH$_2$CH$_2$(CO)HisTrpAlaValDAlaHisDProΨPhe-NH$_2$
$\qquad$ 3.0×10^{-10} $\quad$ 1.0×10^{-9} $\quad$ 4.7×10^{-10}

13 (F$_3$C-phenyl)-CH$_2$CH$_2$(CO)HisTrpAlaValDAlaHisDProΨPhe-NH$_2$
$\qquad$ 3.5×10^{-10} $\quad$ 1.0×10^{-9} $\quad$ 1.6×10^{-10}

Table 1 continued

	A^a	B^b	C^c
14 [structure: phenyl with CF$_3$]—CH$_2$(CO)HisTrpAlaVal$\underline{D}$AlaHis$\underline{D}$ProΨPhe-NH$_2$	3.3×10^{-7}	nt	1.2×10^{-8}
15 [structure: F$_3$C phenyl]—CH$_2$(CO)HisTrpAlaVal$\underline{D}$AlaHis$\underline{D}$ProΨPhe-NH$_2$	2.6×10^{-7}	nt	8.4×10^{-9}
16 [structure: F$_3$C phenyl]—CH$_2$(CO)HisTrpAlaVal$\underline{D}$AlaHis$\underline{D}$ProΨPhe-NH$_2$	2.0×10^{-7}	nt	3.7×10^{-9}
17 [structure: phenyl with NH$_2$]—CH$_2$(CO)HisTrpAlaVal$\underline{D}$AlaHis$\underline{D}$ProΨPhe-NH$_2$	1.3×10^{-8}	1.0×10^{-10}	2.0×10^{-9}
18 [structure: phenyl]—C(CH$_3$)$_2$CH$_2$(CO)HisTrpAlaVal$\underline{D}$AlaHis$\underline{D}$ProΨPhe-NH$_2$	1.8×10^{-8}	1.0×10^{-8}	1.2×10^{-9}
19 [structure: isoquinoline]—(CO)HisTrpAlaVal$\underline{D}$AlaHis$\underline{D}$ProΨPhe-NH$_2$	4.0×10^{-9}	nt	5.7×10^{-10}
20 [$\underline{D}$-[naphthyl]—CH(NH$_2$)CH$_2$(CO)HisTrpAlaVal$\underline{D}$AlaHis$\underline{D}$ProΨPhe-NH$_2$	4.5×10^{-8}	nt	7.7×10^{-9}
21 [structure: phenothiazine N-CH$_2$]—CH$_2$(CO)HisTrpAlaVal$\underline{D}$AlaHis$\underline{D}$ProΨPhe-NH$_2$	2.9×10^{-8}	nt	1.9×10^{-9}
22 [structure: tetrahydronaphthalene]—(CO)HisTrpAlaVal$\underline{D}$AlaHis$\underline{D}$ProΨPhe-NH$_2$	4.0×10^{-9}	3.0×10^{-10}	7.9×10^{-11}
23 [structure: phenyl cyclopropane]—"(CO)HisTrpAlaVal$\underline{D}$AlaHis$\underline{D}$ProΨPhe-NH$_2$	2.0×10^{-9}	3.0×10^{-11}	4.9×10^{-11}
24 [structure: phenyl]—CH$_2$CH$_2$(CO)HisTrpAlaVal$\underline{D}$AlaHis$\underline{D}$ProΨ($\underline{D}$2Nal)-NH$_2$	4.0×10^{-8}	5.0×10^{-9}	2.0×10^{-9}

[a]Mitogenic antagonism (M); [b]S3T3 cells binding (M); [c]S3T3 membrane binding (M)

References

1. Coy, D.H., Taylor, J.E., Jiang, N.-Y., Kim, S.H., Wang, L.-H., Huang, S., Moreau, J.-P., Gardner, J.D. and Jensen, R.T., J. Biol. Chem., 264 (1989) 14691.

Preparation of an iodinated bradykinin analogue with selectivity for a picomolar, high-affinity bradykinin binding site

C. Liebmann[1], R. Bossé[2] and E. Escher[2]

[1]*Institute of Biochemistry and Biophysics, Friedrich-Schiller-University, Philosophenweg 12, D-07743 Jena, Germany*
[2]*Department of Pharmacology, University of Sherbrooke, Sherbrooke Québec, J1H 5N4, Canada*

Introduction

The pharmacology of bradykinin (BK) receptors has been characterized extensively in binding studies by using [2,3-prolyl-3,4-^{3}H(N)]BK (^{3}H-BK) [1, 2] as well as ^{125}I-[I-Tyr^0]BK [3] or ^{125}I-[I-Tyr^8]BK [4]. However, for several tissues there are contradictory results postulating either a single BK binding site [4] or two binding sites, one with picomolar and one with nanomolar affinity constants [1, 2], respectively. The comparison among these studies is relatively difficult because of the variations between the experimental conditions and because of the lack of selective, high-affinity, radiolabeled ligands which might discriminate between the described multiple binding site. Here, we report the synthesis of ^{125}I-[p-I-Phe^5]BK (^{125}I-BK), an analogue with higher specific radioactivity and higher selectivity towards the high-affinity site compared to ^{3}H-BK. We have characterized the BK binding sites from the guinea pig ileum smooth muscle membranes with this new radioligand and the corresponding non-radioactive precursor. We have further compared the binding properties of ^{125}I-BK with those of ^{3}H-BK on the same tissue.

Results and Discussion

In order to obtain a labeled BK with higher specific activity than ^{3}H-BK, we have prepared iodinated BK by the direct introduction of ^{125}I into the para position of Phe^5 of the BK molecule. A control synthesis with p-I-Phe in the respective position was prepared. After solid-phase peptide synthesis with the BOC-TFA-HF strategy and cleavage of the peptide-resin ester, the crude peptides were purified by gel filtration, followed by reversed-phase (C$_{18}$) chromatography. Correct peptide structure was confirmed by FAB-MS and the purity by HPLC and TLC. The p-NO$_2$-Phe containing peptide was modified by catalytic hydrogenation, followed by diazotation and incubation in the presence of freshly precipitated nascent Cu together with 1 mCi of Na^{125}I (12 h at 4°C). After reversed-phase HPLC purification of the buffered reaction mixture, the radioactive peak was collected and an aliquot was co-eluted with the non-radioactive peptide. A perfect co-elution of ^{125}I-BK and I-BK was obtained in two different solvent systems (0.05% TFA and 0.01% HCl). The specific radioactivity was estimated using displacement curves with unlabeled I-BK on guinea pig ileum membranes and varied between 1000 and 1500 Ci/mmol.

Table 1 *Binding parameters of the radioligands ^{125}I-BK and ^3H-BK as well as of the unlabeled precursor I-BK in guinea pig ileum membranes*

Parameter[a]	^{125}I-BK	^{3}H-BK
K_D (pM)	$K_H = 3; K_L = 192$	$K_H = 24; K_L = 200$
B_{max} (fmol/mg protein)	$B_H = 22; B_L = 245$	$B_H = 75; B_L = 265$
IC_{50}-values (pM)		
BK	$K_H = 17; K_L = 375$	$K_H = 10; K_L = 2200$
I-BK	$K_H = 0.3; K_L = 1100$	$K_H = 5; K_L = 3100$

[a]H and L denote the high and the low affinity site, respectively. K_D and B_{max} were calculated using the LIGAND program. All results are the means of three independent experiments in duplicate determinations. The inter-assay variability was less than 10%.

As shown in Table 1, the new tracer displayed enhanced affinity and selectivity for the low picomolar, high affinity binding site (K_H) compared with ^{3}H-BK. The use of ^{125}I-BK as probe for BK receptors has enabled us to confirm the existence of two separate BK binding sites in guinea pig ileum smooth muscle membranes and will permit a better characterization of putative BK receptor subtypes.

Acknowledgements

This work was supported by grants from the Canadian Medical Research Council (CRMC) and by the Deutsche Forschungsgemeinschaft.

References

1. Manning, D.C., Vavrek, R., Stewart, J.M. and Snyder, S.H., J. Pharmacol. Exp. Ther., 237 (1986) 504.
2. Liebmann, C., Offermanns, S., Spicher, K., Hinsch, K.-D., Schmittler, M., Morgat, J.L., Reissmann, S., Schultz, G. and Rosenthal, W., Biochem. Biophys. Res. Commun., 167 (1990) 910.
3. Odya, C.E., Goodfriend, T.L. and Pena, C., Biochem. Pharmacol., 29 (1980) 175.
4. Tousignant, C., Guillemette, G., Barabe, J., Rhaleb, N.-E., Regoli, D., Hiley, C.R. and Stocket, J.-C., Can. J. Physiol. Pharmacol., 69 (1991) 818.

Monocytes augment the contractile potency of endothelin-1 via a thromboxane independent mechanism

H.I. Magazine[1], C.A. Bruner[2] and T.T. Andersen[3]

[1]Queens College and The Graduate School of the City University of New York, Department of Biology, Flushing, NY 11367, U.S.A Albany Medical College, [2]Pharmacology & Toxicology, [3]Biochemistry & Molecular Biology, Albany, NY 12208, U.S.A.

Introduction

Intimal migration of peripheral blood monocytes (PBM) is an early event in atherogenesis. PBM may contribute to the alterations in vascular responsiveness observed in atherosclerotic vessels by secretion of potent vasoactive mediators such as thromboxane (TxA_2). Since macrophages (Mø) release increased levels of TxA_2 following stimulation with the vasoconstrictor peptide endothelin-1 (ET-1) [1], we examined the ability of PBM to alter vascular contractile responses to ET-1.

Results and Discussion

The contractile potency of endothelin-1 (ET-1) was assessed by addition of ET-1 in 1/4-log increments to organ baths containing strips of guinea pig carotid artery in the presence and absence of PBM. The contractile potency of ET-1, $EC_{50} = 2.28 \pm 0.21$ nM, was increased markedly, in the presence of PBM, $EC_{50} = 0.46 \pm 0.06$ nM (Figure 1). Addition of PBM alone did not induce contraction. Mø have been reported to synthesize ET-1 [2]. Therefore, we examined the supernatant of ET-1 stimulated PBM for a vasoactive, transferable factor. Addition of supernatant from stimulated PBM did not result in contraction greater than that attributable to the stimulating dose of ET-1 alone (Figure 2). Thus, release of a transferable factor, such as ET-1, could not account for the increase in ET-1 potency. Treatment of strips and PBM with dazoxiben (TxA_2 synthetase inhibitor) or SQ 29,548 (TxA_2 receptor antagonist) did not diminish the effect of PBM on ET-1 potency, suggesting that the effect is independent of TxA_2 production.

PBM increased the vascular contractile potency of ET-1 via a TxA_2 independent mechanism. Although additional work is required to elucidate the actual mechanism by which ET-1 potency is altered, these data suggest that accumulation of monocytic cells at sites of inflammation may potentiate the vasoconstrictor effects of ET-1.

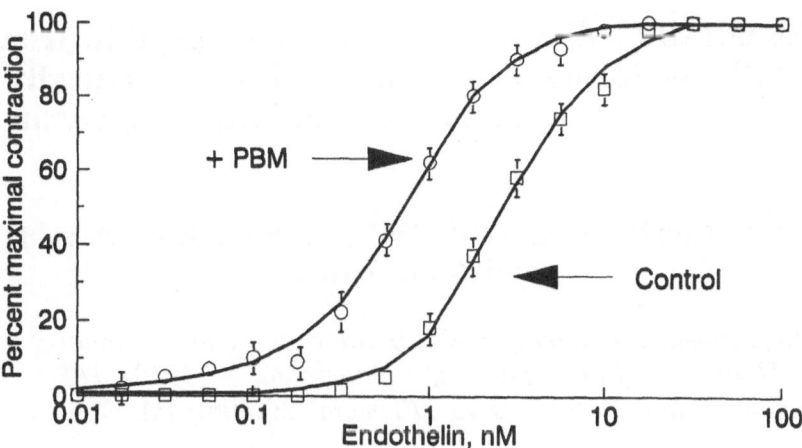

Fig 1. *Effect of PBM on vascular contraction induced by ET-1.*

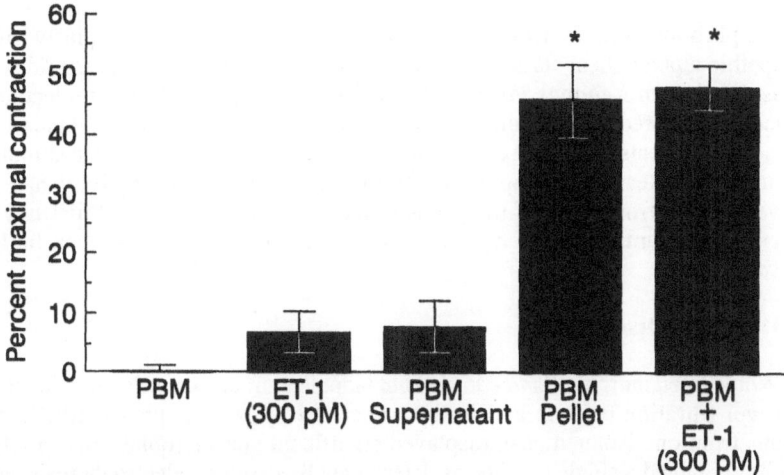

Fig. 2. *A transferable vasoactive factor is not produced by ET-1 stimulated PBM. Addition of stimulated PBM to the vascular strips resulted in a 300 pM ET-1 concentration in the organ bath. PBM + ET-1 represents ET-1-stimulated PBM that have not been separated into supernatant and pellet fractions.*

Acknowledgements

Supported by a grant from the City University of New York Awards Program, PSC-CUNY24 and American Heart Association 89-025G.

References

1.   Ninomiya, H., Yu, X.Y., Hasegawa, S. and Spannhake, E.W., Prostaglandins, 43 (1992) 401.
2.   Ehreinreich, H., Anderson, R.W., Fox, C.H., Rieckmann, P., Hoffman, G.S., Travis, W.D., Coligan, J.E., Kehrl, J.H. and Fauci, A.S., J. Exp. Med., 172 (1990) 1741.

Ranalexin: A novel antimicrobial peptide from bullfrog (*Rana catesbeiana*) skin, structurally related to the bacterial antibiotic, polymyxin

D.P. Clark[1], S. Durell[1], W.L. Maloy[2], K.U. Prasad[2] and M.A. Zasloff[2]

[1]*Department of Pathology and Laboratory Medicine, Hospital of the University of Pennsylvania, Philadelphia, PA 19104, U.S.A.*
[2]*Magainin Pharmaceuticals Inc., Plymouth Meeting, PA 19462, U.S.A.*

Introduction

Amphibian skin is rich in antimicrobial molecules, particularly of the amphipathic alpha helical class. At least fifteen different antimicrobial peptides have been isolated from *Xenopus laevis* and *Bombina* sp. skin [1]. These molecules are produced and stored in specialized dermal structures called granular glands which release their contents onto the external surface of the frog upon adrenergic stimulation or injury [1]. Besides Xenopus and Bombina, antimicrobial activity has been observed in skin from frogs of the genus *Rana* [2]. Our objective in this study is to characterize the antimicrobial activity observed in skin of the American bullfrog, *Rana catesbeiana*.

Results and Discussion

Whole, metamorphic *Rana catesbeiana* tadpole extracts were size fractionated on a P-30 gel filtration column and further separated by reverse-phase HPLC. Only fractions from one isolated peak displayed significant antimicrobial activity. These fractions contained a single peptide, as determined by capillary electrophoresis, amino acid analysis and protein sequence analysis. The amino acid sequence of the isolated peptide is shown in Figure 1.

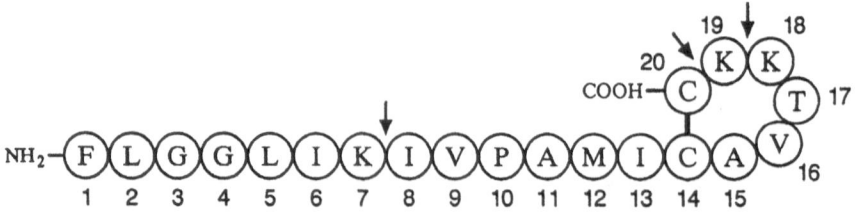

Fig. 1. The amino acid sequence of ranalexin.

Because of the identification of two cysteine residues in the peptide, it was necessary to determine if they were disulfide bonded. Using a combination of FAB mass spectrometry and trypsin digestion, the presence of a single intramolecular disulfide bond was confirmed. The expected molecular mass of ranalexin with a single intramolecular disulfide bond matched the mass measured by FAB mass spectrometry (2103 Da) (Figure 1). Trypsin digestion resulted in two major fragments (748 and 1375 Da). This is consistent with a single cleavage at Lys-7, apparently there is no cleavage at the Lys residues in the disulfide loop. To further establish that the observed mass spectrometry results were due, in part, to a disulfide bond, ranalexin was reduced and then subjected to trypsin digestion. It was found that this reduction reduced the mass of the larger tryptic peptide from 1375 Da to 1146 Da due to the loss of Cys-20 and Lys-19, indicating that once the disulfide loop is reduced tryptic cleavage can occur at Lys-19 and Lys-18.

To determine the antimicrobial spectrum of this novel peptide and its analogs, minimal inhibitory concentrations were determined on gram positive and gram negative bacteria. As shown in Table 1, synthetic ranalexin displayed antimicrobial activity against the gram positive bacterium *Staphylococcus aureus* as well as the gram negative bacteria *Escherichia coli* and *Pseudomonas aeruginosa*. Analogs of ranalexin with amino acid deletions were synthesized to determine which structural features were important for antimicrobial activity. It was found that truncation of ranalexin by deletion of one to four amino terminal amino acids eliminated antimicrobial activity (analogs Del 1, Del 1-2, Del 1-3, and Del 1-4 in Table 1). Deletion of the carboxy terminal cysteine also impaired antibiotic activity (analog Del-20 in Table 1).

Table 1 *Minimal inhibitory concentrations of synthetic ranalexin and several analogs*

		Minimal	Inhibitory	Concentration (μg/ml)
Peptide Sequence	Analog	S. aureus	E. coli	P. aeruginosa
FLGGLIKIVPAMICAVTKKC	Ranalexin	4	32	128
LIKIVPAMICAVTKKC	Del 1-4	256	256	256
GLIKIVPAMICAVTKKC	Del 1-3	>256	>256	>256
GGLIKIVPAMICAVTKKC	Del 1-2	256	>256	>256
LGGLIKIVPAMICAVTKKC	Del 1	256	256	>256
FLGGLIKIVPAMICAVTKK	Del 20	128	256	128

While the primary amino acid sequence of ranalexin is unique, it does share remarkable structural similarity with the polymyxins, a class of membrane-active antibiotics which were first isolated from the bacterium *Bacillus polymyxa* in 1947. The polymyxins are amphipathic molecules consisting of a decapeptide with a hydrophobic fatty acid attached to its amino terminus (Figure 2). Seven of the ten amino acids are configured into a cationic ring structure through a covalent bond. Ranalexin is similar in that it contains a cationic heptapeptide ring and an amino terminus composed of six hydrophobic amino acids in place of a fatty acid. The length of the fatty acid in polymyxin is similar to that of the six amino terminal amino acids of ranalexin configured in an alpha helix. Comparison of the heptapeptide ring of ranalexin and polymyxin reveals conserved residues.

Specifically, Thr-17, Lys-18 and Lys-19 of ranalexin occupy identical positions within the heptapeptide ring as Thr-7, Dab-8 and Dab-9 in polymyxin D (Figure 2). These similarities suggest that ranalexin and the polymyxins may share a common mechanism of antimicrobial action.

A

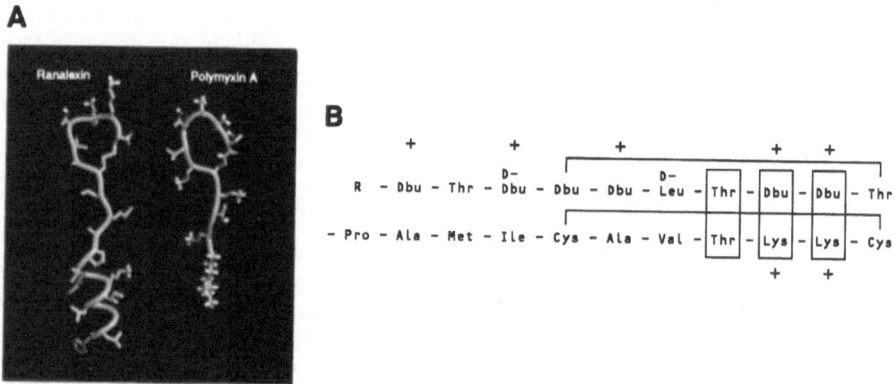

B

Fig. 2. Structural and primary sequence comparison of ranalexin and polymyxin D.

We have isolated a novel antimicrobial peptide from the American bullfrog *R. catesbeiana*, which is responsible for antimicrobial activity observed in its skin. The presence of an intramolecular disulfide bond distinguishes it from previously isolated amphibian antimicrobial molecules. The structural similarity of ranalexin to the membrane-active antibiotic polymyxin suggests that it has a similar mechanism of action.

Acknowledgements

This work was supported by a grant from the G. Harold and Leila Y. Mathers Charitable Foundation.

References

1. Bevins, C.L. and Zasloff, M., Annu. Rev. Biochem., 59 (1990) 395.
2. Cevikbas, A., Toxicon, 16 (1978) 195.

Synthesis of bradykinin analogs containing N-benzylglycine

A.R. Mitchell[1], J.D. Young[2], F. Ghofrani[3] and J.M. Stewart[4]

[1]Lawrence Livermore National Laboratory, Livermore, CA 94550, U.S.A., [2]UCLA, Los Angeles, CA 90024-1737, U.S.A. and Skyline Peptides, Alameda, CA 94501, U.S.A., [3]UCLA, Los Angeles, CA 90024-1737, U.S.A. and [4]University of Colorado Medical School, Denver, CO 80262, U.S.A.

Introduction

N-Benzylglycine (Bzl-Gly) is an achiral structural isomer of phenylalanine that is useful as an amino acid replacement in SAR studies. The first incorporation of this secondary aromatic amino acid into a biologically active peptide was reported by us using bradykinin (BK: Arg-Pro-Pro-Gly-Phe-Ser-Pro-Phe-Arg). [Bzl-Gly⁷]-BK was prepared and shown to be a potent BK-agonist [1]. The synthesis and bioassay of more potent BK-agonists, reported here, allowed the design of the first BK-antagonists containing Bzl-Gly [2]. Our concept of substituting a single naturally occurring amino acid with its achiral structural isomer (Figure 1) has recently been used by others to replace all of the naturally occurring amino acids (proline excepted) with N-alkyl glycine derivatives for the preparation of N-substituted glycine oligomers or "peptoids" [3, 4].

Results and Discussion

The [Bzl-Gly]-BK analogs (Table 1) were synthesized by SPPS using a Boc-benzyl strategy on PAM resins as described earlier [1]. The peptides were purified by HPLC to homogeneity and gave the expected AAA. FABMS verified the expected molecular masses. Bioassays show very low activity for peptides **5** and **7** (Table 1) in isolated smooth muscle but showed striking depressor activity in the rat following IV administration. Peptide **6** possesses 50% BK depressor potency with the blood pressure response reaching a maximum in 1.5-2 min (BK, 30 seconds) and a recovery to normal requiring 10 min (BK, 1 min). The slow response behavior is similar to that reported for cyclic BK [5].

Vavrek and Stewart [6] have shown that potent BK-agonists can be converted to BK-antagonists by replacement of Pro with D-Phe in position 7. [Bzl-Gly⁶]-BK, [Bzl-Gly⁸]-BK and [Bzl-Gly⁶,⁸]-BK were modified by insertion of D-Phe at position 7 to give peptides **8**, **9**, and **10** (Table 1). Peptide **9** showed weak antagonist activity in the rat uterus bioassay. Peptide **10** showed no agonist or antagonist effects in the standard blood pressure bioassay (bolus injection of BK+10) but there was a very slow recovery to the usual response to BK after several administrations of

Table 1 *Bioassay results of bradykinin analogs[a]*

Peptide	Smooth Muscle (%)		Rat Blood Pressure (%)		
	Rat U.	GP L.	IA	IV	Destruc.
(1) BK	100	100	100	100	99
(2) [Bzl-Gly^7]-BK	27	36	97	337	98
(3) [D-Phe^7]-BK[a]	1	b	2	4	36
(4) [Phe^7]-BK[c]	0.1	0.1	0	0	86
(5) [Bzl-Gly^6]-BK	0	0	180	3150	40
(6) [Bzl-Gly^8]-BK	0.2	0	10	50^d	40
(7) [Bzl-Gly^{6,8}]-BK	0.01	0	180	4040	60
(8) [Bzl-Gly^6,D-Phe^7]-BK	0.2	0	0	0	N.D.
(9) [Bzl-Gly^8,D-Phe^7]-BK	0.3	b	0	0	N.D.
(10) [Bzl-Gly^{6,8},D-Phe^7]-BK	0.3	0	e	e	N.D.

[a] See reference [6] for methods of assay. Abbreviations: Rat U., rat uterus contraction;
GP L, guinea pig ileum contraction; IA, intraaortic; IV, intravenous; Destruc., pulmonary destruction; N.D., not determined.
[b] Antagonist; pA$_2$ = 5.7. *[c]* Stewart and Vavrek [7]. *[d]* Blood pressure response atypical, see text.
[e] Antagonist, see text.

peptide **10**. The bioassay was modified by infusion of peptide **10** at various rates (1, 2, and 5 mg/min) followed by frequent challenges with standard doses of BK. This assay showed the peptide to be an antagonist of moderate potency with a long lag to onset of action and a long time to complete recovery [2]. These results highlight the utility of achiral structural isomers of naturally occurring amino acids in peptide SAR studies.

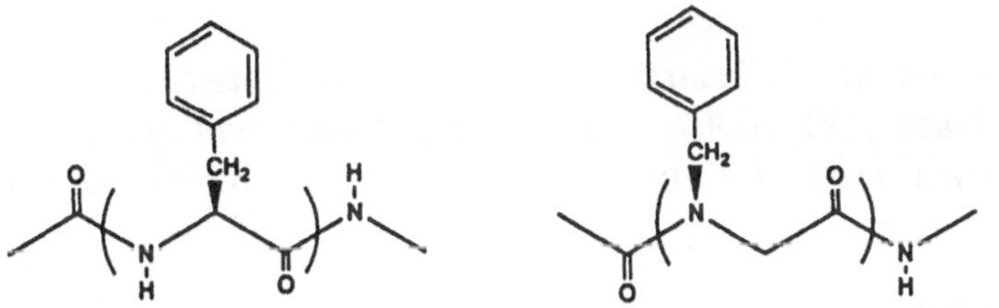

Phenylalanyl residue   **N-Benzylglycyl residue**

Fig. 1. Schematic comparison of peptides containing phenylalanine and N-benzylglycine (Bzl-Gly).

Acknowledgements

We thank K.L. Stevens for some of the synthetic work and Frances Shepperdson for the bioassays. We also acknowledge R.J. Vavrek for informative discussions.

References

1.   Young, J.D. and Mitchell, A.R., In Rivier, J.E. and Marshall, G.R. (Eds.), Peptides: Chemistry, Structure and Biology, Escom, Leiden, The Netherlands, 1990, p.155
2.   Ghofrani, F., Mitchell, A.R. and Young, J.D., Abstract P-323 of the 13th American Peptide Symposium, Edmonton, Alberta, Canada, 1993.
3.˝ Simon, R.J., Kania, R.S., Zuckerman, R.N., Huebner, V.D., Jewell, D.A., Banville, S., Ng, S., Wang, L., Rosenberg, S., Santi, D.V., Cohen, F.E. and Bartlett, P.A., Proc. Natl. Acad. Sci., U.S.A., 89 (1992) 9367.
4.   Zuckermann, R.N., Kerr, J.M., Kent, S.B.H. and Moos, W.H., J. Am. Chem. Soc., 114 (1992) 10646.
5.   Chipens, G.I., Mutulis, F.K., Katayev, B.S., Klusha, V.E., Misina, I.P. and Myshlyriakova, N.V., Int. J. Peptide Protein Res., 18 (1981) 302.
6.   Vavrek, R.J. and Stewart, J.M., Peptides, 6 (1985) 161.
7.   Stewart, J.M. and Vavrek, R.J., In Burch, R.M. (Ed.), Bradykinin Antagonists, Marcel Dekker, New York, U.S.A., 1991, p.51.

Cyclic lactam analogs of the α-factor from *Saccharomyces cerevisiae*

W. Yang[1], X. Rao[1], O. Antohi[1], H.F. Lu[2], A. Mckinney[2],
J.M. Becker[2] and F. Naider[1]

*[1]Department of Chemistry, The College of Staten Island, CUNY,
Staten Island, NY 10301, U.S.A. and [2]Department of Microbiology,
University of Tennessee, Knoxville, TN 37996, U.S.A.*

Introduction

The α-factor (WHWLQLKPGQPMY) from *Saccharomyces cerevisiae* is an important model compound for understanding cell-cell signaling in eukaryotes. Previous studies indicated the existence of a Type II β-turn spanning residues 7 through 10, and suggested that a folded conformation might be involved in the biological activity of the pheromone [1]. To test this hypothesis we prepared a series of cyclic analogs of α-factor containing side chain lactam constraints with variable ring size. We describe here the synthesis and biological activities of 8 cyclic analogs and a structural analysis of corresponding model tetrapeptides using ^{1}H NMR, CD and a Monte Carlo minimum energy conformational search.

Results and Discussion

The chemical structures of the cyclic lactam analogs of α-factor are shown in Figure 1. The peptides were synthesized by SPPS using orthogonal Boc, Fmoc and Ofm protecting groups. On resin cyclization was performed using the BOP reagent to generate the lactam rings. Corresponding linear peptides with free ω amino and β or γ carboxyl groups were also synthesized as controls. All peptides were purified to near homogeneity (>99%) and were characterized by FABMS, amino acid analysis,

Trp-His-Trp-Leu-Gln-Leu-NHCHCO-Pro-Gly-NHCHCO-Pro-Nle-Tyr

$$\underset{NH \quad\text{————————}\quad CO}{\underset{(CH_2)_m \qquad\qquad (CH_2)_n}{}}$$

*	IC(42):	m = 4; n = 2	IIC41:	m = 4, n = 1
	IC32:	m = 3; n = 2	IIC31:	m = 3; n = 1
	IC22:	m = 2; n = 2	IIC21:	m = 2; n = 1
	IC12:	m = 1; n = 2	IIC11:	m = 1; n = 1

* This is a D-residue

Fig. 1. Structure of the cyclic lactam analogs of α-factor.

430

TLC and reversed-phase HPLC. The overall yields of the purified analogs were 5-20%. Model compounds mimicking the central portion of the cyclic lactam analogs were synthesized by SPPS using a similar strategy.

Table 1 *Biological activity against S. cerevisae RC629*

Peptide analog	Minimum amount causing growth inhibition (mg/disk)
[Nle^{12}]α-factor	0.05
IC(42)	>10
IC32	10
IC22	0.5
IC12	0.5
IIC41	>10
IIC31	10
IIC21	5
IIC11	>50

IC22, the most potent cyclic analog of α-factor, is still 10 times less active than [Nle^{12}]α-factor (Table 1). Comparison of the activities among these cyclic analogs indicates that peptides with Glu in position 10 (n = 2 in Figure 1) have higher activity than those with Asp (n = 1 in Figure 1), suggesting that a carbonyl group in the γ position of residue 10 is very important for biological activity. The binding of the cyclic analogs to the α-factor receptor was tested by competition with tritiated [Nle^{12}]α-factor. Interestingly, IC(42), which is virtually inactive, is a good competitor reducing the binding of [Nle^{12}]α-factor by 50% at a concentration of 10^{-5} M. In contrast IC22, which is over 20 times more active than IC(42), has 20 fold lower affinity for the receptor. We conclude that proper spatial orientation of the N-terminal residues is important for triggering the biological response but not for receptor binding.

The temperature coefficients (dδ/dT) of several model tetrapeptides in DMSO-d$_6$ are listed in Table 2. In Tetra II and Tetra III, the NH of Glu has a very low dδ/dT value, providing some evidence for a 4→1 hydrogen-bonded β-turn structure. Analysis of the ROESY (500ms) spectrum of Tetra III revealed a strong NOE between the NH of Gly^3 and the αCH of Pro^2, characteristic of a Type II β-turn structure [2]. The CD spectrum of Tetra III showed a strong positive band (mean residue ellipticity ~190,000 deg.cm^2.dmol^{-1}) at 202-204 nm, a much weaker negative band (mean residue ellipticity ~20,000 deg.cm^2.dmol^{-1}) at 227-230 nm and a cross over point at 221 nm. These features are consistent with the properties of a Type II β-turn [3]. The lowest energy conformation of Tetra III reveals a set of dihedral angles close to an ideal Type II β-turn and a hydrogen bond between NH of Glu^4 and CO of DABA^1 in this model compound.

Table 2 *Temperature coefficients for model tetrapeptides in DMSO-d_6 (Δ ppb/°C)*

Tetrapeptide	αNH (1)	ωNH (1)	αNH (3)	αNH (4)
Ac-K-P-G-E-NH$_2$ (Tetra I)	-7.02	-5.89	-3.14	-4.80
Ac-Orn-P-G-E-NH$_2$ (Tetra II)	-7.44	-2.28	-6.76	-0.90
Ac-(DABA)-P-G-E-NH $_2$ (Tetra III)[c]	-7.44	-5.71	-7.57	+1.80
Ac-(DABA)-P-G-D-NH$_2$ (Tetra VII)[c]	-9.30[a] -6.92[b]	-5.00[a] -3.48[b]	-5.56[a] -5.02[b]	-6.80[a] -5.26[b]

a) Values correspond to cis isomer   b) Values correspond to trans isomer
c) DABA - α,γ-Diamino Butyric Acid

In conclusion, the conformational analysis suggests that Tetra III assumes a Type II β-turn in DMSO-d_6 and water. This model compound corresponds to the center of IC22, the most potent cyclic analog of α-factor. The results of these studies support the previous hypothesis, that a Type II β-turn is an important structural feature for the biological activity of α-factor. The conformational analyses on the cyclic analogs and other model compounds are currently under investigation.

Acknowledgements

These studies were supported by grants from the National Institutes of Health (GM22086 and GM22087).

References

1. Xue, C., Eriotou-Bargiota, E., Miller, D., Becker, J.M. and Naider, F., J. Biol. Chem., 264 (1989) 19161.
2. Wüthrich, K., NMR of Proteins and Nucleic Acids, John Wiley & Sons, New York, U.S.A., 1986, p.126.
3. Woody, R.W. (Ed.), Peptides, Polypeptides and Proteins, John Wiley & Sons, New York, U.S.A., 1974, p.338.

Cloning of a human B_2 bradykinin receptor: Functional expression and pharmacological characterization in *Xenopus* oocytes and mammalian cell lines

E.A. Novotny, M.A. Connolly, S. Chakravarty, S. Lu,
D.E. Wilkins, D.L. Bednar, J.R. Connor,
D.J. Kyle and T.M. Stormann

Scios Nova Inc., Baltimore, MD 21224, U.S.A.

Introduction

The recent cloning of a rat B_2 bradykinin receptor [1] opened up the possibility of isolating a clone for the human B_2 receptor using a homology cloning strategy. Recognizing that being able to screen potential bradykinin antagonists against the human receptor would help in avoiding the problem of pharmacological differences between species, we pursued the isolation of a human B_2 bradykinin receptor clone, and its expression in eukaryotic systems.

Results and Discussion

We have isolated a full length cDNA encoding the human B_2 bradykinin receptor which is identical in sequence to that published [2], assembled an appropriate eukaryotic expression construct, and isolated a stable CHO/K cell line expressing bradykinin receptors at a density approximately 40-fold higher than that seen in guinea pig ileum [3].

The expressed human bradykinin receptors have been demonstrated to bind ^{3}H-bradykinin and the B_2 selective receptor antagonist, ^{3}H-NPC 17731 with high affinity (K_D = 190 pM and 47 pM, respectively) in a saturable fashion. This binding can be competitively inhibited by other B_2 receptor antagonists and bradykinin. The use of frozen membranes allows for the preparation of a large batch of membranes with the same protein content and the assay of multiple novel compounds on a series of successive days. Figure 1 illustrates the SAR of several hydroxyproline ether containing peptides on the cloned human B_2 receptor. The IC_{50} of each antagonist is as follows: NPC 17975, 2.1 pM (n=2); NPC 17974, 4.0 pM (n=2); NPC 17731, 12 pM (n=2); NPC 17761, 19.5 pM (n= 2); NPC 18407, 560 pM (n=3); NPC 18413, 4,400 pM (n=3); NPC 18410, 25,000 pM (n=3). Electrophysiological studies done in our lab using *Xenopus* oocytes injected with *in vitro* synthesized RNA showed that the cloned B_2 receptor is functional. Maximal responses to bradykinin were seen with 100 nM bradykinin. Following a ten minute wash with buffer, a second application of 100 nM bradykinin evoked no response, indicating that the receptors were desensitized. This response to 100 nM bradykinin could be blocked with 500 nM NPC 17731 in a reversible fashion.

433

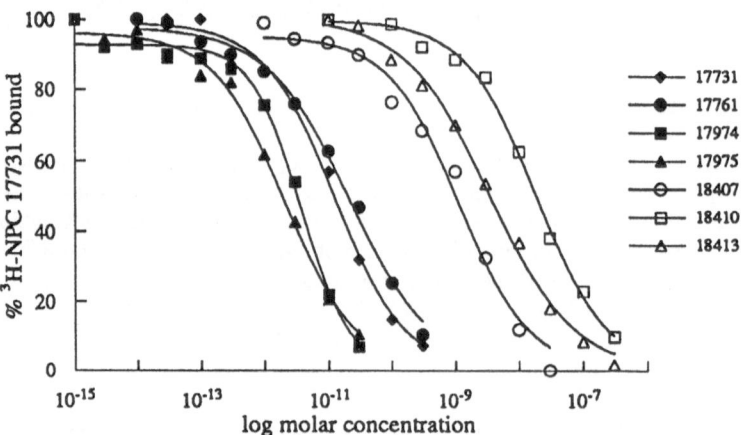

DArg⁰-Arg¹-Pro²-Hyp³-Gly⁴-Phe⁵-Ser⁶-X⁷-Oic⁸-Arg⁹				
X =	(propyl ether)	(thiophenyl)	(isobutyl ether)	(phenylpropyl ether)
Trans	17731	17761	17975	17974
Cis	-------	18407	18413	18410

Fig. 1. *Chemical structures and competition binding of selective B₂ antagonists. The chart illustrates the position 7 cis and trans isomers of these hydroxyproline ether compounds. Competition binding was done using CHO/K cells stably expressing the human B₂ receptor.*

The isolation of the cDNA for the human B_2 bradykinin receptor and its functional expression give us the security of screening with the human receptor which may have different pharmacological properties as compared to guinea pig. We feel that these binding assays and functional studies should improve the efficiency of our primary screening and facilitate our development of new, potent B_2 receptor antagonists that will hold up in clinical trials.

References

1. McEachern, A.E., Shelton, E.R., Bhakta, S., Obernolte, R., Bach, C., Zuppan, P., Fujisaki, J., Aldrich, R.W. and Jarnagin, K., Proc. Natl. Acad. Sci. U.S.A., 88 (1991) 7724.
2. Hess, J.F., Borkowski, J.A., Young, G.S., Strader, C.D. and Ransom, R.W., Biochem. Biophys. Res. Commun., 184 (1992) 260.
3. Burch, R., DuPont Biotech Update, 7 (1992) 127.

Superactive lanthionine-enkephalins

G. Ösapay, S. Wang, D.D. Comer, A. Toy-Palmer, Q. Zhu and M. Goodman

*Department of Chemistry, University of California at San Diego
La Jolla, CA 92093, U.S.A.*

Introduction

Through rational design of opioid ligands as mimetics for the biologically active conformations of endogenous peptides, new insight can be obtained about major opioid receptor sub-types. For this reason a large number of opioid molecules have been prepared and probed for structure-activity relationships. Structural modifications of enkephalins in our laboratory have resulted in numerous potent peptide opioids that have different selectivity for the mu and delta receptors.

As one approach we incorporated a thioether bridge between the side chains of putative alanine residues in positions 2 and 5. The ring of this lanthionine cyclopeptide is one atom smaller than the corresponding disulfide analog. Our initial synthesis of lanthionine-enkephalins led to the molecule of [D-Ala$_L$^{2}, Ala$_L$^{5}]-EA (DA$_L$^{2}LA$_L$^{5}EA, 1), where Ala$_L$ denotes each of the lanthionine amino acid ends linked by a monosulfide bridge [1]. This led to a molecule with a very high analgesic activity at subnanomolar levels. The analog exhibited an increase in metabolic stability as compared to the disulfide analog [2]. On the basis of these results, we prepared other diastereomers of lanthionine enkephalins for biological assays and conformational studies.

Results and Discussion

Two new synthetic methods have been developed and used to prepare the target lanthionine enkephalins. In both cases the ring formation was carried out on solid supports (MBHA-resin and Kaiser-oxime resin).

In a simulation of the biosynthetic pathway of lanthionine formation a **Michael-addition reaction** was used to form the thioether bridge. The peptide chain was assembled on an MBHA-resin containing Dha^2 and Cys^5 residues in the positions of the desired Ala$_L$ units. During the sulfide-bridge formation an unexpected stereoselectivity of the Michael-addition took place resulting in DA$_L$^{2}LA$_L$^{5}EA, probably because of the steric hindrance of the solid support.

Another route for the synthesis of lanthionine peptides involves the **Peptide Cyclization on an Oxime Resin** (PCOR) method [3, 4]. The peptide chain was prepared on the oxime resin, and cyclization reaction took place by a nucleophilic attack of a free amino group on the ester carbonyl of the C-terminal oxime group. The product simultaneously cleaved from the resin. After coupling the Tyr residue to the N-terminus, the protecting groups were removed and the product was purified by RP-HPLC.

Structure-Activity   Relationships

Biological activities of lanthionine-enkephalins have been determined both in the *in vivo* (rat hot plate tests) and *in vitro* assays (GPI and MVD). In parallel experiments, the disulfide analogs have also been tested.

Our new lanthionine-enkephalins show increased bioactivity both *in vivo* (ED_{50} = 0.1 nmol) and *in vitro* assays (Table 1). The lanthionine constraints do not enhance selectivity [IC_{50} (MVD)/IC_{50} (GPI)].

Table 1 *Inhibitory potency and selectivity in GPI and MVD bioassays, rat hot plate test in vivo and enzymatic degradation*

Compounds	GPI IC_{50}[nM]	MVD IC_{50}[nM]	MVD/GPI	*In vivo* test ED_{50} [nmol]	Biodegradation $t_{1/2}$[min]
$DA_L{}^2LA_L{}^5EA$	0.62	0.54	0.88	0.11	1220
$DA_L{}^2DA_L{}^5EA$	1.67	1.67	1.0	0.11	-
DC^2LC^5EA	1.51	0.76	0.5	0.24	330
LC^2LC^5EA	inactive	inactive	-	inactive	-
Leu^5-E	246	11.4	0.05	>100	30
Morphine	58.6	644	11	15	-

A comparison of biological activity of DC^2LC^5EA, $DA_L{}^2LA_L{}^5EA$ and DP^2DP^5E reveals enormous differences. All activity of DP^2DP^5E (100 mg/kg) is lost by 60 min; the disulfide-bridged analog (0.3 mg/kg) retains partial activity, while the lanthionine-enkephalin (0.3 mg/kg) retains almost full activity over the same time period. A direct study of enzymatic degradation (enzyme mixture from rat brain homogenate) demonstrates a significant difference in the stability of three compounds, the Leu^5-enkephalin, the disulfide-bridged enkephalinamide and the lanthionine-enkephalinamide. The lanthionine-enkephalinamide has a half-life at least four times that of the corresponding disulfide analog and forty times that of Leu^5-enkephalin.

Conformational Analysis

NMR spectroscopy and molecular modeling were used to determine the backbone conformations for the lanthionine and disulfide cyclized enkephalin analogs. The results showed similar γ-turns about the Phe^4 residue in both the cystine and lanthionine cyclized enkephalins.

A systematic search of side chains was carried out with the Small Molecule Modeling Tool Version 1.0 module in INSIGHT II (Biosym Technologies, Inc.).

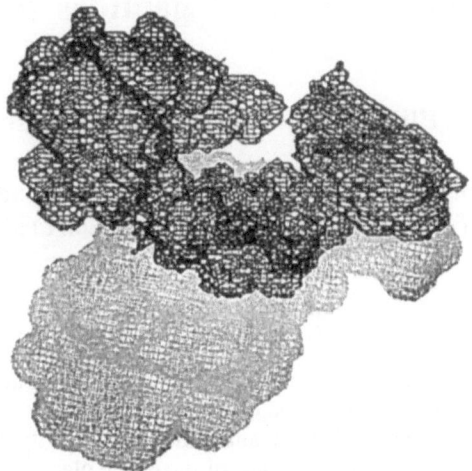

Fig. 1. Comparison of superimposed phenylalanine residues and subsequent exclusion volumes of three superactive and one inactive sulfur-bridged enkephalin analogs (see text).

In Figure 1, the conformations of three biologically superactive compouds $DA_L{}^2LA_L{}^5EA$, $DA_L{}^2DA_L{}^5EA$, and DC^2LC^5EA and the inactive LC^2LC^5EA are represented. The phenyl groups of the residues of position 4 are essentially superimposed allowing the backbone rings and the Tyr^1 groups to sweep out volume elements. In this analysis it is interesting to note that in the active analogs the phenol groups of the Tyr^1 residue prefer an array on the same side of the backbone rings as the phenyl groups. In the inactive analog the phenolic group of Tyr^1 is found on the opposite side.

Acknowledgements

This work was supported by grants NIH DA-05539 and NIH DK-15410. The authors thank Dr. Tony Yaksh (Department of Anesthesiology, UCSD) for the *in vivo* assays, and Dr. Peter W. Schiller (Clinical Research Institute of Montreal) for the *in vitro* tests of the compounds.

References

1. Polinsky, A., Cooney, M.G., Toy-Palmer, A., Ösapay, G. and Goodman, M., J. Med. Chem., 35 (1992) 4185.
2. Ösapay, G., Wang, S., Shao, H. and Goodman, M., In Yanaihara, N. (Ed.), Peptide Chemistry 1992, Proceedings of the 2nd Japan Symposium on Peptide Chemistry, Escom, Leiden, The Netherlands, 1993, p.152.
3. Ösapay, G. and Taylor, J.W., J. Am. Chem. Soc., 112 (1990) 6043.
4. Ösapay, G., Bouvier, M. and Taylor, J.W., In Villafranca, J.J. (Ed.), Techniques in Protein Chemistry II, Academic Press, Inc., San Diego, CA, U.S.A., 1991, p.221.

Neurohypophyseal hormone analogues and blood coagulation: How successful can a design of "antihemophilic" peptides be?

V. Pliška and S. Neuenschwander

Department of Animal Science, Swiss Federal Institute of Technology, ETH Zürich, CH-8092 Zürich, Switzerland

Introduction

Vasopressin and its structural analogue deamino-D-arginine-vasopressin (dDAVP) exhibit a dual effect on blood coagulation in humans: the enhancement of blood clotting factor VIII (F-VIII) [1] ("antihemophilic activity"), and the activation of fibrinolytic system *via* enhanced release of the tissue plasminogen activator (tPA) [2]. In the overall coagulation response to a drug, the two effects are mutually antagonistic. In spite of this, dDAVP has been introduced as an F-VIII releasing drug in mild and medium severe forms of coagulopathies [1] like hemophilia A or von Willebrand disease. Its great advantages are viral safety and low costs of this life-long therapy. Potential need of new antihemophilic substances (failing of dDAVP response, development of immune incompatibility against dDAVP in patients, design of substances without side effects of dDAVP, etc.), and missing knowledge of the mechanism of action, require further effort in this area.

Results and Discussion

Activity estimates. Antihemophilic (AH) and plasminogen activating (PA) potencies were expressed as integrals of the time-response curves for F-VIII and tPA, respectively, within the first 90 (AH) and 60 min (PA), as percent of the corresponding integrals for dDAVP. Table 1 summarizes published values for different species, and new values from our laboratory [3,4] for peptides interacting specifically with individual classes of neurohypophyseal hormone receptors (see Table 1) [5]. Responses to 1 μg/kg b.w. doses of the peptide tested and of dDAVP were compared. Standard values of antidiuretic (AD), vasoconstrictory (BP), *in vitro* uterotonic (UT) and milk ejecting (GB) activities were taken from [6]. AD and BP for peptides 3, 4, 5, 7 and 11 were unpublished data by Dr. P. Melin.

Receptor selectivity. Selectivity of a neurohypophyseal peptide interaction with a particular receptor subtype V_1, V_2 or OT (oxytocin-type) was expressed by means of the three standard activities AD, BP and UT as a "selectivity index": $SV_1 = BP/(BP + AD + UT)$; $SV_2 = AD/(BP + AD + UT)$; $SOT = UT/(BP + AD + UT)$. The peptides were then grouped as follows (Table 1): 1. V_1, V_2 or OT selective peptides with the corresponding selectivity $\geq 2/3$; 2. $\{V_1V_2\}$ peptides for which $0.5 \leq SV_1/SV_2 \geq 2$, and $SV_1 \& SV_2 \geq 1/3$; 3. $\{V_1V_2OT\}$ peptides for which ratios of any two selectivity indices SA_i, SA_j are $0.5 \leq SA_i/SA_j \geq 2$, and $SV_1 \& SV_2 \& SOT \geq 1/3$.

Table 1 *Antihemophilic (AH) and plasminogen activating (PA) potencies of neurohypophyseal hormone analogues[a]*

		AH	PA				AH	PA
V$_2$ peptides					**V$_2$ peptides**			
1	dDAVP	100	100		7	[Gly^5]dDAVP	0^v	nd
2	DAVP	29^v	nd		8	carba-6-dDAVP	41^v	103^c
3	[phe^3]dDAVP	26^v	nd				96^p	
4	[trp^3]dDAVP	0^v	nd		9	[orn^8]dVP	0^c	nd
5	[Ail^4]dDAVP	88^v	nd		10	[MeArg^8]dAVP	3^p	nd
6	[Val^4]DAVP	61^n	169^n		11	desGly^9-AVP	0^v	nd
V$_1$ peptides					**{V$_1$V$_2$} peptides**			
12	[Phe^2,Orn^8]VT	33^n	71^n		14	AVP	33^n	44^c
13	[Gly^7]dDAVP	0^v	nd		15	LVP	54^m	46^m
								39^c
					16	Gly-Gly-Gly-LVP	nd	6^c
{V$_1$V$_2$OT} peptides								
17	AVT	0^p	35^c		**OT peptides**			
					18	OT	31^n	27^n
								0^c
nonspecific antagonists					19	[Thr^4,Gly^7]OT	32^n	5^n
20	[Mep^1,Tyr(Me)^2]DAVP	30^n	nd					
21	[Mpp^1,tyr(Me)^2,Val^4]AVP	48^n	0^n					

[a]AH and PA in percent of dDAVP. Abbreviations: AVP, arginine vasopressin; AVT, arginine vasotocin; DAVP, [arg^8] vasopressin; dDAVP, deamino-D-arginine vasopressin; dVP, deamino-vasopressin; LVP, lysine vasopressin; OT, oxytocin; VT, vasotocin. Amino acids according to IUPAC-IUB; Xyz stands for L- xyz for D-enantiomers; Ail, *allo*-isoleucine; Mep, β-mercapto-β,β-diethylpropionic acid; Mpp, β-mercapto-β,β-cyclopentamethylenepropionic acid; Orn, ornithine. References (species): ^{c}Cash et al. [8] (human); ^{m}Mannucci et al. [9] (human); ^{n}Neuenschwander and Pliška, unpublished (sheep); ^{p}Prowse et al. [10] (human); ^{v}Vilhardt and Barth [11] (dog). Abbreviations: nd, not determined.

Relationships between blood clotting and standard potencies. Due to the data inconsistency (see above), correlations of AH and PA with standard potencies and with the corresponding selectivity indices were evaluated by a robust correlation analysis PROGRESS [7] based on the least median of squares method. In order to account for putative nonlinearities, significance of the Spearman's rank correlation coefficient was tested. The results indicate that i) solely the antidiuretic potency (determined in rat) shows significant correlations with AH and PA (determined in the species indicated in Table 1), ii) this correlation is linear, and that iii) selectivity indices SV$_1$, SV$_2$ and SOT show no significant effects upon AH and PA. We conclude that AH and PA are in some way determined by V_2-types of interactions. However, it would be far-fetched to state that these responses are directly initiated on a V_2 receptor type; rather, we assume a linkage to a physiologic mechanism started there.

Peptides belonging to the $\{V_1V_2\}$ and to OT groups (naturally occurring peptides AVP, LVP, OT) were detected as outliers by the PROGRESS routine. When the V_2-group alone was subjected to the correlation analysis, the multiple correlation coefficient and the variance explained by the regression model employed, were considerably higher.

Using the same methods, we have investigated the interdependence of AH and PA. Pearson's coefficient of linear correlation is not highly significant ($p<0.1$) but the rank correlation between them clearly exists ($p<0.05$). At present, the size of the sample does not allow conclusions about the type of relationship between the two.

Activity predictions for AH and PA. Considering these findings, a design of "specific" antihemophilic and plasminogen activating peptides may meet severe difficulties. First, despite their loose correlation, it may turn difficult to separate AH and PA. Second, antidiuretic activity probably remains the most dominating side effect. It seems likely that substances possessing the activity spectrum of dDAVP will remain therapeutics of choice in the future.

Acknowledgements

Supported by the Swiss National Science Foundation, ETH Zürich and SANDOZ Foundation Basel. Newly assayed peptides were donated by Dr. M. Manning, Toledo, OH, U.S.A.

References

1. Mannucci, P.M., Ruggeri, Z.M., Pareti, F.I. and Capitanio, A., Lancet i (1977) 869; Lancet ii (1977) 1171.
2. Mannucci, P.M., Thrombosis Research, 4 (1974) 539.
3. Neuenschwander, S., Lang, E., Heiniger, J. and Pliška, V., In Thorn, N.A., Vilhardt, H. and Treiman, M. (Eds.), Proceedings of the Fourth International Conference on the Neurohypophysis, Oxford University Press, U.K., 1989, p.162.
4. Heiniger, J., Kissling-Albrecht, L., Neuenschwander, S., Rössli, R. and Pliška, V., Br. J. Pharmacol., 94 (1988) 279.
5. Manning, M. and Sawyer, W.H., J. Receptor Res., 13 (1993) 195.
6. Hruby, V. and Smith, C.W., In Udenfriend, S. and Meienhofer, J. (Eds.), The Peptides, Analysis, Synthesis, Biology, Vol. 8, Academic Press, New York, NY, U.S.A., 1987, p.77.
7. Rousseeuw, P.J., Journal of the American Statistical Association, 79 (1984) 871.
8. Cash, J.D., Gader, A.M.A. and Da Costa, J., Br. J. Haematol., 27 (1974) 363.
9. Mannucci, P.M., Åberg, M., Nilsson, I.M. and Robertson, B., Br. J. Haematol., 30 (1975) 81.
10. Prowse, C.V., Sas, G., Gader, A.M.A., Cort, J.H. and Cash, J.D., Br. J. Haematol., 41 (1979) 133.
11. Vilhardt, H. and Barth, T., J. Receptor Res., 11 (1991) 233.

SAR studies on MSI-103, a repeat heptamer peptide sequence exhibiting antimicrobial activities

K.U. Prasad, M.M. Brasseur, S.M. French, M.T. MacDonald, R.J. White, D.L. MacDonald, C.J. Messler and W.L. Maloy

Magainin Pharmaceuticals Inc., Plymouth Meeting, PA 19462, U.S.A.

Introduction

A repeat heptamer peptide sequence (KIAGKIA)$_3$-NH$_2$, MSI-103 (I) exhibiting good antimicrobial activities was derived from structure-activity relationship (SAR) studies [1] on PGLa (Peptide between Glycine and Leucine as amide), a 21 amino acid long peptide with the sequence GMASKAGAIAGKIAKVALKAL-NH$_2$, extracted from the granular glands of African clawed frog *Xenopus laevis* [2, 3]. When represented on a helical wheel it is obvious that MSI-103 like PGLa is amphipathic in nature in that all the basic hydrophilic amino acids are on one side of the helix and the hydrophobic amino acids are on the other side of the helix. It is proposed [4] that PGLa (which can be the case with MSI-103 also), like other antibacterial peptides is membrane active and dissipates membrane potential by forming transmembrane pore like structures involving several molecules. MSI-103 exhibits a wide spectrum of antibacterial activities against gram positive (*Staphylococcus aureus*) and gram negative (*E. coli* and *Pseudomonas aeruginosa*) organisms and it does not lyse human red blood cells, which is an index of the specificity of these antimicrobial peptides. In our continuous efforts to improve the antibacterial activity and the specificity of MSI-103 various analogs are synthesized and their biological activities determined and some of those studies are presented here.

Results and Discussion

Variation of chain length: In order to determine the optimum length that is necessary to exhibit full biological activities, analogs II-VII (Table 1) are synthesized. There is a lowering in the antibacterial activities as the chain length is decreased. When the length is increased (VI-VII), improved biological activities are observed, but there is also an increase in hemolysis (loss of specificity). Thus it appears that at least 21 amino acids (which will be needed to span the lipid bilayer as an α-helix) are necessary to have the antibacterial activities.

Isoleucine modifications: Isoleucine with β-branching in the side chain is not a very favorable amino acid to form α-helical structures, especially when it occurs at i and i+4 positions along the helical axis, as is the case with MSI-103. When Leu with no β-branching is substituted, to improve the helix forming ability the resulting

Table 1 *Antimicrobial Activities*

| | MIC (µg/ml) | | | % Hemolysis |
	S.aureus	E.coli	P.aeruginosa	(100µg/ml)
MSI-103				
I KIAGKIAKIAGKIAKIAGKIA-NH$_2$	8-16	4-8	64-128	0
II _IAGKIAKIAGKIAKIAGKIA-NH$_2$	16-32	4	128	0
III __AGKIAKIAGKIAKIAGKIA-NH$_2$	128	16	128	0
IV ___GKIAKIAGKIAKIAGKIA-NH$_2$	64	8	64	0
V _____KIAGKIAKIAGKIA-NH$_2$	256	64-128	128-256	0
VI KIA(KIAGKIA)$_3$-NH$_2$	4	4	8	12
VII (KIAGKIA)$_4$-NH$_2$	4	4	16	45
VIII (KLAGKLA)$_3$-NH$_2$	16	8	32	0
IX (KVAGKIA)$_3$-NH$_2$	256	32	256	0
X (KIAGKVA)$_3$-NH$_2$	64-128	32	128	0
XI (KFAGKFA)$_3$-NH$_2$	32	4-16	32	0
XII (KFAGKIA)$_3$-NH$_2$	32	8-16	128	0
XIII (KIAGKFA)$_3$-NH$_2$	32	8-16	64	0
XIV (KJAGKJA)$_3$-NH$_2$	>256	>256	>256	83
XV (KJ*AGKJ*A)$_3$-NH$_2$	>256	>256	>256	33
XVI (KF*AGKF*A)$_3$-NH$_2$	128	128	256	1
XVII (KUAGKUA)$_3$-NH$_2$	32	32	64	71
XVIII (KUAGKIA)$_3$-NH$_2$	4-8	2-4	16-32	61
XIX (KIAGKUA)$_3$-NH$_2$	32	32	64	32
XX (KXAGKXA)$_3$-NH$_2$	4-8	4	32-64	8
XXI (KXAGKIA)$_3$-NH$_2$	16	4-8	64	5
XXII (KIAGKXA)$_3$-NH$_2$	8	2-4	32	0
XXIII (KZAGKZA)$_3$-NH$_2$	128-256	32	128	0
XXIV (KZAGKIA)$_3$.NH$_2$	128	8	128	0
XXV (KIAGKZA)$_3$-NH$_2$	128	16	128	0
XXVI (KA*AGKA*A)$_3$-NH$_2$	>256	>256	>256	0

MIC, minimum inhibitory concentration; J, homophenylalanine; j*, phenylglycine; F*, p.NH$_2$-Phe; U, cyclohexylalanine; X, norleucine; Z, norvaline; A*, α-aminobutyric acid

analog, VIII, retained the antimicrobial activities. For comparison, analogs IX-X are synthesized with Val (with β-branching) which is known to destabilize the helical structure, and as expected, the two analogs are considerably less active. Hydrophobic and aromatic amino acid, Phe, with no β-branching is substituted for Ile and all analogs (XI-XIII), as expected, showed good antimicrobial activities. Analogs are synthesized positioning the aromatic ring either away (XIV) or closer (XV) to the backbone of the molecule. To our surprise, complete loss of activities is observed for both the analogs. The increased hemolysis for (XIV) may be explained by the overall increase in the hydrophobicity of the molecule, but the loss of antibacterial activities is unexpected. Substitution of Ile with the hydrophilic amino acid, p-amino-Phe, (analog XVI) resulted in the loss of activities. This can be reasoned by the fact that the hydrophilic amino acid now occurs on the hydrophobic side of the amphipathic helical structure. Saturation of the aromatic ring (Cha analogs XVII XIX) still retained the antimicrobial activities, but once again there is an increase in hemolysis. This is in accordance with the earlier observation that increased hydrophobicity led to increased hemolysis. Analogs XX-XXVI are synthesized using amino acids with no branching in the side chain. Nle (which is more hydrophobic than Ile) substituted analogs (XX-XXII) showed good antimicrobial activities and also increased hemolytic values at 500 µg/ml concentration (not shown here). Analogs XXIII-XXVI which are less hydrophobic overall, exhibited low antimicrobial and low hemolytic activities. From these studies it can be concluded that the overall hydrophobicity of the molecule plays an important role in eliciting antimicrobial and hemolytic properties.

Acknowledgements

We thank Dr. Meenakshi L. Rao for assisting in preparing the posters and manuscript and Carol A. Billetta and Clara M. Stewart for typing the manuscript and posters.

References

1. Prasad, K.U., Brasseur, M.M., French, S.M., MacDonald, M.T., White, R.J., MacDonald, D.L., Messler, C.J. and Maloy, W.L., Presented at the European Peptide Symposium, September 13-19, 1992, Abs. # P310.
2. Gibson, B.W., Poulter, L., Williams, D.H. and Maggio, J.E., J. Biol. Chem., 261 (1986) 5341.
3. Andreu, D., Aschauer, H., Kreil, G. and Merrifield, R.B., Eur. J. Biochem., 149 (1985) 531.
4. Juretic, D., Studia Biophysica, 138 (1990) 79.

Cross-sectional molecular shapes of amphipathic α helixes control membrane stability

E.M. Tytler[1], R.M. Epand[2], G.M. Anantharamaiah[1]
and J.P. Segrest[1]

[1]Department of Medicine, UAB Medical Center, Birmingham,
AL 35294, U.S.A., and [2]Department of Biochemistry,
McMaster University, Hamilton, Ontario, L8N 3Z5, Canada

Introduction

Amphipathic α helixes vary drastically in their effects on membrane bilayer stability [1]. Apolipoprotein (class A) amphipathic helices are membrane-active and are postulated to act as detergents by virtue of their cross-section being wedge shaped [2]. Low molecular weight polypeptides containing lytic (class L) amphipathic helixes, such as mastoparan, are also membrane-active but have biological activities opposite to or distinct from class A amphipathic helices. We used synthetic model peptides to find the structural basis for these differences in activity. Using computer analysis [3] of naturally occurring class A and class L amphipathic helices, we designed two archetypical model peptides called 18A (DWLKAFYDKVAEKLKEAF) and 18L (GIKKFLGSIWKFIKAFVG).

Results and Discussion

Analogs of these two peptides, incorporating substitutions or modifications of interfacial or basic residues, had the following effects: Class A peptides stabilized bilayer structure, reduced leakage from dye entrapped lipid vesicles and erythrocytes and inhibited lysis induced by class L peptides. Class L peptides destabilized bilayer structure in model membranes and erythrocytes. The ability of class L analogs to lyse membranes and induce inverted lipid phases was reduced by either decreasing the bulk of an interfacial residue, increasing the angle subtended by the polar face, or increasing the bulk of the basic residues. The ability of the class A analog to stabilize bilayer structure and inhibit erythrocyte lysis by class L peptides was enhanced by methylation of the Lys residues. Class L peptides also increased significantly the binding of apolipoprotein A-I to erythrocyte membranes. Molecular modeling indicated that class A and class L amphipathic helices have reciprocal wedge shapes in cross-section.

These results can be explained by a model (Figure 1) that we term the reciprocal wedge hypothesis. By analogy to the reciprocal effects of phospholipid shapes on membrane structure [4], we propose that the wedge shape of class A helices stabilizes membrane bilayers while the inverted wedge shape of class L helices destabilizes membrane bilayers, and, thus, one class will neutralize the effect of the other class on

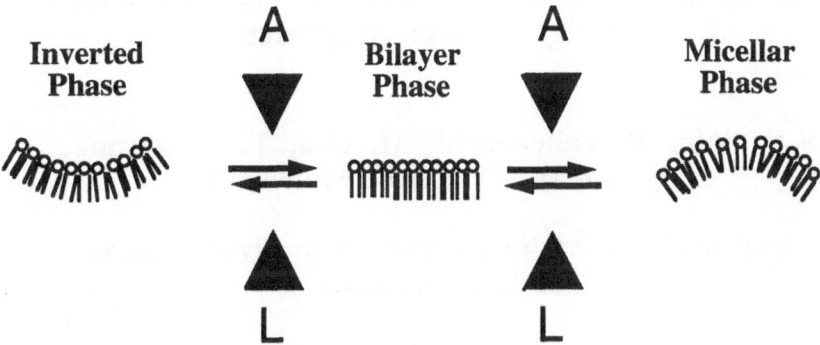

Fig. 1. Schematic representation of the postulated effects of cross-sectional shape of amphipathic α helixes on phospholipid structure.

membranes via reciprocal steric effects. We suggest that this reciprocal wedge concept has broad biological implications.

References

1. Segrest, J.P., de Loof, H., Dohlman, J.G., Brouillette, C.G. and Anantharamaiah, G.M., Proteins, 8 (1990) 103.
2. Small, D.M., In Kreisberg, R.A. and Segrest, J.P., (Eds.), Plasma Lipoproteins and Coronary Artery Disease, Blackwell Scientific Publications, Boston, MA, U.S.A., 1992, p.57.
3. Jones, M.K., Anantharamaiah, G.M. and Segrest, J.P., J. Lipid Res., 33 (1992) 287.
4. Cullis, P.R. and De Kruijff, B., Biochim. Biophys. Acta, 559 (1979) 399.

The synthesis and biological activity of EGF-family related peptides

S.Y. Shin, T. Takenouchi, M. Uga, T. Yokoyama,
N. Kobayashi and E. Munekata

*Institute of Applied Biochemistry, University of Tsukuba,
Tsukuba 305, Japan*

Introduction

Epidermal growth factor (EGF) is a regulatory peptide which controls proliferation and differentiation of various cells. Recently, new members of EGF family peptide, amphiregulin (AR) [1], heparin-binding EGF like growth factor (HB-EGF) [2] and schwannoma-derived growth factor (SDGF) [3] have been isolated and characterized from conditioned medium of human breast carcinoma (MCF-7), human macrophage-like U-937 cells and rat schwannoma cells, respectively. These novel growth factors are glycosylated polypeptides which are processed by the proteolytic cleavage of larger transmembrane precursors. These growth factors compete with ^{125}I-EGF for binding to the EGF receptor (EGF-R) on A431 epidermoid carcinoma or vascular smooth muscle cells and they stimulate the proliferation of EGF-response cells. However, it is unclear whether the biological activities of these growth factors are mediated through the only EGF-R which possesses intrinsic tyrosine kinase and autophosphorylation activity or some unknown cell surface receptors. In contrast to EGF and TGF-α, these growth factors contain distinct hydrophilic N-terminal domains that have been suggested as potentional heparin binding region. The EGF-like domain that has striking high sequence homology to EGF and TGF-α is involved in their C-terminus. Most amino acid residues implicated in binding to the EGF-R and inducing mitogenic stimulation of normal cells are conserved in this EGF-like domains. Until now, very little is known about the role of the EGF-like domain in mitogenic activities of these EGF family peptides. Thus, in this study, to investigate whether the only EGF-like domain of these EGF family growth factors have growth-promoting activity in normal cells or not, human AR (44-84), human HB-EGF (44-86) and mouse SDGF (38-80) corresponding to this domain were synthesized and their biological response was assessed in mitogenesis assays.

Results and Discussion

All peptides were synthesized manually by stepwise solid phase method using hydroxymethylphenoxy(HMP)-resin and Fmoc chemistry. The following base (piperidine) stable protecting groups were employed for side chain protection: Asp(OtBu), Glu(OtBu), Asn(Trt), Gln(Trt), His(Trt), Lys(Boc), Arg(Mtr), Ser(tBu) and Tyr (tBu). Peptide chain elongation was carried out by double coupling protocol using DCC/HOBt, BOP/HOBt and HBTU/HOBt method. The protected-peptide resins of each peptides was treated with TFA or TMSBr/TFA in the presence of

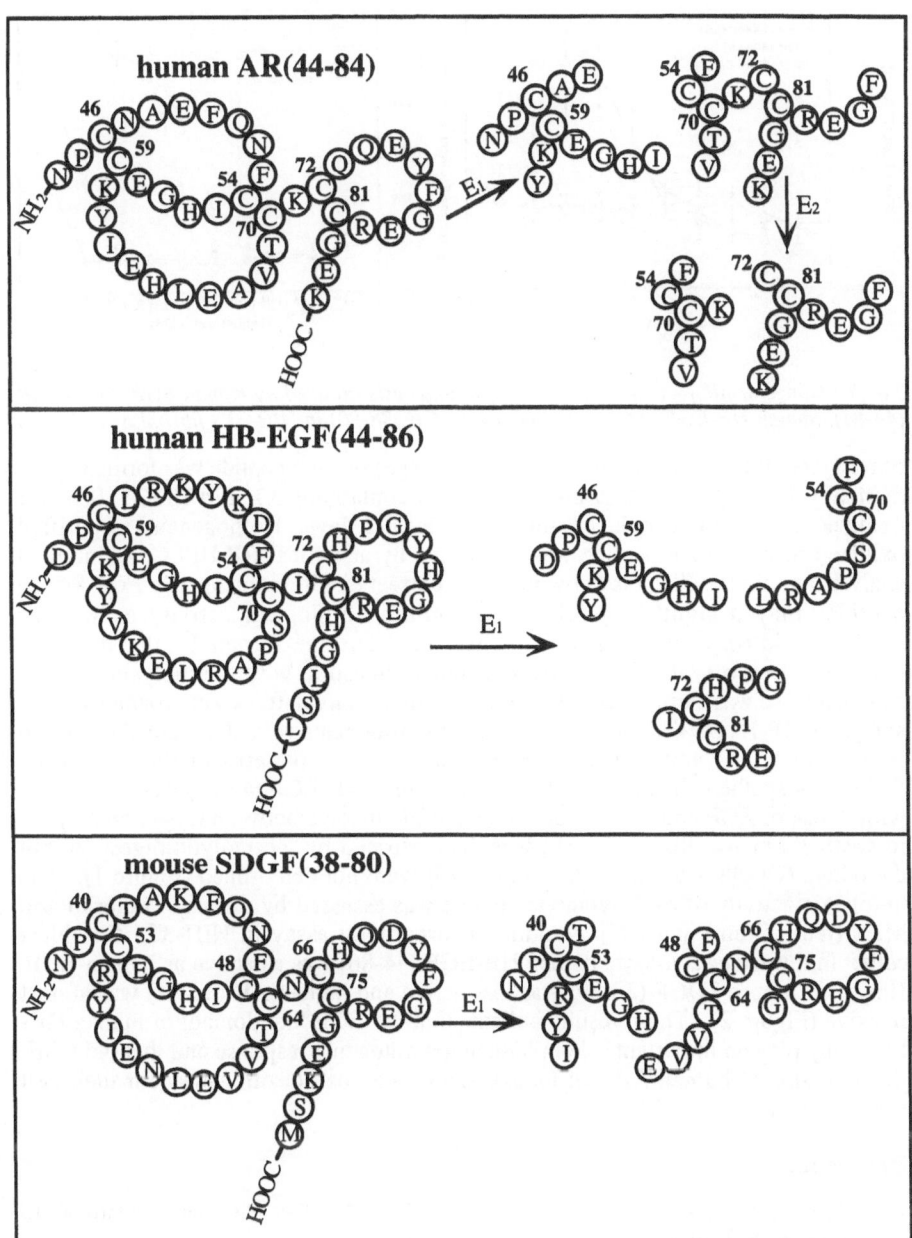

Fig. 1. Primary structure of human AR(44-84), human HB-EGF(44-86) and mouse SDGF(38-80) (left). Amino acid sequences of cystine-containing fragments determined after enzymatic digestion of human AR(44-84), human HB-EGF(44-86) and mouse SDGF(38-80)(right). E1 and E2 represent thermolysin and lysyl endopeptidase, respectively.

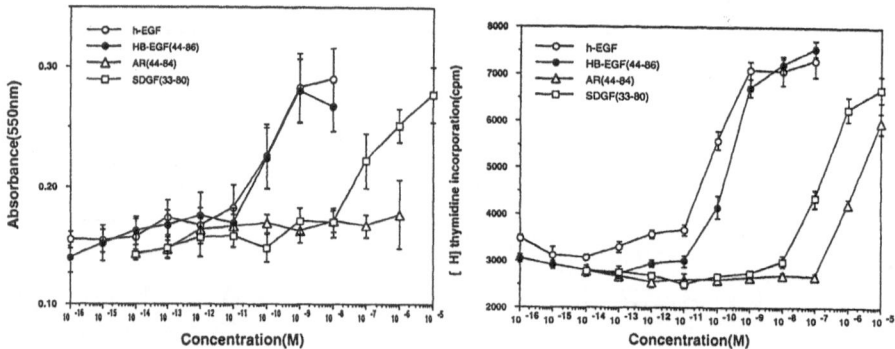

Fig. 2. Concentration - response curves of mitogenesis induced by human EGF, human AR (44-84), human HB-EGF (44-86) and mouse SDGF (38-80) in NIH-3T3 fibroblast cells.

scavengers. Intramolecular three disulfide linkages of each peptide was formed by air oxidation in 0.1 M Tris-HCl buffer (pH 8.0) containing 0.1 mM EDTA, 0.3 mM oxidized and 0.15 mM reduced glutathione for 2-3 days. Homogeneity of purified oxidized form of each peptide was confirmed by analytical RP-HPLC, amino acid analysis and FAB-MS. It is known that the correct disulfide linkages in EGF family peptides play a significant role in determining biological activity as well as stabilizing intrinsic conformation. Thus, to determine intramolecular disulfide topology of each synthetic peptide, enzymatic digestion with thermolysin or lysyl endopeptidase was carried out. The enzymatic digestion mixtures were fractionated by analytical RP-HPLC and the cystine-containing fragments were determined by amino acid analysis and amino acid microsequencing. The patterns of three disulfide linkages in synthetic human AR (44-84) and human HB-EGF (44-86) were consistent with those of EGF and TGF-α, and the disulfide linkage between Cys40 and Cys53 in synthetic mouse SDGF (38-80) was demonstrated by thermolytic digestion, but the others (Cys48, Cys64, Cys66 and Cys75) were not determined (Figure 1). The mitogenic activity of each synthetic peptide was assessed by means of colorimetric MTT (tetrazolium) and [^{3}H]thymidine incorporation assay in NIH-3T3 fibroblast cells. In mitogenesis assays, human HB-EGF (44-86) was as active as human-EGF. However, mouse SDGF (38-80) was less active and human AR (44-84) was almost inactive (Figure 2). These results indicate that the EGF-like domain of human HB-EGF may play an important role in inducing a mitogenic response and that the EGF-like domains of human AR and mouse SDGF may be not obligatory for their cell growth activity.

References

1. Shoyab, M., Plowman, G.D., McDonald, V.L., Bradley, J.G. and Todaro, G.J., Science, 243 (1989) 1074.
2. Higashiyama, S., Abraham, J.A., Miller, J., Fiddes, J.C. and Klagsbrun, M., Science, 251 (1991) 936.
3. Kimura, H., Fischer, W.H. and Schubert, D., Nature, 348 (1990) 257.

Bradykinin antagonists containing unusual amino acids show long-lasting action *in vivo*

L. Gera[1], V. Srivastava[1]*, F.W. Dunn[2] and J.M. Stewart[1]

[1]*Department of Biochemistry, University of Colorado Medical School, Denver, CO 80262 and* [2]*Abilene Christian College, Abilene, TX 79601, U.S.A.*

Introduction

Bradykinin, a nonapeptide having the structure:

Arg[1]-Pro[2]-Pro[3]-Gly[4]-Phe[5]-Ser[6]-Pro[7]-Phe[8]-Arg[9] (BK)

is released from its precursor kininogens by the action of kallikreins and exerts its biological effects through specific receptors to trigger various physiological and pathological processes such as pain, inflammation, contraction of smooth muscle and lowering of blood pressure. Extensive attempts to develop a competitive bradykinin antagonist were unsuccessful until 1984 when the first bradykinin antagonists, [D-Phe[7]]-BK and D-Arg-[Hyp[3]-Thi[5,8]-D-Phe[7]]-BK were reported by Vavrek and Stewart [1, 2]. Despite their considerable potency, these antagonists were degraded *in vivo* and have short durations of action. BK is inactivated *in vivo* principally by cleavage at 7-8 and 8-9 bonds. Structural modifications of BK antagonists by several groups including the authors' laboratory led to development of second generation BK antagonists by incorporating unusual amino acids, conformationally constrained and stable to enzymatic degradation, in the C-terminal part of the peptides. We now report a series of BK antagonists showing long duration of action on rat blood pressure that contain unusual amino acids in position 5, 7 and 8, such as cyclopentylglycine (Cpg), octahydroindole-2-carboxylic acid (Oic), 2,3-methanophenylalanine (E and Z-isomers), p-chloro-2,3-methanophenylalanine(Z-isomer), 2,4-methanoproline (MPIV), allylglycine (Alg) and thienylalanine (Thi).

Cpg   CpPhe   MPIV   Alg

The peptides were synthesized by standard solid phase synthesis methods [3], using Boc chemistry and BOP or TBTU coupling agents, on a Beckman 990 automatic synthesizer. They were purified by countercurrent distribution and HPLC (where necessary), characterized by amino acid analysis, mass spectrometry, TLC and paper electrophoresis.

---

*Present Address: Cortech Inc., 6840 N. Broadway, Denver, CO 80221, U.S.A.

The bradykinin antagonists were assayed *in vitro* on isolated rat uterus (RUT) and guinea pig ileum (GPI), and *in vivo* on rat blood pressure (RBP) by standard assay methods. Inhibition potencies on smooth muscles are expressed as pA_2 values of Schild [4]. The *in vivo* effect of bradykinin antagonists on blood pressure in the anesthetized rat were determined as described by Stewart et al. [5]. The antagonists produce inhibition of the hypotensive action of bradykinin when administered as bolus admixture with BK or as an infusion, but potency in this assay was not quantitated precisely.

Results and Discussion

The bradykinin antagonist D-Arg-[Hyp^3-Thi^5-D-Cpg^7-Cpg^8]-BK, (B8820) has high potency in these assays, and its effect on blood pressure lasts for 180 min. following an infusion (Table 1). Replacement of Phe with Cpg at position 5, keeping D-Cpg at 7 and L-Cpg at 8 (B8838), reduces RUT and GPI activity, and this analog does not show long term inhibition on rat blood pressure following an infusion. Cpg at position 5 in combination with D-Tic^7-Oic^8 (B9076) gives antagonists with significant activity. Incorporation of cyclopropane phenylalanine, especially its E-isomer, at position 8 (B8844) in the [D-Phe^7] BK antagonist, shows enhancement in pA_2 value and a prolonged action lasting 45 min following an infusion in rat blood pressure. Incorporation of 2,4-methanoproline at either position was not useful for increasing activity.

Table 1 *Structures and activities of bradykinin antagonists*

NUMBER	STRUCTURE	BIOLOGICAL ACTIVITIES		
		RUT	GPI	RBP(min)
NPC349	D-Arg-[Hyp^3-Thi^{5,8}-D-Phe^7]-BK	I(6.5)	I(6.4)	I(LT,16)
B7890	D-Arg-[Hyp^3-Thi^5-D-Phe^7-Cpg^8]-BK	I(7.0)	I(6.5)	I(LT,50)
B8820	D-Arg-[Hyp^3-Thi^5-D-Cpg^7-Cpg^8]-BK	I(7.9)	I(6.6)	I(LT,180)
B8838	D-Arg-[Hyp^3-Cpg^{5,8}-D-Cpg^7]-BK	I(7.4)	I(7.0)	I(LT,38)
B8844	D-Arg-[Hyp^3-Thi^5-D-Phe^7-CpPhe(E)^8]-BK	I(7.0)	I(6.9)	I(LT,45)
B8862	D-Arg-[Hyp^3-Thi^5-D-Cpg^7-Tic^8]-BK	I(6.7)	I(6.0)	I
B8868	D-Arg-[Hyp^3-Thi^5-DAlg^7-Tic^8]-BK	I(7.2)	I(7.2)	I
B8878	D-Arg-[Hyp^3-Thi^5-D-Cpg^7-CpPhe(E)^8]-BK	I(7.1)	I(7.1)	I(LT,35)
B8976	D-Arg-[Hyp^3-Thi^5-D-Phe^7-pClCpPhe(Z)^8]-BK	I(5.5)	I(6.0)	
B8994	D-Arg-[MPIV^3-Cpg^5-D-Cpg^7-Cpg^8]-BK	I(7.2)	I(6.5)	I
B9070	D-Arg-[Hyp^3-Thi^5-D-Phe^7-pCl-D-Phe^8]-BK	I(5.5)	0	0
B9076	D-Arg-[Hyp^3-Cpg^5-D-Tic^7-Oic^8]-BK	I(6.8)	I(7.5)	I(LT,140)

I(LT): Long term inhibition of rat blood pressure (RBP) response by intraaortic infusion of antagonist at 0.1 mg/min; Antagonist activity on smooth muscles is given as the pA_2 determined on 4-10 tissues; RUT: rat uterus; GPI: guinea pig ileum; I: inhibition on RBP; not quantitated.

Acknowledgements

This work was supported by grants from the U.S. NIH (grant HL-26284) and from Nova Pharmaceutical Corp. We thank Prof. Charles H. Stammer for providing cyclopropane amino acids, Robin Reed and Robert Binard for assistance with chemistry, and Frances Shepperdson for the assays.

References

1. Vavrek, R.J. and Stewart, J.M., Peptides, 6 (1985) 161.
2. Stewart, J.M. and Vavrek, R.J., In Burch, R.M. (Ed.), Bradykinin Antagonists, Marcel Dekker, New York, NY, U.S.A., 1991, p.51.
3. Stewart, J.M. and Young, J.D., Solid Phase Peptide Synthesis, 2nd Ed., Pierce Chemical Co., Rockford, IL, U.S.A.
4. Schild, H.O., Brit. J. Pharmacol., 2 (1947) 189.
5. Ryan, J.W., Roblero, J. and Stewart, J.M., Biochem. J., 110 (1968) 795.

Structure-activity relationships of positions 7, 8, and 9 in LHRH antagonists with respect to histamine release, LH, and testosterone suppressions

R.E. Swenson[1], N.A. Mort[1], F. Haviv[1], E.N. Bush[1], G. Diaz[1], G. Bammert[1], N. Rhutasel[1], A. Nguyen[1], J. Leal[1], V. Cybulski[1], J. Knittle[2] and J. Greer[1]

[1]*Pharmaceutical Products Division, Abbott Laboratories, and*
[2]*TAP Pharmaceuticals Inc., Abbott Park, IL 60064, U.S.A.*

Introduction

The major safety problem of LHRH antagonists was related to the degranulation of mast cells and release of histamine [1]. SB-75 is a known LHRH antagonist in phase II clinical trials [2], which is administered sc bid, because of its short duration of action. In the histamine release (HR) rat peritoneal mast cells assay SB-75 has an ED_{50} of 1.1 µg/ml [3]. Our goal in this study was to improve the potency and safety profile of SB-75. Haviv *et al.* reported [3] the *in vitro* activity and LH suppression *in vivo* of NMeTyr^5-SB-75, **2,** in comparison with the parent, **1**. Recently we tested this antagonist **2** in intact rat by infusion for three days at a dose of 3.6 µg/kg/hr. This compound was far more effective than SB-75 (Figure 1) and therefore was selected as a lead in the following SAR study of positions 7, 8, and 9, designed to improve the safety profile.

Results and Discussion

LHRH antagonists 1-18 were synthesized by SPPS, and tested *in vitro* for rat pituitary LHRH receptor binding (pK_I), inhibition of LH release from rat pituitary cells (pA_2) and HR (ED_{50}) [3]. Selected compounds were also tested *in vivo* for suppression of testosterone (T) in intact rats by infusion. N-methylation of position 7 of NMeTyr^5-SB-75, **4,** increased the HR ED_{50} by 10-fold over SB-75. Likewise, introduction of bulkier side groups such as Val^7, **6,** or t-Butylglycine^7, **5,** increased the HR ED_{50} by 3- and 6-fold respectively. Substitution of position 8 with NMeArg^8, **7,** resulted in a 20-fold, improvement in HR ED_{50}, although a substantial loss in other *in vitro* activities was observed. This remarkable enhancement in HR ED_{50} value as a result of backbone conformational restriction is novel and unique. Previous attempts to suppress HR activity focused on decreasing basicity and reducing hydrophobicity [1]. The 3-fold improvement in safety of the HArg^8 analog, **8,** indicates the four carbon side chain length at position 8 is preferred for safety. Replacement of Arg^8 with Lys(Isp)^8 **9** also improved the safety profile 15-fold, possibly suggesting that changing the character of the basic residue is beneficial.

The Lys(Isp)^8 analog **9** was also very effective *in vivo* in suppressing T (Figure 1). Substitution of NMeTyr^5-SB-75, **2** with Ala^9 produced an analog 12 with similar *in vitro* activity and only a moderate improvement in HR ED$_{50}$. Substitution of azetidinecarboxylic acid^9, **16**, or pipecolic acid^9, **15**, did not have any significant effect on *in vitro* potency and safety indicating that conformational restriction at this site is neither beneficial nor detrimental. The poor binding of NMeLeu^9, **11**, suggests that the steric environment at this site is limited. The two best modifications with highest HR ED$_{50}$ values were the aminobutyric acid^9, **13**, and the 4-hydroxyproline^9, **14**, analogs. The transposition of positions 7 and 8 yielded compound **18**, which maintained *in vitro* potency and had an almost 10-fold improvement in safety [4].

Table 1 *In vitro activities of LHRH antagonists*

NAcD2Nal^1-D4ClPhe^2-D3Pal^3-Ser^4-NMeTyr^5-DCit^6-Leu^7-Arg^8-Pro^9-DAla^{10}-NH$_2$

Compound	X	pK$_I$ [a]	pA$_2$ [b]	ED$_{50}$ [c]
1	Tyr^5,Leu^7,Arg^8,Pro^9 SB-75^d	9.93	10.45	1.11
2	as above NMeTyr^5-SB-75^a	10.87	10.80	1.65
3	Ile^7,Arg^8,Pro^9	10.58	11.25	4.53
4	NMeLeu^7,Arg^8,Pro^9	10.59	11.02	11.45
5	tBugly^7,Arg^8,Pro^9	10.32	11.85	6.93
6	Val^7,Arg^8,Pro^9	10.67	11.15	3.26
7	Leu^7,NMeArg^8,Pro^9	8.75	8.14	23.86
8	Leu^7,Harg^8,Pro^9	10.21	11.10	3.52
9	Leu^7,Lys(Isp)^8,Pro^9	9.39	11.40	15.00
10	Leu^7,Arg^8,NMeAla^9	10.47	11.05	0.61
11	Leu^7,Arg^8,NMeLeu^9	9.49	9.30	0.53
12	Leu^7,Arg^8,Ala^9	10.35	10.90	3.20
13	Leu^7,Arg^8,Abu^9	9.96	10.50	6.05
14	Leu^7,Arg^8,Pro(OH)^9	10.33	10.95	4.99
15	Leu^7,Arg^8,Pipec^9	10.36	11.05	1.39
16	Leu^7,Arg^8,Aze^9	10.53	11.15	1.62
17	Leu^7,Arg^8,Tic^9	9.82	10.45	1.68
18	Arg^7,Leu^8,Pro^9	10.18	10.29	10.53

[a] pK$_1$ = the negative log of the equilibrium dissociation constant in the rat pituitary receptor binding assay. [b] pA$_2$ = the negative logarithm of the concentration of antagonist that requires 2-fold higher concentration of agonist to release LH from rat pituitary cells. [c] ED$_{50}$ = the effective dose of antagonist that gives 50% of maximal release of histamine from rat peritoneal mast cells. Abu = amino butyric acid. Pro(OH) = 4-hydroxyproline. Pipec = pipecolic acid. Aze = azetidinecarboxylic acid. Tic = 1,2,3,4-tetrahydro-3-isoquinolinecarboxylic acid. [d] Ref 3.

These results show for the first time that the propensity of LHRH antagonists to release histamine can be decreased solely as the result of a backbone restriction caused by N-methylation of position 8 or by increasing the side chain bulk at position 7. This finding also indicates that HR can be suppressed not only by varying the physicochemical properties of antagonists, but also by restricting conformation. Substitution of NMeTyr⁵-SB-75, **2**, with Lys(Isp)⁸ yielded antagonist 9, which was very efficacious both *in vitro* and *in vivo* (Figure 1).

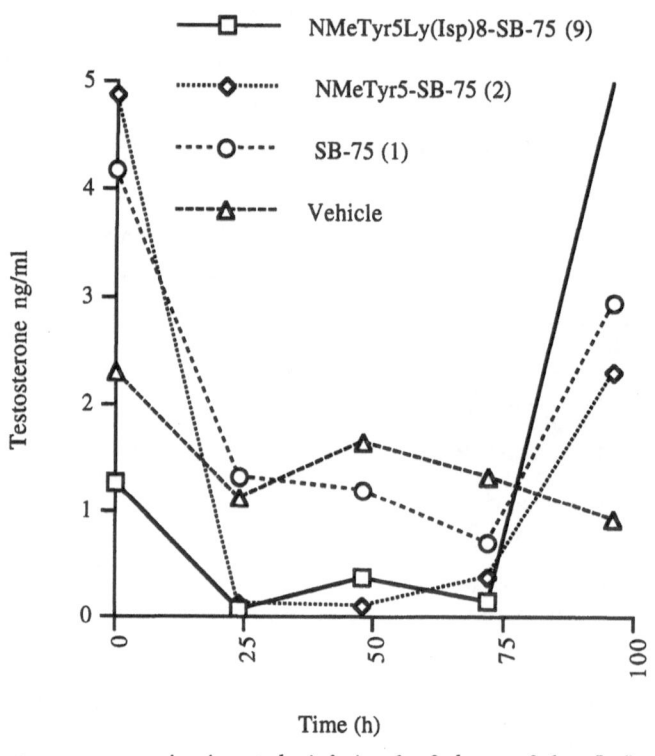

Time (h)

Fig. 1. Testosterone suppression in rats by infusion for 3 days at 3.6 µg/kg/hr

References

1. Karten, M.J. and Rivier, J.E., Endocr. Rev., 7 (1986) 44.
2. Bokser, L., Bajusz, S., Groot, K. and Schally, A.V., Proc. Natl. Acad. Sci. U.S.A., 87 (1990) 7100.
3. Haviv, F., Fitzpatrick, T.D., Nichols, C.J., Swenson, R.E., Mort, N.A., Bush, E.N., Diaz, G., Nguyen, A.T., Holst, M.R., Cybulski, V.A., Leal, J.A., Bammert, G., Rhutasel, N.S., Dodge, P.W., Johnson, E.W., Cannon, J.B., Knittle, J. and Greer, J., J. Med. Chem., 36 (1993) 928.
4. Floret, G., Mahan, K. and Majewski, T., J. Med. Chem., 35 (1993) 636.

Anti-HIV activity and conformational analysis of a novel synthetic peptide, T22 ([Tyr^{5,12}, Lys^7]-polyphemusin II)

H. Tamamura[1], H. Nakashima[2], M. Masuda[1], A. Otaka[1],
S. Funakoshi[1], T. Murakami[2], N. Yamamoto[2]
and N. Fujii[1]

[1]*Faculty of Pharmaceutical Sciences, Kyoto University, Sakyo-ku, Kyoto 606, Japan*
[2]*Department of Microbiology, Tokyo Medical and Dental University, School of Medicine, Yushima, Bunkyo-ku, Tokyo 113, Japan*

Introduction

Tachyplesins I and II and polyphemusins I and II are antimicrobial peptides isolated from the acid extracts of the hemocytes of the Japanese horseshoe crab (*Tachypleus tridentatus*) and the American horseshoe crab (*Limulus polyphemus*) [1, 2]. Tachyplesin I was also shown to have antiviral activity against human immunodeficiency virus type 1 [3]. However, its cytotoxicity is relatively strong. Problems in reducing the cytotoxicity as well as in potentiating the anti-HIV activity has remained to be further improved. In the present study, we synthesized more than 20 peptides through modification of tachyplesin or polyphemusin (Figure 1) and evaluated their abilities to suppress the replication of HIV *in vitro*. Among these peptides, we found that T22 ([Tyr^{5,12}, Lys^7]-polyphemusin II) showed strong anti-HIV activity and relatively low cytotoxicity. Thus, we determined the solution structures of T22 by NMR to disclose the relationship between the conformation and the expression of anti-HIV activity.

Results and Discussion

Table 1 summarizes anti-HIV activity of the native tachyplesins and their analogs. The anti-HIV activity was evaluated on the basis of protection of virus-induced cytopathogenicity in MT-4 cells by the MTT method [4]. Tachyplesin I, tachyplesin II, polyphemusin I and polyphemusin II inhibited HIV-1-induced cytopathic effects (CPE). However, their antiviral activity was only seen at the concentration of 3 to 17 times lower than their cytotoxic concentration. The results of structure-activity relationship study of T5-T10 indicated that Trp^2 and the major disulfide ring are essential to the inhibition of viral replication. T11, T14 and T15 strongly inhibited HIV-1-induced CPE. The anti-HIV activity of T12 or T13 was weaker than that of T11, indicating the importance of the N-terminal Arg residue. Especially, 50% effective concentration of T22 was 0.008 μg/ml. Its selectivity index (ratio of CC_{50} to EC_{50}) was 6700. This value was comparable to that of AZT.

T11, T14, T15 and T22 have two repeats of Cys-Tyr-Arg-Lys-Cys and two disulfide bonds. This unique structure seems to be closely related with high anti-HIV activity.

tachyplesin I	NH_2-K-W-C-F-R-V-C-Y-R-G-I-C-Y-R-R-C-R-$CONH_2$
tachyplesin II	NH_2-R-W-C-F-R-V-C-Y-R-G-I-C-Y-R-K-C-R-$CONH_2$
polyphemusin I	NH_2-R-R-W-C-F-R-V-C-Y-R-G-F-C-Y-R-K-C-R-$CONH_2$
polyphemusin II	NH_2-R-R-W-C-F-R-V-C-Y-K-G-F-C-Y-R-K-C-R-$CONH_2$
T5	NH_2-W-C-F-R-V-C-Y-R-G-I-C-Y-R-R-C-R-$CONH_2$
T6	NH_2-C-F-R-V-C-Y-R-G-I-C-Y-R-R-C-R-$CONH_2$
T7	NH_2-K-F-C-F-R-V-C-F-R-G-I-C-F-R-R-C-R-$CONH_2$
T8	NH_2-K-W-C-F-R-V-C-F-R-G-I-C-F-R-R-C-R-$CONH_2$
T9	NH_2-K-W-C-F-R-V-A-Y-R-G-I-A-Y-R-R-C-R-$CONH_2$
T10	NH_2-K-W-A-F-R-V-A-Y-R-G-I-A-Y-R-R-A-R-$CONH_2$
T11	NH_2-R-W-C-Y-R-K-C-Y-R-G-I-C-Y-R-K-C-R-$CONH_2$
T12	NH_2-W-C-Y-R-K-C-Y-R-G-I-C-Y-R-K-C-R-$CONH_2$
T13	NH_2-A-W-C-Y-R-K-C-Y-R-G-I-C-Y-R-K-C-R-$CONH_2$
T14	NH_2-R-W-C-Y-R-K-C-Y-K-G-I-C-Y-R-K-C-R-$CONH_2$
T15	NH_2-R-W-C-Y-R-K-C-Y-K-G-F-C-Y-R-K-C-R-$CONH_2$
T22	NH_2-R-R-W-C-Y-R-K-C-Y-K-G-Y-C-Y-R-K-C-R-$CONH_2$

Fig. 1. Structure of tachyplesin I, its isopeptides and their analogs.

In order to understand the mechanism of the expression of strong anti-HIV activity, knowledge of the solution structure of T22 was required. The solution structure of T22 was investigated using NMR. From the information of NOE, the coupling constants, the amide proton exchange rates and the chemical shifts, we concluded that the secondary structure of T22 resembles that of tachyplesin I [5]. T22 is composed of an antiparallel β-sheet structure with a type II β-turn [Figure 2]. Although the difference of sequence between natural polyphemusin II and T22 was only three amino acid residues, T22 showed two hundred times stronger anti-HIV activity than polyphemusin II. The side chains of these three amino acids protrude in the same direction from the plane of the β-sheet. This side of the plane may be related to the expression of high activity, and the replacement of the three amino acid residues (Phe[5]-Tyr, Val[7]-Lys, Phe[12]-Tyr) is thought to affect the interaction with the target material, which is not identified yet. The increase of positive charges may also influence the interaction. Taking its high anti-HIV activity and high selectivity index into consideration, it is concluded that T22 has the potency to become an attractive candidate compound for the chemotherapy of HIV infection. The results from this conformation analysis will aid us in disclosing the reason for the high activity of T22, and possibly lead to rational designs of drugs possessing higher activity.

Table 1 *Anti-HIV activity of tachyplesin I, its isopeptides and their analogs*

Compound	EC_{50} (µg/ml)	CC_{50} (µg/ml)	S.I. (CC_{50}/EC_{50})
tachyplesin I	8.4	29	3
tachyplesin II	6.6	36	5
polyphemusin I	5.9	34	6
polyphemusin II	1.9	33	17
T5	8.8	37	4
T6	>100	38	<1
T7	>100	31	<1
T8	5.4	27	5
T9	5.8	31	5
T10	>500	109	<1
T11	0.35	39	113
T12	6.9	206	30
T13	6.9	300	44
T14	0.18	52	300
T15	0.087	53	600
T22	0.008	54	6700
dextran sulfate	0.76	>1000	>1316
AZT	0.0014	2.89	2038

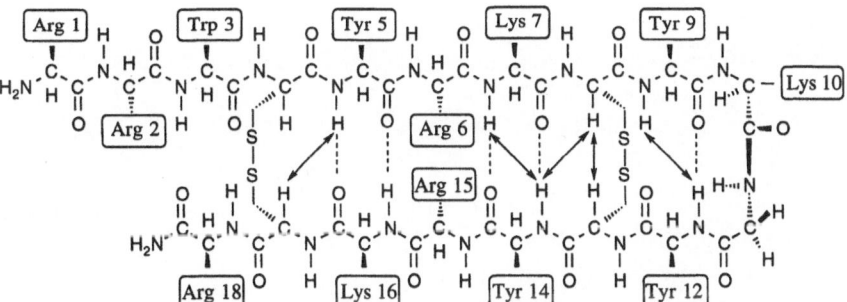

Fig. 2. *Schematic representation of the structure of T22 deduced from the information from NMR analysis. The observed interstrand NOE connectivities between NH and CαH protons are shown by arrows. The hydrogen bonds deduced from slowly exchanging amide protons are exhibited by dashed lines.*

References

1. Miyata, T., Tokunaga, F., Muta, T., Iwanaga, S., Niwa, M., Takao, T. and Shimonishi, Y., J. Biol. Chem., 263 (1988) 16709.
2. Miyata, T., Tokunaga, F., Yoneya, T., Yoshikawa, K., Iwanaga, S., Newa, M., Takao, T. and Shimonishi, Y., J. Biochem., 106 (1989) 663.
3. Morimoto, M., Mori, H., Otake, T., Ueda, N., Kunita, N., Niwa, M., Murakami, T. and Iwanaga, S., Chemotherapy, 37 (1991) 206.
4. Pauwels, R., Balzarini, J., Baba, M., Snoeck, R., Schols, D., Herdewijn, P., Desmyter, J. and De Clercq, E., J. Virol. Methods, 20 (1988) 309.
5. Kawano, K., Yoneya, T., Miyata, T., Yoshikawa, K., Tokunaga, F., Terada, Y. and Iwanaga, S., J. Biol. Chem., 265 (1990) 15365.

Highly active atrial natriuretic factor (ANF) analogues, lacking the exocyclic N- and C-termini

H. Thøgersen[1], P. Faarup[1], N.L. Johansen[1], B.F. Lundt[1], K. Madsen[1], J.U. Weis[1], H. Echner[2] and W. Voelter[2]

[1]Novo Nordisk A/S, Pharmaceuticals Research, Novo Nordisk Park, DK-2760 Maaloev, Denmark, [2]Abteilung für Physikalische Biochemie des Physiologisch-chemischen Instituts der Universität Tübingen, Hoppe-Seyler-Strasse 4, D-7400 Tübingen, Germany

Introduction

In our search for a potent reduced-size ANF analogue we guided our synthetic work on models for the bioactive conformation of ANF which initially was derived from the spectroscopic evidence about the conformation of mainly ANF(5-28) (1) in solution and subsequently refined in accordance with additional information from biological activities of our synthetic ANF analogues.

Results and Discussion

In the presence of lipid bilayers, CD [1] and FT-IR [2] spectroscopic data show that ANF(5-28) mainly adopts a β-sheet-like or extended structure, some turn structure and only little if any α-helical structure. We proposed that ANF(7-28) upon receptor-binding might adopt an extended structure with β-turns at $Asp^{13}Arg^{14}$ and $Cys^{23}Asn^{24}$. These turns segregate the hydrophobic and hydrophilic residues (except for Ser^{19}) at opposite faces of the ANF ring and pack the hydrophobic residues of the otherwise extended C-terminus against this hydrophobic surface (Figure 1A). The suggested packing of the C-terminus led to the synthesis of 4-6 (Table 1) for which we propose that the essential Arg^{27} [3] might be mimicked by a replacement of Gly^{20} by Arg or D-Arg. The most potent analogue 4 maintains ~25% of the potency of 1 in, both, a rabbit renal artery assay and a rat diuretic assay; and is almost 100 fold more potent than 3 on the renal artery.

Furthermore, we found that Arg^4 could be mimicked by a replacement of Ala^{17} by Arg leading to the synthesis of 7 and 8 which showed improved renal activity and unchanged diuretic properties as compared to 4 and 5 (Table 1). The spatial proximity proposed between Arg^4 and Ala^{17} and between Arg^{27} and Gly^{20} suggested a particular folding of 2 at the receptor. A refined model for the bioactive conformation of 7 (Figure 1B) was derived from the proposed folding of 2 by removal of the exocyclic termini and proper mutations in the ring structure. Based on this model, we synthesized the potent analogues 9 and 10 (Table 1) with additional truncations in the Gly^{22},Cys^{23},Cys^7 segment. 9 and 10 are both more potent than the endogenous peptide hormone and are to the best of our knowledge the most active ANF analogues of this size known today.

459

Table 1 *Relative biological activities of ANF analogues*

ANF analogue	rabbit renal artery	rat diuresis
1 ANF(5-28)	1	1
2 ANF(4-28)	5	1
3 Ac-ANF(7-23)-NH$_2$	0.003	0.1
4 [D-Arg^{20},Phe^{21}]Ac-ANF(7-23)-NH$_2$	0.2	0.3
5 [Arg^{20},Phe^{21}]Ac-ANF(7-23)-NH$_2$	0.05	0.04
6 [Arg^{20}]Ac-ANF(7-23)-NH$_2$	0.03	-
7 [Arg^{17},D-Arg^{20},Phe^{21}]Ac-ANF(7-23)-NH$_2$	1.8	0.2
8 [Arg^{17},Arg^{20},Phe^{21}]Ac-ANF(7-23)-NH$_2$	0.3	0.04
9 [D-Cys^7,Arg^{17},D-Arg^{20},Phe^{21},Cys^{22}]-Ac-ANF(7-22)-NH$_2$	10	1
10 [Mercaptopropionyl^7,Arg^{17},D-Arg^{20},Phe^{21},Cys^{22}]-ANF(7-22)-NH$_2$	16	1

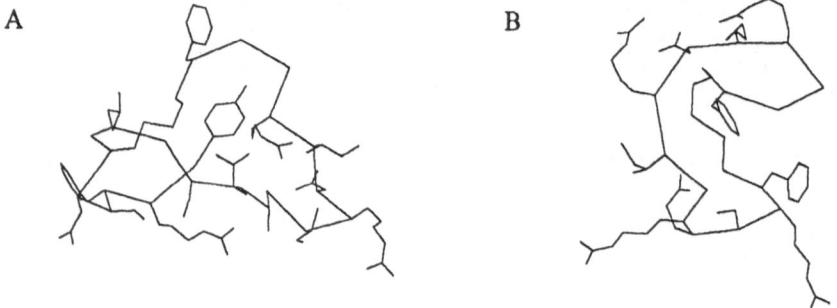

A   B

Fig. 1. *Hypothetical models for the bioactive conformation of Ac-ANF(7-28) (A) and of [Arg^{17},D-Arg^{20},Phe^{21}]Ac-hANF(7-23)-NH$_2$ (B). 50 ps molecular dynamics simulation at 300 K and energy minimization has been applied on both models to remove unfavorable intra-molecular interactions.   The peptide backbones are shown as Cα-traces.   All hydrogen atoms are invisible.*

References

1.   Epand, R.M. and Stahl, G.L., Int. J. Protein Res., 29 (1987) 238.
2.   Surewicz, W.K., Mantsch, H., Stahl, G.L. and Epand, R.M., Proc. Natl. Acad. Sci. U.S.A., 84 (1987) 7028.
3.   Nutt, R.F., Ciccarone, T.M., Brady, S.F., Colton, C.D., Paleveda, W.J., Lyle, T.A., Williams, T.M., Veber, D.F., Wallace, A. and Winquist, R.J., In Marshall, G.R. (Ed.), Peptides, Proceedings of the 10th American Peptide Symposium, Escom, Leiden, The Netherlands, 1988, p.444.

Synthetic approach to the identification of the antibacterial domain of bactenecins, Pro/Arg-rich peptides from bovine neutrophils

R. Gennaro[1], M. Scocchi[2], B. Skerlavaj[1], A. Tossi[2] and D. Romeo[2]

[1]Department of Biomedical Sciences and Technologies, University of Udine, I-33100 Udine [2]Department of Biochemistry, Biophysics and Macromolecular Chemistry, University of Trieste, I-34127 Trieste, Italy

Introduction

We have isolated two highly cationic peptides, named bactenecins, which represent a new family of Pro/Arg-rich antibiotics [1]. They are termed Bac5 and Bac7 from their respective molecular weight of 5 and 7 kDa, and both efficiently kill various gram negative and gram positive bacteria *in vitro*. These peptides are stored as inactive proforms in the so called large granules of bovine neutrophils, and are activated by proteolytic removal of the pro region by elastase [2, 3]. Their sequence is characterized by a highly cationic N-terminal portion, followed by frequent repeats of the Arg-Pro-Pro-X (Bac5) or Pro-Arg-Pro-X (Bac7) motifs, where X are hydrophobic residues (Figure 1). Moreover, in Bac7, these motifs are arranged in three tandemly repeated tetradecamers from residues 15 to 56 [4]. This unusual primary structure prompted us to investigate which characteristics (e.g., charge distribution, repeats) were necessary for the antibacterial activity. To this end, we chemically synthesized a series of fragments covering the whole sequence of both peptides (Figure 1), and tested their antibacterial and membrane permeabilizing activities.

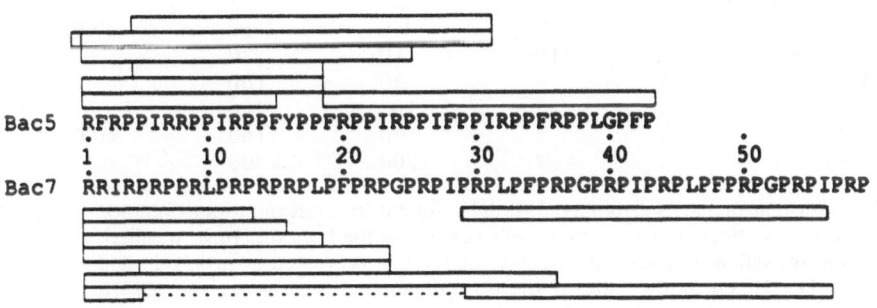

Fig. 1. Sequences of Bac5 and Bac7 and schematic representation of synthesized fragments.

461

Results and Discussion

A series of fragments of Bac5 and Bac7 were synthesized by the solid phase method, using Fmoc-amino acid derivatives, and purified by RPHPLC. Amino acid analysis and mass spectrometry indicated that the peptides had the expected composition and were obtained with a high degree of purity and excellent yield.

The antibacterial activity of the fragments was compared to that of the parent peptides using the minimal inhibitory concentration assay (MIC), which determines the minimal concentration at which no bacterial growth is observed. We tested the peptides against several gram negative and gram positive bacteria, including *Escherichia coli, Bacillus megaterium* (Table 1), *Salmonella typhimurium, Serratia marcescens* and *Staphylococcus epidermidis* (not shown).

Table 1 *Antibacterial and membrane permeabilization activities of Bac5, Bac7 and related synthetic peptides*

PEPTIDE	# of residues	positive[a] charges	MIC (μM)[b]		permeabilization[c]
			E. coli	*B. megaterium*	
Bac5	43	10	1	2	0.44
Arg+1-31	32	9	0.5	2	0.36
1-31	31	8	1.0	20	0.21
1-25	25	8	1.0	50	0.12
1-18	18	6	40	160	
1-15	15	6	80	>160	
4-31	28	6	>100	>100	
4-18	15	4	>160	>160	
19-43	25	5	>200	>200	<0.01
Bac7	59	18	0.5	1	0.33
1-35	35	12	0.25	0.5	0.25
1-23	23	10	0.5	2.0	0.09
1-23am	23	10	1.0	2.0	0.20
ac1-23am	23	9	2.0	10	<0.08
1-23am (all-D)	23	10	2.0	5	
1-18	18	9	1.0	10	<0.05
1-4+29-56	32	10	10	80	
1-15	15	8	20	160	
29-56[d]	28	7	>100	>100	
5-23	19	7	100	>160	<0.05
1-13	13	7	>200	>200	

[a]Includes that of the N-terminus; [b]Minimal inhibitory concentration, obtained by serial dilution of peptides into microtiter wells containing the bacteria, 16 hr incubation at 37°C and determination of the minimal concentration that prevented growth; repeated at least 3 times [1]; [c]Given as the ratio of the time required to obtain a given absorbance after addition of 50 μg/ml of peptide to that required to reach the same absorbance after sonication (100% permeabilization); [d]Identical to Bac7(15-42); am = amidated; ac = acetylated.

The results show that the highly cationic N-terminal portion of Bac5 or Bac7 is essential for antibacterial activity and, given its presence, a minimum length of 18-20 residues is required. Thus, removal of only four residues (RRIR) from the N-terminus of Bac7(1-23) markedly reduces its activity [see Bac7(5-23)], while the shorter fragment 1-18, conserving the RRIR sequence, shows an activity comparable to that of Bac7. Similarly, removing RFR from the N-terminus of Bac5(1-31) virtually eliminates its activity [see Bac5(4-31)]. The relatively long C-terminal fragments Bac5(19-43) and Bac7(29-56) are inactive, but the activity of the latter can be partially recovered by adding the RRIR motif to its N-terminus [see Bac7(1-4 + 29-56)]. The importance of positive charge is further emphasized by the fact that addition of a single Arg residue to the N-terminus of Bac5(1-31) doubles its activity [see Bac7(Arg+1-31)], while N-terminal acetylation of the Bac7(1-23am) causes a slight reduction of its activity. We have also found that in the case of Bac7(1-23am), the all-D enantiomer maintains a comparable activity. Apart from pointing to a means of increasing resistance to proteolytic degradation, this also indicates that the mode of action of these peptides is not likely to be receptor based.

Bac5 and Bac7, as many other antibacterial peptides, were found to permeabilize both the inner and outer membrane of *E. coli* [5]. The ability of the synthetic peptides to permeabilize the inner membrane was tested using a lactose permease-deficient strain (ML-35) of *E. coli*, which constitutively produces β-galactosidase in the cytoplasm. The kinetics of permeabilization can be followed by monitoring the hydrolysis of the substrate *o*-nitrophenyl-β-D-galactopyranoside, accessible to the enzyme only upon inner membrane disruption. The activity of several peptides is reported in Table 1 as the ratio of the time required to obtain a given absorbance after the addition of peptide, to that after sonication (100% permeabilization).

These experiments confirm the trends observed for antibacterial activity, *i.e.* the presence of the highly cationic N-terminal portion of the peptides and a minimum length of 18-20 residues is required. Moreover it appears that the kinetics of permeabilization is directly dependent on fragment length. For example, permeabilization velocity decreases in the order Bac7>Bac7(1-35)>Bac7(1-23). Interestingly, amidation of Bac7(1-23) increases the velocity to a value comparable to that of Bac7(1-35).

In conclusion, by synthesizing a set of peptides of different length, covering the whole sequence of Bac5 or Bac7, we have pinpointed those factors which determine their antibacterial activity. This information is being used in ongoing studies on the structure of these peptides, their mode of interaction with bacterial membranes, and in designing peptides for biotechnological uses.

References

1.   Gennaro, R., Skerlavaj, B. and Romeo, D., Infect. Immun., 57 (1989) 3142.
2.   Zanetti, M., Litteri, L., Gennaro, R., Horstmann, H. and Romeo, D., J. Cell Biol., 111 (1990) 1363.
3.   Scocchi, M., Skerlavaj, B., Romeo, D. and Gennaro, R., Eur. J. Biochem., 209 (1992) 589.
4.   Frank, R.W., Gennaro, R., Schneider, K., Przybylski, M. and Romeo, D., J. Biol. Chem., 265 (1990) 18871.
5.   Skerlavaj, B., Romeo, D. and Gennaro, R., Infect. Immun., 58 (1990) 3724.

Russtoxin: A new family of two-component phospholipase A_2 toxins from Russell's vipers

I.-H. Tsai, P.-J. Lu and Y.-C. Su

Institute of Biological Chemistry, Academia Sinica, Taipei, Taiwan 10798, Republic of China

Introduction

The venom components and antivenin of Russell's viper have been considered as medically important because the snake caused more deaths than any other snake in Southeast Asia and India [1, 2]. The major symptoms of envenomation by Russell's vipers are consumptive coagulation and neurotoxic signs [3-5]. We have previously found that the phospholipase A_2 (PLA$_2$) from the venom of *Vipera russelli (V. r.) formosensis* was a neurotoxic heterodimer RV-4:RV-7 (1:1) [5], designated as russtoxin hereafter. The venoms of the other Russell's vipers were found containing many PLA$_2$ isoforms [5, 6], but their amino acid sequences have not been studied. The objective of this study is to identify the presence of russtoxin in other species of Russell's vipers, and analyze how the acidic subunit of russtoxin affects the biologic function of the basic subunit.

Results and Discussion

We compared venom proteins of four species of Russell's vipers: *V. r. formosensis* from Taiwan, *V. r. pulchella* from Sri Lanka, *V. r. russelli* from India and Pakistan, and *V. r. siamensis* from Thailand. Each of these venoms was separated into three major fractions by Sephadex G100 gel filtration column. The PLA$_2$s found in the major peak were further purified by C_{18} RP-HPLC (Figure 1). After confirmation of their purity by both C_8-RP-HPLC re-chromatography and SDS-PAGE, the purified PLA$_2$s were subjected to automated protein sequencing up to the 40-50 residues. Based on their sequences the PLA$_2$s could be categorized into several groups, Table 1 lists two of the sequence groups. We found that the venoms of *V. r. russelli* and *V. r. siamensis* contained russtoxin subunits similar to those obtained from *V. r. formosensis*. R3 and S4 were very similar if not identical to F7 (i.e. RV-7 [5]); R2 and S2 were also similar or identical to F4 (i.e .RV-4 [5]) except that the HPLC-elution times of R2 and S3 differed slightly from those of F4 and S2 (Figure 1).

The enzymatic activity of R2 was slightly higher than F4, while both F7 and R3 showed much lower catalytic activities (only about 1/150 those of F4 or R2). The complexes of R2/R3 (1:1) and F4/F7 (1:1) had a similar LD$_{50}$ (i.v.) of 0.15 μg/g mice, and at 0.3-0.6 μg/ml they effectively blocked the contraction of chick biventer cervicies tissue [5]. Therefore, Russtoxin-like molecules exist in all of the three Russell's vipers studied. In contrast, the venom of *V. r. pulchella* contains a different

464

set of PLA$_2$ isoforms. This conclusion was also confirmed by immunochemical analysis using antisera prepared from purified F4 and F7 as antigens.

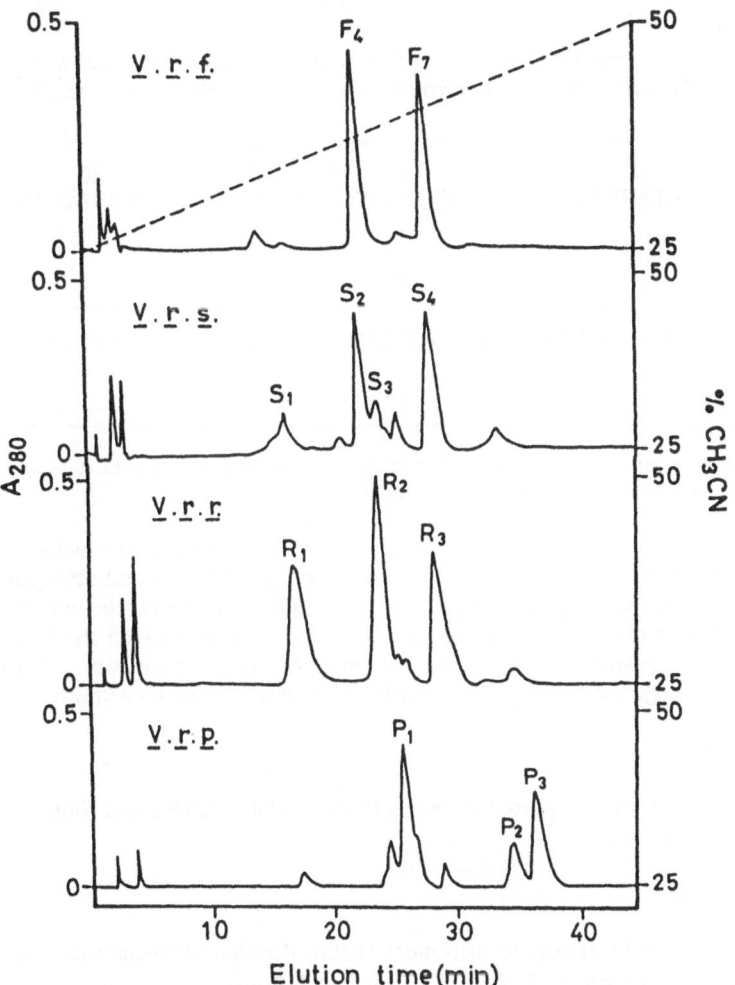

Fig. 1. RP-HPLC of the PLA$_2$-containing fractions from Sephadex G-100. A Chemcosorb C$_{18}$-RP column (ODS-H, 5 μm, 1 x 25 cm) was used to isolate the PLA$_2$ isoforms from the four viper species. Solvent A was 0.07% TFA in H$_2$O, solvent B was 0.07% TFA in CH$_3$CN, and the elution was by a linear gradient 25-50% of solvent B over 36 min at a flow rate of 2 ml/min.

R2 alone was much less neurotoxic toward the chick tissue than F4 but the 1:1 complex of R2 with R3 or F7, was as toxic as the F4/F7 complex. We therefore determined the amino acid sequence of R2 by endoproteinase fragmentation followed by automated N-terminal sequencing of the fragments. The complete amino acid sequence of R2 differed from that reported for F4 [5] by only 5 substitutions at the

C-terminal region: the substitution in R2 with respect to F4 are: E116R, K125Y, and E131K besides two semiconserved changes of S124Y and R129K. Among these, K125Y probably is an important factor responsible for the low neurotoxicity of R2 since region 119-128 has been suggested to be a part of the neurotoxic site [5].

Table 1   *The N-terminal amino acid sequences of russtoxin-related PLA$_2$s from Russell's vipers: V. r. formosensis, V. r. russelli and V. r. siamensis*

```
PLA2          10        20        30        40        50
F4    NLFQFARMINGKLGAFSVWNYISYGCYCGWGGQGTPKDATDRCCFVHDCC
R2
S2
S3             DA QE    F   K
----------------------------------------------------------------
F7    NLFQFGEMILEKTGKEVVHSYAIYGCYCGWGGQGRAQDATDRCCFVHDCC
R3
S4
```

Abbreviations of PLA$_2$ were as in Figure 1. In both groups of the PLA$_2$s only the first sequence and the residues different from the first are shown.

The acidic subunits of russtoxin (F7 or R3) inhibited the enzymatic activity of the basic subunit (F4 or R2) toward phospholipid substrates and also protected it against inactivation by *p*-bromophenancyl bromide. However, the acidic subunits increased the lethality and neurotoxicity of the basic subunits in the chick-neck tissue [5]. It is suggested that the acidic subunit protects the basic subunit from non-specific binding, and facilitates its targeting for certain neuronal receptors.

Acknowledgements

This study was supported by grants from Academia Sinica and National Science Council, Taiwan, R.O.C.

References

1.   Tu, A.T. (Ed.), Handbook of Natural Toxins, Reptile and Amphibian Venoms, Vol. 5, Marcel Dekker, New York, NY, U.S.A., 1991, p.3.
2.   Woodhams, B.J., Wilson, S.E., Xin, B.C. and Hutton, R.A., Toxicon, 28 (1990) 427.
3.   Lee, C.Y., J. Formosan Med. Assoc., 47 (1948) 65.
4.   Chippaux, J. P., Williams, V. and White, J., Toxicon, 29 (1991) 1279.
5.   Wang, Y.M., Lu, P.J., Ho, C.L. and Tsai, I.H., Eur. J. Biochem., 209 (1992) 635.
6.   Jayanthi, G.P. and Gowda, T.V., Toxicon, 26 (1989) 257.

Glucagon antagonist design based on replacement of active site triad residues

C.G. Unson, A. Sadarangani, K. Fitzpatrick, C.-R. Wu and R.B. Merrifield

The Rockefeller University, 1230 York Avenue, New York, NY 10021, U.S.A.

Introduction

Glucagon is a 29-residue peptide hormone, which together with insulin is responsible for maintaining basal glucose concentrations critical for survival. It is a prominent member of a family of peptides that includes secretin and VIP, with which it shares remarkable sequence similarity. Not surprisingly, the glucagon receptor sequence deduced from its cDNA [1] also bears resemblance to a subset of G protein-coupled receptors whose striking feature is a large extracellular domain which presumably contains the ligand binding site. The structural criteria by which these receptors recognize their ligands, as well as the events that trigger the subsequent generation of the transmembrane signal are gradually being defined by structure function analyses of the hormone, utilizing solid phase peptide synthesis. Our previous work has established that Asp^9 and His^1 are important residues for transduction of the glucagon response [2, 3]. Extensive investigation of all four serine residues in the sequence led to the observation that Ser^{16} also contributed significantly to adenylate cyclase activation [4, 5]. Thus, we have proposed that, upon binding, the resulting hormone-receptor complex acquires serine protease-like activity, and that the His^1, Asp^9, and Ser^{16} residues of glucagon constitute an active site triad that mediates a proteolytic event that triggers the signaling cascade. The hydroxyl groups of Ser^2 and Ser^8 provide a binding determinant so that substitution led to decreases in receptor binding and consequently, to decreased activity. Interestingly, Ser^{11} has an influence on binding. Replacement of Ser^{11} by a more hydrophobic residue such as Ala or Thr increased binding 3 to 5 fold, while the effect on signal transduction was minimal [4, 5]. Utilizing data from these structure function studies, we designed and synthesized a series of nine new analogs that incorporated at least two and as many as four replacements of the putative active site residues. Some of the derivatives are the most potent glucagon antagonists to date.

Results and Discussion

Nine replacement analogs of glucagon amide were synthesized by the solid phase peptide method on an Applied Biosystems synthesizer. We designed these derivatives based on the presumption that substitution of at least two of the active site residues should lead to inactive analogs that might still bind to the glucagon receptor. In addition, we combined these changes with Ala^{11} and a deletion of histidine at position 1. All of the peptides were assayed for their receptor binding affinity and potency in the activation of adenylate cyclase.

Table 1 *Position 1, 9, 11,& 16 replacement analogs*

	Analog of glucagon amide	Membrane binding, %	Adenylate cyclase activity		
			% Relative potency[a]	$(I/A)_{50}$[b]	pA_2[c]
	Glucagon amide	100	15		
1	Nle^9Ala^{16}	56.2	.038	3.2	8.2
2	$Ala^{11}Ala^{16}$	100	1.34	-	-
3	$Ala^{11}Asn^{16}$	49	1.1	-	-
4	$Ala^{11}Gln^{16}$	60.3	12	-	-
5	$Nle^9Ala^{11}Ala^{16}$	26.3	<0.0012	3.8	8.8
6	des-$His^1Nle^9Ala^{16}$	100	<0.001	2.14	8.4
7	des-$His^1Ala^{11}Ala^{16}$	28.2	<0.007	19.1	7.0
8	des-$His^1Ala^{11}Asn^{16}$	22.4	.014	9.5	8.0
9	des-$His^1Nle^9Ala^{11}Ala^{16}$	100	<0.0043	0.85	8.4

[a]Relative potency is the ratio (x100) of the concentration of natural glucagon to the concentration of analog which gives 50% of maximum response of analog.
[b]$(I/A)_{50}$, the inhibition index, is the ratio of inhibitor concentration to agonist concentration when the response is reduced to 50% of the response of agonist in the absence of inhibitor.
[c]pA_2 is the negative logarithm of the concentration of inhibitor that reduces the response to 1 unit of agonist to the response obtained from 0.5 unit of agonist.

In general, as predicted, Table 1 shows that all of the active-site replacement analogs retained very good receptor binding affinities ranging from 22% to 100%. Any two modifications of the $His^1 Asp^9 Ser^{16}$ triad, as in Nle^9Ala^{16}, analog **1** reduced the relative activity in the adenylate cyclase assay to 2500 times less than the natural hormone, or 375 times less than glucagon amide. Analogs **2-4** that combined an Ala^{11} substitution with a Ser^{16} replacement had partial agonist activity and thus were not glucagon antagonists. Not unexpectedly, a deletion of position-1 histidine improved the inhibitory potential of the analogs. Notably, when His^1 was deleted from analogs **2** and **3**, the resulting des-His^1 derivatives, analogs **7** and **8** were converted to glucagon antagonists. Des-$His^1Nle^9Ala^{16}$, analog **6** bound better and was a slightly improved antagonist than the His^1-containing analog **1**. Interestingly, analog **5** proved to be the most potent glucagon antagonist reported, with an inhibition index (I/A_{50}) of 3.8 and a pA_2 value of 8.8. The des-His^1 derivative, analog **9** has by far the best inhibition index, 0.85 of any reported antagonist.

The contribution of position 11 to the binding affinity of the analogs is less clear. Previous analyses suggested that Ser^{11} appears to fit into a binding groove, since a more hydrophobic functional group actually enhanced the hydrophobic interaction. [Ala^{11}]glucagon amide for example, bound 500% as well as glucagon itself [4, 5]. Analogs **5** and **9** do not have improved binding potencies compared to the corresponding analogs **1** and **6** which have not incorporated alanine at position 11. However, notwithstanding the role of Ala^{11}, all four analogs are potent antagonists with pA$_2$ values ranging from 8.2 to 8.8.

The observation that replacement of at least two of the active site triad residues resulted in antagonists underscores our contention that the three residues, His^1, Asp^9, and Ser^{16} are indeed involved in the activation step, and that they must interact in some way. In close parallelism with a charge relay system, we considered the possibility that Ser^{16} behaved as a nucleophile in a serine protease-like pathway where His^1 accepts the hydroxyl proton and is in turn stabilized by Asp^9. Our idea is bolstered by the observation that cAMP production by the glucagon adenylate cyclase system was inhibited by serum protease inhibitors, although at a relatively high concentration [4, 5]. Nevertheless, the ability to design potent antagonists based on the putative active site triad supports our finding that the three residues are indeed responsible for activation and not binding, and act in a cooperative manner. Furthermore, it is a widely accepted notion that glucagon is very sensitive to alteration in its natural sequence, since the entire peptide is required for binding and full agonist activity. In our study, we have been able to design analogs with up to four points of modification at positions 1, 9, 11 and 16, which are still recognized by the glucagon receptor and are able to compete effectively with natural glucagon for receptor sites. These results present new possibilities for the design of glucagon antagonists.

References

1. Jelinek, L.J., Lok, S., Rosenberg, G.B., Smith, R.A., Grant, F.J., Biggs, S., Bensch, P.A., Kuijper, J.L., Sheppard, P.O., Sprecher, C.A., O'Hara, P.J., Foster, D., Walker, K.M., Chen, L.H.J., Mckernan, P.A. and Kindsvogel, W., Science, 259 (1993) 1614.
2. Unson, C.G., Macdonald, D., Ray, K., Durrah, T.L., and Merrifield, R.B., J. Biol. Chem., 266 (1991) 2763.
3. Unson, C.G., Macdonald, D. and Merrifield, R.B., Arch. Biochem. Biophys., 300 (1993) 747.
4. Unson, C.G. and Merrifield, R.B., Proc. Natl. Acad. Sci. U.S.A., 1994, in press.
5. Merrifield, R.B. and Unson, C.G., In Du, Y.C., Tam, J.P. and Zhang, Y.S. (Eds.), Peptides: Biology and Chemistry (Proceedings of the 1992 Chinese Peptide Symposium), Escom, Leiden, The Netherlands, 1993, p.251.

Replacement of D-amino acids with L-amino acids in the Cecropin A-Melittin hybrid, all-D CA(1-13)M(1-13)NH$_2$, produces anomalous effects on antibacterial activities

D. Wade[1,2], H.G. Boman[3], S.A. Mitchell[2] and R.B. Merrifield[2]

[1]Protein Engineering Network of Centres of Excellence, 720 Heritage Medical Research Centre, University of Alberta, Edmonton, Alberta, T6G 2S2, Canada, [2]The Rockefeller University, 1230 York Avenue, New York, NY 10021-6399, U.S.A. and [3]Department of Microbiology, Stockholm University, S-106 91 Stockholm, Sweden

Introduction

Cecropin A is an antibacterial peptide isolated from the hemolymph of silkmoth pupae (review [1]), and melittin is a peptide toxin isolated from the venom of the honeybee [2]. The solid-phase synthesis, and antibacterial and hemolytic activities of the cecropin A-melittin hybrid peptide, all-L CA(1-13)M(1-13)NH$_2$, were reported by Boman, et al. in 1989 [3], and those of its D-enantiomer by Wade, et al. in 1990 [4]. A two-dimensional ^{1}H-NMR (2D-NMR) structure for the all-L hybrid was reported by Sipos et al. in 1991 [5] (Figure 1). The all-L and all-D hybrids had increased antibacterial and decreased hemolytic activities with respect to the all-L and all-D parent peptides, cecropin A [CA(1-37)NH$_2$] and melittin [M(1-26)NH$_2$]. In addition, the antibacterial and hemolytic activities of the all-L and all-D enantiomeric pair were very similar against both Gram-positive and Gram-negative bacteria, and erythrocytes (Table 1), indicating that bioactivity did not involve interaction with chiral receptors.

K-W-K-**L-F-K-K-I-E-K-V-G**-Q-G-I-**G-A-V-L-K-V-L-T-T-G-L**-CONH$_2$
        α-helix   α-helix

Fig. 1. Primary structure of the cecropin A-melittin hybrid, CA(1-13)M(1-13)NH$_2$. The N-terminal half of the peptide consists of residues 1-13 of cecropin A, and the C-terminal half consists of residues 1-13 of melittin [3]. 2D-NMR experiments [5] indicate that residues 4-12 of the cecropin A portion and 3-13 of the melittin portion are α-helical (underlined), and separated by a short region of flexibility.

Results and Discussion

We synthesized an analog of the all-D hybrid in which D-Ile residues were replaced with L-Ile, at position 8 of the CA(1-13) segment and position 2 of the M(1-

470

13) segment, and compared the antibacterial and hemolytic activities of the analog with those of the all-D peptide. The analog had reduced activities (Table 1) against one Gram-negative (3-fold reduced) and all Gram-positive organisms (2-50 fold reduced), especially against *Staphylococcus aureus* (SaC1).

We also synthesized single replacement analogs of the all-D hybrid which contained an L-Ile replacement at either position 8 of the CA(1-13) segment or position 2 of the M(1-13) segment, and compared their activities with those of the all-D hybrid and the double replacement analog (Table 1). Surprisingly, the activities of both single replacement analogs were identical, and not significantly different from the activities of the all-D hybrid.

Circular dichroism (CD) spectra of the single and double replacement analogs, in 20% hexafluoroisopropanol-water solution, showed that the single replacement analogs had the same content of α-helix as the all-D hybrid, 70%, whereas the double replacement analog had only 55% α-helix.

In addition, we replaced both Ile's of the all-D hybrid with D-Leu, a less expensive isomer, and found that this analog had slightly better bioactivity than the parent peptide, especially against *Staph. aureus* (6-fold more active).

A half-D, half-L analog, D[CA(1-13)]-L[M(1-13)]NH$_2$, was also synthesized. The antibacterial activities of this analog, with respect to those of the all-L and all-D hybrids, were mixed: activities were 17-27 fold improved against *E. coli* (D21), 2-5 fold reduced against *P. aeruginosa* (OT97), 7-10 fold improved against *B. subtilis* (Bs11), and 28-113 fold reduced against *S. aureus* (SaC1).

None of the new peptides caused hemolysis.

Table 1 *Lethal and lytic concentrations (μM) of peptides*

Peptide Amide:	Gram-Neg. [a]		Gram-Positive[a]			SRC
	D21	OT97	Bs11	SaC1	Sp1	
L[CA(1-37)][b]	0.2	1	3	>300	5	>200
D[CA(1-37)][b]	0.3	0.8	3	>300	2	>300
L[M(1-26)][b]	0.8	3	0.2	0.2	0.5	4
D[M(1-26)][b]	1	2	0.4	0.1	0.9	2
L[CA(1-13)M(1-13)][b]	0.5	1	0.7	2	1	>200
D[CA(1-13)M(1-13)][b]	0.8	2	1	8	0.8	500
L-Ile[8]-D[CA(1-13)M(1-13)]	0.8	3.6	0.9	6.7	0.9	>210
D[CA(1-13)]-L-Ile[2]-D[M(1-13)]	0.5	3.2	0.9	8.9[c]	0.6	>188
L-Ile[8]-D[CA(1-13)]-L-Ile[2]-D[M(1-13)]	0.9	7	2	>400	4	>400
D[Leu[8]-CA(1-13)-Leu[2]-M(1-13)]	0.5	1.1	0.5	1.3	0.8	>195
D[CA(1-13)]-L[M(1-13)]	0.03	4.8	0.1	226	0.5[c]	700

[a]D21, *Escherichia coli*; OT97, *Pseudomonas aeruginosa*; Bs11, *Bacillus subtilis*; SaC1, *Staphylococcus aureus*; Sp1, *Streptococcus pyogenes*; SRC, sheep red blood cells. Small values for D21, OT97, Bs11, SaC1, and Sp1 indicate good antibacterial activities, and a small value for SRC indicates that the peptide is hemolytic. Two-fold differences in antibacterial activities are considered to be within the range of experimental error. [b]Data from Ref. 4. [c]Flat concentration dependence; values difficult to interpret.

The results for D[CA(1-13)]-L-Ile^2-D[M(1-13)]NH$_2$ can be rationalized by noting that replacement occurs in the flexible interhelical region, which may reduce a disruptive effect on the C-terminal α-helix. The results obtained with the L-Ile[8] analog were surprising because replacement of a D- with an L-amino acid at position 8 would be expected to disrupt the α-helix. Andreu et al. [6] showed that replacing Ile[8] in cecropin A with helix disrupting Pro, caused 2-50 fold reductions in antibacterial activities against Gram-negative and Gram-positive bacteria. However, Sipos et al. [5] found that the N-terminal α-helix of CA(1-13)M(1-13)NH$_2$ was less stable than either the corresponding N-terminal α-helix of cecropin A [7] or the C-terminal α-helix of CA(1-13)M(1-13)NH$_2$. Therefore, substituting an L- for a D-amino acid in this helix may not be very disruptive, or, alternatively, this helix may simply be less important than the C-terminal α-helix for bioactivity of the hybrid.

The secondary structure of the half-D, half-L analog would be expected to change from a left-handed α-helix to a right-handed α-helix in the center of the flexible region of the peptide, thereby minimizing any disruptive effect on the helices. However, the antibacterial activities of this analog were mixed with respect to those of the replacement analogs, which makes it difficult to speculate about possible structural reasons for differences in activity.

In conclusion, the reduction in antibacterial activities and content of α-helix seen with the double replacement hybrid cannot be accounted for by either of the single replacements, and may be due to a combination of the effects of both replacements; a possible cooperative effect. Molecular modeling studies are in progress to determine if there is any obvious structural basis for these phenomena.

Acknowledgements

We thank A. Boman (Stockholm Univ.) and E. Merrifield (Rockefeller Univ.) for assistance with antibacterial assays, Dr. H. Steiner (Stockholm Univ.) for the computer program used to calculate lethal concentration values, and R. Luty (Univ. of Alberta) for CD analyses.

References

1.   Boman, H.G. and Hultmark, D., Ann. Rev. Microbiol., 41 (1987)103.
2.   Habermann, E. and Jentsch, J., Z. Physiol. Chem., 348 (1967) 37.
3.   Boman, H.G., Wade, D., Boman, I.A., Wåhlin, B. and Merrifield, R.B., FEBS Lett., 259 (1989) 103.
4.   Wade, D., Boman, A., Wåhlin, B., Drain, C.M., Andreu, D., Boman, H.G. and Merrifield, R.B., Proc. Natl. Acad. Sci. U.S.A., 87 (1990) 4761.
5.   Sipos, D., Chandrasekhar, K., Arvidsson, K., Engström, Å. and Ehrenberg, A., Eur. J. Biochem., 199 (1991) 285.
6.   Andreu, D., Merrifield, R.B., Steiner, H. and Boman, H.G., Biochemistry, 24 (1985) 1683.
7.   Holak, T.A., Engström, Å, Kraulis, P.J., Lindeberg, G., Bennich, H., Jones, T.A., Gronenborn, A.M. and Clore, G.M., Biochemistry, 27 (1988) 7620.

Molecular basis for prokaryotic specificity of magainin-induced lysis

D.E. Walker, E.M. Tytler, M.N. Palgunachari,
G.M. Anantharamaiah and J.P. Segrest

*Departments of Medicine and Biochemistry, UAB Medical Center,
Birmingham, AL 35294, U.S.A.*

Introduction

Membrane active peptides from frog skin secretion, such as the magainins (MG), that specifically lyse prokaryotic cells [1] are class L (lytic) amphipathic α helixes [2]; other class L peptides from bee and wasp stings, such as mastoparan, lyse eukaryotic as well as prokaryotic cells [3]. Since eukaryotic cell membranes contain cholesterol and bacterial cell membranes do not, interaction between the lytic peptide and cholesterol may be involved with the specificity of the lytic activity of these peptides. Recent studies have shown that some magainins adopt an α helical structure in the presence of phospholipid and a β structure in the presence of phospholipid and cholesterol [4]. Computer analysis [5] of naturally occurring class L amphipathic helixes showed that there is a conserved Glu residue on the polar/non polar interface of the helix of antibiotic peptides which is not present in the peptide venoms. This negatively charged Glu residue has the potential to interact with cholesterol found in eukaryotic membranes. We have also observed that the bulk of amino acid residues on the polar/non polar interface on the opposite side of the helix affects the hemolytic activity of these peptides [6]. We used this information in the design model peptide analogs to study the molecular basis for prokaryotic specificity of peptide-induced lysis.

Results and Discussion

Three peptides, $18L_{MG}$ (LGSIWKFIKAFVGGIKKF), $[E^{14}]18L_{MG}$ and $[G^5, E^{14}]18L_{MG}$ were synthesized [7] and examined for hemolytic and bactericidal activity (Table 1). The rank order hemolytic activity was $18L_{MG} > [E^{14}]18L_{MG} > [G^5,E^{14}]18L_{MG} = MG-2$; the rank order bactericidal activity was $18L_{MG} > [G^5,E^{14}]18L_{MG} > [E^{14}]18L_{MG} > MG-2$ (Table 1). Although the hemolytic activities of both $[G^5,E^{14}]18L_{MG}$ and MG-2 are <1%, $[G^5,E^{14}]18L_{MG}$ has 4-fold greater bactericidal activity than MG-2. These results indicate that a hemolytic, bactericidal class L peptide can be converted to a non-hemolytic, bactericidal peptide by two changes in the hydrophobic face: addition of Glu and reduction of "steric bulk".

We have shown that the presence or absence of an interfacial Glu residue in these peptides determines the specificity of its lytic activity, perhaps by forming H-bonds with cholesterol in eukaryotic cell membranes. The bulk of interfacial amino acid residues on the opposite side of the helix determines the degree of eukaryotic cell lysis.

Table 1 *Hemolytic and antibacterial activity of peptides*

Peptide	% Hemolysis[a]	C_l (μM)[b]
$18L_{MG}$	44	0.03
$[E^{14}]18L_{MG}$	11	3.3
$[G^5,E^{14}]18L_{MG}$	<1	1.4
Magainin-2 (MG-2)	<1	5.6

[a]Lysis of human erythrocytes was determined after incubation at 37°C for 10 min at a peptide concentration of 30μM.
[b]Lethal concentrations (C_l) against *E.coli* D31 were determined on agarose plates [8].

References

1. Bevins, C.L. and Zasloff, M., Ann. Rev. Biochem., 59 (1990) 395.
2. Segrest, J.P., de Loof, H., Dohlman, J.G., Brouillette, C.G. and Anantharamaiah, G.M., Proteins, 8 (1990) 103.
3. Argiolas, A. and Pisano, J.J., J. Biol. Chem., 258 (1983) 13697.
4. Jackson, M., Mantsch, H.H. and Spencer, J.H., Biochemistry, 31 (1992) 7289.
5. Jones, M.K., Anantharamaiah, G.M. and Segrest, J.P., J. Lipid Res., 33 (1992) 287.
6. Tytler, E.M., Epand, R.M., Anantharamaiah, G.M. and Segrest, J.P., this volume.
7. Anantharamaiah, G.M., Methods Enzymol., 128 (1986) 627.
8. Hultmark, D., Engström, A., Andersson, K., Steiner, H., Bennich, H. and Boman, H.G., EMBO J., 2 (1983) 571.

Ac-Gly-Arg-Gly-Asp-Ser-OH, an RGD-peptide, induces endothelium-dependent relaxation in isolated rabbit aorta

L. Torday[1], M. Zarándi[2], G.E. Balogh[1], J. Pataricza[1], J.G. Papp[1] and B. Penke[2]

[1]Department of Pharmacology and [2]Department of Medical Chemistry, Albert Szent-Györgyi Medical University, Szeged, Dóm tér 12., P.O.B. 115, H-6701, Hungary

Introduction

Platelet adhesion to the subendothelium of a damaged vessel wall and the subsequent platelet aggregation are critical steps in the beginning of uncontrolled platelet deposition on thrombogenic surfaces. Such impairments of the vessel wall caused by, for example, atherosclerosis may provoke vascular occlusions resulting in myocardial infarction or cerebral stroke. In the growth of thrombus the fibrinogen cross-linkings between platelets mediated via platelet fibrinogen receptor glycoprotein IIb/IIIa complex (gpIIb/IIIa) were shown to be essential [4]. To block this step monoclonal antibodies against gpIIb/IIIa and the army of analogues of fibrinogen α-chain mimicking Arg-Gly-Asp(RGD)-containing peptides were developed [1, 2, 3].

Ac-Gly-Arg-Gly-Asp-Ser-OH (GRGDS) is a pentapeptide representing the family of RGD-peptides powerfully inhibits platelet aggregation induced by ADP and by other platelet agonists.

This peptide is not only an antifibrinogen agent but it can block other RGD-mediated receptor-ligand interactions, e.g. the binding of fibronectin, vitronectin, von Willebrand factor, thrombospondin, and disintegrins to their own receptor sites [6].

Although systematically administered RGD-peptide analogues successfully inhibit thrombotic events *in vivo* investigated in animal models, and therefore, they may have therapeutical applications. There are no data available in the literature how RGD-peptides influence the tone of the vessel wall. In order to obtain information about the possible effect of an RGD-peptide on vascular tissue we investigated the vasoactive properties and the mechanism of action of GRGDS.

Results and Discussion

New Zealand rabbits with 1.5-2.5 kg body weight from either sex were used. After injection of 500 IU/kg heparin into the marginal ear vein, the rabbits were killed by blowing on the head. The chest was opened immediately, the heart was cut out to achieve exsanguination and the thoracic aorta was excised. When the vessel was carefully cleaned from the surrounding connective tissue, it was cut with blades into rings of 5 mm in width. Before equilibration process the rings were stored in

100 ml organ bath (freshly made Krebs-Henseleit (K-H) solution, pH 7.4, 37°C, continuously bubbling with a gas mixture containing 95% O_2 and 5% CO_2).

During the 45 min incubation periods the rings were mounted in a recording chamber with a volume of 2 ml of K-H solution continuously bubbling with the above-mentioned gas mixture and a resting tension of 10 mN was applied. In order to achieve perfect equilibration, the K-H solution was changed every 15 min and the resting tension was adjusted continuously to 10 mN.

In some cases the endothelium was removed by rubbing with a glass rod covered by a cotton swab.

In the second part of our investigations 30 min preincubation of rings was performed with 30 µM L-N^G-nitro-arginine (LNNA), the most powerful nitric-oxide synthase inhibitor prior to addition of GRGDS.

Before beginning the experiments the rings were suspended in recording chambers under similar conditions as described above and were preconstricted by 2.5×10^{-7} M phenylephrine (Phe). The endothelium was considered suitable, if the amplitude of relaxation caused by 2.5×10^{-7} M acethylcholine (Ach) was at least 50% of the magnitude of contraction induced by Phe. The denuded rings were contracted after Ach administration. After washing a new Phe constriction was performed, and GRGDS was added when the steady state had been accomplished.

The isometric tension was measured with a force transducer (Experimetria, Hungary) connected to a bridge amplifier (Mikromed, Hungary) and displayed on a pen-recorder (KUTESZ, Type 175, Hungary).

Student's t test for unpaired observations was used for statistical evaluation of the data. The IC_{50} was calculated by fitting the logistic function.

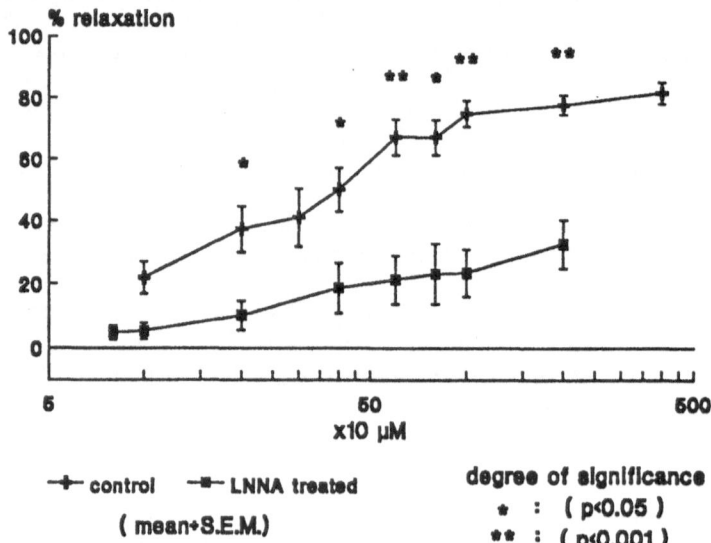

Fig. 1. Endothelium dependent relaxation caused by GRGDS on rabbit aortic rings (n = 5-13).

1. When the effect of GRGDS on endothelium intact rabbit aortic rings preconstricted with Phe was studied, the pentapeptide showed a dose-dependent

vasorelaxant activity in the concentration range of 10^{-5} to 4×10^{-4} M. The measured maximum relaxation amounted to $81.48 \pm 3.44\%$ (n=13, p<0.001). The IC_{50} of GRGDS was $2.74 \pm 0.53 \times 10^{-5}$M.

2.   The responsiveness of denuded rings to GRGDS was also studied. In such preparations no relaxation was observed with GRGDS up to a concentration of 4×10^{-4} M.

3.   Furthermore, the effect of LNNA on GRGDS induced relaxation was studied. 3×10^{-5} M LNNA decreased the maximum vasorelaxant effect of GRGDS to $33.69 \pm 7.49\%$ (n=5). On the other hand this effect of LNNA was completely reversed by adding 10^{-4} M L-arginine.

Results are shown in Figure 1.

The data of the present study indicate that the pentapeptide GRGDS is a powerful endothelium dependent vasodilator, and its action is mediated probably via the nitric-oxide pathway.

It was published previously that this peptide was successfully tested in order to prevent reocclusion of dog coronaries after streptolysis [5, 7]. Our data suggest that the EDRF released by GRGDS administered in platelet inhibitory concentration may contribute to keep open the recanalized vessel not only via the potentiation of the platelet inhibitory effect of GRGDS but through a direct vasodilating action [8].

In this phase of investigations we do not know the exact nature of the endothelial GRGDS receptor. There are also no published data available concerning the vascular action of other RGD-peptides. Undoubtedly, in order to obtain more information about the new endothelial RGD-receptor we need further investigations.

The antifibrinogen pentapeptide Ac-GRGDS-OH is a potent endothelium dependent vasodilator.

Acknowledgements

Special thanks to Ervin Lukács for his work about the management of the computer workstation and to Katalin Nagy for her labour to synthesize and purify the peptide.

References

1.   Alig, L., Edenhofer, A., Hadváry, P, Hurzeler, M., Knopp, D., Muller, M., Steiner, B., Trzeciak, A. and Weller, T., J. Med. Chem., 35 (1992) 4393.
2.   Samanen, J., Ali, F., Romoff, T., Calvo, R., Sorenson, E., Vasko, J, Storer, B., Berry, D., Benett, D., Strohsacker, M., Powers, D., Stadel, J. and Nichols, A., J. Med. Chem., 34 (1991) 3114.
3.   Gold, H.K., Gimple, L.W., Yasuda, T., Leinbach, R.C., Werner, W., Holt, R., Jordan, R., Barger, H., Collen, D. and Coller, B.S., J. Clin. Invest., 86 (1990) 651.
4.   Shattil, S.J., Hoxie, J.A., Cunningham, M. and Brass, L.F., J. Biol. Chem., 260 (1985) 11107.
5.   Roux, S.P., Tschopp, T.B., Kuhn, H., Steiner, B. and Hadváry, J., Pharm. Exp. Ther., 264 (1993) 501.
6.   Rouslahti, E., J. Clin. Invest., 87 (1991) 1.
7.   Shebuski, R.J., Berry, D.E., Bennett, D.B, Romoff, T., Storer, B.L., Ali, F. and Samanen, J., Thromb. Haemostasis, 61 (1989) 183.
8.   Bult, H., Fret, H.R.L., Van den Bossche, R.M. and Herman, A.G., Br. J. Pharmacol., 95 (1988) 1308.

The design and synthesis of antimicrobial peptides based on the principle of hydrophobic moments

L. Zhong[1], R.J. Putnam[1], C.W. Johnson, Jr.[2] and A.G. Rao[1]

[1]Department of Biotechnology Research, Pioneer Hi-Bred International, Inc., 7250 NW 62nd Avenue, Johnston, IA 50131, U.S.A., [2]Department of Biochemistry and Biophysics, Oregon State University, Corvallis, OR 97331, U.S.A.

Introduction

In recent years there have been many reports suggesting that the mechanism of action of several antibacterial peptides may well involve their ability to lyse cells through perturbation of the membrane bilayer [1, 2]. Although there is no sequence homology amongst many of these peptides, they all appear to be characterized by a common structural feature that is critical to their antimicrobial activity i.e. amhipathic α-helices. The amphipathy of a segment in a polypeptide chain can be quantitated through the hydrophobic moment algorithm developed by Eisenberg [3]. Using this approach, Eisenberg noted that a significant proportion of the known antimicrobial peptides could be identified by their typically high hydrophobic moment <u_h> and low hydrophobicity <H> values. We have used the algorithm to screen the vast number of sequences in the protein data base to locate segments that possess the hydrophobic moment properties associated with the known antimicrobial peptides. Here, we describe the antimicrobial and other properties of peptide derivatives of a 14-residue segment derived from murine perforin [4].

Results and Discussion

The 14-residue peptide derived from murine perforin (Table 1) comprises residues 191 through 204 of the mature protein. To facilitate intrinsic fluorescence measurements, **Phe** at position 13 was replaced with **Trp**. Additional substitutions included **Asp** to **Lys** (position 1), **Phe** to **Leu** (position 2), **Pro** to **Ala** (position 10), **Asn** to **His** (position 12) and **Asn** to **Lys** at position 14. The resulting peptide, **P3**, was more basic than the wild type peptide (Table 1). Three other derivatives were also made to examine the effect of increasing the hydrophobicity and the amphipathy on the properties of **P3** (data not shown). Peptide **P3** in aqueous solution adopted a random conformation. However, in the presence of acidic liposomes (phosphatidylcholine:phosphatadic acid, 3:1), a predominantly α-helical structure was induced (Figure 1A). This ability to form an α-helical structure was lost in peptide **P3S**, a scrambled derivative of **P3** with a low hydrophobic moment. All peptides exhibited similar antifungal activity against two plant pathogens (Table 1) albeit with a significantly reduced MCIC value in the case of P3S. However, the persistence of the activity suggests that, at least in the case of fungal cells, the juxtaposition of basic charges may play a significant role in addition to the amphipathy. In antibacterial assays the activity of P3S against several g- and g+

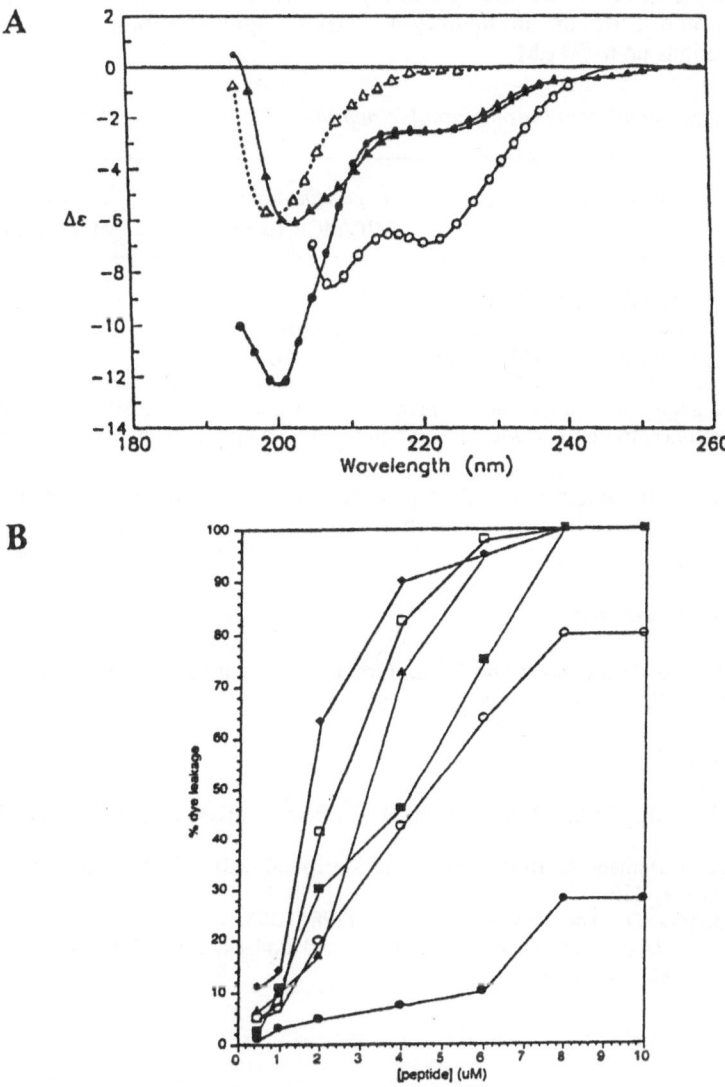

Fig. 1A. *CD spectra of peptides in acidic liposomes (PC:PA, 3:1) or PBS, pH 7.4.* ● *P3 in PBS,* ○ *P3 in liposome,* △ *P3S in PBS,* ▲ *P3S in liposome.*

Fig. 1B. *Lipolytic activity of peptides. Samples were prepared by mixing different concentrations of peptides with 0.1 mg/ml acidic liposomes containing catboxyfluorescein, and incubating for 10 min at room temperature.* □ *PS,* ◆ *P3S,* ▲ *P3,* ◆ *Q3,* ■ *P10,* ○ *Q4.*

bacteria were significantly reduced (data not shown). The membranolytic action of the peptides was evidenced by their ability to leach carboxyfluorescein from acidic liposomes encapsulating the dye (Figure 1B). Significantly, the lytic activity of the scrambled peptide P3S, was markedly reduced. Furthermore, the peptides demonstrated little or no hemolytic activity against rabbit erythrocytes at concentrations up to 50 μM.

Table 1 *Antifungal data for amphipathic peptides*

Peptide	Sequence $(u_h)^a$	*F. graminearum* MIC/MCIC(μg/ml)	*F. moniliforme* MIC/MCIC(μg/ml)
Perf.	DFKKALRALPRNFN(0.80)		
P3	KLKKALRALARHWK(0.78)	10/20	5/40
P3S	KAHWLRLKALAKRK(0.21)	40/>80	15/>80

$^a u_h$ is the averaged hydrophobic moment for the whole sequence; MIC is the minimum inhibitory concentration and MCIC is the minimum complete inhibitory concentration.

Our results suggest that it is possible to design a library of antimicrobial peptides by identifying segments in the protein sequence data base that have hydrophobic moment values comparable to the other known antimicrobial peptides.

Acknowledgements

We thank Tracy Rood and Susan Grant for the antifungal assays and Jeannine Lawrence for CD measurements.

References

1. Zasloff, M., Martin, B. and Chien, H.-C., Proc. Natl. Acad. Sci. U.S.A., 85 (1988) 910.
2. Fink, J., Boman, A., Boman, H.G. and Merrifield, R.B., Int. J. Peptide Protein Res., 33 (1989) 412.
3. Eisenberg, D., Ann. Rev. Biochem., 53 (1984) 595.
4. Lowrey, D.M., Aebischer, T., Olsen, K. and Podack, E.R., Proc. Natl. Acad. Sci. U.S.A., 86 (1989) 247.

Session VI
Peptide Hormones/Neuropeptides

Chairs: **Maurice Manning**
Medical College Ohio
Toledo, Ohio, U.S.A.

and

Antonio C.M. Paiva
Escola Paulistade Medicina
São Paulo, Brazil

TIPP opioid peptides: Development of extraordinarily potent and selective δ antagonists and observation of astonishing structure-intrinsic activity relationships

P.W. Schiller, T.M.-D. Nguyen, G. Weltrowska,
B.C. Wilkes, B.J. Marsden, R. Schmidt,
C. Lemieux and N.N. Chung

Laboratory of Chemical Biology and Peptide Research
Clinical Research Institute of Montreal, 110 Pine Avenue West
Montreal, Quebec, H2W 1R7, Canada

Introduction

Receptor-selective opioid antagonists are of interest as pharmacological tools and as potential therapeutic agents. In recent years, significant progress has been made in the development of both peptide and non-peptide antagonists with improved selectivity for each of the three major opioid receptor types (μ, δ, κ) (for a recent review, see ref. [1]). Among various reported δ antagonists, the enkephalin analog N,N-diallyl-Tyr-Aib-Aib-Phe-Leu-OH (ICI 174864) shows considerable δ receptor selectivity but only moderate potency [2], whereas the δ-selective non-peptide antagonist naltrindole is highly potent [3]. However, naltrindole also displayed significant antagonist potency against μ and κ agonists in the guinea pig ileum (GPI) assay (K_e values of 29 nM and 46 nM, respectively). Recently, we reported the discovery of a new class of opioid peptide-derived δ antagonists that contain a tetra-hydroisoquinoline-3-carboxylic acid (Tic) residue in the 2-position of the peptide sequence [4]. The two prototype antagonists were the tetrapeptide H-Tyr-Tic-Phe-Phe-OH (TIPP) and the tripeptide H-Tyr-Tic-Phe-OH (TIP). TIPP showed high antagonist potency against various δ agonists in the mouse vas deferens (MVD) assay (K_e = 3-5 nM), high δ receptor affinity (K_i^δ = 1.22 nM) and extraordinary δ selectivity (K_i^μ/K_i^δ = 1410). Furthermore, TIPP displayed no μ or κ antagonist properties in the GPI assay at concentrations as high as 10 μM. In comparison with TIPP, TIP was a somewhat less potent and less selective δ antagonist. Whereas both TIPP and TIP were stable in aqueous buffer solution (pH 7.7) for at least six months, these peptides underwent slow, spontaneous Tyr-Tic diketopiperazine formation with concomitant cleavage of the Tic-Phe peptide bond in DMSO and MeOH [5]. Here we describe a number of TIPP analogs with increased δ antagonist potency, with further improved δ selectivity and with structural characteristics that prevent spontaneous or enzymatic degradation. Furthermore, we systematically examined the effect of halogenation at the 3'-position of Tyr[1] in TIPP on the intrinsic activity ("efficacy").

Results and Discussion

Replacement of the Phe[3] residue with 3-(2'-naphthyl)alanine (2-Nal) or of the Phe[4] residue with *p*-nitrophenylalanine [Phe(pNO$_2$)] produced TIPP analogs that retained about the same antagonist potency as the parent peptide in the MVD bioassay but showed a 3-fold improvement in δ receptor selectivity in the opioid receptor binding assays (Table 1). Analogs containing tryptophan or homo-phenylalanine (Hfe) in place of Phe[3] were 2 and 7 times more potent than TIPP as antagonists, respectively. These compounds displayed δ receptor affinities in the subnanomolar range as well as further improved δ selectivities. Two other analogs with subnanomolar δ antagonist potency resulted from substitution of the N-terminal tyrosine residue with either N$^\alpha$-methyltyrosine [Tyr(NMe)] or 3',5'-dimethyltyrosine (Dmt). In comparison with naltrindole, [Tyr(NMe)[1]]TIPP was about 50% more potent as antagonist against the δ agonist [D-Ala[2]]deltorphin I in the MVD assay and 500 times more δ-selective in the receptor binding assays. [Dmt[1]]TIPP showed unprecedented δ antagonist potency (K$_e$ = 0.152 nM), being about 4 times as potent as naltrindole.

To exclude the possibility of spontaneous peptide degradation via Tyr-Tic diketopiperazine formation we prepared TIPP- and TIP analogs that contain a reduced peptide bond between the Tic[2] and Phe[3] residues. The resulting pseudopeptide

Table 1 *Antagonist potencies and opioid receptor affinities of TIPP analogs*

Compound	K$_e$ [nM][a]	K$_i$ $^\mu$[nM][b]	K$_i$ $^\delta$[nM][b]	K$_i$ $^\mu$/K$_i$ $^\delta$
H-Tyr-Tic-Phe-Phe-OH	2.96	1720	1.22	1410
H-Tyr-Tic-Phe-OH	12.6	1280	9.07	141
H-Tyr-Tic-Trp-Phe-OH	1.65	1790	0.301	5950
H-Tyr-Tic-2-Nal-Phe-OH	3.07	6330	1.31	4830
H-Tyr-Tic-Hfe-Phe-OH	0.408	1990	0.277	7180
H-Tyr-Tic-Phe-Phe(pNO$_2$)-OH	2.79	2890	0.703	4110
Tyr(NMe)-Tic-Phe-Phe-OH	·0.436	13400	1.29	10400
H-Dmt-Tic-Phe-Phe-OH	0.152	141	0.248	569
H-Tyr-Ticψ[CH$_2$-NH]Phe-Phe-OH	2.58	3230	0.308	10500
H-Tyr-Ticψ[CH$_2$-NH]Phe-OH	9.05	10800	1.94	5570
Naltrindole	0.636	3.86	0.182	21.2
DPDPE	-	943	16.4	57.5
[D-Ala[2]]deltorphin II	-	3930	6.43	611
[Leu[5]]enkephalin	-	9.43	2.53	3.73

[a]Determined against [D-Ala[2]]deltorphin I in the MVD assay. [b]Binding assay based on displacement of [^{3}H]DAMGO (μ-selective) and [^{3}H]DSLET (δ-selective) from rat brain membrane binding sites.

analogs, H-Tyr-Ticψ[CH$_2$-NH]Phe-Phe-OH (TIPP[ψ]) and H-Tyr-Ticψ[CH$_2$-NH]Phe-OH (TIP[ψ]), showed slightly improved δ antagonist potency and significantly enhanced δ receptor selectivity as compared to their respective parent peptides.

TIPP[ψ] displayed unprecedented δ selectivity (K_i^μ/K_i^δ = 10 500), being 180 times and 17 times more δ-selective than the δ agonists DPDPE and [D-Ala²]deltorphin I, respectively. In enzymatic degradation studies based on incubation of 10^{-4} M peptide solutions with rat brain membrane suspensions at 37°C TIPP[ψ] was found to be completely stable against enzymatic degradation, whereas the parent peptide TIPP underwent slow degradation ($t_{1/2}$ = 22 h). All TIPP analogs listed in Table 1 had insignificant affinity for κ receptors (K_i^κ > 1 μM). Furthermore, at concentrations up to 10 μM, none of the TIPP-derived antagonists described here showed μ or κ antagonist properties in the GPI assay or any agonist effect in the MVD assay. These novel δ antagonists are likely to find wide use as pharmacological tools in opioid research.

The results of binding studies performed with [^{125}I]TIPP suggested that substitution of an iodine atom at the 3'-position of Tyr¹ in TIPP had turned the δ antagonist into a δ agonist [6]. Indeed, [Tyr(3'-I)]TIPP behaved as a full agonist in the MVD assay (Figure 1, Table 2) and its effect was antagonized by TIPP (K_e = 11 nM). Corresponding iodination of the Tyr residue in TIP and TIPP[ψ] did not result in agonism (data not shown) and, therefore, it appears that the astonishing conversion observed with TIPP may be due to an overall conformational effect rather than to a direct, local effect of the iodine substituent. Interestingly, substitution of a bromine or chlorine atom at the 3'-position of Tyr¹ in TIPP produced partial agonists with

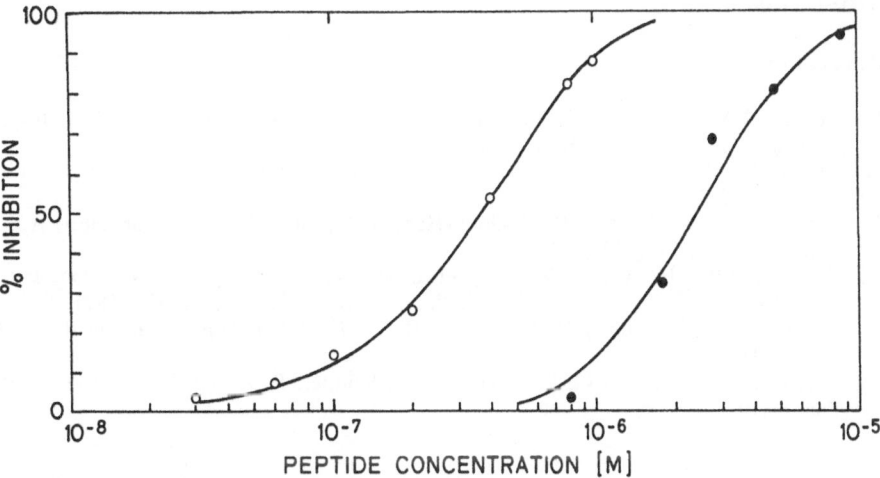

Fig. 1. Antagonism of [Tyr(3'-I)¹]TIPP by TIPP in MVD assay. Log dose-response curve of [Tyr(3'-I)¹]TIPP in absence (O) and in presence (●) of TIPP (100 nM).

respective "efficacies" (e) of 0.16 and 0.12, whereas the Tyr(3'-F)-analog was again a pure antagonist. Thus, systematic substitution of halogen atoms beginning with iodine and in the order of the periodic table produced a progressive decrease in intrinsic activity and a concomitant increase in affinity at the δ receptor (Table 2).

Conformational studies of this series of halogenated TIPP analogs can be expected to provide new insight into structure – intrinsic activity relationships.

Table 2 *δ Opioid agonist/antagonist properties of halogenated TIPP analogs*

| Analog | MVD assay | | Binding[a] |
	Agonist/antagonist	Potency (nM)	K_i^δ (nM)
H-Tyr(3'-I)-Tic-Phe-Phe-OH	full agonist (e =1.0)	IC50 = 97	24.2
H-Tyr(3'-Br)-Tic-Phe-Phe-OH	partial agonist (e = 0.16)	K_e = 24-32	3.62
H-Tyr(3'-Cl)-Tic-Phe-Phe-OH	partial agonist (e = 0.12)	K_e = 19-20	3.00
H-Tyr(3'-F)-Tic-Phe-Phe-OH	antagonist	K_e = 6-13	1.62
H-Tyr-Tic-Phe-Phe-OH	antagonist	K_e = 3-6	1.22

[a]Displacement of [^{3}H]DSLET from rat brain membrane binding sites.

Acknowledgements

This work was supported by grants from the MRCC (MT-5655) and NIDA (DA-04443).

References

1. Schiller, P.W., In Herz, A. (Ed.), Opioids I, Handbook of Pharmacology, Vol. 104/I, Springer-Verlag, Berlin, FRG, 1992, p.681.
2. Cotton, R., Giles, M.G., Miller, L., Shaw, J.S. and Timms, D.S., Eur. J. Pharmacol., 31 (1984) 281.
3. Portoghese, P.S., Nagase, H., MaloneyHuss, K.E., Lin, C.-E. and Takemori, A.E., J. Med. Chem., 34 (1991) 1715.
4. Schiller, P.W., Nguyen, T.M.-D., Weltrowska, G., Wilkes, B.C., Marsden, B.J., Lemieux, C. and Chung, N.N., Proc. Natl. Acad. Sci. U.S.A., 89 (1992) 11871.
5. Marsden, B.J., Nguyen, T.M.-D. and Schiller, P.W., Int. J. Peptide Protein Res., 41 (1993) 313.
6. Lee, P.H.K., Nguyen, T.M.-D., Chung, N.N., Schiller, P.W. and Chang, K.-J., Mol. Pharmacol., in press.

A new potent and highly selective, long lasting, peptide based Neurokinin A antagonist: Rational design of MEN 10627

V. Pavone[1], A. Lombardi[1], C. Pedone[1], L. Quartara[2] and C.A. Maggi[2]

[1]Centro Interdipartimentale di Ricerca sui Peptidi Bioattivi,
Via Mezzocannone 4, I-80134 Napoli, Italy
[2]A. Menarini Industrie Farmaceutiche Riunite,
Via Sette Santi 3, I-50131 Firenze, Italy

Introduction

Tachykinins (TKs) are a family of peptides that exert their biological actions by stimulating three distinct receptors (termed NK-1, NK-2 and NK-3). Among the three natural mammalian TKs, substance P (SP) displays the highest affinity for the NK-1 receptor, Neurokinin A (NKA) for the NK-2 receptor and Neurokinin B (NKB) for the NK-3 receptor [1].

In order to design highly selective peptides, which are able to interact specifically with NK-2 receptor and not with NK-1 and NK-3 receptors and acting either as agonists or antagonists, it is essential both to have some structural information on the bioactive three-dimensional structure of the ligand and a well-defined and rigid scaffold on which the chemical groups, directly interacting with the receptor, could be attached.

At the beginning of our study, some information was available on the biological activity of NKA analogues acting either as agonists or antagonists, but very little was known on the structural requirements for the interaction with NK-2 receptors. A highly reliable structure activity relationship, based on NMR studies, was reported [2] on a series of NKA (4-10) analogues containing different residues in position 8. The most active and selective analogue [β-Ala8]NKA (4-10) showed at least two well-defined structural domains. Two β-turns type I involving residues 5-8 and 8-10 (including the amide terminal group) were proposed.

We were the first to propose and to demonstrate [3], that two consecutive β amino acids, when incorporated in *cyclo*-tetra-peptides of 14-membered ring size, are capable to force the remaining pair of α-amino acids to adopt a β-turned structure. We have also successively proposed that this structural domain can easily be incorporated in larger sequences [4].

We disclose and report here the structure of a new potent, highly selective, long lasting, peptide based NKA antagonist.

Results and Discussion

MEN 10627 consists of two *cyclo*-tetra-peptides fused together, each containing a β-turn domain, namely *cyclo*-(Met-Asp-Trp-Phe-Dap-Leu)*cyclo*(2β-5β), (Dap = 2,3 di-amino propionic acid) as shown in Figure 1.

MEN 10627 was designed on the amino acid sequence of other known NKA antagonists [5]. The initial hypothetical structure was approximately modelled on a Silicon Graphics work station using the known structure of *cyclo*-hexa-peptides [6] and successively minimized to ascertain its stability using the Discover program. The overall shape of the molecule remained almost unchanged. A view of the molecular model obtained with this procedure is reported in Figure 1.

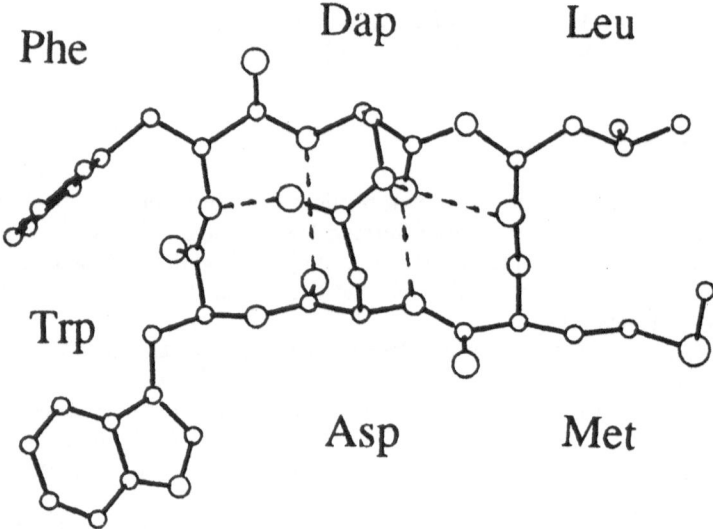

Fig. 1. Molecular model of MEN 10627. Intramolecular hydrogen bonds are indicated as broken lines.

The peptide was synthesized by the solid phase method, using Boc chemistry on a Boc-Leu-OCH$_2$-Pam resin. Side chains of Asp and Dap residues were protected as OFm and Fmoc derivatives, respectively. Treatment with piperidine and cyclization (using PyBop) was performed, before coupling the Met residue. After completion of peptide assembling on the resin and removal (with HF) of the mono-cyclic peptide, the second cyclization step was performed in diluted DMF solution using PyBop. The overall yield, with respect to the initial resin substitution, of the HPLC purified bi-cyclic peptide was 27%.

Preliminary structural analysis in solution, using NMR spectroscopy, and in the solid state, using X-ray diffraction techniques, fully confirms the hypothetical model.

In vitro biological activity of MEN 10627 was determined on the following bioassay: guinea pig ileum (GPI) for NK-1 receptor, rabbit pulmonary artery (RPA) and hamster trachea (HT) for NK-2 receptors, and rat portal vein (RPV) for NK-3 receptor. Conditions of the bioassays were as described previously [7].

MEN 10627 showed very high affinity for NK-2 receptor: pKB values on the RPA and HT were 8.1 and 10.1, respectively. This high potency corresponds also to a remarkable selectivity towards NK-1 and NK-3 receptors, as MEN 10627 was inactive on RPV (at 1 μM) and showed very low potency on GPI (pKB = 6.1).

The *in vivo* activity of MEN 10627 was also investigated in urethane anesthetized rats for its ability to inhibit contraction of the urinary bladder following i.v. administration of the NK-2 selective agonist [β-Ala8]NKA (4-10) (1 nmol/Kg i.v.). MEN 10627 (3-30 nmol/kg i.v.) inhibited in a dose-dependent manner the response to the NK-2 agonist. The activity of MEN 10627 was long lasting, a significant inhibition being observed up to 3 hours from i.v. administration of a 30 nmol/Kg dose.

MEN 10627 is the prototype of a new class of cyclic peptide-based NK-2 receptor antagonists. Owing to its high potency and long lasting activity *in vivo* MEN 10627 and its analogues are suitable candidates for clinical testing in humans.

With the present work we also demonstrate that peptides as drugs can be developed if it is possible to design and synthesize an analogue that will adopt a unique well-determined three-dimensional structure, corresponding to the bioactive conformation. The conformational rigidity increases potency, specificity and especially increases the half life of the molecule in the living body.

References

1. Maggi, C.A., Patacchini, R., Rovero, P. and Giachetti, A., J. Auton. Pharmacol., 13 (1993) 23.
2. Saviano, G., Temussi, P.A., Motta, A., Maggi, C.A. and Rovero, P., Biochemistry, 30 (1991) 10175.
3. Pavone, V., Lombardi, A., Yang, X., Pedone, C. and Di Blasio, B., Biopolymers, 30 (1990) 189.
4. Pavone, V., Lombardi, A., D'Auria, G., Saviano, M., Nastri, F., Paolillo, L., Di Blasio, B. and Pedone, C., Biopolymers, 32 (1992) 173.
5. McKnight, A.T., Maguire, J.J., Elliott, N.J., Fletcher, A.E., Foster, A.E., Tridgett, R., Williams, B.J. and Iversen, L.L., Br. J. Pharmacol., 104 (1991) 355.
6. Gierasch, L.M., Rockwell, A.L., Thompson, K.F. and Briggs, M.S., Biopolymers, 24 (1985) 117.
7. Patacchini, R., Maggi, C.A., Rovero, P., Regoli, D., Drapeau, G. and Meli, A., J. Pharmacol. Exp. Ther., 250 (1989) 678.

Hydrogen bonding patterns in X-ray diffraction studies of neuropeptides

J.L. Flippen-Anderson, C. George and J. Deschamps

Laboratory for the Structure of Matter, Code 6030, Naval Research Laboratory, Washington, DC 20375-5341, U.S.A.

Introduction

In the years between 1977 and 1989 thirteen crystal structure studies on enkephalin and its analogs were published. We have recently completed X-ray studies on 7 additional enkephalin analogs including the linear peptides DTLET [1], DADLE [2] and N,N-diallyl-O-t-butyl-Tyr-Aib-Aib-Phe-Leu [3] and the cyclic peptides DPDPE [4], L- and D-Ala DPDPE [5] and Tyr-D-Cys-Phe-DPen [6]. With these new structures in hand, we have begun to look at hydrogen bonding patterns within this family of neuroactive peptides. For this presentation, we concentrate on the acyclic peptides that exhibit folded conformations in the solid state. This provides 14 unique molecules, from studies on 3 tetrapeptides, 6 pentapeptides and 1 hexapeptide [1-3, 7-12].

Results and Discussion

Overall, these peptides exhibit two different folded conformations: A single type I′ β-bend (Figure 1 - left) and a double β-bend, consisting of a type III and a type I turn (Figure 1 - right). Peptides with more than one molecule in the asymmetric unit all have the same general conformation although there are differences in their intermolecular interactions and even some differences in their intramolecular hydrogen bonding.

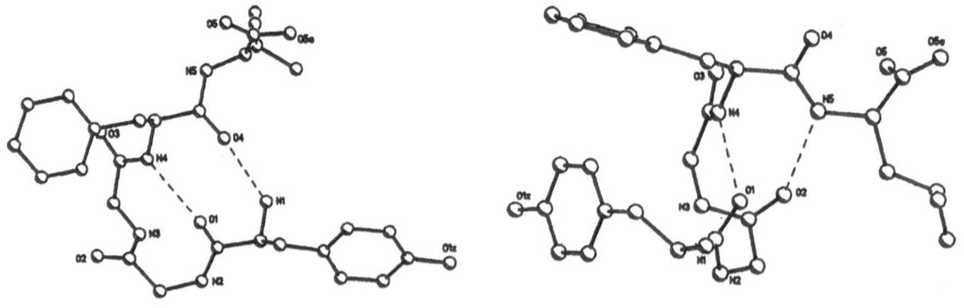

Fig. 1. The results of two independent X-ray studies [8, 12] on native Leu-enkephalin showing the 2 different folded conformations seen in the solid state.

Three of the peptides exhibit a double β bend conformation (Figure 2): Leu-enkephalin [2], an N,N-diallyl Leu-enkephalin analog in which residues 2 and 3 have been replaced by Aib [3] and a tetrapeptide analog of Met-enkephalin, Aib-Aib-Phe-Met [9]. This is the only occurrence of a Met-enkephalin analog which exhibits a folded conformation in the solid state.

Fig. 2. Superposition of double β – bend structures showing the similarities in conformation for the three molecules.

One of the tetrapeptides, TGGP, [7] and the 4 remaining pentapeptides [8-11] (8 unique molecules) exhibit a single type I′ β-bend. All 9 molecules have an N4•••O1 intramolecular H-bond. In a typical type I′ β-bend there is a second intramolecular H-bond, between N1 and O4, and this is seen in 6 of the 9 molecules. TGGP [7] does not form the second intramolecular H-bond nor does one of the 2 unique molecules of [11] even though the N•••O distance (3.23 Å) is of the proper order. However, in both cases, the β-bend is stabilized by a second connection in which both N1 and O4 hydrogen bond to a water molecule. DADLE [2] also does not quite fit the pattern in that the second H-bond forms between N1 and O5, resulting in a somewhat more open conformation.

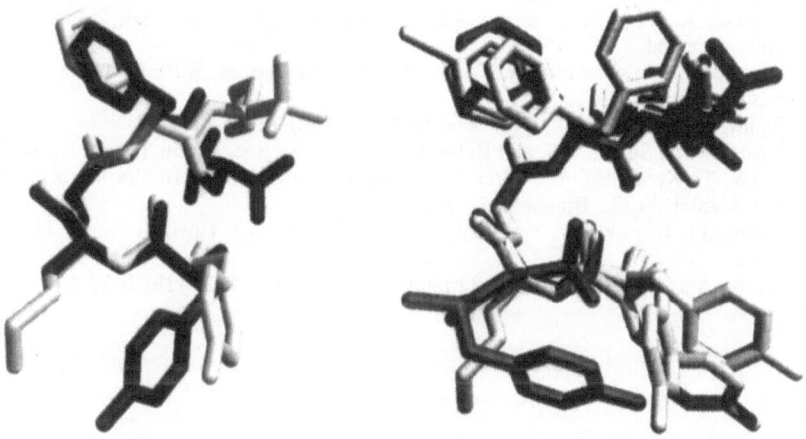

Fig. 3. Type I′ structures. Left: DADLE [2] superimposed on another enkephalin analog with a D-residue at position 2 [11]. Right: DTLET [1] superimposed on four type I′ enkephalins [7, 8, 10, 11] clearly indicating the involvement of the D-Thr side chain in the 'pseudo' β-bend.

The real outlier is the hexapeptide DTLET [1] which does not form a true β-bend in that the torsions $\chi 1$ and $\chi 2$ of the hydrophilic D-Thr^2 side chain, instead of main-chain ϕ,ψ angles define the bend. We have, therefore, classified, this conformation as a 'pseudo' β-bend. DTLET also has only one intramolecular H-bond (O2$^\gamma$•••N4) but there is a second stabilizing connection through a water molecule which links O2$^\gamma$ and O4.

All the peptides contain co-crystallized solvent molecules. Seven of the compounds contain only co-crystallized water (varying from 1 - 6 molecules of water per peptide molecule in the asymmetric unit). Of these, each participates in at least 3 peptide - water interactions with up to as many as 9 for the double β-bend enkephalin [12] and 11 for DTLET [1]. The N atoms most commonly bound to water are those on residues 1 and 2. The O atoms most commonly bound to water are the Oζ of Tyr1, the C=O of residues 2 and 4 and the C-O$^-$ of the terminal residue. These patterns persist regardless of the length of the peptide.

Acknowledgements

This work was supported by ONR and NIDA.

References

1. Flippen-Anderson, J.L., George, C., Deschamps, J., Ward, K.B. and Houghten, R., Int. J. Peptide Protein Res., in press.
2. Carroll, F.I., Brine, G.A., Boldt, K.G., Sawyer, D.K., Moreland, C.G., Flippen-Anderson, J.L., George, C. and Deschamps, J., in preparation.
3. Flippen-Anderson, J.L., George, C., Deschamps, J. and Brine, G., in preparation.
4. Hruby, V.J., Collins, N., Flippen-Anderson, J.L., George, C. and Cudney, B., in preparation.
5. Haaseth, R.C., Flippen-Anderson, J.L., George, C., Collins, N., Kover, K.E. and Hruby, V.J., in preparation.
6. Lomize, A., Flippen-Anderson, J.L., George, C. and Mosberg, H., J. Am. Chem. Soc., accepted.
7. Fournie-Zaluski, M., Prange, T., Pascard, C. and Roques, B.P., Biochem. Biophys. Res. Commun., 79 (1977) 1199.
8. Smith, D.G. and Griffin, J.F., Science, 199 (1978) 1214.
9. Prasad, B.V., Sudha, T.S. and Balaram, P., J. Chem. Soc. Perkin Trans., 1 (1983) 41.
10. Ishida, T., Kenmotsu, M., Mino, Y., Inoue, M., Fujiwara, T., Tomita, K., Kimura, T. and Sakakibara, S., Biochem. J., 218 (1984) 677.
11. Stezowski, J., Eckle, E. and Bajkusz, S., J. Chem. Soc. Chem. Comm., 11 (1985) 681.
12. Aubry, A., Sakarellos, C. and Marraud, M., Biopolymers, 28 (1989) 27.

A comparative study on the physical and conformational state of double-tailed lipo-gastrin and -CCK derivatives

L. Moroder[1], R. Romano[1], J. Winand[2], J. Christophe[2], W. Guba[3], D.F. Mierke[3] and H. Kessler[3]

[1]Max-Planck-Institut für Biochemie, Martinsried, Germany
[2]Department of Biochemistry and Nutrition, Université Libre de Bruxelles, Bruxelles, Belgium and [3]Organisch- Chemisches Institut, TU München, Garching, Germany

Introduction

Structural studies on membrane-bound peptides are hampered by the different conformational equilibria established in the partition process between water and a lipid phase. Using human gastrin (HG) and cholecystokinin (CCK) as model peptides, two lipophilic adducts, i.e. (2R,S)-1,2-dimyristoyl-3-mercaptoglycerol/N^{α}-maleoyl-β-alanyl-[Nle^{15}]-HG-(2-17) (=DM-HG) [1] and (2R,S)-1,2-dimyristoyl-3-mercaptoglycerol/N^{α}-maleoyl-β-alanyl-[Thr^{28},Nle^{31}]-CCK-(25-33) (=DM-CCK) [2], were synthesized in order to induce a drastic shift in the distribution of the peptides in favor of the lipid phase as resulting from interdigitation of the double-tailed lipo-moieties with the membrane bilayers.

Results and Discussion

The two peptide hormones HG and CCK were derivatized at their N-termini, since modifications at this sequence position usually does not affect their bioactivities. The structure of the two lipo-peptides was expected to induce in aqueous media a behavior similar to that of phospholipids, i.e. low solubility and spontaneous formation of vesicles. This, however, should not prevent partitioning of the lipo-peptides in favor of natural membranes, if the relative fluidity of their vesicles is sufficient to guarantee a net intervesicular lipid transfer.

Dynamic light scattering and electron microscopy showed that DM-HG aggregates into unilamellar spherical vesicles, which upon extrusion form a stable monodispersed system [1]. Conversely, vortexing or sonication of DM-CCK leads to a highly polydispersed system of vesicles even after extrusion [2]. This reflects a surprisingly strong effect of the peptide head groups on the lipid packing despite their sequence homology. Since no phase transition (T_c) was detected for DM-HG and DM-CCK by high sensitivity differential scanning calorimetry above 5˚C, the vesicles are in the liquid state. This low T_c of the lipo-peptide vesicles was expected to facilitate their net transfer to phosphatidylcholine (DPPC and DMPC). In fact, statistical insertion of DM-HG into DPPC vesicles occurs quantitatively upon 1.5 h incubation of the two vesicle populations above the T_c of DPPC, whereas the transfer of DM-CCK to DMPC vesicles is rapid, already observed in the first scan (30°C/h), i.e. even below the T_c of the DMPC, and formation of one DMPC domain (T_c=24.72°C) containing statistically inserted DM-CCK and two differently enriched DM-CCK domains (T_c =16.32° and 19.87°C) takes place. Addition of Ca^{2+} to this system affects the T_c of the two CCK-rich domains (shift by 5°C) more than that of the DMPC-rich domain (shift by 3°C) confirming the presence of differently populated domains. This also suggests a higher affinity of the CCK-rich domains for Ca^{2+}, a fact which could be of physiological relevance since CCK is known to stimulate Ca^{2+} release from reserves on cell membranes.

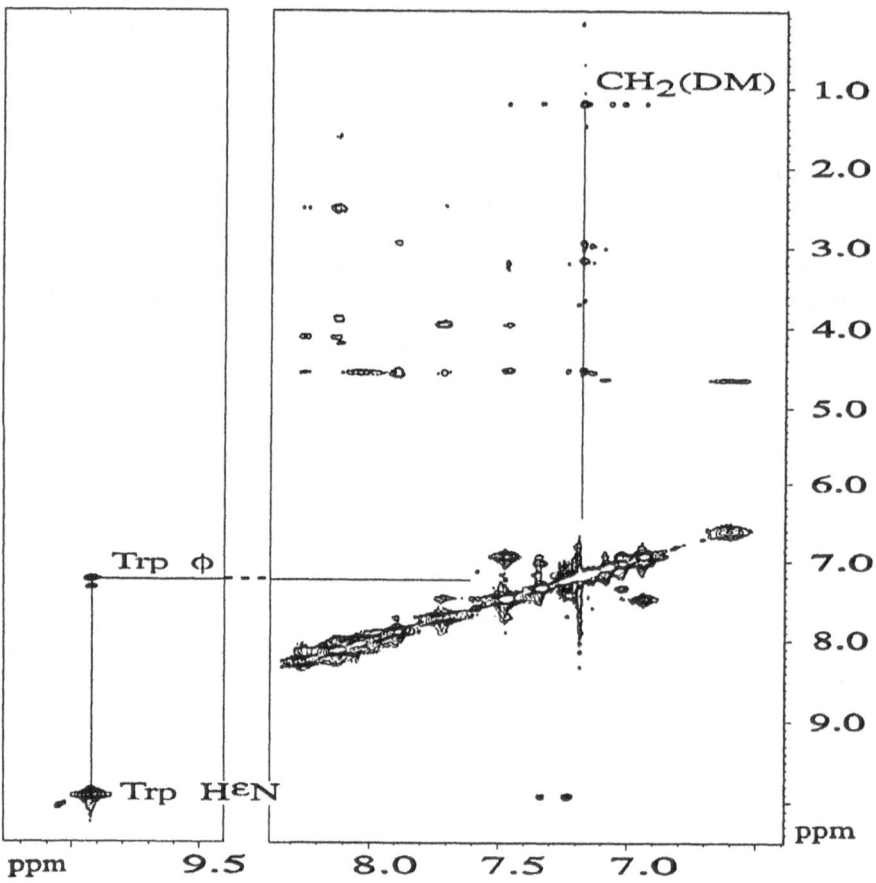

Fig. 1. Expanded portion of a NOESY (mixing time 50ms) illustrating the NOE between the aromatic side chain of Trp and the alkane chain of the dimyristoyl moieties.

Merging of DM-HG with DMPC vesicles is accompanied by a conformational transition from partially ordered to a predominantly unordered structure of the HG moiety as monitored by CD [3]. Thus incorporation of DM-HG into DMPC bilayers apparently prevents contacts of the peptide head group with more hydrophobic compartments of the membrane and exposes most of the HG randomly coiled to the bulk water. Conversely, the CD of CCK in pure DM-CCK and in mixed DM-CCK/DMPC vesicles is very similar with a strong negative maximum located in the 214-215 nm range which could reflect β-type structures possibly resulting from peptide-peptide associations. Additionally, the Trp fluorescence is blue-shifted in DM-CCK by 5-6 nm and in DM-CCK/DMPC vesicles by 22 nm, and the accessibility of Trp to I-quencher is strongly reduced. Thus the C-terminus of the CCK moiety seems to be inserted into more hydrophobic compartments of the bilayers and this would require a chain reversal which can occur at Thr-Gly in agreement with the preferred conformation of [Thr[28],Nle[31]]-CCK-(25-33) in cryomixtures [4]. According to this model, the Arg-Asp-Tyr(SO$_3$H) portion would act as ionized head group exposed to the water phase. In ^{1}H-NMR measurements (600 MHz, 310K) on DM-CCK magnetization through bonds via COSY and TOCSY experiments could not be observed because of extremely broad resonances. Therefore a complete resonance assignment was not possible. However, the characteristic spin pattern of Trp allowed for the unambiguous assignment of an NOE between the side chain of Trp and the DM-alkane chains (Figure 1) confirming chain reversal in the peptide with insertion of its C-terminus into inner compartments of the bilayer whereby onset of β-sheet type structures would also explain the formation of DM-CCK clusters in DMPC bilayers.

Since a net intervesicular transfer of DM-HG and DM-CCK to phosphatidylcholine was confirmed, it should also occur with natural membranes. In fact, DM-HG retains a receptor affinity in AR4-2J cells which is only 7-fold lower than that of the native HG [1] and DM-CCK exhibits in pancreatic acini a 5.7-fold reduced receptor affinity compared to [Thr[28],Nle[31]]-CCK-(25-33). These values suggest a common factor which could possibly be a reduced two-dimensional migration rate of the lipo-peptide monomers in natural bilayers as induced by the di-myristoyl moiety. On the other hand, the built-in lipid affinity is affecting differently the biopotency of DM-HG and DM-CCK in characteristic assays. Whilst for DM-HG a 5-fold reduced potency with slightly enhanced efficacy was determined in the [^{14}C]-aminopyrine accumulation assay with parietal cells [3], the potency of stimulating amylase release from pancreatic acini by DM-CCK was 80 times lower, but its efficacy was identical to that of [Thr[28],Nle[31]]-CCK-(25-33).

Although lipo-derivatization was found to modify the bioactivities of HG and CCK, intervesicular transfer is occurring with model bilayers and the results indicate that at the level of the collisional event with lipid bilayers, DM-HG and DM-CCK exhibit a significantly different behaviour in their mode of display on the lipid-water interphase and of interaction with inner compartments of membranes.

References

1. Romano, R., Musiol, H.-J., Dufresne, M., Weyher, E. and Moroder, L., Biopolymers, 32 (1992) 1545.
2. Romano, R., Bayerl, T.M. and Moroder, L., Biochim. Biophys. Acta., 1151 (1993) 111.
3. Romano, R., Dufresne, M., Prost, M.-C., Bali, J.-P., Bayerl, T.M. and Moroder, L., Biochim. Biophys. Acta, 1145 (1993) 235.
4. Moroder, L., D'Ursi, A., Picone, D., Amodeo, P. and Temussi, P.A., Biochim. Biophys. Res. Commun., 190 (1993) 741.

First neuromedin B receptor antagonists are based on somatostatin octapeptide analogues

D.H. Coy[1], J. Taylor[2], J-P. Moreau[2], N-Y. Jiang[1],
M. Orbuch[3], J. Mrozinski[3], S. Mantey[3] and R.T. Jensen[3]

[1]Peptide Research Laboratories, Department of Medicine, Tulane
University Medical Center, New Orleans, LA 70112, U.S.A.,
[2]Biomeasure, Inc., Milford, MA 01757, U.S.A.,
[3]Digestive Diseases Branch, NIH, Bethesda, MD 20892, U.S.A.

Introduction

Neuromedin B (NMB), G-N-L-W-A-T-G-H-F-M-NH_2, is a bombesin/gastrin releasing peptide-related peptide possessing unique receptors with vastly different tissue distributions relative to the latter [1]. Although many extremely potent types of competitive receptor antagonists have been developed for bombesin/GRP [2], to date none have retained much affinity for NMB receptors. Recently, we have screened a library of somatostatin (SS) analogues for binding to a number of different receptor preparations, including 5 recently discovered SS receptor subtypes [3] and non-SS receptors such as bombesin/GRP and NMB. Two main classes of SS analogues were examined in the present study - the cyclic octapeptides related to octreotide (peptide 1, Table 1) and lanreotide (peptide 2, Table 1) and the recently developed linearized versions of the octapeptides [3] (peptides 5 and 6). In terms of their affinities to SS receptors on the various transfected cell types, many interesting SARs have been discovered [3]. The cyclic octapeptides exhibited particularly high affinity for SSTR2

Table 1 *Comparison of binding affinities of key cyclic and linear SS octapeptides for 5 SS receptors on transfected cells and NMB receptors on rat olfactory bulb*

| | IC_{50}(nM) | | | | | |
| | SS Receptor | | | | | NMB |
Peptide	1	2	3	4	5	
SS	0.1	0.28	0.07	0.86	1.2	>10000
1. DPheCysPheDTrpLysThrCysThrol	>1000	2.24	2.6	0.57	>1000	–
2. DNalCysTyrDTrpLysValCysThrNH$_2$	>1000	1.60	4.9	0.10	>1000	802
3. DPheCysTyrDTrpLysValCysNalNH$_2$	>1000	0.002	57	0.19	252	245
4. DNalCysTyrDTrpLysValCysNAlNH$_2$	>1000	4.2	181	5.2	102	43
5. DPhePheTyrDTrpLysValPheDNalNH$_2$	>1000	>1000	0.02	43	158	>1000
6. DPhePhePheDTrpLysThrPheThrNH$_2$	23	32	0.42	0.002	18	>1000

and SSTR4 but no or very little affinity for SSTR1 and SSTR5. The linear peptide 5 surprisingly exhibited high affinity only for SSTR3 and appears to be a relatively selective ligand for this receptor. Similarly, linear peptide 6 exhibited very high affinity for SSTR4 and was the only peptide other than full sequence SS peptides to exhibit any affinity for SSTR1 [Terry Reisin, personal communication].

Results and Discussion

Several of the cyclic octapeptides have previously been found to have high affinity for μ-opiate receptors [4] and, indeed, it has proved possible [5] to completely dissociate these properties from traditional SS effects. It was, therefore, logical to screen our extensive library of these peptides for binding to other peptide receptors. It was found that certain cyclic octapeptides bound to NMB receptors on rat olfactory bulb membranes (Table 1). The most potent of these was the highly hydrophobic sequence, *D-Nal-Cys-Tyr-D-Trp-Lys-Val-Cys-Nal-NH$_2$* (peptide 3, Table 1), which bound with a Ki of 43 nM, with significant affinity also being found with *D-Nal-Cys-Tyr-D-Trp-Lys-Nal-Cys-Thr-NH$_2$* (Ki 85 nM) (not shown). These analogues also bound significantly to NMB receptors transfected into Balb 3T3 cells. Peptide 3 displayed a Ki of 216 nM (Figure 1, left panel) in this system, did not stimulate ^{3}H-IP levels in these cells, but inhibited NMB-stimulated ^{3}H-IP release (Figure 1, right panel) as expected for a pure NMB receptor antagonist. The peptides exhibited little affinity for GRP receptors.

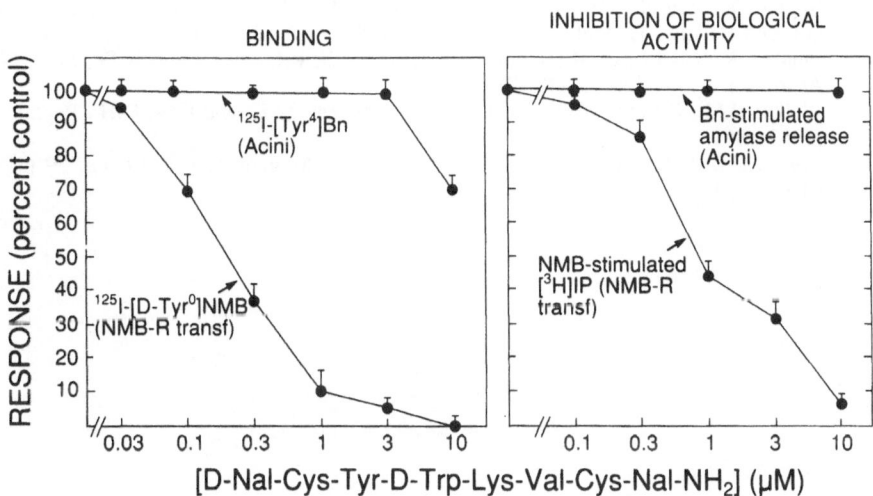

Fig. 1. *Ability of D-Nal-Cys-Tyr-D-Trp-Lys-Val-Cys-Nal-NH$_2$ to displace labeled Bn or NMB from rat pancreatic acinar cells or NMB receptor transfected Balb 3T3 cells (left panel) and to inhibit either Bn-stimulated amylase release or IP release, respectively from the same cell systems (right panel).*

Preliminary SAR studies thus far indicate that, as for SS receptors, the D-Trp and Lys residues, centered in the type II' β-bend region of these molecules [6], are critical but, additionally, aromatic residues at the termini of the chain and/or in position 6 must be present. A combination of Nal substitutions to give *D-Nal-Cys-Tyr-D-Trp-Lys-Nal-Cys-Nal-NH₂*, contrary to expectations, resulted in a lowering of binding affinity.

Although many of the NMB-receptor binding analogues thus far discovered also have high affinity for SS type 2 (pituitary) receptors, it is expected that dissociation of binding affinities will be possible. Since the physiological functions of NMB are presently poorly understood, selective and even more potent NMB antagonists should be useful investigatory tools. Considering that in the past SS octapeptides have also provided conformationally restrained ligands for the μ-opioid peptide receptor system, it appears that the folded motif in these molecules, which has been well characterized by 2D NMR studies and is also a common characteristic of other peptide ligands, could well provide a structural template for agonists/antagonists of other peptide receptors as well.

Acknowledgements

This research was supported in part by NIH grant CA-45153.

References

1. Wada, E., Way, J., Shapira, H., Kusano, K., Lebacq-Verhayden, A.M., Coy, D.H., Jensen, R.T. and Battey, J., Neuron, 6 (1991) 421.
2. Jensen, R.T. and Coy, D.H., Trends Pharm. Sci., 12 (1991) 13.
3. Raynor, K., Murphy, W.A., Coy, D.H., Taylor, J.E., Moreau, J-P., Yasuda, K., Bell, G.I. and Reisine, T., Mol. Pharm., 43 (1993) 838.
4. Maurer, R., Gaehwiler, B.H., Buescher, H.H., Hill, R.C. and Roemer, D., Proc. Natl. Acad. Sci. U.S.A., 79 (1982) 4815.
5. Walker, J.M., Bowen, W.D., Akins, S.T., Hemstreet, M.K. and Coy, D.H., Peptides, 8 (1987) 869.
6. Verheyden, P.M.F., Coy, D.H. and Van Binst, G., Magnet. Reson. Chem., 29 (1991) 612.

The effect of D-amino acid substitutions on the interaction of neuropeptide Y-(18-36) analogues with hydrophobic surfaces

M.I. Aguilar[1], S. Mougos[1], E. Lazoura[1], J. Boublik[2], J. Rivier[3] and M.T.W. Hearn[1]

[1]Department of Biochemistry and Centre for Bioprocess Technology, Monash University, Wellington Rd, Clayton, Victoria 3168, Australia, [2]Baker Medical Research Institute, Commercial Rd Prahran, Victoria 3181, Australia, and [3]The Clayton Foundation Laboratories for Peptide Biology, The Salk Institute, La Jolla, CA 92138, U.S.A.

Introduction

The ability to study the interactive behaviour of peptides at hydrophobic surfaces represents a powerful approach to characterizing the molecular properties which control the orientation and binding of peptides with different surfaces. In the present study, the relative stability and interactive behaviour of a series of analogues of Neuropeptide Y (NPY) were studied by reversed phase liquid chromatography (RP-HPLC) [1].

Results and Discussion

The peptides comprised a series of 16 analogues of NPY-(18-36) which differ by a single substitution of a D-amino acid for the L-isomer at various sequence positions. A number of chromatographic parameters which describe the interactive contact area and binding affinity have been evaluated. The RP-HPLC retention data of these peptides was analyzed in terms of the following expression:

$$\log \overline{k} = \log k_o - S\,\overline{\psi}$$

The parameters S and $\log k_o$ are related to the hydrophobic contact area of the peptide and the affinity of this contact region for the hydrophobic ligands [1, 2]. In order to study the effect of structural substitutions within NPY(18-36) on its interaction with hydrophobic surfaces, retention data were obtained for the series of 16 NPY(18-36) analogues. The conformation of peptides in solution was manipulated by measuring the interactive behaviour of the peptides over a range of temperatures.

If a peptide interacts without any specific orientation, D-substitutions would not be expected to exert any influence on the interactive properties of these NPY analogues. Figures 1(a) and (b) show plots of $\log \overline{k}$ versus $\overline{\psi}$ for the set of NPY(18-36) D-substituted analogues separated on an n-butylsilica (C4) sorbent at 25°C. These plots demonstrate that there are very specific orientation effects associated with the interaction of these peptides with the hydrophobic ligand. The plots shown in Figure 1 were used to derive the corresponding S and $\log k_o$ values.

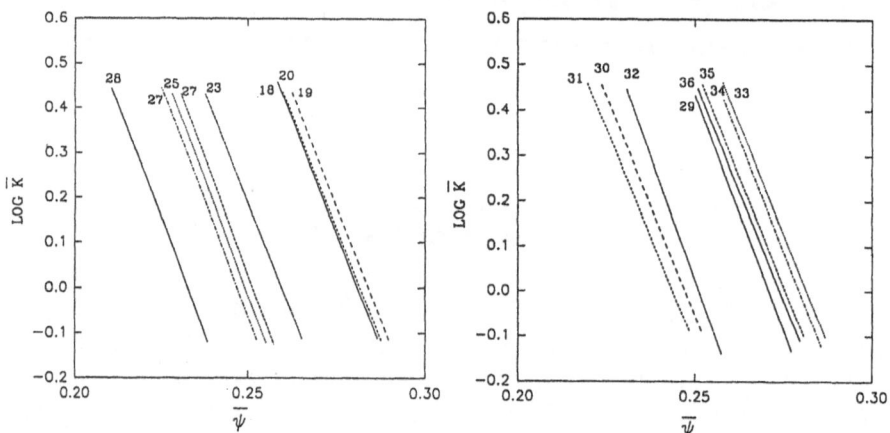

Fig. 1. Plots of log $\bar{k}$ versus $\bar{\psi}$ NPY(18-36) analogues separated on a C4 sorbent with acetonitrile as the organic solvent at 25°C. Chromatographic conditions are described elsewhere [1].

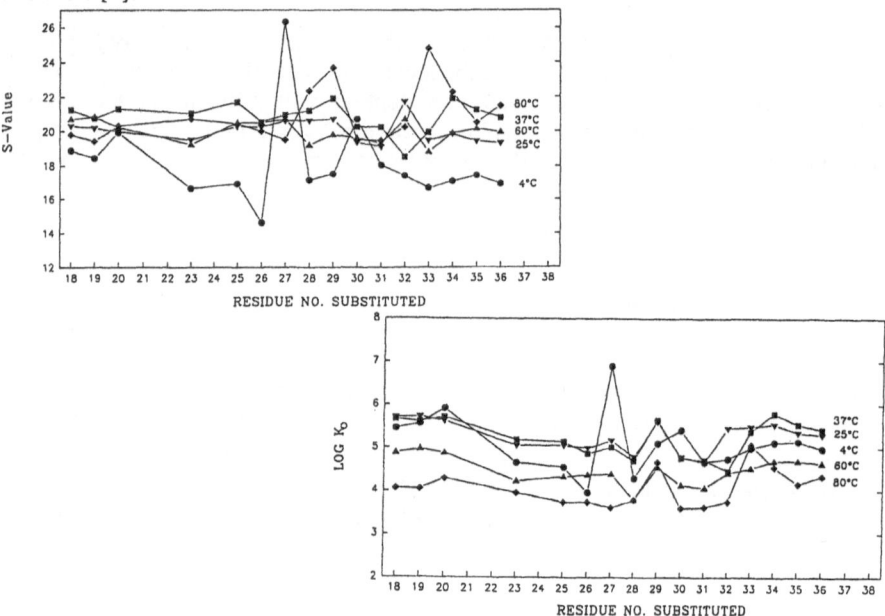

Fig. 2. Plots of (a) S and (b) log k_0 versus position of the D-amino acid substitution in NPY(18-36) analogues separated on a C4 sorbent with acetonitrile as the organic solvent.

Figures 2(a) and (b) show plots of the S and log k_0 values for each D-substituted NPY(18-36) analogue plotted against the residue position for experiments carried out with a C4 ligand and acetonitrile as the organic solvent modifier. At 4°C there was a small decrease in both parameters when D-substitutions were made between amino acid residues Ala(18) and His(26), compared to the values determined for the corresponding α-L-NPY(18-36). Much larger fluctuations in the magnitude of the S and log k_o values were evident when D-substitutions were made between amino acid residues Tyr(27) and Ile(31). This behaviour was then followed by essentially constant S and log k_o values for the D-substitutions in the C-terminal positions.

At the higher temperatures of 25°C, 37°C and 60°C, the S and log k_o values were similar in magnitude to those observed at 4°C and there were only small fluctuations in these parameters for the various D-substituted NPY(18-36) analogues. At 80°C, with the exception of small increases in the S values for the analogues D-Ile(28), D-Asn(29) and D-Arg(33), the S values were similar to the values determined at 25°C, while the log k_o values were consistently lower.

It is apparent from these data that the substitutions by D-amino acids in the central region of the NPY(18-36) peptide cause significant changes in the interactive behaviour of these peptides with the C4 ligand. These results were also similar to those which were evident with the C8 ligand [1]. However, these changes in retention properties occurred at a lower temperature with the C4-sorbent, which suggests that the longer *n*-octyl ligands may be able to more effectively stabilize the secondary structure of the peptides relative to the shorter more rigid n-butyl ligand. Furthermore, the most striking changes in the S and log k_o values evident for the NPY(18-36) analogues eluted from the C8 ligands corresponded to a large decrease in these parameters when the D-substitution occurred at amino acid residues Ile(28) and Asn(29). In contrast, a large increase in these parameters was observed with the C4 ligand for these two D-substituted analogues. These results indicate that the extent of the secondary structure adopted by these analogues is different with these two stationary phase ligands again demonstrating the important role of the ligand in the retention process.

To further characterize the interactive behavior of the NPY(18-36) analogues, the putative hydrophobic binding domain was defined. If residues Arg(19)-Q(34) are constrained into an α-helix (by analogy with the proposed structure of NPY(1-36) [3]) and the resulting structure subjected to energy minimization, the hydrophobic contact region can be characterized by use of hydrophobicity coefficients [4]. For the parent peptide NPY(18-36), the hydrophobic residues which form a continuous surface with a C4 ligand are Tyr(20), Tyr(21), Leu(24), Tyr(27), Ile(28), Ile(31), Gln(31) and Tyr(36). However, this region of the peptide is severely disrupted for the analogue D-Tyr(27) and the residues Tyr(27) and Gln(34) do not form part of a continuous hydrophobic patch. Overall, the results indicate that NPY-(18-36) adopts a significant degree of secondary structure in the presence of hydrophobic surfaces, and that this structure can be severely disrupted by the presence of the D-amino acids at residue positions Tyr(27)-Thr(32). The results of chromatographic measurements such as those presented here can supplement data derived from spectroscopic analyses and biological activity measurements to further characterize structure-function relationships.

References

1. Aguilar, M.I., Mougos, S., Boublik, J., Rivier, J. and Hearn, M.T.W., J. Chromatogr., 646 (1993) 53.
2. Purcell, A.W., Aguilar, M.I. and Hearn, M.T.W., J. Chromatogr., 593 (1992) 103.
3. Mierke, D.F., Durr, H., Kessler, H. and Jung, G., Eur. J. Biochem., 206 (1992) 39.
4. Wilce, M.C.J., Aguilar, M.I. and Hearn, M.T.W., J. Chromatogr., 632 (1993) 11.

Rapid, large-scale synthesis and solubility studies of the amyloid β protein of Alzheimer's disease

E.P. Berger, K.C. Rasmussen, P.H. Weinreb and P.T. Lansbury, Jr.

Department of Chemistry, Massachusetts Institute of Technology, 18-443, 77 Massachusetts Avenue, Cambridge, MA 02139-4307, U.S.A.

Introduction

The amyloid β protein found in the Alzheimer's afflicted brain, β1-42, was synthesized using a fragment-coupling scheme developed in our lab [1]. Fragment condensation greatly simplifies the purification of insoluble peptides. Each of the protected fragments was purified by RPHPLC before the final synthesis, eliminating any single amino acid deletions at an early stage. The final product was then purified by size-exclusion chromatography. We have modified the synthesis to eliminate impurities and to increase the overall yield of the full-length peptide. In addition, syntheses of several β protein analogues were achieved using the same fragments in this improved fragment-coupling strategy.

We have also investigated the mechanism of amyloid fibril formation by the β protein. Understanding the pathway that leads to fibril formation may lead to an understanding of the pathology of Alzheimer's disease (AD). We are particularly interested in the role that the C-terminus plays in fibril formation. The rates of aggregation of several β proteins, differing in the lengths of their C-termini, were measured and found to differ dramatically. Based on these results, a mechanism for amyloid formation is proposed. This mechanism, nucleation-dependent aggregation, may have implications in the cause of amyloid formation in AD [2].

Results and Discussion

Two major impurities were present in the original synthesis of β1-42 [1]. The first impurity resulted from the attachment of the C-terminal fragments to the resin (Figure 1) and was eliminated by using a modified procedure for the synthesis of the C-terminus and a different resin (PAM). The second impurity occurred during the deprotection step. The problem of benzylated impurities was solved by an altered HF deprotection protocol that employed twice the amount of *p*-cresol scavenger.

In addition, several practical improvements were made. Modifications to the syntheses of three fragments (β(1-17), β(18-25) and β(26-33)) resulted in an increase in both the yield and purity of the peptides. The new route to the synthesis of the fragment β(26-33), consisting of an additional fragment coupling, considerably simplified its purification and improved its yield. In addition, the development of a unique purification procedure for β(18-25) significantly increased the recovery of this fragment. This procedure employed the temporary removal of the N-terminal protecting group (Boc) before purification.

Finally, several analogues of the β protein were made using the new strategy. β1-40, β1-39, and the 3 kDa β proteins, β17-42, β18-42, and β17-40, were synthesized using the same fragments. The high purity of β17-42 (greater than 90%) demonstrated the success of our new methodology in the synthesis of the β protein.

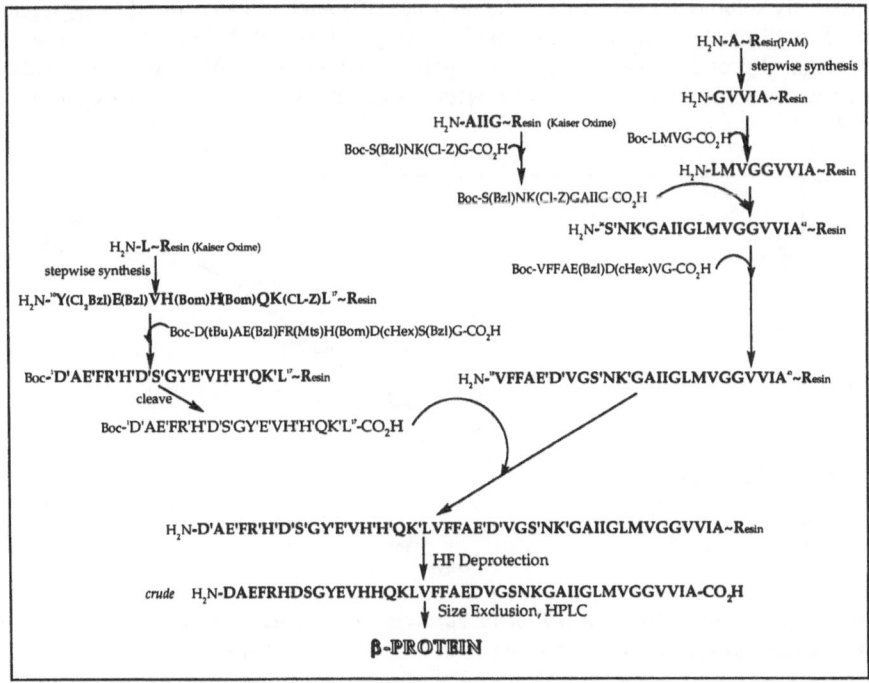

Fig. 1. *Synthetic Scheme of β1-42. Fragment couplings were used to eliminate single amino acid impurities, which simplifies the purification step.*

Dramatic differences were observed in the aggregation rates of the β proteins (Figure 2). Concentrated peptide solutions in DMSO (50 µL) were added to a phosphate buffer at pH 7.4 (950 µL). The rate of aggregation was measured by a turbidity assay using a UV spectrometer at 400 nm. Peptides with shorter C-termini (β1-39 and β1-40) appeared soluble for a week before they began to aggregate, while the longer C-terminal variant (β1-42) began to aggregate immediately. The lag time before aggregation observed with β1-39 and β1-40 may be due to the slow formation of a nucleus that is required for the growth of the fibril. This nucleation-dependent mechanism for amyloid formation was further supported by evidence that a kinetically soluble peptide (i.e., β1-40) can be seeded (caused to aggregate faster) by small amounts of preformed fibrils of either itself or of the kinetically insoluble peptides (i.e., β1-42) [2].

Our synthetic methodology allows us to make numerous β protein variants (β1-42, β1-39, β17/18-42, and β17-40) whose impurities can be identified and eliminated without the use of reverse-phase HPLC. Results of studies of peptides synthesized via this fragment-coupling methodology are much more reproducible than those found with stepwise-synthesized peptides. In addition, the ease of purification increases the yield as well as the purity of the protein.

The aggregation studies suggest the importance of the C-terminus of the β protein in determining the rate of aggregation and the role that an exogenous seed may play in the aggregation of kinetically soluble β protein. A nucleation-dependent mechanism of aggregation may mean that the "normal" β protein (β1-40) is kinetically soluble in the brain and the presence of a nucleus can cause it to aggregate and form fibrils. This nucleus could be in the form of fibrils of the possibly abnormal β protein, β1-42, which aggregates much more readily. Moreover, other proteins found in the plaques may be acting as nuclei or seeds for the aggregation of β1-40.

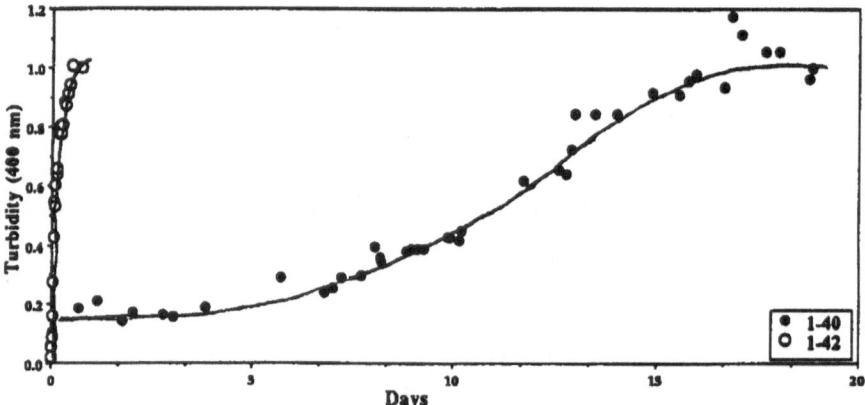

Fig. 2. Aggregation Curves of β Protein Variants. The turbidity of the peptide solutions was measured at 400 nm. Graphs were normalized to the final turbidity.

Acknowledgements

This work was supported by the National Institutes of Health (AG08470-04), the Camille and Henry Dreyfus Foundation, the Sloan Foundation, and the National Science Foundation (Presidential Young Investigator Award; contributions from Parke-Davis, Monsanto, and Hoechst-Celanese). P. L. is the Firmenich Assistant Professor of Chemistry. E. B. was supported by a graduate fellowship from Eli Lilly.

References

1.   Hendrix, J.C., Halverson, K.J. and Lansbury, P.T., Jr., J. Am. Chem. Soc., 114 (1992) 7930.
2.   Jarrett, J.T., Berger, E.P. and Lansbury, P.T., Jr., Biochemistry, 32 (1993) 4693.

Potent bombesin-like agonists and antagonists: Structure/activity at the COOH- and NH$_2$-termini

J.V. Edwards[1], B.O. Fanger[1], L.R. McLean[1],
E.A. Cashman[1], S.R. Eaton[1],
J.L. Krstenansky[1] and J.J. Knittel[2]

[1]East Marion Merrell Dow Research Institute, 2110 Galbraith Rd.,
Cincinnati, OH 45215, U.S.A., and [2]College of Pharmacy,
University of Cincinnati, Cincinnati, OH 45267-0004, U.S.A.

Introduction

The bombesin-like peptides are amphibian skin peptides which possess mitogenic and secretory activities including growth stimulation of small cell lung carcinoma (SCLC) through an autocrine loop mechanism. These peptides are elevated in the serum of SCLC patients. The goal of this research is to develop effective bombesin antagonists for the treatment of neoplastic diseases such as lung, breast, and prostate cancer. Several approaches have been reported to be effective in the design and synthesis of bombesin antagonists. Those studies focus on deletion or modification of the COOH-terminal residues. We report here three groups of peptides with backbone and side chain substitutions which address the receptor binding and activation properties of the active COOH-terminal sequence.

Results and Discussion

Four types of synthetic modifications were introduced into bombesin-like sequences. These include the following: 1) ψ[CH$_2$S] and ψ[CH$_2$N(CH$_3$)] substituted at the penultimate amide bond; 2) conformers of dehydrophenylalanine at the COOH-terminus; 3) COOH-terminal deletion with NH$_2$-terminal acylation and 4) a related cyclic lactam analog thereof. The ψ[CH$_2$S] substitution was accomplished as previously reported [1]. The ψ[CH$_2$N(CH$_3$)] substitution was made according to the solid phase reductive alkylation method of Sasaki and Coy [2]. Incorporation of dehydrophenylalanine was accomplished through preparation of Boc-Leu-Δ^ZPhe-OMe, and subsequent fragment coupling. Photoisomerization of **8** to **7** gave conversion of the Δ^Z to the Δ^E conformer. This was accomplished by photolysis of **8** in a similar manner to that of Nitz et al. [3]. The conversion was monitored by analytical RP-HPLC. The N-acyl[D-Ala^{11}]bombesin(7-13)amide analogs were synthesized on solid phase. The cyclic analog was prepared on Kaiser oxime resin with concomitant cyclization and resin deblocking [6].

A variety of amide bond surrogates have been studied [4] at the penultimate position of bombesin since Coy's finding that Leu$^{11}\psi$[CH$_2$NH]Leu^{14}bombesin is a potent antagonist. In this regard we have shown that ψ[CH$_2$S] yields a potent antagonist [5]. However, $\psi[CH_2N(CH_3)]$ substituted at the identical amide bond position, **6**, yields an agonist.

Aromatic substitutions at the COOH-terminal residue have resulted in highly potent receptor antagonists. To probe for the preferred side chain conformer at the COOH-terminus Δ^ZPhe and Δ^EPhe were substituted. The ϕ and ψ angles of dehydrophenylalanine are significantly restricted and the double bond restricts c-rotation to planar cis and trans positions. The Δ^ZPhe-containing analog, **8**, is 10-fold more potent than Δ^EPhe, **7**, although both conformers retain significant binding affinity. In addition both conformer analogs demonstrated partial agonist/antagonist activity.

The N-acyl analogs, **1-4**, were synthesized to probe a potential lipophilic binding site in the bombesin receptor. It was found that octyl(D-Ala[11])bombesin(7-13) amide is a potent antagonist. To further assess the relation of this lipophilic requirement to a folded conformation, aminooctanoic acid was placed at the NH_2-terminus and the analog cyclized to the COOH-terminus yielding **5**. Significant activity was retained upon cyclization. These results suggest that the octyl analog, **2**, is active in a folded conformation.

Table 1   *Comparison of receptor affinities in competitive binding and PI turnover in mouse pancreas [5]*

Peptides	Binding K_d	PI turnover Ag.	Antag[a]
1 N$^\alpha$-Acetyl--Gln-Trp-Ala-Val-D-Ala-His-Leu-NH$_2$	69	-	+
2 N$^\alpha$-Octyl-Gln-Trp-Ala-Val-D-Ala-His-Leu-NH$_2$	5.0	-	+
3 N$^\alpha$-Lauryl-Gln-Trp-Ala-Val-D-Ala-His-Leu-NH$_2$	320	-	+
4 N$^\alpha$-Palmityl-Gln-Trp-Ala-Val-D-Ala-His-Leu-NH$_2$	350	-	+
5 cyclo[Aoa-Gln-Trp-Ala-Val-D-Ala-His-Leu-]	51	ND	ND
6 Phe$^8\psi$[CH$_2$N(CH$_3$]Leu^{9}litorin	0.56 & 25	+	-
7 Ac-D-Phe-Gln-Trp-Ala-Val-Gly-His-Leu-Δ^EPhe-OCH$_3$	18	+/-	-
8 Ac-D-Phe-Gln-Trp-Ala-Val-Gly-His-Leu-Δ^ZPhe-OCH$_3$	1.7	+/-	+
9 Litorin, pGlu-Gln-Trp-Ala-Val-Gly-His-Phe-Met-NH$_2$	0.11	+	-

[a]Abbreviations: (Ag) Agonist; (Antag) Antagonist.

A comparison of the conformations of the bombesin-like agonists and antagonists was made by CD spectroscopy in TFE (Figure 1). Agonists and antagonists reported here give different CD spectra, however both appear to adopt ordered conformations. Interestingly we have found that the $\psi[CH_2N(CH_3)]$ substituted analog has a similar CD spectrum to that of litorin (5). This suggests a correlation of CD-observed ordered structure to agonist/antagonist activity.

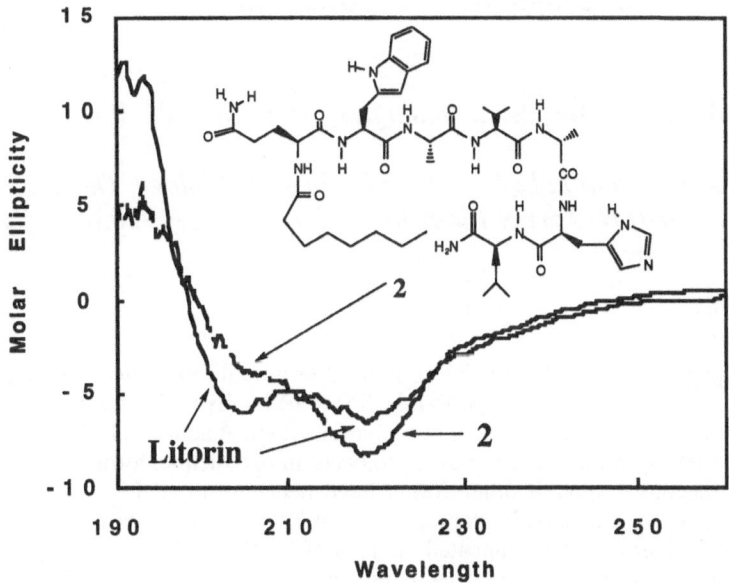

Fig. 1. CD spectra of litorin and analog 2 in trifluoroethanol.

This work examined backbone and side chain modifications having a pronounced effect on receptor binding and activation. The correlation of CD structures to agonist/antagonist activity suggests a role of ordered structure in receptor binding and activation. The conformation of the COOH-terminal phenyl ring significantly influences receptor binding and activation. However, deletion of the COOH-terminal residue in combination with an NH_2-terminal octyl group results in significant retention of antagonist receptor affinity. This effect may be mediated by a folded conformation as evidenced by the octyl containing cyclic peptide.

References

1. Spatola, A.F. and Edwards, J.V., Biopolymers, 25 (1986) 229.
2. Sasaki, Y. and Coy, D.H., Peptides, 8 (1986) 119.
3. Nitz, J.N., Shimohigashi, Y., Costa, T., Chen, H.-C. and Stammer, C.H., Int. J. Peptide Protein Res., 27 (1986) 522.
4. Jensen, R.T. and Coy, D.H., Trends Pharm. Sci., 12 (1991) 13.
5. Edwards, J.V., McLean, L.R., Wade, A.C., Eaton, S.R., Cashman, E.A., Hagaman, K.A. and Fanger, B.O., Int. J. Peptide Protein Res., in press.
6. Osapay, G., Profit, A. and Taylor, J.W., Tet. Lett., 31 (1990) 6121.

High affinity, truncated, cyclic and branched analogs of Neuropeptide Y

D.A. Kirby, S.C. Koerber and J.E. Rivier

Clayton Foundation Laboratories for Peptide Biology, The Salk Institute, 10010 N. Torrey Pines Road, La Jolla, CA 92037, U.S.A.

Introduction

Neuropeptide Y (NPY) is a 36 residue C-terminally amidated polypeptide belonging to the pancreatic polypeptide-fold (PP-fold) family which has been shown to be involved in many physiological and pharmacological actions [1, 2]. Specific receptors for neuropeptide Y are presumed to occur in two distinct forms, designated Y_1 and Y_2 receptors, though other forms have been proposed [3]. Distinction between the two types is derived from differential binding properties of C-terminal fragments and other centrally truncated analogs of NPY. While it was previously shown that the amidated C-terminal segment was essential for triggering biological activity [4], and residues in the N-terminal portion were essential for potent interaction of NPY with the Y_1 receptor by stabilizing the proximity of the N- and C-termini [5], it was postulated that the residues composing the central core of NPY contained only features of structural significance (i.e. helicity, hydrophobicity, intramolecular stability) and therefore were not essential for direct receptor interaction. The use of molecular modeling suggested several possible options for constructing analogs of NPY in which the central segment, residues 10-17, was removed. We have previously reported the binding properties of centrally truncated and internally constrained analogs of NPY [6, 7], ultimately leading to the production of analogs with high selectivity for the Y_2 receptor. In continuation of this study, we report here the synthesis and structure-activity relationships of a series of linear and cyclic deletion analogs based on des-AA$^{10\text{-}17}$[Cys^{7,21}]-NPY, an analog that maintained highest affinity to the Y_1 receptor in our previous investigations.

Results and Discussion

All peptides were synthesized manually using the Boc-strategy, cleaved by anhydrous HF and purified by HPLC. Cyclic compounds were obtained by air oxidation. In previous SAR studies, we noted that analogs having central residues 6-24 deleted produced high selectivity for the Y_2 type receptor without loss of affinity. However, since many of the biological effects of NPY appear to be mediated by the Y_1 receptor, it was our desire to instead construct truncated analogs which retained high affinity for the Y_1 receptor. Circular dichroism (CD) spectroscopy indicated that the deletion of residues 10-17 (**2**), a region expected to form an amphipathic α-helix, indeed failed to demonstrate a strong negative

Table 1 *Peptide sequences and binding affinities*

Cpd	Sequence	Y_1	Y_2
	1 5 10 15 20 25 30 35		
1	YPSKPDNPGEDAPAEDLARYYSALRHYINLITRQRY	2.0	0.3
2	YPSKPDNPG ARYYSALRHYINLITRQRY	77	2.5
3	Ac-YPSKPDNPk(ε-Ac) ARYYSALRHYINLITRQRY	87	1.8
4	Ac-YPSKPDNPk(ε-A-Ac) ARYYSALRHYINLITRQRY	96	1.0
5	YPSKPDCPG ARYCSALRHYINLITRQRY	5.0	1.3
6	YPSKPDCPGk(ε-Ac) ARYCSALRHYINLITRQRY	6.0	3.3
7	Ac-YPSKPDCPGk(ε-L-Ac) ARYCSALRHYINLITRQRY	12	5.5
8	YPSKPDcPK(ε-Ac) ARYCSALRHYINLITRQRY	160	5.4
9	YPSKPDcPk(ε-Ac) ARYCSALRHYINLITRQRY	28	5.2

[a]Small letters indicate D amino acids, *Ac* = acetyl; bridge position represented by underline. [b]Binding values expressed as K_i (in nM), where cell lines were derived from human neuroblastoma cells, SK-N-MC (Y_1 type) and SK-N-BE2 (Y_2 type).

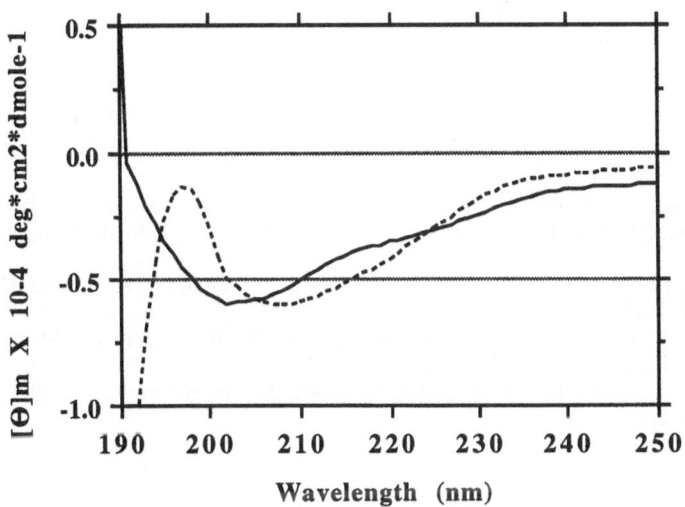

Fig. 1. Circular dichroism spectrum of compounds 2 (----) and 5 (—) in 0.01 M phosphate buffer, pH 7.4; data collected was average of four scans recorded on Aviv Model 62DS spectropolarimeter (Aviv Assoc., Lakewood, NJ) using 0.5 mm cuvettes.

ellipticity at 222 nm (Figure 1). The presence of an intramolecular constraint between residues 7 and 21 resulted in an analog (**5**) with higher affinity to the Y_1 receptor than the linear (**2**), while slightly decreasing the affinity to the Y_2 receptor. The fact that a single modification results in opposing binding properties suggests that each receptor type recognizes specific, yet different, pharmacophores. Despite a 15-fold increase in affinity at the Y_1 receptor over the linear deletion analog, the constrained analog (5) failed to give evidence of secondary structural integrity as measured by CD (Figure 1). Since the constrained analog (5) bound with high affinity, it is therefore suspected that essential information needed for enhancement of Y_1 receptor recognition is independent of the central α-helical segment, but rather it is the spatial alignment of the two termini that contributes most to high affinity binding to the Y_1 receptor. Indeed, if the role of the central segment was merely to act as a template to present the peptide binding site, then the extention of a branch from the turn region of an optimized analog would not be expected to affect binding. When an acetylated D-lysine was substituted at position 10 in a constrained analog (6), binding to the Y_1 receptor was only slightly reduced. However, if the N-terminus was also acetylated (7), binding was reduced by two fold. We have previously found that acetylation of the N-terminus of native NPY reduced affinity to the Y_1 receptor by 10-fold (unpublished results). Branch extension in corresponding linear analogs (3 and 4) likewise did not produce a deletrious effect on binding when compared to the parent deletion analog (2), again suggesting that the central segment is not a critical element of peptide/receptor interaction.

Acknowledgements

This work was supported by NIH grant HL-41910 and the Hearst Foundation. We thank Dr. Anthony Craig for mass spectral analysis and Duane Pantoja for technical assistance.

References

1. Tatemoto, K., Proc. Natl. Acad. Sci. U.S.A., 79 (1982) 5485.
2. Grey, T. and Morley, J., Life Sciences, 38 (1986) 389.
3. Wahlstedt, C., Yanaihara, N. and Håkanson, R., Reg. Peptides, 13 (1986) 307.
4. Boublik, J.H., Scott. N.A., Brown, M.R. and Rivier, J.E., J. Med. Chem., 32 (1989) 597.
5. Forest, M., Martel, J., St-Pierre, S., Quirion, R. and Fournier, A., J. Med. Chem., 32 (1990) 1615.
6. Kirby, D., Koerber, S., Craig, A., Feinstein, R., Delmas, L., Brown, M. and Rivier, J., J. Med. Chem., 36 (1993) 385.
7. Reymond, M., Delmas, L., Koerber, S., Brown, M. and Rivier, J., J. Med. Chem., 35 (1992) 3653.

Synthesis and aggregation of C-terminal analogs of β-amyloid

B.F. McGuinness, P.H. Weinreb and P.T. Lansbury, Jr.

Department of Chemistry, Massachusetts Institute of Technology,
77 Massachusetts Avenue, Cambridge, MA 02139-4307, U.S.A.

Introduction

The Alzheimer's afflicted brain is characterized by the presence of extracellular plaques which consist largely of a 39-43 amino acid protein known as β-amyloid. A synthetic peptide corresponding to β-amyloid's C-terminus (β34-42; NH_2-^{34}LMVGGVVIA^{42}-COOH) has been previously shown to form highly insoluble amyloid fibrils *in vitro* [1]. We have been examining the structure of this portion of the protein under the hypothesis that this region determines the kinetics of aggregation of β-amyloid *in vivo*. Rotational resonance solid state NMR has located a cis amide bond between residues G37 and G38 in the fibrillar aggregate of β34-42 [2]. A number of truncated and sequence variant analogs of this region have now been synthesized and characterized for their ability to form amyloid fibrils by electron microscopy (EM), FTIR, and congo red staining. Furthermore, analogs which explore the cis amide conformational bias at G37-G38 have been synthesized. Each of the synthetic analogs formed fibrils as determined by EM and exhibited similar characteristic β-sheet bands in the amide I region of the FTIR spectrum. However, differences in the kinetics of aggregation of these various analogs stress the importance of the β-branched nature of the C-terminal residues as well as the conformational flexibility of the G37-G38 amide bond.

Results and Discussion

Peptides were synthesized manually using standard solid phase protocols. The ester isostere was incorporated by reaction of bromoacetic anhydride [3] with NH_2-VVIA-PAM resin followed by overnight treatment with the cesium salt of Boc-Gly. The full depsi analog was then completed by standard Boc chemistry. The pyrrole isostere was synthesized via the method of Abell *et al.* [4], protected as its Fmoc derivative, and incorporated into the sequence of β34-42 on the HMPB-MBHA resin (1% TFA cleavage).

The kinetics of aggregation of the peptide analogs were followed utilizing a turbidity assay [5]. Aggregation was initiated by the addition of a DMSO solution (90 μL) of the analog to a stirred cuvette containing buffer (910 μL; 100 mM NaCl, 10 mM NaH_2PO_4, pH = 7.4). The turbidity of the resultant 375 μM solution was then measured over time at 400 nm. Final solubilities of the peptides were determined by amino acid analysis of the filtrate (0.22 μm filter) obtained after one week stirring.

Table 1 *Activity results of β-amyloid analogs*

β34-42 Analogs	Thermodynamic Solubility ($\pm$10%)
1 NH$_2$-LMVGGVVIA-COOH	26μM
2 NH$_2$-LMVGGVNvaLA-COOH	25μM
3 NH$_2$-LMVGΨ(CO$_2$)GVVIA-COOH	19μM
4 NH$_2$-LMVGΨ(NMeCO)GVVIA-COOH	27μM
5 NH$_2$-LMVΨ(C$_4$N)VVIA-COOH	8μM
6 NH$_2$-LMV-C$_6$H$_4$-VVIA-COOH	4μM

Table 1 lists some of the analogs of β34-42 synthesized. All of these analogs formed fibrils which were visualized via EM. In addition, the FTIR spectra of these analogs were characterized by the presence of a strong amide I band at ~1630 indicative of β-sheet structure. However, these methods are not sensitive enough to distinguish detailed structural differences in the solid state of these peptides.

The thermodynamic solubility of the first four entries in Table 1 (peptides **1** - **4**) was similar. However, each of these peptides exhibited differences in aggregation kinetics when studied by the turbidity assay. Previously, similar peptides have been shown to aggregate via a nucleation-dependent mechanism [5]. The natural sequence, β34-42, aggregated immediately under the conditions of this assay (Figure 1). However, the Nva40/Leu41 analog, which conservatively decreases the β-branched nature of the C-terminus, exhibited a lag time before fibril growth. Hence, the β-branching of residues V40 and I41 restricts the conformational freedom of this portion of the peptide, thus increasing the rate of the critical nucleation step.

As previously mentioned, the G37-G38 amide bond assumes a cis conformation in the solid state [1]. Replacement of this amide with its ester isostere yielded a peptide (**3**) which aggregated with a much faster growth rate. Rotation about the ester bond should be faster than rotation about the amide bond. This suggests that the fibril growth rate is influenced by the rate of cis-trans isomerization about this bond.

Introduction of cis amide mimics at the 37-38 position was next explored. The pyrrole analog (**5**) aggregated faster than the natural sequence, as would be predicted if the preferred aggregating conformation was cis (data not shown). The more conformationally restricted p-aminobenzoic acid derivative (**6**), however, exhibited a lag time in this assay. Again, although these peptides display different rates of aggregation, their thermodynamic solubilities are similar. Further characterization of the effect of cis amide mimics on the kinetics of aggregation is underway. One point is clear, however: the final thermodynamic solubility of the peptide analog (Table 1) is not directly related to its aggregation kinetics.

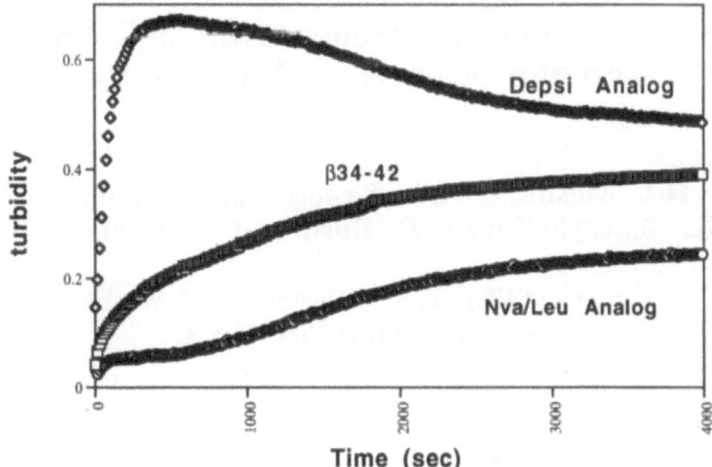

Fig. 1. Turbidity assay of β34-42 ((1), LMVGGVVIA); the Nva/Leu analog ((2), LMVGGVNvaLA); and the depsi analog ((3), LMVG$\Psi(CO_2)$GVVIA). The concentration of each peptide was 375 µM.

Differences in the kinetics of aggregation of the various analogs in this study stress the importance of both the β-branched nature of the C-terminal residues and the conformational flexibility of the G37-G38 amide bond. We believe further study of similar peptide analogs will eventually aid in the design of an inhibitor of amyloid aggregation as a potential therapeutic for Alzheimer's disease.

Acknowledgements

This work was supported by the National Institutes of Health (AG08470-04), the Camille and Henry Dreyfus Foundation, the Sloan Foundation and the National Science Foundation (Presidential Young Investigator Award; contributions from Parke-Davis, Monsanto, and Hoechst-Celanese). P.T.L. is the Firmenich Assistant Professor of Chemistry. B.F.M. is supported by an NIH postdoctoral fellowship (GM14044-03).

References

1. Halverson, K.J., Fraser, P.E., Kirshner, D.A. and Lansbury, P.T. Jr., Biochemistry, 29 (1990) 2639.
2. Spenser, R.G.S., Halverson, K.J., Auger, M., McDermott, A.E., Griffin, R.G. and Lansbury, P.T. Jr., Biochemistry, 30 (1991) 10382.
3. Robey, F.A. and Fields, R.L., Anal. Biochem., 177 (1989) 373.
4. Abell, A.D., Hoult, D.A. and Jamieson, E.J., Tetrahedron Lett., 33 (1992) 5381.
5. Jarrett, J.T., Berger, E.P. and Lansbury, P.T. Jr., Biochemistry, 32 (1993) 4693.

Cyclic deltorphin analogues with high δ opioid receptor affinity and selectivity

H.I. Mosberg, H.B. Kroona, J.R. Omnaas, K. Sobczyk-Kojiro, P. Bush and C. Mousigian

*College of Pharmacy, University of Michigan,
Ann Arbor, MI 48109, U.S.A.*

Introduction

The heptapeptides deltorphin I and II, Tyr-D-Ala-Phe-Xxx-Val-Val-GlyNH$_2$, where Xxx = Asp or Glu for deltorphin I and II, respectively, are naturally occurring opioids isolated from frog skin which have been shown to exhibit high selectivity for the δ type of opioid receptor [1]. This high selectivity is due largely to the anionic side chain of residue 4 which diminishes μ but not δ receptor affinity [2], while high δ binding affinity is dependent upon an intact C-terminal tripeptide [3]. It has been proposed that this tail plays a conformational role by inducing a β-turn involving residues 3-6 [4]. We have previously suggested [5] that a similar conformational element exists in a series of cyclic, δ selective tetrapeptide opioids, exemplified by Tyr-c[D-Cys-Phe-D-Pen] (JOM-13), which we described prior to the discovery of the deltorphins [6]. In JOM-13, which, like the deltorphins has a Tyr-D-Yyy-Phe N-terminal sequence, the disulfide cyclization has a similar conformational effect as the tripeptide tail in the deltorphins and results in the high δ binding affinity displayed by this tetrapeptide. In an effort to better define the role of C-terminal conformation on δ selectivity and affinity, we have examined a series of cyclic heptapeptide and related truncated, pentapeptide deltorphin analogs and report these results here.

Results and Discussion

Binding affinities, K_i, for the deltorphin analogs at μ (displacement of [^{3}H]DAMGO) and δ (displacement of [^{3}H]DPDPE) receptors in guinea pig brain membrane preparations are summarized in Table 1. Binding affinities for κ receptors were also assessed, but none of the analogs displayed significant affinity for this receptor. As seen in Table 1, extending the C-terminus of JOM-13 with the deltorphin tripeptide tail has no effect on δ affinity, but slightly improves μ affinity, consistent with the observation that an anionic side chain in residue 4 of deltorphin enhances δ selectivity by interfering with μ binding. In JOM-13 this role is played by the C-terminal carboxylate which is absent in the extended structure, **1**, and consequently μ affinity of the latter increases.

Table 1 *Opioid receptor binding profiles of cyclic deltorphins*

Peptide Analog	Binding K_i (nM)		
	DAMGO[a]	DPDPE	$K_i(\mu)/K_i(\delta)$
Tyr-*D*-Cys-Phe-*D*-PenOH (JOM-13)	107	1.79	60
Tyr-*D*-Ala-Phe-Glu-Val-Val-GlyNH$_2$ (Delt II)	1310	2.69	487
Tyr-*D*-Ala-Phe-Asp-Val-Val-GlyNH$_2$ (Delt I)	677	1.73	391
Tyr-*D*-Met-Phe-His-Leu-Met-AspNH$_2$ (Denk)	266	3.17	83.9
1 Tyr-*D*-Cys-Phe-*D*-Pen-Val-Val-GlyNH$_2$	70	3.0	23.3
2 Tyr-*D*-Cys-Phe-*L*-Pen-Val-Val-GlyNH$_2$	41	3.3	12.4
3 Tyr-*D*-Cys-Phe-Glu-*D*-Pen-Val-GlyNH$_2$	967	17.8	54.3
4 Tyr-*D*-Cys-Phe-Glu-*L*-Pen-Val-GlyNH$_2$	225	4.18	53.8
5 Tyr-*D*-Cys-Phe-Glu-*D*-PenNH$_2$	50.4	2.33	21.6
6 Tyr-*D*-Cys-Phe-Glu-*L*-PenNH$_2$	45.5	6.42	7.1
7 Tyr-*D*-Cys-Phe-His-*L*-Pen-Nle-AspNH$_2$	21.1	3.56	5.9
8 Tyr-*D*-Cys-Phe-His-*D*-Pen-Nle-AspNH$_2$	756	3.06	247
9 Tyr-*D*-Cys-Phe-His-*D*-PenNH$_2$	29.4	0.58	51

DAMGO=[^{3}H][*D*-Ala^2, NMePhe^4, Gly^5-ol]enkephalin;DPDPE=[^{3}H][*D*-Pen^2, *D*-Pen^5]-enkephalin

    Interestingly, the residue 4 diastereomer, **2**, is essentially equivalent to **1** in both affinity and selectivity. In the heptapeptides **3** and **4** and the corresponding truncated pentapeptides **5** and **6** a carboxyl side chain in residue 4 is reinstated via a 2-5 disulfide cyclization scheme. All four analogs maintain good δ affinity, however δ selectivity is reduced relative to deltorphin II, especially for the truncated peptides. Heptapeptides **7** and **8** utilize a similar 2-5 cyclization approach in analogs related to a third δ selective frog skin derived heptapeptide, dermenkephalin (Denk) [7]. Both analogs display similar δ affinity as Denk, however, while the *L*-Pen^5 diastereomer is virtually nonselective, the *D*-Pen^5 isomer, **8**, is equally selective with Denk, itself. Finally, the C-terminal truncated analog **9**, exhibits reduced selectivity relative to heptapeptide **8**, however its δ affinity is ca. 6-fold higher.

    The results presented above support the premise that the C-terminal tails of the deltorphins and dermenkephalin contribute to the δ receptor affinity of these peptides primarily by a conformational effect and that in the cyclic analogs described here, this conformational role is served by the cyclization. Consequently, extending the C-terminus of the cyclic tetra- and pentapeptide structures yields no additional binding attraction for the δ receptor, and, in fact, sometimes leads to a slight reduction in δ affinity (e.g., **3** vs. **5** and **8** vs. **9**). The extended structures do, however, display slightly improved δ selectivity which can be attributed to the consistently lower μ affinity of the heptapeptides.

Acknowledgements

This study was supported by grants DA 03910 and DA 00118 from the National Institute on Drug Abuse (NIDA). H.B. Kroona was supported by a NIDA Postdoctoral Training Grant, DA 07268.

References

1. Erspamer, V., Melchiorri, P., Falconieri-Erspamer, G., Negri, L., Corsi, R., Severini, C., Barra, D., Simmaco, M. and Kreil, G., Proc. Natl. Acad. Sci. U.S.A., 86 (1989) 5188.
2   Lazarus, L.H., Salvadori, S., Santagada, V., Tomatis, R. and Wilson, W.E., J. Med. Chem., 34 (1991) 1350.
3. Melchiorri, P., Negri, L., Falconieri-Erspamer, G., Severini, C., Corsi, R., Soaje, M., Erspamer, V. and Barra, D., Eur. J. Pharmacol., 195 (1991) 201.
4. Balboni, G., Marastoni, M., Picone, D., Salvadori, S., Tancredi, T., Temussi, P.A. and Tomatis, R., Biochem. Biophys. Res. Commun., 169 (1990) 617.
5. Mosberg, H.I. and Porreca, F., NIDA Research Monograph Series, in press.
6. Mosberg, H.I., Omnaas, J.R., Smith, C.B. and Medzihradsky, F., Life Sci., 43 (1988) 1013.
7. Mor, A., Delfour, A., Sagan, S., Amiche, M., Pradelles, P., Rossier, J. and Nicolas, P., FEBS Lett., 255 (1989) 269.

A new model for the development of Alzheimer's disease

B. Penke, K. Soós, J. Varga, E.Z. Szabó and J. Márki-Zay

Department of Medical Chemistry, A. Szent-Györgyi Medical University, H-6723 Szeged, Hungary

Introduction

The amyloid A4, or β-peptide ("β-amyloid") is a major component of extracellular amyloid deposits that are typical in Alzheimer's disease. According to the amyloid cascade hypothesis [1] the pathological route for the disease is most likely the following: β-amyloid deposition → τ-phosphorylation and formation of neurofibrillary tangles → cell death.

β-amyloid peptides proved to be neurotoxic [2], Kowall and Yankner found that the ß-amyloid 1-40 is a potent neurotoxin in the brain and its effects can be blocked by Substance-P. However, according to Mitsuhashi [3], various β-amyloid peptides do not interact with the cell culture Substance-P receptors.

It is not understood why β-amyloid peptides have high neurotoxicity, and whether Substance-P has any protecting role against the neurotoxic effect of β-amyloids.

In our previous experiments [4] β-amyloid 1-42 peptides definitely proved to be neurotoxic in cell cultures. Our aim was to investigate why β-amyloid peptides are neurotoxic, and what is the first event which triggers the death of neurons and starts the development of Alzheimer's disease.

Results and Discussion

Human β-amyloid peptide 1-42, and two analogs ([Gln]^{22}-β-1-42, [Gln]^{15}-β-1-42) as well as rat amyloid β-1-42 were synthesized with standard Boc-chemistry on MBHA resin. After purification the peptides were stored under different conditions:

   a) in solution (50 µg/ml concentration) at 20°C, 4°C, or -20°C
   b) in the form of lyophilized powder

Peptide purity was regularly checked with RP-HPLC. We have found that all β-1-42 peptides are unique: they show an enormously high affinity for aggregation: the higher the concentration of the aqueous solution the higher the aggregation rate. The β-1-42 peptide monomers cannot be stored as lyophilized powder even at -20°C, after a few weeks a mixture of aggregates can be found in the place of the original homogeneous peptide. The best method for the storage of β-amyloid peptides is to dissolve them in aqueous trifluoroethanol (1:1 v/v) in a concentration of 20 μg/ml, and store the solution at -20°C.

All β-amyloid peptides proved to be toxic in 18-day old rat brain cortical cell cultures containing different types of neurons and glial cells. The neurotoxicity of short peptide sequences (e.g. the β-31-35 pentapeptide) is similar to β-1-42.

Electron microscopic studies have shown that β-amyloid peptides (even the short sequences) impair calcium homeostasis and increase the intraneuronal calcium concentration.

In other experiments we have found that β-amyloid peptides do not interact with the Substance-P receptors; these results support those of Mitsuhashi's investigations.

A New Alzheimer's Model. According to literature data, the amyloid precursor protein (APP) is a heterogeneous membrane glycoprotein. An extracellular protease (secretase) cleaves the polypeptide chain between amino acids 612 and 613 ("normal cleavage") resulting in water soluble polypeptides. In our new model the same enzyme might be responsible for the "alternative cleavage" of APP, which gives water-insoluble peptides (β-amyloids) containing 38-43 amino acid residues. These peptides are neurotoxic and can kill neurons in two different ways:

1.   The peptide chains aggregate and precipitate among the neurons, deteriorating axons and dendrites. Precipitation of other proteins on the surface of the growing β-amyloid aggregate results in formation of amyloid plaques.

2.   The amyloid peptides may interfere with the functioning Ca^{2+} channels. The increase of the intracellular Ca^{2+} level shifts the enzyme equilibrium of the neurons: some enzymes (e.g. protein kinase C) are activated, other enzymes are more or less inhibited. The phosphorylation of APP in "abnormal" position results in conformational changes and facilitates the "alternative" cleavage of APP. More β-amyloid peptides are formed causing an accelerating cell death. Neurons possessing a higher concentration of Ca^{2+}-binding proteins could survive. (This might be the reason why some neurons are not deteriorated in the first stage of Alzheimer's disease and cholinergic neurons die first.)

In the dying neurons neurofilaments and neurofibrillary tangles are formed from τ-proteins and also from other cell proteins. In the final stage the whole protein content of the cell precipitates and the cell dies. (Only <u>one</u> component of these precipitates is the τ-protein.) The increased Ca^{2+} level results in formation of highly phosphorylated proteins, however, this event must not be overestimated: this is only a consequence and not the reason for the disease.

If cell membranes are intact, the "alternative" cleavage (and the formation of β-amyloid peptides) is not preferred. However, highly reactive free oxygen radicals can oxidize the unsaturated fatty acids of the cell membrane resulting in the formation of neurotoxic amyloid peptides. The released peptides attack other neurons causing the increase of cell death in the brain. The whole process is a biological "chain reaction": a positive feed-back resulting in accelerating neuronal death, formation of amyloid plaques and Alzheimer's disease. Free radicals could react with the vitamin E content of the cell membrane, so vitamin E could possess a protective effect preventing membrane deterioration and the formation of neurotoxic amyloid peptides.

According to our experiments, the triggering event of Alzheimer's disease is the formation and aggregation of neurotoxic amyloid peptides. Alzheimer's disease is polyetiological because the release of these peptides could have various reasons (e.g. Down-syndrome, mechanical injury of the neurons in professional fighters, decreased activity of the immune system, hereditary reasons in the familiar Alzheimer's disease, etc.). Protection of membranes with antioxidants (vitamin E and C) might be useful in preventing or to stop the disease.

References

1. Hardy, J.A. and Higgins, G.A., Science, 256 (1992) 184.
2. Kowall, N.W., Beal, M.F., Busciglio, J., Duffy, L.K. and Yankner, B.A., Proc. Natl. Acad. Sci. U.S.A., 88 (1991) 7247.
3. Mitsuhashi, M., Akitaya, T., Turk, C.W. and Payan, D.G., Brain Res., 11 (1991) 177.
4. Penke, B., Soós, K., Szabó, E.Z., Márki-Zay, K.J., Pákáski, M. and Kása, P., In Schneider, C.H. and Eberle, A.N. (Eds.), Peptides 1992 (Proceedings of the 22nd European Peptide Symposium), ESCOM, Leiden, The Netherlands, 1993, p.792.

Study on P05, a new Leiurotoxin I scorpion toxin

J.-M. Sabatier[1], H. Zerrouk[2], H. Darbon[3], K. Mabrouk[1], A. Benslimane[2], H. Rochat[1], M.-F. Martin-Eauclaire[1] and J. Van Rietschoten[1]

[1]Laboratoire de Biochimie, CNRS URA 1455, Faculté de Médecine Nord, Bd Pierre Dramard, 13916 Marseille Cédex 20, France;
[2]Institut Pasteur du Maroc, Laboratoire de Purification des Protéines, 1, Place Charles Nicolle, B.P. 120, Casablanca, Morocco;
[3]Laboratoire de Cristallographie et Cristallisation des Macromolécules Biologiques, CNRS URA 1296, Faculté de Médecine Nord, Bd Pierre Dramard, 13916 Marseille Cédex 20, France

Introduction

Most polypeptide animal toxins are highly active ligands that interfere with the function of specific ion channels. One such toxin, Leiurotoxin I from the scorpion *Leiurus quinquestriatus hebraeus* has been reported to act as a blocker of small-conductance Ca^{2+}-activated K^+ channels in various cell types [1]. Leiurotoxin I possesses binding and physiological properties similar to those of the bee venom toxin apamin. Recently, a toxin, termed P05, was isolated from the venom of the scorpion *Androctonus mauretanicus mauretanicus*. It is both structurally and functionally related to leiurotoxin I with 87% sequence identity (Figure 1) [2]. By analogy with leiurotoxin I, the α-amidated form of P05 (sP05-NH$_2$) was chemically synthesized by the solid phase method. The disulfide bridge pairings of P05 had been mapped by both enzymatic and partial acidolytic cleavages of the synthetic product P05-NH$_2$. The conformational properties of the toxin were assessed by circular dichroism analysis of sP05-NH$_2$ and molecular dynamics. The binding properties of P05 were compared with those of several structural C-terminal carboxyl-amidated analogs, and apamin [3].

Results and Discussion

sP05-NH$_2$ was synthesized on MBHA using optimized Boc/benzyl chemistry. After HF cleavage, the crude peptide was partially purified by preparative MPLC, oxidized by exposure to air, and purified to 98% homogeneity by semi-preparative HPLC. Synthesis of the two Arg-substituted analogs ([Lys$_6$,Lys$_7$] sP05-NH$_2$ and [Leu$_6$,Leu$_7$] sP05-NH$_2$) was also achieved.

All the synthetic peptides, native P05 and apamin, were tested for inhibition of [125I] apamin binding to rat brain synaptosomes. sP05-NH$_2$ competed with [125I] apamin and had a much higher affinity for the binding site. The binding of [125I]-apamin was totally abolished at 10^{-13} M concentration of sP05-NH$_2$ when the

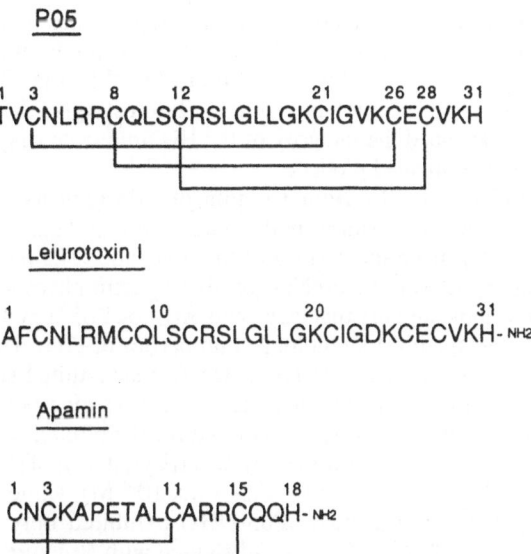

Fig. 1. Primary structures (one-letter code) of P05, leiurotoxin I, and apamin. Established (apamin) or deduced (P05) disulfide bridge pairings are indicated by plain lines.

competing molecules were added together, or when sP05-NH$_2$ was preincubated with the synaptosomes before addition of [^{125}I] apamin. These experiments suggest that sP05-NH$_2$ either binds irreversibly or dissociates much more slowly than apamin. Native P05 and analogs of sP05-NH$_2$ also competed with [^{125}I] apamin for binding to synaptosomes, with half-effects ($K_{0.5}$) obtained at the concentrations of 6 x 10^{-12} (apamin), 2 x 10^{-11} (P05), 2 x 10^{-10} (diiodo-sP05-NH$_2$), 2 x 10^{-9} ([Lys$_6$,Lys$_7$] sP05-NH$_2$) and 10^{-7} M ([Leu$_6$,Leu$_7$] sP05-NH$_2$). *In vivo*, the synthetic peptides, except [Leu$_6$,Leu$_7$] sP05-NH$_2$, caused neurotoxic and lethal effects in mice with clinical symptoms identical to those induced by native P05 or apamin. The LD$_{50}$ per mouse were 12 (apamin), 20 (P05, sP05-NH$_2$), 80 (diiodo-sP05-NH$_2$), and 200 ng ([Lys$_6$,Lys$_7$] sP05-NH$_2$). Peptide [Leu$_6$,Leu$_7$] sP05-NH$_2$ was inactive at the dose of 10 mg.

To investigate conformational properties of sP05-NH$_2$, disulfide pairings were established after enzymatic digestion and partial acid hydrolysis. The disulfides were mapped between Cys$_3$-Cys$_{21}$, Cys$_8$-Cys$_{26}$, and Cys$_{12}$-Cys$_{28}$. A model for P05 was obtained by molecular dynamics using the positions of the disulfide bridges of sP05-NH$_2$ and the secondary structure of leiurotoxin I. P05 appears to be mainly composed of a double-stranded antiparallel β-sheet (from Leu$_{18}$ to Val$_{29}$) linked to an α-helix (from Arg$_6$ to Gly$_{16}$) by disulfides (Cys$_8$-Cys$_{26}$ and Cys$_{12}$-Cys$_{28}$) and to an extended fragment (from Thr$_1$ to Leu$_5$) by disulfide (Cys$_3$-Cys$_{21}$). Circular dichroism analysis of sP05-NH$_2$, [Lys$_6$,Lys$_7$] sP05-NH$_2$, and [Leu$_6$,Leu$_7$] sP05-NH$_2$ were performed at various pH values or percentages of TFE. The spectra obtained for each peptide under different conditions were superimposable, confirming the rigidity of their structures. Remarkably, the P05 polypeptide backbone from the model was found to be very similar to that of particular stretches of CsE V3, AaH II, and AaH IT.

Pharmacological experiments on rat brain synaptosomes indicated that the C-terminal carboxyl-amidated analog of P05 (sP05-NH$_2$) binds irreversibly, a novel property for scorpion toxins. This result indicates that the C-terminal His is implicated in the toxin to receptor interaction. Interestingly, the specific binding properties of sP05-NH$_2$ need the integrity of the His imidazole ring, as shown by the different activity of the iodinated product.

From the models, the N-terminal region of P05 appears to be structurally similar to the C-terminus of apamin: both contain an α-helical core with a pair of Arg residues protruding from the surface of the molecule The pair of Arg residues (Arg$_{13}$-Arg$_{14}$) was reported to be crucial for high apamin pharmacological activity [4]. These observations suggest that Arg$_6$ and Arg$_7$ of P05 may be essential, or at least contribute, to the apamin-like biological properties of P05. To investigate their importance in the bioactivity of sP05-NH$_2$, two Arg-substituted sP05-NH$_2$ analogs were designed and synthesized. Pharmacological assays clearly showed that these residues, or at least positive charges, are required for the biological activity of sP05-NH$_2$. Analogs whose Arg are substituted by Lys ([Lys$_6$,Lys$_7$] sP05-NH$_2$) had lower activities than sP05-NH$_2$ (10% toxicity; K$_{0.5}$ = 2 x 10^{-9} M), while those whose Arg are substituted by Leu ([Leu$_6$,Leu$_7$] sP05-NH$_2$) exhibited only residual activity (toxicity <0.2%; K$_{0.5}$ = 10^{-7} M). These results agree with structure-activity analyses of apamin.

Structure-activity analysis of the new natural analog of leiurotoxin I, P05, has shown the following: 1) the C-terminal carboxyl-amidation of His induces a considerable strengthening of the toxin to receptor interaction, giving to sP05-NH$_2$ the properties of an irreversible, apamin-sensitive, Ca^{2+}-activated K$^+$ channel blocker, ii) the two Arg residues, presumably located in an α-helix (like apamin), and the C-terminal His are involved in the interaction with the receptor. Modeling of P05 suggests that the three residues are located on the same side of the molecule. The overall data strongly suggest the importance of the common α-helical core for activity of this group of pharmacologically-related toxins. Interestingly, structure-activity relationship studies of long scorpion toxins (60 to 70 residues, four disulfide bridges) indicate that, in this case, the α-helical structure is not involved in pharmacological activity, but more probably in stabilizing an active conformation of a solvent-exposed hydrophobic region. This indicates that further studies on scorpion toxins are needed to understand the precise molecular basis of the toxin-receptor recognition.

References

1. Abia, A., Lobaton, C.D., Moreno, A. and Garcia-Sancho, J., Biochim. Biophys. Acta, 856 (1986) 403.
2. Zerrouk, H., Mansuelle, P., Benslimane, A., Rochat, H. and Martin-Eauclaire, M.-F., FEBS Lett., 320 (1993) 189.
3. Sabatier, J.M., Zerrouk, H., Darbon, H., Mabrouk, K., Benslimane, A., Rochat, H., Martin-Eauclaire, M.-F. and Van Rietschoten, J., Biochemistry, 32 (1993) 2763.
4. Vincent, J.-P., Schweitz, H. and Lazdunski, M., Biochemistry, 14 (1975) 2521.

Synthesis and pharmacological characterization of a cyclic β-casomorphin analog with mixed μ agonist/δ antagonist properties

R. Schmidt[1], D. Vogel[2], C. Mrestani-Klaus[2], W. Brandt[2], K. Neubert[2], N.N. Chung[1], C. Lemieux[1] and P.W. Schiller[1]

[1]*Laboratory of Chemical Biology & Peptide Research, Clinical Research Institute of Montreal, Montreal, Quebec, H2W 1R7, Canada*
[2]*Institute of Biochemistry, Martin Luther University, D-06099 Halle, Germany*

Introduction

The conformational flexibility of linear opioid peptides is most likely the reason for the lack of their specificity toward the different opioid receptor classes, because conformational adaptation to the various receptor topographies is possible. Conformational restriction of peptides may often result in improved receptor selectivity, as indicated by the opioid activity profiles of conformationally constrained cyclic opioid peptide analogs with agonist activity [1, 2]. β-Casomorphin-5 (Tyr-Pro-Phe-Pro-Gly), a peptide with opioid properties, can be generated from the milk protein β-casein by proteolytic fragmentation. The cyclic peptides H-Tyr-c[-D-Orn-Phe-Pro-Gly-] (1) and H-Tyr-c[-D-Orn-Phe-D-Pro-Gly-] (2) represent β-casomorphin-5 analogs showing high potency in receptor binding assays and bioassays *in vitro* and high antinociceptive activity [3, 4]. Analog 1 exhibits considerable preference for μ receptors over δ receptors, whereas peptide 2 is highly potent but relatively non-selective. In the present study we describe the syntheses and opioid activity profiles of corresponding analogs containing 1- or 2-naphthylalanine (Nal) in place of the Phe[3] residue.

Results and Discussion

The cyclic β-casomorphin analogs were synthesized by means of conventional solution methods, using N^α-Boc, N^ω-Z and -ONb protection, and mixed anhydride couplings. Diphenylphosphoryl azide was used for the cyclization reaction between the 2-position side chain amino function and the C-terminal COOH group under optimized conditions with cyclization yields >80% [5]. The N-terminal unprotected peptides were purified by RP-HPLC and were characterized by mass spectrometry, amino acid analysis and one- and two-dimensional ^{1}H-NMR techniques. The pharmacological profile of these peptides were studied *in vitro* using the GPI and MVD bioassays and receptor binding assays based on displacement of the tritiated opioid receptor ligands DAGO (μ–selective) and DSLET (δ–selective) from rat brain membrane binding sites (Table 1).

Table 1 *Opioid activities and receptor affinities of cyclic β–casomorphin-5 analogs*

	compound	GPI	MVD	DAGO	DSLET
		IC_{50}, nM		K_i, nM	
1	Tyr-c[-D-Orn-Phe-Pro-Gly-]	13.40	69.90	3.99	1,234
2	Tyr-c[-D-Orn-Phe-D-Pro-Gly-]	2.14	4.89	0.88	13.17
3	Tyr-c[-D-Orn-2-Nal-Pro-Gly-]	1,224	>10,000	81.29	2,137
4	Tyr-c[-D-Orn-2-Nal-D-Pro-Gly-]	384	antagonist	5.89	17.15
5	Tyr-c[-D-Lys-2-Nal-D-Pro-Gly-]	609	antagonist	17.26	62.62
6	Tyr-c[-D-Orn-1-Nal-D-Pro-Gly-]	14.94	29.89	3.45	23.56
	[Leu⁵] enkephalin	246	11.40	9.43	2.53

Substitution of the Phe[3] residue in compound **1** by 2-Nal, resulting in H-Tyr-c[-D-Orn-2-Nal-Pro-Gly-] (**3**), produced a 100-fold potency drop in both the GPI- and the MVD assay and corresponding decreases in μ and δ receptor affinities in comparison with compound **1**. The low opioid activity in the MVD bioassay could be antagonized with the δ receptor selective antagonist TIPP [6], indicating that the effect was mediated by δ receptors.

The diastereoisomer, H-Tyr-c[-D-Orn-2-Nal-D-Pro-Gly-] (**4**), displayed only 7 times lower μ receptor affinity than **2**, but 180 times lower potency in the GPI assay. It retained an affinity for the δ receptor ($K_i^δ$ = 17 nM) similar to that of the relatively non-selective δ agonist **2**, but showed no agonist effect at concentrations up to 50 μM in the MVD assay. Analog **4** turned out to be a moderately potent antagonist against the δ agonists [Leu⁵]enkephalin, [D-Ala²]deltorphin I and DPDPE with K_e values of 268 nM, 202 nM and 233 nM, respectively. Thus, like the recently reported tetrapeptide TIPP-NH₂ [6], analog **4** represents another example of a mixed μ agonist/ δ antagonist. Such compounds are thought to have potential as analgesics that may not produce tolerance and dependence [7]. Increasing the ring size through insertion of another methylene group in the 2-position side chain (compound **5**) resulted in decreased μ and δ receptor affinity, in correspondence with diminished agonist activity in the GPI assay and decreased antagonist potency against [Leu⁵]enkephalin (K_e = 603 nM) in the MVD assay.

Interestingly, replacement of 2-Nal[3] by 1-Nal in **4** led to a compound (**6**) with restored agonist activity in the MVD assay and with only slightly reduced potencies in both bioassays as compared to **2**.

The solution structure of the casomorphin analogs **2** and **4** was studied by performing 1- and 2-dimensional ^{1}H-NMR experiments in DMSO-d_6. Peptides **2** and **4** each showed only one conformer with all-*trans* peptide bonds and a similar backbone conformation, characterized by a Tyr^{1}CO···NH-2-Nal^3(or Phe^3)- and an Orn^2-CO···NH^8-Orn^2 hydrogen bond. For both compounds a close proximity between the aromatic moiety of the 3-position residue and the pyrrolidine ring of the D-Pro residue was established on the basis of ROESY experiments. These results suggest that the antagonist properties of analog **4** may not be due to a difference in its overall conformation as compared to compound **2**, but rather may be a direct effect of the naphthyl moiety *per se* preventing proper alignment of the peptide such as required for receptor activation.

Substitution of the Phe^3 residue in the cyclic β-casomorphin-5 analog **2** by 2-naphthylalanine led to a compound with mixed μ agonist/δ antagonist properties. Since the corresponding compound with the 1-Nal^3 showed restored agonist activity in the MVD, the detailed conformational comparison of the peptide analogs **4**, **5** and **6** may provide insight into the structural requirements of the δ opioid receptor for ligand binding and signal transduction.

Acknowledgements

This work was supported by grants from NIDA (DA-04443), MRCC (MT-5655) and the DFG.

References

1. Schiller, P.W., In Ellis, G.P. and West, G.B. (Eds.), Progress in Medicinal Chemistry, Vol. 28, Elsevier, Amsterdam, The Netherlands, 1992, p.301.
2. Schiller, P.W. and DiMaio, J., Nature, 279 (1982) 74.
3. Schmidt, R., Neubert, K., Barth, A., Liebmann, C., Schnittler, M., Chung, N.N. and Schiller, P.W., Peptides, 12 (1991) 1175.
4. Rüthrich, H.-L., Grecksch, G., Schmidt, R. and Neubert, K., Peptides, 13 (1992) 483.
5. Schmidt, R. and Neubert, K., Int. J. Peptide Protein Res., 37 (1991) 502.
6. Schiller, P.W., Nguyen, T.M.-D., Weltrowska, G., Wilkes, B.C., Marsden, B.J., Lemieux, C. and Chung, N.N., Proc. Natl. Acad. Sci. U.S.A., 89 (1992) 11871.
7. Abdelhamid, E.E., Sultana, M., Portoghese, P.S. and Takemori, A.E., J. Pharmacol. Exp. Ther., 258 (1991) 299.

In vivo activities of conantokin-G analogs

J.L. Torres[1,4], E. Mahé[4], J-F. Hernandez[2,4], C. Miller[4], L.J. Cruz[3], B.M. Olivera[3] and J.E. Rivier[4]

[1]Unit for Protein Chemistry and Biochemistry, C.I.D.-C.S.I.C., 08034 Barcelona, Spain, [2]Institut de Biologie Structurale, Laboratoire d'Enzymologie Moleculaire, 41 av. des Martyrs, 38027 Grenoble, Cedex 1, France, [3]Dept. of Biochemistry, University of Utah, Salt Lake City, UT 84112, U.S.A. and [4]The Clayton Foundation Laboratories for Peptide Biology, The Salk Institute, La Jolla, CA 92037, U.S.A.

Introduction

Conantokins are the first family of peptides known to interact with the N-methyl-D-aspartate (NMDA) receptor [1]. They elicit quite different behavioral symptomatology in neonatal and adult mice (sleep-like state and hyperactivity, respectively [1]). The natural conantokins purified and characterized to date from *Conus geographus* and *Conus tulipa* venom, respectively are represented in Figure 1.

Peptide	Amino acid sequence			
	1	5	11	15
Conantokin-G	GE γγ	LQγNQ	γ LI Rγ	KSN*
Conantokin-T	GE γγ	YQKML	γ NL Rγ	AEVKKNA*

Fig. 1. Primary structure of naturally occurring conantokins. The symbol γ represents γ-carboxyglutamate and *indicates an amidated carboxy terminus.

In order to understand the functional biology of NMDA receptors, ligands which discriminate between different molecular forms of the NMDA receptor are required. The conantokins are promising ligands for further development in this respect [2]. In this paper, the synthesis and biological evaluation of a series of analogs of conantokin-G with substitution in one conserved residue (position 2) and seven variable residues (positions 5, 6, 9, 11, 12, 16 and 17) are reported.

Results and Discussion

Conantokin-G analogs were synthesized by standard Fmoc solid-phase protocols. For the cleavage, a seven hour treatment with TFA/thioanisole/ water/EDT/CH_2Cl_2 (40:8:1:2:49) was successful in releasing the fully deprotected peptide chain. The crude peptides were purified by preparative RP-HPLC using triethylamine phosphate (TEAP) buffer/CH$_3$CN at pH 2.25 and a final desalting step with 0.1% trifluoroacetic acid/CH$_3$CN. On analytical RP-HPLC, the conantokin-G analogs eluted as one major peak with several small phobic peaks.

The analogs were tested for biological activity (sleep in 12-14 days old mice, hyperactivity in 21-28 days old mice) after i.c. injection. Since both glycine and glutamic acid are involved in the NMDA receptor mediated response, conserved Gly[1] and Glu[2] were believed to be important for conantokin-G activity. Consequently, the lack of activity of [Gln[2]]-G was expected.

All analogs were tested by i.c. injection in mice at 30 and 200 pmoles/g body weight. The effects of Ala substitutions on the variable loci (Figure 1) were evaluated (Table 1).

Table 1 *Biological activity of alanine-substituted conantokin-G analogs*

Conantokin	Dose	Activity in mice	
	pmoles/g	12-14 days old	21-28 days old
Conantokin-G	30	sleep	hyperactivity
[Ala[5,6,9]]G	30 200	sleep sleep	no effect slight hyperactivity
[Ala[11,12]]G	30 200 500	no effect no effect no effect	no effect no effect no effect
[Ala[16,17]]G	30	sleep	hyperactivity
[Ala[11]]G	200	sleep	no effect
[Ala[12]]G	200	no effect	no effect
[Ala[11,12,16,17]]G	200	no effect	no effect

It is clear that substitutions at positions 16 and 17 (for Ser and Asn) do not appear to affect biological activity significantly. A single Ala substitution for Leu[11], or multiple Ala substitutions for Leu[5],Gln[6], Gln[9] yielded active analogs with significantly decreased potency. In contrast, any analog with an Ala substitution for Ile[12] was completely inactive in the concentration range tested. Thus, of the seven variable positions, it appears that Ile[12] plays a specific functional role.

It should also be noted that although by the sleep assay, many of these analogs are quite potent compared to the native peptide, the hyperactivity induced in mature mice seems more sensitive to substitutions at the variable positions. Thus, the

[Ala^{11}]conantokin-G analog which is potent in inducing sleep in young mice has lost much of its activity for inducing hyperactivity in older mice. [Ala^{12}]conantokin-G is inactive at any age. Since the biological tests applied are *in vivo* tests, differences in biological activity do not necessarily represent changes in primary interactions between the peptides and their receptor targets (presumably types of NMDA receptor). In any case, these results demonstrate that structural changes in positions 11 and 12 of conantokin-G clearly influence its activity in a way related with the animal's developmental state. Curiously the IRγ and LRγ sequences of conantokin-G and -T respectively, resemble the LRE sequence of S-laminin. This sequence appears to be involved in neural adhesion and the formation of the synapse [3]. The interaction of conantokins with neural membranes through a sequence related with synapse formation would add to the picture of their action on the NMDA receptor and its environment in an age dependent way.

The presence of a glutamic acid residue in position 2 of conantokin-G is crucial for its activity. This is in agreement with conantokins interacting with glutamate receptors, probably of the NMDA type. On the other hand, many of the non-conserved amino acids in the sequence seem not to affect biological activity. Surprisingly, non-conserved position 12 appears to be crucial. Analogs such as [Ala^{12}]conantokin-G may prove to discriminate between receptor (probably NMDA) subtypes or states.

Acknowledgements

This research was supported by NIH grants GM 22737, DK 26741, NS 27219. J.L.T. was supported by a NATO Advanced Research Fellowship. We thank Jeff Abbot and Tom Leitner for technical assistance with the bioassays.

References

1. Olivera, B.M., Rivier, J., Clark, C., Ramilo, C.A., Corpuz, G.P., Abogadie, F.C., Mena, E.E., Woodward, S.R., Hillyard, D.R. and Cruz, L.J., Science, 249 (1990) 257.
2. Hammerland, L.G., Olivera, B.M. and Yoshikami, D., Eur. J. Pharmacol., 226 (1992) 239.
3. Hall, Z.W. and Sanes, J.R., Cell, 72, Neuron, 10 (Suppl.) (1993) 99.

Structure activity relationship studies on kaliotoxin, a peptidic K+ channel blocker

J. Van Rietschoten[1], R. Romi[1], M. Crest[2], F. Sampieri[1],
H. Zerrouk[1], P. Mansuelle[1], G. Jacquet[2], M. Gola[2],
H. Rochat[1] and M.-F. Martin-Eauclaire[1]

[1]Laboratoire de Biochimie, CNRS URA 1455, Faculté de Médecine
Nord, Bd Pierre Dramard, 13916 Marseille Cédex 20, France
[2]Laboratoire de Neurobiologie, CNRS, 31, chemin Joseph Aiguier,
13402 Marseille Cédex 09, France

Introduction

Scorpion venoms contain a variety of peptidic toxins that have been characterized as specific ion channel blockers. In recent studies, a new class of inhibitors has been discovered that interfere with high affinity on several types of potassium channels. From the venom of the Moroccan scorpion *Androctonus mauretanicus mauretanicus* (*Amm*) a toxic peptide, called kaliotoxin (KTX), has been isolated and studied for its activity on neuronal BK-Type Ca++ activated K+ channel [1]. Based on its amino acid sequence, KTX belongs to the family of three other toxins, named noxiustoxin (NTX), charybdotoxin (ChTX), and iberiotoxin (IbTX). Synthesis of KTX has been undertaken possessing either a free carboxylate or an amidated C-terminal end in order to complete the structural knowledge of this peptide, essentially the nature of the C-terminus and the pattern of the disulfide bridges. Non identity of the synthetic peptides (37 amino acid residues) compared to natural KTX prompted a structure revision.

Radioiodination of synthetic KTX(1-37) allowed binding experiments to be performed in competition with different selective toxins.

Shorter fragments based on the 26-32 sequence of KTX have been studied. They include a major conserved cluster of amino acids among the four homologous toxins and expressed competitive antagonistic properties against KTX.

Results and Discussion

The published sequence (37 amino acid residues) with either a free C-terminal carboxylate or an amidated C-terminus -KTX(1-37) and KTX(1-37)NH$_2$- have been synthesized using the Boc solid phase strategy on an Applied Biosystems automated peptide synthesizer. After air oxidation of the cysteines, the peptides were purified by MPLC on reverse phase column and found homogeneous by HPLC and amino acid analysis. However, neither of the two synthetic peptides did coelute with natural KTX when injected simultaneously. Additional structural studies were then performed to compare synthetic KTX(1-37) and natural KTX, especially mass spectrometry experiments and sequencing of C-terminal peptides. The results indicated that the exact sequence of KTX comprises an additional C-terminal lysyl residue, which was undetected in the previous work. KTX (38 amino acid residues) has been synthesized and found identical to natural KTX in all tests. Assignment of the three disulfide bridges has been performed on synthetic peptide by identifying the fragments obtained after pronase proteolysis. The structure is homologous to that of ChTX and IbTX.

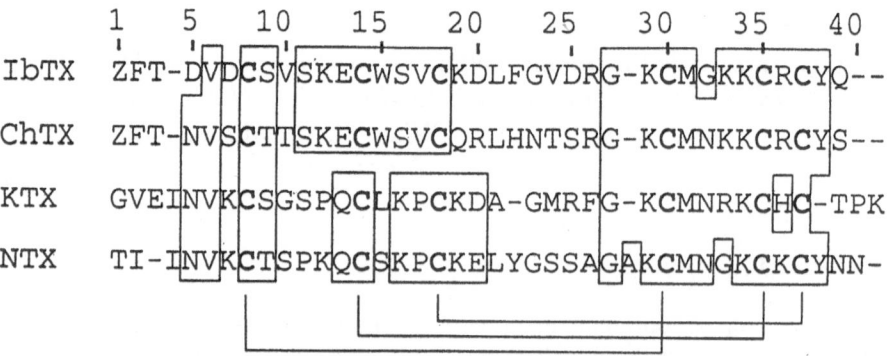

Fig. 1. Amino acid sequences of IbTX, ChTX, KTX and NTX. The toxins have been aligned using the clustal program. The numbers indicate the positions of the amino acid residues in this alignment. Homologous residues are boxed with the cysteine residues in bold. The disulfide bondings are those determined for the three toxins IbTX, ChTX and KTX.

Biological characterization of the synthetic peptides showed that all three KTX(1-37), KTX(1-37)NH$_2$, and KTX expressed approximately the same toxic activity when injected intracerebroventricularly into mice (LD$_{50}$ of 6 to 9 pmoles per 20 g mouse). Blockage experiments on whole-cell molluscan KCa current showed that KTX(1-37) is the most active (2 nM) compared to KTX(1-37)NH$_2$ (4 nM) and KTX (8 nM), indicating that the C-terminal lysine has no favourable effect on the interaction of KTX with the receptor.

Binding competition experiments on rat brain synaptosomal membranes between ^{125}I-KTX(1-37) and KTX, KTX(1-37), and KTX(1-37)NH$_2$ yielded identical IC$_{50}$ values of 10 pM. In this same test, other toxins, ChTX, dendrotoxin (DTX), and MCD peptide gave IC$_{50}$ values of 8 nM, 8 pM, and 1 nM, respectively, indicating that, in this preparation, KTX behaved differently than ChTX and bound in the closeness to or at the DTX binding site. DTX is known as a specific blocker of the voltage-dependent K$^+$ channels. IbTX did not interfere with KTX binding.

Two short peptides have been synthesized, corresponding to the 26-32 and 25-35 sequences of KTX (or positions 27-34 and 26-37 on the aligned toxins in Figure 1) with cysteine residues replaced by the isosteric analogue aminobutyric acid residue, Abu^{28}-KTX(26-32) or H-Gly-Lys-Abu-Met-Asn-Arg-Lys-NH$_2$ and Abu^{28,33,35}-KTX(25-35) or H-Phe-Gly-Lys-Abu-Met-Asn-Arg-Lys-Abu-His-Abu-NH$_2$. Devoid of KTX biological activity, the two peptides competed with ^{125}I-KTX(1-37) in the binding test at an IC$_{50}$ of 8 µM. The hypothesis that these peptides could exert an antagonist effect was confirmed by experiments where 100μM peptide inhibited the KTX effect by 80% on the blockage activity on the *Helix* KCa current. Also in vivo, when co-injected (0.5 μmole) with KTX into mice, the toxic activity of KTX was diminished significantly (a factor 2) and the appearance of envenomation symptoms was delayed from 1 min to 20 min. The 26-32 sequence is very homologous for the four toxins of this family (see Figure 1).

The 26-32 sequence of KTX could represent a low affinity binding site for this class of toxins that permits a first non selective interaction, other part of the toxins, at the N- and C-terminus, could then play a complementary role for the high affinity binding and also for the selectivity of the type of potassium channel.

Acknowledgements

R. Gaillard, T. Brando, M. Alvitre and B. Ceard are acknowledged for their skillful technical expertise.

References

1.   Crest, M., Jacquet, G., Gola, M., Zerrouk, H., Benslimane, A., Rochat, H., Mansuelle, P. and Martin-Eauclaire, M.-F., J. Biol. Chem., 267 (1992) 1640.

Structure-activity correlation in peptide hormones based on regiospecific interaction with Ca²⁺ in the lipid milieu

V.S. Ananthanarayanan

Department of Biochemistry, McMaster University,
Hamilton, Ontario, L8N 3Z5, Canada

Introduction

The determination of the bioactive structures of peptide hormones is rendered difficult by their exhibiting a large ensemble of flexible conformations in water where most structural data have been gathered. Fairly ordered structures are, however, obtained in the lipid medium as many recent studies indicate. In the physiological situation, the hormones encounter millimolar levels of Ca^{2+} in the extracellular aqueous phase. In the absence of a compact folded structure and specific chelating groups, the hormones would not be expected to bind Ca^{2+} appreciably in this phase. However, a totally different picture would prevail at the lipid-water interface or in the lipid interior where these molecules assume a folded structure and interact with their membrane receptors. In this scenario, we expect the hormones to bind extracellular Ca^{2+} and undergo a conformational change as an initial step in the signal transduction process [1]. Here, we summarize our recent studies on the interaction of several peptide hormones with Ca^{2+} in lipid-mimetic solvents and highlight some common characteristics observed in their Ca^{2+}-interacting regions.

Results and Discussion

Table 1 shows the sequences of the hormones studied here (with C-terminal fragments in bold) and also "homologous" sequences of certain other hormones (see below).

Table 1 *Sequences of hormones: (A) this study; (B) "homologous" hormones*

A. Hormone	Sequence	B. Hormone	Sequence
Substance P	RPKPQQ**FFGLM**	Met enkephalin	YGGFM
Bombesin	ZQRLGNQWA**VGHLM**	Gastrin [12-17]	YGWMDF
Glucagon	HSQGTFTSDYSKYLD-**SSRAQDFVQWLMNT**	GRP [20-27]	WAVGHLM
		CCK [27-31]	YMGWM

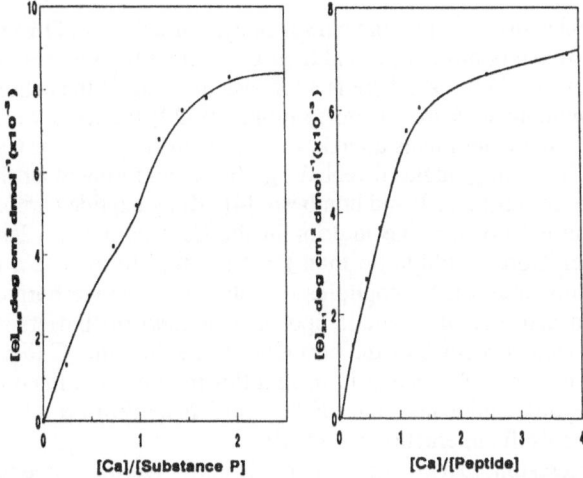

Fig. 1. *Ca²⁺-binding of substance P (left) and its [7-11] fragment (right) in TFE.*

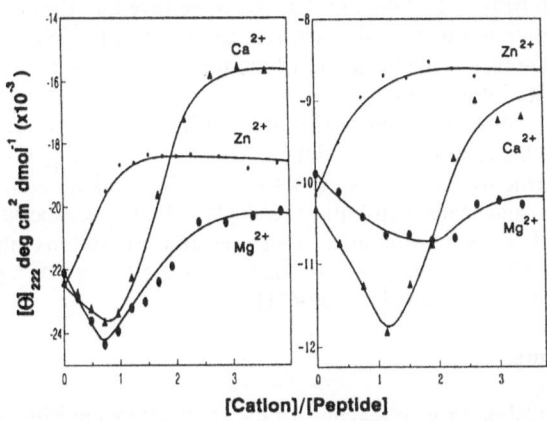

Fig. 2. *Cation-binding of glucagon (left) and its [19-29] fragment (right) in TFE.*

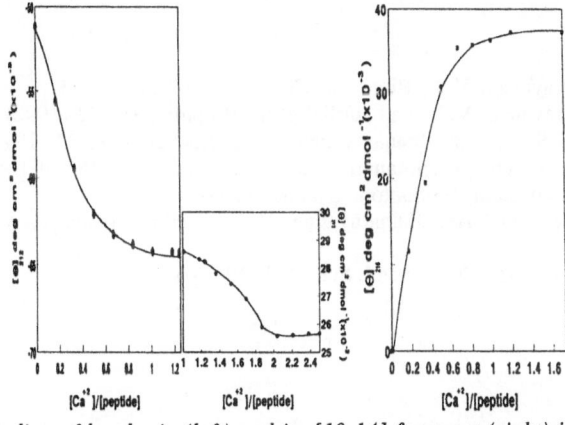

Fig. 3. *Ca²⁺-binding of bombesin (left) and its [10-14] fragment (right) in TFE.*

Figures 1-3 show Ca^{2+} binding curves in trifluoroethanol (TFE) for substance P [2], glucagon [3] and bombesin [4] and their C-terminal fragments monitored by CD spectral changes. In each case, there is a biphasic binding of the cation. In glucagon, both the Ca^{2+}-binding sites are contained in the [19-29] fragment while in substance P and bombesin, the C-terminals contain the higher affinity site for Ca^{2+}.

Molecular modeling studies revealed significant conformational changes induced by Ca^{2+}-binding in substance P and bombesin [4]. Many peptide carbonyls in the free hormone are turned towards the interior in the Ca^{2+} complex. This, in essence, makes the latter more amphiphilic than the Ca^{2+}-free form. As a result, peptide hormones would act as Ca^{2+} ionophores as data on the above hormones [2-4] and insulin [5] in suspensions of synthetic liposomes in *aqueous* buffers show. We find the Ca^{2+}-translocating regions of the hormones to be the same C-terminal parts that bind the ion in the nonpolar media; in insulin this region is confined to the B-chain. We may thus regard peptide hormones to be *"lipid-dependent Ca^{2+}-binding proteins"* exhibiting regiospecific interaction with Ca^{2+}.

A most interesting fact emerges when one notes that the above regions are also the functionally important parts of these hormones involved in signal transduction (see data cited in Refs. 2-5). We are thus led to believe that the so-called "message" part [6] of a peptide hormone is, in fact, its Ca^{2+}-interacting "domain". This domain is distinct from the rest of the hormone which forms the "address" part containing the receptor specificity information.

Finally, we observe a compositional similarity in terms of the presence of hydrophobic residues and the Met residue in the Ca^{2+}-interacting C-terminal domains of the above hormones as well as others such as Met enkephalin, gastrin, gastrin-releasing peptide and cholecystokinin (Table 1). We expect the latter hormones to interact with Ca^{2+} in these domains. Our present observations amplify our earlier proposal that the Ca^{2+}-binding leads to the bioactive structure of a peptide hormone as recognized by the lipid-bound receptor [1].

Acknowledgements

I thank my laboratory colleagues for data collection and MRC Canada for grant support.

References

1.   Ananthanarayanan, V.S., Biochem. Cell. Biol., 69 (1991) 93.
2.   Ananthanarayanan, V.S. and Orlicky, S., Biopolymers, 32 (1992) 1765.
3.   Brimble, K.S. and Ananthanarayanan, V.S., Biochemistry, 32 (1993) 1632.
4.   Saint-Jean, A. and Ananthanarayanan, V.S., Abstract P465 at the 13th American Peptide Symposium, Edmonton, Canada (1993).
5.   Brimble, K.S. and Ananthanarayanan, V.S., Biochim. Biophys. Acta, 1105 (1992) 319.
6.   Schwyzer, R., Ann. N.Y. Acad. Sci., 297 (1977) 3.

A single substitution confers antagonistic properties in NPY

A. Balasubramaniam, S. Sheriff, M. Stein, J.E. Fischer and W.T. Chance

Division of G.I. Hormones, Department of Surgery, University of Cincinnati College of Medicine and VAMC, Cincinnati, OH 45267-0558, U.S.A.

Introduction

Neuropeptide Y (NPY), a 36 residue peptide amide isolated originally from porcine brain [1], occurs in higher concentrations in mammalian brain than any other peptides isolated to date [2]. A number of studies have shown that NPY is localized in hypothalamic regions known to control the feeding behaviour, and that central administration of NPY induces a robust feeding effect in rats [3]. These observations, and the findings that hypothalamic NPY levels and/or the NPY mRNA are elevated in obese rats [4] and decreased in anorectic tumor-bearing rats [5] suggest that the NPY sequence could be modified for therapeutic use. Since nearly the entire sequence of NPY is required to elicit feeding responses [6], we synthesized a series of full length analogs NPY substituting D-Trp/D-Trp(CHO) in the C-terminal receptor binding region and screened their effects on isoproterenol stimulated rat hypothalamic membrane adenylate cyclase (AC) activity. These investigations have led to the identification of [D-Trp^{32}]NPY as a potent antagonist of NPY in rat hypothalamus.

Results and Discussion

Although [D-Trp^{34}]NPY and [D-Trp^{36}]NPY, and the corresponding formylated derivatives inhibited the AC activity, [D-Trp^{32}]NPY and [D-Trp(CHO)^{32}]NPY did not exhibit any intrinsic activity. Moreover, while [D-Trp(CHO)^{32}]NPY exhibited lower receptor affinity, [D-Trp^{32}]NPY inhibited ^{125}I-NPY binding with a potency comparable to that of intact NPY. It is the high receptor affinity and the complete loss of intrinsic activity that suggested that [D-Trp^{32}]NPY may be an antagonist of NPY in rat hypothalamus. Consistently, the presence of 30 and 300 nM of [D-Trp^{32}]NPY shifted the inhibitory dose-response curve of NPY in rat hypothalamus AC activity parallel to the right increasing the IC$_{50}$ value from 0.18 nM to 4.0 and 40.0 nM, respectively. This antagonism was specific to the NPY receptor, since [D-Trp^{32}]NPY exhibited no effect on the inhibitory hypothalamic AC activity of serotonin. Since intrahypothalamic injection of 1 μg of NPY has been shown to elicit a robust feeding response in rats [3], we also tested the effect of [D-Trp)^{32}]NPY on NPY induced feeding. [D-Trp^{32}]NPY (1 or 10 μg) did not exhibit significant stimulatory effect on feeding. Although 1 μg of [D-Trp^{32}]NPY failed to exhibit antagonist effect, 10 μg of [D-Trp^{32}]NPY significantly attenuated the 1 hr

535

feeding response induced by 1 μg of NPY. Thus, [D-Trp^{32}]NPY also behaved as an antagonist of NPY in in vivo models. Other substitutions such as D-Phe, Hyp or even D-2-Nal, a Trp-mimicking residue, at position 32 resulted in agonist activity, suggesting that there are strict structural requirements to induce antagonistic properties to NPY. To determine the structural changes associated with D-Trp substitution, we did modeling studies based on the X-ray structure coordinates of the homologous peptide, avian pancreatic polypeptide (APP) [7]. These investigations suggested that substitution of D-Trp at position 32 induced a type II' β-turn at that position causing the C-terminal to fold back over the PP-fold, rather than extending away from the helix, as it does in APP crystal structure. It appears therefore that the modified conformation in the C-terminal region is responsible for the loss of agonist activity, resulting in a potent antagonist. Furthermore, stabilization of this β-turn through cyclization may result in more potent antagonistic compounds.

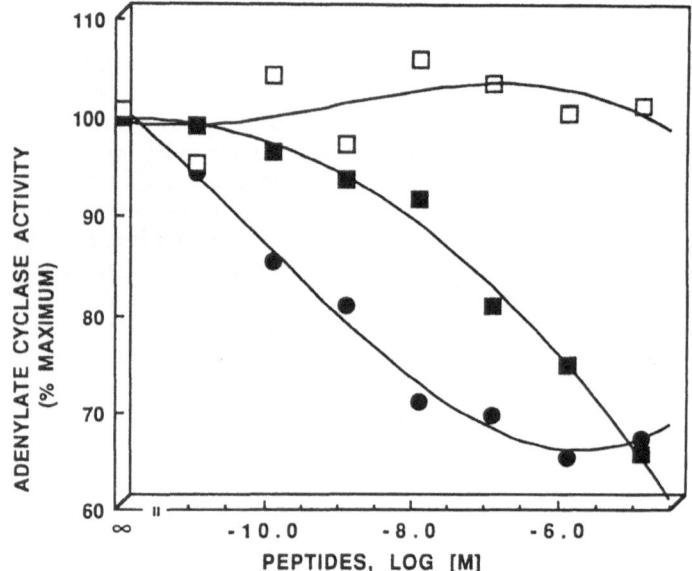

Fig. 1. *Effects of increasing concentrations of [D-Trp^{32}]NPY (Σ), and NPY in the absence (●) and presence (■) of 300 nM [D-Trp^{32}]NPY on isoproterenol stimulated adenylate cyclase activity of rat hypothalamic membranes.*

In summary, we have shown that [D-Trp^{32}]NPY is a competitive antagonist of NPY in both in vitro and in vivo models. Compounds based on [D-Trp^{32}]NPY may have potential clinical application since NPY has already been implicated in the pathophysiology of feeding disorders.

Acknowledgements

Supported by NIH (GM47122), AICR (91B54) and VA grants.

References

1. Tatemoto, K., Proc. Natl. Acad. Sci. U.S.A., 79 (1982) 5485.
2. Allen, Y.S., Adrian, T.E., Allen, J.M., Tatemoto, K., Crow, T.J., Bloom, S.R. and Polak, J.M., Science, 221 (1983) 877.
3. Chance, W.T., Sheriff, S., Foley-Nelson, T., Fischer, J.E. and Balasubramaniam, A., Peptides, 10 (1989) 1283.
4. Sanacora, G., Kershaw, M., Finkelstein, J.A. and White, J.D., Endocrinology, 127 (1990) 730.
5. Chance, W.T., Sheriff, S., Zhang, F., Kalmonpunpour, M., Fischer, J.E. and Balasubramaniam, A., Proc. Soc. Neurosci., (1990) 774.
6. McLaughlin, C.L., Tou, J.S., Rogan, G.J. and Baile, C.A., Phys. Behav., 49 (1991) 521.
7. Blundell, T.L., Pitts, J.E., Tickle, I.F., Wood, S.P. and Wu, C.W., Proc. Natl. Acad. Sci. U.S.A., 78 (1981) 4175.

Amylin(1-23)-NH$_2$ and [Anb^{2,7}]amylin (1-23)-NH$_2$ are potent agonists of amylin

A. Balasubramaniam[1], S. Sheriff[1], L.R. McLean[2], M. Stein[1], G. Hod[1], W.T. Chance[1] and J.E. Fischer[1]

[1]Division of G.I. Hormones, Department of Surgery, University of Cincinnati College of Medicine, VAMC, Cincinnati, OH 45267, U.S.A. and [2]Marion Merrell Dow Research Institute, Cincinnati, OH 45215, U.S.A.

Introduction

Amylin is a 37-residue peptide amide with a single disulfide bond originally isolated from the amyloid rich pancreas of type II diabetic and insulinoma patients [1]. It has now been established that amylin is a normal component of the pancreas, co-stored and co-secreted with insulin from the islet β-cell granules [2]. Amylin inhibits insulin secretion [3] as well as insulin stimulated peripheral glucose uptake and glycogen synthesis by the activation of glycogen phosphorylase and inhibition of glycogen synthase [4]. *In vivo* studies indicate that amylin also antagonizes the inhibition of hepatic glucose output by insulin [5]. Therefore, it appears that alterations in the plasma levels of both insulin and amylin may have deleterious effects on glucose metabolism. These observations suggest that amylin peptides may have useful clinical applications in type I and II diabetes. Structure-activity studies were therefore carried out to identify the minimum sequence requirements for activity, and to develop selective agonists and antagonists of amylin. These studies were performed using the *in vitro* model systems, HEPG2 [6] and C$_2$C$_{12}$ [7] cell lines, established in our laboratory to study amylin receptors and their effects on glucose uptake.

Results and Discussion

Several N-terminal, central and C-terminal fragments of amylin were synthesized and purified according to the procedure reported by us for amylin [8]. The structural integrity of these peptides were confirmed by amino acid and mass spectral analyses. Both rat and human amylin inhibited insulin-stimulated glucose uptake by C$_2$C$_{12}$ cells in a dose-dependent manner with comparable potencies. Although central and C-terminal fragments exhibited little or no effect on glucose uptake, N-terminal fragments, (1-23)-NH$_2$, of both human and rat amylin inhibited insulin-stimulated glucose uptake by muscle cells in a dose-dependent manner, with potencies comparable to that of intact hormones. These findings suggest that the N-terminal region might constitute the active site of amylin. Structure-activity studies revealed that cysteine residues are not crucial, because [Anb^{2,7}]rat amylin(1-23)NH$_2$ also had a potency comparable to that of intact amylin in inhibiting the glucose uptake. CD studies indicated that human amylin was largely α-helical (98%) in trifluoroethanol (TFE) while rat peptide exhibited 52% α, 5% β and 43% random structures. (1-23)-NH_2 fragments of both rat and human amylin as well as the analog, [Anb^{2,7}]rat amylin(1-23)-NH_2 exhibited > 85 % α-helical structures in TFE. The intact peptides and the N-terminal fragments were, however, less structured in water. These observations suggest that the overall conformation of these fragments is not dependent on the disulfide bond, and that the central amphipathic helical region of amylins may be crucial for activity as in calcitonin peptides. We therefore synthesized s. calcitonin(1-22)-NH_2 and a hybrid peptide linking amylin(1-7) with the amphipathic helical region, 8-22, of s. calcitonin. Although these peptides also exhibited predominantly α-helical (> 75%) structures, they failed to have any effect on glucose uptake. Therefore, it appears that certain residues in the central amphipathic helical region of amylin are important for activity.

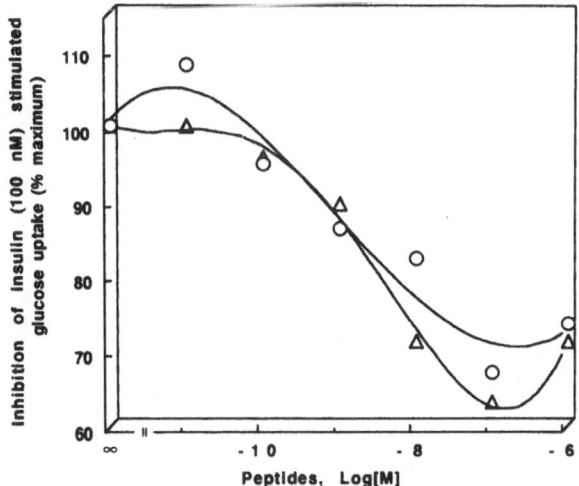

Fig. 1. *Effects of human amylin (O) and human amylin(1-23)-NH_2 (Δ) on insulin (100 nM) stimulated glucose uptake by C_2C_{12} cells.*

In vivo experiments in fasted rats revealed that bolus doses (I.V.) of human amylin(1-23)NH_2 and [Anb^{2,7}]rat amylin(1-23)NH_2 significantly lowered the plasma glucose levels. Furthermore, human amylin(1-23)NH_2 significantly lowered the hyperglycemia induced by amylin. This finding is not surprising because amylin not only affects peripheral glucose uptake, but also hepatic glucose output as well as insulin secretion which are excluded in *in vitro* models. However, further investigations are required to determine why N-terminal analogs of amylin behave as agonists and antagonists in *in vitro* and *in vivo* systems, respectively.

References

1. Cooper, G.J.S., Willis, A.C., Clark, A., Turner, R.C., Sim, R.B. and Reid, K.B.M., Proc. Natl. Acad. Sci. U.S.A., 84 (1987) 8628.
2. Cooper, G.J.S., Day, A.J., Willis, A.C., Roberts, A.N., Reid, K.B.M. and Leighton, B., Biochim. Biophys. Acta, 1014 (1989) 247.

3. Ohsawa, H., Kanatsuka, A., Yamaguchi, T., Makiuo, H. and Yoshida, S., Biochem. Biophys. Res. Commun., 160 (1989) 961.
4. Deems, R.O., Deacon, R.W. and Young, D.A., Biochem. Biophys. Res. Commun., 174 (1991) 716.
5. Molina, J.M., Cooper, C.J.S., Leighton, B. and Olefsky, J.M., Diabetes, 39 (1990) 260.
6. Sheriff, S., Fischer, J.E. and Balasubramaniam, A., Peptides, 13 (1992) 1193.
7. Sheriff, S., Fischer, J.E. and Balasubramaniam, A., Biochim. Biophys. Acta, 1136 (1992) 219.
8. Balasubramaniam, A., Renugopalakrishnan, V., Stein, M., Fischer, J.E. and Chance, W.T., Peptides, 12 (1991) 919.

Cyclic analogs of [MeTyr^1,Ala^{15},Nle^{27}]GRF(1-29)-NH$_2$ with high potencies *in vitro*

L.A. Cervini, A. Corrigan, C.J. Donaldson, S.C. Koerber, W.W. Vale and J.E. Rivier

Clayton Foundation Laboratories for Peptide Biology, The Salk Institute, 10010 N. Torrey Pines Road, La Jolla, CA 92037, U.S.A.

Introduction

Accumulating physicochemical data now support the hypothesis that growth hormone-releasing factor (GRF) preferentially assumes an α-helical structure as its bioactive conformation [1]. Earlier studies with linear analogs suggested that enhancements in potency were due to increased amphiphilicity and conformational bias [2, 3]. Recent modifications incorporate cystine [4-6] and lactam [7, 8] bridges along the sequence of GRF(1-29)-NH$_2$ in attempts to constrain secondary structure. Felix et al. have synthesized analogs that contain i-(i + 4) lactam bridges between residues 4-8, 8-12, 12-16, and 21-25 and show high biological potencies [8]. We modified the C-terminus by constraining residues 25-29 through a disulfide linkage in rat GRF(1-29)-NH$_2$ [6]. Our preliminary study of this region has now expanded to include lactam bridging of various i-(i + 4) ring sizes, amide bond shift and i-(i + 3) bridging. The effect of the structural modifications on conformation is also addressed through analysis by circular dichroism spectroscopy.

Results and Discussion

The four possible D/L bridgehead combinations of cyclo(25-29)[MeTyr^1, Ala^{15},D/LCys^{25},Nle^{27},D/LCys^{29}]rGRF(1-29)-NH$_2$ were synthesized. All were found to be nearly equipotent to the assay standard which is itself ca. 13 times less potent than the parent compound [MeTyr^1,Ala^{15},Nle^{27}]-hGRF(1-29)-NH$_2$ (**1**). Based on the observation that **3** was the most potent analog, others with varying lactam ring sizes were synthesized. The side chain of DAsp in position 25 was linked to the ω-amino function of Dpr (**6**), Dbu (**7**), Orn (**8**) and Lys (**9**) (re. Xaa) in position 29 of cyclo(25-29)[MeTyr^1,Ala^{15},DAsp^{25},Nle^{27},Xaa^{29}]-hGRF(1-29)-NH$_2$. The biological data in Table 1 show that as the lactam ring size increases, the relative potency increases until a maximum is reached with the 19-membered lactam **8** which has a potency 17 times greater than that of the assay standard. The potency then decreases by 50% upon ring enlargement in **9**. Since the location of the amide bond in the lactam may play a role in bioactivity, we chose the most potent lactam analog **8** and replaced the bridgehead residues while keeping the ring size constant. Specifically, DGlu^{25} replaced DAsp^{25} and Dbu^{29} replaced Orn^{29} which shifted the amide bond by one methylene group. The resulting analog **10** showed a 35% decrease in potency from the original analog **8**. Bridging a shorter backbone segment with an i-(i + 3) lactam resulted in an 18-membered ring analog (**11**) that was nearly equipotent to the most potent i-(i + 4) analog **8**. A decrease in the ring size to 17-membered **12** also decreased the potency by >50%. Referring to our earlier SAR study [9] of the D-amino acid and alanine scans, all sites 25, 28 and 29 tolerated both D- and Ala-substitutions which targeted this region for structural manipulation.

Table 1 *In vitro biological potency of cyclic i-(i + 4) and i-(i + 3) GRF analogs with different ring sizes*

GRF Analog		Ring size	Relative potency[a]
	hGRF(1-40)-OH Standard		1.0
1	[MeTyr^1,Ala^{15},Nle^{27}]-hGRF(1-29)-NH$_2$		13 (8-20)
	cyclo(25-29)-rGRF(1-29)-NH$_2$ substitutions		
2	[MeTyr^1, Ala^{15}, Cys^{25}, Nle^{27}, Cys^{29}]	17	1.1 (0.7-1.7)
3	[" " DCys^{25} " Cys^{29}]	17	1.9 (1.3-2.9)
4	[" " Cys^{25} " DCys^{29}]	17	0.5 (0.3-0.8)
5	[" " DCys^{25} " DCys^{29}]	17	1.5 (0.9-2.4)
	cyclo(25-29)-hGRF(1-29)-NH$_2$ substitutions		
6	[MeTyr^1, Ala^{15}, DAsp^{25}, Nle^{27}, Dpr^{29}]	17	1.4 (0.6-3.6)
7	[" " DAsp^{25} " Dbu^{29}]	18	4.2 (2.4-7.5)
8	[" " DAsp^{25} " Orn^{29}]	19	17 (7-37)
9	[" " DAsp^{25} " Lys^{29}]	20	8.4 (5.1-13)
10	[" " DGlu^{25} " Dbu^{29}]	19	11 (5.6-20)
	cyclo(25-28)-hGRF(1-29)-NH$_2$ substitutions		
11	[MeTyr^1, Ala^{15},Glu^{25}, Nle^{27}, Lys^{28}]	18	14 (7-28)
12	[" " Glu^{25} " Orn^{28}]	17	6.6 (3.4-12)

[a]Numbers in parentheses represent 95% confidence limits.

CD spectra were generated for each of the i-(i + 4) lactam analogs **6-10** and the most potent i-(i + 3) analog **11** and were compared to that of the parent linear compound **1**. Since [Nle^{27}]hGRF(1-29)-NH$_2$ has been shown to become maximally α-helical at approximately 30% TFE [10], we anticipated differences in helix-forming tendencies at the halfway point of 15% TFE. Spectral analysis revealed no linear correlation between the parameters of helicity, randomness, or β-sheet character and biological potency or ring size in either solvent. However, all analogs lost random character and concomitantly gained α-helical character in 15% TFE except for the i-(i + 3) analog **11** which showed little change in randomness. Further analysis showed that the linear analog **1** was more random with no β-sheet character in aqueous solvent, whereas the cyclic analogs were less random and displayed more β-sheet character. Also, the most potent analog **8** had the largest percentage increase in helicity from water to 15% TFE. The least potent analog **6** with the i-(i + 4) 17-membered ring showed the lowest amount of α-helix and the highest amount of β-sheet character in 15% TFE.

Earlier studies [1, 7, 8] showed lactam ring sizes of 20 and 21 atoms to be optimal for retention or gain in biological potency. In contrast, we saw maximal potencies for 19-membered i-(i + 4) and 18-membered i-(i + 3) lactam ring analogs. Although subtle differences in CD were observed among the analogs, the 18- through 20-membered lactam rings at the C-terminus did not impede the adoption of helical structure nor did they drastically reduce potencies. These results also corroborate modeling studies [11] which showed that the bridges used in analogs **7-12** are compatible with an α-helix. Smaller lactam ring sizes are sufficient for high potency at the C-terminus whereas in other regions of the peptide the side-chain (or ring) bulk may also be required.

Acknowledgements

Research was supported by NIH Grant DK 26741 and the Hearst Foundation. We thank C. Miller, R. Kaiser, D. Pantoja and R. Galyean for their excellent technical assistance and Dr. A. Craig for mass spectral analysis.

References

1. Fry, D.C., Madison, V.S., Greeley, D.N., Felix, A.M., Heimer, E.P., Frohman, L., Campbell, R.M., Mowles, T.F., Toome, B. and Wegrzynski, B.B., Biopolymers, 32 (1992) 649.
2. Kaiser, E.T. and Kézdy, F.J., Science, 223 (1984) 249.
3. Campbell, R.M., Lee, Y., Rivier, J., Heimer, E.P., Felix, A.M. and Mowles, T.F., Peptides, 12 (1991) 569.
4. Coy, D.H., Murphy, W.A., Lance, V.A. and Geiman, M.L., Peptides, 7, Suppl. 1 (1986) 49.
5. Sato, K., Hotta, M., Kageyama, J., Chiang, T.-C., Hu, H.-Y., Dong, M.-H. and Ling, N., In Ueki, M. (Ed.), Peptide Chemistry 1988, Protein Research Foundation, Osaka, Japan, 1989, p.85.
6. Rivier, J.E., Rivier, C., Koerber, S.C., Kornreich, W.D., de Miranda, A., Miller, C., Galyean, R., Porter, J., Yamamoto, G., Donaldson, C.J. and Vale, W., In Smith, J.A. and Rivier, J.E. (Eds.), Peptides: Chemistry, Structure and Biology (Proceedings of the Twelfth American Peptide Symposium), Escom, Leiden, The Netherlands, 1992, p.33.
7. Felix, A.M., Heimer, E.P., Wang, C.T., Lambros, T.J., Fournier, A., Mowles, T.F., Maines, S., Campbell, R.M., Wegrzynski, B.B., Toome, V., Fry, D. and Madison, V.S., Int. J. Peptide Protein Res., 32 (1988) 441.
8. Felix, A.M., Wang, C.-T., Campbell, R.M., Toome, V., Fry, D.C. and Madison, V.S., In Smith, J.A. and Rivier, J.E. (Eds.), Peptides: Chemistry, Structure and Biology (Proceedings of the Twelfth American Peptide Symposium), Escom, Leiden, The Netherlands, 1992, p.77.
9. Cervini, L., Galyean, R., Donaldson, C.J., Yamamoto, G., Koerber, S.C., Vale, W. and Rivier, J.E., In Smith, J.A. and Rivier, J.E. (Eds.), Peptides: Chemistry, Structure and Biology (Proceedings of the Twelfth American Peptide Symposium), Escom, Leiden, The Netherlands, 1992, p.437.
10. Clore, G.M., Martin, S.R. and Gronenborn, A.M., J. Mol. Biol., 191 (1986) 553.
11. Koerber, S. and Rivier, J., unpublished results.

Conformational differences between ovine and human CRF

M. Dathe[1], D. Zirwer[2], K. Gast[2], M. Beyermann[1], E. Krause[1] and M. Bienert[1]

[1]Research Institute of Molecular Pharmacology, D-10315 Berlin,
[2]Max-Delbrück-Centrum of Molecular Medicine, D-13125 Berlin,
Germany

Introduction

Corticotropin releasing factor (CRF) is a species specific 41-residue neuropeptide [1]. The peptides of ovine (o-CRF) and human (h-CRF) origin differ in six amino acid positions but differences in their physical-chemical properties in solution have not been described. Also, statistical analysis (Chou Fasman) predicts comparable potency of secondary structure formation with helical regions in the central and C-terminal parts of the chains in both o-CRF [2] and h-CRF [3].

Preliminary results which revealed a different behaviour of the two peptides in solution initiated a series of experiments to elucidate the reasons for the structural differences in aqueous environment and to investigate the properties of the peptides in trifluoroethanol (TFE), a structure inducing solvent, by CD-spectroscopy and dynamic light scattering.

Results and Discussion

The increase of peptide concentration in acid aqueous solution from 10^{-5} to 10^{-3} mol/L has only a minor influence on the CD spectral characteristics of o-CRF. The amount of helix in the peptide remains less than 20%. In h-CRF solutions the increasing concentration encourages helical spectral characteristics and the amount of helix was determined to reach 45% in the 1 mM peptide solution (Figure 1). Ionic interactions do not play a role in the conformation of o-CRF. However, the helical content of h-CRF varies with solution pH. It is low in the case of negative or positive charge and reaches a maximum in the range of the isoelectric point of h-CRF (Figure 2).

These findings lead to the assumption that o-CRF exists mainly as a monomer, while h-CRF forms an intermolecularly stabilized structure which is influenced by ionic forces. By dynamic light scattering investigations of o-CRF in aqueous solutions at concentrations lower than 10^{-4} mol/L, a particle distribution with a mean particle radius R of 1.2 nm and an apparent molecular weight, M_{app}, of about 4000 were found. These values correspond to the monomer. The broad distribution at 10^{-3} mol/L reflects an association equilibrium of mainly dimers with a M_{app} of about 9000. h-CRF exists at the lowest investigated peptide concentration in an association equilibrium with predominantly dimers. In a 1 mM solution the distribution becomes narrower. The main particle radius was found to be 2.4 nm and the apparent molecular weight was found to be 19500. The particle distribution consists mainly of tetramers (Figure 3).

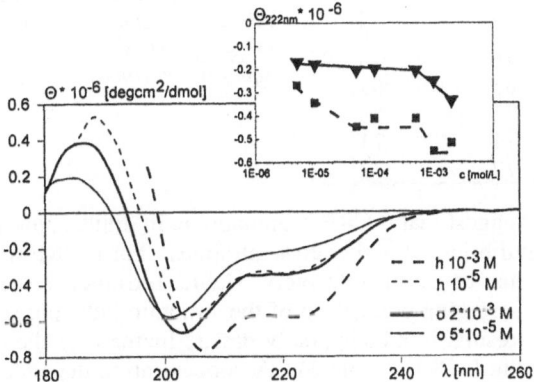

Fig. 1. CD spectra of o-CRF (full line, ▼) and h-CRF (dotted line, ■) in H_2O, pH4. Inset: Concentration dependence of Θ of the two peptides at $\lambda = 222nm$.

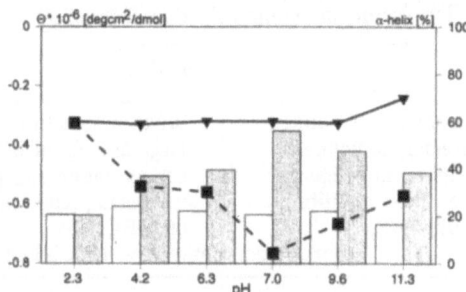

Fig. 2. Molar ellipticity Θ of o-CRF (▼) and h-CRF (■) in H_2O (c=2x10^{-5} mol/L) at 222 nm as a function of pH. The amount of α-helix of the peptides (Δ, o-CRF; shaded box, h-CRF) was estimated from different methods (Θ_{222nm} [4], CONTIN [5], VARSEL - least squares algorithms), pooled and averaged.

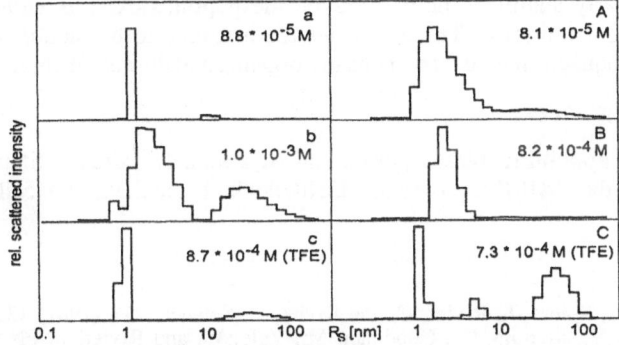

Fig. 3. Particle size distribution of o-CRF (left) at a concentration of 8.8x10^{-5} mol/L a; 1.0x10^{-3} mol/L b and h-CRF (right) at 8.1x10^{-5} mol/L A; 8.1x10^{-4} mol/L B in H_2O. c and C give the distributions in TFE. R_s - mean particle radius.

TFE (%)	o-CRF % helix	h-CRF % helix
0	5	23
10	4	20
20	46	46
40	65	69
60	72	66
80	71	71
100	77	70

Table 1

The amount of α-helix of o-CRF and h-CRF ($c=2x10^{-5}$ mol/L) in TFE/H_2O mixtures, estimated from different methods (Θ_{222nm} [4], CONTIN [5], VARSEL - least squares algorithms), pooled and averaged.

The results suggest that o-CRF is monomeric in diluted solutions, but exists in an association equilibrium at higher concentrations. For h-CRF we could show that the peptide seems to form relatively stable tetramer structures at higher concentrations. Increasing percentage of the structure inducing solvent TFE in the peptide solutions results in a comparably drastic increase of the helical content in o-CRF as well as in h-CRF (Table 1). A concentration dependence of the molar ellipticity is not observed in this system. The particle radius R of about 1 nm of the main distribution of the peptides in TFE corresponds to the peptide monomer (Figure 3). Thus, under structure inducing solvent conditions the two peptides show identical properties. The results are summarized as follows:

Conditions	Properties	
	o-CRF	h-CRF
aqueous $10^{-6}<c[M]<10^{-4}$	monomer, less defined secondary structure independent of charge	broad association equilibrium, large amount of helix dependent on concentration and peptide charge
aqueous $c[M]>10^{-4}$	association equilibrium of oligomers	tetramers predominate with an amount of helix >45%
TFE	monomer, content of helix about 70%	

The induction of the helical structure in the investigated CRF peptides in aqueous solution is the result of intermolecular interactions which are more pronounced in the human sequence than in o-CRF. Under structure inducing conditions more than 70% of the amino acid residues of the two peptides form an intramolecularly stabilized helix. Maybe the peptides exist as monomers under physiological conditions. The molecules are assumed to be random coiled in the extracellular aqueous medium and helically organized at the cell membrane.

Acknowledgements

We are grateful to W.C. Johnson and A. Toumadje, Oregon State University, Corvallis for the VARSEL program. H. Nikolenko is thanked for excellent technical assistance.

References

1. Vale, W., Spiess, J., Rivier, C. and Rivier, J., Science, 213 (1981) 1394.
2. Pallai, P.V., Mabilia, M., Goodman, M., Vale, W. and Rivier, J., PNAS, 80 (1983) 6770.
3. Krause, G., FMP, personal information.
4. Chen, Y.J., Yang, J.T. and Martinez, H.M., Biochemistry, 11 (1972) 4120.
5. Provencher, S.W. and Glöckner, J., Biochemistry, 20 (1981) 33.

Bradykinin analogs via cyclization of side chains and modified backbone

G. Greiner[1], J. Jezek[1], L. Seyfarth[1], B. Müller[1],
C. Liebmann[1], I. Paegelow[2] and S. Reissmann[1]

[1]Institute of Biochemistry and Biophysics, Friedrich-Schiller-
University, Philosophenweg 12, D-07743 Jena, Germany
[2]Institut of Pharmacology and Toxicology, University Rostock,
Schillingallee 70, D-18057 Rostock, Germany

Introduction

Cyclic bradykinin analogs were synthesized in order to study the influence of conformational restrictions on the biological activity and to suppress the enzymatic degradation [1, 2]. To study the influence of the N-terminal part on the hormone receptor interaction we prepared different types of cyclic bradykinin analogs. Besides the classical side chain to side chain cyclization we synthesized also side chain to backbone and backbone to backbone analogs.

Results and Discussion

The peptides synthesized in this study (see Table 1) were prepared by SPPS using a combination of Boc/Bzl and Fmoc/Bu^t strategy. For condensation reagents see Figure 1. We used qualitative Kaiser test monitoring throughout all synthetic steps. Peptides with $\psi(CH_2-NH)$ bonds were prepared in accord with Coy et. al. [3]. The cyclizations (except peptide no. I) were done by BOP or TBTU. Final deprotection and detachment from the resin was made with TFMSA, TFA, thioanisol and EDT. Purification was done on Biogel P2 column and then by RP-HPLC. The peptides obtained were characterized by AAA and FABMS. In comparison with literature [4, 5], we used N to N backbone cyclization. The most problematic building block Fmoc-N(CH_2COOBu^t)'Phe'-OH (XIV) was obtained by alkylation of H-Phe-OBzl with $BrCH_2COOBu^t$ (Ag_2O; DMF) to give $HN(CH_2COOBu^t)$'Phe'-OBzl (XI). Catalytic hydrogenolysis (Pd/C) of XI in AcOH afforded $HN(CH_2COOBu^t)$'Phe'-OH (XII). Compound XII was reacted with N,O-bis(trimethylsilyl)-acetamide to give $(CH_3)_3SiN(CH_2COOBu^t)$'Phe'-O-Si$(CH_3)_3$ (XIII). In situ reaction of XIII with Fmoc-Cl and workup with $KHSO_4$ afforded the key building block XIV. All peptides with exception of VIII afforded [M+H]$^+$ in accord with theory. The synthesis of peptide VII proceeded well including condensation of Fmoc-N(CH_2COOBu^t)'Phe'-OH. The following steps were more difficult and prolonged reaction time or double couplings were necessary. It was not possible to monitor the incorporation of Fmoc-Arg(Mts), because the resin bound

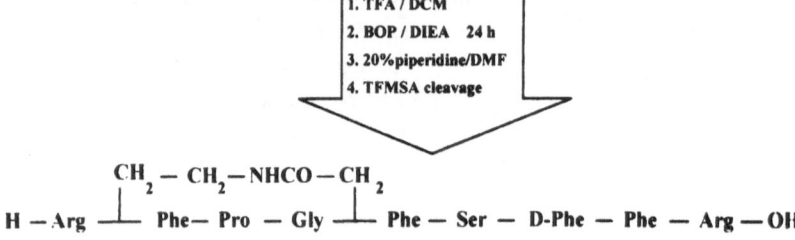

Cyclization on the Solid Support

Fig. 1. Backbone - backbone cyclization of bradykinin analog.

Table 1 *Biological activities of cyclic peptide analogs*

No	Peptides	Antagonistic Activity (pA₂)	
		RUT	GPI
I	Cys-Arg-Pro-Pro-Gly-Phe-Cys-D-Phe-Phe-Arg	7,2	0
II	Lys-Arg-Pro-Pro-Gly-Phe-Glu-D-Phe-Phe-Arg	5,8	0
III	⌐——— Phe-Phe ———⌐ Lys-Arg-Pro-Pro-Gly-Phe-Glu-D-Phe-Phe-Arg	0	0;*
IV	⌐——— Arg-Arg ———⌐ Lys-Arg-Pro-Pro-Gly-Phe-Glu-D-Phe-Phe-Arg	0	0;*
V	⌐——— Glu-Glu ———⌐ Lys-Arg-Pro-Pro-Gly-Phe-Glu-D-Phe-Phe-Arg	5,8	0;*
VI	⌐——————— COCH₂ Lys-Arg-Pro-Pro-Gly⌐Phe-Ser-D-Phe-Phe-Arg CH₂-CH₂-NHCOCH₂	0	0;*
VII	Arg⌐Phe-Pro-Gly⌐Phe-Ser-D-Phe-Phe-Arg CH₂-CH₂-NH——————⌐	6,2	0;*
VIII	Arg⌐Phe-Pro-Gly-Phe-Glu-D-Phe-Phe-Arg CH₂-CH₂-NH₂	n.d.	n.d.
IX	⌐Phe-Pro-Gly-Phe-Glu-D-Phe-Phe-Arg CH₂-CH₂-NH₂	n.d.	n.d.

* potentation; n.d. not determined

Boc-Glyψ(CH$_2$-NH) peptide gave a negative Kaiser test. The synthesis of VIII proceeded well including condensation of phenylalanine in position 2. In order to improve the incorporation of Boc-Glyψ(CH$_2$-NH), we used a 6 fold excess of Boc-Gly aldehyde. In this case, unfortunately, double reductive alkylation [6, 7] took place and after final deprotection we obtained peptide IX. The disulfide bridged analog I has the highest RUT activity. Expansion of the bridge size decreases this activity with the exception of compound V. The side chain to backbone cyclic analog VI has no antagonistic activity in comparison with the corresponding linear peptide.

Acknowledgements

This work was supported by the Bundesministerium für Forschung und Technologie FRG (0319992A).

References

1. Mutule, I., Mutulis, F., Myshliakova, N., Veveris, M., Golubeva, V., Porunkevich, E., Ratkevich, M., Strazda, G., Klusa, V., Bergmann, I., Sekacis, I., Grigoryeva, V., Sulima, A. and Chipens, G., Bioorg. Khim., 16 (1990) 1465.
2. Marshall, G.R., Kaczmarek, K., Kataoka, T., Plucinska, K., Skeean, R., Lunney, B., Taylor, C., Dooley, D., Lu, G., Panek, R. and Humblet, C., In Giralt, E. and Andreu, D., (Eds.), Peptides 1990, Escom, Leiden, The Netherlands, 1991, p.594.
3. Hocart, S.J. and Coy, D.H., In Epton, R. (Ed.), Innovation and Perspectives in Solid Phase Peptide Synthesis: Peptides, Polypeptides and Oligonucleotides, SPCC, Birmingham, U.K., 1990, p.413.
4. Gilon, C., Halle, D., Chorev, M., Selinger, Z., Goldshmith, R. and Byk, G., In Giralt, E. and Andreu, D. (Eds.), Peptides 1990, Escom, Leiden, The Netherlands, 1991, p.404.
5. Gilon, C., Halle, D., Chorev, M., Selinger, Z. and Byk, G., Biopolymers, 31 (1991) 745.
6. Harbeson, S.L., Shatzer, S.A., Le, T.B. and Buck, S.H., J. Med. Chem., 35 (1992) 3949.
7. Cushman, M., Oh, Y., Copeland, T.D., Oroszlan, S. and Snyder, S.W., J. Org. Chem., 56 (1991) 4161.

New potent selective linear and cyclic bradykinin agonists and antagonists

S. Reissmann

Institute of Biochemistry and Biophysics, Friedrich-Schiller-University, Philosophenweg 12, D-07743 Jena, Germany

Introduction

Bradykinin, an endogenous linear nonapeptide hormone with the amino acid sequence Arg-Pro-Pro-Gly-Phe-Ser-Pro-Phe-Arg, is involved in a variety of physiological and pathophysiological processes.

Because of the pathophysiological role of bradykinin, its antagonists are of great interest. Vavrek and Stewart [1] developed the first antagonist in 1984, where the key sequence alteration with respect to bradykinin was the replacement of L-Pro[7] with D-Phe or other D-aromatic amino acids. Since this initial discovery many laboratories have attempted to develop more potent antagonists [2-5].

Our interest was to determine the structural and conformational requirements for antagonistic activity and tissue selectivity of bradykinin antagonists. We synthesized and studied analogs with replacements of D-Phe at position 7 by sterically hindered or more hydrophobic amino acids, with replacements of the proline residues in position 2 and 3 by N-methyl phenylalanine, and cyclic analogs with different bridges, starting from the side chains or from the backbone.

Results and Discussion

The replacement of D-Phe at position 7 in bradykinin antagonists by C^α, C^β and ring substituted amino acids decreases, in most cases, the potency probably by steric hindrance [6]. Only 2-methyl-, 2,5-dimethyl phenylalanine and 2,5-dimethyl cyclohexyl alanine (II) enhance the antagonistic activity, alone or in combination with other hydrophobic amino acids at position 8. The possibility of replacing D-Phe at position 7 by p-benzoyl-D-phenylalanine (Bpa) (IV) without a decrease in potency is of importance for developing a photolabelled bradykinin. Furthermore it is possible to replace D-Phe by 3-iodo-D-tyrosine (V). From the standpoint of structure-activity considerations the substitution with Bpa shows that bulky ring substitutions like 2,3,4,5,6-pentamethyl phenylalanine (III) prevents the interaction with the receptor binding site whereas the planar p-benzoyl moiety seems to interact with a hydrophobic pocket. N-methyl and N-alkyl phenylalanines in the sequence restrict the conformational flexibility of the backbone. In position 7 D-NMePhe (VI), converts antagonists into potent agonists [6]. The more bulky residues in D-N-alkyl amino acids destroy the activity. Analogs with D-NMePhe at position 7 (VII) give the highest tissue selectivity for agonists (RUT/GPI=10000).

Table 1 *Agonistic and antagonistic activities*

No	Peptides	Activity	
		RUT	GPI
I	[DPhe^7]-BK	-	5.6
II	[DLCha(2,5Me)^7]-BK	-	6.0
III	[DLPhe(2,3,4,5,6Me)^7]-BK	-	-
IV	Tyr(I)-DArg-[Bpa^7]-BK	7.98	6.85
V	[Tyr(I)^7,Oic^8]-BK	7.13	6.69
VI	[DNMePhe^7]-BK	136%	0.4%
VII	DArg[Hyp^3,Thi^5,DNMePhe^7]-BK	86%	0.006%
VIII	[LNMePhe^2]-BK	35%	4%
IX	Lys-Arg-Pro-Pro-Gly-Phe-Glu-DPhe-Phe-Arg	5.8	0
X	Arg-Phe-Pro-Gly-Phe-Ser-DPhe-Phe-Arg	6.13	-
	(CH$_2$)$_2$NHCOCH$_2$		

By replacement of proline residues at position 2 and 3 by D- and L-NMePhe only [L-NMePhe^2]-BK (VIII) has moderate agonist activity on rat uterus. The most striking finding with this analog is the antagonistic activity on guinea pig lung strip. Thus this analog represents a new type of antagonist for B$_2$ receptors, without any replacement at position 7. Furthermore this analog inhibits at a concentration of 10^{-13} M cytokine secretion from mononuclear cells.

Based on conformational calculations using molecular mechanics and dynamics we synthesized different types of cyclic bradykinin antagonists to confirm the supposed conformation of two turns, one in the N-terminal part and a second in the C-terminal part; to develop conformationally stabilized potent antagonists and to study the influence of different types of cyclization on the biological activity.

In the first series of analogs we used the linear starting sequence [DPhe^7]-bradykinin. In this relatively weak antagonist the influence of conformational constraints should be easier to observe than in highly potent analogs, where the C-terminal sequence binds strongly with the receptor.

The replacement of serine in position 6 by glutamic acid retains the biological activity. Cyclization of the side chains from Lys^0 and Glu^6 by a lactam bridge (IX) provides an antagonist with nearly the same activity as the linear peptide and as [DPhe^7]-BK itself.

In contrast to the studies on melanotropin by Hruby [7], modifications of the intramolecular bridges by positively charged, negatively charged or hydrophobic amino acids destroy the antagonistic activity. On the contrary these cyclic compounds potentiate the bradykinin action.

To obtain cyclic analogs with minimal changes in the side chains we synthesized analogs with bridges starting from the backbone. Using the results with [L-NMePhe^2]-BK we replaced the proline residue at position 2 by phenylalanine and started the bridge at the amide nitrogen. A very interesting fact of this series of cyclic bradykinin antagonists is the relative high activity of the backbone-backbone cyclized analog (X), an analog with minimal side chain modifications. In most cases the cyclization seems to change the tissue selectivity. Contrary to [DPhe^7]-BK the active antagonists have a higher potency on rat uterus, some are only active on RUT.

Molecular dynamics and molecular mechanics calculations show a good agreement between the conformational shape of [DPhe^7]-BK and the cyclic bradykinin with a lactam bridge between the side chains of lysine at position 0 and glutamic acid at position 6 (X). Thus, the activities of the cyclic antagonists confirm the proposed β-turn in the N-terminal part.

Acknowledgements

The work was performed with financial support from the Bundesministerium für Forschung und Technologie (BEO 21/19992A) and from Hoechst AG.

References

1. Vavrek, R.J. and Stewart, J.M., Peptides, 6 (1985) 161.
2. Regoli, D., Rhaleb, N.E., Dion, S. and Drapeau, G., Trends Pharmacol. Sci., 11 (1990) 156.
3. Cheronis, J.C., Whally, E.T., Nguyen, K.T., Eubanks, S.R., Allen, L.G., Duggan, M.J., Loy, S.D., Bonham, K.A. and Blodgett, J.K., J. Med. Chem., 35 (1992) 1563.
4. Lembeck, F., Griesbacher, T., Eckhardt, M., Henke, St., Breipohl, G. and Knolle, J., Br. J. Pharmacol., 102 (1991) 297.
5. Kyle, D.J., Martin, J.A. and Burch, R.M., J. Med. Chem., 34 (1991) 2649.
6. Reissmann, S., Greiner, G., Schwuchow, C., Pineda, L.F., Seyfarth, L., Paegelow, I., Liebmann, C. and Wiesmüller, K.-H., In Schneider, C.H. and Eberle, A.N. (Eds.), Peptides 1992, ESCOM, Leiden, The Netherlands, 1993, p.695.
7. Sharma, S.D., Nikiforovich, G.V., Jiang, J., Castrucci, A.M.L., Hadley, M.E. and Hruby, V.J., In Schneider, C.H. and Eberle, A.N. (Eds.), Peptides 1992, Escom, Leiden, The Netherlands, 1993, p.95.

High yield synthesis of pseudopeptide analogs of cardiac NPY receptor antagonist, NPY (18-36)

A. Balasubramaniam, M. Stein, S. Sheriff and J.E. Fischer

Division of G.I. Hormones, Department of Surgery, University of Cincinnati College of Medicine, Cincinnati, OH 45267-0558, U.S.A.

Introduction

Neuropeptide Y (NPY) exhibits negative inotropic effects on both isolated hearts [1] and myocytes [2]. Plasma NPY levels are elevated in patients with congestive heart failure (CHF) [3]. These findings suggest that cardiac NPY receptor antagonists may have therapeutic value in CHF. Our previous investigations directed towards these goals resulted in the characterization of cardiac NPY receptors [4] and the demonstration that NPY(17-36) is a physiological antagonist of NPY in rat cardiac membranes [5]. We also reported that NPY(18-36) can competitively antagonize the inhibitory effects of NPY on the adenylate cyclase activity [6] and contractility [7] of rat cardiac membranes and/or myocytes. However, NPY(18-36) has been reported to exhibit agonist activity in other *in vitro* [8] and *in vivo* [9] systems. To minimize these problems, we decided to synthesize NPY(18-36) analogs with the pseudobond (-CH$_2$-NH-) because introduction of the pseudobond has often been shown to generate peptide analogs with increased receptor affinity, selectivity and/or half life [10]. Furthermore, we chose to incorporate the pseudobonds in between residues 30/31, 31/32, and 32/33 because the conformational changes in this receptor binding region were expected to impart dramatic changes in the properties of NPY(18-36). Although we encountered initial difficulties, introduction of a protecting group for the secondary amine in the pseudobond enabled the synthesis of these compounds in good yield and good purity.

Results and Discussion

Elegant investigations by Fehrentz and Castro [11] and subsequently by Sasaki and Coy [12] have shown that: I. optically pure Boc-AA-CHO can be synthesized in high yields by the LiAlH$_4$ reduction of the N-methoxy-N-methylamides of Boc-AA-OH, and II. Boc-AA-CHO can be directly coupled to the α-NH$_2$ group of the peptide resin by reductive alkylation using NaBH$_3$CN (Scheme 1). However, the possibility of branching at the secondary amino group prevents general applicability of this method especially for the synthesis of long peptides containing a pseudobond in the C-terminal region as in NPY(18-36). Therefore, we investigated the possibility of capping the secondary amino group with Tos, Z or (2-Cl)Z groups which could be simultaneously removed during the final HF cleavage to obtain free peptides. Initial investigations revealed that the treatment of the peptide resin containing the

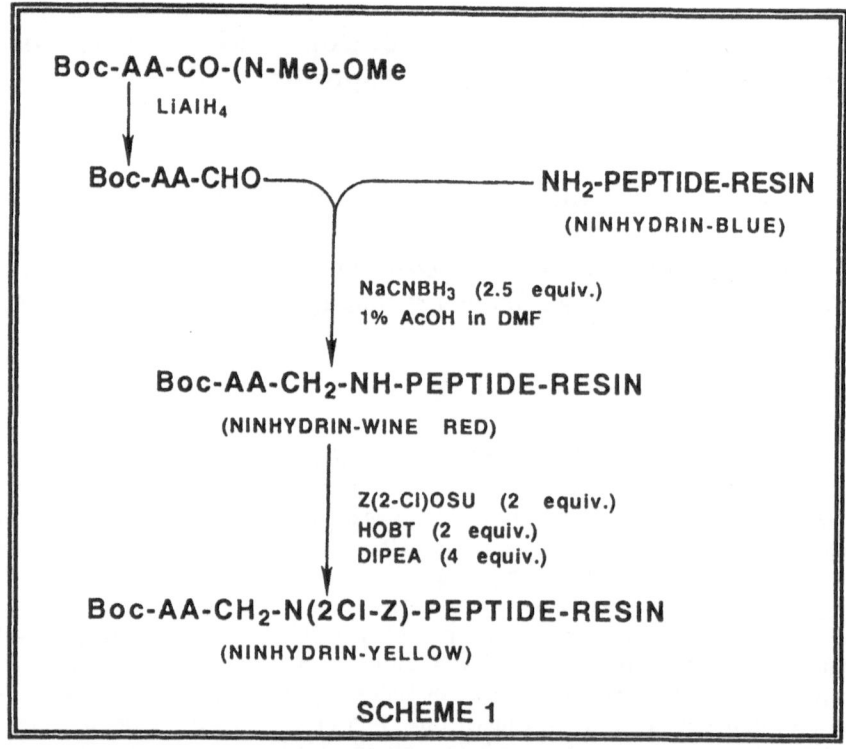

SCHEME 1

pseudobond with Tos-Cl or Z-Cl (2 equiv.) and DIPEA (4 equiv.) resulted in the complete capping of the secondary amine within 30 min. The red wine color of ninhydrin with the secondary amine turned yellow at the end of capping. However, the known lability of the Z-group during repeated acidolysis to remove the N-α-Boc group, and the apparent resistance of the Tos group attached to secondary amine to HF, led us to choose (2-Cl)Z for the capping of the secondary amine group. This was introduced by treating the peptide resin with the pseudobond with (2-Cl)Z-OSU (2 equiv.), HOBT (2 equiv.) and DIPEA (4 equiv.) for 30-60 min. This reaction also blocks the α-NH$_2$ group which has not reacted with Boc-AA-CHO. Analytical RPLC of the three pseudopeptide analogs of NPY(18-36) synthesized by this strategy revealed that the crude peptides contained >65% of the target compounds. These peptides, ($\psi^{30\text{-}31}$)NPY(18-36) (IC$_{50}$=6.00 nM), ($\psi^{31\text{-}32}$)NPY(18-36) (1.00 nM) and ($\psi^{32\text{-}33}$)NPY(18-36) (0.56 nM), exhibited greater affinity and selectivity to cardiac NPY receptors than NPY(18-36) (126 nM). In summary, we have shown that long peptides with pseudobonds in the C-terminal region can be synthesized by solid phase method in high yields by protecting the secondary amine of the pseudobonds. This strategy could now be applied for the synthesis of the pseudopeptide analogs of other peptide hormones with C-terminal active sites.

Acknowledgements

This work was supported in part by a NIH grant GM47122.

References

1. Balasubramaniam, A., Grupp, I., Matlib, M.A., Benza, R., Jackson, R.L., Fischer, J.E. and Grupp, G., Reg. Pept., 6 (1988) 289.
2. Piper, H.M., Millar, B.C. and McDermott, B.J., Arch. Pharmacol., 338 (1989) 333.
3. Maisel, A.S., Scott, N.A., Motulsky, J.H., Michel, M.C., Boublik, J.H., Rivier, J.E., Ziegler, M., Allen, R.S. and Martin, R.B., Am. J. Med., 86 (1989) 43.
4. Balasubramaniam, A., Sheriff, S., Rigel, D.F. and Fischer, J.E., Peptides, 11 (1990) 545.
5. Sheriff, S. and Balasubramaniam, A., J. Biol. Chem., 267 (1992) 4680.
6. Balasubramaniam, A. and Sheriff, S., J. Biol. Chem., 265 (1990) 14724.
7. Millar, B.C., Weis, T., Piper, H.M., Weber, M., Borchard, U., McDermott, B.J. and Balasubramaniam, A., Am. J. Physiol., 261 (1991) H1727.
8. De Quidt, M. and Emson, P.C., Neurosci., 18 (1986) 545.
9. Boublik, J., Scott, N., Taulane, J., Goodman, M., Brown, M. and Rivier, J., Int. J. Pept. Protein Res., 33 (1989) 11.
10. Jenson, R.T. and Coy, D.G., Trends Pharmacol. Sci., 12 (1991) 13.
11. Fehrentz, J.A. and Castro, B., Synthesis, (1983) 676.
12. Sasaki, Y. and Coy, D.H., Peptides, 8 (1987) 119.

Structure-function relationships in human parathyroid hormone: The essential role of amphiphilic α-helix

W.K. Surewicz, W. Neugebauer, L. Gagnon, S. MacLean, J.F. Whitfield and G. Willick

Institute for Biological Sciences, National Research Council of Canada, Ottawa, Ontario, K1A OR6, Canada

Introduction

Human parathyroid hormone (hPTH) plays an important physiological role by regulating the resorption and deposition of calcium in bone. Although the full-size hormone is 84 residues long, all the information required to elicit major biological responses, such as activation of adenylate cyclase or stimulation of membrane-bound protein kinase C, is contained within the biologically active 1-34 region [1]. In the previous report [2], we have identified a sequence, between residues 21 and 34, that can form an amphiphilic α-helix. Furthermore, biological data [3] have indicated that this sequence contains the entire domain responsible for stimulation of membrane bound PKC. To further test the role of amphiphilic helix in biological activities of the hormone, we have designed and synthesized two classes of hPTH(20-34) and hPTH(1-34) analogues: class one peptides which were predicted to have greatly reduced propensity to form amphiphilic helices, and class II peptides of increased propensity to form amphiphilic structures. Here, we report the correlations between the structural properties of these analogues and their biological activities.

Results and Discussion

Figure 1 shows helical wheel projections for the 21-34 region of hPTH and of the hormone analogues used in this study. The distribution of polar and hydrophobic residues in the original peptide is fairly regular. Therefore, if this peptide were to form an α-helical structure, the polar and apolar residues would be segregated on opposite sides of the cylindrical helix (top panel). The amphiphilic character of this helix should be further improved in the mutant peptide obtained by substituting Lys-27 with the more hydrophobic Leu (left panel). Another analogue of the hormone was designed by swapping Leu-24 with Lys-26. Swapping of these two amino acids altered distribution of hydrophobic and polar residues, resulting in the structure of greatly diminished potential to form amphiphilic helix (right panel).

A notable feature of many peptide hormones with propensity to form amphiphilic helix is their ability to interact with phospholipid surfaces. Such an anisotropic environment, which is believed to mimic the properties of a hydrophobic region of the receptor, facilitates folding of these peptides into

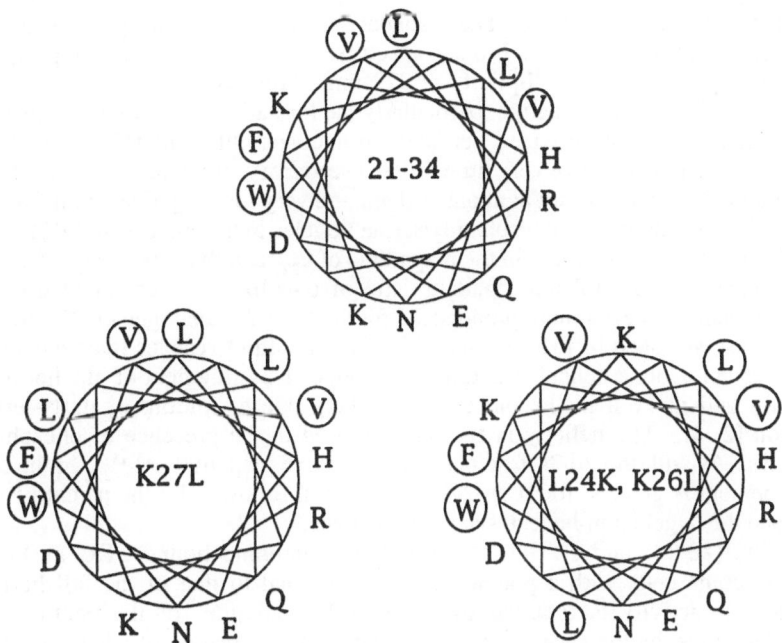

Fig. 1. *Helical wheel projections of hPTH(21-34) and analogues. Hydrophobic residues are circled.*

Table 1   *Circular dichroism and biological activities of hPTH analogues*

Peptide	$-\theta_{222} \times 10^{-3}$		PKC	Adenylate cyclase(C_{50}) (nM)
	Buffer	+POPS[a]		
hPTH-(20-34)	3.2	17.1	+	inactive
[L[27]]-hPTH(20-34)	6.5	24.1	+	inactive
[K[24],L[26]]-hPTH(20-34)	1.7	0.2	-	inactive
hPTH-(1-34)	6.3	13.8	+	14
[L[27]]-hPTH(1-34)	11.6	16.7	+	3
[K[24],L[26]]-hPTH-(1-34)	3.8	6.2	-	inactive

[a]   Palmitoyl-oleoyl-phosphatidylserine.
[b]   Expressed as peptide concentration (C_{50}) inducing 50% of maximum observed production of [3H]-cAMP.

amphiphilic helical structures. The secondary structure of various peptide analogues was assessed by circular dichroism spectroscopy. In Table 1 we report, as an empirical parameter, the ellipticity at 222 nm. More negative values of θ_{222} are indicative of increased structure, particularly the formation of an α-helix. Like many other small and medium-sized peptides, in aqueous buffer hPTH(20-34) and the substituted analogues have very little ordered secondary structure, as indicated by the minimum in the CD spectra at about 200 nm and very low negative ellipticity at 222 nm (Table I). Addition of phosphatidylserine vesicles to the solution of hPTH(20-34) results in a dramatic increase in the magnitude of θ_{222} and the appearance of a second minimum at about 208 nm, characteristic of α-helix. The extent of this lipid-induced helicity is even more pronounced for the [Leu-27] analogue of hPTH(20-34). In contrast, swapped hPTH(20-34) still binds to the lipid vesicles (data not shown), but does not adopt a helical structure. The longer 1-34 fragment of the hormone is slightly helical even in the buffer, most likely due to folding of its N-terminal portion [2, 4]. The helicity is further increased in the presence of phospholipid vesicles. As with the 20-34 fragments, the negative magnitude of θ_{222} in the lipidic environment is greatest for [Leu-27]-hPTH(1-34), followed by the parent hPTH(1-34), and the largely non-helical swapped analogue.

PTH(20-34) and PTH(1-34) stimulate membrane bound PKC in 17/2 rat osteosarcoma cells with a potency essentially equal to that of the full hormone. While this activity is retained by the Leu-27 analogues of the peptides, it is completely lost by the swapped variants (Table 1). Activation of adenylate cyclase requires almost all of the 1-34 sequence. Notably, the adenylate cyclase stimulatory activity of hPTH(1-34) is considerably increased upon substitution of Lys-27 with Leu (as indicated by a lower concentration of the latter peptide required to produce similar response), whereas it completely disappears in the swapped analogue of the hormone (Table 1).

In conclusion, the present data shed new light on structure-function relationships in hPTH. These data clearly demonstrate the importance of conformational properties of hPTH and indicate that the amphiphilic helical secondary structure within the 20-34 region is critically important for biological activities of the hormone.

References

1. Potts Jr., J.T., Tregear, G.W., Keutmann, H.T., Niall, H.D., Sauer, R., Deftos, L.J., Dawson, B.F., Hogan, M.L. and Aurbach, G.D., Proc. Natl. Acad. Sci. U.S.A., 68 (1971) 63.
2. Neugebauer, W., Surewicz, W.K., Gordon, H.L., Somorjai, R.L., Sung, W. and Willick, G., Biochemistry, 31 (1992) 2056.
3. Jouishomme, H., Whitfield, J.F., Chakravarthy, B., Durkin, J.P., Gagnon, L., Isaacs, R.J., MacLean, S., Neugebauer, W., Willick, G. and Rixon, R.H., Endocrinology, 130 (1992) 53.
4. Klaus, W., Dieckmann, T., Wray, V., Schomburg, D., Wingander, E. and Mayer, H., Biochemistry, 30 (1991) 6936.

Synthesis and biological activity of *Bombyx* eclosion hormone

M. Takai, I. Umemura, H. Hayashi, Y. Shibanaka and N. Fujita

International Research Laboratories, Ciba-Geigy Japan Limited, 10-66, Miyuki-cho, Takarazuka 665, Japan

Introduction

Eclosion hormone (EH) is a neuropeptide which plays a crucial role in insect ecdysis behavior. The peptide was isolated from lepidopteran species, *Bombyx mori* and *Manduca sexta. Bombyx* EH consists of 62 amino acid residues and includes three intramolecular disulfide bonds (Figure 1) [1]. In order to study the chemical and biological aspects of *Bombyx* EH, it was necessary to prepare highly purified peptide in large quantities. The synthesis was carried out by both a liquid phase method applying maximum protection-HF procedure and a solid phase method applying Fmoc-strategy. Using the synthetic *Bombyx* EH, we investigated the mechanism of EH-mediated signal transduction in the silkworm abdominal ganglia.

Results and Discussion

The polypeptide chain was assembled from 11 segments as indicated in Figure 1. All the coupling reactions were carried out in N-methylpyrrolidone (NMP) or a mixture of NMP and DMSO using the WSCI-HOBt method to afford the fully protected peptide. All the protecting groups, except Acm, were removed by the HF method [HF/p-cresol/ 1,2-butanedithiol:80/5/15 (V/V/V), -5 °C, 60 min]. The thus obtained hexa-Acm peptide was purified by gel-filtration on Sephadex G-50 followed by RP-HPLC and then treated with $Hg(OAc)_2$ in 5% AcOH to remove the Acm groups. The disulfide bridges were formed by random air oxidation in the presence of GSH/GSSG. The homogenous peptide was obtained after several purification steps by using gel-filtration and preparative RP-HPLC. We have also tried to synthesize EH by a conventional solid phase method applying Fmoc-strategy. The purity of the intermediate 53-62, 43-62, 33-62, 23-62 and 13-62, was checked after deprotection. A major side product formed during the peptide chain elongation was found to lack Ile residue at position 46. However, this side reaction was avoided by the double coupling technique. The final peptide either synthesized by the liquid phase or the solid phase method was characterized by amino acid analysis, sequence analysis, mass analysis and RP-HPLC. The peptide mapping of a thermolysin digest of the synthetic peptide gave multiple peaks on RP-HPLC. The structures of the cystine-containing fragments were determined by sequence and mass analyses. The results indicate that the amino acid sequence and the disulfide pairings of the synthetic peptides are identical to that of *Bombyx* EH. *In vivo* studies showed that the synthetic EH induced eclosion behavior with an ED_{50} value of 0.1-0.2ng/animal, a value very similar to that of the native EH (Figure 2a).

These results suggest that the peptides obtained by both synthetic methods were identical to that proposed for the natural *Bombyx* EH. Furthermore, the linear EH (hexa-Acm EH) did not induce the eclosion behavior, suggesting that the disulfide bonds are important for exhibiting biological activity.

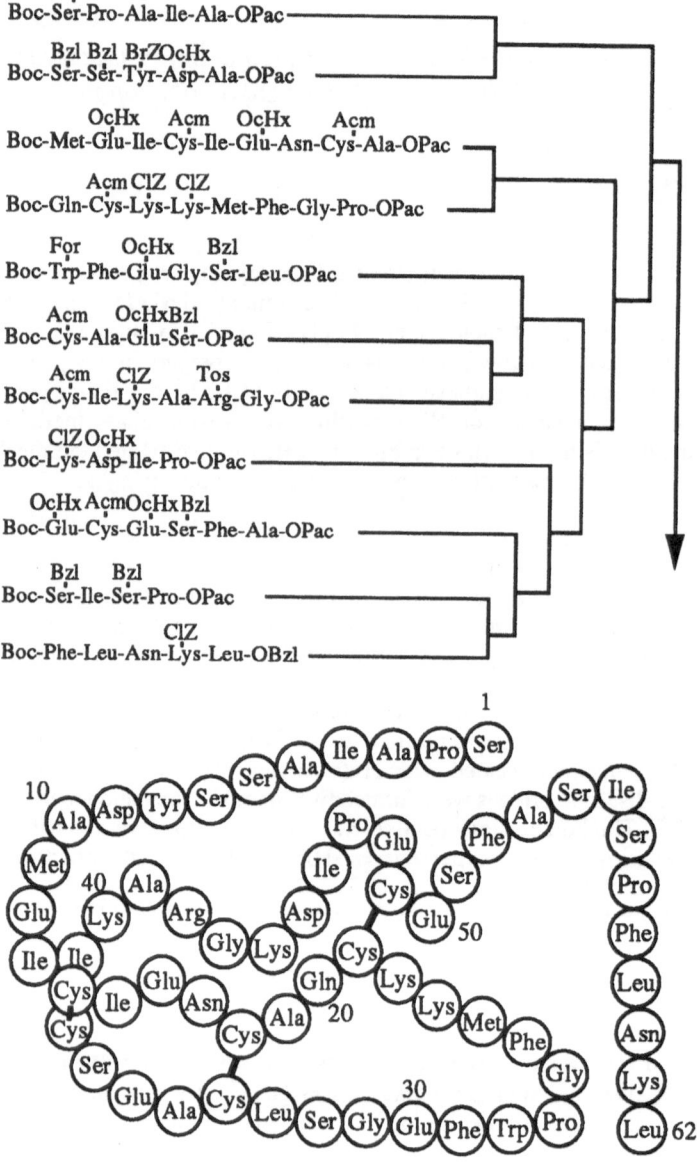

Fig. 1. Synthetic scheme and structure of Bombyx EH.

Using the synthetic EII, we studied the mechanism of EH mediated-signal transduction. Consequently, we found a new biochemical aspect, namely the EH induced-phosphatidylinositol hydrolysis (PtdIns hydrolysis), in the abdominal ganglia. Incubation of the ganglia from silkworm pharate adults with the EH led to an increase in the formation of inositol 1,4,5-trisphosphate (Figure 2b). The EH induced-PtdIns hydrolysis occurred in a dose-dependent manner and was abolished by phospholipase C inhibitors. This kind of EH response developed in parallel to the EH induced-eclosion behavior during the silkworm adult development. In consideration of previous results which show a crucial role of cGMP as a second messenger in the EH mediated-signal transduction [2], we hypothesized a cross talk between cGMP and inositol 1,4,5-trisphosphate in the EH signaling cascade.

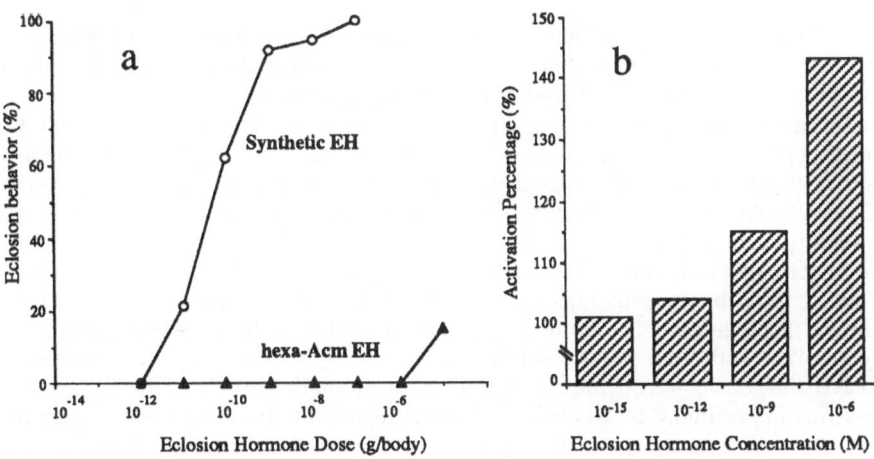

Fig. 2. Dose dependency of eclosion behavior (a) and IP_3 increase (b) by EH.

Acknowledgements

We are grateful to Ms. M. Tanaka for amino acid and sequence analyses, and Mr. A. Kusai of JEOL LTD. for mass analyses. We also thank Dr. A. James for reading the manuscript.

References

1. Kono, T., Nagasawa, H., Kataoka, H., Isogai, A., Fugo, H. and Suzuki, A., FEBS Lett., 263 (1990) 358.
2. Shibanaka, Y., Hayashi, H., Okada, N. and Fujita, N., Biochem. Biophys. Res. Commun., 80 (1991) 881.

Design and synthesis of highly water soluble LHRH antagonists

Z.-P. Tian, Y.-L. Zhang, P.J. Edwards and R.W. Roeske

Department of Biochemistry and Molecular Biology, Indiana University School of Medicine, IN 46202-5122, U.S.A.

Introduction

After the isolation of luteinizing hormone-releasing hormone (LHRH) from porcine hypothalamic extracts in 1971, more than 3500 LHRH antagonists have been synthesized with the goal of finding a better therapeutic agent for the treatment of endocrine-related diseases (e.g., prostate cancer and breast cancer) and for contraception. Significant progress has been made in increasing the anti-ovulatory activity (AOA) of LHRH antagonists, with the development of Nal-Glu [1], Antide [2] and Azaline B [3] as examples. But all these types of antagonists have low water solubility, and are typically administered to experimental rats by dissolving in corn oil or 1% aqueous DMSO. Some of the analogs have a very strong tendency to form a gel at the subcutaneous injection site, which discourages their further study for clinical application. Therefore it is highly desirable to have LHRH antagonists synthesized with high water solubility as well as high AOA. In fact a number of LHRH antagonists were synthesized bearing a hydrophilic O-(α-L-rhamnopyranosyl)-D-serine in position 6 by König et al. [4] and reported to have high water solubility and good AO activity. We report here our studies on the synthesis and characterization of LHRH antagonists with high water solubility and high anti-ovulatory activity.

Results and Discussion

Early structure-activity relationship studies have suggested that the side chain of amino acid residue 6 is not involved in receptor binding. Therefore this position was chosen as the modification site where a variety of highly hydrophilic moieties (carbohydrate, polyethylene glycol and polyhydroxy compounds) were attached through either the ε-amino group of D-lysine or the γ-carboxylic group of D-glutamic acid residue, with the common sequence shown below.

$$Ac\text{-}D\text{-}Nal\text{-}D\text{-}Cpa\text{-}D\text{-}Pal\text{-}Ser\text{-}Tyr\text{-}D\text{-}AA^6\text{-}Leu\text{-}Lys(N^\varepsilon\text{-}iPr)\text{-}Pro\text{-}D\text{-}Ala\text{-}NH_2}$$

This was achieved by a post-synthesis solid phase modification method in which Boc-D-Glu(Fm)-OH or Boc-D-Lys(Fmoc)-OH was initially incorporated during the peptide chain assembly (Boc/Bzl chemistry). Then the fluorenylmethoxy-based side chain protecting group was selectively removed using 50% piperidine in DMF and the newly released carboxyl or amino group was modified. Peptides were released from the MBHA resin by HF in the presence of anisole with simultaneous removal of the side chain protecting groups. Analogs were purified by preparative HPLC and characterized by amino acid analysis and plasma desorption mass spectrometry.

This synthetic strategy works very well for the synthesis of most of the analogs, but there are cases when this method does not work. It was found that DMSO was a useful solvent for solid phase coupling of polyhydroxy acids or amines when DCM, DMF, and NMP failed to dissolve the compounds. For the synthesis of the analog carrying a taurine moiety (1), this strategy failed because taurine was insoluble in organic solvents and could not be coupled to the resin-bound carboxyl group. This problem was circumvented by directly incorporating Boc-D-Glu(γ-NHCH$_2$CH$_2$SO$_3$H)-OH into the sequence. The taurine derivative was synthesized by reacting Boc-D-Glu(α-Bzl)-OH with taurine using the mixed anhydride method followed by hydrogenolysis to remove the α-benzyl protection. We had to sacrifice yield in the taurine coupling by mixed anhydride since we had no alternative except to use water to dissolve taurine in the reaction. In the case of coupling tris(hydroxymethyl)-aminomethane (Tris, 3) to the resin-bound carboxyl group where three hydroxyl groups are competing with one amino group for attachment, an ester linkage was formed along with the desired amide bond formation. But the ester could be easily converted to the desired amide by dissolving the crude peptide in aqueous acetonitrile with the presence of triethylamine (probably through a 5-membered intramolecular rearrangement). For the synthesis of analog 8 in which a carbohydrate moiety was attached to the D-lysine side chain, the hydroxyl groups were protected by a ketal when coupling to the resin-bound ε-amino group, and the ketal was removed at the acidic extraction step using 50% aqueous acetic acid.

Table 1 *Characteristics of LHRH antagonists with side-chain modified D-Glu^6(-X)*

Entry	X	RT (min)[a]	AOA[b]
1	-NHCH$_2$CH$_2$SO$_3$H	13.0	1/8 @ 5.0 μg
2	N(CH$_2$CH$_2$OH)$_2$	14.4	6/8 @ 1.0 μg
3	-NHC(CH$_2$OH)$_3$	13.7	2/8 @ 1.0 μg
4	-N⌒O	15.8	6/8 @ 1.0 μg
5	-O(CH$_2$CH$_2$O)$_7$CH$_3$	16.2	7/8 @ 2.0 μg
6	-NH(CH$_2$CH$_2$O)$_2$H	15.7	3/8 @ 2.0 μg

[a]RT is retention time recorded on HPLC in min. Column: Vydac C18 (250 mm x 4.6 mm), Linear gradient: 36%B to 42%B in 20 min (solvent A: 0.1% TFA in H$_2$O and solvent B: 0.09% TFA in 70% aq. MeCN).
[b]AOA is anti-ovulatory activity expressed by number of rats ovulated/number of rats tested at a dose given in micrograms per rat.

High performance liquid chromatography (HPLC) was used to evaluate the hydrophilicity of the LHRH antagonists synthesized (Tables 1 and 2). The retention time is well correlated with the structure. But from our study, it was found that water solubility of peptides could not be exactly expressed by its retention time (hydrophilicity) measured using a reversed phase (C18) HPLC column. For example,

5 bearing a polyethylene glycol modification has the highest water solubility, but its retention time is among the longest. 2 has a diethanolamine modification and 8 has a carbohydrate moiety. The two peptides have the same retention time, but 8 is much more soluble in water than 2. We also found that there was no solid evidence that the gel formation of LHRH antagonists could be related to its hydrophilicity.

Table 2 *Characteristics of LHRH antagonists with side-chain modified D-Lys^6(-Y)*

Entry	Y	RT (min)[a]	AOA[b]
7	-COCH$_2$CH$_2$COOH	15.5	9/10 @ 1.0 µg
8	-CO(C$_5$H$_9$O$_5$), gulonyl	14.4	4/8 @ 1.0 µg
9	-COCH$_2$OH	16.6	7/8 @ 1.0 µg
10	-COCH$_2$CN	16.6	8/8 @ 1.0 µg
11	-COCONH$_2$	16.5	4/8 @ 1.0 µg
12	-COCH$_3$	16.9	2/10 @ 1.0 µg

[a,b]See Table 1.

All the LHRH antagonists synthesized (Tables 1 and 2) were assayed for their anti-ovulatory activities (AOA). Generally, analogs with modified D-lysine in position 6 have higher activities than those with modified D-glutamic acid in position 6, except the one with D-glu(Tris)^6 (3). The activity differences between the two groups of antagonists may partially be attributed to the length difference of the side chains between D-lysine and D-glutamic acid. Analogs (entries 1 and 7) with a negatively charged residue in position 6 give low AO activities. Modification by Tris and gulonic acid give the most active antagonists, 75% and 50% inhibition of ovulation at a dose of 1 µg in rats.

Acknowledgements

This work was supported by Contract N01-HD-1-3102 from the Contraceptive Development Branch, NICHD.

References

1. Rivier, J.E., Porter, J., Rivier, C.L., Perrin, M., Corrigan, A., Hook, W.A., Siraganian, R.P. and Vale, W.W., J. Med. Chem., 29 (1986) 1846.
2. Ljungqvist, A., Feng, D.-M., Tang, P.-F. L., Kubota, M., Okamoto, T., Zhang, Y., Bowers, C.Y., Hook, W.A. and Folkers, K., Biochem. Biophys. Res. Commun., 148 (1987) 849.
3. Rivier, J., Porter, J., Hoeger, C., Theobald, P., Craig, A.G., Dykert, J., Corrigan, A., Perrin, M., Hook, W.A., Siraganian, R.P., Vale, W. and Rivier, C., J. Med. Chem., 35 (1992) 4270.
4. König, W., Sandow, J., Jerabek-Sandow, G. and Kolar, C., In Jung, G. and Bayer, E. (Eds.), Peptides 1988 (Proceedings of the 20th European Peptide Symposium), Walter de Gruyter & Co., Berlin, Germany, 1989, p.334.

Structure-activity relationships of LHRH antagonists: Incorporation of positively charged Nᵖʸ-alkylated 3-D-pyridylalanines

Y.L. Zhang, Z. Tian, M. Kowalczuk, P. Edwards and R.W. Roeske

Department of Biochemistry and Molecular Biology, Indiana University School of Medicine, Indianapolis, IN 46202-5122, U.S.A.

Introduction

In designing LHRH antagonistic decapeptides, a cluster of hydrophobic aromatic residues at the N-terminus combined with a positively charged residue in position 6 and a basic residue in position 8 is known to result in high antiovulatory (AO) potency [1]. However, the same structural feature can also, unexpectedly, trigger a strong tendency for the cells to release histamine as a side reaction in vivo which has already prevented a number of antagonistic analogs from their clinical application [2]. Among these analogs, the conventional choice for getting a positive charge in position 6 is the incorporation of a basic residue (D-Arg, D-Lys or Nᵋ-isopropyl-D-Lys) which can be protonated under physiological conditions. We report a method to retain the positive charge in position 6 but eliminate its basicity by an incorporation of Nᵖʸ-alkylated 3-(3'-pyridyl)-D-alanine which holds a permanent and relatively shielded positive charge as an alkyl pyridinium. A series of decapeptides containing different kinds of Nᵖʸ-alkyl-3-(3'-pyridyl)-D-Ala (or alkyl-D-Pal) in position 6 has been prepared to investigate the shielding effect related to the size of the alkyl group. Some structure-activity relationships of other designed methyl-Pal containing peptides are also discussed.

Results and Discussion

N^{α}-Boc-Nᵖʸ-alkylated 3-(3'-pyridyl)-D-alanines with alkyl groups of different sizes were prepared from Boc-3-(3'-pyridyl)-D-alanine via either direct alkylation with corresponding alkyl halide or a catalytic process carried out by using relevant alkyl iodide and silver(I) oxide as a catalyst [3]. Solid phase synthesis with Boc/Bzl protecting protocol and DCC/HOBt coupling strategy were employed in an ABI-431A automatic peptide synthesizer for the preparation of all the alkylPal-containing peptides.

The Nᵖʸ-alkyl-D-Pal⁶ containing antagonists, biological data of which are shown in Table 1, were designed in a general sequence of Ac-D-Nal-D-Cpa-D-Pal-Ser-Tyr-D-(Nᵖʸ-alkyl)Pal-Leu-(Nᵋ-iPr)Lys-Pro-D-Ala-NH₂ in order to basically satisfy the structural requirements for high antiovulatory activity based on the reported results [1, 4-6]. Evidently, this sequence has provided the peptides (entries 1-4) with a sufficient AO activity and, simultaneously, a reduction in histamine release (HR).

The shielding effect to the pyridinium positive charge caused by the N^{py}-alkyl group has been observed by a decrease of HR by some three-fold when it is changed from a less hindered carbon (methyl or benzyl) to a hindered carbon (isopropyl). Generally, the introduction of an alkyl pyridinium instead of a protonated amine in position 6 may reduce HR by about forty (entries 1 and 3) to a hundred-fold (entry 4) compared with 'Nal-Arg' (entry 5). It seems that the structural necessity in position 6 for high AO activity should be a positively charged side chain rather than a basic residue.

Table 1   *Biological data comparison among 'Nal-Arg' and N^{py}-alkyl-D-Pal^6 containing LHRH-antagonists in a general sequence of Ac-D-Nal-D-Cpa-D-Pal-Ser-Tyr-D-(N^{py}-alkyl)Pal-Leu-(N^{ε}-iPr)Lys-Pro-D-Ala-NH$_2$*

Entry	Alkyl group	Antiovulatory activity[a]	Histamine release ED_{50} (μg/ml)[b]
1	-CH$_3$	1/10 @ 1.0 μg	7.1
2	-CH$_2$CH$_2$CH$_2$CH$_3$	1/8 @ 1.0 μg	not done
3	-CH$_2$C$_6$H$_5$	0/8 @ 1.0 μg	6.0
4	-CH(CH$_3$)$_2$	0/10 @ 1.0 μg	17.0
5 ('Nal-Arg')		0/10 @ 1.0 μg	0.17

[a] Ovulated/total experimental rats at a dose in microgram.
[b] In vitro test using rat mast cells.

Table 2   *Characteristics of other N^{py}-methyl-Pal containing LHRH antagonists*

Entry	Other residues in a sequence of [Ac-D-Nal^1, D-Cpa^2, ... , D-Ala^{10}]LHRH	AO activity	HR ED_{50} (μg/ml)
6	D-Pal^3, MePal^5, D-Trp^6	1/10 @ 0.5 μg*	2.5
7	D-Pal^3, MePal^5, D-Trp^6, iPrLys^8	7/10 @ 0.5 μg	4.3
8	D-Pal^3, iPrLys^5, D-MePal^6, iPrLys^8	9/10 @ 0.5 μg	not done
9	D-Pal^3, MePal^5, D-iPrLys^6, iPrLys^8	9/10 @ 0.5 μg	7.9
10	D-Pal^3, D-MePal^6, (N^{ε}-triMe)Lys^8	5/10 @ 0.5 μg*	not done
11	D-Pal^3, MePal^5, D-Tyr^6, (N^{ε}-triMe)Lys^8	2/10 @ 1.0 μg	5.6

*0/10 @ 1.0 mg.

Some other MePal containing antagonists have also been synthesized for structural interests (data shown in Table 2). Entries 10 and 11 were designed for consolidating the knowledge of residue-transposition between positions 5 and 6 [4] and the influences based on a basicity elimination in position 8 (N^{ε}-trimethylLys was used instead of the traditional Arg or N^{ε}-iPrLys). The more hydrophobic residue D-Trp compared with D-Tyr in position 6 with a positively charged residue (MePal) in position 5 enhances AO activity but no reduction in histamine release is obtained (entry 6). There is a loss of AO activity by substituting Arg for iPrLys in position 8 with MePal[5] and D-Trp[6] (entries 6 and 7). Our data suggest a strong preference for a single positive charge in either position 5 or 6. The trials to pose two positive charges in positions 5 and 6 caused a big drop in AO activity (entries 8 and 9).

Abbreviations: AO, antiovulatory; Cpa, 3-(4'-chlorophenyl)alanyl; HR, histamine release; iPrLys, N^ε-isopropyllysyl; LHRH, luteinizing hormone releasing hormone; MePal, 3-(1'-methyl-3'-pyridyl)-D-alanyl; Nal, 3-(1'-naphthyl)alanyl; Pal, 3-(3'-pyridyl)-alanyl; py, pyridine.

Acknowledgements

We thank the National Institutes of Health, U.S.A. for financial support (Contract No. N01-HD-1-3102).

References

1. Coy, D.H., Horvath, A., Nekola, M.V., Coy, E.J., Erchegyi, J. and Schally, A.V., Endocrinology, 110 (1982) 1445.
2. Rivier, J., Rivier, C., Perrin, M., Porter, J. and Vale, W.W., In Vickery, B.H., Nestor, Jr., J.J. and Hafez, E.S.E. (Eds.), LHRH and Its Analogs, MTP Press, Lancaster, U.K., 1984, p.11.
3. Zhang, Y.L., Tian, Z., Kowalczuk, M., Edwards, P. and Roeske, R.W., Tetrahedron Lett., 34 (1993) 3659.
4. Roeske, R.W., Chaturvedi, N.C., Rivier, J., Vale, W., Porter, J. and Perrin, M., In Deber, C.M., Hruby, V.J. and Kopple, K.D. (Eds.), Peptides: Structure and Function, Proceedings of the Ninth American Peptide Symposium, Pierce Chemical Co., Rockford, IL, U.S.A., 1985, p.561.
5. Roeske, R.W., Chaturvedi, N.C., Hrinyo-Pavlina, T., Kowalczuk, M., In Vickery, B.H., Nestor, Jr., J.J. and Hafez, E.S.E. (Eds.), LHRH and Its Analogs, Part 2, MTP Press, Lancaster, U.K., 1987, p.17.
6. Tian, Z.P., Zhang, Y.L., Kowalczuk, M., Hrinyo-Pavlina, T., Edwards, P. and Roeske, R., In Du, Y.C., Tam, J.P. and Zhang, Y.S. (Eds.), Peptides: Biology and Chemistry, (Proceedings of the 1992 Chinese Peptide Symposium), Escom, Leiden, The Netherlands, 1993, p.45.

Session VII

Peptide Inhibitors/
Peptide-Receptor Interactions

Chairs: Irwin M. Chaiken

SmithKline Beecham
King of Prussia, Pennsylvania, U.S.A.

and

Arno F. Spatola

University of Louisville
Louisville, Kentucky, U.S.A.

Characterization of peptide receptors

M.J. Brownstein

Laboratory of Cell Biology, National Institute of Mental Health,
Bethesda, MD 20892, U.S.A.

In the last two decades the number of biologically active peptides isolated and sequenced has increased dramatically. Synthetic peptides, their derivatives, or non-peptide agonists and antagonists have been used for radioreceptor assays and physiological or pharmacological studies. These studies have provided convincing evidence that actions of the peptides were mediated by receptors located on the plasma membranes of target cells. Broadly speaking there are three types of peptide receptors: 1) ligand-gated tyrosine kinases (e.g., the receptors for insulin or nerve growth factor); 2) ligand-gated guanylyl-cyclases [1] (e.g., the atrial natriuretic factor receptor); and 3) G-protein coupled receptors. The first two types of receptors above have single membrane spanning segments; the latter type has multiple membrane spanning segments.

There are three families of receptors that operate through trimeric G-proteins. (For a description of G-protein diversity, see reference 2.) Two of these families are relatively small: the VIP/secretion family [3] and the metabotropic glutamate family [4] of receptors. The third family of G-protein coupled receptors - all of which are structurally related to opsins - is very large [5]. To date no peptide-gated ion channels analogous to the nicotinic cholinergic, gamma-aminobutyric acid (GABA), or N-methyl-D-aspartate (NMDA) receptors have been discovered, but sooner or later such a receptor may be found.

I shall devote the remainder of this short review to a description of the neurohypophysial hormone receptors, because this will allow me to make some remarks about the properties of G-protein coupled receptors. Mammals typically have two posterior pituitary hormones: Lysine- or arginine-vasopressin and oxytocin. (Thirteen related peptide hormones are found in other species [6].) These hormones are made by brain cells and released into the bloodstream to act on peripheral organs. In addition they are used by neurons in the central nervous system as neurotransmitters and by certain peripheral cells as "paracrine" mediators. The actions of vasopressin and oxytocin are mediated by receptors which are members of the opsin receptor superfamily.

Mammals have three types of vasopressin receptors (V1a, V1b and V2) and at least one type of oxytocin receptor. Complementary DNAs encoding each of these have been isolated [7-11]. The V1a receptor is responsible for vasopressin's pressor (i.e., vasoconstricting) effect. In the rat large numbers of V1a receptors are found on hepatocytes and cause glycogenolysis. V1a receptors are also found in several areas of the central nervous system. The V1b receptor is present on ACTH-producing cells in the anterior pituitary. Along with CRF, vasopressin elicits ACTH secretion. Oxytocin, released in response to suckling, causes milk let-down in the breast. It is potent in contracting uterine smooth muscle as well, and is thought to be important in parturition. Activation of V1a, V1b, or oxytocin receptors triggers a cascade of events culminating in the mobilization of intracellular calcium by inositol 1,4,5-trisphosphate and the activation of protein kinase C (PKC) by diacylglycerol. Activation of intracellular kinases such as PKC results in phosphorylation of diverse proteins including some that affect gene transcription [12]. Thus, calcium-mobilizing receptors have both short- and long-term effects on cells.

Stimulation of the V2 receptor activates adenylate cyclase resulting in an increase in intracellular cyclic AMP. In cells that comprise the distal collecting ducts of the kidney the increased cyclic AMP opens a pore resulting in greater reabsorption of water. This effect accounts for vasopressin's other name, antidiuretic hormone. People who inherit a defective gene for the V2 receptor cannot concentrate their urine. They have to drink large volumes of water to make up for this handicap, and they have a high risk of becoming dehydrated should they be unable to obtain or drink enough fluid. Most, if not all, of the mutations in the V2 receptor gene discovered thus far have seriously flawed the receptor [13-16]. One can imagine, however, a mutation that alters agonist preference. It is possible that a "new" agonist could be designed for such a receptor and used to treat patients.

While the primary structures of many receptors are now known, their conformations are not. Bacterial rhodopsin has served as a model for these receptors [17]. This molecule is activated by photoisomerization of a covalently bound retinal molecule just as mammalian opsins are, but its amino acid sequence is not at all similar to that of the opsins. Furthermore, it does not interact with a G-protein. Therefore, one might question its usefulness as a structural template for the opsin related receptors. It has served as a useful first approximation however. From its two-dimensional crystal structure people have inferred that the G-protein coupled receptors should have seven membrane-spanning α-helical segments. In the case of the vasopressin receptors, the last five transmembrane domains have proline and/or glycine residues which interrupt the helices in which they are found. Viewed from above the membrane spanning domains are tightly packed in a circle or crude spiral and form a central cleft that includes the ligand binding domain. In the cases of the ligands examined to date, residues on several of the helices participate in binding agonists. (Antagonist compounds, especially organic-chemical peptide antagonists, seem to bind to a site that overlaps but is not identical with the agonist site).

The receptors should be viewed as highly dynamic structures. α-helical domains oscillate on either side of "hinge" (proline and glycine) residues. Receptor occupancy results in an active metastable conformation being favored over an inactive one. How this, in turn, causes the G-protein subunits to dissociate from one another is still a matter of speculation. It is evident, though, that the cytoplasmic loops are the sites of interaction of receptors with their G-proteins. In particular, the third cytoplasmic loop seems to play an especially important role in defining which G-protein α-subunit the receptors will bind. Not unexpectedly, the third cytoplasmic loops of the V1a, V1b, and oxytocin receptors are quite similar. The corresponding region of the V2 receptor is structurally different.

Cell lines that express cloned receptors - especially human receptors - have already begun to be used by scientists in pharmaceutical companies to identify novel peptide agonists and antagonists. Design of non-peptide ligands remains rather empirical. One hopes that a better understanding of the relationship between the structure and function of receptors will make this process a more rational one.

References

1. Garbers, D.L., New Biologist, 2 (1990) 499.
2. Bourne, H., Sanders, D. and McCormick, F., Nature, 349 (1991) 117.
3. Schoepp, D.D. and Conn, P.J., Trends Pharmacol. Sci., 14 (1993) 13.
4. Masu, M., Tanabe, Y., Tsuchida, K., Shigemoto, R. and Nakanishi, S., Nature, 349 (1991) 760.
5. Seeman, P., Receptor Tables Vol. 1 Receptor Amino Acid Sequences of G-linked Receptors, University of Toronto, Toronto, Ontario, 1992.
6. Acher, R., Reg. Peptides, 45 (1993) 1.
7. Morel, A., O'Carroll, A.-M., Brownstein, M.J. and Lolait, S.J., Nature, 356 (1992) 523.
8. Kimura, T., Tanizawa, O., Mori, K., Brownstein, M.J. and Okayama, H., Nature, 356 (1992) 526.
9. Lolait, S.J., O'Carroll, A.-M., McBride, O.W., Konig, M., Morel, A. and Brownstein, M.J., Nature, 357 (1992) 336.
10. Birnbaumer, M., Seibold, A., Gilbert, S., Ishido, M., Barbaris, C., Antaramian, A., Brabet, P. and Rosenthal, W., Nature, 357 (1992) 333.
11. Lolait, S.J., O'Carrol, A.-M. and Brownstein, M.J., unpublished data.
12. Kennedy, M., Trends Neurosci., 12 (1989) 417.
13. Merendino, J.J., Spiegel, A.M., Crawford, J.D., O'Carrol, A.-M., Brownstein, M.J. and Lolait, S.J., N. Engl. J. Med., 328 (1993) 1538.
14. Par, Y., Metzenberg, A., Das, S., Jing, B. and Gitschier, J., Nat. Genet., 2 (1992) 103.
15. Rosenthal, W., Seibold, A., Antaramian, A., Lonergan, M., Arthus, M.F., Hendy, G.N., Birmbaumer, M. and Bichet, D.G., Nature, 359 (1992) 233.
16. van den Ouweland, A.M.W., Dreesen, J.C.F.M., Verdijk, M., Knowers, N.V.A.M., Monnens, L.A.H., Rocchi, M., van Oost, B.A., Nat. Genet., 2 (1992) 99.
17. Trumpp-Kallmeyer, S., Hoflack, J. and Bruinvels, A., Trends Pharmacol. Sci., 14 (1993) 7.

Analysis of the interaction between insulin and its receptor by means of covalent and non-covalent techniques

D. Brandenburg[1], W. Thevis[1], D. Glasmacher[1],
J. Pirrwitz[1], M. Fabry[1], E. Schaefer[2] and L. Ellis[2]

[1]Deutsches Wollforschungsinstitut, D-52062 Aachen, Germany,
[2]Institute of Biosciences and Technology, Texas A & M University,
Houston, TX 77030, U.S.A.

Introduction

Detailed information on the interaction of ligands and receptors is a prerequisite for understanding the molecular mechanisms of hormone action. While peptide chemistry has provided many suitable analogues and derivatives of hormones, access to receptors has been limited. Fortunately, the situation is changing. Soluble insulin receptor ectodomain (ED), $\alpha-\beta'-\beta'-\alpha$, which contains the complete α-subunit and truncated extracellular β-subunit [1], has already become available in milligram quantity. Hence, photoaffinity labelling can be carried out on a new level, and physical studies of the hormone-receptor complex under reversible conditions become feasible. With ED, we could recently identify a second insulin binding domain of the receptor [2]. We now report on two novel photoreactive insulins as well as spin-labelled analogues [3] and their application to probe the ED under irreversible and reversible conditions.

Results and Discussion

Photoaffinity labelling. Since aryl azides, which are generally used to label peptide hormones [4], give only moderate incorporation into receptors, we are currently testing perfluorinated aryl azides [3] and recently 4-benzoylphenylalanine (Bpa), for which encouraging labelling of the complete insulin-receptor has been reported [5]. The photoreactive analogues were prepared by enzyme-catalyzed incorporation of Bpa-amide into position B30 of insulin (**P1**) and B26 of des-(B27-B30)-insulin (**P2**). They are homogeneous in RP-HPLC and exhibit binding affinities to IM-9 lymphocyte receptors of 60% (**P1**) and 47% (**P2**). Both labelled ED specifically. UV-induced incorporation of **P1** into ED yielded up to 35% of the covalent complex. Because the photo-activatable group is located very closely to the postulated binding region of the insulin receptor [6], **P2** appears to be an active site label useful for binding site analysis.

Spin labelling. Electron spin resonance (ESR) spectroscopy directly detects the presence of free radicals and gives information on the structure and motional dynamics at the site of one or more unpaired electrons [7]. Stable organic nitroxide radicals, which are used as reporter groups for biomolecules, give a spectrum

showing 3 sharp, well resolved hyperfine lines. It is very sensitive to
immobilization, which results in characteristic line-broadening. The powerful
method has so far found little application with peptide hormones. The interaction of
spin-labelled EGF and its receptor has recently been studied [8].

    We have prepared two spin labelled insulins with the 2,2,5,5-tetramethyl-3-
carboxy-pyrrolidine-1-oxyl residue in position B1 or B29 (Figure 1 **B1, B29**) via N-
protected insulin intermediates. The shortened analogue **B25**, lacking amino acids
B26-B30, contains a C-terminal 2,2,6,6-tetramethyl-4-amino-piperidine-1-oxyl
residue. It was obtained by trypsin-catalyzed semisynthesis. The ESR spectra are
shown in Figure 1. Characteristic differences in the high-field bands and different
rotational correlations times τ_R indicate that the spin label is subject to partial
immobilization, which is most pronounced for **B1**, intermediate for **B25**, and least
for **B29**.

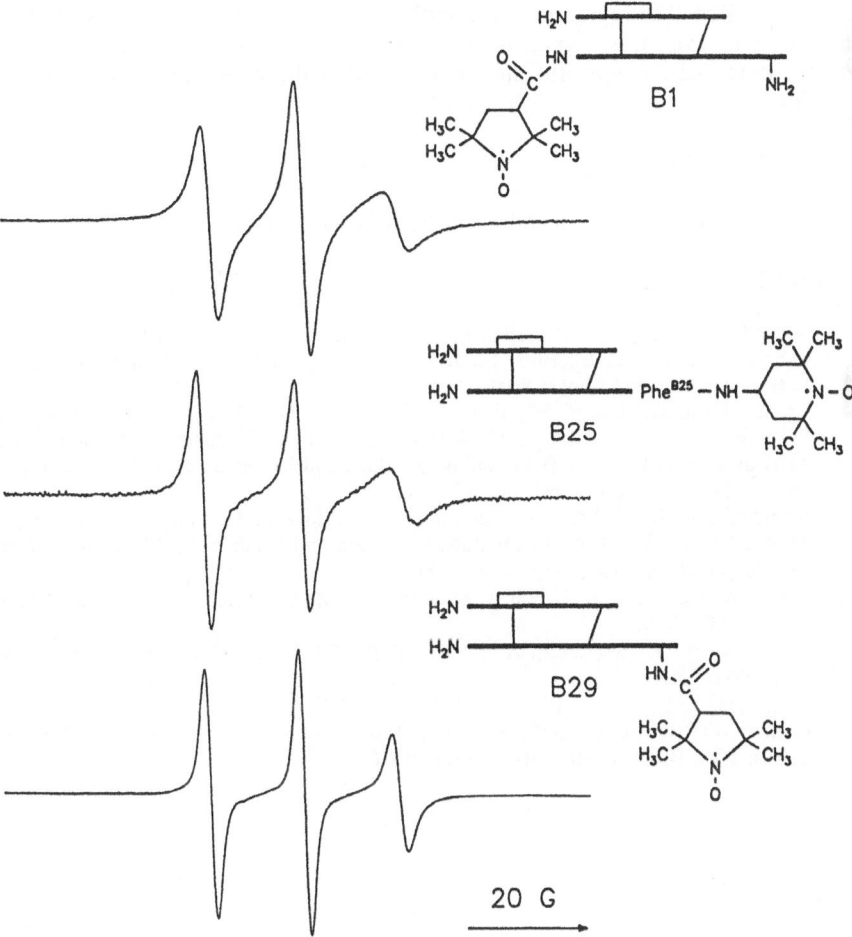

Fig. 1. Structures and ESR spectra of spin labelled insulins.

The interaction of two spin labelled insulins, **B1** and **B25**, and ED was studied by ESR spectroscopy at pH 7.8 and concentrations of 10^{-5} M ligand in a series of preliminary experiments. The spectral changes observed were different for the two insulins. The relative rotational correlation times, corrected for the increase of molecular mass due to binding to the receptor (MW approx. 350,000), was not significantly different for free and receptor-bound **B1**. Thus, if any, there appears to be only little involvement of position B1 in receptor contacts. In contrast, the corresponding values for **B25** were 3.5 ns for free, and 5.2 ns for bound insulin. This reflects a specific immobilization of the label and thus indicates a more intensive receptor interaction at this site. The signals observed were reversible, i.e. increasing amounts of non-labelled insulin abolished the spectral differences and regenerated the spectrum of free **B25**.

Although as yet preliminary, the results show that spin-labelling and ESR spectroscopy can yield valuable information on insulin binding to its receptor. The specific immobilization with **B25**, but not **B1**, would be in line with the hypothesis [6] that the B24-B25 region interacts with the receptor. This is confirmed by the photoaffinity labelling experiments. Our ESR findings are a first direct demonstration of site-specific interactions within the reversible insulin/receptor complex.

Acknowledgements

We thank C. Diaconescu and J. Tenelsen for bioassays.

References

1.   Schaefer, E., Siddle, K. and Ellis, L., J. Biol. Chem., 165 (1992) 19288.
2.   Fabry, M., Schaefer, E., Ellis, L., Kojro, E., Fahrenholz, F. and Brandenburg, D., J. Biol. Chem., 267 (1992) 8950.
3.   Brandenburg, D., Fabry, M., Pirrwitz, J., Strack, U., Thevis, W., Wittkamp, A., Schaefer, E., Ellis, L., Keana, J.F.W. and Cai, S.X., In Schneider, C.H. and Eberle, A. (Eds.), Peptides 1992 (Proceedings of the 22nd European Peptide Symposium), Escom, Leiden, The Netherlands, 1993, p.111.
4.   Brandenburg, D., Fabry, M., Schumacher, F., Strack, U. and Wedekind, F., In Tschesche, H. (Ed.), Modern Methods in Protein and Nucleic Acid Research, Walter de Gruyter, Berlin, Germany, 1990, p.305.
5.   Shoelson, S.E., Lee, J., Lynch, C.S., Backer, J.M. and Pilch, P.F., J. Biol. Chem., 268 (1993) 4085.
6.   Murray-Rust, J., McLeod, A.N., Blundell, T.L. and Wood, S.P., BioEssays, 14 (1992) 325.
7.   Millhauser, G.L., Trends Biochem. Sci., 17 (1992) 448.
8.   Faulkner-O'Brien, L.A., Beth, A.H., Papayannopoulos, I.A., Anjaneyulu, P.S.R. and Staros, J.V., Biochemistry, 30 (1991) 8976.

Regulation of signal transduction pathways by peptide toxins

C.F.B. Holmes[1], M. Craig[1], T.L. McCready[1],
M.P. Boland[1], J.F. Dawson[1], D.Z.X. Chen[1],
K. Wang[1], H. Klix[1], H.A. Luu[1], J. Magoon[2],
M. O'Connor-McCourt[2] and R.J. Andersen[3]

[1]Medical Research Council Group in Protein Structure and Function,
Department of Biochemistry, University of Alberta, Edmonton, Alberta,
Canada, and [2]Biotechnology Research Institute, National Research
Council, Montreal, Quebec, Canada, and [3]Department of Chemistry,
University of British Columbia, Vancouver, British Columbia, Canada

Introduction

Reversible protein phosphorylation on serine, threonine and tyrosine by protein
kinases and phosphatases is widely accepted as a principal mechanism by which
eukaryotic cells respond to extracellular signals [1]. Several aquatic compounds,
including okadaic acid (OA) polyether fatty acids and microcystin/nodularin cyclic
peptide hepatotoxins, have been identified that are potent and specific inhibitors of
the catalytic subunits of type-1 and -2A protein phosphatases (PP-1c and PP-2Ac),
two of the major serine/threonine PPases in eukaryotes [2-5]. OA and the
microcystins are powerful tumor promoters, although the molecular mechanism
underlying this effect is unknown [6, 7]. The chemical feature that characterises the
heptapeptide microcystins and pentapeptide nodularins is the presence of an unusual
C_{20} β-amino acid Adda ([2S, 3S, 8S, 9S]-3-amino-9-methoxy-2,6,8-trimethyl-10-
phenyldeca-4,6-dienoic acid) [8]. Most microcystins differ in the nature of two
variable L-amino acids indicated by suffix letters (i.e., L=leucine, R=arginine) and in
the absence of methyl goups on D-erythro-β-methyl aspartic acid (D-MeAsp) and/or
N-methyldehydroalanine (Mdha) residues [9].

Results and Discussion

A sensitive liquid chromatography (LC)/capillary electrophoresis (CE)-linked
PPase bioassay was used to identify and isolate six new microcystin-like peptides
from freshwater cyanobacteria and marine sponges (Table 1) [10-12]. All of these
peptides are potent inhibitors of PP-1c/PP-2Ac (IC_{50} = 0.06-0.4 nM) and as
hydrophobic as OA, based on reversed-phase LC elution times. The predominant
protein hyperphosphorylated in rat-1 fibroblasts in response to microcystin-LR was
identified as a 23 kDa phosphoseryl protein (PP23). PP23 was largely unaffected by
OA (a 100-fold less potent PP-1c inhibitor than microcystin-LR) but was strongly
hyperphosphorylated in these cells in response to inhibitor-1 peptide (a specific PP-
1c inhibitor) and epidermal growth factor (EGF, Figure 1). These data indicate that
PP23 is a potential substrate for PP-1c *in vivo* and may play a role in EGF-
stimulated signal transduction pathways.

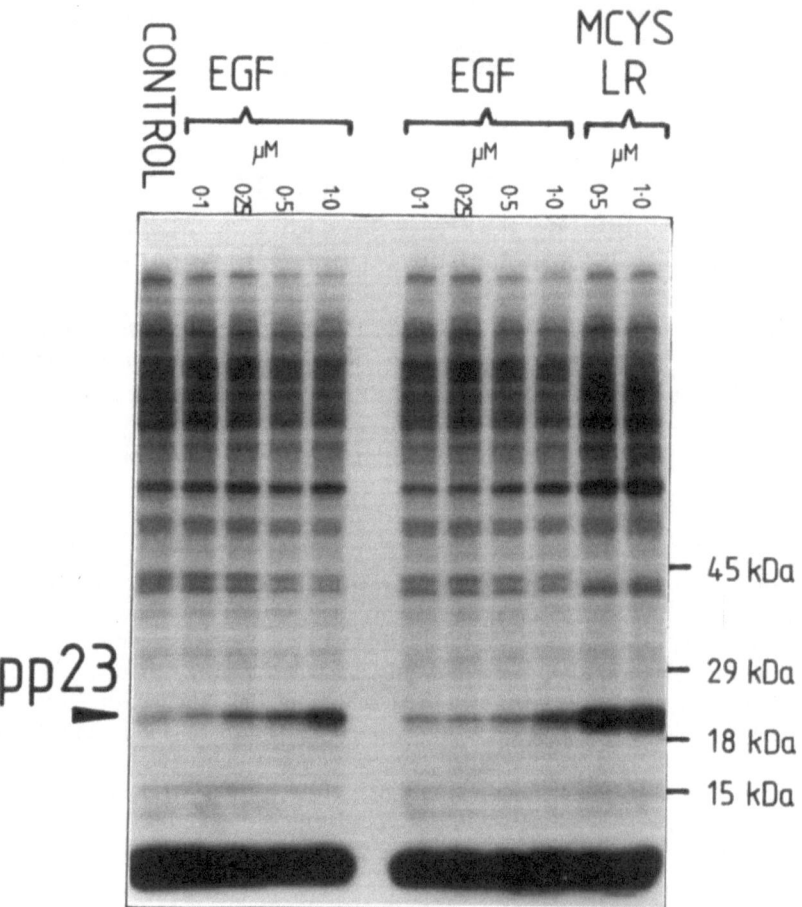

Fig. 1. Identification of a 23 kDa phosphoseryl protein as the major protein hyperphosphorylated in digitonin-permeabilised rat-1 fibroblasts in response to epidermal growth factor (EGF) and microcystin-LR (MCYS-LR). Phosphorylation and treatment of cells with EGF and MCYS-LR was carried out using a modified protocol from Eicholtz et al. [14]. Rat-1 fibroblasts were washed in permeabilisation buffer (PB) comprising: KCl (120 mM), NaCl (30 mM), $MgCl_2$ (1 mM), KH_2PO_4 (1 mM), sodium PIPES (10 mM, pH 7.4), EDTA (1 mM), $CaCl_2$ (0.037 mM). To start phosphorylation assays, fibroblasts were incubated with PB containing: digitonin (Sigma, 40 µM), γ-^{32}P-ATP (Amersham, 100 µCi per gel lane) and protease inhibitor cocktail (PMSF, 1 mM; leupeptin, 20 µg/ml; aprotinin, 20 µg/ml; trypsin inhibitor, 20 µg/ml, benzamidine, 20 mM). Peptides/toxins were dissolved in dimethyl formamide:PB (1:20) and incubated with rat-1 fibroblasts for 15 min at 25°C. Treated cells were subjected to SDS-PAGE (4-12% gradient gels autoradiographed using Kodak X-OMAT film) to reveal ^{32}P-radiolabelled proteins. Lane 1 is a control without toxin/EGF, lanes 2-5 and 6-9 represent treatment with increasing concentrations of two different preparations of EGF (Collaborative Research), lanes 10 and 11 represent treatment with MCYS-LR. Extraction and partial amino acid analysis of hyperphosphorylated gel proteins was carried out as described in Haystead et al. [13].

Table 1 *Structures of novel hydrophobic cyclic peptide toxins*

Peptide	Structure	IC_{50} [nM]vs PP-1c
Microcystin-LV	cyclo(D-Ala-L-Leu-D-MeAsp-L-Val-Adda-D-Glu-Mdha)	0.3
Microcystin-LM	cyclo(D-Ala-L-Leu-D-MeAsp-L-Met-Adda-D-Glu-Mdha)	0.1
Microcystin-LL	cyclo(D-Ala-L-Leu-D-MeAsp-L-Leu-Adda-D-Glu-Mdha)	0.1
Microcystin-LF	cyclo(D-Ala-L-Leu-D-MeAsp-L-Phe-Adda-D-Glu-Mdha)	0.4
Nodularin-V	cyclo(D-MeAsp-L-Val-Adda-D-Glu-Mdha)	0.06
Nodularin-I	cyclo(D-MeAsp-L-Ile-Adda-D-Glu-Mdha)	<0.1

Using a combination of protein purification and PCR cloning, PP-1 was also identified as the predominant PPase in the OA-producing marine dinoflagellate *Prorocentrum lima* and overexpression of this enzyme may allow *P. lima* to counterbalance the effects of a powerful tumor promoter (OA). We recently established for the first time that the microcystins are prevalent throughout the marine environment. Increasing evidence suggests that many cyanobacteria and dinoflagellates combine to produce PPase inhibitors with heterogeneous potencies, effective against signal transduction pathways in higher eukaryotes. The study of protein kinases and PPases in marine eukaryotes will be facilitated by a novel non-radioactive CE-based assay which allows resolution of phospho- and dephosphopeptides in crude tissue extracts. It will now be of interest to examine the effects of microcystins on PP-1 activity in marine eukaryotes and to identify PP23 and its associated protein kinase(s).

References

1. Hubbard, M.J. and Cohen, P., Trends Biochem. Sci., 18 (1993) 172.
2. Cohen, P., Holmes, C.F.B. and Tsukitani, Y., Trends Bio. Sci., 15 (1990) 98.
3. Bialojan, C. and Takai, A., Biochem. J., 256 (1988) 283.
4. MacKintosh, C., Beattie, K., Klumpp, S., Cohen, P. and Codd, G., FEBS Lett., 264 (1990) 187.
5. Yoshizawa, S., Matsushima, R., Watanabe, M.F., Harada, K.I., Ichihara, A., Carmichael, W.W. and Fujiki, H., J. Cancer Res. Clin. Oncol., 116 (1990) 609.
6. Nishiwaki-Matsushima, R., Ohta, T., Nishiwaki, S., Suganuma, M., Kohyama, K., Ishikawa, T., Carmichael, W.W. and Fujiki, H., J. Cancer Res. Clin. Oncol., 118 (1992) 420.
7. Suganuma, M., Fujiki, H., Suguri, H., Yoshizawa, S., Hirota, M., Nakayasu, M., Ojika, M., Wakamatsu, K., Yamada, K. and Sugimura, T., Proc. Natl. Acad. Sci., U.S.A., 85 (1988) 1768.
8. Rinehart, K.L., Harada, K.I., Namikoshi, M., Chen, C., Harvis, C.A., Munro, M.H.G., Blunt, J.W., Mulligan, P.E., Beasley, V.R., Dahlem, A.M. and Carmichael, W.W., J. Am. Chem. Soc., 110 (1988) 8557.
9. Carmichael, W.W., J. Applied Bacteriol., 72 (1992) 445.
10. Boland, M.P., Smillie, M.A., Chen, D.Z.X. and Holmes, C.F.B., Toxicon, 31 (1993) 1393.
11. Craig, M., McCready, T.L., Luu, H.A., Smillie, M.A., Dubord, P. and Holmes, C.F.B., Toxicon, 31 (1993) 1541.
12. DeSilva, E.D., Williams, D.E., Andersen, R.J., Klix, H., Holmes, C.F.B. and Allen, T.M., Tetrahedron Lett., 33 (1992) 1561.
13. Haystead, T.A.J., Weil, J.E., Litchfield, D.W., Tsukitani, Y., Fischer, E.H. and Krebs, E.G., J. Biol. Chem., 265 (1990) 16571.
14. Eicholtz, T., Alblas, J., Van Overveld, M., Hoolenaar, W. and Ploegh, H., FEBS Lett., 261 (1990) 147.

Design of potent and selective herpes simplex virus ribonucleotide reductase inhibitors that mimic the C-terminus of the enzyme's small subunit

N. Moss, R. Déziel, P. Beaulieu, A.-M. Bonneau, R.L. Krogsrud, M. Liuzzi, R. Plante and Y. Guindon

Bio-Méga Boehringer Ingelheim Research Inc., 2100 Cunard Street, Laval, Québec, H7S 2G5, Canada

Introduction

Herpes simplex virus (HSV)-encoded ribonucleotide reductase (RR) catalyzes the conversion of ribonucleoside diphosphates into the corresponding 2'-deoxy derivatives required for viral DNA synthesis. HSV-RR is comprised of two distinct homodimeric subunits, the association of which is required for enzymatic activity [1]. The C-terminus of the small subunit (R2) is known to play a critical role in this association, and the nonapeptide H-Tyr-Ala-Gly-Ala-Val-Val-Asn-Asp-Leu-OH, which corresponds to the nine C-terminal amino acids of the small subunit, inhibits HSV RR with an IC_{50} of 38 μM [2, 3]. This nonapeptide was selective for HSV-RR over mammalian RR but failed to inhibit HSV growth in tissue culture. We have conducted an extensive structure activity investigation using the nonapeptide as a lead in the hope of identifying inhibitors more potent against HSV-RR and also effective at preventing viral replication inside the cell.

Results and Discussion

The relative potency of the inhibitors is evaluated in an enzyme assay [4], which directly measures the ability of the compounds to inhibit the conversion of radiolabeled CDP to dCDP, and a solid phase competition binding assay [5], which measures the ability of the compounds to compete with a radiolabeled tracer (inhibitor) for binding to immobilized HSV-RR large subunit (R1). The binding assay has the advantage of being sensitive for compounds having IC_{50}s greater than 0.001 μM while the enzyme assay is limited to compounds having IC_{50}s greater than 0.1 μM. The binding assay proved crucial for the ranking of the potent inhibitors shown in Table 1.

The four N-terminal amino acids of the nonapeptide are not optimal for inhibitory potency as they can be replaced effectively with a 3-phenylpropionyl group. The resulting inhibitor (compound 1) is five times more potent than the nonapeptide (IC_{50} = 25 μM in our binding assay). One of the most significant increases in inhibitor potency is obtained by alkylating the asparagine side chain nitrogen (cf. compound 1 and 2). The best substitution we have found so far, i.e., replacing the NH_2 with a pyrrolidine, boosts activity more than a hundred times. Compound 2 is shown as a point of reference to illustrate the effect of individually incorporating what we have so far found to be the best substitutions at each of the other four amino acid positions (see compounds 3 to 7).

Table 1 *Ribonucleotide reductase inhibitors - structure activity*

Compound		IC_{50} Binding Assay (µM)	EC_{50} Cell Culture Assay	
			HSV-2 (µM)	HSV-1 (µM)
1		5		
2		0.034		
3		0.010		
4		0.005		
5		0.002		
6		0.006		
7		0.043		
BI-LD-598		0.0002	120	120
BI-LD-733		0.0004	50	35

Although compounds 2 to 7 all have low nanomolar potencies, none were effective at inhibiting viral replication in tissue culture at concentrations below 1000 μM. However, when all of the optimized modifications were combined into one molecule (BI-LD-598), we obtained the first subunit association inhibitor to have efficacy in tissue culture. BI-LD-598, which has subnanomolar activity in the binding assay, inhibits the growth of both HSV-1 and HSV-2 in serum starved BHK cells (EC$_{50}$ = 120 μM). Tissue culture efficacy can be improved by reducing the C-terminal carboxyl to an alcohol (cf. BI-LD-598 and 733). Although this modification lowers inhibitor potency in the binding assay (cf. compound 6 and 7), it seems reasonable that the decreased hydrophilicity of BI-LD-733 favors cell penetration and consequently tissue culture efficacy.

In order to provide support that HSV RR is inhibited inside the cell, we developed an assay to measure the ability of our compounds to lower the pools of deoxyribonucleotides in virally infected serum starved BHK cells [6]. Of the compounds shown in the table, only BI-LD-598 and 733 effectively lower the deoxyribonucleotide pools. These compounds reduce the concentration of dATP, dCTP, and dGTP by 50% at a concentration of ~32 μM and ~7 μM respectively. Antiviral activity is thus associated with a reduction of deoxyribonucleotides inside the cell.

In conclusion, compounds BI-LD-598 and 733 are greater than 50,000 times more potent than the lead nonapeptide and more than 100 times more effective at binding to the large subunit of HSV-RR than the natural ligand, the small subunit. These compounds inhibit viral replication in tissue culture and in addition do not inhibit human ribonucleotide reductase at concentrations up to 1000 μM. We believe that the discovery of these compounds constitutes an important step towards the development of novel and selective antiherpetic drugs.

References

1.   Ingemarson, R. and Lankinen, H.J., Virology, 156 (1987) 417.
2.   Dutia, B.M., Frame, M.C., Subak-Sharpe, J.H., Clarke, W.N. and Marsden, H.S., Nature, 321 (1986) 439.
3.   Cohen, E.A., Gaudreau, P., Brazeau, P. and Langelier, Y., Nature, 321 (1986) 441.
4.   Gaudreau, P., Paradis, H., Langelier, Y. and Brazeau, P., J. Med. Chem., 33 (1990) 723.
5.   Krogsrud, R.L., Welchner, E., Scouten, E. and Liuzzi, M., Anal. Biochem., in press.
6.   Garrett, C. and Santi, D.V., Anal. Biochem., 99 (1979) 327.

Potent, selective inhibitors of human gelatinase and their potential in the treatment of tumour metastases

J.R. Morphy[1], T.S. Baker[1], N.R.A. Beeley[1], M. Birch[1],
B.A. Boyce[1], S. Chander[1], M. Cockett[1], T. Crabbe[1],
A.J.P. Docherty[1], D. Eaton[1], I. Hart[2], P. Hynds[1],
J. Leonard[1], B. Mason[1], A. Mountain[1], T.A. Millican[1],
J. O'Connell[1], J. Porter[1], S. Tickle[1],
F.W. Willenbrock[1] and N. Willmott[1]

*[1]Division of Chemistry and Division of Biology, Celltech Ltd.,
216 Bath Road, Slough, SL1 4EN, U.K., and [2]CRF, P.O. Box 123,
Lincolns Inn Fields, London, WC2A 3PX, U.K.*

Introduction

Gelatinase-A and gelatinase-B are members of the matrix metalloproteinase family of enzymes [1] which includes the collagenases, stromelysins and matrilysin. We and others [2] have been able to show by in situ hybridisation experiments using radiolabelled oligonucleotide probes specific for gelatinase mRNA that the gelatinase gene is turned on in the vicinity of many tumour tissues. When sections of tissue surgically removed from human breast cancer patients are subjected to immunohistochemistry using a gelatinase-A specific monoclonal antibody the staining patterns reveal increasing amounts of gelatinase-A according to the severity of disease. Examination of cytosols generated from such tumour tissue on a substrate gel shows that the active forms of both gelatinase-A and B are present. In addition, we have carried out a series of experiments *in vivo* with C127 cells which have been transfected with the gelatinase-A gene and selected for their ability to secrete gelatinase-A [3]. When these cells are injected into Balb/C nu/nu mice and the mice examined for the presence of lung nodules four weeks later, a direct correlation between numbers of nodules and amounts of gelatinase secreted is observed. C127 cells transfected with vectors alone or collagenase and stromelysin genes produced only background levels of lung nodules. All of this constitutes evidence for the involvement of gelatinase-A in the process of tumour metastasis and selective inhibitors may offer therapeutic potential in preventing this process.

Results and Discussion

Having access to a panel of the pure human forms of gelatinase-A, stromelysin and collagenase has allowed us to examine a series of hydroxamic acid derivatives for their potency and selectivity *in vitro*. Optimisation of a series containing isobutyl in the P_1' position led to potent inhibitors such as **1** which lacked selectivity. Based on substrate knowledge [4] we then introduced an aromatic residue into the P_1' position, optimisation of which led to the highly potent and selective phenyl propyl analogue **2**. The P_3' position tolerates a wide range of substituents but variations which culminated in the arylsulphonamide series proved to be particularly interesting as in **3**. We then returned to the P_1' group for a fine tuning exercise of which **4** and **5** are examples with K_i's against gelatinase-A of < 0.010 nM and selectivities over collagenase of up to 80,000. The majority of these analogues were synthesised via the Evan's chiral auxiliary [5] route which is illustrated for the synthesis of **2**. Thus formation of the desired chiral auxiliary derivative followed by deprotonation with sodiumhexamethyldisilylazide and alkylation with t-butylbromoacetate gave exclusively the diastereoisomer shown as measured by NMR. Removal of the chiral auxiliary with lithium hydroperoxide gave the optically pure carboxylic acid. Standard peptide coupling followed by cleavage of the t-butyl group afforded the carboxylic acid. The best results for formation of the hydroxamic acid were obtained with ethylchloroformate as activating agent and trimethylsilylhydroxylamine. The carboxylic acid intermediates shown are also potent inhibitors, for example, the carboxylic acid corresponding to **4** has a potency of 1 nM vs. gelatinase-A and comparable selectivity figures to the hydroxamic acid.

SAR's of HYDROXAMIC ACID INHIBITORS.

K_i in nM (relative activity)

	P_1'	P_2'	P_3'	GL-A	SL	CL
1				0.33 (1)	25.3 (76)	7.8 (23.6)
2				0.062 (1)	8.3 (134)	203 (3270)
3				0.025 (1)	5.92 (237)	150 (6000)
4				< 0.010 (1)	2.98 (298)	329 (32900)
5				< 0.010 (1)	1.08 (108)	790 (79000)

Early analogues appeared to have poor pharmacokinetics when examined for their ability to inhibit the formation of lung nodules induced by gelatinase-A secreting C127 cells in Balb/C nu/nu. mice. However, members of the arylsulphonamide series possessed good half lives at 4 mg/kg i.v. in normal mice.

A range of analogues have been evaluated in a simple model which assesses the inhibition of gelatinase-A *in vivo*. The duration of action of the sulphonamide derivatives has been confirmed and we have recently optimised the above series to obtain orally active analogues. In a syngeneic model of tumour metastasis in mice, inhibition of the formation of lung nodules was observed with selected inhibitors both i.v. and orally.

We have demonstrated that gelatinase is a key enzyme in the process of tumour metastasis and have identified the structural requirements for potent and selective inhibitors. Several inhibitors were efficacious in animal models of metastasis, some possessing good duration of action and oral bioavailability.

References

1. Docherty, A.J.P., O'Connell, J., Crabbe, T., Angal, S. and Murphy, G., Tibtech, 10 (1992) 200.
2. Pyke, C., Ralfkiaer, E., Huhtala, P., Hurskainen, T., Dano, K. and Tryggvason, K., Cancer Res., 52 (1992) 1336.
3. Cockett, M., Birch, M., Crabbe, T., Murphy, G., Hart, I., Morphy, J.R., Millican, A., Beeley, N. and Docherty, A.J.P., in preparation.
4. Nagase, H., Okada, Y., Suzuki, K., Enghild, J.J. and Salvesen, G., Biochem. Soc. Trans., 19 (1991) 715.
5. Evans, D.A., Ennis, M.D. and Mathre, D.J., J. Am. Chem. Soc., 104 (1982) 1737.

Investigation of the influence of N^{α}-substituted amino acid on the platelet fibrinogen receptor antagonist activity of cyclic disulfide peptides

F.E. Ali, D. Bennett, R. Calvo, S.M. Hwang, A. Nichols, D. Shah, J. Vasko, A. Wong, C.K. Yuan and J.M. Samanen

Departments of Medicinal Chemistry, Biomolecular Discovery, Cellular Biochemistry and Pharmacology, SmithKline Beecham Pharmaceuticals, King of Prussia, PA 19406-0939, U.S.A.

Introduction

The conformationally constrained analogues cyclo-S,S-[Ac-Cys-Arg-Gly-Asp-Pen]-NH$_2$, **2** and cyclo-S,S-[Mba-Arg-Gly-Asp-Man], **3**, (Mba/Man = 2-mercaptobenzoyl/2-mercaptoaniline disulfide) have been described as potent platelet aggregation inhibitors and fibrinogen receptor antagonists [1, 2]. The ten fold enhancement in potency and receptor affinity associated with replacing (N^{α}-Me)Arg for Arg as in cyclo-S,S-[Ac-Cys-(N^{α}-Me)Arg-Gly-Asp-Pen]-NH$_2$, **4** (SK&F 106760) and cyclo-S,S-[Mba-(N^{α}-Me)Arg-Gly-Asp-Man], **5** (SK&F 107260), directed our attention to investigate the effect of other N^{α}-substituents on Arg, and other amino acids within the cyclic structure, on platelet aggregation inhibition potency and receptor affinity.

Results and Discussion

We have generated potent fibrinogen (Fg) receptor antagonists from a small linear peptide, Ac-RGDS-NH$_2$ **1**, by utilizing conformational constraints and increasing lipophilicity, through enclosing -RGD- into a cyclic disulfide bearing the Ac-Cys/Pen-NH$_2$ tether to give **2**, and substituting the Mba/Man tether for the Ac-Cys/Pen-NH$_2$ tether in **2** to give **3**, Table 1. Substituting (N^{α}-Me)Arg for Arg gave rise to analogues **4** and **5**, with significant enhancement in platelet aggregation inhibition (potency) and Fg receptor antagonist (affinity). The effect of alpha amine methylation on potency has also been examined for Gly and Asp within the framework of the cyclic peptide **2**. The corresponding (N^{α}-Me)Gly and (N^{α}-Me)Asp peptides **6** and **7** displayed decreased activity compared to **2**. Similar modification on peptide **4** gave **10** and **11** with drastic loss of potency and affinity compared to **4**. These results indicate that, contrary to the Arg residue, N^{α}-methylation on Gly or Asp is unfavorable. These substitutions may impose an undesirable influence upon peptide conformation, may promote unfavorable receptor interaction or eliminate a favorable hydrogen bond interaction. The probable solution conformations of **4** have been determined by ^{1}H NMR and molecular modeling [3]. The conformation determined for Gly and Asp in **4** fall within the manifold of conformations of

(N^α-Me) amino acid residues in peptides [4]. Thus, N^α-methylation of these residues may not perturb conformation. Instead, N^α-methylation of Gly or Asp in **4** may introduce a negative steric interaction, or loss of a favorable receptor interaction. The role of the N^α-methyl group on Arg was probed with the (N^α-Et) analogue **8** and the (N^α-Bzl) analogue **9**. Peptide **8** displayed enhanced potency over **2** and comparable potency to **4**, while **9** displayed diminished potency relative to **4** or **8** and

Table 1   *Structure activity of N^αsubstituted amino acid modifications in peptides 2 and 3*

No.		Plat. Agg. IC_{50} (μM)[a]	Binding K_i (μM)[b]	K_i (μM)[c]
1	Ac-RGDS-NH$_2$	91.3±0.1	4.2	37± 0.28
2	cyclo(S,S)-[Ac-CRGD-Pen]-NH$_2$	4.12±0.6	--	
3	cyclo(S,S)-[Mba-RGD-Man]	0.29±0.09	0.027	--
4	cyclo(S,S)-[Ac-C(N^α-Me)RGD-Pen]-NH$_2$ (SK&F 106760)	0.36±0.04	0.058±0.02	0.175±0.025
5	cyclo(S,S)-[Mba-(N^α-Me)RGD-Man] (SK&F 107260)	0.09±0.02	0.0021±0.006	0.004±0.0002
6	cyclo(S,S)-[Ac-CR(N^α-Me)GD-Pen]-NH$_2$	73.4±8.2	--	1.25±0.061
7	cyclo(S,S)-[Ac-CRG(N^α-Me)D-Pen]-NH$_2$	136.7±21.9	--	25.3±1.2
8	cyclo(S,S)-[Ac-C(N^α-Et)RGD-Pen]-NH$_2$	0.9±0.15	0.065	--
9	cyclo(S,S)-[Ac-C(N^α-Bzl)RGD-Pen]-NH$_2$	7.74±1.78	>100	6.38±0.1
10	cyclo(S,S)-[Ac-C(N^α-Me)R(N^α-Me)GD-Pen]-NH$_2$	39.5±6.0	--	15.1±0.62
11	cyclo(S,S)-[Ac-C(N^α-Me)RG(N^α-Me)D-Pen]-NH$_2$	23.4±8.7	--	7.04±0.03
12	cyclo(S,S)-[Ac-CGRGD-Pen]-NH$_2$	11.4±2.0	0.61±0.13	3.79±0.02
13	cyclo(S,S)-[Ac-C(N^α-Me)GRGD-Pen]-NH$_2$	1.2±0.71	0.077±0.0	0.740±0.005
14	cyclo(S,S)-[Ac-CG(N^α-Me)RGD-Pen]-NH$_2$	0.36±0.05	--	0.135±0.015
15	cyclo(S,S)-[Ac-C(N^α-Me)G(N^α-Me)RGD-Pen]-NH$_2$	0.15±0.03	0.02±0.01	0.017±0.001
16	cyclo(S,S)-[Ac-C(N^α-Bzl)G(N^α-Me)RGD-Pen]-NH$_2$	0.29±0.06	0.43±0.059	0.052±0.006
17	cyclo(S,S)-[Ac-CP(N^α-Me)RGD-Pen]-NH$_2$	24.4±9.81	>10	15.45±2.3
18	cyclo(S,S)-[Mba-(N^α-Me)GRGD-Man]	3.46±1.1	0.143±0.072	--
19	cyclo(S,S)-[Mba-(N^α-Me)G(N^α-Me)RGD-Man]	0.102±0.043	0.0245±0.006	--
20	cyclo(S,S)-[Mba-P(N^α-Me)RGD-Man]	1.01±0.44	--	0.82±0.03

[a]   Inhibition of platelet aggregation in canine platelet-rich plasma induced by ADP.
[b]   Inhibition of ^{125}I fibrinogen binding to purified GPIIb/IIIa isolated from human platelets, reconstituted in liposomes, no statistical limits indicate only one determination.
[c]   Inhibition of ^3H-SK&F 107260 binding to purified GPIIb/IIIa.

comparable to **2**. Thus, a certain degree of steric bulk in the N^α-substituent on Arg can be tolerated by the receptor. The solution conformations of **4** and **5** [3] do not support the promotion of a cis-amide by the N^α-Me group since neither displayed a cis-amide. The conformations of **4** and **5** do suggest that the N^α-Me group adds conformational constraint to these molecules. The methylation of Arg could enhance potency furthermore, by removing an unfavorable receptor interaction or introducing a favorable receptor interaction arising from: a) increasing the electronegativity of the amide oxygen or b) promoting a positive lipophilic receptor interaction.

In our initial search the cyclic hexapeptide **12** displayed activity comparable to the pentapeptide **2**. We examined the effect of N^α-alkylation on the intercalated amino acid in **12**. The $(N^\alpha$-Me)Gly **13** displayed enhanced potency and affinity relative to **12** or **2**. The $(N^\alpha$- Me)Arg peptide **14**, displayed enhanced potency over **12** and **13**, and comparable potency to **4**. The doubly N^α-methylated peptide **15**, displayed superior activity to either **13** or **14**. The $(N^\alpha$-benzyl)Gly peptide **16** displayed no drastic loss, however the Pro analogue **17** produced a dramatic loss in potency. A similar intercalation was carried out in **3** and **5** to give **18-20** with a drop of activity in **18** and **20** and no dramatic loss in **19**, despite decrease in receptor affinity relative to **5**. These results indicate that the enhancement of potency obtained by N^α-methylation is dependent on the specific amino acid (Arg) and not on its position in the sequence or the size of the ring.

In conclusion, the importance of an $(N^\alpha$-Me)Arg in the cyclic -RGD-peptides to Fg-receptor antagonist and inhibition of platelet aggregation activity has been demonstrated.

References

1. Samanen, J.M., Ali, F.E., Romoff, T., Calvo, R., Sorenson, E., Vasko, J., Storer, B., Berry, D., Bennett, D., Strohsacker, M., Powers, D., Stadel, J. and Nichols, A., J. Med. Chem., 34 (1991) 3114.
2. Ali, F.E., Samanen, J.M., Calvo, R., Romoff, T., Yellin, T., Vasko, J., Powers, D., Stadel, J., Bennett, D., Berry, D. and Nichols, A., In Smith, J.A. and Rivier, J.E. (Eds.), Peptides: Chemistry and Biology, Escom, Leiden, The Netherlands, 1992, p.761.
3. Kopple, K.D., Baures, P.W., Bean, J.W., D'Ambrosio, C.A., Peishoff, C.E. and Eggleston, D.S., J. Am. Chem. Soc., 114 (1992) 9615.
4. Manavalan, P. and Momany, F., Biopolymers, 19 (1980) 1943.

Inhibition of Interleukin-1β converting enzyme by peptide derivatives

I. Fauszt, E. Széll, K. Németh, M. Patthy and S. Bajusz

Institute for Drug Research, P.O. Box 82, H-1325 Budapest, Hungary

Introduction

Interleukin-1β (IL-1β)-converting enzyme (ICE) has been identified [1] as a cysteine protease which cleaves the IL-1β precursor (pIL-1β) at Asp^{116}-Ala^{117} to yield mature IL-1ß, an important mediator of inflammation. ICE also cleaves pIL-1β at Asp^{27}-Gly^{28}.

$$\overbrace{\qquad\qquad}^{\text{IL-1β}}$$

pIL-1β: Met-...-Ala-Asp-↓Gly-Pro-...-His-Asp-↓-Ala-Pro-...........-Ser
      1   27   28   116   117   269

To develop reversible inhibitors for ICE, peptide aldehydes and peptide substrates having the scissile bond as keto-methylene isostere were derived from the ICE cleavage sites. Trans-epoxysuccinyl (Eps) peptides, analogues of E-64 (Eps-Leu-Agm), were prepared to find irreversible inhibitors of ICE.

Results and Discussion

Compounds involved in this study are listed in Table 1. Peptide aldehydes **1-4** were prepared by stepwise elongation of β-tert-butyl-L-aspart-1-al semi-carbazone, Asp(O^{t}Bu)SC, followed by successive removal of the SC and ^{t}Bu. (An alternative synthesis of an analogous peptidyl L-aspart-1-al has been published recently [2].) For the synthesis of substrate analogues **5** and **6** the key intermediate, Z-Asp(O^{t}Bu)^{k}Gly was prepared analogously to [3]. Eps peptides **7-9** were obtained by stepwise elongation.

ICE inhibiting activity of the peptides was determined on THP_1 cells stimulated by LPS and silica. The release of IL-1β from the cells was measured by ELISA.

It has long been known that peptide aldehydes are reversible inhibitors of serine and cysteine proteases [4], and that potent and selective aldehyde inhibitors can be constructed for such proteases, e.g., thrombin, from the P_n-P_1 portions of their native substrates [5]. In view of these findings tripeptide aldehyde **1** and pentapeptide aldehydes **2-4** were derived from the IL-1β fragments P_3-P_1 and P_5-P_1, respectively. Of these **2** was the most inhibitory indicating the significance of residues P_4 and P_5 and that of His at P_2.

Substrate analogues **5** and **6** were less inhibitory than the aldehydes. Thus the presence of P_1'-P_4' does not seem to help the binding of **5** and/or the keto-methylene isostere group may be less reactive than the aldehyde.

E-64, Eps-Leu-Agm, is an irreversible inhibitor of most cysteine proteases and seems highly specific for this class of enzymes [6]. ICE showed resistance to E-64, which was explained by its unique substrate specificity [1]. To obtain ICE-inhibiting analogues the Leu and Agm of E-64 have been replaced with residues of pIL-1β in consideration of the possible binding modes of E-64. It is very likely that the Leu-Agm of E-64 binds to the S_1'-S_2' or S_2'-S_3' subsites of the enzymes [6]. Thus Leu and Agm as substrate subsites correspond to P_1' and P_2' or P_2' and P_3', respectively.

Based on the crystal structure of E-64-papain complex, it has been also supposed that E-64 occupies the S sites of the enzyme [7]. To comply with the side chain requirements for these three binding modes **7**, **8** and **9** have been prepared.

Table 1 *ICE inhibitors derived from P_5-P_5' of pIL-1β*

No.	Peptide and its relation to pIL-1β sequence Ala-Tyr-Val-His-Asp↓Ala-Pro-Val-Arg-Ser P_5 P_4 P_3 P_2 P_1 P_1' P_2' P_3' P_4' P_5'	ICE inhibition[a] IC_{50} μM	%
1	Z-Val-His-Asp-H	10	-
2	Eoc-Ala-Tyr-Val-His-Asp-H	4	-
3	Eoc-Ala-Tyr-Val-Ala-Asp-H	7	-
4	Eoc-Ala-Tyr-Val-Gly-Asp-H	10	-
5	Z-Tyr-Val-His-Asp^{k}Gly-Pro-Val-Arg-NH$_2$[b]	400	29
6	Ac-Tyr-Val-Ala-Asp^{k}Gly-NH-CH$_3$	-	20
7	Eps-Ala-dcPro	-	NI
8	Eps-Pro-dcVal	-	NI
9	dcVal←Ala←Eps	-	NI

[a]Determined on THP$_1$ cells by using ELISA for measuring IL-1β release. % Inhibition of peptides at 10 μM. NI, no inhibition.
[b]Asp^{k}Gly, keto-methylene isostere of Asp-Gly; Eps, L-trans-epoxysuccinyl; dcAA descarboxyamino acid; ←, reversed direction of the peptide bond (NH-CO). Neither of them are inhibitory, which may indicate that ICE is not a cysteine protease.

Based on the ICE sensitive region of pIL-1β potent peptide aldehyde inhibitors of ICE can be prepared, e.g., Eoc-Ala-Tyr-Val-His-Asp-H inhibits IL-1β release from THP$_1$ cells with an IC_{50} of 4 μM. Substrate analogues having the scissile bond as keto isostere are much less active. E-64 analogues derived from pIL-1β are inactive. This latter finding may show that ICE is not a cysteine protease as classified [1, 8], but rather a serine protease which contains a thiol group near the active site whose integrity is necessary for enzyme activity. In our opinion this view is supported by recent findings [8] that ICE contains the active-site consensus sequence of Ser proteases (Gly-X-Ser-X-Gly) at 287-291 as well as a nearby thiol, that of Cys[285], which is sensitive to alkylation by peptidyl diazoketone.

References

1. Black, R.A., Kronheim, S.R. and Sleath, P.R., FEBS Lett., 247 (1989) 386.
2. Chapman, K.T., Bioorg. Med. Chem. Lett., 2 (1992) 613.
3. García-Lopez, M.T., González-Muñiz, R. and Harto, J.R., Tetrahedron Lett., 29 (1988) 1577.
4. Aoyagi, T., Miyata, S., Nanbo, M., Kojima, F., Matsuzaki, M., Ishizuka, M., Takeuchi, T. and Umezawa, H., J. Antibiot., 22 (1969) 558.
5. Bajusz, S., Barabas, E., Szell, E. and Bagdy, D., In Walter, R. and Meienhofer, J. (Eds.), Peptides: Chemistry, Structure and Biology, Ann Arbor Sci. Publ. Inc., Ann Arbor, MI, U.S.A., 1975, p.603.
6. Rich, D.H., In Barrett, A.J. and Salvesen, G. (Eds.), Proteinase inhibitors, Elsevier, Amsterdam, 1986, p.161.
7. Yamamoto, D., Ohishi, H., Ishida, T., Inoue, M., Sumiya, S. and Kitamura, K., Chem. Pharm. Bull., 38 (1990) 2339.
8. Thornberry, N.A., Bull, H.G., Calaycay, J.R., Chapman, K.T., Howard, A.D., Kostura, M.J., Miller, D.K., Molineaux, S.M., Weidner, J.R., Aunins, J., Elliston, K.O., Ayala, J.M., Casano, F.J., Chin, J., Ding, G.J.-F., Egger, L.A., Gaffney, E.P., Limjuco, G., Palyha, O.C., Raju, S.M., Rolando, A.M., Salley, J.P., Yamin, T.-T., Lee, T.D., Shively, J.E., MacCross, M., Mumford, R.A., Schmidt, J.A. and Tocci, M.J., Nature, 356 (1992) 768.

Thrombin-inhibiting decapeptides deduced from the C-terminus of hirudin

H.E.J. Bernard[1], H.W. Höffken[1], W. Hornberger[2], K. Rübsamen[2] and B. Schmied[1]

[1]BASF AG, Central Research Laboratory, D-67056 Ludwigshafen, Germany, and [2]Knoll AG, Department of Angiology, D-67008 Ludwigshafen, Germany

Introduction

The serine protease thrombin is an important component of the blood coagulation cascade. Selective thrombin inhibitors are of interest as anticoagulant drugs for the treatment and prophylaxis of diseases like vein thrombosis, pulmonary embolism and arterial thrombosis. The catalytic site of thrombin cleaves peptide sequences at Arg-AA bonds. In addition thrombin also interacts with substrates via a so called anion binding exosite which is distinct from the active site and is responsible for the specificity of the enzyme.

Peptides derived from the C-terminus of hirudin have been shown to inhibit thrombin by binding to the non-catalytic exosite of the enzyme. The decapeptide hirudin$_{56-65}$ has been recognized as the minimal structural unit for inhibitory action in earlier studies [1]. The therapeutic utility of presently known hirudin$_{56-65}$ derivatives is limited by low "in vivo" activities and their short duration of action.

Results and Discussion

Based on the 3D-structure of the thrombin-hirudin complex [2] and molecular modeling methods, various series of N-acylated decapeptide analogs of hirudin$_{56-65}$

$$\text{H-Phe-Glu-Glu-Ile-Pro-Glu-Glu-Phe(4-OSO}_3\text{H)-Leu-Gln-OH}$$

were synthesized. In the course of this work variations of each position of the native sequence have been evaluated.

Peptide synthesis was carried out on solid supports using Boc- or Fmoc-chemistry. Unnatural amino acids, e.g., Boc-L-p-sulfomethyl-phenylalanine (Boc-Smp) were synthesized via alkylation of (2R)-(-)-2,5-dihydro-3,6-dimethoxy-2-isopropyl-pyrazine [3].

Concentrating first on **position 63** (Table 1), we found that the chemically labile tyrosine-O-sulfate can be replaced by 4-sulfomethyl-phenylalanine. This replacement results in an increase of inhibitory potency and may be expected to favourably influence the kinetic properties of the corresponding derivatives.

Table 1   *Variations of position 63 (thrombin time assay)*

Suc-Phe-Glu-Glu-Ile-Pro-Glu-Glu-AA-Leu-Gln-OH	EC_{100} [μmol/l]
Smp	0.27
Phe(4-OSO$_3$H)	0.96
Phe(4-COOH)	1.08
Phe(4-SO$_3$H)	1.09
Tyr(3-NO$_2$)	1.45
Tyr	5.64

Thrombin time assay (TT): The EC_{100}-value was determined as concentration of test compound leading to 100% prolongation of clotting time of human plasma as compared to control.

Further optimization confirmed that as **N-terminus** the succinyl group Suc (des-amino-Asp) is a good choice. Its carboxylic acid function forms a salt bridge with Arg_t^{73} and Lys_t^{149} (AA_t = AA in thrombin). Tyr in **position 56** forms an additional hydrogen bond to the thrombin backbone and is slightly better than Phe which only interacts with Phe_t^{34}. Replacement of Ile in **position 58** by Pro, Hyp or Glu gave inhibitors with nearly equal potencies. Whereas Glu is able to form a salt bridge to Arg_t^{77}, Pro and Hyp lead to a more rigid structure. The side chains of both amino acids **59** and **64** reach into the same hydrophobic pocket; combination of Ile[59] and Cha[64] gives the best interactions. Substitution of Pro in **position 60** by

Table 2   *Comparison of LU 58463 with known antithrombotics (exosite and bifunctional inhibitors)*

	TT EC_{100} [nmol/l]	PA[a] IC_{50} [nmol/l]	AVS[b] ED_{15} [mg/kg]	$t_{1/2}$[c] [min]
LU 58463	80	67	0.49	57
Hirugen [5]	5,100			
MDL 28050 [6]	776	348	28.8	
r-Hirudin	22	2	1.42	35
Hirulog-1 [7]	110	32	2.38	6

[a]Platelet aggregation (PA) was measured with platelet-rich human plasma.
[b]Arteriovenous shunt model (AVS): Thrombus formation was evoked by a glass capillary inserted into a shunt between the right carotid artery and left jugular vein of anaesthetized rats. ED_{15} was determined as the dose of compound required to cause a 15 min prolongation of the shunt patency.
[c]Plasma concentration after intravenous administration ($t_{1/2}$) in anaesthetized rats was determined by comparing the thrombin time of venous blood drawn at different time intervals with a thrombin time standard curve.

Hyp leads to an increase in potency by a factor of about 5. None of the replacements investigated at **positions 57, 61, 62, 65** resulted in compounds more potent than the native sequence.

By combining the results of the systematic variation of each position of hirudin$_{56-65}$ we were led to the synthesis of

$$\text{Suc-Tyr-Glu-Pro-Ile-Hyp-Glu-Glu-Smp-Cha-Gln-OH} \quad \textbf{LU 58463}$$

The "in vitro" activity of LU 58463 in the thrombin time assay is in the same range as r-hirudin. The X-ray structure of the thrombin-LU 58463 complex shows that the sulfonate group interacts - like the sulfate in hirugen [4] - with the phenolic group of $\text{Tyr}_t{}^{76}$, with a backbone nitrogen of $\text{Ile}_t{}^{82}$ and, via a water molecule, with the carbonyl groups of $\text{Gln}_t{}^{80}$ and $\text{Arg}_t{}^{77}$.

Evaluation of LU 58463 in several animal models of thrombosis demonstrated its high activity in vivo. The biological data (Table 2) indicate that this newly developed inhibitor is equipotent or superior to other known thrombin inhibitors.

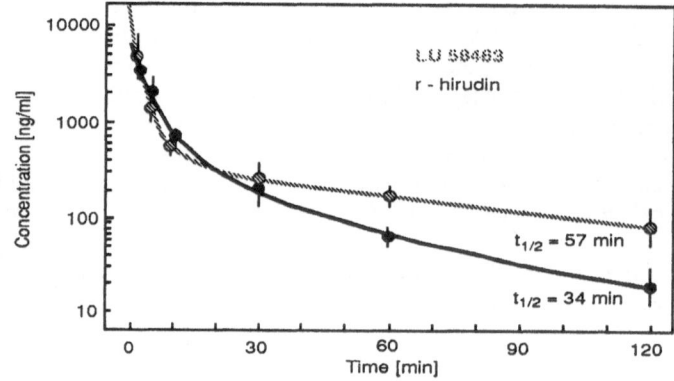

Fig. 1. Kinetics of LU 58463 and r-hirudin in plasma after intravenous administration of 1 mg/kg to anaesthetized rats (n = 5).

References

1. Mao, S.J.T., Yates, M.T., Owen, T.J. and Krstenansky, J.L., Biochemistry, 27 (1988) 8170.
2. Rydel, T.J., Tulinsky, A., Bode, W. and Huber, R., J. Mol. Biol., 221 (1991) 583.
3. Schöllkopf, U., Tetrahedron, 39 (1983) 2085.
4. Skrzypczak-Jankun, E., Carperos, V.E., Ravichandran, K.G., Tulinsky, A., Westbrook, M. and Maragonore, J.M., J. Mol. Biol., 221 (1991) 1379.
5. Maraganore, J.M., Adv. Appl. Biotechnol. Ser., 11 (1990) 103.
6. Krstenansky, J.L., Broerma, R.J., Owen, T.J., Payne, M.H., Yates, M.T. and Mao, S.J.T., Thromb. Haemost., 63 (1990) 208.
7. Maraganore, J.M., Bourdon, P., Jablonski, J., Ramachandran, K.L. and Fenton, J.W., Biochemistry, 29 (1990) 7095.

Inhibition of HIV-1 protease by its C- and N-terminal peptides

J. Franciskovich and J. Chmielewski

*Department of Chemistry, Purdue University,
West Lafayette, IN 47907, U.S.A.*

Introduction

HIV protease is a critical enzyme in the replication cycle of the human immunodeficiency virus. The protease is responsible for the posttranslational cleavage of the *gag* and *gag/pol* gene products which are the precursors to structural proteins p17, p24 and p9/p6 which make up the virion core, and the enzymes reverse transcriptase, integrase, and the protease itself. As such, virions arising from transfection with a provirus mutated in the protease-coding region contained unprocessed viral proteins and were non-infectious [1]. The crucial role of this protease in the processing of HIV proteins and the production of infective virus particles has made it a prime target for drug design [2].

HIV-1 protease is a 99 amino acid residue protein which assembles into a homodimeric structure [3]. Dimerization of the protease generates the catalytic center of the enzyme (containing two active aspartate residues) and also the substrate binding pocket. The dimeric nature of HIV-1 protease offers a new mode of enzyme inhibition based on blocking homodimer assembly or disrupting the dimeric interface.

The N- and C-terminal regions of HIV-1 protease are the main area of dimeric overlap within the protease. These sequences interdigitate to form a four-stranded β-sheet structure which accounts for more than 50% of the interfacial region of the dimeric protease (Figure 1).

Results and Discussion

Since the N- and C-terminal sequences of HIV protease are involved in dimer formation, the addition of peptides corresponding to these sequences could inhibit HIV protease by interfering with dimerization. We evaluated the ability of peptides corresponding to the N- and C- termini of HIV-1 protease (with a Ser mutation at residue 95) to inhibit protease activity [4]. We synthesized peptides 1-3 corresponding to the last 4-6 amino acid residues and the first 5-7 residues of HIV-1 protease. The peptides were synthesized by a solid phase procedure using the Merrifield resin modified with a *p*-alkoxybenzylalcohol linker [5]. The inhibitory effect of peptides 1-3 on HIV-1 protease activity was evaluated using a fluorogenic substrate assay developed by Toth and Marshall [6]. We found that with the C-terminal peptides maximum inhibition was observed with pentapeptide 1b (Table 1).

This corresponds well with the C-terminal region involved in the β-sheet interface of the protease which spans from residue 99 to 95.

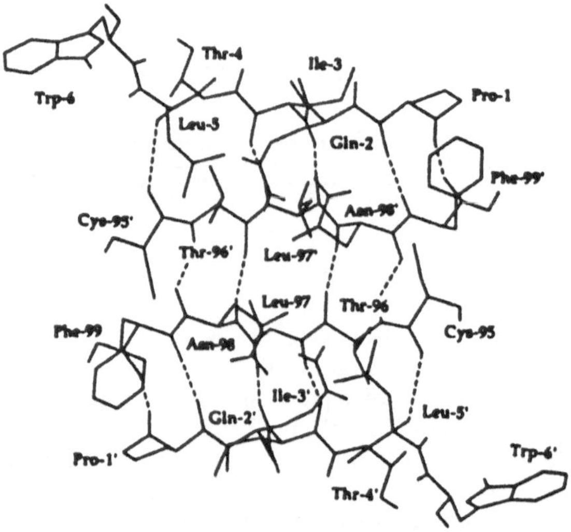

Fig. 1. The four-stranded β–sheet structure found at the N- and C-termini of the HIV protease homodimer (dashed lines indicate interstrand hydrogen bonds).

The N-terminal portion of HIV-1 protease extending from residues 1 through 6 forms half of the four-stranded β-sheet structure. The hexapeptide **2b** is a significantly better inhibitor than the shorter pentapeptide and the longer heptapeptide, which corresponds well with the maximum length N-terminal sequence involved in the β-sheet structure (Table 2). We have also explored the effect of blocking the N-terminus in these peptides, which is involved in a salt bridge with the carboxylate of Phe(99) in the homodimer. As expected, there was a small decrease in inhibition with peptides **3a** and **3b** but the hexapeptide was still quite active. Peptide **3c**, however, had improved inhibition upon acetylation, which may be due to H-bonding interactions between the acetyl-amide and C-terminal carboxylate, in addition to hydrophobic interactions with the Phe(99) sidechain.

Table 1   *Inhibition of HIV-1 protease with C-terminal fragments corresponding to the dimerization interface of HIV-1 protease*

	C-terminal Peptides	% Inhibition[a]
1a	AcNH-Thr-Leu-Asn-Phe-OH	3.9
1b	AcNH-Ser-Thr-Leu-Asn-Phe-OH	37.2
1c	AcNH-Gly-Ser-Thr-Leu-Asn-Phe-OH	4.6

[a]At 100 μM concentration.

Table 2   *Inhibition of HIV-1 protease with N-terminal fragments corresponding to the dimerization interface of HIV-1 protease*

	N-terminal Peptides	% Inhibition[a]
2a	NH-Pro-Gln-Ile-Thr-Leu-OH	5.6
2b	NH-Pro-Gln-Ile-Thr-Leu-Trp-OH	39.9
2c	NH-Pro-Gln-Ile-Thr-Leu-Trp-Gln-OH	1.3
3a	AcN-Pro-Gln-Ile-Thr-Leu-OH	-1.4
3b	AcN-Pro-Gln-Ile-Thr-Leu-Trp-OH	28.1
3c	AcN-Pro-Gln-Ile-Thr-Leu-Trp-Gln-OH	26.0

[a]At 100 μM concentration of peptides.

In conclusion, we have demonstrated that N- and C-terminal peptides of the dimerization interface of HIV-1 protease can act as HIV-1 protease inhibitors. In systematically varying the length of the peptide sequences, definite trends were obtained in which it was determined that peptide **1b** of the C-terminus and peptide **2b** of the N-terminus had the maximum inhibitory activity towards HIV-1 protease. Interestingly these sequences correspond exactly to the full length β-strands in the four-stranded, β-sheet region of the protease, which lends support to the belief that these peptides may be interfering with the dimerization interface. Although the inhibition obtained with these peptides is not large, they may act as good starting points for the design of novel dimerization inhibitors of HIV-1 protease.

Acknowledgements

We thank Richard Mueller and Kathryn Houseman of Searle Pharmaceutical for obtaining the IC$_{50}$'s of our peptides with HIV-1 protease and the Monsanto Company for financial support.

References

1.   Kohl, N.E., Emini, E., Schlief, W., Davis, L., Heimbach, J., Scolnick, E. and Sigal, I., Proc. Natl. Acad. Sci. U.S.A., 85 (1988) 4686.
2.   (a) Norbeck, D.W. and Kempf, D.J., Annu. Rep. Med. Chem., 36 (1991) 141; (b) Meek, T.D. and Dreyer, G.B., Ann. N.Y. Acad. Sci., 616 (1990) 41; (c) Korant, B.D., Ann. N.Y. Acad. Sci., 616 (1990) 252; (d) Scharpe, S., De Meester, I., Hendriks, D., Vanhoff, G., Van Sande, M. and Vriend, G., Biochimie, 73 (1991) 121; (e) Kent, S., Protein Eng., 4 (1990) 1.
3.   Wlodawer, A., Miller, A., Jaskolski, M., Sathyanarayana, B., Baldwin, E., Weber, I., Selk, L., Clawson, L., Schneider, J. and Kent, S., Science, 245 (1989) 616.
4.   (a) Franciskovich, J., Houseman, K., Mueller, R. and Chmielewski, J., Bioorg. Med. Chem. Lett., 3 (1993) 765; (b) Zhang, Z., Poorman, R., Maggiora, L., Heinrikson, R. and Kezdy, F., J. Biol. Chem., 266 (1991) 15591; (c) Schramm, H.J., Nakashima, H., Schramm, W., Wakayama, H. and Yamamoto, N., Biochem. Biophys. Res. Commun., 179 (1991) 847.
5.   Wang, S.S., J. Am. Chem. Soc., 95 (1973) 1328.
6.   Toth, M.V. and Marshall, G.R., Int. J. Pept. Protein Res., 36 (1990) 544.

Endothelin antagonists: The rational design of combined and ET_B receptor subtype selective antagonists

W.L. Cody[1], A.M. Doherty[1], J.X. He[1], P.L. DePue[1], L.A. Waite[1], S.J. Haleen[2], D.M. LaDouceur[2], M.A. Flynn[2], K.M. Welch[2] and E.E. Reynolds[2]

Departments of [1]Chemistry and [2]Pharmacology, Parke-Davis Pharmaceutical Research Division, The Warner-Lambert Company, 2800 Plymouth Road, Ann Arbor, MI 48016, U.S.A.

Introduction

The endothelins (ETs) and sarafotoxins (SRTXs) comprise a family of potent vasoconstricting peptides that contain 21-amino acids arranged in a unique bicyclic motif formed by disulfide bridges between cysteines located in positions 1-15 and 3-11 [1]. All members of this family possess a hydrophobic C-terminal hexapeptide terminating in an L-tryptophan with a free carboxylate that is necessary for high binding affinity to isolated endothelin receptor subtypes (ET_A and ET_B) and for vasoconstrictor activity [1]. We have utilized the C-terminal hexapeptide of endothelin-1 (Ac-His[16]-Leu-Asp-Ile-Ile-Trp[21], (I)) to develop functional antagonists of ET-1 [2-4]. A D-amino acid scan of compound I revealed that the D-His[16] analogue (II) enhanced binding (5 to 20-fold) to both ET_A and ET_B receptor subtypes versus (I). The D-Phe[16] substitution (III) led to a modest enhancement in binding affinity over (II) [2, 3]. In addition, we have reported [3, 4] that substitution of beta, beta-disubstituted hydrophobic D-aromatic amino acids in position 16 of the C-terminal hexapeptide led to peptides with high binding affinity to both receptor subtypes (Ac-D-Dip[16]-Leu-Asp-Ile-Ile-Trp[21] (IV) (PD 142893), Dip = 3,3-diphenylalanine [5]). Several antagonists of ET-1 that are selective for the ET_A receptor have been reported in the literature [6-8], but PD 142893 and related analogues are combined receptor antagonists. Compound (VI) blocked ET-1 stimulated arachidonic acid release (AAR) with a potency consistent with its binding in tissues that express either the ET_A or rat ET_B receptor. In addition, PD 142893 has been shown to block the *in vitro* vasoconstrictor activity of ET-1 and SRTX-6c with pA_2 values of 6.6 (ET_A) and 6.3 (ET_B), respectively [3, 4]. In examining the role of positions 17, 18, 19, and 20 to binding affinity and receptor subtype selectivity, we have found compounds selective for the ET_B receptor subtype by incorporating aromatic amino acids, specifically phenylalanine.

Results and Discussion

Starting with compound **IV** (PD 142893), phenylalanine was substituted in positions 18, 19 and 20. In the case of positions 18 and 19 the phenylalanine substitution led to compounds that were selective for the rat ET_B receptor (compounds **V** and **VI**). However, phenylalanine was not tolerated at position 20 with the resulting compound (**VII**) having only micromolar affinity for either receptor subtype. In vitro, Ac-D-Dip[16]-Leu-Asp-Phe-Ile-Trp[21] (**VI**) was able to block SRTX-6c stimulated vasoconstriction in the rabbit pulmonary artery (ET_B) with a pA_2 value of 6.4, but only had a pA_2 value of 5.3 in the rabbit femoral artery (ET_A). Surprisingly, even though **V** had binding affinities similar to **VI**, it showed no in vitro activity in either tissue. Compound **VI** is a functional antagonist of ET-1 that is selective for the rat ET_B receptor albeit only slightly.

Table 1 *Binding affinities and biochemical activities of C-terminal ET-1 analogues*

Peptide[c]	Binding Assay[a]		Biochemical Assay[b]	
	ET_A	ET_B	AAR_A	AAR_B
I	>50	>50	ND[d]	ND
II	9.5	10.0	3.2	ND
III	2.8	3.3	3.1	ND
IV	0.04	0.06	0.07	0.02
V	0.67	0.03	1.3	0.85
VI	0.15	0.06	0.13	0.054
VII	4.0	7.5	>10	ND
VIII	0.10	0.33	0.42	2.7
IX	1.0	6.0	0.45	ND
X	7.0	5.2	>10	ND

[a] Binding data in rabbit renal artery vascular smooth muscle cells (VSMC) (ET_A) or rat cerebellar membranes (ET_B) (micromolar IC_{50} values).

[b] Inhibition of ET-1 stimulated AAR (micromolar EC_{50} values) in rabbit renal artery VSMC (ET_A) or CHO cells expressing recombinant rat ET_B receptors.

[c] (**I**) Ac-His-Leu-Asp-Ile-Ile-Trp; (**II**) Ac-D-His-Leu-Asp-Ile-Ile-Trp; (**III**) Ac-D-Phe-Leu-Asp-Ile-Ile-Trp; (**IV**) Ac-D-Dip-Leu-Asp-Ile-Ile-Trp.2Na+ (PD 142893); (**V**) Ac-D-Dip-Leu-Phe-Ile-Ile-Trp; (**VI**) Ac-D-Dip-Leu-Asp-Phe-Ile-Trp; (**VII**) Ac-D-Dip-Leu-Asp-Ile-Phe-Trp; (**VIII**) Ac-D-Dip-Leu-Asp-Ala-Ile-Trp; (**IX**) Ac-D-Dip-Leu-Asp-Glu-Ile-Trp; (**X**) Ac-D-Dip-Leu-Asp-Lys-Ile-Trp.

[d] ND = Not Determined.

In order to determine the structural parameters of endothelin antagonists that contribute to ET_B receptor selectivity we incorporated neutral (Ala, **VIII**), acidic (Glu, **IX**), and basic (Lys, **X**) amino acids into position 19 of Ac-D-Dip[16]-Leu-Asp-Ile-Ile-Trp[21] (**XI**). Although, the Ala substitution maintained high nanomolar binding affinity, all receptor selectivity was lost. Both the Glu and Lys substitutions led to compounds with only micromolar affinity for each receptor subtype and no selectivity. Similar results were obtained for the substitution of nonaromatic amino acids in positions 17 and 18 (data not shown). This data suggests that the rat ET_B receptor may have an additional/auxiliary hydrophobic binding pocket that can accommodate a phenylalanine residue. Apparently, such hydrophobic binding pockets may not be present or important for binding to the ET_A receptor.

The incorporation of beta, beta-disubstituted hydrophobic D-aromatic amino acids (Dip) in position 16 of the C-terminal hexapeptide of ET-1 led to potent functional antagonists at both the ET_A and ET_B receptor subtypes. In addition, the incorporation of phenylalanine in positions 18 (**V**) and 19 (**VI**) of Ac-D-Dip[16]-Leu-Asp-Ile-Ile-Trp[21] imparts selectivity for the rat ET_B receptor. In fact, Ac-D-Dip[16]-Leu-Asp-Phe-Ile-Trp[21] (**VI**) is a functional antagonist of ET-1 stimulated vaso-constriction *in vitro*. Selective ET_B receptor antagonists may help determine the physiological and/or pathophysiological role of ET-1.

References

1. Doherty, A.M., J. Med. Chem., 35 (1992) 1493.
2. Doherty, A.M., Cody, W.L., He, J.X., DePue, P.L., Leonard, D.M., Dunbar, J.B., Hill, K.E., Flynn, M.A. and Reynolds, E.E., Bioorg. Med. Chem. Lett., 3 (1993) 497.
3. Cody, W.L., Doherty, A.M., He, J.X., DePue, P.L., Rapundalo, S.T., Hingorani, G.A., Major, T.C., Panek, R.L., Dudley, D.D., Haleen, S.J., LaDouceur, D.M., Hill, K.E., Flynn, M.A. and Reynolds, E.E., J. Med. Chem., 35 (1992) 3301.
4. Cody, W.L., Doherty, A.M., He, J.X., DePue, P.L., Topliss, J.G., Haleen, S.J., LaDouceur, D.M., Flynn, M.A., Hill, K.E. and Reynolds, E.E., Med. Chem. Res., 3 (1993) 154.
5. Chen, H.G., Beylin, V.G., Leja, B. and Goel, O.P., Tetrahedron Lett., 33 (1992) 3293.
6. Spinella, M.J., Malik, A.B., Everitt, J. and Andersen, T.T., Proc. Natl. Acad. Sci. U.S.A., 88 (1991) 7443.
7. Ihara, M., Noguchi, K., Saeki, T., Fukuroda, T., Tsuchida, S., Kimura, S., Fukami, T., Ishikawa, K., Nishikibe, M. and Yano, M., Life Sci., 50 (1992) 247.
8. Sogabe, K., Nirei, H., Shoubo, M., Nomota, A., Henmi, K. and Notsu, Y., Jap. J. Pharm., 58 (1992) 105P.

Conformational scan of bombesin/GRP reveals new position 11 receptor antagonists

D.H. Coy[1], M. Neya[1], N-Y. Jiang[1], J.E. Mrozinski[2], S.A. Mantey[2] and R.T. Jensen[2]

[1]*Peptide Research Laboratories, Department of Medicine, Tulane University Medical Center, New Orleans, LA 70112, U.S.A.,* [2]*Digestive Diseases Branch, NIH, Bethesda, MD 20892, U.S.A.*

Introduction

There are several design approaches which yield highly potent bombesin (Bn) antagonists. All result from alterations in the geometry of the C-terminal pair of amino acids or complete deletion of the C-terminal residue [1]. In the present study, further methods for producing receptor antagonists were explored *via* the introduction of conformationally restrictive amino acids. (D)-F-Q-W-A-V-G-H-L-L-NH_2 (peptide I, Table 1) is a potent nonapeptide agonist analogue of bombesin/gastrin releasing peptide (GRP) (Ki 7 nM for rat pancreatic acinar cells, Table 1) and was used for the present study in which its amino acids were replaced with the side-chain tethered cyclic amino acids Pro (Ala analog), Tic (Phe analog), Tee (2,3,4,5-tetrahydro-β-carboline-4-COOH) (Trp analog) or Tip (4,5,6,7-tetrahydro-1H-imidazo [4,5-c]pyridine-6-COOH, also known as spinacin) (His analog):

Tic   Tcc   Tip

Results and Discussion

Replacement of D-Phe in position 1 (equivalent to position 6 in Bn) or His in position 2 (normally present in GRP) with L-Tic and L-Tip, respectively, had little effect on binding affinity to rat pancreatic acinar cells or ability to release amylase from them. However, in the more conformationally sensitive center of the chain, severe loss of affinity was observed with Tcc replacements for Trp^3 and Tic replacements for Ala and Val in positions 4 and 5, respectively. Substitution of Gly[6] with D-Phe surprisingly, in view of full potencies of previously synthesized D-Ala[6] analogs [2], almost eliminated binding affinity (K_i 2 μM) and produced a partial agonist possessing weak antagonist properties. L-Phe[6] retained significant affinity (K_i 333 nM), albeit far lower than the parent peptide I.

Table 1   *Abilities of various Bn analogs to stimulate or inhibit release of Bn-stimulated pancreatic amylase release and to bind to rat pancreatic acinar cells*

Peptide Added	Ag/Ant	GRP Receptor (Pancreatic Acini)	
		EC_{50}/IC_{50}(nM)	K_i(nM)
Bn	Ag	0.06 ± 0.01	4 ± 1
I	Ag	0.08 ± 0.1	7 ± 1
[D-Tic[1]]	Ag	0.14 ± 0.01	6 ± 1
[L-His[2]]	Ag	0.3 ± 0.1	36 ± 10
[L-Tip[2]]	Ag	0.87 ± 0.10	78 ± 10
[L-Tec[3]]	Ag	5.9 ± 1.0	426 ± 117
[L-Tic[4]]	P.Ag (30% max)	2990 ± 444	>30000
[L-Tic[5]]	Ag	1430 ± 270	>30000
[L-Tic[6]]	Ag	>10000	9370 ± 700
[D-Tic[6]]	Ant	520 ± 65	265 ± 16
[L-Phe[6]]	Ag	333 ± 76	>30000
[D-Phe[6]]	P.Ag (15% max)	5400 ± 330	1990 ± 330
[L-Pro[6]]	Ag	690 ± 250	1980 ± 100
[D-Pro[6]]	Ant	478 ± 79	134 ± 18

The L-Tic[6] analog had no observable affinity (K_i > 10 μM), however, the D-Tic[6] analog exhibited a K_i of 265 nM. This analog was devoid of effects on amylase release and further studies showed it to be a new receptor antagonist of Bn/GRP with an IC_{50} of 520 nM. The L-Pro[6] analog was an agonist (K_i 690 nM), however, as in the case of the D-Tic analog, the D-Pro[6] peptide was an antagonist (K_i 134 nM, IC_{50} 478 nM). This position, together with His[7], forms part of a putative [3,4] type II β-bend region in the postulated receptor binding conformation of Bn agonists and antagonists in which the imidazole side-chain may constitute the active site. For instance, D-Phe substitution of His results in weak receptor antagonists [5] and replacement of His with certain amino acids greatly lowers agonist potencies whilst having little effect on antagonist potencies [2]. The present results demonstrate that position 6 side chain size has dramatic effects on biological properties, particularly when they are locked into position in the tethered amino acids. Molecular modeling of the standard H-bond stabilized type II β-bend encompassing the Val-Gly-His-Leu sequence suggests that the D-Phe, D-Pro, D-Tic analogs (but not the L-Phe, L-Tic and L-Pro analogs) must have a side chain projecting into the same region of space as the imidazole group on His[7] (Figure 1) and we propose that this could possibly prevent activation of Bn/GRP receptor by the latter, but not the general binding of the analog.

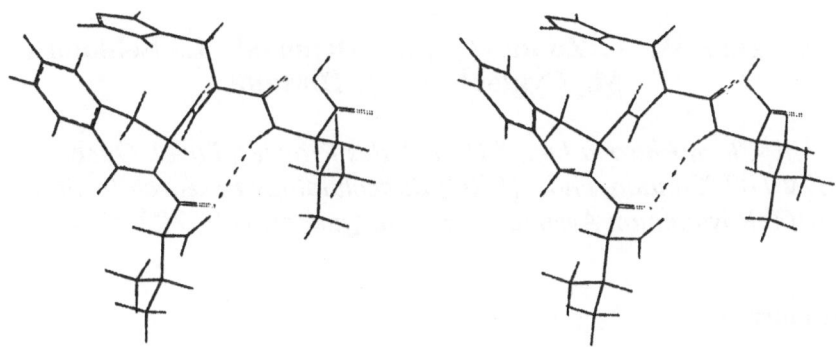

Fig. 1. Stereo molecular model of -Val-D-Tic-His-Leu- sequence displayed in a type II β-bend conformation. Note the projection of the D-Tic and His side chains into the same area of space. Dotted line represents standard H-bond.

Conformational effects induced by the substitution of constrained amino acids into a potent Bn(6-14) agonist were examined. The most interesting results included retention of virtually full potency with N-terminal substitutions and the generation of new Bn receptor antagonists by incorporation of D-Pro and D-Tic in Bn position 11. The latter effect seems to provide further support to a putative folded conformation for receptor-bound Bn and the participation of His^{12} in the receptor activation process. Incorporation of these modifications into existing restrained cyclic structures may further help to elucidate the conformational properties of Bn agonist/antagonists via NMR and molecular modeling studies.

Acknowledgements

This research was supported in part by NIH grant CA-45153.

References

1.   Jensen, R.T. and Coy, D.H., Trends Pharm. Sci., 12 (1991) 13.
2.   Coy, D.H. and Jensen, R.T., Growth Factors, Peptides and Receptors, In Moody, T. (Ed.), Plenum, New York, NY, U.S.A., 1993, p.161.
3.   Coy, D.H., Heinz-Erian, P., Jiang, N-Y., Sasaki, Y., J. Taylor, J., Moreau, J-P., Wolfrey, W.T., Gardner, J.D. and Jensen, R.T., J. Biol. Chem., 263 (1988) 5056.
4.   Coy, D.H., Jian, N-Y., Kim, S.H., Moreau, J-P., Lin, J.T., Frucht, H., Qian, J-M., Wang, L-W. and Jensen, R.T., J. Biol. Chem., 25 (1991) 16441.
5.   Heinz-Erian, P., Coy, D.H., Tamura, M., Jones, S.W., Gardner, J.D. and Jensen, R.T., Am. J. Physiol., 252 (1987) G439.

Structure-based design of small thrombin anion binding exosite inhibitors

M. Tarazi[1], A. Zdanov[2], J.M. Demuys[1], L. Leblond[1],
M. Cygler[2] and J. DiMaio[1]

*[1]BioChem Pharma Inc., 531 blvd. des Prairies, Laval, Quebec,
H7V 1B7, Canada, and [2](NRC) Biotechnology Research Institute,
6100 Royalmount Avenue, Montreal, Quebec, H4P 2R2, Canada*

Introduction

Thrombin-mediated proteolysis of fibrinogen to fibrin and the enzyme's interaction with specific receptors on platelets and endothelial cells are key processes in the formation and dissolution of a haemostatic plug at a site of vascular injury. The biochemical events manifested by thrombin are abrogated by hirudin or related inhibitors such as hirulog and hirutonin peptides [1, 2] (Figure 1), which are being developed for various indications of thrombotic disorders. The specificity of these inhibitors is secured by the presence of a common COOH-terminal tail whose sequence [N-D-G-D^{55}-F-E-E-I-P-E^{61}-E-Y-L-Q^{65}] is complementary to and binds within the putative *"fibrinogen anion-recognition exosite"* (FRE) of thrombin. We have solved the crystal structure of the complex between human α-thrombin (h-thrombin) and two hirutonin peptides. Hirutonin-6 [(R)F-P-AOGD-SP2-D-F-E-P-I-P-L (Figure 1)], which embodies a truncated FRE segment, inhibits thrombin with a K_i= 3.5 nM despite its reduced size. On the basis of the interactions within the enzyme's FRE, we investigated the effects of strategic modifications on the antithrombin activity of the new truncated exosite binding motif.

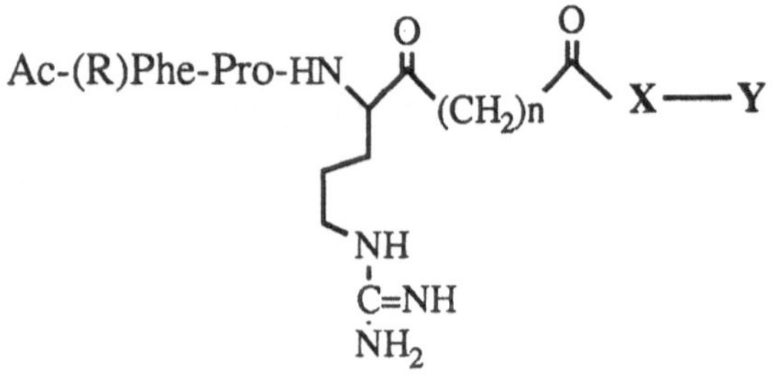

hirutonin-2	n= 2	X = hir49-54	Y = hir55-65
hirutonin-6	n= 4	X = [NHCH$_2$CH=CHCH$_2$]$_2$	Y = D-F-E-P-I-P-L

Fig. 1. *Structures of Hirutonin-2 and Hirutonin-6. Hirutonin-2: n=2, X=hir 49-54, Y=hir 55-65. Hirutonin-6: n=4, X=[NHCH$_2$CH=CHCH$_2$CO]$_2$, Y=D-F-E-P-I-P-L.*

Table 1 *Inhibitory concentrations required to double fibrinogen clotting time*

	Exosite Recognition sequence	IC_{50} μM
I	Ac-Asp-Phe-Glu-Glu-Ile-Pro-Glu-Glu-Tyr-Leu-Gln	1.2
II	Ac-Asp-Phe-Glu-Glu-Ile-Pro-**Leu**	60
III	Ac-Asp-Phe-Glu-**Pro**-Ile-Pro-**Leu**	32
IV	Ac-Asp-Phe-Glu-**Pro**-Ile-Pro-**Glu**	70
V	Ac-Asp-Phe-Glu-**Pro**-Ile-Pro-**Pgy**	35
VI	Ac-Asp-Phe-Glu-**Pro**-Ile-Pro-**Tyr**	18
VII	Succinyl-Phe-Glu-**Pro**-Ile-Pro-**Tyr**	10
VIII	Ac-Asp-Phe-GluΨ[CH=CH]Gly-Ile-Pro-**Leu**	>500
IX	Ac-Asp-Phe-GluΨ[COCH$_2$]Gly-Ile-Pro-**Leu**	>500
X	Ac-Asp-Phe-GluΨ[COHCH$_2$]Gly-Ile-Pro-**Leu**	>700
XI	Ac-Asp-Phe- HN (COOCH$_3$) ... Ile-Pro-Leu (COOH, O)	>500
XII	Ac-Asp-Phe-Glu–**Gly**-Ile-Pro-Leu	300
XIII	Ac-Asp-(S)**Act**-Glu–Pro-Ile-Pro-Leu	20
XIV	Succinyl-(S)**Tic**-Glu–Pro-Ile-Pro-Leu	77
XV	Succinyl-(S)**Nal**-Glu–**Gly**-Ile-Pro-Leu	140

Abbreviations: **Act**, 2-amino-2-carboxy tetralin; **Tic**, 2-carboxy tetrahydro-isoquinoline; **Nal**, 2-naphthylalanine.

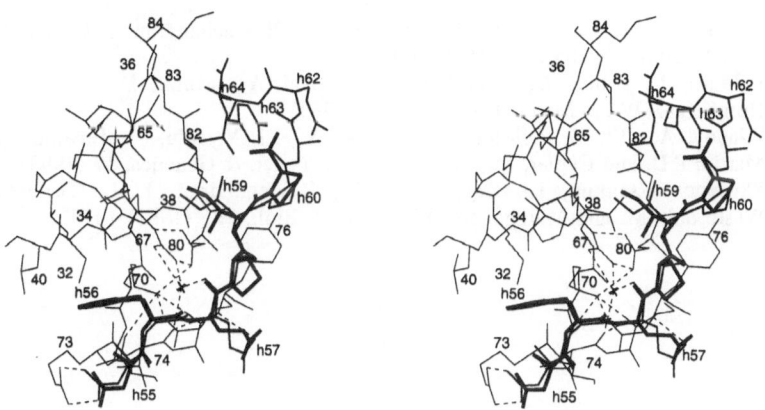

Fig. 2. *Bound conformation of hirutonin-6 (bold) and hirutonin-2 (thin) in the h-thrombin FRE.*

Results and Discussion

Figure 2 shows the conformations of the portion of hirutonin-6 (bold) and hirutonin-2 (thin line) bound to the h-thrombin fibrinogen recognition exosite (FRE). Residues corresponding to hir55-60 comprise the minimum core required to bind to the putative FRE. The backbone conformation from Asp^{55} to Pro^{60} is similar for both hirutonin peptides and is fully extended across the FRE surface. Hirutonin-6 terminates with a leucine residue in position 61 and therefore the 3_{10} helical reverse turn characteristic of hirutonin-2 and other C-terminal hirudin peptides [4] is lacking. The electron density for Leu^{61} was not clearly defined (<B>=70 Å^2) in the crystal and its orientation is not shown in Figure 2. Nevertheless, the data in Table 2 indicates that antithrombin activity is preserved and enhanced in truncated FRE inhibitors terminating with a hydrophobic residue. As an explanation for this finding, this residue could compete for the locus normally occupied by the greasy cluster formed by Pro^{60}, Tyr^{63} and Leu^{64} in the native hirudin sequence.

The carbonyl group of Glu^{57} in both inhibitors is oriented toward the enzyme surface and is hydrogen bonded to a water molecule (WAT84) which participates in a hydrogen bonding array involving enzyme residues Arg^{67}, Glu^{80}, Lys^{70} and Thr^{74}. In order to probe the relevance of this interaction to thrombin inhibition, the modifications shown in Table 1 were made to the amide bond connecting Glu^{57}-Glu^{58}. Neither alkene, ketoethylene, hydroxyethylene or methoxycarbonyl functions were compatible replacements. However, the single point mutation $Glu^{58}Gly$ caused a similar potency drop suggesting an important conformational role for the amino acid residue in position 58 which is not simulated by the above dipeptidyl mimetic functions. Interestingly, the same mutation to the full length FRE inhibitor (55-65) caused only a twofold drop in antithrombin activity.

Table 1 also shows results of modifications to Phe^{56} which, in the bound state, is buried in a hydrophobic pocket lined by side chains of Met^{32}, Phe^{34} and Leu^{40} of the enzyme. Among the selected modifications shown, only a (S)-2-amino-2-carboxy tetralin residue effectively enhanced inhibitory activity.

References

1. Maraganore, J.M., Bourdon, P., Jablonski, J., Ramachandran, K.L. and Fenton, J.W. II, Biochemistry, 29 (1990) 7095.
2. DiMaio, J., Gibbs, B., Lefebvre, J., Konishi, Y., Munn, D., Yue, S.Y. and Hornberger, W., J. Med. Chem., 35 (1992) 3331.
3. Zdanov, A., Wu, S., DiMaio, J., Konoshi, Y., Li, Y., Wu, X., Edwards, B.F.P., Martin, P.D. and Cygler, M., Proteins, Struct. Funct. & Genetics, 17 (1993).
4. Skrzypczak-Jankun, E., Carperos, V.E., Ravichandran, K.G., Tulinsky, A., Westbrook, M. and Maraganore, J.M., J. Mol. Biol., 221 (1991) 1379.

A rational approach for the design of mechanism-based inactivators for zinc proteases

S.S. Ghosh[1], O. Levy[2] and S. Mobashery[3]

[1]*Life Sciences Research Laboratory, Baxter Diagnostics Inc., San Diego, CA 92121, U.S.A.,* [2]*SIBIA Inc., La Jolla, CA 92037, U.S.A. and* [3]*Wayne State University, Detroit, MI 84202, U.S.A.*

Introduction

A number of physiologically important zinc metalloproteases, such as angiotensin-converting enzyme (ACE), neutral endopeptidase 24.11 (NEP), collagenase and renal dipeptidase, are targets for inhibition for the regulation of their biological action. These proteases share similarities in their mechanism of action with the well-studied carboxypeptidase A (CPA) [1], the prototypic member of this family of enzymes. While traditional strategies for inhibitor design have been based on competitive inhibition approaches, we have developed a rationale for the mechanism-based inactivation of such proteases. Based on reports that CPA and ACE catalyze the stereospecific deprotonation of ketonic substrate analogues [2, 3], we describe a mechanism-based inactivation approach for the model zinc protease, CPA, and demonstrate the application of the design principle for the irreversible inactivation of NEP and ACE.

Results and Discussion

(R)-2-Benzyl-5-cyano-4-oxo-pentanoic acid (**1a**) and N-cyanoacetyl-L-phenylalanine (**1b**) were designed as mechanism-based inactivators for CPA, based on the premise that enzyme-initiated deprotonation/isomerization would lead to the formation of reactive ketenimine intermediates which would trap an active site nucleophile. CPA-catalyzed α,β-elimination of N-(3-chloropropionyl)-L-phenylalanine (**1c**) was envisioned to furnish N-(acrolyl)-L-phenylalanine as the electrophilic inactivating species.

1a	X=CH₂	Y=CN
1b	X=NH	Y=CN
1c	X=NH	Y=CH₂Cl

2a	X=CH₂
2b	X=NH

Compounds **1a-c** [4, 5] irreversibly inactivated CPA in a time-dependent manner, and the kinetic parameters measured for the inactivation process showed that compound **1a** is a very effective inactivator of CPA (Table 1). The compounds partition between turnover and inactivation, and the high partition (k_{cat}/k_{inact}) ratios for **1b** and **1c** reveal that hydrolysis to L-phenylalanine predominates over the inactivation process. Attempts to characterize the turnover products from compound **1a** have thus far been unsuccessful. Experiments with ^{14}C-labeled **1a** and **1b** provided a 1:1 stoichiometric ratio of inactivator and CPA upon inactivation. N-(Acrolyl)-L-phenylalanine was independently shown to be an affinity inactivator, providing additional evidence for the proposed mechanism of action of compound **1c**.

Table 1 *Mechanism-based inactivation for CPA*

Inactivator	k_{inact} (min^{-1} x 10^{-2})	K_m (mM)	k_{cat}/k_{inact}
1a	8.3 ± 0.3	4.9 ± 0.4	28 ± 3
1b	3.5 ± 0.3	4.2 ± 0.4	1180 ± 40
1c	3.9 ± 0.2	1.4 ± 0.1	1680 ± 60

The classical active-site model for zinc proteases [6] was used as a template for the design of N-[(R)-2-benzyl-5-cyano-4-oxo-pentanoyl]-L-phenylalanine (**2a**) and N-[N-(cyanoacetyl)-L-phenylalanyl]-L-phenylalanine (**2b**) as inactivators for ACE [7] and NEP [8]. The following structural motifs were deemed necessary for binding at the active sites and for inactivation: (i) a C-terminal carboxylate group for ion pairing with an active-site basic residue, (ii) a hydrophobic P'1 residue based on the preference for aromatic or large hydrophobic residues in the S'1 subsites of these enzymes, and (iii) incorporation of an α-cyanomethyl moiety for the requisite enzyme-catalyzed formation of the electrophilic ketenimine intermediates.

Table 2 *Mechanism-based inactivation for NEP and ACE*

Inactivator	NEP			ACE[a]		
	k_{inact} (min^{-1})	K_m (mM)	k_{cat}/k_{inact}	k_{inact} (min^{-1})	K_m (mM)	k_{cat}/k_{inact}
2a	0.071	5.1	1340	—	—	—
2b	0.022	2.4	4700	0.08	10.5	8300

[a] Compound **2a** does not activate ACE even at 10 mM concentrations.

The results of the kinetic analysis of inactivation of NEP and ACE by compounds **2a** and **2b** are presented in Table 2. The kinetic parameters for the inactivation of the two enzymes by the peptidic compound **2b** are quite comparable. Treatment of compound **2b** as a competitive inhibitor for NEP hydrolysis of the substrate, dansyl-D-Ala-Gly-pNO2-Phe-Gly, and ACE hydrolysis of N-[(2-furyl)acrolyl]-Phe-Gly-Gly gave K_i values of 17 μM and 65 μM, respectively. Compound **2a** is, remarkably, not a mechanism-based inactivator for ACE and also binds poorly to the enzyme (K_i >10 mM) as a reversible inhibitor. The selective inactivation of NEP by compound **2a** and its K_i value of 14 μM indicate that the presence of a penultimate amide bond is not a critical requirement for reversible binding to the NEP active site or for the chemistry of inactivation.

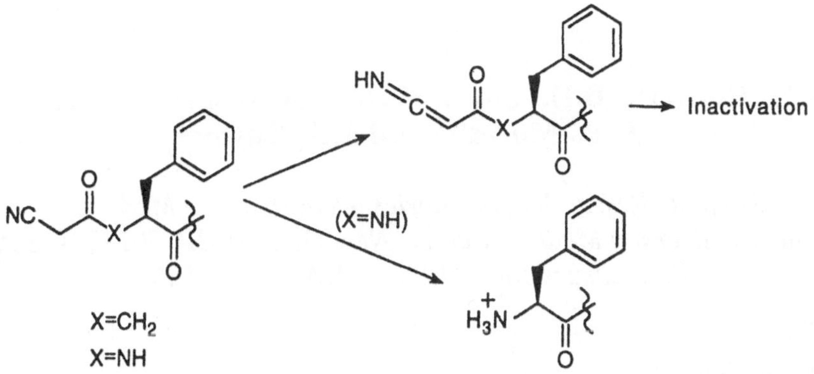

The deprotonation reaction described in the above examples appears to be a general feature of zinc proteases. Recently, this characteristic was also exploited for designing mechanism-based inactivators for renal dipeptidase [9]. We propose that inactivation of CPA, NEP and ACE with the cyanomethyl derivatives involves enzymic deprotonation, initiated either by a promoted water attack or by an active-site basic residue, followed by rearrangement to a ketenimine (see above scheme). This reactive intermediate then traps an active-site nucleophile, resulting in enzyme inactivation. Our studies with the dipeptide derivative, **2a**, and its ketomethylene analogue, **2b**, suggest that the NEP active site is more flexible than the ACE active site in accommodating the substitution of the penultimate NH group with a methylene functionality. This feature thus provides a motif for the design of potent mechanism-based inactivators which selectively differentiate between NEP and ACE.

Acknowledgements

This work was supported by funds from SIBIA, Inc. and an SBIR grant (1 R43 HL46629-01) from the National Heart, Lung, and Blood Institute.

References

1.   Christianson, D.W. and Lipscomb, W.N., Acc. Chem. Res., 22 (1989) 62.
2.   Sugimoto, T. and Kaiser, E.T., J. Am. Chem. Soc., 101 (1979) 3946.
3.   Spratt, T.E. and Kaiser, E.T., J. Am. Chem. Soc., 106 (1984) 6440.
4.   Mobashery, S., Ghosh, S.S., Tamura, S.Y. and Kaiser, E.T., Proc. Natl. Acad. Sci. U.S.A., 87 (1990) 578.
5.   Ghosh, S.S., Wu, Y.-Q. and Mobashery, S., J. Biol. Chem., 266 (1991) 8759.
6.   Ondetti, M.A., Rubin, B. and Cushman, D.W., Science, 196 (1977) 441.
7.   Ghosh, S.S., Said-Nejad, O., Roestamadji, J. and Mobashery, S., J. Med. Chem., 35 (1992) 4175.
8.   Levy, O., Taibi, P., Mobashery, S. and Ghosh, S.S., J. Med. Chem., 36 (1993) 2408.
9.   Wu, Y.-Q. and Mobashery, S., J. Med. Chem., 34 (1991) 1914.

Alternative bridging groups for proline in endothelin pentapeptide receptor antagonists

S.J. Hocart[1], D.H. Coy[1], W.J. Rossowski[1], M. Neya[1], J.-P. Moreau[2] and J.E. Taylor[2]

[1]Peptide Research Laboratories, Department of Medicine,
Tulane University Medical Center, New Orleans, LA 70112, U.S.A.,
[2]Biomeasure Inc., Milford, MA 01757, U.S.A.

Introduction

The type A endothelin (ET-1) receptor antagonist, *cyclo(D-Trp-D-Asp-Pro-D-Val-Leu)* (BQ-123; analog I) [1], has been the subject of several NMR studies and appears to adopt a solution conformation consisting of a type II β-bend involving the D-Val-Leu-D-Trp-D-Asp tetrapeptide and a γ' turn centered around the Pro residue [1-3]. In the present study we have examined the ability of several cyclic, side chain-constrained amino acids to replace Pro in this structural environment.

Results and Discussion

The parent compound (I) and the following peptide analogs were prepared by solid phase peptide synthesis on *p*-nitrobenzophenone oxime resin with Boc chemistry and cleaved by *in situ* cyclization:

II	Tic	1,2,3,4-tetrahydroisoquinoline-3-carboxylic acid
III	Tcc	2,3,4,5-tetrahydro-b-carboline-4-carboxylic acid
IV	Tip	4,5,6,7-tetrahydro-1H-imidazo[4,5-c]pyridine-6-carboxylic acid
V	Oic	3aS,7aS-octahydroindole-2-carboxylic acid

The structures of these amino acids are given in Figure 1. All the peptides gave the expected amino acid analyses and were judged to be pure by HPLC and TLC. The yields of peptide from the syntheses were low and the crude material contained several side products including the diketopiperazine which formed during the cyclization. The side reactions occurred in the following order: Tip>Tcc>Tic>Pro. This spontaneous degradation has been reported recently with Tic in an unrelated tetrapeptide [4]. The biological activity of each analog was assessed in two assay systems. The binding affinity (Ki) of each peptide for ET-1 type A receptors was measured using A10 smooth muscle cells and the inhibition (ED_{50}) of ET-1 (1 nM) induced contraction of rat uterus smooth muscle was measured using young (150 g B.W.) virgin Wistar rats [5]. The results obtained are shown in Table 1.

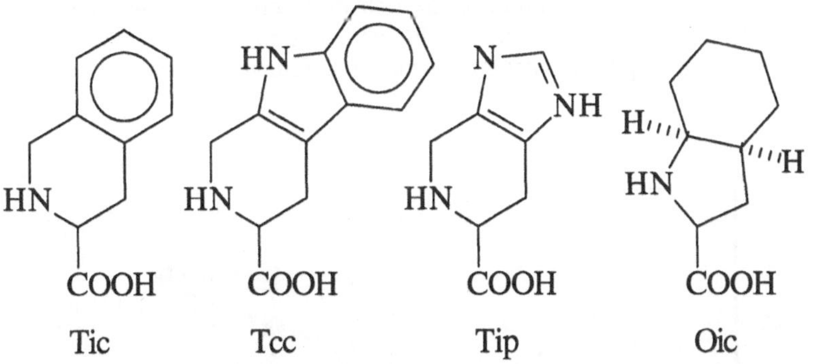

Fig. 1. Structure of the constrained amino acids used as replacements for proline.

The parent peptide I exhibited high affinity (Ki 7.9 ± 2.9 nM) for ET-1 type A receptors present on A10 smooth muscle cells and, at a concentration of 0.16 μM, completely blocked contractions of isolated rat uterus produced by 1 nM endothelin (Table 1 and Figure 2). All the analogs were able to completely block smooth muscle contractions with similar potencies to I except the Tip-containing analog IV which surprisingly exhibited full agonist contracting activity at a 10^{-6} M concentration. This could be blocked by prior addition of I and thus appeared to be caused by activation of endothelin receptors. None of the peptides exhibited significant affinity for endothelin type B receptors (not shown).

Table 1 *Analog receptor affinities for endothelin type A receptors and ED_{50} values in rat uterus smooth muscle assay*

Analog	Ki (nM)	ED_{50} (μM)
I	7.9 ± 2.9	0.16 ± 0.06
II	5.7 ± 1.5	0.87 ± 0.04
III	15 ± 4.0	0.25 ± 0.017
IV	44 ± 24	1.75 ± 0.4
V	16 ± 9.5	0.22 ± 0.02

This is, as far as we are aware, the first instance where a small antagonist structure has been converted into an agonist analog with activity comparable to the parent natural compound. In analog IV, the substitution of Tip introduces an imidazole ring into a crucial folded region of the pentapeptide, and we feel that this is probably mimicking His present in position 16 of endothelin itself. In published studies by others, replacement of this His residue in C-terminal hexapeptides with aromatic D-amino acids has also produced antagonists [6], so that His could be the active center of endothelin.

It appears that cyclic amino acids with more complex side chains than Pro perform well in this and perhaps other bridging situations and that they offer a unique method for introducing functional groups into constrained peptides.

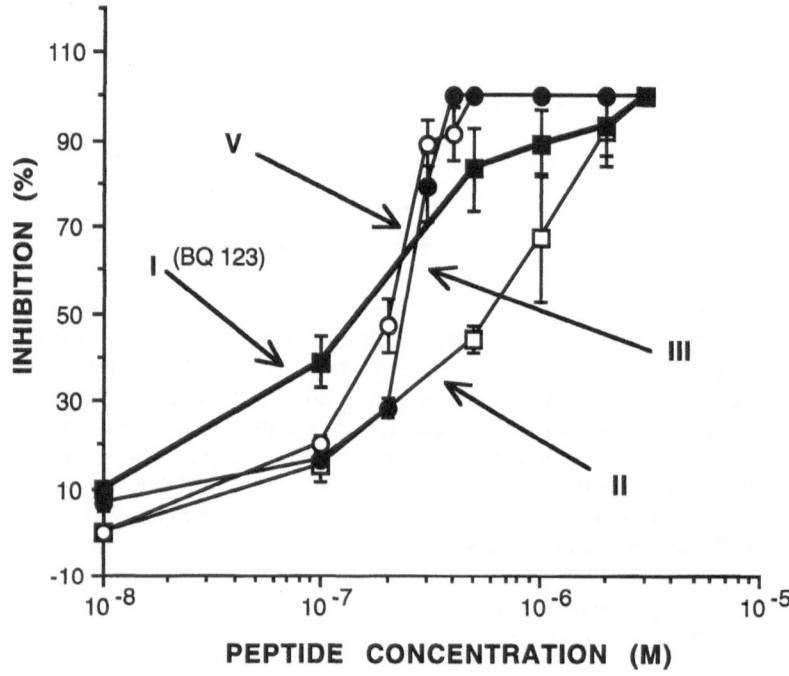

Fig. 2. *Inhibition of endothelin-1-induced rat uterus smooth muscle contraction by analogs and BQ-123.*

References

1. Atkinson, R. and Pelton, J.T., FEBS Lett., 296 (1992) 1.
2. Krystek, S.R., Jr., Bassolino, D.A., Bruccoleri, R.E., Hunt, J.T., Porubcan, M.A., Wandler, C.F. and Anderson, N.H., FEBS Lett., 299 (1992) 255.
3. Reily, M.D., Thanabal, V., Omecinsky, D.O., Dunbar, Jr., J.B., Doherty, A.M. and DePue, P.L., FEBS Lett., 300 (1992) 136.
4. Marsden, B.J., Nguyen, T.M.-D. and Schiller, P.W., Int. J. Peptide Protein Res., 41 (1993) 313.
5. Kozuka, M., Ito, T., Hirose, S., Takahashi, K. and Hagiwara, H., Biochem. Biophys. Res. Commun., 159 (1989) 317.
6. Cody, W.L., Doherty, A.M., He, J.X., DePue, P.L., Rapundalo, S.T., Hingorani, G.A., Major, T.C., Panek, R.L., Dudley, D.T., Haleen, S.J., LaDouceur, D., Kill, K.E., Flynn, M.A. and Reynolds, E.E., J. Med. Chem., 35 (1992) 3301.

Design of octadecapeptide loop structures by modeling the active-site region of barley serine proteinase inhibitor-2

F. Jean[1], A. Basak[1], H. Dugas[2], N.G. Seidah[3], M. Chrétien[4] and C. Lazure[1]

[1]Neuropeptides Structure and Metabolism Laboratory, Institut de Recherches Cliniques de Montréal, 110 Pine Avenue West, Montréal, Québec, H2W 1R7, Canada, [2]Department of Chemistry, University of Montréal and J.A. de Sève Laboratories of [3]Biochemical and [4]Molecular Neuroendocrinology, I.R.C.M., 110 Pine Avenue West, Montréal, Québec, H2W 1R7, Canada

Introduction

Biologically active peptides, including hormones, are often biosynthesized from inactive precursor molecules following cleavage by specific proteinases with a marked preference for pairs of basic residues [1]. Molecular biology techniques allowed identification of numerous candidate enzymes, known as pro-hormone convertases (PC). These enzymes are Ca^{2+}-dependent serine proteinases and possess a subtilisin-like catalytic domain [2]. To better understand their mode of action, two members of that family, PC1 (also called PC3) and furin, were expressed as vaccinia virus recombinants and characterized [3, data not shown].

Recent studies on the design of proteinase inhibitors revealed that small cyclic peptides can mimic efficiently the active-site region of larger native inhibitors. Indeed, Leatherbarrow and Salacinski [4] have successfully prepared an 18-residue cyclic loop related to part of the 83-residue barley serine proteinase inhibitor-2 (BSPI-2) known to inhibit subtilisin. Considering the similarity of the PC catalytic region to subtilisin, this cyclic structure was chosen as a starting point for preparing novel PC inhibitors.

Here, we describe the synthesis, purification and characterization of three mutated forms of the BSPI-2 reactive loop as well as preliminary results concerning their reactivities against PC1 and furin.

Results and Discussion

Three mutated forms were prepared which respectively contain one and two mutations. The first one (**I**) substitutes the normally occurring Met_{59} by an Arg residue (Figure 1). The second one (**II**) in addition to the Arg_{59} contains another Arg at position 58 in agreement with the P_1 and P_2 basic amino acid specificity of the PCs. The third one (**III**) contains Arg_{59} but, in addition, the Ile_{56} was changed into an Arg in view of the reported preference of PC1 and furin for such residues at P_1 and P_4 positions [2, 3]. Other substitutions with respect to the native BSPI-2 were done as previously described [4]. The envisioned structures were modeled and minimized using the Macromodel V3.5 software. Introduction of the single (Figure 1) or the double Arg substitutions has minimal effect on the overall conformation while the Arg side chains are predicted to protrude from the loop rendering them accessible to the enzymes.

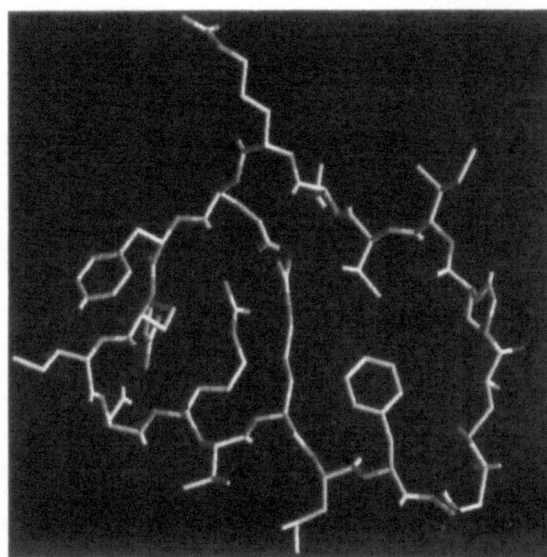
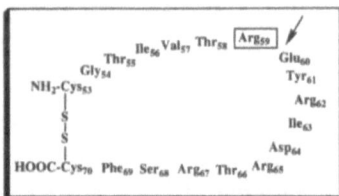

Fig. 1. Model structure based on X-ray data [5] and peptide sequence of the octadecapeptide I (numbering is based on the corresponding BSPI-2 sequence).

The synthetic peptides were prepared using FMOC chemistry on an ABI-431A peptide synthesizer with HMP-resin. The sequence **I** (giving side-chain protecting groups) synthesized was: C(X)-G-T(tBu)-I-V-T(tBu)-R(Pmc)-E(tBu)-Y(tBu)-R(Pmc)-I-D(tBu)-R(Pmc)-T(tBu)-R(Pmc)-S(tBu)-F-C(X) where X represents a Trt or an Acm group. Both **II** and **III** were prepared after splitting the resin at cycle 12. Deprotection and cleavage was achieved by two treatments with reagent K, as described [3], with the Trt-derivatives or with TFA:phenol:water (v/v 90:5:5) for the Acm-derivatives. Following purification by RP-HPLC, all peptides were characterized by amino acid composition and ion spray/FAB-mass spectrometry (MS).

Recovery of the cyclic peptides led to numerous problems. Indeed, cyclization of **I** and **II**-Trt-derivatives in their resin-bound forms [6] led to formation of multimers based on MS data. Similarly, treatment of the purified Trt-peptides, as suggested [7], yielded minimal amounts of the expected monomer. Finally, as shown in Figure 2 for **I**, treating the **I** and **III**-Acm derivatives with methanolic iodine solution [7] yielded the desired product in good yields. Both amino acid composition and MS-data agree with the expected values. The presence of the disulfide bridge was confirmed by examining the chromatographic behavior of the peptides following treatment with either DTT or mercuric acetate; in both cases, the linear peptides are eluting later than their cyclic counterparts.

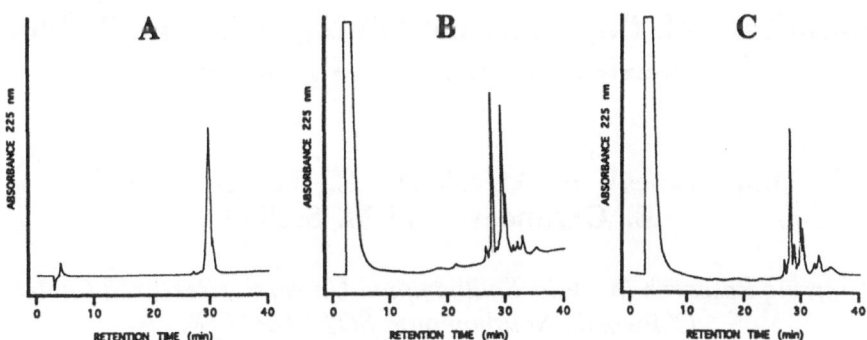

Fig. 2. *RP-HPLC analysis of iodine oxidation to yield cyclic I. Conditions were as described in [3]. (A) T=0; (B) 4 equivalents I_2, 1 h; (C) 8 equivalents I_2, 2 h.*

Incubation of the linear Acm protected-**I** and -**III** with either recombinant PC1 or furin showed that **III** behaves as substrate for furin and PC1 whereas **I** under identical conditions, was almost entirely left intact; similar results with smaller P_1 and P_4 Arg-containing substrates were obtained previously confirming their importance for proper recognition by the enzymes. Furthermore, the cyclic **I** derivative proved to be an inhibitor of PC1 when assayed with small fluorogenic substrates [3] whereas the cyclic **III** peptide did not inhibit PC1.

In conclusion, it does appear that this approach could prove useful in developing related inhibitors for convertases. In any case, more studies on similar derivatives are required in order to assess their usefulness.

Acknowledgements

The authors would like to thank Ms. Dany Gauthier for expert technical assistance and the MRC for financial support (PG 11474). C.L. is a "chercheur-boursier" of the FRSQ and F.J. is the recipient of a studentship from the FCAR and the University of Montréal.

References

1. Lazure, C., Seidah, N.G., Pélaprat, D. and Chrétien, M., Can. J. Biochem. Cell Biol., 61 (1983) 501.
2. Seidah, N.G., Day, R., Marcinkiewicz, M., Benjannet, S. and Chrétien, M., Enzyme, 45 (1992) 271
3. Jean, F., Basak, A., Rondeau, N., Benjannet, S., Hendy, G.N., Seidah, N.G., Chrétien, M. and Lazure, C., Biochem. J., 292 (1993) 891.
4. Leatherbarrow, R.J. and Salacinski, H.J., Biochemistry, 30 (1991) 10717.
5. McPhalen, C.A. and James, M.N.G., Biochemistry, 26 (1987) 261.
6. Seidel, C., Klein, C., Empl, B., Bayer, H., Lin, M. and Batz, H.G. In Giralt, E. and Andreu, D. (Eds.), Peptides 1990, Escom, Leiden, The Netherlands, 1991, p.236.
7. Kamber, B., Hartmann, A., Eisler, K., Riniker, B., Rink, H., Sieber, P. and Rittel, W., Helv. Chim. Acta, 63 (1980) 899.

Inhibitors of thrombin containing a ketomethylene isostere of the P_1 - P_1' bond

D.M. Jones[1], B. Atrash[1], A.C. Teger-Nilsson[2], E. Gyzander[2] and M. Szelke[1]

[1]Ferring Research Institute, Southampton University Research Centre, Chilworth, Southampton SO1 7NP, U.K.
[2]Astra Hässle AB, S-431 83 Mölndal, Sweden

Introduction

In recent years much attention has been focused on inhibitors of thrombin as potential anticoagulants [1]. As part of our programme directed at the development of enzyme inhibitors based on transition-state mimics [2, 3], we discovered in the early 1980s that fragments of human fibrinogen Aα containing a ketomethylene isostere [4] at P_1 - P_1' were potent inhibitors of thrombin [5]. More recently, ketomethylene dipeptide isosteres incorporated into fragments of hirudin have also been shown by DiMaio and co-workers to yield effective thrombin inhibitors [6]. We describe here the synthesis, structure-activity relationships and *in vivo* anticoagulant properties of novel ketomethylene analogues based on human fibrinogen.

Results and Discussion

Suitably protected Arg^KGly was synthesised by the method described in our patent [5] and shown in Figure 1. An analogous route to the naturally occurring ketomethylene dipeptide isostere H-Arg^KPhe-OH was reported by Umezawa et. al. in 1983 [7] and a more general one for X^KY by García-López et. al. in 1988 [8]. Contrary to the observations of the latter authors, we found that ca. 20-30% racemisation occurred at the Arg α-carbon during synthesis which appeared to be general in other ketomethylene isostere syntheses (unpublished observation). However, the diastereomers were readily separated by chromatography on silica gel prior to final deprotection. All inhibitors were homogeneous by RP-HPLC and their structures were confirmed by amino acid analyses and FAB-MS.

$$Boc\text{-}Arg(Z)_2OH \;\; \xrightarrow{\text{(i), (ii), (iii)}} \; Boc\text{-}Arg(Z)_2CH_2Br$$

$$\xrightarrow{\text{(iv)}} Boc\text{-}Arg(Z)_2CH_2CH(CO_2Tce)_2 \;\; \xrightarrow{\text{(v), (vi)}} \; Boc\text{-}Arg(Z)_2{}^KGly\text{-}OH$$
*

Fig. 1. Synthesis of Boc-Arg(Z)$_2{}^K$Gly-OH. Reaction conditions (i) NMM, IBC, -10°C, THF; (ii) CH_2N_2/Et_2O, 1h; (iii) HBr-EtOAc, -15°C, 15'; (iv) $H_2C(CO_2Tce)_2$, NaH,THF, 0°C, 2h; (v) Zn-HOAc, 3h; and (vi) Reflux, toluene, 3h.
*20-30% racemisation observed; Tce = 2,2-Trichloroethyl.

Table 1 shows a representative selection of inhibitors synthesised together with their K_i values for hTh, IC_{50} TT and ED. Structure **1** corresponds to the natural P_3 - P_3' sequence of human fibrinogen Aα and **2** contains the P_3 - P_2 replacement discovered by Bajusz [9].

Table 1 *Thrombin inhibitors based on Arg^{K}Gly*

No.	Th P_3 P_2 $P_1' \downarrow P_2'$ P_3'	K_i vs. hTh[b] nM	IC_{50}TT[c] nM	ED[d] min
1	H-Gly- Val-Arg-Gly-Pro-Arg-OH			
2	H-**DPhe-Pro**-Arg-Gly-Pro-OH	443,000		
3	H-DPhe-Pro-**Arg^{K}Gly**-Pro-**NHEt**[a]	1,000	1,000	36
4	H-DPhe- Pro-Arg^{K}Gly-**Prl**	1,700	1,900	nd
5	H-**DCha**-Pro-Arg^{K}Gly-Prl	150	210	22
6	H-DCha-Pro-Arg^{K}Gly-**Phe-NHEt**	10	24	11
7	H-DCha-Pro-Arg^{K}Gly-**NHCH(Me)Ph** ^(R)	20	41	12
8	H-DCha-Pro-Arg^{K}Gly- **Thi**	12	26	12
9	H-DCha-Pro-Arg^{K}Gly-**Phe-Agm**	2.5	9.5	12
10	H-DCha-Pro-**Lys^{K}Gly**-Phe-NHEt	520	480	22

[a]Abbreviations: $\underline{K}$ = -COCH$_2$- in place of -CONH-; Cha = 2-cyclohexylalanine; Prl = pyrrolidide; Agm = agmatine; hTh = human thrombin; TT = thrombin-induced clotting time in human plasma; Thi = N-1,2,3,4-tetrahydroisoquinoline; nd = not determined; [b]K_i values determined using chromogenic substrate S-2238 (Kabi) and Dixon plots. [c]Concentration of inhibitor that doubles the TT. [d]Effect duration, in vivo (rats): arbitrarily defined here as time interval between maximum TT and reduction to 1/3 of maximum TT.

In rats, inhibitor **3** showed a bioavailability of ~5% after intra-duodenal administration and was metabolically stable. Inhibitor **3** exhibited a significant beneficial effect in a model of venous thrombosis in the rat. Details of these investigations will be published separately.

In summary:
1. Potent inhibitors (K_i = 10^{-7} - 10^{-8} M) of human thrombin have been developed from the P_3 - P_3' fragment of human fibrinogen Aα by introducing (a) a ketomethylene isostere of the scissile Arg-Gly bond; introducing a Lys side-chain in place of the Arg side-chain at P_1 leads to a 50-fold loss of activity; (b) DCha at P_3; and (c) Phe or an aromatic amide at P_2'.
2. These inhibitors significantly prolonged thrombin time in human plasma.
3. They had duration of action in rats of 12 - 36 min.

Acknowledgements

We thank Astra Hässle AB and Ferring AB for financial support, and Dr. Kurt-Jürgen Hoffman for the pharmacokinetic studies.

References

1. Jakubowski, J.A., Smith, G.F. and Sall, D.J., Ann. Rep. Med. Chem., 27 (1992) 99.
2. Szelke, M., Leckie, B., Hallett, A., Jones, D.M., Sueiras, J., Atrash, B. and Lever, A.F., Nature, 299 (1982) 555.
3. Szelke, M., Jones, D.M., Dudman, P.A., Atrash, B., Sueiras-Diaz, J. and Leckie, B.J., In Rivier, J.E. and Marshall, G.R. (Eds.), Peptides: Chemistry, Structure and Biology, Escom, Leiden, The Netherlands, 1990, p.78.
4. Sharpe, R. and Szelke, M., G.B. Patent 1,587,809 (priority date 15th March 1977) and corresponding foreign patents.
5. Szelke, M. and Jones, D.M., Eur. Patent 118,280 (G.B. priority date 4th March 1983).
6. DiMaio, J., Gibbs, B., Lefebre, J., Konishi, Y., Munn, D. and Yue, S.Y., J. Med. Chem., 35 (1992) 3331.
7. Umezawa, H., Nakamura, T., Fukatzsu, S., Aoyagi, T. and Tatsuta, K., J. Antibiot., 36 (1983) 1787.
8. García-López, M.T., González-Muñíz, R. and Harto, J.R., Tetrahedron Lett., 29 (1988) 1577.
9. Bajusz, S., Barabás, E., Széll, E. and Bagdy, D., In Walter, R. and Meienhofer, J. (Eds.), Peptides: Chemistry, Structure and Biology, Ann Arbor Science Publishers, Ann Arbor, MI, U.S.A., 1975, p.603.

The promising anti-HIV agent kynostatin (KNI)-272: A highly selective and super-active HIV protease inhibitor containing allophenylnorstatine

Y. Kiso[1], T. Mimoto[1,2], J. Imai[1], S. Kisanuki[1,2], H. Enomoto[1], N. Hattori[2], K. Takada[1], K. Akaji[1], S. Kageyama[3] and H. Mitsuya[3]

[1]Departments of Medicinal Chemistry and Pharmaceutics, Kyoto Pharmaceutical University, Kyoto 607, Japan, [2]Nikko Kyodo Co. Ltd., Toda 335, Japan, and [3]National Cancer Institute, Bethesda, MD 20892, U.S.A.

Introduction

Human immunodeficiency virus (HIV) codes for an aspartic protease known to be essential for retroviral maturation and replication. The HIV protease can recognize Phe-Pro and Tyr-Pro sequences as the virus specific cleavage sites. These features provided a basis for the rational design of selective HIV protease-targeted drugs for the treatment of AIDS. Based on the substrate transition state, we designed a novel class of HIV protease inhibitors [1-4] containing allophenylnorstatine (Apns: (2S, 3S)-3-amino-2-hydroxy-4-phenylbutyric acid) with a hydroxymethylcarbonyl (HMC) isostere [5]. Having identified the tripeptide derivative KNI-102 (Z-Asn-Apns-Pro-NHBu^t) as a lead compound, we undertook a study of lead optimization (Figure 1). The conformationally constrained tripeptide (KNI-174; Noa-Asn-Apns-Dmt-NHBu^t) exhibited remarkable increase in potency. Further structure-activity relationship study considering the penetration across cell membrane and the behaviour *in vivo* led to the promising protease active site-targeted anti-HIV agent kynostatin (KNI)-272 (iQoa-Mta-Apns-Thz-NHBu^t).

Results and Discussion

Combinations of each preferred side chain led to highly potent HIV protease inhibitors, such as KNI-174 (Ki=0.0068 nM), KNI-225 and KNI-170. These compounds exhibited potent antiviral activities against HIV-1 in CD4$^+$ ATH8 cells.

The behaviors of compounds *in vivo*, such as penetration across the cell membrane and non-specific adsorption in blood, are important factors for the *in vivo* antiviral activity. Therefore, considering the subtle balance of lipophilicity-hydrophilicity and molecular size, we incorporated the 5-isoquinolinyloxyacetyl (iQoa) moiety at the P$_3$ position and combined it with each preferred side chain. Such modifications resulted in the super-active HIV protease inhibitors KNI-227 (iQoa-Mta-Apns-Dmt-NHBu^t; Ki=0.0023 nM) and KNI-272 (Ki=0.0055 nM) [4, 6]. Both compounds exhibited potent antiviral activities against clinical HIV-1 isolates in phytohemagglutinin-stimulated peripheral blood mononuclear cells (PHA-PBM) assay (Table 1). These antiviral activities against clinical HIV-1 isolates were more than 10-fold potent compared to a C_2 symmetric protease inhibitor, A-77003 [7]. As expected from the protease active site-targeted antiviral mechanism, these also exhibited potent antiviral activities (IC_{50}=0.01 µM for both compounds) against AZT-insensitive clinical HIV-1 isolates.

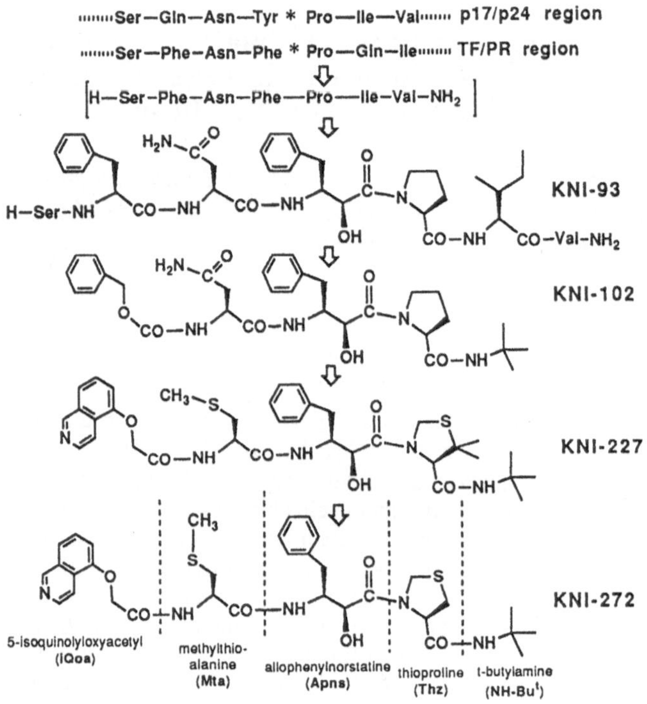

Fig. 1. Design of HIV protease inhibitors.

Also, KNI-227 and KNI-272 showed potent antiviral activities against the infectivity and cytopathic effect of HIV strains, including HIV-1$_{LAI}$, HIV-1$_{RF}$, HIV-1$_{MN}$ and HIV-2$_{ROD}$, as tested in CD4$^+$ ATH8 cells. The 50% inhibitory concentrations (IC_{50}) of KNI-227 against these strains were 0.1, 0.02, 0.03 and 0.13 µM, while those of KNI-272 were 0.1, 0.02, 0.04 and 0.14 µM, respectively. From the viewpoint of the action mechanism, the active site-targeted HIV protease inhibitors have reason to exhibit activities against a wide spectrum of HIV strains, including HIV-2.

KNI-272 exhibited excellent enzyme selectivity, with practically no inhibition of other aspartic proteases such as human plasma renin (IC_{50}>100,000 nM), porcine pepsin (IC_{50}>80,000 nM) and bovine cathepsin D (IC_{50}>100,000 nM). On the other hand, IC_{50} values of KNI-227 were >100,000 nM(renin), 13,000 nM(pepsin) and 23,000 nM(cathepsin D). Although the P$_1$' dimethylthioproline (Dmt) residue remarkably enhanced the protease inhibitory activity due to the conformational constraint and hydrophobicity of the bulky dimethyl group, enzyme selectivity of KNI-227 was a little lower than KNI-272 with P$_1$' thioproline (Thz). This may be explained by the fact that Dmt residue is regarded as a hybrid-type of Thz and Val, and that Apns-Thz is a closer mimic to the virus specific Phe-Pro sequence than Apns-Dmt.

Table 1 *Antiviral activity against clinical HIV-1 isolates in PHA-PBM^a*

Compound	Structure P3 P2 P1 P1' P2' Phe -Asn-Phe -Pro -Ile	IC_{50} (μM)	IC_{90} (μM)	TC_{50} (μM)	TC_{50}/IC_{50}
KNI-154	Noa-Asn-Apns-Thz-NHBu^t	0.14	0.24	77	550
KNI-174	Noa-Asn-Apns-Dmt-NHBu^t	0.13	0.23	32	246
KNI-170	Noa-Msa-Apns-Dmt-NHBu^t	0.03	0.05	19	633
KNI-225	Noa-Mta-Apns-Dmt-NHBu^t	0.03	0.06	8	267
KNI-227	iQoa-Mta-Apns-Dmt-NHBu^t	0.01	0.04	49	4900
KNI-272	iQoa-Mta-Apns-Thz-NHBu^t	0.01	0.04	>80	>8000

$^a IC_{50}$ values were determined based on levels of p24 gag protein production. Abbreviations: Noa=1-naphthyloxyacetyl, Apns=allophenylnorstatine, Thz=L-thiazolidine-4-carboxylic acid, Dmt=L-5,5-dimethylthiazolidine-4-carboxylic acid, Msa=L-methanesulfonylalanine, Mta=L-methylthioalanine, iQoa=5-isoquinolyloxyacetyl.

Interestingly, a relatively low-lipophilic and small-sized tripeptide derivative, KNI-272, combined with iQoa moiety and Thz residue, exhibited highly potent antiviral activities and low cytotoxicity (TC_{50}>80μM). After intraduodenal (i.d.) administration in rats [8], bioavailabilities of KNI-174, KNI-227 and KNI-272 were 5.37%, 5.90% and 42.3% respectively. As compared with other HIV-protease inhibitors, the bioavailability of KNI-272 was improved.

In conclusion, ease in the synthetic procedure, the HIV protease-specific inhibition and potent antiviral properties as well as favorable cytotoxicity profile warrant further studies in the direction of clinical application of KNI-272 as an oral anti-HIV drug.

References

1. Mimoto, T., Imai, J., Tanaka, S., Hattori, N., Takahashi, O., Kisanuki, S., Nagano, Y., Shintani, M., Hayashi, H., Sakikawa, H., Akaji, K. and Kiso, Y., Chem. Pharm. Bull., 39 (1991) 2465.
2. Mimoto, T., Imai, J., Tanaka, S., Hattori, N., Kisanuki, S., Akaji, K. and Kiso, Y., Chem. Pharm. Bull., 39 (1991) 3088.
3. Mimoto, T., Imai, J., Kisanuki, S., Tanaka, S., Hattori, N., Takahashi, O., Katoh, R., Yumisaki, T., Sakikawa, H., Akaji, K. and Kiso, Y., In Suzuki, A. (Ed.), Peptide Chemistry 1991, Protein Res. Found., Osaka, Japan, 1992, p.395.
4. Mimoto, T., Imai, J., Kisanuki, S., Enomoto, H., Hattori, N., Akaji, K. and Kiso, Y., Chem. Pharm. Bull., 40 (1992) 2251.
5. Iizuka, K., Kamijo, T., Harada, H., Akahane, K., Kubota, T., Umeyama, H., Ishida, T. and Kiso, Y., J. Med. Chem., 33 (1990) 2707.
6. Kageyama, S., Mimoto, T., Murakawa, Y., Nomizu, M., Ford, H., Shirasaka, T., Gulnik, S., Erickson, J., Takada, K., Hayashi, H., Broder, S., Kiso, Y. and Mitsuya, H., Antimicrob. Agents Chemother., 37 (1993) 810.
7. Mitsuya, H., Yarchoan, R., Kageyama, S. and Broder, S., FASEB J., 5 (1991) 2369; Kageyama, S., Weinstein, J.N., Shirasaka, T., Kempf, D.J., Norbeck, D.W., Plattner, J.J., Erickson, J. and Mitsuya, H., Antimicrob. Agents Chemother., 36 (1992) 926.
8. Kiriyama, A., Mimoto, T., Kiso, Y. and Takada, K., Biopharm. Drug Dispos., 14 (1993) 199.

Tetrapeptide based inhibitors of p21[ras] protein farnesyl transferase

K. Leftheris, T. Kline, W. Lau, L. Mueller, V.S. Goodfellow, M.K. DeVirgilio, Y.H. Cho, C. Ricca, S. Robinson, V. Manne and C.A. Meyers

Bristol-Myers Squibb Pharmaceutical Research Institute, Princeton, NJ 08540, U.S.A.

Introduction

The protein farnesyl transferase (FT) farnesylates p21[ras] on the Cys residue of the C-terminal consensus sequence referred to as a CA_1A_2X box [1]. This post-translational modification is required for plasma membrane association and function of normal and transforming ras activity. Transformed ras proteins are implicated in a number of human cancers including colon, pancreatic and lung carcinomas. Therefore, selective inhibition of FT could lead to a new class of potent and selective anticancer agents. CVFM is a very potent inhibitor of FT (IC_{50} = 37 nM). Unlike known tetrapeptide substrate inhibitors containing aliphatic residues at the A_2 position (CVVM, CVIM), CVFM appears not to be farnesylated; i.e., it is a true inhibitor. To better define the structural requirements for both the substrate and inhibitor peptide classes, analogs containing backbone amide modifications as well as replacements at the A_1 and A_2 positions were explored. Minimized energy conformations of selected compounds were measured to determine any conformational preferences.

Results and Discussion

Four series of compounds were investigated (see Table 1). These include compounds containing A_2 modifications, A_1 turn-inducing residues, methyleneamine amide bond isosteres and N-methyl derivatives. Peptides containing A_1 and A_2 replacements were assembled on a Wang resin using BOP/HOBt couplings and N^αFmoc-protected amino acids. The methyleneamine-containing peptides were prepared via the N,O-dimethylhydroxamate using the methods of Castro [2] and Borch [3]. N-methyl containing peptides were assembled using previously described methods [4-6]. The Monte Carlo conformational searches of CVFM, CAibFM and CV(N-Me)FM were carried out in MACROMODEL using the AMBER force field. Minimizations were determined in vacuo using a distance dependent dielectric of 4r, an electrostatics cutoff of 12 Å, a van der Waals cutoff of 7 Å, and a hydrogen bonding cutoff of 4 Å. The *in vitro* inhibition assay was carried out essentially as described [7] and was adapted to TCA precipitation in microtiter plates. The TLC procedure used for the substrate assay was taken from a previously described method [8].

Table 1 *Peptide-based inhibitors of p21^{ras} protein farnesyl transferase*

Structure	IC$_{50}$ (nM)	Substrate
CVFM	37	no
CVVM	68	yes
CVPhgM	696	yes
CVHphM	120	yes
CAibFM	400	no
CAcc^{6}FM^a	1200	no
Cψ[CH$_2$NH]VFM	12	no
Cψ[CH$_2$NH]VVM	29	yes
CVFψ[CH$_2$NH]M	2220	no
N$^{\alpha}$Me-CVFM	420	yes
C(N$^{\alpha}$Me)VFM	250	no
CV(N$^{\alpha}$Me)FM	360	no
CVF(N$^{\alpha}$Me)M	1400	no
C(N$^{\alpha}$Me)VVM	90	NT
CV(N$^{\alpha}$Me)VM	120	NT
CVV(N$^{\alpha}$Me)M	6750	NT

[a]Abbreviations: Acc^6 - aminocyclohexylcarboxylic acid, Aib - aminoisobutyric acid, Hph - homophenylalanine, Phg - phenylglycine.

As shown in Table 1, distance between the aromatic ring at the A$_2$ position and the backbone is critical for nonsubstrate inhibition. Placing the aromatic ring closer or farther away from the backbone (as in CVPhgM or CVHphM) results in a less potent inhibitor that *reverts to a substrate*. Two turn-inducing replacements at the A$_1$ (i + 1) position were investigated to explore the possibility of a turn as the preferred conformation. These compounds were weak nonsubstrate inhibitors. The methylene-amine amide bond replacement between Cys and Val for both CVVM and CVFM *conferred improved potency* over the progenitor peptides. This is a very critical discovery since it is the first step toward developing a proteolytically stable inhibitor. This same modification between Phe and Met led to a very poor inhibitor, suggesting that either the Phe-Met amide carbonyl is involved in a critical hydrogen bond or a rigid geometry is required at that position. N-methyl peptides were explored exhaustively in CAAX tetrapeptides. For the CVFM series, this modification in any position generally led to a weaker nonsubstrate inhibitor. Interestingly, methylating the N-terminus of CVFM generated a substrate. In the CVVM series, methylation was well tolerated at both the A$_1$ and A$_2$ positions. These findings suggest there are critical interactions required for nonsubstrate (CVFM-like) inhibitor binding which may be less necessary for substrate inhibition. In both the CVVM and the CVFM series, N-methylating the C-terminal methionine led to very weak inhibition, suggesting that the A$_2$-X amide may play a similar role (i.e., through one or several critical interactions) in both substrate and inhibitor peptides.

Monte Carlo (MC) conformational searches of CVFM, CAibFM and CV(N-Me)FM were carried out to determine if these compounds adopted a conformational preference. For each peptide, the NH-amide bonds were kept in the transconfiguration and were not varied during the MC searches. However, as the N(Me)-amide for CV(N-Me)FM could exist in the trans or the cis configuration, two separate MC searches were carried out for this compound with the amide bond fixed in either the trans or cis configurations. The lowest energy conformer for all three peptides was found to be extended. A comparison of the conformational profile of the three peptides suggests that CVFM is a relatively flexible molecule. Substitution of Val with Aib induces more flexibility into the first two residues due to increased flexibility at A_1. However, N-methylation of Phe reduces the flexibility of A_1 and A_2 and also reduces the torsional variation of the Phe sidechain. This results in an overall reduction of the flexibility of the peptide. Both substitution with Aib and N-methylation led to a loss of activity. For CAibFM, these effects could be due to 1) greater conformational flexibility which is entropically unfavorable or 2) steric hindrance from the α-methyl group (the Acc^6 analog is less active than Aib). For CV(N-Me)FM, loss of activity could be from 1) cis-trans isomerization, 2) loss of amide NH for hydrogen bonding, 3) steric occlusion from the N-methyl. Calculations suggest that CAibFM does promote turn formation. Type I and II β turns were found that were 7 kcal higher in energy than the lowest energy conformer. For CVFM, type I and II β turns were 10.3 kcal and 11.1 kcal higher in energy than the lowest energy conformer. These findings suggest that although turns can exist, they are unlikely to be preferred in these compounds.

In conclusion, no definitive active conformational structure has yet been identified for potent peptide-based inhibitors of $p21^{ras}$ FT. As we identify active inhibitors, we will continue to probe secondary structural motifs as a means to identify active conformations.

References

1. Reiss, Y., Goldstein, J.L., Seabra, M.C., Casey, P.J. and Brown, M.S., Cell, 62 (1990) 81.
2. Fehrentz, J. and Castro, B., Synthesis, (1983) 676.
3. Borch, R.F., Bernstein, M.D. and Durst, H.D., J. Am. Chem. Soc., 93 (1971) 2897.
4. Freidinger, R.M., Hinkle, J.S., Perlow, D.S. and Arison, B.H., J. Org. Chem., 48 (1983) 77.
5. Diago-Meseguer, J. and Palomo-Coll, A.L., Synthesis, (1980) 547.
6. Yamashiro, D., Aanning, H.L., Branda, L.A., Murti, V.V.S. and du Vigneaud, V., J. Am. Chem. Soc., 90 (1968) 4141.
7. Manne, V., Roberts, D., Tobin, A., O'Rourke., E., De Virgilio, M., Meyers, C., Ahmed, N., Kurz, B., Resh, M., Kung, H.-F. and Barbacid, M., Proc. Natl. Acad. Sci. U.S.A., 87 (1990) 7541.
8. Goldstein, J.L., Brown, M.S., Stradley, S.J., Reiss, Y. and Gierasch, L.M., J. Biol. Chem., 266 (1991) 15575.

Bestatin-derived analogs inhibit *E. coli* methionine aminopeptidase

S. Lim and D.H. Rich

School of Pharmacy and Department of Chemistry,
University of Wisconsin-Madison, Madison, WI 53706, U.S.A.

Introduction

Methionine aminopeptidase (MAP) catalyzes the co-translational removal of amino-terminal methionine from proteins. *E. coli* MAP contains two divalent cobalt ions which are not replaceable by Zn^{2+}, Mg^{2+} or Mn^{2+} [1]. Although this unique cobalt protease enzyme is structually similar to the Zn metalloprotease, bovine lens leucine aminopeptidase (blLAP), in that the analogous binuclear metals are separated by the same distance (2.9 Å) [2], the two aminopeptidases do not share sequence homology nor 3-dimensional structural similarity.

Bestatin, a slow- and tight-binding inhibitor of LAP, was postulated to bind in the active site of LAP with its α-amino group and hydroxyl group coordinated to the zinc ion [3], which was confirmed by the X-ray structure of bestatin complexed to blLAP [4]. Bestatin thus may be a natural mimetic of a reaction pathway intermediate for the aminopeptidase-catalyzed hydrolysis of substrate.

We report here the design and synthesis of the first tight-binding inhibitor of *E. coli* MAP. Our strategy for inhibitor design follows that developed for designing inhibitors of other enzymes by inserting the reaction pathway mimetic found in one natural enzyme inhibitor into the target substrate structure [5]. *E. coli* MAP substrate specificity is quite different from that of zinc aminopeptidases. MAP strongly favors substrates having a small and uncharged amino acid (Gly, Ala, Ser, Pro, Val, Thr and Cys) at the P_1' position. Only L-Met or L-Nle are tolerated at the N-terminus of substrate, and the enzyme seems to utilize only tripeptides or longer peptides. We synthesized several bestatin analogs, **1 - 6**, which contain Nle-Ala side chains in place of Phe-Leu of bestatin. The Nle side chain of bestatin analogs is asssumed to bind in the S_1 subsite and Ala side chain then binds in the S_1' subsite.

Results and Discussion

Bestatin is a poor inhibitor of *E. coli* MAP (K_i >1 mM), as is amastatin, a tetrapeptide analog of bestatin (K_i = 188 μM). These results could be explained by steric factors, because MAP requires a small amino acid in S_1', and both bestatin and amastatin have a large amino acid at P_1'. Accordingly, we designed our analogs to more closely mimic the MAP substrate.

Table 1 *Inhibition of* E. coli *MAP by bestatin analogs*

Compounds	X	C_2	C_3	K_i (μM)
1	OH	S	S	212
2	OH	S	R	748
3	NHNH-Dns	RS[a]	S	119
4	NHNH-Dns	S	R	5.6
5	Leu-Val-Phe-OMe	RS[a]	S	17.7
6	Leu-Val-Phe-OMe	S	R	0.0236[b]

[a] The ratio of R:S is 4:1.
[b] Slow-binding inhibitor (10 min. preincubation at 30°C).

Compounds **1** - **6** were synthesized by a stepwise solution strategy [6], and assayed by the HPLC method with H-Nle-Ala-NHNH-Dns as substrate, an adaptation of the method of Lin and Wart [7].

The results shown in Table 1 indicate that the length of peptide chain and the configuration at C_2 and C_3 are critical factors for the activity. Dipeptide analogs **1** and **2** are weak inhibitors, slightly more potent than bestatin. Substrate analogs **3** and **4**, where the hydrazide dansyl group probably binds in the S_2' pocket, are better inhibitors than the dipeptide analogs. The introduction of the hydrazide dansyl group decreased K_i 130-fold in the case of (2S,3R)-3-amino-2-hydroxyheptanoic acid (AHHpA) derivatives (**2** vs **4**). Pentapeptide analogs **5** and **6** are even better inhibitors than the dipeptide analogs. A decrease in K_i of 32,000-fold is seen with the (2S, 3R) configuration at AHHpA moiety (**2** vs **6**).

Slow-binding inhibition of aminopeptidases is characteristic of bestatin and its analogs. Among compounds in Table 1, only the pentapeptide analogs containing (2S,3R)-configuration (**6**) show slow-binding behavior.

In conclusion, a pentapeptide analog of bestatin (**6**) was shown to be the first slow, tight-binding inhibitor of *E. coli* MAP reported to date. The stereochemistry of the amino-substituted carbon atom is the same as found in the natural product bestatin but reversed (D-configuration) from that found in substrate (L-configuration).

Acknowledgements

This work was supported by a grant from NIH (GM 40092). S.L. is grateful for an American Peptide Society Travel Award.

References

1. Ben-Bassat, A., Bauer, K., Chang, S.Y., Myambo, K., Boosman, A. and Chang, S., J. Bacteriol., 169 (1987) 751.
2. Roderick, S.L. and Matthews, B.M., Biochemistry, 32 (1993) 3907.
3. Takita, T., Nishizawa, R., Saino, T., Suda, H., Aoyagi, T. and Umezawa, H., J. Med. Chem., 20 (1977) 510.
4. Burley, S.K., David, P.R. and Lipscomb, W.N., Proc. Natl. Acad. Sci. U.S.A., 88 (1991) 6916.
5. Wiley, R.A. and Rich, D.H., Medicinal Research Reviews, Vol. 13, No. 3, John Wiley & Sons, Inc., New York, NY, U.S.A., 1993, p.327.
6. Rich, D.H., Moon, B.J. and Boparai, A., J. Org. Chem., 45 (1980) 2288.
7. Lin, W.-Y. and Wart, H.E.V., Biochemistry, 27 (1988) 5054.

Structure-based design and synthesis of inhibitors of a drug resistant mutant of HIV-1 protease

P. Majer[1], S.K. Burt[1], S. Gulnik[1], D.D. Ho[2] and J.W. Erickson[1]

[1]Structural Biochemistry Program, PRI/DynCorp,
National Cancer Institute, Frederick, MD 21702, U.S.A.,
[2]Aaron Diamond AIDS Research Center, NYU School of Medicine,
New York, NY 10016, U.S.A.

Introduction

A number of antiviral agents have been developed for treatment of AIDS that target the HIV-1 protease, an enzyme essential for the processing of viral proteins and virion maturation. The HIV-1 protease is a highly specific enzyme such that its primary sequence is highly conserved among differing viral strains. Recent studies [1] revealed the emergence of HIV-1 strains in vitro with mutated protease that exhibit resistance towards inhibitors. Recently [1] we examined HIV-1 resistance to the pseudo C_2 symmetric inhibitor A-77003 [2] (Figure 1) by growing NL4-3, a virus derived from an infectious molecular clone, in MT-4 cells in the presence of subinhibitory concentrations of A-77003. Considerable increase of ID_{50} was observed after 14-19 passages (10 to 35 times). The amino acid sequences of mutant HIV-1 protease regions were determined using PCR, cloning into M13 phage followed by nucleotide sequencing of multiple clones. The most abundant mutation was an arginine to glutamine substitution at position 8 in the protease sequence (R8Q). This mutant was subsequently cloned, expressed and used for the present study. The main goal of this study was to design inhibitors that are equally active towards both the wild type (WT) and R8Q mutant enzymes.

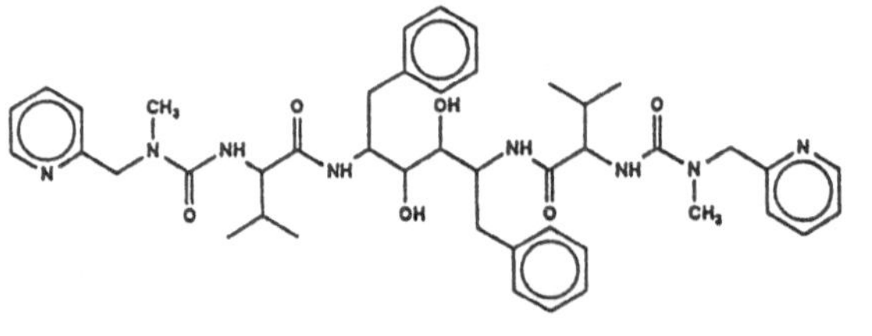

	Ki [nM]	
	WT	R8Q
	0.012	0.6

Fig. 1. C_2 pseudosymmetric inhibitor A-77003 [2].

Results and Discussion

a)   *Synthesis of inhibitors*

The inhibitor core unit (2,5-diamino-1,6-diphenyl-3,4-hexandiol) was synthesized by McMurry's pinacol coupling of Boc-phenylalaninal [3], that was obtained from Boc-phenylalaninol by oxidation with periodinane [4]. The SRSS isomer (pseudomesoform) was isolated by chromatography and used for synthesis of inhibitors. Coupling steps were performed using TBTU/HOBt/DIEA method.

The N,N'- unsymmetrically disubstituted urea for synthesis of inhibitor (IV) has been prepared by a newly developed method based on subsequent addition of different amino components to a solution of triphosgene. The reactive intermediate of this reaction (more likely amidoylchloride than isocyanate) is apparently less reactive than phosgene (triphosgene) itself. This enables one to distinguish substitution of the first and the second chlorine atom in the formal phosgene molecule.

$$\text{1eq. } R_1\text{-NH}_2.\text{HCl } + \text{ 1/3eq. Cl}_3\text{COCOOCCl}_3$$

$$1. \quad \downarrow \text{ 2.2 eq. } iPr_2NEt$$

$$[\, R_1\text{-NH-CO-Cl} \,] \quad + \quad \text{1eq. } R_2\text{-NH}_2.\text{HCl}$$

$$2. \quad \downarrow \text{ 2.2 eq. } iPr_2NEt$$

$$R_1\text{-NH-CO-NH-R}_2$$

Fig. 2. Synthesis of unsymmetrical urea based on subsequent addition of amino components to the solution of triphosgene.

b)   *Activities of inhibitors*

The R8Q mutation strongly affects the S_3 (S_3') subsite of the protease binding pocket. From the crystal structure of the A-77003/protease complex, it is known that the pyridine groups interact closely with the Arg8 (resp. Arg108) residue. The planes of the interacting pyridine ring and guanidinium group are approximately parallel and stacked at a distance of 3.4-3.6 Å. We have modelled the R8Q mutant into the crystal structure of the A-77003/protease complex. Glutamine makes less extensive van der Waals contacts with the pyridinium groups. In addition, the strength of the interaction should be reduced by both the reduced positive charge of the Gln side chain and the greater distance dependence of the amide dipolar interaction with the pyridine ring. With these facts in mind, we designed a series of four inhibitors modified in the critical P_3(P_3') positions. The truncated inhibitors I, II and III were designed to prevent interactions with Arg in position 8 thus making inhibitor binding independent of this residue. The obtained K_i values (Figure 3) revealed that we succeeded in the first part of the task, i.e., to achieve similar activity towards both the mutant and wild type proteases.

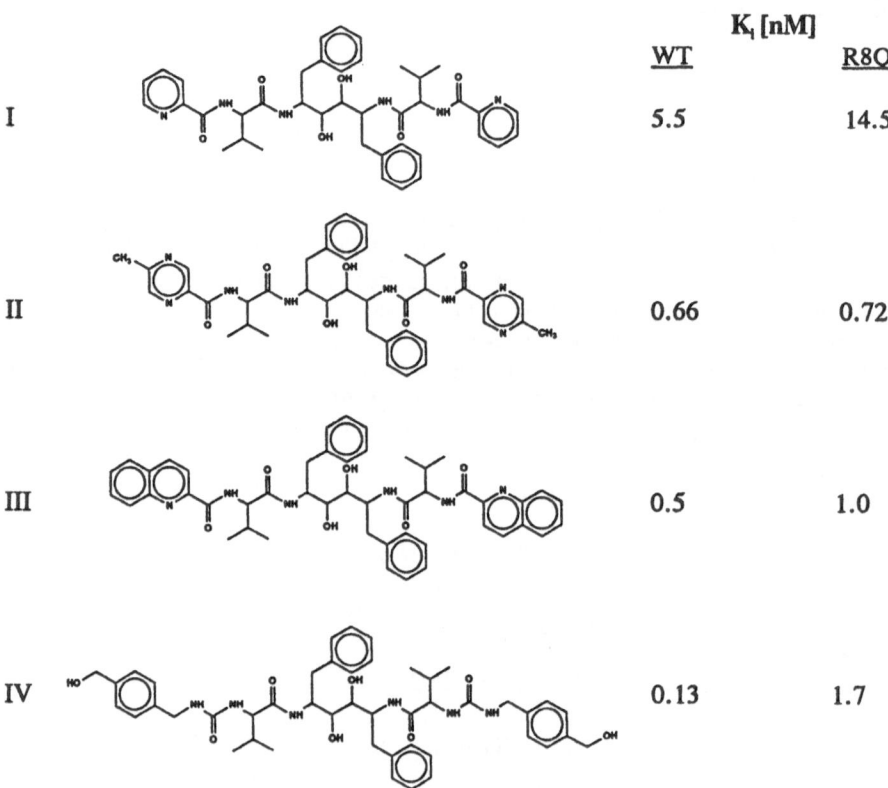

		K_i [nM]	
	WT		R8Q
I	5.5		14.5
II	0.66		0.72
III	0.5		1.0
IV	0.13		1.7

Fig. 3. Inhibition constants of P_3 modified inhibitors.

The K_i's ratio (WT/mutant) decreased from 1/50 for A-77003 to about 1/1 for inhibitor II. However, we did not succeed in reaching the deep subnanomolar activity found in the case of WT protease and A-77003. The inhibitor IV has been designed to preserve the beneficial stacking interaction of the aromatic ring with Arg of WT protease and also to make stabilizing hydrogen bonds with Gln8 of the R8Q mutant. Although its activity towards WT protease was better than that of truncated inhibitors, activity towards R8Q mutant remained at the nanomolar level.

References

1. Ho, D.D., Toyoshima, T., Kempf, D.J., Norbeck, D., Erickson, J.W. and Singh, M.K., manuscript in preparation.
2. Kempf, D.J., Marsh, K.C., Paul, D.A., Knigge, M.F., Norbeck, D.W., Kohlbrenner, W.E., Codacovi, L., Vasavanonda, S., Bryant, P., Wang, X.C., Wideburg, N.E., Clement, J.J., Platner, J.J. and Erickson, J., Antimicrob. Agents Chemother., 35 (1991) 2209.
3. Kempf, D.J., Sowin, T.J., Dohert, E.M., Hannick, S.M., Codacovi, L., Henry, R.F., Green, B.E., Spanton, S.G. and Norbeck, D.W., J. Org. Chem., 57 (1992) 5692.
4. Dess, D.B. and Martin, J.C., J. Org. Chem., 48 (1983) 4156.

Synthesis and structure-activity studies of SH2 binding peptides containing hydrolytically stable analogs of O-phosphotyrosine

A. Otaka[1], M. Nomizu[1], M.S. Smyth[1], S.E. Shoelson[2], R.D. Case[2], T.R. Burke, Jr.[1] and P.P. Roller[1]

[1]Laboratory of Medicinal Chemistry, DTP, DCT, NCI, National Institutes of Health, Bethesda, MD 20892, U.S.A., [2]Joslin Diabetes Center and Department of Medicine, Brigham and Women's Hospital, Harvard Medical School, Boston, MA 02215, U.S.A.

Introduction

Protein tyrosine kinases (PTKs) are enzymes that provide critical functionality for a variety of cellular signal transduction processes [1]. Specific tyrosine-containing sequences in PTKs, such as PDGF receptor, are autophosphorylated upon ligand binding. These phosphotyrosyl residues (pTyr) in turn are recognized by proteins containing *src*-homology 2 (SH2) domains. For example, the PDGF receptor forms complexes with phosphatidylinositol-3-kinase (PI3-kinase), PLC-γ1, and GTPase-activating protein. Modulation of SH2 mediated mitogenic signal transduction processes may provide a useful approach for anti-cancer therapy. Phosphonomethyl phenylalanine (Pmp) [2] can serve as a phosphatase resistant pTyr mimetic in peptides, and such peptides have been found to inhibit SH2 binding interaction [3]. In the present study, we utilized Pmp analogs bearing substitutions at the phosphono α-methylene carbon in solid-phase peptide synthesis (SPPS) of SH2 inhibitory peptides and evaluated the structure-activity relationship of these peptides for the inhibition of PI3 kinase (p85) binding.

Results and Discussion

The peptide Tyr(751)-Val-Pro-Met-Leu represents one of the auto-phosphorylation sequences of PDGF receptor-β, and it is a known SH2 binding site for PI3-kinase [4]. Sets of linear and cyclic glycine extended peptides were synthesized with incorporation of Pmp analogs (Figure 1). An Fmoc / t-Bu protection strategy was used for the incorporation of Pmp, HPmp (hydroxyPmp), and FPmp (monofluoroPmp) residues into the peptide [5]. For incorporation of F_2Pmp (difluoroPmp), the t-Bu protection was found to be too unstable, while ethyl side chain protection proved to be satisfactory [6]. For the syntheses of Pmp, HPmp, FPmp and pTyr-containing linear peptides, an Fmoc-based SPPS was utilized which combined piperidine-mediated N-deprotection and DIPCDI / HOBt-mediated coupling on 4-[(2',4'-dimethoxyphenyl) hydroxymethyl]phenoxy (super acid-labile Rink) resin.

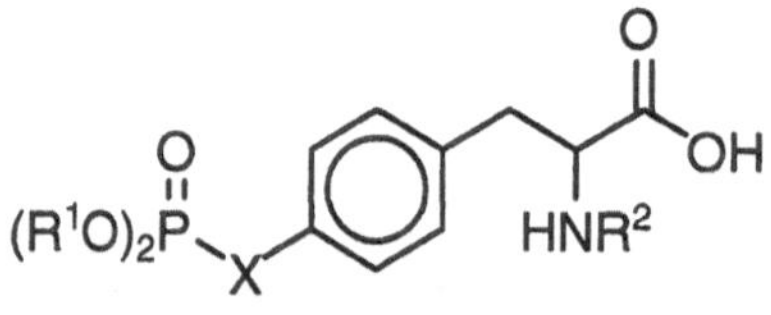

	X	R^1	R^2
pTyr	O	Me	Boc or Fmoc
Pmp	CH$_2$	t-Bu	Fmoc
HPmp	CHOH	t-Bu	Fmoc
FPmp	CHF	t-Bu	Fmoc
F$_2$Pmp	CF$_2$	t-Bu	Fmoc
F$_2$Pmp	CF$_2$	Et	Boc or Fmoc

Fig. 1. Structures of phosphatase resistant pTyr analogs.

Treatment of the protected peptide resins with TFA-thioanisole-ethanedithiol (EDT) gave diastereomeric mixtures of L- and D-Pmp analog-containing peptides. Separation of D/L-mixtures was accomplished via HPLC on an ODS column and absolute configurations were assigned by digestion with aminopeptidase-M. Deprotection of the Tyr(OPO$_3$Me$_2$)-containing peptide resin was accomplished using 1 M trimethylsilyl bromide (TMSBr)-thioanisole / TFA, *m*-cresol, EDT [7,8]. For the preparation of the F$_2$Pmp-containing peptides, Boc-based SPPS on Merrifield resin or Fmoc-based SPPS on super acid-labile Rink resin was employed. Treatment of the F$_2$Pmp(OEt)$_2$-containing peptide resin with 1 M trimethylsilyl trifluoro-methanesulfonate (TMSOTf)-2 M dimethylsulfide (DMS) / TFA, *m*-cresol, EDT (100 : 4 : 20, v/v; 4°C, 30 min; r.t., 2 hr) yielded a diastereomeric mixture of completely deprotected L-F$_2$Pmp- and D-F$_2$Pmp-containing peptides. The TMSOTf-mediated deprotection method [9] afforded the best results for removing the Et groups on the F$_2$Pmp residue. The requisite linear side-chain protected peptides for the synthesis of cyclic peptides incorporating pTyr and F$_2$Pmp were prepared using the Fmoc method on p-alkoxybenzylalcohol (Wang) resin. In this synthesis, to prevent partial piperidine-mediated deprotection of Me and Et groups on Tyr(OPO$_3$Me$_2$) or F$_2$Pmp(OEt)$_2$, respectively, these derivatives were introduced into the N-terminal position of the peptide as Boc-derivatives or as a Boc-protected dipeptide unit. Treatment of the protected resins (e. g. Boc-Gly-F$_2$Pmp(OEt)$_2$-Val-Pro-Met-Leu-O-Wang resin) with TFA-H$_2$O yielded Tyr(OPO$_3$Me$_2$)- and D/L-F$_2$Pmp(OEt)$_2$-containing linear peptides. After HPLC separation of D- and L-containing diastereomers, the linear peptides were cyclized using diphenylphosphorylazide (DPPA)/N-methylmorpholine (NMM) in DMF. Final deprotection was achieved with TMSBr- or TMSOTf-mediated methods as mentioned above. The Pmp-containing cyclic peptide was obtained using a synthetic approach analogous to the method for the synthesis of the F$_2$Pmp-containing cyclic peptide, except that 1% TFA / CH$_2$Cl$_2$ was used to cleave the t-Bu phosphonate protected linear peptide from super acid-labile Rink resin and TFA was used to deprotect the t-Bu groups. The various phosphonopeptides were evaluated in competitive binding assays [10] to the C-terminal SH2 domain of p85 PI3-kinase subunit. Data are shown in Table 1. The relative affinities of Pmp analog-containing peptides are: pTyr = F$_2$Pmp > FPmp >Pmp > HPmp.

Table 1 *PI3-kinase binding inhibition of synthetic SH2 inhibitory peptides*

Synthetic Peptides	$ED_{50} \pm SD$ (μM)
G-(L-Tyr-OPO$_3$H$_2$)-V-P-M-L	0.17 ± 0.02
G-(L-Pmp)-V-P-M-L	0.98 ± 0.09
G-(L-HPmp)-V-P-M-L	3.30 ± 0.5
G-(L-FPmp)-V-P-M-L	0.40 ± 0.03
G-(L-F$_2$Pmp)-V-P-M-L	0.17 ± 0.02
G-(L-Tyr)-V-P-M-L	> 500
G-(D-F$_2$Pmp)-V-P-M-L	34 ± 9.6
G-(L-Pmp)-V-P-Nle-L	1.6 ± 0.2
Cyclic[G-(L-Pmp)-V-P-M-L]	5.2 ± 0.7
Cyclic[G-(L-F$_2$Pmp)-V-P-M-L]	$2.24 \pm 0.40*$
Cyclic[G-(L-Tyr-OPO$_3$H$_2$)-V-P-M-L]	$1.01 \pm 0.15*$
Cyclic[G-(D-Pmp)-V-P-M-L]	156 ± 30

*preliminary results

The good affinity of F$_2$Pmp containing peptides is attributed both to its low pKa2 value [11] and the hydrogen bond forming ability of F atoms. Met at pTyr + 3, which has been thought to be critical for binding to SH2 of PI3-kinase, can be replaced by a Nle residue. The conformationally more rigid cyclic peptides possess significant affinity and provide a useful lead for peptidomimetic SH2 inhibitor development.

References

1. Carpenter, G., FASEB J., 6 (1992) 3283.
2. Burke, Jr., T.R., Russ, P. and Lim, B., Synthesis, (1991) 1019.
3. Domchek, S.M., Auger, K.R., Chatterjee, S., Burke, Jr., T.R. and Shoelson, S.E., Biochemistry, 31 (1992) 9865.
4. Fantl, W.J., Escobedo, J.A., Martin, G.A., Turck, C.W., Delrosario, M., McCormick, F. and Williams, L.T., Cell, 69 (1992) 413.
5. Burke, Jr., T.R., Smyth, M.S., Nomizu, M., Otaka, A. and Roller, P.P., J. Org. Chem., 58 (1993) 1336.
6. Burke, Jr., T.R., Smyth, M.S., Otaka, A. and Roller, P.P., Tetrahedron Lett., 34 (1993) 4125.
7. Fujii, N., Otaka, A., Sugiyama, N., Hatano, M. and Yajima, H., Chem. Pharm. Bull., 35 (1987) 3880.
8. Kitas, E.A., Knorr, R., Trzeciak, A. and Bannwarth, W., Helv. Chim. Acta, 74 (1991) 1314.
9. Fujii, N., Otaka, A., Ikemura, O., Akaji, K., Funakoshi, S., Hayashi, Y., Kuroda, Y. and Yajima, H., J. Chem. Soc., Chem. Commun., (1987) 274.
10. Piccione, E., Case, R.D., Domchek, S.M., Hu, P., Chaudhuri, M., Backer, J.M., Schlessinger, J. and Shoelson, S.E., Biochemistry, 32 (1993) 3197.
11. Smyth, M.S., Ford, Jr., H. and Burke, Jr., T.R., Tetrahedron Lett., 33 (1992) 4137.

Topographically designed peptidomimetic inhibitors of human renin: Incorporation of novel, tethered $P_1 \rightarrow P_3$ side chain functionalization

M. Plummer, J. Hamby, E. Lunney, G. Hingorani, C. Humblet, B. Batley and S. Rapundalo

Departments of Chemistry and Pharmacology, Parke-Davis Pharmaceutical Research, Division of the Warner-Lambert Company, 2800 Plymouth Road, Ann Arbor, MI, 48106, U.S.A.

Introduction

Since the initial work by Boger and co-workers [1], a cyclohexylmethyl group has been the standard P_1 sidechain in a variety of transition-state based renin inhibitors. Recently, cyclic pseudopeptidyl inhibitors of the aspartyl proteinase pepsin have been described [2] in which the sidechains of P_1 and P_3 are covalently linked. In contrast, we chose to append the P_3 sidechain functional group directly to the P_1 side chain, thereby eliminating the P_3 sidechain to backbone linkage. Molecular modeling studies suggested that the proposed novel transition-state isosteres could satisfy the hydrophobic requirements of both the S_1 and S_3 binding pockets, lending support for their synthesis and incorporation into peptidomimetic renin inhibitors.

Results and Discussion

A flexible and convergent synthesis of a variety of transition state isosteres, **1**, with alkyl and aryl groups appended from the 4-position of the P_1 cyclohexylmethyl group was desired. Retrosynthetically, we envisioned that these groups could be appended to the cyclohexyl ring through the 4-keto derivative **2**; which, in turn, might be derived from the diol, **3**. The diol should be available following known procedures [3] from N-Boc-O-Bzl-Tyr (Scheme 1). In practice, the key intermediate **2** was synthesized in 17% overall yield from the commercially available protected amino acid. The 4-keto functionality was found to react smoothly with organolithium and Wittig reagents providing the novel transition state isosteres in 30-60 percent yield after functional group deprotection. Coupling of these isosteres, employing standard carbodiimide chemistry, with N-(4-morpholino-sulfonyl) Gly-Alg [4] yielded the fully elaborated renin inhibitors **4-9**.

Docking experiments were based on a human renin model using the conserved interaction scheme observed for the aspartic proteinase inhibitors [5]. Substitution of a 1-naphthyl group on C(4) of the P_1 cyclohexyl ring of the transition state isostere resulted in a structure that remained compatible with the enzyme cleft region. The appended 1-naphthyl group extended into the S_3 region normally occupied by the side chain of the P_3 Phe. A conformational analysis of the 4-(1-naphthyl) cyclohexyl P_1

Scheme 1

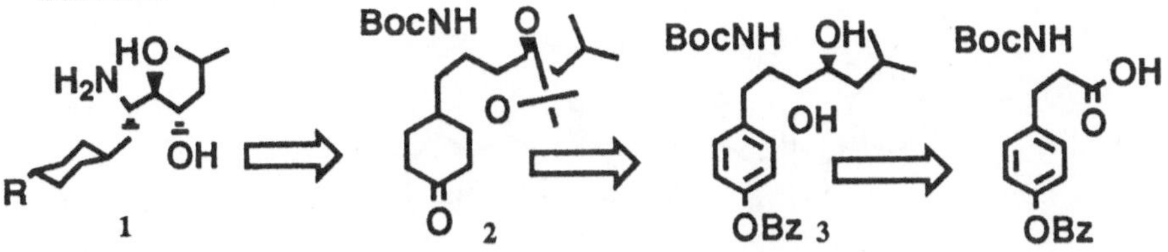

sidechain run without the confines of the cleft indicated that the bound conformation was a valid conformer.

The inhibitors were tested against monkey plasma renin according to a previously described method [6]. The binding affinities were compared to two renin inhibitors, 4 and 5, to evaluate any advantage offered by tethering a group from the cyclohexyl P_1 moiety into the S_3 binding pocket (Table 1). The most potent inhibitor, that containing the 1-naphthyl appendage, 9, has a binding affinity of 11 nM. This affinity is a seven-fold improvement over the 82 nM binding affinity of the P_3 glycine standard, 5, but is still less potent than the highly potent P_3 Phe containing parent, 4. The loss of potency observed with 9 may be partially attributed to the increased flexibility in the P_3 Gly residue relative to the P_3 phenylalanine moiety in 4. Inhibitor 6 containing a smaller phenyl substituent appended to C(4) of the cyclohexyl ring binds with 61 nM affinity and also appears reasonable based on modeling studies. Although the phenyl substituent in 6 does not seem to fully occupy the S_3 pocket, the group does bind in the vicinity of the hydrophobic phenylalanine side chains located at the base of the cleft in the enzyme binding site model. Insertion of a methylene spacer into the phenyl derivative 6 provides 8 with a benzyl tether on C(4) of the cyclohexane ring. The binding affinity of the derivative 8 is 202 nM, demonstrating that minor structural variations can have a significant impact on binding affinity.

The key intermediate 2 has proven to be a versatile and readily available starting material for a variety of novel P_1-P_1' transition state mimetics. Although the increase in binding affinity attributed to the incorporation of a tether is modest (compare 9 with 5), the ability to extend from a transition state mimetic directly to relatively distant binding pockets without involving the peptide backbone is seen as a new and promising strategy for designing smaller topographically based enzyme inhibitors.

Table 1 *Inhibitor structure and activity*

No.	R	R'	Binding affinity IC$_{50}$ (nM)
4	Benzyl	Cyclohexyl	0.2
5	H	Cyclohexyl	82
6	H		60.5
7	H		18.5
8	H		202
9	H		11

References

1. Boger, J., Payne, L.S., Perlow, D.S., Lohr, N.S., Poe, M., Blaine, E.H., Ulm, E.H., Schorn, T.W., LaMont, B.I., Lin, T.-Y., Kawai, M., Rich, D.H. and Veber, D.F., J. Med. Chem., 38 (1985) 1779.
2. Szewczk, Z., Rebholz, K.L. and Rich, D.L., Int. J. Pept. Protein Res., 40 (1992) 233.
3. For the first synthesis of dihydroxyethylene isosteres see: Luly, J.R., BaMaung, N., Soderquist, J., Fung, A.L.K., Stein, H., Kleinert, H.D., Marcotte, P.A., Eagan, D.A., Bopp, B., Meritis, I., Bolis, G., Greer, J., Perun, T.J. and Plattner, J.J., J. Med. Chem., 31 (1988) 2264.
4. Repine, R.T., Himmelsbach, R.J., Hodges, J.C., Kaltenbronn, J.S., Sircar, I., Skeean, R.W., Brennan, S.T., Hurley, T.R., Lunney, E., Humblet, C.C., Weishaar, R.E., Rapundalo, S., Ryan, M.J., Taylor, D.G., Jr., Olson, S.C., Michniewicz, B.M., Kornberg, B.E., Belmont, D.T. and Taylor, M.D., J. Med. Chem., 34 (1991) 1935.
5. Blundell, T.L., Cooper, J.B., Foundling, S.I., Jones, D.M., Atrash, B. and Szelke, M., Biochemistry, 26 (1987) 5585.
6. All inhibitors were tested against monkey plasma renin according to a previously described method: Haber, E., Koerner, T., Page, L.B., Kliman, B. and Purnode, A., J. Clin. Endocrinol., 29 (1969) 1349.

Structure-activity and modeling studies of tryptophan-modified peptidyl antagonists of endothelin

J.V.N. Vara Prasad[1], W.L. Cody[1], X.-M. Cheng[1], A.M. Doherty[1], P.L. DePue[1], J.B. Dunbar, Jr.[1], K.M. Welch[2], M.A. Flynn[2], E.E. Reynolds[2] and T.K. Sawyer[1]

Departments of [1]Chemistry and [2]Pharmacology, Parke-Davis Pharmaceutical Research Division, The Warner-Lambert Company, 2800 Plymouth Road, Ann Arbor, MI 48106, U.S.A.

Introduction

The endothelins (ETs) are a family of peptides that are highly potent vasoconstrictors and possess a variety of additional biological activities [1, 2]. These peptides act through at least two distinct receptor subtypes termed ET_A and ET_B [3]. A D-amino acid scan of Ac-ET-1_{16-21} and further SAR studies have advanced potent and non-selective ET_A/ET_B antagonists such as PD 142893 [4]. Recently, the cyclic pentapeptide BQ-123 and a linear tripeptide, FR 139317, were reported [5, 6] as potent and selective ET_A antagonists. A common structural feature of ET-1, BQ123, FR 139317 and PD-142893 is a Trp residue (L or D stereochemistry), which is critical to the SAR of each of these peptidyl ligands.

Results and Discussion

Two series of chimeric heptapeptides were designed from FR 139317 and compound 2 and each showed µM affinity for the ET_A receptor (Table 1, **3 - 6**). However, only chimeric heptapeptides **5** and **6** containing the N-terminus of FR 139317, showed both ET_A and ET_B receptor binding. These results indicated that the C-terminus of PD 142893 is particularly important for ET_B binding.

Replacement of D2Pyr by 2Pyr (**7**) in FR 139317 resulted in a 10-fold loss in binding affinity at the ET_A receptor. Deletion of the C-terminal D2Pyr residue (**8**) showed only a 10-fold decrease in ET_A receptor binding affinity. Other modifications such as the substitution of Aze by Chx or Chp, DTrp(Me) by DTrp and D3Pyr by DPhe also did not significantly alter ET_A binding affinity (**9-11**). Interestingly, the modification of DPhe to DDip, (**12**), although it did not alter ET_A receptor binding, did significantly increase the ET_B receptor binding (>100-fold). Similarly, backbone modification of DTrp by αMeDTrp resulted in a 100-fold increase in ET_B receptor binding affinity, albeit ET_A receptor binding was decreased 10-fold (**13**). Both backbone modification (αMeDTrp) and β-branching at the C-terminus (DDip) resulted in potent binding to both the ET_A and ET_B receptors (**15**). Furthermore,

Table 1 SAR of Peptidyl Antagonists of ET-1

	Peptide							Binding Assay[a] (IC$_{50}$, µM)		Biochemical Assay (IC$_{50}$, µM)	Selectivity ET$_A$ vs. ET$_B$
								ET$_A$	ET$_B$	AAR$_A$	
1.	Ac-	DHis-	Leu-	Asp-	Ile	Ile	Trp[a]	9.0	8.1	3.2	~1
2.	Ac	DTrp	Leu	Asp	Ile	Ile	Trp[a]	0.13	1.3	0.45	~10
PD 142893	Ac	DDip	Leu	Asp	Ile	Ile	Trp[a]	0.030	0.050	0.07	~1
BQ-123			Cyclo(DVal	Leu	DTrp	DAsp	Pro)	0.012	>10	0.027	>800
FR 139317			Aze	(U)	Leu	DTrp(Me)	D2Pyr	0.002	10	0.026	~5000
3.	Ac	DTrp	Leu	Asp	Ile	DTrp	D3Pyr	1.5	>10	ND	>5
4.	Ac	DTrp	Leu	Asp	Ile	DTrp	D3Pyr	2.2	>10	ND	>5
5.	Chx(U)Leu	DTrp	Leu	Asp	Ile	Trp		0.2	0.43	0.25	~2
6.	Chx(U)Leu	DTrp	D3Pyr	Asp	Ile	Trp		0.225	5.05	0.52	>10
7.			Aze	(U)	Leu	DTrp(Me)	2Pyr	0.058	>10	0.1	>100
8.			Aze	(U)	Leu	DTrp(Me)		0.35	>10	ND	>30
9.			Chx	(U)	Leu	DTrp	D3Pyr	0.028	>10	0.170	>100
10.			Chx	(U)	Leu	DTrp	DPhe	0.027	>10	ND	>100
11.			Chp	(U)	Leu	DTrp	DPhe	0.019	>10	0.057	>100
12.			Chx	(U)	Leu	DTrp	DDip	0.037	1.8	0.045	~30
13.			Chx	(U)	Leu	αMeDTrp	D3Pyr	0.107	0.155	0.170	~1
14.			Chx	(U)	Leu	αMeDTrp	3Pyr	3.2	7.0	ND	~2
15.			Chx	(U)	Leu	αMeDTrp	DDip	0.030	0.050	0.283	~1

[a] See Ref. 4.; ET$_A$: Rabbit renal artery vascular smooth muscle cells; ET$_B$: Rat cerebellar membranes; AAR$_A$: Inhibition of ET-1 stimulated arachidonic acid release in the same tissue as the binding assay; Dip: 3,3-diphenylalanine ; ND: not determined; Aze: azepine;(U): urea linkage; 2Pyr: 2-pyridylalanine; 3Pyr: 3-pyridylalanine Chx: cyclohexyl; Chp: cycloheptyl.

compounds **13-15** are non-selective antagonists in contrast to FR 139317 which is 5000-fold more selective for the ET_A receptor in our assay systems.

An analysis of a proposed conformation of FR 139317 (based on the hypothesis that BQ-123 [7] and FR 139317 occupy the same binding site) relative to BQ-123 suggested a reasonable overlay of the two peptides (FR 139317/BQ123) based on superpositioning Leu/Leu, DTrp(Me)/DTrp, Aze/DVal and DPyr/AspPro. Based on a 3-D model for FR 139317, we systematically modified the ligand template to that of compound **15**. These studies suggest that the αMeDTrp substitution may induce a turn conformation that orients the DDip, backbone COOH and hydrophobic side chain groups to provide favorable interactions at both ET receptors.

The tripeptide ET_A receptor selective antagonist FR 139317 was systematically modified to yield a series of compounds which were potent ET-1 antagonists at both ET_A and ET_B receptors. Specifically, backbone α-methylation of the Trp and modification at the C-terminus by Dip resulted in the discovery of several non-selective antagonists, **13-15** of ET-1.

References

1. Yanagisawa, M., Kurihara, H., Kimura, S., Tomobe, Y., Kobayashi, M., Mitsui, Y., Yazaki, Y., Goto, K. and Masaki, T., Nature (London), 332 (1988) 411.
2. Doherty, A.M., J. Med. Chem., 35 (1992) 1493.
3. Arai, H., Hori, S., Aramori, I., Ohkubo, H. and Nakanishi, S., Nature (London), 348 (1990) 730.
4. Cody, W.L., Doherty, A.M., He, J.X., DePue, P.L., Rapundalo, S.T., Hingorani, G.A., Major, T.C., Panek, R.L., Dudley, D.T., Haleen, S.J., LaDouceur, D.M., Hill, K.E., Flynn, M.A. and Reynolds, E.E., J. Med. Chem., 35 (1992) 3301.
5. Ishikawa, K., Fukumui, T., Nagase, T., Fujita, K., Hayama, T., Niyama, K., Mase, T., Ihara, M. and Yano, M., J. Med. Chem., 35 (1992) 2139.
6. Sogabe, K., Nirei, H., Shoubo, M., Nomota, A., Henmi, K. and Notsu, Y., Japan J. Pharm., 58 (1992) 105.
7. Reily, M.D., Thanabal, V., Omecinsky, D.O., Dunbar, Jr., J.B., Doherty, A.M. and DePue, P.L., FEBS Lett., 300 (1992) 136.

Modifications to angiotensin II (AII) peptide analogs that reduce AT_1 receptor affinity have little to no effect upon AT_2 receptor affinity

J. Samanen, N. Aiyar and R. Edwards

Departments of Medicinal Chemistry and Pharmacology, SmithKline Beecham Pharmaceuticals, King of Prussia, PA 19406, U.S.A.

Introduction

[(pNH$_2$)Phe^6]-AII [1] was the first reported AT_2-selective peptide. Rat peripheral vasculature displays the AT_1 receptor preferentially, whereas rat cerebral artery displays the AT_2 receptor [2]. Thus, AII analog potency determined in a peripheral vascular smooth muscle assay, e.g., rabbit aorta, reflects analog affinity and efficacy at the AT_1 receptor. In contrast, AII peptide analog in vivo antihypertensive potency may reflect analog affinity and efficacy at both the AT_1 and AT_2 receptors, since the analog in circulation may pass through tissues expressing one or both receptors. Two isoforms AT_{1A} and AT_{1B} were recently discovered in rat tissues [3]. Most previous AII structure-activity studies utilized smooth muscle assays, e.g., rabbit aorta (RB-A), which displays only the AT_1 receptor. To begin to sort out the different structural requirements for the AT_1 and AT_2 receptors, we have reevaluated previously published peptide analogs in the bovine adrenal cortex (B-AC) for the study of AT_1 receptor affinity and bovine ovary (B-O) and cerebellum (B-C) for the study of AT_2 receptor affinity [4]. Rat mesenteric artery (R-M), employed in early experiments, was replaced with B-AC.

Results and Discussion

The rank ordering of peptides in Table 1 primarily by RB-A does not completely parallel B-AC activity, perhaps due to species differences. None of the peptides, furthermore, display AT_1 selectivity. A variety of modifications to [Sar^1]-AII and [Sar^1,Ile^8]-AII in the critical positions Arg^2, Tyr^4, and His^6 reduce AT_1 receptor affinity but not AT_2 receptor affinity, as evidenced in the AT_2-selective analogs **4, 7, 11**, and **15**. AT_2 selectivity can be demonstrated in other positions as well. Thus, the AII analog structural requirements for AT_2 receptor affinity are remarkably different from AT_1 requirements. Within the bovine AT_2-containing tissues, however, analog selectivities are apparent: **14** and **24** prefer the ovary over brain, whereas **13** and **17** prefer brain over ovary (suggesting AT_2 receptor subtypes?). In vitro activity (RB-A) and in vivo activity (R-BP) fall within 1 order of magnitude of AT_1 receptor affinity in the R-M or the B-AC although species differences may account for exceptions, especially **8** which displays high affinity in all bovine tissues. Subtle modifications can alter selectivity, e.g., Ile to (OMe)Thr decreases affinity for B-O in **14** to **13**. The para substituents of [(pX)Phe^4]-AII

Table 1 *In vitro and in vivo activities[a] vs. receptor affinities[b]*

	Agonist Activity In Vitro % A II[c]	Antagonist Activity In Vitro Rabbit Aorta pA$_2$[d]	Antagonist Activity In Vivo Rat BP ID$_{50}$ (ng/rat/m)[e]	Binding ^{125}I-A II to Bovine Tissue (AT$_1$) Adr.Cort. IC$_{50}$ (nM)	(AT$_2$) Ovary IC$_{50}$ (nM)	(AT$_2$) Brain IC$_{50}$ (nM)
1, [Sar^1]-A II	100, 190^c			3.0	1.2*	1.02*
2, [Sar^1,Ile^8]-A II	0	9.1	10	3.0	0.5	3
Position Two:						
3, [des-Asp^1,D-(NMe)Ala^2,Ile^8]-A II	0	7.1	25	537	11.2	
4, [Sar^1,Cit^2,Ile^8]-A II	0	6.2		1220	80	5.0
Position Four:						
5, [Sar^1,(pNO$_2$)Phe^4]-A II	pa^f	pa^f		1350	200	58
6, [Sar^1,(pOEt)Phe^4]-AII	0	pa^f	150	2200	1600	
7, [Sar^1,(pF)Phe^4]-AII	0	7.5	150	6000	10	5.5
8, [Sar^1,Phe^4,Ile^8]-AII	0	8.5	80	1.2*	2.2	1.2
9, [Sar^1,(pNO$_2$)Phe^4,Ile^8]-AII	0	7.1	1000	3000	110	26
10, [Sar^1,(pOMe)Phe^4,Ile^8]-AII	0	7.1	100	150	30	59
11, [Sar^1,(pF)Phe^4,Ile^8]-AII	0	6.8	750	8000	8	43
Position Six:						
12, [Sar^1,(pOEt)Phe^4,Ile^8]-AII	0	<6.0	>1000	6000	3800	178
13, [Sar^1,(OMe)Thr,^{5}Thi^6,Ile^8]-AII	0	7.3	30	300	400	2.0
14, [Sar^1,Thi^6,Ile^8]-AII	0	7.5	100	950	0.5*	13
15, [Sar^1,Trp^6,Ile^8]-AII	0	7.4	100	700	1.0	0.11

analogs 5-12 have a dramatic influence upon selectivity, e.g., p-H in 8 vs. p-F in 11. Further structure-affinity studies will be required to determine the full extent of the different requirements for AT$_1$ and AT$_2$ receptors.

	Agonist Activity In Vitro % AII[c]	Antagonist Activity		Binding ^{125}I-AII to Bovine Tissue		
		In Vitro Rabbit Aorta pA$_2$[d]	In Vivo Rat BP ID$_{50}$ (ng/rat/m)[e]	(AT$_1$) Adr.Cort. IC$_{50}$ (nM)	(AT$_2$) Ovary IC$_{50}$ (nM)	(AT$_2$) Brain IC$_{50}$ (nM)
Position Six, continued:						
16, [Sar^1,(N^{3}Me)His^6,Ile^8]-AII	0	6.75	100	30	3	9.5
17, [Sar^1,Phe^6,Ile^8]-AII	0	6.5	100	300	100	2.0
18, [Sar^1,Lys^6,Ile^8]-AII	0	<6.0		>10^5	300	263
Position Eight:						
19, [Sar^1,D-(αMe)Phe^8]-AII	0	9.7	10	32.8 / 1.3 (RM)	1.65	
20, [Sar^1,D-Phe^8]-AII	0	9.0	12.5	1.3 (RM)		
21, [Sar^1,Thr^8]-AII	0	8.9	15	24 / 4.4 (RM)	2.5	0.8
22, [Sar^1,Thi^8]-AII	pa[f,h]	8.6	2.5	10 / 0.11 (RM)	38	18
23, [Sar^1,Ala^8]-AII	0	8.6	15	3.4 (RM)[g]		
24, [Sar^1,His^8]-AII	0	7.2	750	120 / 26 (RM)	3.0	264
25, [Sar^1,D-Phe^8]-AII-NH$_2$	0	7.5	12.5	300 / 2000 (RM)	500	509
26, [D-Cys^1,Cys^8]-AII	0	<6.0		1387 (RM)		

[a]Activities described in previous references in Ref. [5]; [b]Described in [4]; All IC$_{50}$ values (n=3) are within ± 0.2, or ± 0.30 for analogs *; [c]In vitro % agonist activity rabbit aorta rel. AII; [d]pA$_2$ = -log IC$_{50}$ [5]; [e]Inhibition of AII induced rat blood pressure increase, employing 250 g rats [5]; [f]Partial agonist activity, difficult to quantitate; [g]RM = rat mesentery; [h]Partial agonist activity seen above 100 ng/ml.

References

1. Speth, R.C. and Kim, K.H., Biochem. Biophys. Res. Commun., 169 (1990) 997.
2. Wong, P.C., Chiu, A.T., Duncia, J.V., Herblin, W.F., Smith, R.D. and Timmermans, B.M.W.M., Trends Endocrin. Metab., 3 (1992) 211.
3. Kitami, Y., Okura, T., Marumoto, K. and Wakamiya, R., Biochem. Biophys. Res. Commun., 188 (1992) 446.
4. Aiyar, N., Griffin, E., Edwards, R., Weinstock, J., Samanen, J. and Nambi, P., Pharmacol., 46 (1993) 1.
5. Samanen, J., Cash, T., Narindray, D., Brandeis, E., Adams, W., Weideman, H., Yellin, T. and Regoli, D., J. Med. Chem., 34 (1991) 3036.

Inhibition of thrombin by derivatives of the dipeptide aspartic acid-amidinophenylalanine

W. Stüber, R. Koschinsky, C. Kolar, M. Reers,
G. Dickneite, D. Hoffmann, J. Czech, K.-H. Diehl
and E.-P. Pâques

*Research Laboratories, Behringwerke AG, P.O. Box 1140,
35501 Marburg, Germany*

Introduction

The serine proteinase thrombin plays an essential role in the coagulation pathway. Thus, thrombin is involved in many thrombotic disorders and it is often desirable to control thrombin activity *in vivo*. Thrombin activity can be blocked by active site directed inhibitors. Due to the similarities of the active sites of related serine proteinases, such as trypsin and plasmin, potency and specificity play major roles in a successful treatment of coagulatory disorders. It is thought that competitive inhibitors are particularly useful as drug candidates. Widely known are MD-805 (2R,4R-4-methyl-(N^2-(3-methyl-1,2,3,4-tetrahydro-8-quinolinyl)) sulphonyl-L-arginyl-2-piperidine carboxylic acid) and NAPAP (β-naphthyl-sulfonylglycyl-D,L-4-amidinophenylalanyl-piperidide). However these compounds display drawbacks, such as limited efficacy *in vivo*, short duration of action and side effects *in vivo*. We report herein on the development of new and highly potent thrombin inhibitors with improved characteristics. For this purpose we optimized the inhibitor structure of NAPAP employing molecular modelling and synthesized and characterized a series of inhibitor molecules. An important starting point for molecular modelling was the elucidation of the thrombin structure [1, 2]. Based on the X-ray crystal structure data of low molecular weight thrombin inhibitors, rational drug design was possible.

Results and Discussion

In order to overcome the shortcomings of NAPAP (**1**) we introduced an acidic moiety aiming at improved tolerability and half-life *in vivo*. It was known from literature that the introduction of an acidic moiety into basic molecules improved the above characteristics [3]. The most potent compound in this series, β-naphthyl-sulfonylglycyl-D,L-4-amidinophenylalanine-L-proline displayed a Ki of 510 nM. Thus, we looked for possible sites for the introduction of acidic functionalities into NAPAP gaining compounds without a loss of antithrombotic activity. Initially, we introduced carboxylic functions at the piperidine group. However, this led to an undesired loss of antithrombotic potency reflected by the Ki. We found that the substitution of Gly by Asp led to compounds with acceptable characteristics. The synthesis of **7** is a representative outline for the other compounds (Figure 1).

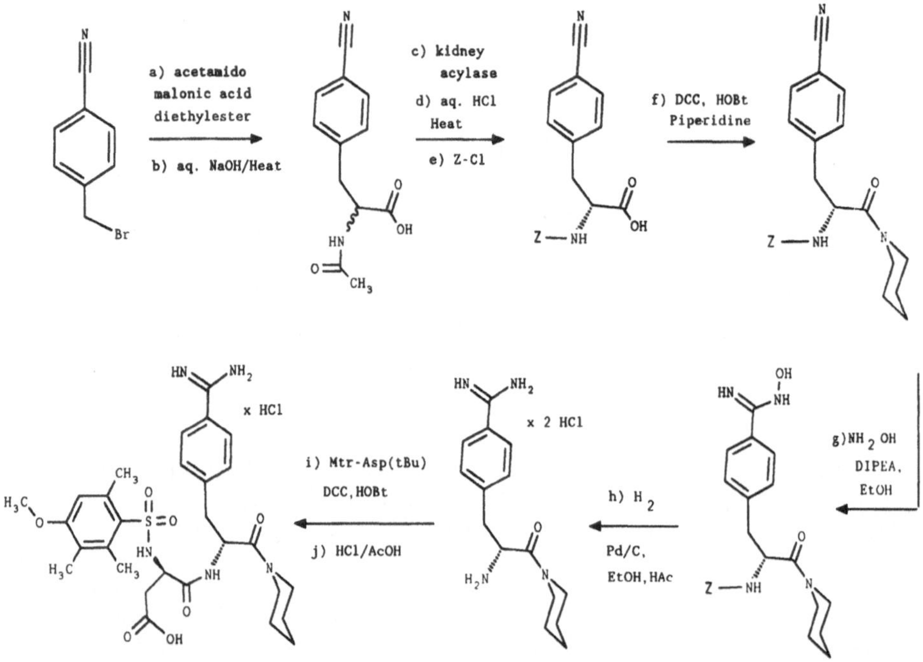

Fig. 1. Outline of the synthesis of compound 7. Other compounds shown in Table 1 have been synthesized similarly. For every single inhibitor, steps a) to h) have been the same yielding the intermediate 4-amidinophenylalanylpiperidide. The final inhibitors have been synthesized by coupling individually protected amino acid derivatives (data not outlined).

In comparison to NAPAP, **2** displayed only a slightly higher Ki against thrombin. It could be demonstrated that the introduction of aspartic acid instead of Gly reduced also the influence of the inhibitors on blood pressure and histamine release. Our next goal was to enhance the antithrombotic potency of **2**. Our CAMD studies pointed out that the β-naphthylsulphonyl group did not fit optimally into the active site of thrombin. We found that the sulphamide portion of this inhibitor type had to be optimized. Results are compiled in Table 1.

The data in Table 1 support the importance of the derivatization pattern of the phenylsulphamide portion. It can be summarized that the 4-methoxy-2,3,6-trimethylphenylsulphonyl group exhibited the most potent contribution to the binding forces. We found that the antithrombotic potency *in vitro* and *in vivo* could be enhanced by derivatization or substitution of the carboxylic moiety of Asp. Even an exchange of the carboxylic group by a sulphonic acid moiety (compound **10**) yielded an improved Ki against thrombin. The most potent compound was a β-glycosylated structure (**12**) with a Ki of 40 pM. The data in Table 1 demonstrate that improved inhibition of thrombin had only little influence on trypsin inhibition. Hence, the inhibitors **7,10,11,12** displayed a high selectivity. This was also confirmed for similar enzymes tPA and plasmin (data not shown). Thrombin

Table 1 *Structure activity relationship of thrombin inhibitors*

No.	Structure[a,b]	Inhibition constants Ki(nM)	
		Thrombin	Trypsin
1	β-naphthylsulphonyl-Gly-D,L-Adf-pip(NAPAP)	11	499
2	β-naphthylsulphonyl-L-Asp-D,L-Adf-pip	20	505
3	4-methoxy-2,5-dimethylphenylsulphonyl-L-Asp-D,L-Adf-pip	279	582
4	4-methoxy-2,3,5-trimethylphenylsulphonyl-L-Asp-D,L-Adf-pip	103	618
5	4-methoxy-2,6-dimethylphenylsulphonyl-L-Asp-D,L-Adf-pip	54	270
6	4-methoxy-2,3-dimethylphenylsulphonyl-L-Asp-D,L-Adf-pip	30	271
7	4-methoxy-2,3,6-trimethylphenylsulphonyl-L-Asp-D-Adf-pip	2.4	312
8	4-methoxy-2,3,6-trimethylphenylsulphonyl-L-Asp-D,L-Adf-pip	19	631
9	4-methoxy-2,3,6-trimethylphenylsulphonyl-L-Glu-D,L-Adf-pip	33	2952
10	4-methoxy-2,3,6-trimethylphenylsulphonyl-L-Cya-D-Adf-pip	0.8	118
11	4-methoxy-2,3,6-trimethylphenylsulphonyl-L-Asp(GABA)-D-Adf-pip	0.8	1449
12	4-methoxy-2,3,6-trimethylphenylsulphonyl-L-Asn ((β-glucopyranosyl)uronate)-D-Adf-pip	0.04	174

[a]Adf = 4-amidinophenylalanine, [b]pip = piperidine.

inhibitors with Ki values <10 nM (**7,10,11,12**) have been tested for further *in vitro* and *in vivo* characteristics. It turned out that these compounds were highly stable in body fluids and influenced strongly the coagulation parameters TT and aPTT. In *in vivo* thrombus models, i.e. vena cava thrombosis, stenosis of arteria carotis and DIC, these inhibitors demonstrated their high efficacy. This data suggests that the most potent inhibitors in the above series may serve as useful antithrombotic drugs.

References

1. Bode, W., Mayr, I., Baumann, U., Huber, R., Stone, S.R. and Hofsteenge, J., EMBO J., 8 (1989) 3467.
2. Bode, W., Turk, D. and Stürzebecher, J., Eur. J. Biochem., 193 (1990) 175.
3. Voigt, B., Stürzebecher, J., Wagner, G. and Markwardt, F., Pharmazie, 43 (1988) 412.

Design, synthesis and study of a selective cyclopeptidic mechanism-based inhibitor of human thrombin

M. Wakselman[1], J.P. Mazaleyrat[1], R.C. Lin[1], J. Xie[1], B. Vigier[1], A.C. Vilain[2], S. Fesquet[2], N. Boggetto[2] and M. Reboud-Ravaux[2]

[1]CNRS-CERCOA, 94320 Thiais, France, [2]Lab. Enz. Mol. Fonct., Inst. J. Monod, Univ. Paris VII, 75251 Paris Cedex 05, France

Introduction

Thrombin plays a central regulatory role in haemostasis, wound healing and various diseases such as arterial and venous thrombosis. If the extent of thrombin generation and activity in prethrombic situations can be inhibited, then antithrombic efficacy will result [1].

In order to obtain selective suicide substrates of serine proteinases, we have previously introduced a latent functionality into cyclopeptides related to good substrates of these enzymes. Inactivation of α-chymotrypsin [2] or trypsin-like proteinases [3, 4] has already been observed. These cyclopeptides possess a P'_1 functionalized *meta*-aminobenzoyl residue (m-aB(CH$_2$)X) which constitutes a latent electrophilic moiety and a P_1 amino acid residue displaying a good affinity for the S_1 primary binding site of the target enzyme (Schechter and Berger notation [5]).

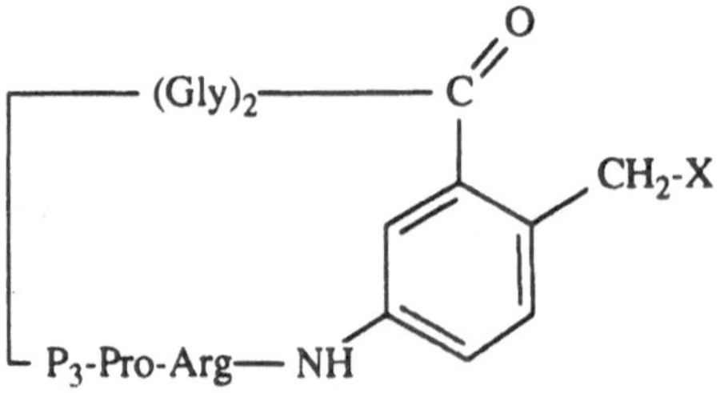

1a: P_3= D-Phe, 1b: L-Phe, 1c: Gly; X= H

2: P_3= D-Phe, X= S$^+$Me$_2$

Fig. 1. Structure of the cyclopeptides 1a-c and 2.

Selective enzymic cleavage of the P_1-P'_1 anilide bond should result in the formation of an acyl-enzyme and an aminobenzoic derivative having a good sulfide leaving group (when X=$^+$SR$_1$R$_2$). Then substitution at the benzylic position by an active site nucleophile should lead to enzyme inactivation, by a dissociative mechanism involving a quinoniminium methide intermediate. During the life-time of the acyl-enzyme, diffusion of the reactive electrophile out of the active site should be prevented by the presence of a peptidic linkage. When P_1=Arg and the peptide link was a tetraglycyl chain, an interesting selectivity for the inhibition of urokinase was observed, without a noticeable inactivation of t-PA [4].

In thrombin substrates and inhibitors, acylation of the D-Phe-Pro-Arg sequence levels down the favorable effect of the D configuration of the P_3 residue [6]. In the aim of obtaining good substrates of thrombin, a series of c(P_3-Pro-Arg-m-aB(CH_3)-Gly_2) cyclopeptides 1 (Figure 1), in which P_3=Gly, L-Phe or D-Phe, has now been synthesized and studied as substrates. Then, the sequence leading to the best cyclopeptidic substrate has been retained for the synthesis of the corresponding c(P_3-Pro-Arg-m-aB(CH_2X)-Gly_2) inhibitor 2.

Results and Discussion

The cyclopeptides 1a-c and 2 were prepared by solution peptide synthesis. Diglycine ethyl ester was first coupled to 2-methyl-5-nitrobenzoyl or 2-(phenoxymethyl)-5-nitrobenzoyl chloride. Then, reduction of the nitro substituent followed by elongation of the peptide chain gave a compound which was cyclized by the azide method (Figure 2). The sulfonium salt 2 was obtained from the corresponding phenyl ether c(D-Phe-Pro-Arg-m-aB(CH_2-OC_6H_5)-Gly_2) by treatment with Me_2S/CF_3CO_2H.

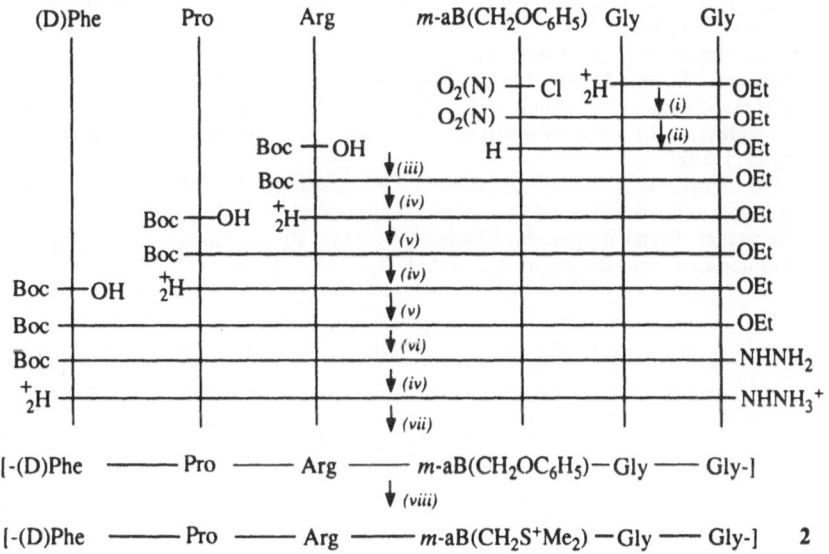

Fig. 2.   Synthesis of the cyclopeptide 2. (i) Et_3N/DMF (ii) H_2/PtO_2 (iii) DCC/pyridine/DMF (iv) TFA/CH_2Cl_2 (v) EtOCOCl/NMM (vi) H_2N-NH_2 (vii) 1. RONO/H^+/DMF/-40° 2. dilution (DMF) 3. iPr_2NEt (viii) Me_2S/TFA.

Analysis of the influence of the nature of the amino acid at the P_3 position of the protio compounds **1a-c** (X=H) showed that the D-Phe-Pro-Arg sequence considerably increases the rate of the thrombin catalyzed hydrolysis: k_{cat}/K_M=1.6 x 10^5 $M^{-1}s^{-1}$ (**1a**: P_3=D-Phe), 1,780 $M^{-1}s^{-1}$ (**1b**: P_3=L-Phe) and 370 $M^{-1}s^{-1}$ (**1c**: P_3=Gly) at pH 7.5 and 25°C. Therefore **1a** was used as a core structure for the synthesis of the inactivator. The obtained functionalized cyclic peptide **2** inactivated human thrombin (k_{inact}/k_i=3,500 $M^{-1}s^{-1}$ at pH 7.5 and 25°C) through a process fulfilling the kinetic criteria expected for an enzyme-activated inhibition. The inactivation is remarkably selective: **2** has no effect on bovine trypsin or chymotrypsin and human factor Xa, and very poor effect on human urokinase and plasmin (k_{inact}/k_i<0.1 $M^{-1}s^{-1}$).

The model study on the cyclopeptides **1a-c** (X=H) demonstrated that the sequence D-Phe-Pro-Arg, even when included in a cyclic structure, considerably enhanced the efficiency of the thrombin-catalyzed hydrolysis. The corresponding cyclopeptide **2** (X=$^+$SMe$_2$) was a highly selective inactivator of human thrombin. The present study suggests that cyclopeptides possessing a latent electrophilic function, like compound **2**, can be tailored in order to obtain selective inhibitors of therapeutic interest.

Acknowledgements

We thank the ADIR Society for financial support of this project.

References

1. Fenton II, J.W., Ofosu, F.A., Moon, D.G. and Maraganore, J.M., Blood Coagul. and Fibrin., 2 (1991) 69.
2. Wakselman, M., Mazaleyrat, J.P., Xie, J., Montagne, J.J., Vilain, A.C. and Reboud-Ravaux, M., Eur. J. Med. Chem., 26 (1991) 699.
3. Reboud-Ravaux, M., Vilain, A.C., Boggetto, N., Maillard, J., Favreau, C., Xie, J., Mazaleyrat, J.P. and Wakselman, M., Biochem. Biophys. Res. Commun., 178 (1991) 352.
4. Wakselman, M., Xie, J., Mazaleyrat, J.P., Boggetto, N., Vilain, A.C., Montagne, J.J. and Reboud-Ravaux, M., J. Med. Chem., 36 (1993) 1539.
5. Schechter, I. and Berger, A., Biochem. Biophys. Res. Commun., 27 (1967) 157.
6. Izquierdo, C. and Burguillo, F.J., Int. J. Biochem., 21 (1989) 579.

Improved ionic interactions of the hirudin C-terminal (55-65) analogs with thrombin

S.-Y. Yue, Z. Szewczuk and Y. Konishi

Biotechnology Research Institute, National Research Council of Canada, 6100 Royalmount Ave., Montreal, Quebec, H4P 2R2, Canada

Introduction

Hirudin is the most potent thrombin inhibitor (Ki ~ 10^{-14}M) and its C-terminal fragment 55-65, Ac-DFEEIPEEYLQ-OH ($IC_{50} = 1.7$ μM), binds to the Fibrinogen Recognition Exosite (FRE) of thrombin, one of the most important structural features that is responsible for the high selectivity of thrombin functions in the blood coagulation system. The previous study suggested that three residues, Phe^{56}, Glu^{57}, and Ile^{59}, are the most important for the binding affinity [1]. While the FRE of thrombin is also called the 'anion binding exosite' because of the many basic residues present in this region, there is only one acidic residue, Glu^{57}, showing the crucial role in the hirudin C-terminal side. Recently, we demonstrated that branching an Asp residue to both Suc^{55} and Glu^{57} side chains of a Glu^{58} and Lys^{61} cyclic analog of hirudin 55-65, [Asp-γ-Suc^{55}, Asp-δ-Glu^{57}], cyclo-[Glu^{58}, Lys^{61}] $hirudin^{55-65}$ ($IC_{50} = 0.21$ μM), improved the potency 8-fold [2]. In this report, we focus on the analysis of ionic interactions at the hirudin Glu^{57} binding site of a truncated hirudin sequence (hirudin 55-62).

Results and Discussion

Table 1 shows the hirudin 55-62 analogs prepared in this study and their inhibition of fibrin clot formation. The IC_{50} values presented in the table were estimated as the inhibitor concentration required to double the clotting time relative to the control at the enzyme concentration of 0.10 NIH unit/mL. Since the N-terminal part of hirudin 55-65 plays a more important role than the C-terminal part [1], we used a truncated hirudin C-terminal form, hirudin 55-62, instead of the entire tail of the segment (hirudin 55-65), in this study.

In the first six analogs listed in Table 1, the effect of branching an Asp at the Glu^{57} side chain was studied with three different hirudin 55-62 forms. These forms include P432, the original hirudin sequence; P434, Glu^{58} replaced by Pro, a hirudin variant that was reported to have a higher potency for the entire tail [3]; and P436, the cyclic form between the side chains of Glu^{58} and Lys^{61} that also showed an improved potency over the original sequence [4]. All the Asp branched analogs, P433, P435, and P437, showed more than a 2-fold improvement in potency compared to their counterpart respectively in all three forms. It seems that the

Table 1 *Inhibition of fibrin clot formation*

Peptide	Sequence (Hirudin 55-62)[a]	IC$_{50}$ [μM]
P432	Ac-D F-Glu-E I P E E-OH	71.6 ± 2.2
P433	Ac-D F-Glu-E I P E E-OH Asp	31.4 ± 2.4
P434	Ac-D F-Glu-P I P E E-OH	29.5 ± 0.2
P435	Ac-D F-Glu-P I P E E-OH Asp	12.5 ± 1.5
P436	Ac-D F-Glu-E I P K E-OH	45.4 ± 2.1
P437	Ac-D F-Glu-E I P K E-OH Asp	18.2 ± 1.2
P439	Ac-D F-Orn-E I P K E-OH Suc	> 500
P440	Ac-D F-Lys-E I P K E-OH Suc	> 500
P457	Ac-D F-Glu-P I P K E-OH Gly	55.5 ± 0.4
P458	Ac-D F-Glu-P I P K E-OH β-Ala	147 ± 2.0

[a] All branched amino acids have a carboxyl terminal group.

substitutions of the 58th hirudin residue did not affect the binding affinity of the Asp residue branched to Glu^{57}. While more rigid structures, a cyclization between Glu^{58} and Lys^{61}, or Pro^{58} substitution showed clear benefits for the potency of the truncated forms (hirudin 55-62), this structural feature seems less important for the entire length of the exosite inhibitor, hirudin 55-65. The same cyclization or replacing Glu^{58} by a Gly residue have little effect on potency.

Based on the X-ray data of the hirugen-thrombin complex, the hirudin Glu^{57} side chain interacts only with Arg^{77A} (chymotrypsin numbering), via a water molecule, at the thrombin exosite (another interaction is with Arg^{75} of a symmetric thrombin molecule in the crystal form) [5]. Branching an Asp residue to the side chain of Glu^{57}, increased the length of the side chains and introduced one more carboxyl group to the side chain. Molecular graphics analysis suggested that both carboxyl groups of the branched Asp are close to the guanidine groups of the thrombin residues Arg^{75} and Arg^{77A}. Therefore, further analogs were prepared to analyze the contributions of these carboxyl groups.

One Asp residue can be considered as a combination of a Gly and a β-Ala. Both amino acids were separately branched to the Glu^{57} residue as P457 and P458. These analogs showed a significant decrease in the potency, by 4- and 11-fold, compared with P435. This means that the combination of the two carboxyl groups of the branched Asp has some contributions to the binding. We might think that linking these amino acids at the Glu^{57} side chain should be conformationally more flexible than linking Asp. The high flexibility should be also one reason for the low inhibition of two other analogs, P439 and P440, where a succinic acid was branched to either an Orn or Lys residue at the 57th position of hirudin. If we compare P458 and P439, which have a similar side-chain length, there are two structural differences. The first one is the different substitutions at the 58th position, proline versus the cyclization, which could only introduce a moderate variation of the potency. The major structural difference is a reversed amide bond at the middle of the side chain for P439 (and P440). This observation suggested that the 'normal' amide bond, which links the Asp residue to Glu[57], may also contribute to the binding potency.

In the protein-ligand interactions, the side chain-side chain ionic pair might show a moderate contribution to the binding due to their conformational flexibility and the compensation with water molecules. However, the intermolecular electrostatic interactions may have a long-range effect to direct the ligand for an initial recognition of binding sites. This might be the reason why increasing the number of the negative charges of the hirudin (55-62) analogs, which may be helpful to recognize the 'anion binding exosite', can improve the binding affinity.

This study demonstrated that extending the length and increasing the number of negative charges at the hirudin Glu[57] position can improve the binding affinity, and that a rigid structure at the 58th position in hirudin can also increase the potency.

Acknowledgements

We thank Jean Lefebvre and Bernard F. Gibbs for their excellent technical assistance.

References

1. Yue, S.-Y., DiMaio, J., Szewczuk, Z., Purisima, E.O., Ni, F. and Konishi, Y., Protein Eng., 5 (1992) 77.
2. Szewczuk, Z., Gibbs, B.F., Yue, S.-Y. and Konishi, Y., In Schneider, C.H. and Eberle, A.N. (Eds.), Peptides 1992 (Proceedings of the 22nd European Peptide Symposium), Escom Science Publishers B.V., Leiden, The Netherlands, 1993, p.555.
3. Krstenansky, J.L., Broersma, R.J., Owen, T.S., Payne, M.H., Yates, M.T. and Mao, S.J., Thromb. and Haemo., 63 (1990) 208.
4. Szewczuk, Z., Gibbs, B.F., Yue, S.-Y., Purisima, E.O. and Konishi, Y., Biochemistry, 31 (1992) 9132.
5. Skrzypczak-Jankun, E., Carperos, V.E., Ravichandran, K.G., Tulinsky, A., Westbrook, M. and Maraganore, J.M., J. Mol. Biol., 221 (1991) 1379.

Synthetic peptides for the development of pertussis toxin receptor-targeted drugs

P. Chong[1], L.D. Heerze[2], G.D. Armstrong[2] and M.H. Klein[1]

[1]Connaught Centre for Biotechnology Research,
1755 Steeles Avenue West, Willowdale, Ontario, M2R 3T4, Canada,
[2]Department of Medical Microbiology and Infectious Diseases,
University of Alberta, Edmonton, Alberta, T6G 2H7, Canada

Introduction

Recent studies suggest that glycolipids and glycoproteins play an important biological role in the attachment of bacteria as well as bacterial toxins to oligosaccharidic receptors expressed on host cell membranes. Pertussis toxin (PT) is a virulence factor produced by the organism *Bordetella pertussis* that binds to glycoprotein receptors with a terminal sialyllactosamine moiety. PT is a classical A-B type toxin, and its lectin-like activity is an intrinsic property of its B-oligomer which consists of four different subunits (S2 to S5). Although a serological or cell-mediated response to PT that correlates with protection against pertussis disease has not been identified, extensive evidence indicates that PT is an important component of any defined whooping cough vaccine [1]. The majority of research efforts aimed at characterizing the functional domains of PT have centered around the ADP-ribosyl transferase activity of its A subunit. There is very little information on the protein sequences contributing to the lectin-binding domains of PT. Using synthetic peptides corresponding to selected segments of S2 and S3 [2], we have shown that PT may contain more than one carbohydrate binding domain in subunits S2 and S3. S2 might bind to nonsialyated glycolipids while S3 could preferentially bind to gangliosides [2, 3]. To determine which amino acid sequences are involved in the lectin-like binding domain(s) of PT, 23 peptides covering the entire sequences of the four B-oligomer subunits (S2-S5) were synthesized. Each peptide was then screened for its ability to (1) bind to fetuin or ceruloplasmin and (2) inhibit biotinylated-PT binding to fetuin or ceruloplasmin.

Results and Discussion

All pertussis peptides were first tested for their reactivities to biotinylated fetuin using a solid-phase immunoassay as previously described [2]. Peptides corresponding to residues 78-98 and 123-154 of S2; 78-108, 103-127 and 149-176 of S3; 1-34 and 68-103 of S4 reacted strongly with biotinylated fetuin. These peptides also bound to biotinylated ceruloplasmin. The binding of these peptides to fetuin was found to be specific. As shown in Figure 1, synthetic peptides (residues 78-98 and 123-154 of S2; 78-108, 103-127 and 149-176 of S3; and 68-103 of S4) were capable of inhibiting the binding of biotinylated fetuin to PT in a competitive binding inhibition assay [2].

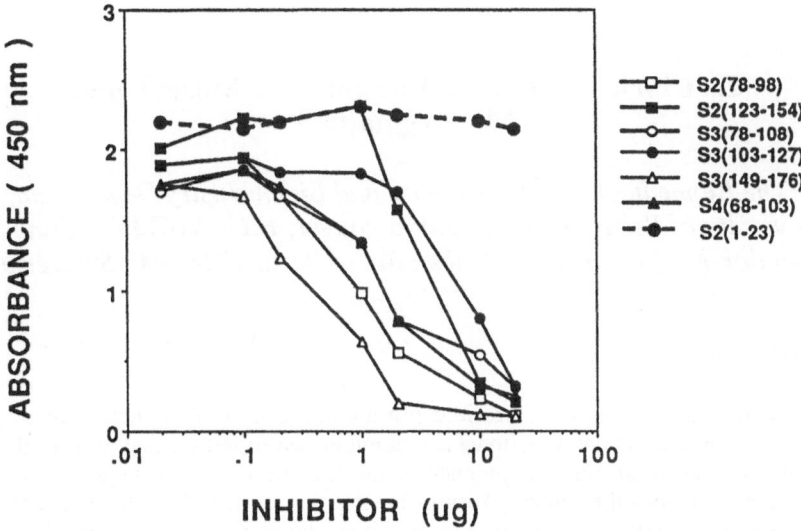

Fig. 1. *Competitive inhibition of biotinylated fetuin binding to PT-coated plate by PT peptides. The experimental protocol has been previously described [2].*

These results are in agreement with our early studies that rabbit antisera raised against peptides (residues 78-98 of S2, 78-108 and 148-176 of S3) were capable of neutralizing the toxicity of PT in the CHO cell clustering assay [4]. Furthermore, that most of the peptides tested contain tyrosine and lysine residues. Earlier results obtained with iodinated PT, have revealed the functional importance of tyrosine residues in its lectin-like activity [5]. In addition, PT analogs with mutations in S2 (PT-S2$^{(91-93)\Delta}$, PT-S2$^{(Y102A,Y103A)}$) exhibited reduced binding to fetuin. Similarly, PT analogs with mutations in S3 or S4 (PT-S3$^{(Y82A)}$, PT-S3$^{(91-93)\Delta}$, PT-S4$^{(Y4A)}$ and PT-S4$^{(Y21A)}$) retained less than 50% of PT toxicity in the CHO cell assay [6]. Therefore, the peptides identified in this study may represent lead compounds for the design of pertussis toxin receptor-targeted drugs.

References

1. Zealey, G., Loosmore, S., Yacoob, R. and Klein, M., Vaccine Res., 1 (1992) 413.
2. Heerze, L.D., Chong, P.C.S. and Armstrong, G.D., J. Biol. Chem., 267 (1992) 25810.
3. Saukkonen, K., Burnette, N.W., Mar, V.L., Masure, R. and Tuomeanen, E.I., Proc. Natl. Acad. Sci. U.S.A., 89 (1992) 118.
4. Chong, P., Zobrist, G., Sia, C., Loosmore, S. and Klein, M., Infect. Immun., 60 (1992) 4640.
5. Armstrong, G.D., Howard, L.A. and Peppler, M.S., J. Biol. Chem., 263 (1988) 8677.
6. Loosmore, S., Zealey, G., Cockle, S., Boux, H., Chong, P., Yacoob, R. and Klein M., Infect. Immun., 61 (1993) 2316.

Interleukin-8 structure-activity relationships solved by use of synthetic hybrids

I. Clark-Lewis[1], B. Dewald[2], B. Moser[2] and M. Baggiolini[2]

[1]The Biomedical Research Centre and Biochemistry Department, University of British Columbia, Vancouver, B.C., V6T 1Z3, Canada, [2]Theodor-Kocher Institute, University of Bern, CH-3000, Switzerland

Introduction

Interleukin-8 (IL-8) is a 72 residue pro-inflammatory protein that stimulates the chemotactic migration and functional activation of neutrophil leukocytes [1]. IL-8 is one of a subfamily of nine human chemokines that are related in sequence and are distinguished from other chemokines in having the motif C X C for the first two cysteines. The NMR structure indicates that it is a homodimer which is stabilized by formation of a 6-stranded β sheet (3 strands from each subunit) and by the overlying α-helices [2]. The monomer consists of: a highly disordered N-terminal region (residues 1-6), a flexible loop 7-18, a 3_{10} helical turn (residues 19-22), β strand 1 (residues 23-29), an atypical turn (30-35), β strand 2 (36-42), a type I β turn (43-46), β strand 3 (47-51), a type I β turn (52-55), and a C-terminal α helix (56-72) [2].

IL-8 is associated with a large number of acute and chronic inflammatory diseases suggesting that IL-8 antagonists could be therapeutically useful. To design antagonists, information regarding the structural regions that determine both receptor binding, and also receptor triggering, is required. We have acquired this information by studying IL-8 structure-activity relationships. Appropriately designed analogs were chemically synthesized using automated tBoc/Bzl chemistry on an Applied Biosystems 430A synthesizer. After low-high HF deprotection, the crude product was folded by air oxidation and purified by reverse phase HPLC. Multiple techniques including electrospray mass spectrometry, isoelectric focussing and amino analysis were used to characterize the product and verify authenticity [3].

Results with truncation analogs demonstrated that 3 residues, Glu-4, Leu-5 and Arg-6 (termed the ELR motif) in the flexible N-terminal region were essential for function [4]. Single amino acid substitution analogs demonstrated that all three amino acids were highly sensitive to modification, with the arginine being the most sensitive. Many of the modified analogs were found to be receptor antagonists with IC_{50}s of 0.3-2 μM [5]. However the ELR motif was not sufficient for activity because the ELR peptide and several cyclic and non-cyclic peptides containing the ELR motif did not bind to receptors. The most likely interpretation of these results is that the ELR motif of IL-8 is critical for receptor binding and activation but that other structural features are also necessary. This was further investigated by design of substitution analogs and a hybrid approach.

Results and Discussion

The hypothesis that additional structural features of IL-8 other than ELR are necessary was supported by two analogs which had either the 7-34 or 9-50 disulfide bridge eliminated by conversion of the respective cysteines to α-aminobutyric acid, and which were inactive. This strongly indicated the importance of the compact tertiary structure for activity. Analogs with individual substitutions covering the 1-35 region were designed to identify residues/regions that, like the ELR motif, were highly sensitive to modification and by implication, involved in receptor binding. No such regions were identified using this approach.

To identify which structural features are functionally important, a hybrid approach was used. The aim was to convert the inactive related protein, IP10 [6], into a hybrid protein with IL-8-like function by splicing in the minimal amount of IL-8 sequences. In general, for this approach to be effective, the surrogate protein (in this case IP10) must be related but have low overall sequence similarity. IP10 has 17 residues that are identical in IL-8, and 14 of these are predicted to be involved in maintaining the structure. The more solvent exposed regions were the focus of our structure-function studies and the hybrids were carefully designed to maintain the IP10 core structure.

An IP10 hybrid containing IL-8 residues 4-23, 30-35 and 44-49 was found to be equivalent in activity to the fully active control, IL-8, 4-72, thus indicating that these sequences were sufficient for conferring receptor function. The question remained: how much of this IL-8 structure is necessary? Additional hybrids demonstrated that the 30-35 β turn is essential, but the 44-49 β turn was not required.

Two hybrids addressed the role of the N-terminal loop and 3_{10} turn. A hybrid with IL-8 4-10, 16-23, and 30-35 had extremely low activity, >100-fold lower than IL-8, 4-72, thus indicating the importance of residues 11-15. A hybrid with IL-8, 4-16 and 30-35 had detectable activity but this was 100-fold lower than IL-8,4-72. Therefore, the 3_{10} turn is also critical.

It is interesting to speculate as to the role of the loop and 3_{10} turn. The N-terminal loop, residues 10-18, is highly variable in the other IL-8-related proteins, and they also vary in potency in IL-8 assays. Residues 11-15 can be substituted individually without substantially affecting function. (However, the results show that the region is essential.) Neither these substitutions nor sequence comparisons suggest a clear explanation for the observation that the region is essential. The most likely explanation is that the loop region is directly involved in binding to the receptor but is not directly involved in receptor triggering and binding, as is clearly the case with the ELR region. Thus the loop region was less sensitive to modification than ELR but was still required for binding. Understanding the detailed role of this region will require further analogs coupled with structural studies. The 3_{10} turn is also important; however, its role could be structural and deviation from the IL-8 3:10 turn could also affect the 11-18 loop. For example, the F21L analog of IL-8 has a significant reduction in activity and Phe-21 is involved in aromatic interactions with Tyr-13, Phe-17 and Trp-56.

The role of the 30-35 turn may also be structural as the conformation of this turn will affect the 7-34 disulfide bridge. This disulfide anchors the essential ELR motif. Furthermore from substitution analogs, His-33 and Ala-35 are not essential and it is probable that Gly-31, Pro-32 and Cys-34 have structural roles. Even though these residues could directly bind receptors as well as determine the ELR position, our hypothesis is that the role of the 30-35 turn is conformational.

In summary, we have demonstrated that the ELR motif is the most critical region for receptor activation, although the tertiary structure and N-terminal loop are also essential for function. The β turn that encompasses residues 30-35, and, via the cysteine 7-34 disulfide tethers the ELR motif, is also essential. Evaluating the precise roles of these regions will require detailed structural information on the analogs and further modifications to IL-8 aimed at fine mapping the functional determinants.

References

1.   Baggiolini, M. and Clark-Lewis, I., FEBS Lett., 307 (1992) 97.
2.   Clore, G.M., Appella, E., Yamada, M., Matsushima, K. and Gronenborn, A.M., Biochemistry, 29 (1990) 1689.
3.   Clark-Lewis, I., Moser, B., Walz, A., Baggiolini, M., Scott, G.J. and Aebersold, R., Biochemistry, 30 (1991) 3128.
4.   Clark-Lewis, I., Schumacher, C., Baggiolini, M. and Moser, B., J. Biol. Chem., 266 (1991) 23128.
5.   Moser, B., Dewald, B., Barella, L., Schumacher, C., Baggiolini, M. and Clark-Lewis, I., J. Biol. Chem., 268 (1993) 7125.
6.   Dewald, B., Moser, B., Barella, L., Schumacher, C., Clark-Lewis, I. and Baggiolini, M., Immunol. Lett., 32 (1992) 81.

Modification of the address sequence of dynorphin A that produces highly potent δ-selective peptide agonists

N. Collins[1], J.-Ph. Meyer[1], T. Zalewska[2], H.I. Yamamura[2], P. Davis[2], F. Porreca[2] and V.J. Hruby[1]

[1]Department of Chemistry and [2]Department of Pharmacology, University of Arizona, Tucson, AZ 85721, U.S.A.

Introduction

Dynorphin A (Dyn A) is the putative endogenous ligand for the κ opioid receptor. However, in the mammalian brain, Dyn A is highly potent at all three of the commonly accepted opioid receptors (κ:μ:δ IC_{50}'s 0.07 : 1.08 : 6.99 nM respectively) and thus is potentially a good lead for the design of selective ligands for all three receptors. To date there are several potent analogues of Dyn A showing moderate selectivity for κ and/or μ receptors, but none have been found with δ selectivity. Dyn A is postulated to be comprised of a "message" sequence ([Leu⁵]enkephalin; residues 1-5) that produces the opioid response, and an "address" that targets the message sequence to the κ receptor (residues 6-17, or in Dyn A_{1-11} residues 6-11) (Figure 1) [1]. Notably [Leu⁵]enkephalin carrying no address sequence is somewhat δ selective. We therefore wished to locate and modify the regions of the address sequence that determine selectivity to produce highly potent δ opioid ligands.

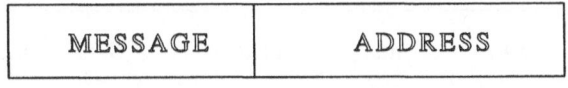

Tyr¹ Gly Gly Phe Leu Arg⁶ Arg⁷ Ile Arg Pro Lys¹¹

Fig. 1. Dyn A_{1-11} displaying message and address regions.

Results and Discussion

Analogues of Dyn A_{1-7}, Dyn A_{1-8} and Dyn A_{1-11} were prepared by standard Merrifield solid phase methods in which Arg⁶ and Arg⁷ were replaced with Nle. While DynA$_{1-7}$-NH₂ (1) was κ selective, [Nle⁶,⁷]Dyn A_{1-7}-NH₂ (2) and [Nle⁶,⁷,⁸]Dyn A_{1-8}-NH₂ (3) exhibited 10-fold and 100-fold selectivity, respectively, for δ versus κ receptors in GPB binding assay (Table 1). These findings were reflected in the GPI and MVD bioassays (Table 2). While extension of the Dyn A address sequence in the analogue [Nle⁶,⁷,⁸]Dyn A_{1-11}-NH₂ (4) altered the selectivity ratios slightly, binding at the δ receptor was still favored and potency at all opioid receptors was enhanced in the GPB.

657

Table 1 *Binding affinities (IC_{50}, nM) in guinea pig brain (GPB)*

Compound	κ^a	μ^b	δ^c	κ/δ	μ/δ
1	0.15	2.55	23.2	0.01	0.11
2	271		10.7	25.3	
3	1014		20.5	49.5	
4	24.2	33.5	2.25	10.8	14.9
5	1367	4489	160	8.54	28.1
6	1615	1356	9.55	169	142
7	174	202	1.17	149	173

[a]vs. [^{3}H]U-69593, [b]vs. [^{3}H]PL-17 and [c]vs. [^{3}H](p-Cl)DPDPE.
1: $DynA_{1-7}$-NH_2, **2**: [$Nle^{6,7}$]$DynA_{1-7}$-NH_2, **3**: [$Nle^{6,7,8}$]$DynA_{1-8}$-NH_2,
4: [$Nle^{6,7,8}$]$DynA_{1-11}$-NH_2, **5**: [$DPen^{2,5}$, $Nle^{6,7,8}$]$DynA_{1-11}$-NH_2,
6: [$DPen^2$,Pen^5,$Nle^{6,7,8}$]$DynA_{1-11}$-NH_2, **7**: [$DPen^2$, Cys^5, $Nle^{6,7,8}$]$DynA_{1-11}$-NH_2.

Table 2 *Inhibitory potencies (IC_{50}, nM) in mouse vas deferens (MVD) and guinea pig ileum (GPI) bioassays*

Compound	MVD			GPI		
	IC_{50}	δ shift[a]	κ shift[b]	IC_{50}	μ shift[c]	κ shift[b]
1	30.7	n.s.	11.0	5.65	n.s.	42.8
2	32.0	21.9	n.s.	640	n.s.	7.5
3	72.7	25.8	n.d.	17,549	n.d.	n.d.
4	93.2	9.6	n.d.	335	n.s.	n.d.
5	>1,000	n.d.	n.d.	22,725	n.d.	n.d.
6	8.23	13.8	n.s.	5,210	n.d.	n.d.
7	0.84	29.0	n.s.	106	7.4	n.d.

[a]vs. ICI 174864 (1000 nM, δ antagonist), [b]vs. nor-BNI (10 nM, κ antagonist) and [c]vs. CTAP (1000 nM, μ antagonist). n.s.: no significant shift was observed. n.d.: not determined.

This group has previously reported that the cyclic peptides [D-Pen^2,D-Pen^5]enkephalin (DPDPE) and [D-Pen^2,L-Pen^5]enkephalin (DPLPE) are highly potent δ selective ligands [2]. We therefore synthesized chimeric hybrid analogues of Dyn A$_{1-11}$ containing DPDPE in the message 1 - 5 positions and the δ selective modified Dyn A address sequence in positions 6 - 11. While [D-Pen^{2,5},Nle^{6,7,8}]Dyn A$_{1-11}$-NH$_2$ (**5**) was moderately δ selective it displayed weak binding and analgesia. However substitution of D-Pen^5 with L-Pen produced a highly potent δ selective ligand (**6**) indicating that an L residue is essential in position 5 of extended sequences. Replacement of L-Pen^5 with L-Cys provided a ligand (**7**) with increased potency in the GPB and strong δ opioid activity in the MVD (IC$_{50}$ 0.8 nM).

In modifying the address sequence of Dyn A$_{1-11}$ to alter the selectivity of this peptide hormone between the opioid receptors we have demonstrated that the address sequence can in fact be subdivided into two regions. An address proper or "selector" region responsible for selectivity is located at positions 6 and 7 (e.g., Arg^6-Arg^7 in the case of κ selectivity); while residues 8-11 define a "potentiator" region [3] which increases potency at all three opioid receptors (see Table 1, peptides **1**, **2**, **3** and Figure 2). As we have shown here, proper choice of message sequence and selector region in Dyn A analogues can in principle be utilized to produce potent and selective ligands for all three (κ, μ and δ) opioid receptors.

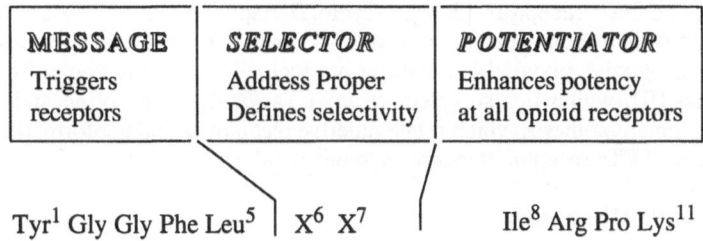

Fig. 2. Breakdown of address into selector and potentiator regions.

References

1. Schwyzer, R., Ann. N.Y. Acad. Sci., 297 (1977) 3.
2. Mosberg, H.I., Hurst, R., Hruby, V.J., Gee, K., Yamamura, H.I., Galligan, J.J. and Burks, T.F., Proc. Natl. Acad. Sci. U.S.A., 80 (1983) 5871.
3. Schwyzer, R., Karlanganis, G. and Lang, V., In Ananchenko, S.N. (Ed.), Frontiers of Bioorganic Chemistry and Molecular Biology, Pergamon Press, New York, U.S.A., 1980, p.277.

Endothelin ETa receptor antagonist pharmacophore hypotheses

T. Cordova, M. Hassan, J.C. Hempel, S.C. Koerber, E.R. Vorpagel and A.T. Hagler

Biosym Technologies Inc., San Diego, CA 92121, U.S.A.

Introduction

Endothelin-1 (ET-1) is a 21 residue peptide hormone with potent cardiovascular effects. Two types of receptors are known for endothelin: ETa and ETb [1, 2]. In this study we investigate the conformational preferences of three structural classes of competitive ET-1 antagonists at the ETa receptor and compare them to define putative shared pharmacophores. The first compound modeled, a linear hexapeptide [Ac-D-Dip^1-Leu^2-Asp^3-Ile^4-Ile^5-Trp^6] (Antagonist I), is a functional antagonist at both ETa and ETb receptors [3, 4]. Cyclo[D-Asp^1-Pro^2-D-Val^3-Leu^4-D-Trp^5] (Antagonist II) is a cyclic pentapeptide which is selective for the ETa receptor [5, 6]. The third antagonist modeled is a natural product steroid, myriceron caffeoyl ester (Antagonist III), which is also selective for ETa [7, 8]. We define and evaluate pharmacophore hypotheses which relate putative receptor bound conformations of the three classes of ETa receptor antagonists to one another.

Results and Discussion

Databases of energetically accessible conformations of each antagonist were generated using high temperature molecular dynamics and systematic search strategies. Pharmacophore hypotheses are defined by three dimensional patterns of descriptor centers (structural features critical for activity) that are shared by the conformers of the antagonists included in the databases. Since only certain features of each of the three classes of antagonists are reported to be critical for activity [3-8], we focus on side chain and backbone carbonyl descriptors for the peptide antagonists. Pharmacophore hypotheses were identified using an expert system which, given the database of conformers for each bioactive molecule, is targeted to identify shared structural patterns that may be associated with biological activity (APEX-3D) [9-10]. This approach provides a 3D representation of the three molecules in which the individual conformers are overlaid according to the pharmacophore hypothesis.

To simplify the problem, we first matched the cyclic pentapeptide (Antagonist II) to the steroid and then in a second step matched the hexapeptide (Antagonist I) to the steroid. We define a matching hypothesis in which the descriptor centers are divided into classes (hydrophobic side chains, aromatic rings, carboxylic acids, H-bond acceptor/donor sites, carbonyls) and any member of a given descriptor center class for one molecule may be matched to any member of that descriptor class in another molecule. Multiple pharmacophore hypotheses were defined and evaluated.

Using the appropriate conformer overlay, we determined the molecular volume of each antagonist, the volume common to all three molecules (common volume), the union of volumes shared by any two antagonists (shared volume), and the volume which is occupied by any antagonist but not shared (excluded volume). Pharmacophore hypotheses for which the common and shared volume is maximal and the excluded volume is minimal are ranked highest and the associated conformations of the three antagonists are putative bioactive conformations. An optimal hypothesis (Figure 1) identified in this manner matches the side chains of Asp[1] of the pentapeptide and Asp[3] of the hexapeptide to the carboxylic acid moiety of the steroid, the carbonyls of Leu[4] of the pentapeptide and of Dip[1] in the linear hexapeptide to the ester carbonyl in the steroid, and the side chain of Val[3] of the pentapeptide to C4 in ring A in the steroid, and one of the aromatic rings in Dip[1] in the hexapeptide to the aromatic ring side chain in the steroid.

Antagonist I : linear hexapeptide

$$[\text{Ac-D-Dip}^1\text{-Leu}^2\text{-Asp}^3\text{-Ile}^4\text{-Ile}^5\text{-Trp}^6]$$

Antagonist II: cyclic pentapeptide

$$\text{cyclo } [\text{D-Asp}^1\text{-Pro}^2\text{-D-Val}^3\text{- Leu}^4\text{-D-Trp}^5]$$

Antagonist III: myriceron caffeoyl ester

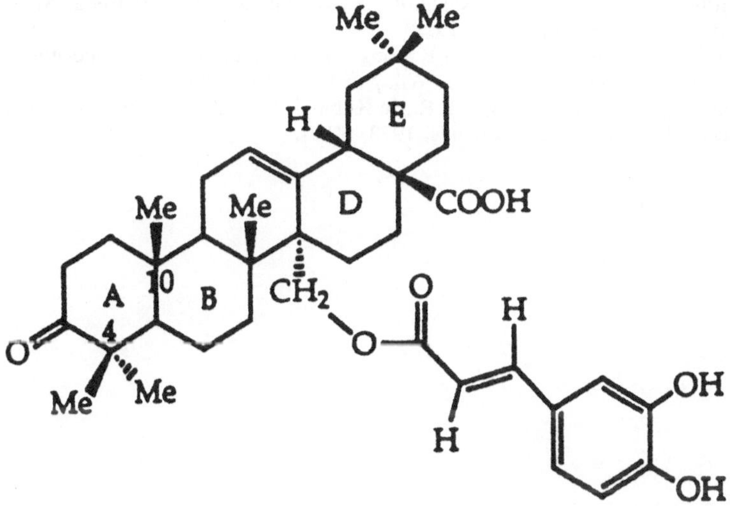

Fig. 1. *In one of the optimal overlaps we identified, the side chains of Asp[1] of the pentapeptide and Asp[3] of the hexapeptide are matched to the carboxylic acid moiety of the steroid, the carbonyls of Leu[4] of the pentapeptide and Dip[1] of the hexapeptide are matched to the ester carbonyl in the steroid, the side chain of Val[3] of the pentapeptide is matched to C4 in the steroid and one of the aromatic rings in Dip[1] of the hexapeptide is matched to the aromatic ring side chain in the steroid.*

Putative pharmacophore hypotheses defined for ETa receptor antagonists will be used as a starting point for the design of novel compounds in the search for new antagonists. The strategy used in the identification of pharmacophore hypotheses is general and can be used for series of bioactive molecules. When constrained analogs are included, the number of possible pharmacophores is reduced.

References

1. Arai, H., Hori, S., Aramori, H., Ohkubo, H. and Nakanishi, S., Nature, 348 (1990) 730.
2. Sakurai, T., Yanagisawa, M., Takuwa, Y., Miyasaki, H., Kimura, S., Goto, K. and Masaki, T., Nature, 348 (1990) 732.
3. Cody, W.L., Doherty, A.M., He, J.X., DePue, P.L., Rapundalo, S.T., Hingorani, G., Major, T.C., Panek, R.L., Dudley, D.T., Haleen, S.J., LaDouceur, D.M., Hill, K.E., Flynn, M.A. and Reynolds, E.E., J. Med. Chem., 35 (1992) 3303.
4. Doherty, A.M., Cody, W.L., He, X., DePue, P.L., Leonard, D.M., Dudley, D.T., Rapundalo, S.T., Hingorani, G.P., Panek, R.L., Major, T.C., Hill, K.E., Flynn, M.A. and Reynolds, E.E., 203rd National Meeting of the American Chemical Society, San Francisco, CA, U.S.A. (1992) MEDI 174.
5. Ihara, M., Noguchi, K., Saeki, T., Fukuroda, T., Tsuchida, S., Kimura, S., Fukami, T., Ishikawa, K., Nishikibe, M. and Yano, M., Life Sci., 50 (1992) 247.
6. Ishikawa, K., Fukami, T., Nagase, T., Fujita, K., Hayama, T., Niiyama, K., Mase, T., Ihara, M. and Yano, M., J. Med. Chem., 35 (1992) 2139.
7. Fujimoto, M., Mihara, T., Nakajima, S., Ueda, M., Nakamura, M. and Sakurai, K., FEBS Lett., 305 (1992) 41.
8. Konoide, T., Hayashi, T., Sakurai, K., Tojo, T., Yasuda, F., Nakamura, M., Fujimoto, M., Mihara, S., Nakajima, S., Matsumura, S. and Ueda, M., Structure Activity Symposium, Japan (1992).
9. Golender, V. and Rozenblit, A.B., Logical and Combinatorial Algorithm for Drug Design, Research Studies Press, Wiley & Sons, 1983, p.8, p.45.
10. Golender, V.E. and Vorpagel, E.R., In Kubinyi, H. (Ed.), 3D-QSAR in Drug Design, Escom, Leiden, The Netherlands, 1993, p.137.

Structure-function relationship between guanylin and the *E. coli* heat-stable enterotoxin

B. Carpick and J. Gariépy

Department of Medical Biophysics, University of Toronto and The Ontario Cancer Institute, 500 Sherbourne Street, Toronto, Ontario, M4X 1K9, Canada

Introduction

The heat-stable enterotoxins are a group of small homologous peptides elaborated by enterotoxigenic strains of bacteria [1]. They are collectively responsible for a large proportion of worldwide cases of secretory diarrhea in human and animals. These enterotoxins, abbreviated ST I (or ST_A), are known to bind to receptors located on the brush border surface of intestinal cells [2], and to cause an elevation of intracellular cGMP levels [3-5]. The binding of the enterotoxin to its receptor is coupled to the activation of a guanylate cyclase. cGMP acts as the intracellular second messenger causing the eventual onset of diarrhea. Recently, a naturally occurring peptide termed guanylin, was isolated from rat jejunum and found to activate a particulate form of guanylate cyclase present on the human colonic carcinoma cell line, T_{84} [6]. Guanylin was able to displace the binding of ^{125}I-labeled ST I to receptors on the surface of T_{84} cells. This 15-amino acid long peptide is homologous in sequence to a region of ST I, abbreviated ST Ib(6-18), that codes for its receptor binding and enterotoxigenic properties [7-9] (Figure 1). As a consequence of structural and functional similarities between guanylin and ST I, one would expect guanylin to cause diarrhea in mammals at a concentration relative to ST I that parallels its ability to inhibit the binding of radiolabeled ST I to intestinal cells. In this study, we report that guanylin is not enterotoxic and propose that ST I enterotoxins may represent long-lived superagonists of guanylin.

Results and Discussion

We report the synthesis of guanylin and a second guanylin analogue, termed N^9P^{10}guanylin, which is a closer homologue of ST Ib (Figure 1), particularly within the central region of the ST Ib molecule (residues 11-14). This turn region was previously found to be the most important region of ST Ib in terms of biological activity and is particularly sensitive to amino acid substitutions [9]. The objectives of this study were thus twofold: to quantitatively compare the biological activities of guanylin and ST Ib(6-18) using established in vitro and in vivo assays and to compare the structural domain of guanylin to that of ST I in order to evaluate if guanylin represents an appropriate peptide scaffold for the design of potent ST I antagonists.

663

	1	2	3	4	5	6	7	8	9	10	11	12	13	14	15	16	17	18	19
ST Ib	Asn	Ser	Ser	Asn	Tyr	Cys	Cys	Glu	Leu	Cys	Cys	Asn	Pro	Ala	Cys	Thr	Gly	Cys	Tyr
ST Ib(6-18)						Cys	Cys	Glu	Leu	Cys	Cys	Asn	Pro	Ala	Cys	Thr	Gly	Cys	
Guanylin			Pro	Asn	Thr	Cys	Glu	Ile	Cys	Ala	Tyr	Ala	Ala	Cys	Thr	Gly	Cys		
N^9P^{10}guanylin			Pro	Asn	Thr	Cys	Glu	Ile	Cys	Ala	Asn	Pro	Ala	Cys	Thr	Gly	Cys		

Fig. 1. *The amino acid sequences of the E. coli heat-stable enterotoxin ST Ib and the three peptides prepared for this study; ST Ib(6-18), guanylin and N^9P^{10}guanylin. Numbering refers to the position of the amino acids in relation to the ST Ib sequence. ST Ib(6-18) corresponds to the biologically active domain of ST Ib [7-9]. Regions of sequence homology are boxed. The substitution $Leu^9 \rightarrow Ile$ represents a conserved substitution observed in other ST I sequences. All peptides were synthesized as described elsewhere [9].*

Table 1 *Biological properties of ST Ib(6-18), guanylin and N^9P^{10}guanylin*

Peptide	IC_{50} (M)[a]	ED_{50} (mole)[b]	GC_{50} (M)[c]
ST Ib(6-18)	3.0×10^{-8}	2.5×10^{-12}	10^{-8}
Guanylin	10^{-6}	$>3.0 \times 10^{-8}$	7.0×10^{-7}
N^9P^{10}guanylin	6.0×10^{-5}	3.2×10^{-10}	6.3×10^{-6}

[a]IC_{50}, peptide concentration required to inhibit 50% of the specific binding of radiolabeled Y^4ST Ib(4-18) to rat villus cells; [b]ED_{50}, peptide dose required to cause a half-maximal increase in gut-to-carcass weight ratio after an oral administration of the analogue to infant mice; [c]GC_{50}, peptide concentration required to cause a half-maximal activation of intestinal guanylate cyclase.

The binding of guanylin analogues to the ST I receptor on rat small intestinal villus cells was assessed by measuring their ability to compete with radiolabeled ST I for receptor sites on enterocytes [7, 9]. As shown in Table 1, synthetic guanylin displaced, in a concentration-dependent manner, the binding of ^{125}I-Y^4ST Ib(4-18) to receptors on rat intestinal cells with an IC_{50} value 30-fold higher than ST Ib(6-18). The peptide N^9P^{10}guanylin, on the other hand, was ~ 2000-fold less active than ST Ib(6-18) in the binding assay (Table 1). The capacity of ST I analogues to cause intestinal fluid accumulation in infant mice [10] correlates well with their ability to bind to rat villus cells [7, 9]. Guanylin was unable to cause diarrhea in mice except at concentrations 4 orders of magnitude higher than ST Ib(6-18) (Table 1). In contrast, N^9P^{10} guanylin was only 125-fold less active than ST Ib(6-18) in this experiment (Table 1). If a single class of ST I receptors existed that was mechanistically linked to the onset of diarrhea, one would expect guanylin to act as an antagonist and block diarrhea induced by ST I. Experimentally, guanylin was unable to act as an antagonist since diarrhea was observed in all suckling mice even at a molar ratio of guanylin to ST Ib(6-18) of 10^2 (10^{-9} mole guanylin to 10^{-11} mole ST Ib(6-18)) (data not shown). Both guanylin analogues were assayed for their ability to activate rat intestinal brush border guanylate cyclase in relation to ST Ib(6-18). Results presented in Table 1 indicate that the pattern of activation of guanylate cyclase for the three peptides mirrors their pattern of binding affinities for the ST I receptor. Structural differences between guanylin and ST I enterotoxins may explain their differences in causing fluid accumulation. The simplest model would view ST I as a long-lived superagonist of guanylin, and show that processing or clearance of ST I from the guanylin receptor is not as efficient as in the case of guanylin causing a chronic activation of the guanylate cyclase. The tyrosine-alanine segment present in the sequence of guanylin is the only site absent in ST I enterotoxins which represent a substrate region susceptible to protease digestion (putative chymotryptic site; Figure 1). Preliminary experiments indicate that guanylin's ability to bind to the rat villus ST I receptor is lost within minutes of exposure to chymotrypsin while the binding of N^9P^{10}guanylin as well as ST I(6-18) to intestinal cells remains unaffected by such a treatment even after 3 hours. Thus the susceptibility of guanylin to chymotrypsin digestion suggests that the conserved Asn-Pro sequence present in all ST I enterotoxins increases dramatically their biological half-life and dictates their enterotoxicity.

Acknowledgements

This work was supported by a grant (AI26152) from the National Institutes of Health.

References

1.   Thompson. M.R. and Giannella, R.A., J. Recep. Res., 10 (1990) 97.
2.   Gariépy, J. and Schoolnik, G.K., Proc. Natl. Acad. Sci. U.S.A., 83 (1986) 483.
3.   Field, M., Graf, L.H., Jr., Laird, W.J. and Smith, P.L., Proc. Natl. Acad. Sci. U.S.A., 75 (1978) 2800.
4.   Giannella, R.A. and Drake, K.W., Infect. Immun., 24 (1979) 19.
5.   Hughes, J.M., Murad, F., Chang, B. and Guerrant, R.L., Nature, 271 (1978) 755.
6.   Currie, M.G., Fok, K.F., Kato, J., Moore, R.J., Hamra, F.K., Duffin, K.L. and Smith, C.E., Proc. Natl. Acad. Sci. U.S.A., 89 (1992) 947.
7.   Gariépy, J., Judd, A.K. and Schoolnik, G.K., Proc. Natl. Acad. Sci. U.S.A., 84 (1987) 8907.
8.   Shimonishi, Y., Hidaka, Y., Koizumi, M., Hane, M., Aimoto, S., Takeda, T., Miwatani, T. and Takeda, Y., FEBS Lett., 215 (1987) 165.
9.   Carpick, B.W. and Gariépy, J., Biochemistry, 30 (1991) 4803.
10.   Gyles, C.L., Can. J. Comp. Med., 43 (1979) 371.

Probing the steric tolerance of the insect hypertrehalosemic hormone receptor to develop an effective photoaffinity probe

T.K. Hayes, F.L. Nails and M.M. Ford

*Institute of Biosciences and Technology, Department of Entomology,
Texas A&M University, College Station, TX 77843, U.S.A.*

Introduction

Hypertrehalosemic hormone (HTH) stimulates the insect fat body to synthesize and release trehalose into the hemolymph for use as an energy source by other tissues. It was isolated from whole head extracts of the cockroach *Blaberus discoidalis* and was found concentrated in the major insect neurosecretory gland (the corpora cardiaca) located at the base of the brain [1]. The peptide is a member of the adipokinetic hormone family (AKH) and possesses many of the conserved family traits [2]. It is a blocked decapeptide with a C-terminal amide, has a pGlu residue at the N-terminus and two aromatic residues at positions 4 and 8, and has no charge (Table 1). Other members of the AKH family stimulate the release of other metabolites from the insect fat body (i.e., lipid and amino acids), stimulate heart contractions, or, when found in crustaceans, stimulate pigment concentration in the epidermis. This report will concentrate on the description of initial structure-activity studies and how this information is being used to develop a photoaffinity label to study the properties of HTH receptors.

Results and Discussion

Structure-activity studies have been used to gain an impression of how HTH binds and activates its receptor. Analogs have been prepared by standard tBoc chemistry on a Milligen-Biosearch 9500 instrument. After purification by RP-HPLC, the correct synthesis and homogeneity of each peptide was confirmed by analytical HPLC, amino acid analysis and mass spectrometry. The ability for each analog to stimulate trehalose production was evaluated with an in vivo assay and often confirmed with an in vitro assay using fragments of insect fat body tissue in temporary organ culture [3].

Previous structure-activity studies have defined many of the important aspects of how biological information is encoded in the structure of HTH. Single amino acid replacement analogs helped determine that the most critical side chains were pGlu[1], Phe[4], and Trp[8] [4]. Further, the ring nature of the pGlu seems important and the aromaticity of the Phe and Trp is required [5]. The double ring structure of the Trp is also needed for full potency.

Table 1 *Comparison of HTH analog structure and potency values from the hypertrehalosemic assay*

Peptide	ED$_{50}$(pmole injected)
pQVNFSPGWGTamide　　(HTH)	1
Ac-PSFNVGPVNFSPGWGTamide	2

SINGLE D-AA REPLACEMENT ANALOGS:

Peptide	ED$_{50}$(pmole injected)
pQVNFSPGWGTamide[a]	10
pQVNFSPGWGTamide	350
pQVNFSPGWGTamide	NA
pQVNFSPGWGTamide	150
pQVNFSPGWGTamide	NA
pQVNFSPGWGTamide	30
pQVNFSPNWGTamide	1
pQVNFSPGWGTamide	NA
pQVNFSPGWATamide	90
pQVNFSPGWGTamide	9

BOLTON-HUNTER & Tyr REPLACEMENT ANALOGS:

Peptide				ED$_{50}$(pmole injected)
BPVN	F	SP G	WGTamide	>1000
pQYN	F	SP G	WGTamide	100
pQVN	Y	SP G	WGTamide	2
pQVN (Y*)	SP G		WGTamide	NA
pQVN	F	SP Y	WGTamide	NA
pQVN	F	SP (KB)	WGTamide	NA
pQVN	F	SP G	YGTamide	30
pQVN	F	SP G	WGYamide	100

TRI-Ser TETHER ANALOGS:

Peptide		ED$_{50}$(pmole injected)
AcSSSPVNFSPGWGTamide		185
pQVNFSPXWGTamide	(Ac-SSSK)	NA
pQVNFSPXWGTamide	(Ac-SSSK)	1
pQVNFSPGWGXamide	(Ac-SSSK)	60
pQVNFSPGWGXamide	(Ac-SSSK)	60
pQVNFSPGWGTSSSamide		100

[a]Abbreviations:　pQ represents pyroglutamic acid; Ac represents a N-acetyl moiety; Underlined amino acids differ from native HTH; Double underlined amino acids are in the D-configuration; NA means not active; B represents the Bolton-Hunter reagent; (KB) represents the Bolton-Hunter reagent attached to the epsilon amino group of Lys; (Y*) represents an iodinated Tyr; X represents the attachment site for a branched peptide chain through the epsilon amino group of Lys (the sequence of the branch is shown in parentheses).

A series of analogs where a single amino acid was replaced with its D-enantiomer were compared with the bioassay system to gain initial indications where important conformational elements might be located or induced for receptor interaction (Table 1). The loss of activity or potency around the aromatic amino acids confirms the importance of these residues towards the biological activity of HTH. One explanation for the loss of potency and activity within the N-terminal pentamer of HTH is that a chiral repetitive structure may be disrupted in the analog when it is bound by the receptor(s). An examination of HTH in detergent micelles by CD and NMR indicates that the N-terminal pentamer of HTH has tendencies to form a β-strand. A special N-terminally extended analog of HTH has near full activity and potency (Table 1). That analog was designed to have an additional turn and strand to promote the formation of a small β-sheet and thereby stabilize the postulated β-strand for the native hormone.

The relatively high potency of the HTH analogs that have D-amino acids in positions 6 or 7 is a first indicator that a turn centered around the Pro-Gly sequence may be important for the biological activity of the hormone. The inactivity of the single amino acid deletion analogs of HTH corresponding to positions 4 through 8 is consistent with emphasizing the importance of the spacing between the critical aromatic amino acids. That spacing would contain the suspected β-turn. The design of a photoaffinity label to better study the properties of HTH receptors must include a site for radioiodination. However, attempts to either incorporate a site for iodination or to iodinate acceptable sites resulted in analogs that were inactive or had unacceptably low potencies (Table 1). Tri-Ser tether analogs were designed to have hydrophilic probes to explore for channels out of the likely receptor sites (Table 1). The aim of additional analogs is to attach radioiodination and photocrosslinking sites outside the region of HTH that is critical for receptor interaction. The construction of a branched chain analog at position 7 in the D-configuration is the best candidate to continue this approach. The acceptance of the large hydrophilic D-tether and rejection of the large L-tether is consistent with a turn localized around the Pro-Gly sequence when HTH interacts with its receptor(s). Active analogs with potential as photoaffinity probes have been prepared. However, additional progress is needed to develop more soluble analogs after radioiodination.

Acknowledgements

We are grateful for the support of this research by NIH grant RO1 NS20137.

References

1. Hayes, T.K., Keeley, L.L. and Knight, D.K., Biophys. Biochem. Res. Comm., 140 (1986) 674.
2. Hayes, T.K. and Keeley, L.L., J. Comp. Physiol., 160 (1990)187.
3. Hayes, T.K. and Keeley, L.L., Gen. Comp. Endocr., 57 (1985) 246.
4. Ford, M.M., Hayes, T.K. and Keeley, L.L., In Marshall, G.R. (Ed.), Peptides Chemistry and Biology, Escom, Leiden, The Netherlands, 1988, p.653.
5. Hayes, T.K., Ford, M.M. and Keeley, L.L., In Masler, P. and Borkovec, A.B., (Eds.), Insect Neurochemistry and Neurophysiology, Humana Press, New York, NY, U.S.A., 1990, p.247.

Analogues of cholecystokinin$_{26\text{-}33}$ selective for B-type CCK receptors possess δ opioid receptor agonist activity in vitro and in vivo: Evidence for similarities in CCK-B and δ opioid receptor requirements

V.J. Hruby[1], S.N. Fang[1], T.H. Kramer[2], P. Davis[2],
D. Parkhurst[2], G. Nikiforovich[1], L.W. Boteju[1],
J. Slaninova[2], H.I. Yamamura[2] and T.F. Burks[2]

*Departments of [1]Chemistry and [2]Pharmacology, University of Arizona,
Tucson, AZ 85721, U.S.A.*

Introduction

It has been known that central and peripheral administration of sulfated cholecystokinin$_{26\text{-}33}$ (CCK-8) can cause analgesia [1]. The mechanism is unknown, but postulated to be an allosteric interaction between CCK-B and opioid receptors. Recently we have reported [2] that certain CCK analogues that bind selectively to CCK-B receptors and δ opioid receptors can produce analgesia. We report further evidence in support of this observation, and suggest that CCK-B and δ opioid receptors have overlapping topographical structural requirements for their agonist ligands.

Results and Discussion

The syntheses of the CCK-8 analogues described here were accomplished using solid phase synthetic methods and purification methods similar to those previously reported [3]. Purity of the final products was established by reversed phase HPLC, TLC, FAB-MS and amino acid analysis.

Binding affinities of ligands for CCK-A and CCK-B receptors were determined [2, 3] using guinea pig pancreatic and guinea pig cortex membranes, and Bolton-Hunter labeled ^{125}I-CCK-8 and [^{3}H]SNF-8702, respectively. Binding assays were performed as previously outlined. Opioid receptor binding assays were performed using guinea pig whole brain membranes. The radiolabelled ligands used to label μ, δ and κ receptors respectively were [^{3}H]CTOP, [^{3}H][4'-Cl-Phe^4]DPDPE and [^{3}H]U69,593. Mouse hot plate analgesic experiments were performed using standard procedures using a 55°C hot plate and a 30 sec cut off for detection of nociception.

Table 1 summarizes binding data for CCK-8 (sulfated) and six CCK-8 analogues that possess high CCK-B receptor potency and selectivity, as well as δ opioid agonist potency (mouse vas deferens; MVD) and hot plate potency. The analogues

Table 1   *Binding affinities, bioassay, and analgesic data for CCK-8 and selected cholecystokinin analogues*

Compd No.	Binding Affinity (IC$_{50}$ in nM)				IC$_{50}$,nM	A$_{50}$,nM
	μ	δ	CCK-B	CCK-A	MVD	Hot Plate
CCK-8	>20,000[a]	>50,000[a]	0.46[a]	0.11[a]	I.A.[b]	40%
1	3,200[a]	1,040[a]	1.01[a]	960[a]	850	1.3
2	N.D[c]	~1,000	3.3	8,000	130	8.6
3	34,000[a]	1,020[a]	2.07[a]	2,100[a]	2,600	230
4	~20,000[a]	570[a]	0.086[a]	4.0[a]	900	2.4
5	N.D.	N.D.	1.4	4,500	17	1.4
6	702[a]	29[a]	0.79[a]	>10,000[a]	71	0.55

[a]From Ref. [3]; [b] I.A. = inactive; [c] N.D. = not determined.

tested were: Asp-Tyr-N-MeNle-Gly-Trp-N-MeNle-Asp-Phe-NH$_2$ (**1**); Asp-Tyr(SO$_3$H)-Ile-Gly-Trp-Nle-Asp-Phe-NH$_2$ (**2**); Asp-Tyr-t-L-β-MePhe-Gly-Trp-N-MeNle-Asp-Phe-NH$_2$ (**3**); (SO$_3$H)Asp-Tyr(SO$_3$H)-t-L-β-MePhe-Gly-Trp-N-MeNle-Asp-Phe-NH$_2$ (**4**); Asp-Tyr-N-MeNle-Gly-Trp-N-MeNle-Asp-p-NO$_2$Phe-NH$_2$ (**5**); and Asp-Tyr-D-Phe-Gly-Trp-N-MeNle-Asp-Phe-NH$_2$ (**6**). Careful statistical analysis of all results was performed, but are not provided for clarity. Kappa receptor binding potencies were weaker than 15,000 nM.

All analogues have high potency at the CCK-B receptor (about 0.1 nM to 3 nM) and, with the exception of **4**, low potency at the CCK-A receptor (1,000 to >10,000 nM). Nearly all compounds have very weak μ binding affinity, compounds **1** to **4** have modest binding to δ receptors (500-1,000 nM IC$_{50}$) and **6** binds well (29 nM IC$_{50}$). In the in vitro δ receptor assay (MVD) **1-6** showed variable potency (17 to 2,600 nM). All MVD activities were reversed by the δ antagonist ICI 174,864, but not by the μ antagonist CTAP. In the hot plate analgesic assay, i.c.v. administration demonstrated quite potent activity except for CCK-8 and analogue **3**. The analgesic activity was unaffected by the CCK-B antagonist L365,260 or the potent μ antagonist CTAP. Thus the analgesic activity of the CCK-8 analogues **1-6** appears to be directly due to its binding to the δ opioid receptor and not to the μ or CCK-B receptor. The exact mechanism of action is under further investigation.

These results led us to examine a possible similarity in the structural or topochemical requirements for δ and CCK-B receptors. Previous studies using NMR, molecular mechanics calculations, topographical constraints in the Phe[4] and Tyr[1] positions [4, 5] and calculations of low energy conformations, suggest a topographical model for the "bioactive conformation" of the δ opioid receptor ligand [D-Pen^2, D-Pen^5]enkephalin (DPDPE) (Figure 1) [6]. Recently, using a similar approach, we have suggested a "bioactive conformation" for CCK-8 at CCK-B receptors (Figure 1) [7]. Examination of these proposed bioactive conformations demonstrates topographical similarity of the surfaces containing the Phe and Tyr moieties (Figure 1). The rather bulky Trp residue in the CCK-8 conformation on the same surface as the other aromatic side chains may explain the somewhat different potencies of the two ligands at δ receptors. These results suggest that at least in part δ opioid receptors and CCK-B receptors have overlapping stereostructural and topographical requirements involving these aromatic side chain residues.

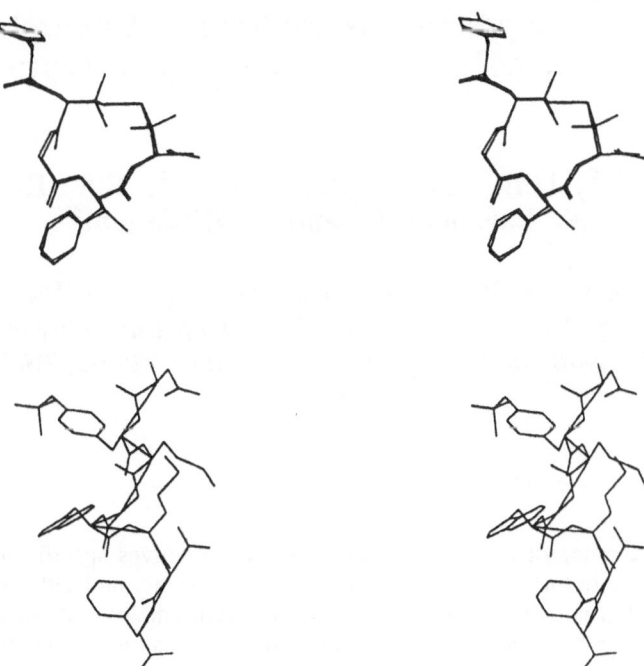

Fig. 1. *Stereoviews of proposed bioactive conformations of DPDPE and [(2S,3S)β-MePhe⁴]DPDPE (top) and CCK (bottom) at δ opioid and CCK-B receptors.*

Acknowledgements

This work was supported by grants from the U.S. Public Health Service DK 36289 and National Institute of Drug Abuse DA 04248.

References

1. See Baber, N.S., Dourish, C.T. and Hall, D.R., Pain, 39 (1989) 307 for a review.
2. Slaninova, J., Knapp, R.J., Wu, J., Fang, S., Kramer, T., Burks, T.F., Hruby, V.J. and Yamamaura, H.I., Eur. J. Pharmacol., 200 (1991) 195.
3. Hruby, V.J., Fang, S., Knapp, R., Kazmierski, W., Lui, G.K. and Yamamura, H.I., Int. J. Pept. Protein Res., 35 (1990) 566.
4. Hruby, V.J., Toth, G., Gehrig, C.A., Kao, L.-F., Knapp, R., Lui, G.K., Yamamura, H.I., Kramer, T.H., Davis, P. and Burks, T.F., J. Med. Chem., 34 (1991) 1823.
5. Toth, G., Russell, K.C., Landis, G., Kramer, T.H., Fang, L., Knapp, R., Davis, P., Burks, T.F., Yamamura, H.I. and Hruby, V.J., J. Med. Chem., 35 (1992) 2384.
6. Nikiforovich, G.V., Hruby, V.J., Prakash, O. and Gehrig, C.A., Biopolymers, 31 (1991) 941.
7. Nikiforovich, G. and Hruby, V.J., Biochem. Biophys. Res. Commun., 194 (1993) 9.

Peptide-carbohydrate interactions: Understanding bacterial adherence to host cell receptors

K.K. Lee[1,5], H.B. Sheth[5], R.T. Irvin[1,5], R.S. Hodges[2,5], W. Paranchych[3] and O. Hindsgaul[4]

Departments of [1]Medical Microbiology & Infectious Diseases, [2]Biochemistry, [3]Microbiology, and [4]Chemistry, University of Alberta, and [5]SPI Synthetic Peptides Inc., Edmonton, Alberta, T6G 2H7, Canada

Introduction

The adherence of bacteria to host cells often involves specific interactions between receptors on cell surfaces and bacterial adhesins, molecules that confer attachment properties [1]. Pathogenic bacteria often employ adhesins that have lectin-like properties and mediate binding to glycoconjugate receptors [2, 3]. *Pseudomonas aeruginosa* employs a pilus adhesin, a filamentous structure that is a polymer of pilin subunits, to mediate attachment to epithelial cell surfaces [4]. We have previously demonstrated that the C-terminal disulfide-looped region of the *Pseudomonas* pilin structural subunit confers the receptor-binding property to the pilus [5, 6]. In the present studies, we show that pili from *P. aeruginosa* strain PAK and synthetic peptides corresponding to the C-terminal region of the pilin PAK sequence possess lectin-like properties.

Results and Discussion

Previous studies have demonstrated that PAK and PAO pili bound to the glycosphingolipid, βGal(1-3)βGalNAc(1-4)βGal(1-4)βGlc-ceramide (asialo-GM$_1$), but not to monosialo-GM$_1$ in direct and competitive ELISAs [7]. It has been suggested that the minimal carbohydrate receptor sequence consists of βGalNAc(1-4)βGal [2]. Using synthetic βGalNAc(1-4)βGal-BSA conjugates as immobilized receptors, PAK pili was observed to bind to these disaccharides in a concentration-dependent manner but no significant binding to a control that consisted of βGlcNAc(1-3)βGal(1-4)βGlc-BSA was observed (Figure 1). These lectin-like properties were specific as the free synthetic disaccharides inhibited PAK pili binding to the immobilized conjugated receptors. A maximal inhibition of 30% was obtained at the highest concentration of the disaccharide hapten esters employed (Table 1). The low level of inhibition may be due to the multivalency of pilus adhesin (~5 binding domains are displayed at the tip of the pilus) and clustering of receptors on the solid phase.

We have previously shown that synthetic peptides corresponding to the receptor-binding domain of PAK and PAO pilins (region 128-144) bound to asialo-GM$_1$ [7]. We have used a synthetic peptide, N$^\alpha$-acetylated PAK(128-144)ox-OH, in competitive binding assays and found that it inhibited PAK pili binding to βGalNAc(1-4)βGal-BSA by 40% at the highest concentration of competing peptides (Table 2). This indicated that residues 128-144 of the *Pseudomonas* pilin were sufficient to mediate binding to the minimal carbohydrate receptor structure.

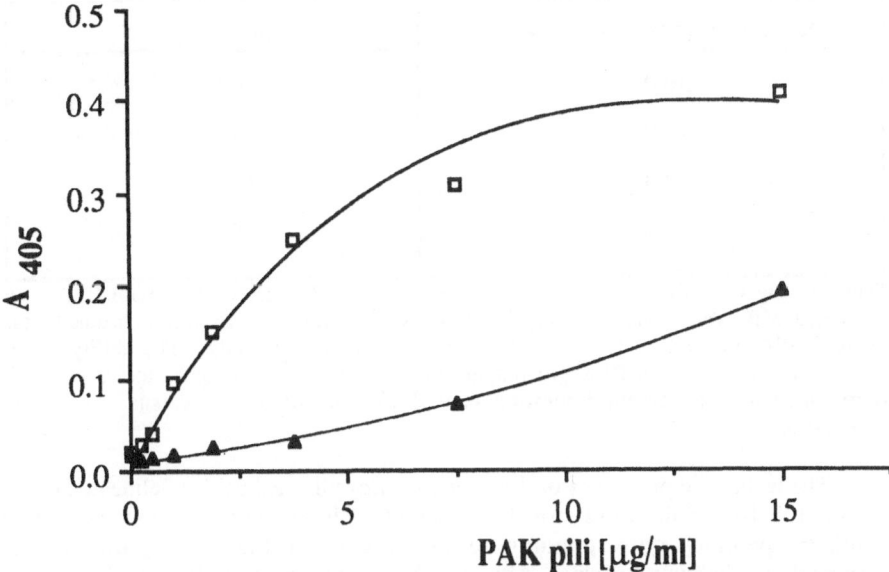

Fig. 1. Binding of P. aeruginosa PAK pili to βGalNAc(1-4)βGal-BSA (□) and to βGlcNAc(1-3)βGal(1-4)βGlc-BSA (▲). The glycoconjugates were coated onto microtiter wells (5 μg/ml) and incubated for an hour at 37°C with purified pili that have been diluted with PBS, pH 7.4, containing 0.05% (wt/vol) BSA. Bound PAK pili were quantitated using ELISA protocols with an anti-PAK pili monoclonal antibody. Absorbance readings (A_{405}) were recorded.

Table 1   Specificity of βGalNAc(1-4)βGal for P. aeruginosa PAK pili demonstrated by competitive binding assays with immobilized βGalNAc(1-4)βGal-BSA[a]

βGalNAc(1-4)βGal [μM]	% Inhibition
18.5	29.4
1.85	28.9
0.185	4.7
0.0185	0.5

[a]The binding of PAK pili (5.0 μg/ml) to immobilized βGalNAc(1-4)βGal-BSA was competed with free βGalNAc(1-4)βGal hapten esters. PAK pili was quantitated using ELISA protocols with an anti-PAK pili monoclonal antibody.

Table 2   *The ability of synthetic peptides corresponding to the receptor binding region of PAK pilin to compete with PAK pili for βGalNAc(1-4) βGal receptors in a competitive binding assay[a]*

N^α-AcPAK (128-144)ox-OH [μM]	% Inhibition
30.0	39.8
10.0	18.2
3.3	10.8
1.1	8.3

[a]The binding of PAK pili (2.5 μg/ml) to immobilized βGalNAc(1-4)βGal-BSA was competed with N^α-acetylated PAK(128-144)ox-OH peptides. PAK pili was quantitated using ELISA protocols with an anti-PAK pili monoclonal antibody. The ability of the peptides to compete with PAK pili binding to the immobilized disaccharide receptors is represented as the percent inhibition caused by the competitor versus absence of competitor.

The lectin-like properties of the *Pseudomonas* pilus adhesin is delineated to the region 128-144 of the pilin structural subunit. These studies demonstrated that synthetic peptides corresponding to this region of the PAK pilin possess carbohydrate-binding properties. Although the true nature of the *P. aeruginosa* pilus receptors on cell surfaces are still under investigation, the application of peptides representing receptor-binding domains may be useful in the purification of these receptors and in understanding the interactions involved in bacterial adherence to glyco-receptors.

Acknowledgements

The financial support of the Canadian Bacterial Diseases Network and Protein Engineering Network Centers of Excellence, the Medical Research Council of Canada and the Canadian Cystic Fibrosis Foundation is gratefully acknowledged.

References

1.   Beachey, E.H., J. Infect. Dis., 143 (1981) 325.
2.   Karlsson, K.-A., Annu. Rev. Biochem., 58 (1989) 309.
3.   Krivan, H.C., Roberts, D.D. and Ginsburg, V., Proc. Natl. Acad. Sci. U.S.A., 85 (1988) 6157.
4.   Prince, A., Microb. Pathog., 13 (1992) 251.
5.   Irvin, R.T., Doig, P., Lee, K.K., Sastry, P.A., Paranchych, W., Todd, T. and Hodges, R.S., Infect. Immun., 57 (1989) 3720.
6.   Lee, K.K., Doig, P., Irvin, R.T., Paranchych, W. and Hodges, R.S., Mol. Microbiol., 3 (1989) 1493.
7.   Lee, K.K., Sheth, H.B., Wong, W.Y., Sherburne, R., Paranchych, W., Hodges, R.S., Lingwood, C.A., Krivan, H. and Irvin, R.T., Mol. Microbiol., (1994), in press.

Ile and Val residues are helix-promoters in transmembrane segments

C.M. Deber and S.-C. Li

Division of Biochemistry Research, Research Institute, Hospital for Sick Children, Toronto, M5G 1X8, and Department of Biochemistry, University of Toronto, Toronto, Ontario, M5S 1A8, Canada

Introduction

The bulky aliphatic amino acids Ile, Leu and Val are the three residues most frequently found in transmembrane (TM) segments of integral membrane proteins [1]. In single span membrane proteins, they account for nearly 60% of the total amino acid composition of the TM domain [2]. Despite their similar hydrophobicity, which makes them suitable membrane-anchors, these residues, nevertheless, differ in their propensity in forming various types of secondary structures in globular proteins. While Leu is one of the best α-helix formers, the branched side chains of Ile and Val tend to destabilize an α-helix, and prefer the β sheet conformation. Therefore, the established structural propensity of Ile and Val in globular proteins as β-sheet promoters appears to conflict with their frequent occurrence within membrane protein TM domains, which are generally assumed to be α-helical [3]. Our working hypothesis is that the secondary structure of an amino acid is a function of its molecular environment (sequence context and environmental medium), and the helical propensity of Ile and Val may be greatly enhanced in the helix-favoring environment of membrane bilayers in order to satisfy the structural requirements for TM domains.

Results and Discussion

We are testing this hypothesis by constructing model peptides having the prototypical sequence NH_2-SKSK-**AXA-AXA-W-AXA**-KSKSKS-OH, where X = L, I, or V. The resulting peptides are designated ALA, AIA, and AVA, respectively, according to the triad repeat of their hydrophobic stretches (bolded above; Table 1). It was expected that these peptides would represent condensed versions of single span membrane proteins, in which Leu, Ile, Val and Ala residues similarly constitute up to 70% of their TM domains [2]. The single Trp residue was incorporated into the peptide sequence to facilitate monitoring peptide interactions with their environments.

Circular dichroism (CD) spectroscopy was employed to study the conformational behavior of these peptides under both aqueous and membranous conditions. As shown in Figure 1 (a), peptide conformation in H_2O is consistent with what may be predicted from primary sequences, e.g., helicity of peptide ALA > peptide AIA > peptide AVA. However, this pattern is dramatically altered when these peptides are dissolved in a vesicle suspension of 3 mM dimyristoylphosphatidylglycerol (DMPG). In DMPG, as displayed in Figure 1 (b), all three peptides adopt significant helical conformation essentially indistinguishable from one another. This result suggests that β-branched Ile and Val residues, as well as Leu, are comparable helical promoters in membrane environments, with their helical propensities governed by their hydrophobicities.

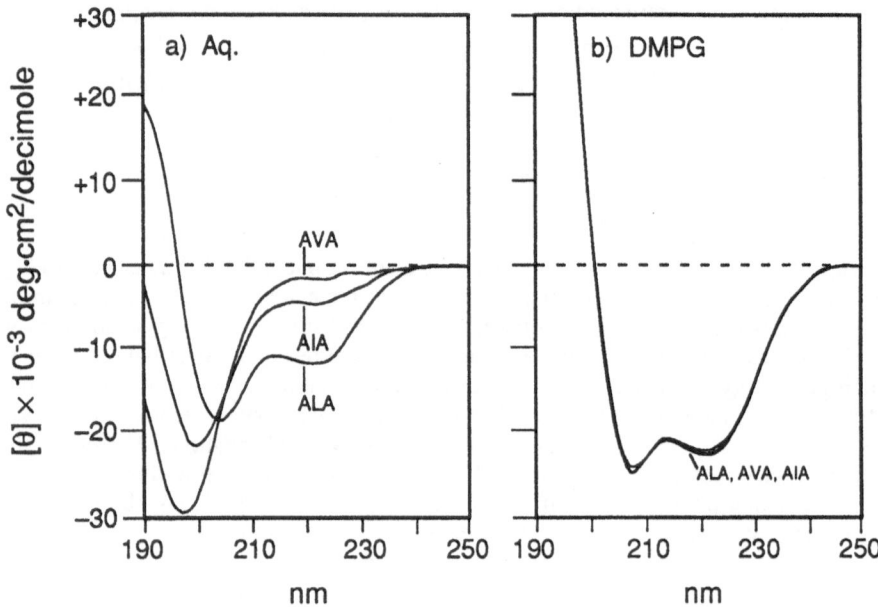

Fig. 1.   CD spectra of peptides ALA, AIA, and AVA in (a) aqueous buffer, pH 7.0, and (b) lipid vesicle suspensions of 3 mM DMPG.   Synthesis and characterization of peptides were as described [3,4]. Small unilamellar vesicles (SUV) were prepared according to [5]. CD spectra were recorded on a Jasco J-720 spectrometer from freshly prepared peptide samples (30 μM).   Spectra shown were averaged over 6 scans for each peptide with background subtracted.

This notion is further supported by Trp fluorescence studies. Although the hydrophobic segments in these peptides are too short to span the membrane bilayer, large "blue shifts" of λ_{max} for Trp fluorescence emission (Table 1), as these peptides are transferred from aqueous buffer to DMPG vesicles, indicate that extensive hydrophobic interactions exist between the peptide mid-segments and the vesicle membrane. Moreover, the extent of such interactions, which is proportional to peptide hydrophobicities, is likely to be the same for all three peptides as can be appreciated from their similar λ_{max} values in DMPG. Fluorescence experiments therefore provide supporting evidence that the CD patterns observed in Figure 1(b) are a genuine reflection of comparably high helical propensities for Ile, Val and Leu residues in the membrane.

Table 1 *Sequences and fluorescent properties of peptides NH$_2$-SKSK-AXA-AXAW-AXA-KSKSKS-OH*

Peptide	Hydrophobic segment[a]	λ_{max} of Trp fluorescence (+/-1 nm)[b]	
		in aq. buffer	in DMPG vesicles
ALA	-ALA-ALA-W-ALA-	354	337
AIA	-AIA- AIA -W-AIA-	354	338
AVA	-AVA-AVA-W-AVA-	355	338

[a] Sequence of the central portion of the peptide bolded in the title.
[b] λ_{max} is the wavelength at which fluorescence emission is maximum. Trp emission spectra were recorded on a Hitachi F-400 fluorescence spectrometer at 37°C. Excitation wavelength was 280 nm. Reported λ_{max} values are an average of at least 3 independent measurements in each case.

Conformational behavior of Ile and Val in membrane-mimetic materials has previously been studied using peptides with triad repeat of AXG (X=L, I, or V) [3-5]. While results from these studies grossly agree with the present observations, Ile and Val residues were found to be slightly inferior to Leu in promoting helices in some lipid vesicle suspensions [5]. Thus, the helical propensity of an amino acid is not only a function of its environment, but a function of its sequence context as well. The structural versatility of Ile and Val residues, which can act both as β-sheet promoters in aqueous media and α-helix promotors in membranes, may be of significance in membrane protein assembly and/or functioning.

Acknowledgements

This work was supported, in part, by grants to C.M.D. from the Natural Sciences and Engineering Research Council of Canada (NSERC) and the Medical Research Council of Canada (MRC). S.-C.L. holds a University of Toronto Open Studentship.

References

1. Deber, C.M., Brandl, C.J., Deber, R.B., Hsu, L.C. and Young X.Y., Arch. Biochem. Biophys., 251 (1986) 68.
2. Landolt-Marticorena, C., Williams, K.A., Deber, C.M. and Reithmeier, R.A.F., J. Mol. Biol., 229 (1993) 602.
3. Li, S.-C. and Deber, C.M., Int. J. Pept. Protein Res., 40 (1992) 243.
4. Li, S.-C. and Deber, C.M., FEBS Lett., 311 (1992) 217.
5. Li, S.-C. and Deber, C.M., In Schneider, C.H. and Eberle, A.N., (Eds.), Peptides 1992, Proceedings of the 22nd European Peptide Symposium, Escom, Leiden, The Netherlands, p.93.

Discrimination between bradykinin receptor subtypes in the rat uterus

C. Liebmann and S. Nawrath

Institute of Biochemistry and Biophysics, Friedrich-Schiller-University, Philosophenweg 12, D-07743 Jena, Germany

Introduction

The isolated rat uterus (RUT) is very sensitive to bradykinin (BK), responds with a rapidly reversible contraction and serves, therefore, as one of several model systems of kinin agonists and antagonists with the B_2 kinin receptor [1]. The myotropic effect of BK in the whole uterus seems to be due to a direct action in the myometrium and an indirect action via the endometrium (sensitive to indomethacin) that depends on the release of prostaglandins [2]. Because uterine contraction is primarily dependent on the presence of extracellular Ca^{2+}, then Ca^{2+}-influx via receptor-operated Ca-channels is presumably the most important source of the increase of intracellular Ca^{2+} concentration which is necessary for contraction [3]. Nevertheless, BK-induced stimulation of phosphoinositide turnover and subsequent release of Ca^{2+} from internal stores may be of importance [4]. A third component by which BK might bring about contraction of RUT is the modulation of sensitivity of the contractile proteins [3]. The binding of [^{3}H]-BK to rat [5] and bovine [6] uterine membranes has been found to be heterogeneous displaying two binding sites. The question is how these binding sites may be involved in the BK-induced uterine contraction. Here we summarize the hitherto known data and propose a hypothetical molecular model of a dual BK action in the rat uterus.

Results and Discussion

Binding studies with [^{3}H]BK in rat myometrial membranes constantly reveal two binding sites: a high-affinity site (K_H) with an affinity of approximately 20 pM and a low-affinity site (K_L) with 0.5-1.0 nM. Both sites also occur when the binding was measured under physiological conditions but the detectable binding site concentrations are drastically reduced in high-ionic strength buffers. Furthermore, both binding sites are found independent of the hormonal state of the RUT. In the presence of Gpp[NH]p the affinity of the K_H-site is shifted to the right by about one order of magnitude but not converted into the K_L-site. Thus, the K_H-site does not represent an interconvertible conformation of K_L but seems to be a separate G protein-coupled binding site. In contrast to the G protein activation via the K_H-site, under our experimental conditions a G protein stimulation via the K_L-site is only detectable after treatment with estrogen. It may be assumed, therefore, that the functional activity of K_L-coupled G proteins depends on the hormonal state. The classical B_2 receptor antagonist, DArg[Hyp^3,Thi^{5,8},D-Phe^7]BK, binds to both sites

[7], whereas another highly potent antagonist, HOE 140, selectively blocks the K_L-site [8]. Thus, both binding sites could be classified as different B_2 receptors.

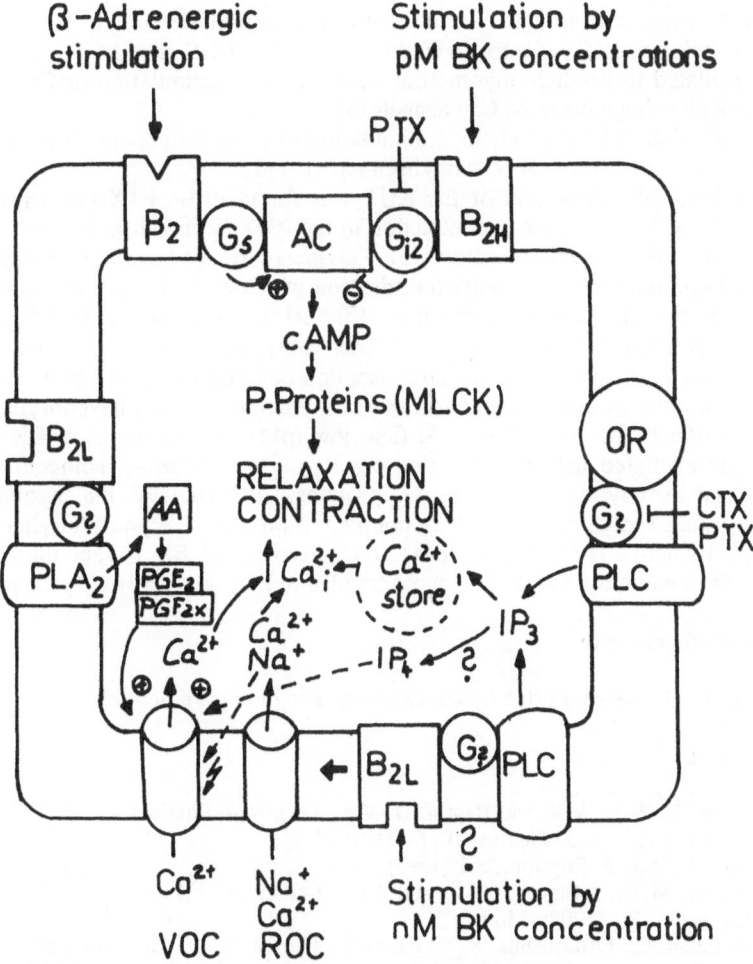

Fig. 1. *Proposed hypothetical mechanisms of bradykinin-induced contraction of the rat uterus via a dual signal transduction pathway. Abbreviations: $\beta_2 = \beta_2$-adrenoceptors; B_{2H} = high-affinity bradykinin B_2 receptor; B_{2L} = low-affinity bradykinin B_2 receptor; OR = oxytocin receptor; G_s = stimulatory G protein; G_{i2} = inhibitory G protein of the G_{i2} type, pertussis toxin (PTX)-sensitive; $G_?$ = unknown type of G protein, PTX-insensitive; PLC = phospholipase C; PLA_2 = phospholipase A_2; AC = adenylate cyclase; VOC = voltage-operated Ca-channel; ROC = receptor-operated ion channel; IP_3 = inositol 1,4,5-trisphosphate; IP_4 = inositol 1,3,4,5-tetrakisphosphate; AA = arachidonic acid; PG = prostaglandin; MLCK = myosin light-chain kinase.*

The putative B_2 receptor subtypes B_{2H} and B_{2L} (with high and low affinity for BK, respectively) are obviously acting via different signal transduction pathways:

- B_{2H} receptors activate PTX-sensitive G proteins of the G_{i2} type [5] and inhibit myometrial adenylate cyclase activity [9].

- B_{2L} receptors stimulate (endometrial) PLA_2 activity leading to the release of arachidonic acid and prostaglandin production. PGE_2 and $PGF_{2\alpha}$ have been postulated to produce myometrial contraction by stimulation of Ca^{2+}-influx through voltage-operated Ca-channels [3].

- In addition, a PTX-insensitive stimulation of phosphoinositide breakdown in cultured rat myometrial cells has been reported [4].

The BK-induced contraction of the RUT was found to be PTX-insensitive. In contrast, the oxytocin-induced contraction in the RUT is known to be mediated by IP_3 formation which is PTX-sensitive [3]. Obviously, BK and oxytocin use different types of G proteins in their signal transduction which mediate uterine contraction. Concerning the BK-induced contraction of the RUT, we assume a dual regulatory pathway involving both putative B_2 receptor subtypes. The RUT has autonomic innervation and is relaxed via β_2-adrenoceptors coupled to G_s/adenylate cyclase. cAMP is known to cause relaxation of rat myometrium via phosphorylation of MLCK and/or by lowering Ca^{2+}. At first, low (pM) concentrations of BK inhibit adrenergic-mediated uterine relaxation via B_{2H} receptors which compete for the myometrial acdenylate cyclase by G_{i2} proteins. In that way, the myometrium becomes more sensitive because less contractile proteins are phosphorylated and, thus, inactivated. Then, higher (nM) concentrations of BK trigger directly the myometrial contraction via B_{2L} receptors coupled to PTX-insensitive G proteins.

Acknowledgements

This work was supported by the Deutsche Forschungsgemeinschaft.

References

1. Barabe, J., Park, W.K. and Regoli, D., Can. J. Physiol. Pharmacol., 53 (1975) 345.
2. Whalley, E.T., Br. J. Pharmacol., 64 (1978) 21.
3. Wray, S., Am. J. Physiol., 264 (1993) C1.
4. Tropea, M.M., Munoz, C.M. and Leeb-Lundberg, F.L.M., Can. J. Physiol. Pharmacol., 70 (1992) 1360.
5. Liebmann, C., Offermanns, S., Spicher, K., Hinsch, K.-D., Schnittler, M., Morgat, J.L., Reissmann, S., Schulz, G. and Rosenthal, W., Biochem. Biophys. Res. Commun., 167 (1990) 910.
6. Leeb-Lundberg, F.L.M. and Mathis, S.A., J. Biol. Chem., 265 (1990) 9621.
7. Liebmann, C., Schnittler, M., Stewart, J.M. and Reissmann, S., Eur. J. Pharmacol., 199 (1991) 363.
8. Liebmann, C., Nawrath, S., Ludwig, B. and Paegelow, I., Eur. J. Pharmacol., 235 (1993) 183.
9. Liebmann, C. and Reissmann, S., Biomed. Biochim. Acta, 49 (1990) 1231.

Structure-activity studies of the [Leu13]motilin-(1-14) pharmacophore

M.J. Macielag[1], M.S. Marvin[1], T.L. Peeters[2],
R. Dharanipragada[1], I. Depoortere[2], J.R. Florance[1],
R.A. Lessor[1] and A. Galdes[1]

[1]Ohmeda PPD, 100 Mountain Avenue, New Providence, NJ 07974,
U.S.A., [2]Gut Hormone Laboratory, University of Leuven,
Gasthuisberg, B-3000 Leuven, Belgium

Introduction

Motilin (Mot) is a linear polypeptide of 22 amino acid residues (H-Phe[1]-Val-Pro-Ile- Phe[5]-Thr-Tyr-Gly-Glu-Leu[10]-Gln-Arg-Met-Gln-Glu[15]-Lys-Glu-Arg-Asn-Lys [20]-Gly-Gln-OH) originally purified from the mucosa of the small intestine of hogs [1]. Although the precise physiological role of motilin remains controversial, it is generally accepted that the hormone is involved in the regulation of fasting gastrointestinal motor activity through stimulation of receptors on gut smooth muscle cells and enteric neurons. Exogenous administration of motilin to healthy humans accelerates gastric emptying, decreases intestinal transit time, and initiates phase III of the migrating motor complex. In addition, motilin infusion to diabetic patients stimulates the emptying of solids and liquids in even the most refractory cases of gastroparesis [2].

A previous study in our laboratories demonstrated that the N-terminal tetradecapeptide of motilin, [Leu13]Mot-(1-14), retained most of the binding affinity and biological activity of the full molecule [3]. Alanine and D-amino acid scans of this bioactive sequence were then used to establish that Phe[1], Val[2], Ile[4] and Tyr[7] are residues involved in receptor contact [4]. In the present study, we have examined several series of analogs of [Leu13]Mot-(1-14) in which structural features of these key amino acid residues have been systematically modified. The goal of this investigation is to define the physicochemical basis for the high affinity interaction between motilin and its receptor in order to design more potent peptide agonists.

Results and Discussion

Peptides were synthesized by SPPS on PEG-PS or PepSyn KA supports using Fmoc continuous flow methods. Following TFA cleavage, the crude peptides were purified by RPHPLC using dual preparative Vydac C_{18} columns in series. Structures were confirmed by quantitative AAA and FABMS. Receptor binding affinity (pIC_{50}) was determined by displacement of [(^{125}I)Nle[13]]-motilin from rabbit antral smooth muscle membranes [5]. Contractility (pEC_{50}) was measured in a tissue bath assay employing rabbit duodenal smooth muscle strips [6].

Table 1 *Potency of [Leu^{13}]motilin-(1-14) analogs in binding and in contractility experiments*

H-Phe^1-Val^2-Pro-Ile^4-Phe-Thr-Tyr^7-Gly-Glu-Leu-Gln-Arg-Leu-Gln-OH

#	Analog	pEC$_{50}$	pIC$_{50}$	#	Analog	pEC$_{50}$	pIC$_{50}$
1	[Leu^{13}]Mot-(1-14)	7.55	8.36		Position 2 modifications:		
	Position 1 modifications:			16	Leu	7.29	8.45
2	p-F-Phe	7.47	8.24	17	Ile	7.74	8.46
3	p-Cl-Phe	7.47	8.47	18	Cha	7.57	8.58
4	p-I-Phe	7.46	8.52	19	Phe	6.98	8.37
5	Trp	7.41	8.65		Position 4 modifications:		
6	1-Nal[a]	8.18	8.42	20	Leu	7.29	8.45
7	2-Nal	7.64	8.53	21	Val	6.53	7.32
8	Cha[a]	7.59	8.51	22	t-BuGly	7.17	7.55
9	p-NO$_2$-Phe	6.37	7.25	23	Cha	7.78	8.56
10	Pal[a]	5.47	7.25	24	Phe	7.48	8.62
11	Leu	6.81	7.82		Position 7 modifications:		
12	Phg[a]	5.68	6.50	25	Phe	7.27	8.20
13	D-Phg	6.35	6.61	26	p-MeO-Phe	7.11	7.87
14	N-AcPhe	6.75	7.59	27	p-Cl-Phe	6.62	6.86
15	des-amino-Phe	6.02	6.96	28	2-Nal	7.16	7.72
				29	Cha	6.31	6.76
				30	Trp	7.27	8.53

[a]3-(1-Naphthyl)alanine, Nal; 3-Cyclohexylalanine, Cha; 3-(3-Pyridyl)alanine, Pal; Phenylglycine, Phg.

The results from the study of Phe^1 modified analogs (Table 1) reveal that the receptor does not discriminate between a number of structurally diverse aromatic and aliphatic side chains. In particular, potency is largely unaffected by substitution of the aromatic ring (**2-4**), an increase in steric bulk (**5-7**), or loss of aromaticity (**8**). However, introduction of more hydrophilic moieties like p-NO$_2$Phe (**9**), Pal (**10**), and Leu (**11**) or shortening the length of the side chain (**12, 13**) is detrimental to in vitro biological activity. Acetylation of the N-terminal amino group (**14**) or substitution of this functionality by hydrogen (**15**) significantly reduces potency.

Modification of Val^2 and Ile^4 is readily tolerated as long as the lipophilic nature of the side chain is maintained. The critical dependence of biological activity on hydrophobic interactions is exemplified by the Val^4 analog (**21**), which differs from the parent peptide by a single methylene unit but is 10-fold less potent in receptor binding and tissue bath assays. Increasing hydrophobicity by introducing the cyclohexylmethyl side chain (**18, 23**), however, affords only modest increases in potency. Incorporation of the sterically encumbered t-BuGly^4 residue (**22**) significantly affects receptor binding affinity, possibly through distortion of the conformation of the peptide backbone.

In contrast to the other pharmacophoric elements, the Tyr^7 side chain exhibits limited tolerance for structural modification. Although replacement of the p-OH group with hydrogen (**25**) only slightly affects potency, substitution of the aromatic ring with other functional groups like 4-methoxy- (**26**) or 4-chloro- (**27**) is incompatible with high levels of bioactivity. The diminished potency of the 2-Nal (**28**) and Cha (**29**) analogs presumably reflects the importance of excluded volume and $\pi-\pi$ interactions, respectively, in the high affinity binding of motilin to its receptor. Only the Trp^7 analog (**30**), the side chain of which contains both an aromatic ring and a hydrogen bond donor, exhibits potency comparable to the parent peptide in both receptor binding and tissue bath assays.

The structure-activity relationships of the Phe^1, Val^2, and Ile^4 side chains of [Leu^{13}]Mot-(1-14) suggest that the motilin receptor contains a substantial hydrophobic pocket which is capable of accomodating a wide variety of aromatic and aliphatic moieties without loss of binding affinity. The results from the study of Tyr^7 replacements indicate that effective ligand-receptor interaction is dependent on a delicate balance between electronic, steric, and hydrogen bonding effects. Although dramatic increases in potency have not as yet been realized, it is clear that the design of high affinity peptide agonists to the motilin receptor must incorporate the following structural features: 1) a basic N-terminal amino group; 2) hydrophobic residues at positions 1, 2, and 4; and 3) π-electron density and possible hydrogen bond donation from the side chain of residue 7.

References

1. Brown, J.C., Cook, M.A. and Dryburgh, J.R., Gastroenterology, 62 (1972) 401.
2. Peeters, T.L., Muls, E., Janssens, J., Urbain, J.L., Bex, M., Van Cutsem, E., Depoortere, I., De Roo, M., Vantrappen, G. and Bouillon, R., Gastroenterology, 102 (1992) 97.
3. Macielag, M.J., Peeters, T.L., Konteatis, Z.D., Florance, J.R., Depoortere, I., Lessor, R.A., Bare, L., Cheng, Y.-S. and Galdes, A., Peptides, 13 (1992) 565.
4. Peeters, T.L., Macielag, M.J., Depoortere, I., Konteatis, Z.D., Florance, J.R., Lessor, R.A. and Galdes, A., Peptides, 13 (1992) 1103.
5. Bormans, V., Peeters, T.L. and Vantrappen, G., Regul. Pept., 15 (1986) 143.
6. Depoortere, I., Peeters, T.L., Matthijs, G., Cachet, T., Hoogmartens, J. and Vantrappen, G., J. Gastroint. Mot., 1 (1989) 150.

Chimeric amino acids in cyclic GnRH antagonists

A. Olma[1,3], G.V. Nikiforovich[1], B. Nock[2] and G.R. Marshall[1]

Center for Molecular Design and Departments of [1]Molecular Biology and Pharmacology and [2]Psychiatry, Washington University School of Medicine, St. Louis, MO 63110, U.S.A., [3]Institute for Organic Chemistry, Politechnika, Lodz, Poland

Introduction

The search for an appropriate antagonist of gonadotropin releasing hormone (GnRH, LHRH), Glp-His-Trp-Ser-Tyr-Gly-Leu-Arg-Pro-Gly-NH$_2$, as a contraceptive has led to the synthesis and testing of thousands of analogs. As yet, no peptidomimetics with appropriate properties have been developed. As an exercise in peptidomimetic design [1], GnRH antagonists provide an excellent test case. In order to define the receptor-bound conformation of potent GnRH antagonists such as Antide [2], N-Ac-D-Nal-D-Cpa-D-Pal-Ser-Lys(Nic)-D-Lys(Nic)-Leu-ILys-Pro-D-Ala-NH$_2$, 13 cyclic analogs were prepared with disulfide bridges between residues 4 and 8, 4 and 9, 5 and 8, 5 and 9, and 4 and 10. The use of sidechain cyclization to help constrain the backbone conformation is a common approach, but conformational analysis has suggested [3] that little effect on the backbone is achieved in most cases due to the size and flexibility of the sidechains used. The presence in Antide of a proline residue in position nine led to the incorporation of chimeric proline analogs [4] with a functionalized sidechain, *cis*- (MPc) and *trans*-4-mercaptoproline (MPt) to provide a cyclization functionality while preserving the conformational constraints associated with proline. Both Hcy and Cys were substituted in positions 4, 5, 8, 9 and 10 in order to vary the ring size. Two novel compounds were designed based on models of the receptor-bound conformation of GnRH agonist and antagonist developed by Nikiforovich and Marshall [5]. GnRH antagonists possess low-energy structures with the spatial arrangement of the residues in the N-terminal tripeptide similar to that in the conformer, which was found to be common for GnRH agonists. These studies suggested new approaches to the design of GnRH antagonists, namely by mimicking this specific arrangement of the N-terminal tripeptide, or by cyclization to residue 4 from residue 8 in which a D-amino acid has been substituted.

Results and Discussion

Synthesis of peptides - The linear precursors were synthesized using SPPS using sidechain protection of pMeBzl for Mpt and Mpc and Acm for Cys and Hcy. Couplings used either BOP or TBTU with the addition of HOBt. Oxidations were carried out with either I$_2$ in 25% HOAc or DMSO. The analogs were purified on preparative HPLC and characterized by AAA, FABMS and analytical HPLC.

Additional analogs with amide bonds between residues 4 (Asp) and 10 (Dpr) were also prepared as controls. The cyclic pentapeptide, *cyclo*(Gly-D-Nal-D-Cpa-D-Pal-Gly), was prepared by SPPS, cleaved by HF and cyclized in solution.

Table 1 *Selected cyclic analogs of GnRH and their biological activities (AOA assay, 0/8 = no animals ovulated out of 8 tested)*

Amino acid substitutions and AOA results
4 to 8
[Ac-D-Nal^1,D-Cpa^2,D-Pal^3,Cys^4,Lys(Nic)^5,D-Lys(Nic)^6,Cys^8,D-Ala^{10}] inactive at 50 mg
[Ac-D-Nal^1,D-Cpa^2,D-Pal^3,Cys^4,Tyr^5,D-Arg^6,Leu^7,D-Cys^8,D-Ala^{10}] 0/8 @ 25 mg
[Ac-D-Nal^1,D-Cpa^2,D-Pal^3,Hcy^4,Tyr^5,D-Arg^6,Leu^7,D-Cys^8,D-Ala^{10}] 4/8 @ 50 mg
[Ac-D-Nal^1,D-Cpa^2,D-Pal^3,Cys^4,Tyr^5,D-Arg^6,Leu^7,D-Hcy^8,D-Ala^{10}] 2/8 @ 25 mg
[Ac-D-Nal^1,D-Cpa^2,D-Pal^3,Hcy^4,Tyr^5,D-Arg^6,Leu^7,D-Hcy^8,D-Ala^{10}] 4/8 @ 50 mg
[Ac-D-Nal^1,D-Cpa^2,D-Pal^3,Hcy^4,Lys(Nic)^5,D-Lys(Nic)^6,Hcy^8,D-Ala^{10}] inactive at 50 mg
4 to 9
[Ac-D-Nal^1,D-Cpa^2,D-Pal^3,Cys^4,Lys(Nic)^5,D-Lys(Nic)^6,ILys^8,Mpt^9,D-Ala^{10}] inactive at 50 mg
[Ac-D-Nal^1,D-Cpa^2,D-Pal^3,Cys^4,Lys(Nic)^5,D-Lys(Nic)^6,ILys^8,Mpc^9,D-Ala^{10}] 4/8 at 50 mg
[Ac-D-Nal^1,D-Cpa^2,D-Pal^3,Hcy^4,Lys(Nic)^5,D-Lys(Nic)^6,ILys^8,Mpt^9,D-Ala^{10}] inactive at 50 mg
[Ac-D-Nal^1,D-Cpa^2,D-Pal^3,Hcy^4,Lys(Nic)^5,D-Lys(Nic)^6,ILys^8,Mpc^9,D-Ala^{10}] inactive at 50 mg
5 to 9
[Ac-D-Nal^1,D-Cpa^2,D-Pal^3,Hcy^5,D-Lys(Nic)^6,ILys^8,Mpc^9,D-Ala^{10}] 5/8 at 50 mg
[Ac-D-Nal^1,D-Cpa^2,D-Pal^3,Ser^4,Cys^5,Lys(Nic)^6,Leu^6,ILys^8,Hcy^9,D-Ala^{10}] inactive at 50 mg
cyclo(Gly-D-Nal-D-Cpa-D-Pal-Gly-)   4/8 @ 500 mg

Both competitive binding and antiovulatory activity (AOA) were measured. While the trends were similar, considerable differences were apparent in the data. Compounds retaining the highest binding affinity contained cycles between residues 5 and 8, 4 and 10, and 5 and 9 in accord with previous studies on cyclic GnRH analogs, but were relatively inactive in the AOA assay. A novel analog with a cyclic between residues 4 and 8, c[Ac-D-Nal^1, D-Cpa^2, D-Pal^3, Cys^4, Tyr^5, D-Arg^6, Leu^7, D-Cys^8, D-Ala^{10}]-GnRH, designed based on the models of agonists and antagonists [5] showed complete inhibition of ovulation at 50 mg. The cyclic pentapeptide showed 50% inhibition of ovulation at a dose of 500 mg also consistent with the conformational hypothesis.

Using the proline of residue 9 as a point for cyclization to either residue 4 or 5 by incorporation of the chimeric amino acids, Mpt and Mpc, did not enhance the activity. c[Ac-D-Nal[1], D-Cpa[2], D-Pal[3], Cys[4], Lys(Nic)[5], D-Lys(Nic)[6], ILys[8], Mpc[9], DAla[10]]-GnRH and c[Ac-D-Nal[1], D-Cpa[2], D-Pal[3], Hcy[5], D-Lys(Nic)[6], ILys[8], Mpc[9], D-Ala[10]]-GnRH showed 50% inhibition in the AOA assay at 50 mg. Conformational analyses on the impact of these modifications offer further insight into recognition by GnRH receptors.

Acknowledgements

The authors thank the NIH (Contract NO1-HD-1-3104) for partial support and acknowledge use of the W. U. Mass Spectroscopy Resource (NIH grant RR00945). Thanks to Dr. Marvin Karten and the Contraceptive Development Branch of NICHD for the AOA assay results.

References

1.   Marshall, G.R., Tetrahedron, 49 (1993) 3547.
2.   Ljungqvist, A., Feng, D.-M., Hook, W., Shen, Z.-X., Bowers, C. and Folkers, K., Proc. Natl. Acad. Sci. U.S.A., 85 (1988) 8236.
3.   Kataoka, T., Beusen, D.D., Clark, J.D., Yodo, M. and Marshall, G.R., Biopolymers, 32 (1992) 1519.
4.   Plucinska, K., Kataoka, T., Cody, W.L., He, J.X., Humblet, C., Lu, G.H., Lunney, E., Major, T.C., Panek, R.L., Schelkun, P., Skeean, R. and Marshall, G.R., J. Med. Chem., in press.
5.   Nikiforovich, G.V. and Marshall, G.R, In Schneider, C.H. and Eberle, A.N. (Eds.), Peptides 1992, (Proceedings of the 22nd European Peptide Symposium), Escom, Leiden, The Netherlands, 1993, p.541.

Chimeric amino acids in cyclic bradykinin analogs: Evidence for receptor-bound turn conformation

K. Kaczmarek[1,3] K.-M. Li[1], R. Skeean[2], D. Dooley[2], C. Humblet[2], E. Lunney[2] and G.R. Marshall[1]

[1]Departments of Molecular Biology and Pharmacology, Washington University School of Medicine, St. Louis, MO 63110, U.S.A., [2]Parke-Davis Pharmaceutical Research Division, Warner-Lambert Co., Ann Arbor, MI 48105, U.S.A., and [3]Institute for Organic Chemistry, Politechnika, Lodz, Poland

Introduction

The receptor-bound conformation of peptide hormones can be determined by analysis of the effects of chemical modification with conformationally constrained amino acids and dipeptide analogs [1]. Cyclization through sidechains offers an additional method of introducing constraints on possible backbone conformations and relative positioning of sidechains during receptor interaction. Little work has been published on the receptor-bound conformation of bradykinin (BK), Arg-Pro-Pro-Gly-Phe-Ser-Pro-Phe-Arg. We had explored [2] substitution of a methyl group for the α-proton at positions 4, 5 and 8. [D-Ala[4]]-BK was essentially inactive while [MeF[5]]-BK retains slight activity (1-4%). [MeF[8]]-BK retained 30% activity and showed reduced pulmonary inactivation. The enhanced activity of [dehydroPhe[5]]-BK observed by Fisher et al. [3], led to the idea that a β-turn at position 5 may be recognized at the receptor as the proclivity of dehydroamino acids to induce β-turns has been well documented [4-6]. Analysis of structure-activity studies of BK, Arg-Pro-Pro-Gly-Phe-Ser-Pro-Phe-Arg, shows sensitivity to substitution of a proton by a methyl group of the entire tripeptide segment, Gly-Phe-Ser, at either α-carbon or amide nitrogen, suggesting a tight turn in apposition to the receptor.

Conformational analysis [7] of cyclic tripeptides shows that Ac-cyclic-S,S-[Cys-AA-Cys]-NHCH$_3$ limits the backbone somewhat, but that larger rings resulting from homocysteine substitution for Cys results in possible backbone conformations resembling those of the linear, non-cyclic peptide. A chimeric amino acid in which conformational properties are combined with sidechain requirements for recognition, or introduction of constraints, was appropriate. The presence in bradykinin of three proline residues flanking the proposed turn site led to cyclic BK analogs in which chimeric proline analogs [8] with a functionalized sidechain, cis-(MPc) and trans-4-mercaptoproline (MPt), cis-(M^{3}Pc) and trans-3-mercaptoproline (M^{3}Pt), and cis-(APc) and trans-4-amino-proline (APt) were used.

Results and Discussion

Synthesis of cyclic bradykinin analogs

The solid-phase syntheses were performed with Boc-amino acids with the sidechain protecting groups: benzyl for Ser, tosyl for Arg, 4-methoxy- or 4-methyl-benzyl for Cys and mercaptoprolines. Boc-amino acids were coupled by means of BOP, TBTU or diisopropylcarbodiimide in DMF in the presence of HOBt and DIEA. Completion of the reaction was checked by the ninhydrin test, or in the case of the proline or mercaptoproline amino group, a 2% solution of bromophenol blue in N,N-dimethylacetamide was successfully used. Subsequently, the peptides were cleaved from the resin by the action of HF in the presence of anisole (9:1) for 1 h at 0°C, dissolved in water, and lyophilized. The APt and Apc analogs were prepared by SPPS coupling Boc-4-amino(Fmoc)-Pro at position three and Boc-Asp(OFm) at position six. After the peptide was assembled, the base-labile sidechain protecting groups were removed, and the peptide cyclized, cleaved and purified. Crude peptides were oxidized in 95% acetic acid at r.t. by dropwise addition of 0.01 M iodine solution in glacial acetic acid with stirring until a yellow color persisted for 5 min. Then the reaction solution was lyophilized and the crude cyclic products purified by RP-HPLC. In the case of [D,L-M^3pt^2,Mpc^7]-BK, oxidation gave three products (ratio approximately 4:1:1), which were separated by HPLC. FAB-MS spectra showed the first peak (HPLC) to be cyclic monomer, the next two to be cyclic dimers. After chymotrypsin hydrolysis of the Phe^5-Ser^6 amide bond, the antiparallel nature of the first cyclic dimer was confirmed as it showed only one peak on HPLC (t_R=10.89 min, 5-50%B, t=25 min, v=1 mL/min, identical with the product from enzymatic hydrolysis of cyclic monomer), presumably the disulfide-bridged peptide Arg-M^3pt-Pro-Gly-Phe to Ser-Mpc-Phe. Chymotrypsin digestion of the second isolated dimer gave two peptides (t_R=11.20 and 11.90 min), presumably the disulfide-linked dimers of Arg-M^3pt-Pro-Gly-Phe and Ser-Mpc-Phe, confirming the parallel nature of this dimer.

The cyclic bradykinin analogs were assayed with a neuronal bradykinin receptor binding assay. Over 30 BK analogs with disulfide cycles from 2-6, 3-6, 2-7 and 3-7 were prepared and significant recognition in almost all cases was retained with high binding affinity (13%) shown by c[MPt^3,Cys^6]-BK among others, implying that the postulated turn conformation of BK at the receptor is probable.

688

One analog, c[Mpt³,Mpc⁷]-BK, in which two chimeric proline residues were constrained by a disulfide bond, retained significant activity (2.4%) as well. In another case, both the parallel (0.1%) and antiparallel dimer (0.04%) retained more binding activity than the cyclic monomer, c[M³pt²,Mpc⁷]-BK (0.02%). This has also been described for AII dimers [8].

In conclusion, introduction of constrained amino acids into bradykinin led to a hypothesis that a turn centered on residues 4 and 5 was the biologically relevant conformation. Introduction of chimeric proline analogs which allowed cyclization without loss of the conformational constraints of the proline ring retained significant activity at the B_2 receptor consistent with this hypothesis. The analogs will be useful in conformational analysis of the three-dimensional recognition requirements for bradykinin.

Acknowledgements

The authors thank the National Institutes of Health (NIH GM24483) for partial support of this research. They acknowledge use of the facilities at the Washington University Mass Spectroscopy Resource (NIH grant RR00945).

References

1. Marshall, G.R., Tetrahedron, 49 (1993) 3547.
2. Turk, J., Needleman, P. and Marshall, G.R., J. Med. Chem., 18 (1975) 1139.
3. Fisher, G.H., Berryer, P., Ryan, J.W., Chauhan, V. and Stammer, C.H., Arch. Biochem. Biophys., 211 (1981) 269.
4. Bach, A.C., II and Gierasch, L.M., Biopolymers, 25 (1986) S175.
5. Imazu, S., Shimohigashi, Y., Kodama, H., Sakaguchi, K., Waki, M., Kato, T. and Izumiya, N., Int. J. Pept. Protein Res., 32 (1988) 298.
6. Chauhan, V.S., Sharma, A.K., Uma, K., Paul, P.K.C. and Balaram, P., Int. J. Pept. Protein Res., 39 (1987) 126.
7. Kataoka, T., Beusen, D.D., Clark, J.D., Yodo, M. and Marshall, G.R., Biopolymers, 32 (1992) 1519.
8. Plucinska, K., Kataoka, T., Cody, W.L., He, J.X., Humblet, C., Lu, G.H., Lunney, E., Major, T.C., Panek, R.L., Schelkun, P., Skeean, R. and Marshall, G.R., J. Med. Chem., in press.

The importance of the aromatic ring quadrupole for receptor interactions of the position 4 side chains of angiotensin-II analogues

G.J. Moore[1], R.C. Ganter[1], J.M. Matsoukas[2], J. Hondrelis[2], G. Agelis[2], K. Barlos[2], S. Wilkinson[3], J. Sandall[3] and P. Fowler[3]

[1]Department of Medical Biochemistry, University of Calgary, Calgary, Alberta, T2N 4N1, Canada, [2]Department of Chemistry, University of Patras, Patras, Greece, and [3]Department of Chemistry, University of Exeter, Devon EX4 4QD, England

Introduction

The role of the octapeptide angiotensin II (ANG II, Asp-Arg-Val-Tyr-Ile-His-Pro-Phe) in blood pressure regulation and in hypertension derives largely from its contractile action at vascular smooth muscle receptors [1, 2]. Structure-activity studies on ANG II have demonstrated that methylation [3] or deletion [4, 5] of the Tyr hydroxyl group of ANG II produces antagonists, implicating the Tyr OH group in receptor activation. Chemical reactivity studies [6], conformational studies by NMR [7, 8] and tyrosinate fluorescence studies [9] on ANG II analogues have suggested that intramolecular interaction of the Tyr OH with the His imidazole and C-terminal carboxylate of ANG II, resulting in a charge relay system (CRS) analogous to that found at the active site of serine proteases [10], is an important element of ANG II action at its receptors. The CRS triad of interacting groups creates a tyrosinate anion pharmacophore (tyranophore) which has a role in activating/triggering the receptor consequent to binding of the ligand [9]. In the present study, we investigated substituents for the Tyr[4] residue of [Sar[1]]ANG II in order to delineate aspects of this pharmacophore which provide for its receptor triggering properties. In particular, the aromatic amino acids Nal, Pal, DL-Phg(4-F), Phe(4-F), Phe(F_5) and His were investigated.

Results and Discussion

[Sar[1] Nal[4]]ANG II, [Sar[1] Pal[4]]ANG II, [Sar[1] DL-Phg(4-F)[4]]ANG II, [Sar[1] Phe(4-F)[4]]ANG II, [Sar[1] Phe(F_5)[4]]ANG II and [Sar[1] His[4]]ANG II had agonist activities of 4.5%, 7%, < 0.1%, 0.2%, 1% and 0.6%, respectively. All peptides investigated were devoid of measurable antagonist activity except [Sar[1] Phe(4-F)[4]]ANG II (pA_2 = 7.7). These findings illustrate that for ANG II analogues containing an aromatic amino acid other than Tyr at position 4, ligand binding and agonist activity are not dependent on the electronegativity or dipole moment of the aromatic ring, or on the ability of the 4 ring substituent to accept a proton, although there appears to be a dependence on the ring quadrupole (Table 1). *Ab initio* calculations (10) based on aromatic ring multipoles suggest that the binding of ANG II analogues is associated with an interaction of the out of plane axis of the position 4 aromatic ring quadrupole with a receptor-based group. In addition, intramolecular interactions providing for the conformation of the ligand as it approaches its receptor appear to have a role in determining agonist *versus* antagonist activity.

Table 1　*Electrostatic properties and biological activities of aromatic acids in position 4 of [Sar¹]ANG II analogues*

AROMATIC RING	$O^{\ominus}$	OCH_3	F	(phenyl)	F_4 (tetrafluoro)
ELECTRONEGATIVITY[1]	+0.07	-0.27	+0.06	0	+0.86
DIPOLE MOMENT[2]	+5.6	-1.4	+1.6	0	+1.6
QUADRUPOLE MOMENT[3]	+6.8	-6.9	-4.6	-6.8	+5.7
AGONIST ACTIVITY (%)[4]	100	<0.1	0.2	10	<0.1
ANTAGONIST ACTIVITY (pA₂)[5]	–	7.7	7.7	7.7	<5

[1]Hammett factor (σ). [2]Debyes. [3]Atomic units; values given are for the axis perpendicular to the ring and were calculated as described previously [10]. For phenoiate the overriding electrostatic property is the charge (-1), which reduces all other considerations (multipoles, etc.) to insignificance. [4]Relative to ANG II. [5]pA₂ is the negative logarithm of the concentration of antagonist required to reduce the response to an ED50 dose of ANG II to that of ED50/2 dose.

References

1.　Kanagy, N.L., Pawloski, C.M. and Fink, G.D., Am. J. Physiol., 259 (1990) R102.
2.　Chatziantoniou, C., Daniels, F.H. and Arendshorst, W.J., Am. J. Physiol., 259 (1990) F372.
3.　Scanlon, M.N., Matsoukas, J.M., Franklin, K.J. and Moore, G.J., Life Sci., 34 (1984) 317.
4.　Vine, W.H., Brueckner, D.A., Needleman, P. and Marshall, G.R., Biochemistry, 12 (1973) 1630.
5.　Matsoukas, J.M., Goghari, M.H., Scanlon, M.N., Franklin, K.J. and Moore, G.J., J. Med. Chem., 28 (1985) 780.
6.　Moore, G.J., Int. J. Pept. Protein Res., 26 (1985) 469.
7.　Matsoukas, J.M., Bigam, G., Zhou, N. and Moore, G.J., Peptides, 11 (1990) 359.
8.　Matsoukas, J.M., Yamdagni, R. and Moore, G.J., Peptides, 11 (1990) 367.
9.　Turner, R.J., Matsoukas, J.M. and Moore, G.J., Biochim. Biophys. Acta, 1065 (1991) 21.
10.　Fowler, P.W. and Moore, G.J., Biochem. Biophys. Res. Commun., 153 (1988) 1296.

Neuropeptide receptors: Molecular characterization of receptors for somatostatin and vasotocin

D. Richter[1], W. Meyerhof[1], I. Wulfsen[1], S. Mahlmann[1], J. Heierhorst[1], C. Schönrock[1] and K. Lederis[2]

[1]*Institut für Zellbiochemie und klinische Neurobiologie, Universität Hamburg, UKE, Martinistrasse 52, 2046 Hamburg, Germany and* [2]*Department of Pharmacology and Therapeutics, University of Calgary Faculty of Medicine, 3330 Hospital Drive N.W., Calgary, Alberta, T2N 4N1, Canada*

Introduction

Recent years have seen significant progress in the elucidation of the sequences of a number of neuropeptide receptors by the application of recombinant DNA technology. In general, their structures exhibit seven putative membrane-spanning domains resembling those for other receptors of the guanine nucleotide binding protein (G-protein) coupled family [1]. Here we discuss the properties of two neuropeptide receptors, namely those for mammalian somatostatin (SST) and fish vasotocin (VT). SST is a multifunctional peptide which, for instance, blocks the release of pituitary and pancreatic peptide hormones. VT is the structural counterpart of the mammalian antidiuretic hormone vasopressin.

Results and Discussion

Somatostatin receptors (SSTR)

Receptor binding studies have revealed the existence of at least two subtypes of SSTR; one mediates its activity via GTP-binding proteins by inhibiting adenylate cyclase and by modulating Ca^{2+} currents, the other additionally potentiates a delayed rectifier potassium current. To date, molecular cloning techniques have revealed the presence of five different SSTR subtypes (SSTR1-5; Table 1). The mRNAs encoding four of these have been localized in the brain and also in peripheral organs of mammals; mRNA for the fifth subtype, which has a high preference for binding SST-28, an N-terminally extended form of the tetradecapeptide SST-14, is found predominantly in the hypophysis.

The SSTR subtypes contain several potential N-linked glycosylation sites in the N-terminal extracellular portion of the molecules, a number of conserved cysteine residues in the first and second extracellular loops which presumably form disulfide bridges, and several consensus sites for phosphorylation [5]. As indicated in Figure 1 there is a rather high sequence identity between the SSTRs and the recently cloned delta opiate receptor [6, 7] which suggests that they evolved from a common ancestor.

Table 1 *Properties of cloned rat somatostatin receptors*

	Mol.W. (kDa)	a.a. residues	affinity (EC$_{50}$, nM)[a]				signal transduction
			SST14	SST28	SMS	MK 648	
rSSTR1	43.4	391	1.1[b]	2.0[b]	600[b]	–	?
rSSTR2	41.2	369	0.26	0.15	0.53	n.d.	ion channels[c]
rSSTR3	47.0	428	0.89	0.24	35.0[b]	n.d.	cAMP ↓[b]
rSSTR4	42.1	384	4.5	7.5	–	–	cAMP ↓
rSSTR5	48.0	383	2.6	0.087	0.19	7.3	cAMP ↓

[a]The data are taken from Refs. 2-5; [b]refers to our unpublished data, [c]to ref. 3. Relative affinities for the individual subtypes are depicted but these cannot be compared for different receptors because of the different experimental conditions that were used.

Vasotocin receptors (VTR)

Clones encoding several VTRs have been isolated from white sucker (*Catostomus commersoni*) brain cDNA and genomic libraries [8]. Expression in frog oocytes by injecting cRNA for one of the cDNAs shows that the encoded VTR is coupled to the IP$_3$/calcium second messenger pathway and hence functionally related to the V$_1$ vasopressin receptor from mammals. Sequence analysis reveals that the VTR is also structurally related to the mammalian V$_1$-receptor (Figure 1). Conservation is particularly obvious in the second and third extracellular domains, both of which are thought to contain possible hormone recognition sites.

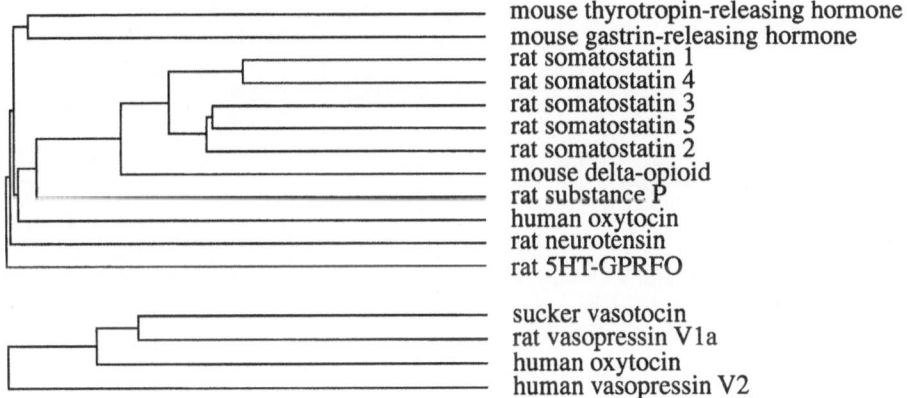

mouse thyrotropin-releasing hormone
mouse gastrin-releasing hormone
rat somatostatin 1
rat somatostatin 4
rat somatostatin 3
rat somatostatin 5
rat somatostatin 2
mouse delta-opioid
rat substance P
human oxytocin
rat neurotensin
rat 5HT-GPRFO

sucker vasotocin
rat vasopressin V1a
human oxytocin
human vasopressin V2

Fig. 1. A dendrogram derived from the sequence of five somatostatin receptor subtypes and those of other neuropeptide and neurotransmitter receptors.

There is considerable sequence identity between sucker and salmon VTRs and the corresponding receptor from *Xenopus laevis*. Studies using a mutant VTR construct modified at the N-terminus (18 amino acid (a.a.) residues are replaced by a randomized sequence of 12 a.a. residues) revealed significant changes in ligand specificities and affinities. This suggests that the N-terminus of the VTR plays an essential role in the process of ligand-receptor interaction.

References

1. Richter, D., Meyerhof, W., Buck, F. and Morley, S.D., In Seifert, G. (Ed.), Cell Receptors, Current Topics in Pathology, Molecular Biology of Receptors for Neuropeptide Hormones, Vol. 83, Springer Verlag, Berlin, Germany, 1991, p.117.
2. Meyerhof, W., Paust, H.-J., Schönrock, C. and Richter, D., DNA and Cell Biol., 10 (1991) 689.
3. Bell, G.I. and Reisine, T., Trends Neurosci., 16 (1992) 34.
4. O'Carroll, A.-M., Lolait, S.J., König, M. and Mahan, L.C., Mol. Pharmacol., 42 (1992) 939.
5. Meyerhof, W., Wulfsen, I., Schönrock, C., Fehr, S. and Richter, D., Proc. Natl. Acad. Sci. U.S.A., 89 (1992) 10267.
6. Kieffer, B.L., Befort, K., Gaveriaux-Ruff, C. and Hirth, C.G., Proc. Natl. Acad. Sci. U.S.A., 89 (1992) 12048.
7. Evans, C.J., Koith, D.E., Morrison, H., Magendzo, K. and Edwards, R.H., Science, 258 (1992) 1952.
8. Mahlmann, S., Meyerhof, W., Hausmann, H., Heierhorst, J., Schönrock, C., Zwierst, H., Lederis, K. and Richter, D., Proc. Natl. Acad. Sci. U.S.A., 91 (1994) 1342.

Thrombin receptor antagonists derived from "tethered ligand" agonist peptides

R.M. Scarborough, W. Teng, J.W. Rose, V. Alves, A. Arfsten and M.A. Naughton

COR Therapeutics, Inc., South San Francisco, CA 94080, U.S.A.

Introduction

Thrombin is an important enzyme regulator of hemostasis and plays critical roles in thrombotic and inflammatory responses to vascular injury. In addition to the procoagulant actions of thrombin, this enzyme mediates a variety of cellular responses in platelets and other vascular cells through activation of the thrombin receptor, a novel member of the G-protein coupled receptor family [1]. A unique mechanism wherein the thrombin receptor is proteolytically cleaved at a specific site within the N-terminal domain of the receptor affording a "tethered ligand" which leads to activation of the receptor has been demonstrated [2, 3]. Peptides derived from the "tethered ligand" domain can also behave as full agonists of the receptor [2]. Structure-activity relationships for the minimal length pentapeptide agonist sequence, Ser-Phe-Leu-Leu-Arg-NH_2, have revealed the importance of the N-terminal amino group and the Phe^2 side chain for receptor activation [4]. Using this information, a series of novel thrombin receptor antagonists have been prepared from agonist sequences in which the amino-terminal ammonium group has been removed.

Results and Discussion

The initial attempts to convert "tethered ligand" agonist peptides into thrombin receptor antagonists were suggested by the observations that the minimal length sequence for agonists required a free α-amino group to display activity [4]. Synthesis of the peptide, desamino-Ser-Phe-Leu-Leu-Arg-NH_2, by solid-phase peptide synthesis resulted in a compound which displayed weak agonist activity in a platelet aggregation assay (EC_{50}=920 μM) without any apparent antagonist activity. However, a more conveniently prepared series of desamino compounds in which the amino terminal Ser is replaced by the 3-mercaptopropionyl-group (Mpr) did afford the first class of receptor antagonists which were evaluated in a convenient 96-well format platelet aggregation assay [5] using either a-thrombin or the agonist peptide, Ser-Phe-Leu-Leu-Arg-Asn-Pro-NH_2 for receptor activation (Table 1). Extension of the peptide sequence carboxyl-terminal to the Arg^5 residue and substitution of more optimized residues discovered in studies of "tethered ligand" agonist sequences [6] within the remainder of the peptide sequence enhanced the inhibitory activity for this class of antagonists. The most potent peptides in this series displayed IC_{50}'s in the 15-25 μM range. Higher concentration of some antagonists (>500 μM) stimulated platelet microaggregate formation which may be a result of partial agonist activity for this series of analogs.

Table 1 *"Tethered ligand" antagonists of the thrombin receptor*

No.	Analog	Inhibition of Platelet Aggregation[a]	
		Thrombin IC_{50} (μM)	Agonist Peptide IC_{50} (μM)
1	[desamino-Ser^1]-Phe-Leu-Leu-Arg-NH$_2$	NA	NA
2	Mpr-Phe-Leu-Leu-Arg-NH$_2$	300	ND
3	Mpr-Phe-Cha-Cha-Arg-Asn-Pro-Asn-Asp-Lys-Tyr	25-50	15-25
4	Mpr-Phe-Cha-Cha-Arg-Lys-Pro-Asn-Asp-Lys-NH$_2$	25	25
5	nBu-N-Phe-Cha-Cha-Arg-Lys-Pro-Asn-Asp-Lys-NH$_2$	15-20	15-20
6	Me-N-(n-pentyl)-Phe-Cha-Cha-Arg-Lys-NH$_2$	30	30
7	(n-pentyl)$_2$-N-Phe-Cha-Cha-Arg-Lys-NH$_2$	30	30
8	(cyclohexylmethyl)$_2$-N-Phe-Cha-Cha-Arg-Lys-Pro-Asn-Asp-Lys-NH$_2$	5	7.5

[a]Agonist concentrations: 300-600 pM a-thrombin; 3-5 μM SFLLRNP-NH$_2$; NA = not active as an antagonist; ND = not determined.

The ability of compounds to antagonize both thrombin and agonist peptide induced aggregation in these assays supports the concept that these analogs are binding at a site within the receptor which is identical or nearly identical to the site where the "tethered ligand" domain and agonist peptides bind. In addition, the compounds do not inhibit the catalytic activity of thrombin nor do they block other platelet agonists such as collagen or calcium ionophore-induced platelet aggregation responses. Interestingly, we found that analog **3** could also inhibit g-thrombin-induced platelet aggregation (IC_{50}=20 μM), suggesting that aggregation responses to both a- and g-thrombin presumably occur through identical platelet receptors.

An additional series of antagonists in which the N-acyl-group attached to the Phe^2-residue was altered to afford mono- or di-N-alkyl-Phe^2-containing analogs were also examined. Compounds in this series were prepared by reductive alkylation of the appropriate aldehyde with the corresponding α-amino group of the Phe residue in peptides attached to solid-phase resins using NaCNBH$_3$ in acidified DMF [7, 8]. The mono- and di-N-alkyl-Phe^2-analogs required at least one alkyl group larger than methyl to display antagonist activity. Modest antagonist activity was observed with either C$_4$ or C$_5$ chains but was significantly enhanced when a cyclohexylmethyl group was one of the N alkyl groups of the N-terminal Phe residue (Analog **8**). Again, partial agonist activity could be seen at higher concentrations for some of these analogs.

Previous observations with "tethered ligand" agonist sequences from the thrombin receptor have implicated the critical importance of the free ammonium group for receptor activation [4]. We therefore proposed that one potential way of obtaining thrombin receptor antagonists was to modify the amino terminal ammonium group while retaining other important side chain interactions of the "tethered ligand" agonist sequences. This approach has yielded two classes of novel receptor antagonists which appear to antagonize both thrombin and agonist peptide-induced responses in platelets. This data is consistent with the hypothesis that these ligands bind to a site within the body of the receptor that is the same as the domain to which the cleaved "tethered ligand" binds, resulting in receptor activation. Although these initial antagonists display only modest antagonist activity in inhibiting platelet activation, additional improvements to the activity of these analogs and removal of the partial agonist activity of analogs should afford agents with significant utility in studying the physiological and pharmacological conse-quences of thrombin receptor antagonism.

References

1. Coughlin, S.R., Vu, T.-K.H., Hung, D.T. and Wheaton, V.I., J. Clin. Invest., 89 (1992) 351.
2. Vu, T.-K.H., Hung, D.T., Wheaton, V.I. and Coughlin, S.R., Cell, 64 (1991) 1057.
3. Vu, T.-K.H., Wheaton, V.I., Hung, D.T., Charo, I. and Coughlin, S.R., Nature, 353 (1991) 674.
4. Scarborough, R.M., Naughton, M.A., Teng, W., Hung, D.T., Rose, J.W., Vu, T.-K.H., Wheaton, V.I., Turck, C.W. and Coughlin, S.R., J. Biol. Chem., 267 (1992) 13146.
5. Fratantoni, J.C. and Poindexter, B.J., Am. J. Clin. Pathol., 94 (1990) 613.
6. Coughlin, S.R. and Scarborough, R.M., European Patent Application WO 92/14750 (1992).
7. Coy, D.H., Hocart, S.J. and Sasaki, Y., Tetrahedron, 44 (1988) 835.
8. Hocart, S.J., Nekola, M.V. and Coy, D.H., J. Med. Chem., 31 (1988) 1820.

Selective photoaffinity labeling of type 2 receptors of angiotensin II

G. Servant, R. Bosse, G. Guillemette and E. Escher

Department of Pharmacology, Faculty of Medicine, University of Sherbrooke, Sherbrooke, Québec, J1H 5N4, Canada

Introduction

The recent development of non-peptide angiotensin II (Ang) receptor antagonists has led to the identification of at least two Ang receptor subtypes. One subtype exhibiting a high affinity for DuP 753 (a non-peptide Ang antagonist [IC_{50} of ~20 nM]) has been termed AT_1. The other subtype exhibiting a high affinity for PD 123319 (also a non-peptide ligand [IC_{50} of ~30 nM]) has been termed AT_2 [1]. The molecular properties of the AT_1 receptor have been extensively investigated and it is now well established that this receptor and its subtypes belong to the super family of G protein coupled receptors with 7 transmembrane domains [2, 3]. The molecular and physiological properties of the AT_2 receptor are much less known, and, since a clear physiological role is still not defined, cloning and isolation efforts have not succeeded so far. Earlier AT_2 photoaffinity labeling attempts with ^{125}I-[Sar^1, Phe(N$_3$)^8]Ang (^{125}I-Ang-N$_3$), the photoaffinity label which permitted successful AT_1 labeling, were however unsuccessful and a new photolabeled peptide was prepared: ^{125}I-[Sar^1, Bpa^8]Ang (Bpa=p-benzoylphenylalanine). This compound was incorporated with a high yield of 70% into the AT_2 receptor [4]. We therefore conducted labeling experiments with AT_2 bearing tissues from several sources: Human myometrium, R3T3 cells (murine fibroblast), and PC12 cells (rat pheochromocytoma) and compared them to AT_1-labeling experiments.

Results and Discussion

Photoaffinity labeling experiments on AT_1 tissue (bovine adrenal cortex membranes, ^{125}I-Ang-N$_3$) produced after solubilization and SDS-PAGE a specifically labeled band at 58 kDa (Figure 1, lane 2) which was prevented by co-incubation of either Ang or L-158,809 (a highly selective non-peptide AT_1 antagonist) but not by PD 123319, all at 1 µM. AT_2-bearing membranes were photolabeled in the presence of ^{125}I-Ang-Bpa and specifically labeled bands were obtained after SDS-PAGE at 68 kDa (human myometrium, Figure 1, lane 4), at 95 kDa (R3T3, Figure 1, lane 6) and at 130 kDa (PC12, Figure 1, lane 8). All AT_2 labeling was suppressed in the presence of Ang (1 µM) or of PD 123319 (1 µM) but not by L-158,809 (1 µM). Solubilized-photolabeled membranes were treated with Glycopeptidase-F (PNGase-F) and also subjected to SDS-PAGE. All experiments, regardless of tissue origin, resulted in similar protein sizes around 32 kDa (Figure 1. AT_1: adrenal, lane 1; AT_2: myometrium, lane 3; R3T3, lane 5; and PC12, lane 7). Time-course deglycosylation assured absence of protease activity and residual glycosylation· Incubation of solubilized labeled mymometrium membranes with PNGase-F (3 U/ml) were interrupted at 5, 15, 30 and 60 min,16 h and 24 h. The initial, diffuse band at 68 kDa disappeared gradually within 1 h, after 16 h, a predominant and narrow band at 36 kDa was visible, and after 24 h another band at 32 kDa became visible. A high strength digestion with PNGase-F (60 U/ml) showed at 1 h this 32 kDa band as predominant with traces at 36 kDa remaining. Further exposure up to 60 h revealed a single 32 kDa band with no higher or lower molecular weight protein present.

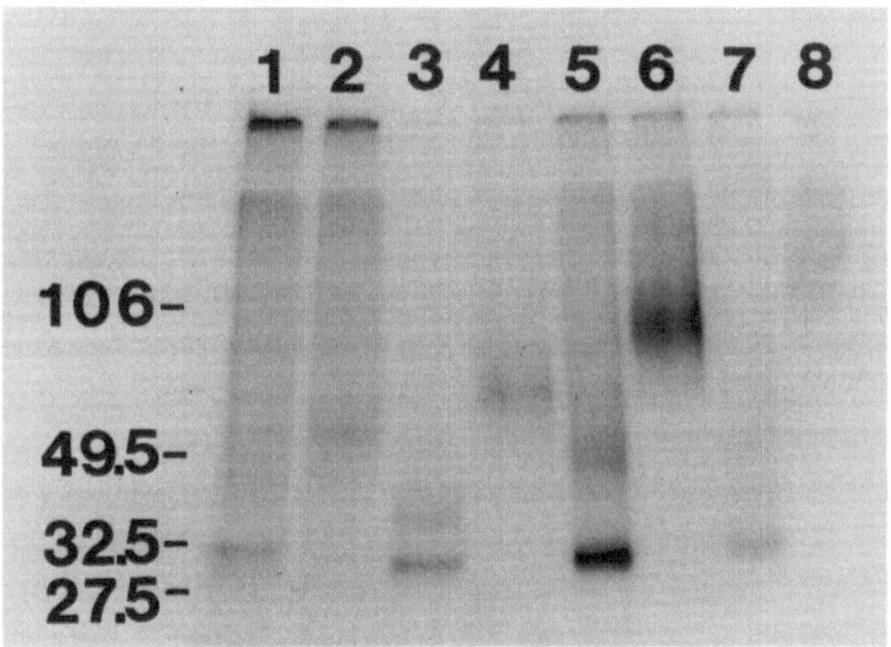

Fig. 1. SDS-PAGE analysis of denatured-photolabeled Ang receptors. AT_1 receptors from bovine adrenals were photolabeled with ^{125}I-Ang-N_3, solubilized and incubated in the presence (lane 1) or absence (lane 2) of PNGase-F (3 U/ml). AT_2 receptors from human myometrium, R3T3 cells and PC12 cells were photolabeled with ^{125}I-Ang-Bpa, solubilized and incubated in the presence (lane 3: human myometrium; lane 5: R3T3 cells; lane 7: PC12 cells) or absence (lane 4: human myometrium; lane 6: R3T3 cells; lane 8: PC12 cells) of PNGase-F (3 U/ml).

We report here the use of a new photoactivatable probe (^{125}I-Ang-Bpa) for the photoaffinity labeling of Ang receptors. This probe is very efficient for the covalent labeling of AT_2 receptors. SDS-PAGE analysis revealed heterogeneity in the size of AT_2 receptors varying from 68 to 130 kDa for the AT_2 receptor of human myometrium and PC12 cells, respectively. An intermediate value was obtained for the AT_2 receptor of R3T3 cells with an M_r close to 100 kDa. Dudley et al. [5], using a chemical cross-linking technique, found a similar M_r of 100 kDa for the AT_2 receptor of R3T3 cells. Pucell et al. [6], using the same approach, found a M_r of 79 kDa for the AT_2 receptor of rat ovarian granulosa cells. These results strongly suggest that there are not only species differences but also tissue differences in the molecular structure of AT_2 receptors. Interestingly, deglycosylation of AT_2 receptors of human myometrium, R3T3 cells and PC12 cells resulted in the production of at least two discernable populations of proteins, one of them having a common M_r of 32 kDa. The higher molecular weight proteins are the result of partial deglycosylation since longer periods of incubation with higher amounts of endoglycosidase cause their disappearance, leaving only the protein migrating at an M_r of 32 kDa. The deglycosylation of the AT_1 receptor of bovine adrenal cortex reduced its M_r from 58 kDa to the same 32 kDa value. Carson et al. [7], using the same endoglycosidase, evaluated a similar M_r of 34 kDa for the deglycosylated form of the AT_1 receptor of bovine adrenal cortex. The theoretical molecular weight of the AT_1 receptor deduced from its sequence has been evaluated at 41 kDa [2, 3]. The gap existing between this value and the one obtained by Carson et al. [7] and our group could be explained by a partial specific proteolysis in the C-terminal part of the receptor occurring during the membrane preparation or the different incubations performed for the labeling and the deglycosylation. Since specific proteolysis of the receptor to 32 kDa may occur [8], it would be premature to conclude anything about the similarity existing between the M_r of deglycosylated AT_1 and AT_2 receptors. The cloning and sequencing of the AT_2 receptor will most certainly help to understand this phenomenon. In conclusion, the results reported here demonstrate that variations in carbohydrate content contribute to the physical heterogeneity of AT_2 receptors in individual target tissues.

Acknowledgements

This work was supported by grants from the Medical Research Council of Canada.

References

1. Bumpus, F.M., Catt, K.J., Chiu, A.T., de Gasparo, M., Goofriend, T., Husain, A., Peach, M.J., Taylor, D.G. and Timmermans, P.B.M.W.M., Hypertension, 17 (1991) 720.
2. Murphy, T.J., Alexander, R.W., Griendling, K.K., Runge, M.S. and Bernstein, K.E., Nature, 351 (1991) 233.
3. Sasaki, K., Yamano, Y., Bardhan, S., Iwai, N., Murray, J.J., Hasegawa, M., Matsuda, Y. and Inagami, T., Nature, 351 (1991) 230.
4. Servant, G., Boulay, G., Bossé, R., Escher, E. and Guillemette, G., Mol. Pharmacol., 43 (1993) 677.
5. Dudley, D.T., Hubbell, S.E. and Summerfelt, R.M., Mol. Pharmacol., 40 (1991) 360.
6. Pucell, A.G., Hodges, J.C., Sen, I., Bumpus, F.M. and Husain, A., Endocrinology, 128 (1991) 1947.
7. Carson, M.C., Leach Harper, C.M., Baukal, A.J., Aguilera, G. and Catt, K.J., Mol. Endocrinol., 1 (1987) 147.
8. Boyd, N.D., White, C.F., Cerpa, R., Kaiser, E.T. and Leeman, S.E., Biochemistry, 30 (1991) 336.

Cationized deltorphin analogues: Potent and selective ligands for delta opiate receptors

S.D. Sharma[1], M. Shenderovich[1], R. Horvath[2],
P. Davis[2], S. Weber[2], F. Porreca[2], T. Davis[2],
H.I. Yamamura[2] and V.J. Hruby[1]

*Departments of [1]Chemistry and [2]Pharmacology,
University of Arizona, Tucson, AZ 85721, U.S.A.*

Introduction

Introduction of additional positively charged amino acids in peptides acting on membrane bound receptors has been demonstrated to cause significant enhancement in their biological effects [1, and Sharma et al., this volume]. This may be the result of a more pronounced cationic character of these derivatives causing an increased affinity for cell membranes [2]. Incorporation of these additional positive charges in a peptide requires the identification of a site(s) in the molecule at which their incorporation does not compromise its ability towards receptor recognition, selectivity, and signal transduction. In the present study cationic analogues of an opioid peptide selective for delta opiate receptors, Tyr-$\underline{D}$-Ala-Phe-Glu-Val-Val-Gly-NH_2 (deltorphin-II), were synthesized and studied for their biological characteristics so as to identify such site(s) in the molecule. It is well established that the deltorphins, like all other opioid peptides, require a cationic center at their N-terminus as one of the absolute requirements for them to be recognized by their receptors. It has also been shown earlier that in enkephalins this cationic center can be removed further from the N-terminus by one amino acid residue [3]. Based on these considerations various cationized deltorphin-II analogues with a generic structure: $(Xxx)_n$-Tyr-$\underline{D}$-Ala-Phe-Glu-Val-Val-Gly-NH_2 (Xxx is a dibasic amino acid, and n is 1-6) were synthesized and evaluated for their biological activities. This study has further helped us to design a series of cyclic deltorphin analogues in which the tyrosine residue is also a part of the cycle.

Results and Discussion

The biological activity results on these analogues in rat brain receptor binding assays and peripheral functional assays (Table 1) demonstrated that all the analogues retained their selectivity for δ receptors. This clearly suggests that the structural changes at the N-terminus of the molecule have no bearing on the selectivity of the peptides. They, however, do affect the potency. The analogues having one or more additional amino acid(s) at the N-terminus (e.g. **1-6**, **9**, and **11**) were still quite potent in the MVD assay and were also good binders. A comparison of the activity profiles of analogues **4**, **11**, and **12** strongly suggests that it is the α-NH_2 functionality (not the ϵ-NH_2) that is the all-important cationic center required for interaction in this series of peptides with the opioid receptors. This is also suggested by the low activities of **7**, **8**, and **10** in comparison with their corresponding linear analogues **1**, **3**, and **9**. Further, it was very interesting to note that the biological activities of **1-10** under assay conditions in which proteolytic enzyme inhibitors were excluded tend to approach that exhibited by deltorphin-II under similar conditions (data not shown). This suggests that all the basic amino acid residue(s) inducted at the N-terminus in these analogues tend to cleave off by the enzymes at the receptor site to liberate a molecule of deltorphin-II. These results suggest the possibility of creating prodrugs specifically for the delta opiate receptors in approaches to target them through the blood-brain-barrier.

Table 1 *Biological activities (IC_{50}, nM) of deltorphin analogues*

	Analogue	Radioreceptor binding		Peripheral assay	
		[³H]pCl-DPDPE	[³H]CTOP	MVD	GPI
	Y-a-F-E-V-V-G-NH₂	0.73	1700	0.35	15000
1.	R-Y-a-F-E-V-V-G-NH₂	14.9	>80μM[1]	8.54	6885
2.	R-R-Y-a-F-E-V-V-G-NH₂	8.89	11113	6.73	2405
3.	K-R-Y-a-F-E-V-V-G-NH₂	20.7	>80μM[2]	8.27	15659
4.	K-Y-a-F-E-V-V-G-NH₂	25.2	37459	7.9	9958
5.	K-K-Y-a-F-E-V-V-G-NH₂	38.8	>80μM[3]	7.01	23146
6.	R-K-Y-a-F-E-V-V-G-NH₂	29.4	60394	2.33	12728
7.	⌐R-Y-a-F-E-V-V-G-NH₂	1704	39257	1712	>60μM[4]
8.	⌐K-R-Y-a-F-E-V-V-G-NH₂	7659	73545	2366	>60μM[5]
9.	R-K-R-K-R-K-Y-a-F-E-V-V-G-NH₂	31.8	13360	8.77	19660
10.	⌐R-K-R-K-R-K-Y-a-F-E-V-V-G-NH₂	94.3	7409	28.6	>25μM[6]
11.	(Nle⁰)-Y-a-F-E-V-V-G-NH₂	10.4	12171	0.46	5420
12.	(NH₂-(CH₂)₅-CO)-Y-a-F-E-V-V-G-NH₂	642	>80μM[6]	138.2	>60μM[4]
13.	K-Y-a-F-E-V-V-G-NH₂	524	>10μM[2]	502.7	>50μM[3]
14.	K-Y-a-F-E-V-V-G-NH₂	3900	>80μM[7]	1064.6	>50μM[7]

Observed inhibition at this concentration: [1]44%, [2]48%, [3]40%, [4]0%, [5]22%, [6]19%, and [7]20%.

Based on these studies and to explore the stereochemical features of the cationic center with respect to the message sequence **13** and **14** were investigated. Geometrical shape comparison of a low energy conformer of **13** generated by ECEPP

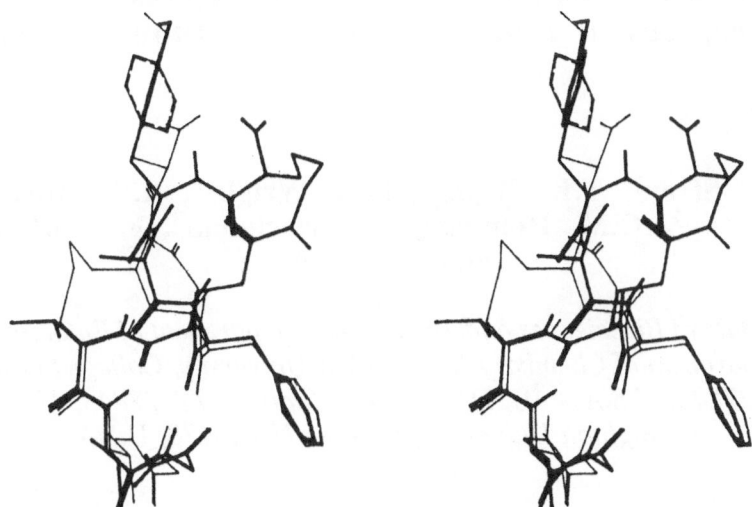

Fig. 1. Geometrical shape comparison of low energy conformers of **13** (bold line) and [D̲-Cȳs², Cȳs⁵]deltorphin-I (thin line).

force field [4] with a structure suggested earlier for Tyr-D̲-Cȳs-Phe-Asp-Cȳs-Val-Gly-NH$_2$, a potent deltorphin-I analogue [5], exhibited excellent overlap of these two analogues (Figure 1) both in respect to the backbone and the side chains of Tyr and Phe residues (rms = 0.6 Å at all common C^α and C^β atoms). However, **13** was 600 fold and 2185 fold less potent than [D̲-Cȳs², Cȳs⁵]deltorphin-I in δ-receptor binding MVD assay respectively, thereby suggesting that it presents a different topography at the receptor. A switch in chirality of the lactam bridge (e.g. **14**), however, still suggests that **13** represents a favorable chirality. It, therefore, appears that a definitive lactam ring size might be crucial for stabilizing the bioactive conformation as has been reported earlier in the case of cyclic lactam analogues of melanotropin [6]. Investigations in this direction are continuing. This research was supported by USPHS and NIDA.

References

1. Sharma, S.D., Nikiforovich, G.V., Jiang, J., Castrucci, A.M.L., Hadley M.E. and Hruby, V.J., Peptides 1992, In Schneider, C.H. and Eberle, A. (Eds.), Escom, Leiden, The Netherlands, 1993, p.95.
2. Sargent, D.F. and Schwyzer, R., Proc. Natl. Acad. Sci. U.S.A., 83 (1986) 5774.
3. Morley, J.S., Annu. Rev. Pharmacol. Toxicol., 20 (1980) 81.
4. Momany, F.A., McGuire, R.F., Burgess, A.W. and Scheraga, H.A., J. Phys. Chem., 79 (1975) 2361.
5. Misicka, A., Nikiforovich, G., Lipkowski, A.W., Horvath, R., Davis, P., Kramer, T.H., Yamamura, H.I. and Hruby, V.J., Bioorg. Med. Chem. Lett., 2 (1992) 547.
6. Al-Obeidi, F., Castrucci, A.M.L., Hadley, M.E. and Hruby, V.J., J. Med. Chem., 32 (1989) 2555.

The interaction of Culekinin peptides with receptors in isolated Malpighian tubules from mosquitoes

A.A. Strey[1], T.K. Hayes[1], M.S. Wright[1], R.W. Meola[1],
J. Kelly[2], G.M. Holman[3], R.J. Nachman[3], F. Clottens[3]
and D. Petzel[4]

[1]*Institute of Biosciences and Technology, Department of Entomology,*
[2]*Department of Chemistry, Texas A&M University, College Station,*
TX 77843, U.S.A., [3]*USDA, College Station, TX 77843, U.S.A.,*
[4]*Creighton University, Omaha, NE 68178, U.S.A.*

Introduction

Culekinins, members of the leucokinin (LK) family of insect neuropeptides, were isolated from the mosquito *Culex salinarius* (CDPI-NPFHSWGamide; CDPII-NNANVFYPWGamide; CDPIII-WKYVSKQKFFSWGamide). The leucokinins (LK) are a family of multifunctional neuropeptides found in at least four insect species which share a conserved C-terminal pentamer (FX(S/P)WGamide) [1, 3]. Culekinins and other leucokinin family members stimulate a Cl⁻ dependent depolarization of the transepithelial voltage and fluid secretion from Malpighian tubules (*Aedes aegypti*) [2]. The effect of several leucokinin analogs on depolarization of the transepithelial voltage of *Aedes aegypti* Malpighian tubules was examined to begin to gain insight into structure-activity relationships. Identification of core sequences and important side chains were necessary for the design of photoaffinity probes to begin to study the mosquito receptor system.

Results and Discussion

A core LK fragment (FYSWGamide) was examined closely with a series of analogs that contained a single Ala replacement per analog (Figure 1). Replacement of any of the three aromatic side chains in the LK fragment resulted in a large potency loss for the respective analog. The order of importance for these side chains for depolarizing Malpighian tubules was Phe^1>Trp^4>Tyr^2. Examination of other naturally occurring core pentapeptides from the LK family suggested that the achetakinin pentamer, FXPWGamide (isolated from *Acheta domestica*) was preferred over the standard LK pentamer, FXSWGamide. Additionally, phenylalanine was the preferred aromatic amino acid in the "X" position of the LK pentamer.

A photoaffinity analog (PAL) containing a photoreactive amino acid, 4-benzoylphenylalanine (Bpa) and a D-Tyr to the N-terminus of the receptor interactive C-terminal pentamer [(dA)(dY)(Bpa)KFFSWGamide] has been prepared. The PAL successfully stimulated the depolarization of mosquito Malpighian tubules even after

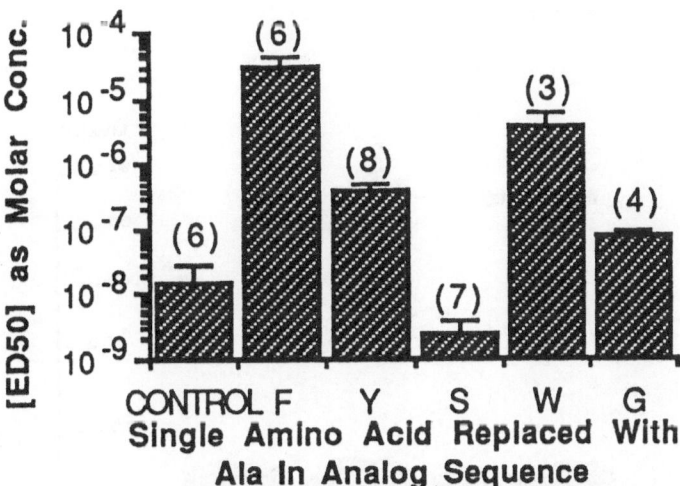

Fig. 1. ED₅₀s of single Ala replacement analogs for a common leucokinin core that stimulated depolarization of mosquito Malpighian tubule TEV. TEV-Transepithelial voltage.

iodination. Photolysis (350 nm) of the iodinated PAL with *Aedes* Malpighian tubule membranes resulted in the specific labeling of a single 60 kDa membrane protein (Figure 2). An excess (10 nmol) of unlabeled analog (FFSWGamide) protected the 60 kDa protein labeling suggesting specific binding. The photoaffinity label can be used to search for receptor-like binding proteins in other tissues. For example, the photolysis reaction was also carried out with cricket Malpighian tubule membranes. Two cricket membrane proteins were specifically labeled (Figure 3).

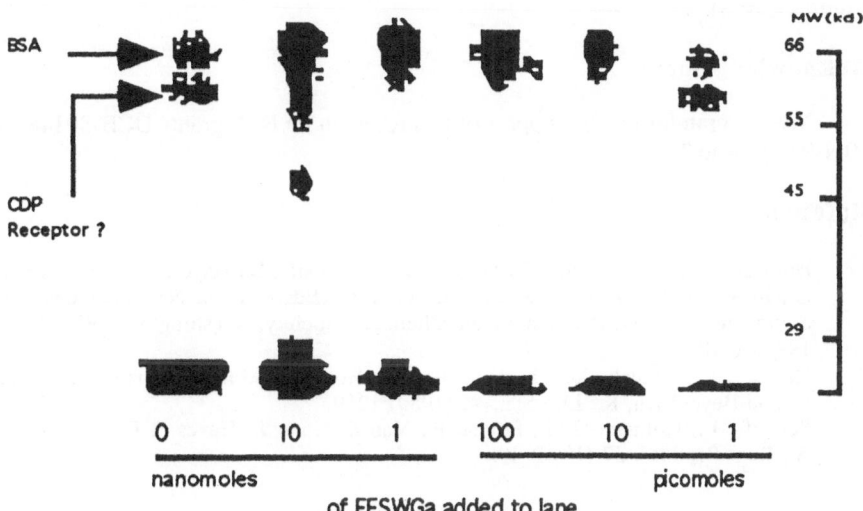

Fig. 2. SDS-PAGE of mosquito Malpighian tubule membrane preparations.

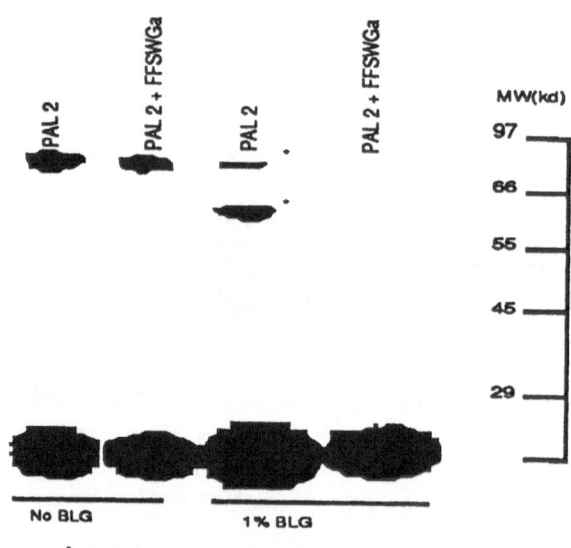

* Achetakinin Receptors ?

Fig. 3.   SDS-PAGE of cricket Malpighian tubule membrane preparations.   PAL-photoaffinity label.   BLG-βLactoglobulin.

An important point for successful labeling was the addition of relatively large amounts of exogenous protein (i.e., 1% w/v) to the medium during membrane preparation and incubation. β-Lactoglobulin (BLG) proved to be more useful than BSA for this purpose. For example, the low molecular weight of BLG meant that this protein did not interfere with SDS-PAGE analysis as did the BSA (compare Figures 2 & 3).

Acknowledgements

We are grateful for the support of this research by NSF grants DCB-89148 and IBN-9209884 to TKH.

References

1.   Holman, G.M., Nachman, R.J., Wright, M.S., Schoofs, L., Hayes, T.K. and DeLoof, A., In Menn, J.J., Kelly, T.J. and Masler, E.P. (Eds.), Insect Neuropeptides, ACS Symposium Series 453, American Chemical Society, Washington, DC, U.S.A., 1991, p.40.
2.   Hayes, T.K., Pannebecker, T., Hinckley, D., Holman, G.M., Nachman, R.J., Petzel, D. and Beyenbach, K., Life Sci., 44 (1989) 1259.
3.   Schoofs, L., Holman, G.M., Proost, P., Van Damme, J., Hayes, T.K. and DeLoof, A., Reg. Peptides, 37 (1992) 49.

Peptides

Chemistry, Structure and Biology

Peptides

Chemistry, Structure and Biology

Proceedings of the Thirteenth American Peptide Symposium
June 20-25, 1993, Edmonton, Alberta, Canada

Edited by

Robert S. Hodges
Department of Biochemistry
University of Alberta
Edmonton, Alberta, T6G 2H7, Canada

and

John A. Smith
Department of Pathology
University of Alabama at Birmingham
Birmingham, Alabama 35233-7331, U.S.A.

ESCOM ▪ Leiden ▪ 1994

CIP-Data Koninklijke Bibliotheek, Den Haag

Peptides

Peptides: Chemistry, Structure and Biology: Proceedings of the Thirteenth American Peptide Symposium, June 20-25, 1993, Edmonton, Alberta, Canada/ed. by Robert S. Hodges and John A. Smith. - Leiden : ESCOM. - Ill.
With index, ref.
Subject headings: Peptides/Proteins.

ISBN 90-72199-19-7 (hardbound)

Published by:

ESCOM Science Publishers B.V.
P.O. Box 214
2300 AE Leiden
The Netherlands

Preface

The Thirteenth American Peptide Symposium was held in the Edmonton Convention Centre, Edmonton, Alberta on June 20-25, 1993. This Symposium was held under the auspices of the American Peptide Society and the University of Alberta. Compared to previous symposia, this meeting was the largest in terms of attendance (1447), and truly an international Symposium with participants from 32 countries and more than 30% of participants from outside North America. It was a pleasure to see the large number of first-time participants at the Symposium. The scientific program consisted of plenary lectures, poster presentations, workshops and exhibits. Of the 762 presentations (58 plenary lectures involving 14 sessions; 3 workshops involving 13 additional speakers; and 691 poster displays) the Program Committee selected 372 articles for publication in the Proceedings Volume. These manuscripts were selected on the basis of originality and scientific significance.

As with past symposia, the Thirteenth Symposium continued to show the exponential growth of the use of peptides in the diverse fields of medical science.

We began the Symposium with an exciting lecture on "Recollections of the Past 40 Years in Peptide Chemistry" by Dr. Bruce Merrifield, Nobel laureate, which focused on past scientists and their contributions that have paved the way to what is now a mushrooming field of peptide chemistry in all aspects of biology. Sessions I and II and 189 posters discussed Synthetic and Analytical Methods, the cornerstone of our field. It is interesting to observe the continuing improvements to solution and solid-phase peptide chemistry such as new supports, coupling agents and optimization procedures, fragment condensation methods and enzymatic synthesis. The development of new instrumentation (*e.g.*, capillary electrophoresis, on-line HPLC with electrospray mass spectrometry, micro-peptide sequencing and ultra-fast HPLC) is aiding, and will continue to aid greatly, the peptide chemist in purification and characterization of peptides.

Session III dealt with the exciting developments in designing peptide mimetics, cyclic and constrained peptides and secondary structure mimetics. Session IV discussed the synthesis of glycopeptides and lipopeptides, the role glycosylation plays on the biological properties of synthetic peptides, and the role of lipids as delivery systems for peptides.

Sessions V, VI and VII and 230 posters discussed the area of Biologically Active Peptides/Neuropeptides/Peptide Hormones/Peptide Inhibitors/Peptide Receptor Interactions. This area continues to be of fundamental importance, not only for understanding mechanism of action, but also in the development of new potent agonists and antagonists in the pharmaceutical sector as well.

Antibiotics are facing a losing battle with the resurgence of antibiotic resistant pathogens, which is predicted to be "nothing short of a medical disaster" in the 1990s and beyond. Session VIII titled Peptide Vaccines and Immunology again emphasized the important contributions that peptides are making in the field of immunology and the increasing importance of peptides to immunologists, microbiologists and virologists in the development of synthetic peptide vaccines against bacterial and viral pathogens.

Session X emphasized the important breakthroughs that are occurring in peptide delivery; strategies to enhance intestinal permeability of peptides; protease-resistant peptide analogs for oral delivery; and delivery of peptides into the central nervous system by sequential metabolism. In addition, the use of peptides as diagnostics, a field in its infancy, was included.

Two sessions, IX and XI, discussed the area of Conformational Analysis and the use of Computational Biochemistry to study peptide structure. Session XII, Peptide Macromolecular Interactions, went to the next step in complexity with the study of the interactions of peptides with biological macromolecules, the important first steps in signal transduction and understanding ligand-receptor interactions.

Session XIII focused our attention on the rapidly developing field of using Peptide Libraries as screening tools in research and drug discovery. Novel methods of synthesis and strategies of using these libraries are being developed. Libraries of peptides, free in solution, bound to the solid support or being generated by phage can contain cyclic peptides, unnatural amino acids, and colored and fluorescent markers. We look forward to the exciting new developments that this technology will bring in the coming years in terms of unique peptide chemistry and application.

Session XIV on *de novo* design represented the inroads peptide chemistry is bringing to understanding protein folding and stability. The future will bring small chemically synthesized protein molecules with novel enzymatic and binding properties to the forefront of the biotechnology industry.

The Symposium introduced three workshops, which provided participants with an opportunity to familiarize themselves at the basic level of an area outside their own expertise: Workshop I, Approaches and Advances in Peptide Synthesis, Purification and Analysis; Workshop II, An Introduction to NMR Spectroscopy of Peptides; and Workshop III, An Introduction to Energy Minimization, Molecular Dynamics, Molecular Modelling and Conformational Analysis of Peptides. These workshops ran simultaneously and the workshop concept was an overwhelming success and will continue at future meetings.

This year's Alan E. Pierce Award recipient, Dr. Victor J. Hruby, was recognized for his outstanding contributions to the chemistry and biology of peptides, especially the design of peptides, pseudopeptides and peptidomimetics to understand the relationship of structure and biological activity. His fascinating lecture illustrated the wide-ranging contributions Victor has made to the field of peptide chemistry and, combined with his outstanding scientific and personal qualities, clearly demonstrated his deserving position among the previous eight awardees.

The use of the Edmonton Convention Centre and the success of this venue indicates the need for this type of facility to provide an atmosphere which can deliver the high quality of scientific lectures, poster presentations, workshops and exhibits in an effective and efficient manner to such a large group of participants. The large lounge area in the center of the exhibit/poster hall provided an excellent atmosphere for discussion and exchange of information.

This meeting was organized solely with the use of volunteers and the success of the meeting was directly related to their dedication and hard work. As Chairman, I am forever grateful to the Department of Biochemistry personnel including technologists, graduate students and postdoctoral fellows from my laboratory. A special thanks goes to Dr. Colin Mant, who as Conference Manager contributed to all aspects of the meeting from fund raising to managing the exhibition and without his help I could not have brought the Symposium to Edmonton; Colleen Iwanicka, who as Financial Administrator diligently handled and maintained all financial transactions; Janice MacDonald, who as Social Director made this aspect of the meeting a success and for many who were involved in the post conference tour of the Canadian Rockies an unforgettable experience; and lastly Janet Wright who as head of Office Administration and Dawn Lockwood together organized the daily correspondence, created the Symposium database required to correspond with all conference attendees, kept track of adjudicated abstracts, typed all printed material and lastly, literally corrected, typed and reformatted the Proceedings Volume as camera ready, altogether an enormous task.

To members of the Program Committee who assisted in organizing the scientific program, who volunteered to organize the workshops and who evaluated all abstracts and selected papers for publication in the Proceedings Volume, your hard work and commitment is greatly appreciated. To the Chairmen who stimulated discussion and kept the meeting on schedule, we are indebted.

We are especially grateful to the generous support by the Benefactors, Sponsors, Donors and Contributors (see the following pages). Such support is critical to the quality and success of the American Peptide Symposium. We wish to express our special thanks to Dr. Alan E. Pierce and Pierce Chemical Company for providing funds for the Alan E. Pierce Award and partial support for the Awards Banquet and symposium photographer, and to the sponsors of the American Peptide Society Travel Awards (Applied Biosystems Inc., Millipore and Star Biochemicals) which enabled young scientists from around the world to attend the Symposium.

We congratulate the Student Affairs Committee of the American Peptide Society who again organized the Job Fair and the Student Poster Competition which facilitate career development of young scientists and reward their individual collective achievements.

As Chairman, I would like to acknowledge the Department of Biochemistry, Faculty of Medicine, and the University of Alberta for their support of my efforts to hold the Thirteenth American Peptide Symposium and The Alberta Heritage Foundation for Medical Research for providing travel funds to speakers at the Symposium.

Lastly, the Chairman would like to acknowledge his wife, Phyllis, who understands the challenges and workload involved in this endeavour.

Robert S. Hodges
John A. Smith

Thirteenth American Peptide Symposium

Edmonton Convention Centre, Edmonton, Alberta, Canada
June 20-25, 1993

THE AMERICAN PEPTIDE SOCIETY

OFFICERS

Charles M. Deber	President	University of Toronto
Jean E. Rivier	President-Elect	The Salk Institute
Arthur M. Felix	Secretary	Hoffmann-La Roche Inc.
John A. Smith	Treasurer	University of Alabama at Birmingham

COUNCILLORS

Irvin M. Chaiken	SmithKline Beecham
David H. Coy	Tulane University
Bruce W. Erickson	University of North Carolina
Ralph F. Hirschmann	University of Pennsylvania
Maurice Manning	Medical College of Ohio
Garland R. Marshall	Washington University School of Medicine
Arno E. Spatola	University of Louisville
James P. Tam	The Vanderbilt University
Daniel F. Veber	Merck Sharp and Dohme Research Laboratories
Janis D. Young	Skyline Peptides

CHAIRMAN - THIRTEENTH AMERICAN PEPTIDE SYMPOSIUM

Robert S. Hodges	University of Alberta

PROGRAM COMMITTEE

Murray Goodman	University of California at San Diego
Robert S. Hodges	University of Alberta
Randy T. Irvin	University of Alberta
Pravin T. Kaumaya	Ohio State University College of Medicine
Cyril M. Kay	University of Alberta
Rachel E. Klevit	University of Washington
Gilles Lajoie	University of Waterloo
Garland R. Marshall	Washington University School of Medicine
Ruth F. Nutt	Merck Sharp and Dohme Research Laboratories
Jean E. Rivier	The Salk Institute
Peter W. Schiller	Clinical Research Institute of Montreal
Bhagirath Singh	University of Alberta and University of Western Ontario
John A. Smith	University of Alabama at Birmingham
Brian D. Sykes	University of Alberta
James P. Tam	The Vanderbilt University

ORGANIZING COMMITTEE - UNIVERSITY OF ALBERTA

Robert S. Hodges   Symposium Chairman
Colin T. Mant   Conference Manager
Colleen Iwanicka   Financial Administrator
Janice E. MacDonald   Social Director
Janet Wright   Administrative Assistant
Morris R. Aarbo
T.W. Lorne Burke
Cyril M. Kay
Dawn Lockwood
Gail Redmond
Brian D. Sykes

LOCAL VOLUNTEERS - UNIVERSITY OF ALBERTA

Christi Andrin
Sherron Becker
Jim Black
Lori Burke
Rose Caday
Michael Carpenter
Irene Church
Len Daniels
Beth-Anne Exham
Rod Gagne
Devon Husband
Shari Kasinec
Terri Keown
Xin Liu
Vicki Luxton
Joyce MacDonald
Jane Miller
Oscar D. Monera

Jack Moore
Sai Ming Ngai
Stacy Oare
Bob Parker
Paul Semchuk
Terry Sereda
Cindy Shaughnessy
Sue Smith
Natalie Strynadka
Kathy van Denderen
Gary Van Domselaar
Jennifer Van Eyk
Iain Wilson
Wah Wong
Shirley Woywitka
Peter Wright
Mae Wylie

The Thirteenth American Peptide Symposium greatly appreciates the support and generous financial assistance of the following organizations:

BENEFACTORS

Alberta Economic Development and Tourism
Alberta Heritage Foundation for Medical Research
Applied Biosystems
Bachem California
Mallinckrodt Specialty Chemicals Company
Millipore Corporation
Novabiochem/Calbiochem
Pierce Chemical Company
Protein Engineering Network of Centres of Excellence, Canada
Sigma Chemical Company
U.S. Army Medical Research Acquisition Activity

SPONSORS

Abbott Laboratories
Advanced ChemTech
Amgen, Inc.
Bachem Bioscience, Inc.
Bachem Feinchemikalien AG
Biogen, Inc.
Hoffmann - La Roche, Inc.
Propeptide
Vydac/The SEP/A/RA/TIONS Group, Inc.

DONORS

BioChem Pharma, Inc.
Biomeasure Incorporated
Boehringer Mannheim GmbH
Bristol-Myers Squibb Company
ESCOM Science Publishers B.V.
Glaxo
Hewlett-Packard
Hoechst-Roussel Pharmaceuticals, Inc.
Merck Research Laboratories
Orpegen
Peninsula Laboratories, Inc.
Peptide Institute / Protein Research Foundation
Peptides International
Peptisyntha and Cie, SNC
Pfizer

Senn Chemicals AG
Sterling Winthrop, Inc.
Syntex Research
Synthetech, Inc.
The DuPont Merck Pharmaceutical Company
The Upjohn Company
UCB-Bioproducts S.A.
Wyerst-Ayerst Research

CONTRIBUTORS

Alberta Tourism, Parks and Recreation
Astra Pharma, Inc.
Bio-Mega, Inc.
Ciba-Geigy Canada, Inc.
CSPS
Cytogen
Edmonton Convention and Tourist Authority
Immunobiology Research Institute
Lilly Research Laboratories
Sandoz Canada, Inc.
Sandoz Pharmaceuticals Corporation
Tanabe Research Laboratories, USA, Inc.
The Peptide Laboratory, Inc.
Warner-Lambert
Zeneca Pharmaceuticals

Alan E. Pierce Award

(Sponsored by Pierce Chemical Company)

The recipient is an individual who has made outstanding contributions to techniques and methodology in the chemistry of amino acids, peptides and proteins.

1993	Victor J. Hruby	University of Arizona Tucson, AZ, U.S.A.
1991	Daniel Veber	Merck Sharp and Dohme Research Laboratories, West Point, PA, U.S.A.
1989	Murray Goodman	University of California - San Diego San Diego, CA, U.S.A.
1987	Choh Hao Li	University of California - San Francisco San Francisco, CA, U.S.A.
1985	Robert Schwyzer	Swiss Federal Institute of Technology Zürich, Switzerland
1983	Ralph Hirschmann	Merck Sharp and Dohme Rahway, NJ, U.S.A.
1981	Klaus Hofmann	University of Pittsburgh School of Medicine, Pittsburgh, PA, U.S.A.
1979	Bruce Merrifield	The Rockefeller University New York, NY, U.S.A.
1977	Miklos Bodanszky	Case Western Reserve University Cleveland, OH, U.S.A.

American Peptide Symposia

Symposium		Chair(s)	Location
First	1968	Saul Lande Yale University, New Haven Boris Weinstein University of Washington, Seattle	Yale University New Haven, CT, U.S.A.
Second	1970	F. Merlin Bumpus Cleveland Clinic, Cleveland	Cleveland Clinic Cleveland, OH, U.S.A.
Third	1972	Johannes Meienhofer Harvard Medical School, Boston	Children's Cancer Research Foundation, Boston, MA, U.S.A.
Fourth	1975	Roderich Walter University of Illinois Medical Center Chicago	The Rockefeller University and Barbizon Plaza Hotel New York, NY, U.S.A.
Fifth	1977	Murray Goodman University of California-San Diego	University of California- San Diego San Diego, CA, U.S.A.
Sixth	1979	Erhard Gross National Institutes of Health Bethesda	Georgetown University Washington, DC, U.S.A.
Seventh	1981	Daniel H. Rich University of Wisconsin-Madison	University of Wisconsin-Madison Madison, WI, U.S.A.
Eighth	1983	Victor J. Hruby University of Arizona, Tucson	University of Arizona Tucson, AZ, U.S.A.
Ninth	1985	Kenneth D. Kopple Illinois Institute of Technology Chicago Charles M. Deber University of Toronto, Toronto	University of Toronto Toronto, Ontario, Canada
Tenth	1987	Garland R. Marshall Washington University School of Medicine, St. Louis	Washington University St. Louis, MO, U.S.A.

Symposium		Chair(s)	Location
Eleventh	1989	Jean E. Rivier The Salk Institute for Biological Studies, La Jolla	University of California- San Diego San Diego, CA, U.S.A.
Twelfth	1991	John A. Smith Massachusetts General Hospital Boston	Massachusetts Institute of Technology, Cambridge, MA, U.S.A.
Thirteenth	1993	Robert S. Hodges University of Alberta, Edmonton	Edmonton Convention Centre Edmonton, Alberta, Canada

Abbreviations

Abbreviations used in the proceedings volume are defined below:

α-AE	α-amidating enzyme	ADE	atrial peptide degrading enzyme
A₂bu	2,4-diaminobutyric acid	ADR	adriamycin
A₂pr	2,3-diaminopropionic acid; *see* Dpr	AEC	3-amino-9-ethylcarbazol
AA, aa	amino acids	Agm	4-guanidino-butyl-amine (descarboxy arginine)
AAA	amino acid analysis		
Aab	3-aminomethyl-4-aminobutanoic acid	AFP	antifreeze protein
αAT	α-antitrypsin	AGSP	atrial granule serine proteinase
Ab	antibody		
Aba	2-aminobutyric acid	Aha	7-aminoheptanoic acid
ABTS	2,2'-azido-bis(3-ethyl-benzthiazoline sulfonic acid)	AHPBA	3-amino-2-hydroxy-4-phenylbutanoic acid; phenyl-norstatine
ABZ	aminobenzoic acid		
AC	adenylate cyclase	AHPPA	4-amino-3-hydroxy-5-phenylpentanoic acid
AC₂O	acetic anhydride		
ACE	angiotensin-converting enzyme	Ahx	aminohexyl
ACh	acetylcholine	Aib	α-aminoisobutyric acid
ACHPA	4-amino-3-hydroxy-5-cyclohexylpentanoic acid	Aic	2-aminoindan-2-carboxylic acid
AChR	acetylcholine receptor	AIDS	acquired immune deficiency syndrome
Acm	acetamidomethyl		
ACN; Acn	acetonitrile		
AcOH	acetic acid	Alloc, Aloc	allyloxycarbonyl
ACP	acyl carrier protein	Allyl	allyl ester
ACSA	adenylate cyclase-stimulating activity	AMD	actinomycin D
		Amf	aminophenylalanine; *see* Aph
Acsc	aminocyclopentane-carboxylic acid	AMP	aminomethyl-piperidine
ACT	Advanced ChemTech, Inc.	AMPA	aminomethylphenyl-acetic acid
AcT	Nα-acetyltransferase	ANF	atrial natriuretic factor
ACTH	corticotropin	Ang II, AII	angiotensin II
AD	Alzheimer's disease	Ang, ANG	angiotensin; angiotensinogen
Ada	adamantyl		

ANP	atrial natriuretic peptide	BN, Bn	bombesin
Anq	anthraquinone	BnPeOH	2,2-[bis(4-nitro-phenyl)]-ethanol
AO	antiovulatory	Boc	tert-butyloxycarbonyl
Aoc	1-azabicyclo[3.3.0]-2-carboxylic acid	Boc-ON	2-tert-butyloxy-carbonylamino-2-phenylacetonitrile
AOGO	5-amino-4-oxo-8-guanidinooctanoic acid	BOI	2-(benzotriazol-1-yl)-oxy-1,3- dimethyl-imidazolinium
AP	aminopeptidase		
APC	antigen presenting cell	Bom	benzyloxymethyl
APG	azidophenyl glyoxal	BOP	benzotriazolyloxy tris-(dimethylamino) phosphonium hexafluoro-phosphate
Aph	aminophenylalanine		
APM	aminopeptidase M		
Apo	apolipoprotein		
APP	avian pancreatic polypeptide		
		BOP-Cl, BopCl	bis(2-oxo-3-oxazo-lidinyl) phosphinic chloride
APY	anglerfish peptide YG		
AR	adrenergic receptor		
ARC	AIDS related complex	BPA	benzylphenoxyacet-amidomethyl
Asa	azidosalicylic acid		
ASF	African swine fever	Bpa	benzoylphenylalanine
Asu	aminosuberic acid	Bpo	D-α-benzoyl-penicilloyl
AT	antithrombin		
Atc	2-aminotetralin-2-carboxylic acid	Bpoc	biphenylpropyloxy-carbonyl
ATIII	antithrombin III	BPTI	bovine trypsin inhibitor
ATR	attenuated total internal reflection		
		bR	bacterial rhodopsin
ATZ	anilinothiazolinone	Br$_2$Dmb	3,5-bis(bromo-methyl)benzoate
AVP	arginine-8-vasopressin		
AZT	3'-azido-3'-deoxy-thymidine; zidovudine	BroP	bromo tris(dimethyl-amino)phosphonium hexafluorophosphate
		BSA	bovine serum albumin
b	bovine	Bt	biotinoyl
Bab	3,5-bis(2-aminoethyl) benzoic acid	BTD	bicyclic β-turn dipeptide
Bal	β-alanine	BTU	*O*-benzotriazolyl-*N,N,N′,N′*-tetra-methyluronium hexafluorophos-phate
BAP, βAP	beta amyloid protein		
BBB	blood brain barrier		
BGG	bovine gamma globulin		
BHAR	benzhydrylamine resin	Bum	tert-butyloxymethyl
BHI	biosynthetic human insulin	BU^t, ^{t}BU	tert-butyl
		Butaz	1,2-diphenyl-pyrazolidine-3,4-dione
Biot	biotin		
BK	bradykinin		
BME	β-mercaptoethanol	BW	body weight

Bzl	benzyl	ClZ	2-chlorobenzyloxy-carbonyl
c-3-PP	cis-3-propyl-L-prolyl	Cle	cycloleucine (1-amino-1-carboxyl-cyclopentane)
CA	chemical acetylation		
cAMP	cyclic adenosyl mono-phosphate	CM	chloromethyl; carboxymethyl
CAP	core amyloid peptide		
CAT	chloramphenicol acyl transferase	CNS	central nervous system
Cbz, Z	carbobenzoxy; benzyloxycarbonyl	COSY	correlated NMR spectroscopy
CCD	countercurrent distribution	Cpa	4-chlorophenylalanine
		CPD, CP	carboxypeptidase
CCK	cholecystokinin	CPF	caerulin precursor fragment
CD	circular dichroism		
CD	complement domain	CPMAS	cross-polarization/ magic angle spinning
cDNA	complementary DNA		
CE	carbetocin; capillary zone electro-phoresis; *see* CZE		
		CR	chain recombination
		CRF	corticotropin releasing factor
CEC	cation exchange chromatography		
		CRP	C-reactive protein
CFA	complete Freunds adjuvant	CsA	cyclosporin A
		CT	carboxy terminus; calcitonin; chymotrypsin; cholera toxin
CGRP	calcitonin gene-related peptide		
CgTx	conotoxin		
CHA	cyclohexylamine	CTAB	cetyl trimethyl ammonium bromide
Cha	cyclohexylalanine		
CHAPS	3-[(3-cholamido-propyl)-dimethyl-ammonio]-1-pro-pane-sulfonate		
		CTL	cytotoxic T-lympho-cytes
		CTMS	chlorotrimethylsilane
cHex, cHx	cyclohexyl	Ctp	chloroacetyl-trypto-phan
CHF	congestive heart failure		
		CVAP	cerebrovascular amyloid peptide
CHO	Chinese hamster ovary; aldehyde		
		CVS	cardiovascular system
		Cyp	cyclophilin
ChTX	charybdotoxin	CZE	capillary zone electrophoresis
CID	chemically ionized desorption; collision-induced dissociation		
		DA	D/Ala substitution factor
CINC	cytokine-induced neutrophil chemoattractant	Dab	diaminobutyric acid
		DABCYL	4-dimethyl-amino-azobenzene-4'-sulfonyl chloride
CLA	cyclolinopeptide A		

DABITC	4-dimethylamino-phenyl-4'-isothio-cyanate	DIEA	diisopropylethylamine
DADLE	[D-Ala^2,D-Leu^5] enkephalin	DIP	4,7-diphenyl phenanthroline
DAGO	[D-Ala^2,N-MePhe^4,Gly^5-ol] enkephalin	DIPCDI	diisopropylcarbo-diimide; *see* DIC
Dah	1,6-diaminohexane	DKP	[Asp^{B10},Lys^{B28},-Pro^{B29}]-insulin;
Dap	diaminopimelic acid		diketopiperazine
Dat	desamino tyrosine	DLPS	dilauroylphosphatidyl-serine
DBU	1,8-diazabicyclo [5.4.0]-undec-7-ene	DMA	dimethylacetamide
Db$_z$g	dibenzylglycine	DMAP	dimethylamino-pyridine
DCCI, DCC	dicyclohexyl-carbodiimide	DMBA	9,10-dimethyl-1,2-benzathracene
DCHA	dicyclohexylamine	DMBHA	2',4'-dimethoxybenz-hydryl amine
DCI	3,4-dichloro-isocoumarin	DMF	dimethylformamide
DCM	dichloromethane	Dmp, DMP	dimethylphosphinyl
Dcp	dichlorophenyl	DMPC	dimyristoylphospha-tidylcholine
DCU	dicyclohexylurea		
DDDA	2,9-diamino-4,7-dioxadecanedioic acid	DMPG	dimyristoylphospha-tidylglycerol
Dde	N-(1-(4,4-dimethyl-2,6-dioxocyclo-hexylidene)ethyl)	DMPSE	dimethylphenylsilyl ethane
		DMPTU	1-dimethyl-3-phenyl-2-thioures
DDQ	dichlorodicyano-quinone	DMS	dimethyl sulfide
DEAE	diethylaminoethanol	DMSO	dimethyl sulfoxide
DEAM	diethylacetamido-malonate	Dmt-OH	2,2-dimethyl-L-thiazolidine-4-carboxylic acid
Deg	diethylglycine	Dncp	2,4-dinitro-6-carboxy-phenyl
DEPC	diethylphosphoro-cyanidate	DNP	dinitrophenyl
Dφg	diphenylglycine	Dns	dansyl
DG/SA	distance geometry/ simulated annealing	DOACl	dimethyl-dioctadecyl ammonium chloride
Dha	dehydroalanine	DOC	deoxycholate
Dhc	S-(2,3-dihydroxy-propyl)cysteine	DOPC	dioleoyl-*sn*-glycero-phosphocholine
DHO	dihydroorotic acid	Dpa	β,β-diphenylalanine
DHP	dihydroxypropyl	Dpa	diphenylalanine
DIBAL	diisobutyl aluminium hydride	DPBT	diphenylphosphoryl-benzoxazolthione
DIC	N,N'-diisopropyl-carbodiimide	DPCDI	diisopropyl-carbodiimide;

	see DIC
DPDPE	cyclo[DPen²-DPen⁵]enkephalin
Dpg	dipropylglycine
DPI	despentapeptide insulin
DPP	dipeptidyl peptidase
DPPA	diphenylphos-phorylazide
DPPC	dipalmitoyl phosphatidyl-choline
DPPG	dipalmitoyl phosphatidyl-glycerol
DPP-IV	dipeptidylpeptidase-IV
Dpr	2,3-diaminopropionic acid
DPTU	*N,N'*-diphenylthioures
DQF	double quantum focused
DSB	4-(2,5-dimethyl-4-methylsulfinyl-phenyl)-4-hydroxy-butanoic acid
DSP	dimethylsulfonium methyl sulfate
DTC	dimeric tripeptide chemoattractants
Dtc	*S,S*-dimethyl-thiazolidine-4-carboxylic acid
DTH	delayed type hyper-sensitivity
DTNB	dithiobis(2-nitro-benzoic acid)
DTPA	diethylenetriamine pentaacetic acid
Dts	dithiasuccinoyl
DTT	dithiothreitol
Dyn	dynorphin
e	eel; equine
EA	ergotamine
EAE	experimental autoimmune encephalomyelitis

EDAC, EDC	1-(3-dimethyl-amino-propyl)-3-ethyl-carbodiimide hydrochloride
EDRF	endothelium-derived relaxing factor
EDTA	ethylenediamine-tetraacetic acid
EGF; EGFR	epidermial growth factor; epidermial growth factor receptor
EI	epidermal cell inhibitor
EIAV	equine infectious anemia virus
ELAB	enantiomer labeling
ELISA	enzyme-linked immunosorbent assay
Enk	enkephalin
EP	endorphin
EPNP	1,2-epoxy-3-(*p*-nitro-phenoxy)propane
ESI-MS	electrospray ionisation mass spectrometry
ESR	electron spin resonance
ET	endothelin
Et₃N	triethylamine; *see* TEA
EtA	α-ethylalanine
Etm	ethyloxymethyl
FAA	fatty amino acid
FAB-MS	fast atom bombardment mass spectrometry
Farn	farnesyl
FeLV	feline leukemia virus
Fg	fibrinogen
FGF	fibroblast growth factor
FID	flame ionization detector
FITC	fluorescien isothiocyanate
Flg	fluorenylglycine

Fm, fm	fluorenylmethyl		GnRH	gonadotropin-releasing hormone
FMDV	foot-and-mouth disease virus		GPI	guinea pig ileum
FMOC, Fmoc	9-fluorenylmethoxy-carbonyl		GRF, GHRH	growth hormone-releasing factor
Fpa	4-fluorophenylalanine		GRP	gastrin releasing peptide
FPLC	fast protein liquid chromatography		GS	gramicidin S
FRET	fluorescence resonance energy transfer		GSH	reduced glutathione
			GSSG	oxidized glutathione
FSH	follicle stimulating hormone; follitropin		GTP	guanosine triphosphate
			GvH	graft versus host
FTIR	Fourier transform infrared		GVIA	conotoxin G VIA
			h	human
GA	gramicidin A		Hat	6-hydroxy-2-aminotetralin-2-carboxylic acid
GABA	gamma aminobutyric acid			
GAL	galanin		hBP	human serum binding protein
GAP	growth associated protein; gonadotropin-releasing hormone associated protein		HBTU	*O*-benzotriazolyl-*N,N,N',N'*-tetramethyl-uronium hexa-fluorophosphate
GC	gas chromatography			
gCSF	granulocyte-colony stimulating factor		HBV	hepatitis B virus
			HC	heparin cofactor II
GDA	glutaraldehyde		hCG	human chorionic gonadotropin
GEMSA	guanidino ethyl-mercaptosuccinic acid		hCGRP	human calcitonin gene-related peptide
GH	growth hormone		hCys, Hcys	homocysteine
GHRH, GRF	growth hormone releasing hormone		HEL	hen egg lysozyme
			HEL	human erythroleukemia
GHRP	growth hormone releasing peptide		Hepes	*N*-[2-hydroxy-ethyl]piperazine-*N'*2-ethanesulfonic acid]
GITC	2,3,4,6-tetra-*O*-Ac-β-D-glucopyranosyl isothiocyanate			
Gla	D-galactopyranosyl; gamma carboxy-glutamic acid		HF	hydrogen fluoride
			Hfa	homophenylalanine
			HFBA	heptafluorobutyric acid
GLP	glucagonlike peptide		hGH	human growth hormone
GM	gramicidin M; [Phe[9,11,13,15]] gramicidin A		HHM	humoral hypercalcemia of malignancy
Gn, Gu	guanidine			

HI	hemoregulatory cell inhibitor	HR	histamine release
HIC	hydrophobic interaction chromatography	HR, hr	human recombinant
		HSA	human serum albumin
		Hse	homoserine
HILIC	hydrophilic interaction chromatography	HSPS	high speed peptide synthesis
HIV	human immuno-deficiency virus	HSV	herpes simplex virus
		Htc	7-hydroxytetrahydro-isoquinoline-3-carboxylic acid
HIV-PR	human immuno-deficiency virus protease		
		HTE	hamster trachea epithelial cell
HLE	human leukocyte elastase	HTLV	human T-cell leukemia virus
HMP	hydroxymethyl-phenoxyacetic acid; hydroxymercapto-propionic acid	HUB	hamster urinary bladder
		HUVEC	human umbilical vein endothelial cell
HMP	hydroxymethyl-phenoxymethyl	Hyp	hydroxyproline
		Hz	hertz
HMPA, HMPT	hexamethylphosphoric triamide	ICE	interleukin-1 beta converting enzyme
HNE	human neutrophil elastase	i.c.v.	intra-cerebro-ventricular
hNP	human neutrophil peptide	i.m.	intramuscular
		i.v.	intravenous
HOBt	N-hydroxybenzo-triazole	IC	inhibitory concentration
HODhbt	hydroxyoxodihydro-benzotriazine	ICAM	intracellular adhesion molecule
HONp	nitrophenol	ICE	interleukin convertase
HOOBt	hydroxyoxodihydro-benzotriazine	IEC	ion-exchange chromatography
HOSu	N-hydroxysuccinimide	IEF	isoelectric focusing
HOTic	7-hydroxy-1,2,3,4-tetrahydroquinoleic-3-carboxylic acid	IFNα	interferon α
		Ig	immunoglobulin
HPA	hypothalamic-pituitary-adrenal axis	IGF	insulin-like growth factor
		IL	interleukin
HPI	human proinsulin	ILys	lysine(N$^\varepsilon$-isopropyl)
HPLC	high performance liquid chromatography	im	imidazole
		in	indole
		INEPT	insensitive nuclei enhancement by pulse transfer
Hpp	3-(4-hydroxyphenyl)-propionyl		
HPSEC	high performance size exclusion chromatography	Ing	indenylglycine
		IP	inositol phosphate

IR	infrared; insulin receptor	MBP	myelin basic protein
IRMA	immunoradiometric assay	MBS	*m*-maleimidobenzoyl-*N*-hydroxy-succinimide ester
IU	international units	MCH	melanin concentrating hormone
KLH	keyhole limpet hemocyanin	MCPBA	*m*-chloroperbenzoic acid
KP	[Lys^{B28},Pro^{B29}]-insulin	MCPS	multiple constrained peptide synthesis
		MD	molecular dynamics
LDH	lactate dehydrogenase	Me	methyl
LDTOF	laser desorption time-of-flight	Mea	2-mercaptoethylamine
		MeA	α-methylalanine
LEC	ligand-exchange chromatography	MeCN	acetonitrile
LFA	lymphocyte associated antigen	MeNTI	*N*-methyl noroxy-morphindole
		MeOH	methanol
LH	luteinizing hormone; lutropin	MHC	major histocompati-bility complex
LHRH	*see* GnRH	MIC	minimal inhibitory concentration
LPC	lauroylphosphoryl-choline	MIR	main immunogenic region
LPH	lipotropin		
LPS	lipopolysaccharide	Mls	minor lymphocyte stimulating gene
LSIMS	liquid secondary ion mass spectrometry	MMC	migrating motor complex
LVP	lysine-8-vasopressin		
		Mom	methyloxymethyl
m	murine; messenger	MoMuLV	Moloney murine leukemia virus
MAb	monoclonal antibody	Mot	motilin
Man	2-mercaptoanaline	Mpg	3-methoxypropyl-glycine
MAP, MAp	membrane-anchored protein; multiple antigen peptide; mean arterial pressure	MPM	p-methoxyphenyl-methyl ester
		Mpr	3-mercaptopropionyl
		Mpr	mercaptopropionic acid
MAPs	macromolecule-associated proteins	MS	mass spectrometry
MARCKS	myristolated alanine-rich C kinase substrate	MSH	melanocyte stimulating hormone; melanotropin
MAS	magic angle spinning		
Mba	2-mercaptobenzoic acid	Msob	methylsulfinylbenzyl
		Msz	methylsulfinylbenzyl-oxycarbonyl
Mbh	methoxybenzhydryl		
MBHA	methylbenzhydryl-amine		

Mtr	4-methoxy-2,3,6-trimethylbenzene-sulfonyl	NVOC	nitroveratryloxy-carbonyl
Mts	mesitylene sulfonyl	o	ovine
MuLV	murine leukemia virus	OEt	ethyl ester
MxAn	mixed anhydride	Oic	2,3,4,5,6,7,8-octa-hydroindole-2-carboxylic acid
Nal	2-naphthylalanine		
Nbb	nitrobenzamidobenzyl	OMe	methyl ester
Nbb	*o*-nitrobenzamido-benzyl	OMP	outer membrane protein
NBD	7-nitrobenz-2-oxa-1,3-diazole	ONb	*o*-nitrobenzyl
		OPA	*o*-phthaldialdehyde
NBS	*N*-bromosuccinimide	OSu	*O*-succinimide ester
NCp	nucleocapsid protein	OT	oxytocin
NCS	neocarcinostatin	OTf	*O*-triflate
NEM	*N*-ethyl maleimide	OVA	ovalbumin
NFT	neurofibrallary tangles	OVLT	organum vasculosum laminate terminalis
Nic	nicotinoyl		
NIDD	non-insulin dependent diabetes	Ox	oxazolidines
		OXT	oxytocin
NIS	*N*-iodosuccinimide		
NK	neurokinin	P_3CSS	macrophage activator tripalmitoyl-S-glycerylcysteinyl-serylserine
NM	neuromedin		
NMB	neuromedin B		
NMDA	*N*-methyl-D-aspartate		
NMM	*N*-methylmorpholine	PA	parent antagonist
NMP	*N*-methyl-pyrrolidinone	PAB	*p*-alkoxybenzyl
		PAC, Pac	phenacyl
NMR	nuclear magnetic resonance	PAF, Paf	*p*-aminophenylalanine
		PAGE	polyacrylamide gel electrophoresis
NMT	*N*-myristoyl transferase	PAL	photoaffinity labeling
NOE	nuclear Overhauser effect	Pal	3-pyridylalanine
		Paloc	3-(3-pyridyl)allyloxy-carbonyl
NOESY	nuclear Overhauser enhanced spectroscopy	PAM	phenylacetamido-methyl
NP	neutrophil peptide; neurophysin	Pas	6,6-pentamethylene-2-aminosuberic acid
Npp	nitrophenyl-pyrazolinone	Pbf	2,2,4,6,7-pentamethyl-dihydrobenzofuran-5-sulfonyl
NPY	neuropeptide Y		
Npys	3-nitro-2-pyridyl-sulfenyl	PBM	peripheral blood monocytes
NT	N-terminus; amino terminus; neurotensin	PBS	phosphate-buffered saline
		PCOR	peptide cyclization on oxime resin

PCP	phencyclidine	Pmc	2,2,5,7,8-penta-methylchroman-6-sulfonyl
PDB	phorbol 12,13-dibutyrate		
PD-MS	plasma desorption mass spectrometry	Pmp	3,3-pentamethylene-3-mercaptopropionic acid
PEG	polyethylene glycol		
Pen	penicillamine	Pmp	phosphonomethyl-phenylalanine
PEPS	polystyrene-grafted polyethylene film	pNA	*p*-nitroaniline
PFC	plaque forming cell	PND	principal neutralizing determinant
Pfp	pentafluorophenyl ester	PON	periodically oscillating neuron
PG	proteoglycan		
PGF	proteoglycan growth factor	POPS	palmitoyl-oleoyl-phosphatidylserine
Pgl	n-pentylglycine	PP	pancreatic polypeptide
PGL[a]	peptide glycine leucine amide	PPA	n-propylphosphoric anhydride
Phaa	phenylacetic acid	PPE	porcine pancreatic elastase
PHBT	polymeric hydroxy-benzotriazole		
PHF	paired helical filaments	PPIase	peptidyl-prolyl isomerase
Phi	4-iodophenylalanine	PPL	porcine pancreatic lipase
Phpa	3-phenylpropanoic acid	Ppt	diphenylphosphino-thionyl
pI	isoelectric point	PQ	paraquat
Pic	picolinoyl	Pqt	3-(1'-methyl-4,4'-bipyridinium-1-yl)propyl
Pilot	peptide identification and lead optimization		
		Pra	propargylglycine
Pin	β-pineyl	PRL	prolactine
Pip	pipecolinyl; piperidine	PRP	platelet-rich plasma
Pipes	piperazine-*N,N'*-bis[2-ethanesulfonic acid]	PT	pertussis toxin
		PTFCE	polytrifluorochloro-ethylene
PITC	phenylisothiocyanate	PTH	phenylthiohydantoin; parathyroid hormone
Piv	pivaloyl		
Piz	piperazic acid	PTK	protein tyrosine kinase
PK	protein kinase		
PLA$_2$	phospholipase A$_2$	Ptm	phenyloxymethyl
PLP	poly-L-proline	PTPase	protein-tyrosine phosphatase
PMA	phorbol myristate acetate		
		PTZ	phenothiazine
PMB	polymyxin B	Ptz	3-(10-phenothiazinyl)-propanol
pMBHA	4-methylbenz-hydrylamine		
		PVA	polyvinyl alcohol
		PVDF	polyvinylidene fluoride

PVN	paraventricular nuclei	Sar	sarcosyl; sarcosine
PyBOP	(benzotriazolyl)-*N*-oxy-pyrrolidinium phosphonium hexafluoro-phosphate	SCAL	safety catch amide linkage
		SCC	short circuit current
		SCLC	small cell lung carcinoma
PYL[a]	peptide tyrosine leucine amide	SEC	size exclusion chromatography
PYY	peptide tyrosine-tyrosine	SEM	standard error of the mean
		SH	sulfhydryl
QUIS	quisqualate	SHMT	serine hydroxymethyl transferase
recDNA	recombinant DNA	SLE	systemic lupus erythematoius
REDOR	rotational echo double resonance	SMPS	simultaneous multiple peptide synthesis
RGD	Arg-Gly-Asp fibrinogen binding sequence	SP	substance P
		SPCL	synthetic peptide combination libraries
RIA	radioimmunoassay		
RMSD, rmsd	root mean square deviation	SPPS, SPS	solid phase peptide synthesis
RNase	ribonuclease	SRIF	somatostatin
ROE	rotating frame nuclear Overhauser effect	SRP	signal recognition peptide
ROESY	rotating frame nuclear Overhauser enhanced spectroscopy	ssDNA	single stranded DNA
		ST	heat stable enterotoxin
		Sta	statin
ROS	rat osteosarcoma cells	Su	succinimide
RP	reversed phase	Suc	succinoyl
RP-HPLC (RPC)	reversed-phase high performance liquid chromatography	SWM	sperm whale myoglobin
		t-3-PP	*trans*-3-propyl-L-prolyl
s	salmon		
s.c.	subcutaneous	Tacm	*S*-trimethyl-acetamidomethyl
SA	symmetrical anhydrides	TAP	tick anticoagulant peptide
SAH	*S*-adenosylhomo-methionine	TASP	template-assembled synthetic protein
SAM	*S*-adenosylmethionine		
SAMBHA	(4-succinylamido-2',2',4'-trimethoxy)benz-hydryl amine	TBDMSCl	tert-butyldimethylsilyl chloride
		TBS	tert-butyldimethylsilyl
		TBTA	tert-butyl-2,2,2-tri-chloroacetamide
SAP	serum amyloid protein		
SAR	structure-activity relations	TBTU	*O*-(benzotriazol-l-yl)-*N,N,N',N'*-tetra-

	methyluronium tetrafluoroborate	TMSE	β-trimethylsilyl ethane
^tBu	tert-butyl	TMSOTf	trimethylsilyl trifluoromethane-sulfonate
Tca	trichloroacetamide		
TCEP	tris(2-carboxyethyl) phosphine	Tn	troponin
TCR	T lymphocyte antigen receptor	TNF	tumor necrosis factor
TCS	trypsin-catalyzed semi-synthesis	TPA	12-*O*-tetradecanoy-phorbol-13-acetate; tissue plasminogen activator
TCT	tracheal cytotoxin		
TEA	triethylamine	TPI	triose phosphate isomerase
TEAP	triethylamine phosphate	TPK	tyrosine-specific phosphate kinase
TEDOR	transferred echo double resonance	TPTU	1,1,3,3-tetramethyl-2-(2-oxo-1(2H)-pyridyl)uronium tetrafluoroborate
Teoc	trichloroethyloxy-carbonyl		
TEP	triethylphosphite	TPyClU	1,1,3,3-bis(tetra-methylene)-chlorouronium tetrafluoroborate
TFA	trifluoroacetic acid		
TFE	trifluoroethanol		
TFM	trifluoromethyl	TR	time resolved
TFMSA	trifluoromethane-sulfonic acid	TR-COSY	transferred rotational correlated NMR spectroscopy
TGF	transforming growth factor		
Th	thiazolidines	TRH	thyrotropin releasing hormone
THF	tetrahydrofuran		
Thi	*see* Dtc	Tris	tris(hydroxymethyl)-aminomethane
THTP	tetrahydrothiophene		
Thz	thiazolidine carboxylic acid	TRNOE	transferred nuclear Overhauser effect
Tic, Tiq	1,2,3,4-tetrahydro-quinoleic-3-carboxylic acid	Trt	trityl
		TSH	thyroid stimulating hormone
TicOH	7-hydroxy-1,2,3,4-tetrahydro-quinoleic-3-carboxylic acid	TT	tetanus toxoid
		TxA_2	thromboxane A_2
TLC	thin layer chromatography	UK	urokinase
		UNCA	urethane protected α-amino acid N-carboxy anhydride
TM	transmembrane		
Tm	melting temperature		
TMBD	tetramethylbenzadine	UNCA's	urethane N-carboxy-anhydrides
Tmob	trimethoxybenzyl		
TMP	3,4,7,8-tetramethyl-phenanthroline	UV	ultraviolet
TMS	trimethylsilyl	VCD	vibrational circular dichroism
TMSCN	trimethylsilyl nitrile		

Vly	valeryl	XAL	5-(9-aminoxanthen-2-oxy)valeric acid
VMH	ventro medial hypothalamus		
Vn	vitronectin	Z, Cbz	carbobenzoxy; benzyloxycarbonyl
VNA	virus neutralizing antibody		
VSMC	vascular smooth muscle cells	Ψ	pseudo
WSCI	water soluble carbodiimide		

Contents

Contents

Contents

Contents

Session II: Synthetic and Analytical Methods

Contents

Session III: Peptide Mimetics

Session IV: Glycopeptides/Lipopeptides

Contents

Session V: Biologically Active Peptides

Contents

Contents

Session VI: Peptide Hormones/Neuropeptides

Contents

Session VII: Peptide Inhibitors/Peptide-Receptor Interactions

Contents

Contents

Session VIII: Peptide Vaccines and Immunology

Session IX: Conformational Analysis

Contents

Session X: Peptide Pharmaceuticals/Diagnostics and Peptide Delivery

Session XI: Computational Biochemistry

Session XII: Peptide Macromolecular Interactions

Contents

Session XIII: Peptide Libraries

Contents

Session XIV: *De Novo* Design of Peptides and Proteins

Contents

Indexes

Session VIII
Peptide Vaccines and Immunology

Chairs: Vadim T. Ivanov
Shemyakin Institute of Bioorganic Chemistry
Moscow, Russia

and

Conrad H. Schneider
Institute of Clinical Immunology
Bern, Switzerland

A novel role for the immunogenic peptides in the assembly and expression of class II MHC molecules on B lymphocytes

B. Singh, B.J. Rider, B. Agrawal, Q. Yu and E. Fraga

Department of Microbiology and Immunology, University of Western Ontario, London, Ontario, N6A 5C1 and Department of Immunology, University of Alberta, Edmonton, Alberta, T6G 2H7, Canada

Introduction

Antigen presenting cells (APC's) process and present antigens in the form of peptides in association with class II MHC molecules for the activation of CD4+ T lymphocytes as shown in Figure 1. Peptide binding has been shown to be critical for the expression of MHC class I. MHC class II α and β chains form a compact molecule before surface expression. Recent evidence suggests that peptide binding may influence compact state formation and stabilize α and β chain association. TA3 cells, a murine B cell hybridoma, express both I-A^d and I-A^k. We have reported that TA3 cells slowly lose I-A^d expression in culture while I-A^k and B220 expression remain constant [1]. This decrease may therefore be due to the lack of peptide ligand saturation of I-A^d molecules. We show here that addition of selected peptides can restore I-A^d expression within 24 hrs in a dose- and affinity-dependent fashion. We have explored the possible mechanism for the variation in I-A^d expression on TA3 cells.

Results and Discussion

The invariant chain (Ii) blocks the peptide groove binding while the $\alpha\beta$ complex is in the endoplasmic reticulum (ER). Once released the Ii chain is degraded and peptides are able to bind and stabilize the complex which is then exported and displayed on the APC's surface for immune recognition [2].

We have used a low I-A^d expressing clone, TA3.11, which, when incubated with I-A^d binding antigens, regained the expression of I-A^d. Figure 2 shows that ovalbumin (OA) increased I-A^d surface expression more than one of its synthetic OA peptides 323-339. Other I-A^d restricted peptides, (EYA)$_5$ (E5) and EYK(EYA)$_4$ (K4), also upregulated TA3.11 I-A^d surface expression. Control antigen hen egg lysozyme (HEL) which is not I-A^d restricted did not increase the expression of I-A^d.

As expected, recombinant interleukin 4 (rIL-4) also increased I-A^d surface expression on TA3.11 cells but interferon γ (IFNγ) downregulated expression [3]. When rat concavalin A supernatant (rCAS), containing both IL-4 and IFNγ, was used there was no effect on I-A^d expression. It seems that these cytokines cancel each other's effects.

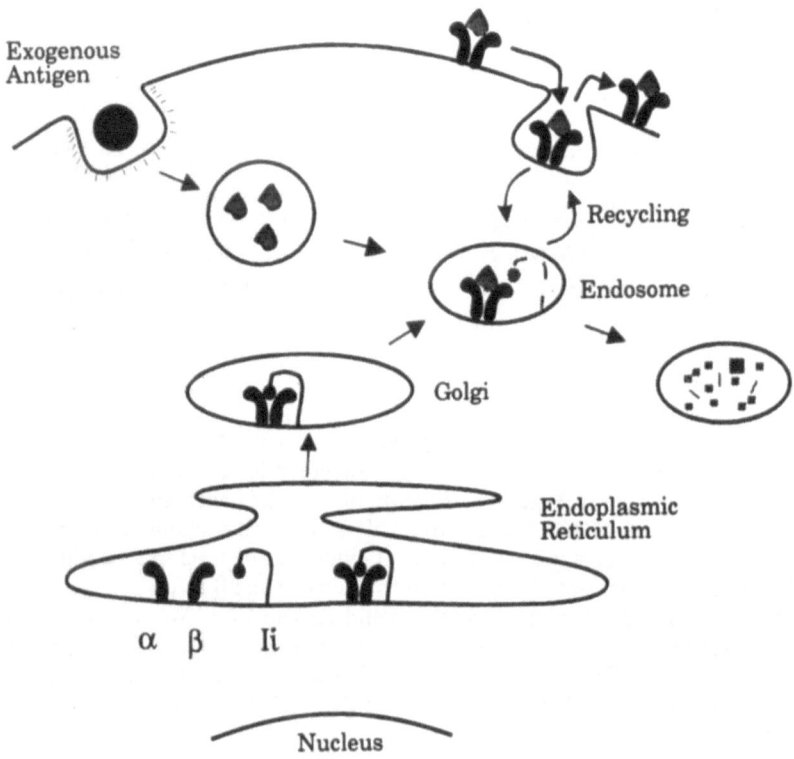

Fig. 1. MHC class II α and β complex formation and antigen processing and presentation. α, β, and invariant (Ii) chains are assembled in the endoplasmic reticulum. The Ii chain blocks peptide binding to the αβ complex until it reaches the endosomal stage where the Ii chain is degraded, allowing exogenous endocytosed peptide to bind. Peptide binding may stabilize this compact state, ensuring transport to and expression on the cell surface.

Titrations of rIL-4 and OA or its 323-339 peptide added simultaneously did not show synergy for I-A^d expression upregulation.

TA3.26, a high I-A^d expressing subclone, did not show these effects. Northern blots of TA3.11 and TA3.26 suggest that these subclones contain the same amounts of mRNA for I-A. Therefore, the difference in I-A^d expression is not because of different levels of transcription but due to the difference in the assembly of class II chains induced by the I-A^d binding antigens.

These results suggest that peptide ligand alone can influence the expression of class II molecules on the surface of APC's presumably by stabilizing the compact state formation of αβ chain complex. Our results explain why low MHC class II expressing APC's could present antigen to T cells.

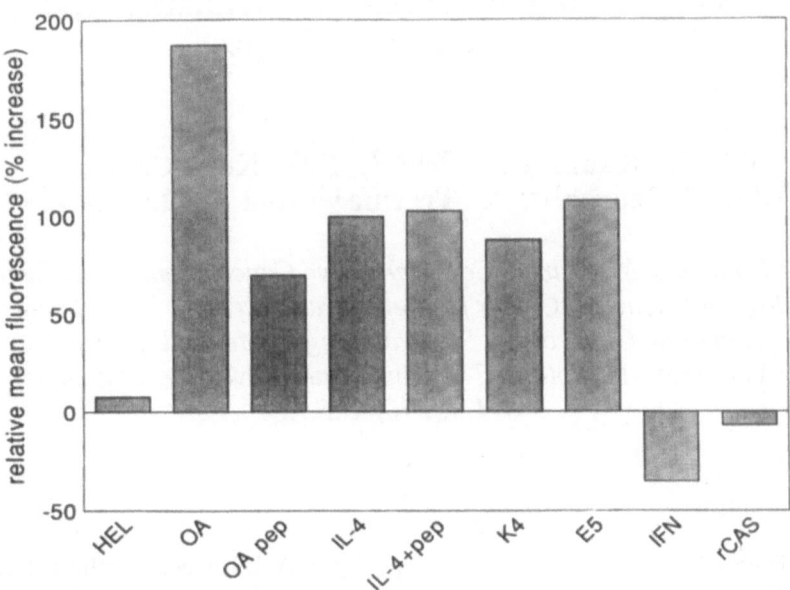

Fig. 2. Regulation of MHC class II I-A^d surface expression on TA3.11 cells. 10^6 TA3.11 cells were incubated with various antigens and/or cytokines for 24 hrs, washed and then incubated with 100 µl MKD6 supernatant for 30 min, washed and then incubated with 100 µl 1/50 fluorescein conjugated goat anti-mouse Ig for 30 min. Cells were then resuspended in PBS before FACS analysis. HEL, OA, OA peptide, K4, and E5 were used at 10 µM. IL-4, IFNY, and rCAS were used at 400 U/ml, 100 U/ml, and 10% respectively.

References

1. Agrawal, B., Fraga, E., Kane, K. and Singh, B., J. Immunol., 152 (1994) in press.
2. Germain, R.N. and Margulies, D.H., Annu. Rev. Immunol., 11 (1993) 403.
3. Glimcher, L.H. and Kara, C.J., Annu. Rev. Immunol., 10 (1992) 13.

A universal HTLV-I template vaccine incorporating cytotoxic, neutralizing and promiscuous epitopes

P.T.P. Kaumaya[1,2,3,5,6,7], S.F. Kobs-Conrad[2], A.M. DiGeorge[1,5], A. Trevino[4,8] and M. Lairmore[4,8]

[1]The College of Medicine, [2]Comprehensive Cancer Center, [3]College of Biological Sciences, [4]Center for Retrovirus Research, Departments of [5]Obstetrics & Gynecology, [6]Microbiology, [7]Medical Biochemistry, [8]Veterinary Pathology, The Ohio State University, Columbus, OH 43210, U.S.A.

Introduction

Human T-cell lymphotropic virus type 1 (HTLV-I) has been implicated as the causative agent of adult T-cell leukemia and a chronic degenerative myelopathy [1]. The immunopathogenesis of HTLV-I infection and disease is poorly understood despite a large body of information concerning the molecular biology of the virus. Even though several immunodominant epitopes of the HTLV-I envelope gp46 have been mapped [2, 3] that induce both humoral and cellular immune responses, the identification and characterization of these epitopes which elicit a protective immune response remains a central obstacle to the development of a vaccine against the virus infection. Some epitopes have elicited antibodies which block HTLV-I mediated cell fusion but have not elicited a protective immune response. In order to further delineate the immunogenicity, antigenicity and immunoprotective capacity of HTLV-I gp46 envelope protein, we have selected two known regions encoding a neutralizing epitope (SP2, 86-107) [4] and an overlapping cytotoxic and B-cell epitope (SP4a, 190-209) [2]. We recently developed strategies for bypassing haplotype restricted immune responses through the design of chimeric constructs incorporating B-cell epitopes with "promiscuous" T-cell epitopes, as well as the concurrent and parallel assembly of multiple and individual epitopes on a multivalent template, and have demonstrated their utility for rational vaccine design [5, 6]. In the work described here, we have designed and synthesized HTLV-I chimeras of SP2 and SP4a with a "promiscuous" measles virus (MVF) T-cell epitope. Additionally, in order to develop a universal HTLV-I vaccine able to elicit optimal B-cell, helper T-cell and cytotoxic responses, we have constructed two multivalent vaccines comprising two "promiscuous" T-cell epitopes from MVF and tetanus toxoid (TT), with either the neutralizing epitope SP2 alone, or with SP2 in combination with SP4a on a multivalent β sheet template.

Results and Discussion

Both chimeric constructs SP4-MVF and MVF-SP2 produce antipeptide antibodies in rabbits which are cross-reactive with the immunogen, the individual B-cell epitopes and the MTA-1 recombinant protein (in the case of SP4-MVF). Our studies in three inbred strains of mice (C3H/HeJ, H-2^k; BALB/c, H-2^d; and C57BL/6, H-2^b) show that all mice (5/5, or 4/4) responded to the chimeric immunogens with moderate titers (6400-12800) in the case of MVF-SP2, and with relatively high titers (>12800) in the case of SP4-MVF. The high reactivity of SP4-MVF to the immunogenic sequence paralleled its reactivity to the MTA-1 recombinant protein (kindly provided by Dr S. Foung, Stanford University Blood Bank, Palo Alto, CA) in all strains. Modest titers to individual peptide may be due to inappropriate presentation on ELISA plates.

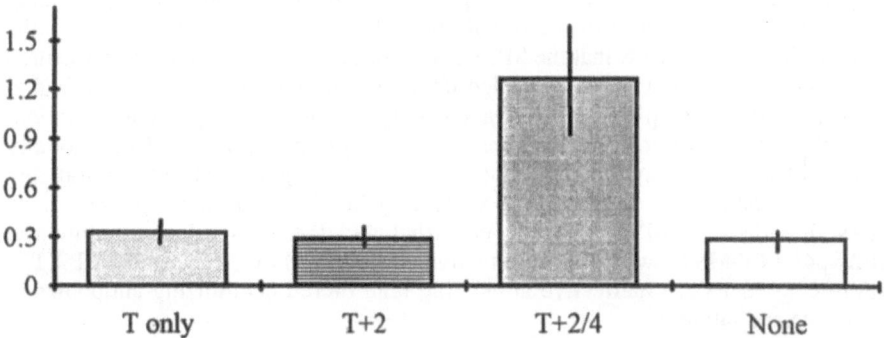

Fig. 1. Whole virus ELISA of immunized rabbits. Rabbits were immunized with template constructs (e.g., T + 2/4 = rabbits immunized with template with SP2 and SP4a). Serum (1:20) was tested at one week after 2nd immunization (mean 0.0+/- 5.E.M., n=5).

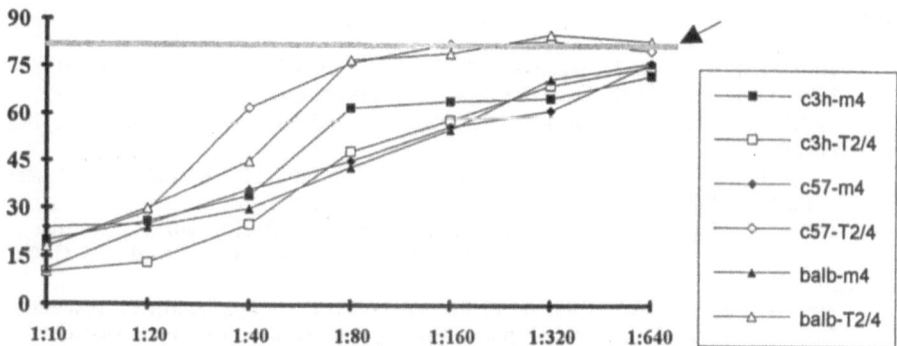

Fig. 2. Syncytia inhibition assay of mice strains SP4a. Mice strains were immunized with either MVF + SP4a (e.g., c3h - m4) or beta template with SP2 and SP4a (e.g., c3h - T2/4). Syncytia for 4 x 10x fields. Arrow = syncytia in wells with prebleed serum samples.

Mice (5/5) from all three strains (C3H/HeJ, BALB/c and C57BL/6) immunized with the SP2/Sp4a combination template constructs elicited high titers of antibodies (>12800) to both the immunogen and the recombinant protein, while once again minimal immune reactivity was observed when the SP2 epitope was incorporated into the template design. The antibodies recognized native forms of HTLV-I envelope proteins in whole virus ELISA, radioimmunoprecipitation, and immunofluorescence assays. In ELISA to whole virus preparations three out of the six rabbits immunized with SP2/SP4a template and SP4-MVF (Figure 1) showed high positive reactivity and the same animals recognize the surface of infected cells (1:40 - 1:320 titers) by immunofluorescence. Serum samples from immunized mice and rabbits were further tested in syncytia inhibition assay (Figure 2) to determine the ability of elicited antibodies to prevent HTLV-I mediated cell fusion. Antibodies produced in response to the SP4a-MVF and SP2/SP4a template constructs significantly inhibited HuT102 (HTLV-I) induced syncytia induction in human osteosarcoma cells.

The linear form of the SP2 epitope was ineffective at eliciting neutralizing antibodies in the presence of the promiscuous T-cell epitopes. This result can be easily rationalized in view that the SP2 site encodes a conformational loop-structured motif as predicted by two predictive algorithms and consequently antibodies raised to the sequential SP2 epitope do not adequately mimic the proper conformation. Further work in progress to engineer the SP2 epitope into a conformationally dependent sequence may provide an alternative to eliciting high affinity neutralizing antibodies, and provide an efficacious vaccine against HTLV-I infection. Our results show that the SP4-MVF chimeric construct and the multivalent combination SP2/Sp4a (TT-MVF) were highly effective while the MVF-SP2 and the SP2(TT-MVF) construct were ineffective at eliciting high titered neutralizing antibodies in both mice and rabbits.

Acknowledgements

This work was supported by the National Cancer Institute (NIH/NCI) grant CA16058 and an Elsa Pardee Fnd. grant 726778 to PTPK and ML.

References

1. Hinuma, Y., Nagata, K., Hanooka, M., Nakai, N., Matsumoto, T., Shirakawa, K. and Miyoshi, I., Proc. Natl. Acad. Sci. U.S.A., 78 (1981) 6476.
2. Palker, T., Tanner, M., Scearce, R., Clark, M. and Hayes, B., J. Immunol., 142 (1989) 971.
3. Lairmore, M.D., Rudolph, D.L., Roberts, B.D., Dezzutti, C.S. and Lal, R.B., Cancer Lett., 66 (1992) 11.
4. Palker, T., Riggs, E., Spragion, D., Muir, A., Scearce, R., Randall, R., McAdams, M., McKnight, A., Clapham, P., Weiss, R. and Haynes, B., J. Virol., 66 (1992) 5879.
5. Kaumaya, P.T.P., Kobs-Conrad, S., Seo, Y.H. and DiGeorge, A.M., In Schneider, C.H. and Eberle, E.N. (Eds.), Peptides 1992, Escom, Leiden, The Netherlands, 1993, p.139.
6. Kaumaya, P.T.P., Seo, Y.H., Kobs, S., Feng, N., Sheridan, J. and Stevens, V.C., J. Mol. Recog., 6 (1993) 81.

Synthetic peptides as mini antibodies

**M. Sällberg[1,2], M. Levi[2], U. Rudén[2], P. Pushko[3],
V. Bichko[3], L.O. Magnius[2], A. Tsimanis[3] and B. Wahren[2]**

*[1]Department of Clinical Virology, F 69, Huddinge University Hospital,
S-141 86 Stockholm, Sweden, [2]Department of Virology, The National
Bacteriological Laboratory, S-105 21 Stockholm, Sweden, [3]Institute of
Molecular Biology, Latvian Academy of Sciences, Krustpils str 53A,
Riga, Latvia*

Introduction

The complementary determining regions (CDRs) of antibodies are responsible
for the specificity of the antibody. X-ray crystallography has shown that the three
CDRs of the variable (V) region of the heavy chain and the three CDRs of the V
region of the light chain may all have contact with the epitope in an antigen-antibody
complex [1]. Also residues in the conserved β-sheet forming sequences, the
framework regions (FR), spacing the CDRs, can add to the binding [1]. Single
peptides corresponding to the CDRs of mAbs to various antigens have been shown
to mimic the recognition by the respective mAb [2]. We have recently found that a
peptide corresponding to the CDRH3 of a mAb, specific for the V3 region of human
immunodeficiency virus-1 (HIV-1), neutralizes the virus in vitro [3].

The HBcAg has been immunologically well characterized, but the main
recognition site for human HBc specific antibodies which is known to be
discontinuous [4] has not been identified. The HBc specific mAbs C1-5 and 35/312
recognize neighbouring linear epitopes [5, 6] and inhibit the binding of human
antibodies to HBcAg (anti-HBc). The amino acid sequence of the V region of mAb
C1-5 of the heavy and light chains has recently been revealed [7]. We were therefore
interested to analyze whether synthetic peptides, corresponding to the CDRs of the
mAb C1-5 HBc specific for HBcAg, were able to bind HBcAg, thereby allowing a
detailed characterization of the paratope.

Results and Discussion

The C1-5 mAb was raised against recombinant HBcAg, and was found reactive
to a peptide covering residues 74 to 90 of HBcAg (Figure 1a). However, the binding
of C1-5 to peptide 74-90 could only be inhibited by particulate HBcAg, suggesting a
discontinuous determinant as the main target (Figure 1b). The HBc mAb 35/312
(Behringwerke AG) recognizes a linear epitope at residues 77 to 83 of the HBc
sequence [5, 6], and HBc mAb 7/275 (Behringwerke AG) recognizes a yet unknown
determinant within the HBc sequence [5, 6]. Synthetic peptides were synthesized
according to a method for multiple peptide synthesis using Fmoc protected amino
acid esters [8]. Peptides were synthesized corresponding to the six CDRs of the

heavy and light chains of mAb C1-5 [7], three cyclized CDRH2 peptides, and a set of substitution peptide analogues of CDRH2 where each amino acid within the CDRH2 sequence was sequentially substituted by Ala or Gly.

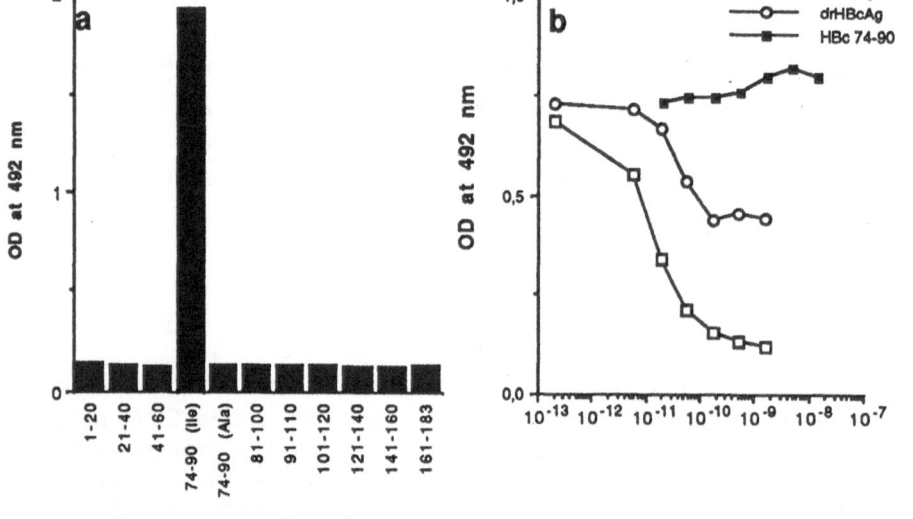

Fig. 1. *Mapping of the recognition site for mAb C1-5 using synthetic peptides covering the HBcAg sequence (a). Fig. 1b shows inhibition of mAb C1-5 binding to peptide HBc 74-90, by peptide HBc 74-90, by denaturated recombinant HBcAg (drHBcAg), and by rHBcAg, added in solution.*

All peptides were analyzed by reverse phase HPLC using a Pep-S 5μ column (Pharmacia, Uppsala, Sweden), run with a gradient from 0% to 60% CH$_3$CN against water containing 0.1% trifluoroacetic acid. If the purity was less than 80%, peptides were purified by the same procedure using a slower gradient. Three peptides, in which two residues were substituted by Cys, were cyclized on the resin using oxidation by iodine for 24 hours at room temperature. The Cys residues were placed on each side of the predicted β-turn between positions 52a and 54 of the CDRH2 sequence.

Peptides were coated to microtiter wells (Nunc 96F Certificated; Nunc, Roskilde, Denmark) at various concentrations in 0.05 M sodium carbonate buffer, pH 9.6, at +4°C overnight. Excess peptides were removed by washing with PBS containing 0.05% Tween 20. The CDR peptide-coated plates were assayed for binding using rHBcAg, subtype ayw with an Ile at position 80. The rHBcAg was diluted from 2 μg/ml to 0.156μg/ml, added in 100 μl portions, and incubated with the peptides for 60 minutes at +37°C. Excess rHBcAg was removed by washing, and bound rHBcAg was indicated by horseradish peroxidase labelled mAb to HBcAg (Clone 231; Behringwerke AG). In general, each reaction differing more than seven SD from the mean of the negative control was regarded as reactive. In the inhibition experiments, the cut off was set so that a reaction giving <50% of the non-inhibited reaction was regarded as a significant inhibiting concentration (IC_{50}).

The CDRH2 peptide of mAb C1-5 was the only peptide capable of capturing recombinant HBV core antigen (Figure 2). Two μg or more of this CDR peptide significantly bound rHBcAg. The CDRH2 binding of HBcAg was dependent on the amount of added HBcAg (data not shown). None of the other heavy chain CDR peptides or the light chain CDR peptides showed any reproducible HBcAg binding.

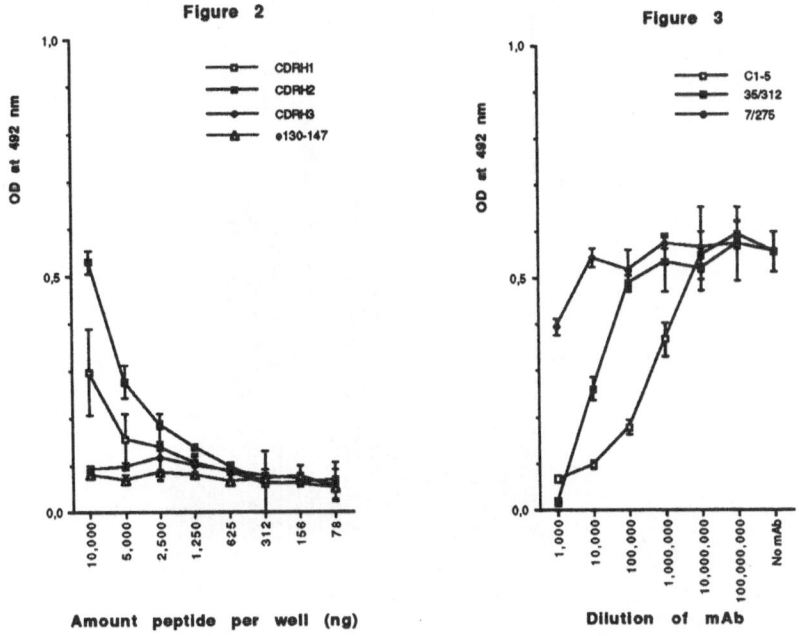

Figure 2   **Figure 3**

Amount peptide per well (ng)   Dilution of mAb

Fig. 2. *Testing of the CDRH peptides immobilized to microplate wells for ability to capture rHBcAg (100ng/well) using the EIA procedure. The peptide e130-147 corresponds to residues 130-147 of HBc/eAg and served as negative control.*

Fig. 3. *Ability of HBc mAbs C1-5, 35/312, and 7/275 to inhibit the binding of rHBcAg (100ng/well) to CDRH2 peptide coated to the wells of microtiter plates.*

The binding of rHBcAg to CDRH2 could be inhibited by addition of the mAbs C1-5 and 35/312 (Figure 3). MAb 7/275 showed no significant inhibition of the rHBcAg-CDRH2 binding (Figure 3). All three CDRH2 peptides with an artificial disulphide bridge showed a more efficient binding to rHBcAg than the linear peptide, indicated by the fact that 8-52 times more mAb was necessary to displace the cyclized peptide binding to rHBcAg. None of the CDRH2 linear or cyclized peptide analogues gave reproducible inhibition patterns when tested in solution. The assaying of the substitution analogues coated to microtiter plates (3μg/well) is given in Figure 4. As shown, the residues most essential for binding of CDRH2 to HBcAg are Val[51], Ser[52a], Ser[52c], Phe[53], and Gly[65]. Thus, the main paratope of CDRH2 would have the sequence Val[51] -x- Ser[52a] -x- Ser[52c] - Phe[53]. Also shown is that substitution of residues 55, 58, 61, and 62 by Ala increases the binding, indicating that a reduction in the number of charged residues in the carboxy terminal half of the CDRH2 peptide enhances the reactivity of the amino terminal paratope. The same enhancement occurs when substituting residues 56 and 60 by Gly.

In conclusion, we believe that this HBcAg and our HIV-1 based mini antibody systems provide a tool for epitope-paratope interaction studies at the single amino acid level, and can give further information about the binding mechanisms of antibodies.

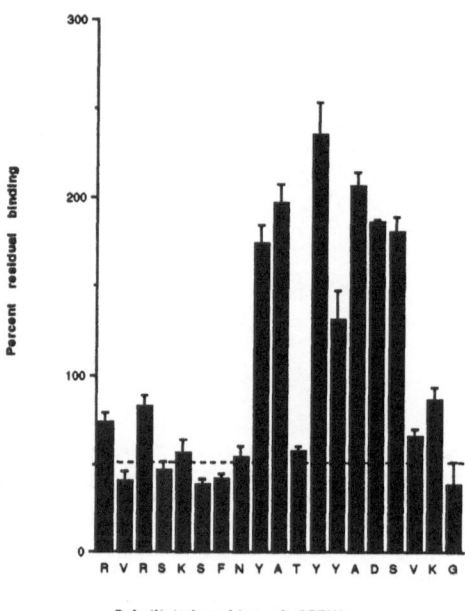

Fig. 4. Assaying of ability of the substitution peptide analogues of CDRH2 to bind rHBcAg. Each residue was sequentially substituted by Ala, except the two original Ala residues, which were substituted by Gly. Values are given as the percent residual binding of each substitution peptide analogue as compared to the peptide with the original sequence (dotted line).

References

1. Amit, A.G., Maruzzia, R.A., Phillips, S.E.V. and Poljak, R.J., Science, 233 (1986) 747.
2. Williams, W.V., Guy, R., Rubin, D.H., Robey, F., Myers, J.N., Kieber-Emmons, T., Weiner, D.B. and Greene, M.I., Proc. Natl. Acad. Sci. U.S.A., 85 (1988) 6488.
3. Levi, M., Sällberg, M., Rudén, U., Herlyn, D., Maruyama, H., Wigzell, H., Marks, J. and Wahren, B., Proc. Natl. Acad. Sci. U.S.A., 90 (1993) 4374.
4. Ferns, R.B. and Tedder, R.S., J. Med. Virol., 19 (1986) 193.
5. Salfeld, J., Pfaff, E., Noah, M. and Schaller, H., J. Virol., 63 (1989) 798.
6. Sällberg, M., Rudén, U., Magnius, L.O., Harthus, H.P., Noah, M. and Wahren, B., J. Med. Virol., 33 (1991) 248.
7. Skrivelis, V., Steinberg, Y., Bichko, V., Gren, E. and Tsimanis, A., Scand. J. Immunol., 37 (1993) 637.
8. Sällberg, M., Rudén, U., Magnius, L.O., Norrby, E. and Wahren, B., Immunol. Lett., 30 (1991) 59.

Conformational constraints of neutralizing epitopes from a major antigenic area of the human respiratory syncytial virus fusion (F) glycoprotein

D. Andreu[1], J.A. López[2], C. Carreño[1], G. Taylor[3] and J.A. Melero[2]

[1]Department of Organic Chemistry, University of Barcelona, E-08028 Barcelona, Spain, [2]Department of Molecular Biology, CNMVIS, Majadahonda, E-28220 Madrid, Spain, and [3]AFRC Institute for Animal Health, Compton, Nr. Newbury, Berks RG16 ONN, U.K.

Introduction

A major conserved antigenic area has been recently located between residues 260-275 of human respiratory syncytial virus (RSV) F glycoprotein [1]. Mutants selected with neutralizing mAbs had amino acid changes at residues 262, 268 and 272. While the vicinity of these amino acids and the reactivity of the mAbs with the F_1 subunit in western blots could suggest that the antibodies recognized sequential epitopes, most epitopes on this area were not reproduced by short (20-residue) synthetic peptides. In addition, amino acid changes at positions distant from 260-275 significantly affected mAb binding.

Results and Discussion

In order to explore the conformational requirements of the F_1 epitopes, a series of overlapping peptides of increasing length and incorporating the essential 265-275 region were synthesized (Figure 1). These peptides were used as ELISA antigens against mAbs and polyclonal (rabbit hyperimmune and human convalescent) sera. The largest peptide, F215-275, was reactive with most mAbs and rabbit antisera. The smaller (41-residue) peptide F235-275 was significantly less reactive and F255-275 was recognized only by a single mAb and very few sera. Peptides F215-234, F215-254 or F235-254 did not react at all. Thus, most mAb and polyclonal antibody epitopes were only reproduced in 41- and 61-residue peptides that spanned the 255-275 region of F_1. Some of these epitopes were only present in F215-275. It became clear, considering the length of this peptide, that structural features other than primary structure played a role in antibody recognition.

Previous work had shown [1, 3] that this antigenic area is highly resistant to trypsin digestion, a fact suggestive of a particular conformation. The susceptibility of the above peptides to digestion with increasing amounts of agarose-bound trypsin was examined by HPLC and evaluated as an indication of their capacity to reproduce the conformation adopted by the homologous sequences in native F protein. Despite a high content of trypsin-cleavage sites (Figure 1), F215-F275 was significantly more resistant to proteolysis than F235-275 and F255-275. The possible contribution of preferential conformations to the immunoreactivity of the peptides was then examined by CD in aqueous solution. The spectra of F235-275 and F255-275 were very similar and typical of aperiodic structures. In contrast, F215-275 showed minimal ellipticity values at 205 and 222 nm and a positive maximum at 190 nm, suggesting that it could adopt an ordered conformation in solution that was related to its increased trypsin resistance. Secondary structure prediction of the F protein favors the presence of two α-helices at the region included in peptide F215-275 (Figure 1). These aspects will be further investigated by means of ongoing NMR experiments. The conformational dependence of antibody binding was also confirmed by the fact that treatment of F215-275 with SDS previous to ELISA led to loss of several epitopes. Several other epitopes were also lost upon incorporation of F215-275 to KLH, presumably affecting peptide conformation, providing further evidence for discontinuous epitopes in the F_1 region.

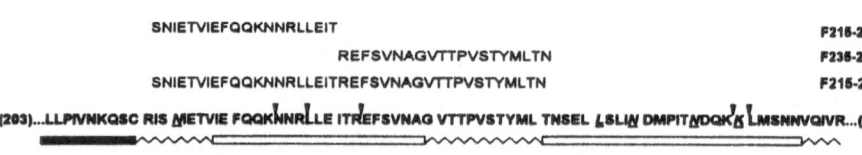

SNIETVIEFQQKNNRLLEIT	F215-234
REFSVNAGVTTPVSTYMLTN	F235-254
SNIETVIEFQQKNNRLLEITREFSVNAGVTTPVSTYMLTN	F215-254

(203)...LLPIVNKQSC RIS METVIE FQQKNNRLLE ITREFSVNAG VTTPVSTYML TNSEL LSLIN DMPITNDQKK LMSNNVQIVR...(282)

SNIETVIEFQQKNNRLLEITREFSVNAGVTTPVSTYMLTNSELLSLINDMPITNDQKKLMS	F215-275
REFSVNAGVTTPVSTYMLTNSELLSLINDMPITNDQKKLMS	F235-275
SELLSLINDMPITNDQKKLMS	F255-275

Fig. 1. Amino acid sequence (residues 203-282) of the F_1 subunit from Long RSV (boldface). Changes in mAb escape mutants [1] are shown in underlined italics. Predicted secondary structure: (open bar) α-helix, (wavy line) β-sheet, (solid bar) random coil. Trypsin susceptible sites are shown by inverted filled triangles. Synthetic peptides containing the 255-275 sequence are shown at the bottom. Other synthetic peptides are shown above the sequence.

Mice immunized with peptide F215-275 developed high titers of antibody to peptides F215-275, F235-275, F255-275, F215-234 and F215-254, three weeks after the last immunization. In contrast, none of the mice immunized with F235-275 or F255-275 developed anti-peptide antibodies, even after three boosts. Despite the high anti-peptide response induced by F215-275, mouse sera reacted poorly with RSV. Furthermore, there was no evidence of protection against infection in mice immunized with any of the peptides, five days after challenge with RSV. It is worth mentioning that the anti-peptide titer was observed after the first inoculation of F215-275, whereas anti-RSV antibodies were only detected after the fourth inoculation of the peptide.

References

1. Arbiza, J., Taylor, G., López, J.A., Furze, J., Wyld, S., Whyte, P., Stott, J., Wertz, G., Sullender, W., Trudel, M. and Melero, J.A., J. Gen. Virol., 73 (1992) 2225.
2. García-Barreno, B., Palomo, C., Peñas, C., Delgado, T., Pérez, P. and Melero, J.A., J. Virol., 63 (1989) 925.
3. López, J.A., Peñas, C., García-Barreno, B., Melero, J.A. and Portela, A., J. Virol., 64 (1990) 927.

Immunological relevance of the RGDL sequence of foot-and-mouth disease virus

I.S. Novella[1], B. Borrego[2], M.G. Mateu[2], E. Domingo[2], E. Giralt[1] and D. Andreu[1]

[1]Department of Organic Chemistry, University of Barcelona, E-08028 Barcelona, Spain, and [2]Center for Molecular Biology, CSIC-UAM, E-28049 Madrid, Spain

Introduction

The main antigenic site A of FMDV is a flexible loop on VP1 exposed on the virus surface. B-cell epitopes on site A (positions 138-150 in isolate C-S8c1) map at residues surrounding a conserved RGD tripeptide (positions 141-143) [1-3] thought to be a cell-attachment site [4]. Serotype C, most O and some type A viruses also show a conserved Leu at position 144. Frequent amino acid substitutions occur within site A of field and laboratory FMDV variants, including antibody-escape mutants. The effect of such replacements on the reactivity of FMDV with MAbs has been faithfully reproduced by synthetic peptides including the relevant substitutions [2, 5]. The extremely variable positions 138-140 and 146-150 seem to attract immune responses which might be detoured from RGDL. The practical absence of mutants in this RGDL region is a significant hurdle to assess its contribution to the immunogenicity and antigenicity of site A. We have approached this problem by evaluating the effect of changes at each of the RGD tripeptide positions on the interaction of synthetic peptides with MAbs. The immune response of rabbits and guinea pigs to synthetic peptides composed of RGDL tandem repeats has also been examined.

Results and Discussion

To explore the possible contribution of the RGD tripeptide to the interaction of site A with antibodies we replaced each of the three residues by Ala (Table 1). Binding of the substituted peptides with MAbs mapping at continuous epitopes within site A [3] was measured by competition ELISA. Tolerance to Ala replacement was greatest for Arg-141 and lowest for Asp-143 (Table 2). Amino acids other than Ala, particularly those resulting from point RNA mutations and leading to gross side chain and/or charge modification, were also tested, showing larger alterations in MAb recognition. In particular, Glu at position 142 led to loss of all epitopes probed, an alteration as drastic as that caused by His-146 → Arg [3] (peptide HR, Tables 1 and 2). Thus, each residue in RGD has an important contribution to the recognition of site A by antibodies.

Since the RGD sequence is involved in the recognition of antigenic site A by antibodies, we considered the possibility that synthetic peptides based on such a

Table 1 *Synthetic peptides related to the RGD sequence of the FMDV loop*

Peptide[a]	Amino acid sequence[b]	Peptide/carrier ratio[c]	
		KLH	BSA
RGD	ASAR**GD**LAHLTTTHARHLP	-	-
AGD	ASA**A**GDLAHLTTTHARHLP	-	-
RAD	ASAR**A**DLAHLTTTHARHLP	-	-
RGA	ASARG**A**LAHLTTTHARHLP	-	-
KGD	ASA**K**GDLAHLTTTHARHLP	-	-
SGD	ASA**S**GDLAHLTTTHARHLP	-	-
RED	ASAR**E**DLAHLTTTHARHLP	-	-
RGK	ASARG**K**LAHLTTTHARHLP	-	-
RGE	ASARG**E**LAHLTTTHARHLP	-	-
HR[d]	ASARGDLA**R**LTTTHARHLP	-	-
RGDL₃AXA	(C)**ARGDLAXARGDLAXARGDLA**	2775	18
RGDL₃AA	(C)V**ARGDLA..ARGDLA..ARGDLA**	2413	15
RGDL₃A	(C)**RGDLA....RGDLA.... RGDL**	3003	15
RGDL₃X	(C)**RGDLX.... RGDLX.....RGDL**	1130	15
RGDL₃	(C)**RGDL........RGD .... RGDL**	3504	16

[a]Prepared as described in [5]. All peptides gave correct MW values (± 2 Da) by PDMS.
[b]Underlined: changes from original sequence of site A (peptide RGD). X= Ahx.
[c]Conjugation through additional Cys residue (C) via MBS. Molar peptide/carrier ratios determined by AAA.
[d]Reproducing the His-146 → Arg mutation; see [3].

Table 2 *Antigen-competition ELISA with RGD-substituted peptides*

Peptide	Monoclonal antibody[a]						
	SD6	4C4	6D11	7FC12	7AH11	7JD1	7CA11
RGD	+	+	+	+	+	+	+
AGD	+	+	-	+	+	+	+
RAD	+	+	+	-	-	-	+
RGA	-	-	-	-	-	+	+
SGD	+	-	-	-	-	-	+
KGD	-	-	-	-	-	±	+
RED	-	-	-	-	-	-	-
RGK	-	-	-	-	-	+	+
RGE	-	-	-	-	-	+	+
HR	-	-	-	-	-	-	-

[a]Monoclonal antibodies have been described [3, 6]. +, -, and ± indicate, respectively, more than, less than, and close to 30% inhibition of Mab binding by a 20-fold molar excess of the competing peptide against plate-bound RGD peptide.

sequence could induce neutralizing antibodies which may define a broader antigenic spectrum than peptides involving variable residues around RGD. We synthesized tandem repeats of the basic unit RGDL because Leu-144 is strictly conserved in all type C viruses sequenced to date and in nearly all type O isolates. The RGDL repeats were separated by different spacer residues as shown in Table 1. Anti-peptide antibody titers in rabbits and guinea pigs were generally high. All sera reacted with heterologous RGDL-containing antigens, though to a much lesser extent than with the homologous peptide, suggesting a low proportion of antibodies with RGDL as the sole target. As expected, the first boost improved anti-peptide antibody levels, though further inoculations had no effect. Anti-RGDL tandem peptide sera from rabbits and guinea-pigs were tested for their ability to neutralize C-S8c1 FMDV. Neutralization levels, at their highest, were 2 to 3 orders of magnitude lower than those attained with peptides reproducing the full site A sequence. Neutralizing activity in rabbits was higher for peptides including more amino acids of the consensus C-S8c1 sequence. Guinea pig sera gave better results when shorter peptides were used. Another difference between the two animal systems was the effect of including Ahx as spacer: in rabbits, no differences were found for peptides with or without Ahx, while in guinea pigs neutralizing activity was completely lost when this amino acid was present. In addition to C-S8c1, serotypes O_1K and A_5W were also tested for cross-neutralization by the anti-peptide sera. Again, low but significant neutralization was noted with several of the tested sera. It is also noteworthy that mutant HR [3] was neutralized at least as efficiently as its wild-type counterpart C-S8c1. These results point to a relaxation of the specificity of the immune response against FMDV when sequences responsible for serotype specificity are excluded from the peptide immunogens.

In most animals, neutralization values dropped significantly with the number of boosts. To study this phenomenon, IgGs and IgMs from sera after 1 and 3 inoculations were fractioned and assayed for their neutralizing ability. Significant neutralization activity was only found in the IgM fractions of sera from the first immunization. This suggested a possible suppression of those anti-peptide antibodies that may also be directed to cellular RGD sequences.

In conclusion, our results show that the RGD tripeptide is an important determinant of the interaction of antigenic site A with antibodies, and that a significant anti-peptide and FMDV-neutralizing response can be obtained with peptides based on RGD. However, a potent neutralizing anti-viral response necessitates the highly variable residues surrounding the RGD. Appropriate combinations of peptides representing most residues of site A, along with tandem repeats of the conserved residues around the RGD core may be useful in the formulation of anti-FMDV synthetic vaccines with broader antigenic spectra than formulations assayed so far.

Acknowledgements

Work supported by CICYT (PB91-0051-C02-01; PB91-0266).

References

1. Domingo, E., Mateu, M.G., Martínez, M.A., Dopazo, J., Moya, A. and Sobrino, F., In Kurstak, E., Marusyk, R.G., Murphy, S.A. and Van Regenmortel, M.H.V. (Eds.), Applied Virology Research, Vol. 2, Plenum Press, New York, NY, U.S.A., 1990, p.233.
2. Mateu, M.G., Martínez, M.A., Rocha, E., Andreu, D., Parejo, J., Giralt, E., Sobrino, F. and Domingo, E., Proc. Natl. Acad. Sci. U.S.A., 86 (1989) 5883.
3. Mateu, M.G., Martínez, M.A., Capucci, L., Andreu, D., Giralt, E., Sobrino, F., Brocchi, E. and Domingo, E., J. Gen. Virol., 71 (1990) 629.
4. Fox, G., Parry, N.R., Barnett, P.V., McGinn, B., Rowlands, D. and Brown, F., J. Gen. Virol., 70 (1989) 625.
5. Carreño, C., Roig, X., Cairó, J., Camarero, J., Mateu, M.G., Domingo, E., Giralt, E. and Andreu, D., Int. J. Peptide Protein Res., 39 (1992) 41.
6. Mateu, M.G., da Silva, J.L., Rocha, E., de Brum, D.L., Alonso, A., Enjuanes, L., Domingo, E. and Barahona, H., Virology, 167 (1988) 113.

The identification by chemical synthesis of a conformational epitope on merozoite surface protein 1 of *P. falciparum* malaria

J.C. Calvo[1], M. Perkins[2] and A.C. Satterthwait[1]

[1]Department of Molecular Biology, The Scripps Research Institute, La Jolla, CA 92037, U.S.A. and [2]Rockefeller University, New York, NY 10021-6399, U.S.A.

Introduction

Malaria is a parasitic disease where the parasite undergoes a complex life cycle in human and mosquito hosts. While the infective form of malaria is the sporozoite, it is the merozoite that proliferates in red blood cells and causes the symptoms of the disease that has become a recent focus of attention. In particular, the cysteine rich C-terminal fragment of merozoite surface protein 1 (MSP-1) has become a vaccine candidate since monoclonal and polyclonal antibodies to it can inhibit merozoite proliferation. However, epitopes within this fragment have gone unidentified since antisera to the native protein show no reactions with the reduced fragment indicating that all epitope(s) are conformational in character [1]. We have taken advantage of a new method for folding peptides [2] to identify a conformational epitope on the C-terminal domain of MSP-1.

Results and Discussion

Initially, peptides (20-mers) displaced every ten amino acids were synthesized for the two allelic forms of the C-terminal domain of *P. falciparum* MSP-1 [3] and coupled to maleimide-activated BSA for screening rabbit antiserum R294. R294 was raised against MSP-1 purified from the FCR3 strain of *P. falciparum*. IFA titers of R294 against merozoites are low (1/100). No reaction of R294 was observed with the 20-mers in ELISA. These results indicated that if polyclonal antibodies were formed against the C-terminal domain, they were formed against conformational epitopes.

Since the cysteine pattern in the MSP-1 C-terminal domain indicated two tandem epidermal growth factor (EGF)-like domains and since NMR structures for EGF-like proteins are similar, cyclic MSP-1 peptides corresponding to the solvent exposed B-loop β-hairpin of a prototypical EGF-like domain were synthesized by replacing predicted main chain hydrogen bonds with a hydrazone hydrogen bond covalent mimic on the solid support [2].

R294 shows a positive reaction with cyclic peptides corresponding to the β-loop of the EGF-2 domain (Figure 1), no reaction with the corresponding linear peptide, probable reaction with the cyclic peptide corresponding to the EGF-1 β-loop

...

and no reaction with a control loop or preimmune serum. Reaction of the EGF-2 loop with R294 has been confirmed in competition ELISA.

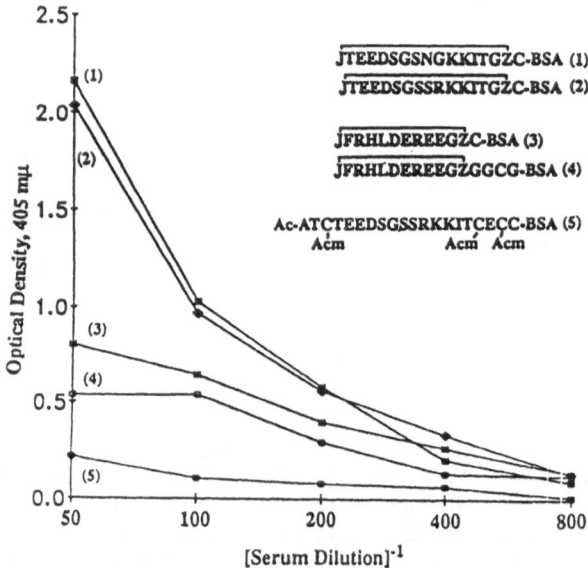

Fig. 1. Titers of R294 serum against modeled EGF-2 B-loops from K-1 (1) and FC-27 (2) P. falciparum strains and the EGF-1 B-loop (3,4) and linear (5) MSP-1 peptides.

The important observation is that large improvements in antigenicity can be achieved by folding peptides with a covalent hydrogen bond mimic. A related example is also reported [4]. This provides a method for identifying epitopes that are missed in peptide screens. Although conformational epitopes are often considered to be discontinuous, these results demonstrate that continuous epitopes or an epitope residing within a short amino acid sequence can be conformational. More are anticipated and one can now explore this phenomenon further with folded peptides.

Acknowledgements

This research was supported by the Agency for International Development Contract No. DPE-5979-A-1035-00. This is publication 8128-MB from The Scripps Research Institute.

References

1. Chang, S.P., Gibson, H.L., Chun, T.L., Barr, P.J. and Hui, G.S.N., J. Immunol., 149 (1992) 548.
2. Chiang, L.-C., Cabezas, E., Calvo, J.C. and Satterthwait, A.C., this volume.
3. Daly, T.M., Burns, J.M. and Long, C.A., Mol. Biochem. Parasitol., 52 (1992) 279.
4. Stura, E., Kaslow, D. and Satterthwait, A.C., this volume.

Efficient continuous-flow SPPS of immunogenic lipophilic MAPS: Application of orthogonal Dde amino-protection

W.C. Chan[1], B.W. Bycroft[1], D.J. Evans[1] and P.D. White[2]

[1]Department of Pharmaceutical Sciences, University of Nottingham, University Park, Nottingham NG7 2RD and [2]Calbiochem Novabiochem Ltd., Highfields Science Park, Nottingham NG7 2QJ, England

Introduction

In recent years, the realization that both the lipopeptide adjuvant Pam_3Cys and multimeric presentation of peptide antigenic determinants can significantly enhance the immunogenicity of such epitopes [1] has prompted considerable interest in synthetic vaccines. Tam and co-workers [2] recently showed that the multimeric antigenic constructs with built-in lipo-adjuvant, termed lipophilic MAPS, appear to be obvious candidates as efficacious vaccines. However, the total chemical synthesis of lipophilic MAPS presents considerable problems, particularly the lack of versatility of current synthetic procedures. Present SPPS protocols invariably employ the expensive and preformed reagent Fmoc-Lys(Pam_3Cys)-OH. We describe herein an efficient SPPS protocol which exploits the orthogonality of our recently reported novel amino-protecting group, N-(1-(4,4-dimethyl-2,6-dioxocyclo-hexylidene)ethyl) (Dde) [3] within the Sheppard Fmoc/continuous-flow methodology [4].

Results and Discussion

Automated SPPS was carried out on either PepSynthesizer 9050 or NovaSyn Crystal, and acylations were accomplished with Fmoc-X_{aa}(R)-OH : HOBt : HBTU : DIPEA in 4 times excess for 60 to 120 minutes. Both Pam_3Cys and Pam were incorporated manually by DIC : HOBt activation chemistry; we encountered considerable problems with uronium or phosphonium (e.g. HBTU, PyBOP) activation chemistries due to DMF insolubility of the palmitoyl-uronium and -phosphonium intermediates. Following completion of synthetic assembly, all peptidic constructs were cleaved/deprotected by TFA, 88.5 : thioanisole, 2.5 : EDT, 3.0 : TIS, 1.0 : H_2O, 5.0%. The synthesized lipophilic MAPS immunogens have the general schematic structure:

$$\Uparrow$$

Peptide $\Longleftarrow$ Lys ⌐   LipoX ⌐

Antigen $\Longleftarrow$ Lys — Lys — Ser — Ser — Lys – β-Ala - NH$_2$

$$\Downarrow \qquad \Delta \ \text{LipoX} = \text{H}, \text{Pam}_3\text{Cys}, \text{Pam}_2\text{Dap}$$

The highly efficient and flexible Fmoc/continuous-flow SPPS protocol which we have established first involves the synthesis of the resin bound Dde-Lys(Fmoc)–ßAla–. The Lys ε-amino group following deprotection by 20% piperidine/DMF was acylated by the activated Pam$_3$Cys. Nα-Dde was then removed by 2% hydrazine/DMF; UV monitoring at 270 (λ_{max}) or 290 nm for the Dde-hydrazine cyclic adduct 3,6,6-trimethyl-4-oxo-4,5,6,7-tetrahydro-1*H*-indazole indicates complete deprotection after 5 minutes. The lipopeptide was then elaborated to afford the resin-bound core lipopeptide Lys–Ser(OtBu)–Ser(OtBu)–Lys(Pam$_3$Cys)–ßAla–(**1**). This material when treated with TFA : TES : H$_2$O yielded the desired lipopeptide (LD-MS, MH$^+$ found 1411.3, requires 1410.8) in purity of over 95% by RP-HPLC. Alternatively, in a similar manner, the resin bound Dde-Lys–ßAla– was first elaborated to Dde-Lys(Pam$_2$Dap)–ßAla– (using Fmoc-Dap(Fmoc)-OH, m.p. 210-213°C, Dap = DL-diaminopropionic acid; and palmitic acid), and then secondly to Lys–Ser(OtBu)–Ser(OtBu)–Lys(Pam$_2$Dap)–ßAla– (**2**); TFA treatment again afforded the desired lipopeptide (LD-MS, MH$^+$ found 1081.4, requires 1081.6) in good purity. This facile route clearly demonstrates the versatility of the described SPPS protocol which will allow the *in situ* incorporation of any desired lipophilic moieties. Analogues of the resin-bound core lipopeptide have been synthesized utilizing Fmoc-Lys(Dde)-OH, Fmoc-Orn(Dde)-OH (m.p. 113-115°C) or Boc-Lys(Dde)-OH (m.p. 67-69°C).

Using **1** and **2**, and following acylation with Fmoc-Lys(Fmoc), the synthesis of lipophilic MAPS immunogens was exemplified by the simultaneous tetrameric construction of a dodecapeptide antigen ("RNSQFVALMPTA", herpes simplex virus ribonucleotide reductase 1-(959-970)). As a result of steric hindrance, double couplings were found to be required for certain amino acid residues. The assembled lipophilic immunogens were cleaved/deprotected from the resin, precipitated by diethyl ether, and then dialyzed for 24 hours. Analysis of the crude product by SDS-PAGE and amino acid analysis demonstrated the presence of the desired materials, which are at present undergoing comparative immunological evaluations. A further advantage of our described protocol is the ability to undertake total SPPS of di-epitopic lipophilic MAPS *via* e.g. the acylation of **1** with Fmoc-Lys(Dde). We believe that this will lead to the synthesis of very exciting and potentially efficacious vaccines for application in both animal and human therapies.

References

1. (a) Jung, G., Wiesmuller, K., Becker, G., Buhring, H. and Bessler, W., Angew. Chem. Int. Ed. Engl., 24 (1985) 872; (b) Tam, J.P., Proc. Natl. Acad. Sci. U.S.A., 85 (1988) 5409.
2. Defoort, J.P., Nardelli, B., Huang, W., Ho, D.D. and Tam, J.P., Proc. Natl. Acad. Sci. U.S.A., 89 (1992) 3879.
3. Bycroft, B.W., Chan, W.C., Chhabra, S.R. and Hone, N.D., J. Chem. Soc., Chem. Commun., (1993) 778.
4. Arshady, R., Atherton, E., Clive, D.L. and Sheppard, R.C., J. Chem. Soc., Perkin Trans I, (1981) 529.

Mapping of the immunodominant B- and T-cell epitopes of the outer membrane protein P6 of *H. influenzae* type b using overlapping synthetic peptides

P. Chong, O. James, Y.-P. Yang, C. Sia, B. Tripet, Y. Choi and M. Klein

*Connaught Centre for Biotechnology Research,
1755 Steeles Avenue West, Willowdale, Ontario M2R 3T4, Canada*

Introduction

Haemophilus influenzae type b (Hib) infection is a major cause of bacterial meningitis, epiglottitis, cellulitis and pneumonia in young children. Current *Haemophilus* capsular polysaccharide (polyribosyl ribitol phosphate, PRP) conjugate vaccines protect against Hib infection, but they do not protect against other invasive typeable and non-encapsulated strains which are a common cause of otitis media in children. Recent studies [1, 2] indicate that antibodies raised against the outer membrane protein P6 are protective in the infant rat model of bacteremia, and a P6-specific monoclonal antibody was shown to have bactericidal activity in vitro against both Hib and non-typeable strains. Therefore, the use of P6 or its immunodominant epitopes as both additional immunogens and carriers for PRP represents a promising strategy to develop new conjugate Hib vaccines with enhanced protective ability and T-cell priming capability. Thus, the purpose of this study was to map the antigenic determinants of P6 using overlapping synthetic peptides, and assess their immunogenicity for possible inclusion in a cross-protective synthetic *H. influenzae* vaccine.

Results and Discussion

To identify the B-cell epitopes of P6, rabbits, rats, guinea pigs and mice were immunized with chromatographically purified P6 emulsified in Freund's adjuvant. After two immunizations, all animals generated a strong P6-specific antibody response as judged by both ELISA and immunoblots. The synthetic P6 peptides were tested for their reactivities with the various anti-P6 antisera in peptide-specific ELISAs as previously described [3]. *Bordetella pertussis* peptides were used as negative controls. Two immunodominant linear B-cell epitopes were mapped to residues 73-96 and 109-134 of the mature P6 sequence.

Peptides containing T-cell epitopes were characterized by their ability to stimulate the proliferation of mouse (Balb/c and C57BL/6 strains) T-lymphocytes primed with P6 in a standard in vitro T-cell proliferation assay [4]. Four synthetic peptides corresponding to residues 19-41, 35-58, 73-96 and 109-134, respectively, were found to stimulate the proliferation of primed T-cells from Balb/c mice, whereas only peptide P6-4 (residues 54-77) induced a good proliferative response of P6-primed T-lymphocytes from C57BL/6 mice. These results indicate that peptides 73-96 and 109-134 contain both T- and B-cell epitopes. The identification of these immuno-dominant B- and T-cell epitopes represents a first step towards the rational design of a potentially cross-protective Hib vaccine.

Acknowledgements

We thank M. Flood, W. Williams, W. Xu-Li, and T. Olivier for their excellent technical assistance. This work was partially supported by IRAP grant# CA103-9-1423 from the National Research Council Canada.

References

1. Granoff, D.M. and Munson, Jr., R.S., J. Infect. Dis., 153 (1986) 448.
2. Murphy, T.F., Nelson, M.B., Dudas, K.C., Mylotte, J.M. and Apicella, M.A., J. Infect. Dis., 152 (1985) 1300
3. Chong, P., Sydor, M., Wu, E., Zobrist, G., Boux, H. and Klein, M., Mol. Immunol., 28 (1991) 239.
4. Sia, D.Y. and Chou, J.L., Scand. J. Immunol., 26 (1987) 683.

Immune responses of seven different "promiscuous" T-cell epitopes on chimeric peptide vaccine design

A.M. DiGeorge[1,4], B. Wang[1,4], S.F. Kobs-Conrad[2] and P.T.P. Kaumaya[1,2,3,4,5,6]

[1]The College of Medicine, [2]Comprehensive Cancer Center, [3]College of Biological Sciences, Departments of [4]Obstetrics and Gynecology, [5]Medical Biochemistry, [6]Microbiology, The Ohio State University, Columbus, OH 43210, U.S.A.

Introduction

In the design of synthetic vaccines "promiscuous" T-cell epitopes may provide a means to circumvent the problem of MHC restricted responses. We have reported the use of a "promiscuous" epitope (sequence 580-599, TT) [1] which bypassed strain differences associated with a B/T-cell epitope from the protein antigen lactate dehydrogenase C_4 (LDH-C_4) [2]. Recently, we also reported the use of a measles virus epitope (MVF, sequence 288-302) [3] and another tetanus toxoid epitope (TT$_2$, sequence 830-844) [4] in association with the same B-cell epitope of LDH-C_4 in producing high titered, high affinity antibodies. In this report, we have completed an in-depth study on these epitopes as well as additional "promiscuous" T-cell epitopes derived from (1) Human hepatitis B surface antigen, HBsAg (HBV, sequence 19-33) [5]; (2) P. vivax (CSP, sequence 317-336) [6]; and (3) tetanus toxoid (TT$_3$, sequence 947-967) [7]. The aim of this comparative study is to identify one or more optimal "promiscuous" T-cell epitopes for incorporation into a multivalent template that will not require an unwieldy number of peptides to be universally effective.

Results and Discussion

Peptides were designed and synthesized as previously reported [2, 4, 8] and were purified by reverse phase HPLC. The immune responses of the 7 chimeric constructs in several inbred strains of mice [C3H/HeJ, B10.BR, (H-2^k); BALB/c, B10.D2, (H-2^d); C57BL/6, C57BL/10, (H-2^b)] are summarized in Figure 1. In the responder strain C3H/HeJ (H-2^k), all the chimeric constructs elicited high titer antipeptide antibodies for the respective immunogenic sequence and the native protein antigen LDH-C_4. Haplotype restricted immune responses associated with the LDH epitope were bypassed in several strains of mice depending on the choice of the "promiscuous" T-cell epitope. Thus, the tetanus toxoid (TT) epitope was able to overcome the H-2^k restriction in H-2^b strains but was ineffective in H-2^d strains of mice. The tetanus toxoid (TT$_2$) epitope was able to provide T-cell help in 5 out of 6

Haplotypes	H-2b				H-2k				H-2d			
Strains	C57BL/6		C57BL/10		C3H/HeJ		B10.BR		B10.D2		BALB/c	
Immunogens	Peptide	Protein	Peptide	Protein	Peptide	Protein	Peptide	Protein	Peptide	Protein	Peptide	Protein
a1TT	++	++	+++	+++	+++	+++	+++	+++	+	+	--	--
aNTT	+++	+++	+++	+++	+++	+++	+++	+++	+	+	--	--
aNTT2	+++	+++	--	--	+++	+++	+++	+++	+++	+++	++	++
aNTT3	+++	+++	**	**	+++	+++	**	**	**	**	+++	++
aN-HBV	++	++	**	**	+++	+++	**	**	**	**	+++	+++
aN-MVF	++	++	+	+	+++	+++	+++	+++	+++	+++	+++	+++
MVF-aN	--	--	**	**	+++	+++	**	**	**	**	+++	+++
TT3-aN	+++	+++	**	**	+++	+++	**	**	**	**	+++	+++
CSP-aN	--	**	**	**	--	**	**	**	**	**	+++	**

— Negative   +++ High titers   ++ Medium titers   + Low titers   ** Not tested

Fig. 1. Immune responsiveness of various chimeric "promiscuous" constructs in inbred mice strains. Results are based on direct ELISA titers to native protein (determined as the dilution giving an absorbance of 0.2 above the blank) and are a compilation of pooled individual sera at the secondary + 1 week bleed. Responses are tabulated according to titers: high (>30,000), medium (approx. 4000-10,000), and low titers (<400).

strains tested while TT_3 was consistently effective in providing help to the B-cell epitope in all strains tested. In the case of measles virus (MVF) H-2^d strains of mice produced high titer protein-reactive antibodies and low titers was evidenced in H-2^b strains. In rabbits high antibody titers (>32,000) for native LDH-C_4 were obtained in the secondary response for 6 out of 8 immunogens tested and moderate titers (5000-15000) for two immunogens ($\alpha_1 TT$ and $\alpha_N TT$). In conclusion, these results provide several leads in defining the necessary T-cell epitopes (TT_3, TT_2, MVF) that may represent the best method for ensuring adequate T-cell stimulating in an outbred population with a heterogeneous MHC make up.

Acknowledgements

This work was supported by an NIH grant AI25790 to PTPK.

References

1. Ho, P.C., Mutch, D.A., Winkel, K.D., Saul, A.J., Jones, G.L., Soran, T.J. and Rzepczyk, C.M., Eur. J. Immunol., 20 (1990) 477.
2. Kaumaya, P.T.P., Feng, N., Kobs-Conrad, S., VanBuskirk, A.M. and Sheridan, J.F., In Smith, J.A. and Rivier, J. (Eds.), Peptides: Chemistry and Biology, Escom, Leiden, The Netherlands, 1992, p.883.
3. Partidos, C.D. and Steward, M.D., J. Gen. Virol., 71 (1990) 2099.
4. Seo, Y.H., Kobs-Conrad, S. and Kaumaya, P.T.P., In Schneider, C.H. and Eberle, A.N. (Eds.), Peptides 1992, Escom, Leiden, The Netherlands, 1993, p.139.
5. Schad, V.C., Garman, R.D. and Greenstein, J.L., Sem. in Immunol., 3 (1991) 217.
6. Fern, J. and Good, M.F., J. Immunol., 148 (1992) 907.
7. Panina-Bordignon, P., Tan, A., Termijtelen, A., Demotz, S., Corradin, G. and Lanzavecchia, A., Eur. J. Immunol., 19 (1989) 2237.
8. Kaumaya, P.T.P., Seo, Y.H., Kobs, S., Feng, N., Sheridan, J. and Stevens, V., J. Mol. Recog., 6 (1993) 81.

Peptides as inhibitors of enveloped virus formation

J. Gorka[1], A. Loewy[2], N.C. Collier[2], J. Mo[2] and M.J. Schlesinger[2]

[1]Howard Hughes Medical Research Institute and [2]Department of Molecular Microbiology, Washington University School of Medicine, St. Louis, MO 63110, U.S.A.

Introduction

We have initiated a program to determine if short peptides with sequences corresponding to those domains of virus-encoded polypeptides that participate in virus specific protein-protein interactions can act as antiviral agents for control of viral infections. In this brief report we describe results obtained from a set of peptides that correspond in sequence to a part of the influenza virus hemagglutinin protein (HA) that is postulated to participate in the final stages of assembly and release of this virus from infected cells.

Influenza virus HA is a transmembranal glycoprotein that forms a trimeric "spike" on the surface of the influenza virus particle and functions at three critical stages in the infectious cycle of this virus: (1) it binds to sialic acid residues on host cells, (2) it participates as an essential component in a membrane fusion that brings viral genomes into the host cell cytoplasm, and (3) its carboxy terminal 10 amino acids is a specific attachment site for virus components during assembly and release of virus from cells. It is this latter activity that we sought to inhibit with peptides.

Results and Discussion

Initial studies with a 10-amino acid peptide that precisely matched the carboxy terminus of the WSN strain of HA blocked release of virus particles by >90% during a 2 h period at levels of 50-200 µM [1]. At these levels, there was <20% inhibition of particle release by two other unrelated enveloped viruses, Sindbis and vesicular stomatitis. Based on these results, a series of peptides were synthesized and tested in order to establish the minimum size and the essential components of the sequence.

The syntheses were performed utilizing t-Boc chemistry on an Applied Biosystems Model 403A peptide synthesizer. Peptides were purified by reversed-phase HPLC and subjected to purity assessment techniques previously described [2]. They were lyophilized and stored at 4°C. For application to virus-infected cells, peptides were weighed and dissolved in deionized H_2O at 10 mg/ml. They were added to the media of virus-infected cells at levels ranging from 50 to 200 µg/ml. The titer of infectious WSN influenza was determined immediately after addition of the peptide and after 2 h of virus replication by plaque titration on confluent cultures of Madin-Darby bovine kidney cells. Typically there were about 10^4 PFU/ml at the start of the inhibitory procedure and this level rose to about 10^6 in the 2 h period of infection. An inhibitory peptide blocked this increase by >90%, whereas a change <50% from the untreated sample was considered to be a non-inhibitory peptide.

The results obtained with 16 peptides are summarized in Table. 1.

Table 1   *Effect of peptide composition on inhibition of influenza and Sindbis viruses[a]*

Peptide Composition	Influenza Virus	Sindbis Virus
N.G.S.L.Q.C.R.I.C.I-NH$_2$	+	-
N.G.S.L.Q.C.R.I.C.-NH$_2$	-	-
G.S.L.Q.C.R.I.C.I - "	+	-
S.L.Q.C.R.I.C.I.- "	+	-
L.Q.C.R.C.I. - "	+	+/-
L.Q.C.R.-C.I. - "	-	-
L.Q.C.R.I.C.I.- OH	+	+/-
Q.C.R.I.C.I "	-	+/-
Q.C.R.I.C.I- NH$_2$	+	+
Ac-Q.C.R.I.C.I.- "	+	+/-
T.C.R.I.C.I-. "	+	+
G.C.R.I.C.I- "	+	+
Y.C.R.I.C.I- "	+	+
I.R.C.N.I.C.I- "[b]	+	-
M.R.C.T.I.C.I- "[c]	+	+/-
I.C.I.R.C.Q.L.[d]	-	-

[a]Results tabulated were obtained with 200 μg/ml peptide; (+) are those with >90% inhibition; (+/-) are those with ~70% inhibition; (-) are those with <50% inhibition.
[b,c]Sequence of carboxy termini of influenza serotypes HA3 and HA7, respectively.
[d]Retro-inverted peptide; all D-amino acids utilized.

Earlier data indicated that substitution of the two cysteines by glycine destroyed inhibitory activity [3]. Removal of the carboxy-terminal isoleucine as well as the internal isoleucine also blocked inhibition (Table 1). In contrast, the amino terminus could be truncated to a 6-amino acid peptide that retained antiviral activity. However, unlike the larger peptides, the 6-mer inhibited the release of infectious Sindbis virus particles.

We had postulated that the specific inhibition by the 10-mer on influenza virus assembly occurred by the peptide competing with the influenza virus HA cytoplasmic domain for binding sites on the nucleoprotein core or matrix protein of influenza. The molecular basis for inhibition by the smaller, less specific peptides is still unclear.

References

1. Collier, N.C., Knox, K. and Schlesinger, M.J., Virology, 183 (1991) 769.
2. Gorka, J., McCourt, D.W. and Schwartz, B.D., Pept. Res., 2 (1989) 376.
3. Schlesinger, M.J., Collier, N.C., Loewy, A. and Gorka, J., Arch. Virol., in press.

Synthesis and evaluation of novel lipopeptide constructs for *in vivo* induction of virus-specific cytotoxic T-lymphocyte mediated response

B. Déprez[1], H. Gras-Masse[1], F. Martinon[2], E. Gomard[2], J.P. Lévy[2] and A. Tartar[1]

[1]Université Lille II, URA CNRS 1309, Institut Pasteur de Lille, 59019 Lille, France, and [2]Immunologie et Oncologie des maladies rétrovirales, INSERM U152 Institut Cochin de Génétique Moléculaire, 75014 Paris, France

Introduction

Generation of CTL is a critical component of the immune response to intracellular pathogens. Recently, the possibility of inducing *in vivo* a virus-specific cytotoxic T-cell response using a cytotoxic T-cell epitope from the nucleoprotein of the influenza virus associated to a lipopeptidic adjuvant P3CSS represented a decisive advance in the field of synthetic vaccines: thus, chemically defined, low-molecular weight synthetic antigens offered an interesting alternative to the use of replicating viruses in vaccination, and appeared as a possible new approach to the induction of cell-mediated immunity. Recently, we have shown that an early detectable and virus specific CTL-mediated response could be induced by using a hexadecapeptide (V3 312-327 = IRIQRGPGRAFVTIGK from the *env* protein of HIV-1 LAV/BRU) modified in the C-terminal position by an amino acid possessing a saturated 14 carbon aliphatic side chain, administered in water medium, without adjuvant [1].

Here we describe the synthesis of several other covalent modifications of the same peptide by simple lipidic amino acids (Figure 1), and their evaluation in terms of ability to induce a virus-specific CTL response.

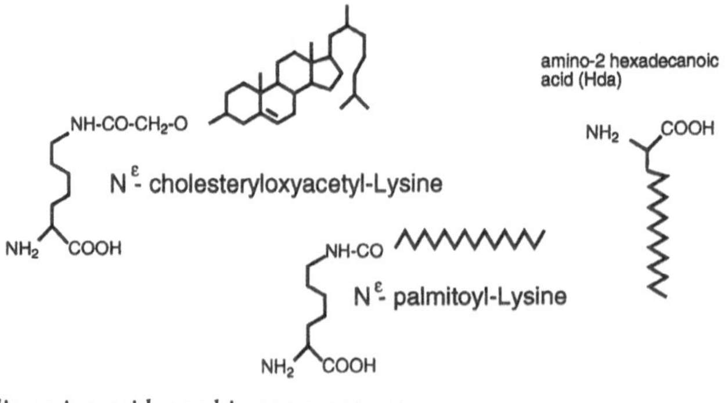

Fig. 1. *Lipidic amino acids used in our constructs.*

Results and Discussion

P3CSS was a kind gift provided by G. Jung P3CSS- and cholesteryl-modified analogs were synthesized using the Fmoc-tBu chemistry. All other peptides were synthesized using the "Boc-benzyl strategy" in an ABI 430A peptide synthesizer. Peptides possessing a carboxylic C-terminal end were synthesized on N-t-Boc-Gly Pam resin, while peptides possessing a carboxamide C-terminal end were synthesized on MBHA resin.

Early detectable, MHC class I-restricted CTL, capable of specifically lysing endogeneously expressing antigen target cells (i.e., P815{H-2^d, DBA/2}, infected by a vaccinia virus recombinant for the *env* gene (Vac-*env*)), can be isolated from BALB/c (H-2^d) mice immunized with Vac-*env*. We evaluated the efficiency of *in vivo* priming with the different constructs according to two criteria: a short delay before detection of cytolytic activity (before the 18th day of culture of the splenocytes, i.e., after only two *in vitro* stimulations), and the ability of the CTL to lyse specifically virus-infected target cells, as previously described [1]. Results of CTL induction obtained with the different constructs, without adjuvant, are summarized in Table 1.

Table 1 *Results of CTL induction obtained with different constructs*

N-term / Cterm		positif/total
H-	-OH	0/5
P$_3$CSS-	-OH	2/2
H-	-Hda-NH$_2$	16/17
H-Hda-	-Hda-NH$_2$	1/3
H-Hda-	-OH	6/10
(N^{a,e}-dipalmitoyl)Lys-	-Hda-NH$_2$	0/3
H-	-(N^{e}palmitoyl)Lys-NH$_2$	7/9
H-	-(N^e-cholesteryloxyacetyl)Lys-	2/2
(N^a-Ac, (N^e-cholesteryloxyacetyl)Lys-	-OH	2/2

Modification by the P3CSS lipopeptide induces an early detectable and virus-specific CTL response. We also show that an efficient CTL response can be induced by using as immunogens the same peptide modified in the C- or N-terminal position by a lipidic amino acid possessing no intrinsic immunoadjuvant properties. C-terminal modification appears as a relatively constant means to obtain efficient immunogens (in terms of CTL induction), while N-terminal modification appears more problematic. Introduction of a second lipidic amino acid may impair the immunogenicity of a relatively short (16 residues) peptide.

Reference

1. Martinon, F., Gras-Masse, H., Boutillon, C., Chirat, F., Deprez, B., Guillet, J.G., Gomard, E., Tartar, A. and Levy, J.P., J. Immunol., 149 (1992) 3416.

Heterologous expression of the adherence domain of *Pseudomonas aeruginosa* PAK pilin as carboxy- and amino-terminal fusion proteins in *Escherichia coli*

H. Hahn, P.M. Lane-Bell and W. Paranchych

Department of Microbiology, University of Alberta, Edmonton, Alberta, T6G 2E9, Canada

Introduction

Pseudomonas aeruginosa (PA) is a primary pulmonary pathogen in patients with cystic fibrosis and other immunosuppressive illnesses [1]. Initial infection involves pili-mediated bacterial attachment to epithelial cells. The adherence domain of PA pili has been shown to reside in a disulfide loop (DSL) region, 12 to 17 amino acids in length, found at the C-terminus of the pilin protein [2, 3]. These adherence domains bind to a common set of receptors, but amino acid differences in these regions result in different affinities for mammalian cells [4]. One method to study the pili-receptor binding is by examination of the 3D structure of the DSL region alone and in complex with the receptor through NMR techniques. To generate the required amounts of unlabelled and ^{13}C- or ^{15}N-labelled peptide, we have chosen a recombinant DNA approach. Here, we present data on the production of the DSL-peptide from PA strain PAK as N-terminal and C-terminal fusion proteins in *Escherichia coli*.

Results and Discussion

Figure 1 shows the two constructed expression vectors containing the sequence encoding the DSL region of PAK pilin. The expression vector pLB201 carries a tripartite fusion between the β-galactosidase (βG) gene, a segment of the chicken proα-2 collagen gene containing several potential collagenase cleavage sites, and the PAK DSL sequence under the control of the thermally inducible lambda promoter. The N-terminally fused peptide can be released from the resultant fusion protein by cleavage with collagenase. Plasmid pHE311 contains the glutathione-*S*-transferase (GST) gene behind the chemically inducible *tac* promoter. The peptide attached to the C-terminus of GST can be released by cleavage with factor Xa. Upon either thermal or chemical induction of the gene expression, the βG and GST fusion proteins were produced as about 33% and 20-25% of the total protein content of the *E. coli* cells, respectively.

The DSL-βG fusion protein (M_r ~123 kDa) was obtained in the amount of 20-25 mg/L culture after 40% ammonium sulfate precipitation and subsequent affinity column chromatography on *p*-aminobenzyl 1-thio-β-D-galactopyranoside-agarose, resulting in a calculated peptide yield of 0.4-0.5 mg/L. The major protein purification was achieved by salt fractionation, and not in the affinity chromatography step as expected. Also, 50% of the βG protein was lost during the affinity purification. The DSL-GST fusion protein (M_r ~30 kDa) was purified directly from the crude extract by affinity chromatography on S-linked reduced glutathione-agarose beads, providing 10-15 mg/L of the recombinant protein which corresponds to 1-1.5 mg of DSL-peptide. Western blot analysis showed that both fusion proteins contain the DSL adherence domain (Figure 2). Cleavage of the DSL-βG fusion protein was achieved with *Achromobacter* collagenase resulting in a single βG band of lower M_r on SDS-PAGE.

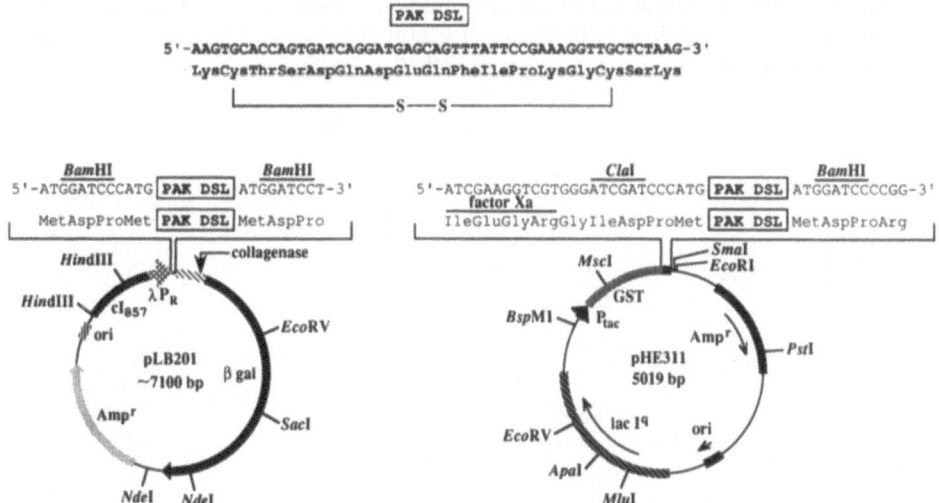

Fig. 1. *Plasmid vectors for the expression of PAK DSL-peptide as N- and C-terminal fusion proteins. The DSL-peptide, including some flanking amino acids, was synthesized in the form of two complementary 63-base oligonucleotides. The resultant double stranded oligonucleotide was then cloned by standard DNA techniques into the BamHI site of the expression vector pDS100 [5], giving plasmid pLB201. The same oligonucleotide was cloned in the right orientation in the BamHI site of the polylinker from expression vector pGEX-3X [6]. To obtain a correct in-frame fusion, the BamHI site at the fusion site was enzymatically filled-in and re-ligated. This resulted in the addition of 4 nucleotides encoding an aspartic acid residue. Correct orientation and in-frame fusion was confirmed for both constructs by DNA sequencing and SDS-PAGE analysis of the fusion proteins.*

However, a nested set of peptides, as a result of the multiple collagenase cleavage sites, was found on HPLC purification. Further incubation with collagenase did not result in continued cleavage to a single peptide species in contrast to previous reports [7]. This may be due to the reduced efficiency with which collagenase cleaves polypeptides. In comparison, the DSL-GST fusion protein was purified in a single step to homogeneity as shown on silver-stained SDS-PAGE. Quantitative cleavage of the fusion protein with factor Xa was obtained either with purified DSL-GST or with the fusion protein still bound to the affinity gel. The advantage of the latter is that only the peptide is released from the column while the

GST moiety remains bound to the matrix, providing an easy and efficient separation of the two species.

These data clearly demonstrate that the PA DSL-peptides can be successfully expressed as either an N- or C-terminal fusion protein in high yields in the expression systems examined here. Both fusion proteins can be easily purified by affinity chromatograhpy. The βG fusion system, however, suffers from relatively low peptide yields due to the large subunit size of the carrier protein and an incomplete proteolytic cleavage by collagenase. The GST fusion system, on the other hand, produces lower amounts of protein, but the smaller size of GST provides much higher yields of the desired peptide. Since this fusion protein can be cleaved efficiently by factor Xa while the protein remains bound to the affinity gel, the entire purification process is reduced to a single chromatography step. Based on these results, GST fusions appear to be the most useful system for future cost-effective synthesis of mg amounts of pure isotopically labelled PA DSL-peptides.

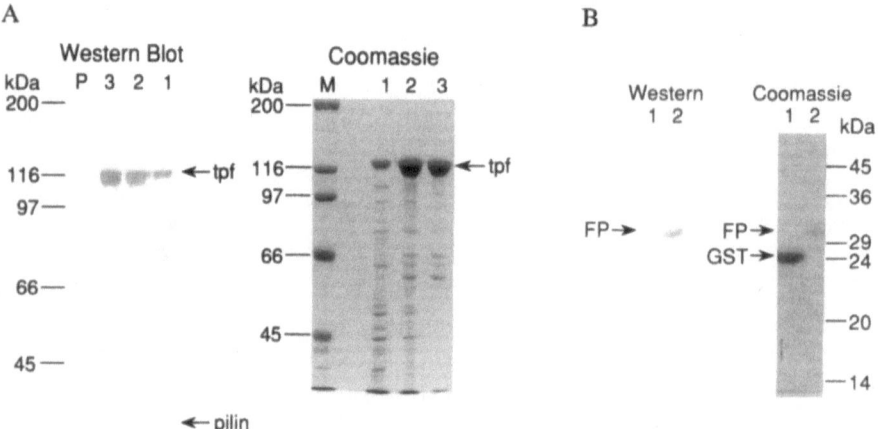

Fig. 2. Immunoblot of DSL-βG and DSL-GST fusion proteins. (A) SDS-PAGE and Western blot analysis of β-galactosidase tripartite fusion protein (tpf) purification. Samples were separated in 6.5% PAA gels and were either stained with Coomassie blue or the proteins were electrophoretically transferred to an Immobilon-P membrane and standard immuno-detection was done with a peptide-specific antibody in a 1:10000 dilution. Lanes contain samples of affinity purified fusion protein (3), 40% $(NH_4)_2SO_4$ fraction (2), and crude extract prepared from E. coli harbouring plasmid pLB201 (1). Lanes: P, 2.5 µg pilin as control; M, standard proteins. (B) Western blot analysis of DSL-GST fusion protein (FP). Purified DSL-GST and GST were separated in a 12.5% SDS-PAA gel. The protein bands were visualized either by Coomassie staining or by immunochemical techniques as above. Lanes: 1, GST; 2, DSL-GST.

Acknowledgements

We thank Dr. R. Irvin for providing antibodies and Dr. R.S. Hodges and L. Burke for HPLC assistance. This work was supported by the Medical Research Council of Canada, the Canadian Bacterial Diseases Network, and by the Alberta Heritage Foundation for Medical Research (P.L.-B.).

References

1. Pier, G.B., J. Infect. Dis., 151 (1985) 575.
2. Sastry, P.A., Pearlstone, J.R., Smillie, L.B. and Paranchych, W., Can. J. Biochem. Cell Biol., 63 (1985) 284.
3. Doig, P., Todd, T., Sastry, P.A., Lee, K.K., Hodges, R.S., Paranchych, W. and Irvin, R.T., Infect. Immun., 56 (1988) 1641.
4. Irvin, R.T., Doig, P.C., Sastry, P.A., Heller, B. and Paranchych, W., Micro. Ecol. in Health and Dis., 3 (1990) 39.
5. Moskaluk, C. and Bastia, D., Proc. Natl. Acad. Sci. U.S.A., 85 (1988) 1826.
6. Smith, D.B. and Johnson, K.S., Gene, 67 (1988) 31.
7. Germino, J. and Bastia, D., Proc. Natl. Acad. Sci. U.S.A., 81 (1984) 4692.

Utilization of carbohydrate-binding synthetic peptide sequences as novel antiinflammatory agents

L.D. Heerze[1], N. Wang[2], R.H. Smith[2] and G.D. Armstrong[1]

1Department of Medical Microbiology and Infectious Diseases, University of Alberta, Edmonton, Alberta, T6G 2H7, Canada and 2Carbohydrate Research Program, Department of Biotechnology, Alberta Research Council, Edmonton, Alberta, T6H 5X2, Canada

Introduction

Several plant lectins such as those from *Sambucus nigra* (SNA), *Maackia amurensis* (MAL) and wheat germ (WGA) have similar oligosaccharide binding specificities as pertussis toxin (PT), a virulence factor produced by the organism *Bordetella pertussis*, the etiological agent of whooping cough [1]. All these lectins recognize sialic acid containing oligosaccharide sequences. An important group of mammalian sialic acid-specific lectins, called selectins play a crucial role in immune responses (including inflammatory responses) by promoting cell-cell contact or extravasation of leukocytes. Specific carbohydrate receptors for selectins have been determined to contain similar oligosaccharide sequences as those which are recognized by PT and the three plant lectins. Moreover, we, as well as others, have recently determined functional similarities between PT and the mammalian selectins [2, 3]. We have also recently demonstrated that peptide sequences derived from the S2 subunit of PT were able to bind specifically to sialic acid containing oligosaccharides and inhibit the binding of PT and the plant lectins [4]. Since these PT derived peptides have the ability to bind to sialylated oligosaccharide structures and inhibit lectin binding, these peptides in principle should be able to antagonize selectin mediated inflammatory activity. If selectin binding activity can be altered by the use of the inhibitory peptide sequences, then they have the potential to control inflammation. These types of compounds will lead to the development of novel and more effective antiinflammatory agents.

In this report we evaluated a series of synthetic peptides derived from sequences found in PT, which had the ability to bind to sialic acid containing oligosaccharide sequences and inhibit lectin binding, for their potential to bind to the putative oligosaccharide selectin receptors. Those peptide sequences which had the potential to bind these oligosaccharide sequences were screened for the potential to inhibit selectin mediated inflammatory responses.

Results and Discussion

The peptide sequences that were chosen for further binding studies were sequences found in the S2 subunit of PT and corresponded to amino acids 9-23

(ACS2P1) and 18-23 (SPYGRC). Both of these peptides contained within their sequences a short 6-amino acid span (SPYGRC) which displayed good homology to a sequence found in WGA (SQYGHC) that was found to be a component of the sialic acid binding site as determined by X-ray crystallography [5]. Since these peptides have been shown to play a major role in binding sialic acid, biotinylated derivatives of these peptide sequences were screened for the ability to react with a panel of synthetic BSA glycoconjugates including the sialyl Lewis X (SLe^x) and the sialyl Lewis A (SLe^a) oligosaccharide sequences, which have been identified as putative receptors for selectins. Table 1 shows that both peptides can bind to a panel of sialic acid containing glycoconjugates with variable affinities relative to control binding to fetuin.

Table 1 *Binding of ACS2P1-biotin and SPYGRC-biotin to BSA Conjugates**

Carbohydrate Structure of BSA Conjugate	Common Name	Binding Relative to Fetuin of ACS2P1-b (n = 3)	Binding Relative to Fetuin of SPYGRC-b (n = 3)
αNeuAc(2-3)βGal(1-4)βGlcNAc-BSA	SLacNAC	1.17 ± 0.12	1.20 ± 0.07
αNeuAc(2-3)βGal(1-3)βGlcNAc-BSA	SLe^c	0.70 ± 0.04	0.97 ± 0.12
αNeuAc(2-3)βGal(1-4)βGlcNAc-BSA	SLacNAC	0.94 ± 0.02	1.22 ± 0.05
αNeuAc(2-3)βGal(1-4)βGlcNAc-BSA (1-3) αFuc	SLe^x	0.84 ± 0.03	1.20 ± 0.04
αNeuAc(2-3)βGal(1-3)βGlcNAc-BSA (1-4) αFuc	SLe^a	0.84 ± 0.14	1.03 ± 0.08

*Experiments were done by coating 50 µg/ml BSA-conjugate or fetuin and probed with 0.1 µg ACS2P1-biotin (ACS2P1-b) or 0.5 µg of SPYGRC-b for 1 h at room temperature as described previously [4]. The binding of biotinylated peptide to fetuin is set at a value of 1.0.

The results in Table 1 indicate that both biotinylated peptide sequences, ACS2P1 (2275) and SPYGRC (2283) were able to bind to the oligosaccharides SLe^x and SLe^a, and thus could serve as antagonists for a selectin mediated inflammatory response. To assess the potential of these peptides as mediators of a DTH inflammatory response, an *in vivo* animal model system was employed which measures the degree of mouse footpad swelling in Balb/c mice after re-exposure to an antigen [6]. The peptides ACS2P1 and SPYGRC as well as two control peptides that contain randomized sequences of ACS2P1 (RHQGQSPIYGEQPCT, 2295) and SPYGRC (GSYRPC, 2294) were used as potential inhibitors of footpad swelling.

The oligosaccharide SLe^x was utilized as a positive control. Seven days after immunization with an antigen (ovalbumen), groups of 10 mice were footpad-challenged with ovalbumin. Five hours after re-challenge each group was treated with 100 µg of peptide or SLe^x in PBS while control groups received only PBS. The effect of each peptide or carbohydrate on reducing mouse footpad swelling was determined after 24 h. The results shown in Figure 1 indicate that the peptides ACS2P1 (2275) and SPYGRC (2283) were as effective at reducing a DTH response in mice as an equal amount of SLe^x oligosaccharide; a compound which is presently being considered as a new antiinflammatory drug. Control peptides 2294 and 2295 which contain randomized sequences were found to be ineffective at reducing inflammation indicating that the hexapeptide sequence (SPYGRC) found in peptides 2275 and 2283 must play an important role in preventing selectin binding to its receptor thus reducing a DTH inflammatory response.

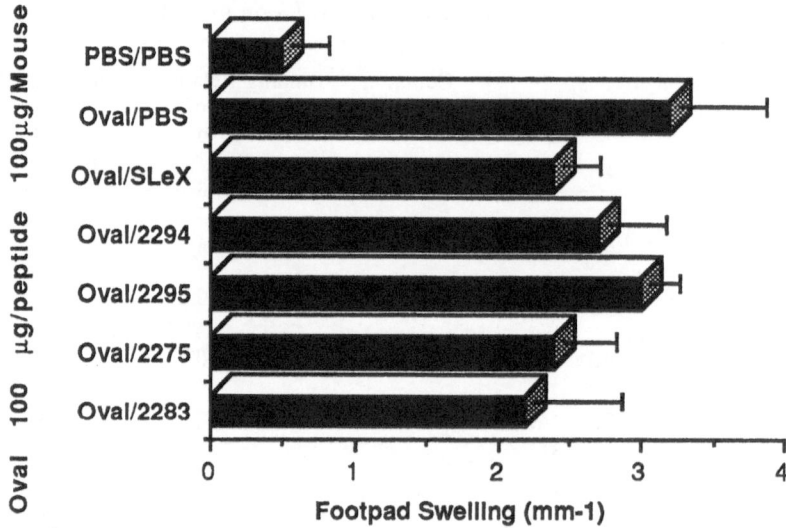

Fig. 1. Effect of peptides on an ovalbumin-induced DTH response.

The utilization of carbohydrate binding peptides to block selectin binding provides a novel approach for the development of alternative antiinflammatory agents. We have identified small amino acid sequences which can serve as lead compounds for the development of antiinflammatory drugs. New peptide sequences are presently being synthesized and examined in various animal model systems.

References

1. Heerze, L.D. and Armstrong, G.D., Biochem. Biophys. Res. Commun., 172 (1990) 1224.
2. Smith, R.H., Heerze, L.D. and Armstrong, G.D., unpublished data.
3. Tuomanen, E., Mar, V. and Burnette, W.N., 92nd General Meeting of the American Society of Microbiology, New Orleans, LA, U.S.A., May 26-30, 1992, Poster B-16.
4. Heerze, L.D., Chong, P.C.S. and Armstrong, G.D., J. Biol. Chem., 267 (1992) 25810.
5. Wright, C.S., J. Mol. Biol., 215 (1990) 635.
6. Smith, R.H. and Ziola, B., Immunology, 58 (1986) 245.

McPC603 V$_{II}$(1-115): Synthesis and folding of immunoglobulin variable regions

L.M. Martin, K.S. Rotondi and R.B. Merrifield

The Rockefeller University, New York, NY 10021, U.S.A.

Introduction

We seek to chemically synthesize small model antibodies based on the framework of the murine myeloma protein McPC603, an IgA immunoglobulin whose three-dimensional structure is known [1]. To better understand the required sequences for antibody-antigen interaction we have chemically synthesized the N-terminal VH(1-68) and VH(1-115) of the α-heavy chain. Our previous synthesis of VH(1-68) [2] demonstrated the feasibility of a stepwise synthesis of such a large fragment using the PAM resin. The present synthesis of VH(1-115) includes the 22-98 disulfide bond that is thought to be necessary for the correct folding of the heavy chain. This VH fragment also contains the three hypervariable loops, which are responsible for the specificity of the antibody and also contains the framework region of the M603 idiotope. Using these synthetic fragments, and the natural light chains isolated from the mouse, we explored domain association, folding, and stoichiometry by affinity chromatography and capillary electrophoresis.

```
VH(1-115):   EVKLVESGGGLVQPGGSLRLSCATSGFTFS
             DFYMEWVRQPPGKRLEWIAASRNKGNKYTT
             EYSASVKGRFIVSRDTSQSILYLQMNALRA
             EDTAIYYCARNYYGSTWYFDVWGAG
```

Fig. 1. Sequence of the N-terminal of McPC603 α–heavy chain showing ninhydrin/amino acid analysis monitoring points during the synthesis in bold type.

Results and Discussion

Both VH(1-68) and VH(1-115) were synthesized with Boc-protection via a linear stepwise solid phase approach on a PAM resin. We find that the linear stepwise synthetic approach for hydrophobic proteins of this size is very efficient, due to solubilization of the growing peptide chains by the resin. By using amino acid analysis to carefully monitor the synthesis at positions marked in italics (Figure 1), we identified a general problem spot in the synthesis of antibodies at the cysteine couplings, which were overcome by using HOBT esters in *N*-methylpyrrolidinone. Significant problems in the synthesis of large β-barrel proteins arise not during the synthesis, but in the isolation and purification steps. Denatured or incorrectly folded fragments aggregate and subsequently become insoluble. RP-HPLC of the denatured products, once thought impossible, even with trifluoroacetic acid, improves dramatically with the addition of stronger ion-pairing reagents such as pentanesulfonic acid (Figure 2) [3].

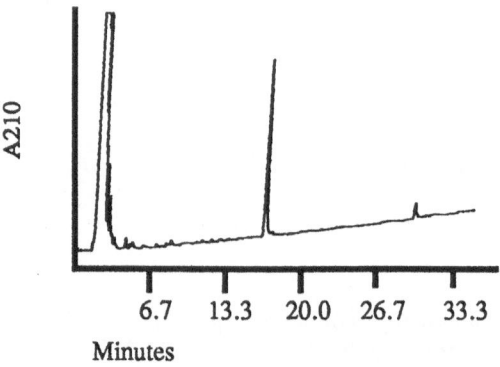

Fig. 2. C18 HPLC on VH(1-115). Air oxidized sample, 4.5 M guanidine HCl, 20 mM borate pH 8.5. 5-45% acetonitrile, 1%/min, in 5 x 10⁻³M pentanesulfonic acid.

Capillary electrophoresis below the isoelectric point (pH 9), also improves on addition of these reagents. Above the pI, a coated capillary (Supelco, C1) gives sufficient resolution (Figure 2). The crude synthetic product contains two minor impurities by CE, that are not seen by HPLC. The charge on the impurities is close to that of the main product at pH 9.5 and 7.

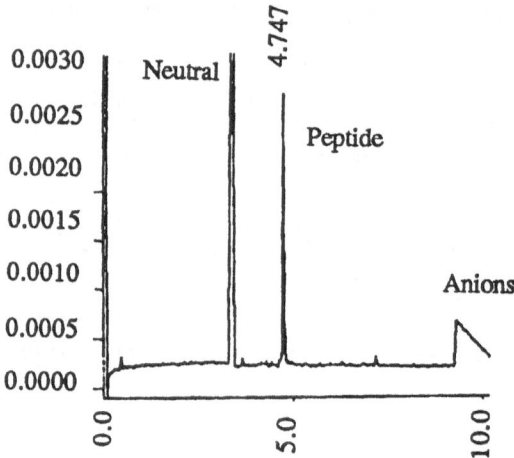

Fig. 3. CE of VH in 20 mM borate pH 9.98. C1 column 214 nm 30kV, 72 cm capillary, 52 cm to detector.

The folding and solubility of the product was measured in guanidine HCl (Figure 4). At higher pH's (8.0 and 10.0), a discontinuity may reflect the ability of the protein to form a disulfide bond. Solubilities were verified by amino acid

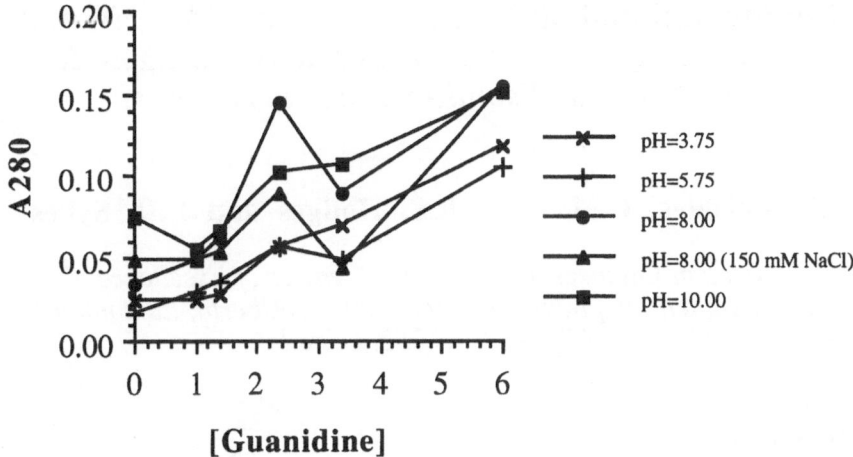

Fig. 4. Guanidine titration of chemically synthesized VH(1-115) in different buffers monitored by UV spectroscopy and amino acid analysis. Aliquots of a 50% acetic acid extract of the crude peptide were diluted into buffers, then denatured with guanidine HCl.

analysis. We found that correct folding of the synthetic VH proteins relies on the adoption of a stable secondary structure such as a β-barrel, that formation of a disulfide bond may be necessary, and that correct folding is facilitated by renaturation in the presence of a template. Removal of denaturant at physiological pH without the complimentary light chain gave insoluble particles.

Analysis of sequence homology between VH(1-68) and the second half of VH(1-115) led us to suspect that dimers of VH(1-68) may effectively mimic the longer chain and recombine with the light chain to form active hybrids. These hybrids might have similar properties to the native structure and provide yet another variation on the β-barrel structural motif. This would also explain a difference between our results and those on the product from genetic engineering [4].

The hapten of McPC603, phosphocholine (PC), is a small, negatively charged organic substrate. The hybrid synthetic and natural recombination products were isolated by affinity chromatography on a column with covalently linked PC. Two recombination products with slightly different affinities were isolated and will be further characterized. Further studies are underway to fully characterize the stoichiometry and structure of these recombinants.

References

1. Satow, Y., Cohen, G.H., Padlan, E.A. and Davis, D.R., J. Mol. Biol., 190 (1986) 593.
2. Smith, J.A. and Rivier, J.E. (Eds.), Peptides (Proceedings of the Twelfth American Peptide Symposium), Escom, Leiden, The Netherlands, 1992, p.849.
3. Hancock, W.S., Bishop, C.A., Meyer, L.J., Harding, D.R.K. and Hearn, M.T.W., J. Chrom., 161 (1978) 290.
4. Glockshuber, R., Schmidt, T. and Plückthun, A., Biochemistry, 31 (1992) 1270.

Conformational differences between the *cis* and *trans* proline isomers of a synthetic antigen from the pilus of *Pseudomonas aeruginosa*

C. McInnes, C.M. Kay, R.S. Hodges and B.D. Sykes

Protein Engineering Network of Centres of Excellence and Synthetic Peptides Inc., University of Alberta, Edmonton, Alberta, T6G 2S2, Canada

Introduction

Pseudomonas aeruginosa is an opportunistic pathogen which affects burn, cystic fibrosis and cancer patients [1]. A potential anti-adhesion vaccine from the C-terminal region of the pilin of *P. aeruginosa* strain K (PAK) has been studied by 1- and 2-dimensional NMR techniques in order to gain insight into the process of antibody-antigen recognition. The study of the pilin of this and other strains of this bacterium should allow a greater understanding of the cross-reactivity of antibodies specific for this region and facilitate the development of an effective anti-adhesion vaccine. We have shown that a 17 residue disulfide bridged peptide (PAK 128-144) KCTSDQDEQFIPKGCSK corresponding to the receptor binding domain of *P. aeruginosa* exists as two isomers in solution and in this paper demonstrate the differences between the two forms.

Results and Discussion

1 and 2 dimensional NMR spectra of PAK 128-144 oxidized indicated that two isoforms of the peptide exist in aqueous solution in a ratio of 3:1. Due to well resolved cross-peaks of the two forms in COSY, TOCSY and NOESY spectra, a complete set of resonance assignments [2] were obtained for both. Through the assignment of characteristic NOEs, the two isoforms were shown to arise as a consequence of *cis/trans* isomerization of the I138-P139 amide bond. The two isoforms were demonstrated to differ markedly in their solution structure through comparison of chemical shift and temperature coefficient data. Several residues in the *cis* isoform exhibit considerable differences in their NH and α proton chemical shifts with respect to the *trans*. These variations do not occur exclusively in the region of the *cis* amide bond, suggesting that conformational factors play a role in stabilization of the minor isoform, however, this might be expected since *cis/trans* isomerization of proline is not an entirely localized event [3]. Examination of NH proton temperature coefficients for the two forms also gives insight into the conformational differences that occur. The temperature dependence of the NH chemical shift is an indication of intramolecular hydrogen bonding in a peptide or protein [4]. The *trans* isoform has putative H-bonds to the amides of F137 and C142 while the *cis* form apparently has H-bonds to S131, E135, Q136 and C142. The

different hydrogen bonding in the *cis* isomer indicates that its 3-D structure differs significantly from that of the *trans* form and may even be more rigid due to the greater number of H-bonds. The differences in the chemical shifts are consistent with the variation in hydrogen bonding between the two isoforms. The residues which apparently change their H-bonding from *cis* to *trans* are those which show marked differences in their chemical shift. Distance and torsion angle restraints have been obtained for the *trans* isomer and used to generate an ensemble of solution structures which is shown in Figure 1. This was not possible for the *cis* isomer due to the limited number of NOEs observed. The conformational ensemble of the *trans* form shows the presence of two β-turns between residues 134-137 and 139-142 in the region of highest definition in the molecule. The epitope of the monoclonal antibody PK99H has been shown to consist of residues 134-140 of PAK 128-144 [5]. The presence of the β-turn between D134 and F137 is consistent with other immunogenic peptides which have shown an innate tendency to adopt secondary structure in the epitope region. The *cis* isoform apparently lacks the β-turn between residues 134 and 137 since the characteristic H-bond of F137 is not present and H-bonds are present at E135 and Q136 indicating that the turn is disrupted. These results show that the two isoforms differ markedly in their 3-D structure and thus may also differ significantly in their biological activity.

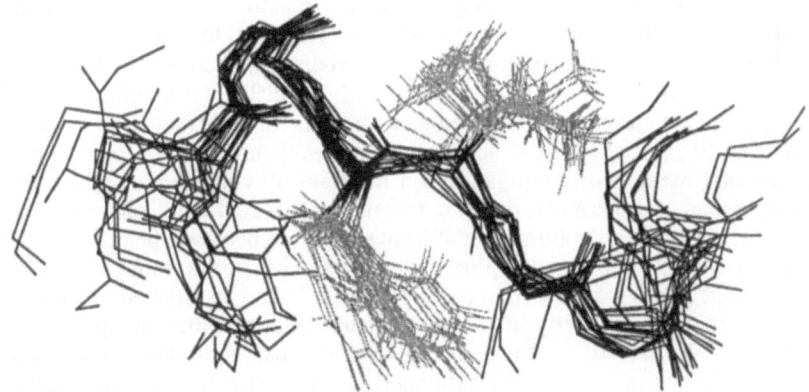

Fig. 1. Conformational ensemble for PAK 128-144 trans. The backbone of residues 133-142 and the sidechains of F137 and I138 are displayed.

Acknowledgements

This work has been supported by the PENCE which is funded by the Government of Canada. We thank Drs. F.D. Sönnichsen and J.J. Wang for helpful advice.

References

1. Rivera, M. and Nicotra, M.B., Am. Rev. Respir. Dis., 126 (1982) 833.
2. Wüthrich, K., NMR of Proteins and Nucleic Acids, John Wiley, New York, NY, U.S.A., 1986.
3. Grathwohl, C. and Wüthrich, K., Biopolymers, 20 (1981) 2623.
4. Rose, G.D., Gierasch, L.M. and Smith, J.A., Adv. Protein Chem., 37 (1985) 1.
5. Wong, W., Irvin, R., Paranchych, W. and Hodges, R.S., Prot. Sci., 1 (1992) 1308.

A novel approach to a synthetic malaria vaccine using the multiple antigen peptide system

J.C. Spetzler, C. Rao and J.P. Tam

Department of Microbiology and Immunology, A-5119 MCN, Vanderbilt University, Nashville, TN 37232-2363, U.S.A.

Introduction

Malaria infects 200 million human beings each year with a fatality rate of 2-3% and is a major health problem in developing countries. To help solve this problem, a viable vaccine candidate is needed for clinical trials. A cysteine-rich domain at the carboxyl terminus of the merozoite surface protein (MSP-1) [1] is the most promising antigen for a vaccine that protects against infection challenge in primate and rodent models. The antigen belongs to the three-disulfide epidermal growth factor (EGF) family based on the alignment of the six cysteines. Our strategy is to allow the EGF-like domain to fold and form well-defined disulfide bonds before conjugation to the MAP core matrix [2]. We have recently accomplished the synthesis of MSP-1 antigen [3] and are currently developing new methods for its conjugation to the MAP core matrix. Because the MSP-1 antigen has formed disulfide bonds, we cannot use the traditional method of conjugating via the thiol alkylation of cysteine with a haloacetyl group. Furthermore, there are few methods available for such site-specific conjugation in the selective formation of a covalent bond between the peptide antigen and the core-matrix. A rational approach would make use of the conjugation reaction mediated through a weakly basic nitrogen nucleophile and an aldehyde under acidic aqueous conditions, whereby all the side chain functionalities are protected by protonation. We describe here a model reaction that validates our concept of attaching a folded protein to a MAP core matrix, which can result in a chemically and structurally well-defined synthetic vaccine with a MW exceeding 25,000 daltons and which is technically difficult to achieve in stepwise MAP synthesis.

Results and Discussion

Our approach to assembling unprotected peptide antigens on a scaffold utilizes the reaction between a phenyl hydrazine and an alkyl aldehyde to form a hydrazone bond. The model peptide, SSQFQIHGPR, was synthesized by the solid phase method using Boc chemistry. The peptide was capped with a 4-Boc-hydrazinobenzoyl (Boc-Hob) group via DCC/HOBt at the N-terminal. Hob is a suitable choice for this approach because it contains a weak nitrogen base and can act as the only nucleophile in acidic pH even in the presence of unprotected side chains. The aldehyde attached to the core matrix was generated from a mild periodate oxidation of the amino terminal serine residue by utilizing the 2-amino alcohol moiety of the unprotected N-terminal serine that is sensitive to periodate oxidation. A tetravalent Ser-MAP core was

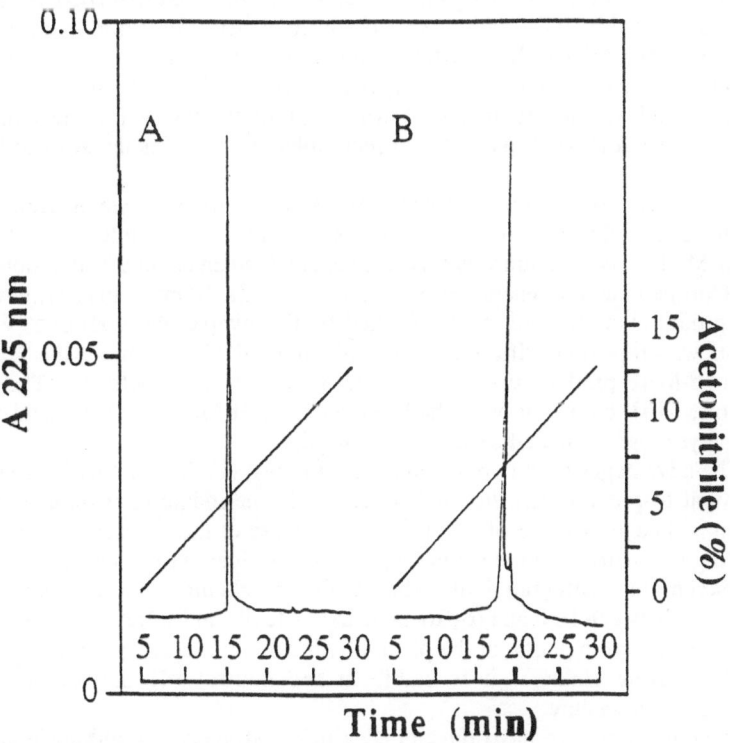

Fig. 1. *Analytical C_{18} reverse-phase HPLC profiles of Ser_4-Lys_2-Lys-βAla-OH (A) and after oxidation by periodate to $(HCOCO)_4$-Lys_2-Lys-βAla-OH (B).*

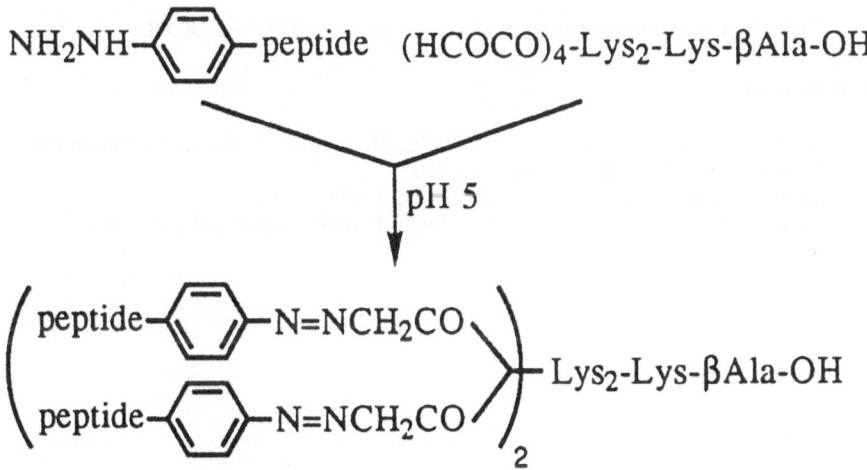

Fig. 2. *Conjugation between the α-oxoacyl-MAP core and the Hob-peptide.*

synthesized on solid phase using Fmoc chemistry. (Fmoc-Ser(tBu))$_4$-Lys$_2$-Lys-Ala-resin was cleaved by TFA and directly oxidized at pH 7 for 5 min to an α-oxoacyl derivative of MAP (HCOCO-MAP) by sodium periodate. The reaction was quenched by ethylene glycol and purified by C$_{18}$ reverse-phase HPLC (Figure 1). Due to its instability and sensitivity to air, it was used immediately for the conjugation reaction, α-oxoacyl-MAP gave the correct molecular mass as determined by FAB-MS.

The conjugation (Figure 2) was carried out at pH 5, in which Hob obtained from the TFA deprotection of Boc-Hob is the only nucleophile to react with α-oxoacyl-MAP. The reaction was very efficient and, when using a 2-fold molar excess of the Hob-peptide, the reaction was completed within 10 min. Furthermore, only a single product was formed, as determined by C$_4$ reverse-phase HPLC. When the reaction was repeated with 1 or 1.5 equivalent of Hob-peptide, the tetravalent hydrazone-MAP product was completed in 5 h and 1 h, respectively. The peptide-hydrazone-MAP conjugate was characterized by ESI-MS and amino acid analysis. Both analyses agreed with the expected composition.

The advantages of our approach are as follows: (1) the Ser-MAP core matrix is convenient to prepare; (2) the mild oxidation by periodate to α-oxoacyl-MAP is efficient and generally free of side reactions because of the absence of side chains in the MAP core matrix; (3) the Hob group can be considered to be an amino derivative and conveniently integrated into the synthetic scheme in solid phase; (4) the conjugation between Hob and α-oxoacyl-MAP is facile and selective; and (5) because of the aromatic conjugation in the phenyl hydrazone linkage, and unlike the normal alkyl hydrazone linkage, it is stable at the neutral pH that is used in the immunization procedure.

In conclusion, we have developed a novel strategy for linking unprotected peptide antigens to the MAP core matrix to produce structurally and chemically defined macromolecules for vaccine purposes.

Acknowledgements

This work was in part supported by a grant by USPHS AI 28701.

References

1. Mackay, M., Goman, M., Bone, N., Hyde, J.E., Scaife, J., Certa, U., Stummenberg, H. and Bujard, H., EMBO J., 4 (1985) 3823.
2. Tam, J.P., Proc. Natl. Acad. Sci. U.S.A., 85 (1988) 5409.
3. Spetzler, J.C., Rao, C. and Tam, J.P., Int. J. Peptide Protein Res., in press.

Directionality in "promiscuous" T-cell epitope selection on colinear B- and T-cell constructs

B. Wang, S.F. Kobs-Conrad and P.T.P. Kaumaya

College of Medicine (Departments of Obstetrics & Gynecology, and Medical Biochemistry) and the Comprehensive Cancer Center, The Ohio State University, Columbus, OH 43210, U.S.A.

Introduction

The development of chimeric constructs containing a T-cell and B-cell epitope for the rational design of future vaccines represents an effective alternative to the use of undefined carrier protein conjugates [1]. The mechanism of directional help in T-cell dependent B-cell responses elicited by chimeric constructs is poorly understood, but is dependent on a number of factors: peptide processing by antigen presenting cells (APC), the presence of T-cells of appropriate specificity, affinity of the T-cell epitope for MHC class II molecules and antigen mimicry. There are a number of examples that show the influence of epitope orientation on the immunogenicity of chimeric synthetic peptides and the affinity of antibodies induced by them [2-4]. The aim of this study is to explore the mechanisms of the interplay involved in the pairing of B-cell and T-cell epitopes within a single synthetic peptide.

Results and Discussion

A B-cell epitope (α_N) derived from the protein antigen mouse lactate dehydrogenase C_4 (LDH-C_4), was colinearly synthesized with a "promiscuous" T-cell epitope from either tetanus toxoid (TT$_3$, sequence 947-967) or the fusion protein of measles virus (MVF, sequence 288-302). The four constructs were synthesized such that the T-cell epitopes were at the amino terminus (α_NTT$_3$ and α_NMVF) or the carboxyl terminus (TT$_3\alpha_N$ and MVFα_N) relative to the B-cell epitope. The four chimeric peptides were synthesized on a Milligen/Biosearch 9600 synthesizer using FMOC/t-butyl synthetic strategy, BOP/HOBt in situ activation coupling protocols on a MBHA resin (0.6 mmol/g, 4-hydroxymethyl phenoxyacetyl linker). Peptides were deprotected and cleaved from the resin with TFA and were purified (>95% homogeneity) by reverse-phase semi-preparative HPLC.

We studied the immune responses of these constructs by immunizing New Zealand white female rabbits three weeks apart and collecting sera at weekly intervals during the primary, secondary and tertiary response. Since all constructs produced high titered antisera specific for the immunogen and the native protein LDH-C_4, the orientation of the B-cell epitope (α_N) relative to the promiscuous T-cell epitope [TT3 (data not shown) or MVF] did not affect the immunogenicity of these peptides. In order to delineate how orientation of the epitopes affects specificity and affinity of antibody recognition, competitive inhibition ELISA were carried out. Our results show that all of the chimeric constructs irrespective of epitope orientation were good inhibitors of anti-chimera antibody binding to the native protein on the plate (Figure 1).

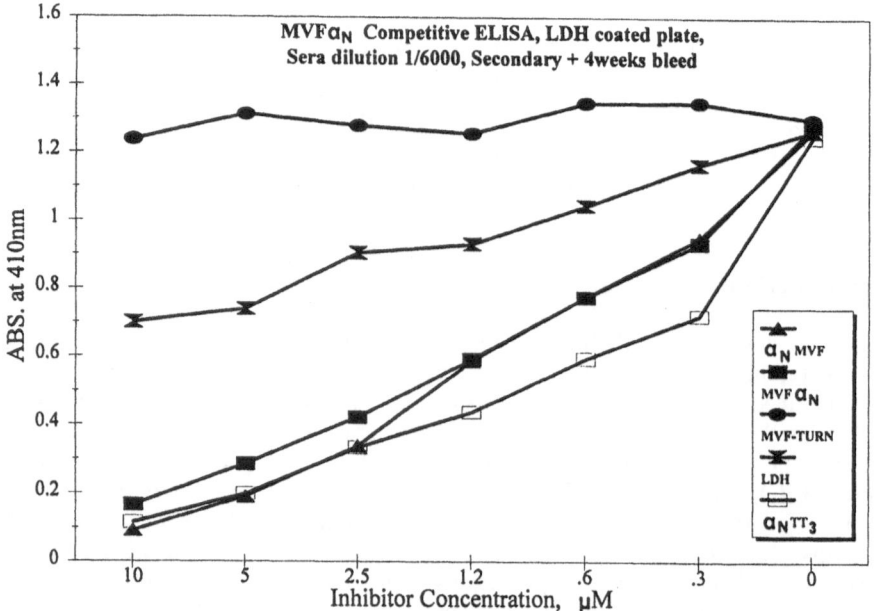

Fig. 1. Inhibitors α_NMVF, MVFα_N, α_NTT3, LDH-C_4 and MVF-Turn competing for the binding of anti-MVFαN antibodies to native protein (LDH-C_4) coated plates.

The chimeric antibodies are specific for the α_N epitope of LDH-C_4 indicating that the B-cell epitope conformation in the chimeric peptides is mimicking the structure of the native antigen protein. Thus, the influence of epitope orientation on the ability of the various peptide constructs to inhibit antibody binding to LDH-C_4 cannot be easily distinguished. In conclusion, although chimeras with opposite orientation were slightly less effective inhibitors than chimeras with the same orientation in inhibiting antibody binding to the native protein, the immunogenic capacity of individual chimeras was not impaired by choice or selection of epitope orientation. *(This work was supported by an NIH grant AI25790 to PTPK.)*

References

1. Kaumaya, P.T.P., Feng, N., Kobs-Conrad, S., Seo, Y.H., VanBuskirk, A.M. and Sheridan, J.F., In Smith, J.A. and Rivier, J. (Eds.), Peptides: Chemistry and Biology, Escom, Leiden, The Netherlands, 1992, p.883.
2. Partidos, C., Stanley, C. and Steward, M., Mol. Immunol., 29 (1992) 651.
3. Cox, J.H., Ivanyi, J., Young, D.B., Syred, A. and Francis, M.J., Eur. J. Immunol., 18 (1988) 2015.
4. Partidos, C., Stanley, C. and Steward, M., Eur. J. Immunol., 22 (1992) 2675.

Session IX
Conformational Analysis

Chairs: Evaristo Peggion
University of Padua
Padua, Italy

and

Ken Kopple
SmithKline Beecham Pharmaceuticals
King of Prussia, Pennsylvania, U.S.A.

Glu-mediated dimerization of the transmembrane region of the oncogenic *neu* protein

S.-C. Li[1], C.M. Deber[1] and S.E. Shoelson[2]

[1]Division of Biochemistry Research, Hospital for Sick Children, Toronto, Ontario M5G 1X8, and Department of Biochemistry, University of Toronto, Toronto, Ontario M5S 1A8, Canada, [2]Joslin Diabetes Center and Department of Medicine, Harvard Medical School, Boston, MA 02215, U.S.A.

Introduction

Receptor dimerization upon ligand binding has been suggested as a general mechanism for receptor kinase activation [1]. While ligands may directly participate in or facilitate dimerization of receptors by inducing corresponding conformational changes, the transmembrane (TM) regions of the receptors may also play an active role in this process. For example, a single mutation of Val^{664} to Glu within the putative TM domain of the *neu* protein, p185, dimerizes and constitutively activates this protein kinase, and consequently renders it oncogenic. It has been hypothesized that dimerization of the oncogenic *neu* protein may be mediated by the TM Glu residue [2].

Results and Discussion

We have adopted a peptide approach to address the above problem (see also [3]). Two 23-residue peptides were synthesized of amino acid sequence H$_2$N-Arg-Ala-Ser-Pro-Val-Thr-Phe-Ile-Ile-Ala-Thr-Val-X-Gly-Val-Leu-Leu-Lys-Arg-Arg-Arg-Gln-Lys- OH, where **X = Val** or **Glu**. These two peptides, designated *neu*(TM)-V and *neu*(TM)-E, respectively, represent major portions of the putative TM segments of the proto-oncogenic and oncogenic forms of the corresponding *neu* proteins.

Circular dichroism (CD) spectra of these peptides in water, and upon titration with sodium dodecylsulfate (SDS) are displayed in Figure 1. While both peptides undergo conformational transitions from random structure in aqueous buffer to largely α-helix in SDS (pH 7.0), the helicity of peptide *neu*(TM)-V apparently decreases as the SDS concentration is increased from 2.5 to 10 mM (Figure 1B). This phenomenon may reflect the relatively stronger aggregational tendency for peptide *neu*(TM)-V in SDS solutions at neutral pH, as is further suggested by its decreased absolute ellipticity in 10 mM SDS compared with peptide *neu*(TM)-E (Figure 1A). Thus, SDS molecules may act as counter-ions to the positively-charged ends of peptide *neu*(TM)-V, and promote its hydrophobic self-association. Parallel SDS-PAGE gels run at pH 8.8 (Figure 2A) reveal that peptide *neu*(TM)-V (monomer MW 2608) assumes largely helical dimers along with some highly aggregated β-sheet species (noted at the top of gel A), whereas peptide *neu*(TM)-E exists predominantly as diffuse monomers. Ionized TM Glu side chains may mitigate against dimerization of *neu*(TM)-E because of charge-charge repulsion, but this peptide could, in principle, form dimers when the Glu side chain is protonated (neutral). This proved to be the case, as evidenced by SDS gels run at a pH below the pK value (4.4) of the Glu side chain; Figure 2B demonstrates that both peptides *neu*(TM)-V and *neu* (TM)-E form dimers (~5 kDa) at pH 3.5. Although running SDS gels under acidic conditions has not been widely reported, the well-resolved separation of the molecular weight standards in Figure 2B suggests that the results are valid.

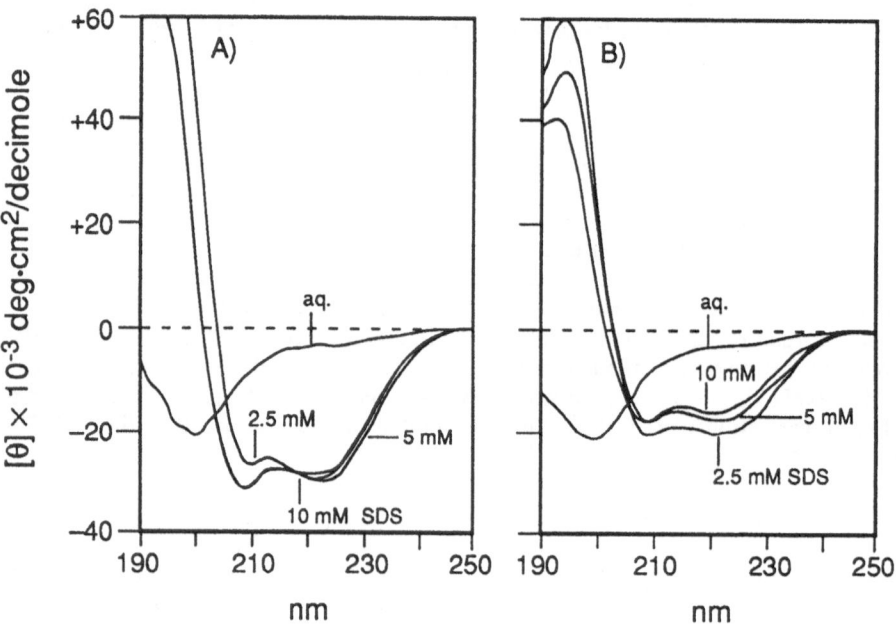

Fig. 1. CD spectra of peptides (A) neu(TM)-E (30 μM) and (B) neu(TM)-V (30 μM) upon titration with SDS at the indicated concentrations. Methods for peptide synthesis and CD measurements were as described [3]. Peptide sequences are given in the text.

The *in vitro* tendency toward self-association of peptide *neu*(TM)-V in the membrane-mimetic environment of SDS micelles may not necessarily be representative of the situation in intact p185 proteins, where close packing of TM segments may generally be hindered by the bulky periplasmic and cytoplasmic domains of the protein. As well, wild type TM residues can interact hydrophobically with themselves or surrounding lipids, and in principle, equilibrate between the two. The present results suggest that when the Glu side chain is protonated - as may occur in the physiological, non-polar environment of a membrane-spanning segment - this residue can also mediate the dimerization of the (oncogenic) *neu* protein in the membrane, likely by H-bonding between pairs of (protonated) Glu side chains and/or between Glu side chains and accessible (Ala?) neighboring main chain C=O groups

[2]. However, in contrast to the Val dimer, the site of interaction in the Glu dimer is four side chain bonds distant from the main chain; such an interaction may not require as close an approach of peptide backbones, yet may be sufficient to bring the corresponding cytoplasmic protein kinase domains into contact. Once formed, such Glu dimers may be more tightly associated than corresponding Val dimers, since their interaction involves the attractive energies of two H-bonds in an otherwise low dielectric local environment. Such relative inability of Glu dimers to dissociate would then provide, in part, an underlying basis for the mechanism of receptor constitutive activation.

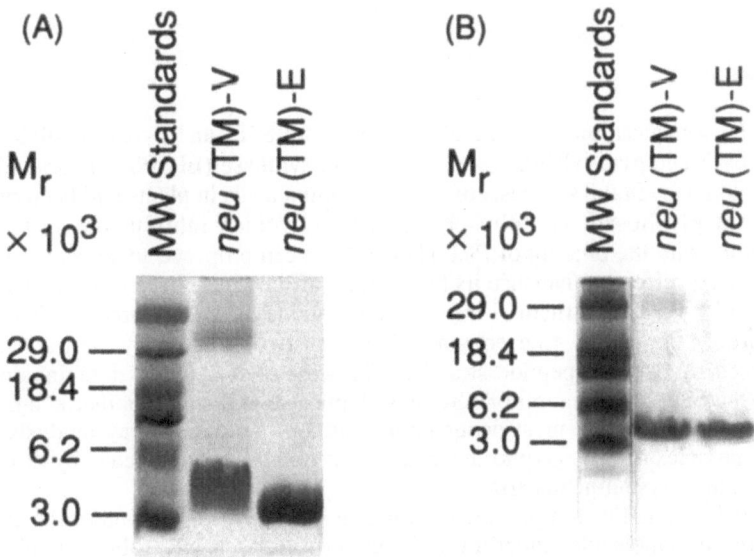

Fig. 2. SDS-PAGE gel analysis of peptides at (A) pH 8.8, and (B) pH 3.5. Lanes are as indicated above each gel.

Acknowledgements

This work was supported, in part, by grants to C.M.D. from the National Cancer Institute of Canada, and to S.E.S. from the National Institutes of Health. S.-C.L. holds a University of Toronto Open Studentship.

References

1.   Yarden, Y. and Schlessinger, J., Biochemistry, 26 (1987) 1434.
2.   Sternberg, M.J.E. and Gullick, W.J., Nature, 339 (1989) 587.
3.   Li, S.-C. and Deber, C.M., FEBS Lett., 311 (1992) 217.

Solid-state NMR determination of the geometry of substrate and inhibitor bound to EPSP synthase

D.D. Beusen, L.M. McDowell, A. Schmidt, E.R. Cohen and J. Schaefer

Center for Molecular Design and Department of Chemistry,
Washington University, St. Louis, MO 63130, U.S.A.

Introduction

The condensation of shikimate 3-phosphate (S3P) and pyruvylenolphosphate (PEP) by 5-enolpyruvylshikimate-3-phosphate synthase (EPSPS) to form EPSP is an essential step in the synthesis of aromatic amino acids in plants and bacteria. The herbicide glyphosate (N-[phosphonomethyl]glycine) inhibits the enzyme by competing with the binding of PEP [1] and has been proposed to act as a transition state analog. Efforts to enhance its micromolar affinity have been of limited success, hindered by a lack of structural information to guide the design process. The X-ray structure of the 46 kD free enzyme [2] shows two structurally similar domains connected by two polypeptide strands. The wide cleft between domains does not define the bound orientation of S3P and glyphosate and suggests that a significant conformational change must occur upon binding. To date, X-ray analysis of the ternary complex of S3P, glyphosate, and the enzyme has been impeded by difficulties in generating crystalline material.

When crystalline samples are unavailable for X-ray analysis and/or the motional properties of a molecule render it unsuitable for solution NMR, solid-state NMR can provide structural information otherwise unattainable. Rotational Echo-DOuble-Resonance (REDOR) NMR allows interatomic separation to be determined based on the r^3 distance dependence of the dipolar coupling [3]. In this magic-angle spinning experiment, signal echoes that normally form during each rotor cycle for one nucleus are attenuated by rotor-synchronized π pulses applied to the other nucleus. The reduction in signal intensity is a function of the dipolar coupling of the nuclei involved. In this study, we report REDOR NMR analyses of the EPSPS ternary complex which measure distances within glyphosate, between glyphosate and S3P, and between each ligand and the enzyme. These distances, along with information from chemical modification and site-directed mutagenesis studies, have been used to generate a model for the bound geometry of S3P and glyphosate.

Results and Discussion

EPSPS was purified from cultures of an *E. coli* strain (Monsanto Co.) engineered to overproduce the enzyme. Samples for NMR analysis (~150 mg) were prepared by lyophilization of a buffered dilute solution of the enzyme, S3P, and glyphosate [4]. Bound ^{31}P chemical shifts were the same in the lyophilized powder and solution, confirming that the structure of the binding site is preserved in the lyophilized enzyme.

REDOR experiments with specifically labelled ^{13}C-glyphosate in which the ^{31}P signal was dephased by ^{13}C yielded glyphosate intramolecular distances of 4.92 ± 0.04 Å (C1-P) and 4.04 ± 0.05 Å (C2-P). Dephasing by surrounding natural abundance ^{13}C was treated as a single, "mean" spin and applied as a correction to the observed dephasing. A systematic conformational search of glyphosate identified four families of conformations consistent with these distances, of which two are similar to X-ray structures reported for glyphosate [5, 6]. All four conformations could be superimposed on a carbocation transition-state model for PEP.

These experiments and others using ^{15}N-glyphosate also yielded inter-ligand distances: 6.2 ± 0.2, 6.4 ± 0.1, 7.1 ± 0.3, and 6.3 ± 0.3 Å between the S3P phosphorus and C1, C2, N, and C3 of glyphosate, respectively. Restrained molecular dynamics simulations of the two ligands in the absence of the enzyme did not eliminate any of the four glyphosate conformations and did not define the orientation of glyphosate with respect to S3P. These simulations did reveal that the closest approach of the S3P oxygen at position five to the glyphosate nitrogen is >5.0 Å. One model for glyphosate inhibition proposes hydrogen bonding between these two atoms to form a complex similar in geometry to the tetrahedral intermediate formed by EPSPS from S3P and PEP [7]. The simulations reveal that this mode of interaction is unlikely and also explain the unexpectedly low affinity of a cosubstrate of S3P and glyphosate [8].

REDOR experiments using [6-^{15}N]lysine-EPSPS in which the ^{15}N signal was dephased by ^{31}P suggested that 3 of the 17 lysine sidechains in EPSPS are near glyphosate and S3P. A TEDOR [9] experiment established that two of the lysines are 4.0 ± 0.3 Å from the phosphorus of glyphosate while one is 3.7 ± 0.4 Å from the S3P phosphorus. Another REDOR experiment with [6-^{15}N]lysine-EPSPS and [1-^{13}C]glyphosate revealed that one of the 3 lysines was 4.3 ± 0.3 Å from C1 of glyphosate. Proximity of S3P and glyphosate phosphorus to two arginines and histidine was discovered in experiments using [U-^{15}N]EPSPS and [^{15}N]histidine-EPSPS, respectively.

Chemical protection and site-directed mutagenesis experiments have implicated Lys22 [10, 11], Lys340 [12], Lys411 [13], and Arg27 plus another unidentified arginine [14] in the binding of S3P and glyphosate to EPSPS. These residues are located in the cleft of the EPSPS X-ray structure. S3P and glyphosate were docked into the crystal structure, and molecular dynamics simulations were performed on the complex using the glyphosate intramolecular and glyphosate/S3P inter-ligand distances as restraints. A restraint from Arg27 to S3P C1 was inferred from the crystal structure and additional restraints consistent with the REDOR NMR results linked Arg100 and Arg124 to S3P phosphorus; Lys22 to S3P phosphorus and glyphosate C1; and Lys340 and Lys411 to glyphosate phosphorus. The resulting orientation of glyphosate with respect to S3P is shown in Figure 1. Further REDOR studies of the ternary complex will test this model and provide insights into the mechanism by which glyphosate inhibits EPSPS.

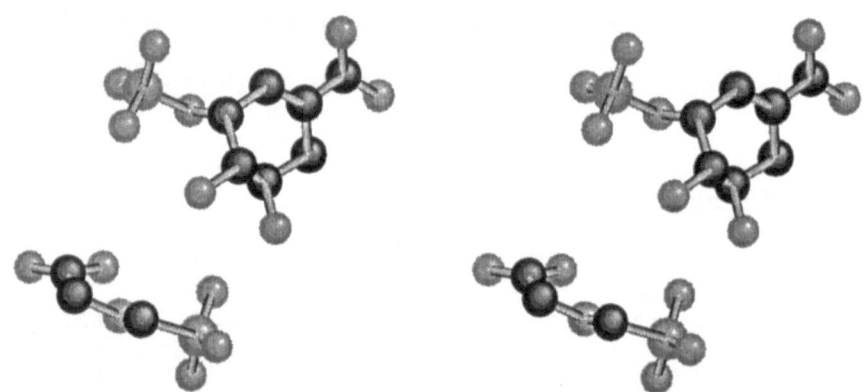

Fig. 1. Stereo view of the model of S3P and glyphosate bound to EPSP synthase consistent with solid-state REDOR NMR results. Hydrogens are omitted for clarity and O, P, and N atoms are shown in gray.

Acknowledgements

This work was supported by grants from the National Science Foundation and the National Institutes of Health (General Medicine). Monsanto Company supplied X-ray coordinates for EPSPS and Jeremy Evans provided [U-^{15}N]EPSPS.

References

1. Steinrucken, H.C. and Amrhein, N., Eur. J. Biochem., 143 (1984) 351.
2. Stallings, W.C., Abdel-Meguid, S.S., Lim, L.W., Shieh, H.-S., Dayringer, H.E., Leimgruber, N.K., Stegeman, R.A., Anderson, K.S., Sikorski, J.A., Padgette, S.R. and Kishore, G.M., Proc. Natl. Acad. Sci. U.S.A., 88 (1991) 5046.
3. Pan, Y., Gullion, T. and Schaefer, J., J. Magn. Reson., 90 (1990) 330.
4. Christensen, A.M. and Schaefer, J., Biochemistry, 32 (1993) 2868.
5. Knuuttila, P. and Knuuttila, H., Acta Chem. Scand. B., 33 (1979) 623.
6. Smith, P.H. and Raymond, K.N., Inorg. Chem., 27 (1988) 1056.
7. Anderson, K.S., Sikorski, J.A. and Johnson, K.A., Biochemistry, 27 (1988) 7395.
8. Marzabadi, M.R., Font, J.L., Gruys, K.J., Pansegrau, P.D. and Sikorski, J.A., Bioorg. Med. Chem. Lett., 2 (1992) 1435.
9. Hing, A.W., Vega, S. and Schaefer, J., J. Magn. Reson., 96 (1992) 205.
10. Huynh, Q.K., Kishore, G.M. and Bild, G.S., J. Biol. Chem., 263 (1988) 735.
11. Huynh, Q.K., Bauer, S.C., Bild, G.S., Kishore, G.M. and Borgmeyer, J.R., J. Biol. Chem., 263 (1988) 11636.
12. Huynh, Q.K., J. Biol. Chem., 265 (1990) 6700.
13. Huynh, Q.K., Arch. Biochem. Biophys., 284 (1991) 407.
14. Padgette, S.R., Smith, C.E., Huynh, Q.K. and Kishore, G.M., Arch. Biochem. Biophys., 266 (1988) 254.

The structure of the synthetic *M.tuberculosis* GroES protein

P. Mascagni[1], A.W.E. Chan[1], A.R.M. Coates[2], G. Fossati[1], P. Giuliani[1] and P. Lucietto[1]

[1]*Department of Peptide Chemistry, Italfarmaco Research Centre, Lavoratori 54, Cinisello B., 20092 Milan, Italy;*
[2]*St. George's Hospital Medical School, London, U.K.*

Introduction

Heat-shock proteins (hsp) belong to a family of highly conserved proteins whose MWs range from about 10 to 110 kD and are of general importance for protein folding and assembly. The members of the hsp60 sub-family (GroEL) have been shown to mediate the folding of many proteins *in vivo* and *in vitro* [1]. These so-called chaperones [2] are high MW complexes, consisting of 14, 60 kD subunits arranged in two stacked heptameric rings [3]. In the presence of ATP, GroEL forms a complex with the co-chaperone hsp10 (GroES), also a heptameric ring of identical 10 kD subunits. Binding of GroES to GroEL is required for full function in protein folding.

The molecular mechanism of chaperone action is still poorly understood and 3D structures of GroEL and GroES are not yet available.

After having shown that synthetic *E. coli* GroES assembles spontaneously into an active heptamer [4], we have now progressed to study the structure of GroES molecules. Since the dimensions of the latter make solution NMR unfeasible, we have used chemical synthesis coupled to CD spectroscopy, chromatography and molecular modeling to study the aggregation and structural properties of several *M. tuberculosis* hsp10 fragments and correlate them to those of the full-length protein.

Results and Discussion

The GroES protein (1-99) and its shorter fragments (26-99, 51-99, 59-99 and 75-99) were chemically synthesized using t-BOC chemistry and purified by RP-HPLC and ion-exchange chromatography. The immunological properties of the full-length protein were shown to superimpose with those of a sample of recombinant material.

At concentrations larger than 500 µg/ml synthetic and recombinant full-length protein, the 26-99, 51-99 and 59-99 fragments but not the 75-99 fragment aggregated to tetramers in contrast with the predicted heptameric structure based upon all other hsp10 molecules studied to date. At concentrations of about 10 µg/ml the protein was mainly dimeric while all other fragments remained as tetramers. Thus, one of the protein aggregation motifs was contained in the C-terminus half.

To predict the protein secondary structure CD spectroscopy was then used. In aqueous solution (pH 7.3) the CD spectrum of the 59-99 fragment had a large β contribution (minimum at 218 nm). The β contribution was only partly maintained in the 51-99 fragment and progressively decreased with increasing the chain length. In fact, the full-length protein had a spectrum which others have attributed to the so-called "random-coil" structure. To induce conformational changes which would allow the determination of secondary structure elements, CD studies were conducted as a function of pH, temperature and addition of organic solvents (MeOH, TFE and AcCN). Varying the pH from 2 to 10 did not change substantially the spectrum of the protein while it reinforced the β contribution in the shorter fragments at acidic pHs. Deconvolution of the 59-99 spectrum at pH 3 using two independent methods gave between 21% and 35% β structure with little or no α-helix contribution depending on the method used. The next step involved titration with organic solvents of the aqueous solutions containing either the hsp molecule or the shorter fragments. This induced conformational changes, summarized as follows: (i) the CD spectra in 70% MeOH were of the α+β type in all cases except for the full-length protein at high concentration and the 59-99 fragment whose spectra were of the α/β type; (ii) similar trends to that observed with MeOH were obtained using TFE and AcCN, although conformational changes occurred for different organic solvent concentrations.

To evaluate whether addition of MeOH had any effect on the aggregation states of the peptides as well as their conformation, similar titration studies were repeated and followed by size exclusion chromatography. This indicated that in 70% MeOH and for a peptide concentration of 500 µg/ml both full-length protein and fragments were dimeric. When the peptide concentrations were reduced to either 100 or 10 µg/ml, the protein became monomeric while the fragments changed their aggregation state from tetrameric in 100% buffer to dimeric in 70% MeOH. In conclusion, the above CD experiments when coupled to the S.E. chromatography results indicated that the GroES protein was characterized as having three different conformational states: a "random coil" conformation in aqueous solution, an α+β conformation at low concentration and in 70% MeOH and finally an α/β conformation at higher protein concentration and in 70% MeOH.

Before attempting a rationalization of these findings in terms of 3D models, sequence and structure homology studies were performed. These revealed that the 59-99 fragment had about 40% sequence homology with the 45-83 region of the MHC II 1 chain. This region has been proposed to adopt the same four β stranded conformation found at the base of MHC class I peptide binding groove. This and the CD results which had shown that the 59-99 fragment had a predominantly β structure in aqueous solution were used to model the peptide on MHC antigens.

Furthermore, inspection of the protein data bank revealed a second possible sequence homology, that between hsp10 and the 76 amino acid long stress protein ubiquitin. Interestingly, ubiquitin, whose 3D structure has been solved, gave a CD spectrum both in aqueous solution and in buffer-MeOH mixtures identical to that of monomeric hsp10 in 70% MeOH.

Based upon these observations, it was felt possible to explain the three conformational states detected for the hsp10 molecule:

1.   The higher concentration CD spectrum (α/β) in 70% MeOH was reminiscent of the structure of the 59-99 fragment in aqueous solutions. The structure of dimeric hsp10 may therefore correspond to that of the MHC II peptide binding

domain. Although the homology between the C-terminus half of GroES and MHC II might be coincidental, recent speculations, evolutionary relating the latter to the more ancient hsp molecules [5] support this conclusion.

2.   The overall sequence homology with ubiquitin and the fact that the latter has a CD spectrum identical to monomeric hsp10 were used to model the second conformational state on the published 3D structure of ubiquitin. The model obtained was consistent with the aggregation properties of the protein and its fragments. Thus, in the proposed ubiquitin-like hsp structure the β strand at the C-terminus (86-90) is H-bonded to another β strand, that between residues 2 and 7. In the shorter fragments (51-99 and 26-99) which lack the N-terminus sequence, the intramolecular interaction between the two strands is replaced by an intermolecular one, that between the C-terminus strands of two different molecules thus generating the dimeric structures detected by SE chromatography.

3.   Finally, the so-called "random coil" conformation detected in aqueous solutions could be explained either with the existence of conformational equilibria or more convincingly, as indicating that hsp10 is a compact molecule with defined secondary elements but lacking a well defined tertiary structure. A corollary to this hypothesis would be that hsp molecules adopt a "random coil" conformation to properly carry out their biological function which requires binding to different polypeptide chains. Our recent findings indicating that synthetic Rattus Norvegicus GroES forms heptameric structures with the expected chaperoning activity but with a CD spectrum of the "random coil" type support this conclusion.

References

1.   Ellis, R.J., Nature, 328 (1987) 378.
2.   Ellis, R.J., Van der Vies, S.M. and Hemmingsen, S.M., Biochem. Soc. Symp., 55 (1989) 145.
3.   Saibil, H., Dong, Z., Wood, S. and Auf der Mauer, A., Nature, 353 (1991) 25.
4.   Mascagni P., Tonolo, M., Ball, H.L., Lim, M., Ellis, R.J. and Coates, A.R.M., FEBS Lett., 286 (1991) 201.
5.   Parham, P., Trends Biochem. Sci., 16 (1991) 357, and references therein.

Defining the active conformation of gonadotropin releasing hormone (GnRH) through design and conformational analysis of constrained GnRH analogs

J. Rizo[1], R.B. Sutton[1], J. Breslau[1], S.C. Koerber[2], J. Porter[2], J.E. Rivier[2], A.T. Hagler[3] and L.M. Gierasch[1]

[1]Department of Pharmacology, University of Texas Southwestern Medical Center, Dallas, TX 75235-9041, U.S.A.
[2]The Clayton Foundation Laboratories for Peptide Biology, The Salk Institute, La Jolla, CA 92307, U.S.A.
[3]Biosym Technologies Inc., San Diego, CA 92121, U.S.A.

Introduction

GnRH is a linear decapeptide that regulates ovulation and spermatogenesis by stimulating the secretion of the gonadotropins by the pituitary. In order to define the conformational requirements for GnRH antagonist activity, we are applying a multidisciplinary approach that involves the design and synthesis of progressively more constrained analogs, in combination with analysis by nuclear magnetic resonance (NMR) and computational techniques. Our ultimate goal is to design peptidomimetics that could be used as non-steroidal contraceptive agents and for treatment of sex steroid dependent pathologies. Our initial analysis of a marginally potent, monocyclic (1-10) analog (1, Table 1) led to the design of monocyclic (4-10) analogs with high antagonist activity [1]. Analysis of one of these antagonists (2, Table 1) by NMR and molecular dynamics showed that the (4-10) cycle adopts a β-hairpin conformation with a Type II' β turn around residues 6-7 [2, 3]. In this conformation, the side chains of residues 5 and 8, as well as the tail formed by residues 1-3, are located on the same side of the molecule, suggesting new bridging possibilities to design more constrained GnRH antagonists. Dicyclic (4-10/5-8) antagonists with high potency have been obtained [4], and structural analysis of one of these compounds (3, Table 1) revealed a very similar conformation to that found for the parent monocyclic analog 2 [5]. Here we report the conformational analysis of another dicyclic (4-10/5-8) antagonist (4, Table 1) and of the first dicyclic compound with a 1-8 bridge that displays a significant GnRH antagonist activity (5, Table 1).

Results and Discussion

The conformational behavior of 4 in $CDCl_3/DMSO-d_6$ 2:1 (v/v) was studied by a combination of NMR and computational techniques. A total of 70 restraints deduced from nuclear Overhauser effects (NOEs), coupling constants and temperature

Table 1 *Biological potency of constrained GnRH antagonists*

	compound	AOA[a]
1	cyclo(1-10)[Δ^3Pro^1,D-pClPhe^2,D-Trp^3,D-Trp^6, NMeLeu^7,βAla^{10}]GnRH	1000 (5/8)
2	cyclo(4-10)[Ac-Δ^3Pro^1,D-pFPhe^2,D-Trp^3, Asp^4,D-2Nal^6,Dpr^{10}]GnRH	10 (2/10)
3	dicyclo(4-10/5-8)[Ac-D-2Nal^1,D-pClPhe^2,D-Trp^3, Asp^4,Glu^5,D-Arg^6,Lys^8,Dpr^{10}]GnRH	5 (2/10)
4	dicyclo(4-10/5,5'-8)[Ac-D-2Nal^1,D-pClPhe^2,D-Pal^3, Asp^4,Glu^5(Gly),D-Arg^6,Dbu^8,Dpr^{10}]GnRH	5 (2/8)
5	dicyclo(1-8/4-10)[Ac-D-Glu^1,D-pClPhe^2,D-Pal^3, Asp^4,Arg^5,D-Nal^6,Lys^8,Dpr^{10}]GnRH	100 (3/6)

[a]AOA-antiovulatory assay in the rat: dosage in μg (rats ovulating/total rats).

coefficients of amide protons were incorporated into simulated annealing and molecular dynamics calculations to obtain structures compatible with the NMR data. A representative structure is shown in Figure 1. Rather than forming a Type II' β turn around residues 6-7, which has been proposed to be essential for the biological activity of GnRH, **4** adopts a Type I' β-turn conformation around these residues. In addition, a Type II β turn is formed around residues 5-6. The tail formed by residues 1-3 is flexible, as was observed for **2** and **3**, but is oriented outwards with respect to the 4-10 ring. These results show that, despite the dicyclic nature of **3** and **4**, the 4-10 backbone is able to adopt at least two distinctly different conformations, and suggest that the presence of a turn around residues 6-7, rather than the type of turn, is important for GnRH antagonist activity. Our results also stress the necessity to define the active conformation in the N-terminal residues.

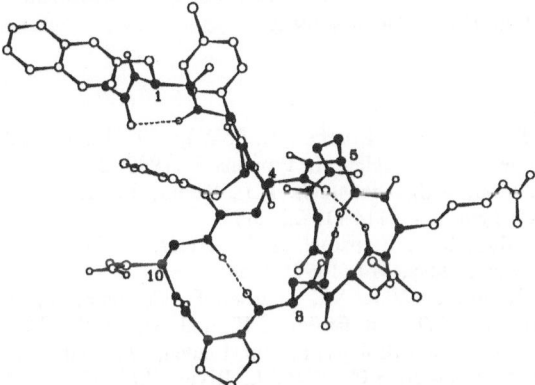

Fig. 1. *A representative structure of the dicyclic (4-10/5,5'-8) analog **4**. Only heavy atoms and amide protons are shown. Solid circles correspond to atoms of the backbone and of the bridging side chains. Dashed lines represent hydrogen bonds.*

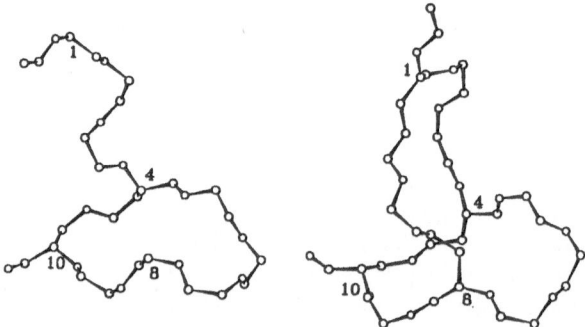

Fig. 2. Comparison of the backbone conformations of the monocyclic (4-10) analog 2 (left) and of the dicyclic (1-8/4-10) analog 5 (right). Only N and C atoms of the backbone and the bridging side chains are shown.

Using the same methodology outlined above, we have initiated the analysis of compound **5**, where a 1-8 bridge constrains the conformation of the N-terminus. The NMR data observed for this compound indicate strong similarity between its conformation and that adopted by the parent monocyclic analog **2**, including the Type II' β-turn conformation around residues 6-7 (Figure 2). A number of NOEs between residues 4 and 8 suggest that the conformation of **5** is highly rigid. The bridge between the N-terminus and the Lys[8] side chain is located above the 4-10 ring, on the same side as the Arg[5] side chain. A conformation similar to a Type II β turn is observed around residues 1-2. This feature could be important for biological activity, as it was observed previously in the monocyclic (1-10) analog **1** [6], and was suggested to be populated in **2** [2]. We are currently optimizing the potency of the dicyclic (1-8/4-10) series of GnRH antagonists. Conformational analysis of the most potent compounds, in combination with the information obtained for the dicyclic (4-10/5-8) antagonists, should lead to a more complete definition of the structural requirements for GnRH antagonist activity.

References

1. Struthers, R.S., Tanaka, G., Koerber, S., Solmajer, T., Baniak, E.L., Gierasch, L.M., Vale, W., Rivier, J. and Hagler, H., Proteins, 8 (1990) 295.
2. Rizo, J., Koerber, S.C., Bienstock, R.J., Rivier, J., Hagler, A.T. and Gierasch, L.M., J. Am. Chem. Soc., 114 (1992) 2852.
3. Rizo, J., Koerber, S.C., Bienstock, R.J., Rivier, J., Gierasch, L.M. and Hagler, A.T., J. Am. Chem. Soc., 114 (1992) 2860.
4. Rivier, J.E., Rivier, C., Vale, W., Koerber, S., Corrigan, A., Porter, L., Gierasch, L.M. and Hagler, A.T., In Rivier, J.E. and Marshall, G.R. (Eds.), Peptides: Chemistry, Structure and Biology, Escom, Leiden, The Netherlands, 1990, p.33.
5. Bienstock, R.J., Koerber, S.C., Rizo, J., Rivier, J.E., Hagler, A.T. and Gierasch, L.M., In Smith, J.A. and Rivier, J.E. (Eds.), Peptides: Chemistry and Biology, Escom, Leiden, The Netherlands, 1992, p.262.
6. Baniak, E.L. II, Rivier, J.E., Struthers, R.S., Hagler, A.T. and Gierasch, L.M., Biochemistry, 26 (1987) 2642.

The interactive behaviour of amphipathic peptides with hydrophobic surfaces

A.W. Purcell, M.T.W. Hearn and M.I. Aguilar

Department of Biochemistry and Centre for Bioprocess Technology, Monash University, Wellington Rd, Clayton, Victoria 3168, Australia

Introduction

The development of interactive modes of chromatography to investigate the physicochemical nature of peptide and protein surface interactions has advanced considerably over the past decade. In the present study, the interactive behaviour of a series of peptides designed to adopt amphipathic helical structure has been investigated using reversed phase liquid chromatography (RP-HPLC).

Results and Discussion

Parameters which describe the hydrophobic contact area and the solute affinity for the stationary phase ligands were determined for the P-series of peptides listed in Table 1 over a temperature range of 5-85°C. The retention of peptides in RP-HPLC can be described by the following expression:

$$\log \bar{k} = \log k_o - S \bar{\psi}$$

The parameters S and $\log k_o$ were determined from plots of $\log \bar{k}$ *versus* $\bar{\psi}$ and are related to the hydrophobic contact area of the peptide and the affinity of this contact region for the hydrophobic ligands [1]. The influence of ligand hydrophobicity was assessed by comparison of peptide retention on C4 and C18 stationary phases.

Table 1 *Peptide sequences*

Peptide	Sequence
P1	H_2N-KSEEQLA-COOH
P2	H_2N-KSEEQLAKSEEQLA-COOH
P3	H_2N-KSEEQLAKSEEQLAKSEEQLA-COOH
P4	H_2N-KSEEQLAKSEEQLAKSEEQLAKSEEQLA-COOH

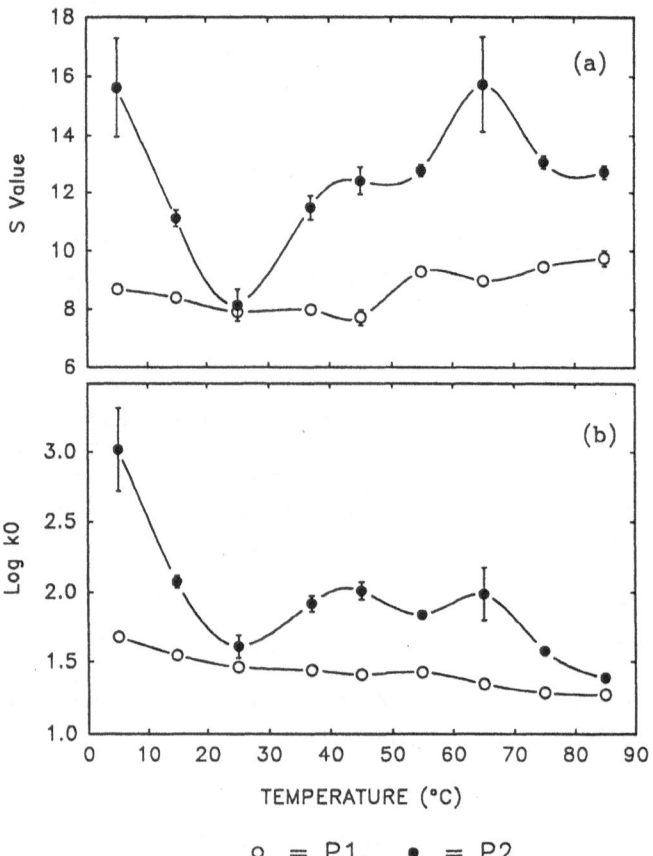

Fig. 1. The dependence of S and log k_o on temperature for P1 and P2 separated on a C18 sorbent.

Figures 1 and 2 show the plots of S and log k_o *versus* temperature for the P-peptides separated on an *n*-octadecylsilica (C18) sorbent with acetonitrile as the organic solvent. It can be seen that the dependence of the parameters becomes more complex as peptide length is increased. The results correlate with the CD analysis of these peptides in acetonitrile and trifluoroethanol which indicate that a significant degree of α-helix can be induced with P3 and P4 in hydrophobic environments. The changes in the S and log k_o values may thus correspond to the induction of helical structure during the interaction of these peptides with hydrophobic ligands. The results also demonstrate that ligand-induced effects can stabilize peptide structure depending on the chromatographic residence time and peptide length.

The chromatographic bandwidths also permit the kinetics of conformational interconversions to be investigated in terms of time-dependent changes in the conformation and molecular composition of the hydrophobic contact area established between the peptide and the chromatographic ligands [2]. Additional information on

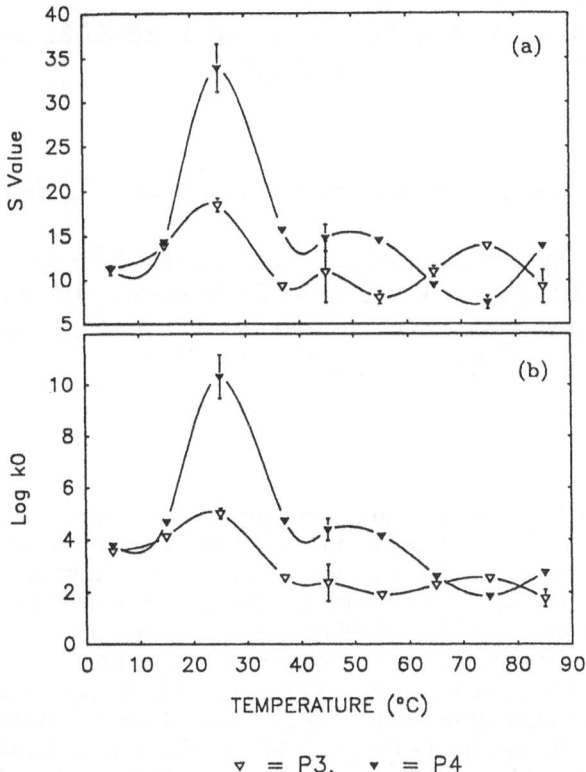

$$\triangledown = P3, \quad \blacktriangledown = P4$$

Fig. 2. The dependence of S and log k_o on temperature for P3 and P4 separated on a C18 sorbent.

the interactive dynamics of the P-peptides was thus obtained from analysis of bandwidth dependencies over the different chromatographic conditions. For P1 and P2 planar dependencies of bandwidth were observed which is consistent with these peptides chromatographing as random coils. For P3 and P4 large increases in bandwidth were observed at discrete temperatures and gradient times which correspond to the experimental conditions at which maxima in the S and log k_o values were observed. These observations indicate that the structure of these longer peptides was undergoing conformational transitions during the chromatographic timescale. Furthermore, these changes in bandwidth were seen at lower temperatures on the C4 sorbent compared to the C18 sorbent, demonstrating that the longer more hydrophobic ligand acts to stabilize the secondary structure.

These studies have provided further insight into the contribution of peptide secondary structure to peptide retention behaviour and the role of hydrophobic forces in the stabilization of peptide structures in RP-HPLC.

References

1.   Purcell, A.W., Aguilar, M.I. and Hearn, M.T.W., J. Chromatogr., 593 (1992) 103.
2.   Purcell, A.W., Aguilar, M.I. and Hearn, M.T.W., Anal. Chem., 65 (1993) 3038.

How do magainins interact with neutral and acidic lipids?

J. Blazyk[1], J. Hua[1], A.W. Hing[2] and J. Schaefer[2]

[1]*Chemistry Department, Molecular and Cellular Biology Program, and College of Osteopathic Medicine, Ohio University, Athens, OH 45701, U.S.A. and* [2]*Chemistry Department, Washington University, St. Louis, MO 63130, U.S.A.*

Introduction

Magainin 2, a 23-amino acid cationic peptide, was isolated by Zasloff in 1987 from the skin of the female African clawed frog, *Xenopus laevis*. This peptide possesses antimicrobial activity against a broad range of organisms, is relatively nonhemolytic and has the potential to form a highly amphiphilic α-helical conformation with a large hydrophobic moment and an overall length of about 30 Å. It soon became clear that the bactericidal activity of magainin is associated with its ability to disrupt bacterial membranes. Current evidence indicates that magainin may form cation-selective channels [1]. Also, results from ^{15}N NMR suggest that magainin may lie parallel to the plane of the bilayer as α-helical rods at the membrane surface [2]. In contrast, Williams et al. [3] found that while magainins induce leakage of calcein and carboxyfluorescein from unilamellar liposomes, they do not readily penetrate the bilayer surface subsequent to binding, but cause a disordering of the fatty acyl chains of acidic phospholipids and, at high peptide concentrations, can solubilize lipid bilayers. Based upon this evidence, these workers dispute the channel-forming theory and propose that magainins may disrupt membrane structure by a mechanism similar to detergents.

Results and Discussion

FT-IR spectroscopy was used to study lipid and peptide structure. Using the symmetric C–H stretching band near 2850 cm^{-1} to monitor the fluidity of the lipid fatty acyl chains, an ordering effect is observed in both DPPC and dipalmitoyl-phosphatidylglycerol (DPPG) which is proportional to peptide concentration. In each case, a shift in T_m is observed at molar lipid to peptide ratios below 100:1. For DPPC, T_m increases to >45°C at a lipid to peptide ratio of 20:1, but does not change at higher levels of peptide. In contrast, at lipid to peptide ratios below 10:1, the T_m for DPPG decreases and the fatty acyl chains become more fluid. In order to assess the conformation of magainin 2 amide at a lipid to peptide ratio of 20:1, the resolution of the amide I band (after appropriate subtraction using the pure lipid in D$_2$O buffer) was enhanced by calculating the second derivative. These band shapes are essentially identical to those obtained by Fourier self-deconvolution. For each, the major component occurs near 1650 cm^{-1} with a smaller component near 1635

cm^{-1}. The 1650 cm^{-1} band is generally attributed to α-helical secondary structure, while the 1635 cm^{-1} band may indicate β-sheet (or possibly 3_{10}-helix) structure. These results are similar to those reported by Jackson et al. [4]. From these data, several qualitative observations can be made. For both DPPC and DPPG, the peptide appears to be predominantly, but certainly not completely, α-helical. Also, the ratio of the 1650 to 1635 cm^{-1} components increases when the lipid undergoes a phase change from the gel to liquid crystal state, which may indicate greater α-helical structure in a fluid lipid environment. Further IR studies are continuing with magainin 2 amide and other more potent antimicrobial peptides.

We are using a new solid-state NMR technique developed by Schaefer and coworkers called rotational-echo double-resonance (REDOR) [5, 6] for the direct measurement of dipolar coupling between stable-isotope-labeled nuclei in the same molecule or between labels in different molecules. From the intensity of dipolar coupling, interatomic distances can be estimated. In the experiment shown here, magainin 2 containing ^{15}N-labeled Ala at position 9 was added at a 1:5 molar ratio to DPPG, which has a phosphorus atom in the polar headgroup. The experiment was performed at –50°C to minimize motional averaging. For the ^{31}P-containing lipid and ^{15}N-labeled peptide, observable dipolar coupling may be expected if the distance between the two nuclei is less than ~5 Å. In the REDOR experiment, the ^{31}P rotational spin echoes may be prevented from reaching full intensity by inserting two ^{15}N π pulses per rotor period, with the P-N dipolar coupling determining the level of dephasing. The difference between the ^{15}N-pulsed ^{31}P spectrum and one where no ^{15}N π pulses are applied measures the extent of ^{31}P-^{15}N coupling, which is inversely proportional to the distance between these nuclei. A loss of intensity in the ^{15}N-pulsed ^{31}P spectrum would indicate that at least some of the ^{15}N-Ala in magainin is very close to the polar headgroup of DPPG. Although the results of this experiment showed no detectable dipolar coupling, this does not rule out the possibility that the peptide resides at the bilayer surface. Complications arise from the possibility of motional averaging (even at –50°C) and the short distance for which coupling can be measured between these two nuclei. Another problem stems from the excess of DPPG, since even if most or all of the magainin molecules were interacting with DPPG polar groups, only a relatively small population of the ^{31}P atoms in the lipids might be expected to be close enough to ^{15}N-Ala in magainin to experience dipolar coupling. If the average interatomic distance were not substantially less than 5 Å, no difference signal could be expected. Future REDOR experiments will measure an intramolecular ^{13}C-^{15}N distance between amino acids separated by four residues to test for α-helical structure when the peptide is bound to lipid bilayers.

References

1. Cruciani, R.A., Barker, J.L., Durell, S.R., Raghunathan, G., Guy, H.R., Zasloff, M. and Stanley, E.F., Eur. J. Pharmacol., 226 (1992) 287.
2. Bechinger, B., Zasloff, M. and Opella, S.J., Biophys. J., 62 (1992) 12.
3. Williams, R.W., Starman, R., Taylor, K.M.P., Gable, K., Beeler, T., Zasloff, M. and Covell, D., Biochemistry, 29 (1990) 4490.
4. Jackson, M., Mantsch, H.H. and Spencer, J.H., Biochemistry, 31 (1992) 7289.
5. Gullion, T., Poliks, M.D. and Schaefer, J., J. Mag. Res., 80 (1988) 553.
6. Gullion, T. and Schaefer, J., J. Mag. Res., 81 (1989) 196.

The conformation of amphipathic peptides in lipid complexes

R.J. Cushley, A.S. Tracey, Q. Zhong, W.D. Treleaven and G. Wang

Department of Chemistry, Simon Fraser University, Burnaby, British Columbia, V5A 1S6, Canada

Introduction

The lipid-associating domains of the exchangeable human serum apolipoproteins A, C and E are purported to form a series of amphipathic α-helical structures. The hydrophobic face of the helix is proposed to be in contact with the phospholipid side chains in the lipoprotein complex. The hydrophilic face is exposed to the aqueous phase and the charged side chains may then form associations with the zwitterionic phospholipid headgroups [1]. Only the lipid-bound protein is active physiologically, i.e. binds to receptors or activates enzymes. To test the amphipathic helix motif we prepared two peptides which we studied in complexes with perdeuterated dodecylphosphocholine (DPC-d_{38}) by 2-dimensional NMR. The peptides are: **1)** a segment (residues 35-53; SAKMREWFSETFQKVKEKL) of the lipid associating domain of apolipoprotein C-I, and **2)** the lipid associating peptide LAP-20 (VSSLLSSLKEYWSSLKESFS) [2]. LAP-20 in lipid complexes was found to activate the enzyme LCAT to approximately 65% the extent that apolipoprotein A-I (M_r=28,083) did. Both peptides are excellent amphipathic helical structures based on predictive methods and helical wheel projections.

Results and Discussion

NMR spectra (400 and 600 MHz) were run at 25°C (LAP-20) or 37°C (C-I peptide). 2-Dimensional DQF-COSY, TOCSY and NOESY spectra were collected in pure phase absorption mode using TPPI. From the 2D NMR spectra H^α-protons were assigned which were confirmed by NOE connectivities. H^α secondary shifts (Figure 1) compare an amino acid residue in an ordered structure and in random coil, with upfield shifts characteristic of a helical conformation and downfield shifts of a β-structure [3, 4].

From Figure 1 LAP-20 was shown to adopt a helix in DPC-d_{38} complexes except for residues at the C- and N-terminal ends. C-I peptide, which is predicted to be a better amphipathic helix than LAP-20, appeared to adopt a two domain structure. The C-terminal end formed a helical structure of about two turns, starting at residue 10. This was confirmed by NOE connectivities. H^N_i–H^N_{i+1} NOE cross peaks were measured for approximately ten residues of C-I peptide in DPC-d_{38} complexes. As well, we found consecutive NOE cross peaks of H^α_i–H^β_{i+3} and H^N_i–H^N_{i+2} for residues 11 through 18. The overall NOESY cross peak patterns suggested that the C-terminal end of the peptide adopted a loosened helix, which spanned about nine or ten residues. A second helix indicated in Figure 1, spanning residues 7 - 9, was not supported by the NOE connectivity data which indicated the N-terminal half of the molecule was unstructured in complexes. According to the amphipathic helix theory, the entire length of such a peptide should bind to lipids.

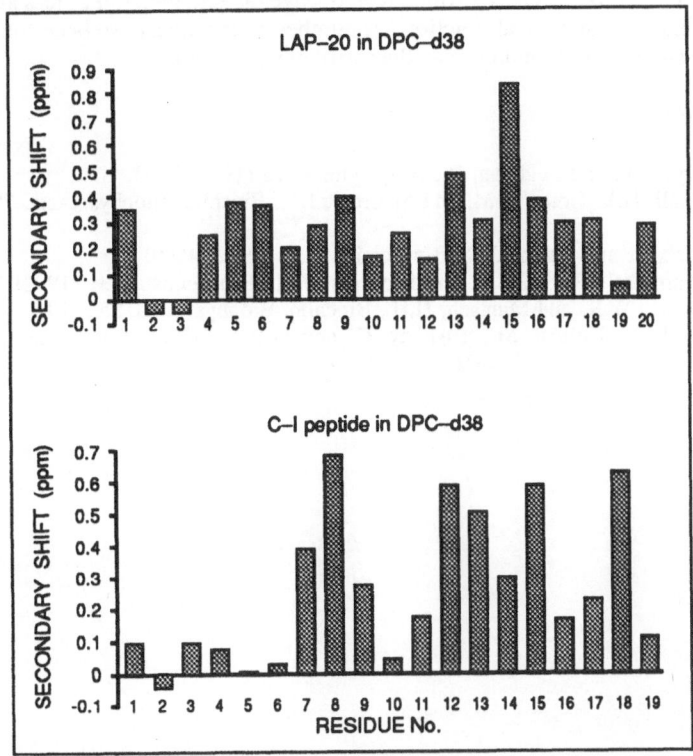

Fig. 1. H^α secondary shifts of peptides in lipid complexes. Upfield is positive.

The 2D NMR results for LAP-20 in DPC-d$_{38}$ complexes were compared with estimates of secondary structure determined by deconvolution of CD and FTIR spectra. FTIR of LAP-20/DPC-d$_{38}$ micellar complexes were deconvoluted and subjected to curve-fitting with variable Gaussian-Lorentzian band shapes. Protein secondary structures were determined from the amide I region [5]. Using CD spectra of LAP-20/DPC-d$_{38}$ complexes peptide secondary structures were estimated from the convex constraint analysis (CCA) deconvolution method [6] and LINCOMB, a least squares fit to a reference set, obtained from CCA. Results: α-helix; 75% (NMR), 23% (IR) and 65% (CD); β-structure 0% (NMR), 37% (IR) and 5% (CD); turns/random: 25% (NMR), 40% (IR) and 30% (CD). CD is supposed to over-estimate helical content but in phosphocholine micellar complexes this does not appear to be the case. All three techniques predict significant amounts of "less ordered" structures (turns, random) which is similar to what we observed with C-I peptide in DPC-d$_{38}$ by 2D NMR. IR is supposed to overestimate β-structure. A band at 1638 cm^{-1} in the IR spectrum, ≈2/3 the intensity of the α-helical band, falls within the "normal" range for β-structure so that appears to be the case with LAP-20 in complexes. Binding to lipids does not shift IR bands from their normal range.

Results of using small peptides to model apolipoprotein behavior have been erratic. Our study indicated that some amphipathic regions of the human serum apolipoproteins may, in fact, serve a dual function. That is, only part of the segment binds to the lipoprotein surface which would leave the remainder, the "less structured" part, available for biological function. A number of enzymes have been found where the active site is less structured than the surrounding scaffold.

References

1. Segrest, J.P. and Feldman, R., Biopolymers, 16 (1977) 2053.
2. Pownall, H.J., Gotto, A.M. and Sparrow, J.T., Biochim. Biophys. Acta, 793 (1984) 149.
3. Saudek, V. and Pelton, J.T., Biochemistry, 29 (1990) 4509.
4. Wishart, D.S., Sykes, B.D. and Richards, F.M., Biochemistry, 31 (1992) 1647.
5. Surewicz, W.K. and Mantsch, H.H., Biochim. Biophys. Acta, 952 (1988) 115.
6. Perczel, A., Hollósi, M., Tusnàdy, G. and Fasman, G.D., Protein Eng., 4 (1991) 669.

Interaction of the replication primer tRNA^{Lys} with the HIV-1 nucleocapsid protein NCp7: Structural properties of zinc finger motifs monitored by fluorescence measurements

H. de Rocquigny[1], Y. Mely[2], C.Z. Dong[1], D. Ficheux[1], D. Gérard[2], J.L. Darlix[3], M.C. Fournié-Zaluski[2] and B.P. Roques[2]

[1]U 266 INSERM, URA 1500 CNRS, UFR des Sciences Pharmaceutiques et Biologiques, 4 avenue de l'Observatoire, 75007 Paris, France, [2]UA 491 CNRS, Université Louis Pasteur, 67401 Illkirch, France, [3]LaboRetro, Ecole Normale Supérieure, 69000 Lyon, France

Introduction

In the core of the human immunodeficiency virus type 1 (HIV-1), the diploid genomic RNA is tightly associated with the 72 amino acid nucleocapsid protein designated NCp7 (Figure 1). This protein is composed of two successive zinc finger domains of the type $CX_2CX_4HX_4C$ surrounded by basic rich sequences. *In vitro*, NCp7 activates viral RNA dimerization and promotes annealing of the replication primer tRNA^{Lys} to genomic RNA [1, 2]. This annealing process is the first step of the initiation of reverse transcription and therefore is of importance for the synthesis of the provirus.

NCp7

To investigate the interaction between tRNA and NCp7, studies were performed using commercially available tRNA^{Phe} and natural and synthetic tRNA^{Lys}. The protein-RNA interaction was monitored by the fluorescence quenching of tryptophan residues found in NCp7 and NCp7 derived peptides [3].

Preliminary experiments were performed by titrating the two separate zinc fingers with tRNA^{Phe}, one of these being (13-30)NCp7 where Phe^{16} was replaced by a Trp residue. To characterize the role of the zinc finger in this interaction, a series of NCp7 derived peptides was synthesized: *S^{15}, S^{18}, S^{28}, S^{36}, S^{39}, S^{49} hexa acetamidomethyl-NCp7* (AcmNCp7) in which the six cysteine residues cannot bind the zinc and **CCHC boxes deleted-NCp7** (1-72)NCp7 where the zinc fingers were replaced by a glycine-glycine dipeptide. Both peptides were compared to native NCp7. Moreover, to probe further the structural relevance of the CCHC box, a third peptide **C^{23}(13-64)NCp7** was synthesized in which His^{23} was substituted by a cysteine residue. This modification has been demonstrated by NMR spectroscopy to induce an overall change of peptide conformation [4]. This peptide was titrated with tRNA and compared with (13-64)NCp7.

Results and Discussion

Reverse titration was performed by following the fluorescent behavior of tryptophan residues of NCp7 and NCp7 derived peptides during the addition of small aliquots of tRNA to a given peptide concentration. Binding isotherms were analyzed using the McGhee and Von Hippel [5] equation for non-cooperative binding. Moreover, the method of Record et. al. was used to differentiate the electrostatic and the non-electrostatic forces which stabilize the complex [6].

Table 1 *tRNA binding properties of NCp7 and NCp7 derived peptides*

	[NaCl] (mM)	K (M^{-1})	n	RNA
W^{16}(13-30)NCp7	0	1.2 ($\pm$0.2) 10^4	3 ($\pm$0.5)	tRNAPhe
(34-51)NCp7	0	4.5 ($\pm$0.5) 10^3	2.9 ($\pm$0.7)	tRNAPhe
NCp7	100	2 ($\pm$0.2) 10^6	7.5 ($\pm$0.5)	Synthetic tRNA^{Lys}
NCp7	500	1.2 ($\pm$0.3) 10^4	6.5	Synthetic tRNA^{Lys}
Acm-NCp7	100	5.6 ($\pm$0.7) 10^5	5.6 ($\pm$0.5)	Synthetic tRNA^{Lys}
(13-64)NCp7	100	1.5 ($\pm$0.3) 10^6	7.6 ($\pm$0.5)	Synthetic tRNA^{Lys}
C^{23}(13-64)NCp7	100	1.5 ($\pm$0.4) 10^5	7.5 ($\pm$0.3)	Synthetic tRNA^{Lys}

Using the fluorescent properties of tryptophans, we have monitored the binding of tRNA^{Phe} to the two separate zinc finger motifs of NCp7. In both peptides, the binding of tRNA^{Phe} induced an almost complete quenching of Trp fluorescence. This phenomenon suggests that the quenching corresponds to the stacking of the indole moiety of Trp to tRNA^{Phe} bases [7]. This non-electrostatic interaction is clearly the main driving force stabilizing the interaction. Nevertheless, increasing NaCl concentrations led to reduction of (34-51)NCp7 affinity. The affinity of (13-30W^{16})NCp7 for tRNA^{Phe} is three fold higher than the (34-51)NCp7 C-terminal domain, which we assume is due to an additional electrostatic interaction, probably involving basic residues (Lys^{14} and/or Arg^{29}) (Table 1). Interestingly, these amino acids were found to be crucial for "in vitro" annealing activities of NCp7 [2].

To investigate the role of the structure NCp7 [8] for the interaction with tRNA, a series of peptides was synthesized and compared to native NCp7 or to (13-64)NCp7. This latter peptide is as active as NCp7 for the *in vitro* annealing of tRNA to genomic RNA [2].

The affinities of Acm-NCp7 and (1-72)NCp7 for the tRNA were one and two orders of magnitude less, respectively, compared to native NCp7 (Table 1). These results show that the presence and the folding of each CCHC box around a zinc atom is of importance for the stable interaction of NCp7 with tRNA. To support this data, C^{23}(13-64)NCp7 possessing a CCCC box was synthesized. This peptide was titrated by tRNA^{Lys} and the results show a loss of affinity (one order of magnitude) compared to (13-64)NCp7 (Table 1). In fact, the characterization of C^{23}(13-64)NCp7 by NMR spectroscopy showed that the CCCC box had a different conformation leading to a drastic modification of the structure of the mutant peptide [4].

In conclusion, the three dimensional structure of individual CCHC boxes induces a specific conformation of NCp7, critical for recognition of target RNA by correct exposition of basic and aromatic residues. Experiments are now currently underway to extend these studies to genomic RNA and to develop a test for screening molecules aimed at inhibiting tRNA recognition by NCp7.

References

1. Darlix, J.L., Gabus, C., Nugeyre, M.T., Clavel, F. and Barre-Sinoussi, F., J. Mol. Biol., 216 (1990) 689.
2. de Rocquigny, H., Gabus, C., Vincent, A., Fournié-Zaluski, M.C., Roques, B.P. and Darlix, J.L., Proc. Natl. Acad. Sci. U.S.A., 89 (1992) 6472.
3. de Rocquigny, H., Ficheux, D., Gabus, C., Fournié-Zaluski, M.C., Darlix, J.L. and Roques, B.P., Biochem. Biophys. Res. Commun., 180 (1991) 1010.
4. Demene, H. and Roques, B.P., in preparation.
5. McGhee, J.D. and von Hippel, P.H., J. Mol. Biol., 86 (1974) 469.
6. Record, Jr., M.T., Lohman, T.M. and de Haseth, P., J. Mol. Biol., 107 (1976) 145.
7. Mely, Y., Sorinas-Jimeno, M., de Rocquigny, H., Jullian, N., Morellet, N., Roques, B.P. and Gerard, D., Biophys. J., in press.
8. Morellet, N., Jullian, N., de Rocquigny, H., Maigret, B., Darlix, J.L. and Roques, B.P., EMBO J., 11 (1992) 3059.

Structure-activity relationship of small cyclic RGD peptides as inhibitors of platelet aggregation

M. Eguchi[1], G.B. Crull[1], Q. Dong[1], I. Ojima[1] and B.S. Coller[2]

[1]Department of Chemistry, and Division of Hematology;
[2]School of Medicine, State University of New York at Stony Brook,
Stony Brook, NY 11794, U.S.A.

Introduction

Three-dimensional structures of biologically active peptides are related to their biological function at the molecular level. The precise 3D structure of a peptide in solution can be determined using multidimensional NMR techniques combined with computational methods, such as constrained distance geometry search and molecular dynamics [1, 2]. In our study on the structure-activity relationship (SAR) of small cyclic RGD peptides as inhibitors of platelet aggregation, we found that the inhibitory activity depends on the ring size and its adjacent sequence. In this report, we focus on Phe-Pro sequence in small cyclic RGD peptides, such as cyclo[Arg-Gly-Asp-Phe-Pro] and cyclo[Arg-Gly-Asp-(D)Phe-Pro].

Results and Discussion

Protected linear peptides were prepared on solid support using a commercially available methoxy-alkoxybenzyl alcohol resin. The cyclization of these linear peptides was carried out using DPPA as the coupling reagent followed by deprotection with TFA/thioanisole [3]. Table 1 summarizes the platelet aggregation inhibitory activity of cyclic RGD peptides. The replacement of Phe with (D)Phe in the cyclic peptides exerts a remarkable effect on the inhibitory activity.

The NMR data show that most of the small cyclic RGD peptides are conformationally rigid in water or DMSO, displaying a well-defined geometry of the RGD sequence. Possible backbone structures consistent with the experimental NOE data were identified by means of SYBYL™ 5.5 and 6.0 programs on a Silicon Graphics Iris workstation. Figure 1 exemplifies the possible solution structures of cyclo[Arg-Gly-Asp-Phe-Pro] (**I**, IC_{50} 8.8 µM) and cyclo[Arg-Gly-Asp-(D)Phe-Pro] (**II**, IC_{50} >200 µM) on the basis of 2D NOE data acquired with a 600 MHz NMR spectrometer. The significant difference between these structures is the spatial orientation of aspartyl and arginyl side chains. In the structure of **I**, the side chains point toward the outside of the ring system, while these side chains are perpendicular to the ring in the structure **II**. The observed difference in the conformation appears to be responsible for the marked difference in the inhibitory activity.

Table 1 *Platelet aggregation inhibitory activity of cyclic RGD peptides*

Peptide	IC_{50} $(\mu M)^a$
Cyclo[Arg-Gly-Asp-Phe]	12
Cyclo[Arg-Gly-Asp-Phe-Pro]	8.8
Cyclo[Arg-Gly-Asp-(*D*)Phe-Pro]	>200
Cyclo[Arg-Gly-Asp-Phe-Pro-Gly]	>200
Cyclo[Arg-Gly-Asp-(*D*)Phe-Pro-Gly]	9.2
Cyclo[Arg-Gly-Asp-Phe-Pro-Ala-Gly]	3.7
Cyclo[Arg-Gly-Asp-(*D*)Phe-Pro-Ala-Gly]	>200
Arg-Gly-Asp-Phe	17

[a]Freshly prepared platelet-rich plasma with ADP as the initiator was employed.

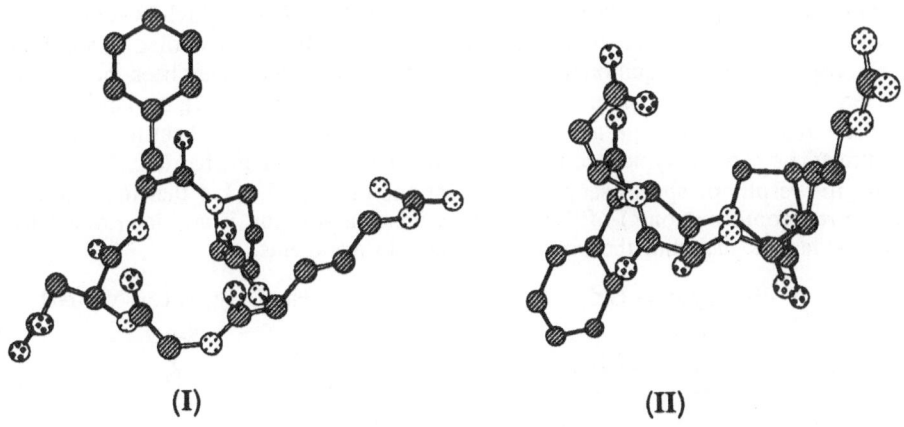

(I)   **(II)**

Fig. 1. *Molecular model of cyclo[Arg-Gly-Asp-Phe-Pro] (I) and cyclo[Arg-Gly-Asp-(D)Phe-Pro] (II).*

References

1. Kopple, K.D., Baures, P.W., Bean, J.W., D'Ambrosio, C.A., Hughes, J.L., Peishoff, C.E. and Eggleston, D.S., J. Am. Chem. Soc., 114 (1992) 9615.
2. McDowell, R.S. and Gadek, T.R., J. Am. Chem. Soc., 114 (1992) 9245.
3. Sugita, T., Hamada, Y. and Shioiri, T., Tetrahedron Lett., 28 (1987) 2251.

Solution conformations of the cyclic Cys=Cys disulfide structural motif

C. Garcia-Echeverría[*] and D.H. Rich

*School of Pharmacy and Department of Chemistry,
University of Wisconsin-Madison, 425 North Charter St.,
Madison, WI 53706, U.S.A.*

Introduction

In 1968, Chandrasekaran and Balasubramanian [1] concluded from energy calculations that ring closure by a disulfide bond in an eight membered ring (R)-cysteinyl-(R)-cysteine dipeptide is possible only when the peptide amide bond adopts a non-planar *cis* conformation. Since then the cyclic [(R)-cysteinyl-(R)-cysteinyl] structural motif (abbreviated in this text as Cys=Cys) has been studied in solution and observed to be an equilibrating mixture of conformers, which has been interpreted in terms of amide bond conformers [2] or disulfide bond rotamers [3], the latter due to a preferred dihedral angle around +/-90° with a barrier of 7-14 kcal mol^{-1} that is intermediate between typical carbon-carbon bond and amide bond barriers. We report here the results of spectroscopic investigations on peptides **1-4** that indicate the preferred conformation(s) of the Cys=Cys dyad in solution is/are determined by several factors, including the amino and carboxyl substituents.

N$^\alpha$-Boc-Cys-Cys-OX
S---S
1a, X = Bzl
1b,   Me

H-Phe-Lys-Cys-Cys-Val-NH$_2$
S---S
2

N$^\alpha$-Boc-Ala-Cys-Cys-OBzl
S---S
3

N$^\alpha$-Boc-Cys-Cys-Ala-OBzl
S---S
4

Results and Discussion

Syntheses of peptides **1, 3**, and **4** were carried out in solution, while peptide **2** was synthesized manually on solid phase using N$^\alpha$-Fmoc amino acids in a standard protocol. The cyclic peptides were obtained by simultaneous deprotection and oxidation of the linear S-acetamidomethyl-protected precursors with iodine in methanol (**1,3,4**) or AcOH-H$_2$O (4:1, v:v; **2**) under high dilution conditions.

The conformations of peptides **1-4** in d$_6$-DMSO were studied using NMR

---

[*]Present address: Pharmaceutical Division, Ciba-Geigy Ltd., CH-4002, Basel, Switzerland.

spectroscopy. Two sets of signals in a 85:15 ratio were observed in the ^{1}H-NMR spectra of peptide 2. The ratio is close to that reported by Lawrence et al. [2], but we cannot assign our ratio to isomerization of the Cys=Cys amide bond because the $^\alpha$H-protons of both cysteines have identical chemical shifts. Compounds 1b and 4 exist in d_6-DMSO as a mixture of several conformations, which interconvert slowly on the NMR time scale. These complex NMR data suggest that simultaneous isomerization of the amide and disulfide bonds occurs for these two peptides. If this is the case, the four possible conformations: trans/M; trans/P; cis/M; and cis/P would be present in solution. The ^{1}H-NMR spectra of peptides 1a and 3 were found to contain only one set of signals, which can be due to the presence of a predominant conformation or to an averaged conformation on the NMR time scale as a result of a rapid equilibrium between several conformations. We used high-field 2D-NMR spectroscopy to determine the conformation of the Cys=Cys amide bond of 1a and 3. A *cis* peptide amide bond is characterized by the spatial proximity between $^\alpha$H(i)-$^\alpha$H(i+1), and NH(i)-$^\alpha$H(i+1). Analysis of the 2D-ROESY spectra (t_m=200 ms) revealed a short distance between the NH of Cys^i and the NH of Cys^{i+1}, which is characteristic of a *trans* amide bond. No cross-peaks were observed between $^\alpha$H Cys^i/$^\alpha$H Cys^{i+1} or NH Cys^i/$^\alpha$H Cys^{i+1}.

To further determine the conformation of peptide 1a in solution, we carried out circular dichroism (CD) studies and molecular mechanics calculations. The CD spectrum of peptide 1a in TFE showed a weak positive band near 225 nm and a weak negative band close to 202 nm. The CD data agree with the NMR experiments in indicating a fairly rigid structure. This can be seen by the insensitivity of the spectra to temperature variation from 278 to 328K. The spectrum showed a broad CD negative n to σ^* band with a maximum at 265 nm. Following the quadrant rule [4], a negative long-wavelength CD band can be ascribed to a right-handed disulfide with *transoid* conformation or to a left-handed disulfide with a *cisoid* conformation.

Systematic conformational search on peptide 1a was performed with the SYBYL program on an Indigo system (Silicon Graphics). The sulfur-sulfur bond was "opened", allowing the other bonds within the ring to rotate. The permitted variation in the distance between the ring closure atoms was 0.2 Å, and the permitted variation in the valence angles about the ring closure atoms was 10°. Two starting conformations were used, which included a *cis* and a distorted *trans* amide bond. Energy minimizations were performed using the Powell method until the energy difference between successive iterations differed by less than 0.05 kcal/mol. Ten conformations were found within 8 kcal/mol of the global minimum conformation (Table 1). Among them, b-c and e-j possess a distorted *trans* amide bond while a and d adopt a non-planar *cis* conformation. A disulfide bond of P-helical conformation is observed in a, c, f, and j, while the remainder present an M-helical disulfide conformation. *Trans* conformation c is in agreement with the network of ROEs observed in d_6-DMSO, and with the negative CD band of the first n to σ^* disulfide bond transition. A similar distorted *trans* amide bond has been reported in the eight-membered ring cyclic disulfide compounds, (4R)-hexahydro-7,7-dimethyl-6-oxo-1,2,5-dithiazocine-4-carboxylic acid, and phenylacetyl-L-cysteinyl-D-penicill-amine cyclic disulfide methyl ester [5]. From these publications and the results reported here, we conclude that rigid conformations containing only non-planar *cis* amide bonds cannot be assumed for the Cys=Cys structural motif when designing conformationally restricted peptides. The low barriers to *cis/trans* isomerization of the amide bond and to disulfide interconversion will depend on the sequence of the entire peptide. Our results are consistent with previous reports [5] except that *cis/trans* interconversion is not determined only by the stereochemistry of the cysteinyl-derived residues.

Table 1 *Amide bond torsion angles for low energy conformations of cyclo (cysteinyl-cysteinyl) derivatives*

Conformations	ω	χ	Energy (kcal/mol)
a	3.0	103.7	12.16
b	157.4	- 96.0	12.50
c	-157.7	96.4	12.74
d	-3.7	- 103.4	12.97
e	- 158.8	- 112.7	15.18
f	150.9	83.6	18.04
g	- 150.4	- 100.0	19.66
h	- 146.7	- 72.6	19.88
i	149.8	- 100.8	19.98
j	-149.5	97.1	20.20

Acknowledgements

C.G.-E. acknowledges a postdoctoral fellowship from Ministerio de Educación y Ciencia (Spain) and the financial support by Ciba-Geigy (Switzerland) to attend this meeting. We thank S. Hohmbeck and Prof. L. Lerner (Department of Chemistry, University of Wisconsin-Madison) for the 2D-ROESY spectra.

References

1. Chandrasekaran, R. and Balasubramanian, R., Biochim. Biophys. Acta, 188 (1969) 1.
2. Sukumaran, D.K., Prorok, M. and Lawrence, D.S., J. Am. Chem. Soc., 113 (1991) 706.
3. Veber, D.F., In Smith, J.A. and Rivier, J.E. (Eds.), Peptides: Chemistry and Biology (Proceedings of the Twelfth American Peptide Symposium), Escom, Leiden, The Netherlands, 1991, p.3.
4. Woody, R.W., Tetrahedron, 29 (1973) 1273, and references cited therein.
5. a) Baxter, R.L., Glover, S.S.B., Gordon, E.M., Gould, R.O., McKie, M.C., Scott, A.I. and Walkinshaw, M.D., J. Chem. Soc. Perkin Trans. I, (1988) 365; b) Cumberbatch, S., North, M. and Zagotto, G., J. Chem. Soc. Chem. Commun., (1993) 641.

Conformational analysis of peptides from the high affinity receptor for IgE

R.C. Thomas[1], G.J. Anderson[1], I. Toth[1], M. Zloh[1], D. Ashton[2] and W.A. Gibbons[1]

[1]Pharmaceutical Chemistry, The School of Pharmacy, U.L. 29-39 Brunswick Sq., London, WC1N 1AX, U.K., and [2]Physical Chemistry, Wellcome Research Foundation, Beckenham, Kent, U.K.

Introduction

Aggregation of IgE complexed IgE receptors by multivalent antigens triggers a host of complex cellular reactions [1] which result in the clinical symptoms of allergy. Blank and co-workers [2] proposed the topographical model shown on the next page wherein a functional receptor molecule is thought to be an $\alpha\beta\gamma_2$ tetramer with each subunit containing extracellular, transmembrane and cytoplasmic domains.

We have synthesized peptides corresponding to portions of receptor subunits and investigated their conformation using a variety of spectroscopic techniques. The conformational information obtained for the individual domain peptides should shed light on the conformation of the receptor as a whole. Perturbation circular dichroism (CD) and Fourier transform infrared (FTIR) have been used in combination to assess the main conformational components for the peptide corresponding to amino acids 200-222 of the alpha subunit (IERAC23). Preliminary data has also been obtained for the peptide corresponding to residues 198-243 of the beta subunit (IERBC46) and a peptide corresponding to residues 69-106 of the alpha subunit (IERAE38). IERAC23 corresponds to the C-terminal cytoplasmic domain of the alpha subunit. IERBC46 corresponds to the C-terminal cytoplasmic domain of the beta subunit and IERAE38 to an extracellular portion of the alpha subunit.

Both cytoplasmic domain peptides exhibited CD spectra which were dominated by helical conformers in nonpolar solvents and which were characteristic of classic 'random coil' in aqueous media. Solvent titrations between nonpolar and polar solvents typically resulted in the disappearance of helical conformers.

The reduced intensities of the bands compared to those of polypeptides adopting a single conformation [3] together with nonlinear plots, at single wavelength in the spectrum indicated that both domain peptides adopted more than one conformation in all solvents screened.

For IERAC23 in water higher salt concentrations, pH and temperature favoured a small (<6%) helical component whilst lower temperatures favoured a conformer characterized by a weak positive band around 215 nm and a stronger negative band in the region of 195 nm. In ethanediol/water mixtures both of these conformers appeared to be stabilized relative to a third at lower temperatures as indicated by the constant ratio of the 222 and 205 nm bands. Both wavelength analysis and the low intensity bands indicated that more than one conformer was present at low temperatures in 67% ethanediol/water.

Single bands were seen in the Amide I' region of FTIR spectra during a methanol/water solvent perturbation study of 1 mM IERAC23. The FTIR results identified β turn conformers which had not been characterized directly by CD and indicated that three main conformers were present (turns, helix and extended structure) whose relative populations depended on solvent. CD spectra of IERAC23 obtained in the same solvent mixtures as the FTIR spectra were then simulated using weighted sums of reference spectra corresponding to the components identified by FTIR. Woody's theoretical β turn spectra [4] were used as reference spectra for β turns. The reference spectrum used for helix was poly-L-lysine in 67% ethanediol/water at pH 11. Poly-L-lysine in 67% ethanediol/water at -80°C was used as the reference spectrum for extended structure of all types [5].

Table 1 *Summary of spectral parameters of the Amide I' band obtained during a methanol/water titration*

% Methanol	λ_{max}/cm^{-1}	$v_{1/2}$/cm^{-1}	Deconvolution Bands λ_{max}/cm^{-1} ($f_{integrated\ intensity}$)			
90	1653	33	1667 (0.16)	1655 (0.53)	1638 (0.16)	1627 (0.15)
80	1653	43	1669 (0.18)	1650 (0.49)	1640 (0.20)	1632 (0.13)
75	1650	45	1673 (0.24)	1651 (0.53)		1633 (0.23)
50	1651	37	1672 (0.21)	1651 (0.55)		1633 (0.24)
25	1648	40	1670 (0.2)	1657 (0.27)	1644 (0.34)	1632 (0.19)
0	1643	42	1666 (0.3)		1643 (0.73)	

The CD and FTIR results for IERAC23 were consistent with three major conformational components, helical, β turn and extended. The stability of each was dependent on the mode of perturbation. The data obtained for IERBC46 suggested similar conformational preferences but these were not reflected by the extracytoplasmic peptide, IERAE38. The techniques employed could not distinguish between three conformers on one molecule or three interconverting molecules. ^{1}H NMR studies of IERAC23 indicated that whilst some segments of IERAC23 existed in discrete conformers, conformational averaging occurred in other segments [6].

References

1. Kinet, J.-P. and Metzger, H., In Metzger, H. (Ed.), Fc Receptors and the Action of Antibodies, American Society for Microbiology, Washington, DC, U.S.A., 1990, p.239.
2. Blank, U., Ra, C., Miller, L., White, K., Metzger, H. and Kinet, J.-P., Nature, 337 (1989) 187.
3. Manning, M. and Woody, R.W., Biopolymers, 31 (1991) 569.
4. Woody, R.W., In Blout, E.R., Bovey, F.A., Goodman, M. and Wotan, N. (Eds.), Peptides, Polypeptides and Proteins, Wiley, New York, NY, U.S.A., 1974, p.338.
5. Drake, A.F., Siligardi, G. and Gibbons, W.A., Biophys. Chem., 31 (1988) 143.
6. Thomas, R.C., Anderson, G.J., Toth, I., Zloh1, M., Ashton, D. and Gibbons, W.A., 11th International Meeting on NMR Spectroscopy, University College of Swansea, Wales, 4-9 July 1993.

Investigation of the calcium binding site in the N-terminal EGF-like domain of blood coagulation factor IX

Y. Gong[1], J.P. Tam[2], A. Pardi[3] and W.V. Sweeney[1]

[1]Department of Chemistry, Hunter College, CUNY, 695 Park Avenue, New York, NY 10021, U.S.A., [2]Department of Microbiology and Immunology, Vanderbilt University, A5321 Medical Centre North, Nashville, TN 37232, U.S.A., and [3]Department of Chemistry and Biochemistry, University of Colorado, Boulder, CO 80309-0215, U.S.A.

Introduction

Several EGF-like domains are known to exhibit calcium binding, including the N-terminal EGF-like domains of blood coagulation factors IX and X. The solution structure of the calcium-loaded N-terminal EGF-like domain of bovine factor X (45-86) (FX-EGFn) was reported recently [1]. Based on this structure, calcium ligands were identified. In this work a complementary effort is made to identify residues involved in the calcium binding by human factor IX (46-87) (FIX-EGFn). pH titrations were performed in the presence and absence of calcium, monitored using 2D-NMR techniques. The calcium dependence of the chemical shifts in these titration curves are useful to diagnose which residues are involved in calcium ligation. For example, the aspartic β-protons report sidechain calcium binding, while the amide proton chemical shift of the Nth amino acid is useful to diagnose backbone carbonyl binding to calcium in the (N-1) amino acid. Apparent differences in calcium binding between FIX-EGFn and FX-EGFn will be discussed.

Results and Discussion

The 2D-NMR spectra of synthetic human factor IX (45-87) were acquired in H_2O, D_2O in unbuffered solution at various pH values. Spectra were acquired at a range of proton frequencies, including 400 MHz (JEOL), 500 MHz (Varian), and 600 MHz (Bruker). Spectra of the calcium loaded protein were acquired in a solution containing 20mM $CaCl_2$. The 2D-NMR data were processed using a VAX station 3100 or IBM RS6000 computer with the FTNMR or FELIX programs (Hare Research). In the following discussion, a sequential amino acid numbering system of 1 to 43 will be used for human factor IX (45-87). For purposes of comparison, a numbering of 2 to 43 will be used for bovine factor X (45-86) to align the homologous peptide sequences in the two peptides.

FIX-EGFn has two anti-parallel beta sheets between residues 16-20, 25-29 and 32-34, 40-42 [2, 3]. There are beta turns at residues 21-24 and 35-39. In general, the chemical shifts of the residues for both the calcium-loaded and apo forms are very similar at pH 4.2, suggesting that little or no calcium binding occurs at this pH, as expected. Even at higher pH the chemical shifts of most residues are nearly calcium independent, suggesting that the conformation of the peptide does not change significantly as a result of calcium binding.

Amide protons of all the residues except Tyr-1, Pro-11, and Pro-30 were detected for both Ca^{2+} and apo forms at a range of pH values. At pH 5.8, where calcium binding is expected, significant calcium dependence of the chemical shift is observed for the amide protons of Asp-5 and Cys-7, and to a lesser extent, for those of Gln-6 and Asp-21. The chemical shifts of the beta protons of Asp-5 show some calcium dependence, while effectively none is observed for the gamma protons of Gln-6 or the beta protons of Asp-3, Asp-20, and Asp-21.

In FX-EGFn calcium was found to be coordinated [1] by the two backbone carbonyls of Gly-4 and Gly-21, and the sidechains of Gln-6, β-hydroxyaspartic acid-20, and possibly Asp-3.

Our data are consistent with backbone carbonyl coordination to calcium by Gly-4 in FIX-EGFn. However, the chemical shift of the amide proton of Asp-21 is calcium dependent, while that of Ile-22 is not, suggesting that the backbone carbonyl of Asp-20 rather than Asp-21 binds calcium in FIX-EGFn. The chemical shifts of the beta protons of both Asp-20 and Asp-21 are calcium independent, suggesting that there is no sidechain involvement by either aspartic acid. Further, the calcium independence of the Gln-6 gamma proton chemical shifts, in concert with the calcium dependence of the Cys-7 amide proton chemical shift, strongly indicates that Gln-6 binds calcium by its carbonyl backbone oxygen rather than its sidechain in FIX-EGFn. No evidence was found for calcium coordination by Asp-3. Thus, based on this NMR data, it appears likely that factor X and factor IX bind calcium in somewhat different motifs.

Acknowledgements

The authors gratefully acknowledge Linda Huang and Yan Yang for synthesis of human factor IX (45-87), and Mike Blumenstein for helpful discussions.

References

1. Selander-Sunnerhagen, M., Ullner, M., Persson, E., Teleman, O., Stenflo, J. and Drakenberg, T., J. Biol. Chem., 267 (1992) 19642.
2. Huang, L.H., Cheng, H., Pardi, A., Tam, J.P. and Sweeney, W.V., Biochemistry, 30 (1991) 7402.
3. Baron, M., Norman, D.G., Harvey, T.S., Handford, P.A., Mayhew, M., Tse, A.G.D., Brownlee, G.G. and Campbell, I.D., Protein Sci., 1 (1992) 81.

Structural analysis of synthetic peptides corresponding to domains of a voltage-gated potassium channel protein

P.I. Haris, B. Ramesh and D. Chapman

Department of Protein and Molecular Biology,
Royal Free Hospital School of Medicine, University of London,
Rowland Hill Street, London, NW3 2PF, U.K.

Introduction

Ion channels are integral membrane proteins which catalyze the ion fluxes across biological membranes. In recent years many advances have been made towards our understanding of these proteins through the combined application of electrophysiological and molecular biological techniques. However, as with other membrane proteins, very little is known about the structure of ion-channel proteins and there are no high resolution X-ray structures available. The various problems that limit the application of X-ray crystallography and NMR spectroscopy to the study of membrane proteins also apply to ion-channel proteins. In addition, voltage-gated ion-channel proteins are not available in sufficient quantities for structural analysis. In the light of these problems, one approach which seems particularly promising for structural studies of ion-channel proteins is the use of synthetic polypeptides. This is possible as genes have been cloned for a number of ion channels and these have been useful for identification of local regions of the protein sequence that are involved in different channel properties such as voltage dependence of gating, ion selectivity etc. [1-4]. Hence, these functionally important domains/motifs can be chemically synthesized and their structure and function characterized by spectroscopic and electrophysiological techniques. We are using this strategy for the study of voltage-gated potassium channels in an attempt to gain an understanding of the structural organization of this protein in biological membranes. Here, our recent study on the ion-selective pore of the Drosophila Shaker potassium channel is described.

Results and Discussion

Voltage-gated potassium channels are tetrameric membrane proteins [1-4] with intracellular N- and C-termini, and six predicted transmembrane α-helices (called S1, S2, S3, S4, S5 and S6). Recently, a hydrophobic region between S5 and S6 has been identified (called H5 or SS1-SS2) as the pore forming region of the potassium channel [1-3]. We have synthesized a peptide corresponding to the pore region (residues 431-449 of the Drosophila shaker voltage-gated potassium channel [4]).

The amino acid sequence of the H5 peptide synthesized is as follows:

CH₃CO-**DAFWWAVVTMTTVGYGDMT**-CONH₂

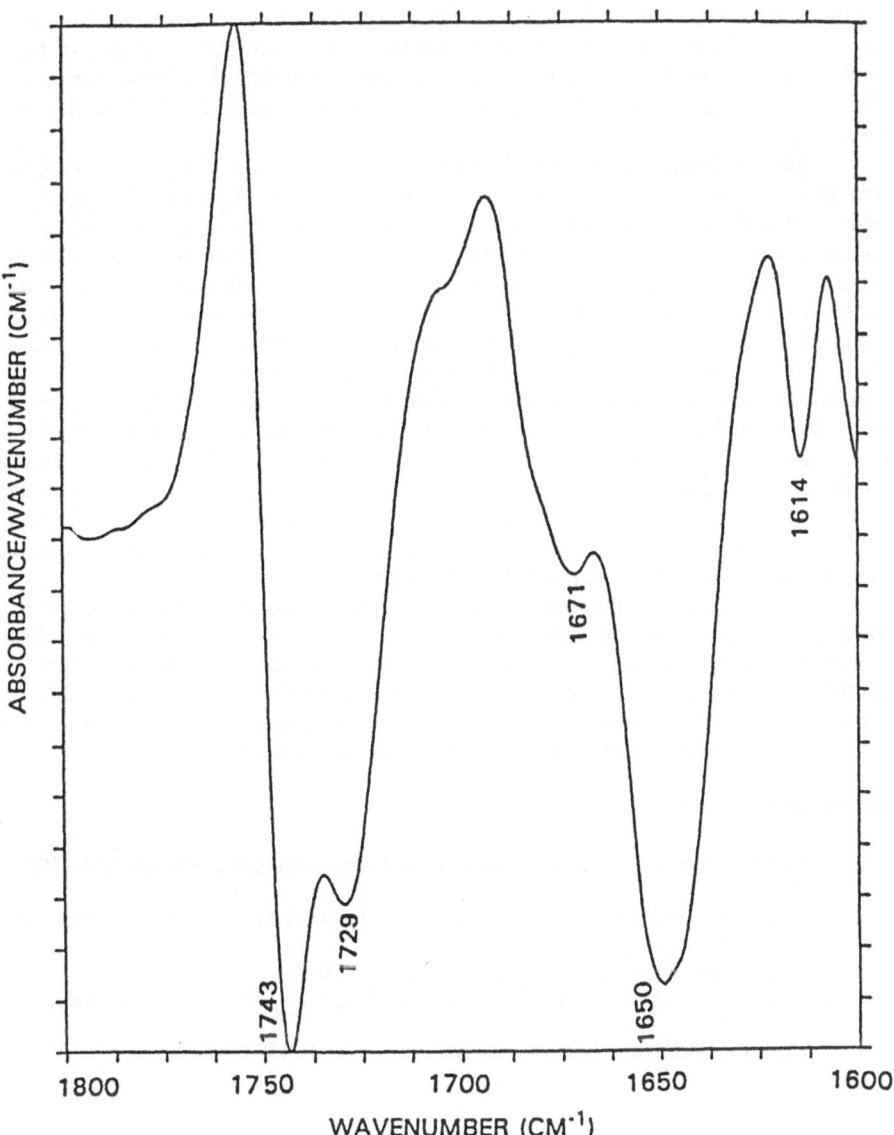

Fig. 1. *FTIR second-derivative spectrum of the H5 peptide in LPC micelles in* ^2H_2O *buffer at 30°C. The amide I maximum at 1650cm⁻¹ indicates the presence of α-helical structure [5]. The bands at 1734cm⁻¹ and 1729cm⁻¹ are due to the C=O vibration of the lipid ester groups.*

It is suggested that four H5 regions make up the channel ion-selective pore, hence a tetrameric version of this sequence was also synthesized by linking together four H5 peptides via a lysine core. The H5 peptides were synthesized using an automated peptide synthesizer and were purified on a reverse-phase HPLC column. Further characterization of the peptides were carried out by amino acid analysis and the correct molecular weight was confirmed by mass spectrometry. Spectra of both peptides were recorded for samples in lysophosphatidylcholine (LPC) micelles and in dimyristoylphosphatidylcholine bilayers using a Perkin-Elmer 1750 FTIR spectrometer [5].

Figure 1 shows the second-derivative spectrum of the H5 peptide in LPC micelles. The amide I maximum is centered at 1650cm^{-1}. This band frequency is consistent with the presence of α-helical structure [5]. Similar results were obtained for the tetrameric form of the H5 peptide. As can be seen from the FTIR spectrum presented in Figure 1, there are no bands in the region 1630-1640cm^{-1} to indicate the presence of any β-sheet structure. This result differs from the suggestion that the H5 segment forms a ß-hairpin type structure within the membrane [1-3]. It is possible that the H5 peptide forms an α-helical structure as an isolated peptide but forms a ß-hairpin structure when part of the total protein structure. Nevertheless, it is clear from our study that the sequence of amino acids that line the ion-selective pore of the potassium channel show a high propensity to form an α-helical structure in a lipid environment. If the H5 segment is indeed located in a hydrophobic region within the native channel protein, surrounded either by the lipid hydrocarbon chains or by the hydrophobic side chains of the transmembrane helices, then our experimental results indicate that it may adopt such an α-helical conformation.

This study provides the first experimental evidence for an α-helical structure for the sequence corresponding to the ion-selective pore of the voltage-gated potassium channels. Further studies are in progress in order to determine the structure of synthetic peptides corresponding to the predicted transmembrane helices (S1-S6) of this channel protein. These studies are aimed towards providing information on the structural arrangement of the channel protein in the lipid membrane.

References

1. Yellen, G., Jurman, M.E., Abramson, T. and MacKinnon, R., Science, 251 (1991) 939.
2. Hartman, H.A., Kirsch, G.E., Drewe, J.A., Taglialatela, M., Joho, R.H. and Brown, A.M., Science, 251 (1991) 942.
3. Yool, A.J. and Schwarz, T.L., Nature, 349 (1991) 700.
4. Tempel, B.L., Papazian, D.M., Schwarz, T.L., Jan, Y.N. and Jan, L.Y., Science, 237 (1987) 770.
5. Haris, P.I. and Chapman, D., Trends Biochem. Sci., 17 (1992) 228.

Propensity for helix formation by the dipropylglycyl residue (Dpg): Crystal structure of Boc-Aib-Ala-Leu-Ala-Leu-Dpg-Leu-Ala-Leu-Aib-OMe

I.L. Karle[1] and P. Balaram[2]

[1]Laboratory for the Structure of Matter, Washington, DC 20375,
U.S.A. and [2]Molecular Biophysics Unit, Indian Institute of Science,
Bangalore 560 012, India

Introduction

It has been demonstrated that the inclusion of one or very few α-amino-isobutyric (Aib) residues restricts linear peptides to helical conformations almost exclusively [1, 4]. Crystal structure analyses of numerous peptides containing the Aib residue have shown that short peptides form a 3_{10}-helix, while longer peptides (generally 10 or more residues) form an α-helix [3, 4]. By comparison, there is considerably less structural information about the conformational preferences for disubstituted glycine residues when the side chains are longer than methyl, that is, diethyl glycine (Deg), dipropyl glycine (Dpg) and dibutyl glycine (Dbg). Currently available structural data, concerning a number of peptides with 5 or fewer repeating Deg or Dpg residues, show fully extended backbones with C_5 type hydrogen bonding [5]. The potential usefulness of dialkyl glycine residues for constraining conformation in the synthesis of enzyme-resistant agonists and antagonists prompted us to synthesize and examine by x-ray diffraction the structures of peptides with various lengths containing a single Dpg or Dbg residue. Our first structural results on three peptides show that Dpg and Dbg have a strong propensity for helix formation.

Results and Discussion

Crystal structure analyses by x-ray diffraction have been completed for the following penta- to decapeptides:

  I.   Boc-Leu-**Dbg**-Val-Ala-Leu-OMe
  II. Boc-Leu-**Dpg**-Leu-Ala-Leu-Aib- OMe
  III. Boc-Aib- Ala- Leu-Ala-Leu-**Dpg**-Leu-Ala-Leu-Aib-OMe

Each peptide is composed of at least 40% Leu residues and each peptide has one Dpg or Dbg residue within the body of the sequence. Peptide I does not have any Aib residues. The backbones in peptides I and II form 3_{10}-helices with fully extended side chains, while the longer backbone in III forms an α-helix with six 5 → 1 hydrogen bonds and fully extended side chains. The conformation of peptide III is shown in Figure 1. The average values for the torsional angles φ and ψ in residues 1-8 are -61°±2° and -39°±6°, respectively. The only distortion in the α-helix occurs at the penultimate residue Leu[9] where ϕ and ψ values are -110° and +1°. The Dpg[6] residue does not cause any distortion.

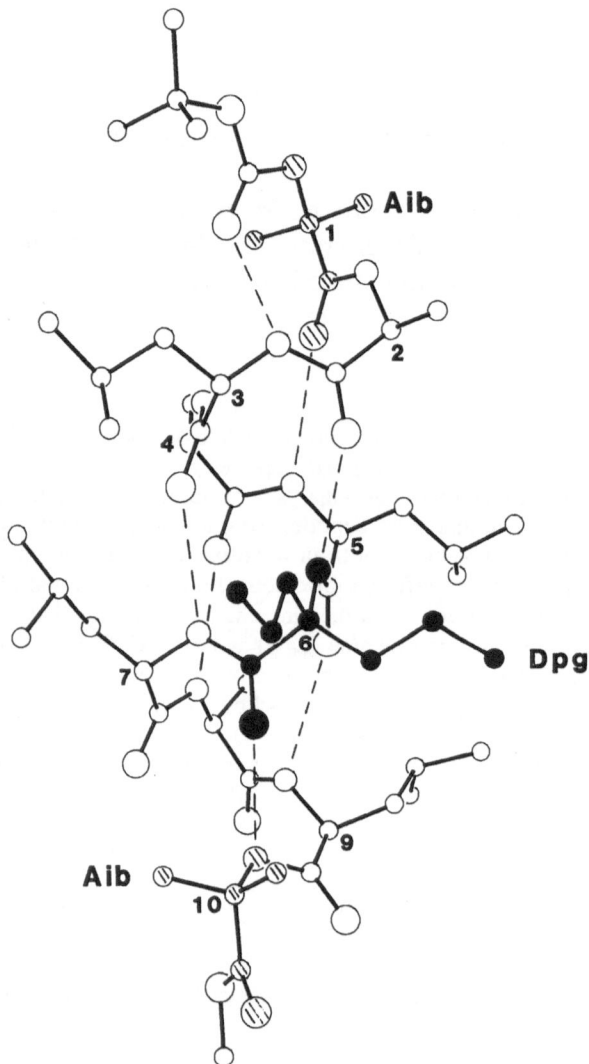

Fig. 1. The α-helical conformation for Boc-Aib-Ala-Leu-Ala-Leu-Dpg-Leu-Ala-Leu-Aib-OMe found in the crystalline state by x-ray diffraction. The Dpg residue with two prolyl side-chains is shown with black spheres. Aib residues at both ends of the peptides are shown with striped spheres. The dashed lines represent NH•••OC hydrogen bonds.

These experiments show that the introduction of a single $C^{\alpha,\alpha}$–dialkyl residue into a sequence of 5-10 residues, with or without additional Aib residues, results in folding the backbone into a 3_{10}- or α-helix; in contrast to earlier experiments with shorter sequences [6] and with multiple Deg residues in which the backbones were found to be fully extended [5]. Further, the bulkiness of two propyl or butyl chains substituted on C^α atoms is accommodated very well without distorting a helical backbone whether the bulky residues are incorporated near an end or in the middle of a sequence.

Acknowledgements

This research was supported in part by the National Institutes of Health Grant GM30902, by the Office of Naval Research and by a grant from the Department of Science and Technology, India.

References

1. Marshall, G.R. and Bosshard, H.E., Circ. Res., 30/31 (Supppl. II) (1972) 143.
2. Balaram, P., Proc. Indian Acad. Sci., 93 (1984) 703.
3. Toniolo, C., Bonora, G.M., Bavoso, A., Benedetti, E., di Blasio, B., Pavone, V. and Pedone, C., Biopolymers, 22 (1983) 205.
4. Karle, I.L. and Balaram, P., Biochemistry, 29 (1990) 6747.
5. Benedetti, E., di Blasio, B., Pavone, V., Pedone, C., Toniolo, C. and Crisma, M., Biopolymers, 32 (1992) 453.
6. Dentino, A.R., Raj, P.A., Bhandary, K.K., Wilson, M.E. and Levine, M.J., J. Biol. Chem., 266 (1991) 18460.

NMR and computational evidence that high affinity bradykinin receptor antagonists adopt C-terminal β–turns

D.J. Kyle[1], P.R. Blake[2], D. Smithwick[2], L.M. Green[2], J.A. Martin[1], J.A. Sinsko[1] and M.F. Summers[2]

[1]*Scios Nova Inc., Baltimore, MD 21224, U.S.A.*
[2]*Department of Chemistry and Biochemistry, University of Maryland, Baltimore County, Baltimore, MD 21228, U.S.A.*

Introduction

For some time there has been speculation that high affinity peptide bradykinin receptor antagonists adopt C-terminal β-turns [1-3]. To study this proposal, we prepared three tetrapeptides, each corresponding to the four C-terminal amino acids of the three most potent, second generation antagonists yet reported [3-5]. The specific tetrapeptides are (I) Ser-D-Phe-Oic-Arg, (II) Ser-D-Tic-Oic-Arg, and (III) Ser-D-Hype(trans propyl)-Oic-Arg.

Results and Discussion

Conformational searches performed on models corresponding to each tetrapeptide led to the discovery of local energy minima, from which representative structures were compared with those determined via distance geometry/simulated annealing calculations which incorporated NMR distance constraints determined at 600 MHz in aqueous solution at neutral pH. Excellent correlation between the calculated and experimentally determined structures was observed (Figure 1). The predominant conformation observed for each tetrapeptide contained a 1 - 4 hydrogen bond, characteristic of β-turn geometry. The amide proton chemical shift of Arg[4] in each tetrapeptide was independent of varying pH (~ 4.4 to ~ 7.7) further confirming the involvement of that proton in a hydrogen bond. Thirty minimum penalty NMR-derived structures corresponding to the three tetrapeptides are shown in Figure 2.

Overall, the methods of NMR and computational chemistry independently produced similar conformations for each of these peptides. Taken together, the data supports the hypothesis that high affinity bradykinin receptor antagonists adopt a C-terminal β-turn.

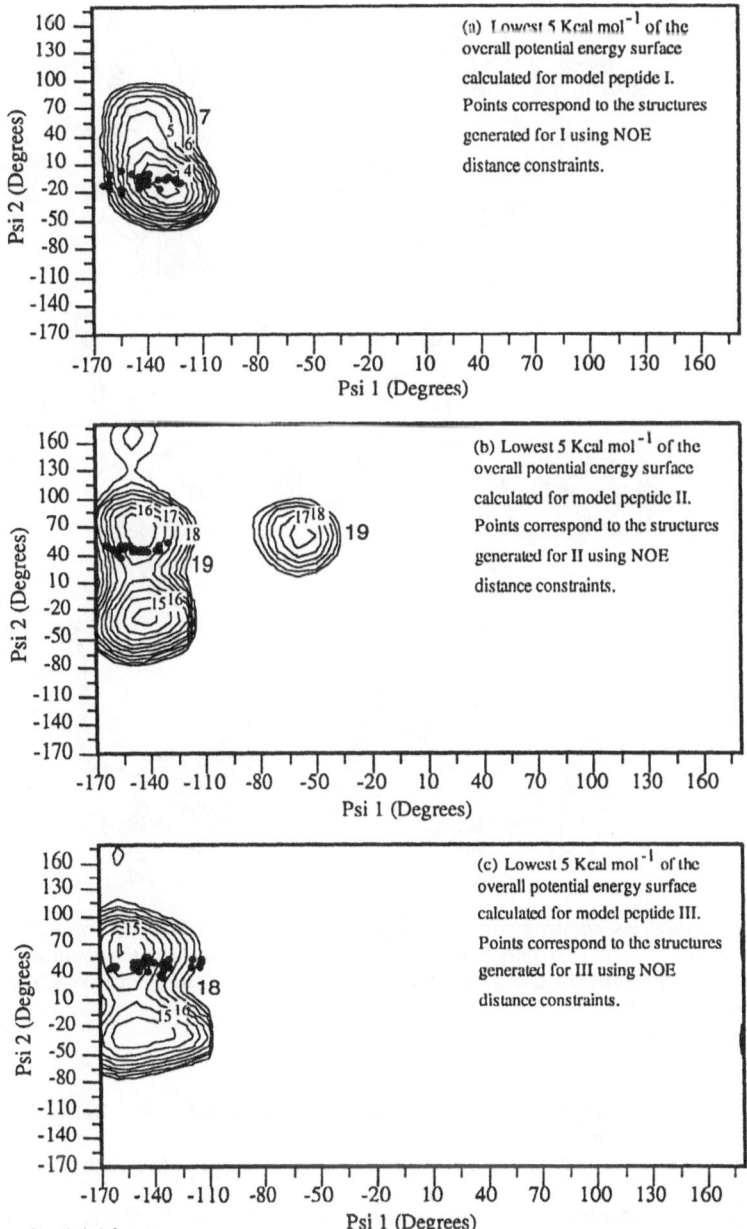

Fig. 1.   Lowest 5 kcal mol⁻¹ of the calculated overall potential energy surface for (a) model I, (b) model II, and (c) model III.   The contour interval is 0.5 kcal mol⁻¹, and for each the highest (outermost) and lowest contour energy values are labeled.   Superimposed on the contour plots are values for ψ_{i+1} and ψ_{i+2} from each of the 30 structures generated from the NMR data for (a) peptide I, (b) peptide II, and (c) peptide III.

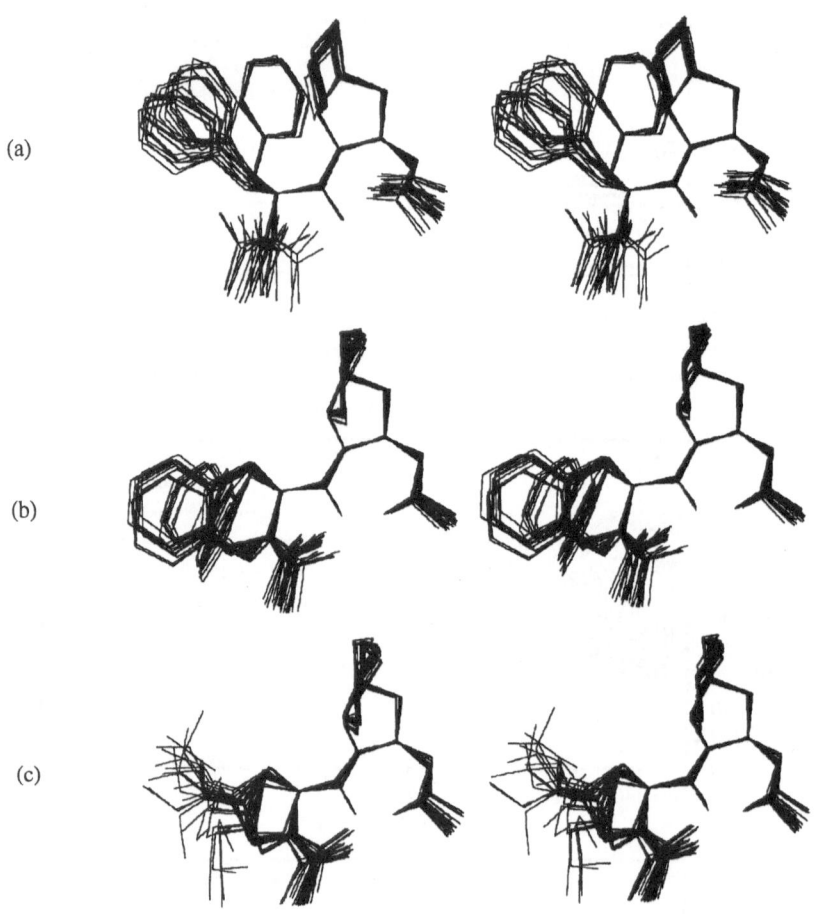

(a)

(b)

(c)

Fig. 2. Stereoviews showing the 30 best-fit superpositions of C, Cα, and N atoms of residues 2 and 3 of (a) tetrapeptide I, (b) tetrapeptide II, and (c) tetrapeptide III. All structures were generated via distance geometry/simulated annealing with NOE constraints incorporated.

References

1. Burch, R.M. (Ed.), Bradykinin Antagonists: Basic and Clinical Research, Marcel Dekker, New York, NY, U.S.A., 1990, p.131.
2. Kyle, D.J., Martin, J.A., Farmer, S.G. and Burch, R.M., J. Med. Chem., 34 (1991) 1230.
3. Kyle, D.J., Martin, J.A., Burch, R.M., Carter, J.P., Lu, S., Meeker, S., Prosser, J.C., Sullivan, J.P., Togo, J., Noronha-Blob, L., Sinsko, J.A., Walters, R.F., Whaley, L.W. and Hiner, R.N., J. Med. Chem., 34 (1991) 2649.
4. Hock, F.J., Wirth, K., Albus, U., Linz, W., Gerhards, H.J., Wiemer, G., Henke, St., Breipohl, G., Knoig, W., Knolle, J. and Scholkens, B.A., Brit. J. Pharmacol., 103 (1991) 769.
5. Wirth, K., Hock, F.J., Albus, U., Linz, W., Alpermann, H.G., Anagnostopoulos, H., Henke, St., Breipohl, G., Knoig, W., Knolle, J. and Scholkens, B.A., Brit. J. Pharmacol., 103 (1991) 774.

Conformational transitions around phosphorylation sites of human and bovine tau protein often indicate transformations characteristic of tau found in the paired helical filaments of Alzheimer's disease

E. Lang[1,2], G.I. Szendrei[1] and L. Otvos, Jr.[1]

[1]The Wistar Institute of Anatomy and Biology, 3601 Spruce Street, Philadelphia, PA 19104, U.S.A., and [2]Teacher Training Faculty, L. Eotvos University, H-1055 Budapest, Hungary

Introduction

Among the hallmark neuropathological lesions of Alzheimer's disease are the intracellular neurofibrillary tangles (NFTs). Tangles represent dense accumulations of ultrastructurally distinct paired helical filaments (PHFs), characterized by β-pleated sheets at the molecular level. The major constituent of PHF, the low molecular weight microtubule-associated protein τ is, however, a highly elastic molecule in the healthy brain. τ appears to be abnormally hyperphosphorylated in PHF [1]. We studied the conformation of synthetic mid-sized peptides corresponding to various regions of τ in order to understand how the normal protein folds into the β-pleated sheets as a consequence of phosphorylation of certain serine residues.

Results and Discussion

Table 1 lists the synthetic peptides and phosphopeptides, their location in τ, and the conformation of the protein fragments as detected by circular dichroism (CD) and computer-assisted molecular modelling techniques. The selection of the peptide panel is explained as follows: T1NM is the minimal epitope for monoclonal antibody (mAb) Tau-1 [2], which recognizes τ in normal human τ, but not τ in PHF. The recognition pattern of one of the most commonly used PHF-specific mAbs, AT8, was reported to be complementary with Tau-1. T3P is the epitope of the other preferred anti-PHF mAb, PHF-1 [3]. PHF-1 does not cross-react with peptide T3. *In vitro* phosphorylation of Ser^{416} of τ by the Ca^{2+}/calmodulin-dependent kinase (camk) makes the protein long and stiff. mAb Tau-2 was raised against bovine τ, was reported to recognize a conformational epitope [4], and stained τ was found in NFTs, but not normal human τ. As seen in Figure 1A, the three non-phosphorylated human sequences (which are not involved in PHF), and the human analog of Tau 2BS (Tau 2H), assume different types of β-turns or α-helices. Secondary structural prediction proposes the presence of β-pleated sheets in T3 and Ac-PRH-camk, but this extended structure becomes apparent only after phosphorylation (Figure 1B). Ac-PRH-camkPh is entirely in β-pleated sheets in trifluoroethanol (TFE), and the weak shoulder of T3P at 220 nm in TFE indicates the

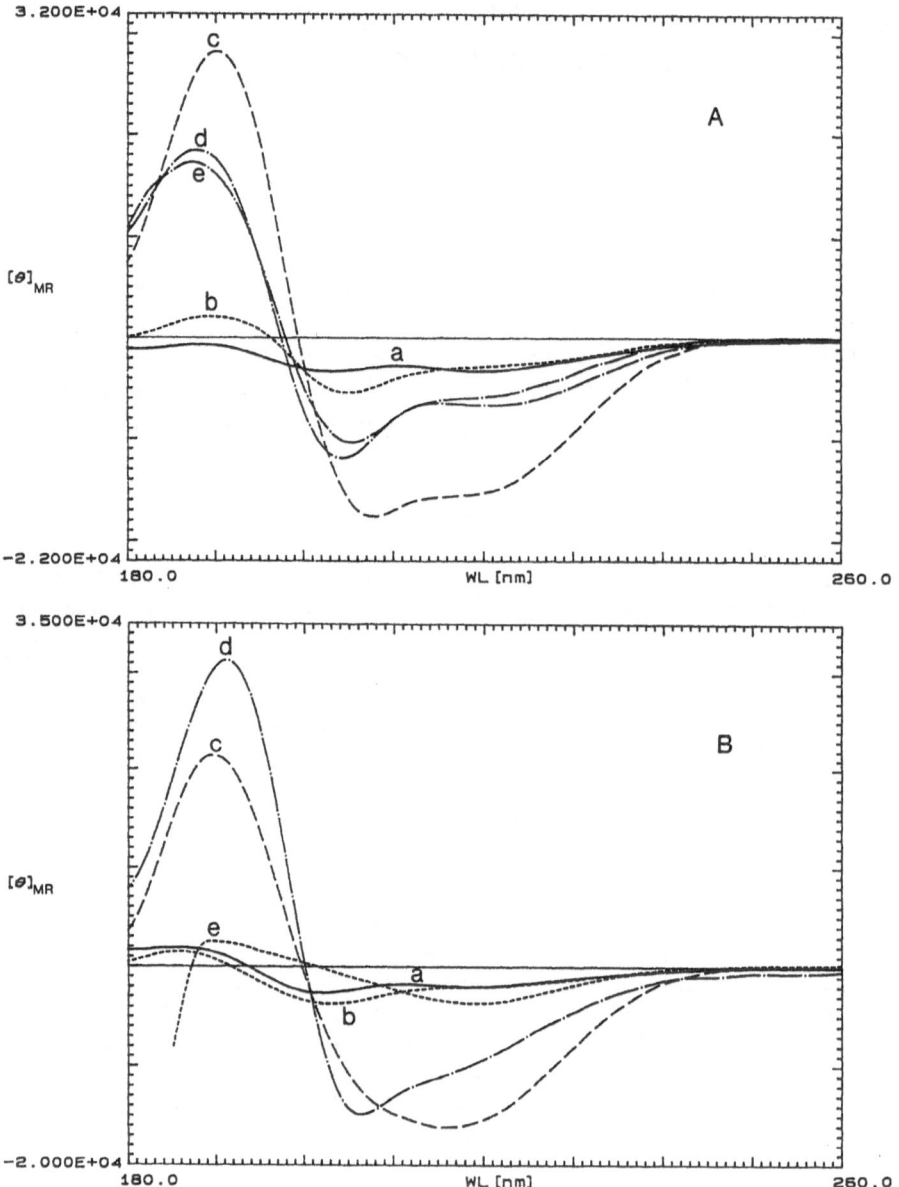

Fig. 1. CD spectra of synthetic τ fragments. The peptides and the used solvents are as follows: Panel A: curve a (continuous line) is T1NM in 75% aqueous TFE; curve b (dotted line) is T3 in TFE; curve c (dashes) is Ac-PRH-camk in TFE; curve d (dots and dashes) is Tau 2BSPh in TFE; curve e (dots and dashes) is Tau 2H in TFE. Panel B: curve a (continuous line) is T1NM 202P in 75% aqueous TFE; curve b (dotted line) is T3P in TFE; curve c (dashes) is Ac-PRH-camkPh in TFE; curve d (dots and dashes) is Tau 2BS in TFE; curve e (dotted line) is T3P in 90% aqueous ethanol. CD spectra were taken on a Jasco J720 instrument at room temperature. Peptide concentration was 0.5 mg/ml.

Table 1 *Synthetic peptide and phosphopeptide fragments of τ proteins*

Peptide	Sequence[a]	Origin	Cross-reacts with		Confor-mation
			normal τ	PHF-τ	
T1NM	GDRSGYSSPGSPG	human 192-204	+	-	turns
T1NM 202P	GDRSGYSSPGS*PG		-	+	turns
T3	GAEIVYKSPVVSGD	human 389-402	+	-	turns
T3P	GAEIVYKS*PVVSGD		-	+	turn/extended
Ac-PRH-camk	PRHLSNVSSTGSIDMVD	human 405-421	?	?	turn/helix
Ac-PRH-camkPh	PRHLSNVSSTGS*IDMVD		?	?	extended
Tau 2BS	AGIGDTSNLEDQAA	bovine 95-108	-	+	extended
Tau 2BSPh	AGIGDTS*NLEDQAA		?	?	turns
Tau 2H	AGIGDTPSLEDEAA	human 106-109	+	-	turns

[a] * denotes the phosphorylated serine residues.

presence of the β-structure that is fully manifested in ethanol. T1NM is not predicted to form β-pleated sheets, and indeed remains in turns after phosphorylation (T1NM 202P). In contrast, Tau 2BS assumes a high percentage of extended structure before phosphorylation, and this can be correlated with its immunological cross-reactivity with NFTs isolated from humans. Phosphorylation of Tau 2BS (the phosphopeptide is Tau 2BSPh) results in a conformation very similar to the genuine human sequence of this region, which lacks the β-pleated sheet structure.

First, it appears that the PHF-like state of τ fragments prefers, but does not necessarily require, an extended secondary structure. Second, phosphorylation of this protein may overcome interspecies conformational gaps.

Acknowledgements

This work was supported by NIH grant AG 10670.

References

1.   Lee, V.M.-Y., Balin, B.J., Otvos, L., Jr. and Trojanowski, J.Q., Science, 251 (1991) 675.
2.   Szendrei, G.I., Lee, V.M.-Y. and Otvos, L., Jr., J. Neurosci. Res., 34 (1993) 243.
3.   Lang, E., Szendrei, G.I., Lee, V.M.-Y. and Otvos, L., Jr., Biochem. Biophys. Res. Commun., 187 (1992) 783.
4.   Lang, E. and Otvos, L., Jr., Biochem. Biophys. Res. Commun., 188 (1992) 162.

Stable helicity in linear growth hormone releasing factor analogs

S.R. Lehrman, M.E. Lund, E.L. Ulrich, K.A. Farley and W.M. Moseley

The Upjohn Company, Biotechnology Development and Performance Enhancement Research, Kalamazoo, MI 49001, U.S.A.

Introduction

Substitution of alanine for glycine at position 15 enhances the stability of secondary structure in bovine growth hormone releasing factor (bGRF) [1]. Other changes in peptide sequence may also stabilize the secondary structure of this peptide. This is of interest since it has been suggested that the stabilization of the bGRF helix enhances its receptor binding affinity [2].

Enhancements in peptide helical stability are expected when the amino acids that have higher helix forming propensity are added to the peptide chain [3]. In addition, substitutions that enhance amphiphilicity when a peptide forms a helix also encourage this type of secondary structure. In this report, we examine the effect of leucine substitutions at positions 9 and 18 on helicity and bioactivity.

Results and Discussion

We attempted to enhance the helical stability of bGRF by replacing Ser^9 or Ser^{18} with leucine residues. These substitutions are expected to enhance the helix forming propensity of the primary structure and the amphiphilicity of the resulting helix. The helical content of these peptides was determined by measuring their circular dichroic spectra in trifluoroethanol (TFE) and aqueous buffers. TFE, which is well-known for its ability to induce helical structure, has been shown recently to provide a meaningful estimate of helical propensity in protein fragments [4]. Circular dichroism studies of Leu^{27}, $Leu^{9,27}$, $Leu^{18,27}$, and $Leu^{9,18,27}$ bGRF(1-29)NH$_2$ peptides, as a function of TFE concentration, indicate that the helical content of these peptides decreases in the following order: $Leu^{9,18,27} > Leu^{18,27} > Leu^{9,27} > Leu^{27}$ (Table 1). These data are consistent with the helical content of these peptides (0.5 mg/mL) in 5 mM NH$_4$HCO$_3$ (pH 7.2). These effects are more marked than the helix stabilization that is observed on replacement of Gly^{15} with Ala. The latter peptide forms 60% as much helix as the leucine analogs discussed above (data not shown). 2D ^{1}H NMR NOESY spectra of the Leu^9 analogs define helical regions extending from residues 5 to 10, and, 16 to 27. These regions of stable helicity are absent in the Leu^{27} analog.

It has been suggested that stabilization of helical secondary structure in bGRF may enhance the affinity of this hormone for the bGRF receptor [2]. This suggestion is primarily based, in part, on bioassay data obtained with Ala^{15} analogs of bGRF. Since multicyclic helical analogs of bGRF vary in their bioactivity, the benefits of stabilizing secondary structure may be restricted to specific peptide regions [5]. We find that leucine substitutions at positions 9 and 18 do not significantly alter growth hormone release in steers, relative to Leu[27] (Table 1). Since these structural modifications likely alter several processes *in vivo*, resolution of this issue will likely require direct binding studies performed with isolated receptor.

Table 1   *Conformational and biological characteristics of Leu[9]-, Leu[18]-, and Leu[9,18]-bGRF(1-29)NH₂ analogs*

Compound	MRE$_{222nm}$ (TFE)	MRE$_{222nm}$ (H$_2$O)	Relative bioactivity (%)[a]
[Leu[27]]-GRF(1-29)NH$_2$	-23,900	-3,520	–
[Leu[9,27]]-GRF(1-29)NH$_2$	-27,950	-5,700	117
[Leu[18,27]]-GRF(1-29)NH$_2$	-30,700	-8,300	87
[Leu[9,18,27]]-GRF(1-29)NH$_2$	-38,150	-16,500	78

[a]Bioactivities are expressed relative to the Leu[27] analog. These bioactivities are not statistically significant.

In summary, substitution of leucine for Ser[9] and Ser[18] of bGRF stabilizes the helicity of this peptide. This effect has been observed in TFE- and water-containing buffers. Leucine substitutions at positions 9 and 18 stabilize bGRF helicity more than is observed for the Gly[15]→Ala substitution. NMR data suggest that stable helicity is induced in distinct peptide regions.

References

1.   Bewley, T.A., Brovetto-Cruz, J. and Li, C.H., Biochemistry, 8 (1969) 4701.
2.   Fry, D.C., Madison, V.S., Bolin, D.R., Greeley, D.N., Toome, V. and Wegrzynski, B.B., Biochemistry, 28 (1989) 2399.
3.   Chou, P.Y. and Fasman, G., Annu. Rev. Biochem., 47 (1978) 251.
4.   Lehrman, S.R., Tuls, J.L. and Lund, M.E., Biochemistry, 29 (1990) 5590.
5.   Fry, D.C., Madison, V.S., Greeley, D.N., Felix, A.M., Heimer, E.P., Frohman, L., Campbell, R.M., Mowles, T.F., Toome, V. and Wegrzynski, B.B., Biopolymers, 32 (1992) 649.

Fine-tuned conformation of dithioacylpapain intermediates: Insights from resonance Raman spectroscopy

M. Kim, W. Neugebauer, G.I. Birnbaum and P.R. Carey

Institute for Biological Sciences, National Research Council of Canada, Ottawa, Ontario, K1A 0R6, Canada

Introduction

Resonance Raman (RR) spectroscopy offers a powerful method of providing structural and dynamical information on functioning, catalytically active enzyme-substrate complexes. Although the RR spectra of the complexes, e.g., dithioacyl-papain intermediates, can be recorded on a rapid time scale under essentially physiological conditions [1], interpretation of the RR data is not always unambiguous [2]. The intermediates have at least four rotationally permissive bonds about the P_1 glycine linkages and the cysteine-25 side chain, defined by the (ϕ', ψ') and (χ_1, χ_2) torsion angles, respectively, in the enzyme's active site [3].

$$RC(=O) \text{–} NH \text{–} CH_2C(=S)S \text{–} CH_2 \text{–} Papain$$
$$\phi' \quad \psi' \qquad\qquad \chi_2 \quad \chi_1$$

A detailed knowledge of these chemical bonds is essential for a comprehensive understanding of enzyme catalysis at the molecular level. Recently, we derived a new set of structure-spectra correlations for model compounds such as N-acylglycine ethyl dithioesters, $R\text{-}C(=O)NHCH_2C(=S)SCH_2CH_3$, by combining X-ray structural parameters with resonance Raman frequencies for single crystals of the above ethyl dithioesters. This provided a quantitative value for the ψ' torsion angle of N-benzoylglycine dithioacylpapain intermediates, which is larger than any ψ' torsion angle so far observed for N-acylglycine ethyl dithioesters [4]. In this presentation, we discuss the possible mechanistic significance of the increase in ψ' in the enzyme's active site.

Results and Discussion

One of the main features revealed by X-ray crystallography is that the structural parameters in the central $\text{-}C(=O)NHCH_2C(=S)SCH_2CH_3$ skeleton are correlated with the ψ' (NC-CS(thiol)) torsion angle. The structural changes, involving bond lengths (e.g., NC-CSS), bond angles (e.g., N-C-C) and torsion angles (e.g., NC-CS(thiol)), are mainly ascribed to the non-bonded N···S attractive interaction (1,4-interaction). At the same time, RR bands, sensitively reflecting the structural changes, were found for single crystals of the above dithioesters. Thus we can set up structure-spectra correlations for several selected Raman bands of the ethyl dithioesters. These correlations allow us to probe the conformation about the P_1 glycine linkages of N-benzoylglycine dithioacylpapain to give $(\psi', \phi') = (+15°, -105°)$. Importantly, the estimated value of $+15°$ for ψ' in the acylenzyme is very different from those of the corresponding model compounds, N-benzoylglycine ethyl dithioesters, and is outside the ψ' torsion angle range so far observed for N-acylglycine ethyl dithioesters. In all likelihood, enzyme-substrate contacts, e.g., hydrogen bonding and hydrophobic interactions, bring about the increase in the ψ' in the enzyme's active site. Since the direction of the hydrogen bonding interactions involving the amide group of the substrate is nearly orthogonal to the rotation axis of the NC-CSS single bond, we suggest that the value of the ψ' torsion angle in the active site is regulated, at least in part, through these hydrogen bonding interactions. The increase in ψ' in the active site represents a small but significant departure of ψ' in the direction of ψ' for the transient tetrahedral intermediate for deacylation, which is believed to be close to the deacylation transition state on the reaction pathway, and must be driven in this direction by the enzyme-substrate contacts. In addition, this fine-tuned conformation serves to decrease the N···S non-bonded interaction in the enzyme's active site and reduce its stabilizing effect on the acylenzyme energy.

At the moment, the structure-spectra correlations are being extended using model compounds, $R\text{-}C(=O)NHCH_2C(=S)SCH_2CH(CH_3)NHC(=O)CH_3$, which more closely resemble the cysteine linkages of cysteine-25.

Acknowledgements

We are grateful to Dr. Rosemary C. Hynes for her assistance in the X-ray crystallographic analysis of N-(p-methylbenzoyl)glycine ethyl dithioester.

References

1.   Storer, A.C., Murphy, W.F. and Carey, P.R., J. Biol. Chem., 254 (1979) 3163.
2.   Carey, P.R., In Spiro, T.G. (Ed.), Biological Application of Raman Spectroscopy, Vol. 2, John Wiley, New York, NY, U.S.A., 1987, p.303
3   Kim, M. and Carey, P.R., J. Mol. Struct., 242 (1991) 421.
4.   Kim, M., Birnbaum, G.I., Hynes, R.C., Neugebauer, W. and Carey, P.R., J. Am. Chem. Soc., 115 (1993) 6230.

Assembly of laminin triple stranded coiled-coil domain

M. Nomizu[1], A. Utani[1], A. Otaka[2], P.P. Roller[2] and Y. Yamada[1]

[1]Laboratory of Developmental Biology, National Institute of Dental Research, and [2]Laboratory of Medicinal Chemistry, DTP, DCT, NCI, National Institutes of Health, Bethesda, MD 20892, U.S.A.

Introduction

Laminin, a large heteromeric glycoprotein specific in the basement membrane, has diverse biological activities including promoting cell adhesion, growth, migration and differentiation, and influencing metastatic potential of tumor cells. Laminin from the mouse Engelbreth-Holm-Swarm (EHS) tumor consists of A, B1 and B2 chains, and has a cruciform shape with one long and three short arms when examined by electron microscopy. Several laminin isoforms have been identified with at least 8 genetically distinct subunits. Merosin (M) is an A chain homolog and is expressed in muscle, kidney, trophoblast and in some other tissues. The three chains are assembled through an α-helical coiled-coil structure in the long arm region of laminin which contains approximately 600 amino acids from each chain [1]. It has been shown that a proteolytic fragment consisting of the carboxy one third of the long arm can be reconstituted to a structure similar to the native molecule after denaturation [2]. More recently, we have demonstrated that a short sequence of the carboxy end of each chain is required to initiate the assembly of the laminin chains (Utani et al., unpublished data). In this paper, we have studied the specific interaction of the laminin chain using short synthetic peptides.

Results and Discussion

The synthetic peptides from B1 and B2 subunits of laminin (31- or 32-, and 51-mers), and M-55 (55-mer) used are listed in Figure 1. M-55 was synthesized by Boc based SPPS methodology using an automated peptide synthesizer. The resulting protected peptide resin was treated by the two-step method [3, 4] to deprotect and cleave the peptide from resin. This consisted of consecutive treatments with trimethylsilyl bromide (TMSBr)-thioanisole and HF procedures. The B1 and B2 peptides were manually synthesized by Fmoc based SPPS methodology followed by 1M TMSBr-thioanisole deprotection procedure [5].

Interactions between the B1, B2 and M peptides were examined by mixing experiments of the components at 10 μM of each peptide in 50 mM phosphate buffer (pH 7.4). After a mixture of various peptide pairs, the resulting disulfide bonded hetero and homo dimers of B1 and B2 were analyzed by HPLC at various time intervals. It was found that in an equimolar mixture of B1-32 and B2-31, and

┌➤**B1-51**   ┌➤**B1-32**
SKLQLLEDLERKYEDNQKYLEDKAQELVRLEGEVRSLLKDISEKVAVYSTC
┌➤**B2-51**   ┌➤**B2-31**
KASDLDRKVSDLESEARKQEAAIXDYNRDIAEIIKDIHNLEDIKKTLPTGC
┌➤**M-55**
VRNLEQEADRLIDKLKPIKELEDNLKKNISEIKELINQARKQANSIKVSVSSGGD

Fig. 1. List of synthetic peptides. B1-51: mouse laminin B1 chain 1714-1764, B2-51: mouse laminin B2 chain 1515-1565 (X=Nle), M-55: merosin 274-328.

separately of B1-51 and B2-51, there was very little selectivity for heterodimer formation. However, when M-55/B1-32/B2-31 or M-55/B1-51/B2-51 mixtures were analyzed, the B1-B2 heterodimer was preferentially formed. The oxidative complexation of the M-55/B1-51/B2-51 mixture was completed within 6 hr (Figure 2), but that of M-55/B1-32/B2-31 mixture required more than 48 hr for completion of the oxidative reaction. When the B1-51-B2-51 and M-55 were mixed in equimolar amounts, the heterotrimer was formed as determined by HPLC on a TSK-G2000SW column. The CD spectra demonstrated that the heterotrimer has significantly higher α-helical content than the heterodimer. The thermal stabilities of the heterodimer and trimer were determined by monitoring changes in the CD spectra (222 nm) at various temperatures. The trimer is considerably more stable (Tm = 61.5°C) than the B1-B2 dimer (Tm = 43°C). This Tm value of the trimer is comparable to the previously reported Tm values of the proteolytic laminin molecule [6]. These results confirm that the laminin A chain (merosin here) is required for the selective formation of the native like heterotrimer consisting of B1 (51-mer), B2 (51-mer) and merosin (55-mer) subunits.

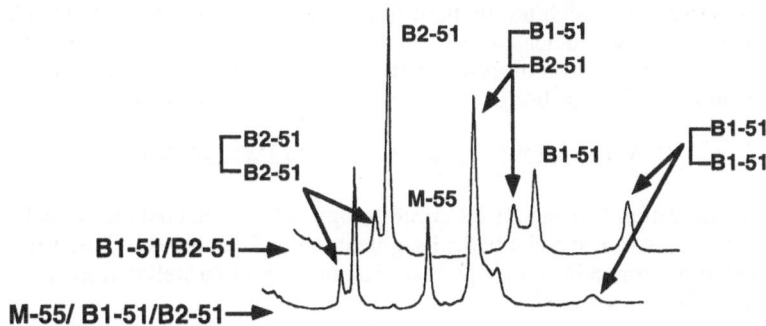

Fig. 2. HPLC analysis of the peptide mixtures incubated at room temperature for 6 hr.

References

1. Beck, K., Hunter, I. and Engel, J., FASEB J., 4 (1990) 148.
2. Hunter, I., Schulthess, T. and Engel, J., J. Biol. Chem., 267 (1992) 6006.
3. Nomizu, M., Inagaki, Y., Yamashita, T., Ohkubo, A., Otaka, A., Fujii, N., Roller, P.P. and Yajima, H., Int. J. Pept. Protein Res., 37 (1991) 145.
4. Nomizu, M., Utani, A., Shiraishi, N., Yamada, Y. and Roller, P.P., Int. J. Pept. Protein Res., 40 (1992) 72.
5. Yajima, H., Fujii, N., Funakoshi, S., Watanabe, T., Murayama, E. and Otaka, A., Tetrahedron, 44 (1988) 805.
6. Paulsson, M., Deutzmann, R., Timpl, R., Dalzoppo, D., Odermatt, E. and Engel, J., EMBO J., 4 (1985) 309.

Structural studies on pro-oxytocin-neurophysin peptides reproducing the dibasic processing site

L. Paolillo[1], M. Simonetti[2], G. D'Auria[1], L. Falcigno[1], M. Saviano[1], T. Carlomagno[1], A. Scatturin[3], C. Di Bello[2] and P. Cohen[4]

[1]*Department of Chemistry, University of Naples, via Mezzocannone 4, 80134 Naples, Italy,* [2]*Institute of Industrial Chemistry, University of Padova, Italy,* [3]*Department of Pharmaceutical Chemistry, University of Ferrara, Ferrara, Italy,* [4]*Groupe de Neurobiochimie Cellulaire et Moléculaire, Université Pierre et Marie Curie, Paris, France*

Introduction

Bioactivation of pro-proteins by limited proteolysis is a general mechanism in the biosynthesis of hormones, receptors and viral protein precursor. The cleavage process occurs at the level of pairs or singlets of basic residues in the pro-forms. It was proposed that the basic residues, which act as signal loci for the proteolytic enzyme are located in flexible structural segments like β-turns and/or loops [1]. The structural parameters which are thought to play a key role in the enzyme-substrate recognition have been elucidated in various synthetic peptides [2]. In this work we present spectroscopic and molecular dynamics data on the whole peptide segment 1-20 (containing a disulfide bridge) of the common pro-OT/Np precursor:

$$C^1\text{-Y-I-Q-N-}C^6\text{-}P^7\text{-L-G-G-K-R-A-V-}L^{15}\text{-D-L-D-V-}R^{20}$$

NMR as well as CD and FT-IR spectroscopy indicate that the molecule has two folded regions in a β-turn and a helical segment in the C-terminus. These results are in line with those found in shorter peptides [2] and point to a well defined molecular architecture of the cleavage site.

Results and Discussion

The CD spectrum of proOT/Np in TFE/H_2O (95/5 v/v) shows a positive band below 200 nm and negative dichroism between 200-250 nm. These features indicate the existence in solution of a conformational equilibrium between aperiodic structures and folded conformations [3].

The FT-IR analysis carried in TFE/D_2O (90/10 v/v) shows an intense band around 1659 cm^{-1} and a weaker band at 1680 cm^{-1}. These bands may be attributed to the existence in solution of an ordered population containing β-turns and α-helical structures.

1D, 2D and 3D NMR experiments were performed at 400, 600 and 750 MHz in solutions of DMSO and DMSO/H$_2$O (80:20). The sequential assignment of proton resonances was accomplished with homonuclear correlation experiments such as TOCSY and NOESY. In DMSO/H$_2$O, as in pure DMSO, sequential NH$_{i+1}$-NH$_i$ and NH$_{i+1}$–αCH$_i$ NOE connectivities have been observed in the C-terminal segment, starting from the Ala[13] to the last Arg[20] and support the hypothesis of α helical-type structure in this region. Nevertheless, no extreme values have been observed for the $^3J_{NH\alpha}$ coupling constants, thus indicating that conformational averaging is taking place for the peptide in solution. It should be noted that a uniform trend is observed by these coupling constants to assume lower values in going from DMSO to DMSO/H$_2$O. This fact can be interpreted as indicative of a higher population of a folded structure in the latter solvent. As far as the N-terminal region is concerned, the observation of sequential NOEs, together with the values of the $^3J_{NH\alpha}$ and the low values of the temperature coefficients in the Cys[1]-Cys[6] ring, indicates the presence of one β-turn in the sequence Tyr[2]-Asn[5], as observed in the solid state structure of the Oxytocin ring [4].

In the middle region of the peptide (Pro[7]-Arg[20]) the resonances of the NHs severely overlap. NH-NH NOE interactions can neither be confirmed nor denied. The NOE αCys[6]-δδ'Pro[7] indicates a trans configuration of the bond Cys[6]-Pro[7]. Further information can be obtained from the 3D-NOESY-HOHAHA spectrum, where NOE NH$_{i+1}$–αCH$_i$ connectivities between the residues Lys[11]-Gly[10] and Arg[12]-Lys[11] have been observed. It should be noted that the Gly[10] NH resonance exhibits a low temperature coefficient value (-2.4 ppb/K in DMSO/H$_2$O).

The results presented in this work come from different spectroscopic methods including CD, FT-IR and NMR. They provide physico-chemical evidence for the existence of an ordered secondary structure at the processing site of a simple prohormone such as pro-OT/Np. The C-terminal region of 1-20pro-OT/Np (Ala[13]-Arg[20]) seems to be organized in a helical structure like the shorter peptides previously examined [2]. Information about the possible existence of a β-turn in the middle region of the peptide chain is obtained from NMR spectra at 750 MHz that point to a β-turn (type I) folding from the Pro[7] residue. The MD simulation, with the insertion of a β-turn (type I) from Pro[7] residue which is similar to that found in shorter peptides, seems fully compatible with the available NMR data.

References

1. Rholam, M., Nicolas, P. and Cohen, P., FEBS Lett., 207 (1986) 1.
2. Paolillo, L., Simonetti, M., Brakch, N., D'Auria, G., Saviano, M., Dettin, M., Rholam, M., Scatturin, A., Di Bello, C. and Cohen, P., EMBO J., 11 (1992) 2399.
3. Rholam, M., Cohen, P., Brack, N., Scatturin, A., Paolillo, L. and Di Bello, C., Biochem. Biophys. Res. Commun., 168 (1990) 1066.
4. Wood, S.P., Tickle, I.J., Treharne, A.M., Pitts, J.E., Mascarenhas, Y., Li, J.Y., Husain, J., Cooper, S., Blundell, T.L., Hruby, V.J., Buku, A., Fischman, A.J. and Wyssbrod, H.R., Science, 232 (1986) 633.

Conformational studies on bombolitin III-related peptides in aqueous solution containing SDS

E. Peggion, R. Battistutta, A. Bisello, S. Mammi and S. Sala

Biopolymer Research Center, Department of Organic Chemistry, University of Padova, Via Marzolo 1, 35131 Padova, Italy

Introduction

Bombolitins are five structurally related heptadecapeptides initially isolated from the venom of a bumblebee [1]. They share the ability to lyse erithrocytes and liposomes and to enhance the activity of phospholipase A_2 (PLA_2). Bombolitins are also able to form amphiphilic α-helices under suitable conditions, a feature which could explain their ability to interact with membranes. In fact, amphiphilicity rather than sequence specificity was suggested to be the main determinant for biological activity [2]. We have recently undertaken a systematic investigation of the conformational properties of bombolitin-related peptides in an attempt to understand the structural basis for biological activity. Here, we report the synthesis and preliminary conformational studies on two analogs of bombolitin III (B-III).

Results and Discussion

Both analogs were synthesized by solid phase methods using FMOC chemistry. After purification, both analogs exhibited a single peak under different HPLC conditions and also in capillary electrophoresis. The peptides were characterized by amino acid analysis, sequence analysis and FAB-MS. All results were consistent with the pure sequences IKIMDILAKLGKVLAHV-NH$_2$ (*all-D* B-III) and VHALVKGLKALIDMIKI-NH$_2$ (retro-B-III).

Similar to other bombolitins [3], these analogs are random coils at 10^{-5} M in water, and their CD patterns are mirror images (data not shown). Like B-III, the conformation of the *all-D* analog depends on concentration and, at 5.72 mM peptide, pH 6.2, the α-helical content is about 55%. Intermolecular aggregation, very likely driven by interactions among the hydrophobic surface of the helices, is at the origin of these results [3].

The results of a titration of retro-B-III with sodium dodecylsulphate (SDS) are reported in Figure 1. Increasing the detergent concentration, there is an increase of ordered structure. At SDS concentration below the c.m.c., the CD pattern suggests that a type II β-turn component is present in the conformer population. Above the c.m.c. of SDS, the α-helical conformation becomes stable and saturation is reached at 1.4 mM SDS, with the equilibrium completely shifted towards the micelle-bound state (Figure 1). The same behavior is observed with *all-D* B-III, with spectra that are the mirror images of those of the L-sequence. We can therefore conclude that the way of interaction of these analogs with detergent micelles is very similar to that of native B-III. *In vitro* biological activity tests (PLA_2-induced release of fatty acid from L-α-phosphatidyl-choline) indicated that both analogs are as active as the native sequence. These results, together with those previously obtained on retro-bombolitin I [4], confirm that neither sequence nor helical sense are the main factors responsible for bioactivity. Rather, the ability to fold into amphiphilic helices seems to be the major determinant of bombolitin activity.

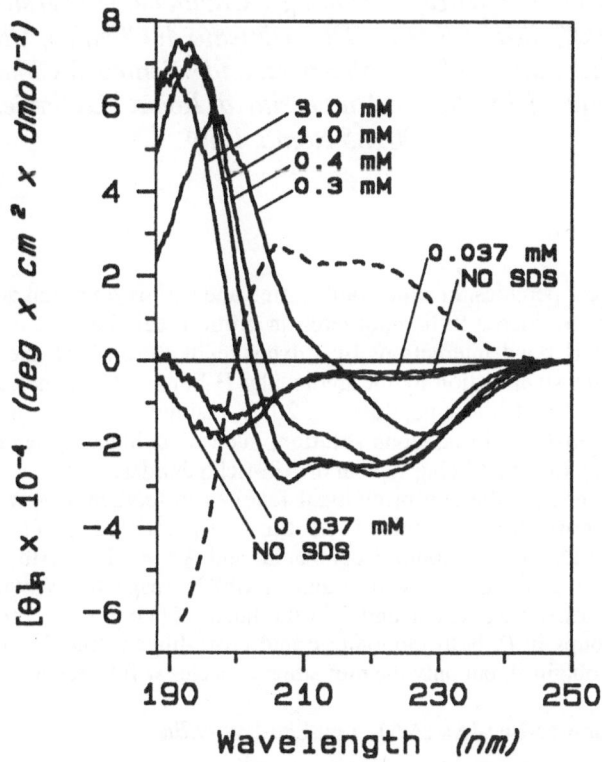

Fig. 1. CD spectra of $3.7x10^{-5}$ M retro-B-III in 10 mM phosphate buffer, pH 5, in the presence of increasing amounts of SDS (indicated). The CD spectrum of all-D B-III, $4.28x10^{-5}$ M and in the presence of 2.85 mM SDS (saturation conditions) is also shown (broken line).

References

1. Argiolas, A.A. and Pisano, J.J., J. Biol. Chem., 260 (1985) 1435.
2. Kaiser, E.T. and Kezdy, F.J., Proc. Natl. Acad. Sci., 80 (1983) 1137.
3. Bairaktari, E., Mierke, D.F., Mammi, S. and Peggion, E., Biochemistry, 29 (1990) 10090 and 10097.
4. Battistutta, R., Bisello, A., Mammi, S. and Peggion, E., In Schneider, C.M. and Eberle, A.N. (Eds.), Peptides 1992 (Proceedings of the 22nd European Peptide Symposium), Escom, Leiden, The Netherlands, 1993, p.525.

Conformational features of sequential peptides probed by fluorescent donor-acceptor pairs

B. Pispisa[1], A. Palleschi[2], M. Venanzi[1] and G. Zanotti[3]

[1]Dipartimento di Scienze e Tecnologie Chimiche, Universita' di Roma "Tor Vergata", 00133 Roma, [2]Dipartimento di Chimica, Universita' di Roma "La Sapienza", 00185 Roma, and [3]Centro di Chimica del Farmaco del CNR, c/o Universita' di Roma "La Sapienza", 00185 Roma, Italy

Introduction

Short linear peptides, in which both amino and carboxyl termini are derivatized, are in general considered to be unordered in protic media because competition for intermolecular H-bond interactions by solvent tends to diminish the advantage of folded structures stabilization by intramolecular H-bonds. Nevertheless, it has been recently reported that linear peptides (of 20-13 residues or less) are able to attain ordered conformations in aqueous solution, such as α-helix, β-bend or 3_{10}-helix structures [1, 2], the latter being typical of Aib-rich peptides.

We present here the conformational features in methanol or water-methanol solution of short, linear peptides, as those represented in (1), where P = protoporphyrin IX, N = 1-naphthylacetic acid, and AA = Ala or Aib. They will be denoted in the following as $P(Ala)_nN$ and $P(Aib)_nN$, respectively, with n = 1, 2, 3 and 4 in the former case, and 1 and 2 in the latter. Owing to the presence of two carboxylic groups in P, both monomeric and cross-linked dimeric protoporphyryl peptides were obtained, but only the monomeric species will be reported here.

$$Boc\text{-}Leu\text{-}Leu\text{-}Lys\text{-}(AA)_n\text{-}Leu\text{-}Leu\text{-}Lys\text{-}OtBu \qquad (1)$$
$$\begin{array}{cc} | & | \\ P & N \end{array}$$

Results and Discussion

The rate of N* quenching should depend on the separation distance between P and N groups, that in a rigid framework is expected to increase by, e.g., 3.8 Å per alanine residue (shortest through-space distance) or 4.4 Å per alanine residue (through-bond distance). The observation that both the fluorescence time decay and quantum yield of naphthalene chromophore [λ_{ex} = 280 nm (N*), λ_{em} = 340 or 630 nm (N* or P*, respectively)] vary with n, but that the trend cannot be simply related to the number of amino acids AA in the peptides strongly suggests that folded (helical) structures form. This agrees with IR spectral results on CD_3OD solution of oligopeptides (1) at concentration ≈0.10-0.30 mM. For instance, in the Ala-based peptides the integrated intensity of the bands at around 3330 cm^{-1} (amide A region) and 1655 cm^{-1} (amide I region), corresponding to intramolecularly H-bonded N-H and C=O groups, respectively, was found to increase dramatically on going from the samples carrying naphthalene or porphyrin molecules only (blanks) to the peptides, although the question as to whether the Ala-based and Aib-based compounds differ in structure could not be clearly answered [3]. However, the finding that exciplex formation in methanol is definitely minor in the former peptides than in the latter strongly suggests that the intramolecularly H-bonded conformations are different. Consistently, structural considerations indicate that the position of the probes linkages in a 3_{10}-helix arrangement, that should be favored by the Aib residues [2], is such as to allow the chromophores to reside more closely to each other than in an α-helical peptide. ^{1}H-NMR spectral patterns in the NH region (5-9 ppm) and preliminary NOE data confirm this conclusion. Therefore, despite the fact that CD spectra of the Ala-based and Aib-based peptides in methanol are not too dissimilar, we are inclined to think that the former compounds are characterized by an H-bonding scheme of i → i+4 (C_{13} type) and the latter by a scheme of i → i+3 (C_{10} type).

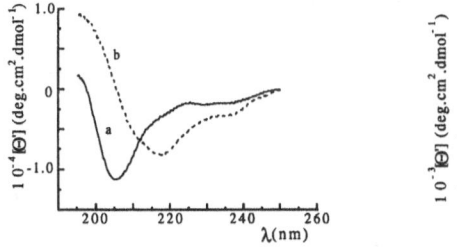 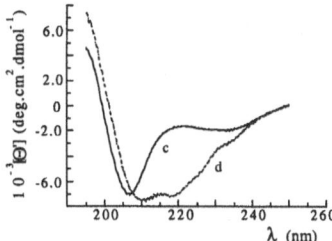

Fig. 1. CD spectra of PAibN (a), PAlaN (b), P(Aib)$_2$N (c), and P(Ala)$_2$N (d) in water-methanol solution (75/25, v/v); concentration ≈ 0.3 mM. Molar ellipticity is on per residue basis.

By contrast, the CD spectra of the two sets of peptides in water-methanol solution exhibit quite different features, as illustrated in Figure 1, and this is a further confirmation of the above conclusion. Theoretical calculations by Woody et al. [4] show that systematic distortions due to the dynamic fluctuation of the α-helical conformation bring about significant changes in the CD spectrum of short linear peptides. For instance, the rotational strength in the nπ* absorption region decreases upon tilting of C=O groups, which may account for the fact that the Ala-based oligopeptides in methanol have a weak rotational strength at around 220 nm. By addition of water, the CD patterns of these peptides change dramatically, now resembling closely those of α-helix (Figure 1). In aqueous medium, hydrophobic interactions between the bulky, apolar N and P groups appear to stabilize the α-helical conformation, provided the main chain is long enough, and to slow down the dynamic distortions of the ordered structure that very likely occur in methanol solution. By contrast, the CD spectra of the Aib-based peptides are only little affected by the presence of water, as one would expect for a 3_{10}-helix which is more tightly bound by intramolecular H-bonds than an α-helix.

References

1. Marqusee, S., Robbins, V.H. and Baldwin, R.L., Proc. Natl. Acad. Sci. U.S.A., 86 (1989) 5286.
2. Basu, G., Bagchi, K. and Kuki, A., Biopolymers, 31 (1991) 1763.
3. Pispisa, B., Venanzi, M., Palleschi, A. and Zanotti, G., J. Mol. Liquids, in press.
4. Manning, M.C., Illangasekare, M. and Woody, R.W., Biophys. Chem., 31 (1988) 77.

Conformational study on a fibronectin-like sequence (250-257) of Leishmania gp63, using ^{1}H-NMR spectroscopy

V. Tsikaris[1], M. Sakarellos-Daitsiotis[1], C. Sakarellos[1], M.T. Cung[2], A.K. Tzinia[3] and K.P. Soteriadou[3]

[1]Department of Chemistry, University of Ioannina, P.O. Box, 45110 Ioannina, Greece, [2]LCPM, CNRS-URA 494, ENSIC-INPL, Nancy, France, and [3]Department of Biochemistry, Hellenic Pasteur Institute, 11521 Athens, Greece

Introduction

The peptide sequence RGD is the principal recognition segment in adhesive proteins associated with biochemical processes such as hemostasis, tumor metastasis and tissue remodeling. It has been demonstrated recently that the IASRYDQL synthetic octapeptide (250-257) of the major surface glycoprotein of Leishmania, gp63, efficiently inhibits parasite attachment to the macrophage receptors *in vitro*, and the SRYD-containing tetrapeptide mimics antigenically and functionally the RGDS segment of fibronectin [1]. The conformational properties of the IASRYDQL octapeptide are now investigated in DMSO solution at pH 5 with the combined use of NMR data and MD simulations, in order to define the structural requirements of the SRYD-moiety for the receptor-mediated event of intracellular parasitism and the design of good vaccine candidates against leishmaniasis.

Results and Discussion

The complete assignment of all proton resonances was based on the combined use of COSY, HOHAHA, ROESY and NOESY experiments and is in agreement with previously reported ^{1}H-NMR data. The temperature coefficient values of the $R^{253}NH$, $D^{255}NH$, $Q^{256}NH$ and $L^{257}NH$ (above -3x10^{-3} ppm/K) indicate that these amide protons are not entirely exposed to the solvent and that they could be involved possibly in intramolecular helical interactions at pH 5. However, the coupling constant values (above 6Hz) between NH and $C^{\alpha}H$ protons exclude this possibility. The NOESY spectrum of the octapeptide showed intense NOE cross peaks in the C-terminal part between the successive amide protons $Y^{254}NH/D^{255}NH$ and $D^{255}NH/Q^{256}NH$. These findings and the absence of strong NOE correlations between the consecutive $Y^{254}C^{\alpha}H/D^{255}NH$ are in favor of a type I β-turn in the $-R^{253}-Y^{254}-D^{255}-Q^{256}-$ part, involving the $Q^{256}NH \rightarrow R^{253}CO$ interaction. Intense NOE connectivities were also found between successive $C^{\alpha}H_i/NH_{i+1}$ and NH_i/NH_{i+1} protons ($A^{251}C^{\alpha}H/S^{252}NH$ and $S^{252}NH/R^{253}NH$ respectively), as well as between $S^{252}C^{\alpha}H/R^{253}NH$. These qualitative NOE data did not allow us to define with certainty the conformational features of the I^{250}-A^{251}-S^{252}-R^{253} N-terminal part.

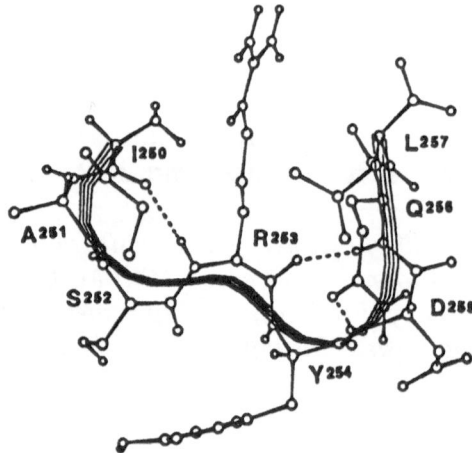

Fig. 1. The lowest energy structure of the IASRYDQL from the constrained molecular dynamics simulations.

With the aim to supplement the NMR data and to refine the reported structure, MD simulations and energy minimizations were carried out using backbone-backbone constrained distances provided from NOE cross-correlations. The torsional angles obtained from the time-averaged MD minimized conformations are in agreement with a type I β-turn in the -R^{253}-Y^{254}- D^{255}-Q^{256}- sequence, with 98% occurrence of the $Q^{256}NH \rightarrow R^{253}CO$ hydrogen bonding interaction, consistent with the NMR data. Moreover the time-averaged MD minimized conformers are also compatible with a type I β-turn in the I^{250}-A^{251}-S^{252}-R^{253} N-terminal part, with 72% occurrence of the $R^{253}NH \rightarrow Ile^{250}CO$ interaction. On the other hand, a ten-membered cycle was stabilized through a hydrogen bonding interaction between $D^{255}NH$ and $Q^{256}C^{8}O$ (79% occurrence), which is in agreement with the low temperature coefficient value of $D^{255}NH$ (Figure 1).

In conclusion, the reported NMR data and MD simulations pointed out that the SRYD-analogue adopts a restricted conformation, which may prove crucial for the receptor-ligand interaction in cellular adhesion.

References

1.   Soteriadou, K.P., Remoundos, M.S., Katsikas, A.K., Tzinia, A.K., Tsikaris, V., Sakarellos, C. and Tzartos, S.J., J. Biol. Chem., 267 (1992) 13980.

Crystallization of a neutralizing malaria immunoglobulin with linear and cyclic peptides

E.A. Stura[1], D.C. Kaslow[2] and A.C. Satterthwait[1]

[1]The Scripps Research Institute, La Jolla, CA 92037, and
[2]NIAID, NIH, Bethesda, MD 20892, U.S.A.

Introduction

Malaria is a parasitic disease that afflicts hundreds of millions of people annually and accounts for millions of deaths among children. This past decade has seen the parasite develop multiple drug resistance and vaccine development has now become a priority. The parasite evolves through several stages in the human and mosquito hosts. Each stage is currently under investigation as a vaccine target [1]. Our goal is to develop a transmission blocking vaccine to be administered in conjunction with drug therapy and as part of a multi-stage vaccine. A major role for this vaccine would be to prevent escape mutants formed in reaction to drugs or other vaccine components from proliferating in the population.

The target of this work is Pfs25, a sexual stage protein of *Plasmodium falciparum* malaria. Monoclonal antibody (Mab) 4B7 which binds to the third epidermal growth factor (EGF)-like domain of Pfs25 completely blocks transmission of the parasite in the mosquito gut [2]. The X-ray crystal structure of this monoclonal complexed with antigens will provide critical information regarding conformation and fine structure specificity for vaccine development.

Results and Discussion

Mab 4B7 binds to an epitope, ILDTSNPVKT, which corresponds to a B loop β-hairpin of a prototypical EGF-like domain. A series of cyclic peptides was prepared by replacing predicted main chain hydrogen bonds (NH...O=CRNH) with a hydrazone covalent hydrogen bond mimic (N-N=CHCH$_2$CH$_2$) (Figure 1). Both linear and cyclic peptides were synthesized on the solid support using described procedures [3].

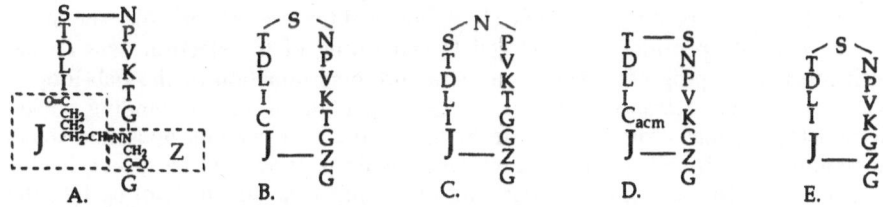

Fig. 1. Cyclic peptides with a hydrazone covalent hydrogen bond mimic.

Table 1 *Relative affinities of anti-Pfs25 monoclonal antibody 4B7 for cyclic peptides (A-E) and corresponding linear peptide F as determined from competition ELISA*

Peptide	Peptide Sequence	Rel. Affinity	
A.	J-**ILDTSNPVKT**-GZG-NH$_2$	+++	Cyclic
B.	**JCILDTSNPVKT**-GZG-NH$_2$	++	Cyclic
C.	J-**ILDTSNPVKTG**GZG-NH$_2$	++	Cyclic
D.	JC$_{acm}$**ILDTSNPVKT**-GZG-NH$_2$	++	Cyclic
E.	J-**ILDTSNPVK**--GZG-NH$_2$	++	Cyclic
F.	**AcGILDTSNPVKTGV**--NH$_2$	+	Linear

The Pfs25 sequence is shown in bold. J and Z are covalently linked to form the hydrazone hydrogen bond mimic [3].

While each of the loops bound with higher affinity to 4B7 Mab than linear peptides, the smaller loop A bound best, reflecting conformational preference of Mab 4B7 (Table 1). We have obtained crystals for the intact IgG complexed with linear peptides which diffract to low resolution and crystals of an elastase released Fab fragment both free and in complex with linear and cyclic peptides. The different peptides have a profound influence on the crystallization of the antibody complexes and appear to control the space group in which crystals grow. The development of crystallization conditions was aided by the screening of the Fab complexed with several loops and linear peptides.

The higher affinity of loops may have aided the crystallization by producing stable complexes. The free Fab and Fab-peptide complexes with cyclic peptides A, D and E nucleate spontaneously. Fab complexes with cyclic peptides B and C do not nucleate spontaneously, but by employing streak seeding and macroseeding [4] with seeds from native crystals, we have produced cocrystals with these peptides. Cocrystals with linear peptide F were obtained by cross-seeding from cocrystals with loop E. The unit cell volume doubles yielding crystals of the seeded complex diffracting to significantly higher resolution (2.7Å). The crystallizing solutions were similar ranging from 10% polyethylene glycol 8,000 for peptide B to 15% for the linear peptide F. The precipitant solutions were buffered with 0.2M imidazole malate from pH 5.5 to 8.0. The optimum crystallization conditions vary from complex to complex within the ranges given above. The different crystal forms diffract from 2.4Å to 3.6Å, depending on the peptide, and are all suitable for X-ray structure determination (Table 2). Data sets have been collected for the free Fab and for all the peptide complexes shown in Table 2 to the diffraction limit of each complex. Molecular replacement solutions have been obtained for free Fab, and for each of the complexes with peptides A, B, C and D and fitting of the electron density and refinement is in progress. The X-ray structure determination of the Fab-loop B complex shows electron density for the cyclic peptide in the Fab binding pocket demonstrating that cross seeding from the native Fab crystals does not select uncomplexed Fab and that complexes with cyclic peptides can be grown in this manner. The data sets provide a basis for determining the structures of each of the peptides in the antigen binding site of the antibody.

Table 2 *Lattice parameters and R_{min} for Fab 4B7 and for Fab-peptide complexes*

Peptide	Space group	a (Å)	b (Å)	c (Å)	α (°)	β (°)	γ (°)	R_{min} (Å)
Free	C2	149.8	73.6	60.7	90.0	96.8	90.0	2.5
A.	$P3_121$	72.1	72.1	135.7	90.0	90.0	120.	3.1
B.	C2	148.6	73.5	60.1	90.0	97.2	90.0	2.8
C.	C2	148.0	73.2	59.7	90.0	97.0	90.0	2.8
D.	C2	71.2	74.2	104.7	90.0	93.4	90.0	3.3
E.	C2	71.1	73.3	104.2	90.0	93.3	90.0	3.3
F.	C2	105.3	77.1	146.1	90.0	95.3	90.0	2.7

Although the crystallization conditions are similar, the resolution and packing arrangement of the Fab, as reflected in the change in space group and in the unit cell dimensions, is affected by changes in the peptide sequence and positioning of the hydrogen bond mimic.

Such structures will help us understand the mechanism for malaria neutralization at the atomic level and should aid in the design of constrained malaria peptides for use with T-cell epitopes in a transmission blocking vaccine.

Acknowledgements

This work was supported by funding from U.S.A.I.D. DPE-5979-A-00-1035-00 and NIH grants GM-38419, GM-46192 and AI-23498. This is publication 8125-MB from The Scripps Research Institute.

References

1. Oaks, S.C. Jr., Mitchell, V.S., Pearson, G.W. and Carpenter, C.J. (Eds.), Malaria Obstacles and Opportunities, National Academy Press, Washington, DC, U.S.A., 1991, p.169.
2. Barr, P.J., Green, K.M., Gibson, H.L., Barhurst, I.C., Quakyi, I.A. and Kaslow, D.C., J. Exp. Med., 174 (1991) 1203.
3. Chiang, L.-C., Cabezas, E., Calvo, J.C. and Satterthwait, A.C., this volume.
4. Stura, E.A. and Wilson, I.A., In Ducruix, A. and Giegé, R. (Eds.), Crystallization of Nucleic Acids and Proteins, IRL Press Limited, Oxford, U.K., 1992, p.99.

Conformational preferences of Magainin2a and PGLa antibiotics in aqueous solution: CD and FTIR spectroscopic detection of secondary structure intermediates

A.E. Shinnar[1,2], L.R. Olsen[3], G.H. Reed[3], J.S. Leigh[4] and M.A. Zasloff[2]

[1]Dept. Biochem. and Biophys., University of Pennsylvania, Philadelphia, PA 19104, U.S.A., [2]Magainin Research Institute, Magainin Pharmaceuticals Inc., 5110 Campus Drive, Plymouth Meeting, PA 19462, U.S.A., [3]Inst. for Enzyme Research and Dept. of Biochem., College of Agriculture and Life Science, University of Wisconsin, Madison, WI 53705, U.S.A., [4]Dept. Radiology, Hospital of the University of Pennsylvania, Philadelphia, PA 19104, U.S.A.

Introduction

Magainin2a (Mgn2a) and PGLa, two amphipathic peptide antibiotics isolated from frog skin, adopt α-helices in the presence of lipid bilayers [1-3]. In aqueous solution, however, they display very low α-helix content unless a helix promoting solvent such as TFE is added [4]. Understanding the details of how these and similar length peptides undergo the conformational transition from an apparently unordered state to α-helical conformation is central to the field of protein folding. In this study, we employ CD and FTIR spectroscopy to explore if Mgn2a and PGLa harbor small populations of secondary structure elements, which might serve as intermediates in the folding process.

Results and Discussion

CD spectra of both Mgn2a and PGLa are dominated by a strong band near 197 nm, which becomes less intense with increasing temperature while a weaker band (210-225 nm) becomes slightly more negative (Figure 1). These temperature dependent spectral changes are reversible and characterized by a sharp isodichroic point at 207.4 nm, suggesting an equilibrium between two states. These properties are similar to poly-L-lysine, in which the strong band is attributed to the "random coil" conformation or to another rigid conformation (RC) of the carbonyl similar to poly-L-proline (II) [5, 6]. CD difference spectra reveal a characteristic band shape with λ_{min} = 217, λ_{max} = 194 nm, which is reminiscent of β-sheet and suggests that a population with properties similar to β-sheet is favored at higher temperature at the expense of the RC population. θ_{197} and θ_{217} do not change significantly over >500-fold concentration range, suggesting that this β-structure is due to intramolecular rather than intermolecular interactions or aggregation. Secondary structure analysis, conducted with LINCOMB [7] and four reference spectra (α-helix, β-sheet, RC [8] and β-turn Type I [9]), shows a consistent decrease in RC with a modest increase in total β-structure (β-sheet plus β-turn).

FTIR spectra show that the amide I' band for Mgn2a and PGLa is dominated by a peak at 1642 cm⁻¹, typical of an RC conformation (Figure 2). Reversible temperature dependent changes in amide I' produce an isosbestic point at 1648 cm⁻¹, suggesting an equilibrium between interconverting species. Fitting of Gaussian peaks to resolution enhanced spectra [10, 11] shows a band at 1675 cm⁻¹ for both peptides, which can be assigned to reverse turns [12]. Difference spectra show that with increasing temperature, there is an absorbance loss at 1637 cm⁻¹ with an accompanying broad absorbance increase from 1650-1670 cm⁻¹. These changes can be interpreted as an increase in loose turns at the expense of RC structure.

CD and FTIR spectra of Mgn2a and PGLa show a temperature dependent equilibrium which at higher temperatures favors a conformation with properties of a β-structure. In addition, the spectra are consistent with a significant percentage of reverse or β-turn. These populations might serve as nascent secondary structures in the folding process.

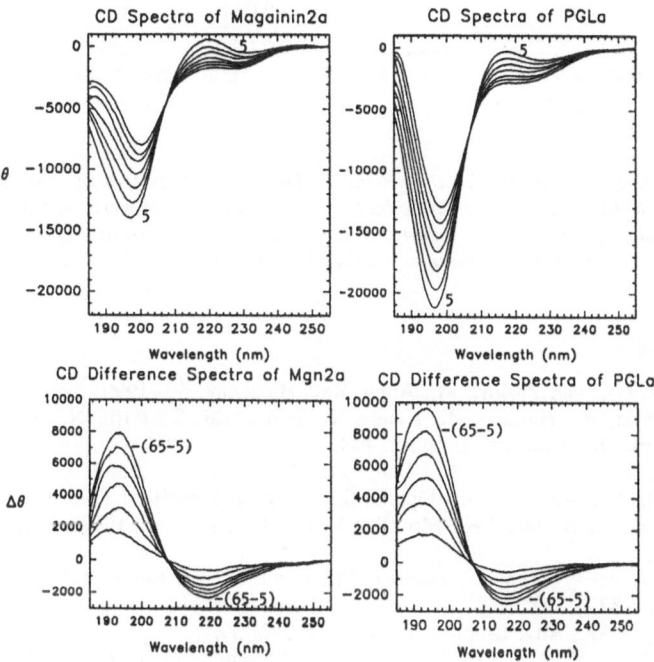

Fig. 1. CD spectra of Mgn2a and PGLa in aqueous solution. CD spectra of the TFA salts of Magainin2a (GIGKFLHSAKKFGKAFVEIMNS_{NH2}) and PGLa (GMASKAGAIAGKIAK VALKAL_{NH2}) in 10 mM phosphate buffer pH 7, 100 mM NaF are shown nested as a function of temperature (5, 15, 25, 35, 45, 55, 65 °C). Difference spectra were generated by subtracting the spectrum at 5 °C from those at higher temperatures.

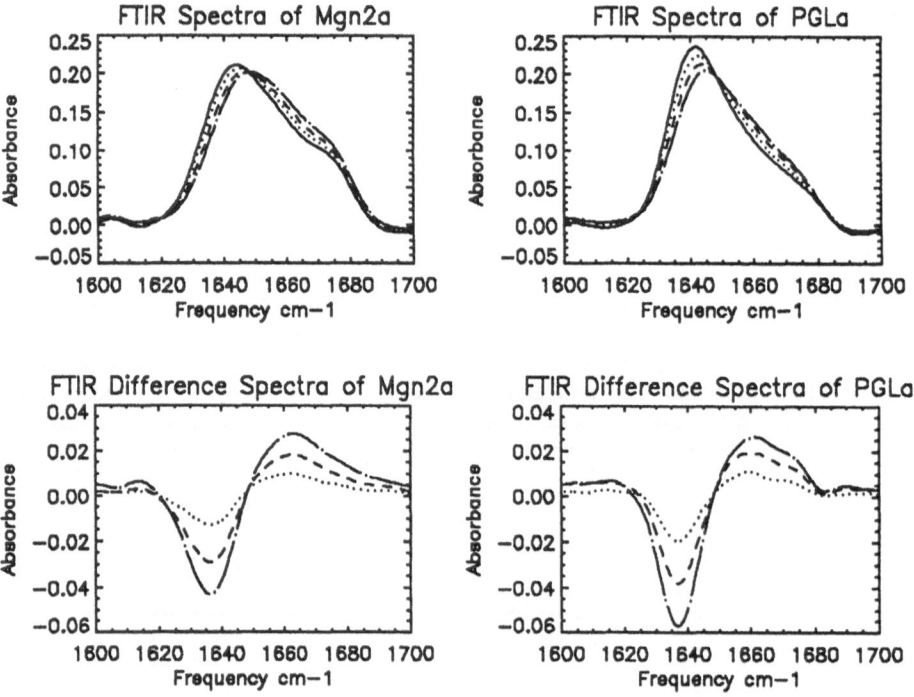

Fig. 2. FTIR spectra of Mgn2a and PGLa in D₂O. FTIR spectra of Mgn2a and PGLa (hydrochloride salts, 5 mM phosphate buffer, pD 7, 4096 scans) were resolution enhanced with linewidths of 19-22 and k factors of 2.0-2.3 [10, 11]. Resulting spectra were fit by non-linear least squares to Gaussian bands. Temperatures are 5 (——), 25 (····), 45 (---), and 65°C (–·–).

References

1.   Bevins, C.L. and Zasloff, M., Annu. Rev. Biochem., 59 (1990) 395.
2.   Matsuzaki, K., Harada, M., Handa, T., Funakoshi, S., Fujii, N., Yajima, H. and Miyajima, K., Biochim. Biophys. Acta, 981 (1989) 130.
3.   Bechinger, B., Kim, Y., Chirlian, L.E., Gesell, J., Neumann, J.-M., Montal, M., Tomich, J., Zasloff, M. and Opella, S.J., J. Biomol. NMR, 1 (1991) 167.
4.   Chen, H.-C., Brown, J.H., Morell, J.L. and Huang, C.M., FEBS Lett., 236 (1988) 462.
5.   Shinnar, A.E., Olsen, L., Shinnar, M., Reed, G. and Leigh, J.S., Biophys. J., 64 (1993) A377.
6.   Woody, R.W., Adv. Biophys. Chem., 2 (1992) 37.
7.   Perczel, A., Hollosi, M., Tusnady, G. and Fasman, G.D., Protein Eng., 4 (1991) 689.
8.   Greenfield, N. and Fasman, G.D., Biochemistry, 8 (1969) 4108.
9.   Brahms, S. and Brahms, J., J. Mol. Biol., 138 (1980) 149.
10.   Kauppinen, J.K., Moffatt, D.J., Mantsch, H.H. and Cameron, D.G., Appl. Spect., 35 (1981) 271.
11.   Byler, D.M. and Susi, H., Biopolymers, 25 (1986) 469.
12.   Prestelski, S.J., Byler, D.M. and Liebman, M.N., Biochemistry, 30 (1991) 133.

The interaction between a DNA-binding SPXX peptide and an A/T rich DNA hairpin

N. Zhou and H.J. Vogel

Department of Biological Sciences, University of Calgary,
Calgary, Alberta, T2N 1N4, Canada

Introduction

Most sequence-specific DNA-binding proteins recognize and bind to their cognate DNA in the major groove. Binding motifs such as the helix-turn-helix, zinc-finger and leucine-zipper and their interactions with DNA have been characterized. The binding of proteins to the minor groove is much less understood. Examples of minor groove DNA binding proteins are histones, DNase I and HU type proteins. Their binding tends to be less sequence-specific and involves interactions largely with backbone phosphate groups, but also with minor groove base groups. Among the various motifs that have been proposed for minor groove binding is an "SPK(R)K(R)" sequence, found repeatedly in DNA-binding "arms" of the histones from sea urchin sperm. A more general "SPXX" sequence was found to occur three times more frequently in gene regulatory proteins than in general proteins [1]. An "SPKK"-rich fragment of sea urchin spermatogenous histone H1 has been shown to be a competitive inhibitor of the drug Hoechst 33258 [2], which is known to interact specifically with the minor groove of A/T rich DNA segments. To further explore the structural requirements for binding of the proposed SPXX motif, we have used a 14-mer peptide MRSRSPSRSKSPMR and an A/T-rich DNA hairpin for a proton and phosphorus 2D NMR study. This peptide contains two SPXX sequences and corresponds to a repeated sequence found in the high molecular weight basic nuclear proteins that are present in the sperm chromatin of winter flounder [3].

Results and Discussion

The oligo-DNA T6C4A6 was characterized by NOESY and TOCSY NMR spectroscopy. Below 30°C it forms a hairpin structure in aqueous phosphate buffer solution (pH 6.5). The stem consists of six A-T base pairs and the four C residues form a loop. Unusual cross-strand NOEs between stem residues were detected, and the first C residue in the loop displayed an abnormal chemical shift thermal melting behavior. These features are similar to a homologous DNA hairpin with a longer stem, which was previously characterized by 2D NMR and restrained molecular dynamics [4]. This hairpin has a bent stem with a slightly narrower minor groove, and the first C in the loop is not stacked but exposed to the solvent. The DNA hairpin used here is expected to have a similar structure. No regular secondary structure can be detected by 2D NMR experiments for the 14-mer peptide.

Upon addition of aliquots of a peptide solution to the DNA sample, the peptide amide signals appear at shifted resonance positions and they are broadened, indicating that binding takes place (Figure 1). Up to a 1:1 DNA-peptide ratio, chemical shift changes only occur for the proton and phosphorus signals of the loop residues, suggesting that under these conditions the peptide binds preferentially to the loop residues. Addition of up to two equivalents of peptide causes changes for four of the T imino and all of the A H2 protons; while the NOESY pattern of the DNA stem residues is not disturbed. The imino protons are involved in the base-pairing of the stem; the A H2 protons are located in the minor groove of the stem. Thus, at the higher ratio, binding of the peptide to the minor groove of the stem takes place. Most of the non-solvent-exchangeable proton signals of the peptide do not display significant changes in their chemical shifts, hence it is unlikely that the peptide undergoes major changes in structure upon binding. Therefore, an extended peptide which interacts with the convex minor groove of the bent stem through some of its amide and side chain groups is consistent with the latter NMR results.

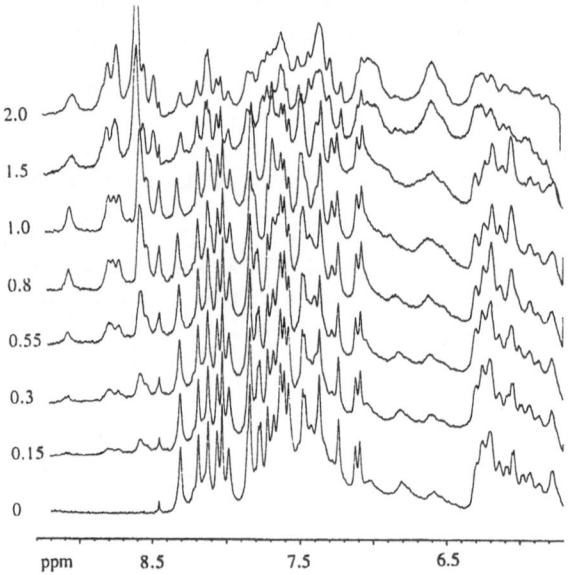

Fig. 1. 500 MHz proton NMR spectra of the DNA hairpin T6C4A6 recorded at 5°C, with different amounts of the peptide MRSRSPSRSKSPMR as indicated.

Acknowledgements

This work was supported by MRC Canada. We wish to thank Drs. P.L. Davies and R.T. Pon for providing the peptide and DNA samples respectively.

References

1. Suzuki, M., J. Mol. Biol., 207 (1989) 61.
2. Churchill, M.E.A. and Suzuki, M., EMBO J., 8 (1989) 4189.
3. Kennedy, B.P. and Davies, P.L., J. Biol. Chem., 260 (1985) 4338.
4. Zhou, N. and Vogel, H.J., Biochemistry, 32 (1993) 637.

Session X

Peptide Pharmaceuticals/Diagnostics and Peptide Delivery

Chairs: Teresa M. Kubiak
The Upjohn Company
Kalamazoo, Michigan, U.S.A.

and

Thomas J. Lobl
Tanabe Research Laboratories, U.S.A.
San Diego, California, U.S.A.

Rational strategies to enhance the intestinal permeability of peptides

R.T. Borchardt

Department of Pharmaceutical Chemistry, The University of Kansas, Lawrence, KS 66045, U.S.A.

Introduction

Through rational drug design, medicinal chemists have synthesized many peptides with novel therapeutic potential. However, one of the major problems in developing these compounds as therapeutic agents is their low oral bioavailability [1]. Low oral bioavailability can result from several factors, including low intestinal mucosal cell permeability, first pass metabolism in the intestinal mucosa and the liver, rapid liver clearance and rapid metabolism in the blood or peripheral tissues. The metabolic liability of peptides has been resolved in part by the incorporation of bioisosteres of the peptide bond. Such peptide mimetics are not susceptible to hydrolysis by peptidases and are thus more metabolically stable. However, medicinal chemists have not achieved this same degree of success in designing peptide mimetics with enhanced mucosal cell permeability and with minimal potential to be cleared by the liver and eliminated in the bile.

This presentation will focus on several strategies employed in our laboratory in an attempt to enhance the intestinal permeability of peptides and peptide mimetics. The strategies to be discussed include: (1) conjugating the peptidic molecules to bile acids that are normally orally absorbed by an intestinal transporter; and (2) altering the lipophilicity and hydrogen bonding potential of peptidic molecules to optimize passive diffusion.

Results and Discussion

One possible strategy to enhance intestinal permeability of peptides and peptide mimetics is the conjugation of these molecules with nutrients that are absorbed by endogenous transporters in the intestine. One transporter that medicinal chemists have tried to exploit for this purpose is the bile acid transporter, which is enriched in the ileal intestinal mucosa [2–4]. This concept is based on the rationale that conjugated and nonconjugated bile acids are rapidly and effectively absorbed in the ileum, and that the essential molecular requirement of bile acids for active transport is the retention of the acidic side chain at position 17 on the D ring. Therefore, our laboratory [5] has evaluated the strategy of targeting this bile acid transporter for the enhanced intestinal permeability of a renin-inhibitory peptide (ditekiren), which represents an important class of potential antihypertensive agents that are reported to have poor oral bioavailability [6]. Ditekiren-cholic acid conjugates and ditekiren-taurocholic acid conjugates were synthesized. Conjugation was through the N-terminus of a ditekiren and the 3-position of the bile acid via a six-carbon spacer. A derivative of ditekiren, containing the spacer without the bile acid moiety, was also synthesized. The ability of these ditekiren derivatives to bind to the bile acid transporter and be transported across an epithelial cell monolayer was evaluated using an *in vitro* model of the intestinal mucosa consisting of Caco-2 monolayers grown on microporous membranes [7]. The bile acid transporter in Caco-2 cells was recently characterized by our laboratory and shown to have properties similar to those of the transporter found in the intestinal mucosa [8]. Both the ditekiren-cholic conjugate (K_i=60±10 μM) and the ditekiren-taurocholic acid conjugate (K_i=19±5 μM) were shown to be potent inhibitors of the apical-to-basolateral transport of [^{14}C]-taurocholic acid ([^{14}C]-TA). These ditekiren-bile acid conjugates at concentrations up to 250 μM had no effect on the diffusion of [^{3}H]-PEG (800–1000), which is a marker of the paracellular pathway. The ditekiren derivative, which lacks the bile acid moiety, had no effect on [^{14}C]-TA transport at concentrations up to 250 μM. When the permeability coefficients of the ditekiren-bile acid conjugates were determined using Caco-2 monolayers, they were shown to be six times less than that of [^{3}H]-PEG (800–1000). The transport of the cholic acid conjugate of ditekiren was also investigated in the perfused rat ileum and its disappearance from the lumen as well as the appearance in the blood outflowing from the mesenteric vein were measured. The ditekiren-cholic acid conjugate was not detected in blood samples taken from the mesenteric vein, while the concentration of the conjugate in intestinal perfusate remained almost constant during the perfusion experiment. These results clearly show that the ditekiren-bile acid conjugates synthesized in this study retain high affinity for the intestinal bile acid transporter, but do not undergo transcellular transport. Studies are currently in progress in an attempt to elucidate whether uptake and/or efflux of these peptide-bile acid conjugates limit their intestinal permeability.

Another pathway which could be exploited to enhance the intestinal absorption of peptides and peptide mimetics is that of passive diffusion. In an attempt to determine which physicochemical parameter (lipophilicity or hydrogen bonding potential) is the major determinant of the intestinal permeability of peptides by passive diffusion, a series of model peptides were synthesized and their permeabilities were determined using an *in situ* perfused rat ileum model [9] and a cell culture model of the intestinal mucosa [10, 11].

The model peptides, which were all blocked on the N-terminal (acetyl, Ac) and the C-terminal (amide, NH$_2$) ends, consisted of D-phenylalanine (F) residues (e.g., AcFNH$_2$, AcFFNH$_2$, AcFFFNH$_2$). To alter the degree of hydrogen bonding potential, the nitrogens of the amide bonds were sequentially methylated [e.g., AcFF(Me)FNH$_2$, AcF(Me)F(Me)FNH$_2$, Ac(Me)F(Me)F(Me)FNH$_2$, Ac(Me)F(Me)F-(Me)FNH(Me)]. These peptides were shown not to be metabolized in the *in situ* perfused rat ileum system or the Caco-2 cell culture system. The results of the transport experiments showed that there were no correlations between the apparent permeability coefficients determined in the *in situ* perfused rat system (Papp values) or the Caco-2 cell culture system (Pmono values) and the octanol-water partition coefficients of the peptides. However, good correlations were observed between the Pmono values for these peptides and their hydrogen bond numbers and the Papp and Pmono values and their partition coefficients in heptane-ethylene glycol and the differences in their partition coefficients between octanol-water and isooctane-water. The latter partitioning systems provide an estimate of hydrogen bonding potential. These results suggest that lipophilicity may not be the major factor in determining the intestinal permeability of peptides and that hydrogen bonding potential may also be a major contributing factor. It should be noted that a good correlation was also observed between the Papp values determined for these peptides in the *in situ* perfused rat ileum model and the Pmono values determined in the *in vitro* cell culture model (Caco-2). These results suggest that permeability values determined in the Caco-2 cell culture model are a good predictor of intestinal permeability of peptides.

The structural features of peptides and peptide mimetics that provide optimal interaction with their pharmacological targets are not necessarily the same structural features that ensure optimal oral bioavailability. Therefore, medicinal chemists in collaboration with drug delivery and drug metabolism scientists need to dedicate more effort to the design of peptide mimetics having structural features which will ensure optimal oral bioavailability. Only through these types of collaborative interactions will it be possible to rationally design peptide mimetics with adequate oral bioavailability to make them useful therapeutic agents.

Acknowledgements

The author's work in this area has been supported by grants from The Upjohn Company, Glaxo, Inc., and the National Institutes of General Medical Sciences.

References

1. Lee, V.H.L. (Ed.), Peptide and Protein Delivery, Marcel Dekker, New York, NY, U.S.A., 1991.
2. Ho, N.F.H., Ann. NY Acad. Sci., 507 (1987) 315.
3. Kramer, W., Wess, G., Schubert, G., Bickel, M., Girbig, F., Gutjahr, U., Kowalewski, S., Baringhaus, K.H., Enhsen, A., Glombik, H., Mullner, S., Neckermann, G., Schulz, S. and Petzinger, E., J. Biol. Chem., 267 (1992) 18598.
4. Mills, C.O. and Elias, E., Biochim. Biophys. Acta, 1126 (1992) 35.
5. Kim, D.C., Harrison, A.W., Ruwart, M.J., Wilkinson, K.F., Fisher, J.F., Hidalgo, I.J. and Borchardt, R.T., J. Drug Targeting, in press.
6. Thaisrivongs, S., Drug News Perspec., 1 (1988) 11.
7. Hidalgo, I.J., Raub, T.J. and Borchardt, R.T., Gastroenterology, 96 (1989) 736.
8. Hidalgo, I.J. and Borchardt, R.T., Biochim. Biophys. Acta, 1035 (1990) 97.
9. Kim, D.C., Burton, P.S. and Borchardt, R.T., Pharm. Res., 10 (1993) 1710.
10. Conradi, R.A., Hilgers, A.R., Ho, N.F.H. and Burton, P.S., Pharm. Res., 9 (1992) 1453.
11. Conradi, R.A., Hilgers, A.R., Ho, N.F.H. and Burton, P.S., Pharm. Res., 9 (1992) 435.

Delivery of peptides into the central nervous system by sequential metabolism

N. Bodor and L. Prokai

*Center for Drug Discovery, College of Pharmacy,
University of Florida, Gainesville, FL 32610-0497, U.S.A.*

Introduction

Intracellular or transcellular transport, i.e., directly through the endothelial cell membrane, is the principal route into and out of the central nervous system (CNS) because of the existence of the blood-brain barrier (BBB) [1]. As a result, the BBB exhibits a low permeability to hydrophilic substances such as peptides that do not have specific transport mechanisms. Enzymes are also present in the BBB [2], and metabolically unstable substances may be rapidly degraded before they could reach the brain tissue (enzymatic BBB). These issues have called for the development of methods for delivering peptides into the CNS. Invasive strategies that involve the implantation of an intraventricular catheter, followed by delivery into the ventricular compartment [3], or the intracarotid infusion of high concentration of osmotically active substances to shrink the brain capillary endothelial cells to open the tight junctions [4] suffer from serious drawbacks [5], and the use of these surgical routes is only justified for some life-threatening conditions. Previous noninvasive strategies involving peptide prodrugs [3] and chimeric peptides [6] have severe limitations and are unable to deliver significant amounts of the desired biologically active peptides into the CNS, since important aspects of the transport through the BBB are overlooked. Although prodrugs may provide an increase in lipophilicity, their peptide character is prevalent, which results in the recognition and rapid degradation by peptidases. Besides, lipid-soluble compounds that are able to cross the BBB can maintain active concentrations in the CNS only if their blood concentrations are maintained at adequately high levels, and the peptide prodrug will be removed rapidly from the brain upon its systemic clearance. The CNS-delivery by chimeric peptides relies on covalently coupling the peptide that is not normally transported through the BBB to a suitable transport vector (insulin, transferrin, cationized albumin, etc.) that undergo carrier- or receptor-mediated transcytosis. However, the physiologically limited transporter capacity (saturable) prevents pharmacologically significant amounts from entering the brain. To deliver peptides into the brain, our enzyme-based strategy [7] places the peptide in a molecular environment through attaching specific functional groups that permit BBB penetration by passive transport, provide protection against peptidases, enhance retention in the CNS, and deliver the biologically active peptide to the site of action.

Results and Discussion

By our design, brain-delivery occurs in the following sequence: (1) the "packaged" molecule crosses the BBB by passive transport, and within the CNS (2) the specific, 1,4-dihydrotrigonellyl targetor is converted to a membrane-impermeable trigonellyl, (3) the biolabile, lipophilic protection is removed and a relatively stable peptide conjugate remains "locked-in," and (4) peptidases process the conjugate to release the peptide.

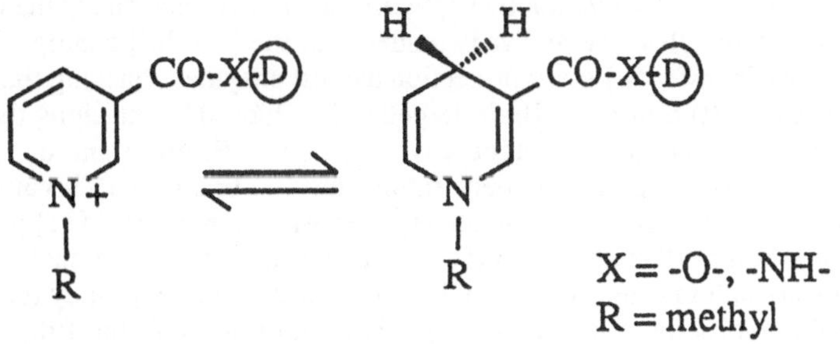

X = -O-, -NH-
R = methyl

Fig. 1. The redox targetor.

The peptide (P) is modified to provide increased lipophilicity through biolabile functional groups which are susceptible to easy removal. The targetor (T), a specific functional group on the molecule, also enhances BBB penetration and, most importantly, can be converted by enzymatic oxidation to a water soluble, lipid insoluble quaternary pyridinium salt (T$^+$). While many moieties may serve such functions, the trigonellinate $\leftrightarrow$ 1,4-dihydrotrigonellinate redox pairs have been the most useful (Figure 1). After covalently linking to nicotinic acid, this derivative is then quaternized to generate the 1-methylnicotinate salt or trigonellinate, and chemically reduced to give the 1,4-dihydrotrigonellinate targetor. This moiety is designed to undergo an enzymatically-mediated oxidation which converts the lipophilic, membrane permeable dihydrotrigonellinate to a hydrophilic, membrane impermeable trigonellinate salt. This conversion occurs ubiquitously. The mechanism of this oxidation is analogous to the oxidation of NAD(P)H, a coenzyme associated with numerous oxidoreductases and cellular respiration [8]. Should oxidation occur in the brain, the polar targetor-peptide conjugate is trapped behind the lipoidal BBB and remains "locked-in" in the CNS. The attachment of the redox targetor alone results in brain-specific delivery for small molecules such as dopamine [10]. However, the targetor will not furnish sufficient increase in lipid-solubility to a peptide, and will only protect against aminopeptidases. The unmodified COOH-terminal part of the molecule will be susceptible to cleavage by BBB exo- and endopeptidases [2].

A bulky and lipophilic moiety (L) attached to the COOH-terminus of the peptide through an ester bond will substantially increase the lipid solubility and also prevents the molecule from being recognized by peptide-degrading enzymes. Cholesteryl esters of amino acids and dipeptides have been chemically stable to be considered as suitable protecting functions [10]. This part of the molecule is labile toward esterase and/or lipase, which permits its removal after delivery.

Coupling the targetor directly to the N-terminal residue may not afford the target peptide because of the low amidase activity of the brain tissue. Therefore, a "spacer" function (S) separating the peptide sequence to be delivered (P) from the targetor part (T) with additional amino acid residue or residues is applied. This portion of the molecule should be selected based on the peptidolytic activity prevalent at the site of the action, so that release of the desired peptide in this reaction is favored over degradation by other peptidases.

We have designed the system for brain-delivery of the analog for Leu-enkephalin (Tyr-Gly-Gly-Phe-Leu, YGGFL) considering several factors. The target peptide is prone to cleavage and deactivation by peptidases at the Tyr^1 end, and at the Gly^3-Phe^4 position. While the cleavage at Tyr is hindered in the [D-Ala^2]-analog, the bulky, lipophilic steroidal ester (L) gives protection against enzymes attacking the carboxy-terminus during BBB transport. Brain-targeting is achieved by attaching the targetor to the N-terminus via an amide-type covalent bond. Esterase and/or lipase can remove the cholesteryl part after penetration into the brain. The involvement of alanine-aminopeptidase in the enkephalinergic transmission in brain [11] justifies the selection of L-Ala as (S) that allows the cleavage of the enkephalin analog from the expected locked-in molecule, T$^+$-AYAGFL. The receptor binding study (competition with [^{3}H]-diprenorphine, a μ-receptor agonist) confirmed that the fully packaged peptide and T$^+$-AYAGFL has low affinity to opioid receptors, as compared to the parent peptide, [D-Ala^2]-Leu-enkephalin. By electrospray ionization mass spectrometry, we have estimated the amount of the "locked-in" compound T$^+$-AYAGFL to be about 600 pmol per gram of tissue 15 min after intravenous (i.v.) administration of the peptide delivery system to rats [7]. In tissue collected 1, 2, and 4 hours after systemic administration, the amount of targetor-peptide conjugate was decreased with approximately 40- to 60-minute half-life. The compound was not detectable in the blood. However, *in vitro* experiments showed that the neutral endopeptidase EC 3.4.24.11 (enkephalinase) is the major degrading enzyme resulting in an inactive T$^+$-AYAG species due to the cleavage at the Gly^3-Phe^4 bond. Carboxypeptidase also cleaves the conjugate by removing the C-terminal (L)-Leu^5 (T$^+$-AYAGF). Dipeptidyl peptidase (EC 3.4.14.5) cleaves the primary product T$^+$-AYAG to yield T$^+$-AY. We have not found YAGFL among the cleavage products, although a small amount may be released from the conjugate in the CNS *in vivo*, as inferred from the slight increase of the latency of the tail-flick response in rats, the measure of the centrally-mediated analgesia.

The undesirable cleavages leading to the loss of central activity have been prevented by the use of the metabolically stable [D-Ala^2]-[D-Leu^5]-enkephalin. By obstructing the deactivating cleavages observed with the [L-Leu^5]-analog, a slow and sustained release of the biologically active peptide (YAGFL) has been observed *in vitro*. Also, a long-lasting and statistically significant increase in the tail-flick latency has been obtained, whereas the unmanipulated enkephalin analog and the partially conjugated peptides showed no effect compared with the animals injected with the vehicle solution. About 20 s increase has been obtained even five hours after i.v. administration, compared to the 10±3 s tail-flick latency of the control animals.

In the centrally active thyrotropin-releasing hormone (TRH) analog Pyr-Leu-Pro-NH$_2$, no free hydroxy or amino groups are present. However, TRH itself is derived from the processing of a precursor polyprotein [11]. The TRH progenitor sequences (QHPG) are flanked by dibasic residues that are typical sites of processing by carboxypeptidase B-like enzymes in polyproteins. The C-terminal glycine functions as an amide donor for the proline by the enzyme peptide glycine alpha-amidating monooxygenase (PAM). Glutamine is the precursor of the N-terminal pyroglutamyl residue. Cyclization of the N-terminal glutamine is catalyzed by a specific enzyme, glutaminyl cyclase. Therefore, we have incorporated a Gln-Leu-Pro-Gly (QLPG) progenitor sequence into the peptide delivery system. The hypothetical "locked-in" precursor T$^+$-AQLPG is, indeed, processed to the prolinamide, as validated *in vitro*, and this reaction by PAM tissue occurs fast in the brain. Thus, the processing to the desired TRH-analog is only dependent on the slow release of Gln-Leu-Pro-NH$_2$, which proceeds similarly to that of the enkephalin analog from the "locked-in" targetor-peptide conjugate. The CNS-delivery of a pharmacologically significant amount of THR analog is evidenced by the profound decrease in the barbiturate-induced sleeping time, the measure of the activational effect on cholinergic neurons, in mice. At equimolar (30 µmol/kg) dose, the i.v. administration of the TRH analog showed only marginal decrease (limited BBB-penetration), while the CDS with (L)-Ala and Ala-Ala spacers has resulted in ca. 30 and 50% reduction in the sleeping time, respectively.

In conclusion, our approach has been the first noninvasive strategy that delivered peptides into the CNS in pharmacologically significant amounts. Our findings underscore the need and importance of controlling transport and metabolism to deliver peptide-based drugs to the brain.

Acknowledgements

This research has been supported by a grant from the National Institute for Aging (1 PO 1 AG10485). The contribution of Drs. W.M. Wu, H. Farag, J.S. Sastry, and J. Simpkins is kindly acknowledged. The important work of X. Ouyang has been made possible by a Graduate Research Assistantship funded by Genentech.

References

1. Brightman, M.W. and Reese, T.S., J. Cell Biol., 40 (1969) 648.
2. Brownlees, J. and Williams, C.H., J. Neurochem., 60 (1993) 793.
3. Pardridge, W.M., Peptide Drug Delivery to the Brain, Raven Press, New York, 1991, p.108
4. Neuwelt, E. and Rapoport, S., Fed. Proc., 43 (1984) 214.
5. Pardridge, W., Endocrine Rev., 7 (1986) 314.
6. Pardridge, W.M., Triguero, D. and Buciak, J.L., Endocrinology, 126 (1990) 977.
7. Bodor, N., Prokai, L., Wu, W.-M., Farag, H., Kawamura, M. and Simpkins, J., Science, 257 (1992) 1698.
8. Hoek, J. and Rydstrom, J., Biochem. J., 254 (1988) 1.
9. Bodor, N. and Simpkins, J., Science, 221 (1983) 65.
10. Shashoua, V.E., Jacob, J.N., Ridge, R., Campbell, A. and Baldessarini, R.J., J. Med. Chem., 27 (1984) 659.
11. Schwartz, J.C., Giros, B., Gros, C., Llorens-Cortes, C., Arrang, J.M., Garbang, M. and Pollard, H., Cephalagia, 7, Suppl. 6 (1987) 32.
12. Jackson, I., Ann. N.Y. Acad. Sci., 553 (1989) 71.

Hydrins, hydroosmotic neurohypophysial peptides involved in neuroendocrine control of amphibian water homeostasis through specific receptors

R. Acher, J. Chauvet, G. Michel and Y. Rouillé

*Laboratory of Biological Chemistry, University of Paris VI,
96 Boulevard Raspail, 75006 Paris, France*

Introduction

Most peptide hormones have a C-terminal α-amidated residue and are excised from a precursor protein at the level of a signal Gly-Lys-Arg sequence. A cascade of four enzymes is involved in the complete maturation: 1) a dibasic endopeptidase cleaving after the pair Lys/Arg-Arg; 2) a carboxypeptidase B-like enzyme (carboxypeptidase E) removing the two basic residues; 3) a glycine monooxygenase converting C-terminal glycine into hydroxyglycine; and 4) an α-amidating dealkylase that splits hydroxyglycine whose amino group becomes the amidated group of the penultimate residue.

Neurohypophysial hormones are nonapeptides linked in their precursors to neurophysins through a Gly-Lys-Arg sequence [1]. The antidiuretic hormone is vasopressin in mammals and vasotocin ((Ile3)-vasopressin) in all non-mammalian tetrapods [1]. Whereas in birds and reptiles the processing of provasotocin is complete and only mature α-amidated vasotocin is found, in anuran Amphibia that must face dehydration because of the particular permeability of their skin, processing intermediates termed hydrins have been identified along with vasotocin.

Results and Discussion

The four processing enzymes have now been located in neurohypophysial secretory granules (the last compartment of the secretory pathway) of the hypothalamoneurohypophysial neurons [1]. In these granules, isolated by differential centrifugation from ox and rat neurointermediate pituitaries, two calcium-dependent endopeptidases displaying predominantly either Lys-Arg or Arg-Arg specificity, but devoid of Lys-Lys or monobasic specificity have been identified [2]. The Lys-Arg enzyme, more abundant, has a pH optimum near 5.5, close to the granule pH (5.8) whereas the Arg-Arg would be more active near pH 7.0 (near the trans-Golgi pH). Two forms of the Lys-Arg endopeptidase, soluble and membrane-associated, have been detected. Partitioning into Triton X-114 suggests that the second form is associated to the membrane through a C-terminal amphiphilic α-helix [2]. Carboxypeptidase E and α-amidating enzyme system have also been identified in the granules by their actions on specific substrates, hippuryl-L-arginine and D-Val-Tyr-Gly, respectively [3].

The differential processing of prohormones can imply any of the four enzymes of the cascade and can be related either to the cell differentiation or to a particular evolutionary adaptation. In amphibians, a down-regulation affects constitutively either carboxypeptidase E (hydrin 1) or the α-amidating system (hydrin 2) in the provasotocin processing [4].

Table 1 *Hydroosmotic activities of amphibian neurohypophysial hormones on frog tissues*

Neurohypophysial Peptide	Skin permeation in vivo		Bladder permeation in vitro (mU/pmol)
	Peptide injected (pmol)	Water uptake (g/100g)	
Saline	–	0.2 ± 0.3	–
Vasotocin	200	7.0 ± 0.9	20
Hydrin 1 (Vasotocinyl-GKR)	200	4.4 ± 0.6	18
Hydrin 2 (Vasotocinyl-G)	200	8.7 ± 0.8	20

Partial reconstitution of the processing *in vitro* has been carried out using purified neurohypophysial secretory granules from ox and rat pituitary glands. Conversion of synthetic vasopressinyl-Gly-Lys-Arg into vasopressinyl-Gly has been quantitatively carried out using carboxypeptidase E from rat granule content at pH 5.5 and with addition of 1 mM $CoCl_2$ [3]. Vasopressinyl-Gly has been converted into mature pharmacologically active vasopressin using the α-amidating enzyme system from the same granules at pHs 6 and 8 with addition of 1 mM $CuCl_2$ and 1 mM ascorbic acid. However, the conversion yield of this step, estimated from the generated pressor activity and absorbance in HPLC, is low (about 6-8%) [3].

In the permanent aquatic *Xenopus laevis*, vasotocinyl-Gly-Lys-Arg (hydrin 1) resulting from a down-regulation of the carboxypeptidase E has been characterized [4]. In 10 European, American, African and Asiatic frogs and toads that are semi-aquatic or terrestrial, vasotocinyl-Gly (hydrin 2) resulting from a down-regulation of the α-amidating enzyme system has been identified [4,5]. Hydrins 1 and 2 have no oxytocic or pressor activities in contrast to vasotocin. They are however at least as active as vasotocin on the water permeation of frog skin and frog urinary bladder (Table 1) but are devoid of frog anti-diuretic activity in contrast to vasotocin. Hydrins 1 or 2 are usually found in molar amounts approximately equal to that of vasotocin. However the molar ratio hydrin 2 to vasotocin reaches 2 in toads able to withstand desiccation through burrowing [5].

It is suggested that water homeostasis of amphibians is submitted to a cooperative hormonal control, hydrins acting mainly on the skin for the water uptake (external water) through specific receptors whereas vasotocin operates mainly on the kidney tubule for reclaiming urine (internal water). Adaptation has been realized in amphibians on the one hand by a differential processing of provasotocin giving two distinct hormones from a single precursor, on the other by diversification of specific receptors located in skin, bladder and kidney nephron.

References

1. Acher, R., Regul. Pept., 45 (1993) 1.
2. Rouillé, Y., Spang, A., Chauvet, J. and Acher, R., Biochem. Biophys. Res. Commun., 183 (1992) 128.
3. Rouillé, Y., Chauvet, J. and Acher, R., Biochem. Int., 26 (1992) 739.
4. Rouillé, Y., Michel, G., Chauvet, M.T., Chauvet, J. and Acher, R., Proc. Natl. Acad. Sci. U.S.A., 86 (1989) 5272.
5. Michel, G., Ouedraogo, Y., Chauvet, J., Katz, U. and Acher, R., Neuropeptides, in press.

Novel peptide prodrugs: Lipid derivatization of protease inhibitor enhances *in vivo* plasma half life

C. Basava[1], L.M. Selk[2], R. Basava[2], M. Gardner[3],
S. Parker[2], D.D. Richman[3] and K.Y. Hostetler[3]

[1]*Novabiochem U.S.A., La Jolla, CA 92037, U.S.A.,*
[2]*Vical Incorporated, San Diego, CA 92121, U.S.A., and*
[3]*University of California San Diego and the VA Medical Center,
La Jolla, CA 92093, U.S.A.*

Introduction

During viral replication, the human immunodeficiency virus (HIV) produces a long gag-pol polyprotein which is cleaved during viral budding into smaller proteins that form the structural elements of the viral progeny. This proteolytic cleavage is catalyzed by a specific HIV aspartyl protease (HIV-PR) and must occur if infectious virions are to be produced [1]. HIV-PR is a 99 amino acid protein and specifically cleaves Phe-Pro or Tyr-Pro sequences of the viral proteins. A number of laboratories are developing inhibitors of this protease based on the small peptide fragments which conformationally resemble the protease binding site but have the P1-P' site replaced by non-hydrolyzable structures as alternate therapies for AIDS. However, such small peptide inhibitors are removed extremely rapidly from the plasma by a variety of mechanisms including cellular uptake and metabolism, filtration by the kidney and urinary excretion or destruction by the proximal tubule cells of the kidney. Furthermore, they lack the ability to target infected cell populations such as macrophages and to penetrate cell membranes. The present study was undertaken to evaluate the effects of lipid conjugation on *in vitro* antiviral activities and *in vivo* pharmacokinetics of small peptide based inhibitors.

Results and Discussion

H-(D-Phe)-(D-α-Nal)-Pip-(α-(OH)-Leu)-Val-NH$_2$ (7169), iBoc-(D-Phe)-(D-α-Nal)-Pip-(α-(OH)-Leu)-Val-NH$_2$ (7140), iBoc-Phe-(D-β-Nal)-Pip-(α-(OH)-Leu)-Val-OH (7194) and iBoc-Phe(4-Br)-(D-β-Nal)-Pip-(α-(OH)-Leu)-Val-OH (7195) were prepared by the Merrifield solid phase method [2] and purified to greater than 95% purity. 1,2-Dipalmitoyl-sn-glycero-phosphatidylethanolamine (DPPE) was reacted with succinic anhydride to give DPPE-succinic acid which in turn was coupled (DCC/HOBt) with 7169 to provide DPPE-succinyl-(D-Phe)-(D-σ-Nal)-Pip-(α-(OH)-Leu)-Val-NH$_2$ (7172). 7194 and 7195 were conjugated (DCC/HOBt) to the amino group of DPPE to provide iBoc-Phe-(D-β-Nal)-Pip-(α-(OH)-Leu)-Val-DPPE (7196) and iBoc-Phe(4-Br)-(D-β-Nal)-Pip-(α-(OH)-Leu)-Val-DPPE (7197), respectively.

7195 and 7197 were catalytically tritiated to provide iBoc-Phe(^{3}H)-(D-β-Nal)-Pip-(α-(OH)-Leu)-Val-OH (^{3}H-7194) and iBoc-Phe(^{3}H)-(D-β-Nal)-Pip-(α-(OH)-Leu)-Val-DPPE (^{3}H-7196), respectively.

Table 1 *Antiviral activities of peptides and their lipid conjugates*

No.	Structure	$IC_{50}(\mu M)$
7140	iBoc-(D-Phe)-(D-α-Nal)-Pip-(α-(OH)-Leu)-Val-NH$_2$	14.0 ± 4.6 (6)
7172	DPPE-succinyl-H-(D-Phe)-(D-α-Nal)-Pip-(α-(OH)-Leu)-Val-NH$_2$	15.3 ± 8.1 (3)
7194	iBoc-Phe-(D-β-Nal)-Pip-(α-(OH)-Leu)-Val-OH	11.5 ± 6.2 (5)
7196	iBoc-Phe-(D-β-Nal)-Pip-(α-(OH)-Leu)-Val-DPPE	9.6 ± 4.6 (5)

Antiviral activities of the peptides were determined by the plaque reduction assay using HIV 1 (lav)-infected HT-64C cells [3].

As shown in Table 1, the peptide inhibitors, 7140 and 7194, and their corresponding prodrugs, 7172 and 7196, showed almost identical antiviral activities in the plaque reduction assay. This data indicates that lipid prodrugs prepared by conjugating DPPE to the N-terminal or C-terminal of small peptides retain full *in vitro* antiviral activity.

Table 2 *Plasma concentrations and T 1/2 of 7194 and its lipid conjugate 7196*

No.	Structure	T 1/2	AUC (nmol.hr/ml)
^{3}H-7194	iBoc-Phe(^{3}H)-(D-β-Nal)-Pip-(α-(OH)-Leu)-Val-OH	0.069 hr	0.317
^{3}H-7196	iBoc-Phe(^{3}H)-(D-β-Nal)-Pip-(α-(OH)-Leu)-Val-DPPE	1.880 hr	24.33

(^{3}H-7194) and (^{3}H-7196) were administered (IV) to mice at 3.0 μmol/kg and plasma levels at various time intervals were measured by radiometric detection. Plasma half lives (T 1/2) and plasma concentrations were calculated based on these determinations.

Pharmacokinetic studies, summarized in Table 2, indicate that the free peptide, ^{3}H-7194, is removed from circulation very rapidly while its C-terminal lipid prodrug (^{3}H-7196) is present in the body for a significantly long period of time. Plasma half life is increased 27 fold by converting the peptide to the lipid prodrug. Also, plasma concentration of the prodrug as measured by the area under the curve is almost 80 fold higher than that for the free peptide. Possible reasons for improved pharmacokinetic behavior of the lipid prodrugs are: metabolism and removal from circulation is slower for the lipid conjugate compared to the free peptide.

Peptide prodrugs prepared by conjugating small peptides to lipid groups appear to exhibit no loss of *in vitro* antiviral activity and improved *in vivo* pharmacokinetic behavior and may form a basis for the design of novel therapeutic agents.

References

1. Kohl, N.E., Emini, E.A., Schleif, W.A., Davis, L.J., Heimbach, J.C., Dixon, R.A., Scolnick, E.M. and Sigal, I.S., Proc. Natl. Acad. Sci. U.S.A., 85 (1988) 4686.
2. Merrifield, R.B., J. Am. Chem. Soc., 85 (1963) 2149.
3. Larder, B.A., Darby, G. and Richman, D.D., Science, 31 (1989) 1731.

Cloning of peptide transport genes from eucaryotes

K. Alagramam[1], J. Perry[1], M. Basrai[1], F. Naider[2] and J.M. Becker[1]

[1]*Department of Microbiology, University of Tennessee, Knoxville, TN 37996, U.S.A., and* [2]*Department of Chemistry, College of Staten Island, City University of New York, Staten Island, New York, NY 10301, U.S.A.*

Introduction

The transport of small peptides across the plasma membrane, a phenomenon shown to occur in a range of procaryotes and eucaryotes, is mediated by specific proteins in an energy dependent fashion. Peptide transport systems have been well studied in prokaryotes, especially *Escherichia coli* (colon bacteria) and *Salmonella typhimurium* where several genes have been cloned and characterized [1, 2]. However very little is known about the genes responsible for peptide transport in eucaryotes. In *Saccharomyces cerevisiae* (baker's yeast) at least three genes are required for peptide transport: *PTR1*, *PTR2* and *PTR3* [3]. Strains carrying a mutation in one of these genes are resistant to the toxic dipeptide alanyl-ethionine and are unable to utilize/transport dipeptides, such as Lys-Leu and His-Leu, in order to satisfy their auxotrophic requirements. We report the cloning of peptide transport genes from *Saccharomyces cerevisiae* and *Candida albicans* by functional complementation of peptide transport mutants of *Saccharomyces cerevisae*. The structure of these genes will provide clues to the understanding of peptide transport in eucaryotes which is essential for rational design of peptide-based drugs.

Results and Discussion

The peptide transport mutants *ptr1* and *ptr2* were used to screen plasmid based genomic libraries of *S. cerevisiae* and *C. albicans*. Transformants were selected by the ability to grow on Lys-Leu containing minimal medium. These transformants were able to utilize other dipeptides in order to satisfy their requirements for certain amino acids and were also sensitive to toxic dipeptides, such as alanyl-ethionine, leucyl-ethionine and leucyl-*m*-fluorophenylalanine, like the wild-type (*PTR*) S288C strain (Table 1). In addition, transformants were able to transport radiolabeled dileucine at levels comparable to the wild type while the mutant(s) or the control strain(s) failed to accumulate the radiolabeled substrate (Table 2). These results indicate the restoration of peptide transport to the relevant mutants by the cloned genes *PTR1* and *PTR2* from *S. cerevisiae* and by a *C. albicans* peptide transport gene.

PTR1 has been localized to the right arm of chromosome VII of *S. cerevisiae* while *PTR2* has been localized to chromosome XI of this yeast. Southern analysis indicates that both are present at unique loci within the *S. cerevisiae* genome. The sequence of the encoded protein indicates that *PTR2* may be related to ABC transporters characterized by multiple membrane spanning domains and ATP-binding sites and to the nitrate transporter of the plant *Arabidopsis thaliana.*

Table 1 *Phenotype of* Saccharomyces cerevisiae *transformants*

Strains	Relevant genotype	Plasmid	Dipeptides[a]			Toxic Dipeptides[b]		
			Leu-Leu	Lys-Lys	Lys-Leu	Ala-Eth	Leu-Eth	Leu-F-Phe
S288C[c]	PTR	None	+	+	+	S	S	S
PB4X-5B	PTR1	None	-	-	-	R	R	R
PB1X-9B	PTR2	None	-	-	-	R	R	R
PB4X-5B	PTR1	YCp50[d]	-	-	-	R	R	R
PB4X-5B	PTR1	pKA2[e]	+	+	+	S	S	S
PB1X-9B	PTR2	YCp50	-	-	-	R	R	R
PB1X-9B	PTR2	pJP9[f]	+	+	+	S	S	S
PB1X-9B	PTR2	YEp24[g]	-	-	-	R	R	R
PB1X-9B	PTR2	pMB3[h]	+	+	+	S	S	S

[a]The growth response is measured in minimal medium containing the dipeptide instead of the required amino acid. The + symbol indicates growth and the - symbol indicates no growth. [b]Sensitivity to toxic dipeptides was measured using disk assay as described by Island et. al. [4]. S indicates sensitivity to toxic effects of dipeptides and R indicates resistance. [c]S288C is wild-type *S. cerevisiae* with a fully functional peptide transport system. [d]YCp50 plasmid is a yeast-*E. coli* shuttle vector which contains the *S. cerevisiae* URA3 selectable marker and is maintained at a low copy number in the host. [e]pKA2 is plasmid YCp50 containing *S. cerevisiae* PTR1 gene. [f]pJP9 is plasmid YCp50 containing *S. cerevisiae* PTR2 gene. [g]YEp24 plasmid is a yeast-*E. coli* shuttle vector which contains the *S. cerevisiae* URA3 selectable marker and is maintained at a high copy number in the host. [h]pMB3 is plasmid YEp24 containing *C. albicans* PTR2 gene.

Table 2 *Transport of L-leucyl-L-[³H]leucyl in* S. cerevisiae *transformants*

Strains[a]	Plasmid [b]	Uptake of dileucine[c] (nmoles/mg cell wt)			
		0.25 min.	1 min.	2 min.	3 min.
S288C	None	0.05	0.28	0.41	0.57
PB4X-5B	None	0.01	0.02	0.02	0.02
PB4X-5B	YCp50	0.02	0.03	0.03	0.04
PB4X-5B	pKA2	0.04	0.25	0.39	0.60
PB1X-9B	None	0.02	0.03	0.02	0.02
PB1X-9B	YCp50	0.03	0.03	0.02	0.02
PB1X-9B	pJP9	0.06	0.26	0.40	0.58
PB1X-9B	YEp24	0.01	0.01	0.02	0.02
PB1X-9B	pMB3	0.04	0.22	0.36	0.42

[a]For relevant genotype refer to Table 1.   [b]For details of the plasmid refer to Table 1. [c]L-leucyl-L-[³H]leucine uptake was determined as described by Basrai et. al. [5].

These findings represent the first eucaryotic genes cloned for transport of small peptides. The characterization of these genes, combined with the cloning of other genes involved in transport of peptides, will provide a better understanding of peptide transport in eucaryotes.

Acknowledgements

These studies were supported by grant BE-39 B/C from the American Cancer Society.

References

1. Payne, J.W., (Ed.), Microorganism and Nitrogen Sources, John Wiley and Sons, Inc., New York, NY, U.S.A., 1980, p.257.
2. Higgins, C.F., Ann. Rec. Cell Biol., 8 (1992) 67.
3. Island, M.D., Perry, J.R., Naider, F. and Becker, J.M., Curr. Genet., 20 (1987) 457.
4. Island, M.D., Naider, F. and Becker, J.M., J. Bacteriol., 169 (1987) 2132.
5. Basrai, M.A. Zhang, H.-L., Naider, F. and Becker, J.M., J. Gen. Microbiol., 138 (1992) 2353.

Ro 25-1553: A potent, metabolically stable vasoactive intestinal peptide agonist

D.R. Bolin, J.M. Cottrell, J. Michalewsky, R. Garippa, N. Rinaldi, M. O'Donnell and W. Selig

Roche Research Center, Hoffmann-La Roche, Inc., Nutley, NJ 07110, U.S.A.

Introduction

Vasoactive intestinal peptide (VIP) is an octacosapeptide of the glucagon/ secretin family. VIP is present in nerve fibers supplying the smooth muscle, blood vessels, and submucosal glands in the trachea, bronchi, and bronchioles [1]. There is significant evidence that VIP may be the endogenous mediator of bronchial smooth muscle relaxation in human airways linking VIP to the disease of asthma [2]. Reports from these laboratories [3-5] have described our work toward determining the solution structure of VIP, identifying the functional elements involved in binding of VIP to its receptor, and establishing the metabolic pathways which function in the pulmonary system. In the present study, we have developed a VIP analog which exhibits high potency and enhanced *in vivo* duration of action.

Results and Discussion

Native VIP is degraded in both human and guinea pig BAL fluid by enzymatic cleavage primarily between Ser^{25} and Ile^{26} [5]. In order to stabilize this site to proteolysis, a series of VIP analogs were synthesized and their degradation halftimes were compared. Incorporation of D-residue or N-alkylated amino acids failed to significantly alter the degradation halftimes. However, formation of a cyclic structure between the side-chains of Lys^{21} and Asp^{25} yielded compounds with enhanced metabolic stability. From SAR studies, the VIP analog Ro 25-1553, Ac-[Glu^8,Lys^{12},Nle^{17},Ala^{19},Asp^{25},Leu^{26},$Lys^{27,28}$,$Gly^{29,30}$,Thr^{31}]-VIP cyclo (Lys^{21} → Asp^{25}), has emerged as a lead compound. This compound was profiled as an airway smooth muscle relaxant and bronchodilator with comparison to two β-agonists, salbutamol and salmeterol, and to native VIP.

The relaxant effects of VIP, Ro 25-1553, salbutamol, and salmeterol on resting tone in isolated guinea pig tracheal smooth muscle was examined. All four test compounds exhibited concentration-dependent relaxations and produced 100% of the maximal relaxation obtainable with isoproterenol. The rank order of potency was Ro 25-1553 > salbutamol > VIP > salmeterol (1:12:53:77) as evidenced by EC_{50} values of 0.3, 3.5, 16 and 23 nM, respectively.

Relaxant experiments were performed on isolated human bronchial tissue from organ donors. Each of the VIP agonists and β-adrenoceptor agonists were comparatively evaluated for their ability to relax histamine-induced contractions of human bronchial smooth muscle. As shown in Figure 1, all four compounds when

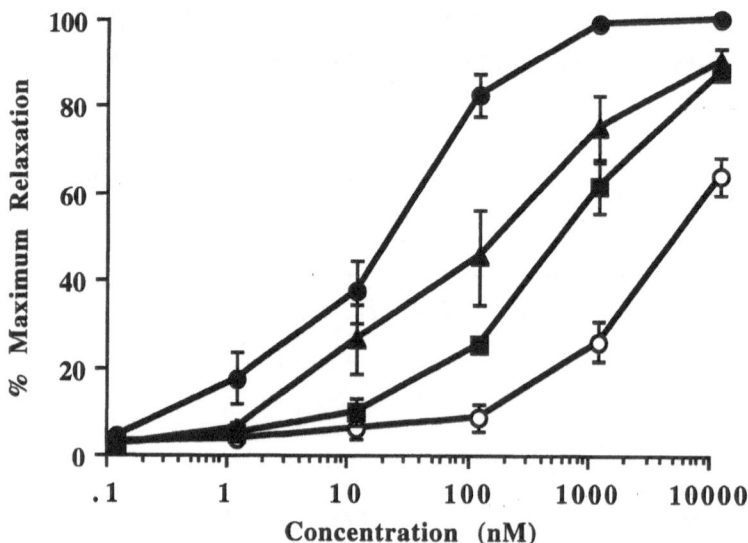

Fig. 1. Cumulative relaxation concentration-response curves (n=6) for Ro 25-1553 (●), salbutamol (▲), salmeterol (■), and VIP (O) in isolated human bronchial smooth muscle precontracted with histamine (0.3 mM). Relaxant responses are expressed as a percentage of the maximum obtainable relaxation.

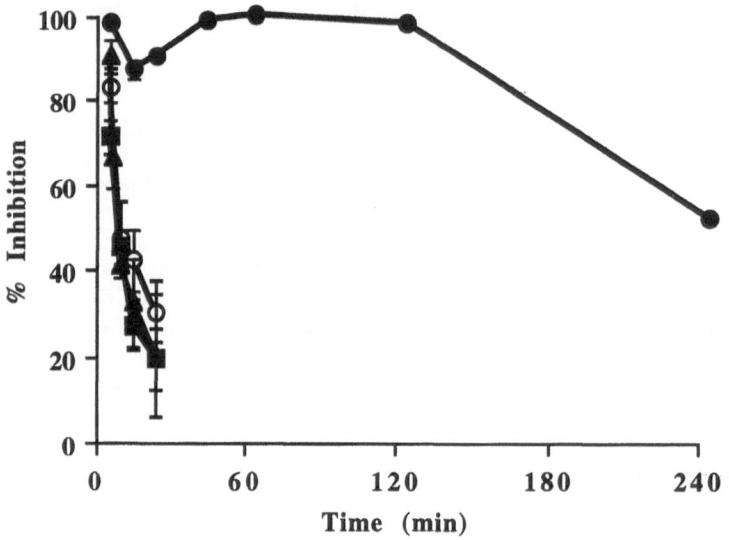

Fig. 2. Comparison of the time course (n=6) of inhibition by Ro 25-1553 (●), VIP (O), salbutamol (▲), and salmeterol (■) of histamine-induced bronchoconstriction in anesthetized guinea pigs. The time between administration of compound (300 µg) and subsequent challenge with histamine (50 µg/kg, i.v.) was varied.

added cumulatively to human bronchus preparations *in vitro* caused concentration-dependent relaxations. Ro 25-1553 showed full intrinsic activity, while salbutamol, salmeterol and VIP showed partial intrinsic activity on human bronchial segments. Indeed, concentrations of VIP as high as 10 μM produced only 62% of the maximal relaxation. On isolated preparations of human bronchial tissue, the rank order of potency was: Ro 25-1553 > salbutamol ≈ salmeterol > VIP (1:29:34:385) as evidenced by EC_{50} values of 20, 570, 680 and 7700 nM, respectively. In human bronchial smooth muscle, Ro 25-1553 is about 385 times more potent than native VIP.

The ability of these compounds to prevent histamine (50 μg/kg)-induced bronchoconstriction in guinea pigs was examined. Upon acute intratracheal instillation, Ro 25-1553, VIP, salbutamol, and salmeterol each caused dose-dependent inhibition of bronchospasms evoked by histamine. Ro 25-1553 was capable of producing nearly 100% inhibition indicative of a full agonist with high efficacy. Comparison of the resultant ED_{50} values show the order of decreasing potency for the four bronchodilators against histamine was: Ro 25-1553 ≈ salmeterol > VIP > salbutamol (1:2:76:470). Ro 25-1553 was markedly superior to native VIP and the two β-agonists as a bronchodilator *in vivo*.

The time courses of bronchodilator activity of Ro 25-1553, VIP, isoproterenol, and salmeterol were compared following administration of a single intratracheally instilled dose of 300 μg. As illustrated in Figure 2, only Ro 25-1553 produced long-lasting inhibition of bronchoconstriction evoked by exogenous histamine. Although maximal pharmacological effects were apparent within 1 min after instillation of all four bronchodilators, statistically significant inhibition was evident with Ro 25-1553 for greater than 240 minutes compared with short durations of action for VIP, salbutamol, and salmeterol (approximately 5 min) in this model.

These studies demonstrate that Ro 25-1553 expresses its bronchodilator activity on VIP receptors in a number of *in vitro* pharmacological assays and this activity extends *in vivo* where it has been shown to be active by intratracheal instillation against a potent spasmogen. If the bronchodilator potency of Ro 25-1553 coupled with its long duration of action translate to the *in vivo* situation in man, this unique VIP-like peptide has the potential to be clinically efficacious in the treatment of bronchospastic diseases, such as asthma.

References

1. Saga, T. and Said, S.I., Trans. Assoc. Am. Physicians, 97 (1984) 304.
2. Matsuzaki, Y., Hamasaki, Y. and Said, S.I., Science, 210 (1980) 1252.
3. Fry, D.C, Madison, V.S., Bolin, D.R., Greeley, D.N., Toome, V. and Wegrzynski, B.B., Biochemistry, 28 (1989) 2399.
4. O'Donnell, M., Garippa, R.J., O'Neill, N.C., Bolin, D.R. and Cottrell, J., J. Biol. Chem., 266 (1991) 6389.
5. Bolin, D.R., Cottrell, J., Michalewsky, J., Garippa, R.J., O'Neill, N., Simko, B. and O'Donnell, M., Biomed. Res. Suppl. 2, 13 (1992) 25.

Angiotensin II - mediation of the myoproliferative response of injured vascular smooth muscle

R. Bourgeois, S. Laporte and E. Escher

*Department of Pharmacology, Faculty of Medicine,
University of Sherbrooke, Sherbrooke, Québec, J1H 5N4, Canada*

Introduction

Transluminal angioplasty of stenosed arteries with balloon catheters is a very important therapeutic procedure e.g. for the prevention of cardiac infarction. Angioplasty and all other vascular remodeling techniques are however beset by a one third incidence of clinically significant restenosis [1], due to proliferation of vascular smooth muscle cells (VSMC) after vascular injury resulting in formation of substantial neointima. This myoproliferative response has been significantly reduced in animal models by administration of Angiotensin Converting Enzyme inhibitors (ACEi) [2], and has been completely suppressed by intralesional infusion of Br_5Ang, a potent non-selective, peptidic angiotensin II (Ang) antagonist [3], establishing thus the essential Ang mediation of this myoproliferative response. Br_5Ang or $[Sar^1, Val^5, Phe(2',3',4',5',6'-Br_5)^8]Ang$ is a peptidic Ang antagonist which has a long duration of action *in vivo* as an anti-hypertensive in the periphery [4], as well as centrally [5]. With the recent development of selective, non-peptide Ang antagonists [6], the tools were available for the determination of the Ang receptor types (AT_1 or AT_2) responsible for the Ang mediation of this stenotic process. After balloon angioplasty of the left carotid artery of male Sprague-Dawley rats, the selective, non-peptide Ang antagonists, losartan (AT_1 selective) and/or PD123319 (AT_2 selective) were continuously infused for two weeks into this artery by Alzet mini-osmotic pumps. The carotid arteries were perfusion fixed, excised, embedded in paraffin and semithin sections (7μm) were prepared. After hematoxylin-eosin coloration, morphometric analysis of neointima formation was performed using a computerized digitizer (program Bioquant, R & M Biometrics).

Results and Discussion

In Figure 1, typical cross-sections of rat carotid arteries 14 days after balloon angioplasty show the great difference between controls having undergone angioplasty (Cont(+)) and controls having undergone no angioplasty (Cont(-)). A highly significant degree of stenosis is present in the Cont(+) group. Also, the superior capacity of Br_5Ang over losartan to prevent neointima formation is shown. Figure 2 is a histogram representing the mean neointimal thickness of the respective experimental groups. There is a significant difference ($p < 0.05$) between the Cont(+) group and all treated groups except for the PD123319 treated group. The ineffectiveness of this AT_2 selective compound to prevent neointima formation leads us to conclude that the action of Ang on this myoproliferation is not mediated via the AT_2 receptor. On the other hand, the intralesional administration of an AT_1 selective antagonist, losartan, significantly reduced the myoproliferative response of VSMC after angioplasty therefore indicating an action through the AT_1 receptor. A combination infusion of PD123319 and losartan produced a neointima reduction identical to that obtained by losartan alone therefore excluding a synergistic action. However, the non-selective, peptidic Ang antagonist, Br_5Ang, reduced the post-angioplastic response to a significantly ($p < 0.05$) greater degree than the reduction obtained with losartan. It is therefore concluded that this VSMC response is mainly but not exclusively AT_1 mediated, that AT_2 mediation has no contribution to this process and that yet unknown, other Ang receptors may contribute. In contrast to this finding, it has been reported [7] that the AT_2 receptor type may be involved in the mediation of this process. These results were obtained by infusing another AT_2 selective compound in the perivascular space. There are, however, two major flaws in this approach: firstly, there was no control group with this infusion method, and secondly, the very high dose used would seem to affect the AT_1 receptor type [8].

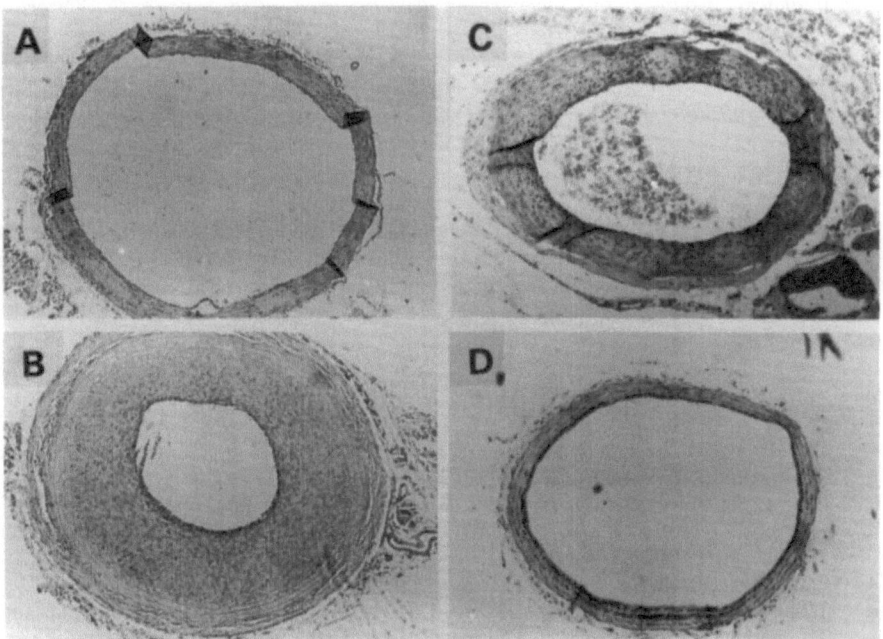

Fig. 1. *Typical cross-sections of rat carotid arteries 14 days after balloon angioplasty.*
A. *Cont(-): group of rats having undergone neither angioplasty nor drug treatment.*
B. *Cont(+): rats receiving no drug treatment after angioplasty.*
C. *Losartan (DUP753): rats receiving 12.5 µg/hr/rat of losartan (i.a.) after angioplasty.*
D. *Rats receiving 5 µg/hr/rat of Br_5Ang (i.a.) after angioplasty.*

The above-presented results suggest that non-selective Ang antagonists are probably the best drug candidates for the prevention of restenosis after vascular remodeling procedures like coronary angioplasty.

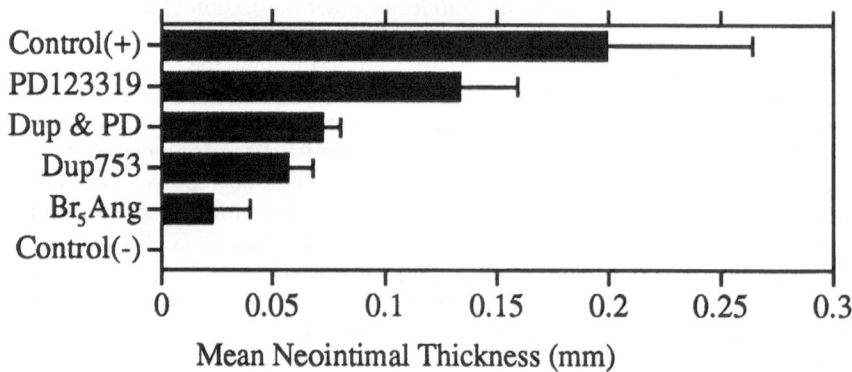

Fig. 2. Histogram representing the mean neointimal thickness (area/perimeter) 14 days after angioplasty (x70). Cont(-): group of rats having undergone neither angioplasty nor drug treatment. Cont(+): rats receiving no drug treatment after angioplasty. Losartan (DUP753): rats receiving losartan after angioplasty. Dup & PD: rats receiving a combination of losartan and PD123319 (same concentrations as each drug alone). PD123319: rats receiving 6.25 µg/hr/rat of PD123319 after angioplasty. Br₅Ang: rats receiving Br₅Ang after angioplasty.

Acknowledgements

This research has been supported by grants from the Canadian Medical Research Council and the Québec Chapter of the Canadian Heart Foundation to E. Escher and by studentships from FCAR (R. Bourgeois) and BioMéga Inc. (S. Laporte).

References

1. Forrester, J.S., Fishbein, M., Helfant, R. and Fagin, J., J. Am. Coll. Cardiol., 17 (1991) 758.
2. Powell, J.S., Clozel, J.-P., Müller, R.K.M., Kuhn, H., Hefti, F., Hosang, M. and Baumgartner, H.R., Science, 245 (1989) 186.
3. Laporte, S. and Escher, E., Biochem. Biophys. Res. Commun., 187, 3 (1992) 1510.
4. Bossé, R., Gerold, M., Fischli, W., Holck, M. and Escher, E., J. Cardiovasc. Pharmacol., 16, Suppl. 4 (1990) 550.
5. Paré, M.-C., Maltais, S. and Escher, E., Regul. Pept., 47 (1993) 81.
6. Wong, P.C., Hart, S.D., Zaspel, A.M., Chui, A.T., Ardecky, R.J., Smith, R.D. and Timmermans, P.B.M.W.M., J. Pharmacol. Exp. Ther., 255 (1990) 584.
7. Janiak, P., Pillon, A., Prost, J.-F. and Vilaine, J.-P., Hypertension, 20 (1992) 737.
8. Criscione, L., Thomann, H., Whitebread, S., de Gasparo, M. Bühlmayer, P., Herold, P., Ostermayer, F. and Kamber, B., J. Cardiovasc. Pharmacol., 16, Suppl. 4 (1990) 556.

Factors which control the transport of cyclic dipeptides across mucosal barriers

T.B. Frigo* and J. Burton

The Unit for Rational Drug Design, University Hospital/Boston University Medical Center, 88 East Newton Street, Boston, MA 02118, U.S.A.

Introduction

Previous research has shown that linear oligopeptides can cross epithelial membranes at high rates [1]. There are, however, few correlations between the rates of absorption of the oligopeptides and their physical properties. As a result, it is not possible to increase the oral activity of a peptide based drug by the methods of rational drug design. To learn which, if any, physical properties are correlated with passage through the gut membrane, a homologous series of six cyclic dipeptides of the form *cyclo*-[Gly-D-Aaa], where the side chain of the Aaa residue varied from H to C_6H_{12}, were synthesized and their rates of passage across membranes of the human gut were measured.

Results and Discussion

Radiolabeled cyclic dipeptides (Table 1) were synthesized by coupling the Boc-D-amino acid with [^{3}H]Gly-OEt using WSCI. Deprotection with TFA and cyclization (NH_3/Et_3N) furnished the cyclic dipeptide with a specific activity of 10 Ci/mol. Chemistry developed by Evans [2] was used to prepare D-Ile(4Me). The permeability coefficient, Pe for each cyclic dipeptide was determined using a modified Ussing chamber, where Pe x [peptide] = rate of passage of the tritium labeled peptide across a T-84 cell monolayer. To control for variability of Pe within the same cell lines, values of Pe are normalized to *cyclo*-(D-Val-Gly). Observed and normalized Pe values are shown in Table 1.

Table 1 *Permeability data, Pe, lipophilicities ($K_{o/w}$, for an octanol/Ringers system) and calculated volumes (from QUANTA) for cyclic dipeptides*

Compound	Observed Pe (nm/s), ± 1 SD	Normalized Pe (nm/s), ± 1 SD	# samp.	$K_{o/w}$	Volume, $Å^3$
cyclo-(Gly-Gly)	5.6 ± 1.9	7.7 ± 2.4	5	0.013	90.6
cyclo-(D-Val-Gly)	10.3 ± 4.4	10.3	27	0.151	136
cyclo-(D-Val(3Me)-Gly)	11.8 ± 2.7	11.8 ± 2.9	4	0.339	147
cyclo-(D-Ile-Gly)	14.8 ± 6.9	13.9 ± 2.9	4	0.501	147
cyclo-(D-Ile(4Me)-Gly)	21.6 ± 7.6	29.0 ± 16.8	6	1.148	162
cyclo-(D-Chg-Gly)	30.6 ± 7.4	48.2 ± 12.5	5	2.089	173

*Current Address: Advanced Magnetics, Inc., 61 Mooney St., Cambridge, MA 02139, U.S.A.

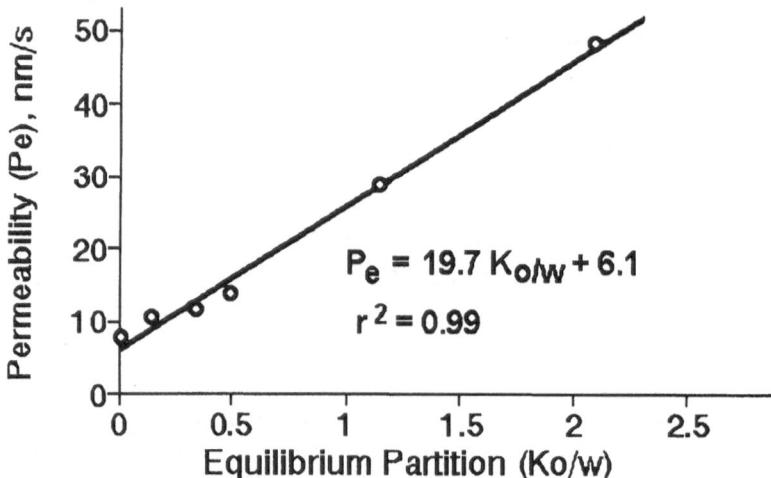

Fig. 1. *Correlation of permeability coefficient and lipophilicity for cyclic dipeptides. The equation shown is for the best fit of the data.*

When the concentration was varied for a single cyclic dipeptide, the flux rates were shown to be linearly related to concentration. The rate of passage (flux) was also independent of direction: mucosal to serosal or vice versa. These observations are consistent with a diffusion controlled absorption mechanism. The integrity of the cell monolayer was monitored by electrical resistance and by [14C]inulin flux. HPLC with a radioactive flow monitor was used to verify that the peptide was transported through the membrane intact.

It is a common understanding that lipophilicity is an important factor for determining how well a drug is absorbed from the gut. Lipophilicities of the cyclic dipeptides were obtained ($K_{o/w}$, Table 1) and range from 0.013 to 2.089. When values of Pe are plotted against $K_{o/w}$, the data are highly correlated ($r^2 = 0.99$, see Figure 1). The simplest interpretation of the data is diffusion into and out of a membrane bilayer as shown in Figure 2 (model 1). The cyclic dipeptide first diffuses into the cell membrane, and then diffuses out to either the serosal or mucosal chamber. This model assumes that the equilibrium between the membrane layer and the outer aqueous phase is the limiting factor for diffusive uptake of the cyclic dipeptide.

An alternative model is shown in Figure 2 (model 2), wherein the cyclic dipeptide penetrates the outer cell membrane, diffuses across the cell in a rate controlling step, and then diffuses out of the cell. The dependence of Pe on factors associated with diffusion through the cell such as volume is important in this model. To estimate the effect of volume on the observed Pe's, the compounds were modeled using the QUANTA molecular mechanics package. The calculated molecular volumes for the minimized structures are given in Table 1. The plot of molecular volume versus permeability gives a poor correlation ($r < 0.6$), supporting the more simple model for absorption of cyclic dipeptides.

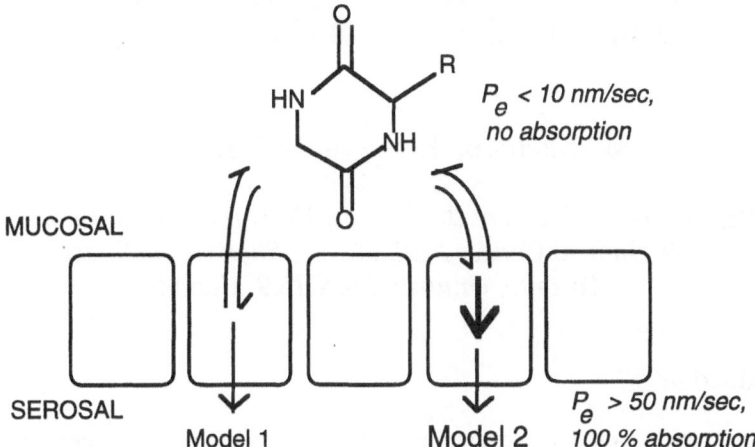

Fig. 2. *Two models for diffusion mediated absorption of cyclic dipeptides across gut epithelial membranes. Model 1, simple diffusion, and model 2, diffusion mediated by intracellular effects.*

Permeability and absorption in humans have been compared for a number of drugs [3]. The data reported can be fitted to an equation (Eq. 1) which shows that for Pe greater than 50 nm/s, absorption is nearly 100%, while for compounds with Pe less than 10 nm/s, the percentage of absorption drops to zero. According to Eq. 1, the cyclic dipeptides reported here range in absorption from 1 - 90%. This modeled dependence of absorption on changes in cyclic dipeptide structure shows how quantitative structure activity relationships can be used to enhance the oral activity of cyclic dipeptides.

$$\% \text{ Absorption} = \frac{100}{1 + 10^{(4.9 - 3.5 \log P_e)}} \qquad \text{Eq. 1}$$

Acknowledgements

The authors thank Dr. J.L Madera of Brigham and Women's Hospital (Boston, MA) for providing valuable expertise in cell culture and flux studies.

References

1. Takaori, K., Burton, J. and Donowitz, M., Biochem. Biophys. Res. Commun., 137 (1986) 682.
2. Evans, D.A. and Britton, T.C., J. Am. Chem. Soc., 109 (1987) 4649.
3. Artursson, P. and Karlsson, J., Biochem. Biophys. Res. Commun., 175 (1991) 880.

Novel radioimmunoconjugates containing a bifunctional metal chelating peptide

K.M. Sheldon, R. Arya and J. Gariépy

Department of Medical Biophysics, University of Toronto, and the Ontario Cancer Institute, 500 Sherbourne Street, Toronto, Ontario, M4X 1K9, Canada

Introduction

One major problem associated with the creation of successful imaging probes and therapeutic reagents has been the practical compromise between introducing a sufficient number of individual metal binding sites on targeting agents and the effects of excessive labelling on the binding affinity of the modified targeting agent for its receptor [1]. Most procedures used to label monoclonal antibodies with radioactive metals rely on the random reaction of activated chelators with available sites on the antibody molecule [2]. Conjugation of large numbers of chelators in this fashion frequently results in a reduction of immunoreactivity of the final product. We have developed a site directed method of producing high specific activity, immunoreactive, radioimmunoconjugates using a bifunctional metal chelating peptide. The peptide scaffold contains several EDTA-like groups and a single functional group that can be conjugated to the carbohydrate moieties located in the Fc portion of the antibody molecule. By defining the site of conjugation we can prevent the loss of immunoreactivity due to modification of the antigen binding domains. Additionally, the functional group of the metal chelating peptide can be modified to enable conjugation to other reactive groups such as the free thiols on Fab'.

Results and Discussion

The concept of using amino acid branching methods [3] and protected metal chelating groups [4] in solid-phase peptide synthesis led us to the design and construction of prototypic peptides such as the one illustrated in Figure 1. The overall approach allows one to dramatically simplify the construction and design of metal-chelating peptides, and permits the rapid creation of an unlimited number of new agents.

To demonstrate the utility of the metal chelating peptide in constructing bioconjugates, we selectively introduced the peptide into the Fc domain of a monoclonal antibody. First a maleimide group was introduced on the peptide by modifying the single amino group near the C-terminus with m-maleimido-benzoyl-N-hydroxysulfosuccinimide ester and the carbohydrate chains on the antibody were oxidized with sodium periodate to create aldehyde groups which were subsequently converted to thiol moieties by reaction with 2-acetamido-4-mercaptobutyric acid hydrazide. The final immunoconjugate was created by reaction of the maleimide substituted peptide with the thiolated antibody. Immunoconjugates produced in this fashion retained their antigen binding ability as determined by ELISA assay.

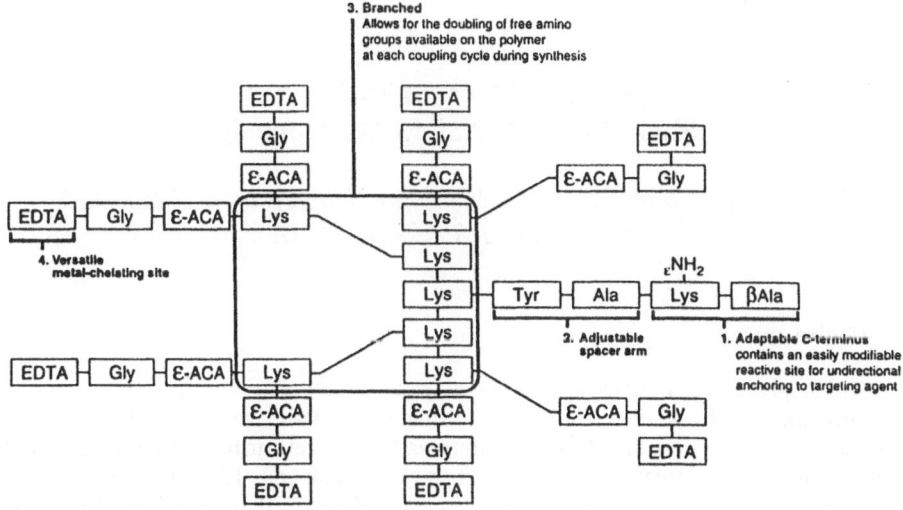

Fig. 1. Structure of the bifunctional branched metal-chelating peptide.

In summary, a synthetic strategy was developed for the rapid design of bifunctional metal-chelating peptides that can be coupled unidirectionally to targeting agents. The overall approach is flexible and permits the incorporation of large clusters of chelating sites such as EDTA without compromising the immuno-reactivity of the targeting agent.

Acknowledgements

This work was supported by a grant from the National Cancer Institute of Canada in association with the Canadian Cancer Society. RA is the recipient of a National Cancer Institute of Canada studentship. KS is the recipient of a postdoctoral fellowship from the George Knudson Foundation.

References

1.   Wang, T.S., Fawwaz, R.A. and Alderson, P.O., J. Nucl. Med., 33 (1992) 570.
2.   Hnatowich, D.J., Layne, W.W., Childs, R.L., Lanteigne, D. and Davis, M.A., Science, 220 (1983) 613.
3.   Tam, J.P., Proc. Natl. Acad. Sci. U.S.A., 85 (1988) 5409.
4.   Ayra, R. and Gariépy, J., Bioconj. Chem., 2 (1991) 323.

[111]Indium labeled laminin peptide fragments as potential diagnostic agents for metastatic cancers

S.S. Jonnalagadda[1], M. Mokotoff[1], D.P. Swanson[1], M.L. Brown[2] and M.W. Epperly[2]

Schools of [1]Pharmacy and [2]Medicine, University of Pittsburgh, Pittsburgh, PA 15261, U.S.A.

Introduction

The glycoprotein laminin is known to promote cell adhesion, spreading, neurite outgrowth, growth and differentiation [1]. In the metastatic evolvement of breast and colon tumors it appears that the expression of a 67 kDa cell surface receptor occurs at levels substantially greater than usually found in normal cells [2]. This high affinity receptor recognizes the specific peptide sequences CDPGYIGSR and YIGSR, which are a part of the B1 chain of laminin [3]. If these peptide sequences could be radiolabeled without altering their binding to the 67 kDa receptor, it might permit the non-invasive early detection and/or differentiation of malignant tumors of the breast and colon via nuclear medicine techniques.

Results and Discussion

Earlier studies [4] showed that the amide derivatives of the laminin peptides, CDPGYIGSR-NH$_2$ and YIGSR-NH$_2$, are significantly more active than their carboxy-terminal forms, therefore we used the amino-terminal group of these peptides on which to attach the chelating functional group diethylenetriaminepentaacetic acid (DTPA), in order to allow the chelation of [111]In. In the case of YIGSR-NH$_2$, we actually added one more amino acid, Gly (i.e. GYIGSR-NH$_2$) to the N-terminus so that the chelating reagent would not have to be attached directly to Tyr. The following peptides, 1-4, were synthesized as agents to which we chelated ^{111}In for potential malignancy diagnosis. In addition, the nonsense peptide 5 was prepared as a negative control. The key intermediates, needed for the preparation of the peptides below, were synthesized by SPPS using 4-methyl-benzhydrylamine (MBHA) as the anchoring resin; specifically, peptides 1-4 required GY(Dcb)IGS(Bzl)R(Tos)-MBHA and C(Meb)D(Bzl)PGY(Dcb)IGS(Bzl)R(Tos)-MBHA, while GAGAGA-MBHA was the precursor to 5. The DTPA monomers 1, 3 and 5 were obtained by reacting their respective key intermediates, while still resin bound, with a mono HOBt-activated ester of DTPA followed by cleavage of all blocking groups with liquid HF.

1	DTPA-GYIGSR-NH$_2$	4	DTPA-(CDPGYIGSR-NH$_2$)$_2$
2	DTPA-(GYIGSR-NH$_2$)$_2$	5	DTPA-GAGAGA-NH$_2$
3	DTPA-CDPGYIGSR-NH$_2$		

Similarly, the DTPA dimers **2** and **4** were obtained by reacting their respective key intermediates with DTPA dianhydride followed by HF cleavage. Peptides **1-5** were purified via preparative HPLC, their purity confirmed by analytical HPLC, and the structures established by amino acid analysis and/or FAB or electrospray mass spectrometry. The labeling of peptides **1-5** was accomplished with ^{111}InCl$_3$ in 0.1 M acetic acid (previously passed through Sigma's chelating resin) at 1.8 mCi/1 nmol for **1, 3** and **4** and at 1 mCi/2 nmol for **2** and **5**. The radiochemical purity of each peptide was assessed by HPLC analysis.

Studies involving the imaging/biodistribution of radiolabeled peptides utilized mice in which a C3H mammary carcinoma had been implanted. The C3H mammary carcinoma line was obtained from Dr. Bernard Fisher (School of Medicine, University of Pittsburgh). Whether the 67 kDa laminin receptor is expressed on the surface of this tumor is not known. C3H mammary tumor cells were implanted into the left thigh of 20 g, 12 week old C3H female mice and allowed to grow (2-4 weeks) until the tumor was significantly noticeable. The animals (n=3/agent/time interval) were injected via the tail vein with 20-25 μCi of test ^{111}In-labeled peptides **1-5** and euthanized at 1, 4, 8, or 24 hours post injection. The animals were subsequently dissected and the tissue distribution of radioactivity quantitated (mean % injected dose/g tissue) using an autogamma scintillation counter. The biodistribution data (Tables 1 and 2) reveal specific uptake of ^{111}In-**2-4** by the tumor, but not of ^{111}In-**1** or ^{111}In-**5**.

Table 1 *Biodistribution of ^{111}In-1-3 in the C3H mammary tumor model*

| Tissue | Percentage injected dose/g tissue | | | | | | | | | | | |
| | ^{111}In-**1** | | | | ^{111}In-**2** | | | | ^{111}In-**3** | | | |
	1hr	4hr	8hr	24hr	1hr	4hr	8hr	24hr	1hr	4hr	8hr	24hr
Brain	0.06	0.02	0.01	0.01	0.17	0.04	0.03	0.02	0.12	0.08	0.07	0.04
Blood	0.73	0.09	0.09	0.10	1.89	0.44	0.55	0.19	1.55	0.88	1.00	0.28
Heart	0.48	0.06	0.05	0.02	1.51	0.35	0.33	0.16	0.96	0.44	0.57	0.26
Lung	0.77	0.11	0.10	0.04	1.87	0.39	0.44	0.26	1.65	0.77	0.96	1.65
Liver	2.17	0.19	0.17	0.12	1.22	0.34	0.37	0.37	1.08	0.71	0.87	0.73
Spleen	0.48	0.13	0.12	0.10	0.81	0.30	0.41	0.46	0.49	0.59	1.01	0.92
Stomach	1.01	0.30	0.07	0.06	1.18	0.12	0.11	0.12	0.72	0.13	0.17	0.14
Gut	6.49	2.72	0.45	0.21	1.92	0.56	0.25	0.32	0.87	0.72	0.54	0.40
Kidney	3.72	1.44	1.32	1.04	5.61	2.43	1.41	2.18	8.25	9.13	9.77	5.91
Ovary	0.90	0.44	0.49	0.09	1.46	0.38	0.31	0.38	1.14	0.67	0.72	0.54
Skin	3.48	0.29	0.11	0.06	1.51	0.37	0.21	0.41	1.19	0.55	0.55	0.36
Muscle	0.72	0.05	0.03	0.02	0.81	0.21	0.17	0.09	0.40	0.12	0.23	0.13
Bone	0.48	0.06	0.05	0.04	0.72	0.14	0.17	0.16	0.61	0.37	0.34	0.40
Tumor	1.42	0.22	0.09	0.09	1.62	0.50	0.55	0.46	1.54	0.99	1.07	0.66

Table 2 *Biodistribution of ^{111}In-4-5 in the C3H mammary tumor model*

Tissue	Percentage injected dose/g tissue							
	^{111}In-4				^{111}In-5			
	1hr	4hr	8hr	24hr	1hr	4hr	8hr	24hr
Brain	0.05	0.04	0.03	0.02	0.07	0.01	0.02	0.00
Blood	2.16	0.05	0.27	0.07	0.44	0.07	0.03	0.03
Heart	0.59	0.23	0.17	0.11	0.40	0.06	0.04	0.03
Lung	0.99	0.23	0.27	0.16	0.49	0.08	0.06	0.04
Liver	0.64	0.28	0.39	0.42	0.37	0.12	0.10	0.10
Spleen	0.42	0.24	0.38	0.49	0.37	0.13	0.10	0.13
Stomach	0.31	0.14	0.92	0.12	0.26	0.23	0.07	0.03
Gut	0.59	0.16	0.59	0.22	0.41	0.44	0.24	0.05
Kidney	13.88	14.77	12.94	11.50	1.81	1.42	1.13	1.09
Ovary	0.64	0.33	0.31	0.61	1.18	0.12	0.11	0.14
Skin	0.33	0.18	0.30	0.15	0.94	0.12	0.09	0.06
Muscle	0.24	0.06	0.09	0.06	0.22	0.04	0.03	0.02
Bone	0.34	0.10	0.18	0.12	0.18	0.04	0.03	0.03
Tumor	1.19	0.45	0.59	0.43	0.83	0.20	0.09	0.08

Biodistribution data reveal specific uptake of ^{111}In-**2**, ^{111}In-**3** and ^{111}In-**4** in the C3H mammary tumors approaching 0.6, 1.1 and 0.6 percent of the administered dose/g, respectively, at 8 hours post injection. At this time interval, tumor-to-blood and tumor-to-muscle radioactivity ratios were 1.0 and 3.3, respectively, for ^{111}In-**2**, 1.1 and 4.6, respectively, for ^{111}In-**3** and 2.2 and 6.5, respectively, for ^{111}In-**4**. Retention of radioactivity in the tumor was longest, nearly 0.7 percent after 24 hours, with ^{111}In-**3** and the tumor-to-blood and tumor-to-muscle ratios were 2.4 and 5.2, respectively. These preliminary data suggest the potential use of ^{111}In-labeled laminin peptide fragments in the early detection of malignant tumors of the breast.

References

1.   Kleinman, H.K., Cannon, F.B., Laurie, G.W., Hassell, J.R., Aumailley, N., Terranova, V.P. and Martin, G.R., J. Cell Biochem., 27 (1985) 317.
2.   Castronovo, V., Colin, C., Claysmith, A.P., Chen, P.H.S., Lifrange, E., Lambotte, R., Krutzsch, N., Liotta, L.A. and Sobel, M.E., Am. J. Path., 137 (1990) 1373.
3.   Graf, J., Iwamoto,Y., Sasaki, M., Martin, G.R., Kleinman, H.K., Robey, F.A. and Yamada, Y., Cell, 48 (1987) 989.
4.   Graf, J., Ogle, R.C., Robey, F.A., Sasaki, M., Martin, G.R., Yamada, Y. and Kleinman, H.K., Biochemistry, 26 (1987) 6896.

Synthesis of [poly(ethylene glycol)]-[fibronectin and laminin related peptide] hybrids and their inhibitory effect on experimental metastasis

K. Kawasaki[1], S. Inouye[1], T. Hama[1], S. Yamamoto[2], N. Okada[2] and T. Mayumi[2]

[1]Faculty of Pharmaceutical Sciences, Kobe-Gakuin University, Ikawadani-cho, Nishi-ku, Kobe 651-21, Japan, and [2]Faculty of Pharmaceutical Sciences, Osaka University, Ymadaoka, Suita-shi 565, Japan

Introduction

Since poly(ethylene glycol) (PEG) is stable, low toxic, low immunogenic, bioinert and soluble in both aqueous and organic solvents, it is noted as a drug carrier. Recently we found that the inhibitory effects of [fibronectin and laminin-related peptide]-[amino-poly(ethylene glycol)] hybrids [Arg-Gly-Asp-aminoPEG (RGD-aPEG) and Tyr-Ile-Gly-Ser-Arg-aminoPEG(YIGSR-aPEG)] on experimental metastasis in mice were more potent than those of RGD and YIGSR [1, 2]. Following these studies, we prepared H-Arg-Glu-Asp-Val-NH2(REDV), REDV-aPEG, H-Glu-Ile-Leu-Asp-Val-NH2(EILDV), cPEG-EILDV (cPEG=carboxy-PEG) and cPEG-YIGSR. REDV and EILDV are sequences of fibronectin and YIGSR is a sequence of laminin.

Results and Discussion

To synthesize REDV-aPEG, Boc-Arg(Tos)-Glu(OcHx)-Asp(OcHx)-Val-OH was prepared by the solution method and coupled with aPEG which was prepared from PEG 6000(MW 7500-9000) according to the procedure reported by Pillai and Mutter [3]. The resulting protected hybrid was treated with HF to give REDV-aPEG, followed by purification by HPLC. Other peptides and hybrids were prepared by the solid-phase method using Boc strategy. cPEG was prepared from PEG monomethyl ether (average MW 5000) by oxidation according to the procedure reported by Ueyama et al [4].

The inhibitory effect of synthetic peptides and hybrids on the experimental metastasis of B16 melanoma BL6 in mice was examined. The inhibitory effect of cPEG-YIGSR was almost equal to that of YIGSR-aPEG. The inhibitory effect of REDV was potentiated by the hybridization with aPEG as shown in Figure 1. Of the synthetic peptides and their hybrids, EILDV exhibited the most potent inhibitory effect.

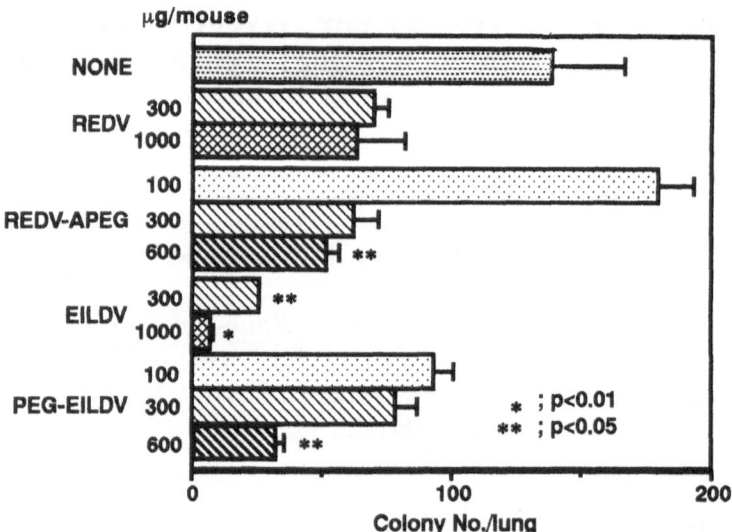

Fig. 1. Inhibitory effect of synthetic peptides and hybrids on the lung metastasis in mice.

EILDV, REDV, and their PEG hybrids exhibited an inhibitory effect on experimental metastasis of B16 melanoma BL6 in mice. The potent inhibitory effect of the hybrid can be roughly explained by the hypothesis that the bulky PEG portion might prevent enzymatic hydrolysis of the peptide portion and stabilize the binding of the peptide portion to its receptor. Koyama reported that subcutaneously injected PEG in mice was distributed in tumor cells in high concentration [5]. This fact also suggests the potent inhibitory effect of the hybrid.

References

1. Kawasaki, K., Namikawa, M., Murakami, T., Mizuta, T., Iwai, Y., Hama, T. and Mayumi, T., Biochem. Biophys. Res. Commun., 174 (1991) 1159.
2. Kawasaki, K., Namikawa, M., Yamashiro, Y., Hama, T. and Mayumi, T., Chem. Pharm. Bull., 39 (1991) 3373.
3. Pillai, V.N.R. and Mutter, M., J. Org. Chem., 45 (1980) 5364.
4. Ueyama, N., Nakata, M. and Nakamura, A., Polymer J., 17 (1985) 721.
5. Koyama, Y., In Inada, Y. (Ed.), The Proceedings of the 6th Symposium on Protein Hybrid, Toin-Gakuin Yokohama University, Yokohama, Japan, 1991, p.5.

N-terminally extended analogs of bGRF with a general formula $[X^{-1},Y^0,Leu^{27}]bGRF(1\text{-}29)NH_2$ as pro-drugs and potential targets for processing by plasma dipeptidylpeptidase IV (DPP-IV)

T.M. Kubiak, R.A. Martin, R.M. Hillman, J.F. Caputo, G.R. Alaniz, W.H. Claflin, D.L. Cleary and W.M. Moseley

The Upjohn Company, Kalamazoo, MI 49001, U.S.A.

Introduction

Recently, we presented the concept of N-terminally extended GRF analogs of $[Leu^{27}]bGRF(1\text{-}29)NH_2$ (**1**) as pro-drugs for the generation of the core peptide **1** in bovine plasma *in vitro* through a DPP-IV-mediated hydrolysis [1]. It has been shown that only GRF analogs with N-terminal extensions specific for DPP-IV recognition generated **1** while other pro-GRFs with a wrong "reading frame" for DPP-IV were not processed to **1** [1]. Also, the apparent $t_{1/2}$ of **1** generated from an extended DPP-IV-cleavable pro-GRF almost doubled as compared to the $t_{1/2}$ of **1** incubated directly in bovine plasma *in vitro* [1].

DPP-IV preferably cleaves dipeptides X-Pro, X-Ala or X-Hyp from the N-termini of unsubstituted peptides or proteins [2]. DPP-IV-assisted hydrolysis rates are dependent upon X. About 20 times faster cleavages were reported for short peptides with Pro at position P_1 than those with Ala [2]. This information prompted us to examine whether pro-GRFs with different N-terminal extensions could be cleaved at different rates by plasma DPP-IV to control the release of the core peptide **1**.

Results and Discussion

Included in this study were pro-GRFs from the $[X^{-1},Y^0,Leu^{27}]bGRF(1\text{-}29)NH_2$ series (prepared by SPPS, Boc-chemistry [3, 4]) having either $Ile^{-1}\text{-}Pro^0$ (analog **2**), $Val^{-1}\text{-}Ala^0$ (analog **3**), $Tyr^{-1}\text{-}Ala^0$ (analog **4**) or $desNH_2Tyr^{-1}\text{-}Ala^0$ (analog **5**) as the N-terminal extension. The main objectives of this study were (i) to compare the course of appearance and degradation of $[Leu^{27}]bGRF(1\text{-}29)NH_2$ (**1**) derived from pro-GRFs with different extensions upon incubation in bovine plasma *in vitro* [1], and (ii) to determine the GH-releasing activity of the N-terminally extended analogs in primary bovine anterior pituitary prime cell cultures *in vitro* [5] and in steers *in vivo* [6].

In agreement with our earlier reports [1, 4, 7], **1** was degraded in bovine plasma *in vitro* via a DPP-IV-dependent mechanism to form $[Leu^{27}]bGRF(3\text{-}29)NH_2$ (**6**). Plasma incubations with analogs **2**, **3** and **4** resulted in sequential cleavages, first of the $X^{-1}\text{-}Y^0$ extension to generate **1**, followed by the removal of $Tyr^1\text{-}Ala^2$ from **1** to form $[Leu^{27}]bGRF(3\text{-}29)NH_2$ (**6**).

Table 1   *N-terminally extended analogs of [Leu^{27}]bGRF(1-29)NH$_2$ (1). GH-releasing activity in primary bovine anterior pituitary cell cultures in vitro[a]*

No.	Extension in peptide 1	Potency[a] (%)	Potency 95% CL[b]	EC$_{50}$ (nm)	Efficacy (%)	Efficacy 95% CL[b]
1	None	100.0	–	0.088	100	–
2	Ile^{-1}-Pro^0	4.5[d]	–	2.15	–	–
3	Val^{-1}-Ala^0	13.0	4.7 - 35	0.70	100.4	86.6-116.1
4	Tyr^{-1}-Ala^0	0.5	0.2 - 2.0	16.43	90.7	74.3-109.7
5	desNH$_2$Tyr^{-1}-Ala^0	3.0	1.0 - 4.4	2.62	68.9	43.0- 98.1

[a]Assay conditions as described [5]. Potencies based on peptides EC$_{50}$'s relative to 1.
[b]95% CL denotes 95% confidence limits.
[c]Efficacy based on the maximum GH-release induced by 1 taken as 100%.
[d]Taken from [1].

The half-lives determined for the removal of N-terminal extensions from analogs 2, 3, and 4 were as follows: 20.4 min (Ile^{-1}-Pro^0), 20.7 min (Val^{-1}-Ala^0) and 23.8 min (Tyr^{-1}-Ala^0). These values do not support the literature data on much faster cleavages of substrates with penultimate Pro than those with Ala [2]. Possibly, in contrast to di- or tri-peptidic substrates, secondary structures of longer peptides could have greater effects on the enzyme binding and cleavage rates. The apparent half-lives determined for 1 derived from 2, 3 and 4 were respectively 174% (43.2 min), 186% (46.2 min) and 155% (38.5 min) of the $t_{1/2}$ of 1 (24.8 min) incubated directly with bovine plasma.

Analog 5 with desNH$_2$Tyr^{-1}-Ala^0 in the extension, an inactive DPP-IV substrate lacking the free N-terminal amino group, was not cleaved by plasma to form 1. Although no metabolites were identifiable under the HPLC conditions used in the study, this extended GRF slowly disappeared from plasma due to other unknown, but non-DPP-IV-related cleavages.

The *in vitro* GH-releasing potencies of the four extended GRF analogs 2, 3, 4 and 5 were in the range of 0.5 to 13% of the potency of 1 (Table 1). Despite low inherent (*in vitro*) potency, 2, 3, and 4 were respectively 121%, 103% and 113% as active as 1 (p>0.05) in releasing serum GH *in vivo* in steers (Figure 1), which suggests that 1 could also have been generated *in vivo* from the DPP-IV-cleavable pro-GRFs. This notion is supported by the fact that the non-DPP-IV cleavable analog 5 released no more GH (56%) than water injected controls (54%) under the same *in vivo* conditions (Figure 1).

In summary, our results confirm the pro-drug concept of N-terminally extended GRF analogs as targets for the DPP-IV mediated release of the core GRF and extension of its $t_{1/2}$ in bovine plasma *in vitro*. However, the DPP-IV cleavable pro-GRFs did not appear to significantly delay nor prolong the analog-induced release of GH *in vivo*. The serum GH profiles in steers treated with these pro-drugs were

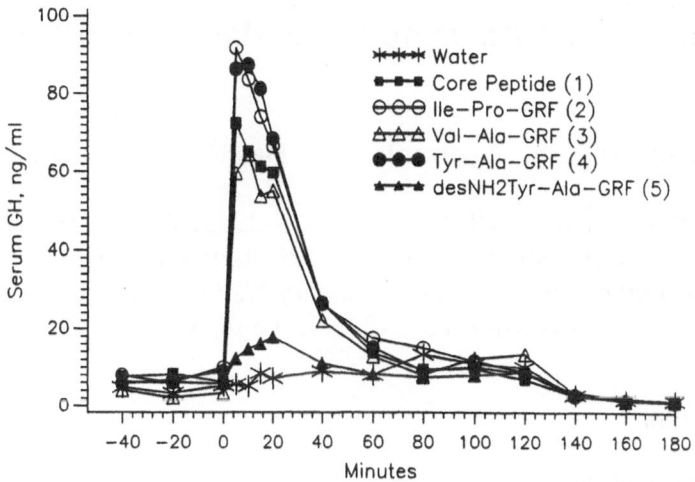

Fig. 1. Profiles of serum GH induced by a single i.v. dose (0.1 nmol/kg) of [Leu^{27}]-bGRF(1-29)NH$_2$ (1) or its analogs 2-5 in steers. Assay conditions as described [6].

similar to those induced by the core peptide despite the analogs' low GH-releasing potency *in vitro*. This suggests the pro-GRFs could have been rapidly processed to the core peptide by blood and/or organ and tissue DPP-IV *in vivo*.

Acknowledgements

We thank Lavern F. Krabill and Marjorie R. Zantello for carrying out the GH RIA assays.

References

1. Kubiak, T.M., Cleary, D.L. and Krabill, L.F., In Schneider, C.H. and Eberle, A.N. (Eds.), Peptides 1992, Escom, Leiden, The Netherlands, 1993, p.739.
2. Walter, R., Simmons, W.H. and Yoshimoto, T., Mol. Cell. Biochem., 30 (1980) 111.
3. Merrifield, R.B., J. Am. Chem. Soc., 85 (1963) 2149.
4. Kubiak, T.M., Friedman, A.R., Martin, R.A., Ichhpurani, A.K., Alaniz, G.R., Claflin, W.H., Goodwin, M.C., Cleary, D.L., Hillman, R.M., Kelly, C.R., Downs, T.R., Frohman, L.A. and Moseley, W.M., J. Med. Chem., 36 (1993) 888.
5. Friedman, A.R., Ichhpurani, A.K., Brown, D.M., Hillman, R.M., Krabill, L.F., Martin, R.A., Zurcher-Neely, H.A. and Guido, D.M., Int. J. Pept. Protein Res., 37 (1991) 14.
6. Moseley, W.M., Alaniz, G.R., Claflin, W.H. and Krabill, L.F., J. Endocrinol., 117 (1988) 252.
7. Kubiak, T.M., Kelly, C.R. and Krabill, L.F., Drug Metab. Dispos., 17 (1989) 393.

Peptide serological classification of HIV-1 infection with support of data algorithms

R. Pipkorn[1,*], S. Ayehunie[2], R. Dash[3] and J. Blomberg[3]

[1]*Replico Medical AB, Storgatan 5, S-23431 Lomma, Sweden*
[2]*Department of Biology, Addis Ababa, Ethiopia*
[3]*Section of Virology, Department of Medical Microbiology,*
Sölvegatan 23, S-223 62 Lund, Sweden

Introduction

Certain synthetic peptides based on various HIV-1 strains show a differential reactivity in enzyme immunoassays with sera from different geographic locations (Africa, India, Europe and North America) [1]. By integrating the reactivity to epitopes which react differentially in various populations a pattern is created which can be used to classify HIV-1 infections. These results can give clues regarding the choice of viral variant to be represented in future HIV-1 vaccines.

Results and Discussion

In the serodendrogram shown in Figure 1, 120 Ethiopian and 46 Swedish HIV-1 positive sera were tested against 6 synthetic full-length (Cysteine to Cysteine) peptides from the third variable domain (V3) of HIV-1 gp 120. Peptides from HIV-1 subtypes A (Z321), B (MN), D (JY1 and a proline substituted derivative of ELI) and unclassified (Z3) were included. The peptide-variable names are written vertically above each column to the right of the dendrogram. To ensure reproducibility and to minimize influences due to differences in the general degree of antibody response of infected individuals a ratio was calculated between the reactivity of each peptide and the average of V3 peptide reactivities (MLV3). Finally, the ratio of the reactivities of Z3 and Z321 was included.

The ranked difference in absorbance between peptide- and mock-coated microtiter wells is shown for each cluster of sera (i.e., one branch in the serodendrogram) as a relative one-figure value, where 9 = maximum, and 0 = minimum of the ranked absorbance difference. The maximum value is shown at the bottom of the Figure. Rank 1.00 is the maximum for all peptide-variables. The reactivity in EIA for each peptide forms one dimension in a 7-dimensional coordinate system. Each serum is represented by a point in this coordinate system. Distances between all points are collected in a distance matrix. Sera with almost identical reactivity patterns are collected in clusters denominated by the identity of the first occurring serum. To distinguish the members of each cluster, Ethiopian members are symbolized with an

---

*Present address: Deutsches Krebsforschungscentrum, Im Neuenheimerfeld 506, D-6900 Heidelberg, Germany.

"E", and Swedish with an "S". Swedish sera cluster in two regions of the serodendrogram, whereas the Ethiopian sera had more variable reactivity patterns. In conclusion, serological similarity analysis and serodendrograms are novel and rational ways of analysing HIV-1 infection which may become useful in large scale epidemiological surveys of HIV-1 antigenic variation at sites selected for vaccine trials.

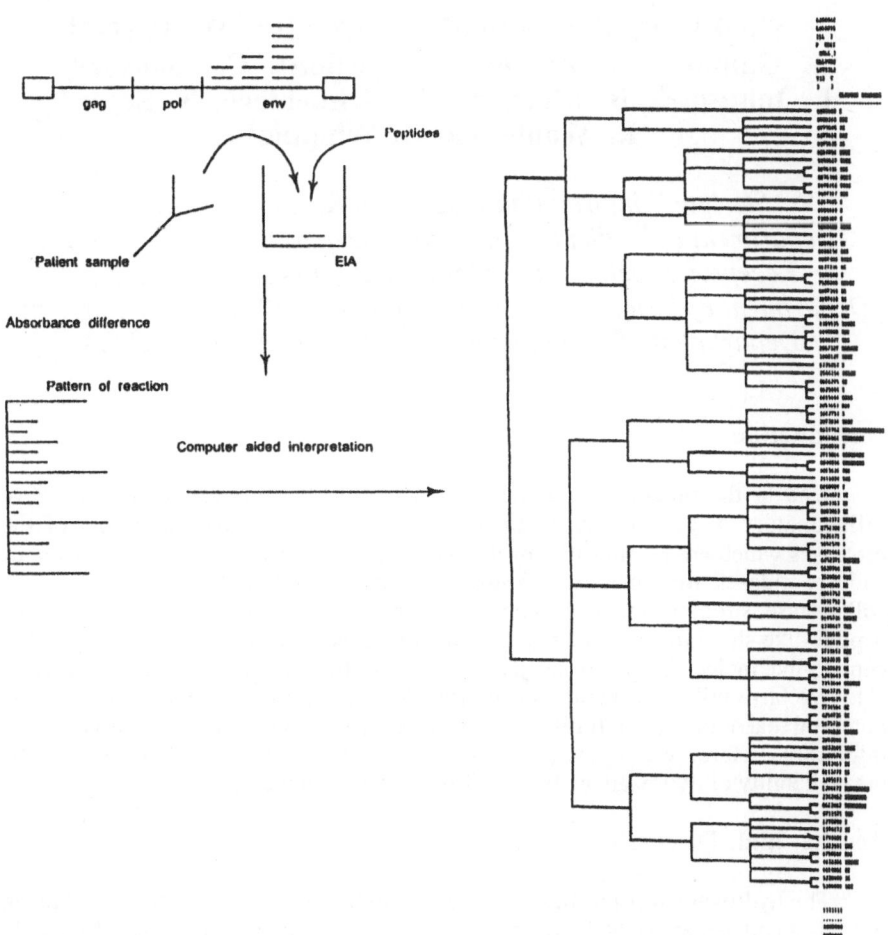

Fig. 1. Serological similarity analysis and a V3 based serodendrogram.

References

1. Blomberg, J., Lawoko, A., Pipkorn, R., Moyo, S., Malmvall, B.-E., Shao, J., Dash, R. and Tswana, S., AIDS, 7 (1993) 759.

Relationship between structure and bio-availability in a series of hydroxamate based metalloprotease inhibitors

J. Singh[1], P. Conzentino[4], K. Cundy[4], W. Earley[1],
J. Gainor[1], C. Gilliam[1], T. Gordon[1], C. Johnson[3],
J. Johnson[3], B. Morgan[1], E. Schneider[2], N. Seger[4],
R. Wahl[2] and D. Whipple[1]

Sterling Winthrop Pharmaceutical Research Division,
[1]Department of Medicinal Chemistry, [2]Department of Enzyme and
Receptor Biochemistry, Collegeville, PA 19426, U.S.A.,
[3]Department of Drug Metabolism, Rensselaer, NY 12144, U.S.A., and
[4]Department of Drug Delivery, Malvern, PA 19355, U.S.A.

Introduction

One of the potential approaches for arresting the progress of degenerative inflammatory disease is the administration of highly specific inhibitors of the proteases which are responsible for the breakdown of cartilage. These endogenous endo-peptidases are collectively known as matrix-metalloproteases (MMPs, e.g., collagenase, stromelysin, etc.). To date, potent inhibitors of MMPs are peptides or peptide-like structures and there is growing evidence that even metabolically stable peptide-like molecules may be subject to extensive first pass loss by biliary excretion [1]. We have utilized a variety of *in vitro*, *in situ* and *in vivo* models to examine potential barriers which limit the bioavailability of hydroxamate based MMP inhibitors. Here, we report on some structural features which influence the bioavailability of hydroxamate based MMP inhibitors in the rat.

Results and Discussion

The hydroxamate **1** has been reported [2] to be a potent inhibitor of the human fibroblast collagenase (HFC). We have found that, although hydroxamate **1** is stable (3-24h) in rat biological fluids, its oral bioavailability in the rat is negligible. Although *in situ* models reveal that the drug is absorbed across rat intestine, no intact **1** escapes the liver. Low levels of **1** were detected in pre-hepatic circulation, but neither the parent hydroxymate **1** nor its major metabolite, the corresponding carboxylate, was detected in post-hepatic circulation. Analysis of bile revealed extensive excretion of unchanged drug and metabolites. In fact, following *iv* administration of 5 mg/kg of **1** to rats with jugular and bile cannulae [3], more than 90% of **1** excreted in the bile in one hour. Moreover, 33% of the total biliary excretion occurred within the first 5 min. The poor pharmacodynamics exhibited by the hydroxamate **1** is probably due to the overall high lipophilicity of this compound. In order to improve the pharmacodynamic properties of the inhibitor, we decided to modify the C-terminal amide portion of the inhibitor **1** which should change the overall physicochemical properties while still maintaining nanomolar enzyme inhibition.

We have shown that the tryptophan analog **2a** (R=Me) is a 3-fold more potent inhibitor [K_i = 2 nM] of HFC than **1**. We have incorporated neutral, polar, cationic and anionic functional groups into various C-terminal amides, **2b-j** (Table 1). These modifications provided analogs with either more or comparable activity against HFC vs. the reference compound **1**.

1   **2a-j**
Ar = (4-OMe)phenyl   3-Indolyl
R = Me   see Table I.

The analogs containing tertiary amines, **2c-e**, improved plasma half life 2-3 fold over the reference compound **1**. More importantly these analogs have considerably reduced biliary excretion. However, the carboxylate analog **2b** behaved like the methyl compound **1** and is essentially completely excreted in bile.

Table 1 *Enzyme inhibition, plasma half life and biliary excretion data*

Compd.	R	Ki vs HFC (nM)	Plasma $t_{1/2}$ (min)	Biliary Excretion (% of total dose[b])	(% as parent)
1	-CH$_3$	6	9.05 ± 2.9	84.1 ± 7.3	43.3 ± 1.1
2 b	-(CH$_2$)$_4$-COOH	1	6.9 ± 2.4	108 ± 18.9	8.8 ± 1.3
2 c	-(CH$_2$)$_2$-N(CH$_3$)$_2$	12	21.0 ± 1.5	18.9 ± 3.7	71.1 ± 2.8
2 d	-(CH$_2$)$_3$-N(CH$_3$)$_2$	3	25.0 ± 4.2	18.1 ± 4.2	12.4 ± 3.2
2 e	-(CH$_2$)$_4$-N(CH$_3$)$_2$	3	28.2 ± 0.6	42.5 ± 3.6	56.5 ± 3.8
2 f	-(CH$_2$)$_2$-O-CH$_3$	5	19.0 ± 4.4	59.2 ± 11.8	21.4 ± 1.7
2 g	-(CH$_2$)$_2$-S-CH$_3$	5	32.4 ± 4.1	27.3 ± 4.3	15.6 ± 2.9
2 h	-(CH$_2$)$_2$-S(O)-CH$_3$	2	7.4 ± 0.7[a]	44.8 ± 10.1	74.3 ± 2.2
2 i	-(CH$_2$)$_2$-SO$_2$-CH$_3$	3	24.0 ± 3.1	67.2 ± 6.1	14.0 ± 4.0
2 j	-(CH$_2$)$_2$-S-(CH$_2$)$_2$-N(CH$_3$)$_2$	5	43.7 ± 7.7	60.8 ± 11.0	42.8 ± 7.1

[a]Mixture of diastereoisomers at sulfur; two peaks in HPLC, so value not accurate.
[b]All compounds were dosed (*iv*) at 5 mg/kg; except compound **2e** was dosed at 2.5 mg/kg.

In a comparable series of tertiary amine **2c**, ether **2f**, and thioether **2g**, surprisingly the thioether gave the longest plasma half life (32 min vs. 21 min and 19 min for the amine and ether analogs, respectively). The longer plasma $t_{1/2}$ of **2g** is also reflected in the lower biliary excretion of the thioether compared to the ether

2f. We have confirmed that the result of the thioether is not due to enzymatic oxidation at sulfur, as authentic sulfoxide **2h**, and sulfone **2i**, gave different results. In addition, the identity of the parent thioether in the biological fluids, plasma and bile, was confirmed by HPLC using the corresponding sulfoxides and sulfone. The analog **2j**, where we have inserted a thioether functionality in the tertiary amine analog **2e**, resulted in an analog which is more potent and has the longest plasma $t_{1/2}$ (44 min). The tertiary amine analog **2e** has been found to be 10% orally bioavailable in rat.

In conclusion, the rational design of C-terminally modified MMP inhibitors has led to analogs that are less susceptible to first pass loss by biliary excretion.

References

1. Boger, J., Bennet, C.D., Payne, L.S., Ulm, E.H., Blaine, E.H., Homnick, C.F., Schron, T.W., LaMont, B.I. and Veber, D.F., Regul. Peptides, 4 (1985) 8.
2. Dickens, J.P., Donald, D.K., Kneen, G. and McKay, W.R., U.S. Patent 4, 599, 361.
3. Details of the rat cannulation and data analysis will be published elsewhere.

A test kit for the determination of blood coagulation factor XIII based on synthetic peptides

W. Stüber, K. Fickenscher and M. Gerken

*Research Laboratories, Behringwerke AG, P.O. Box 1140,
W-3550 Marburg, Germany*

Introduction

Blood coagulation factor XIII (F XIII) is activated by thrombin and plays an important role in coagulation by crosslinking fibrin to insoluble polymers. Additionally F XIII incorporates α_2-antiplasmin and fibronectin into fibrin. F XIII is reported to accelerate wound healing too. For studies and clinical routine analyses several assays for F XIII activity have been described. Most of these tests have to be performed manually. Therefore they are cumbersome, time-consuming and often poorly reproducible. Normally they rely on acid or urea solubility of formed fibrin clots or on the incorporation of labelled amines into natural substrates of F XIII, i.e. casein or fibrin. We developed an assay for the activity of F XIII based on the release of ammonia generated by the F XIII mediated incorporation of glycine ethyl ester into a specific glutamine containing peptide [1]. The test can be carried out on automated routine analyzers or on common photometers. Here we report on the development of a test system with a specific glutamine containing F XIII substrate Leu-Gly-Pro-Gly-Gln-Ser-Lys-Val-Ile-Gly-amide (K9) and a clotting inhibitor Gly-Pro-Arg-Pro-Ala-amide (clot 7).

Results and Discussion

This procedure is carried out directly in a cuvette. The test principle is as follows: a F XIII containing specimen (blood plasma sample) is activated by thrombin and Ca^{2+}. To prevent the formation of a fibrin clot Gly-Pro-Arg-Pro-Ala-amide (clot 7) is added simultaneously. The activated F XIII incorporates Gly-OEt into the Gln of the substrate K9. The continuously released ammonia reacts directly with ketoglutarate. During this process NADH is consumed. The consumption of NADH is detected by the decline of UV absorption at 340 nm. The rate of UV decline is a direct measure for F XIII activity and F XIII content. For this assay two peptides are necessary. One peptide is used to prevent clotting of plasma. We have optimized the potency of Gly-Pro-Arg-Pro, which is known to prevent association of soluble fibrin chains without disturbing the action of thrombin [2]. We found that the elongation of this sequence at the C-terminus led to considerable more potent inhibitors. Particularly useful was Gly-Pro-Arg-Pro-Ala-amide (Table 1). K9 is a substrate that incorporates Gly-OEt via F XIII. We have optimized peptide sequences derived from casein [3, 4] and found K9 (Leu-Gly-Pro-Gly-Gln-Ser-Lys-Val-Ile-Gly-amide) having the most rapid turnover rate of Gly-OEt by F XIII. We synthesized

Table 1 *Structure-activity relationship of clot inhibitors*

Peptide sequence	relative activity (%)[a]
Gly-Pro-Arg-Pro-Ala-amide	100
Gly-Pro-Arg-Pro-Asp-amide	78
Gly-Pro-Arg-Pro-Pro-amide	53
Gly-Pro-Arg-Pro-Val-amide	37
Gly-Pro-Arg-Pro-Gly-amide	28
Gly-Pro-Arg-Pro-Lys-amide	25
Gly-Pro-Arg-Pro-Trp-amide	19
Gly-Pro-Arg-Pro-Ser-amide	17
Gly-Pro-Arg-Pro-Phe-amide	10
Gly-Pro-Arg-Pro-amide	7

[a]50 µl of each peptide (c = 2mg/ml) and 50 µl of thrombin solution (50 IU/ml) were added to 100 µl of plasma. The clotting time was measured in seconds. The most potent compound was set to 100% (449 s).

both peptides by solid phase methods in large amounts. We used TFA labile amide linkers attached to amino methyl polystyrene [5, 6]. The assembly of the peptides was carried out by standard Fmoc-protocols. The coupling reactions were carried out by a 1.5-fold amino acid excess using DIC/HOBt as an activating agent. The batch sizes for K9 were up to 450 g of final peptide resin. The clot inhibitor can be synthesized efficiently by solution methods too. Both peptides were cleaved by TFA/water (97% or 95%, respectively) and were purified by recrystallization employing TFA /tert.-butylmethylether. This procedure yielded substrate and clotting inhibitor pure enough for the above *in vitro* assay.

References

1. Fickenscher, K., Aab, A. and Stüber, W., Thromb. and Haemos., 65 (1991) 535.
2. Laudano, A.P. and Doolittle, R.F., Biochemistry, 19 (1980) 1013.
3. Gorman, J.J. and Folk, J.E., J. Biol. Chem., 259 (1984) 9007.
4. Fesus, L. and Folk, J.E., Clin. Chem., 31 (1985) 2044.
5. Stüber, W., Knolle, J. and Breipohl, G., Int. J. Pept. Protein Res., 34 (1989) 215.
6. Breipohl, G., Knolle, J. and Stüber, W., Int. J. Pept. Protein Res., 34 (1989) 262.

Synthetic peptide-based immunoassay for the detection of antibodies to human T-lymphotropic virus types I and II

P. Su[1], P. Coleman[1], J.W. Morgan[1], J. Payne[2] and F. Aubrit[2]

[1]Genetic Systems Corporation, 6565 185th Avenue NE, Redmond, WA 98052, U.S.A., and [2]Sanofi Diagnostic Pasteur, 3, Bd. Raymond Poincare, B.P. 3, 92430 Marnes La Coquette, France

Introduction

In 1977, Japanese medical officials reported the appearance of adult T-cell leukemia/lymphoma (ATL) in a small cluster of the adult population in southwestern Japan [1]. A retrovirus, human T-Cell Lymphotropic Virus Type I (HTLV-I), was first isolated from patients with ATL in the southeastern part of the United States [2, 3], indicating a possible relationship between the disease and the retrovirus. Collaborative research efforts using a purified HTLV-I antigen in an immuno-precipitation revealed that the Japanese ATL patients' sera reacted with the HTLV-I protein [1, 4]. HTLV-I is usually associated with ATL, but atypical, more benign HTLV-I-positive T-cell malignancies are sometimes found [5]. ATL is characterized by a clinically aggressive course, poor prognosis, hypercalcemia, and visceral involvement [6]. Although HTLV-I requires a long latency period for introduction of ATL, when ATL does develop, it is usually fatal and the survival time of patients is very short, 50% survival being about six months [7].

HTLV-II, first reported in 1982, is associated with T-cell variants of hairy-cell leukemia [8, 9], however, a specific cause-and-effect relationship between HTLV-II and human disease remains unclear. HTLV-II isolates outnumber HTLV-I in the U.S. [10]. Routes of HTLV transmission are similar to that of HIV, including sexual transmission, exposure to contaminated cellular blood components or IV drug abuse, and from infected mother to infant, primarily through breast feeding. HTLV-I and II have 65% nucleotide sequence homology and their encoded protein antigens are closely related [11]. This relationship is especially pronounced in the GAG regions of the viruses.

Methods of identifying antibodies to HTLV-I/II have included enzyme immuno-assays (EIA) based on whole viral lysate, recombinant antigen EIA, and PCR for identification of proviral sequences. This report details the use of synthetic poly-peptide antigens of the ENV region of HTLV-I and HTLV-II for use in an EIA format for the purpose of identifying antibodies to these retroviruses in human serum or plasma samples.

Results and Discussion

Of the total of 449 samples tested, 110 were positive by EIA, western blot, RIPA, PCR, or a combination of the above methods. There were 12 samples which were indeterminate by western blot (i.e., p19 and/or p24 present, and either no gp bands or one of gp68, gp46, or gp21 present), and negative by RIPA. The peptide EIA detected 5 (41.7%) of these indeterminates. Average signal to cutoff ratio with whole viral lysate EIA of the peptide positive samples was 3.8 versus 2.0 for the peptide negative indeterminates. Testing of 327 negative samples, comprised mostly of blood donors, resulted in no repeat reactive samples. Table 1 below summarizes these data.

Table 1 *Performance of the HTLV-I/II peptide EIA*

Sample category	Result	
HTLV-I or HTLV-II Positive Samples (N=110)	110/110	Positive
HTLV-I or HTLV-II Negative Samples (N=327)	0/327	Positive[a]
HTLV-I or HTLV-II Indeterminate Samples (N=12)	5/12	Positive

[a] Indicates repeat reactivity. Five (1.5%) of 327 samples were initially reactive.

Due to the nonspecific nature of recombinant and whole virus-based tests, radio-immunoprecipitation assays (RIPA) have been selected as the most reliable method for HTLV-I/II antibody confirmation. The choice was mainly due to the assay's ability to detect HTLV ENV glycoprotein antibodies with extreme sensitivity. Thus, a screening test based on HTLV-I and HTLV-II ENV proteins, presented as conserved polypeptides, could potentially provide highly reliable results.

Synthetic polypeptide antigens have shown success for detection of antiviral immunoglobulins in HIV-1, HIV-2 and hepatitis C virus. Peptide antigens are prepared based on the identification of epitopes which are both strongly immunogenic and highly conserved from strain-to-strain. Peptide antigens offer the advantage of presenting extremely high densities of immunogenic epitopes in the assay virtually without the presence of contaminating or nonspecific proteins. Mixtures of peptides in immuno-assays, therefore, are much less problematic than whole virus or recombinants.

The peptide-based EIA, with gp21 and gp46 antigens from HTLV-I and HTLV-II, demonstrated the ability to detect 100% of the positive samples tested. Specificity, based on repeat reactive testing, was also 100%. These data show promise for the peptide-based EIA as a valuable screening tool for the detection of antibodies to HTLV-I and II.

References

1. Uchiyama, T., Yodoi, J., Sagawa, K., Takatsuki, K. and Uchino, H., Blood, 50 (1977) 481.
2. Poiesz, B.J., Ruscetti, F.W., Gazdar, A.F., Bunn, P.A., Minna, J.D. and Gallo, R.C., Proc. Natl. Acad. Sci. U.S.A., 77 (1980) 7415.
3. Poiesz, B.J., Ruscetti, F.W., Reitz, M.S., Kalyanaraman, V.S. and Gallo, R.C., Nature, 294 (1981) 268.
4. Kalyanaraman, V.S., Sarngadharan, M.G., Nakao, Y., Aoki, T. and Gallo, R.C., Proc. Natl. Acad. Sci. U.S.A., 79 (1982) 1653.
5. Wong-Staal, F. and Gallo, R., Blood, 65 (1985) 253.
6. Blattner, W.A., Blayney, D.W., Robert-Guroff, M., Sarngadharan, M.G., Kalyanaraman, V.S., Sarin, P.S., Jaffe, E.S. and Gallo, R.C., J. Infect. Dis., 147 (1983) 406.
7. Okochi, K., Sato, H. and Hinuma, Y.A., Vox Sang, 46 (1984) 245.
8. Chen, I.S.Y., McLaughlin, J., Gasson, J.C., Clark, S.C. and Golde, D.W., Nature, 305 (1983) 502.
9. Rosenblatt, J.D., Glode, D.W., Wachsman, W., Giorgi, J.V., Jacobs, A., Schmidt, G.M., Quan, S., Gasson, J.C. and Chen, I.S., N. Engl. J. Med., 313 (1986) 372.
10. Lee, H., Swanson, P., Shorty, V.S., Zack, J.S., Rosenblatt, J.D. and Chen, I.S.Y., Science, 244 (1989) 471.
11. Chen, Y.M.A. and Essex, M., AIDS Res. Hum. Retroviruses, 7 (1991) 453.

Hydrophilic film supports

L. Winther[1], K. Almdal[1], W. Batsberg Pedersen[1], J. Kops[2] and R.H. Berg[1]

[1]Risø National Laboratory, DK-4000 Roskilde, Denmark
[2]The Technical University of Denmark, DK-2800 Lyngby, Denmark

Introduction

We have recently found that polystyrene-grafted polyethylene film (PEPS) [1, 2] is a highly convenient solid support for Merrifield synthesis of multiple peptides. Because of its hydrophobic nature, however, the PEPS film is not wetted by aqueous solutions and, therefore, it is not well-suited for certain biotechnological applications such as diagnostic testing and affinity purification. We now describe the preparation and properties of a hydrophilic version, the PEO-PEPS film, which is equally well-suited for peptide synthesis and, in addition, exhibits a high level of solvation in aqueous solution.

Results and Discussion

The PEO-PEPS film was prepared in the following way: in order to improve its mechanical properties, a polyethylene film was crosslinked by preirradiation with high energy electrons. Subsequent grafting of polystyrene proceeded smoothly on the crosslinked material to obtain a 168 wt % crosslinked PEPS film. Surprisingly, and in contrast with non-crosslinked PEPS films, no inhomogeneities due to bubbles containing occluded homopolymer were visible in the grafted polymer. The crosslinked PEPS film was aminomethylated to obtain a high substitution of ca. 5.5 mmol amino groups per gram film. A Boc-benzhydrylaminopropionic acid linker (BHA) was loaded onto 0.1 mmol amino groups per gram film and the remaining amino groups were coupled with 3,6,9-trioxadecanoic acid to obtain the hydrophilic PEO-PEPS film. The PEO-PEPS film was found to be very hygroscopic, gaining up to 20% in weight at 25°C and 50% relative humidity and, furthermore, it exhibited general solvent compatibility, ranging from toluene, through DMF to water.

In order to demonstrate its usefulness for solid-phase peptide synthesis, we have synthesized human gelsolin fragment (160-169) on the PEO-PEPS film and compared it to that synthesized on the unmodified PEPS film (loaded with 0.3 mmol BHA linker/ gram film). The fully protected Boc-Gln-Arg(Tos)-Leu-Phe-Gln-Val-Lys(Cl-Z)-Gly-Arg(Tos)-Arg(Tos)-BHA-PEO-PEPS film was assembled by a standard solid-phase double coupling protocol employing in situ DIC incorporation as well as some modified procedures for specific amino acids (Boc-Gln was coupled as HOBt ester to minimize the dehydration of amide to nitrile and for coupling of Boc-Phe[163] to Gln[164], the coupling was done with the preformed symmetrical anhydride to minimize pyrolidone carboxylic acid formation). Deprotection and release of the free hGS (160-69) was accomplished with anhydrous HF under standard conditions. Figure 1 shows the HPLC chromatograms of the crude hGS (160-69) obtained from A) PEO-PEPS film and B) PEPS film. The crude hGS-(160-169) was purified and the correct molecular weight was established by fast atom bombardment mass spectrometry [found (calcd), 1285.8 (1285.8)].

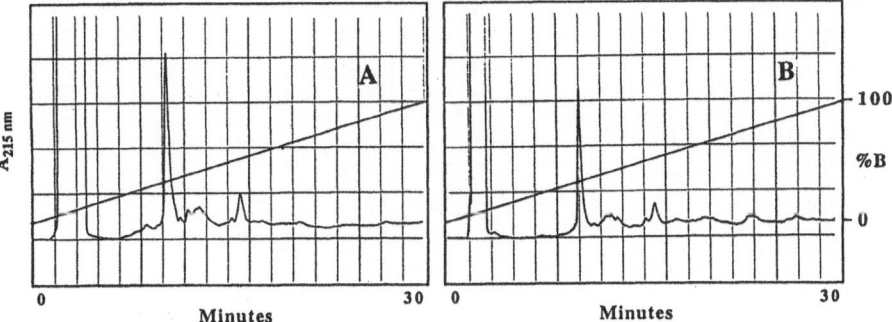

Fig. 1. Analytical HPLC chromatograms of crude hGS (160-169) obtained from (A) the PEO-PEPS film and (B) the PEPS film. Buffer A, 5% $CH_3CN/95\%$ $H_2O/0.0445\%$ TFA; buffer B, 60% $CH_3CN/40\%$ $H_2O/0.0390\%$ TFA; linear gradient, 0-100% of buffer B in 30 min; flow rate, 1.5 ml/min; column, mBONDAPAK^{TM} C_{18} (0.46 x 30 cm).

In conclusion, it has been demonstrated that it is possible to construct a hydrophilic film support, PEO-PEPS film, which is both hygroscopic and at the same time well-suited as a solid support for peptide synthesis in organic solvents.

References

1.   Berg, R.H., Almdal, K., Batsberg Pedersen, W., Holm, A., Tam, J.P. and Merrifield, R.B., J. Am. Chem. Soc., 111 (1989) 8024.
2.   Berg, R.H., Almdal, K., Batsberg Pedersen, W., Holm, A., Tam, J.P. and Merrifield, R.B., In Giralt, E. and Andreu, D. (Eds.), Peptides 1990 (Proceedings of the 21st European Peptide Symposium), Escom, Leiden, The Netherlands, 1991, p.149.

Session XI
Computational Biochemistry

Chairs: Vincent S. Madison
Hoffmann-LaRoche Inc.
Nutley, New Jersey, U.S.A.

and

Isabella L. Karle
Naval Research Laboratory
Washington, District of Columbia, U.S.A.

A critical examination of solvation models for peptides and proteins

M. Hassan[1], J.C. Hempel[1], R.M. Fine[1], S.C. Koerber[1], A. Guaragna[2] and A.T. Hagler[1]

[1]Biosym Technologies, Inc., San Diego, CA 92121, U.S.A.
[2]Farmitalia Carlo Erba, R&D, Computational and Structural Chemistry Group, 20014 Nerviano, Italy

Introduction

The need to include solvation effects in the simulation of biomolecular systems is widely acknowledged, particularly in the case of water and other polar solvents. Water plays a crucial role in determining the structure and conformation of both small and large molecules (peptides, proteins, nucleic acids) and solvation/desolvation energies contribute significantly to binding free energies in ligand/receptor interactions, often determining relative potencies of drug candidates. Solvation effects could be modelled by inclusion of explicit water molecules in the simulations, but this approach is limited by the high computational cost associated with the large number of water molecules needed to simulate the long range interactions acting in these systems and by the complexity of the water-biomolecule interactions, which include electrostatic, hydrophobic and dispersion forces. Because of these limitations, a number of simplified models have been proposed to represent solvation effects when added to classical energy expressions describing *in vacuo* energies. These models include empirical scales obtained by fitting experimental solvation data [1-3], and models that treat the solvent as a continuum media, such as continuum dielectric models based on the Poisson-Boltzmann (PB) equation [4] and models based on a Generalized Born (GB) equation [5, 6]. For solvated peptides and proteins with multiple charged groups, solvation free energies are dominated by the electrostatic component. Continuum dielectric models based on the PB equation have been shown to represent accurately (within the limitations of continuum models) the electrostatic component of solvation for small polar molecules and proteins [7-9]. In this report, we use a PB model to investigate the dependence of electrostatic solvation free energy on conformation, including effects of solvent-accessible surface area, buried charges, charge-charge interactions and molecular shape and we examine other solvation models to determine their ability to reproduce these effects. Finally, we propose a new model based on fitting and parameterization of the results obtained with the PB model. The new model might offer a fast method to couple electrostatic solvation effects to molecular mechanics simulations.

Results and Discussion

Model systems and protocols. The peptide hormone endothelin-1 (ET-1) was selected as model system for these studies. ET-1 has 21 residues, six of which are ionized at neutral pH: N-terminal Cys, Asp 8, Lys 9, Glu 10, Asp 18 and C-terminal Trp, resulting in a net charge of -2. Over 120 different conformations of this molecule, obtained in a previous *in vacuo* conformational search, sample a wide variety of molecular shapes, solvent exposures and relative positions and orientations of the six charged groups. Electrostatic solvation free energies were calculated for each of these conformers using the program Delphi [Biosym Technologies, Inc.]. By selectively turning "on" and "off" the formal charges on these conformers, we were able to investigate the effect on solvation of several conformational parameters, such as solvent-accessible surface area of exposed single charges, distance to the surface for buried charges, distance between charges and overall effect of molecular shape.

Solvation of single charged groups. Examination of electrostatic solvation free energies of individual charged groups (ΔGi) reveals a correlation with their corresponding solvent-accessible surface areas (S), but several observations are noteworthy. First, the relationship between ΔGi and S is not linear, as assumed by empirical scales models, but follows an apparent exponential decay. Second, ΔGi is not a unique function of S; a spread of ± 5 kcal/mol in the correlation between the two variables is typical, suggesting a dependency of ΔGi on other parameters related to molecular shape. Finally, buried charges (with zero solvent accessible surface areas) contribute significantly to ΔGi, with the magnitude of the effect being related to the distance from the charge to the surface. Empirical scales do not take into account the effect of buried atoms. Models based on the GB equation rely on numerical calculation of effective Born radii for each charged atom in the molecule to represent the dependency of ΔGi on solvent exposure and conformation. In our implementation of one GB equation model [6], we found that by adjusting the dielectric offset (an empirical parameter of the model) we could reproduce the results obtained with Delphi to within 10-15% error, but we encounter difficulties in reproducing electrostatic solvation free energies in systems with both exposed and buried charges. In addition, the computational cost of these type of models increases rapidly as the size of the molecule increases.

Effect of charge-charge interactions on solvation. The total electrostatic solvation free energy of a peptide or protein with multiple charged groups is not given by the sum of the electrostatic solvation free energies of the individual charged groups. We found that the relative positions of the charged groups can have a drastic effect on solvation. The magnitude of the electrostatic solvation free energy due to charge-charge interactions (given by the difference between the electrostatic solvation free energy calculated with all the charges present simultaneously and the sum of the electrostatic solvation free energies of the individual charged groups) depends on the magnitude and signs of the charges, the distance among them, their solvent-accessibility and molecular shape. Simple empirical scales models do not have terms for charge-charge interactions and cannot reproduce these effects. Models based on the GB equation include effects of charge-charge interactions on solvation as a pairwise function of effective Born radii and distance between charges.

New model based on parameterization of results obtained with PB model. Functional forms were developed and parameterized to fit the electrostatic solvation free energies obtained with Delphi for the set of ET-1 conformers. The resulting model includes terms to represent solvation of exposed or partially exposed charged groups as a function of solvent-accessible surface area, solvation of buried charges as a function of distance to the surface and effect of charge-charge interactions as a function of distance between charges and solvent-accessible surface areas. The new model reproduces the electrostatic solvation free energies of the ET-1 conformers obtained with Delphi with an average error of less than 5%, although larger errors were observed in cases with close approach between charges. Application of the new model to calculate electrostatic solvation free energies of a set of 27 high-resolution protein structures from the Brookhaven Data Base resulted in similar average errors, less than 5%, indicating that the model might be transferable to systems outside of the training set.

The computational times associated with the application of the new model to peptides and proteins with formal charges are of the same order of magnitude as for empirical scales models and significantly less than the times required by continuum dielectric models based on Poisson-Boltzmann equation or models based on the Generalized Born equation. Further tests are under way to ensure the transferability of the model and to investigate the possibility of coupling it to empirical force field energy expressions representing the energy of the peptide or protein in vacuo, in order to include solvation effects on molecular simulations. The effect on solvation of other electrostatic phenomena such as dielectric and ionic boundary pressures [10], not included in our parameterization of the Poisson-Boltzmann results, will also be investigated.

References

1. Ooi, T., Oobatake, M., Nemethy, G. and Scheraga, H.A., Proc. Natl. Acad. Sci. U.S.A., 84 (1987) 3086.
2. Eisenberg, D. and McLachlan, A.D., Nature, 319 (1986) 199.
3. Wesson, L. and Eisenberg, D., Protein Science, 1 (1992) 227.
4. Sharp, K.A. and Hohig, B., Annu. Rev. Biophys. Chem., 19 (1990) 301.
5. Still, W.C., Tempczyk, A., Hawley, R.C. and Hendrickson, T., J. Am. Chem. Soc., 112 (1990) 6127.
6. Cramer, C.J. and Truhlar, D.G., J. Am. Chem. Soc., 113 (1991) 8305.
7. Jean-Charles, A., Nicholls, A., Sharp, K., Honig, B., Tempczyk, A., Hendrickson, T. and Still, W.C., J. Am. Chem. Soc., 113 (1991) 1454.
8. Gilson, M.K. and Hohig, B., Proteins, 4 (1988) 7.
9. Mohan, V., Davies, M.E., McCammon, J.A. and Pettitt, B.M., J. Phys. Chem., 96 (1992) 6428.
10. Gilson, M.G., Davies, M.E., Luty, B.A. and McCammon, J.A., J. Phys. Chem., 97 (1993) 3591.

Geometry optimization, energetics, and solvation studies on six-membered cyclic peptides, using the programs QUANTA3.3/CHARMm22

F.A. Momany[1], R. Rone[1], H. Kunz[1] and L. Schäfer[2]

[1]Molecular Simulations Inc., 16 New England Executive Park,
Burlington, MA 01803, U.S.A.
[2]Department of Chemistry and Biochemistry, University of Arkansas,
Fayetteville, AR 72701, U.S.A.

Introduction

CHARMm22 parameters, partially derived from *ab initio* studies [1, 2], are used in calculations of two cyclic hexapeptides whose structures are known from x-ray crystallography [3, 4]. In comparing calculated results with x-ray structures, one must consider that the latter are at or near room temperature with full environmental interactions. Energy minimized structures may be considered to be at zero Kelvin. Thus, it is not obvious that by minimizing a cyclic peptide without the crystal environment, one should expect to obtain the exact conformation found in the crystal. However, adding water molecules and carrying out dynamic simulations can provide some modeling information on the stability of these cyclic peptides in solution, and these results in turn can be compared to the crystal data [5-8].

Results and Discussion

All calculations are carried out on a Silicon Graphics Iris 4D/310GTX or IBM RS/6000 using QUANTA3.3 and CHARMm22. The starting conformations of the molecules studied are the x-ray structures [3, 4] with hydrogen atoms added. Atoms are typed and charged according to QUANTA template files. TIP3P waters are placed around the peptides and dynamics is used to allow water reordering. Heating by dynamics (300 K) and subsequent energy minimization is carried out several times to achieve consistent results for solvated molecules. Unit dielectric is used throughout. Peptides studied are cyclic-$(APdF)_2$ and cyclic-$(GPdF)_2$.

The results of the calculations are summarized in Tables 1 and 2. Cyclic-$(APdF)_2$ and cyclic-$(GPdF)_2$ conformational states are searched using peptide flip and dynamics simulations in an attempt to find conformations that are more stable than the x-ray structure. The results of these *in vacuo* searches showed that the Ala containing peptide was most stable close to the x-ray conformation while the glycine containing peptide folds into a lower energy conformation (see Table 2) that is more compact than the x-ray structure and is an asymmetric conformation.

Table 1   *Comparison of calculated and experimental dihedral angles (Phi and Psi) for energy minimized structures of cyclic-(Ala-Pro-DPhe)$_2$*

Residue	Exp.[a]	Vacuo Symm	Solv Symm	Symm Diff	Solv Diff	Vacuo Asym	Solv Asym	Vacuo Diff	Solv Diff
Ala	-157	-156	-160	1	3	-156	-172	1	15
	172	165	163	7	9	168	166	4	6
Pro	-60	-62	-57	2	3	-56	-58	4	2
	123	111	113	12	10	111	118	12	5
DPhe	78	66	64	12	14	61	76	17	2
	9	30	21	21	12	31	9	22	0
Ala	-157	-156	-147	1	10	-153	-156	4	1
	172	165	165	7	7	164	166	8	6
Pro	-60	-62	-66	2	6	-63	-57	3	3
	123	111	114	12	9	109	120	14	3
DPhe	78	66	76	12	2	70	63	8	15
	9	30	19	21	10	26	45	17	36
Avg Diff				9[b]	8[c]			10[d]	8[e]

Energy(kcal/mol)	-112.8	-110.6				-113.5[d]	-107.6		
Binding Energy		-91.3[c]				-91.3	-106.0[e]		

[a]Experimental values from [3]; [b]Comparative values are 45° in [9], and 10° from previous Q/C calculations [8]; [c]The value for the total solvation energy (38 waters, water-water included) is -540 kcal/mol, minimized to a gradient <0.01 kcal/mol/Å. The loss in internal energy of symmetry upon solvation is 2.2 kcal/mol. [d]Asymmetry created by proline up/down ring pucker. [e]The solvation energy is -566 kcal/mol, from 50 ps dynamics, energy minimized as sampled from the last 10 ps. The prolines are symmetrical after dynamics. (N.B. The Phi and Psi angles are listed, respectively, by pairs for each residue.)

Internal geometries calculated in vacuum for both compounds agree well with the x-ray structures. The largest differences between calculated and experimental individual bond lengths are 0.03 Å, with an average value over all bonds of 0.01 Å. Deviations in average bond angle are approximately 1°.

The results described above lead us to conclude that the CHARMm22 force field is effective in modeling peptide structures, energetics, and conformational properties. Solvating these peptides resulted in their internal energy increasing, more in the asymmetric than the symmetric form for both analogs. Further, summing the binding energy and internal energy favored the asymmetric form of these analogs. Thus, one must be aware that these flexible peptides may not be the same in solution as in the crystal environment. These results have ramifications in the design of peptide drugs where the solution structure may be the preferred conformation for rational drug design.

Table 2   *Comparison of calculated and experimental dihedral angles (Phi and Psi) for energy minimized structures of cyclic-(Gly-Pro-DPhe)$_2$*

Residue	Exp.[a]	Vacuo Symm	Solv[b] Symm	Vacuo Diff	Solv Diff	Vacuo Asym[c]	Solv Asym[d]	Vacuo Diff	Solv Diff
Gly	178	177	180	1	2	-65	-68	113	110
	163	-166	-169	31	28	121	-176	42	21
Pro	-64	-51	-63	13	1	-45	-63	19	1
	132	112	115	20	17	118	114	14	18
DPhe	106	123	127	16	21	120	131	14	25
	-14	-28	-31	14	17	-23	56	9	70
Gly	178	177	-174	1	8	-146	90	26	88
	163	-166	166	31	3	-163	179	34	16
Pro	-64	-51	-45	13	19	-58	-56	6	8
	132	112	117	20	15	114	120	18	12
DPhe	106	123	131	17	25	82	131	24	25
	-14	-28	-34	14	20	-74	-113	60	74
Avg Diff				16	16			32	39
Energy(kcal/mol)		-116.1	-113.1			-121.0	-101.3		
Binding Energy			-109.8[e]				-128.1[e]		
RMS dev (all)			0.7[f]						
RMS dev (bb)			0.3[f]						

[a]Experimental values from [4]; [b]Values for solvated energy minimized conformation, after heating to 500 K; [c]Molecule is made asymmetric by different up/down proline ring pucker; [d]After 50 ps of dynamics, a sample structure from the last 10 ps is minimized. Molecule returns to asymmetric form without waters; [e]Interaction energy between 36 waters and the peptide; [f]RMS deviation of (all) heavy atoms included, and RMS deviation of only backbone (bb) heavy atoms. (N.B. The Phi and Psi angles are listed, respectively, by pairs for each residue.)

References

1.   Frey, R.F., Coffin, J., Newton, S.Q., Ramek, M., Cheng, V.K.W., Momany, F.A. and Schafer, L., J. Am. Chem. Soc., 114 (1992) 5369.
2.   Remek, M., Cheng, V.K., Frey, R.F., Newton, S.Q. and Schafer, L., J. Mol. Struct., 235 (1991) 1.
3.   Brown, J.N. and Teller, R.G., J. Am. Chem. Soc., 98 (1976) 7565.
4.   Brown, J.N. and Yang, C.H., J. Am. Chem. Soc., 101 (1979) 445.
5.   Momany, F.A. and Rone, R., J. Comp. Chem., 13 (1992) 888.
6.   Momany, F.A., Klimkowski, V.J. and Schafer, L., J. Comp. Chem., 11 (1990) 654.
7.   Scarsdale, J.N., van Alsenoy, C., Klimkowski, V.J., Schäfer, L. and Momany, F.A., J. Am. Chem. Soc., 105 (1983) 3438.
8.   Momany, F.A., Klimkowski, V.J. and Schäfer, L.L., In Renugopalakrishnan, V., Carey, P.R., Huang, S.G. and Storer, A. (Eds.), Proteins: Structure, Dynamics and Design, Escom, Leiden, The Netherlands, 1991, p.379.
9.   Kitson, D.H., Avbelj, F., Eggleston, D.S. and Hagler, A.J., Ann. N.Y. Acad. Sci., 482 (1986) 145.

Relative free energies of folding and refolding of model secondary structure elements in aqueous solution

A. Tropsha[1], Y. Yan[2], S.E. Schneider[2], L. Li[1], and B.W. Erickson[2]

[1]Laboratory for Molecular Modeling, School of Pharmacy and
[2]Department of Chemistry, University of North Carolina Chapel Hill,
Chapel Hill, NC 27599, U.S.A.

Introduction

The folding of local secondary structure has been a key subject for study of protein folding and for rational protein design. Rational protein design would benefit from the quantitative analysis of sequence-stability relationships of locally folded protein structure. The stability of locally folded conformation may depend on the chirality of composing residues. Experimental analysis of local folding is difficult because short oligopeptides generally do not form stable folded conformations in solution. Molecular simulation allows one to constrain the conformation of a model polypeptide to a desired secondary structure and calculate the effect of the primary sequence on the stability of local folds. Recently, we applied free energy simulations to estimate the sequence dependence and the role of chirality in the stability of type-I and type-I' β-turn [1]. For β-turn models, we studied a chirally representative set of nine dipeptides of the form CH_3CO-L1-L2-$NHCH_3$, where loop residues L1 (i+1) and L2 (i+2) are Gly (G), L-Ala (A), or D-Ala (a). In this paper, we extend our work to include other types of secondary structure, i.e. types II and II' β-turns, right and left-handed helices, right and left-handed β-sheet and extended conformation.

Results and Discussion

The relative free energies of the model dipeptides configured as various typical secondary structure elements were estimated by free-energy simulations (slow growth method). All calculations were performed in explicit water environment using the CEDAR program developed by Prof. J. Hermans at the University of North Carolina. An α-hydrogen atom of a Gly residue at L1 and/or L2 was replaced by a methyl group (thus converting Gly into either L-Ala or D-Ala) and the associated change in free-energy was calculated. The general scheme of molecular replacement is presented in Figure 1. The free energy change along each path was calculated eight times (four forward and four reverse) in order to obtain the mean $\Delta\Delta G°$ and the rmsd of the mean.

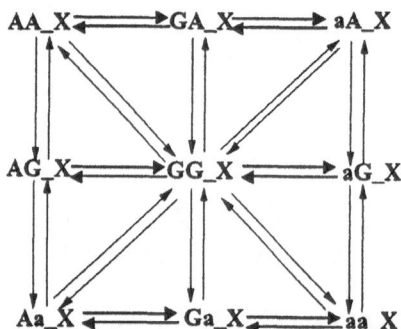

Fig. 1. Replacement paths for interconverting the model dipeptides in various conformations X, where X is right-handed or left-handed helix (H and H', respectively), right-handed or left-handed β-sheet (S and S'), type-I or type-I' (I and I') and type-II or type-II' (II and II') β-turn, and extended (E) conformations.

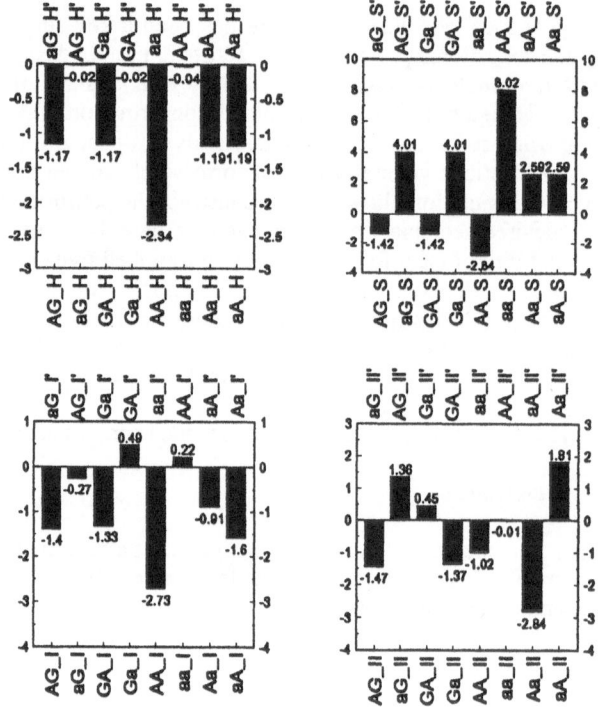

Fig. 2. Calculated relative free energy of folding (vs. GG dipeptide) of model dipeptides from extended conformation into various locally folded conformations. The dipeptides that are mirror images of each other have identical free energies. Each reported value is the difference between ΔG of folding a given dipeptide and GG dipeptide from extended into folded conformation.

Due to molecular symmetry, the free energy of mirror-image backbone conformations of the GG dipeptide, i.e., H and H', S and S', I and I', and II and II' are identical. The same relationship also holds for other pairs of model dipeptides with opposite chirality, i.e. **AA_X/aa_X', GA_X/Ga_X'**, etc. Therefore, the results of free energy calculations for any type of secondary structure X of Figure 1 are directly applicable to X'. Furthermore, the $\Delta G°$ of molecular replacement **AA_X$\Rightarrow$aa_X** has the same meaning as the $\Delta G°$ of refolding **AA_X$\Rightarrow$AA_X'** (because **AA_X'** and **aa_X** are computationally equivalent). Thus molecular replacement calculations of Figure 1 provide estimates of the free energy of refolding of model dipeptides between mirror-image backbone conformations. From these calculations, the following dipeptides were found to have the highest tendency to refold toward a particular conformation X: **AA_I, aa_I', Aa_II, aA_II', AA_H', aa_H', AA_S, and aa_S'**.

The consideration of extended backbone conformation as the model of the unfolded state and the application of the thermodynamic cycle concept [2] afforded calculations of relative free energy of folding of model dipeptides into typical secondary structures (Figure 2). The chirality had a profound effect on the folding propensity of dipeptides, and the strength of the effect was dependent on the local secondary structure. For instance, the DAla-containing dipeptides had either a negligible or slightly negative tendency to form helices and type-I turns, respectively, but strongly disfavored the formation of typical β-sheet.

The molecular replacement calculations of Figure 1 provided the free energy of refolding of model dipeptides between mirror-image conformations. However, from such simulations we cannot estimate the free energy of refolding between symmetry-nonrelated conformations, e.g., helices and β-sheets. In order to address this problem, we devised a novel, peptide-growth simulation approach. By calculating the $\Delta G°$ of changing CH_3CO-$NHCH_3$, which lacks any conformation, into CH_3CO-$GlyGly$-$NHCH_3$ constrained to different main chain conformations, one can estimate the relative free energies of different stable folded conformations of the GG dipeptide. No knowledge or assumption about the nature of the unfolded state is needed. These peptide-growth simulations yielded the following relative free energy values (in kcal/mol): **GG_H**, 0; **GG_S**, 0.4; **GG_II**, 1.2; **GG_I**, 1.7. By combining these data with the results of molecular replacement calculations of Figure 1, the relative free energy of refolding between all possible folded conformations of model dipeptides can easily be obtained.

References

1.   Yan, Y., Tropsha, A., Hermans, J. and Erickson, B.W., Proc. Natl. Acad. Sci U.S.A., 90 (1993) 7992.
2.   Tembe, T.L. and McCammon, J.A., J. Comput. Chem., 8 (1984) 281.

The use of molecular modeling to design highly selective inhibitors of tissue kallikrein

Z. Dong and J. Burton

The Unit for Rational Drug Design, Boston University Medical Center, 88 East Newton Street, Boston, MA 02118, U.S.A.

Introduction

Tissue kallikrein (EC 3.4.21.35) is a serine protease that cleaves low molecular weight kininogen to yield kinins such as kallidin (lysyl-bradykinin). Kinins may be involved in many physiologic functions including the regulation of blood flow in some organs, vascular permeability and pain. Overproduction of kinins is implicated in rheumatoid arthritis, diabetic glomerulonephritis, algesia, and rhinitis. There are many other biologically important serine proteases in the body which are also closely related to trypsin and inhibitors which specifically block tissue kallikrein are not available. Research to clarify the *in vivo* function of tissue kallikrein is hindered by a lack of a selective inhibitor. In addition, use of nonspecific human tissue kallikrein inhibitors as therapeutic agents may be hampered by side-effects caused by the lack of specificity. Specificity has thus become a key issue in the development of tissue kallikrein inhibitors.

An effort has been made to design a highly specific tissue kallikrein inhibitor by modifying the P_1 residue (arginine) in the lead substrate analog inhibitor Pro-Phe-Arg-Ser-Val-Gln-NH_2 (KIBR-11) [1]. Molecular modeling studies leading to a highly specific tissue kallikrein inhibitor (KIZD-06) are reported here.

Results and Discussion

The X-ray crystal structures of porcine pancreatic kallikrein (ßPPK) and bovine trypsin clearly show that the former has a restricted opening to the S_1 subsite [2]. This is mainly because tissue kallikreins, including human urinary kallikrein (HUK) and ßPPK, have an extra Pro[219] in the segment forming the entrance to S_1 when compared with other members of the trypsin-like serine proteases (Table 1). Another unique aspect of the tissue kallikreins is that residue-226 in the S_1 pocket is serine. All other serine proteases whose sequences are known have a glycyl residue at this position. Molecular modeling (Hyperchem) shows that incorporation of L-4-aminophenylalanine (Aph) into KIBR-11 at P_1 should yield specific inhibitors for tissue kallikrein. The bulky benzene ring of Aph can fit through the roomy entrance to the S_1 pocket of tissue kallikrein to form an H-bond with Ser[226]. The Aph residue should not fit through the constricted entry to other members of the trypsin family which also lack an oxygen in the side chain at position-226. In addition, there is a possible hydrophobic interaction between the aromatic ring of Aph and residue-192 in tissue kallikreins. In all other serine proteases checked this residue is more polar and this interaction would not be favored (Table 1). Synthesis of the sequence Pro-Phe-Aph-Ser-Val-Gln-NH$_2$ verified these predictions. KIZD-06 has an improved Ki for human tissue kallikrein (21 vs. 50 μM) and does not block other members of the trypsin family including human plasma kallikrein, human thrombin, human plasmin, human urokinase, or human trypsin (K$_i$ > 5 mM). In addition the inhibitor does not block the clotting cascade. KIZD-06 is the most specific tissue kallikrein inhibitor ever reported.

Table 1 *Comparison of the amino acid sequences in the S$_1$ subsites of various proteases*

Residue →	192	215	216	217	218	219	220	226
βPPK	Met	Trp	Gly	His	Thr	Pro	Cys	Ser
HUK	Val	Trp	Gly	Tyr	Val	Pro	Cys	Ser
HPK	Lys	Trp	Gly	Glu	Gly	....	Cys	Gly
hTrypsin	Gln	Trp	Gly	Asp	Gly	....	Cys	Gly
Throm	Glu	Trp	Gly	Glu	Gly	....	Cys	Gly
Plas	Gln	Trp	Gly	Leu	Gly	....	Cys	Gly
Uro	Gln	Trp	Gly	Arg	Gly	....	Cys	Gly

βPPK, porcine tissue kallikrein; HUK, human tissue kallikrein; HPK, human plasma kallikrein; hTrypsin, human trypsin; Throm, human thrombin; Plas, human plasmin; Uro, human urokinase.

References

1.   Deshpande, M.S. and Burton, J., J. Med. Chem., 35 (1992) 3094.
2.   Chen, Z. and Bode, W., J. Mol. Biol., 164 (1983) 283.

Solution structure of RP71955, a new 21-amino acid tricyclic peptide active against HIV-1 virus

D. Frechet, J.D. Guitton, F. Herman, D. Faucher,
G. Helynck, B. Monegier du Sorbier, J.P. Ridoux,
E. James-Surcouf and M. Vuilhorgne

*Rhône-Poulenc Rorer S.A., Central Research,
13 Quai Jules Guesde, 94403 Vitry-sur-Seine, France*

Introduction

The search for new and effective agents in the treatment of HIV infection and AIDS related complex is a major priority in the pharmaceutical industry. Upon screening for new anti-HIV agents, a secondary metabolite from a strain of actinomycetes, RP71955, has been found to inhibit HIV replication in cell culture.

In this paper, we will focus on the determination of the structure of RP71955. This product has been found to be an original 21 amino acid tricyclic peptide. Its primary structure was determined using an ensemble of spectroscopic techniques, combined with sequencing of fragments obtained by limited chemical hydrolysis. A detailed conformational analysis of RP71955 was performed from NMR derived constraints using distance geometry, restrained molecular dynamics, NOE back calculation and an iterative refinement using a full relaxation matrix treatment.

Results and Discussion

RP71955 could not be sequenced classically using Edman degradation, suggesting that there was no free terminal NH_2. Therefore, the determination of its primary structure had to rely on the results of an ensemble of spectroscopic techniques: FAB/MS, NMR and Raman spectroscopy, supported by amino acid analysis. The sequence of RP71955 was determined by NMR spectroscopy taking into account the fact that for all secondary structure motifs, sequential $H\alpha_i$ - HN_{i+1} NOEs were observed [1]. Based on this information, the following sequence was proposed:

```
    1              10               20
    X L G I G S X N X F A G X G Y A V V V X F W
```

Raman spectroscopy showed the presence of two disulfide bridges suggesting that four of the unknown residues were cysteines. From the results of FAB/MS as well as amino acid analysis, it was shown that the fifth unknown amino acid was D and that the peptide had an internal amide bond. Furthermore, a strong NOE between HN of X1 and the β protons of X9 suggested that X9 was D and that the internal amide bond, was between HN of X1 and the γCO of D9. Therefore, residues 1, 7, 13 and 19 were C. NOEs between Hα of C1 and the 2 Hβ of C13, and between the Hβ of C7 and the Hβ of C19 showed that the disulfide bridges were linking C1 to C13 and

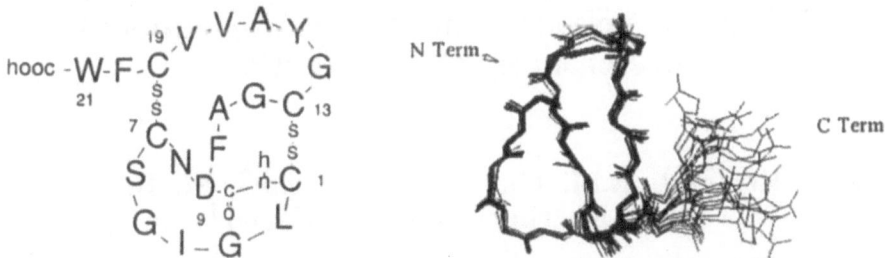

Fig. 1. *RP71955 primary structure.*   Fig. 2. *20 structures of RP71955 are superimposed*

C7 to C19. The sequence was confirmed by sequencing peptide fragments obtained by limited chemical hydrolysis. The primary structure is presented in Figure 1.

About 120 distance constraints were determined using MARDIGRAS [2] and were used for generating 3D structures using DIANA [3]. The best structures were submitted to energy minimization followed by restrained Molecular Dynamics (Discover software from Biosym Technologies). The structures showing the best fit with the experimental data were used for backcalculation of the NOESY spectrum and comparison with the experimental data. The refinement procedure was repeated until the agreement between experimental and calculated data was satisfactory. The final set of distance constraints consisted of 145 distances between isolated proton pairs, 66 distances extracted from the interpretation of overlapping proton pair and 1771 lower distances limits between proton pairs presenting no correlations. A 50 cycles high temperature restrained MD (2 ps at 900 K, 2 ps at 300 K, minimization) was performed on one of the best structure in order to get further insight into the mobility of the structure. The 50 structures obtained were superimposed. The average pairwise RMSD calculated for the main chain of residues 1 to 18 is 0.6 Angstrom. The R factor calculated using CORMA [4] is less than 0.4 Angstrom. Bond angles and solvent accessibility are compatible with experimental results. The structures having the best fit with experimental data were checked for structural quality using Procheck (Oxford Molecular). The structures selected are presented superimposed in Figure 2.

RP 71955 has been found to inhibit HIV-1 induced syncitia formation but not the interaction of GP120 with CD4. Although the mode of action of RP71955 is not elucidated, it is interesting to point out the fact that RP71955 show a 33% sequence identity with residues 608-628 of GP41 [5]. This region of GP41 has been shown to be of critical importance in the assembly of *env* glycoproteins [6]. The activity of RP71955 could be due to an inhibition of GP41 association which seems to be essential for GP41 induced fusion.

References

1. Wüthrich, K., Billeter, M. and Braun, W., J. Mol. Biol., 180 (1984) 715.
2. Borgias, B.A. and James, T.L., J. Magn. Reson., 87 (1990) 475.
3. Güntert, P. and Wüthrich, K., J. Biomol. NMR, 1 (1991) 447.
4. Borgias, B.A., Gochin, M., Kerwood, D.J. and James, T.L., Prog. in Nucl. Mag. Res. Spect., 22 (1990) 83.
5. Andeweg, A.C., Groening, M., Leeflang, P., de Goede, R.E.Y., Osterhaus, A.D.M.E., Tersmette, M. and Bosch, M.L., Aids Res. Human Retrovirus, 8 (1992) 1803.
6. Earl, P.L., Doms, R.W. and Moss, B., Proc. Natl. Acad. Sci., 87 (1990) 648.

Effects of solvation on the conformational preferences of endothelin-1

J.C. Hempel[1], R.M. Fine[1], A. Guaragna[2], M. Hassan[1], S.C. Koerber[1] and A.T. Hagler[1]

[1]Biosym Technologies Inc., San Diego, CA 92121, U.S.A.
[2]Farmitalia Carlo Erba, R&D, Computational and Structural Chemistry Group, 20014 Nerviano, Italy

Introduction

Our goal in modeling endothelin-1 (ET-1) is to investigate the effect of environment on conformational preference for this twenty-one residue peptide. Because ET-1 is a highly charged peptide (six ionizable groups at neutral pH), we investigate the influence of the electrostatic contribution to the solvation energy on conformational preference in water by using a continuum electrostatics model to screen conformers of ET-1 previously generated by *in vacuo* conformational search. In further studies, we are investigating the influence on conformational preference of interactions of ET-1 with individual water molecules in molecular dynamics simulations with explicit waters.

Results and Discussion

An extensive search of conformational space was carried out for ET-1 modeled as a neutral unionized species by the combination of two different techniques *in vacuo*, high temperature molecular dynamics simulations and a Monte Carlo/minimization procedure in torsion angle space. Low energy conformers for ET-1 modeled in this way are compact structures stabilized by interaction of the ring region of ET-1 (defined by disulfide bridges between Cys^1-Cys^{15} and Cys^3-Cys^{11}) with the tail (residues 16-21). The electrostatic contribution to the solvation free energy in aqueous solution was calculated for 140 diverse conformers using the program *DelPhi* (Biosym Technologies, Inc.). In this approach the electrostatic field in and around the molecule is calculated using a finite difference solution to the Poisson-Boltzmann equation [1, 2]. ET-1 (the solute) is treated at an atom level while the solvent is represented by a continuum dielectric medium. In order to calculate the electrostatic polarization energy using *DelPhi*, the conformer is mapped onto a three dimensional grid by assigning values of charge and dielectric constant to grid points. The internal dielectric constant of the molecule is defined as $\varepsilon=2$ in order to take into account the effects of electronic polarizability in the solute. Grid points outside the molecular volume are assigned the dielectric constant of the environment.

The electrostatic contribution to the solvation energy for a given ET-1 conformer is calculated as the difference in reaction field energy for the conformer in water ($\varepsilon = 80$) and vacuum ($\varepsilon = 1$). The total energy is the sum of the internal electrostatic energy for that conformer and the electrostatic polarization energy. The internal coulombic energy was obtained for the 140 conformers with full charges on Asp^8, Lys^9, Glu^{10}, Asp^{18}, the N- and C-terminus, and with partial charges otherwise. *DelPhi* is used to calculate the electrostatic polarization energy using these same charges. The internal electrostatic energy and the electrostatic polarization energy have compensating effects. While the internal coulombic energy favors the close approach of charges of like sign, the electrostatic polarization energy favors the close approach of charges of opposite sign.

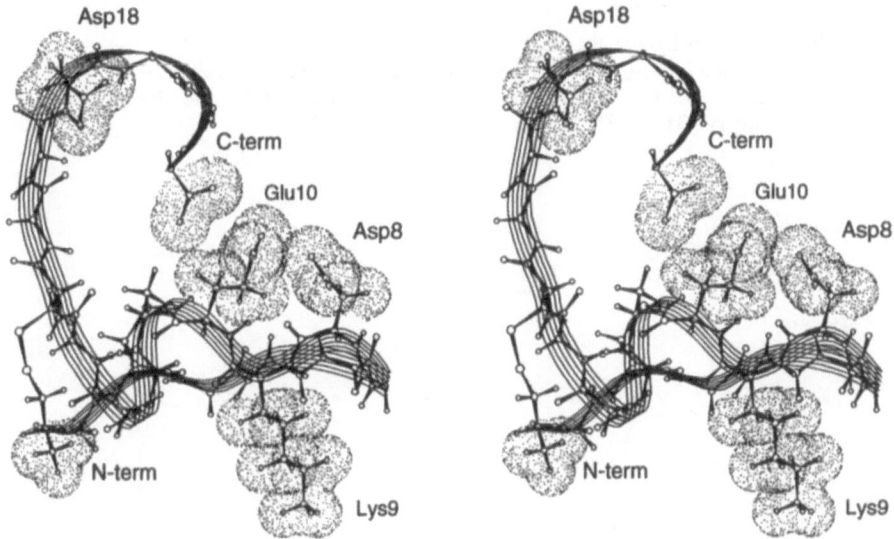

Fig. 1. *ET-1 conformer in this study most favored by the electrostatic contribution to solvation energy in aqueous solution. Note the helical backbone motif and orientation of Asp^8, Lys^9 and Glu^{10} which, along with the placement of the tail, enhances solvation by placing sidechains of like charge "adjacent" in space.*

It is interesting to note that the conformer in this study that is most favored by the electrostatic contribution to the solvation energy in water (Figure 1) is characterized by a helical backbone motif in the region of residues 9-13. Negatively charged sidechains of Asp^8 and Glu^{10} are in close proximity with the positively charged Lys^9 side chain directed away from this region. The tail of ET-1 is folded to place Asp^{18} and the C-terminus well away from Lys^9 and the N-terminus. Solvation is enhanced in this conformation of ET-1 by sidechains of like charge that are "adjacent" in space. Comparison with an NMR model for which coordinates have been reported [3] reveals that the helical backbone conformation of residues 10-13 in Figure 1 is shared with the NMR model. This helical motif has been reported in NMR studies in a variety of solvents by many authors, however, preferred orientations of the tail of ET-1 can be presumed to vary with solvent.

We observe that the electrostatic contribution to the solvation energy substantially alters conformational preferences for the highly charged peptide ET-1 in water from that predicted *in vacuo*. The *in vacuo* conformational search for ET-1 modeled as a neutral unionized species provides a wide range of diverse conformers and the electrostatic contribution to the solvation energy calculated for ET-1 as an ionized species dramatically favors some conformers over others to reveal conformational preferences in water. In this study, the conformer most favored by the electrostatic contribution to the solvation energy, Figure 1, is more than 30 kcal/mol higher in energy as a neutral unionized species than the lowest energy conformer generated for ET-1 as the unionized species.

References

1. Gilson, M. and Honig, B., Nature, 330 (1987) 84.
2. Sharp, K. and Honig, B., Ann. Rev. Biophys. Biochem., 19 (1990) 301.
3. Andersen, N.H., Chen, C., Marschner, T.M., Krystek, S.R. and Bassolino, D.A., Biochemistry, 31 (1992) 1280.

Identification of unified coordinate frameworks for antiparallel and parallel helix to helix packing in peptides and proteins

A.L. Lomize

M.M. Shemyakin Institute of Bioorganic Chemistry,
Russian Academy of Sciences, Moscow, Russia

Introduction

Nearly antiparallel or parallel orientation of adjacent α-helices is the common feature of α-bundle, membrane and α-fibrous proteins. There are strong constraints on the mutual arrangement of the helices as result of close side chain packing like "knobs into holes" [1] or "ridges and grooves" [2] at the helix to helix interfaces. This suggests that some average sets of α-carbon coordinates for the packing classes may be calculated and applied for energy docking of α-helices and design of new α-helical peptides and proteins. These unified coordinate frameworks were identified here by superposition of helix pairs taken from the protein spatial structures represented in the Protein Data Bank.

Results and Discussion

Two groups of helix pairs with left-handed antiparallel and right-handed parallel packing (65 and 49 pairs respectively) were represented by well-separated peaks with maxima at -165° and +20° in the overall distribution of interhelix angles ω for 411 helix pairs from 71 selected protein structures. All long helix pairs from membrane (photoreaction center and bacteriorhodopsin) and 4-α-bundle proteins fall into these groups. The search for the common coordinate frameworks consists of calculation of root mean square differences of α-carbon positions (D) between all helix pairs within each group and choice of the best reference helix pairs for the overall superposition. The differences D were calculated for the overlapped portions of helices (Figure 1) which have lengths of at least 5 residues per helix.

Helix pairs of the antiparallel packing class with different interhelix distances d (d varies from 6.5 to 12.0 Å) can be superimposed well as a result of their longitudinal shift relative to each other (Figure 1). Two helix pairs (19-37, 41-64 and 19-37, 93-108 from the myohemerythrin spatial structure) with different interhelix distances (9.0 and 10.9 Å respectively) were chosen as the best references. 60 helix pairs from the total of 65 have deviations $D \leq 1.2$ Å from at least one of these helix pairs (see Figure 2). A unified framework may be calculated by superposition and averaging of these two reference pairs (as shown in Figure 1) and all other pairs of the same packing class. This framework looks like a pair of long slightly coiled α-helices with variable interhelix distance. The relative position of any two individual antiparallel helices may be defined approximately (with a precision ≤1.2 Å in terms of *D*) by two integer values, which are shifts of the helices within the framework, instead of usual six degrees of freedom.

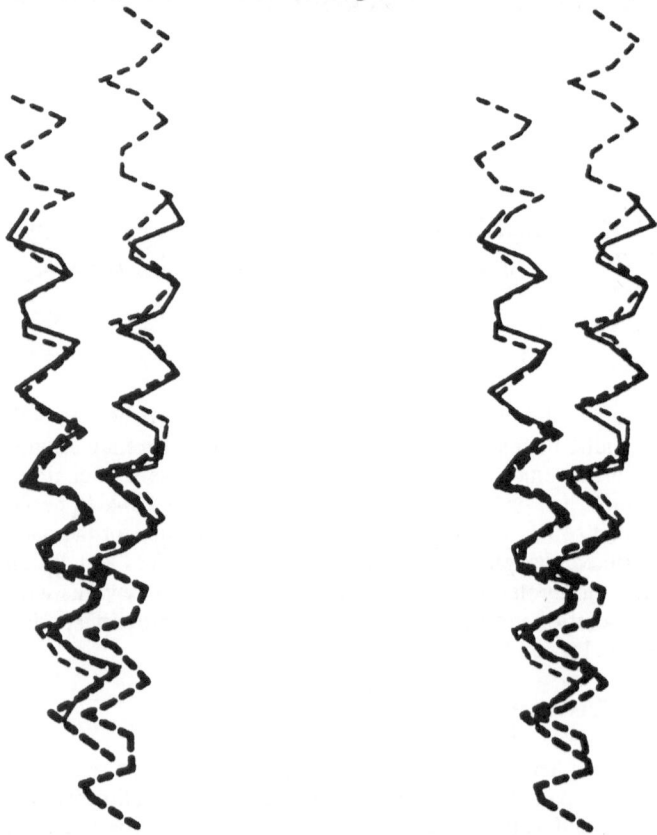

Fig. 1. Superposition of antiparallel helix pairs with different interhelix distances d: helices 171-198 and 226-250 of photoreaction center L subunit (dashed line, d=7.2 Å), 19-37 and 41-64 of myohemerythrin (solid line, d=9.0 Å) and 19-37 and 93-108 of myohemerythrin (bold dashed line, d=10.9 Å).

In a similar manner, the parallel packing class may be represented by two sets of overlapped helix pairs with medium and wide separation of α-helices and the difference $D \leq 1.2$ Å from the correspondent reference pairs: helices 143-166, 261-284 of photoreaction center M subunit ($d=8.5$ Å) and two 74-128 helices of different hemagglutinin subunits of ($d=11.2$ Å). However, a single average coordinate set probably does not exist in that case.

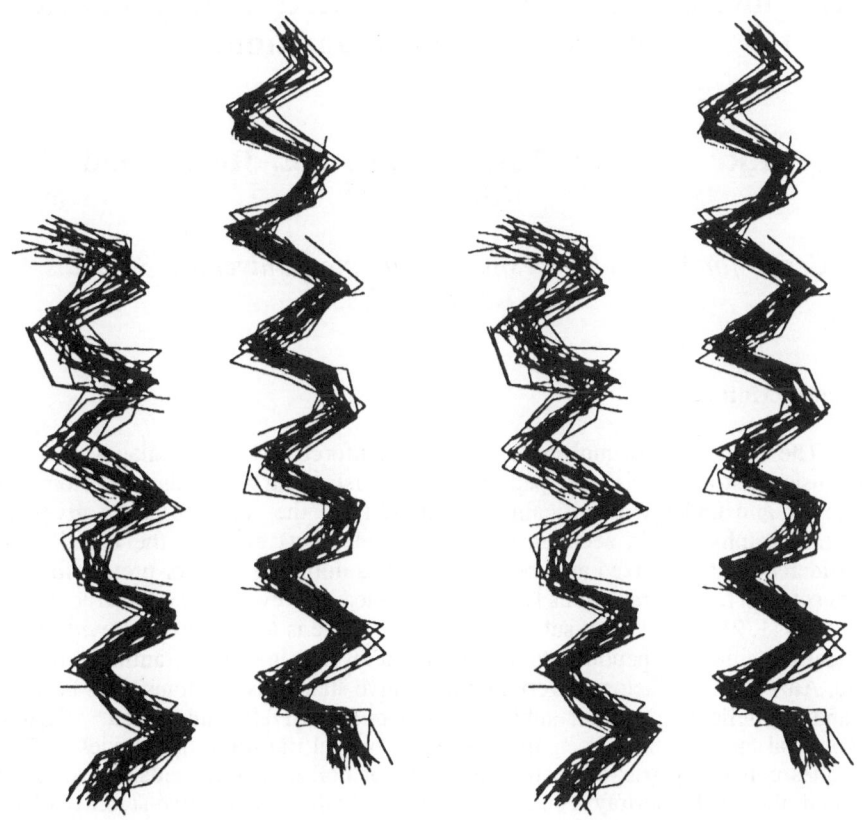

Fig. 2. Superposition of 48 antiparallel helix pairs from 26 protein structures (PDB files, 1PRC, 1BRD, 2MHR, 2CCY, 156B, 2TMV, 2CTS, 2YHX, 2CYP, 3TLN, 1UTG, 3LZM, 2GLS, 6API, 2CPP, 2TAA, 1PAZ, 1PHH, 1BP2, 4PFK, 1GOX, 8ADH, 1INS, 1ECD, 1LH1 and 1MBD) with the reference helix pair 19-37, 41-64 of myohemerythrin. Only areas of overlap of each helix pair with the reference one are shown.

References

1.   Crick, F.H.C., Acta Crystallogr., 6 (1953) 689.
2.   Chothia, C., Levitt, M. and Richardson, D., J. Mol. Biol., 145 (1981) 215.

The molten helix: Low activation energy barrier for helix-helix transitions

G.R. Marshall, M.L. Smythe, S.E. Huston and R.D. Bindal

Center for Molecular Design, Washington University, St. Louis, MO 63130, U.S.A.

Introduction

The static view of molecular structure reinforced by visual images of three-dimensional structures is an impediment to understanding mechanism. Proteins are dynamic and undergo significant excursions from the static structure given by crystallography, while secondary structural elements such as the α-helix are considered relatively fixed and, perhaps, serve as initiation sites for protein folding. Two α to 3_{10}-helical transitions have been reported, however, in crystal structures of enzymes [1, 2] as a result of substrate binding. There is also significant evidence of helical transitions in peptides containing α-methylalanine (MeA, aminoisobutyric acid, Aib) [3]. The factors governing the relative stabilities and transitions between α and 3_{10}-helices are rather subtle and not quantitatively understood. Various theoretical methods have been applied to this helical transition in peptides of 7-12 residues containing α,α-disubstituted (MeA) or normal amino acids. In all cases studied, the results portray a consistent view that while the relative stability of the two helices is influenced by length, composition and environment, there is little or no barrier (<2 kcal/mol) for the transition between the α- and 3_{10}-helix [4, 5].

Results and Discussion

The helix-helix transition was characterized by several approaches to eliminate any methodological bias. The potential of mean force as a function of reaction coordinate was determined initially using the standard simulation procedure of umbrella sampling involving decomposition of the reaction coordinate between the α- and 3_{10}-conformations into overlapping segments, or windows, with simulation of the molecular motions in each segment. The molecular configurations were biased to remain in each window, centered on a value of the reaction coordinate (helix length, or Ψ torsion values), by the use of a harmonic restraining potential. Similar results have been obtained with unconstrained simulations. Table 1 summarizes results from simulations conducted in vacuo and in various solvents (Figure 1) using different force fields. There is a consistently low barrier; helical transitions can be frequently observed if the two helical forms have comparable free energies. A variety of reaction coordinates including distance and torsional restraints were explored to drive the helical transition based on crystal data and molecular mechanics calculations. The values obtained are upper bounds on the actual activation free energy as other transition paths of lower free energy may exist.

Table 1   *Summary of the potential of mean force calculations. A_{310}, A_α, A_{TS} are the Helmholtz free energies of the 3_{10}, α and transition states*

Peptide	Conditions	$A_{310}-A_\alpha$	$A_{TS}-A_{310}$
Ac-MeA$_8$-NMe	Amber united-atom	0.6	<0.2
"	Amber all-atom	-2.8	<0.2
Ac-MeA$_9$-NMe	Amber/OPLS	1.1	0.8
Ac-MeA$_{10}$-NMe	Amber united-atom	2.6	<0.2
"	Amber all-atom	-1.5	<0.2
"	Amber united-atom unrestrained	2.5	<0.2
"	Amber/OPLS	3.2	0.2
"	Amber/OPLS CH$_3$CN	7.2	<0.2
"	Amber/OPLS H$_2$O	7.6	<0.2
NF3,5 [6]	Amber all-atom	1.6	<0.2
NF4,5 [6]	Amber all-atom	1.8	<0.2

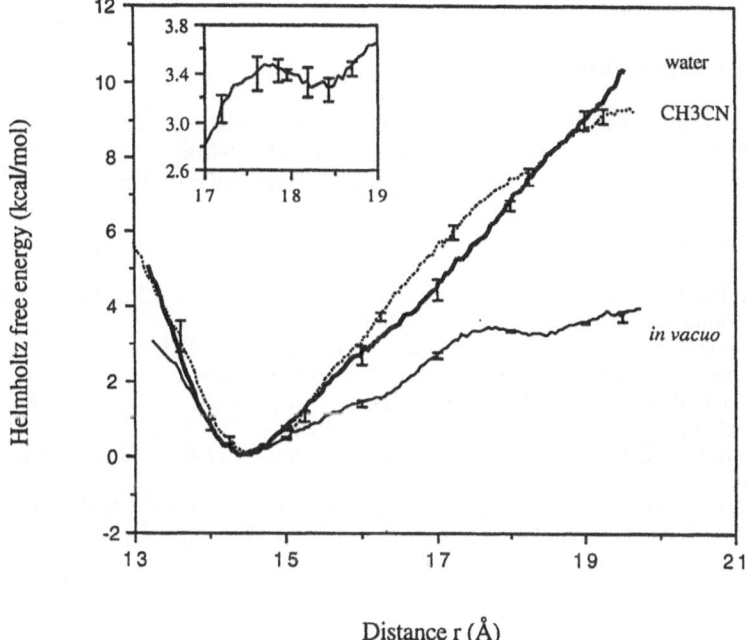

Fig. 1.   *Computed potentials of mean force for the α to 3_{10}-helical transition of Ac-MeA$_{10}$-NHMe in various solvents using the Amber/OPLS force field. The inserted figure is an expansion of the 3_{10}-helical region in vacuo.*

The reaction pathway of Ac-MeA$_8$-NMe was also examined by constrained geometry-optimization with semi-empirical quantum mechanics (AM1). Three reaction paths for the estimation of barrier height for the interconversion of α- to 3_{10}-helices were considered. The simulation of the reaction path with a three-step, sequential use of a distance constraint between C$_\alpha$-atoms of the MeA octamer showed an energy barrier of 2.0 kcal/mol, with the α-helix more stable than the 3_{10}-helix by 0.5 kcal/mol. Single-point aqueous solvation calculations (AMSOL/SM2) further increased the stability of α-helix over 3_{10}-helix by 5 kcal/mol, similar to the effect seen with explicit solvent and MD simulations for the MeA decamer (Figure 1). Alternatively, reaction path calculations by simultaneously driving all eight, or only four, psi-dihedral angles revealed transition barriers of 1.8 and 1.3 kcal/mol, respectively.

The various procedures used to ascertain the height of the barrier for the transition between the α- and 3_{10}-helical conformations yield a consistent result, a very low to non-existent intrinsic barrier for the helical transition ("molten helix"), comparable to the torsional barrier calculated and observed for the rotation of butane. The "molten helix", in which the relative orientations of sidechains and the length between helical residues vary as each helix explores its environment through α- to 3_{10}-helical transitions, may well be a dominant feature of the "molten globule" phase of protein folding when sidechain packing densities are reduced. These results suggest that environmental changes as well as specific sidechain interactions can modulate the relative stability of the two helical forms in proteins and couple chemical to mechanical transformation (helical length of 3_{10}-helix is 1/3 longer than α-helix) at the molecular level.

Acknowledgements

This work was supported by grants from the National Institutes of Health (GM24483) and the Office of Naval Research (N00014-90-J-1393). SEH acknowledges support from NIH Cardiovascular Training Grant No. 5T32Hl07275.

References

1. McPhalen, C.A., Vincent, M.G., Picot, D., Jansonius, J.N., Lesk, A.M. and Chothia, C., J. Mol. Biol., 327 (1992) 197.
2. Gerstein, M. and Chothia, C., J. Mol. Biol., 220 (1991) 133.
3. Karle, I.L. and Balaram, P., Biochemistry, 39 (1990) 6747.
4. Marshall, G.R., Hodgkin, E.E., Langs, D.A., Smith, G.D., Zabrocki, J. and Leplawy, M.T., Proc. Natl. Acad. Sci. U.S.A., 87 (1990) 487.
5. Hodgkin, E.E., Clark, J.D., Miller, K.R. and Marshall, G.R., Biopolymers, 30 (1990) 533.
6. NF3,5 = Ac-MeA$_2$-Nal-MeA-Phe-MeA$_3$-NHCH$_3$ and NF4,5 = Ac-MeA$_3$-Nal-Phe-MeA$_3$-NHCH$_3$, Basu, G., Bagchi, K. and Kuki, A., Biopolymers, 31 (1991) 1763.

The local states method for simulating the free energy of peptides and proteins

H. Meirovitch

Supercomputer Computations Research Institute, Florida State University, Tallahassee, FL 32306, U.S.A.

Introduction

Calculation of the entropy, S and thus the Helmholtz free energy, F by computer simulation has been an important theoretical goal in biophysics for many years [1]. This stems from the fact that F constitutes the criterion of stability, which is useful, for example, in determining the relative binding of different ligands to an enzyme. However, calculation of both, the absolute and relative free energies with computer simulation is difficult. The main problem is that the commonly used methods, the Metropolis Monte Carlo [2] and molecular dynamics (MD) [3, 4] are *dynamical* procedures (i.e., one starts from a conformation that evolves in time) and they do not provide therefore the *value* of the sampling probability, P, of a conformation, which leads to the entropy, S ($S \sim \ln P$). Hence it is important to develop efficient techniques for estimating P. The local states (LS) method, has been proposed by Meirovitch [5] to handle Ising and lattice gas models and has later been extended to polymers and proteins. With this method the value of the sampling probability P can be obtained approximately from the frequencies of occurrence of the so-called local states. The LS method, like procedures that are based on the harmonic or the quasi-harmonic approximations, provides approximately the *absolute* entropy and therefore enables one to obtain differences in the entropy of conformational states with a *significant* structural variance. However, the method, in contrast to these procedures, is not limited to handle only stable states (e.g., an α–helix) but is also applicable to states with *any* chain flexibility, such as the random coil.

Results and Discussion

For simplicity, suppose that the LS method is applied to a linear peptide without side chains (e.g., polyglycine). The molecule is first simulated by MD (or by any Monte Carlo procedure) at a given temperature, T, where every, say 10 femtoseconds the Cartesian coordinates of the instantaneous conformation are stored. Then, each of the stored conformations is expressed in terms of internal coordinates, i.e., the K dihedral angles ϕ, ψ, and ω and the bond angles θ. These angles are ordered according to their position along the backbone and they are denoted by α_k, $1 \leq k \leq K$ (the bond lengths are assumed to be constant). In the next stage, one calculates the ranges $\Delta\alpha_k$, within which the values of α_k lie, $\Delta\alpha_k = \alpha_k(\max) - \alpha_k(\min)$, where $\alpha_k(\max)$ and $\alpha_k(\min)$ are the maximal and minimal values of α_k found in the sample. Then, each $\Delta\alpha_k$ is divided into l equal segments of sizes $\Delta\alpha_{k,l}$

$= \Delta\alpha_k / l$, where l is the discretization parameter; we denote these segments by v_k ($v_k = 1, l$), $1 \leq k \leq K$. Thus, each angle is now defined by the segment to which it belongs and a conformation i is expressed by the particular set of segments ($v_{1(i)}$, ..., $v_{K(i)}$). A local state of α_k is the set of values (v_k, v_{k-1},...v_{k-b}) of the b segments preceding v_k. One can define a transition probability for v_k which depends on the local state and which can be calculated from the frequency of occurrence of certain local states in the sample. Thus, the probability of a conformation is the product of the transition probabilities of its K angles. The lowest approximation for the entropy S is based on $b = 0$ (which means that correlations between successive angles are ignored) and on the assumption that the distribution of angles within the ranges $\Delta\alpha_k$ is homogeneous i.e., the discretization parameter, l, is 1. The transition probability densities are $\rho(a_k, b = 0, l = 1) = 1 / \Delta\alpha_k$ and the approximate entropy is,

$$\bar{S}^A (b=0, l=1) = -k_B \sum_{k=1}^{K'} ln \frac{1}{\Delta\alpha_k} \qquad (1)$$

where k_B is the Boltzmann constant and the bar above S means estimation. Better approximations for $b = 0$ are obtained by increasing l; then, correlations between angles can be taken into account, where the largest value of b used thus far is 3. The approximate probabilities enable one to define upper and lower bounds for the free energy $F^B(b,l)$ and $F^A(b,l)$, respectively; however, F^A can be estimated from significantly smaller samples than F^B, and therefore it is more useful. In fact, recent studies have shown that the correct difference of the free energy between two states can be approximated reliably by the differences, ΔF^A.

Results and Discussion

In an initial study [6] the LS method was applied to a continuum model of decaglycine in vacuo, described by rigid geometry (i.e., constant bond lengths and angles) using the Cornell package ECEPP. Two different Monte Carlo simulations at 100K were carried out, at the α-helical and hairpin regions. It was found that the α-helix is more stable than the hairpin, i.e., its free energy is lower by ΔF=4.0(7) kcal/mole and that the contribution of the entropy to this difference is significant, about 1/3 of that of the energy. Later it has been extended to peptides with side chains simulated by MD (i.e., described by flexible geometry [7]). Thus, a detailed model of the cyclic peptide cyclo-(Ala-Pro-D-Phe)$_2$, based on a full valence force field, was simulated in vacuo and in the crystal and the difference in the chain entropy, $T\Delta S=T[S(\text{vacuo}) - S(\text{crystal})]$ in these two environments was calculated. It was found that $T\Delta S$ =9.4(8) kcal/mol, in a good agreement with calculations based on the harmonic approximation which on the average yielded, $T\Delta S$ =12.3 kcal/mole. Very recently the method has been further developed [8] and applied to a relatively large molecule (188 atoms), the potent "Asp^4-Dpr^{10}" antagonist [cyclo(4/10)-(Ac-D^3-Pro^1-D-pFPhe^2-D-Trp^3-Asp^4-Tyr^5-D-Nal^6-Leu^7-Arg^8-Pro^9-Dpr^{10}-NH$_2$)] of gonadotropin releasing hormone (GnRH). The molecule was simulated in vacuo at T=300K in two conformational states, previously investigated [9], which differ by the orientation of the N-terminal tail, above (tail up, TU) and below (tail down, TD) the cyclic heptapeptide ring. It was found that the values of ΔF^A for various approximations (b,l) have been converged to, $\Delta F^A - F^A(\text{TD})\ F^A(\text{TU}) = 5.6(1)$ kcal/mole, i.e., TD is more stable than TU (The corresponding difference in entropy, $T\Delta S^A = 1.3(2)$ kcal/mole, is equal to the value obtained by the harmonic approximation). Also, as in the previous LS studies the lowest approximation, which has the minimal number of local states$(b=0, l=1)$ also leads to the correct value of ΔF. This is important since the lowest approximation can be applied even to large proteins. The LS method enables one to define the entropy of a part of the molecule and thus to measure the flexibility of this part. Thus, the value for $T[S^A(\text{TU}) - S^A(\text{TD})]$ of the tail alone converged to 2.4(1) kcal/mole, which demonstrates the relatively high flexibility of the tail in the TU state. In order to study the random coil the $\text{Asp}^4\text{-Dpr}^{10}$ analogue and its linear version were simulated by MD at 1000K. In this case a lower bound, of~25 kcal/mole for $T[S(\text{linear})-S(\text{cyclic})]$ was obtained which is the reduction in the entropy caused by the ring closure.

The LS method has reached a state of maturity. It allows the reliable calculation of differences in the total free energy and partial entropies, the handling of the random coil and provision of at least a lower bound for the entropy change upon ring formation. Therefore it can be used for a wide range of applications. For example: One can define entropies and free energies for clusters of peptide conformations with different structures providing thereby a thermodynamic classification of these clusters in addition to the commonly used structural classifications. Such theoretically based classification along with other methodologies may facilitate the identification of the biologically relevant structures in drug design studies. However, the greatest challenge would be to extend the scope of the method to systems that consist of a peptide in solvent. In principle this can be accomplished since the LS method is also applicable to diffusive systems such as water.

Acknowledgements

I would like to acknowledge the Supercomputer Computations Research Institute, which is supported in part by the U.S. Department of Energy through contract No. DE-FC05-85ER2500000.

References

1. Beveridge, D.L. and DiCapua, F.M., Annu. Rev. Biophys., Biophys. Chem., 18 (1989) 431.
2. Metropolis, N., Rosenbluth, A.W., Rosenbluth, M.N., Teller, A.H. and Teller, E., J. Chem. Phys., 21 (1953) 1087.
3. Alder, B.J. and Wainwright, T.E., J. Chem. Phys., 31 (1959) 459.
4. McCammon, J.A., Gelin, B.R. and Karplus, M., Nature, 267 (1977) 585.
5. Meirovitch, H., Chem. Phys. Lett., 45 (1977) 389.
6. Meirovitch, H., Vásquez, M. and Scheraga, H.A., Biopolymers, 26 (1987) 651.
7. Meirovitch, H., Kitson, D.H. and Hagler, A.T., J. Am. Chem. Soc., 114 (1992) 5386.
8. Meirovitch, H., Koerber, S.C., Rivier, J. and Hagler, A.T., Biopolymers, submitted.
9. Rizo, J., Koerber, S.C., Bienstock, R.J., Rivier, J., Gierasch, L.M. and Hagler, A.T., J. Am. Chem. Soc., 114 (1992) 2860.

Solvent simulations and internal motions

D.F. Mierke[1,2] and H. Kessler[1]

[1]Organisch-Chemisches Institut, Technische Universität München,
W-8046 Garching, Germany and [2]Department of Chemistry, Clark
University, Worcester, MA 01610, U.S.A.

Introduction

The use of solvents in molecular dynamics (MD) simulations of peptides has been rather limited. One reason for this is that up till recently only water was available for such calculations, while most of the conformational studies were carried out in DMSO or $CHCl_3$. Here some advantages of using organic solvents in the simulations of peptides are highlighted.

Results and Discussion

Recently, we illustrated the use of DMSO in the simulations of a cyclic hexapeptide [1]. The parameters for $CHCl_3$ have been published by Jorgensen and coworkers [2]. Both of these solvents have advantages over water in the MD simulations of peptides.

DMSO can only act as an acceptor of hydrogen bonds, therefore only a single solvent shell around the peptide is observed. In contrast, water has been reported to form up to four solvent shells and therefore a much greater cut-off distance must be used. With DMSO as the solvent, a random orientation, similar to that of the pure solvent, is obtained at a distance of 7 Å away from the peptide. In contrast, cut-off distances of up to 16 Å may not be sufficient for water simulations [3]. Since the cut-off is smaller, a much smaller periodic system can be used in the DMSO simulations. In addition, for the same volume, DMSO requires only one quarter of the molecules (i.e., the densities are approximately the same while the molecular weight of DMSO is four times larger). With the united atom approach DMSO is treated as four atoms, single-point-charge water is treated as three points.

Very similar conclusions can be drawn from the simulations we have carried out with $CHCl_3$ [4, 5]. There is the presence of only one solvent shell; the $CHCl_3$ interacts specifically with the carbonyl oxygens that are solvent exposed [5]. This is in contrast to DMSO, which forms intermolecular hydrogen bonds explicitly with the amide protons exposed to the solvent [1]. Again, the size of the periodic system and therefore the number of solvent molecules required for the simulation is greatly reduced.

In addition to the great reduction of computer time, there are other questions that can be addressed when the same solvent as used in the NMR examination is used in the MD simulations. One point of interest is internal dynamics, and the effect that such dynamics will have on distances calculated using the standard two-spin approximation, shown below, where r_{ij} and σ_{ij} are the calculated distance and NOE intensity of atoms i and j, respectively, and σ_{std} is the intensity of a pair of protons of known distance, r_{std}. This approximation requires that each of the proton-proton vectors have identical correlation times (i.e., undergoing identical motions). If the MD simulation is assumed to be a good model for the dynamics that are taking place in solution, a good approximation if a good starting structure is used, then the validity of the two-spin approximation can be examined.

$$r_{ij} = r_{std} \left(\frac{\sigma_{std}}{\sigma_{ij}} \right)^{1/6}$$

The method, first illustrated by Karplus and coworkers [6, 7], involves carrying out a long, free (no experimental restraints) MD simulation. From this simulation the correlation times for each proton-proton vector involved in an NOE is then calculated from the simulation using,

$$C(t) = \left\langle \frac{A(t)A(0)}{R_{ij}^3(t)R_{ij}^3(0)} \right\rangle$$

where $A(t)$ is the second order spherical harmonics expressing the orientation of the proton-proton vector, ij, relative to the external, static magnetic field taken as the z axis. The correlation will plateau after a few picoseconds (ps) from the internal motions. This plateau value is equal to the order parameter, S^2, of Lipari and Szabo [8]. The spectral densities, $J(\omega)$, can then be scaled according to the differing motions [6], as shown below.

$$J_{ij}(\omega) = \frac{S^2 \langle r_{ij}^{-6} \rangle}{4\pi} \left[\frac{\tau_c}{1+(\omega\tau_c)^2} \right]$$

We use a protocol of 100-200 ps of restrained (NOEs, coupling constants) MD to create a starting structure. The MD is then continued without the restraints for up to 2 ns (the first 200 ps are to allow the molecule to equilibrate to the lack of restraints and are not used in the calculation). Calculations with cyclosporin A indicate that ignoring the effects of internal dynamics can introduce an error of up to 15%. Similar results have been obtained for antanamide [7].

References

1. Mierke, D.F. and Kessler, H., J. Am. Chem. Soc., 113 (1991) 9466.
2. Jorgensen, W., Briggs, J.M. and Contreras, M.L., J. Phys. Chem., 94 (1990) 1683.
3. Schreiber, H. and Steinhauser, O., Biochemistry, 31 (1992) 5856.
4. Eberstadt, M., Mierke, D.F., Köck, M. and Kessler, H., Helv. Chim. Acta, 75 (1992) 2583.
5. Konat, R., Mierke, D.F., Kessler, H., Kutscher, B., Bernd, M. and Voegeli, R., Helv. Chim. Acta, 76 (1993) 1649.
6. Olejniczak, E.T., Dobson, C.M., Karplus, M. and Levy, R.M., J. Am. Chem. Soc., 106 (1984) 1923.
7. Brüschweiler, R., Roux, B., Blackledge, M., Griesinger, C., Karplus, M. and Ernst, R.R., J. Am. Chem. Soc., 114 (1992) 2289.
8. Lipari, G. and Szabo, A., J. Am. Chem. Soc., 104 (1982) 4546.

A computational modeling of the δ receptor-bound conformation that is similar for cyclic opioid tetrapeptides and δ-selective alkaloids

A.L. Lomize and H.I. Mosberg

*College of Pharmacy, University of Michigan, Ann Arbor,
MI 48109, U.S.A.*

Introduction

We have previously described the conformationally restricted tetrapeptide, Tyr-c(D-Cys-Phe-D-Pen)OH (JOM-13), a high affinity, δ opioid receptor selective ligand [1]. The 11 membered cyclic portion of this peptide can be expected to possess only limited conformational flexibility, however the same restriction does not apply to the exocyclic Tyr residue, nor to the Phe[3] side chain which contain the key elements of the opioid pharmacophore. Consequently, further conformational restriction applied to these conformationally labile parts of the molecule is necessary to identify the binding geometry at the δ receptor. We report here the results of conformational search and molecular mechanics calculations applied to JOM-13 and to analogs in which the Tyr[1] residue is replaced by the more conformationally well defined substitutes: α-MeTyr; 2',6'-dimethyl Tyr (DMT); 1,2,3,4,-tetrahydro-7-hydroxyisoquinoline-3-carboxylic acid (Tic-OH); 6-hydroxy-2-aminoindan-2-carboxylic acid (Hai), 6-hydroxy-2-aminotetralin-2-carboxylic acid (Hat), and *cis*- and *trans*-3-(4'-hydroxy)-phenylproline (*c*-Hpp and *t*-Hpp, respectively).

Results and Discussion

Low-energy conformations calculated for JOM-13 and the related Tyr[1] replacement analogs confirm that the cyclic part of the tetrapeptides has a rigid mainchain structure, but that the exocyclic N-terminal residue and the Phe[3] side chain are flexible. Two isoenergetic conformations of the Cys-Pen disulfide bridge are observed, in agreement with ^{1}H NMR [2] and X-ray crystallography [3] data. All the Tyr[1] modifications have no influence on backbone and SS bridge conformers within the rigid cycle. At the same time, the orientation of the Tyr[1] aromatic ring relative to the rest of the molecule differs from that of JOM-13 for the residue 1 replacements incorporating additional ring structures (c-Hpp, t-Hpp, Hat, Hai and HO-Tic), but remains like the parent peptide for methylated analogs (DMT and α-MeTyr).

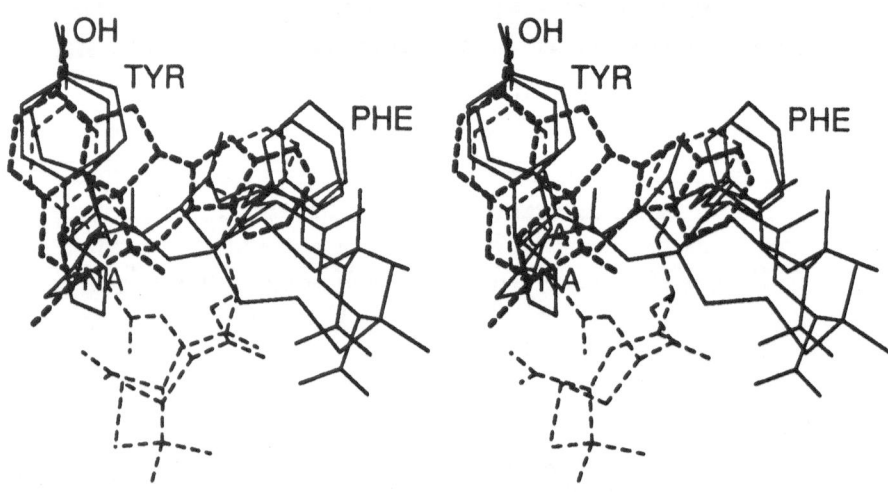

Fig. 1. Superposition of proposed δ receptor bound conformations for JOM-13 analogs with high (c-L-Hpp and t-L-Hpp, solid line) and moderate (L-Hai, dashed line) binding affinities with oxymorphindole spatial structure (bold dashed line). All functionally important points (NA, CA and OH groups and Tyr¹ and Phe³ aromatic rings) are labeled.

All peptides can be classified into three groups according to their binding affinities for the δ opioid receptor: high affinity (residue 1 = Tyr, c-Hpp, t-Hpp and DMT, all with $K_i \sim 1$ nM), moderate affinity (Hai, L- and D-Hat, all with $K_i \sim 10\text{-}20$ nM) and inactive analogs (α-MeTyr and HO-Tic, $K_i > 1000$ nM). The search for δ-bound conformations with the same spatial arrangement of all functionally important groups (N-terminal αNH_3^+ group and the Tyr¹ and Phe³ aromatic rings) leads to the following conclusions. The δ-bound conformations of high and moderate affinity peptides are different outside the rigid tripeptide cycle. The first family of peptides has *trans* conformations of both Tyr¹ and Phe³ sidechains ($\chi^1 \sim 180°$, $\chi^2 \sim 90°$), the second has "morphine-like" ($\chi^1 \sim -90°$, $\chi^2 \sim 0°$) and g+ ($\chi^1 \sim -60°$, $\chi^2 \sim 90°$) conformations of Tyr¹ and Phe³ sidechains, respectively. The torsion angles ψ of Tyr¹ and φ of D-Cys² are different also (ψ, φ = ~ -50°, ~90° and ~180°, ~165° respectively). All peptides of high and moderate affinity have the same orientation of the Tyr¹ aromatic ring plane after superposition of their functionally important groups in the proposed bound conformations. However, the aromatic rings of different peptides are shifted in this plane, especially near their C^δ atoms where two additional methyl groups (i.e., DMT) can be introduced without changing binding affinity. The methyl and methylene substituents on the C^α atom have completely different spatial positions in the inactive α-MeTyr¹ and moderate affinity (L-Hai¹, L-Hat¹ and D-Hat¹) analogs. Presumably, the inactivity of the α-MeTyr¹ analog is due to steric hindrance involving this α-Me at the δ binding site. Both D-Hat and L-Hat can adopt low energy conformers with good overlap of the pharmacophore elements, however, the L-Hai¹→D-Hai¹ replacement results in a shift of the ζOH group position which is the probable reason for the observed loss of binding affinity. The inactivity of the HO-Tic¹ analog may be explained simply by the different structure observed for its tyramine component in which the αNII_3^+ and ζOH groups are located too close together.

The functionally important parts of JOM-13 analogs (the Tyr[1] residue and Phe[3] aromatic ring) in the proposed δ receptor-bound conformations may be superimposed perfectly with correspondent parts (the tyramine moiety and indole ring, respectively) of the δ-selective opioid agonist, oxymorphindole, which has binding affinity similar to the high affinity peptides described here (Fig. 1). All additional methyl groups and aliphatic rings in JOM-13 analogs are overlapped with aliphatic morphine rings and may participate in some additional stabilizing interactions with δ-receptors. These results indicate that the binding modes of δ-selective opioid peptides and alkaloids may be similar.

The present model of the δ receptor-bound conformation for the parent peptide, JOM-13, differs from a previously proposed one [4] by incorporating a definite Tyr[1] side chain conformation (undefined in the previous study), in the orientation of the entire Tyr[1] residue (the ψ torsion angle of Tyr[1] is different) and in the conformation of the disulfide bridge. However, both models have the same backbone structure within the rigid D-Cys-Phe-D-Pen cycle and the same orientation of the Phe[3] side chain ($\chi^1 \sim 180°$).

Acknowledgements

This study was supported by grants DA 03910 and DA 00118 from the National Institute on Drug Abuse (NIDA).

References

1.   Mosberg, H.I., Omnaas, J.R., Smith, C.B. and Medzihradsky, F., Life Sci., 43 (1988) 1013.
2.   Mosberg, H.I. and Sobczyk-Kojiro, K., In Renugopalakrishnan, V., Carey, P.R., Smith, I.C.P., Huang, S.-G. and Storer, A.C. (Eds.), Proteins: Structure, Dynamics and Design, Escom, Leiden, The Netherlands, 1991, p.105.
3.   Flippen-Anderson, J., Naval Research Laboratory, personal communication, 1993.
4.   Nikiforovich, G.V., Hruby, V.J., Prakash, O. and Gehrig, C.A., Biopolymers, 31 (1991) 941.

New leads in angiotensin conformational studies

G.V. Nikiforovich[1], K. Plucinska[2], W.J. Zhang[2],
J.L.-F. Kao[3], R. Panek[4], R. Skeean[4] and
G.R. Marshall[1,2]

[1]Center for Molecular Design, Departments of [2]Pharmacology and
[3]Chemistry, Washington University, St. Louis, MO 63130, U.S.A.
[4]Parke-Davis Pharmaceutical Research Division, Ann Arbor, MI 48105,
U.S.A.

Introduction

The backbone conformer with a β-III-like turn at the Tyr^4-Val^5 residues was earlier proposed as a model for the angiotensin II (Asp^1-Arg^2-Val^3-Tyr^4-Val/Ile^5-His^6-Pro^7-Phe^8, AT) receptor-bound conformation by Marshall et al. (see [1] for final model), and, independently, by Nikiforovich et al., [2, 3]. Recently, several types of cyclic AT analogs with conformational restrictions have been found to possess high affinity towards AT specific receptors. However, the conformations available for these analogs seem to be not consistent with a unique model for the backbone of the AT receptor-bound conformation(s) (see [4] for detailed discussion). In this study, an alternate approach involving the search for a common spatial arrangement of functionally important (pharmacophoric) groups in low-energy conformers of AT and its active analogs has been used for modeling the AT bioactive conformation. It is commonly accepted that four groups are indispensable for optimal triggering of the biological response of AT, which are the aromatic moieties of Tyr^4, His^6 and Phe^8 residues, as well as the C-terminal carboxyl.

Results and Discussion

Energy calculations employing the ECEPP force field were performed for AT, $[(\alpha Me)Phe^4]$-, $[Pro^5]$-, cyclo$[HCys^{3,5}]$- and cyclo$[Sar^1, HCys/Cys^3, Mpt^5]$-AT analogs. Calculation technique involves the build-up procedure for all peptides in question, which yields the sets of low-energy backbone conformers possessing relative energies less than 10 kcal/mol. Geometrical comparison between each AT low-energy conformer vs. every low-energy conformer of its analogs was performed to reveal the mutual spatial orientation of the atomic centers chosen to represent a fragment bearing the pharmacophoric groups of AT molecule, namely, the C^β-atom for Tyr^4 residue, the C^α- and C^β- atoms for Val^5 - Phe^8 residues and the carbon atom of C-terminal carboxyl. One type of possible backbone conformer for AT 4-8 fragment was found to meet the requirement of rms ≤ 1.0 Å when compared to each analog, therefore being a suitable candidate for the receptor-bound conformation of AT. This conformer is presented in Table 1 together with corresponding conformers of the analogs.

Table 1 *Possible models for receptor-bound conformers of AT and its analogs*

Compound	Tyr/(αMe)Phe		Val/Pro/Cys/HCys		His		Pro		Phe
	φ	ψ	φ	ψ	φ	ψ	φ	ψ	φ
AT	-81	-32	-86	-35	-138	83	-60	64	-138
[(αMe)Phe^4]-AT	-58	-31	-87	-40	-132	85	-60	73	-138
[Pro^5]-AT	-107	166	-60	-27	-143	77	-60	73	-138
cyclo[HCys^{3,5}]-AT	-72	-37	-76	-27	-147	101	-60	61	-142
cyclo[Cys^3, Mpt^5]-AT	-130	56	-60	129	52	77	-60	74	-135
NMR Data	-145	60	-60	75	-	-	-	-	-
cyclo[HCys^3, Mpt^5]-AT	-133	70	-60	-13	-145	75	-60	67	-147
NMR Data	-155	75	-60	75	-	-	-	-	-
[D-Tyr^4, Pro^5]-AT	-123	163	-60	-31	-142	77	-60	73	-138

The difference between conformers in Table 1 in dihedral angle values (especially in ψ_4 value), does not affect significantly the spatial location of pharmacophoric groups. Figure 1 depicts an alignment between conformers of AT vs. [Pro^5]- and cyclo[Cys^3, Mpt^5]-AT, which differ in dihedral angle values from the respective AT conformer. It is evident that the Tyr^4 rings can occupy nearly the same spatial location for all three molecules, with very similar spatial positions of the His^6 and Phe^8 side chains, as well as the C-terminal carboxyls. At the same time, the N-terminal parts of both molecules (exemplified in Figure 1 by the Val/Cys^3 residues) are directed quite differently.

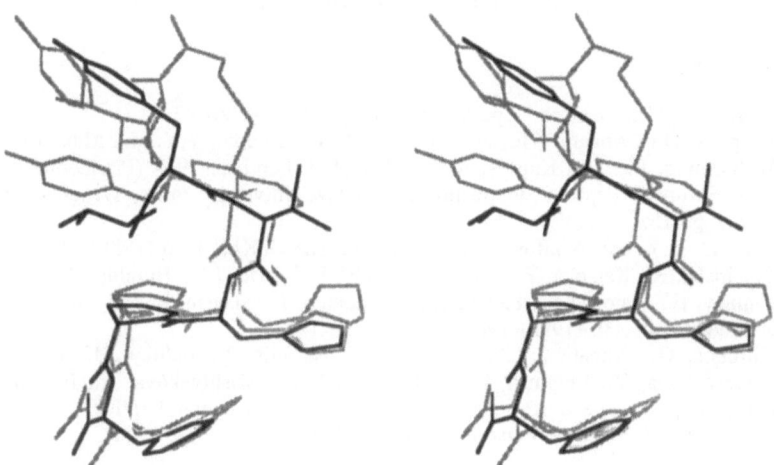

Fig. 1. Stereoview of AT conformer (in bold) overlapped on compatible [Pro^5]- and cyclo[Cys^3, Mpt^5]-AT conformers from Table 1. Only 3-8 fragments are shown. All hydrogen atoms are omitted.

Table 1 also contains the estimated values of dihedral angles for cyclic moieties of cyclo[Sar^1, Cys/HCys^3, Mpt^5]-AT analogs, which were derived from the J(HNC$^\alpha$H) coupling constants values and the NOE's measured by NMR spectroscopy in DMSO solution. Remarkably, they are in an excellent agreement with the models of receptor-bound conformers proposed for these analogs by calculations.

Thus, the problem of the biologically active conformation(s) of AT and its analogs should not be viewed in terms of common backbone structures, i.e., different kinds of turns, peptide chain reversals, etc. We can rather propose that any backbone three-dimensional structure that ensures a common spatial arrangement for the four important functional groups, such as that shown in Figure 1, is a suitable candidate for the receptor-bound conformation of AT agonists. In this sense, the original model predicting the β-III-like turn at the Tyr^4-Val^5 region of AT [1-3] is quite appropriate, but should not be regarded as unique.

Finally, it was suggested based on Figure 1, that the AT conformer in Table 1 could be matched even more closely by an analog, which still contains the Pro^5 residue, but with D-Tyr^4 instead of Tyr^4. The backbone conformation of such an analog might be close to that of [Pro^5]-AT in Table 1, and the D-Tyr^4 aromatic ring would be more close to the Tyr^4 aromatic ring in AT due to inversion of a configuration. This hypothesis was completely confirmed by energy calculations for [D-Tyr^4,Pro^5]-AT (see last line in Table 1). The analog was synthesized and tested for binding to rabbit aorta preparations, showing the IC$_{50}$ value of 42.8 nM. It is noteworthy that [D-Tyr^4]-AT showed a much lower affinity (pD$_2$ value of 5.77), when binding to tissue from rat ascending colon [5]. Thus, we were able to design a new and rather unusual AT analog on the basis of the three-dimensional model for AT recognition proposed in this study.

Acknowledgements

This work was supported in part by the NIH grant GM 24483.

References

1.   Marshall, G.R., Deutsche Apotheker Zeitung, 126 (1986) 87.
2.   Chipens, G.I., Ancan, Y.E., Nikiforovich, G.V., Balodis, Y.Y. and Makarova, N.A. In Siemion, I.Z. and Kupryszewski, G. (Eds.), Peptides 1978 (Proceedings of the 15th European Peptide Symposium), Wroclaw University Press, Wroclaw, Poland, 1979, p.415.
3.   Balodis, Y.Y. and Nikiforovich, G.V., Bioorganich.Khimia, 6 (1980) 865.
4.   Plucinska, K., Kataoka, T., Yodo, M., Cody, W.L., He, J.X., Humblet, C., Lu, G.H., Lunney, E., Major, T.C., Panek, R.L., Schelkun, P., Skeean, R. and Marshall, G.R., J. Med. Chem., 36 (1993) 1902.
5.   Chipens, G., Ancan, J., Afanasyeva, G., Balodis, J., Indulen, J., Klusha, V., Kudryashova, V., Liepinsh, E., Makarova, N. and Mishlyakova, N., In Loffet, A. (Ed.), Peptides 1976 (Proceedings of the 14th European Peptide Symposium), University of Brussels, Bruxelles, 1976, p.353.

Dependence of proton chemical shifts on backbone conformation in proteins and peptides

K. Ösapay and D.A. Case

*Department of Molecular Biology, The Scripps Research Institute,
La Jolla, California 92037, U.S.A.*

Introduction

Proton chemical shifts are sensitive indicators of structural changes in proteins and peptides; the qualitative dependence of Hα and HN chemical shifts on regular secondary structures has been known for some time [1]. Nevertheless, quantitative predictions based on quantum chemistry and empirical analyses [2-4] are still limited. For globular proteins, decomposition of the total shielding into local and long-range effects makes possible the analysis of contributions of distant polar and aromatic groups. For example, our previous model [4] derived from the analysis of 17 proteins uses an empirical equation for conformation-dependent (structural) shifts that are related to differences between the resonance positions in a folded protein and in a "random-coil" polypeptide:

$$\delta_{protein} - \delta_{random\ coil} = \sum \delta_{rc} + \sum \delta_m + \sum \delta_{el} - \delta_{const} \qquad (1)$$

Here, *rc*, *m*, and *el* refer to ring-current, magnetic anisotropy, and electrostatic interactions, respectively. Our parameter fitting resulted in two values for the constant in Eq. 1: 0.041 ppm for side-chain protons and 0.75 ppm for Hα-protons. The simplest interpretation for δ_{const} is that it represents in an averaged way the contributions of peptide groups in the random-coil state, usually taken to be GGXA, for which experimental shifts have been tabulated [5]. Here, we use calculations on dipeptide models to show that this is a reasonable interpretation.

Results and Discussion

Our earlier study [4] did not consider glycine residues. When we apply Eq. 1 for 121 glycine Hα protons -- the average shift for Hα2 and Hα3 is taken -- we find a correlation coefficient of 0.93 between calculated and observed structural shifts with an rmsd (root-mean-square deviation) of 0.29 ppm and a mean error of 0.24 ppm. Mean errors for other residue types are given in Fig. 1, and show that two residues (Pro and Gly) have significant residual shifts. This suggests that special values of δ_{const} be used for these residues, which is plausible because they have unusual Ramachandran plots. Indeed, when a constant of 0.51 ppm is used for glycine, the mean error vanishes, and the rmsd between calculated and observed shifts drops from 0.29 ppm to 0.16 ppm. When $\delta_{const} = 0.51$ ppm is used for proline (instead of

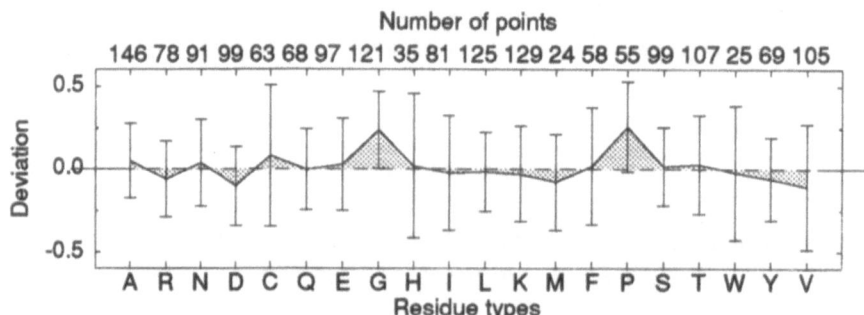

Fig. 1. Deviations between structural shifts calculated by Eq. 1 (for Hα protons in 17 proteins) and experimental $\delta_{protein} - \delta_{random\ coil}$ values. Averages, standard deviations, and numbers of data points are shown for each residue type. Data from Ref. [4].

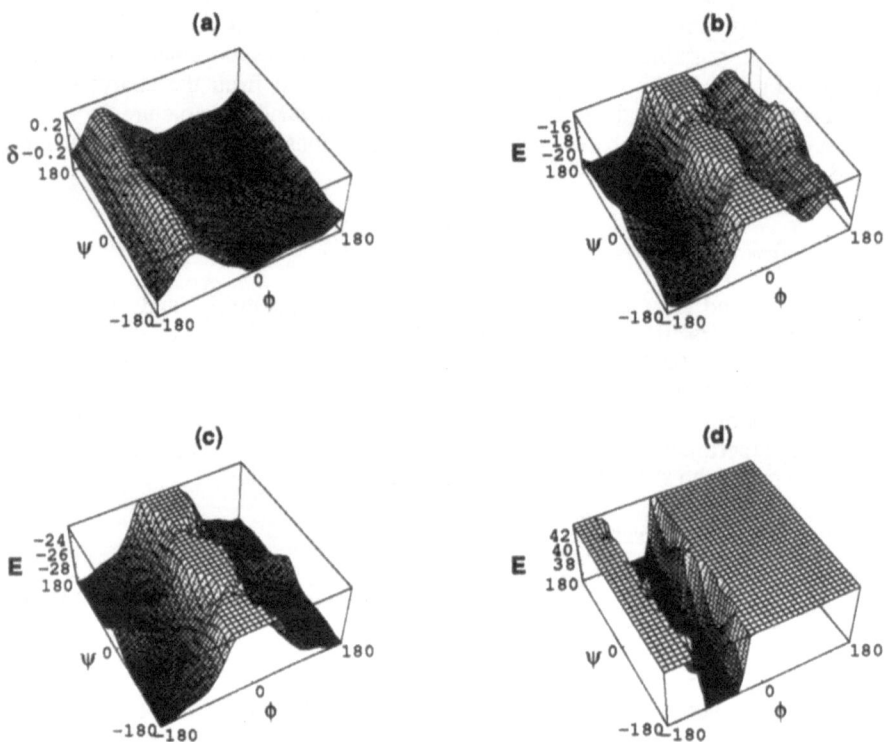

Fig. 2. (a) Structural shifts calculated for a dipeptide as a function of the ϕ and ψ backbone angles. (b)-(d) $\phi - \psi$ energy maps calculated by CHARMM 22 potential and corrected by the MEAD solvent contributions for (b) Ala dipeptide, (c) Gly dipeptide, and (d) Pro dipeptide.

0.75) the rmsd between calculated and observed shifts decreases from 0.37 ppm to 0.24 ppm.

To study the physical origin of δ_{const}, we investigated the conformational effect of two neighboring peptide groups to the Hα proton in acetyl-alanine-N-methylamide and its glycine and proline analogues (the "dipeptide"). The conformational dependence of the calculated Hα structural shift is presented in Figure 2a. As noted earlier [4], the primary dependence is on the ϕ dihedral angle.

The average chemical shift that we estimate for the random-coil conformation (Table 1) was calculated by using a Boltzmann-averaged weighting of the $\phi - \psi$ distribution. The CHARMM 22 potential was used to calculate the adiabatic conformational energy in vacuum (energy minimized with dihedral constraints), and the MEAD continuum dielectric model was applied to calculate solvation corrections (Figures 2b-d) [6,7].

Table 1   *Neighboring peptide contributions in the random-coil state calculated for the Ala, Gly, and Pro dipeptide models and values for δ_{const} as empirically fitted for 17 proteins*

	CHARMM 22 vacuum	CHARMM 22 MEAD solvent	empirically fitted
Ala	0.75	0.83	0.75
Gly	0.54	0.55	0.51
Pro	0.71	0.70	0.51

Our dipeptide calculations support the notion that the empirical constant used in our analysis of chemical shifts mainly arises from neighboring peptide contributions. Thus, differences in distributions of $\phi - \psi$ dihedral angles in the "random-coil" state can explain differences in the empirical constant term. It is suggested that special values be used for glycine and proline residues, which differ significantly from other residues in their allowed ϕ, ψ-ranges.

References

1. Wishart, D.S., Sykes, B.D. and Richards, F.M., Biochemistry, 31 (1992) 1647.
2. Williamson, M.P. and Asakura, T., J. Magn. Res., 91 (1991) 557.
3. DeDios, A.C., Pearson, J.G. and Oldfield, E., Science, 260 (1993) 1491.
4. Ösapay, K. and Case, D.A., J. Am. Chem. Soc., 113 (1991) 9436.
5. Bundi, A. and Wüthrich, K., Biopolymers, 18 (1979) 285.
6. Brooks, B.R., Bruccoleri, R.E., Olafson, B.D., States, D.J., Swaminathan, S. and Karplus, M., J. Comp. Chem., 4 (1983) 187; version CHARMM 22.
7. MEAD program set, Bashford, D., The Scripps Research Institue, La Jolla, CA, U.S.A., 1990.

Elucidation of multiple conformations for flexible peptides from averaged NOE data: Application to *desmopressin*

J. Wang, R.S. Hodges and B.D. Sykes

Protein Engineering Network of Centres of Excellence and Synthetic Peptides Inc., University of Alberta, Edmonton, Alberta, T6G 2S2, Canada

Introduction

Flexible peptides with multiple conformations in solution give rise to one set of averaged NOEs if these multiple conformations are in fast exchange [1]. The elucidation of the multiple conformations for flexible peptides based on averaged NOE data is hindered using current NOE-derived structure generation methods, since one can only construct an 'average' structure [2]. There are several approaches proposed recently to try to solve this problem [1, 2]. We have developed a new method which combines a 'molecular construction' method such as simulated annealing with a 'best fit' method to simulate the multiple conformations for flexible peptides based averaged NOE data. The method actually is a back calculation procedure for analysis of single conformations which was extended to handle multiple conformation systems [3, 4]. Its main hypothesis is that it searches for the widest range of structures whose calculated averaged NOEs best fit the experimental NOE data. The method is demonstrated on a flexible peptide *desmopressin* which contains a six-residue disulfide-bonded loop and a three-residue tail. It is known from our previous work that it exhibits multiple conformations in 20% TFE solution [3, 5]. The results indicate that the procedure successfully generates multiple conformations for *desmopressin* in 20% TFE solution which best fit the experimental NOE data.

Results and Discussion

Two-dimensional ^{1}H-NMR spectra were measured for *desmopressin* in aqueous solution containing 20% TFE at 5°C, pH 5.0. In total 157 NOE distance restraints were generated from a NOESY spectrum with 300 ms mixing time since the NOESY spectra showed a linear buildup of the NOEs up to a mixing time of almost 300 ms. The restraints were grouped into strong, medium and weak which corresponds to distances of 1.8 - 2.7, 1.8 - 3.3 and 2.3 - 5.0 Å and used in the calculation. The new computational procedure is called PEPFLEX-II which is an extension of PEPFLEX [4]. It is an iterative calculation that combines both the "structural generation" methods (simulated annealing) and "best fit" comparisons into a single procedure. The procedure generates structures based on the NOE constraints in each iteration, calculates the RMS deviations between the experimental NOEs and calculated NOEs based upon a proper average for the family of structures, and superimposes the

914

structures generated in each iteration to their averaged structure, and calculates the structural deviations for each structure. It tries to optimize the fit to the experimental data and maximize the structural RMS deviations amongst the family of structures at the same time by modification of the distance restraints. Table 1 shows part of the results of the calculation.

Table 1   *Structural statistics for desmopressin: Total energy (kcal/mol Å^2), number of NOE violations (>0.2Å) and the number of modified NOEs in each iteration*

Iteration	1	2	3	4	5	6
Total energy	98±13	69±7	58±7	43±7	44±9	37±7
NOE$_{viol}$	66	38	28	20	18	16
NOE$_{viol}$/struct.	30±3	9±3	4±2	1±2	2±2	1±1
NOE modified	49	20	20	18	16	0

Based on the criteria for 'goodness' of the NOE-derived structures, the results shown in Table 1 indicate that after each iteration better structures are generated. Finally the approach generates the structures which have minimum of total energy as well as minimum of NOE violations. This can be seen clearly when the procedure calculates the RMS deviations between experimental NOEs and calculated NOEs based on the structures generated by the procedure.

Table 2   *RMS deviations between experimental and calculated NOEs*

Iteration	1	2	3	4	5	6
RMSD$_{NOE}$(all)	0.24	0.22	0.21	0.18	0.17	0.17
RMSD$_{NOE}$(loop)	0.24	0.23	0.21	0.19	0.18	0.18
RMSD$_{NOE}$(tail)	0.22	0.21	0.22	0.15	0.17	0.17

Table 2 indicates that the structures generated from the last iteration fit the experimental NOE data much better than those from the first iteration. However, since we are dealing with a multiple conformation problem, we expect that a good procedure should allow us to cover more conformational space when the calculated NOEs fit the experimental NOEs better, and finally will cover a maximum conformational space when the RMS deviation drops to a minimum.

Table 3 shows that the procedure enables us to search a wider range of conformational space while the calculated NOEs based on the generated structures better fit the experimental NOE data. The procedure includes a cluster analysis after each iteration to determine how many families of structure can be obtained.

The data shown in Table 4 and Table 5 demonstrate that the structures generated from the final iteration can be divided into six families. The structures between families are significantly different (Table 5), but the structures within a given family are well-defined (Table 4).

Table 3   *Structural RMS deviations (Å) for all backbone atoms*[a]

Iteration	1	2	3	4	5	6
$RMSD_{struct}$(all)	0.56±0.41	0.91±0.57	1.25±0.43	1.07±0.45	1.28±0.54	1.46±0.78
$RMSD_{struct}$(loop)	0.34±0.23	0.39±0.38	0.54±0.38	0.69±0.37	0.95±0.25	0.93±0.41
$RMSD_{struct}$(tail)	0.34±0.15	0.55±0.28	0.98±0.14	0.52±0.17	0.46±0.16	0.77±0.10

[a]Structural RMS deviations were determined by atomic RMS differences relative to the mean structure calculated from all structures in the iteration.

Table 4   *Families of structures and the backbone atom structural RMS deviations (Å) of each family in the final iteration*

Families	1	2	3	4	5	6
population (%)[a]	57	13	13	7	7	3
$RMSD_{struct.}$	0.50±0.18	0.78±0.10	0.85±0.29	0.75±0.00	0.49±0.00	----[b]

[a]The population here is the number of structures in the family over the number of total structures generated in the iteration.
[b]The family only contains one structure.

Table 5   *The RMS deviations (Å) between the two families in the final iteration. The upper triangle is only for backbone atoms, the lower triangle is for all atoms of the peptide*

1	1.90	3.57	1.84	2.70	3.89
3.67	**2**	3.55	2.19	2.51	4.21
6.00	6.70	**3**	2.97	2.62	3.06
2.91	3.72	4.83	**4**	1.91	4.44
4.61	4.46	4.03	2.91	**5**	3.88
5.61	5.46	3.99	5.19	5.04	**6**

A new procedure was developed which was applied to *desmopressin*. It was found that based on the averaged experimental NOE data, the procedure could search the maximum conformational space and result in the elucidation of the multiple conformations which best fit the experimental NOE data.

Acknowledgements

We are grateful for the financial support of the Protein Engineering Network of Centres of Excellence. We thank Dr. Frank D. Sönnichsen for helpful discussion.

References

1. Torda, A.E., Scheek, R.M. and Van Gunsteren, W.F., J. Mol. Biol., 214 (1990) 223.
2. Landis, C. and Allured, V.S., J. Am. Chem. Soc., 113 (1991) 9493.
3. Wang, J., Sönnichsen, F.D., Boyko, R., Hodges, R.S. and Sykes, B.D., Techniques in protein chemistry, IV, Academic Press Inc., San Diego, CA, U.S.A., 1993, p.569.
4. Sönnichsen, D.S., Boyko, R, Wang, J. and Sykes, B.D., PENCE, University of Alberta, Program PEPFLEX.
5. Wang, J., Hodges, R.S. and Sykes, B.D., PENCE, Department of Biochemistry, University of Alberta, Edmonton, Canada, T6G 2S2, unpublished data.

Session XII
Peptide Macromolecular Interactions

Chairs: Ann E. Shinnar
University of Pennsylvania
Philadelphia, Pennsylvania, U.S.A.

and

Derek Hudson
Arris Pharmaceutical Corporation
South San Francisco, California, U.S.A.

NMR studies of the structure of peptide complexes that duplicate functional interactions between proteins

G.S. Shaw and B.D. Sykes

Department of Biochemistry and MRC Group in Protein Structure and Function, University of Alberta, Edmonton, Alberta, T6G 2H7, Canada

Introduction

For several years synthetic peptides have been used successfully to study the calcium-binding properties of proteins such as troponin-C (TnC), calmodulin and calbindin D_{9k}. This approach is well-suited for these proteins, since each contains either two or four calcium-binding sites where each site is a contiguous stretch of about 30 amino acids [1]. In addition, high field ^{1}H NMR spectroscopy has proven ideal for monitoring the conformational changes which occur in the peptides upon metal-ion binding. Recently, we have shown that calcium-binding to a 34-residue peptide corresponding to site III of the C-terminal domain of troponin-C (SCIII) not only binds calcium but also undergoes a dimerization to form a two-site peptide homodimer similar in structure to that observed in the x-ray structure of the C-terminal domain of TnC [2]. This important discovery of the dimerization of synthetic calcium-binding peptides is likely a general observation in calcium-binding proteins as two-site homodimers and heterodimers have been noted for peptides derived from TnC [3] and calbindin D_{9k} [4]. In this work we describe a comparison of the three-dimensional structure of the SCIII homodimer derived from NMR data [5] with that of the C-terminal domain of TnC as observed from x-ray crystallographic studies [6].

Results and Discussion

The solutions for the three-dimensional structures of the SCIII homodimer and the C-terminal domain from TnC have been described elsewhere [5, 6]. In this work, all comparisons for the SCIII homodimer are based on the average structure reported which has been energy-minimized to relieve poor van der Waals contacts from the averaging protocol. Figure 1 shows schematic diagrams of the SCIII homodimer and the C-terminal domain of TnC. It is clear that the overall fold of the two structures is very similar. Each structure consists of α-helices between residues 97-106 and 115-122, based on φ and ψ angles, and the analogous helices are present in the symmetrically related portion of SCIII and in site IV in the C-terminal of TnC. In order to examine these overall structural features more closely, Ramachandran plots for both structures have been calculated and are presented in Figure 2. These two plots reinforce the similarity of the structures. In both cases the majority of the residues have φ,ψ angles clustered near φ =-60°, ψ =-60°, with only twelve outlying points. Four of these are found near $\phi = 60°$, $\psi = 30°$ typical for glycine residues. In the SCIII homodimer, two of these points result from G111 and G111' while two result from A109 and A109'. The corresponding residues G111, G147, A109 and N145 are found in similar positions in the Ramachandran plot for TnC. Of the remaining eight points outside the α-helical range, 6 are derived from residues in a β-sheet conformation; Y112, I113, D114 and Y112', I113', D114' in the SCIII homodimer and F112, I113, D114, R148, I149 and D150 in TnC.

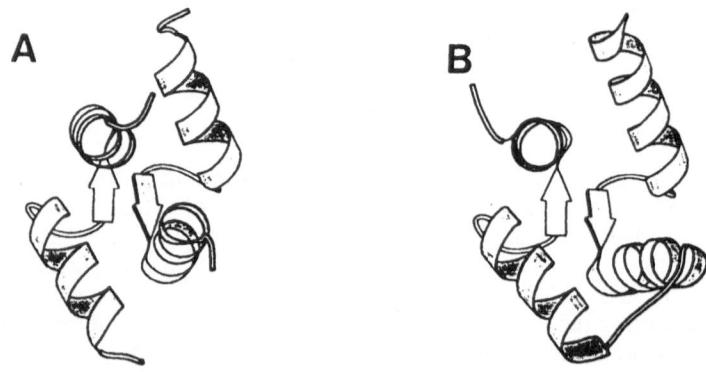

Fig. 1. Ribbon diagrams of the SCIII homodimers (A) and the C-terminal domain of troponin-C (residues 93-162, (B)) created using the program Molscript [7].

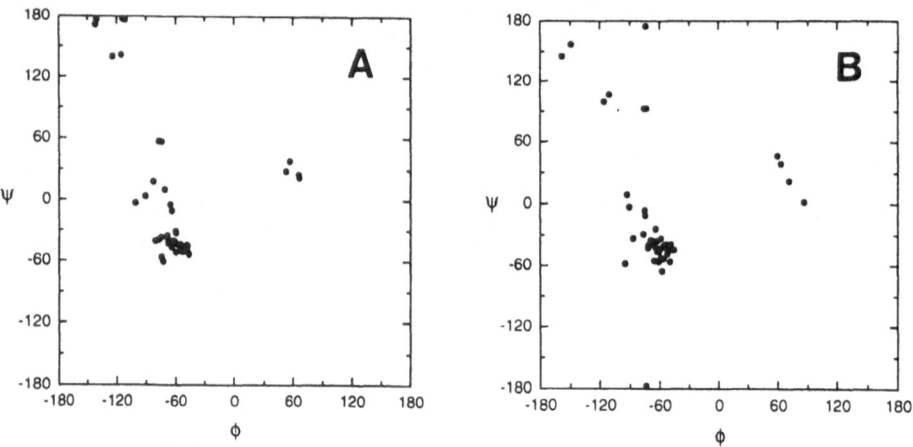

Fig. 2. Ramachandran plots for residues 97-122 and 97'-122' for the SCIII homodimer (A) and analogous residues in the C-terminal domain of troponin-C (97-122 and 133-158, (B)).

In order to compare the ligation of the calcium ions in the SCIII homodimer and the C-terminal domain of TnC, the χ_1 angles for the ligating residues in each were measured (Table 1). Several conclusions can be drawn from the values in this Table. First, the symmetric nature of the SCIII homodimer is reflected in the χ_1 angles for symmetrically related residues. The χ_1 angles for each pair of ligating residues varies by about 1.8° on average. In contrast similar pairs of residues in TnC vary by 3.6° on average. The magnitudes of the χ_1 angles are similar for 8 of the 10 co-ordinating residues for the SCIII homodimer and TnC. The only exception is D114 (and D114') in the SCIII homodimer (χ_1 = 19.3, 19.9°) which are significantly different compared with the corresponding residues D114 (χ_1 = 66.7°) and D150 (χ_1 = 65.2°) in TnC. This interesting observation may be a consequence of the dimer formation in SCIII resulting in an altered χ_1 angle compared to TnC. Alternatively, this angle may be modulated by an intervening water molecule which has been observed between the side chain carboxyl group and the calcium ion in the x-ray structure.

Table 1 χ_1 angles for calcium co-ordinating residues in the SCIII homodimer and C-terminal domain of TnC

Residue	SCIII	C-terminal TnC
106	-175.7	178.8
108	57.6	65.2
110	68.8	72.1
114	19.3	66.7
117	-68.3	-61.7
142	-176.6	176.1
144	57.2	71.9
146	64.6	67.0
150	19.9	65.2
153	-71.3	-66.5

In summary the results described here show the similarities in structure between the SCIII homodimer and the C-terminal domain of TnC and strongly support the usefulness of synthetic calcium-binding peptides as models for the native protein.

References

1. Kretsinger, R.H. and Nockolds, C.E., J. Biol. Chem., 248 (1973) 3313.
2. Shaw, G.S., Hodges, R.S. and Sykes, B.D., Science, 249 (1990) 280.
3. Shaw, G.S., Findlay, W.A., Semchuk, P.D., Hodges, R.S. and Sykes, B.D., J. Am. Chem. Soc., 114 (1992) 6258.
4. Linse, S., Thulin, E. and Sellers, P., Protein Science, 2 (1993) 985.
5. Shaw, G.S., Hodges, R.S. and Sykes, B.D., Biochemistry, 31 (1992) 9572.
6. Herzberg, O. and James, M.N.G., J. Mol. Biol., 203 (1988) 761.
7. Kraulis, P.J., J. Appl. Cryst., 24 (1991) 946.

Use of peptomers to study the binding of HIV-1 to CD4: Effects of polymerization on the biology and chemistry of a synthetic peptide

F.A. Robey[1], T. Harris-Kelson[1], P. Roller[2], M. Cleriqi[3] and D. Batinic[1]

[1]National Institute of Dental Research, [2]DCT of NCI, [3]National Cancer Institute, National Institutes of Health, Bethesda, MD 20892, U.S.A.

Introduction

For many years there has been a growing interest in the simple construction of conformationally constrained analogues of synthetic peptides as part of an effort to rationally design novel drugs and vaccine candidates. A minor contribution to this effort has been the discovery that chloroacetyl and bromoacetyl moieties are stable to many of the reaction conditions and environments used in t-BOC-based peptide synthesis [1-3]. The result of this finding is that chemists now are able to place a chloroacetyl or bromoacetyl moiety at any position in a synthetic peptide and, together with a thiol like that found in cysteine, new cyclic, polymeric, and conjugated peptide-based materials can be formed.

We have applied this technology to the challenge of determining how, on a molecular level, the envelope glycoprotein from HIV-1, gp160, binds to its cellular receptor, CD4. By understanding this mechanism, we may have a chance of having something - a drug or a vaccine - that could block the binding of HIV-1 to a cell and this could lead to preventing intracellular transmission of HIV-1 *in vivo*.

In 1987, Lasky and colleagues determined the region of gp120 that appeared to be most responsible for its high affinity binding to CD4 [4] and several other groups have used various methods to reach some agreement about those amino acids in CD4 which are primarily involved in binding gp120.

From a synthetic peptide approach, no groups have been able to definitively show the Lasky-derived region of gp120 to participate in CD4 binding although in 1991 Reed and Kinzel came very close by demonstrating some minor binding of radiolabeled pentadecapeptides derived from this region to HeLa cells expressing CD4 [5]. However, most noteworthy of the Reed and Kinzel study was their demonstration that this peptide could convert to an α-helical secondary structure under apolar solvent conditions and the suggestion was made that the peptide in gp120 may be in the α-helical form when bound to CD4 [5].

In this presentation, we have combined the haloacetyl chemistry mentioned above to the challenge of understanding HIV-1 infection and we demonstrate that peptomers composed of the CD4 binding region from gp120 - initially described by Lasky et al. [4] and expanded upon by Reed and Kinzel [5] - have enhanced secondary structure and, perhaps because of the enhanced secondary structure, can bind CD4.

Results and Discussion

Using the one letter abbreviations for the amino acids, the peptide under study has the sequence K-I-K-Q-I-I-N-M-W-Q-E-V-G-K-A-M-Y-A. We define a peptomer as a polymer composed of a synthetic peptide that is specifically cross-linked to itself. To make the peptomer from the above-listed peptide, a bromoacetyl group is added to the amino terminus and the C-terminal amino acid is cysteine in the amide form. Polymerization is carried out in pH 7.0 to 7.5 buffer as outlined in [2] and the concentration of the starting material is 5 mg/ml.

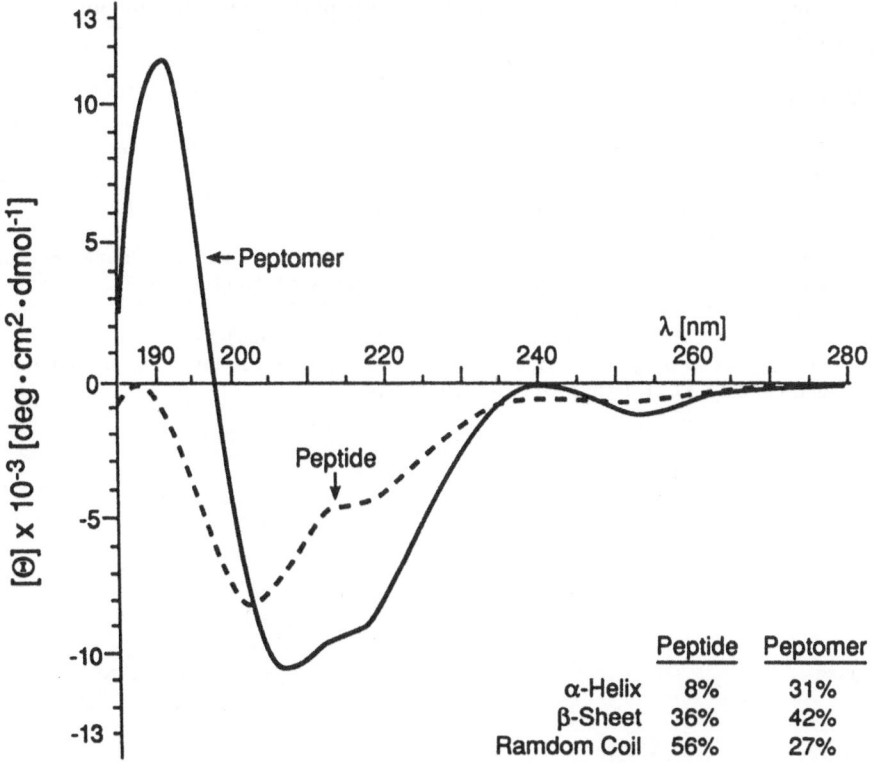

	Peptide	Peptomer
α-Helix	8%	31%
β-Sheet	36%	42%
Ramdom Coil	56%	27%

Fig. 1. *CD Spectra of peptide(419-436) and peptomer(419-436)-MN isolate of HIV-1. Spectra were recorded from 280 to 185 nm on a Jasco Model J-500A/DP-501N CD spectropolarimeter, in Hellma QS cells with 1 mm path length at room temperature. Peptide concentrations were 36 μM in H₂O, pH 5.3 or in 5 mM pH 7.2 sodium cacodylate buffer. The conformations of the peptide and the peptomer were estimated using Provencher's spectral deconvolution program [6] using the experimental CD data.*

The CD spectra of peptomer(419-436) compared with peptide(419-436), both in water is given in Figure 1. From the CD spectra we can conclude that the peptomer contains 31±1% α-helical content compared with 8±7% for the free peptide.

Several tests were then performed to evaluate the binding of the peptomer to CD4 and antibodies derived from HIV(+) serum samples. Most noteworthy was the finding that peptomer immobilized to nitrocellulose and to polystyrene bound CD4 in a dose-dependent fashion and the peptomer also bound antibodies derived from HIV-1-positive serum samples (results not shown). The corresponding peptide was virtually inactive in the above assays.

The current understanding of HIV-1 infection leaves little agreement among researchers. However, the idea that HIV-1 infection might lead to autoimmunity brought on by the immunogenicity of gp160 and the sequence homology of gp160 to MHC Class II beta chain [7], leaves one with the belief that circumventing these problems requires an approach that involves the peptide-containing subunits of the viral envelope protein.

References

1. Lindner, W. and Robey, F.A., Int. J. Peptide Protein Res., 30 (1987) 794.
2. Robey, F.A. and Fields, R.L., Anal. Biochem., 177 (1989) 373.
3. Inman, J.K., Highet, P.F., Kolodny, N. and Robey, F.A., Bioconjugate Chem., 2 (1991) 458.
4. Lasky, L.A., Nakamura, G., Smith, D.H., Fennie, G., Patzer, E., Shimasaki, C., Patzer, D., Berman, P., Gregory, T. and Capon, D., Cell, 50 (1987) 975.
5. Reed, J. and Kinzel, V., Biochemistry, 30 (1991) 4521.
6. Provencher, S.W. and Glockner, J., Biochemistry, 20 (1981) 33.
7. Golding, H., Robey, F.A., Gates, F.T., III, Lindner, W., Beining, P., Hoffman, T. and Golding, D., J. Exp. Med., 167 (1988) 914.

Determinants of specificity in phosphoprotein/ SH2 domain interactions

R. Case[1], E. Piccione[1], M. Chaudhuri[1], G. Gish[2], R. Lechleider[3,4], B. Neel[3,4], T. Pawson[2] and S.E. Shoelson[1,4]

[1]*Research Division, Joslin Diabetes Center, Boston, MA 02215, U.S.A.; [2]Division of Molecular & Developmental Biology, Samuel Lunenfeld Research Institute, Toronto, Ontario, M5G 1X5, Canada; [3]Molecular Medicine Unit, Beth Israel Hospital, Boston, MA 02215, U.S.A.; [4]Department of Medicine, Harvard Medical School, Boston, MA, U.S.A.*

Introduction

Cytoplasmic molecules involved in tyrosine kinase signal transduction are frequently modular, containing discrete catalytic domains and additional protein binding modules termed Src homology (SH) domains. SH2 domains bind directly to phosphorylated Tyr residues in growth factor receptor and non-receptor Tyr kinases, their substrates, and viral transforming proteins [1, 2]. Fortuitously, isolated expressed SH2 domains and small synthetic phosphopeptides retain the capacity to interact with one another with high affinity and specificity [3-7]. In the current study we exploit the capacity of isolated SH2 domains to retain function by using SH2 domain/ phosphopeptide binding assays to directly assess determinants of specificity.

Results and Discussion

SH2 domains of phosphatidylinositol (PI) 3-kinase p85, phospholipase C-γ (PLC-γ), the phosphatase SH-PTP2, and Src itself were expressed as glutathione S-transferase (GST) fusion proteins in *E. coli*. Phosphopeptides synthesized following an Fmoc/tBu strategy using Fmoc-Tyr(O(P(OCH$_3$)$_2$)) have sequences surrounding the numbered Tyr residues of the PDGF receptor: PDGFR740, DGGpYMDMSKDE; PDGFR751, SVDpYVPMLDMK; PDGFR1009, SVLpYTAVQPNE; PDGFR1021, DNDpYIIPLPDPK; the hamster polyoma middle T antigen: *h*mT324, EPQpYEEI-PIYL; and c-Src527: EPQpYQPGENL.

Intact PI 3-kinase binds to multiple proteins which contain phosphorylated YMXM or YVXM motifs [2]. An isolated SH2 domain of PI 3-kinase exhibits the same specificity (Figure 1A) [4]. ED$_{50}$ values for binding of pYM/VXM-peptides PDGFR740 and PDGFR751 are ≈1.0 μM, whereas the PDGFR1021 peptide with a pYIIPL motif binds with 50-fold lower relative affinity (Figure 1A).

In simulated cells, PLC-γ binds to Tyr1021 of the PDGF receptor [8]. When a competition assay was set up with an SH2 domain from PLC-γ, the PDGFR1021 phosphopeptide bound tightly, with an ED$_{50}$ of ≈2.0 μM (Figure 1B). In this case, the YMXM peptide PDGFR740 bound with 50-fold lower affinity to demonstrate an appropriate reversal in specificities between p85 and PLC-γ SH2 domains.

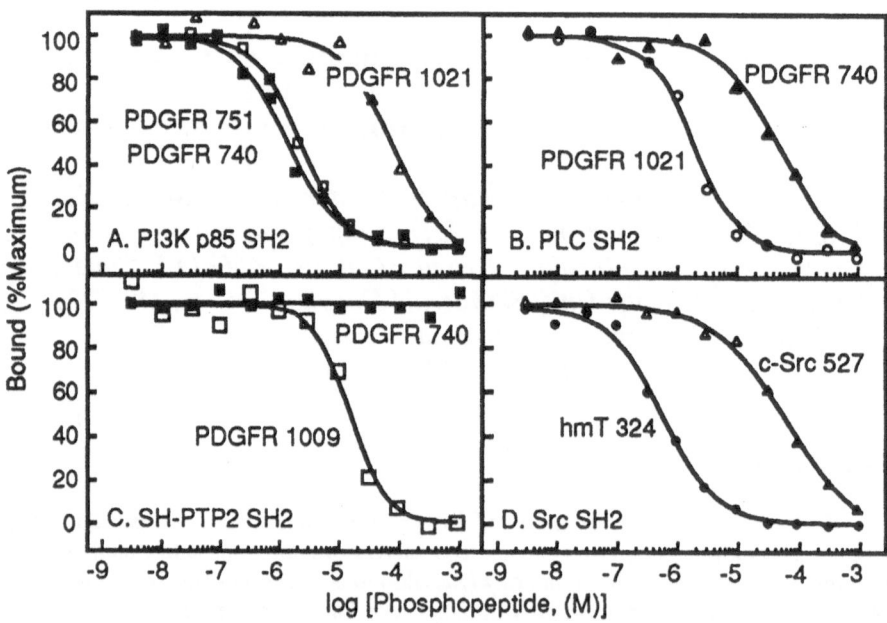

Fig. 1. *SH2 domain binding assays. GST-SH2 domain fusion proteins, tracer amounts of [125I]labeled phosphopeptides, glutathione agarose, and the indicated concentrations of unlabeled phosphopeptides were incubated overnight with shaking. Mixtures were centrifuged and supernatant solutions removed by aspiration. Radioactivity associated with the agarose pellets was used to determine binding [4]. A. Competitive binding to the N-terminal SH2 domain of PI 3-kinase p85 was analyzed using a high affinity IRS-1 derived tracer [4] and unlabeled PDGFR740, PDGFR751 and PDGFR1021. B. Binding to the C-terminal SH2 domain of PLC-γ was analyzed using radiolabeled PDGFR1021 and unlabeled peptides PDGFR1021 and PDGFR740. C. Binding analyses with the N-terminal SH2 domain of SH-PTP2 was with radiolabeled PDGFR1009 and unlabeled PDGFR1009 and PDGFR740. D. Analyses of binding with the SH2 domain of Src were with [125I]labeled hmt324 and unlabeled hmt324 and c-Src527 [5].*

Peptide library studies to map specificity suggested that an SH2 domain of SH-PTP2 might bind tightly to a pYTAV motif [7]. This sequence surrounds Tyr1009 of the PDGF receptor, and was subsequently found to direct interactions between the intact proteins [9]. A phosphopeptide corresponding to this motif, PDGFR1009, binds to an SH2 domain of SHPTP2 with an ED_{50} value of ≈15.0 μM (Figure 1C), whereas peptide PDGFR740 bound with >100-fold lower affinity.

Physiologically relevant, high-affinity interactions with the Src SH2 domain are not understood sufficiently to predict a binding motif, although the phosphorylated C-terminus of c-Src reportedly binds its own SH2 domain [2]. Library studies suggested that pYEEI might bind with high affinity [7], and this motif was found surrounding Tyr324 of the hamster polyoma mT antigen. A corresponding phosphopeptide bound to the Src SH2 domain with high affinity, whereas a peptide corresponding to the c-Src tail sequence bound with 100-fold lower affinity (Figure 1D).

SH2 protein interactions with phosphoproteins can be reconstituted *in vitro* using isolated SH2 domains and phosphopeptides corresponding to phosphorylation sites of growth factor receptor and non-receptor tyrosine kinases and their substrates. Specificity in these interactions is retained. Affinity differences between peptides which bound specifically *vs.* non-specifically ranged from 50- to >100-fold for SH2 domains from four functionally distinct cytoplasmic signaling proteins. The described assay method is generally useful for analyzing SH2 domain specificity.

Acknowledgements

Supported by grants from the NSF (SES), NIH (BGN), NCI-Canada (TP), and a Career Development Award from the Juvenile Diabetes Foundation (SES).

References

1. Pawson, T. and Gish, G.D., Cell, 71 (1992) 359.
2. Cantley, L.C., Auger, K.R., Carpenter, C., Duckworth, B., Graziani, A., Kapeller, R. and Soltoff, S., Cell, 64 (1991) 281.
3. Domchek, S., Auger, K., Chatterjee, S., Burke, T.,Jr. and Shoelson, S.E., Biochemistry, 31 (1992) 9865.
4. Piccione, E., Case, R.D., Domchek, S.M., Hu, P., Chaudhuri, M., Backer, J.M., Schlessinger, J. and Shoelson, S.E., Biochemistry, 32 (1993) 3197.
5. Payne, G., Shoelson, S.E., Gish, G., Pawson, T. and Walsh, C.T., Proc. Natl. Acad. Sci. U.S.A., 90 (1993) 4902.
6. Felder, S., Zhou, M., Hu, P., Urena, J., Ullrich, A., Chaudhuri, M., White, M.F., Shoelson, S.E. and Schlessinger, J., Mol. Cell. Biol., 13 (1993) 1449.
7. Songyang, Z., Shoelson, S.E., Chaudhuri, M., Gish, G., Pawson, T., King, F., Roberts, T., Ratnofsky, S., Lechleider, R.J., Neel, B.G., Birge, R.B., Fajardo, J.E., Chou, M.M., Hanafusa, H., Schaffhausen, B. and Cantley, L.C., Cell, 72 (1993) 767.
8. Rönnstrand, L., Mori, S., Arridsson, A.-K., Eriksson, A., Wernstedt, C., Hellman, U., Claesson-Welsh, L. and Heldin, C.-H., EMBO J., 11 (1992) 3911.
9. Kazlauskas, A., Feng, G.S., Pawson, T. and Valius, M., Proc. Natl. Acad. Sci. U.S.A., in press.

The nature of zinc fingers and conformational behavior of HIV-1 NCp7 are critical for in vitro and in vivo activities: A combined 600 MHz ^{1}H NMR and biological study

H. Déméné[1], N. Morellet[1], N. Jullian[1], C.Z. Dong[1], S. Saragosti[2], J.L. Darlix[3], M.C. Fournié-Zaluski[1] and B.P. Roques[1]

[1]U266 INSERM-D URA D1500 CNRS, Université René Descartes, 4 avenue de l'observatoire, 75006 Paris, France
[2]NSERM U363, ICGM, Hôpital Cochin, 75014 Paris, France
[3]Ecole Normale Supérieure, Laboratoire de Biologie Moléculaire et Cellulaire, 46 avenue de l'Italie, 69634 Lyon Cedex 07, France

Introduction

Mature NCp7 human immunodeficiency virus type 1 (HIV-1) is a 72 amino acid basic protein which contains two CCHC retroviral type zinc finger domains characterized by a high zinc binding affinity. In vitro, NCp7 was shown to promote RNA dimerization and annealing activities, two critical processes for virus infectivity [1]. In vivo, deletion of CCHC boxes caused loss of virion infectivity due to genomic packaging defects. The 3D structure of NCp7, recently elucidated by 600 MHz ^{1}H nuclear magnetic resonance (NMR) and distance-geometry calculations, is characterized by a kink at the Pro[31] level in the basic sequence [29]RAPRKK[34], inducing a spatial proximity of the aromatic residues Phe[16] and Trp[37] of each zinc finger [2]. In order to study the biological relevance of NCp7 structural features, we have on the one hand substituted DPro[31] for Pro[31] in the active (13-64)NCp7 and, on the other hand, Cys[23] for His[23] in the first zinc finger domain of the active (13-64)NCp7. The structures of both peptides were investigated using ^{1}H NMR spectroscopy and their biological activity tested as described [4].

Results and Discussion

The modified peptides were chemically synthesized using the stepwise solid phase synthesis method in Fmoc chemistry. Resonance assignments were made by analysis of the 2-D DQF-COSY, TOCSY and NOESY experiments recorded on a Bruker AMX600 spectrometer.

DPro[31](13-64)NCp7

Comparison of NOESY spectra recorded for native protein and DPro[31](13-64)NCp7 reveals the disappearance in DPro[31](13-64)NCp7 of the bH Phe[16]/H-4 aromatic proton of Trp[37] NOE, but also of the reported NOEs found between the second zinc finger domain and the linker region R[29]APRKK[34]. The stereochemical inversion of P31 therefore rules out the particular relative orientation of the zinc finger domains that is observed in the native protein [3].

Cys[23] (13-64)NCp7

The 1D spectrum of (13-64)NCp7 exhibits deshielded NH and αH resonances, indicating that Cys[23](13-64)NCp7 still binds two equivalents of zinc, but inspection of the proton chemical shifts as well as the NOESY data shows that the modified zinc finger domain, but also the linker region R[29]APRKK[35], undergo drastic conformational changes. NOEs were used as input for DIANA calculations and DIANA conformers were refined using the AMBER package. Figure 1 displays the superposition of the CCCC box and the native CCHC box. The first residues are relatively well superimposed (rmsd= 1.41 Å) but a kink at the Cys[23] level throws the backbone chain ends far away from each other.

The second zinc finger domain, which preserves the same folded conformation in both peptides, is unable to reach the spatial position found for NCp7. Accordingly, the mean distance between aromatic residue centers is increased from 5.1 Å to 16.5 Å.

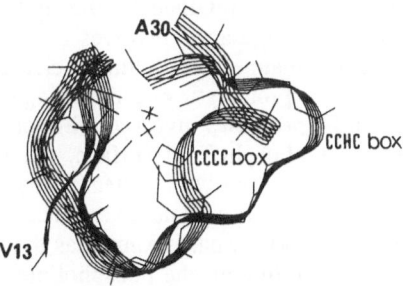

Fig. 1. *Superimposition of alpha carbon chain of the residues from V13 to G22 of native finger (CCHC box) and mutant finger (CCCC box).*

Biological activity

Dimerization of HIV-1 RNA(1-415) reveals that both mutations lead to a drastic decrease in the *in vitro* annealing activities of NCp7. Accordingly, site directed mutagenesis experiments replacing Pro[31] by Leu[31] or His[23] by Cys[23] lead to complete loss of virion infectivity.

The biological inactivity of DPro[31](13-64)NCp7 and Cys[23](13-64)NCp7 is therefore probably due to changes in the relative orientation of the short basic sequences flanking the zinc fingers, thus inhibiting their interactions with nucleotidic sequences. These observations should facilitate the rational design of compounds aimed at inhibiting retrovirus replication.

References

1. Darlix, J.L., Gabus, C., Nugeyre, M.T., Clavel, F. and Barre-Sanoussi, F., J. Mol. Biol., 216 (1990) 689.
2. Morellet, N., Jullian, N., de Rocquigny, H., Maigret, B., Darlix, J.L. and Roques, B.P., EMBO J., 11 (1992) 3059.
3. Morellet, N. and Roques, B.P., submitted.
4. de Rocquigny, H., Gabus, C., Vincent, A., Fournié-Zaluski, M.C. and Roques, B.P., Proc. Natl. Acad. Sci. U.S.A., 89 (1992) 6472.

Ca^{2+} Binding properties of synthetic γ-carboxyglutamic acid containing peptides homologous to protein C

T. Colpitts and F.J. Castellino

Department of Chemistry and Biochemistry, University of Notre Dame, Notre Dame, IN 46615, U.S.A.

Introduction

Mammalian haemostasis is maintained by the highly regulated systems of coagulation and fibrinolysis. Coagulation involves the activation of a number of serine proteases in a sequential manner. The ultimate cleavage, fibrinogen to fibrin, forms the clot. A number of the proteolytic events leading up to production of fibrin are surface dependent; that is, proteases involved will not cleave their physiologic substrates in the absence of phospholipid surface, i.e. platelets. This surface dependent interaction is always found with proteases that contain a number of γ-carboxyglutamic acid residues [1]. These residues are clustered at the amino terminus among the first 38 residues forming what is known as the Gla domain. Its known function is to bind calcium, facilitating the phospholipid interaction. This study investigates both the structural and functional aspects of this region as an independent domain.

Results and Discussion

We have synthesized the 38 residue domain as well as a 48 residue peptide which contains an extra 10 residue α-helix thought to stabilize the folded Gla region. The folding of this region is calcium dependent [2].

The calcium binding data fit a simple model of two types of sites; tight sites with Kd < 300μM, n = 2, and slight negative cooperativity, and weak sites with Kd ~ 1.5mM, n > 5, and positive cooperativity. The conformational change observed upon calcium binding was found to correlate with the weak sites. CD analysis of the peptide at various calcium concentrations shows an increase in α-helical content with a C$_{50}$ of 1.6mM. A schematic is presented in Figure 1 to illustrate the system.

In order to identify the specific Gla residues involved in each type of site, the 48mer peptide was synthesized with individually labelled γ,γ-^{13}C$_2$ Gla. Thus far, four positions have been labelled and titrated. Gla 20 and Gla 14 participate in the weak binding sites. Gla 26 and Gla 16 are involved in the tight sites. Figure 1 shows the chemical shift dependence of Gla 16 on calcium.

The conformations of the folded, calcium bound peptides were similar to protein C. Competition experiments with a monoclonal antibody (JTC1) which recognizes only the folded, calcium bound form of protein C, showed the peptides to have an 8-fold weaker affinity for the antibody. This conformation is required for phospholipid interaction. The calcium dependence for phospholipid binding was the same for protein C as for the peptides, again $C_{50} \sim 0.5 - 1.5mM$. However, the affinity at saturating calcium of protein C for the vesicle surface was two orders of magnitude greater. This strongly suggests that another part of the protein makes contact with the membrane surface.

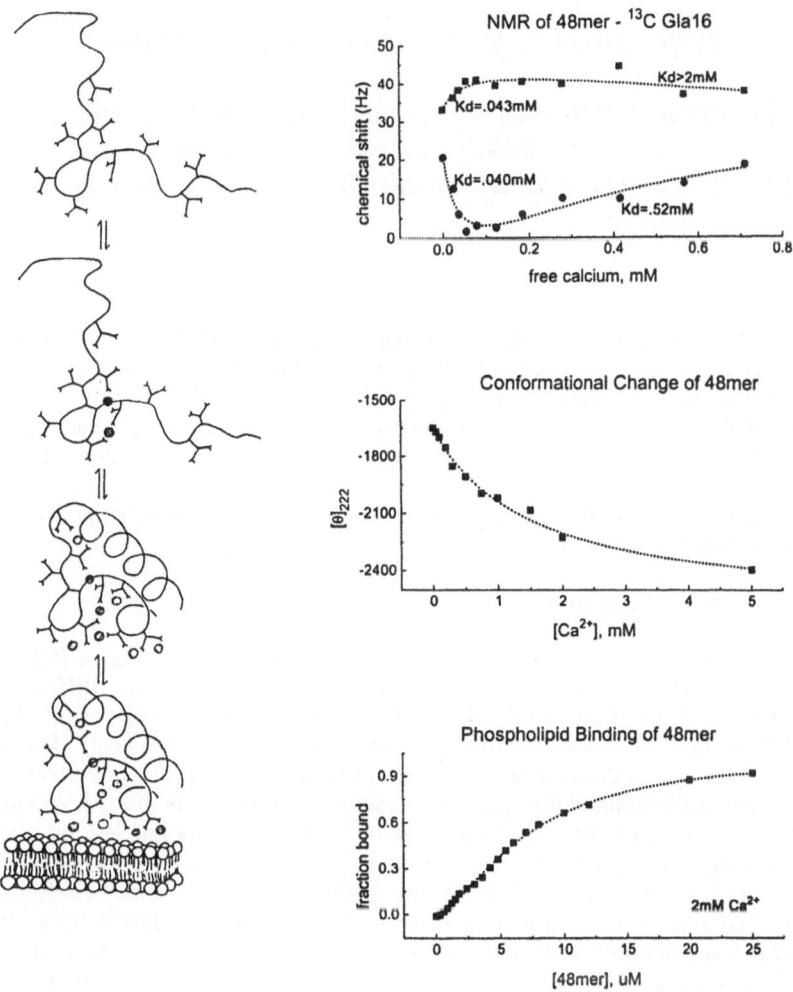

Fig. 1. *Proposed model for calcium binding and folding of the Gla domain.*

References

1.   Furie, B. and Furie, B.C., Cell, 53 (1988) 505.
2.   Colpitts, T. and Castellino, F.J., Biochemistry, 1994, in press.

Recognition of the tripeptide glutathione by cytosolic glutathione S-transferases: An X-ray crystallographic analysis

H.W. Dirr[1,2], P. Reinemer[2] and R. Huber[2]

[1]Department of Biochemistry, University of the Witwatersrand,
PO Wits 2050 Johannesburg, South Africa
[2]Max-Planck-Institut für Biochemie, W-8033 Martinsried, Germany

Introduction

Glutathione (γ-Glu-Cys-Gly), the major nonprotein thiol in cells, participates in a variety of cell functions including the detoxification of toxic electrophilic compounds and the biosynthesis of peptidoleukotrienes. These two functions are mediated through glutathione S-conjugation reactions catalyzed by the supergene family of glutathione S-transferases [1]. The enzymes' active site consists of two distinct regions; a polar G-site for binding glutathione and a hydrophobic H-site for the electrophilic substrate. Features for the recognition and binding of glutathione at the G-site are discussed below.

Results and Discussion

Crystal structures for cytosolic class Pi glutathione S-transferases (GSTP1-1) complexed with glutathione analogues were determined by X-ray diffraction at resolution of 2.8Å (human GSTP1-1·S-hexylglutathione; R-factor = 0.196) and 2.11Å (porcine GSTP1-1·glutathione sulphonate; R-factor = 0.165) [2-4]. Inspection of these structures indicate a network of multiple polar interactions involved in the recognition and binding of the glutathione analogues at the G-site of the homodimeric glutathione S-transferases, as illustrated in Figure 1. Oligonucleotide-directed mutagenesis studies [5] have also confirmed the participation of the G-site residues Tyr7, Arg13, Gln62 and Asp96 in binding glutathione. The extended conformation assumed by the sequestered analogues as well as the strategic interactions closely resemble those features observed for reduced glutathione bound at the G-site of the class Mu isoenzyme [6] and for S-benzyl glutathione bound at the G-site of class Alpha glutathione S-transferase [7]. Surface and electronic complementarity between the protein and glutathione's γ-glutamyl moiety is extensive in support of structure-activity data that this group is the principal binding determinant for the tripeptide. Furthermore, the extent to which the glutathione backbone is sequestered at the G-site through polar interactions indicates tight constraints on thiol substrate recognition and explains the high level of substrate specificity displayed by the glutathione S-transferases for reduced glutathione.

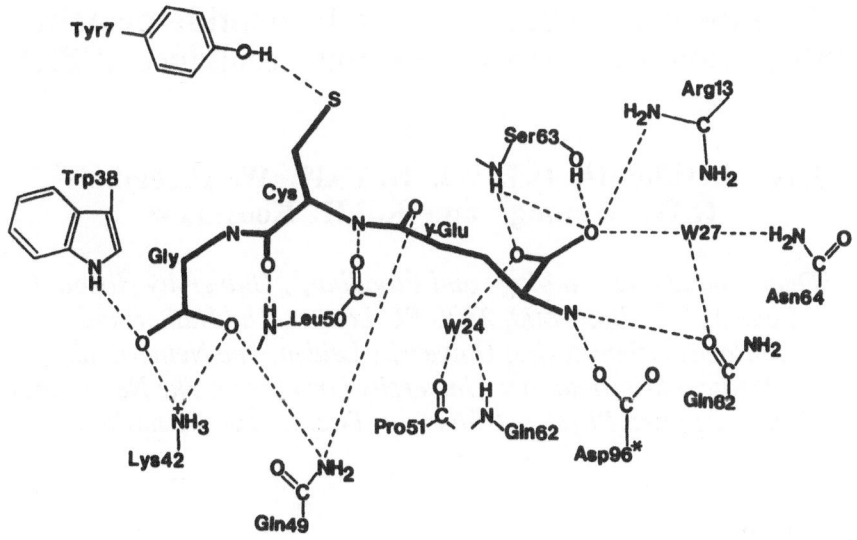

Fig. 1. Schematic representation of the polar interactions between glutathione (thick solid line) and protein moieties at the glutathione-binding site (G-site) of the homodimeric class Pi glutathione S-transferases (Mr 48500). One G-site exists on each subunit and Asp96 (marked with an asterisk), which salt links to glutathione's amino group at the subunit interface, is the only G-site ligand from the adjacent subunit. Dashed lines correspond to hydrogen bonds (<3.5Å). Water-mediated hydrogen bonds with solvent molecules W24 and W27 are also shown.

References

1.   Mannervik, B. and Danielson, U.H., CRC Crit. Rev. Biochem., 23 (1988) 283.
2.   Dirr, H.W., Mann, K., Huber, R., Ladenstein, R. and Reinemer, P., Eur. J. Biochem., 196 (1991) 693.
3.   Reinemer, P., Dirr, H.W., Ladenstein, R., Schäffer, J., Gallay, O. and Huber, R., EMBO J., 10 (1991) 1997.
4   Reinemer, P., Dirr, H.W., Ladenstein, R., Huber, R., LoBello, M., Federici, G. and Parker, M.W., J. Mol. Biol., 227 (1992) 214.
5.   Manoharan, T.H., Gulick, A.M., Reinemer, P., Dirr, H.W., Huber, R. and Fahl, W.E., J. Mol. Biol., 226 (1992) 319
6.   Ji, X., Zhang, P., Armstrong, R.N. and Gilliland, G.L., Biochemistry, 31 (1992) 10169
7.   Sinning, I., Kleywegt, G.J., Cowan, S.W., Reinemer, P., Dirr, H.W., Huber, R., Gilliland, G.L., Armstrong, R.N., Ji, X., Board, P.G., Mannervik, B. and Jones, T.A., J. Mol. Biol., 232 (1993) 192.

An improved method to study peptide-protein interaction by surface plasmon resonance (SPR)

J.W. Drijfhout[1], D.J. v.d. Heuvel[4], W. Bloemhoff[2], G.W. Welling[3] and R.P.H. Kooyman[4]

[1]Dept. Immunohaematology and Bloodbank, University Hospital
Leiden, P.O. Box 9600, 2300 RC Leiden, The Netherlands
[2]Gorlaeus Laboratories, University Leiden, The Netherlands
[3]Dept. Medical Microbiology, University Groningen, The Netherlands
[4]Dept. Applied Physics, University Twente, The Netherlands

Introduction

Application of SPR for investigating interactions between biomolecules is gaining interest because the technique is easy, fast, sensitive and relatively cheap. Moreover, the SPR-technique has shown to be a fruitful concept in the design of various kinds of immunosensors. Methods for immobilization of one of the components to a (gold) layer, a prerequisite for applying SPR, generally result in molecular layers that are randomly immobilized with poorly defined composition. Here we present a method for coupling synthetic peptides directly to the gold, which is easy to perform, highly reproducible and results in an ordered peptide monolayer, of which the precise composition can be accurately tuned.

Results and Discussion

The immobilization method presented here makes use of the high affinity of a gold surface for thiol-functionalities. Since thiol containing peptides are susceptible to (air)oxidation they are produced by SPPS in their S-acetyl-protected form. To induce the formation of an ordered monolayer [1] an alkyl spacer was coupled between the peptide and the thiol function. In order to obtain a monolayer in which the spatial arrangement of the peptide molecules is such that they are able to interact with protein molecules in solution the monolayer was formed applying a mixture of lipopeptide and dilutor. Thus in the last steps of SPPS peptides were elongated with 11-amino-undecanoic acid (Fmoc derivative, Bop/NMM coupling). After Fmoc removal the S-acetylmercaptoacetyl (SAMA) moiety was attached by coupling with the Pfp ester of S-acetylmercaptoacetic acid in the presence of HOBt [2]. S-acetylmercaptoacetyl-11-amino-undecanoic acid was used as a dilutor. Lipopeptide and dilutor (micromolar concentration in PBS) were mixed in various ratios and S-acetyl deprotected with hydroxylamine at neutral pH. By applying such a mixture to the gold surface a strong binding was achieved between the thiol groups and the gold, resulting in a monolayer stabilized by hydrophobic interaction between the aliphatic chains (see Figure 1).

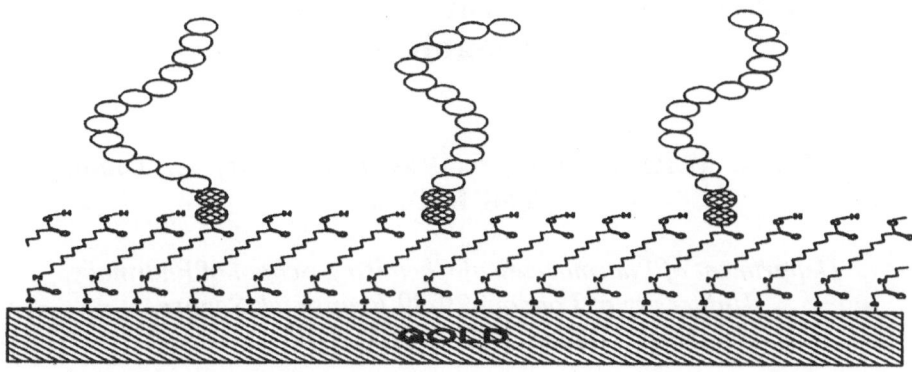

Fig. 1. Schematic representation of the lipopeptide/dilutor mixture forming an ordered monolayer on a gold surface.

As an example we present here the results of SPR studies with a peptide derived from a herpes simplex glycoprotein (9-21) [3] and a monoclonal antibody (A16) which is known to react with this peptide. The binding of 9-21 to A16 proved to be sensitive to the dilutor/lipopeptide ratio. At 3 mol% peptide coverage the binding of the monoclonal was optimal. At this dilution the peptide molecules probably have the proper spacing for allowing interaction with the antibody. With higher peptide loading the peptide molecules are too close together for optimal interaction with the protein, whereas with lower peptide loading not enough protein is bound to cover the surface. At 3 mol% peptide coverage, assuming an occupied area of 21 $Å^2$ per bound thiol compound [4], the average nearest neighbour peptide distance is approximately 100 Å. This is in the order of the dimensions of an antibody molecule. Under these conditions we find a bound antibody/peptide ratio of 0.4, probably caused by the distribution of nearest neighbour distances around the average. The affinity constant of this particular interaction was determined to be 2×10^8 M^{-1} and the detection limit for A16 proved to be 10^{-11} M.

Because of the good reproducibility and high sensitivity the described SPR set-up is very well suited to be used as immunosensor.

References

1. Bain, C.D. and Whitesides, G.M., Angew. Chem. Int. Ed. Engl., 28 (1987) 507.
2. Drijfhout, J.W., Bloemhoff, W., Poolman, J.T. and Hoogerhout, P., Anal. Biochem., 187 (1990) 349.
3. Weijer, W.J., Drijfhout, J.W., Geerligs, H.J., Bloemhoff, W., Feijlbrief, M., Bos, C.A., Hoogerhout, P., Kerling, K.E.T., Popken-Boer, T., Slopsema, K., Wilterdink, J.B., Welling, G.W. and Welling-Wester, S., J. Virol., 62 (1988) 501.
4. Alves, C.A., Smith, E.L. and Porter, M.D., J. Am. Chem. Soc., 114 (1992) 1222.

Investigation of protein kinase substrate recognition

N. Flinn, M.R. Munday, C. Van der Walle, C. Toomey and I. Toth

*Department of Pharmaceutical Chemistry, School of Pharmacy,
University of London, 29-39 Brunswick Square,
London WC1N 1AX, U.K.*

Introduction

Acetyl Co-Enzyme A Carboxylase (ACC) is one of the major enzymes which regulate the rate of mammalian fatty acid biosynthesis. Control of the enzyme is mediated by physiological phosphorylation and inactivation, via a Protein Kinase and dephosphorylation and activation, via a protein phosphatase [1, 2, 3]. Originally it was thought that the protein kinase responsible was cyclic AMP dependent protein kinase (cAMP-PK), however, extensive research has shown the phosphorylation to be carried out by an AMP activated protein kinase (AMP-PK). The major phosphorylation site on ACC is at Ser[79], with its surrounding amino acid sequence being H[73] M R S[76] S M S[79] G L H L V K. The importance of His[73], Met[74] and Arg[75], to the recognition of this sequence by the AMP-PK was assessed. Nine pentadecapeptides were synthesized on solid phase, and the amino acids 73-76 were varied. The ability of the synthetic peptides to act as substrates for the AMP-PK were investigated, using Michaelis-Menten kinetics to provide information regarding binding and activity. Individual protein kinases recognize a distinct amino acid consensus sequence around the reversible phosphorylation site. Some consensus sequences identified by protein kinases are already known. A general requirement is the presence of basic residues N-terminal to the Ser/Thr being phosphorylated. However, the involvement of secondary structure as a possible factor in recognition, cannot be ruled out.

The amino acid sequence of known AMP-PK substrates were studied for any homology about the phosphorylation site, to provide an indication of the residues involved in the recognition of substrates. In enzymes Acetyl-CoA Carboxylase Ser79, Ser1200, Ser1215, Hormone Sensitive Lipase Ser565, Glycogen Synthase Ser7 and HMG-CoA Reductase Ser871, N-terminal to the phosphorylated Ser, basic/polar residues are common at position -3 and -4. Hydrophobic residues are present at -5, particularly Met, while at -6 a ring structure is maintained in all but one of the substrates [4].

The following analogues were synthesized and assayed [5] (Graph 1, Graph 2):

H M R S S M S[79] G L H L V K (R R) CONTROL, H M A R S S M S[79] G L H
L V K (R R), H M A S S M S[79] G L H L V K (R R), H F R S S M S[79] G L H L
V K (R R), W M R S S M S[79] G L H L V K (R R), A M R S S M S[79] G L H L
V K (R R) and P M R S S M S[79] G L H L V K (R R).

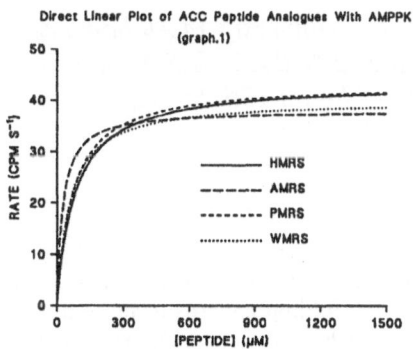

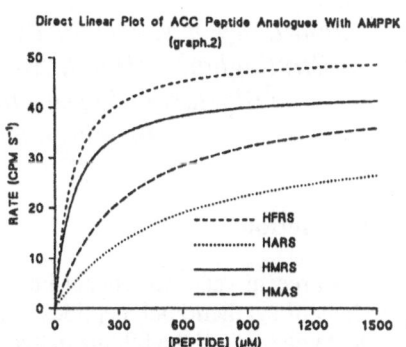

A basic residue such as Arg or Lys seems to be extremely important at positions -3 and -4, indicating the potentiality of a charged interaction between the enzyme binding site and the peptide. At position -5, a hydrophobic residue is required, and primarily one with a large, bulky side chain. Indeed, Phe appears to exhibit the most favourable result, when introduced to this position, suggesting the presence of a hydrophobic pocket in the conformational structure of the binding site within AMPPK. The ring formation at position -6 would appear to be a hindrance to the coupling of the substrate with the enzyme. Surprisingly Ala emerges as the prime candidate to occupy this position, showing that even though altering the ring size produces a slight decrease in the Km value, a small, neutral residue provides an enormous increase in the binding affinity. Thus, knowledge of this consensus sequence could lead to the assembly of potent specific inhibitors of AMPPK which would provide useful information, when used as a pharmacological tool, as to the physiological importance of the enzyme. Also, a specific inhibitor of AMPPK may be of some significance in the treatment of Type I Diabetes Mellitus, by increasing the flux of glucose through the lipid biosynthetic pathways.

References

1. Hardie, D.G., Prog. Lipid Res., 28 (1989) 117.
2. Carling, D., Zammit, V.A. and Hardie, D.G., FEBS Lett., 223 (1987) 217.
3. Munday, M.R., Campbell, D.G., Carling, D. and Hardie, D.G., Eur. J. Biochem., 175 (1988) 331.
4. Aitken, A., Identification of Protein Consensus Sequences, Ellis Horwood, London, U.K., 1990, p.42.
5. Davies, S.P., Carling, D. and Hardie, D.G., Eur. J. Biochem., 186 (1989) 123.

Design and synthesis of poly-tricosapeptides to enhance hydrophobic-induced pKa shifts

D.C. Gowda[1,2], T.M. Parker[1], C.M. Harris[2], R.D. Harris[2]
and D.W. Urry[1]

[1]Laboratory of Molecular Biophysics, The University of Alabama at
Birmingham, VH300, Birmingham 35294-0019, U.S.A. and
[2]Bioelastics Research, Ltd., 1075 South 13th Street,
Birmingham, AL 35205, U.S.A.

Introduction

In recent years significant interest has been generated in studying the *de novo* designs of synthetic polypeptides which mimic the folding patterns of proteins. In order to understand the folding patterns of proteins, it becomes essential to understand the function of ionizable side chains of amino acid residues with shifted pKa values. In this direction, here we report the synthesis and note the partial characterization of a set of protein-based polymers, actually polytricosapeptides, poly[3(GVGVP), 2(GFGFP), (GEGFP)](II); poly[2(GVGVP), 2(GVGFP), (GFGFP), (GEGFP)] (III); poly[GEGFP GVGVP GVGVP GVGVP GFGFP GFGFP] (IV); poly[GEGFP GVGVP GVGFP GFGFP GVGVP GVGFP] (V); and poly[GEGVP GFGFP GFGVP GVGVP GFGFP GVGVP] (VI), all having the same theoretical and essentially the same analytical amino acid compositions. These were designed for the purposes of (1) maximizing hydrophobic-induced pKa shifts by most proximal placing of five Phe residues to a single Glu residue in polymer V being designed on the basis of the working β-spiral structure of poly(GVGVP) [1, 2], (2) using primary structure to achieve proximity of Phe to Glu in polymer IV, and (3) having the same composition but with a disordering of the corresponding six composite pentamers in polymers II and III.

Results and Discussion

In a previous report [3] was described the synthesis of poly[4(GVGVP), GEGVP)] (I) with the side chain function of Glu protected by methyl ester and deblocked by base treatment with NaOH. The synthesis was subsequently repeated by following the same procedure but the same result without the side reaction could not be obtained. Therefore, prior to attempting the synthesis of polytricosapeptides, Glu containing 25 mers were assembled on the solid phase with methyl, benzyl and cyclohexyl ester protection for the Glu side chain in order to choose the best side chain protecting group which would minimize the side reactions of Glu. Only on preparation using the cyclohexyl ester protecting group was the original pKa routinely obtained. Our results support the conclusion that the cyclohexyl ester is a suitable protecting group for the synthesis of peptides containing Glu.

The tricosapeptides IV, V and VI (fixed sequences) were synthesized by the [(5+5+5)+(5+5+5)] fragment coupling strategy in the classical solution methods. The pentamers required for this purpose were synthesized as previously described [4, 5]. In the syntheses, the Boc group was used for N^α-protection and the cyclohexyl group for Glu side chain protection. The C-terminus carboxyl group was protected by the benzyl ester, and its removal was affected by hydrogenolysis using H_2/Pd-C(10%). All coupling reactions and deblocking were achieved by EDCI/HOBt and TFA, respectively. For polymers IV, V and VI, the tricosamer acids were deblocked and 1M solution of each TFA salt was polymerized using EDCI and HOBt in the presence of 1.6 equiv. of NMM as base. After 14 days, the polymers were then each dissolved in water, dialyzed using 3500 mol wt. cut-off dialysis tubing and lyophilized. The Glu side chain protection was deblocked using HF:p-cresol (90:10,v/v) for 1 h at 0°C. The polymers were then dissolved in water, dialyzed using 50kD mol wt. cut-off dialysis tubing and lyophilized. For polymers II and III, the composite pentamers were converted to p-nitrophenyl esters, mixed in the appropriate ratios and polymerized to obtain polymers with disordering of the composite pentamers of polymers IV and V, respectively.

Table 1 *Amino acid composition and experimental pKa values for polymers I through VI*

Amino Acid or pKa	Polymer							
	I		II	III	IV	V	VI	II- VI
	Found	Theoretical	Found	Found	Found	Found	Found	Theoretical
Glu (E)	0.20	0.20	0.18	0.15	0.16	0.15	0.18	0.17
Gly (G)	1.99	2.00	2.10	1.93	2.06	2.01	2.01	2.00
Pro (P)	1.00	1.00	1.00	1.00	1.00	1.00	1.00	1.00
Val (V)	1.79	1.80	0.91	0.99	0.91	0.94	0.93	1.00
Phe (F)	- - - -	- - - -	0.78	0.90	0.86	0.84	0.80	0.83
Experimental pKa	4.3 ± 0.02		4.7 ± 0.05	6.3 ± 0.05	7.7 ± 0.1	8.1 ± 0.1	6.3 ± 0.1	

In the amino acid analyses, Pro is taken as 1.0.

Verification of the syntheses was achieved by amino acid analyses shown in Table 1 and by means of the NMR. The acid-base titrations were carried out at 20° C as previously described using 45 minutes per data point under a nitrogen atmosphere [6], and the experimental acid-base titration curve for an individual run of the polymer VI is given in Figure 1. The normal pKa for the carboxyl moiety of the Glu residue is 4.3. Polymers II, III, IV, V and VI exhibited pKa values for the Glu residues of 4.7, 6.3, 7.7, 8.1 and 6.3, respectively. With only one Glu per 30 residues, the charge-charge repulsion mechanism for pKa shifts is ruled out as the polymers are clearly of a composition for the hydrophobic domain [7]. A 3.8 pH unit shift in pKa represents the largest pKa shift documented for a Glu residue in a simple polypeptide-water system, an apolar-polar repulsive free energy of hydration of 5 kcal/mole, and a significant interaction in protein structure and function. These results were further supported by our recent work on the polymer system, poly[f_V(IPGVG),f_X(IPGXG)] (X=Glu or Asp) where larger pKa shifts were obtained as f_X became less than 0.5 [7, 8].

Fig. 1. Acid-base titration curve for polymer VI and the first derivative. The starting concentration was 60 mg/ml and the temperature was 20°C.

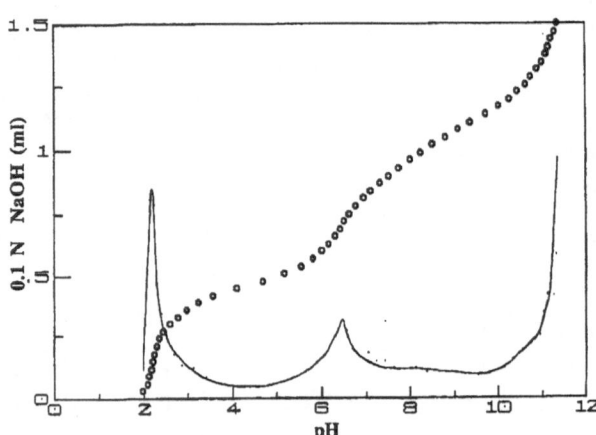

Acknowledgements

This work was supported in part by contracts DAAL03-92-C-0005 (to R.D.H.) from the Department of the Army, Army Research Office, and N00014-89-J-1970 (to D.W.U.) from the Department of the Navy, Office of Naval Research.

References

1. Urry, D.W., Gowda, D.C., Peng, S.Q., Parker, T.M. and Harris, R.D., J. Am. Chem. Soc., 114 (1992) 8716.
2. Urry, D.W., Angew. Chem. Int. Ed. Engl., 32 (1993) 819.
3. Zhang, H., Prasad, K.U. and Urry, D.W., J. Protein Chem., 8 (1989) 173.
4. Williams, D.F. (Ed.), Biocompatibility of Tissue Analogues, CRC Press Inc., Boca Raton, FL, 1985, p.89.
5. Gowda, D.C., Parker, T.M., Harris, R.D. and Urry, D.W., In Basava, C. and Anantharamaiah, G.M. (Eds.), Peptides: Design, Synthesis, and Biological Activity, Birkhäuser, Div. Springer-Verlag, New York, N.Y., U.S.A., in press.
6. Urry, D.W., Peng, S.Q. and Parker, T.M., Biopolymers, 32 (1992) 373.
7. Urry, D.W., Peng, S.Q. and Parker, T.M., J. Am. Chem. Soc., 115 (1993) 7509.
8. Urry, D.W., Peng, S.Q., Parker, T.M., Gowda, D.C. and Harris, R.D., Angew. Chem. Int. Ed. Engl., 32 (1993) 1440.

Synthesis and interaction properties of the DNA binding domain of the sex-determining gene product SRY

F. Heitz, F. Poulat, S. Soullier, B. Calas, R. Bennes
and P. Berta

UPR 9008-U249, CNRS-INSERM, Route de Mende; B.P. 5051,
F-34033 Montpellier Cédex, France

Introduction

The "sex determining region" Y (SRY) gene controls the embryo towards a male development [1]. This gene encodes a 204 residue long protein containing 80 amino acid motif which shows a strong homology with the HMG (High Mobility Group) box [2]. We report here some properties which indicate that a synthetic peptide built of the 80 residues corresponding to this HMG box shows binding properties which can account for the biological activity of the full length protein and compare these properties with these of a deleted analogue.

Results and Discussion

The 80 residue long peptide of sequence DRVKRPMNAF[10]IVWSRDQR RK[20]MALENPRMRN[30]SEISKQLGYQ[40]WKMLTEAEKW[50]PFFQEAQKLQ[60]AM H REKYPNY[70]KYRPRRKAKM[80] called hereafter SRY80 was synthesized by the continuous flow method and it is shown that this peptide, similarly to the full length protein, binds specifically its target nucleotide sequence AATAAAG while that a shortened analogue lacking the 16 N-terminal residues (SRY64) does not [3].

How can such a peptide bind DNA? Owing to the ability of A-T rich polynucleotides to bind the dye (2'[4-hydroxyphenyl]-5-[4-methyl-1-piperazinyl]-2,5'-bi-1H-benzimidazole) or Hoechst 33258 in its minor groove, the fact that this dye could be displaced by the peptide would be a good indication that SRY80 interacts with the polynucleotide through contacts occurring in the minor groove [4]. From the experimental point of view, it is shown that the dye is, indeed, displaced by SRY80 as revealed by a lowering of the fluorescence of the dye. This finding is in agreement with the conclusions reported by van de Wetering and Clevers [5] who concluded to a binding in the minor groove on the basis of binding experiments using combined T-C and A-I substitution.

Is the minor groove the unique binding site? Certain information obtained by CD indicate that the major groove could also be involved in the binding procedure. Indeed, the far UV peptide contribution is modified by the presence of oligonucleotide (Figure 1) in the way that the addition of DNA containing the target sequence induces an increase of the α helical content up to about 20% (Figure 2). As the minor groove is too small to accommodate an α helix, the present observation suggests that the binding of SRY80 to DNA also occurs through the major groove which possesses suitable dimensions to accommodate an α helix [6]. The origin of the selectivity, minor or major groove, is still an open question. It must be mentioned that CD investigations also reveal that some conformational changes of the DNA occur when it binds SRY80. They are mainly characterized by modifications in the base absorption region centered around 280 nm. It is tempting to attribute these structural changes as an early indication of the DNA bending [7].

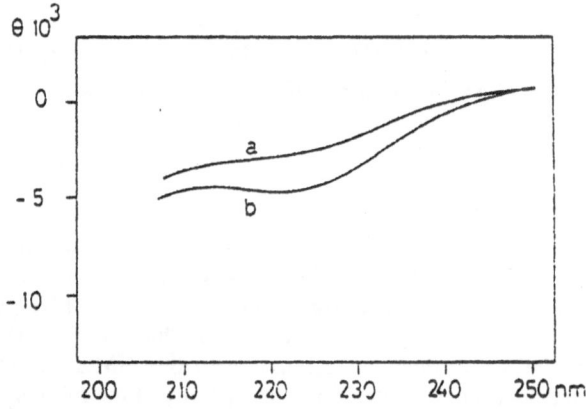

Fig. 1. CD spectra of: a) peptide, and b) peptide + oligonucleotide (the contribution of the oligonucleotide was subtracted and Θ is the ellipticity per amino acid residue).

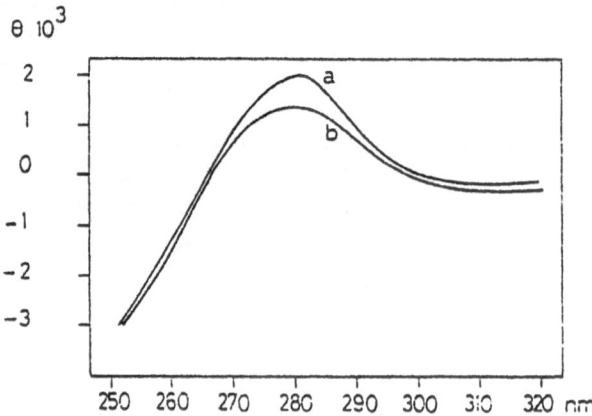

Fig. 2. CD spectra of: a) oligonucleotide, and b) oligonucleotide + peptide (the ellipticity Θ is given per base pair).

Concerning the peptide SRY80 itself, examination of its sequence and structure predictions according to Gibrat et al. [8] suggest the existence of at least three main structural domains. The first one, located at the N-terminus (1-23) is predicted to be in an extended form which could account for the minor groove binding. The remainder of the peptide consists of two α helices (18-23 and 40-65) separated by an unordered domain. It must be noticed that the second helix contains a proline residue in its middle (position 51), the role of which could be to induce a bending of the helix in order to favor the wrapping of the peptide in the major groove of the DNA.

Beside its DNA binding ability, all the properties reported so far for the full length protein are restricted to the synthetic peptide SRY80. The best illustration is found in the fact that, when microinjected into human fibriblast cells, the peptide can translocate to the nucleus while the shortened analogue does not. This suggests that the N-terminus of SRY80 contains a potential nuclear localization signal (NLS).

In conclusion, we have shown that a synthetic peptide with a sequence restricted to 80 residues can display most, if not all, properties of the full length (204 residues) protein. Such a strategy based on the use of synthetic peptides with reduced sequence may provide tools and open a way to get a better understanding of some biological functions mainly in the field of gene regulation.

Acknowledgements

Our thanks are due to Dr. N. Lamb for his help in the microinjection experiments.

References

1. Sinclair, A.H., Berta, Ph., Palmer, M.S., Hawkins, J.R., Griffiths, B.L., Smith, M.J., Foster, J.W., Frischauf, A.M., Lovell-Badge, R. and Goodfellow, P.N., Nature, 346 (1990) 240.
2. Wright, J.M. and Dixon, G.H., Biochemistry, 27 (1988) 576.
3. Poulat, F., Guichard, G., Gozé, C., Heitz, F., Calas, B. and Berta, Ph., FEBS Lett., 309 (1992) 385.
4. Suzuki, M., EMBO J., 8 (1989) 797.
5. van de Wetering, M. and Clevers, H., EMBO J., 11 (1992) 3039.
6. Kissinger, C.R., Liu, B., Martin-Blanco, E., Kornberg, T.B. and Pabo, C.O., Cell, 63 (1990) 579.
7. Ferrari, S., Harley, V.R., Pontiggia, L., Goodfellow, P., Lovell-Badge, R. and Bianchi, M., EMBO J., 11 (1992) 4497.
8. Gibrat, J.-F., Garnier, J. and Robson, B., J. Mol. Biol., 198 (1987) 425.

The binding specificity of kininogen analogues to serine proteases related to tissue kallikrein

G. Lalmanach[1], P. Barmettler[2], K. Tornheim[3] and J. Burton[2]

[1]Laboratoire d'Enzymologie et Chimie des Protéines - URA CNRS 1334, Université François Rabelais, 37032 Tours, France
[2]The Unit for Rational Drug Design, Boston University Medical Center, and [3]The Department of Biochemistry, Boston University School of Medicine, Boston, MA 02118, U.S.A.

Introduction

Tissue kallikrein (EC 3.4.21.35) is a serine protease which releases lysyl-bradykinin (kallidin) from human low molecular weight kininogen under physiologic conditions [1]. The investigation of the exact role of this enzyme requires the development of specific inhibitors which can be used *in vivo*. Protease inhibitors such as aprotinin, benzamidine and the peptidyl chloromethyl ketones inhibit many proteases in addition to tissue kallikrein. The substrate analogue inhibitors however, appear to be reasonably specific for tissue kallikrein [2]. To determine which residues in this sequence are responsible for this specificity, the 21 possible sequences occurring within the heptapeptide SPFFR↓SVQ, which surrounds the C-terminal cleavage site of bovine kininogen, were tested for binding to various trypsin-like proteases. This allowed the energy of interaction of the individual amino acid residues of the kininogen sequence with various serine proteases to be determined [2].

Results and Discussion

The 21 acetyl-peptidyl-amides occurring between Ser^{386} and Gln^{392} of bovine kininogen were synthesized and their ability to inhibit various human serine protease determined using a Biomek 1000 automated laboratory workstation as described previously [5]. The free energy of binding (ΔG) was calculated by using the equation $\Delta G = RT \ln K_i$ (Table 1). The standard deviation for values listed in Table 1 is ± 70 cal/mol.

The peptides show similar binding constants for both tissue kallikreins (β-PPK and HUK) but bind more weakly to HPK and plasmin. Thrombin is poorly inhibited and urokinase does not interact with kininogen analogues (data not shown). The individual contribution of each residue to the total binding energy ($\Delta\Delta G$) appears to be additive for the different enzymes. The combined binding energy for P_2 and P_1 residues is similar (-4 kcal/mol) for the trypsin-like enzymes (except urokinase).

946

Table 1 *Individual energetic contribution of each amino acid in the kininogen sequence*

Subsite Residue		$\Delta\Delta G$ (cal/mol)			
Enzyme $\rightarrow$		ßPPK	HUK	HPK	PL
P_4	Ser	-440	-440	-770	-440
P_3	Pro	-180	-200	+290	+540
P_2	Phe	-850	-660	ND	-630
P_1,	Arg	-2350	-3090	ND	-3450
P_1,	Ser	-1440	-1060	-1350	-130
P_2,	Val	-400	-70	-170	+410
P_3	Gln	-260	-140	-170	-40
TOTAL:		-5920	-5660	-4330	-3740

ßPPK, porcine tissue kallikrein; HUK, human tissue kallikrein; HPK, human plasma kallikrein; PL, plasmin; ND, individual values are not separable. $\Delta\Delta G$ for the combined P_2 and P_1 residues is -2160 cal/mol.

In addition, the affinity for the tissue kallikreins is enhanced by contributions of the P_3 and P_2' residues. For the other enzymes tested the binding energy is diminished by unfavorable interactions at these subsites. This difference is reflected in the tighter binding (10 to 50-fold) of the kininogen sequence to β-PPK and HUK when compared with the other serine proteases. The binding patterns observed with the kininogen analogues suggest that the most important residues for modifications aimed at enhancing the specificity of the inhibitors for tissue kallikrein should be P_3 and P_2'. This approach, combined with molecular modelling studies could be a rational basis to design a selective inhibitor of tissue kallikrein and, in a second step, convert this into a highly efficient drug structure.

References

1. Bhoola, K.D., Figuero, C.D. and Worthy, K., Pharmacol. Rev., 44 (1992) 1.
2. Deshpande, M.S. and Burton, J., J. Med. Chem., 38 (1992) 3094.

Phosphorylated fragments of the epidermal growth factor receptor: Sequence specificity for binding to phospholipase Cγ1 SH2 domains, and for kinase and phosphatase activity

D. Maclean[1], A.M. Sefler[1], D.J. McNamara[1], Z.-Y. Zhang[4], G. Zhu[4], E. Dobrusin[1], A. McMichael[2], D.W. Fry[2], S.J. Decker[3], J.E. Dixon[4], A.R. Saltiel[3], T.K. Sawyer[1] and J.A. Bristol[1]

[1]Departments of Chemistry, [2]Cancer Research, and [3]Signal Transduction, Parke-Davis Research Division, Warner-Lambert Company, Ann Arbor, MI 48105, U.S.A., [4]Department of Biological Chemistry, University of Michigan, Ann Arbor, MI 48105, U.S.A.

Introduction

The C-terminal region of growth-factor receptors (GFRs) is thought to be key to the transduction of signal from these receptors [for review see ref. 1]. Following ligand binding, the inherent kinase activity of the receptor is activated and phosphorylation of several C-terminal tyrosine residues may occur. Intracellular substrates for the activated GFR are sequestered by binding to these phosphotyrosine residues via their SH2 domains. This presumably brings the substrate into proximity with the kinase domain of the receptor, leading to phosphorylation and activation of the substrate. The ensuing cascade of enzyme activity ultimately results in mitogenic or differentiative effects on the cell. The action of intracellular phosphatases reverses these effects. Specificity in these pathways is thought to be the result of the precise sequence of amino acid residues in which the phosphotyrosine is embedded. Consensus patterns for SH2 binding [2], PTPase [3] or kinase [4] substrate potential have been noted, but no peptide series has been used to directly compare the relative specificity of these interactions. We have Ala-scanned a region of the EGF receptor (EGFR) which has been previously identified as being of key importance for binding of phospholipase Cγ (PLCγ) [5] and subjected the resulting series of analogs to analysis in the three assays.

Results and Discussion

Tyr-phosphorylated peptides were prepared as described earlier [6] with minor modifications. Briefly, peptides were assembled by standard Fmoc strategies, with the exception that Tyr was not side-chain protected. On-resin phosphorylation was carried out when required with N,N-diethyl-di-^{t}butylphosphoramidite/tetrazole/THF, followed by ^{t}BuO$_2$H/ CH$_2$Cl$_2$ and cleavage with 95% aq. TFA. Peptides were purified

Table 1 *Binding of Tyr-phosphorylated peptides to PLCγ NC SH2 [10], and kinetics of PTP1-catalyzed dephosphorylation [3]*

		PLCγ NC SH2	PTP1 Phosphatase		
	EGFR$_{988\text{-}998}$ Ala-scan	IC$_{50}$ (μM)	K$_m$ (μM)	k$_{cat}$ (s^{-1})	10^7 x k$_{cat}$/K$_m$ (M^{-1}s^{-1})
IA	DADEpYLIPQQG	20	2.63	75.7	2.88
IIA	AADEpYLIPQQG	7.8	4.94	74.2	1.50
IIIA	DAAEpYLIPQQG	300	5.65	65.8	1.16
IVA	DADApYLIPQQG	17	12.4	54.8	0.442
VA	DADEpYAIPQQG	18	4.05	57.8	1.43
VIA	DADEpYLAPQQG	>1000	4.39	73.8	1.68
VIIA	DADEpYLIAQQG	>1000	5.78	73.4	1.27
VIIIA	DADEpYLIPAQG	22	3.46	79.9	2.31
IXA	DADEpYLIPQAG	8.2	3.88	53.4	1.38
XA	DADEpYLIPQQA	13	2.38	61.4	2.58

Table 2 *Kinetics constants for EGFR-kinase-catalyzed Tyr-phosphorylation [9]*

	EGFR$_{988\text{-}998}$ Ala-scan	K$_m$ (μM)	k$_{cat}$ (s^{-1})	10^7 x k$_{cat}$/K$_m$ (M^{-1}s^{-1})
IB	DADEYLIPQQG	25.6	18.5	0.72
IIB	AADEYLIPQQG	51.2	16.0	0.31
IIIB	DAAEYLIPQQG	10.0	21.6	2.19
IVB	DADAYLIPQQG	40.5	17.6	0.44
VB	DADEYAIPQQG	99.7	9.4	0.09
VIB	DADEYLAPQQG	55.8	19.4	0.35
VIIB	DADEYLIAQQG	30.0	3.3	0.11
VIIIB	DADEYLIPAQG	16.5	25.9	1.57
IXB	DADEYLIPQAG	28.7	25.4	0.89

by reversed-phase HPLC, and the integrity and homogeneity confirmed by analytical HPLC, ^{1}H and ^{31}P NMR, amino acid analysis, and electrospray mass spectrometry.

Table 1 shows the effect of single Ala substitution on the ability of a Tyr-phosphorylated EGFR fragment to block the association of the activated EGFR with a protein construct containing both SH2 domains of PLCγ. Ala is well tolerated in most positions, but substitution of Asp^{990}, and, particularly Leu^{994} and Pro^{995} is deleterious. This effect is presumed to be the result of reduced binding to the SH2 construct. Hydrophobic residues C-terminal to the pTyr residue have been shown to be of general importance in SH2/peptide interactions [2, 7], but the importance of Asp[990], or indeed any residue N-terminal to pTyr has not previously been noted. Table 1 also shows the kinetic constants for the dephosphorylation of analogs IA-XA catalyzed by the mammalian cytoplasmic phosphatase PTP1 [8]. Ala-substitution generally results in smaller effects on PTP1 substrate specificity than it did on PLCγ binding. The major effect arose on replacement of Glu[991] by Ala, reducing the turnover of the resulting peptide [IVB] by almost 10-fold relative to the native sequence [IA]. EGFR-kinase-catalyzed phosphorylation [9] of analogs IB-IXB (Table 2) again shows relatively low specificity in comparison to PLCγ-binding. Leu[993] and Pro[995]-substitution resulted in lower turnover but Asp[990] replacement increased turnover by a factor of three.

In each of these assays we have attempted to study protein-protein interactions by analyzing peptide-protein relationships. Clearly these studies offer merely a simple set of models for the complex processes involved in signal transduction from the EGF receptor. Nevertheless, our data generally supports the hypothesis that specificity in signal transduction results from the primary structure surrounding phosphorylated Tyr residues, and that the binding of SH2 domains may have a particularly important role in molecular recognition and GFR-mediated signal transduction.

References

1. Cadena, D.L. and Gill, G.N., FASEB J., 6 (1992) 2332.
2. Songyang, Z., Shoelson, S.E., Chaudhuri, M., Gish, G., Pawson, T., Haser, W.G., King, F., Roberts, T., Ratnofsky, S., Lechleider, R.J., Neel, B.G., Birge, R.B., Fajardo, J.E., Chou, M.M., Hanafusa, H., Schaffhausen, B. and Cantley, L.C., Cell, 72 (1993) 767.
3. Zhang, Z-Y., Sefler, A.M., Maclean, D., McNamara, D.J., Dobrusin, E.M., Sawyer, T.K. and Dixon, J.E., Proc. Natl. Acad. Sci. U.S.A., 90 (1993) 4446.
4. Geahlen, R.L. and Harrison, M.L., In Kemp, B.E. (Ed.), Peptide and Protein Phosphorylation, CRC Press, Boca Raton, FL, U.S.A., 1990, p.239.
5. McNamara, D.J., Dobrusin, E.M., Zhu, G., Decker, S.J. and Saltiel, A.R., Int. J. Peptide Protein Res., 42(1993) 240.
6. Perich, J.W. and Johns, R.B., Synthesis, (1988) 142.
7. Eck, M.J., Shoelson, S.E. and Harrison, S.C., Nature, 362 (1993) 87.
8. Zhang, Z-Y., Maclean, D., Sefler, A.M., Roeske, R.W. and Dixon, J.E., Anal. Biochem., 211 (1993) 7.
9. Gill, G.N. and Weber, W., Meth. Enzymol., 146 (1987) 82.
10. Zhu, G., Decker, S.J. and Saltiel, A.R., Proc. Natl. Acad. Sci. U.S.A., 89 (1992) 9559.

Photochemical cross-linking between native rabbit skeletal troponin C and benzoylbenzoyl-TnI inhibitory peptide, residues 104-115

S.M. Ngai, F.D. Sönnichsen and R.S. Hodges

MRC Group in Protein Structure and Function, Department of Biochemistry, University of Alberta, Edmonton, Alberta, T6G 2H7, Canada

Introduction

Calcium induced muscle contraction requires the regulatory protein complex consisting of troponin and tropomyosin (TM) [1]. Previous investigations have demonstrated that TnI binds to actin-tropomyosin and inhibits the actomyosin ATPase. The TnI inhibition of the actomyosin ATPase is neutralized when calcium-saturated TnC forms a complex with TnI.

Talbot and Hodges [2, 3] have demonstrated that TnI residues 104-115 (Ip) comprise the minimum sequence necessary for the inhibition of actomyosin ATPase. This synthetic peptide mimicked TnI by possessing tropomyosin specificity [2] and TnC binding capabilities [4, 5], which resulted in neutralizing the Ip induced acto-S1-TM ATPase inhibition [6]. It has been concluded that the Ca^{2+}-dependent switch between muscle contraction and relaxation involves a switch of the TnI inhibitory region between actin-TM and TnC [7].

In the present study, a photoactivatable radioactive TnI peptide (BB-Ip) was synthesized with a benzoylbenzoyl moiety located at the N-terminus by solid-phase methodology. This peptide retained native biological activity and was used to probe the important interactions between TnC and the TnI inhibitory region. Based on our findings and previous crosslinking results together with the X-ray crystallographic structure of TnC [8] and the knowledge of the structure of Ip when bound to skeletal troponin C [9-11], we are able to compute a three dimensional structure of the TnC-Ip complex to demonstrate the corresponding interactions found within the TnC-Ip complex.

Results and Discussion

Our results indicated that the crosslinking of the BBIp to TnC was very specific in either Mg^{2+}- or Ca^{2+}-buffer since only a single cross-linked TnC fragment (residues 154-159) was detected (Figure 1). This gives us a valuable constraint in computing a regio-specific three dimensional structure of the TnC-Ip complex. The resulting structure demonstrated that Cys-98 of TnC is in close proximity to the Ip peptide. It has been shown that Cys-98 can be crosslinked to the TnI inhibitory region [12-15]. Also Phe-105 and Phe-154 of TnC (located in the hydrophobic groove) are participating in the binding (Figure 2). This is in agreement with the observation from fluorescence studies using recombinant TnC (where either Phe-105 or Phe-154 was replaced by Trp), and the two TnC-domains (C-domain where Phe-105 was replaced by Trp, N-domain where Phe-29 was replaced by Trp) that the preferred binding site for the Ip is located in C-terminal domain of TnC [16]. Crosslinking results by Kobayashi *et al.* [17] indicated that TnC in which Ala-57 was replaced by Cys and then modified with 4-maleimidobenzophenone can be crosslinked to native TnI in the region 113-121. With reference to our model of the TnC-Ip structure, we can visualize that if the Ip sequence was extended to include residues 116-121 and the N-domain of TnC is brought into close proximity to the C-domain, residue 57 of TnC would be situated to allow crosslinking to the inhibitory region of TnI.

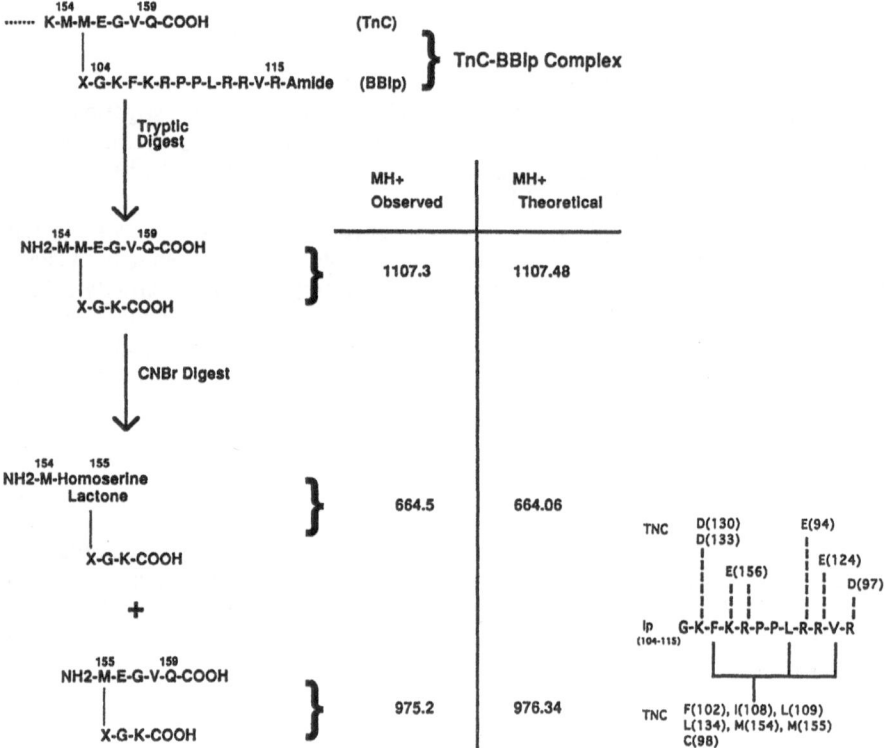

Fig. 1. Summary for characterization of the TnC-BBIp complex (left side). (Results are similar in either the presence or absence of calcium).

Fig. 2. Interactions within the TnC-Ip Complex (right side). Sequence of TnI inhibitory peptide residues 104-115 (Ip) and its interactions with TnC are shown above and below the Ip sequence; solid lines denote hydrophobic interactions whereas broken lines denote electrostatic interactions. Assignments of the corresponding interactions are based on a docked and energy minimized three dimensional TnC-Ip structure.

Wang *et al.* [18] demonstrated that binding of metal cations to the C-domain (sites III and IV) of a mutant TnC altered the environment around the amino acid at position 57 in the N-terminal domain (sites I and II). Grabarek *et al.* [19] and Rosenfeld and Taylor [20] showed that the binding of Ca^{2+} to the N-domain (sites I and II) altered the environment around Cys 98 in the C-terminal domain. These findings indicated that the N and C-terminal domains of TnC are communicating with each other upon the binding of Ca^{2+} or Mg^{2+}. We have demonstrated that the N-terminal TnI peptide (residues 1-40)/Rp when bound to TnC prevents Ip from interacting with TnC [21]. Recently we have shown that it is the C-terminal domain of TnC that strongly interacts with the TnI N-terminal peptide [22]. The switching between the TnI peptides (Rp and Ip) within the hydrophobic groove of TnC C-domain reveals that Rp is a negative effector of the regulatory process. Therefore, there must be a modulating event which governs the release of the TnI inhibitory (Ip) region from TnC by the Rp region of TnI.

Crosslinking studies combined with computer simulations have proposed a three-dimensional structure of the TnC-Ip complex. Though the N-domain of TnC contains the Ca^{2+}- regulatory sites that control muscle contraction, the two most important biologically active regions of TnI (Ip and Rp) bind to the C-domain. This raises the question of how these interactions transmit to the N-domain of TnC.

References

1. Ebashi, S. and Endo, M., Prog. Biophys. Mol. Biol., 18 (1968)123.
2. Talbot, J.A. and Hodges, R.S., J. Biol. Chem., 254 (1979) 3720.
3. Talbot, J.A. and Hodges, R.S., J. Biol. Chem., 256 (1981) 2798.
4. Cachia, P.J., Sykes, B.D. and Hodges, R.S., Biochemistry, 22 (1983) 4145.
5. Cachia, P.J., Gariepy, J. and Hodges, R.S., In Hidaka, H. and Hartshorne, D.J. (Eds.), Calmodulin Antagonist and Cellular Physiology, Ch. 5, Academic Press, New York, N.Y., U.S.A., 1985, p.63.
6. Cachia, P.J., Van Eyk, J.E., Ingraham, R.H., McCubbin, W.D., Kay, C.M. and Hodges, R.S., Biochemistry, 25 (1986) 3553.
7. Van Eyk, J.E. and Hodges, R.S., J. Biol. Chem., 236 (1988) 1726.
8. Herzberg, O. and James, M.N.G., J. Mol. Biol., 203 (1988) 761.
9. Campbell, A.P. and Sykes, B.D., In Hidaka, H. (Ed.), Calcium Protein Signaling: Advance in Experimental Medicine and Biology, Vol. 255, Plenum Press, New York, N.Y., U.S.A., 1989, p.195.
10. Campbell, A.P., Cachia, P.J. and Sykes, B.D., Biochem. Cell Biol., 69 (1991) 674.
11. Campbell, A.P. and Sykes, B.D., J. Mol. Biol., 222 (1991) 405.
12 Chong, P.C.S. and Hodges, R.S., J. Biol. Chem., 256 (1981) 5071.
13. Chong, P.C.S. and Hodges, R.S., J. Biol. Chem., 257 (1982) 2549.
14. Tao, T., Scheiner, C.J. and Lamkin, M., Biochemistry, 25 (1986) 7633.
15. Leszyk, J., Dumaswala, R., Potter, J.D., Gusev, N.B., Verin, A.D., Tobacman, L.S. and Collins, J.H., Biochemistry, 26 (1987) 7035.
16. Smillie, L.B., unpublished results.
17. Kobayashi, T., Tao, T., Grabarek, Z., Gergely, J. and Collins, J.H., J. Biol. Chem., 266 (1991) 13746.
18. Wang, Z., Sarkar, S., Gergely, J. and Tao, T., J. Biol. Chem., 265 (1990) 4953.
19. Grabarek, Z., Leavis, P.C. and Gergely, J., J. Biol. Chem., 261 (1986) 608.
20. Rosenfeld, S.S. and Taylor, E.W., J. Biol. Chem., 260 (1985) 252.
21. Ngai, S.M. and Hodges, R.S., J. Biol. Chem., 267 (1992) 15715.
22. Ngai, S.M. and Hodges, R.S., unpublished data.

Structure-dependent redox potentials of active-site fragments of thiol protein oxidoreductases

S. Rudolph-Böhner, F. Siedler and L. Moroder

Max-Planck-Institut für Biochemie, Au Klopferspitz, 8033 Martinsried, Germany

Introduction

The active sites of thioredoxin-type oxidoreductases are contained within characteristic Cys-X-Y-Cys segments. Their redox potentials reflect the propensity of the bis-cysteinyl-sequence portion for disulfide loop formation. As the main-chain conformation of the active site loop of these proteins is surprisingly similar [1] and only minor changes are observed between the oxidized and reduced state, the more or less relaxed conformation of the disulfide bridge should define the redox properties. In order to assess the influence of the protein packing interactions and of the sequence-dependent intrinsic free energy of formation of the disulfide loops on the redox potentials, the active site fragments 31-38, 10-17, 134-141 and 34-41 of thioredoxin (trx), glutaredoxin (grx), thioredoxin reductase (trr) and PDI were synthesized [2].

Results and Discussion

The bis-cysteinyl-octapeptides Ac-Trp-Cys-Gly-Pro-Cys-Lys-His-Ile-NH$_2$ (trx-His[37]-[31-38], Ac-Gly-Cys-Pro-Tyr-Cys-Val-Arg-Ala-NH$_2$ (grx-[10-17]), Ac-Ala-Cys-Ala-Thr-Cys-Asp-Gly-Phe-NH$_2$ (trr-[134-141] and Ac-Trp-Cys-Gly-His-Cys-Lys-Ala-Leu-NH$_2$ (PDI-[34-41]) differ remarkably in their tendency for intramolecular cyclization by air oxidation, whereby shielding of side-chain charged groups (addition of Gu HC1) and particularly, favoring ordered structures (addition of TFE) is remarkably shifting the product distribution towards monomeric cyclic structures (Table 1). The CD of the reduced peptides is strongly affected by contributions from the aromatic chromophores and therefore difficult to interpret. Nevertheless, the rank order observed in the cyclization experiments seems to agree with the preference of the peptides for ordered conformations both in water and water/TFE mixtures, except for grx-[10-17] which apparently is the most folded in both solvents. The rank order of Table 1 agrees fairly well with the redox potentials of the enzymes, i.e. PDI (-0.11)>grx(-0.23)>trr(-0.25)>trx(-0.27); however, the redox potentials of the peptides determined in glutathione redox buffer [3] are not in line with those of the proteins. This would suggest that the inherent redox potentials of the active sites are significantly affected in the folded proteins by the conformational restraints imposed locally on this portion of the molecule in the oxidized and particularly, in the reduced state. To analyze this aspect, the 3D structure of trr-[134-141] was resolved by NMR and MD simulations. As shown in Figure 1, the conformational characteristics of the active site loop of this class of proteins is fully retained in the excised fragment

Table 1 *Product distribution of air oxidation experiments in various solvents*

peptides (c =2.0·10⁻³M; 20.0°C)	ox.buf.: phosphate buffer (0.1M; pH 7.0)		6M Gu·HCl in ox.buf.		TFE·/ ox.buf. 1:1 (v,v)	
	monom.	oligom.	monom.	oligom.	monom.	oligom.
PDI-[34-41]	93 %	7 %	94 %	6 %	98 %	2 %
trr-[134-141]	84 %	16 %	96 %	4 %	87 %	13 %
grx-[10-17]	65 %	35 %	90 %	10 %	89 %	11 %
trx-His³⁷-[31-38]	29 %	71 %	34 %	66 %	89 %	11 %

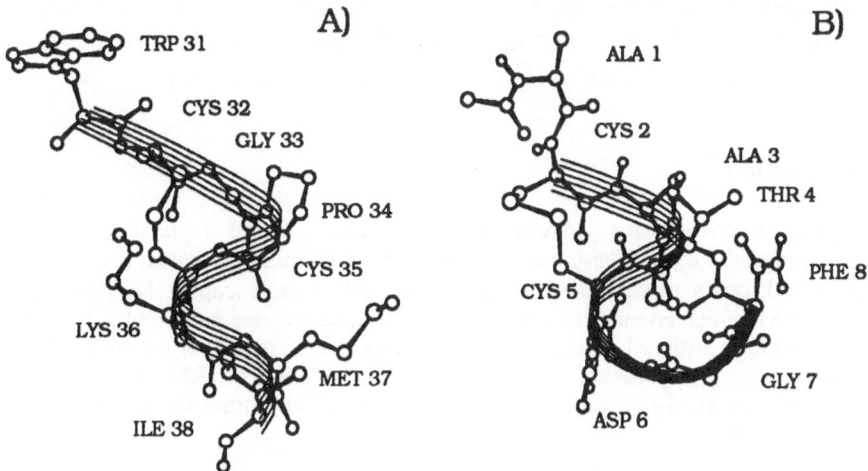

Fig. 1.   Comparison of A) the active site 3D structure of thioredoxin (X-ray) and B) the fragment trr-[134-141] (NMR).

with a β-type "turn III" (3_{10}) involving Ala-Thr-Cys-Asp and with an S-S torsion angle of - 83.3° (> +90° in the proteins). The opposite sign of this torsion angle in the trr-fragment as well as the preliminary observation that in grx-[10-17] the β-type turn is shifted to the N-terminus, i.e. to Cys-Pro-Tyr-Cys, suggest a stronger effect of the interactions in the folded state of the proteins on the conformation of the "functional domains" than expected for a 14-membered loop.

References

1. Eklund, H., Ingelman, M., Söderberg, B.-O., Uhlin, T., Nordlund, P., Nikkola, M., Sonnerstam, U., Joelson, T. and Petratos, K., J. Mol. Biol., 228 (1992) 596.
2. Siedler, F., Rudolph-Böhner, S., Doi, M. and Mororder, L., In Schneider, C.H. and Eberle, A.N. (Eds.), Peptides 1992, Escom, Leiden, The Netherlands, 1993, p.523.
3. Siedler, F., Rudolph-Böhner, S., Doi, M., Musiol, H.-J. and Mororder, L., Biochemistry, in press.

Identification of a binding site for the human immunodeficiency virus type 1 nucleocapsid protein

K. Sakaguchi[1], N. Zambrano[1], E.T. Baldwin[4], B.A. Shapiro[2], J.W. Erickson[4], J.G. Omichinski[3], G.M. Clore[3], A.M. Gronenborn[3] and E. Appella[1]

[1]Lab. of Cell Biology and [2]Laboratory of Mathematical Biology, National Cancer Institute, [3]Laboratory of Chemical Physics, National Institute of Diabetes and Digestive and Kidney Diseases, Bethesda, MD 20892, U.S.A. and [4]Structural Biochemistry, PRI/DynCorp, NCI-FCRDC, Frederick, MD 21702, U.S.A.

Introduction

Type C retroviruses typically contain two copies of the single stranded RNA genome in each virion. These two unspliced RNA molecules are linked together near their 5' ends by the dimer linkage structure. The HIV-1 nucleocapsid (NC) protein p7 (Figure 1A), which contains two sets of a Cys_3His_1 sequence motif and binds zinc to form a unique ordered structure [1], mediates dimerization of retroviral RNA containing the encapsidation sequence Psi and annealing of tRNA to the primer binding site [2-4]. Here, we report that NCp7 specifically binds to and dimerizes a synthetic RNA.

Results and Discussion

The sequences for 8 different isolates of HIV-1 were aligned and 150 nt on either side of the Psi region were chosen for RNA secondary structure prediction. The overwhelming majority of the predicted structures produced a pair of stems and loops separated by single stranded linker (Figure 1B). Stem-Loop 1 contains an unpaired A and the 5' major splice junction which is found within the GGUG loop sequence. This stem-loop is followed by 9 to 13 nt which are palindromic; 2 to 5 adenine nucleotides are followed by a run of 4 to 6 uracil nucleotides. The second stem-loop is shorter than the first and has the loop sequence GGAG. The second loop ends 20 nt 5' to the initiator AUG for the GAG-polyprotein. RNase probing of the synthetic RNA was carried out with either RNase I, V1, A, or T1. The results obtained support the presence of a stem and loop structure as predicted.

As shown in Figure 1C, p7(1-55) binds specifically to the 44-mer RNA with high affinity in presence of zinc ions, yielding two complexes, C1 and C2. Use of radioactive ^{65}Zn showed that both complexes contained metal ion, confirming the presence of peptide in both and proving that they are not just conformers of RNA. Analytical ultracentrifugation experiments indicate that C2 complex is comprised of two RNA molecules and one peptide. Gel shifts performed using a shorter fragment of NCp7 (position 13-51) containing both zinc binding motifs, showed no formation of the C2 complex, while the C1 band was apparent only at high concentration of peptide. Further, neither an N-terminal fragment of NCp7 (position 1-14) nor a different HIV-1 protein, p6, elicited complex formation; addition of zinc ion alone did not yield any complexes. These results suggest that C1 and C2 complex formation requires the N-terminal 55 residues of NCp7 in the presence of zinc. Mutations of Loop 1 or Stem 1 abolished the formation of the C2 complex. No striking changes were found with mutations introduced into Loop 2, Stem 2 or linker. These data show that the predicted stem and loop structure of Stem-Loop 1 plays a critical role in the formation of the NCp7 complex whereas Stem-Loop 2 is dispensable.

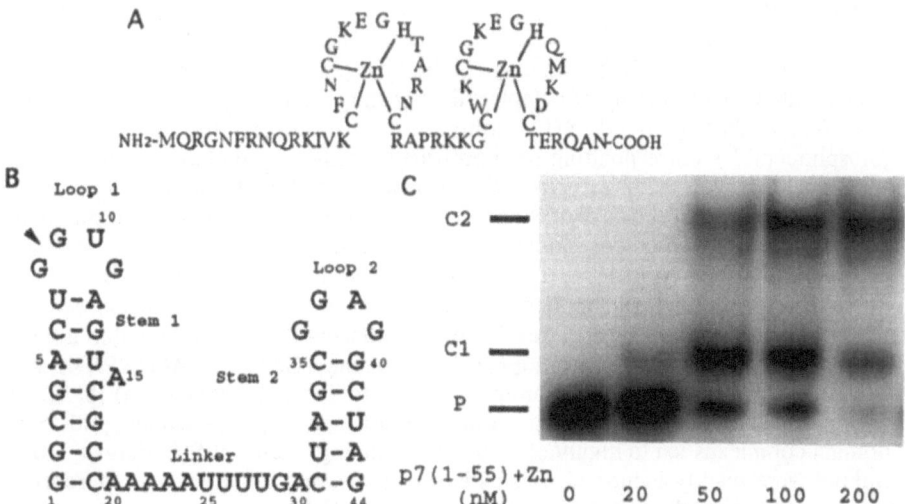

Fig. 1. (A) p7(1-55). (B) Computer-predicted structure of 44-mer. (C) Gel mobility-shift assay of 44-mer RNA with p7(1-55).

Our results show that NCp7 binds to a unique RNA structure within the Psi region; in addition, this structure is necessary for RNA dimerization. We propose that NCp7 binds to the RNA via a direct interaction of one zinc binding motif to Stem-Loop 1 followed by binding of the other zinc binding motif to the second RNA molecule, forming a bridge between the two RNAs.

References

1. Omichinski, J.G., Clore, G.M., Sakaguchi, K., Appella, E. and Gronenborn, A.M., FEBS Lett., 292 (1991) 25.
2. Aldovini, A. and Young, R.A., J. Virol., 64 (1990) 450.
3. Gorelick, R.J., Nigida, S.M., Bess, J.W., Arthur, L.O., Henderson, L.E. and Rein, A., J. Virol., 64 (1990) 3207.
4. de Rocquigny, H., Gabus, C., Vincent, A., Fournié-Zulski, M.C., Roques, B. and Darlix, J.L., Proc. Natl. Acad. Sci. U.S.A., 89 (1992) 6472.

A photoaffinity scan maps regions of an SH2 domain involved in phosphoprotein interactions

K.P. Williams, M. Chaudhuri and S.E. Shoelson

Research Division, Joslin Diabetes Center and Department of Medicine, Harvard Medical School, Boston, MA 02215, U.S.A.

Introduction

Src homology 2 (SH2) domains are modular phosphotyrosine binding pockets found within a wide variety of cytoplasmic signaling molecules [1]. We have shown previously that isolated SH2 domains retain specificity in binding to phosphopeptides corresponding to appropriate motifs surrounding phosphorylated tyrosines in growth factor receptors, their substrates, and viral transforming proteins [2-6]. Moreover, specific phosphopeptide binding to SH2 domains can modulate enzymatic activity with concomitant changes in SH2 domain structure [7]. We have developed a new approach to analyzing protein-protein interfaces, termed photoaffinity scanning, and applied this method to better understand phosphoprotein/ SH2 domain interactions. Each residue except phosphotyrosine (pY) within a tight-binding, IRS-1 derived phosphopeptide sequence (GNGDpYMPMSPKS) was substituted with the photoactive amino acid, benzoylphenylalanine (Bpa) [8, 9]. In addition to the effects on binding affinity, photolysis of Bpa-phosphopeptide/SH2 domain complexes led to highly efficient cross-linking. Sites of SH2 domain cross-linking were readily identified by standard methods and found to map appropriate regions of the phosphopeptide/SH2 domain interface [10].

Results and Discussion

Bpa-substituted phosphopeptides were prepared following an Fmoc/tBu-based synthetic strategy, using Fmoc-Bpa and Fmoc-Tyr(O(P(OCH$_3$)$_2$)) [2, 3, 10]. The N-terminal SH2 domain of phosphatidylinositol (PI) 3-kinase p85 was expressed in *E. coli* as a glutathione S-transferase (GST) fusion protein [4, 7, 10]. Relative affinities between the Bpa-substituted phosphopeptides and the purified SH2 domain were assessed by competition binding assay [2, 3, 10]. Substitutions at -4, -3, -2, -1, +2, +4 and +5, with respect to pY, had little effect on binding affinity. In contrast, Bpa-substitution of Met at the pY+1 and pY+3 positions reduced binding affinity 50- to 100-fold. These findings confirm the importance of Met residues within the YMXM or YVXM recognition motif of PI 3-kinase [3].

For cross-linking studies, the SH2 domain and GST portions of the expressed fusion protein were separated by proteolysis and the SH2 domain was purified by HPLC. Each Bpa-substituted phosphopeptide was incubated with the intact p85 SH2 domain under saturating conditions and irradiated at 350 nm. Cross-linked SH2 domains were retarded during SDS-PAGE, presumably due to addition of both mass and negative charge, and thus readily separated from the corresponding uncross-linked domain (Figure 1). Cross-linking efficiency was high, greater than 50%, for Bpa substitutions at pY-1, pY+1 and pY+4 positions, and these analogues were used for mapping studies. Larger amounts of cross-linked SH2 domains were isolated and proteolytically and/or chemically fragmented. Small cross-linked fragments and cross-link sites were identified by microsequencing, amino acid analysis and electrospray mass spectrometry. The Bpa-1 peptide cross-linked within a tryptic fragment of the SH2 domain corresponding to PI 3-kinase p85 residues 341-346 [11], whereas both Bpa+1 and Bpa+4 peptides cross-linked within a fragment corresponding to p85 residues 410-419.

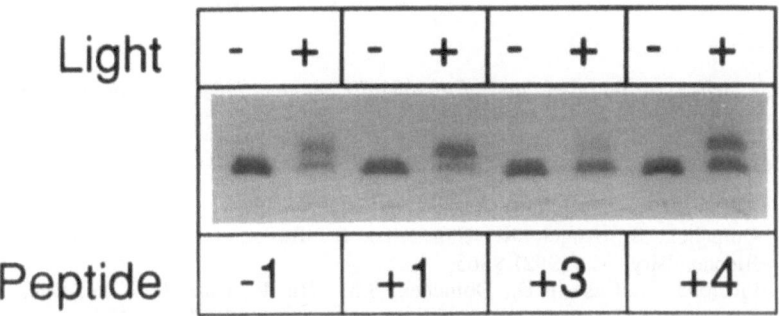

Fig. 1. *SDS-PAGE Separation of SH2 domain preparations following cross-linking with Bpa-substituted phosphopeptides. Cross-linking the SH2 domain results in a shift in apparent molecular weight. High efficiency cross-linking (>50%) was observed with Bpa-1, Bpa+1 and Bpa+4 phosphopeptides.*

Recent structural studies [12-14] allow further interpretation of these findings. The solution structure of the N-terminal p85 SH2 domain in the absence of bound peptide has been determined [12]. Bpa-1 cross-links a position within α-helix A, whereas Bpa+1 and Bpa+4 cross-link the disordered loop C-terminal to α-helix B (Figure 2). These results indicate that in the complex the phosphopeptide sits across the SH2 domain from α-helix A to the loop following α-helix B; peptide residues important for specificity (+1 to +4) interact with the disordered loop. Recently solved crystallographic structures of high-affinity phosphopeptide complexes with Lck and Src SH2 domains [13, 14] allow us to test the accuracy of these assignments. In both structures pY-1 and pY interact with helix A. The side chain of pY+3 makes extensive contacts with two loops, one of which corresponds to the p85 cross-link sites. These findings suggest that phosphopeptides bind p85, Src and Lck SH2 domains in similar configurations and validate the photoaffinity scan method.

In conclusion, photoaffinity scanning provides a new method for analyzing interactive protein surfaces. Bpa can be incorporated anywhere into a peptide or small protein by chemical synthesis. The general approach should prove useful for relating structure and function at many protein-protein interfaces.

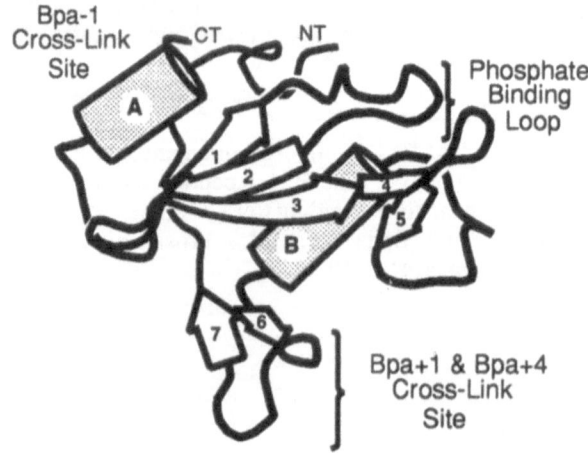

Fig. 2. Schematic representation of the p85 SH2 domain [12]. Sites of cross-linking and the phosphate binding loop are shown.

References

1. Pawson, T. and Gish, G.D., Cell, 71 (1992) 359.
2. Domchek, S., Auger, K., Chatterjee, S., Burke, T., Jr. and Shoelson, S.E., Biochemistry, 31 (1992) 9865.
3. Piccione, E., Case, R.D., Domchek, S.M., Hu, P., Chaudhuri, M., Backer, J.M., Schlessinger, J. and Shoelson, S.E., Biochemistry, 32 (1993) 3197.
4. Felder, S., Zhou, M., Hu, P., Urena, J., Ullrich, A., Chaudhuri, M., White, M.F., Shoelson, S.E. and Schlessinger, J., Mol. Cell. Biol., 13 (1993) 1449.
5. Payne, G., Shoelson, S.E., Gish, G., Pawson, T. and Walsh, C.T., Proc. Natl. Acad. Sci., U.S.A., 90 (1993) 4902.
6. Songyang, Z., Shoelson, S.E., Chaudhuri, M., Gish, G., Pawson, T., King, F., Roberts, T., Ratnofsky, S., Lechleider, R.J., Neel, B.G., Birge, R.B., Fajardo, J.E., Chou, M.M., Hanafusa, H., Schaffhausen, B. and Cantley, L.C., Cell, 72 (1993) 767.
7. Shoelson, S.E., Sivaraja, M., Williams, K.P., Hu, P., Schlessinger, J. and Weiss, M.A., EMBO J., 12 (1993) 795.
8. Kauer, J.C., Erickson-Viitanen, S., Wolfe, H.R. and DeGrado, W.F., J. Biol. Chem., 261 (1986) 10695.
9. Miller, W.T. and Kaiser, E.T., Proc. Natl. Acad. Sci. U.S.A., 85 (1986) 5429.
10. Williams, K.P. and Shoelson, S.E., J. Biol. Chem., 268 (1993) 5361.
11. Skolnik, E.Y., Margolis, B., Mohammadi, M., Lowenstein, E., Fischer, R., Drepps, A., Ullrich, A. and Schlessinger, J., Cell, 65 (1991) 83.
12. Booker, G.W., Breeze, A.L., Downing, A.K., Panayatou, G., Gout, I., Waterfield, M.D. and Campbell, I., Nature, 358 (1992) 684.
13. Eck, M., Shoelson, S.E. and Harrison, S.C., Nature, 362 (1993) 87.
14. Waksman, G., Shoelson, S.E., Pant, N., Cowburn, D. and Kuriyan, J., Cell, 72 (1993) 779.

Probing the functional importance of the amino acid residues in the synthetic peptide actin 1-28 on interaction with troponin I and myosin

J.E. Van Eyk, O.D. Monera and R.S. Hodges

The Medical Research Council Group in Protein Structure and Function, Department of Biochemistry, University of Alberta, Edmonton, Alberta, T6G 2H7, Canada

Introduction

Muscle contraction involves the regulation of the interaction between actin and myosin by the regulatory protein complex, tropomyosin-troponin (TM-Tn). The Tn subunits (TnI, TnT and TnC) interact with each other and work in concert with actin-TM to produce conformational changes which alter the interaction between actin and myosin, ultimately resulting in muscle contraction.

It has been shown that the synthetic peptide, actin 1-28, mimics the interaction between actin and its two target proteins, TnI and myosin [1, 2]. Thus, we have proposed that residues 1-28 of actin are part of the molecular switch involved in the calcium-dependent regulation of muscle. In the absence of calcium, actin residues 1-28 interact with TnI, in particular residues 104-115, which promotes muscle relaxation by preventing actin from interacting with myosin. In contrast, when calcium binds to TnC, residues 104-115 of TnI bind to TnC. This allows actin residues 1-28 to interact with myosin. This in turn increases the myosin ATPase activity, resulting in muscle contraction.

In this report, we have employed two types of synthetic analogs of the actin peptide 1-28 in order to investigate which region(s) of actin, as well as which amino acid residues, are important for biological activity. The first set of analogs had internal sequences either truncated or replaced with a poly-glycine linker, while the second set consisted of single alanine substitutions of selected amino acid residues.

Results and Discussion

The relative binding affinities of the actin peptide analogs for TnI were determined using high performance affinity chromatography using a column derivatized with skeletal TnI. The results indicated that the N-terminal region of actin 1-28 contains the dominate site of interaction with TnI (Figure 1), since the actin peptides 1-7 and 1-14 bound TnI almost as well as actin 1-28 (82, 94 and 102 mM KCl were required to elute the peptides from the column, respectively). Residues 1-4 are most important for the TnI interaction, since the actin 1-28 analogs in which Ala is substituted for any of the first 4 amino acids bound less tightly to the TnI column than actin 1-28 (requiring between 62 and 66 mM KCl for elution).

On the other hand, the N- and C-terminal regions of actin 1-28 are equally important for activation of the myosin ATPase activity (a soluble myosin fragment

HMM was used) (Figure 1). First, the deletion of amino acid residues from either end was detrimental to the ATPase activity (actin peptide 4-28 and 1-24 activated the ATPase activity compared to actin 1-28 by 64%). Secondly, two non-overlapping peptides, actin 1-14 and 18-28, possessed similar activities (50 and 60%, respectively), while actin 8-20 possessed minimal activity (11%). Thirdly, Ala substituted for any single amino acid within the sequence 1-6 or 21-28 dramatically decreased the ATPase activity (for example, D1A actin 1-28 did not activate the HMM ATPase activity, while D25A actin 1-28 inhibited the ATPase activity by 20%). This indicates that the HMM-actin 1-28 interaction is less able to accommodate changes to the amino acid sequence in the N- and C-terminal regions than the interaction between actin 1-28 and TnI.

In addition, the length and composition of the amino acid sequence between the N- and C-terminal regions of actin 1-28 are important for the maximum activation of the HMM ATPase activity, since a mixture of actin 1-14 and 18-28 induced the same degree of activation as 18-28 alone. Residues 8-20 are part of a β-sheet that is buried in the actin molecule [3] and hence should not be available for interaction with HMM. Based on computer modeling, 4 Gly residues were used to link residues 1-7 to 21-28. This analog was less potent than either actin 1-14 or 18-28, requiring approximately 6 times the concentration of peptide to reach the same level of activation. These results suggest that the β-sheet region, residues 8-20 is critical for the correct positioning of the N- and C-terminal regions of actin 1-28. The binding of actin 1-28 to TnI or myosin depends on the conformation of the target proteins and on the ability of the actin residues to adapt to these differences in conformation. Therefore, one can envision that it would require only slight conformational changes in the N-terminal region of actin (residues 1-7) in order to cause the release of this region of actin from the myosin. This would promote the binding of this region of actin to TnI. The sensitivity of the myosin-actin 1-28 interaction means that this region of actin can sense and respond to subtle differences in conformation of myosin. This may represent the fine tuning of the molecular switch that regulates muscle contraction.

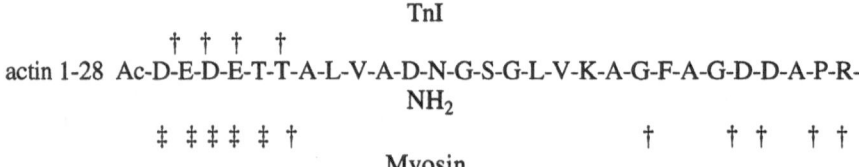

Fig. 1. Amino acid residues of actin 1-28 important for binding and biological activity with TnI (top) and myosin HMM (bottom). ‡ indicates essential residues that if changed to Ala eliminates function, and † indicates less critical residues that when changed to Ala substantially reduce binding or biological activity.

References

1. Van Eyk, J.E., Sönnichsen, F.D., Sykes, B.D. and Hodges, R.S., In Rüegg, J.C., (Ed.), Peptides as Probes in Muscle Research, Springer-Verlag, Heidelberg, Germany, 1991, p.15.
2. Van Eyk, J.E. and Hodges, R.S., Biochemistry, 30 (1991) 11676.
3. Kabsch, W., Mannherz, H.G., Suck, D., Pai, E.F. and Holmes, K.C., Nature, 347 (1990) 37.

Structure determination of the calmodulin binding domain of caldesmon by 2D NMR spectroscopy

M. Zhang and H.J. Vogel

Department of Biological Sciences, University of Calgary, Calgary, Alberta, T2N 1N4, Canada

Introduction

The high molecular weight protein caldesmon (CaD) is a component of the thin filament in smooth muscles. It consists of three domains: an N-terminal myosin-binding domain, an extended central α-helical domain and a C-terminal domain which contains the primary binding sites for tropomyosin, actin and calmodulin (CaM). The long repeating central α-helix gives the molecule an elongated shape. The binding of CaD to actin inhibits the actin-activated ATPase activity, this inhibition can be reversed in a Ca^{2+}-dependent manner by the binding of Ca^{2+}-CaM to CaD. A peptide corresponding to the amino acid residues Gly651 to Ser667 of chicken gizzard CaD has been identified as the CaM binding domain of CaD; this peptide binds to CaM in a Ca^{2+}-dependent manner with a K_d of ≈ 1 μM [1]. In this work, we have studied this 17 residue long synthetic CaM-binding peptide by 2D proton NMR spectroscopy in order to determine the structure of the peptide when it is bound to CaM. We have also studied the involvement of the two globular domains of CaM in the binding of the peptide by homo- and hetero-nuclear 2D NMR spectroscopy.

Results and Discussion

CaD represents a somewhat unique target of CaM because its interaction with CaM is a lot weaker than that observed for the binding of CaM to other target proteins. The CaM-binding sites on the various target proteins of CaM always appear to be contained in a continuous stretch of ≈ 20 amino acid residues. Peptides encompassing these CaM-binding domains usually bind to CaM with nanomolar affinities. This results in slow exchange on the NMR time scale which necessitates the use of isotope editing NMR methods to determine the structure of the bound peptides [2]. However, by taking advantage of the fact that the CaD peptide has a K_d of ≈ 1 μM, it is possible to determine its CaM-bound structure by using the transfer NOE (TRNOE) technique.

The free peptide in H_2O solution shows no regular secondary structure as indicated by standard 2D NMR spectroscopy. The interaction of the peptide with Ca^{2+}-CaM is illustrated in Figure 1A: the free peptide in H_2O solution shows a highly resolved spectrum (bottom trace), and the addition of Ca^{2+}-CaM (molar ratio 1/30) leads to line broadening of the same region of the spectrum (top trace). The line broadening is due to fast exchange of the peptide in the CaM-bound form with the free form. When fast exchange conditions prevail, it is possible to use 2D TRNOE

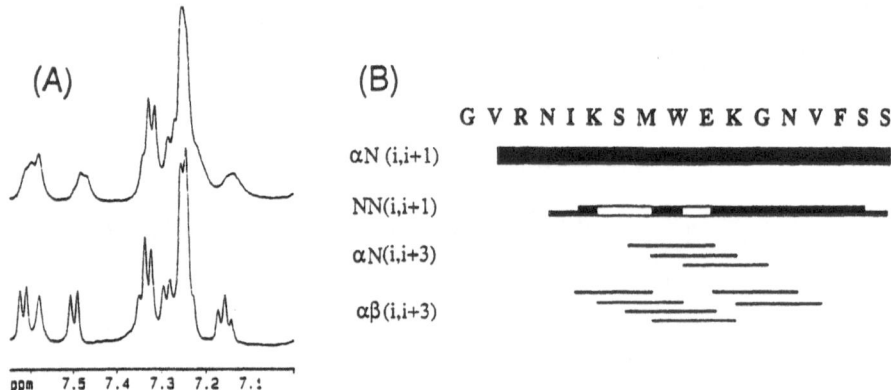

Fig. 1. (A) Aromatic region of the one dimensional ^1H NMR spectrum of the free CaD peptide (bottom trace) and the CaD peptide in the presence of Ca^{2+}-CaM (top trace). The spectra were recorded with 4 mM peptide samples at pH 6.5, 288 K on a Bruker AMX500 NMR spectrometer. (B) Summary of the nOe's observed for the CaD peptide in the TRNOE experiment. The thickness of the lines represents the intensity of the nOe crosspeaks. Open boxes are drawn for nOe's which are probably present but they cannot be observed due to resonance overlap.

NMR to study the structure of the peptide when it is bound to Ca^{2+}-CaM. Figure 2A shows the sequential assignment of the peptide obtained in the presence of 1/30 ratio of Ca^{2+}-CaM. Under these conditions, a substantial number of NN(i, i+1) nOe's can be observed (Figure 2B). Apart from these NN(i,i+1) nOe's, a number of medium range $\alpha\beta$(i,i+3) and αN(i,i+3) nOe's were also observed. The nOe information obtained for the peptide through the TRNOE experiment is summarized in Figure 1B. From the nOe pattern, it can be deduced that the peptide can adopt an α-helical structure from Asn4 to Ser17 in its complex with CaM. The α-helix formed by the peptide has similar basic amphiphilic properties as found for CaM-binding domains from other target proteins. However, the CaM-binding domain in CaD does not have the two characteristic anchoring hydrophobic residues 13 residues apart that are observed for other CaM-binding domains, and this may account for the weaker CaM-binding of CaD as compared to that of myosin light chain kinase [2].

Although the helical region of the CaD peptide in the CaM-bound form is shorter than most other strong CaM-binding peptides, the peptide still appears to bind to both domains of CaM; this could be deduced from titration experiments of CaM with the peptide, where chemical shift changes were observed for amino acid residues from both domains of CaM (data not shown). The titration experiment also indicates that the 17 residue CaD peptide binds to CaM with 1:1 stoichiometry. Using Met methyl ^{13}C labelled CaM, we have been able to show that the two hydrophobic patches in CaM are involved in the binding of the peptide. All the eight Met residues in the two hydrophobic patches appear to be in contact with the peptide, and this result shows the importance of the flexible Met residues in target-binding.

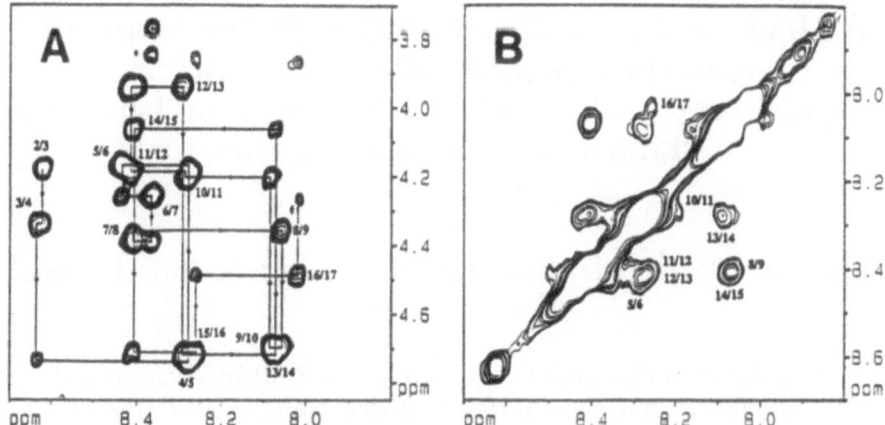

Fig. 2. (A) The fingerprint region of the NOESY spectrum of the CaD peptide in the presence of Ca²⁺-CaM. The sequential assignment is indicated for the interresidue αN(i,i+1) nOe crosspeaks. The mixing time of the experiment was 250 ms. The spectrum was recorded for a 4 mM sample at pH 6.5, 288 K. (B) The amide region of the same NOESY spectrum as in (A). The nOe crosspeaks are labelled by their residue numbers.

By comparison of (¹H,¹⁵N)-HMQC NMR spectra of uniformly ¹⁵N labelled CaM and its complex with the CaD peptide, we have also been able to show that the backbone amides from both domains of CaM undergo chemical shift changes upon complexation of the peptide. Moreover, the central α-helix of CaM also undergoes a conformational change when CaM binds to the CaD peptide. These data suggest that the CaD peptide binds in a similar fashion as other CaM-binding domains, but that the positioning of the two halves of CaM with respect to the peptide may be different.

Acknowledgements

This work is financially supported by the Medical Research Council (MRC) of Canada. HJV is a Scholar of the Alberta Heritage Foundation for Medical Research.

References

1. Zhan, Q., Wong, S.S. and Wang, C.L.A., J. Biol. Chem., 266 (1991) 21810.
2. Ikura, M., Clore, G.M., Gronenborn, A.M., Zhu, G., Klee, C.B. and Bax, A., Science, 256 (1992) 632.

A single amino acid substitution in the troponin I inhibitory peptide alters the calcium responsiveness of reconstituted troponin I-depleted skinned cardiac muscle fibers

J.E. Van Eyk[1], J.D. Strauss[2], R. Weisner[2], R.S. Hodges[1] and J.C. Rüegg[2]

[1]The Medical Research Council Group in Protein Structure and Function, Department of Biochemistry, University of Alberta, Edmonton, Alberta, T6G 2H7, Canada, and [2]II. Physiologisches Institut, Universität Heidelberg, Heidelberg, Germany

Introduction

We have used a 12-residue synthetic peptide mimic of Troponin I (TnI), TnI peptide 104-115 (Ac-G-K-F-K-R-P-P-L-R-R-V-R-amide), to probe the molecular mechanism for Ca^{2+}-dependent regulation of muscle contraction. Previously, using a method for reversibly extracting TnI and Troponin C (TnC) from skinned cardiac muscle fibers, it has been shown that, like TnI, when the synthetic TnI inhibitory peptide (104-115) was reconstituted with skeletal TnC, force development was inhibited and in the presence of Ca^{2+} the inhibition was released [1]. This result indicates that the TnI peptide is able to mimic TnI by switching binding sites from actin to TnC in the presence of Ca^{2+}. Skinned fibers are muscle fibers in which the cell membrane has been permeabilized. This permits precise control of the cytoplasmic environment and allows diffusion of peptides into the intracellular compartment. These fibers are functional in terms of contractility and respond to changes in Ca^{2+} concentrations in much the same manner as living muscle, contracting and relaxing at high and low concentrations, respectively. In skinned cardiac muscle fibers, where the TnI peptides compete with endogenous TnI, the TnI peptide inhibits force development at submaximal Ca^{2+} concentrations while a peptide analog, in which Leu 111 has been replaced by Gly (L111G), stimulates force development [2]. In this report we use TnI-TnC depleted skinned fibers to investigate further the importance of Leu 111 in the regulation of muscle contraction.

Results and Discussion

The single amino acid change from Leu to Gly at position 111 of the TnI inhibitory peptide decreases the effectiveness of the peptide as an inhibitor (Figure 1, panel A), indicating Leu 111 is important for the interaction of TnI inhibitory region with actin-TM. In addition, Pi which decreased force in normal skinned fibers and the TnI-TnC extracted fibers [3], also affected the extent of the L111G analog inhibition (relaxation). Therefore, Pi antagonizes the L111G analog inhibition, suggesting that both Pi and Gly 111 analog affect actin-myosin crossbridges directly.

The Ca^{2+}-dependent development of force (contraction) of fibers reconstituted with the L111G analog-cardiac TnC was shifted with respect to unextracted cardiac fibers and native TnI peptide-cardiac TnC reconstituted fibers (Figure 1, panel B). The 0.16 shift in pCa_{50} value indicates a decrease in the Ca^{2+} affinity of thin filament when reconstituted with cardiac TnC and L111G analog. The skeletal TnC-L111G analog reconstituted fibers display a similar 0.13 shift in the Ca_{50} value (data not shown). This indicates that the single Leu to Gly substitution alters the way in which the inhibitory region of TnI affects Ca^{2+}-dependent contractility of the muscle fiber.

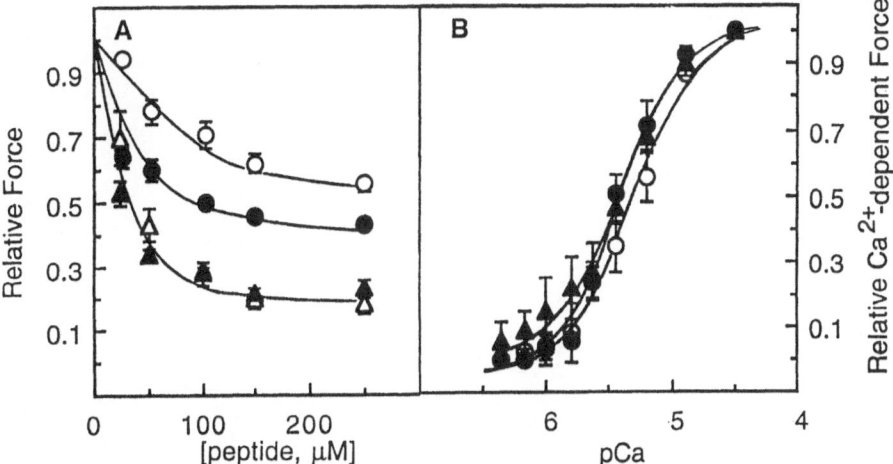

Fig. 1. The Pi and Ca^{2+}-sensitivity of the L111G TnI peptide analog reconstituted fibers. In panel A, the TnI-TnC depleted skinned cardiac fibers [1, 2] were incubated in relaxing solution (-Ca^{2+}) containing increasing quantities of TnI peptide (△, n=4) or L111G analog (○, n=3-5) in the presence (open symbols) or absence (closed symbols) of 5 mM Pi. Force values are relative to the force reached by each fiber following extraction. The decrease in force by Pi was subtracted from the data. In panel B, the Ca^{2+} sensitivity of the muscle fibers prior to extraction (●) and upon reconstitution with L111G analog-cardiac TnC (○, n=4) or the native TnI peptide-cardiac TnC (▲, n=5) is shown. Reconstitution was achieved by incubating the extracted fibers with cardiac TnT (30 μM for 25 min) followed by L111G TnI peptide (250 μM for 200 min) or native TnI peptide (143 μM for 50 min) until maximum inhibition was reached. There is little or no Ca^{2+} sensitivity displayed by these fibers until further incubation with cardiac TnC (80-88 μM for 30 min). The unbound TnC was washed from the fibers prior to the determination of the force/pCa curve. The maximum Ca^{2+}-dependent forces compared to the extracted fibers of the native TnI peptide or L111G analog reconstituted fibers were $83 \pm 8\%$ and $92 \pm 10\%$, respectively.

References

1. Van Eyk, J.E., Strauss, J.D., Hodges, R.S. and Rüegg, J.C., FEBS Lett., 323 (1992) 223.
2. Strauss, J.D., Barth, Z., Van Eyk, J.E. and Rüegg, J.C., Methods: A Companion to Methods in Enzymology, 5 (1993) 281.
3. Strauss, J.D., Zeunger, C. and Rüegg, J.C., Eur. J. Pharmacol., 227 (1992) 437.

Session XIII
Peptide Libraries

Chairs: Günther Jung
Eberhard-Karls-Universität Tübingen
Tübingen, Germany

and

Hossain H. Saneii
Advanced ChemTech
Louisville, Kentucky, U.S.A.

Optimal peptide length determination using synthetic peptide combinatorial libraries

R.A. Houghten, J.R. Appel, S.E. Blondelle, J.H. Cuervo, C.T. Dooley, J. Eichler and C. Pinilla

Torrey Pines Institute for Molecular Studies, San Diego, CA 92121, U.S.A.

Introduction

Synthetic peptide combinatorial libraries (SPCLs) in a variety of formats have recently been presented [1-15]. These synthetic approaches, as well as recombinant approaches, are revolutionizing basic research and drug discovery involving peptides. Such libraries have been successfully used to study antibody/antigen interactions [1-6]; to develop enzyme inhibitors [7, 15], to discover novel antimicrobials [1, 2, 8, 9] and peptides that are potent inhibitors of the binding of opioid and other peptides to their receptors [2, 10-14]. We describe here a previously unreported series of libraries, termed length determination libraries (or sizing libraries for short). These libraries enable the rapid determination of the length of the most effective peptide in a given assay.

Results and Discussion

SPCLs [1], which are not attached to an insoluble support and are therefore available in solution, produce peptide chemical diversities that appear to be usable in virtually any existing *in vitro* or *in vivo* assay system without prior assay modification. This approach, which was first presented by this laboratory [1], uses a format in which two positions of a hexapeptide sequence are individually defined ("O" positions), with the remaining four positions made up of mixtures of amino acids ("X" positions).

The utility of the divide, couple and recombine (DCR, [1, 16, 17]) approach for the preparation of SPCLs begins to falter above a length of four mixture positions when using the 20 most commonly occurring amino acids. This is due to the ever-increasing amounts of resin required to ensure an equal statistical distribution of beads during the "divide" step. This approach is also cumbersome if the defined positions are not located at the N-terminal of the peptides being prepared. Therefore, another strategy was developed for the preparation of peptide resin mixtures, and ultimately SPCLs. In this alternative approach, mixtures of amino acids are used in the coupling step during solid phase synthesis. The advantages of this method are the ease of the synthesis process, as well as the ability to prepare libraries which have more than four mixture positions. Furthermore, the defined positions need not be confined to the N-terminal, but can be located anywhere within the sequence. This synthetic approach was used in the preparation of positional scanning synthetic peptide combinatorial libraries (PS-SPCLs; [3]). The first PS-SPCL prepared

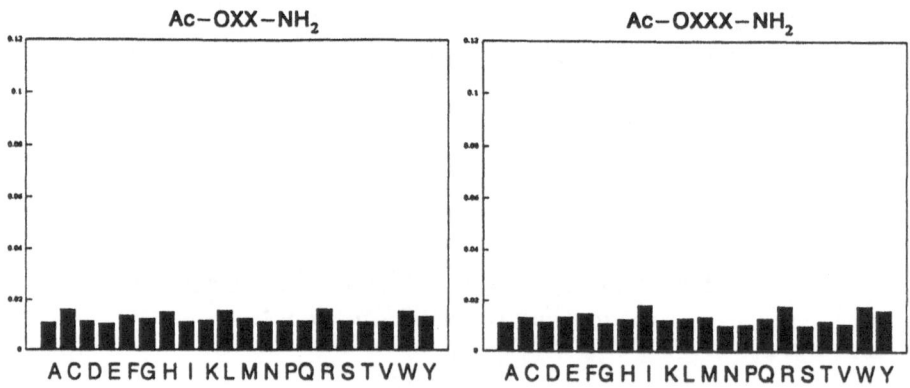

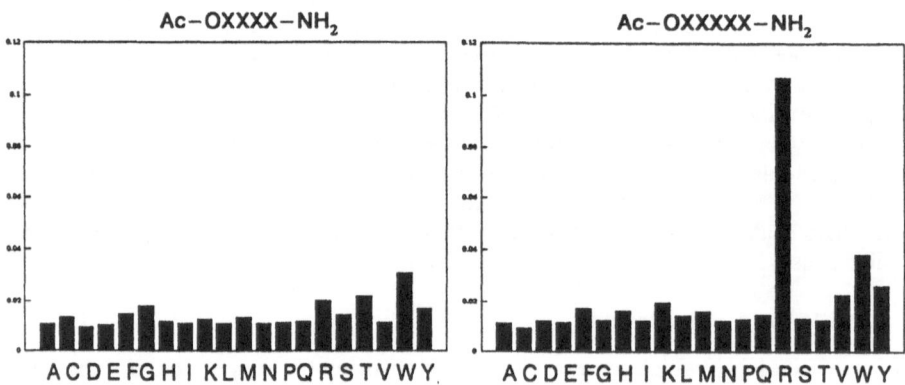

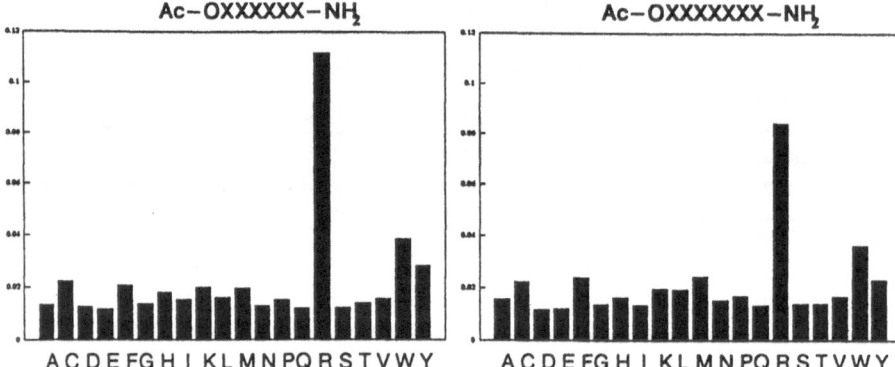

Fig. 1. Varying length and diversity libraries. The libraries vary in length from three residues (Ac-OXX-NH₂; 361 peptides per mixture) to eight residues (Ac-OXXXXXXX-NH₂; 893 x 10⁶ peptides per mixture). The results are expressed as the reciprocal of the IC₅₀ of ³H₂-DAGO binding to crude rat brain homogenates.

consisted of six separate L-libraries, each having a single position defined and the remaining five positions as mixtures (Ac-O_1XXXXX-NH_2, Ac-XO_2XXXX-NH_2, Ac-XXO_3XXX-NH_2, Ac-XXXO_4XX-NH_2, Ac-XXXXO_5X-NH_2, and Ac-XXXXXO_6-NH_2; 120 separate peptide mixtures in total, each composed of approximately 1.9 x 10^6 hexapeptides). This library, and its non-acetylated counterpart, has been used to determine exact peptide sequences recognized by antibodies [3], receptors [3], and enzymes [15].

Every peptide-receptor interaction can be anticipated to be associated with a specific peptide length having the highest binding affinity and/or activity. In order to determine the optimal peptide length associated with any peptide-receptor system, we have prepared six libraries made up of N-acetylated L-amino acid tri- through hexapeptide amides. These six libraries were prepared by both the DCR and the "chemical" mixture approach. These libraries were screened in the opioid receptor assay for their ability to inhibit $^3H_2$-DAGO binding. Figure 1 shows that a length of at least six residues is required for significant inhibition to occur, and also that arginine at the N-terminal position results in the most effective inhibition. These results, in conjunction with an acetylated SPCL [11], were used to determine highly active opioid peptide sequences present in this hexapeptide library. As anticipated from the results found with the sizing SPCL, the most active peptide mixtures contained an arginine in the first position of the acetylated SPCL, Ac-O_1O_2XXXX-NH_2 (Ac-RFXXXX-NH_2, Ac-RYXXXX-NH_2, and Ac-RWXXXX-NH_2). Upon completion of the iterative process, several peptides with the general sequence Ac-RFMWMO-NH_2 were found to be potent inhibitors of $^3H_2$-DAGO binding to the μ receptor. The three peptides, Ac-RFMWMT-NH_2 (IC_{50} = 5nM), Ac-RFMWMR-NH_2 (IC_{50} = 6nM), and Ac-RFMWMK-NH_2 (IC_{50} = 5nM), were also found to be potent μ antagonists in the guinea pig ileum bioassay, and relatively weak antagonists in the mouse vas deferens bioassay. These peptides represent the first instance of new membrane-bound receptor ligands being discovered through the use of peptide libraries. The rapid development of new libraries, both peptide and non-peptide, is expected to facilitate all areas of research, as well as pharmaceutical drug discovery.

Acknowledgements

This work was supported, in part, by Houghten Pharmaceuticals Inc., San Diego, CA, U.S.A.

References

1. Houghten, R.A., Pinilla, C., Blondelle, S.E., Appel, J.R., Dooley, C.T. and Cuervo, J.H., Nature, 354 (1991) 84.
2. Houghten, R.A., Appel, J.R., Blondelle, S.E., Cuervo, J.H., Dooley, C.T. and Pinilla, C., BioTechniques, 13 (1992) 412.
3. Pinilla, C., Appel, J.R., Blanc, P. and Houghten, R.A., BioTechniques, 13 (1992) 901.
4. Appel, J.R., Pinilla, C. and Houghten, R.A., Immunomethods, 1 (1992) 17.
5. Pinilla, C., Appel, J.R. and Houghten, R.A., Gene, 128 (1993) 71.
6. Blake, J. and Litzi-Davis, L., Bioconjugate Chem., 3 (1992) 510.
7. Owens, R.A., Gesellchen, P.D., Houchins, B.J. and DiMarchi, R.D., Biochem. Biophys. Res. Comm., 181 (1991) 402.

8. Houghten, R.A., Blondelle, S.E., and Cuervo, J.H., In Epton, R. (Ed.), Innovation and Perspectives in Solid Phase Synthesis: Peptides, Polypeptides and Oligonucleotides, Intercept Limited, Andover, 1992, p.237.
9. Houghten, R.A., Dinh, K.T., Burcin, D.E. and Blondelle, S.E., In Angeletti, R.H. (Ed.), Techniques in Protein Chemistry IV, Academic Press, Orlando, FL, U.S.A., 1993, p.249.
10. Houghten, R.A. and Dooley, C.T., Biorg. Med. Chem. Lett., 3 (1993) 405.
11. Dooley, C.T., Chung, N.N., Schiller, P.W. and Houghten, R.A., Proc. Natl. Acad. Sci. U.S.A., 90 (1993) 10811.
12. Dooley, C.T. and Houghten, R.A., Life Sciences, 52 (1993) 1509.
13. Hortin, G.L., Staatz, W.D. and Santoro, S.A., Biochem. Int., 26 (1992) 731.
14. Zuckermann, R.N., Kerr, J.M., Siani, M.A., Banville, S.C. and Santi, D.V., Proc. Natl. Acad. Sci. U.S.A., 89 (1992) 4505.
15. Eichler, J. and Houghten, R.A., Biochemistry, 32 (1993) 11035.
16. Furka, A., Sebestyen, F., Asgedom, M., and Dibo, G., Int. J. Pept. Prot. Res., 37 (1991) 487.
17. Lam, K.S., Salmon, S.E., Hersh, E.M., Hruby, V.J., Kazmierski, W.M. and Knapp, R.J., Nature, 354 (1991) 82.

Pilot, a new peptide lead optimization technique and its application as a general library method

R. Cass, M.L. Dreyer, L.B. Giebel, D. Hudson, C.R. Johnson, M.J. Ross, J. Schaeck and K.R. Shoemaker

Arris Pharmaceutical Corporation, Suite 12, 385 Oyster Point Blvd., South San Francisco, CA 94080, U.S.A.

Introduction

We report here on "Pilot" (Peptide Identification & Lead Optimization Technique), which provides spatial array of molecules assembled using conventional chemistry, a simple apparatus, and a sophisticated thin film derivatized support surface which provides display in an unhindered near-aqueous environment.

Results and Discussion

Initial studies for lead optimization used flat polyethylene plates to which Pepsyn K particles were attached with chemically resistant glues. Alternatively, polyamide materials, as gel films, were covalently grafted onto spacer-arm functionalized polyethylene surfaces (provided by a diamino-substituted polyethylene glycol, Jeffamine ED-600, Texaco, as in Figure 1, top and right). A simple slotted block/gasket apparatus was used, with both candidates, to prepare 10 x 10 arrays displaying different alternatives for both H and P within the sequence YGHPQGG (containing the HPQ motif known to bind to streptavidin). Detection with ^{125}I-streptavidin showed no significant binding to any area other than that representing the parent sequence; this binding was reversed in the presence of the natural ligand, (+)-biotin. Parallel experiments using cleavable linkers confirmed that, in all 100 peptides, assembly had occurred as efficiently as on conventional supports, and with no cross-contamination. Although usable as such for lead optimization, neither support was ideal. Dramatic improvement was found with surfaces to which preformed polymers were attached as ultra-thin films. The optimal method, Figure 1, employed water-soluble carbodiimide (EDC) coupling of carboxymethyl dextran (Pharmacia T500) to the previously functionalized polyethylene surfaces. A different system, provided by the Pharmacia BIAcore "chip", can be used for evaluation and testing of ligand/ligate interactions. Use of coarse fritted polyethylene disks (0.25" diameter, 0.125" thick; "tiddly-winks") gave 50-100 nmol loading, allowed high quality peptide synthesis, efficient absorption of radioactivity, and facile and rapid removal of unbound counts by suction washing. Use of ^{125}I labeling with Bolton-Hunter reagent provides sensitive and simple detection by autoradiography, although regeneration of arrays is seldom possible. With ^{35}S and ^{14}C labeling, arrays can be recovered and reprobed numerous times after scintillation counting of individual tiddly-winks pushed out from their holding plate.

975

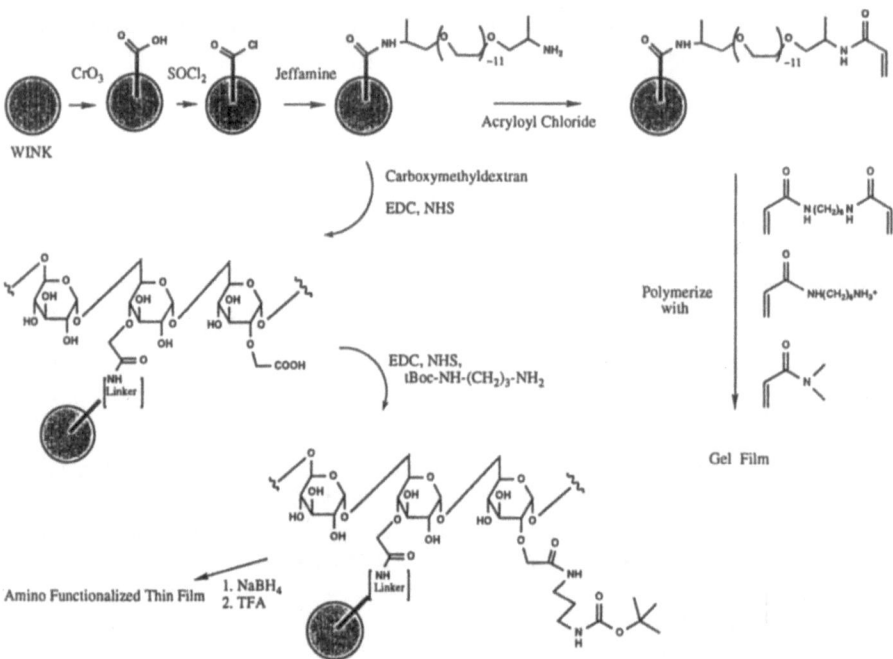

Fig. 1. *Modification of Polyethylene Surfaces used in Pilot experimentation.*

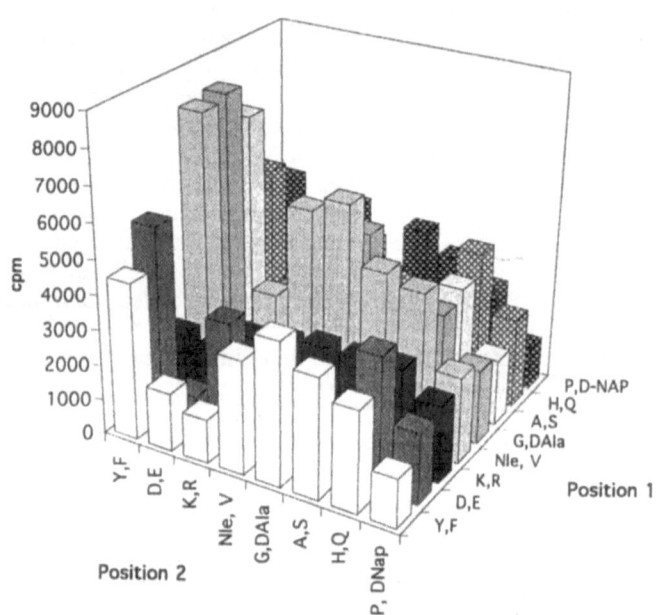

Fig. 2. *Probe of N-terminal two position array with ^{35}S labeled-streptavidin:- linear hexapeptide Pilot study.*

Table 1 *Conditions for equal incorporation of Fmoc- Amino acids from mixtures*

Fmoc-Derivative and Molar Ratio		Fmoc-Derivative and Molar Ratio		Fmoc-Derivative and Molar Ratio	
L-Nle*	1.00	L-Ala*	0.79	D-Nap*	1.50
L-Leu	1.00	L-Ser(tBu)*	1.50	L-Tyr(tBu)*	1.70
L-Val*	1.60	L-His(Trt)*	2.10	L-Phe*	1.00
Gly*	0.60	L-Gln(Trt)*	2.20	L-Asp(OtBu)*	1.40
D-Ala*	0.79	L-Pro*	1.15	L-Glu(OtBu)*	1.20
L-Lys(tBoc)*	1.36	L-Arg(Pmc)*	3.00	L-Thr(tBu)	2.00
L-Asn(Trt)	2.45				

Mixed AA's coupled with 1.25 equiv. HOBt + 1.5 equiv. PyBOP and 1.5 equiv. 0.3 M NMM in DMF after 10 min preactivation; Nap is 3-(2-naphthyl)-alanine, 1-naphthyl derivative couples similarly. *Used in mixture Ω.

For use of Pilot as a general library method, mixtures of Fmoc-amino acids are coupled under conditions which ensure their equal incorporation (Table 1). Sixteen amino acids are incorporated in standard mixture Ω. D-Amino acids couple in similar proportions to L- derivatives. Mixtures may be incorporated at up to 6 positions in a core sequence, and "arrayed" at any other two positions. Individual winks may be processed with conventional synthesizers to provide simulated array components ("pseudo" arrays). Study of non-arrayed mixed winks in the presence and absence of natural ligand indicates whether a full array study will be successful.

A strategy of using arrays and sub-arrays has been used in the streptavidin-biotin system. Figure 2 shows the result of probing, with ^{35}S labeled streptavidin, an 8 x 8 array, in which positions 1 and 2 of a linear hexapeptide were arrayed with 2 amino acid mixtures, and the other 4 positions were uniformly mixed with mixture Ω. The 16 possible combinations from the 4 main areas were prepared, by multiple synthesis, and VY was identified as a strong candidate. The second set of array and sub-array identified VYGF and VYHP, which led on to VYGFRQ and VYHPQF, with numerous other weaker answers. Comparison on Pilot surfaces, and by BIAcore studies, showed the former to be a weak and slow binding species. The concept that these two different sequences might overlap at streptavidin's biotin binding site led to synthesis of the strongly binding combination sequence HPQVYGFRQ. Structural studies and biophysical characterization of this interaction continue. These studies demonstrate a novel technology utilizing simply prepared arrays, optimal display chemistry, and radio-labeling for sensitive detection; and foreshadow its application for receptor mapping, leading to the computer assisted design of new, highly active drug candidates.

Acknowledgements

Thanks are due to R. Arze and J.A. Buettner for assistance with peptide synthesis, and J. Sparrow for helpful advice on formation of grafted polyamide films.

A second dimension in complexity in combinatorial peptide libraries

W.-J. Zhang[1], G.L. Hortin[2,*], R.L. Panek[4], G.H. Lu[4], F.F. Hsu[3] and G.R. Marshall[1]

Departments of [1]Molecular Biology and Pharmacology, [2]Pediatrics, and [3]Medicine, Washington University School of Medicine, St. Louis, MO 63130, and [4]Parke-Davis Pharmaceutical Research Division, Warner-Lambert Co., Ann Arbor, MI 48105, U.S.A.

Introduction

Methods have been developed for the rapid preparation of large numbers of individual synthetic peptides [1, 2] which have improved and greatly simplified earlier techniques. Various strategies in multiple peptide synthesis have been used [3] to generate complex mixtures of peptides, combinatorial peptide libraries, for screening from which a lead compound can be isolated by logical procedures. Alternatively, spatial localization through photolithography allows identification of active sequences by position [4]. As peptide recognition [5] often involves interaction with turn structures, structure-activity studies often indicate discontinuities in sequence.

A second dimension in complexity can be obtained through the introduction of specific functional groups into two sets of peptide mixtures which can then be cross-linked. Our initial test of this concept has utilized disulfide cross-linking and focused on two model systems derived from angiotensin II (Asp-Arg-Val-Tyr-Val-His-Pro-Phe, AT) and bradykinin (Arg-Pro-Pro-Gly-Phe-Ser-Pro-Phe-Arg, BK). For these two peptides, considerable evidence exists [6, 7] that their receptor-bound conformations contain turns centered on residues Tyr-4 and Gly-4 and Phe-5, respectively. In the case of residues which are recognized and which are adjacent because of proximity in adjacent β-sheet structures, incorporation of turn mimetics into linear sequences may be an appropriate strategy. Alternatively, linking two strands together by a disulfide bridge is consistent with β-sheet formation as seen both in proteins [8] and in small peptides [9, 10]. Previous studies [7] had shown that both parallel and antiparallel dimers of AT containing Cys at positions three and five were recognized by AT receptors. Kaczmarek et al. [this volume] have shown similar results for BK dimers.

Results and Discussion

Synthesis of peptide mixtures - Two peptide mixtures were synthesized as fragments of AT to be joined as heterodimers by oxidation:

---

*Current address: Division of Laboratory Medicine, Department of Pathology, University of Alabama Medical School, Birmingham, AL 35233, U.S.A.

Mixture A.   Ac-Cys-(Tyr, Phe, Ala)-Val-His-(Pro, Gly, Phe)-Phe-OH

Mixture B.   Sar-Arg-(Pro, Cle, Val)-(Tyr, Phe, Ala)-Cys-NH$_2$

One (peptide mixture A) of the initial synthetic peptide mixtures which consisted of a six-residue peptide sequence with an acetylated N-terminus and a free carboxyl terminus was prepared using standard SPPS with mixtures of amino acids. The other (peptide mixture B) consisted of a five-residue peptide sequence with a free amino terminus and an amidated C-terminus was synthesized similarly using MBHA resin and Boc protection. Amino acid analysis and HPLC of mixture B as well as FABMS of the resulting dimers indicated that the differences in rates of reaction of the mixtures of activated amino acids gave low yields of Cle incorporation.

Another two peptide mixtures were synthesized as fragments of BK to be joined as heterodimers by oxidation:

Mixture C.   Ac-Cys-(Ser,Phe,D-Phe,Leu)-(Pro,D-Pro,Phe,D-Phe)-Phe-Arg-OH

Mixture D.   NH$_2$-(Arg,Lys,Orn,D-Arg)-Pro-(Pro,MeA,Cle,D-Pro)-Cys-NH$_2$

In peptide mixture D, there were two hindered amino acids, MeA and Cle, at the third position and their coupling rates are quite slow and very different from the other two usual amino acids in the same coupling mixture. For this reason, we employed the divide-react-and-recombine startegy where each part was coupled separately with the individual amino acids (Cle, MeA, D-Pro and Pro); then the four parts were recombined and coupled with Pro. Unfortunately, Cle and MeA were still not completely coupled even after five coupling cycles with different methods. The tripeptide resins were separated again into four portions, and each was coupled with one of the amino acids (Arg, Lys, Orn, D-Arg). Even so, some peptides were less obvious by FABMS due to their relatively low concentrations.

Three methods of disulfide oxidation (I$_2$, and DMSO for AB heterodimer formation, and thalium(trifluoroacetic acid)$_3$ for CD dimer) oxidation were used. Conditions utilizing an excess of one component and kinetic control optimized yield of the mixed disulfides. 73 of the theoretical 81 AB peptides were readily identified by FABMS. The remaining 8 peptides which were observable all contained Cle which coupled poorly. After the oxidation reaction, there were three kinds of dimer mixtures: AA, BB and AB. All CD dimer peaks were readily observed by FABMS. Given the different physicochemical properties of the two parent sets of peptide mixtures A and B, these differences were enhanced by homodimerization and led to easy separation of AA, AB and BB dimer mixtures by HPLC. Control oxidations of mixtures A and mixtures B were performed separately for bioassay and available as standards.

The ability of dimer mixtures AA, AB and BB to inhibit angiotensin II binding to the AT$_1$ receptor (rabbit aorta) was tested:

Dimer Mixture (conc.)	1 x 10^{-5} M	1 x 10^{-6} M
AB	93%	54%
AA	37%	6%
BB	16%	3%

The results imply that one or more of the dimers of the AB mixture binds competitively to the AT receptor at relatively high affinity considering that only one of the 81 peptides in the mixture may possess this property.

A second dimension in complexity can readily be obtained through the introduction of specific functional groups into two sets of peptide mixtures which can then be cross-linked. While the examples focused on the use of the disulfide as the cross-link, such a strategy can easily be adapted. Incorporation of a single Asp or Glu into one peptide mixture and a single Orn or Lys into the other peptide mixture would lead to mixtures which could readily be joined. Another adaptation of this approach is to peptides which are prepared on solid supports and assayed without removal such as the photolithography technique [4]. Use of the disulfide linkage would allow recycling of the support-bound peptides through reduction, washing and combination by oxidation with a different mixture of sulfhydryl-containing peptides.

Acknowledgements

The authors thank the NIH (GM24483) for partial support and acknowledge use of the W.U. Mass Spectroscopy Resource (NIH grant RR00945).

References

1. Geysen, H.M., Immunology Today, 6 (1985) 364.
2. Houghten, R.A., Proc. Natl. Acad. Sci. U.S.A., 83 (1985) 5131.
3. Lam, K.S., Salmon, S.E., Hersh, E.M., Hruby, V.J., Kazmierski, W.M. and Knapp, R.J., Nature, 354 (1991) 82.
4. Fodor, S.P.A., Read, J.L. and Pirrung, M.C., Science, 251 (1991) 767.
5. Marshall, G.R., Curr. Opin. Struct. Biol., 2 (1992) 904.
6. Marshall, G.R., Kaczmarek, K., Kataoka, T., Plucinska, K., Skeean, R., Lunney, B., Taylor, C., Dooley, D., Lu, G., Panek, R. and Humblet, C., In Giralt, E. and Andreu, D. (Eds.), Peptides 1990, ESCOM, Leiden, The Netherlands, 1991, p.594.
7. Plucinska, K., Kataoka, T., Cody, W.L., He, J.X., Humblet, C., Lu, G.H., Lunney, E., Major, T.C., Panek, R.L., Schelkun, P., Skeean, R. and Marshall, G.R., J. Med. Chem., 36 (1993) 1902.
8. Srinivasan, N., Sowdhamini, R., Ramakrishnan, C. and Balaram, P., Int. J. Pept. Protein Res., 36 (1990) 147.
9. Karle, I.L., Flippen-Anderson, J.L., Kishore, R. and Balaram, P., Int. J. Pept. Protein Res., 34 (1989) 37.
10. Balaram, H., Uma, K. and Balaram, P., Int. J. Pept. Protein Res., 35 (1990) 495.

Cyclic peptide libraries

K. Darlak, P. Romanovskis and A.F. Spatola

Department of Chemistry, University of Louisville, Louisville, KY 40292, U.S.A.

Introduction

A number of laboratories have reported many novel methods for the synthesis of peptide libraries—large numbers of linear oligopeptides—which can provide a convenient entry for the rapid screening of biological activities. We wish to report a procedure that is amenable to preparing mixtures of cyclic peptides that are usually regarded as improved lead generators. The strategy involves the concept of resin bound peptide cyclization, developed by Sklyakov and Shashkova, and more recently by Felix, Schiller, and Hruby, among others. It also features the attachment of an amino acid to a solid support through its side chain: an approach designed to mimic solution cyclization procedures with linear peptides.

Results and Discussion

A typical strategy is shown below for the preparation of a series of cyclic pentapeptides. The initial attachment of a trifunctional amino acid is by the aspartyl side chain (with OFm for α-carboxyl protection) and coupling is performed using BOP reagent with racemization suppressant.

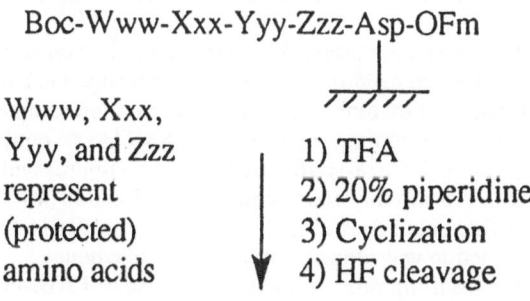

Boc-Www-Xxx-Yyy-Zzz-Asp-OFm

Www, Xxx, Yyy, and Zzz represent (protected) amino acids

1) TFA
2) 20% piperidine
3) Cyclization
4) HF cleavage

Boc chemistry was used for peptide synthesis, cyclization was carried out using BOP or HATU reagents with racemization suppressants (HOBT or HOAT) and strong acid (HF) was used for deprotective cleavage.

The concept of resin-bound cyclization has been previously reported by several groups. The use of side-chain-to-resin linkage combined with resin bound cyclization has been recently reported by Rovero et al. [1]. In order to validate that this synthetic method could be extended to the formation of cyclic peptide libraries, we have prepared a series of increasingly complex mixtures.

In the first, a series of four cyclic pentapeptides (including a "retro-inverso" variant of a recently described cyclic endothelin antagonist [2]) were prepared using the "split resin" strategy [3, 4]. This helps insure that each peptide resin bead will contain only a single cyclic peptide sequence. The FABMS analysis demonstrated that all four expected products were formed. In addition, several other minor products were observed and identified. These included a series of linear peptide piperidineamides, formed presumably from incompletely washed resin following the OFm ester cleavage step. A second series of minor components showed expected molecular masses and amino acid compositions, and were separable from the major isomers by RP-HPLC but not by capillary electrophoresis. These apparently result from some epimerization at the C-terminal aspartyl linkage during cyclization, and presumably may be suppressed by using appropriate racemization preventing methods. The major components were the four expected cyclic analogs.

In a second example, the synthetic protocol was validated through preparation of a set of 16 peptides with general sequence (Xxx-D-Leu-Yyy-D-Pro-Asp), again using the "split-resin" technique to generate a unique sequence on each resin bead. The synthesis gave a mixture of cyclic products after cleavage. FABMS analysis revealed that all expected masses were present. Interestingly, the relative intensities showed the expected dependence on hydrophobicity so that peptides with a Nal, Leu combination were nearly 10 times more abundant in the mass spectrum record than a corresponding Tyr, Gly variant; in contrast, the RP-HPLC profile (and amino acid composition) suggested the expected equimolarity.

In a third example, a more extensive set of cyclic pentapeptides, cyclo(Aaa-Bbb-Ccc-Ddd-Asp), was prepared. The amino acid levels were purposely skewed in order to provide a verifiable and variable ratio using amino acid analysis. A series of six different amino acids was used at each of the four variant positions within the pentapeptide to yield 6 x 6 x 6 x 6 x 1 or 1296 analogs. At least 36 of these final cyclic compounds were designed to contain the Arg-Gly-Asp sequence such that the library products could be tested in appropriate assays for this well-known cell adhesion moiety.

In contrast to earlier syntheses, the resin was extensively washed following OFm deprotection to remove residual piperidine. This resulted in a decreased apparent reaction time, as monitored by ninhydrin testing. Even more dramatic was the effect of the recently described coupling agent HATU (the "Aza" equivalent of HBTU) which also incorporates the aza-HOBt racemization suppressant as described by Carpino [5]. In this presumably more complex cyclization test, in which a large variety of pentapeptide sequences must be cyclized in order to provide a negative ninhydrin test, the reaction was judged complete after one hour.

Analysis of the resulting mixture included the use of FABMS as well as CEEC (capillary electrophoresis with electrochemical detection—designed to identify more electroactive moieties). The molecular weight analysis revealed the presence of both monomeric and dimeric species, the latter presumably occurring with more recalcitrant cyclic-forming sequences. (In this experiment, prolines and glycine residues were not deliberately enhanced to optimize monocyclization yields; analysis of epimerization during coupling is being completed.) Both FABMS and AAA are consistent with the formation of primarily monomeric species containing the expected amino acids. Further analysis of individual products should help in future experimental design to identify the sequence-dependence of cyclization tendencies.

It thus appears feasible to prepare libraries of cyclic peptide mixtures using well-precedented techniques in peptide synthesis. The ring size demonstrated here involves cyclic pentapeptides but examples of both smaller and larger ring sized libraries are being evaluated. We are exploring the use of chemistries that will lead to facile side chain attachment of other amino acids, such as benzyl ether, (Ser, Thr), carbobenzoxy (Lys, Orn), or thioether (Cys) analogs and techniques for preparing deprotected resin-bound sequences.

Additionally, we are evaluating the concept of serial release of cyclic peptides from library mixtures, using a variety of orthogonal linker strategies. This concept has been very well demonstrated using linear peptide libraries [6], but provides a somewhat more challenging set of orthogonality requirements when extended to cyclic peptide mixtures.

Acknowledgements

We thank Dr. Fernando Albericio and Millipore (Bedford, MA) for the samples of HOAT and HATU used in this work and the University of Nebraska for FABMS results. Financial support from NIH GM-33376 is also gratefully acknowledged.

References

1. Rovero, P., Quartara, L. and Fabbri, G., Tetrahedron Lett., 32 (1991) 2639.
2. Ihara, M., Noguchi, K., Saeki, T., Fukuroda, T., Tsuchida, S., Kimura, S. and Fukami, T., Life Sci., 50 (1992) 247.
3. Furka, A., Sebestyen, F., Asgedom, M. and Dibo, G., Int. J. Peptide Protein Res., 37 (1991) 487.
4. Lam, K.S., Salmon, S.E., Hersh, E.M., Hruby, V.J., Kazmierski, W.M. and Knapp, R.J., Nature, 354 (1991) 82.
5. Carpino, L.A., J. Am. Chem. Soc., 115 (1993) 4397.
6. Lebl, M., Lam, K.S., Kocis, P., Krchnak, V., Patek, M., Salmon, S.E. and Hruby, V.J., Peptides 1992, In Schneider, C.H. and Eberle, A.N. (Eds.), Peptides 1992, Escom, Leiden, The Netherlands, 1993, p.67.

New, potent N-acetylated all D-amino acid opioid peptides

C.T. Dooley and R.A. Houghten

Torrey Pines Institute for Molecular Studies,
3550 General Atomics Court, San Diego, CA 92121, U.S.A.

Introduction

Synthetic peptide combinatorial libraries (SPCLs), composed of tens of millions of free peptides, have been shown to be highly effective in the determination of opioid ligands [1, 2]. In these early studies it was shown that methionine and leucine enkephalin could be determined from a library of 52,128,400 hexapeptides. We have also reported the use of an N-acetylated library, composed of the same number of peptides, to determine new peptide ligands for opioid receptors [3]. These peptides, which we have termed Acetalins, bear no sequence homology to known peptides or proteins (Ac-RFMWMO-NH$_2$). Three of these peptides have been shown to bind to μ, δ and K$_3$ opioid receptors, but exhibit negligible activity at K$_1$ and K$_2$ opioid receptors [3]. The peptides, Ac-RFMWMT-NH$_2$, Ac-RFMWMK-NH$_2$, and Ac-RFMWMR-NH$_2$ were shown to behave as antagonists when examined in the guinea pig ileum and mouse vas deferens assays. We have now determined, from a library composed entirely of D-amino acids, individual peptides capable of inhibiting the binding of tritiated (D-Ala^2,MePhe^4,Gly-ol^5)encephalon (DAGO) to the μ receptor.

Results and Discussion

The 400 separate mixtures in this SPCL are represented as Ac-o$_1$o$_2$xxxx-NH$_2$, in which the first two amino acids (o$_1$ and o$_2$) are individually defined, and the last four amino acids consist of equimolar mixtures of 19 D-amino acids (cysteine omitted, glycine included). The library was synthesized as previously described [4]. The library was screened for ability to inhibit the binding of ^{3}H-DAGO to crude rat brain homogenates. Preparation of membranes and assay conditions have been described previously [1]. The three most effective DAGO inhibiting mixtures were Ac-ryxxxx-NH$_2$ (IC$_{50}$ = 8,437 nM), Ac-rwxxxx-NH$_2$ (IC$_{50}$ = 11,686 nM) and Ac-rfxxxx-NH$_2$ (IC$_{50}$ = 14,152 nM). An iterative selection process which sequentially defined the four mixture positions (Ac-o$_1$o$_2$o$_3$xxx-NH$_2$ → Ac-o$_1$o$_2$o$_3$o$_4$o$_5$o$_6$-NH$_2$) was subsequently performed for the two mixtures, Ac-rfxxxx-NH$_2$ and Ac-ryxxxx-NH$_2$. The former mixture ultimately yielded peptides with greater activity and is thus described here in detail. Upon defining the third position with 20 D-amino acids (cysteine included), 10 peptide mixtures inhibited DAGO more effectively than the control mixture, Ac-rfxxxx-NH$_2$ (IC$_{50}$ = 15,673 nM). The mixture Ac-rfwxxx-NH$_2$ (IC$_{50}$ = 1,075 nM) was the most effective at inhibiting ^{3}H-DAGO; with a 14.5-fold increase in activity observed over the control mixture, Ac-rfxxxx-NH$_2$. Seven mixtures had activities greater than Ac-rfwxxx-NH$_2$ when the fourth position was defined. The two most active peptide mixtures found were Ac-rfwwxx-NH$_2$ (IC$_{50}$ = 666 nM) and Ac-rfwixx-NH$_2$ (IC$_{50}$ = 873 nM); they exhibited 4 and 3-fold increases over Ac-rfwxxx-NH$_2$ (IC$_{50}$ = 2,732 nM), respectively. While iterations of both these mixtures have been completed, Ac-rfwixx-NH$_2$ ultimately yielded individual peptides having the greatest activities. Four mixtures were found to be more effective than the control mixture upon defining the fifth position. The most effective mixture was Ac-rfwinx-NH$_2$ (IC$_{50}$ = 235 nM), representing a 2-fold increase in activity over Ac-rfwixx-NH$_2$ (IC$_{50}$ = 559 nM). Of the 20 peptides for which all six positions were defined, three peptides exhibited activity greater than the control mixture, Ac-rfwinx-NH$_2$ (IC$_{50}$ = 559 nM). Ac-rfwink-NH$_2$ (IC$_{50}$ = 16 nM) was the most effective ^{3}H-DAGO inhibiting peptide. Ac-rfwina-NH$_2$ (IC$_{50}$ = 33 nM) and Ac-rfwinr-NH$_2$ (IC$_{50}$ = 34 nM) were the second and third most effective peptides, respectively. We believe this is the first instance of hexapeptides containing all D-amino acids shown to bind with high affinity to an opioid receptor.

Acknowledgements

The authors wish to thank A.M. Prochera, S. Hope and P. Lopez for technical assistance, and the National Institute of Drug Abuse for services provided under contract #271-89-8159. This work was funded in part by Houghten Pharmaceuticals, Inc., San Diego, California.

References

1. Houghten, R.A. and Dooley, C.T., BioMed. Chem. Lett., 3 (1993) 405.
2. Dooley, C.T. and Houghten, R.A., Life Sciences, 52 (1993) 1509.
3. Dooley, C.T., Chung, N.N., Schiller, P.W. and Houghten, R.A., Proc. Natl. Acad. Sci. U.S.A., 90 (1993) 10911.
4. Houghten, R.A., Pinilla, C., Blondelle, S.E., Appel, J.R., Dooley, C.T. and Cuervo, J.H., Nature, 354 (1991) 84.

The use of sub-library kits in a new screening strategy

A. Furka, F. Sebestyen and E. Campian

*Department of Organic Chemistry, Eotvos Lorand University,
P.O. Box 32, H-1518 Budapest 112, Hungary*

Introduction

As a contribution to speeding up the process of drug discovery, a new strategy was introduced in 1988 facilitating discovery of biologically active peptides. Our "portioning-mixing" method [1-3] added two simple operations (dividing the resin into portions before coupling and mixing after coupling) to those used in the Merrifield's solid phase synthesis and, as a result, enabled the user to prepare millions or even billions of peptides (peptide libraries) by building into them the constituent amino acids in all possible combinations. The purpose of this paper is to show how partial libraries can be utilized in screening.

Results and Discussion

Libraries and sub-libraries. First of all, it seems worthwhile to introduce some new definitions. Thus, the name normal library may be suggested for mixtures formed by varying the same number of amino acids in all positions. Standard libraries will represent a special case of the normal libraries where the 20 common amino acids are varied in all positions. Cysteine is mostly omitted in the synthesis. This can be indicated by a "minus C" termination (standard-C library). A special combined library, containing a genetic series of standard libraries will be named genetic library, e.g., the genetic pentapeptide library includes all standard libraries from dipeptides to pentapeptides. The use of non-common amino acids besides the common ones can be indicated by the name: extended library.

Any kind of library – except genetic libraries – can be divided into sub-libraries. They are special partial libraries of parent libraries and can be synthesized without varying the amino acids in one or more positions, while in the rest of the positions they are varied like in the parent library. A non-varied position is occupied by the same amino acid in all peptides. Sub-libraries may be classified according to their order: the first, second, third, etc., order sub-libraries having one, two, three, etc., non-varied positions, respectively.

In the following discussions, if not stated otherwise, the term sub-library means a sub-library of standard libraries. The coupling positions (CP-s) are numbered starting from the C-terminus to express the order of couplings during the synthesis. The amino acids are denoted by one letter symbols.

First order sub-libraries have a single non-varied position. In a first order pentapeptide sub-library shown below, the coupling position No. 2 is non-varied, occupied by glutamic acid, while in the first, third, fourth and fifth positions, all the

20 amino acids are varied.

1	A,C,D,E,F,G,H,I,K,L,M,N,P,Q,R,S,T,V,W,Y
2	E
3	A,C,D,E,F,G,H,I,K,L,M,N,P,Q,R,S,T,V,W,Y
4	A,C,D,E,F,G,H,I,K,L,M,N,P,Q,R,S,T,V,W,Y
5	A,C,D,E,F,G,H,I,K,L,M,N,P,Q,R,S,T,V,W,Y

A short symbol suggested for this type of sub-library shows the non-varied position and the amino acid occupying it: 2E. An additional (optional) figure may be used to indicate the number of residues: 2E5.

There are altogether 100 different first order pentapeptide sub-libraries which form a first order sub-library kit:

1A,1C,1D,1E,1F,1G,1H,1I,1K,1L,1M,1N,1P,1Q,1R,1S,1T,1V,1W,1Y
2A,2C,2D,2E,2F,2G,2H,2I,2K,2L,2M,2N,2P,2Q,2R,2S,2T,2V,2W,2Y
3A,3C,3D,3E,3F,3G,3H,3I,3K,3L,3M,3N,3P,3Q,3R,3S,3T,3V,3W,3Y
4A,4C,4D,4E,4F,4G,4H,4I,4K,4L,4M,4N,4P,4Q,4R,4S,4T,4V,4W,4Y
5A,5C,5D,5E,5F,5G,5H,5I,5K,5L,5M,5N,5P,5Q,5R,5S,5T,5V,5W,5Y

A first order pentapeptide sub-library has 160,000 components. The 1 to 1 mixture of the 20 sub-libraries in any row is equivalent to the standard library (3.2 million components).

The second order sub-libraries are characterized by two non-varied positions. An example of such a pentapeptide sub-library is shown below:

1	A,C,D,E,F,G,H,I,K,L,M,N,P,Q,R,S,T,V,W,Y
2	A
3	L
4	A,C,D,E,F,G,H,I,K,L,M,N,P,Q,R,S,T,V,W,Y
5	A,C,D,E,F,G,H,I,K,L,M,N,P,Q,R,S,T,V,W,Y

The symbol of this sub-library should be either 2A3L or 2A3L5. The number of such CP 2,3 type second order sub-libraries is 400 (part of them are shown below).

2A3A,2C3A,2D3A,2E3A,2F3A,2G3A,   ,2R3A,2S3A,2T3A,2V3A,2W3A,2Y3A
2A3C,2C3C,2D3C,2E3C,2F3C,2G3C,   ,2R3C,2S3C,2T3C,2V3C,2W3C,2Y3C
2A3D,2C3D,2D3D,2E3D,2F3D,2G3D,   ,2R3D,2S3D,2T3D,2V3D,2W3D,2Y3D

2A3W,2C3W,2D3W,2E3W,2F3W,2G3W,,2R3W,2S3W,2T3W,2V3W,2W3W,2Y3W
2A3Y,2C3Y,2D3Y,2E3Y,2F3Y,2G3Y,   ,2R3Y,2S3Y,2T3Y,2V3Y,2W3Y,2Y3Y

Each sub-library has 8000 components. The 1 to 1 mixture of the 400 2,3 type sub-libraries forms a standard library. There are, however, 10 types of second order pentapeptide sub-libraries altogether (or individually) forming a kit: 1,2; 2,3; 3,4; 4,5; (vicinals), 1,3; 1,4; 1,5; 2,4; 2,5; 3,5 (disjuncts).

The domino strategy. It is a characteristic feature of the first order sub-libraries that the sole non-varied position is occupied by the same amino acid in all peptides. For this reason a first order sub-library can be used to detect the presence or absence of an amino acid in the non-varied position of the active peptides. The results of screening experiments done with all components of the first order sub-library kit are expected to show which amino acids occur in the active peptides in the different coupling positions, e.g., 1(A,G,S,T), 2(H,I), 3(K,L,G), 4(W) and 5(M,T,Y).

The found amino acids and their coupling positions define a partial library which is named "occurrence library". These amino acids are considered to be varied in the different positions of this hypothetical library. The occurrence library comprises all active peptides and in addition inactive ones. In the above example, the total number of peptides in the library is only 4x2x3x1x3=72 (instead of the 3.2 million of the parent library).

With a second order sub-library the presence or absence of two amino acids occupying two definite (vicinal or disjunct) positions can be checked, that is, they can be used like dominoes in deducing the sequence of the bioactive peptides. Only a limited number of second order sub-libraries are needed in the experiments. They can be identified knowing the composition of the occurrence library. Second order sub-libraries belonging to both the parent library and the occurrence library are expected to serve equally well in the screening experiments.

The domino strategy is entirely general and involves no condition of any kind concerning the type of peptide libraries and the screening methods.

Acknowledgements

The authors thank the Hungarian Scientific Research Foundation (OTKA, No. T 4192) for financial support.

References

1. Furka, A., Sebestyen, F., Asgedom, M. and Dibo, G., Abstr. 14th Int. Congr. Biochem., Prague, Czechoslovakia, 5 (1988) 47.
2. Furka, A., Sebestyen, F., Asgedom, M. and Dibo, G., Abstr. 10th Int. Symp. Med. Chem., Budapest, Hungary, 1988, p.288.
3. Furka, A., Sebestyen, F., Asgedom, M. and Dibo, G., Int. J. Peptide Protein Res., 37 (1991) 487.

The synthesis and coupling efficiency of 7-hydroxycoumarin-4-propionic acid, a fluorescent marker useful in immobilized substrate libraries

J.A. Gainor, T.D. Gordon* and B.A. Morgan*

Sterling Winthrop Pharmaceutical Research Division, Rensselaer, NY 12144, U.S.A.

Introduction

The assembly of immobilized substrate libraries requires efficient chemistry since the peptide sequences are not removed from the solid support for rigorous analytical analysis following synthesis. The Fmoc strategy of peptide synthesis has been employed in library synthesis with good reliability, however amide formation with carboxylates other than standard protected α-amino acids must be studied carefully. Our first fluorescent marker, 7-hydroxycoumarin-4-acetic acid (Cou), used to quantitatively measure the enzymatic hydrolysis of immobilized substrates exhibited poor coupling efficiencies. We have developed methods to examine coupling efficiency and subsequently improved our fluorescent marker chemistry.

Results and Discussion

The basic construct of our immobilized substrate technology consists of a fluorescent marker at the N-terminus of a peptide sequence attached to a linker comprised of six-aminocaproic acid residues, covalently linked to an aminopropyl anchor on controlled pore glass [1].

Scheme 1

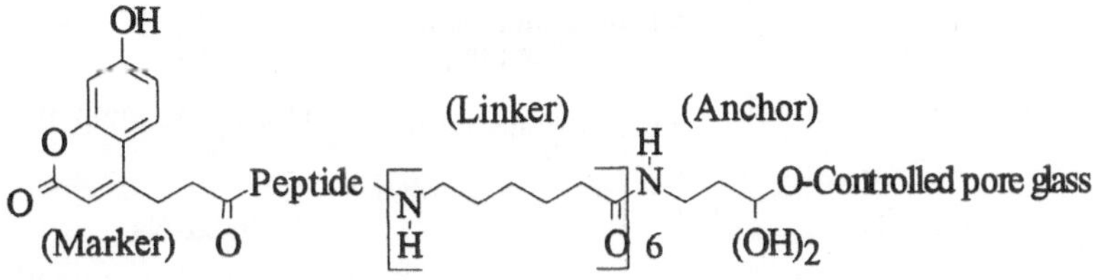

When exposed to enzyme, the peptide sequence is cleaved, releasing a fluorescent fragment into solution. The measured fluorescence is directly correlated with the extent of peptide cleavage.

---

*Current Address: Biomeasure, 27 Maple Street, Milford, MA 01757, U.S.A.

989

We employed commercially available Cou in early experiments. This carboxylic acid was coupled without hydroxyl protection in an automated process. The large extinction coefficient allowed precise measurements even with minute quantities of materials.

On investigation of the solution phase reaction the previously reported [2] decarboxylation of Cou to 7-hydroxy-4-methylcoumarin was found to cause incomplete capping of peptide sequence. The homolog 7-hydroxycoumarin-4-propionic acid (Cop) was developed (Scheme 2) as a more stable analog of Cou which retains nearly identical fluorescence characteristics.

Scheme 2

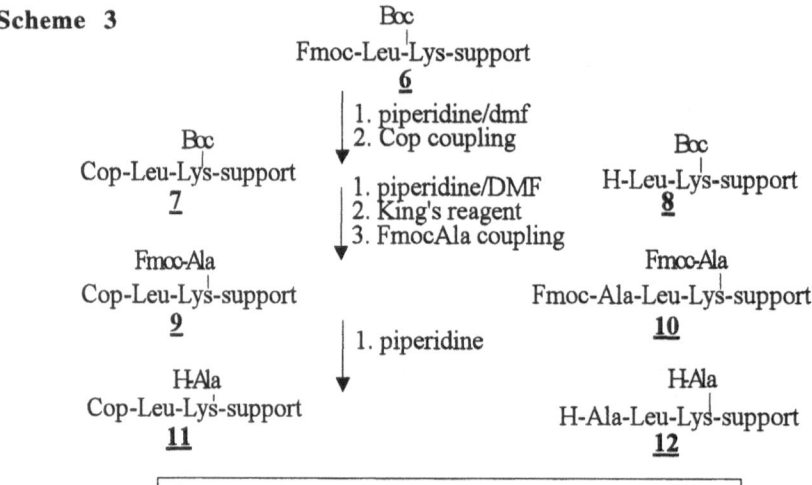

Meldrum's acid **1** was acylated with ethyl succinyl chloride **2** and decarboxylated to yield **3** as previously described [3]. Condensation of the ß-keto ester **3** with resorcinol was achieved in ethanol saturated with HCl at 0°C and left for several days at room temperature tightly stoppered. The crystalline product **4** of the modified Peckman condensation [4, 5] was subject to basic hydrolysis to provide analytically pure Cop **5** after a single recrystallization [6].

Scheme 3

Boc
|
Fmoc-Leu-Lys-support
6

1. piperidine/dmf
2. Cop coupling

Boc
|
Cop-Leu-Lys-support
7

1. piperidine/DMF
2. King's reagent
3. FmocAla coupling

Boc
|
H-Leu-Lys-support
8

Fmoc-Ala
|
Cop-Leu-Lys-support
9

1. piperidine

Fmoc-Ala
|
Fmoc-Ala-Leu-Lys-support
10

H-Ala
|
Cop-Leu-Lys-support
11

H-Ala
|
H-Ala-Leu-Lys-support
12

Support = (Acp)6-aminopropyl-controlled pore glass

The coupling efficiencies of both Cou and Cop were investigated using amino acid analysis and peptide sequencing of products without removal of the peptides from the support. Peptide **6** was synthesized on the controlled pore glass using Fmoc chemistry on an Advanced Chemtech 350 multiple peptide synthesizer. Deprotection of peptide **6** and coupling of Cop provides the capped sequence **7** and, if the marker coupling was less than complete, the side product **8**. The mixture is then treated with TFA to remove the side chain Boc protection and coupled with Fmoc-Ala to yield **9** with possible impurity **10**. Deprotection with piperidine provides final products **11** and **12** which are analyzed on the support.

Amino acid analysis will ideally display equivalent amounts of each amino acid. When coupling is incomplete, the abundance of alanine equals the extent uncoupled. Sequence analysis will indicate that capping with marker was incomplete if leucine is present in the second cycle.

Using these techniques, we have developed a protocol that includes double coupling with a 10-fold excess of Cop. Analysis of products synthesized using this protocol are comparable to results obtained for peptide sequences which were synthesized, purified and then attached to the support as a fragment.

We have described the large scale synthesis of Cop, a chemically stable fluorescent marker useful in immobilized substrate technology. In addition, we have described relatively simple and general techniques for evaluating the coupling efficiencies of carboxylates to a peptide on a solid support. These techniques do not require that the peptide sequence be removed from the support prior to analysis. Using these techniques, we have developed highly reliable coupling protocols for the construction of peptide libraries.

Acknowledgements

We wish to acknowledge Dr. Jasbir Singh and Mr. David Whipple for the amino acid analysis and Ms. Carla Gilliam for the peptide sequencing.

References

1. Ator, M.A., Dankanich, T.C., Echols, M., Gainor, J.A., Gilliam, C.L., Gordon, T.D., Koch, D., Koch, J.F., Kruse, L.I., Morgan, B.A., Krupinski-Olsen, R. and Siahaan, T.J., Proceedings of the 13th American Peptide Symposium, Escom, Leiden, The Netherlands, 1994, this volume.
2. Dey, B.B. and Seshadri, T.R., J. Indian Chem. Soc., 8 (1931) 247.
3. Oikawa, Y., Sugano, K. and Yonemitsu, O., J. Org. Chem., 43 (1978) 2087.
4. Peckman H. and Duisberg, C., Ber., 16 (1893) 2119.
5. Appel, H., J. Chem. Soc., (1935) 1031, and J. Chem Soc., 107 (1915) 1606.
6. Products characterized by elemental analysis, NMR, MS, and TLC.

Peptide reversal on solid supports: A technique for the generation of C-terminal exposed peptide libraries

C.P. Holmes and C.M. Rybak

Department of Chemistry, Affymax Research Institute,
4001 Miranda Ave., Palo Alto, CA 94304, U.S.A.

Introduction

The use of random peptide libraries as a new tool in drug discovery has received considerable recent attention. In addition to the use of soluble libraries, several groups have reported the screening of peptides while still attached to the solid phase synthesis support [1, 2]. Immobilized peptides can be displayed in two orientations: N-terminal exposed and C-terminal exposed, which differ only in which end of the peptide is attached to the support. There are many examples of receptors, enzymes and antibodies which possess inherent preferences for binding to only one end of a peptide chain. Traditional solid phase peptide techniques rely on a C to N strategy of forming the polymer chain, and hence are only compatible with the presentation of N-terminal libraries. The complementary N to C strategy, although known, is believed to be accompanied by racemization of the growing peptide chain. We have begun to explore a technique of *reversing* the peptides on the support, whereby one first creates an N-terminal library, then cyclizes the N-terminus to the support, and one ultimately cleaves the bond anchoring the C-terminus to the support to generate the C-terminal exposed library.

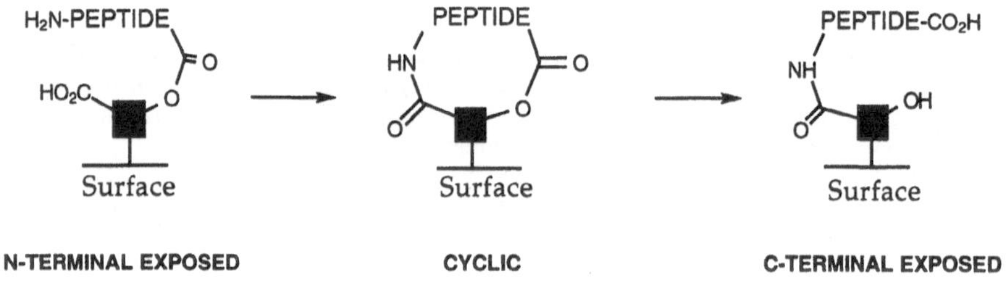

Fig. 1. Peptide reversal on solid supports.

Results and Discussion

The synthesis of a template molecule **6** suitable as a building block to effect the peptide reversal is shown below. We have targeted an amine-to-carboxyl strategy to ring close the peptide chain, and have employed an allyl ester as an orthogonal protecting group for the carboxyl. The alkoxyacetic acid chain is used to attach the template to a solid support. We have incorporated a labile benzylic bond into the tether molecule. This affords the possibility of either the concurrent cleavage of the attachment bond during the removal of the side chain protecting groups upon treatment with TFA, or the stepwise cleavage of the peptide followed by removal of the side chain protecting groups. Since template **6** will have the first amino acid linked through an amide bond, the reversal process generates the peptide as its C-terminal carboxamide. Compound **5** serves as a convenient branch point for exploring other amine protection strategies in addition to the Fmoc strategy we have pursued.

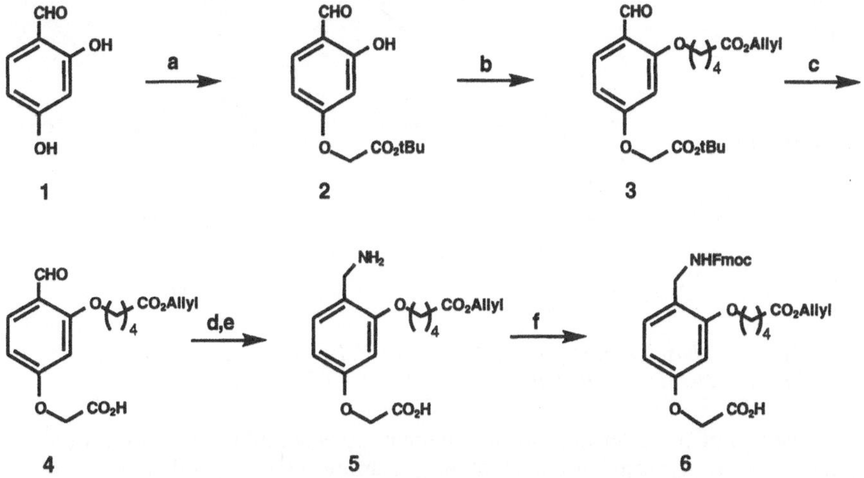

Fig. 2. Synthesis of template molecule. (a) NaH, $BrCH_2CO_2tBu$; (b) K_2CO_3, $Br(CH_2)_4CO_2Allyl$; (c) TFA; (d) H_2NOH, pyridine; (e) Zn, HOAc; (f) Fmoc-Cl, $NaHCO_3$.

We chose to explore the appropriate reaction conditions for the orthogonal deprotections and the cyclization in the solution phase with the template **7**, which is the oxo-glycine analog of **6** and has the carboxyl group used to bind to the support blocked as its t-butyl ester. Conversion of **3** to **7** was accomplished in two steps via sodium borohydride reduction followed by carbodiimide coupling with Fmoc-Gly-OH. Treatment of template **7** with 50% TFA/CH_2Cl_2 for 5 minutes gave complete conversion to Fmoc-Gly-OH, as evidenced by HPLC analysis. This agrees with the known susceptibility of similar cleavable linkers towards TFA [3, 4].

We have found that a two step protocol of generating the cyclized product gives the highest overall yields in this solution phase system. Deprotection of the allyl group with palladium under acidic conditions with acetic acid as scavenger gives >90% yield of acid **8**. The cyclization was most conveniently performed through a one pot, two step procedure of first deprotecting the Fmoc group with 2% DBU in DMF, followed by dilution with DMF to a final concentration of 1 mM and BOP/HOBt treatment to give a 75% yield of **9**. Removal of the Fmoc group with piperidine, followed by cyclization with DCC gave a significant amount of the piperidinyl amide adduct, arising from residual piperidine trapping the activated carboxyl group.

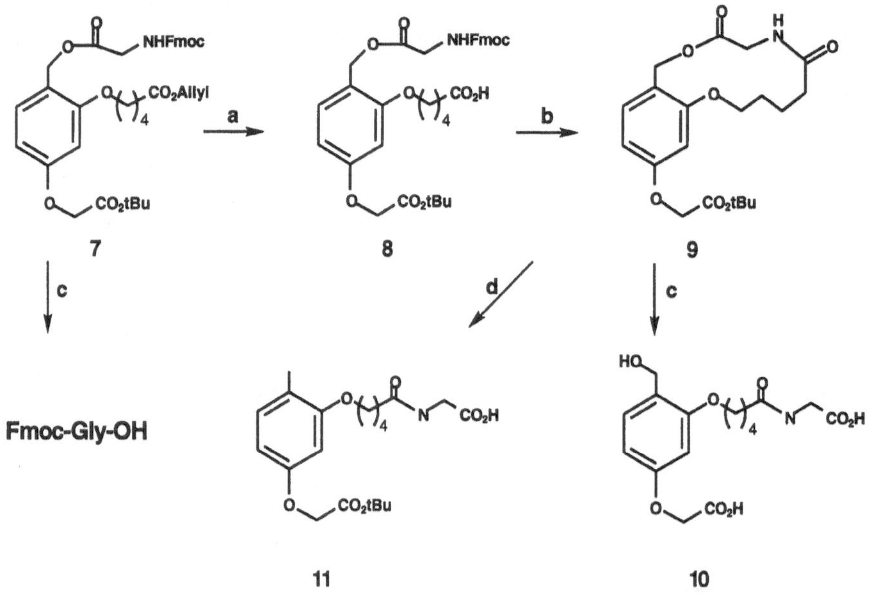

Fmoc-Gly-OH

Fig. 3. Solution phase chemistry of cyclization and cleavage reactions of template. (a) Ph_3P, $Pd(Ph_3P)_4$, HOAc; (b) DBU, then BOP, HOBt; (c) TFA; (d) H_2, Pd/C.

Cleavage of the ester bond to glycine can be accomplished in a multitude of ways. Catalytic hydrogenation, alcoholysis, aminolysis, and other nucleophilic reagents can lead to the generation of acids, esters, amides and the like. We have demonstrated two different types of cleavages. Catalytic hydrogenation of **9** affords cleanly the free acid **11** in 86% yield after chromatography. Treatment of **10** with TFA generates compound **10** in moderate yield, although the *solution* nature of the reaction makes isolation of the product difficult. The application and use of this template to the reversal of peptides and other polymers on solid supports are in progress and will be reported in due course.

References

1. Fodor, S.P.A., Read, J.L., Pirrung, M.C., Stryer, A.T., Lu, A. and Solas, D., Science, 249 (1991) 404.
2. Lam, K.S., Salmon, S.E., Hersh, E.M., Hruby, V.J., Kazmierski, W.M. and Knapp, R.J., Nature, 354 (1991) 82.
3. Sheppard, R.C. and Williams, B.J., J. Chem. Soc. Chem. Comm., (1982) 587.
4. Flörsheimer, A. and Riniker, B., In Giralt, E. and Andreu, D. (Eds.), Peptides 1990 (Proceedings of the 21st European Peptide Symposium), Escom, Leiden, The Netherlands, 1990, p.131.

Investigation of antibody binding diversity utilizing representative combinatorial peptide libraries

W.Y. Wong[1,2], H.B. Sheth[1,3], A. Holm[4], R.T. Irvin[3] and R.S. Hodges[1,2]

[1]*The Protein Engineering Network of Centres of Excellence,*
[2]*Department of Biochemistry, and*
[3]*Department of Medical Microbiology and Infectious Diseases,*
University of Alberta, Edmonton, Alberta, T6G 2H7, Canada
[4]*Centre for Medical Biotechnology, Chemistry Department, Royal*
Veterinary and Agricultural University, Copenhagen, Denmark

Introduction

The advances in solid-phase peptide chemistry have led to the development of various techniques for generating libraries with vast numbers of diverse peptides. These peptide libraries are screened to derive peptide sequences of interest that may be used clinically or used for the development of pharmaceutical compounds. Since peptide diversity of a library constructed with all 20 naturally occurring amino acids is extensive, physical constraints for the synthesis and screening arise and impose serious limitations on the length of peptide that may be examined. To simplify both the synthesis and library screening procedures, a novel approach using representative amino acids instead of all 20 naturally occurring amino acids was investigated. Representative hexapeptide combinatorial libraries were made employing a set of ten amino acids which display the basic physico-chemical properties associated with naturally occurring amino acid side-chains.

These peptide libraries were used to investigate the binding diversity of the murine monoclonal antibody PK99H. This MAb was raised against the pili of the opportunistic pathogen *Pseudomonas aeruginosa* strain K (PAK). The epitope recognized by PK99H has been recently mapped using single alanine substituted peptides and was found to consist of the sequence DEQFIPK [1].

Results and Discussion

The three series of hexapeptide libraries, $Ac-O_1O_2X_3X_4X_5X_6-NH_2$, $Ac-X_1X_2O_3O_4X_5X_6-NH_2$, and $Ac-X_1X_2X_3X_4O_5O_6-NH_2$, were synthesized by Fmoc chemistry using a Multiple Column Peptide Synthesis instrument [2, 3]. Each O_n and X_n were chosen from the following 10 representative amino acids: Ala, Glu (Asp), Phe (Trp, Tyr), Gly, Ile (Val, Met), Lys (His, Arg), Leu (Met), Pro, Gln (Asn) and Ser (Thr) with the representative amino acids given in parentheses. No representative amino acid for Cys was selected. Each series of peptide libraries has

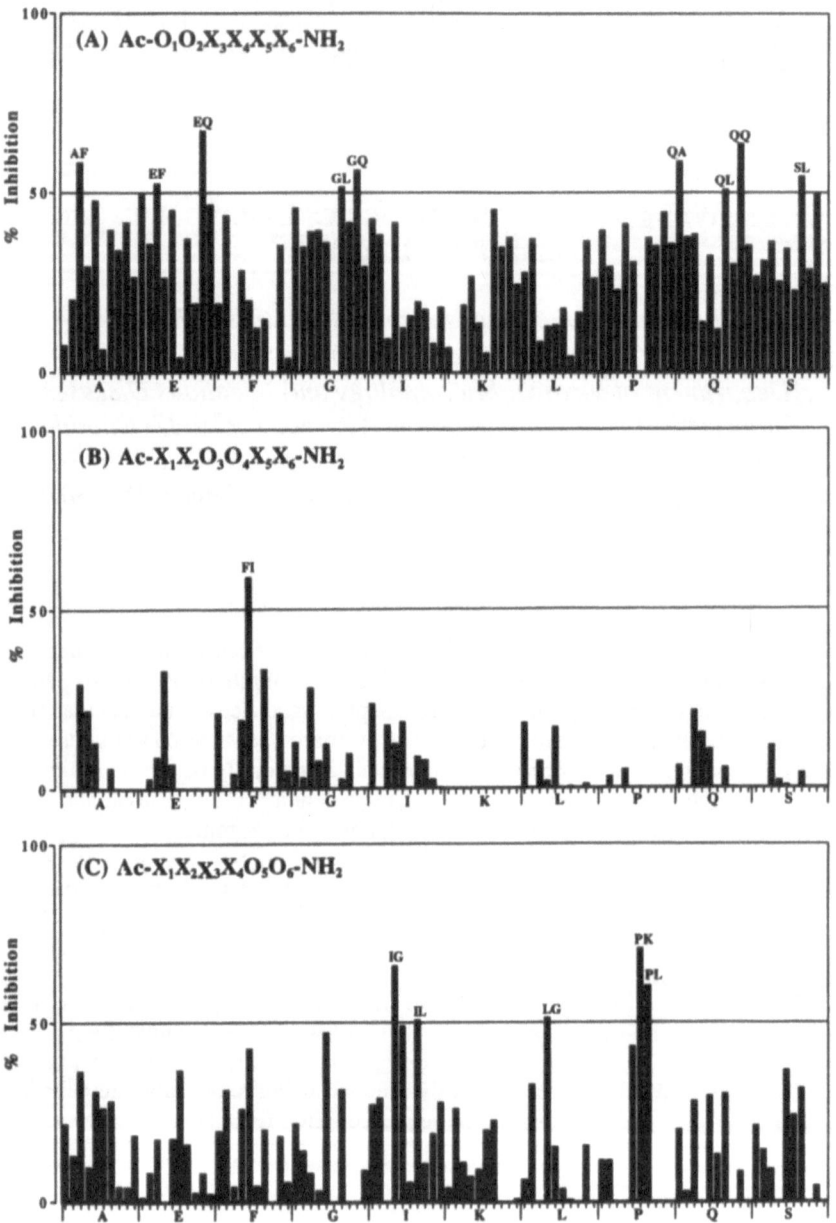

Fig. 1. *Competitive ELISA profiles of the three series of representative peptide libraries. The single letter codes on the abscissa denote the first position (O_n) of the dipeptide sequence O_nO_{n+1}. The order of amino acid residues in the second position (O_{n+1}) was A, E, F, G, I, K, L, P, Q and S. For example, the region denoted A has 10 pools of dipeptide-defined hexapeptide libraries in the following order: AA, AE, AF, AG, AI, AK, AL, AP, AQ and AS.*

two specifically defined positions (O_nO_{n+1}) and thus represents 100 hexapeptide pools with each pool containing approximately 10,000 peptides. The main idea of using a representative library instead of a standard library employing all 20 naturally occurring amino acids is that a representative library can tremendously reduce the number of peptides in the library. This not only simplifies the synthesis procedure but also increases the screening sensitivity. In addition, it also allows the synthesis of a library with longer peptide length. For instance, in a standard octapeptide library, the number of peptides will be 20^8 (25.6 billion), whereas in a representative octapeptide library, the number of peptides will be reduced to 10^8 (0.1 billion).

The peptide libraries were screened by competitive ELISA. Oxidized PAK 17-mer peptide (Ac-KCTSDQDEQFIPKGCSK-OH) corresponding to the C-terminal region of the PAK pilin was adsorbed to the microtitre plate. The ability of the libraries to inhibit the interaction of MAb PK99H to the coated peptide was determined. From the ELISA results (Figure 1), a cutoff at 50% inhibition was used to highlight the significant 'hits' in order to locate those critical residues important for MAb binding. In library (A), nine possible hits were observed, which were in the order of EQ, QQ, QA, AF, GQ, SL, EF, GL and QL. In library (B), only one hit was observed, which was FI. In library (C), five hits were observed, which were PK, IG, PL, LG and IL. The three best dipeptides mapped by these three series of peptide libraries are EQ, FI and PK, respectively. These indeed constitute the native PAK epitope sequence. Presently, in order to verify these ELISA results, a series of synthetic hexapeptides, based on a combination of the best hits from each library, will be made. The binding affinities of these peptides to MAb PK99H will be examined and compared with the native peptide.

The representative combinatorial peptide library is an effective tool in studying the epitope and binding diversity of MAb PK99H. It employs the potential of synthesizing longer peptides in a library without sacrificing the screening sensitivity.

Acknowledgements

We thank Charlotte Holm (Royal Veterinary and Agricultural University) for her excellent technical assistance. This work was supported by grants from the Protein Engineering Network of Centres of Excellence, the Canadian Cystic Fibrosis Foundation, and the Canadian Bacterial Diseases Network.

References

1. Wong, W.Y., Irvin, R.T., Paranchych, W. and Hodges, R.S., Protein Sci., 1 (1992) 1308.
2. Holm, A. and Meldal, M., In Bayer, E. and Jung, G. (Eds.), Peptides 1988, Walter de Gruyter & Co., Berlin-New York, 1988, p.208.
3. Meldal, M., Bisgaard Holm, C., Boejesen, G., Havsteen Jakobsen, M. and Holm, A., Int. J. Pept. Protein Res., 41 (1993) 250.

Experimental limits of the peptide library concept

K.-H. Wiesmüller[1], A.G. Beck-Sickinger[2], H.G. Ihlenfeldt[2], H.A. Wieland[3], K. Udaka[4], P. Walden[4] and G. Jung[2]

[1]*Naturwissenschaftliches und Medizinisches Institut, Gustav-Werner-Str. 3, 72762 Reutlingen, FRG,* [2]*Institut für Organische Chemie, Eberhard-Karls-Universität, 72076 Tübingen, FRG ,* [3]*Preclinical Research, Dr. Karl Thomae GmbH, 88397 Biberach, FRG* [4]*Max-Planck-Institut für Immunbiologie, Corrensstraße 42, 72076 Tübingen, FRG*

Introduction

Peptide libraries are pools of synthetic peptides which can be used as a source of new lead structures for drug development based on peptide receptor interactions. We identified peptide pools binding to MHC class I proteins [1] and these peptide mixtures were synthesized and used for pool characterization by sequencing and electrospray mass spectrometry [2]. Three examples of receptor assays with respect to possible applications of libraries with low and high complexity were selected to estimate the most promising application of peptide mixtures.

Results and Discussion

Receptor binding assay of neuropeptide Y: We investigated the Y1 receptor affinity of NPY [3] admixed to octapeptide mixtures $O_1O_2O_3O_4O_5X_1X_2X_3$ which cover the MHC class I K^b binding motif [2] and are not related to NPY partial sequences. Considerable loss of receptor binding (measured by ^{125}I-BH-NPY displacement) was found for a mixture which contained a tenfold excess of NPY. If we added NPY equimolar to one member of the mixture the receptor affinity of this ligand was blocked. Specificity of the binding was significantly reduced even when NPY was present in the mixture in 100 fold excess. The sublibraries $ITRX_1X_2X_3$-NH_2, Ac-$ITRQX_1X_2$, $ITRQYX$-NH_2, $ITRQYXX$-NH_2 and the C-terminal NPY hexapeptide amide $ITRQRY$-NH_2 were prepared by Fmoc/tBu chemistry (X: L-amino acids except C, W). The C-terminal hexapeptide amide and peptide amide pools were tested for their potency to prevent receptor interaction of NPY in different concentrations and they reduced markedly the affinity to the Y1 receptor. This effect was not influenced by the complexity (5832 or 324) of the mixtures (Figure 1B). Peptide and peptide mixtures itself had no effect on ^{125}I-BH-NPY displacement.

Killer cell activity on MHC class I bound peptides: Killing of target cells preincubated with CTL antigen and peptide mixtures for 30 min at 37°C was investigated in a standard ^{51}Cr release assay [1]. The MHC class I K^b restricted peptide SIINFEKL was added to an octapeptide mixture representing 5832 peptides. Killing of target cells was induced even in a ratio of one molecule SIINFEKL to 174,960 molecules of added peptides. The epitope alone induced 50% killing of target cells at a concentration of 1 pM and in the presence of the mixture 7 pM concentration was necessary for killing. As expected the non-MHC restricted mixture itself had no effect.

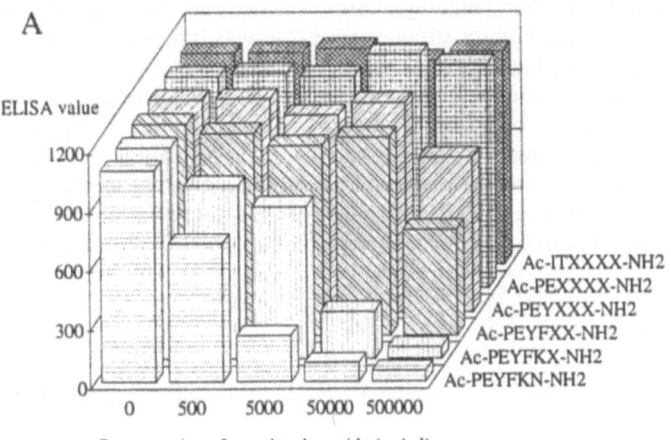

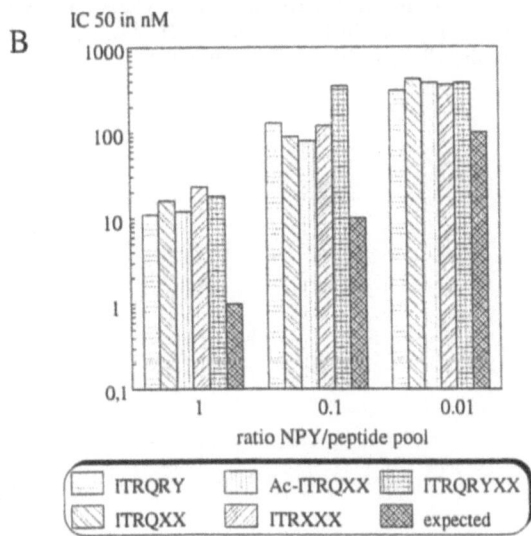

Fig. 1. (A) Competition of B-cell antigen and peptide mixtures with biotinylated antigen biotin-spacer-PEYFKN-NH$_2$ for binding to the mab Dü 162. (B) Inhibition of binding of NPY to Y1 receptor by different peptide pools.

Antigen-antibody interaction: The mab Dü 162 directed against HIV 1-Nef peptide PEYFKN [4] was preincubated (30 min) in streptavidin coated ELISA plates with increasing concentrations of acetylated hexapeptide amide or peptide mixtures carrying partial sequences of the epitope in defined positions. Biotinylated antigen was added (50 ng/well) and reduced binding of the antibody to the streptavidin bound antigen was measured by ELISA. A complex mixture (sublibrary Ac-PEX$_1$X$_2$X$_3$X$_4$-NH$_2$, 104,976 peptides) and the nonrelated sublibrary Ac-ITX$_1$X$_2$X$_3$X$_4$-NH$_2$ were not competing for antibody binding at high concentration (0.5 mg/ml). A significant effect in our optimized biotin-streptavidin assay was induced with peptide mixtures carrying three X-positions (Figure 1A).

In summary, hormone receptor binding was unspecifically influenced by hexapeptide amide and octapeptide mixtures. The 36-peptide amide NPY is blocked for competition at the receptor in the presence of 1440 nonrelated octapeptides. For killer cell activity and antigen antibody interaction the use of mixtures up to three X positions is promising. The presence of unrelated peptides affords higher concentrations of active peptide for induction of a positive signal in immunological assays.

References

1. Falk, K., Rötzschke, O., Stevanović, S., Jung, G. and Rammensee, H.G., Nature, 351 (1991) 290.
2. Stevanović, S., Wiesmüller, K.-H., Metzger, J., Beck-Sickinger, A.G. and Jung, G., BioMed. Chem. Lett., 3 (1993) 431.
3. Beck-Sickinger, A.G., Grouzmann, E., Hoffmann, E., Gaida, W., Van Meir, E.G., Waeber, B. and Jung, G., Eur. J. Biochem., 206 (1992) 957.
4. Spohn, R., Ph.D. Thesis, 1993, University of Tübingen.

An orthogonal partial cleavage approach for solution-phase identification of biologically active peptides from larger chemically synthesized peptide libraries

S.E. Salmon[1], K.S. Lam[1], M. Lebl[2], A. Kandola[1], P.S. Khattri[1], S. Healy[2], S. Wade[2], M. Patek[2], P. Kočiš[2], V. Krchňák[2], D. Thorpe[2] and S. Felder[2]

[1]Arizona Cancer Center, and Department of Medicine, University of Arizona, Tucson, AZ 85724, U.S.A. and [2]Selectide Corporation, 1580 E. Hanley Blvd, Tucson, AZ 85737, U.S.A.

Introduction

Peptide library screening represents a new method to rapidly discover peptide ligands with biological activity, thus greatly facilitating the traditional drug discovery process. We have previously described a synthetic peptide library approach to screen for peptide ligands that bind to specific macromolecular targets (Selectide Technology) [1]. This method is based on the "one-bead one-peptide" concept where each individual peptide-bead expressed only one peptide entity. Because the peptides were not in solution, these libraries could not be used in solution-phase bioassays. We have now broadened the applicability of the technology to permit solution phase assay with releasable peptide libraries [2]. In this system, each bead contains only one peptide entity, but the peptides are attached to each bead via three different types of chemical linkers. Two of these linkers can be cleaved orthogonally and sequentially under very mild but differing chemical conditions. The remaining linker cannot be cleaved under these conditions thereby leaving sufficient peptide on the bead for microsequencing. The peptide library was initially distributed into 96-well plates with 500 beads/well. The peptides were then released into solution phase followed by biological assay. Approximately 100 pmole peptide was released from each bead into solution. Based on volume/well the concentration of each individual peptide in solution was approximately 1µM. The beads from the wells with positive response were then redistributed into individual wells with 1 bead/well. Subsequent release of peptides and biological testing allowed us to physically isolate the bead-of-origin for microsequencing. The general scheme of the methodology is shown in Figure 1.

Results and Discussion

This method has been successfully applied to identification of peptide ligands specific for (i) an anti-β-endorphin monoclonal antibody (MoAb) and (ii) the platelet-derived gp IIb/IIIa membrane fibrinogen receptor.

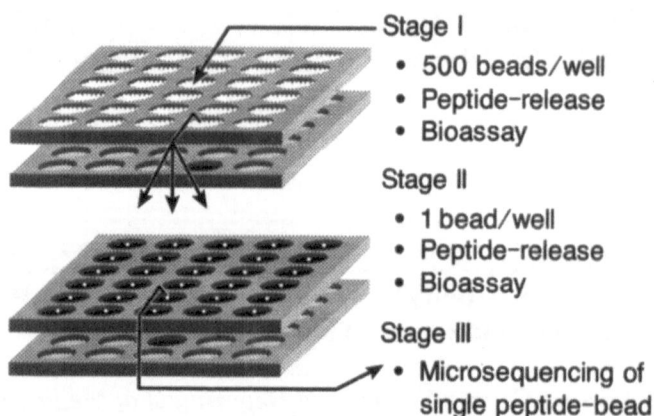

Stage I
- 500 beads/well
- Peptide-release
- Bioassay

Stage II
- 1 bead/well
- Peptide-release
- Bioassay

Stage III
- Microsequencing of single peptide-bead

Fig 1. General scheme of the two-stage release peptide library screening methodology.

A linear tetrapeptide library XXXXG—bead was used in the screening of anti-β-endorphin MoAb (clone 3-E7). This MoAb has been studied extensively [1, 3]. The native epitope is YGGF and the motif identified from library screening was YG_F. The peptides were released into solution and competitive ELISA assay was performed. Four reactive beads were sequenced: YGGF (identified twice), YGVF and YGAF. In another experiment, purified platelet-derived gpIIb/IIIa receptor was used to screen a disulfide cyclic tripeptide library (CXXXC—bead). Again, a competitive ELISA assay was used for screening. Two positive beads were identified both with the following sequence: CRGDC. The RGD sequence is well known to be the native binding motif for gpIIb/IIIa.

These results demonstrate that our two-stage releasable peptide library method can be used to identify binding ligands in solution-phase assays, enabling the easy adaptation of many in vitro bioassays to this methodology. Furthermore, we may now screen for biological endpoint with a functional assay where the target molecule is either not yet purified or even totally unknown. As our libraries are produced synthetically, we have adapted this method to non-peptide chemical libraries which include a coding peptide sequence [4]. The non-peptide chemical structures are first released into solution and screened in a two stage process as described above. The remaining arm on the bead is the peptide coding sequence from which the non-peptide chemical structure can be determined.

References

1. Lam, K.S., Salmon, S.E., Hersh, E.M., Hruby, V.J., Kazmierski, W.M. and Knapp, R.J., Nature (London), 354 (1991) 82.
2. Lebl, M., Patek, M., Kocis, P., Krchnak, V., Hruby, V.J., Salmon, S.E. and Lam, K.S., Int. J. Peptide Protein Res., 41 (1993) 201.
3. Lam, K.S., Hruby, V.J., Lebl, M., Knapp, R.J., Kazmierski, W.M., Hersh, E.M. and Salmon, S.E., Bioorg. &Med. Chem. Letts., 3 (1993) 419.
4. Nikolaiev, V., Stierandova, A., Krchnak, V., Seligmann, B., Lam, K.S., Salmon, S.E. and Lebl, M., Peptide Research, 6 (1993) 161.

Application of Selectide Technology in identifying (i) a mimotope for a discontinuous epitope, and (ii) D-amino acid ligands

K.S. Lam[1], M. Lebl[2], V. Krchňák[2], D.F. Lake[1], J. Smith[1], S. Wade[2], R. Ferguson[2], M. Ackerman-Berrier[2] and K. Wertman[2]

[1]Arizona Cancer Center, and Department of Medicine, University of Arizona, Tucson, AZ 85724, U.S.A. and [2]Selectide Corporation, 1580 E. Hanley Blvd, Tucson, AZ 85737, U.S.A.

Introduction

Selectide Technology is a synthetic peptide library method that is based on the "one-bead one-peptide" concept [1]. The peptides are synthesized on solid-phase beads using a "split synthesis" method [1, 2] resulting in a huge library of peptide beads where each solid-phase bead expresses only one peptide entity [1]. The entire library of peptides (10^6-10^8) is then screened and the reactive bead identified, physically isolated, and microsequenced. Since the Selectide Technology is based entirely on synthetic chemistry, unnatural amino acids and even non-peptide chemical subunits can be used in the construction of the chemical library [3]. In this paper, we report on the application of Selectide Technology in identification of ligands for an anti-insulin monoclonal antibody (MoAb). This murine Ab (Clone AE9D6) recognizes a discontinuous epitope of insulin with a binding constant of 0.01 µM.

Results and Discussion

Table 1 shows the various motifs identified when a linear or cyclic peptide library was screened with the anti-insulin MoAb. Immediately apparent is that several distinct peptide motifs were identified using this parallel approach of peptide library screening. This is in contrast to the convergent approach of screening where an iterative process was used [4], and only one motif was identified. Table 1 also shows that the motif identified is dependent on the length as well as the secondary structure of the library. The binding affinities of most of these ligands identified during the primary screen are relatively low (>10 µM). A sequential screening approach was then applied to one of these motifs, _ W _ _ GF, identified from the primary screen (Table 2). Based on the motif of the primary screen, longer secondary libraries were synthesized and screened under higher stringency. This process was repeated several times until ligands of higher affinity were finally isolated. The binding affinity (IC_{50}) of the best ligand identified (SKQDIWGRGF) thus far was 0.05 µM, 5 fold weaker than that of insulin, the native ligand. The subsequent libraries synthesized, however, need not be linear. We are in fact currently developing libraries where the randomization steps are carried out at a branching arm from the middle of the peptide motif identified from a previous screen. It is hoped that this will generate additional contact points resulting in ligands with affinities higher than that of insulin. In addition, we have also used the anti-insulin MoAb to screen all D-amino acid linear hexa- and octa-libraries. The common motif identified was _ _ _ q _ Gs_G. This motif does not resemble ligands identified from the L-amino acid libraries.

Table 1 *Peptide motifs identified on the primary screen of an anti-insulin MoAb*

Tetra	QNPR	Deca	_ _ _ _W_ _GF
Penta	FNW_ _	Quindeca	_ _ _ _ _ _ _ _ _W_ _GF
	FDW_ _		_FDW_ _ _ _ _ _ _ _ _ _
	_QDPR		_ _ _ _ _ _W_ _GF_ _ _ _
Hexa	_W_ _GF		
	FDW_ _ _	Cyclic 7-mer	CW_ _GF_ _C
	FNW_ _ _		
	_ _QDPR	Cyclic 8-mer	C_F_W_ _GGC
Octa	_ _ _W_ _GF		C_ _ _ _HGVQC
	_ _ _ _ _ _GF		
Nona	_ _ _GF_ _GF	Cyclic 9-mer	CQDI_Y_ _ _ _C

Table 2 *Optimization by sequential screening*

Library		Motif
1°	XXXXXX	_W_ _GF
2°	XXXWXXGF	_ _ _WKYGF, Q_IWG_GF
3°	XXXXWKYGF	NH_(G)WKYGF
	XXQXIWGXGF	S(R/K)Q(D/A)IWG_GF

The V_H and V_L genes of this anti-insulin MoAb were cloned and a single chain Ab has been constructed. This single chain Ab, when compared to the whole native antibody, showed a similar but not identical binding profile to insulin and the various peptide ligands. Work is currently underway to use this single chain Ab and its CDR loop variants to explore antibody-peptide interaction at the molecular level.

References

1. Lam, K.S., Salmon, S.E., Hersh, E.M., Hruby, V.J., Kazmierski, W.M. and Knapp, R.J., Nature (London), 354 (1991) 82.
2. Furka, A., Sebestyen, F., Asgedom, M. and Dib, G., Int. J. Peptide Protein Res., 37 (1991) 487.
3. Nikolaiev, V., Stierandova, A., Krchnak, V., Seligmann, B., Lam, K.S., Salmon, S.E. and Lebl, M., Peptide Research, 6 (1993) 161.
4. Houghten, R.A., Pinella, C., Blondelle, S.E., Appel, J.R., Dooley, C.T. and Cuervo, J.H., Nature (London), 354 (1991) 84.

Streptavidin-peptide interaction as a model system for molecular recognition

K.S. Lam[1], M. Lebl[2], S. Wade[2], A. Stierandová[2], P.S. Khattri[1], N. Collins[3] and V.J. Hruby[3]

[1]Arizona Cancer Center, and Department of Medicine, University of Arizona, Tucson, AZ 85724, U.S.A., [2]Selectide Corporation, 1580 E. Hanley Blvd, Tucson, AZ 85737, U.S.A. and [3]Department of Chemistry, University of Arizona, Tucson, AZ 85724, U.S.A.

Introduction

Peptide library screening has been proven to be a valuable tool in identifying peptide ligands for various macromolecular targets [1]. Our peptide library method (Selectide Technology) is based on the "one-bead one-peptide" concept where each individual bead expresses only one peptide entity [2, 3]. Using this technology, we have identified several peptides with HPQ and HPM motifs that interact specifically with streptavidin [2]. When a histidine depleted library was used, several non-HPQ motifs were identified [3]. When avidin was used as a target macromolecule, four new peptide motifs were identified. One of these motifs, HPYP_, binds to avidin but not to streptavidin. On the other hand HPQ binds to streptavidin but not to avidin [3]. Weber et al. [4] have reported the high resolution X-ray crystallographic structure of streptavidin: FSHPQNT peptide complex. In this paper, we report on our continuing work on using streptavidin as a model system to screen our D-amino acid containing and cyclic peptide libraries.

Results and Discussion

Table 1a shows the structure of the D-amino acid containing peptide ligands isolated when penta-peptide libraries were screened with streptavidin. When an XxXxX penta-peptide library (wherein 'X' represents L amino acids and 'x' represents D amino acids) was screened, three distinct motifs were identified: _ _WpH, _w(F/Y)pH and Y_ _fP. Interestingly, a D-Pro-L-His (_ _ _pH) is present in two of these three motifs. This sequence has some resemblance to that of HPQ or HPM when an all L-library was screened. In contrast, a totally different motif was identified when an all D-amino acid penta-peptide library was screened: wy_ _a. Biotin at 0.1 μM, completely abolishes the binding of streptavidin to these peptide-beads. Furthermore, similar to HPQ and HPM, many of these ligands interact with streptavidin but not with avidin.

Table 1 *Ligands that interact specifically with streptavidin*

a.	XXXXX		XxXxXx		XXXXX	
	HPQ		WkWpH	RwYpH	wyqea (3)	wfrya
	HPM		YgWpH	DwFpH	wyhea	wymel
			QrWpH		wydya	
			LqWpH	YvIfP	wyefa	
			AfWpH	YpFfP	wyfya	
			SyWpH			
b.	CXXXXC		CXXXXXC		CXXXXXXC	
	HHPM		HPQNV	SHPQF	HPQAPK	HPQGPG
	NHPM		HPQNN	LHPQN	HPQAPY	HPQNGG
	QHPM		HPQQV	DHPQN	HPQFAS	HPQNAQ
	LHPM			WHPQN	HPQFAR	HPQVGI
			HPMNA		HPQFPA	HPQSGM
	RHPQ		HPMNP		HPQFPQ	
					HPQGPA	HHPQFP

X = all 18 L amino acids plus glycine, excluding cysteine.
x = all 18 D amino acids plus glycine, excluding cysteine.

We have also screened some cyclic all L-amino acid containing libraries. The results are shown in Table 1b. Disulfide cyclization was accomplished by incubating the library overnight with 20% dimethylsulfoxide. Similar to the linear libraries, HPQ and HPM motifs were both identified. However, HPM was preferred in the cyclic tetrapeptide libraries whereas HPQ was preferred in both cyclic penta- and hexapeptide libraries. Furthermore, the location of this triplet motif with respect to the flanking cysteine residues depends greatly on the ring size.

Work is currently underway to determine the binding affinities of all these ligands. When combining these data with X-ray crystallography and computer modeling, we hope to gain insights into the understanding of peptide-protein and ligand-protein interactions at the molecular level.

References

1. Pavia, M.R., Sawyer, T.K. and Moos, W.H., Bioorg. & Med. Chem. Letts, 3 (1993) 387.
2. Lam, K.S., Salmon, S.E., Hersh, E.M., Hruby, V.J., Kazmierski, W.M. and Knapp, R.J., Nature (London), 354 (1991) 82.
3. Lam, K.S. and Lebl, M., Immunomethods, 1 (1992) 11.
4. Weber, P.C., Pantaliano, M.W. and Thompson, L.D., Biochemistry, 31 (1992) 9350.

Nonsequenceable and/or nonpeptide libraries

M. Lebl[1], V. Krchňák[1], A. Stierandová[1], P. Šafář[1],
P. Kočiš[1], V. Nikolaev[1], N.F. Sepetov[1], R. Ferguson[1],
B. Seligmann[1], K.S. Lam[2] and S.E. Salmon[2]

[1]Selectide Corporation, 1580 E. Hanley Blvd., Tucson, AZ 85737,
U.S.A., [2]Arizona Cancer Center and Department of Medicine,
University of Arizona College of Medicine, Tucson, AZ 85724, U.S.A.

Introduction

The Selectide process consists of three parts: (i) synthesis of a single structure on each bead in a "library"; (ii) screening the library in a solid phase binding assay or in standard solution-phase assays; and (iii) determination of structure on the "active" bead [1]. The use of amino acids for library construction was the first step toward the generation of increasing structural diversity. Libraries based on amino acids incorporate a limited number of building blocks; however, these compounds are easy to synthesize and sequence. Libraries of non-peptidic compounds offer greater diversity of structures, but may require an alternative means of structure identification, such as coding by amino acid sequence for the nonsequenceable building blocks [2].

Results and Discussion

We have combined the simplicity of peptide structures with the diversity available in alternative building blocks besides regular amino acids. The simplest building blocks used to construct the library are given in Figure 1. We have used trifunctional amino acids and modification of a side chain to achieve the structural multiplicity. Amino acids like diaminobutyric acid, aspartic acid, cysteine and/or iminodiacetic acid are the smallest building blocks onto the side chains of which the universe of carboxylic acids, amines or halides (aliphatic, aromatic, heterocyclic) can be attached. To achieve reasonable binding to a neutral acceptor (receptor, antibody, enzyme, nucleic acid...), the appropriate spatial arrangement of the interacting structures must be realized. Linear presentation of amino acid side chains in peptide libraries may not be the optimal format for selection of the best binding structures.

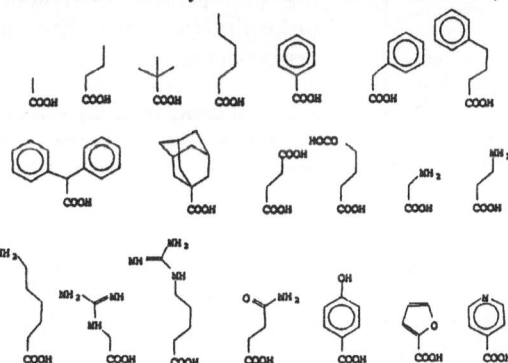

Fig. 1. Blocks used for randomization.

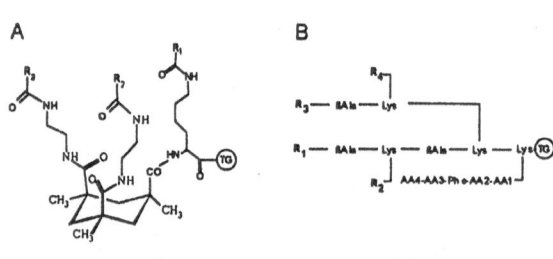

Fig. 2. *Structure of nonpeptide library on nonpeptidic (A) and on peptidic (B) scaffolding.*

The optimal strategy for displaying the interacting structures might be their placement onto a molecular scaffolding which might map the appropriate segments of space. Inter-relationships of the same set of individual building blocks in the scaffolding arrangement can be varied using different scaffolding.

As an example of a small non-peptidic scaffolding, we have built a conformationally constrained scaffolding based on modified Kemp's triacid and constructed a non-peptide library with 20 different carboxylic acids randomized (Figure 2A). Another scaffolding mapping larger conformational space is a simple branched attachment constructed by consecutive coupling of diamino carboxylic acids (Figure 2B). The synthesis of this scaffold required the use of four independent (orthogonal) protecting groups (Tfa, Npys, Fmoc, Ddz). Unnatural building blocks can be combined in the library with standard amino acids. A small library was synthesized with amino acids randomized in position 1, 2 and/or 3 and (i) a set of aromatic amines coupled to the β-carboxyl group of aspartic acid, (ii) a side chain modified iminodicarboxylic acid, (iii) aromatic acids coupled to the side chain of diaminobutyric acid, or (iv) benzylhalides coupled to the side chain of sulfur containing amino acids (cysteine and penicillamine) in position 4. In this case doublets of 6 amino acids were used to code for 36 different building blocks in position 4 which were not directly sequenceable.

A linear library and both scaffold-based libraries were screened in our model systems to determine ligands to anti-β-endorphin monoclonal antibody and streptavidin. Positively reacting beads were subjected to Edman degradation, and the inter-acting structures, deduced from the obtained data are given in Figure 3. These compounds were resynthesized and have shown specific binding (competable by leucine enkephalin or biotin, respectively).

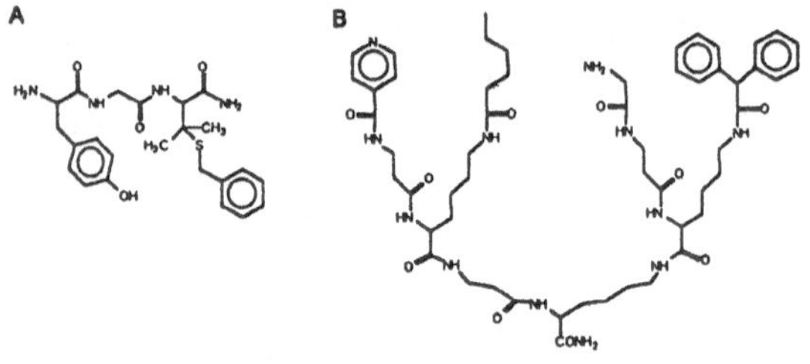

Fig. 3. *Structures interacting with anti-β-endorphin (A) and streptavidin (B) found in the test libraries.*

References

1. Lam, K.S., Salmon, S.E., Hersh, E.M. Hruby, V.J., Kazmierski, W.M. and Knapp, R.J., Nature, 354 (1991) 82.
2. Nikolaiev, V., Stierandová, A., Krchňák, V., Seligmann, B., Lam, K.S., Salmon, S.E. and Lebl, M., Peptide Res., 6 (1993) 161.

Novel concepts in the preparation and use of planar arrays of peptides with unnatural amino acids and cyclic structures

M.H. Lyttle, C.O.A. Berry, H.O. Villar, M.H. Hocker and L.M. Kauvar

Terrapin Technologies, Inc., 750-H Gateway Boulevard, South San Francisco, CA 94080, U.S.A.

Introduction

The simulation of epitope-receptor binding by the use of libraries of immobilized short peptides has generated much interest. The peptide-support linker arm sometimes changes the binding characteristics of short peptides towards their targets, generating misleading results in some systems [1]. Short peptide fragments may also fail to duplicate the desired receptor binding because they lack the conformation imposed on the native fragments by protein secondary and tertiary structure. To address these issues, the affinity of several glutathione analogs [2] immobilized on a cellulose planar array for glutathione-S-transferases (GSTs) was compared with the K_i values obtained with the enzymes and free ligand. Two other arrays containing peptide fragments designed to mimic α-helical portions of protein A, known to be in contact with IgG, were synthesized and tested for their ability to bind IgG. A cyclic loop of 5 residues was moved through the sequence in each array. Allyl side chain based methods [3] were used to generate the library of cyclic structures. A sandwich type assay using alkaline phosphatase suggested differential binding of IgG to one of the arrays. Several of the immobilized cyclic peptides were characterized by AAA and sequence analysis. These results, along with the assay above and computational studies, were used to select a sequence to be made at a larger scale, purified by HPLC and made into an affinity sorbent.

Results and Discussion

Glutathione (GSH) analogs were prepared on amine functionalized cellulose [4] by using Fmoc amino acids and diisopropylcarbodiimide (DIPCDI) mediated coupling. A matrix approach was employed to generate 20 immobilized analogs: five different C-terminal amino acids were coupled to horizontal rows, then four different modified cysteine derivatives were added to the vertical columns, giving 20 different dipeptides. To all of these was added γ-glutamic acid, giving an array of 20 GSH analogs. The immobilized peptides were deprotected with TFA and rinsed. Differential binding affinities of representatives of the major GST isozyme groups A, M, and P to the array were demonstrated by binding these GSTs, followed by enzyme–linked antibody visualization. Table 1 shows the qualitative response of the three GSTs to five of the immobilized analogs compared to the solution phase inhibition of the same GSTs.

Table 1 *Qualitative response of three GSTs to five immobilized GSH analogs, with inhibition values of the free ligands*

Compound	Immobilized			Free inhibition K_is, µM		
	A	M	P	A	M	P
γ-E-C(*S*-4-methylbenzyl)-β-alanine	S	M	M	43	2.1	40
γ-E-C(*S*-benzyl)-β-alanine	S	W	M	360	22	710
γ-E-C(*S*-benzyl)-4-aminobutyric acid	S	W	S	220	67	685
γ-E-C(*S*-benzyl)-phenyl glycine	M	M	M	20	25	0.45
γ-E-C(*S*-benzyl)-G	M	M	S	4.6	2.3	14

S = Strong, M = Medium and W = Weak visualization spot with alkaline phosphatase. Inhibition measured using GSH and chlorodinitro benzene (CDNB) as substrates. Standard amino acid code used. Free analogs > 90 % pure by HPLC. GST isozyme groups A, M and P.

Treatment of immobilized glutathione analogs with the visualization antibody also gave a pattern of variable intensity spots, and this was difficult to subtract from the results with GSTs. Table 1 shows that the inhibition of enzyme by free ligand does not match the pattern observed on the array (lower concentration inhibitory ligand should give stronger binding). Nevertheless, this system could be used to determine the identity of an unknown GST enzyme by pattern recognition applied to the binding profile. When several of these ligands were immobilized at their N-terminus and used as affinity sorbents, the observed binding more closely matched the free inhibition values [5]. In the second planar array system we studied, two peptides were selected from published data [6] showing that these regions of protein A were important for binding to the Fc region of IgG. DEEQRNGFIQSLKD and DQQNAFYEILHLPN were synthesized on cellulose using the same chemistry as above. A loop structure linking carboxy and amino acid side chains five amino acids apart was "walked" through the structure from the second to penultimate residues. Table 2 shows these libraries. **E** and **O** represent side chain linked glutamic acid and ornithine.

Table 2 *Cyclic peptide sequences. Underlined sections denote location of 5 amino acid cyclic structure*

SERIES 1	#	SERIES 2	#
DEEQROGFIQSLKD	1-8	DEQNAOYEILHLPN	2-8
DEEQRNOFIQSLKD	1-7	DQENAFOEILHLPN	2-7
DEEERNGOIQSLKD	1-6	DQQEAFYOILHLPN	2-6
DEEQENGFOQSLKD	1-5	DQQNEFYEOLHLPN	2-5
DEEQREGFIOSLKD	1-4	DQQNAEYEIOHLPN	2-4
DEEQRNEFIQOLKD	1-3	DQQNAFEEILOLPN	2-3
DEEQRNGEIQSOKD	1-2	DQQNAFYEILHOPN	2-2
DEEQRNGFEQSLOD	1-1	DQQNAFYEELHLON	2-1

These arrays were treated with a solution of biotinylated IgG, followed by visualization with avadin alkaline phosphatase and NBT/BCIP [7]. Differential signal intensity was observed with series two, with sequence 2-2 giving the best response, while the peptides in series 1 gave roughly the same signal. Controls of each peptide array without IgG showed high background signal with more nearly uniform intensity. This non-specific binding is a caveat not always recognized with these systems, which influences conclusions that can be drawn with the binding study. Several of these peptides were analyzed directly by AAA and Edman degradation sequencing, without prior cleavage from the cellulose. All of the samples showed similar loading by AAA, however the sequencing data of some of the samples suggested that crosslinking, and not the desired internal cyclization was occurring. This suggests different strain in forming these small rings, depending on the sequence. Sequence 2-2 above gave the best sequencing results, with low PTH signals at the E and O residues involved in the cyclic structures, indicative of a cyclic structure. This sequence was synthesized using a Milligen/Biosearch 9500 synthesizer on 5 g of PEG polystyrene resin, yielding 43 mg of 93% pure cyclic peptide, by HPLC. Synthesis of 2-4 and 2-8 with identical methods gave mixtures of products from which no single peptide could be isolated. Evaluation of purified 2-2 as an affinity ligand for IgG is in progress.

The discrepancy in the binding of GSTs to immobilized GSH analogs versus the free inhibition values questions the use of immobilized peptide libraries in the search for lead inhibitory compounds of GSTs. In the second system, libraries of immobilized cyclic peptides were shown to be empirically useful in the selection of a compound for synthesis at a larger scale. This methodology may therefore be useful, either by planar array or automated multiple peptide synthesis at a small scale, in choosing other cyclic peptide candidates for larger scale synthesis and purification. Background signal with antibody based visualization was a problem in both systems.

Acknowledgements

We thank Jim Kinney at Stanford University Medical Centre for carrying out the sequencing and amino acid analysis.

References

1   Wang, Z. and Laursen, R.A., Pept. Res., 5 (1992) 275.
2.   Lyttle, M.H., Aaron, D.T., Hocker, M.D. and Hughes, B.R., Pept. Res., 5 (1992) 336.
3.   Lyttle, M.H. and Hudson, D., In Smith, J.A. and Rivier, J.E. (Eds.), Peptides, Chem. and Biol. (Proc. 12th APS), Escom, Leiden, The Netherlands, 1992, p.583.
4.   Frank, R., Guler, S., Krause, S. and Lindenmaier, W., In Giralt, E. and Andreu, D. (Eds.), Peptides 1990 (Proc. 21st EPS), Escom, The Netherlands, 1991, p.151.
5.   Kauvar, L.M., In Tew, K., Pickett, C., Mantle, T. and Hayes, J. (Eds.), Structure and Function of Glutathione Transferases, CRC Press, Boca Raton, FL, U.S.A., 1993, p.251.
6   Diesenhofer, J., Biochemistry, 20 (1981) 2361.
7.   5-bromo-4-chloro-3-indolyl-phosphate/nitrobluetetrazolium substrate, Kirkegaard and Perry, Gaithersburg, Maryland.

Immobilized peptide arrays: A new technology for the characterization of protease function

M.A. Ator, S. Beigel, T.C. Dankanich, M. Echols,
J.A. Gainor, C.L. Gilliam, T.D. Gordon*, D. Koch,
J.F. Koch, L.I. Kruse, B.A. Morgan*,
R. Krupinski-Olsen, T.J. Siahaan, J. Singh
and D.A. Whipple

*Sterling Winthrop Pharmaceuticals Research Division,
Collegeville, PA 19426, U.S.A.*

Introduction

Substrate selectivity is a fundamental property of protease action and is of primary importance in the characterization of the protease, in the mechanism-based design of inhibitors, and in the allotment of physiological function. We have sought to design and develop a new technology for the rapid and comprehensive characterization of substrate selectivity by the chemical synthesis and enzymologic evaluation of large numbers of peptides which are attached to a solid-phase matrix: an "immobilized peptide array" (IPA). We decided that this technology should neither require nor presume any preliminary knowledge of the substrate specificity of the protease under study, and that unlike other methods utilizing synthesis of peptide "libraries" [1], our approach should allow timely access to a complete resolution of every component of the array under study.

Results and Discussion

A cornerstone of our strategy was that the action of the protease on the immobilized substrate would release an easily detected species into the solution phase, which could then be readily separated from the solid-phase and quantified. Thus, protease incubation of an N-terminal fluorescence-tagged peptide substrate covalently linked to a solid matrix would release a fluorescence-tagged fragment into the solution phase. This fragment can then be recovered simply by removal of an aliquot of supernatant from the incubation medium. Also, the C-terminal fragment of the cleaved substrate can be characterized by Edman sequencing methods. This strategy thus focuses analysis towards active substrate sequences to a complete resolution of every component of the array under study in two steps: fluorogenic detection followed by sequence analysis. Our first goal was to select an appropriate solid phase for both the chemistry of solid-phase peptide synthesis and enzymology

---

*Current Address: Biomeasure Inc., 27 Maple Street, Milford, MA 01757-3650, U.S.A.

in aqueous media: controlled-pore glass (CPG) functionalized with aminopropyl (Amp) groups seemed to possess the required properties. We decided to include a "spacer" of six aminocaproic acid (Acp) moieties to distance the potential substrate sequence away from the glass allowing access to the protease. After evaluating a number of fluorescent species, we found that the 7-hydroxycoumarin-4-propionoyl group (Cop) was an excellent "tag" for our purpose [2]. At this time, we also systematically investigated the influence of $(Acp)_n$-based linkers on immobilized substrates for thermolysin, and found that optimal linker length was at 7-8 Acps, and appeared to be independent of sequence. Having established the feasibility of IPAs as protease substrates for thermolysin, our next objective was to use the technology to investigate the substrate selectivity of more physiologically-relevant enzymes. The first "application" goal chosen for our new technology was to identify selective substrates for the matrix-metalloproteases collagenase and stromelysin. We wished to achieve a selectivity ratio of 100, that is, we would seek to discover a substrate for collagenase (col) with a $kcat/K_m$ >100 times its $kcat/K_m$ for stromelysin (str), and vice versa.

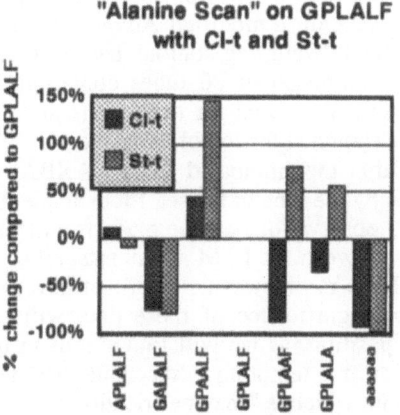

We planned to synthesize the initial peptide probes in a "window" within an alanine matrix "frame". Thus, in the construct Cop-A-A-X-X-X-A-A-$(Acp)_6$-Amp-Cpg, the sequence A-A-X-X-X-A-A is a heptapeptide alanine frame encompassing an X-X-X tripeptide window. In order to determine the size of the frame, and the size and design of the window, we carried out "alanine scans" on the known collagenase substrate GPLALF. The enzyme data for these peptides is shown in Figure 1. The P_3 proline residue was crucial for substrate activity with both col and str. The P_1' leucine was necessary only for mCl-t [3]. The P_2 A for L substitution resulted in a "better" substrate with both enzymes, while alanine substitution at P_4 glycine or P_2' phenylalanine was at least tolerated by each enzyme. With this information in hand, it was clear that a P-X-X-L sequence was necessary for col to recognize a substrate. These data supported our plan to construct a three-volume primary library (Cop-A-A-X-A-X-B-A-$(Acp)_6$-Amp-CPG, Cop-A-A-X-X-A-B-A-$(Acp)_6$-Amp-CPG and Cop-A-A-A-X-X-B-A-$(Acp)_6$-Amp-CPG) in which X is a position utilizing a serial cassette

rack 0003 CopAA(L,M,N,P)AXB A

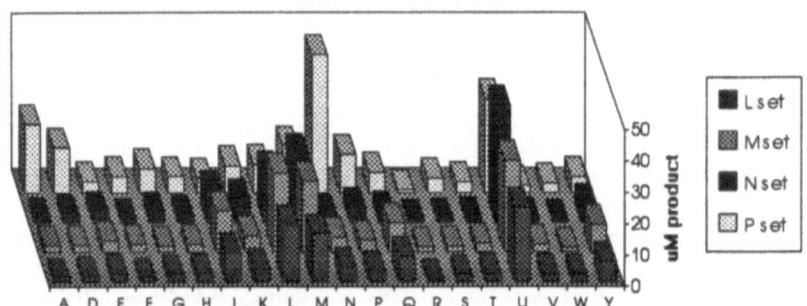

of 20 aminoacids arranged in an array of 20 tubes. The cassette included the 20 coded aminoacids with S-methylcysteine (U) substituted for cysteine. B is a position where the 20 aminoacid cassette is incorporated as an equimolar mixture in a single tube assembled via a protocol similar to that outlined by Furka [4]. Thus each volume consists of 400 tubes (a 20x20 matrix), each tube containing 20 sequences degenerate at B. The A-A-X-A-X-B-A volume will be used to exemplify the technology: this was assembled in five racks on a modified Advanced Chemtech 350 synthesizer utilizing a custom-designed "weighing station" based on a Hewlett Packard Orca robot. Each rack was composed of 80 tubes containing 1600 sequences, plus positive and negative controls. Data for the L,M,N and P sets for col with this volume (AAXAXBA) is shown at the top of this page.

The set is named after the aminoacid $\underline{X}$ AA$\underline{X}$AXBA. The second X position (AAXA$\underline{X}$BA) is defined by the tube in the set, for example the first tube in the L set is AALA$\underline{A}$BA. Several observations can be made from these data: firstly, there is remarkable consistency between the L, M and N sets, all of these have an identical profile with U>L>M>I>Q>Y. However, the P set is different with a profile of N>U>A>D=M>P. The significance of these data will become clearer with a breakout of the P_1'(B) position, as we will find a shift in the site of cleavage for some sequences; however, it is tempting to speculate that the P_3(P) residue imposes a conformational constraint which is "enzyme-friendly" in the case of P_1(N).

The ability to establish B identity by sequence analysis is a cornerstone of IST strategy. B breakout of the AAPANBA tube by one cycle of sequencing shows unambiguously that the "activity" of this tube is due essentially to four residues U>L>I>M. Breakout of this tube by synthesis of the individual components, and enzymologic evaluation confirmed the same four active sequences with essentially the same relative activity. B breakout of AAPA(M,U)BA tubes differs from the AAPANBA in that there is essentially a single residue (M and U respectively) in the first cycle of sequencing. This ambiguity is resolved by a second sequencing cycle. These data suggest that for the AAPA(M,U)BA tubes cleavage occurs predominantly at the A-(M,U) position. The identification of this "frame shift", in which the potential for an M or U residue to occupy a P_1' position can override the dominance of a P_3 directing proline residue, relegating it to the P_2 position, is an example of the power of the IPA methodology in understanding the complexities of protease substrate selectivity.

We are currently continuing the development of IPA technology and extending IPA investigations to other enzymes.

References

1. Meldal, M., In Schneider, C.H. and Eberle, A.N. (Eds.), Peptides 1992, Escom, Leiden, The Netherlands, 1993, p.61; Matthews, D.J. and Wells, J.A., Science, 260 (1993) 1113; Birkett, A.J., Soler, D.F., Wolz, R.L., Bond, J.S., Wiseman, J., Berman, J., and Harris, R.B., Anal. Biochem., 196 (1991) 137.
2. Gainor, J.A., Gordon, T.D. and Morgan, B.A., In Hodges, R.S. and Smith, J.A. (Eds.), Peptides: Chemistry, Structure and Biology (Proceedings of the 13th American Peptide Symposium), Escom, Leiden, The Netherlands, 1994, this volume.
3. mCl-t = mature Collagenase C-terminally truncated.
4. Furka, Á., Sebestyén, F., Asgedom, M. and Dibó, G., Int. J. Pept. Protein Res., 37 (1991) 487.

Identification of an antigenic determinant for the surface antigen of hepatitis B virus using peptide libraries

C. Pinilla[1], J.R. Appel[1], D. Milich[2] and R.A. Houghten[1]

[1]*Torrey Pines Institute for Molecular Studies,*
San Diego, CA 92121, U.S.A.
[2]*The Scripps Research Institute, La Jolla, CA 92037, U.S.A.*

Introduction

Peptide libraries have been shown to identify peptide antigenic determinants accurately and efficiently [1-5]. A more difficult challenge is the identification of antigenic determinants recognized by monoclonal antibodies raised against protein antigens. Using two different peptide libraries and competitive ELISA, we have identified several hexapeptides that inhibit the binding of a monoclonal antibody to the surface antigen of hepatitis B virus (HBsAg). Four of the six residues of the most effective peptides identified have sequence homology with residues 114-117 of HBsAg.

Results and Discussion

The first synthetic peptide combinatorial library (SPCL) consists of two positions defined with each of the 20 L-amino acids (O) and four positions as mixtures (X), and is represented as $Ac-O_1O_2XXXX-NH_2$ [1, 2]. The other peptide library is a positional scanning SPCL (PS-SPCL), which is made up of six individual positional peptide libraries, each one consisting of hexamers with a single position defined and the other positions as mixtures [3]. Both peptide libraries were screened for inhibition of the binding of a monoclonal antibody (mAb) to HBsAg using competitive ELISA.

Upon screening the first SPCL ($Ac-O_1O_2XXXX-NH_2$), the most effective peptide mixtures found to inhibit mAb binding to HBsAg contained either serine or threonine in the first two positions. An iterative screening and synthesis process for the SPCL was carried out to define the remaining mixture positions of $Ac-STXXXX-NH_2$. The amino acids found to be the most effective at positions three and four were threonine and serine, respectively. Leucine and methionine were found to be equally effective at the fifth position, but only leucine was followed. Upon defining the sixth position, the most effective peptides were $Ac-STTSLI-NH_2$ ($IC_{50} = 1.8\mu M$) and $Ac-STTSLM-NH_2$ ($IC_{50} = 2.6\mu M$).

Screening of the PS-SPCL yielded several peptides that effectively inhibited mAb binding to HBsAg. The amino acid residues of the most effective inhibiting peptide mixtures at each position are shown in Table 1A. The 16 individual peptides, representing the combinations of these amino acids, were synthesized and

assayed by competitive ELISA to confirm the results of the PS-SPCL screening
(Table 1B). By comparing the peptide sequences derived from both peptide libraries,
we found that at least four of the six residues exactly matched those found in the
known HBsAg sequence, namely residues 114-117 (-STTS-). Since the relative
affinity of the mAb for Ac-STTSMM-NH$_2$ is nearly 30-fold lower than for HBsAg,
the -STTS- motif probably represents the linear portion of a larger, possibly
discontinuous, antigenic determinant of HBsAg recognized by this mAb.

Table 1 *Peptide combinations derived from the screening of the PS-SPCL*

A

Position:	1	2	3	4	5	6
Amino acid:	S	S, T	T	P, S	A, M	H, M

B

Peptide	IC$_{50}$ (μM)
Ac-STTSMM-NH$_2$	0.390
Ac-SSTSMM-NH$_2$	0.873
Ac-STTSAM-NH$_2$	2.8
Ac-SSTSAM-NH$_2$	17
Ac-STTSMH-NH$_2$	19
Ac-SSTSMH-NH$_2$	81
all others	>100

References

1. Houghten, R.A., Pinilla, C., Blondelle, S.E., Appel, J.R., Dooley, C.T., and Cuervo, J.H., Nature, 354 (1991) 84.
2. Houghten, R.A., Appel, J.R., Blondelle, S.E., Cuervo, J.H., Dooley, C.T., and Pinilla, C., BioTechniques, 13 (1992) 412.
3. Pinilla, C., Appel, J.R., Blanc, P. and Houghten, R.A., BioTechniques, 13 (1992) 901.
4. Appel, J.R., Pinilla, C. and Houghten, R.A., Immunomethods, 1 (1992) 17.
5. Pinilla, C., Appel, J.R. and Houghten, R.A., Gene, 128 (1993) 71.

The *peptide librarian*™:
Fully automated selection and synthesis of peptide libraries

H.H. Saneii[1], J.D. Shannon[1], R.M. Miceli[2], H.D. Fischer[2] and C.W. Smith[2]

[1]Advanced ChemTech, Louisville, KY 40228, U.S.A.
[2]Upjohn Laboratories, Kalamazoo, MI 49001, U.S.A.

Introduction

The aptly named peptide library [1, 2], combinatorial peptide library [3], or portioning mixing method [4], are all manifestations of a strategy to produce mixtures of peptides for rapid screening. This methodology greatly advances peptide-based drug discovery through the synthesis and assay of mixtures of thousands or even millions of peptides at once, rather than one peptide at a time.

The fastest method of generating a large number of peptides by chemical synthesis is by synthesis carried out on solid phase support [5, 6]. A methodology has been reported by Hruby and Furka for the preparation of various peptide libraries and sub-libraries by the solid phase process [7, 8]. Modern automated peptide synthesizers with precise liquid delivery when combined with high quality protected amino acids and mild deprotection methods are quite capable of producing peptides of purity more than sufficient for effective peptide libraries. The libraries can be utilized in the resin-bound form or cleaved to produce a lyophilized peptide library, depending on the requirements of the assay.

The synthesis of a peptide library and of 18 sub-libraries based on a heptapeptide root sequence using a new multiple peptide synthesizer specifically adapted to the fully automated synthesis of combinatorial peptide libraries is described. The effort was undertaken to ascertain how well chemically produced peptide libraries complement phage generated libraries.

Results and Discussion

The synthesis was carried out on an ACT Model 357 *Librarian*™ multiple peptide synthesizer specialized for the fully automated synthesis of combinatorial peptide libraries. The instrument consists of a robotic bench-top chemistry unit controlled by an IBM microcomputer and equipped with two independent robotic arms. Considerable complexity in synthesis parameters can be accommodated, enabling the scientist to program for variables such as sequence length, number of beads per peptide, number of peptides per library and sub-library, and chemical modifications like acetylation, biotinylation, or use of unusual amino acids.

The approach to peptide library synthesis that was employed is one in which a "root peptide" was modified through a specified protocol to produce main libraries and sub-sets ("sub-libraries") of from hundreds to millions of different peptides. The "root peptide" was the basic framework upon which modifications were conducted to produce the desired library and sub-libraries. It consisted of the starting resin or AA-resin plus all the sequence variables including sequence length. Fixed and variable amino acids were defined for each position in the root sequence.

The "root" heptapeptide, $Biotin-Asp-X_2-X_3-Asp-Asp-X_6-Asp-Resin$, thus contains four "constant" amino acids and three variable positions which were further specified as follows:

X_6 is every common L-amino acid except Asp and Cys
X_3 is every common L-amino acid except Lys and Cys
X_2 is every common L-amino acid except Tyr and Cys

Thus, each variable position calls for 18 amino acids. For the cycle specifying X_6, the growing resin was automatically and precisely distributed to 18 individual reactors for coupling steps, then recombined in a larger resin mixing vessel. A sensitive detector contained in one of the robotic arms of the *Librarian*™ enabled the precise distribution of the peptide resin. For cycle X_3 the resin mixture was subsequently redivided into 18 equal amounts back into the smaller reactors, coupled to the next amino acid and recombined into the mixing vessel. When it was desirable to keep a position constant while varying the others [9], (*i.e.*, Fmoc-Asp(OBut) in the experiment described), the coupling and deprotection took place in the mixing vessel itself rather than in the smaller reactors. To maximize coupling efficiency forcing conditions were used, typically DIC/HOBt in six fold excess. After coupling cycle X_2, the peptide resins were not recombined and the final two cycles were completed in the 18 reactors. In this manner 18 sub-libraries containing 324 peptides were completed. After 25 mg of each sub-library was removed from each reactor, the remainder in each reactor was recombined in the mixing vessel to create the main library containing 5,832 peptides. The cycle time was 2-3 hours enabling these hexapeptide libraries to be synthesized in less than a day.

The "root peptide" utilized in this experiment is a derivative of the FLAG peptide/M2 monoclonal antibody epitope system [10]. The FLAG peptide (Asp-Tyr-Lys-Asp-Asp-Asp-Asp-Lys) was originally engineered as a "tag" for immunoaffinity purification of genetically engineered proteins. Using a phage library, a core peptide for recognition by the M2 antibody was determined to be $X_1-Tyr-Lys-X_4-X_5-Asp-X_7-X_8$ [11]. The root peptide was purposely chosen to exclude the core peptide determined by the phage library in order to determine if additional residues could be identified as core residues using a free peptide library. Thus positions 2, 3 and 6 contained specified variables as described above. The result is a library containing 5,832 peptides. At the same time, simply by *not* combining the entire resin back into the main mixing chamber, 18 sub-libraries are already created in which X_2 is individually fixed as a specific residue. The main library can be screened and if it is positive, one is already positioned to determine which X_2 residue is involved.

The resin splitting-recombining approach to the solid phase synthesis of peptide libraries was shown to be capable of efficiently producing libraries and sub-libraries required by scientists engaged in antibody research, drug discovery and other screening activities. This tactic produced evenly distributed mixtures and did not rely on coupling a mixture of amino acids. The procedure utilized a root peptide with constant and variable amino acid positions specified. The *Librarian*™ was sufficiently flexible to permit the use of a variety of chemistries and starting resins and to specify the library or sub-library size, length of peptides, number of resin bead per peptide and other parameters. The efficiency of the system enables the typical library to be synthesized in under a day. With certain positions fixed, practical libraries of peptides greater than heptapeptide size can be synthesized within reasonable time frames and will contain useful concentrations of each peptide.

References

1. Lam, K., Salmon, S., Hersh, E., Hruby, V., Kazmierski, W. and Knapp, R., Nature, 354 (1991) 82.
2. Geysen, H.M., Meloen, R. and Barteling, S., Proc. Natl. Acad. Sci., U.S.A., 81 (1984) 3998.
3. Houghten, R.A., Pinilla, C., Blondelle, S.E., Appel, J.R., Dooley, C.T. and Cuervo, J.H., Nature, 354 (1991) 64.
4. Furka, Á., Sebestyén, F., Asgedom, M. and Dibó, G., Abstr. 14th Int. Cong. Biochem. Prague, Czech Republic, 5 (1988) 47.
5. Schnorrenberg, G. and Gerhardt, H., Tetrahedron, 45 (1989) 7759.
6. Groginsky, C.M. and Saneii, H., Poster #336 at 12th American Peptide Symposium, 1991.
7. Pavia, M.R., Sawyer, T.K. and Moos, W.H., Bioorg. Med. Chem. Lett., 3 (1993) 387.
8. Houghten, R.A. and Dooley, C.T., Bioorg. Med. Chem. Lett., 3 (1993) 405.
9. Lam., K.S., Hruby, V.J., Lebl, M., Knapp, R.J., Dazmierski, W.M., Hersh, E.M. and Salmon, S.E., Bioorg. Med. Chem. Lett., 3 (1993) 419.
10. Hopp, T.P., Prichett, K.S., Price, V.L., Libby, R.T., March, C.J., Cerretti, D.P., Urdal, D.L. and Conlon, P.J., Bio/Technology, 6 (1988) 1204.
11. DeGraaf, M.E., Miceli, R.M., Mott, J.E. and Fischer, H.D., Gene, 128 (1993) 13.

"Difficult" sequence prediction: A requirement for multiple peptide synthesis?

R.M. Valerio, A.M. Bray, H.M. Geysen and N.J. Maeji

Chiron Mimotopes Pty Ltd., 11 Duerdin Street,
Clayton, Victoria, 3168, Australia

Introduction

In recent years, a number of different methods of multiple peptide synthesis (MPS) have been devised for a variety of applications including epitope mapping, systematic screening and generation of so-called peptide libraries comprising thousands of peptides [1]. A forerunner of these MPS approaches, the multipin peptide synthesis procedure of Geysen and coworkers [2], was originally designed for solid phase screening of nmol quantities of pin-bound peptides but has evolved to allow preparation of free peptides with native and other endings at μmol quantities. We have used this new generation multipin approach to prepare thousands of peptides and have recorded information on sequence solubility, purity and difficulty in a database format. An analysis of this data including a retrospective analysis of peptide "difficulty" using two published predictive methods is discussed.

Results and Discussion

Previously, pins were prepared as a single piece rod with a spherical pinhead and radiation grafted with acrylic acid to give a support for solid phase peptide synthesis at the 50-100 nmol scale. To expand the scope of application of the multipin method, we prepared a number of new pins in a detachable crown format with increased surface area for increased peptide loading [3]. In addition, after testing a range of new monomers, we replaced acrylic acid with 2-hydroxyethylmethacrylate ($H_2C=C(CH_3)CO_2CH_2CH_2OH$, HEMA) as the grafted polymer. Radiation grafting of HEMA to polyethylene pins results in higher % polymer grafts and therefore a higher loading capacity. Fmoc-β-alanine was attached to the various pin types by esterification to the HEMA hydroxy groups in order to determine useable loadings and to determine the efficiency of amino acid couplings on the new polymer surface. Loadings of 1-10 μmol of β-alanine were achieved on various size crown-shaped pinheads. Coupling of Fmoc-Arg(Pmc)-OH by PyBOP activation at 0.06M to β-Ala-crowns at loadings from 1-10 μmol was complete within 30 min indicating that the efficiency was comparable to conventional solid phase resins. To assess peptide quality on the new pins, a number of different peptides were assembled on pins with acid labile handles at 7μmol loading using Fmoc-amino acids with BOP couplings. Assembled peptides were deprotected and cleaved from the pins using TFA/scavenger mix within centrifuge tubes. This significantly streamlines this process as the pins are removed after cleavage and the peptides isolated by a series of ether/petrol

˙precipitation steps. We have simultaneously cleaved up to 200 peptides at the 10 µmol scale using this approach. The quality of peptides produced on the HEMA-pins was excellent and comparable to that achieved on the highly efficient Tentagel resins as demonstrated by parallel synthesis of various sequences on both supports. Figure 1 shows the purity distribution obtained from the synthesis of 1545 peptides.

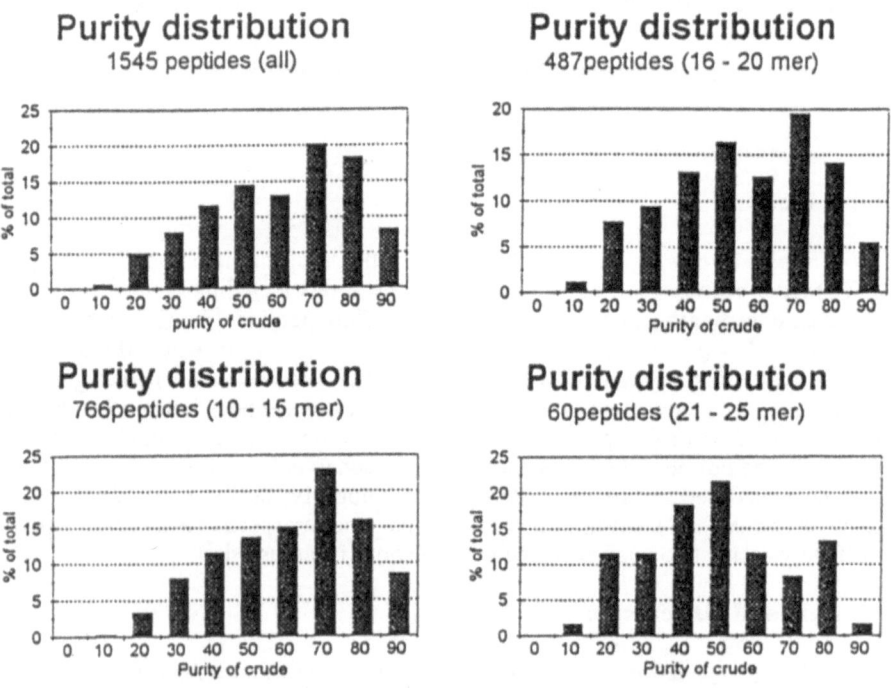

Fig. 1. Purity distribution from the synthesis of 1545 peptides on HEMA-pins.

Analysis of the histogram for all 1545 peptides shows that 60% of the crude peptides prepared were >60% pure which was an excellent result given that peptides up to 35 mers were included. Analysis of purity distribution at specific lengths shows as expected that as peptide length increases, average crude purity drops.

We used two published methods of peptide difficulty prediction [4, 5] to analyze the data we generated from 1545 syntheses retrospectively. Both methods point to intramolecular aggregation caused by ß-sheet hydrogen bonding as a major source of difficult peptide couplings. The first method [4] calculates an average random coil propensity value Pc^* for each peptide which is a measure of the propensity for that peptide to form a random coil structure which is thought to be the optimum conformation for efficient aminoacylation during assembly. The authors found good correlation between the prediction and the actual outcome of a number of syntheses using Boc chemistry on polystyrene resin. We applied the Pc^* parameter to our system which uses Fmoc chemistry on the more polar HEMA polymer and no correlation between the prediction and outcome based on crude purity alone was obtained. In many cases, peptides that were predicted to be very difficult were obtained in high purity. The poor correlation is possibly caused by the different sidechain protected amino acids used in Fmoc chemistry which may have different random coil propensities along with the more polar HEMA support which may help disrupt β-sheet H-bonding and hence aggregation. The second method of difficult peptide prediction [5] uses an aggregation potential parameter Pa to predict peptide difficulty. The authors formulated the values based on results from Fmoc synthesis on polystyrene resin and we were hopeful that a correlation with our results would be obtained. Again, there was no correlation between Pa and our results. Many peptides with a Pa value >1.1 (the value at which a sequence is considered difficult) were obtained in crude purity >80%.

In summary, we have described modifications to the multipin method that allow peptide synthesis at the μmol level. The most important modification appears to be the use of poly-HEMA for synthesis as the quality of peptides produced is equal to that generated on Tentagel resins. We were unable to find a correlation between purity data from 1545 peptide syntheses and two published methods for predicting difficult peptides. This we believe was due in part to the more optimum procedures and the HEMA support we used which produced a better quality peptide than would have otherwise been predicted, which raises the question: Is difficult peptide prediction a requirement for MPS? Based on our data, 63% of all peptides <15 mer were >60% pure. For most MPS applications, peptides <15 mer are used and in order to obtain a binding peptide in an initial screen, >60% purity is sufficient. As a result, for peptides <15 mer, we believe that with the multipin approach, prediction is unnecessary.

References

1.   Jung, G. and Beck-Sickinger, A.G., Angew. Chem. Int. Ed. Engl., 31 (1992) 367.
2.   Geysen, H.M., Rodda, S.J., Mason, T.J., Tribbick, G. and Schoofs, P.G., J. Immunol. Methods, 102 (1987) 259.
3.   Valerio, R.M., Bray, A.M., Campbell, R.A., Maeji, N.J., Margellis, C., Rodda, S.J. and Geysen, H.M., Int. J. Pept. Protein. Res., 42 (1993) 1.
4.   de L. Milton, R.C., Milton, S.C.F. and Adams, P., J. Am. Chem. Soc., 112 (1990) 6039.
5.   Epton, R. (Ed.), Innovation and Perspectives in Solid Phase Synthesis, Intercept Ltd., Andover, U.K., 1992, p.419.

Session XIV
De Novo Design of Peptides and Proteins

Chairs: Cyril M. Kay
University of Alberta
Edmonton, Alberta, Canada

and

Judith C. Hempel
Biosym Technologies Inc.
San Diego, California, U.S.A.

Synthesis of betabellin 14, a 64-residue nongenetic beta protein

Y. Yan and B.W. Erickson

*Department of Chemistry, University of North Carolina,
Chapel Hill, NC 27599-3290, U.S.A.*

Introduction

A betabellin consists of two β sheets packed against each other by the interaction of the hydrophobic face on one side of each sheet. Its amino acid sequence is characterized by the repetition of a 16-residue palindromic pattern (a b c d e f g h h g f e d c b a) in which the exterior residues b, d and f are generally hydrophilic, the interior residues c, e and g are generally hydrophobic, and the adjacent pairs hh and aa are residues L1 and L2 of a tight β turn. Two such patterns per peptide chain could form a 32-residue four-strand antiparallel β sheet. A single disulfide bridge between two such β sheets might stabilize the formation of an eight-strand β barrel in aqueous solution [1-3]. This geometry is most compatible with the presence of type I' β turns. Betabellin 12 and earlier versions had a proline residue at L1 of each β turn. Through molecular dynamics simulation of blocked dipeptides by the slow-growth method, we have shown that LAla-LAla prefers to fold as a type I turn, LAla-DAla as a type II turn, and DAla-LAla as a type II' turn, but that DAla-DAla prefers to fold as a type I' turn [4, 5]. We designed betabellin 14 to test if a 64-residue disulfide-bridged polypeptide having DLys-DAla at the hh positions and DAla-DLys at the aa positions, which should energetically favor the formation of type I' turns, would fold into a stable β protein.

Results and Discussion

In an effort to achieve higher water solubility and folding stability, the design of betabellin 14 involved replacing more than half of the residues of betabellin 12. For example, the net charge of the protein at pH 5 was changed from -2 to +10 by removing most of the Asp and Glu residues and by adding several Lys and His residues. Figure 1 shows the amino acid sequence of betabellin 14.

abcdefghhgfedcbaabcdefghhgfedcba
   * *   * *   * *
HSLTASI**ka**LTIHVQ**ak**TATCQV**ka**YTVHISE

*Fig. 1. Amino acid sequence and the repeated 16-residue palindromic pattern of each of the two 32-residue peptide chains of betabellin 14. The star (*) denotes a D-amino acid.*

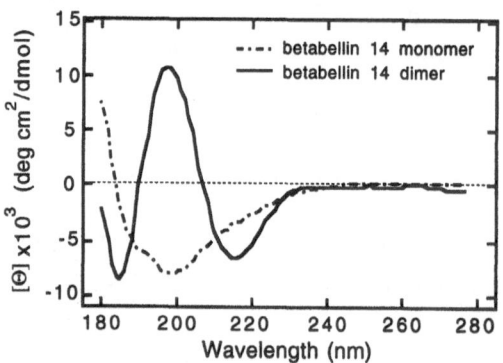

Fig. 2. CD spectra of betabellin 14 and its reduced monomeric form.

This 32-residue peptide chain was assembled by the solid-phase method using Fmoc chemistry and was purified by liquid chromatography (LC) on a column of octyl-silica. The reduced monomer showed the expected molecular mass by electrospray mass spectrometry and an unfolded, "random-coil" conformation by circular dichroic (CD) spectroscopy (Figure 2). Air oxidation of this monomer provided betabellin 14, the 64-residue covalent dimer formed by joining two chains through a single disulfide bond at Cys21. This nongenetic protein, which contained 12 D-amino acid residues, was much more soluble in water than betabellin 12. By reversed-phase LC, betabellin 14 was more hydrophilic than its monomer and much more hydrophilic than betabellin 12. CD spectrophotometry over the range of 180-280 nm showed that betabellin 14 was folded in water at 25°C as a protein having significant β-sheet structure but no detectable α-helical structure (Figure 2). This β structure was maintained at 70°C but was partly unfolded at 85°C (Figure 3).

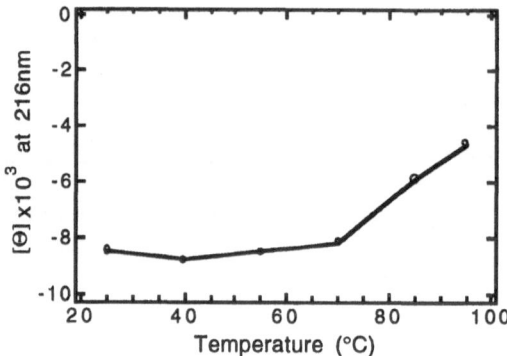

Fig. 3. Thermal denaturation of betabellin 14: Molar ellipticity at 216 nm versus temperature.

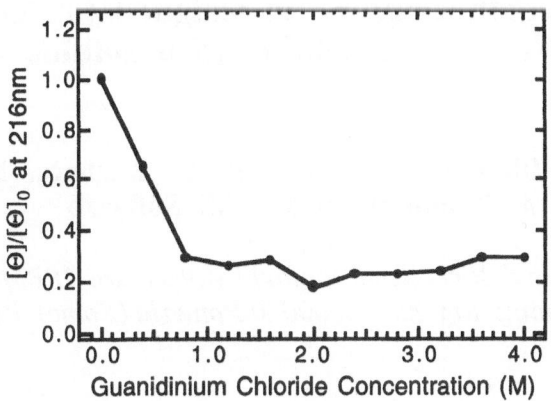

Fig. 4. *Chemical denaturation of betabellin 14: Relative molar ellipticity at 216 nm versus concentration of guanidinium chloride.*

Betabellin 14 was half unfolding by about 0.5 M guanidinium chloride (Figure 4). The thermal and chemical denaturation of this nongenetic protein resembled that of native genetic proteins.

Strikingly, the formation of a single interchain disulfide bond between two unfolded peptide chains induced the resulting covalently bridged polypeptide to fold into a stable β protein. The presence of DLys-DAla and DAla-DLys at the hh and aa positions, respectively, was compatible with betabellin 14 folding as a β protein. The engineering of betabellin 14 represents significant progress in our continuing efforts to design, build, and characterize nongenetic proteins with defined folded structures for use as macromolecular frameworks.

Acknowledgements

This work was supported by NIH research grant GM 42031.

References

1. McClain, R.D., Daniels, S.B., Williams, R.W., Pardi, A., Hecht, M., Richardson, J.S., Richardson, D.C. and Erickson, B.W., In Rivier, J.E. and Marshall, G.R. (Eds.), Peptides: Chemistry, Structure and Biology, Escom, Leiden, The Netherlands, 1990, p.682.
2. McClain, R.D., Yan, Y., Williams, R.W., Donlan, M.E. and Erickson, B.W., In Smith, J.A. and Rivier, J.E. (Eds.), Peptides: Chemistry and Biology, Escom, Leiden, The Netherlands, 1992, p.364.
3. Richardson, J.S., Richardson, D.C., Tweedy, N.B., Gernert, K.M., Quinn, T.P., Hecht, M.H., Erickson, B.W., Yan, Y., McClain, R.D., Donlan, M.E. and Surles, M.C., Biophys. J., 63 (1992) 1186.
4. Yan, Y., Tropsha, A., Hermans, J. and Erickson, B.W., Proc. Natl. Acad. Sci. U.S.A., 90 (1993) 7898.
5. Tropsha, A., Yan, Y., Schneider, S.E., Li, L. and Erickson, B.W., In Hodges, R.S. and Smith, J.A. (Eds.), Peptides: Chemistry, Structure and Biology Escom, Leiden, The Netherlands, 1994, this volume.

Further engineering of a designed protein, the minibody: From an insoluble to a soluble molecule

A. Pessi, E. Bianchi, S. Venturini, G. Barbato, R. Bazzo, A. Tramontano and M. Sollazzo

Istituto di Ricerche di Biologia Molecolare (IRBM)
Via Pontina Km 30.600, 00040 Pomezia (Rome), Italy

Introduction

We have recently reported the design, synthesis and characterisation of the Minibody (MB): a sixty-one residue metal binding protein [1-3]. Using a portion of the heavy chain variable domain of an IgG as a template, we designed a molecule with a β-sheet scaffold of novel fold and two regions corresponding to the hypervariable loops H1 and H2. The aim of the MB design was to transfer into a smaller molecule some desirable properties of the immunoglobulins, namely (a) extreme tolerance to sequence variability in selected regions of the protein (the hypervariable loops) and (b) predictability of the main chain conformation of these regions, directly from the primary structure [4]. The metal binding site was engineered into the molecule in order to explore the possibility of creating functional centres in it. Characterisation of the molecule by CD, size exclusion chromatography and Zn^{++} binding experiments showed the molecule to be folded, globular and able to bind metals [1]: therefore, MB represents the first designed β-protein with a novel fold and a tailored function. The main obstacle to a more detailed characterisation of the protein by NMR or crystallography was its very low solubility in aqueous media. Further engineering of MB was therefore addressed to solve this problem.

Results and Discussion

To this aim, we used two independent strategies. The first one was rational redesign of the β-sheet framework. To select the positions for replacements several factors were taken into account: (i) we excluded the residues predicted to contribute to the stability of the hypervariable loops [1, 4]; (ii) we did not consider substitutions which, according to the model, would result in loss of side chain hydrogen bonds or of the hydrophobic packing; (iii) among the remaining residues, we selected those whose hydrophobic groups were exposed in MB (from here onwards MB1), but had not been changed in the original design to keep alterations of the template IgG framework to a minimum. These residues were substituted with amino acids which, in their most common conformer, would not expose their hydrophobic parts to the solvent. We also took advantage of the higher solvation energy of lysine, in comparison with other charged amino acids, and replaced arginines with lysines whenever possible. Overall, nine residues satisfied the above requirements.

The second strategy was devised as an alternative, since we could not exclude *a priori* that the positions responsible for the stability of the fold would be the same positions that control the solubility of the molecule, this implying that any soluble mutant obtained by changing the β-sheet scaffold would also be destabilized. The strategy consisted in the addition of a solubilizing tail to the protein, without changing the primary structure of the scaffold. Since there are indications that addition of charged residues may enhance solubility in peptides and proteins [5, 6] and we ourselves observed this beneficial effect in synthetic f1 bacteriophage proteins (Pessi *et al.*, unpublished data), we synthesised a MB1 which was elongated with three lysine residues either at the N-terminus (K$_3$-MB1) or at the C-terminus (MB1-K$_3$).

Table 1 *Some features of the mutants used in this study*

Mutants	Source	Mutations[a]	CD (far UV)	Solubility
MB1[b]	SPPS	none	β[c]	10 μM
MB2	SPPS	Q4P-E17D-V19G-R20K-R29K-K34D-E38D-R46K-I48S	random coil	millimolar
MB3	recDNA	Q4P-E17D-V19G-R20K	β[c]	millimolar
MB4	recDNA	E38D-R46K-I48S	β[c]	100 μM
MB5	recDNA	R29K-K34D	β[c]	20 μM
K3-MB1	SPPS	none/+KKK at N terminus	β[c]	millimolar
MB1-K3	recDNA	none/+ KKK at C terminus	β[c]	millimolar

[a]Only those in the framework are considered here; the substitutions are indicated with the equivalent residue in MB1, followed by the number in the sequence and the mutated amino acid;
[b]MB1 is the original molecule [1], with the sequence: RNSQATSGFTFSHFYMEW VRGGEYIAASRHKHNKYTTEYSASVKGRTIVSRDTSQSILYLQ;
[c]The CD spectrum is typical of antiparallel β-sheets (see [1]).

The mutants (Table 1) were produced either by chemical synthesis [1-3] or by recDNA [7]. MB2 is very soluble, but its CD shows a random coil conformation, due either to the presence, among the selected replacements, of a few key residues essential for proper folding, or to the sum of many minor destabilizing distortions, which collectively perturb the structure. To discriminate between these two hypotheses, a set of mutants was prepared in which the mutations of MB2 were divided into three subsets: MB3, MB4 and MB5 contain the first four, the central two and the last three mutations of MB2, respectively. All these mutants are folded, showing a β-sheet-like CD spectrum with bands very similar in position and intensity to MB1; this lends support to the second hypothesis. The mutants display a marked difference in solubility, with only MB3 being in the desired millimolar range. High solubility is also displayed by the mutants with the K$_3$ motif, which both show β-sheet-like CD spectra.

Two of the soluble mutants, MB3 and MB1-K$_3$, were selected for further characterisation. In the same experimental conditions used for MB1 [1] both mutants are monomeric and bind zinc with comparable affinities. Both denaturant-induced unfolding curves (Figure 1A) show a sharp cooperative transition, but with different midpoints (0.74 M and 2.33 M Urea for MB3 and MB1-K_3, respectively) and a different value for the free energy of unfolding extrapolated at zero concentration of denaturant [1] (1.79 and 2.53 kcal/mol, respectively). The values for MB1 are 2.17 M urea and 2.54 kcal/mol. The presence of near UV CD bands (Figure 1B) is a clear indication of tertiary structure; moreover, since the pattern and intensity of the bands are very similar, the aromatic residues should experience the same environment, and therefore the two mutants should have a very similar hydrophobic core.

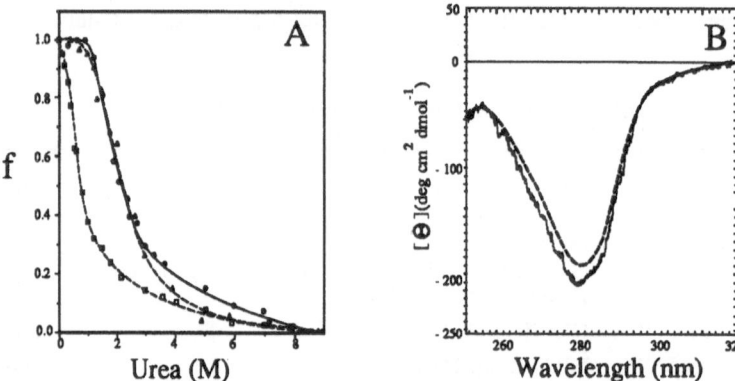

Fig. 1. (A) Denaturant-induced unfolding curve for MB1 (o), MB1-K_3 (▲) and MB3 (▢); f= fraction folded; (B) Near-UV CD spectra of MB1-K_3 (-)and MB3 (- -); conditions as in [1, 7].

The above data, although preliminary, indicate that both strategies were successful, producing soluble mutants with very little alteration in the overall fold. This fold is somehow destabilised in MB3, as shown by the denaturant-induced transition; on the other hand, the only effect of the addition of the lysine tail seems to be on solubility. NMR analysis of these mutants is now in progress.

References

1. Pessi, A., Bianchi, E., Crameri, A., Venturini, S., Tramontano, A. and Sollazzo, M., Nature, 362 (1993) 367.
2. Bianchi, E., Sollazzo, M., Tramontano, A. and Pessi, A., Int. J. Peptide Protein Res., 41 (1993) 385.
3. Bianchi, E., Sollazzo, M., Tramontano, A. and Pessi, A., Int. J. Peptide Protein Res., 42 (1993) 93.
4. Chothia, C., Lesk, A.M., Tramontano, A., Levitt, M., Smith-Gill, S.J., Air, G., Sheriff, S., Padlan, E.A., Davies, D., Tulip, W.R., Colman, P.M., Spinelli, S., Alzari, P.M. and Poljak, J., Nature, 342 (1989) 877.
5. Ford, C.F., Suominen, I. and Glatz, C.E., Protein Expr. Purif., 2 (1991) 95.
6. Keutmann, H.T., Hua, Q.X. and Weiss, M.A., Mol. Endocrinol., 6 (1992) 904.
7. Bianchi, E., Venturini, S., Barbato, G., Bazzo, R., Tramontano, A., Pessi, A. and Sollazzo, M., J. Mol. Biol. (1994), in press.

Design of cytotoxic nuclear targeting peptides

K.M. Sheldon, J.G. Ferguson and J. Gariépy

Department of Medical Biophysics, University of Toronto and the Ontario Cancer Institute, 500 Sherbourne Street, Toronto, Ontario, M4X 1K9, Canada

Introduction

The exposure of cancer cells to cytotoxic drugs initiates a selection process that often leads to the survival and rapid growth of drug-resistant tumor cells [1]. A clinical consequence of this phenomenon is the frequent occurrence of relapse in cancer patients treated with chemotherapeutic agents. Modern approaches to cancer treatment must now focus on searching for new classes of antitumor agents able to circumvent mechanisms of drug resistance. In this report we describe the design of peptide-based agents integrating small peptide signals coding for their endocytosis and compartmentalization inside cells in an effort to deliver toxic or diagnostic moieties to defined intracellular sites.

As an initial step in designing a model cytotoxic agent, *de novo* peptides were constructed that code for three functional domains namely 1) a cytoplasmic translocation domain allowing the peptides to cross the cytoplasmic membrane of live cells, 2) a nuclear localization domain that leads to the accumulation of the peptides in the nucleus of such cells and, 3) a DNA intercalating moiety that can act as the cytotoxic moiety.

Results and Discussion

In order to develop peptide vehicles able to cross the plasma membrane, we exploited the fact that linear chains of polyamino acids such as poly(L-lysine) are rapidly internalized by cells [2]. To facilitate the synthesis of peptides carrying a large but defined number of cationic charges, branched peptides [3] were constructed containing a single C-terminus domain and eight amino-terminal arms (Figure 1). The term "octopeptides" was used to define such peptides in light of their "octopus-like" nature.

In order to carry a DNA intercalator to the nucleus of cells, the model octopeptides were synthesized to include the well characterized nuclear localization signal (NLS) of the SV40 large T antigen [4]. Finally, the DNA intercalator acridine was selected for integration into these octopeptides in view of its appropriate fluorescence properties and its use in the design of potential antitumor drug-peptide conjugates.

In order to test the usefulness of each domain, octopeptides 1, 2, and 3 were synthesized so as to lack one of the three functional domains. Octopeptide 4 contained all three domains (Figure 1b). At a concentration of 1 μM, none of the octopeptides were cytotoxic to CHO cells. Fluorescence microscopy was thus

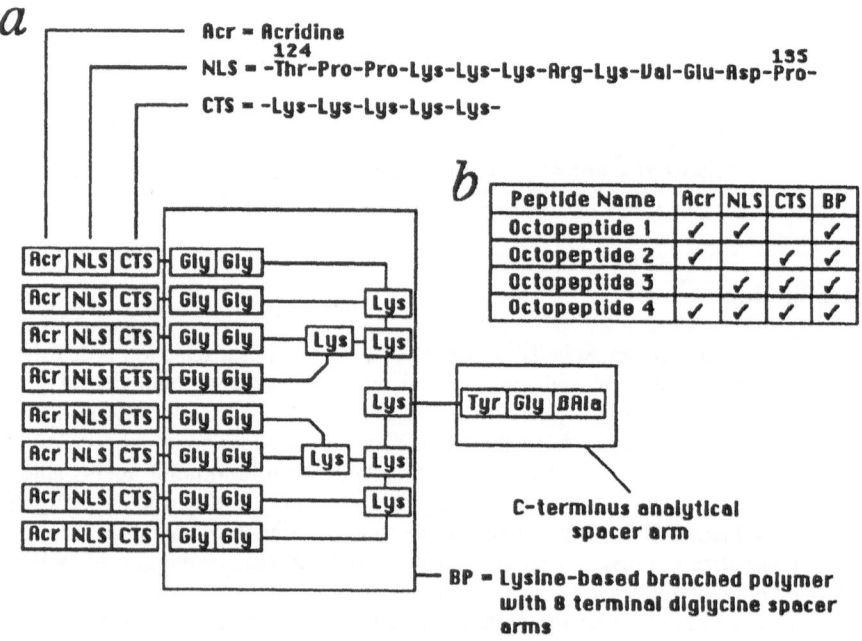

Fig. 1. Primary structure of nuclear-targeting branched peptides (octopeptides). Each peptide is composed of a three-residue long C-terminal region (analytical arm) and 8-N-terminal branches linked together through a branched lysine polymer (BP).

performed at this concentration to characterize the location of each octopeptide in viable CHO cells. Octopeptides 2 and 4 were rapidly internalized by CHO cells in comparison to octopeptide 1 (Figure 2, panels a, b, c), confirming the importance of the CTS domains. After 18 hours, all three octopeptides had entered the cells with only the ones containing NLS domains (Figure 2, panels d and f) being predominantly localized to the nucleus. In the case of octopeptide 2 (Figure 2, panel e), the acridine fluorescence was randomly distributed between cytoplasmic and nuclear compartments in agreement with the recent observation that low molecular weight compounds of less than 60 kD can passively diffuse through the nuclear pore complexes while the nuclear transport of NLS-bearing small molecules is regulated by an active transport mechanism [5].

One can design peptide based agents that are taken up by live cells and rapidly sorted into compartments such as the nucleus. The synthetic flexibility built into the design of octopeptides allows for a range of options in terms of altering their structure, dimensions and biological half-lives.

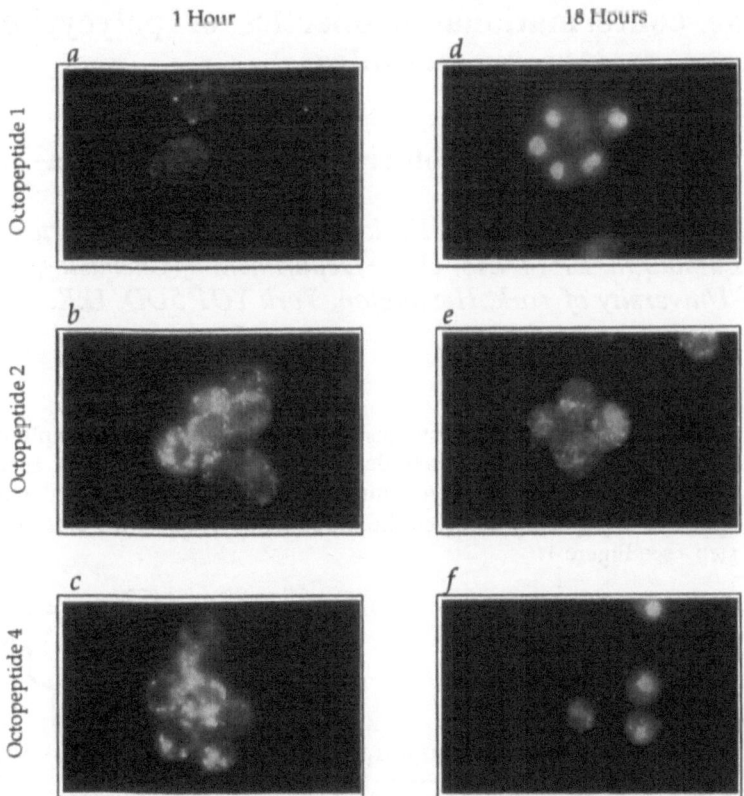

Fig. 2. *Penetration and localization of octopeptides inside live CHO cells as a function of time. The migration of octopeptides bearing acridine groups was monitored by fluorescence microscopy. The efficiency of the cytoplasmic translocation signals was demonstrated by the rapid entry of octopeptides 2 and 4 (panels b,c) into cells as opposed to octopeptide 1 (panel a). The importance of the nuclear localization signals was confirmed by the nuclear accumulation of octopeptides 1 and 4 (panels d and f) but not octopeptide 2 (panel e). Octopeptide 3 lacks acridine.*

Acknowledgements

Supported by a grant from the National Cancer Institute of Canada with funds from the Canadian Cancer Society. K.S is the recipient of a postdoctoral fellowship from the George Knudson Foundation.

References

1. Gottesman, M.M., Cancer Res., 53 (1993) 747.
2. Shen, W.C. and Ryser, H.J.P., Proc. Natl. Acad. Sci. U.S.A., 75 (1978) 1872.
3. Tam, J.P., Proc. Natl. Acad. Sci. U.S.A., 85 (1988) 5409.
4. Kalderon, D., Roberts, B.L., Richardson, W.D. and Smith, A.E., Cell, 39 (1984) 499.
5. Breeuwer, M. and Goldfarb, D.S., Cell, 60 (1990) 999.

The conformational properties of polycyclic peptides

P.D. Bailey[1], I.D. Collier[1] and J.H.M. Tyszka[2]

[1]Department of Chemistry, Heriot-Watt University, Riccarton, Edinburgh, EH14 4AS, U.K. [2]Department of Chemistry, University of York, Heslington, York YO1 5DD, U.K.

Introduction

Whilst there have been many studies on the conformation of simple cyclic peptides, we are the first group to have developed a strategy that gives access to peptides that are constrained by a large number of rings [1]. Initially, our synthesis of these "polycyclic peptides" involved the introduction of all of the cross-links in the final step - see Figure 1.

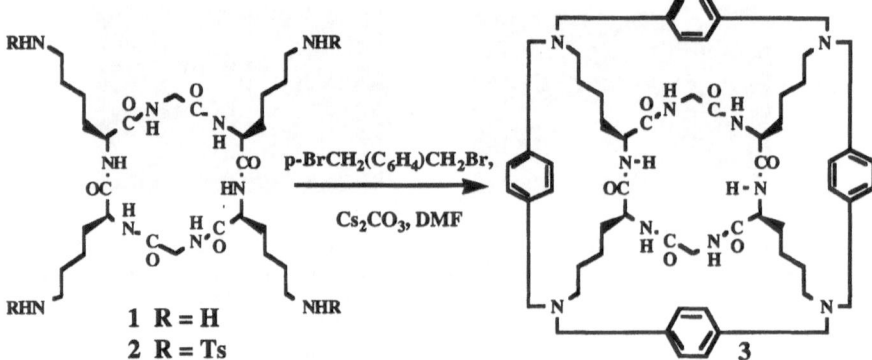

Fig. 1. Formation of the polycyclic peptide 3 from 1 could be achieved by the introduction of all 4 cross-links in the final step.

Of crucial importance is the fact that the cyclic hexapeptide **2** can be obtained directly from Boc-Lys(Ts)-Lys(Ts)-Gly-OH, by cyclodimerisation of the pentafluoro-phenyl ester. In order to improve yields, and allow greater flexibility concerning the structure of the crosslinking units, an alternative strategy was developed (Figure 2), whereby two of the crosslinking units were incorporated at the earlier tripeptide stage.

The polycyclic peptide **3** contains two types of p-xylyl crosslinks: those between adjacent lysyl residues, and those between skipped (non-adjacent) lysyl residues. Therefore there are two possible crosslinked tripeptides **4** and **6**, which when cyclodimerised will give the cyclic isomers **5** and **7** both precursors to the final host molecule **3**. Reaction of the cyclic hexapeptide **2** directly with p-di(bromomethyl)benzene should also produce the same two products. Investigation of the reaction rates and product ratios of these reactions will give important conformational information on both the tri- and hexapeptide precursors **4** - **7**.

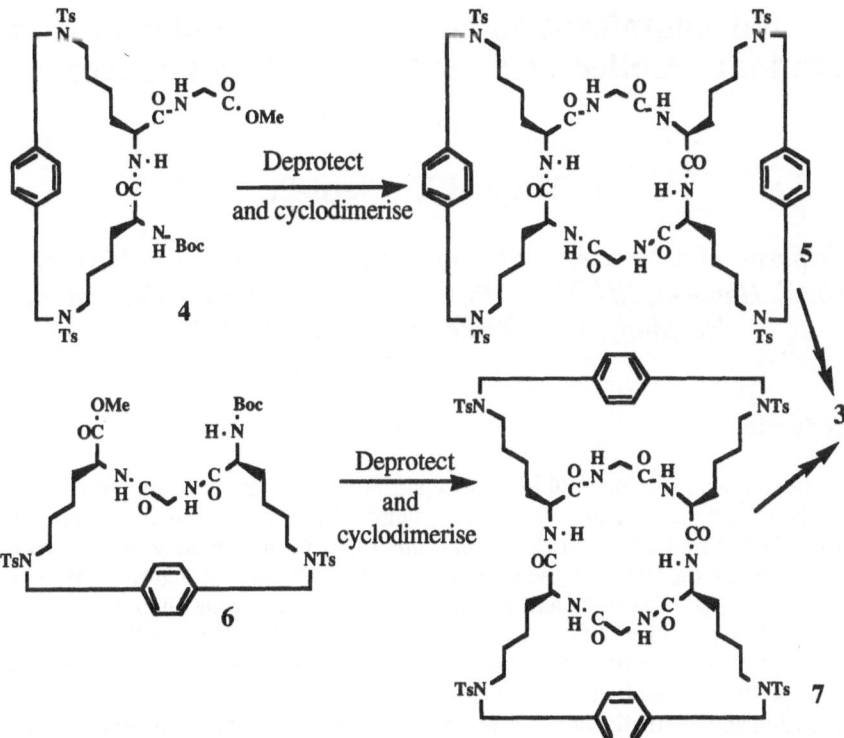

Fig. 2.   Cyclodimerisation of cross-linked tripeptides.

Results and Discussion

The tripeptide precursors **4** and **6** were prepared from the corresponding protected tripeptides by reaction with p-di(bromomethyl)benzene in DMF, with Cs_2CO_3 as base. When **4** was subjected to the deprotection/cyclodimerisation procedure, **5** could be obtained in yields of 70-80%. This is approximately double the yield observed when **2** is prepared by the cyclodimerisation procedure, in which the tripeptide precursor does not possess a cross-link. The formation of **7** has not yet been detected after **6** was subjected to the protection/cyclodimerisation protocol. Finally, when **2** was treated with p-di(bromomethyl)benzene/DMF/Cs_2CO_3, FAB mass spectrometry indicated the formation of **5** and/or **7**; studies are currently underway to determine whether this cross-linking reaction is regio-specific.

The formation of polycyclic peptides *via* cyclodimerisation is much more efficient if adjacent lysyl residues are cross-linked.

Acknowledgements

We thank SERC and Heriot-Watt University for funding.

References

1.   Bailey, P.D., Clarke, D.G.W. and Crofts, G., J. Chem. Soc., Chem. Commun., (1992) 658.

Directed complexation of receptor ectodomains in solution: Coiled-coil mediated subunit assembly

Z. Wu and T. Ciardelli

Department of Pharmacology and Toxicology, Dartmouth Medical School, Hanover, NH 03755-3833 and the Veterans Administrations Hospital, White River Jct., VT 05001, U.S.A.

Introduction

The interleukin-2:interleukin - 2 receptor interaction has been one of the most intensely investigated ligand - receptor systems of the hematopoietic/lymphokine family. Recent findings demonstrate that this system remains an enigma in many respects. Within the last year, the long-standing 3 dimensional structure for IL-2 has been shown to be incorrect and a third cell surface receptor subunit has been identified. The presence of three cell surface subunits (p55, p64 and p75) makes this receptor system unique among the expanding family of hematopoietic receptors. The knowledge of how each subunit functions with respect to ligand capture, signal transmission and internalization is essential for the development of ligand based IL-2 agonists and antagonists. We and others have demonstrated that these subunits must function as heterodimers on the cell surface. A preformed complex of p75 and p55 serves to capture ligand while IL-2 induced association of p75 and p64 is the likely signaling event. The long term goal of this project is to engineer versions of these receptor subunit ectodomains that associate in a directed fashion in solution by exploiting the wealth of knowledge concerning the specificity and stability of coiled - coil (Leu zipper) molecular recognition.

Results and Discussion

Each of the IL-2 receptor subunits can bind ligand independently but with much lower affinity than the heteromeric complexes found on the surface of cells [1]. In order to facilitate our ongoing SAR studies of interleukin-2, we initiated a study to prepare soluble IL-2 receptor subunit complexes designed to associate in a directed fashion in solution and emulate the cell surface complexes. Based on the extensive studies by Hodges [2] and others, we chose idealized coiled - coil heptad repeats as recognition units to mediate solution assembly of the receptor ectodomains in a directed fashion. To test the feasibility of this approach, our first target was the preparation of a p75 homodimeric complex. This protein is composed of the complete p75 ectodomain and, in lieu of the transmembrane and cytoplasmic domains, we substituted 7 heptad repeats of the sequence LEALEKK. This idealized motif is based on the results of Hodges [2] and is designed to form a stable coiled - coil structure. We prepared the genetic constructions for the coiled - coil segment by cloning two large oligonucleotide cassettes, each encoding approximately half of the segment, into a pUC vector. Once complete, we added the cDNA encoding the entire p75 ectodomain 5' to the coiled - coil repeats. Although straightforward in concept, these steps proved difficult in practice. Due to the repetitive nature of the DNA sequences encoding the heptad repeats, we encountered severe problems with recombination. Several clones had to be sequenced before a full length, unmodified version was identified.

For production of these proteins, we chose baculovirus mediated insect cell expression (Max-Bac™, Invitrogen Inc.). This system provides high levels of expression of properly folded and glycosylated proteins. Therefore we subcloned the cDNA encoding the protein (p75cc homo) into the baculovirus expression vector, pBlueBac 2. After complete sequencing, we prepared recombinant virus according to Invitrogen protocols and infected *Spodoptera frugiperda* (Sf9) cells. Expression of p75cc homo exceeded 1mg/L of culture using this system. The protein was purified using affinity chromatography. We determined that ligand affinity (IL-2 column) or immunoaffinity (anti - p75 monoclonal) methods worked equally well for purification, the former demonstrating retention of ligand binding ability by the complex. The purified protein was homogeneous on SDS-PAGE and RP HPLC and N-terminal sequencing confirmed the identity. Figure 1 compares the gel filtration elution of p75cc homo with the p75 ectodomain itself, produced in a similar fashion.

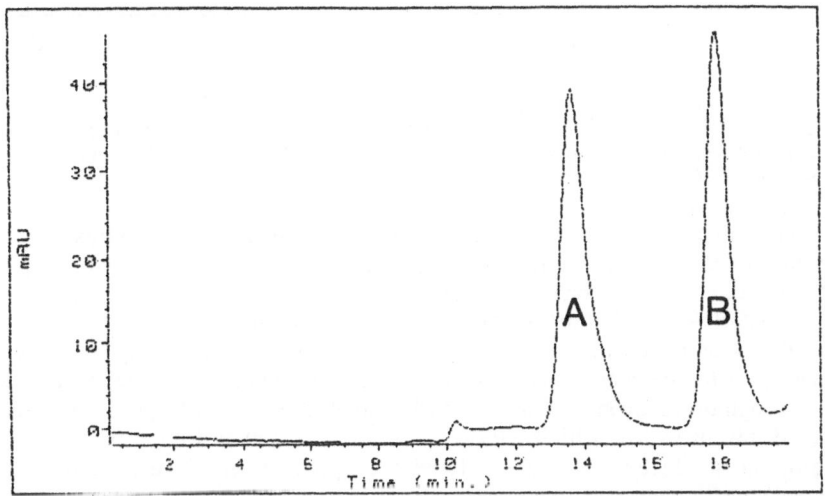

Fig 1. Gel Filtration comparison of A, p75cc homo and B, the p75 ectodomain without the heptad repeats. Chromatogram is a summed overlay of single injections of each sample. Column: Bio-Sil® TSK-250, 300 x 7.5 mm; Buffer: 25 mM Phosphate, pH 7.0 containing 100mM KCl; Flow Rate: 0.5 ml/min.; Absorbance at 215 nm; Sample Concentration 700 - 800 nM based on monomer MW.

The gel filtration results clearly indicate that the addition of the coiled - coil heptad repeats has mediated the formation of a higher MW complex of the p75 ectodomain. That this complex is the result of the coiled-coil helical structure is indicated by comparison of the far UV CD spectrum of the two proteins, Figure 2.

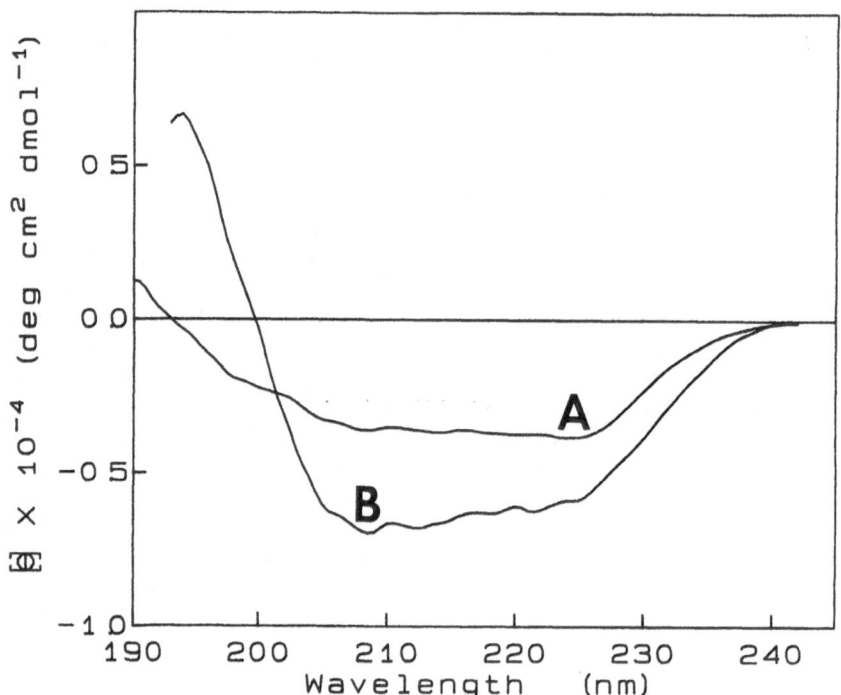

Fig. 2. *Far UV circular dichroism of purified (A) p75 ectodomain and (B) p75 cc homo proteins in 25mM phosphate buffer, pH 7.0, 100mM NaCl. The results are an average of 35-50 scans with buffer baseline subtracted.*

Preliminary deconvolution of the CD spectra confirm the predominant β-sheet structure of the p75 ectodomain (35% β with <2% α-helix). In contrast, p75cc homo contains greater than 13% α-helix or in excess of 75% of the theoretical content based on the addition of 49 residues of helix per monomer.

The feasibility of coiled-coil mediated solution assembly of receptor ectodomains has been demonstrated with our first fusion protein, p75cc homo. This same recognition sequence could be attached to all three IL-2 receptor ectodomains, defined mixtures of which could generate the physiologically significant heterodimers. Instead we have altered the sequence of the heptad repeats to electrostatically favor heteromeric association. This work is in progress.

References

1. Zhou, N.E., Kay, C.M. and Hodges, R.S., J. Biol. Chem., 267 (1992) 2664.
2. Johnson, K., Choi, Y., Wu, Z., Ciardelli, T.L., Granzow, R., Whalen, C., Sana, T., Pardee, G., Smith, K. and Creasey, A., Eur. Cytokine Netw., submitted.

Properties of a potent cationic lytic amphipathic helical peptide

R.M. Epand[1], R.F. Epand[1], W.D. Blackburn[2],
V.K. Mishra[1], E.M. Tytler[1], J.P. Segrest[1] and
G.M. Anantharamaiah[1]

[1]*Department of Biochemistry, McMaster University Health Sciences Centre, 1200 Main Street West, Hamilton, Ontario, L8N 3Z5, Canada,* [2]*V.A. Hospital and the Department of Medicine, University of Alabama at Birmingham Medical Center, Birmingham, AL 35294, U.S.A.*

Introduction

Amphipathic helixes can be classified according to the size and charge distribution of the hydrophilic face [1]. Cationic peptides which form class L amphipathic helixes are lytic to model and biological membranes [1]. The hydrophilic face of this structural motif is characterized by having Lys residues, separated in the middle by a Gly residue [2]. Examples of such peptides are mastoparan and a model peptide, 18L. These peptides lower the bilayer to hexagonal phase transition temperature of dipalmitoleoyl phosphatidylethanolamine [1]. The motif of these peptides can be modified by eliminating the Gly residue which is in the center of the polar face to produce the peptide, K18L, with the sequence KWLLKFYKLVAKLLLKAF. The peptide is fully helical by circular dichroism when it is dissolved in 1:1 (v/v) trifluoroethanol and water. In this conformation the hydrophilic face subtends a narrower angle than class L amphipathic helical peptides. In addition, according to the "snorkel hypothesis" the Lys can allow greater penetration into the membrane by virtue of its long amphipathic sidechain. In the case of K18L, this latter feature would allow the peptide to penetrate further into the membrane than does 18L or mastoparan. We investigated what consequences this had on the interaction of this peptide with model membranes as well as comparing the biological effects of K18L with those of mastoparan and 18L.

Results and Discussion

K18L is several fold more potent than mastoparan or 18L in lowering the bilayer to hexagonal phase transition temperature (T_H) of dipalmitoleoyl phosphatidylethanolamine as measured by differential scanning calorimetry. To our surprise the K18L had an ability to lower this transition temperature by about 5 degrees at a peptide mole fraction of 0.002. This is greater than the effect that hydrocarbons and diacylglycerols have on T_H on a molar basis and similar to their effects on a weight basis [3]. In general, more hydrophobic substances decrease T_H to a greater extent since they expand the hydrophobic volume of the membrane more than they expand the headgroup region. This is obviously not the case with K18L which is not very hydrophobic compared with diacylglycerols. The ability of this peptide to promote the hexagonal phase must reflect the manner in which it intercalates into the membrane. Another membrane property that is increased by the presence of hexagonal phase promoting substances is the formation of structures that give rise to isotropic ^{31}P NMR signals. These structures readily form in suspensions of monomethyldioleoylphosphatidylethanolamine (MeDOPE) and have been associated with increased rates of membrane fusion [4]. The K18L is also much more potent than the other peptides in inducing isotropic ^{31}P NMR signals in suspensions of MeDOPE.

We have shown that a class A amphipathic helix-containing protein, the apolipoprotein A-I, inhibits activation of neutrophils [5]. Generally class L helixes have opposite effects to those containing class A amphipathic helixes [1]. If this antagonism also occurs for neutrophil activation, then K18L should be a more potent peptide than mastoparan or 18L. Indeed this was found to be the case for both lactoferrin release as well as for superoxide production.

The more potent effect of K18L in inducing inverted phases is in accord with a model in which the long sidechain of the Lys residues will allow deeper penetration of the peptide into the bilayer, this effect being greatest when the Lys is opposite the hydrophobic face. K18L was at least twice as active as 18L or mastoparan in inducing leakage in vesicles of MeDOPE, promoting haemolysis and activating neutrophils. These results demonstrate that relatively modest changes in the arrangement of the polar face of cationic, lytic amphipathic helixes can have profound effects on the interactions of these peptides with model and biological membranes.

References

1. Tytler, E.M., Epand, R.M., Anantharamaiah, G.M. and Segrest, J.P., In Hodges, R.S. and Smith, J.A. (Eds.), Peptides: Chemistry, Structure and Biology (Proceedings of the Thirteenth American Peptide Symposium), Escom, Leiden, The Netherlands, 1994, this volume.
2. Segrest, J.P., de Loof, H., Dohlman, J.G., Brouillette, C.G. and Anantharamaiah, G.M., Proteins, 8 (1990) 103.
3. Epand, R.M., Biochemistry, 24 (1985) 7092.
4. Ellens, H., Siegel, D.P., Alford, D., Yeagle, P.L., Boni, L., Lis, L.J., Quinn, P.J. and Bentz, J., Biochemistry, 28 (1989) 3692.
5. Blackburn Jr., W.D., Dohlman, J.G., Venkatachalapathi, Y.V., Pillion, D.J., Koopman, W.J., Segrest, J.P. and Anantharamaiah, G.M., J. Lipid Res., 32 (1991) 1911.

Synthesis and characterization of triple-helical 'mini-collagens'

C.G. Fields, C.M. Lovdahl, A.J. Miles, V.L. Matthias Hagen and G.B. Fields

Department of Laboratory Medicine and Pathology and Biomedical Engineering Center, University of Minnesota, Minneapolis, MN 55455, U.S.A.

Introduction

Collagens are distinguished structurally from other extracellular matrix proteins by their composition of three α chains of primarily repeating Gly-X-Y triplets, which induces each α chain to adopt a left-handed poly-Pro II helix. Three left-handed chains then intertwine to form a right-handed triple-helix. We have developed a generally applicable solid-phase methodology for the synthesis of aligned triple-helical collagen-model peptides (*i.e.*, 'mini-collagens') that can incorporate native collagen sequences [1, 2]. The thermal stabilities of these triple-helical peptides (THPs) are examined here.

Results and Discussion

A generic THP structure is given in Figure 1. The branch introduced at the *C*-terminus of the synthetic peptide is consistent with the natural nucleation of collagen triple-helices from the *C*- to the *N*-terminus. The triple-helix inducing sequence is (Gly-Pro-Hyp)$_8$. THP-1 incorporates residues 531-543 from α1(IV) collagen while THP-2 incorporates residues 1263-1277 from α1(IV) collagen [1, 2]. THP-3 has only the triple-helix inducing sequence [2].

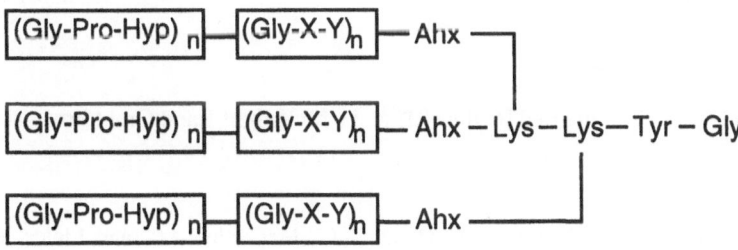

Fig. 1. Generic structure of triple-helical peptides or 'mini-collagens'. Ahx is 6-aminohexanoic acid.

Table 1 T_m and $\Delta H°$ of triple helix $\Leftrightarrow$ coil transitions

Peptide	Hyp (%)	T_m/triplet (°C)	$\Delta H°$/triplet (kcal/mol)
(Pro-Pro-Gly)$_8$	0	0.75	-1.63
(Pro-Pro-Gly)$_{10}$	0	0.83	-1.77
(Pro-Pro-Gly)$_{10}$	0	0.83	-1.72
(Pro-Pro-Gly)$_{20}$	0	1.00	-1.83
THP-2	21	1.50	-2.49
THP-1	22	1.43	-2.30
(Pro-Hyp-Gly)$_{10}$	33	1.93	-3.81
(Pro-Hyp-Gly)$_{10}$	33	1.92	-3.00
(Pro-Hyp-Gly)$_{10}$	33	2.07	-4.13
THP-3	33	1.77	-3.67

The triple-helical melting temperatures (T_m), as evaluated by circular dichroism spectroscopy, were 53, 59, and 42°C for THP-1, THP-2, and THP-3, respectively [1, 2]. T_m/triplet values for our and other [2] THP triple-helical melts could be correlated to Hyp content (Table 1). Although THP-1 contains an interruption in the Gly-X-Y repeat, it has a similar T_m/triplet value as THP-2, which has no Gly-X-Y interruptions. Non-triple-helical inducing regions have substantial detrimental effects on the thermal stabilities of triple-helical peptides due to significant cooperativity of residues [3]. The similar T_m values for THP-1 and THP-2 are probably due to (i) the large number (8) of Gly-Pro-Hyp repeats at the THP N-terminus, providing a cooperative "cluster" which stabilizes the triple-helix, and (ii) the branched Lys-Lys structure, which aligns and entropically stabilizes the C-terminus of the THP.

A good correlation was also found between our and other [2] van't Hoff enthalpies ($\Delta H°$) for THP coil $\Leftrightarrow$ triple-helical transitions and Hyp content (Table 1). An increase in Hyp content results in increasingly negative enthalpies and thus more favorable triple-helix formation. The correlation of THP T_m/triplet and $\Delta H°$/triplet values as a function of Hyp content suggests that these peptides are not destabilized by residue misalignment or "looping" out.

Acknowledgements

This work is supported by the NIH and a McKnight-Land Grant Professorship.

References

1.   Fields, C.G., Mickelson, D.J., Drake, S.L., McCarthy, J.B. and Fields, G.B., J. Biol. Chem., 268 (1993) 14153.
2.   Fields, C.G., Lovdahl, C.M., Miles, A.J., Matthias Hagen, V.L. and Fields, G.B., Biopolymers, 33 (1993) 1695.
3.   Germann, H.-P. and Heidemann, E., Biopolymers, 27 (1988) 157.

A new method for defining dimerization interfaces in proteins

W.M. Kazmierski and J. McDermed

Wellcome Research Laboratories, Research Triangle Park, NC 27709, U.S.A.

Introduction

Gene expression in mammals and other organisms is regulated by, among others, the leucine zipper transcription factor family that includes C/EBP, Fos, Jun, CREB and GCN4 proteins. A prerequisite to DNA recognition by many of the regulatory proteins, including GCN4, is their dimerization. Consequently, molecules able to disrupt the protein dimerization are of potential medicinal interest.

Traditionally, leucine zippers were thought to form via hydrophobic association of the α-helical monomers into a coiled-coil (Figure 1a) [1, 2]. That model was recently challenged by a coiled-coil model based on a periodic hydrophilic-hydrophobic leucine-zipper motif (Figure 1b) [3]. These models are related by a clockwise rotation of both helices. The discrepancy between these models prompted us to develop a simple experimental approach capable to distinguish between the different modes of dimerization in coiled-coil proteins.

Results and Discussion

We synthesized three leucine zipper-containing fragments of GCN4: Ac-[Cys1]-GCN4(1-33)-NH$_2$, (C1); Ac-[Cys2]-GCN4(1-33)-NH$_2$, (C2); and Ac-[Cys5]-GCN4(1-33)-NH$_2$, (C5). In the disulfide trapping experiments 1.2 mg of each peptide was dissolved in 0.3 ml of water and the air oxidation rates were time monitored (Figure 2) by RP-HPLC using a C$_{18}$ stationary phase (Vydac) and 0-45%B/45 min gradient as a mobile phase (A=0.1% TFA in water, B=0.1% TFA in MeCN). HPLC fractions were analyzed by FAB-MS and amino acid analysis, resulting in full and unambiguous characterization of all components.

The cysteines in C1, C2 and C5 were so placed as to provide the distinction between the two leucine zipper models (Figure 1). In C1, both cysteines are placed on **g** and **g'** positions of both helices and the hydrophilic-hydrophobic (Figure 1b) model would allow for a fast disulfide cyclization, while the hydrophobic-hydrophobic (Figure 1a) model would not permit an efficient disulfide bond formation. We observed (Figure 2) a single peak over the period of 150 hr with unchanged retention time corresponding to intact (reduced) C1. In C2, both **a** and **a'** positions face each other in the hydrophobic-hydrophobic coiled-coil (Figure 1a) and are out of phase in the hydrophilic-hydrophobic model (Figure 1b), facilitating fast and slow cyclization rates, respectively. This order is opposite to the expected

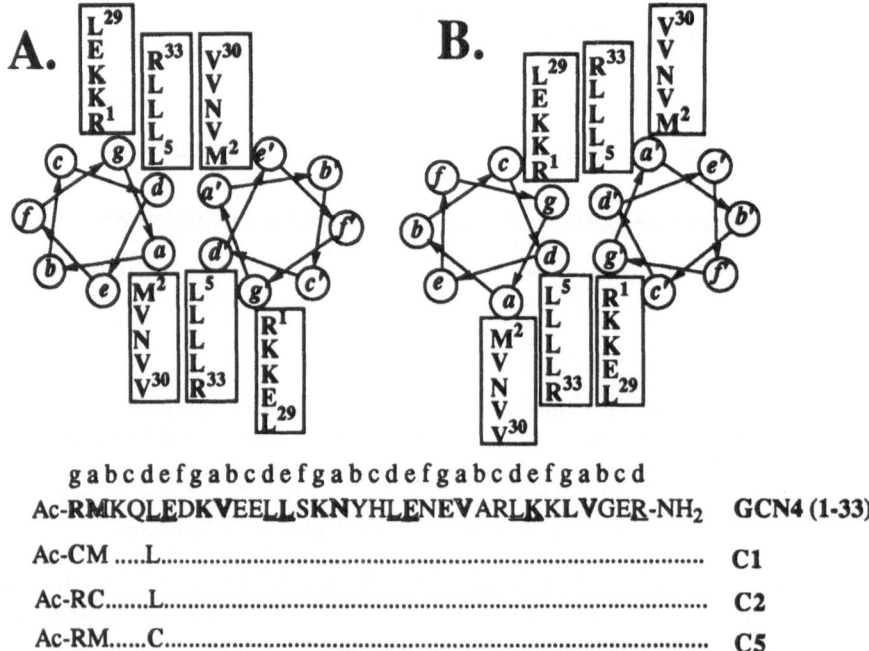

gabcdefgabcdefgabcdefgabcdefgabcd

Ac-RMKQLEDKVEELLSKNYHLENEVARLKKLVGER-NH₂   GCN4 (1-33)

Ac-CM ...L..   C1

Ac-RC.....L..   C2

Ac-RM.....C..   C5

Fig. 1. *The hydrophobic-hydrophobic (a), hydrophilic-hydrophobic (b) leucine zipper models and the sequences of C1, C2, C5.*

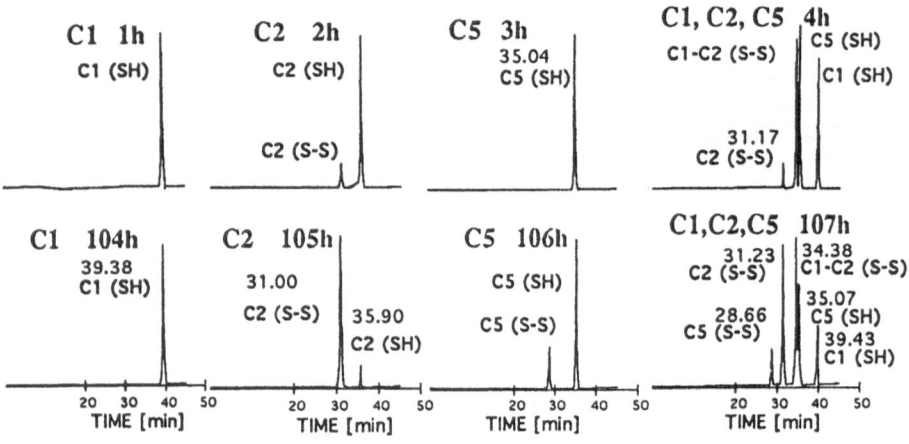

Fig. 2. *Time course of the disulfide oxidation of C1, C2, C5 (separately) and as a mixture.*

cyclization rates for C1. Indeed, we found that within less than 18.5 hrs the reduced form of the C2 peptide is converted to its oxidized form, as confirmed by AAA and FAB-MS. The situation is distinctly different for C5, featuring Cys on positions **d** and **d'**. Both models allow for the disulfide bond formation, as **d** and **d'** face each other in both. Accordingly, we determined that C5 oxidizes slowly with about 30% conversion within 150 hrs. In the experiment involving the mixture of C1, C2 and C5 all of the above observations are manifested: C2 (S-S) forms very fast, followed by a slow rate of C5 (S-S) formation and no detectable C1 (S-S) formation. Additionally, a mixed disulfide C1-C2 appears within 4 hours (Figure 2). An inspection into the models (Figure 1) indicates that C1-C2 disulfide formation is possible in both leucine zipper models. Rates of C2 (S-S) and C5 (S-S) formation are slower than in experiments with C2 and C5 alone, respectively. This is caused by the pseudo-dilution phenomenon as all the leucine-zipper molecules recognize and bind to each other, since the dimer stability is not greatly influenced by a single point mutation. Thus, if C5 binds to either C1 or C2, no cross-linking occurs as a result of mismatched geometry. C5 needs to dissociate and to again associate with another C5 to eventually cross-link, resulting in overall slower cross-linking rates (compared to pure C5).

The experimental evidence clearly establishes that C2 dimerizes and oxidizes very fast. C5 oxidation rate is lower, whereas C1 does not oxidize, even upon extended standing (>150 hr). When these rates are compared with the ones predicted by the models, the hydrophobic-hydrophobic leucine zipper model is strongly supported (Figure 1a) [1, 2]. A recently published X-ray structure of the GCN4 leucine zipper [4] fully supports our conclusions. We have shown, that even in the absence of X-ray data, covalent trapping can be used to determine the mutual orientation of helices in a coiled-coil. We postulate that this approach can define the dimerization interface in biologically important molecules for further drug design.

References

1. O'Shea, E.K., Rutkowski, R., Stafford, W.F. and Kim, P.S., Science, 245 (1989) 646.
2. O'Shea, E.K., Rutkowski, R. and Kim, P.S., Science, 243 (1989) 538.
3. Tropsha, A., Bowen, J.P., Brown, F.K. and Kizer, J.S., Proc. Natl. Acad. Sci. U.S.A., 88 (1991) 9488.
4. O'Shea, E.K., Klemm, J.D., Kim, P.S. and Alber, T., Science, 254 (1991) 539.

Comparative immunogenicity and antigenicity of a cyclized and a zinc finger construct of the β (38-57) hCG loop region

S.F. Kobs-Conrad[1], J.S. Eiden[2], E.A.L. Chung[2],
A.C. Lee[2], J.E. Powell[2], V.C. Stevens[2] and
P.T.P. Kaumaya[1,2,3,4]

*[1]The Comprehensive Cancer Center; Departments of [2]Obstetrics &
Gynecology and [3]Medical Biochemistry, The College of Medicine; and
[4]Department of Microbiology, College of Biological Sciences,
The Ohio State University, Columbus, OH 43210, U.S.A.*

Introduction

We have rationally designed highly engineered synthetic peptides to provide enhanced immunogenicity and specificity for the contraceptive vaccine candidate LDH-C_4 into αα [1] and αβ [2] motifs, and a loop region from the same antigen into a zinc finger peptide motif [3]. Loop-structured and turn motifs have been shown to comprise protein antigenic sites and such segments have traditionally been stabilized via disulfide bonding or chemical cyclization reactions resulting in conformationally rigid structures which as a rule do not accurately mimic the native protein loop. We have demonstrated that the disulfide loop 38-57 of the glycoprotein hormone human chorionic gonadotropin β-subunit (β-hCG) is more immunogenic than the uncyclized peptide, however, with low levels of cross-reactivity to hLH. The aim of this study is to further examine the potential use of the zinc finger peptide motif as an alternative strategy to mimic loop regions.

Results and Discussion

A 12 residue loop sequence of β-hCG was engineered into a consensus zinc finger peptide motif. Of four initial constructs, three formed characteristic zinc finger peptides (two of these had the epitope in the N→C orientation, one with C→N configuration). A promiscuous T-cell epitope (TT$_3$) was colinearly synthesized with one forward (ZF2) and one reverse (ZF4) epitope zinc finger peptide to yield ZF2TT$_3$ and ZF4TT$_3$ immunogens. The zinc finger constructs were synthesized via solid-phase methods using FMOC/t-butyl strategy. The zinc finger peptides were purified to >90% homogeneity (as determined by RP-HPLC and CZE). ZF2, ZF4, ZF2TT$_3$ and ZF4TT$_3$ form 1:1 complexes with Zn^{2+} as determined by absorption spectroscopy of their Co^{2+} complexes (data not shown). CD spectra show α-helical and ß-structural elements in the presence of Zn^{2+}; without metal ion the peptides are in random coil conformation as shown in Figure 1.

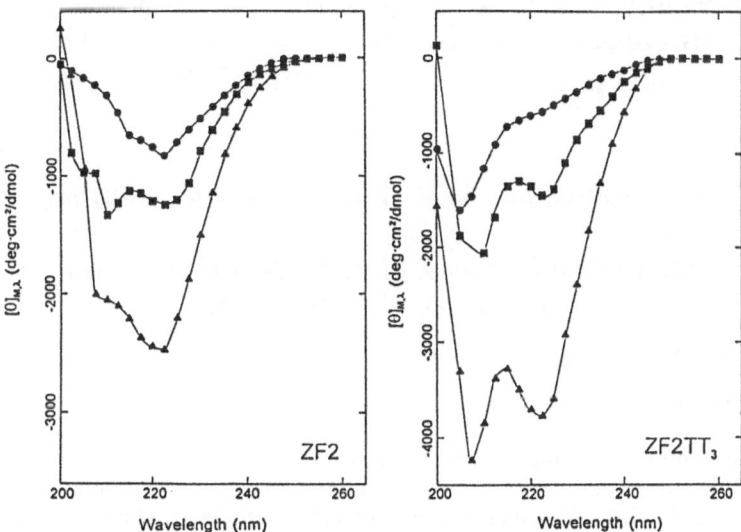

Fig. 1. *CD Spectra of zinc finger construct ZF2 and ZF2TT₃. Peptides (250 µM) were dissolved in 10mM Tris-HCl, pH 8 (●), buffer containing 1 equivalent of ZnCl₂ (■), or buffer containing 1 equiv. ZnCl₂ and 50% TFE (▲). Spectra were recorded under N₂ at 25°C.*

ZF2 and ZF2TT₃ competitively inhibited the binding of labelled whole hCG to anti-native loop antibodies. Thus, in buffered solution, these peptides are able to compete for binding sites to antibodies raised to the cyclized 38-57 peptide in a manner similar to ßhCG(40-52) or native βhCG. However, the reverse sequences ZF4 and ZF4TT₃ did not inhibit the binding of native glycoprotein to the antibody. The ZF2TT₃ and ZF4TT₃ constructs both produced high-titered antipeptide antibodies in rabbits and mice (C3H, C57/BL6 and BALB/c). The antipeptide antibodies were directed towards the zinc finger structure, as evidenced by the cross-reactivity with the individual ZF2 and ZF4 peptides. Antibodies of relatively low titer were produced in mice (BALB/c and C3H/HeJ) immunized with the loop-DT conjugate, and high titer in C57/BL6. On the other hand the chimeric zinc finger constructs elicited high titered antibodies in all three strains. Immune reactivity of the zinc finger constructs for the native hCG was minimal, suggesting that the stability of these constructs in adjuvant is not optimal.

References

1. Kaumaya, P.T.P., Berndt, K., Heindorn, D., Trewhella, J., Kezdy, F.J. and Goldberg, E., Biochemistry, 29 (1990) 13.
2. Kaumaya, P.T.P., VanBuskirk, A., Kobs, S., Goldberg, E. and Pierce, S.K., In Giralt, E. and Andreu, D. (Eds.), Peptides 1990, ESCOM, Leiden, The Netherlands, 1991, p.611.
3. Kobs-Conrad, S., Lee, H., DiGeorge, A.M. and Kaumaya, P.T.P., In Schneider, C.H. and Eberle, A.N. (Eds.), Peptides 1992, Escom, Leiden, The Netherlands, 1993, p.561.

Enhancing the conformational stability of growth hormone via site-directed mutagenesis: Incorporation of a metal binding site

S.R. Lehrman, M.E. Lund and J.R. Shifflett

The Upjohn Company, Biotechnology Development, Kalamazoo, MI 49001, U.S.A.

Introduction

The binding of human growth hormone (hGH) to zinc via His[19], His[22], and Glu[174] leads to the formation of a stable protein dimer [1]. Other growth hormone species (e.g., bovine and porcine) also contain two histidine residues, but at positions 20 and 22. Since these growth hormone homologs bind zinc with low affinity, the relative placement of the His residues appears to be crucial. An analog of bovine growth hormone (bGH) has been reported that contains His residues at positions 19 and 22 [2]. This compound binds zinc with high affinity, and has been purified using immobilized metal affinity chromatography [2]. We report that modifying porcine growth hormone (pGH) in this manner also generates a zinc binding site and that binding of zinc to the modified protein enhances its conformational stability without loss of bioactivity.

Results and Discussion

pGH was modified by replacing Gln[19] with histidine, and His[20] with arginine (Q19H,H20R). These modifications resulted in the placement of two histidine residues at i and i+3 positions of the primary structure. Since this region of the protein forms an α-helix, the histidine sidechains are expected to be separated by a distance of ≈ 5 Å, suitable for zinc binding. Using size exclusion chromatography, we find that this interaction results in the formation of a protein dimer (data not shown).

We find that zinc binding by Q19H,H20R alters its conformational behavior. For example, the equilibrium denaturation behavior of pGH and Q19H,H20R were compared in the presence and absence of zinc. We found that the midpoint of the denaturation curve for Q19H,H20R + Zn increased significantly relative to Q19H,H20R + EDTA (Table 1). In contrast, the denaturation midpoint of pGH was not significantly altered by zinc. In addition, we determined the effect of zinc chelation on the aggregation of partially-denatured pGH and Q19H,H20R. Aggregation of partially-denatured growth hormone is a well-characterized process that can lead to significant loss of protein during purification and handling [3, 4]. Aggregation can be measured ($\pm 10\%$) using a two-step solubility assay. We found that when Q19H,H20R binds zinc there is a significant reduction in the amount of aggregation observed (Table 1). The effect of this modification on rat weight gain activity is also shown. Since the precision of this assay is ⊥ 20% [5], these results indicate that there is no difference in the in vivo bioactivity of these compounds.

Table 1 *Conformational behavior of pGH and Q19H,H20R ± Zn*

Protein	Denaturation midpoint (M)	Relative precipitation (%)	Relative bioactivity (%)
pGH ± Zn	2.6	100	117
Q19H,H20R + Zn	3.0	75	120
Q19H,H20R + EDTA	2.5	97	134

To summarize, in the presence of zinc, the metal-binding analog is more resistant to denaturation and aggregation. In addition, Q19H,H20R is bioequivalent to pGH in a rat weight gain bioassay.

Acknowledgements

We are grateful to J.E. Mott, N. Strakalaitis, J. Gardner, and J. Hoogerheide for their contributions in generating porcine growth hormone and Q19H,H20R. We are also grateful to Paul Elzinga for his comments on evaluation of protein stability.

References

1. Cunningham, B.C., Mulkerrin, M.G. and Wells, J.A., Science, 253 (1991) 545.
2. Suh, S-S., Haymore, B.L. and Arnold, F., Protein Engineering, 4 (1991) 301.
3. Havel, H.A., Kauffman, E.W., Plaisted, S.M. and Brems, D.N., Biochemistry, 25 (1986) 6533.
4. Brems, D.N., Plaisted, S.M., Kauffman, E.W., Lund, M.E. and Lehrman, S.R., Biochemistry, 26 (1987) 7774.
5. Parlow, A.F., Wilhelmi, A.E. and Reichert, L.E., Endocrinology, 77 (1965) 1126.

Fe(pepy)₃: A designed protein with multiple tertiary conformations

M. Lieberman and T. Sasaki

*Department of Chemistry, University of Washington,
Seattle, WA 98195, U.S.A.*

Introduction

Typically, template-assembled synthetic proteins (TASPs) have high secondary structure content, but there is little direct evidence for the presence of native-like folded conformations [1]. We have used an inorganic complex to assemble a three-α-helix bundle protein. This metalloprotein contains a unique reporter group which enables us to probe the steepness of the potential energy surface of the peptide bundle [2].

Results and Discussion

The strong Fe(II) chelating ligand 2,2'-bipyridine was covalently attached to the N' terminus of a 15-residue amphiphilic peptide(AEQLLQEAEQLLQEL-amide) as previously described [3]. The modified peptide (pepy) trimerizes on addition of Fe(II) with a concomitant increase in α-helicity from 35% to 80% at pH 4.8. The ΔG of folding was measured as -2.4 kcal/mole by GnHCl denaturation ([GnHCl]$_{1/2}$=3.4 M, m=0.73). A truncated analog(AEQLLQEL-amide) was also synthesized for comparison.

The iron complex exists as four separable isomers. They were characterized by NMR of the bipyridine region and by CD. The isomers are metastable and interconvert with a half-life of about 8 hours at RT.

The equilibrium isomer ratio of Fe(pepy)₃ is non-statistical. The range of relative energies spanned by the four isomers is only 0.25 kcal/mole. If the peptide is denatured (thermally, by addition of GnHCl, or by altering the pH) before addition of iron, then the isomer ratio approaches the statistical distribution. The truncated analogue showed a nearly statistical distribution of four isomers, presumably due to poor peptide interaction.

CD was used to probe the secondary structure content of the four isomers of Fe(pepy)₃. The Λ-*mer* isomer has a very high helical content (95-100%) and the other three isomers have moderately high helical contents (45-55%).

When an unsymmetrical bidentate ligand such as pepy binds a metal in octahedral coordination, four diastereomers are formed: Δ-*fac*, Λ-*fac*, Δ-*mer*, and Λ-*mer*. The Δ/Λ "propeller" enantiomers could interact diastereomerically with the chiral peptides or with the supercoiling [4] of the helix bundle. In the *facial* isomers, the bases of the helices are all 9 Å apart, while in the *meridional* isomers, the bases of the helices are anchored 9, 9, and 14Å apart. The tertiary structures built on these four templates cannot possibly be the same.

Fe(pepy)$_3$ is a monomeric metalloprotein with four separable tertiary conformations. The system is somewhat similar to ones that contain proline residues which undergo *cis-trans* isomerism and as a result can exist in different conformations [5]. High secondary structure content in this designed protein does not imply the presence of an unique tertiary structure, nor does it correspond to high stability.

The energy differences between the four isomers are small, suggesting that the peptides can fold regardless of what template structure is offered. These results suggest the potential energy surface for this bundle of three parallel α-helices is broad and shallow compared to native proteins.

Acknowledgements

ML thanks the Fannie and John Hertz Foundation for a research fellowship. TS thanks the Petroleum Research Fund for supporting this work.

References

1. a) Mihara, H., Nishiro, N. and Fujimoto, T., Chem. Lett., (1992) 1809. b) Hecht, T. Richardson, J.S., Richardson, D.C. and Ogden, R.C., Science, 249 (1990) 854. c) Mutter, M., Altmann, K.-H. and Vorherr, T., Z. Naturforsch., 41b (1986) 1315. Also see "A *de novo* designed protein shows a thermally induced transition from a native to a molten globule-like state", Raleigh, D.P. and DeGrado, W.F., J. Am. Chem. Soc., 114 (1992) 10079.
2. Sasaki, T. and Lieberman, M., Tetrahedron, 49 (1993) 3677
3. Lieberman, M. and Sasaki, T., J. Am. Chem. Soc., 113 (1991) 1470. See also Ghadiri, M.R., Soares, C.and Choi, C., J. Am. Chem. Soc., 114 (1992) 825, 4000.
4. a) Cohen, C. and Parry, D.A.D., Proteins: Struct., Funct., Genet., 7 (1990) 1. b) Lovejoy, B., Choe, S., Cascio, D., McRorie, D., DeGrado, W.F. and Eisenberg, D., Science, 259 (1993) 1288.
5. See for example Kofron, J.L., Kuzmic, P., Kishore, V., Fesik, S.W. and Rich, D.H., In Smith, J.A. and Rivier, J.E. (Eds.), Peptides: Chemistry and Biochemistry, (Proceedings of the 12th American Peptide Symp.), Escom, Leiden, The Netherlands, 1991, p.785

The design of associating amphiphilic α-helical peptides

R.A. Lutgring, K.S. Lewen and J.A. Chmielewski

Department of Chemistry, Purdue University
West Lafayette, IN 47907, U.S.A.

Introduction

A knowledge of the way in which peptides and proteins assemble is essential for the design of biomolecules with novel functions. One of the simplest self associating systems is the assembly of two or three helical peptides into a coiled-coil motif. The two helix coiled-coil is the basis of the leucine zipper dimerization motif found in many classes of DNA binding proteins [1], whereas the three helix coiled-coil has been identified in proteins such as laminin and thrombospondin [2]. The two helix coiled-coil contains a heptad repeat with a 3-4 hydrophobic repeat. This yields a hydrophobic face within the helix, which dimerizes in a parallel fashion, while packing the sidechains efficiently in a "knobs in holes" manner [3].

Results and Discussion

We have designed a sixteen amino acid residue peptide to dimerize in aqueous solution. The peptide, Leu4, was designed with a minimalistic sequence containing amino acid residues with a high propensity to exist in an α-helix: leucine, glutamate and lysine (Figure 1) [4]. The peptide was designed to be an amphiphilic helix, with all of the hydrophobic leucine residues on one face of the helix and the hydrophilic residues on the other face. The potential for intramolecular salt bridges has been designed into the peptide by positioning glutamate and lysine three or four residues apart whenever possible [5]. The amino terminus has been acetylated and the carboxy

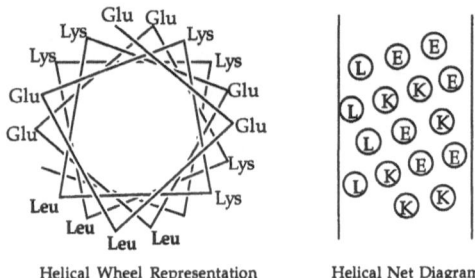

Helical Wheel Representation   Helical Net Diagram

Leu4 Peptide:

AcNH-Glu-Glu-Leu-Glu-Lys-Lys-Leu-Lys-Glu-Leu-Glu-Glu-Lys-Leu-Lys-Lys-NH₂

Fig. 1. The sequence of the minimalistically designed associating amphiphilic α-helical peptide Leu4.

terminus converted into a primary amide to reduce any helix destabilization due to electrostatic repulsion between the charged termini and the helix dipole. Also to this end, glutamate has been positioned at the amino terminus and lysine at the carboxy terminus in the peptide to increase the potential for charge stabilization with the sidechains of these amino acids and the helix dipole [6].

Leu4 was synthesized by standard solid phase methods on a polystyrene support developed by Rink [7], and fluorenylmethyloxycarbonyl (Fmoc) was used as the semipermanent amine protecting group. The peptide was cleaved from the resin with TFA and purified to homogeneity by HPLC. Circular dichroism experiments were performed by measuring the mean residue ellipticity between 190 and 260 nm at a range of peptide concentrations (Figure 2). At a concentration of 119 µM, the spectrum of Leu4 was typical of that of α-helical proteins and the helical content was estimated to be 77% [8]. As the concentration of the peptide was lowered there was a concomitant decrease in the mean residue ellipticity at 222 nm, which is consistent with an aggregation event. The data was analyzed according to various monomer-$n$mer equilibria using a nonlinear regression program (MLAB), and best fits were obtained with both $n=2$ and 3.

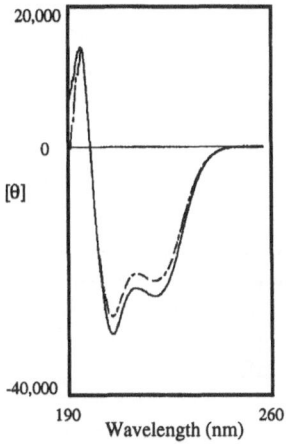

Fig. 2. The CD spectrum of Leu4 at 119 µM and 7 µM.

The apparent molecular weight of Leu4 was determined by size exclusion chromatography on Sephadex G-50. A calibration curve was obtained with known protein standards and the apparent molecular weight of Leu4 was determined to be 5432 by interpolation from the standard curve. By dividing this value by the actual molecular weight an aggregation state of 2.6 was obtained for Leu4.

Crosslinking experiments were performed to determine if Leu4 is in a parallel or antiparallel orientation within the dimer. Two peptides were synthesized containing the core sequence of Leu4, but with a Gly-Gly-Cys sequence added to the N-terminus and a Gly added to the C-terminus of peptide **1**, and a Gly-Gly-Cys sequence added to the C-terminus of peptide **2**. A 1:1 mixture of peptides **1** and **2** was equilibrated and the free cysteines were oxidized. If the peptides had no preference for a parallel or antiparallel arrangement a 2:1:1 statistical mixture of peptides **3**, **4**, and **5** would be obtained, respectively (Figure 3). Upon oxidation, however, a 1:2:2

mixture of peptides **3**, **4**, and **5** was obtained. In aqueous solution, therefore, 80% of the peptides exist as parallel dimers.

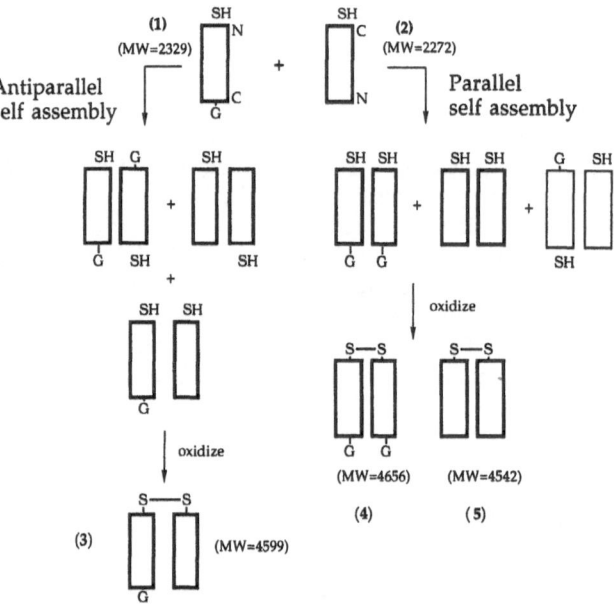

Fig. 3. Crosslinking experiments to determine the orientation of the peptide dimers in aqueous solution.

In conclusion we have prepared the smallest helical peptide reported to date which self assembles into both a two- and three-helix molecule in aqueous solution. A parallel orientation is adopted by 80% of the molecules as determined by crosslinking experiments. Further experiments are underway to design peptides which adopt either two- or three-helix motifs.

References

1. a) Landschulz, W.H., Johnson, P.F. and McKnight, S.L., Science, 240 (1988) 1759;
    b) O'Shea, E.K., Rutkowski, R. and Kim, P.S., Science, 243 (1989) 538;
    c) Adamson, J.G., Zhou, N.E. and Hodges, R.S., Curr. Opin Biotech., 4 (1993) 428.
2. a) Hunter, I., Schulthess, T. and Engel, J., J. Biol. Chem., 267 (1992) 6006;
    b) Sottile, J., Selegue, J. and Mosher, D.F., Biochemistry, 30 (1991) 6556;
    c) Lovejoy, B., Choe, S., Cascio, D., McRorie, D.K., Degrado, W.F. and Eisenberg, D., Science, 259 (1993) 1288.
3. Crick, F.H.C., Acta Crystallogr., 6 (1953) 689.
4. Chou, P. and Fasman, G.D., Ann. Rev. Biochem., 47 (1978) 251.
5. Marqusee, S. and Baldwin, R.L., Proc. Natl. Acad. Sci. U.S.A., 84 (1987) 8898.
6. Shoemaker, K.R., Kim, P.S., York, E.J., Stewart, J.M. and Baldwin, R.L., Nature, 326 (1987) 563.
7. Rink, H., Tetrahedron Lett., 28 (1987) 3787.
8. Lyu, P.C., Sherman, J.C., Chen, A. and Kallenbach, N.R., Proc. Natl. Acad. Sci. U.S.A., 88 (1991) 5317.

Parallel or antiparallel alignment of α-helical chains in two-stranded coiled-coils is controlled by interchain electrostatic interactions

O.D. Monera, N.E. Zhou, C.M. Kay and R.S. Hodges

Department of Biochemistry and the Protein Engineering Network of Centres of Excellence, University of Alberta, Edmonton, Alberta, T6G 2H7, Canada

Introduction

We have previously shown that a stable antiparallel coiled-coil can be formed from a pair of peptides that normally form parallel coiled-coils [1]. However, under benign conditions this antiparallel coiled-coil was not spontaneously formed in a significant amount. Therefore, the objective of this study was to determine what interactions are responsible for controlling parallel or antiparallel orientation of the α-helices in the coiled-coil motif.

Results and Discussion

Two-stranded coiled-coils were formed by disulfide-bond formation between two 35-residue peptides containing five heptad repeats of either the KgLaEbAcLdEeGf or EgLaAbEcLdKeGf sequence. Cysteine residues were included at either position 2 or 33 such that, after disulfide bond formation by air oxidation of appropriate peptide mixtures under benign conditions, 70-residue coiled-coils with either parallel or antiparallel orientation of α-helices were formed. Leucine at either position 16 or 19 of each chain was substituted with Ala in order to bring the stabilities to within the range of Gdn·HCl and urea denaturations. Thus, four disulfide-bridged coiled-coils were formed, namely, parallel coiled-coils with interchain electrostatic attractions (P/A) and repulsions (P/R), as well as antiparallel coiled-coils with interchain electrostatic attractions (A/A) and repulsions (A/R) (Figure 1). Figure 2 shows that under benign conditions the coiled-coils with interchain electrostatic attractions (Figure 2A and 2C) were preferentially formed while those with interchain electrostatic repulsions (Figure 2B and 2D) were not, both for parallel and antiparallel coiled-coils. These results suggest that the interchain electrostatic repulsions prevent the "coming together" of the two chains to form the heterostranded products. This interpretation is consistent with the observation that when the electrostatic repulsions were masked by addition of salts during air oxidation (Figure 2E and 2F) the heterostranded coiled-coils with interchain electrostatic repulsions also became the major products, as expected from random combinations. Circular dichroism spectra of the four disulfide-bridged heterostranded peptides under benign conditions showed that the ratio of the molar ellipticities at 220 nm to those at 208 nm was ~ 1.0, typical of two α-helical chains forming a coiled-coil [2]. In the presence of 50%

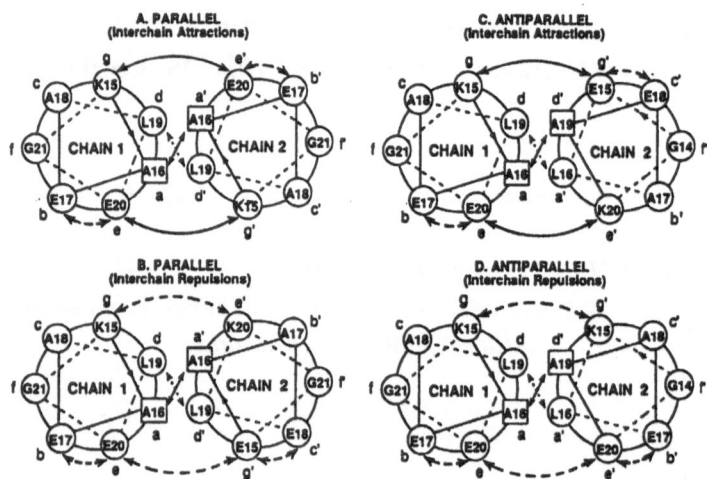

Fig. 1. Cross section diagrams of the third (middle) heptads of the disulfide-bridged coiled-coils showing interchain electrostatic repulsions (dashed arrows) and attractions (solid arrows). In the parallel coiled-coils the direction of chain propagation ($N{\to}C$) goes downward (into the paper for both chains). In the antiparallel coiled-coils, chain 1 goes downward (into the paper) while chain 2 goes upward (towards the reader). The pair of Ala residues are shown in boxes to indicate that they are above (towards the reader) the Leu pairs.

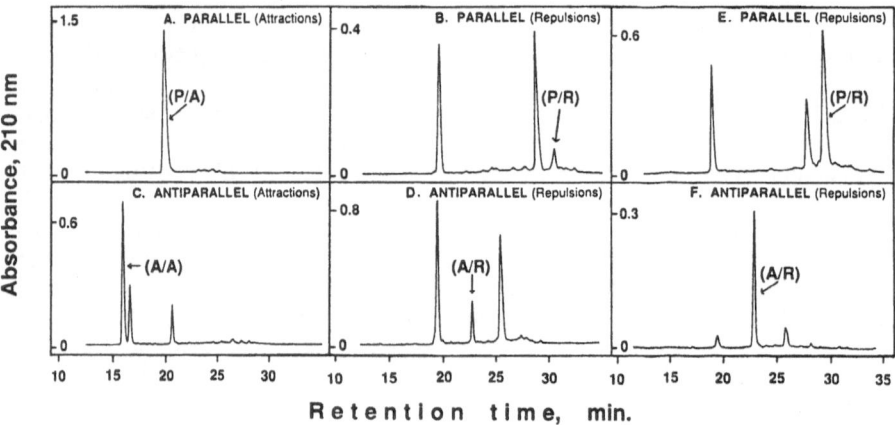

Fig. 2. Reversed-phase HPLC chromatograms of the disulfide-bridged products from the oxidation of a mixture of peptides in 100 mM NH_4HCO_3 pH 8.3 The desired products are highlighted while the other peaks correspond to the parallel homostranded coiled-coils that are formed as oxidized co-products of the reactions. Oxidation was allowed to proceed for about 24 hr by stirring the peptide mixture (0.5 mg of each peptide per ml) in 100 mM NH_4HCO_3, pH 8.3. All peaks correspond to disulfide-bridged (oxidized) products. In E and F the oxidations were carried out in the same buffer system in the presence of mixed salts (0.5 M $NaClO_4$/1M $MgCl_2$).

TFE, which disrupts tertiary and secondary structures, the 220/208 ratio decreased to ~0.9, which is characteristic of non-interacting α-helices [2]. These are strong indications that these disulfide-bridged heterostranded peptides exist as coiled-coils under benign conditions. The Gdn·HCl denaturation data (Table 1) showed that the antiparallel coiled-coils were more stable than the parallel coiled-coils but the type of interchain electrostatic interactions had no or very little effect on the stability. On the other hand, urea and thermal denaturation data showed that those with interchain electrostatic repulsions were less stable than the corresponding coiled-coils with interchain electrostatic attractions. This is probably because Gdn·HCl as a salt masks the interchain electrostatic interactions and therefore measures the stability due to hydrophobic interactions. On the other hand, the uncharged urea preserves the interchain electrostatic interactions and therefore the measured stability reflects the sum of the contributions of hydrophobic, ionic (stabilizing or destabilizing), and possibly other interactions [1].

Table 1 *Ellipticities and stabilities of the three coiled-coil peptide analogs*

Coiled-coil Alignment	Interchain Electrostatic Interactions	$[\theta]_{220}$[a] (deg·cm^2·dmol^{-1})		$[Gdn]_{1/2}$[b] (M)	$[Urea]_{1/2}$[b] (M)	t_m[c] (°C)
		Benign	50% TFE			
Parallel	Attractions	-30,450	-29,060	2.6	3.1	71
Parallel	Repulsions	-28,090	-24,700	2.6	0.8	57
Antiparallel	Attractions	-29,680	-27,600	4.0	4.2	71
Antiparallel	Repulsions	-30,100	-27,800	3.6	1.4	68

[a]$[\theta]_{220}$ is the calculated molar ellipticity of the coiled-coils at 220 nm; [b]The $[Gdn]_{1/2}$ and $[Urea]_{1/2}$ values represent the concentration of denaturant at which 50% of the peptide is unfolded; [c]t_m is the temperature at which 50% of the peptide is unfolded.

This study is the first to demonstrate conclusively that the favored alignment of α-helical chains in coiled-coils is the one that provides interchain electrostatic attractions between oppositely-charged amino acid residues in the "e-g' " and "g-e' " positions of the parallel coiled-coils and the "g-g' " and "e-e' " positions in the antiparallel coiled-coil. The antiparallel coiled-coils were more stable than the parallel coiled-coils, but the overall stability of either coiled-coils is either increased by the stabilizing effect of interchain electrostatic attractions or decreased by the destabilizing effect of interchain electrostatic repulsions. Thus, the order of stability of these coiled-coils was A/A>P/A>A/R>P/R.

References

1.   Monera, O.D., Zhou, N.E., Kay, C.M. and Hodges, R.S., J. Biol. Chem., 268 (1993) 19218.
2.   Lau, S.Y.M., Taneja, A.K. and Hodges, R.S., J. Biol. Chem., 259 (1984) 13253.

De novo design of a coiled-coil stem-loop peptide

D.G. Myszka and I.M. Chaiken

*Department of Molecular Genetics, SmithKline Beecham
Pharmaceuticals, King of Prussia, PA 19406, U.S.A.*

Introduction

We have been interested to use coiled-coil conformational motifs to present α-helical and loop recognition sequences from native proteins and peptides. We have made a 56-residue model polypeptide designed to fold into a stable coiled-coil stem-loop (CCSL) structure (an intramolecular coiled-coil joined at one end by a loop) as shown in Figure 1A. The antiparallel alignment of the coiled-coil is dictated by the sequence of the hydrophobic residues at the coiled-coil interface, which consists of leucine and valine residues in the helical wheel positions d-a' and a-d', respectively (Figure 1C). Glutamic acid and lysine residues are positioned across from each other on the helices in the e-e' and g-g' positions [1] to further align and stabilize the folded conformation through the formation of salt bridges (Figures 1A and C). Serine residues are placed on the exposed b, c and f positions of both helices to increase the solubility of the peptide (Figures 1A and C). The loop segment connecting the α-helices begins and ends with the helix breaking residues glycine and proline (Figure 1B).

Results and Discussion

The CCSL peptide exhibits a typical α-helical circular dichroism spectrum for a coiled-coil peptide in aqueous buffer with high molar ellipticity minima at 222 and 208 nm as shown in Figure 2. The helicity of the 56-residue polypeptide calculated from the observed ellipticity at 222 nm ($-28,000$ deg $cm^2\,dmol^{-1}$) is estimated to be 75% [2] and is independent of peptide concentration. The ellipticity minimum at 222 nm remains unchanged upon the addition of the helix inducing solvent trifluoroethanol (Figure 2) indicating that under these conditions (pH 7.4, 25°C) the peptide is near its maximum α-helical potential. The magnitude of the $[\theta]_{222}/[\theta]_{208}$ ratio in aqueous buffer is 1.02 suggesting the α-helices are stabilized in a coiled-coil conformation [3]. In 50% trifluoroethanol the ellipticity minimum shifts to 205 nm and the $[\theta]_{222}/[\theta]_{208}$ ratio changes to 0.90 indicating single-stranded α-helices [4, 5]. The temperature required to unfold 50% of the α-helical structure, T_m, is greater than 65°C. The gel filtration elution position of the CCSL peptide corresponds to the size of a monomer in solution. Together these results suggest that in aqueous solution the CCSL peptide forms a stable monomeric intramolecular α-helical coiled-coil structure.

The coiled-coil stem loop peptide is a single polypeptide chain that in aqueous solution spontaneously folds into a stable predictable conformation independent of

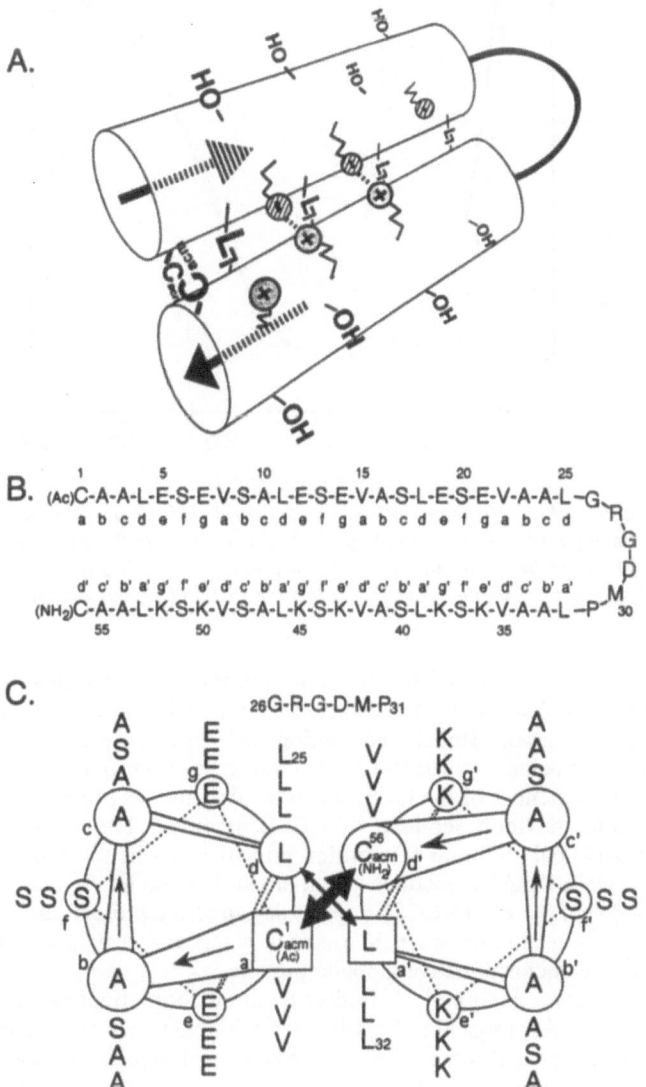

Fig.1. *A: Cartoon of the intramolecular coiled-coil stem-loop (CCSL) peptide. The side chains of the leucine, glutamic acid, lysine and serine residues are represented by L, (-), (+) and OH, respectively. The amino- and carboxy-terminal carboxyamido-cysteine residues are represented by Cacm. B: The 56-amino acid residue sequence of the CCSL peptide. Positions of the heptad repeat are denoted a to g for residues 1-25 and a' to g' for residues 32-56. A six residue loop occupying positions 26-31 joins the two helices. C: Helical wheel representation of the CCSL peptide viewed from the amino-terminus. Leucine residues at position d in the amino-terminal α-helix interact across the interface with residues at position a' in the carboxy-terminal helix.*

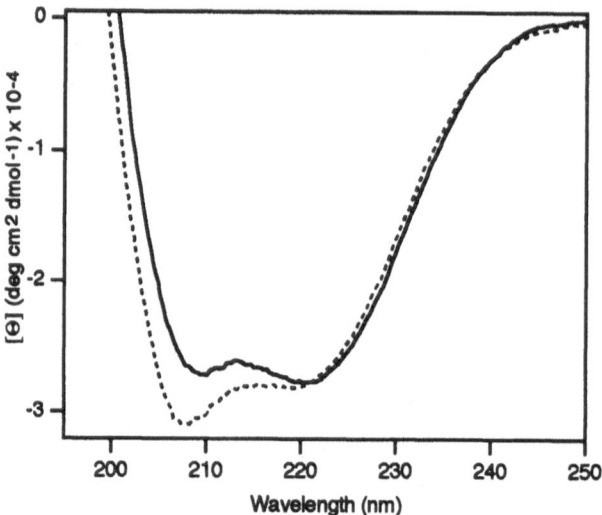

Fig. 2. Circular dichroism spectra of the CCSL peptide (20 μM) in the presence (dashed line) and absence (solid line) of the α-helix inducing solvent, trifluoroethanol (TFE). The buffer was 150 mM sodium chloride, 20 mM phosphate, pH 7.4 (PBS) and the spectra were taken at 25 °C. For the sample containing TFE, the PBS buffer was diluted with TFE (1:1, v/v).

peptide concentration. Therefore, this molecule represents an ideal template into which recognition elements from α-helical surfaces and continuous loop residues of proteins may be inserted. Helical recognition surfaces may be created by inserting the desired amino acids into selected heptad repeat positions *b, c, e, f* and *g* of either one or both of the helices of the CCSL peptide. Continuous protein sequences may be inserted into the loop segment of CCSL peptide. These conformationally constrained α-helical and loop recognition mimetics can be used to study protein-protein and protein-DNA interactions and to produce antibodies against conformation specific peptide epitopes. The CCSL peptide design also represents an ideal molecule for protein folding studies since it assumes a tertiary structure which mimics conformational elements of native globular proteins.

The *de novo* designed coiled-coil stem-loop peptide represents a conformationally stable sequence-simplified scaffold into which binding residues from surfaces of α-helices and loops of native proteins and peptides may be inserted to form mimetic recognition molecules.

References

1. Graddis, T. and Chaiken, I., In Smith, J.A. and Rivier, J.E. (Eds), Peptides: Chemistry and Biology, Escom, Leiden, The Netherlands, 1992, p.360.
2. Chen, Y.-H., Yang, J.T. and Chau, K.H., Biochemistry, 13 (1974) 3350.
3. Zhou, N.E., Kay, C.M. and Hodges, R.S., J. Biol. Chem., 267 (1992) 2664.
4. Cooper, T.M. and Woody, R.W., Biopolymers, 30 (1990) 657.
5. Hodges, R.S., Zhou, N.E., Kay, C.M. and Semchuk, P.D., Peptide Research, 3 (1990) 123.

De novo design of artificial membrane proteins on atropisomeric porphyrin

N. Nishino[1], H. Mihara[2], Y. Tanaka[1], K. Kobata[1] and T. Fujimoto[1]

[1]Department of Applied Chemistry, Faculty of Engineering,
Kyushu Institute of Technology, Tobata, Kitakyushu 804, Japan and
[2]Department of Applied Chemistry, Faculty of Engineering,
Nagasaki University, Nagasaki 852, Japan

Introduction

The *de novo* designs of conjugates of α-helical polypeptide segments with porphyrin ring have been carried out in an attempt to demonstrate various functions, such as enzyme model [1], stable membrane embedding [2], and ion channel [3]. In further evolution of membrane protein models, it is important that a more sophisticated porphyrin compound is designed and combined with appropriate peptide segments bearing other electro- and photo-chemical functions.

In the present study, we employed the bifunctional porphyrin compound, 5,10,15,20-tetrakis(2-amino-5-carboxyphenyl)porphyrin (aminoporphyrinic acid), whose α,α,α,α-atropisomer can combine two different peptide segments on each side of the ring system. An amphiphilic α-helical sequence was designed to be bundled on one side of the porphyrin ring and to help the whole molecule embedded in lipid bilayer membranes. The other side was filled with rather shorter peptides. The porphyrin ring may act as the reaction center of an artificial membrane protein in further design (Figure 1).

Results and Discussion

The aminoporphyrinic acid **1**, a tetraphenylporphyrin with both amino and carboxyl groups, was synthesized as follows. The reaction of 2-nitro-5-methoxycarbonylbenzaldehyde and pyrrole under the conditions of Lindsey [4] gave 5,10,15,20-tetrakis(2-nitro-5-methoxycarbonylphenyl)porphyrin in 37% yield. The nitro groups were reduced to amino groups by the aid of $SnCl_2$ under Rose's conditions [5]. The desired α,α,α,α-isomer of 5,10,15,20-tetrakis(2-amino-5-methoxycarbonylphenyl)porphyrin (aminoporphyrinic acid methyl ester) **2** was successfully separated by silica gel chromatography (benzene/acetone=3/1). The thermal stability of **2** was examined at 80°C in DMF. The half life of α,α,α,α-atropisomer was 4 h, and the activation free energy in atropisomerism was 116 kJ/mol. This value shows that the atropisomer is as stable as 5,10,15,20-tetrakis(2-aminophenyl)porphyrin. The saponification of **2** with NaOH in pyridine gave α,α,α,α-**1** [FAB-MS 851 (M+H)+].

The tetrapeptide, H-Leu-His(Bom)-Leu-Ser(Bzl)-NHCH$_2$CH$_2$OH, was attached to the carboxyl group of **1** with BOP/HOBt in DMF. The amino group was acylated with anhydride of Boc-Gly-OH in TFE/DCM (1/3). The octadecapeptide, Boc-Lys(ClZ)-[Glu(OcHex)-Leu-Leu-Lys(ClZ)-Leu-Leu-Leu-]$_2$-Glu(OcHex)-Leu-Leu-OH, was then condensed to the amino side of the porphyrin ring with WSC/HOBt in TFE/DCM (1/3). After passed through a Sephadex LH-60 column, the protected peptide-porphyrin conjugate was treated with liquid HF. The desired compound was purified by gel-filtration on Sephadex G-50. The mixture of the peptide-porphyrin conjugate with DPPC (50-fold excess) was sonicated to prepare vesicles, which was applied to gel-filtration on Sephadex G-75 (20 mM Tris HCl buffer, pH 7.5). The majority of the protein retained in the fractions containing the DPPC vesicles. This fact suggests that the protein may be stably embedded in the lipid bilayer membrane as illustrated in Figure 1. The spectral characterization of this membrane protein model is under investigation.

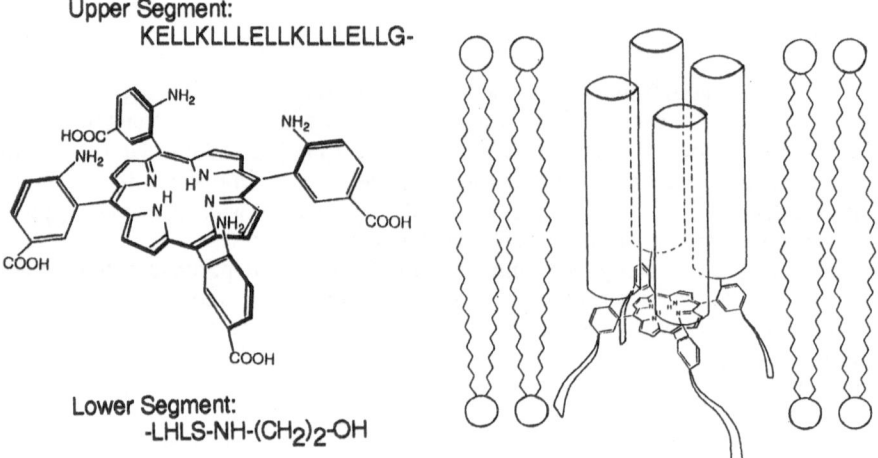

Fig. 1. *Molecular parts and assembled artificial membrane protein model.*

References

1. Sasaki, T. and Kaiser, E.T., J. Am. Chem. Soc., 111 (1989) 380.
2. Mihara, H., Nishino, N., Hasegawa, R. and Fujimoto, T., Chem. Lett., (1992) 1805.
3. Akerfeldt, K.S., Kim., R.M., Camac, D., Groves, J.T., Lear, J.D. and DeGrado, W.F., J. Am. Chem. Soc., 114 (1992) 9656.
4. Lindsey, J.S. and Wagner, R.W., J. Org. Chem., 54 (1989) 828.
5. Lecas, A., Boitrel, B. and Rose, E., Bull. Soc. Chim. Fr., 128 (1991) 407.

Synthesis and characterization of a three-heptad coiled-coil protein

J.E. Rozzelle, Jr. and B.W. Erickson

Department of Chemistry, University of North Carolina,
Chapel Hill, NC 27599-3290, U.S.A.

Introduction

A two-stranded coiled coil consists of two α helices packed against each other by the interaction of a hydrophobic stripe down one side of each helix. Its amino acid sequence is characterized by a repeating seven-residue pattern (abcdefg) in which the a and d residues are generally hydrophobic, the e and g residues are oppositely charged, and the exterior b, c and f residues are hydrophilic [1]. Hodges and co-workers [2] have shown that at least four such heptads per helix are required to form a stable coiled coil in aqueous solution. Interhelical disulfide bridges can stabilize a four-heptad coiled coil [3, 4]. We designed the 44-residue protein P44N to test if two interhelical disulfide bridges could stabilize a three-heptad coiled coil [5].

Results and Discussion

The reduced 22-residue chain of P44N (CDALKSKIATLEYKVASLEAKC) was assembled by the solid-phase method and purified by reversed-phase HPLC. Air oxidation gave a disulfide-bridged monomer, the disulfide-bridged parallel dimer P44N, and disulfide-bridged oligomers. Figure 1A shows that reduced P44N had

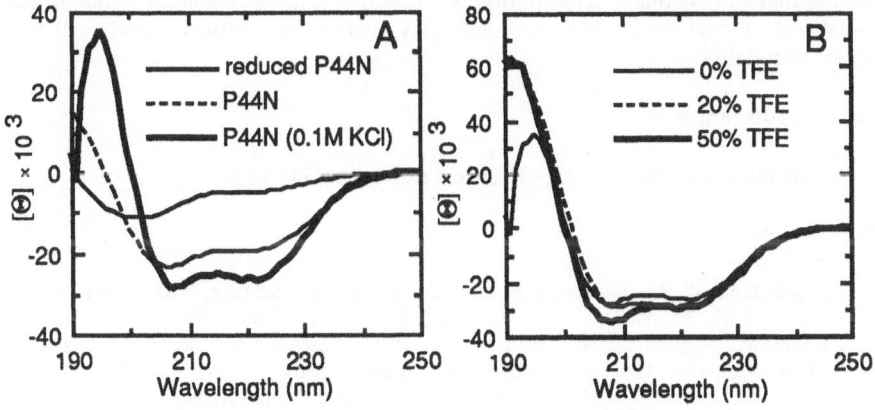

Fig. 1. CD spectra at pH 7 of (A) reduced P44N and of P44N with and without KCl and (B) P44N in 0.1M KCl and varying percentages of trifluoroethanol (TFE).

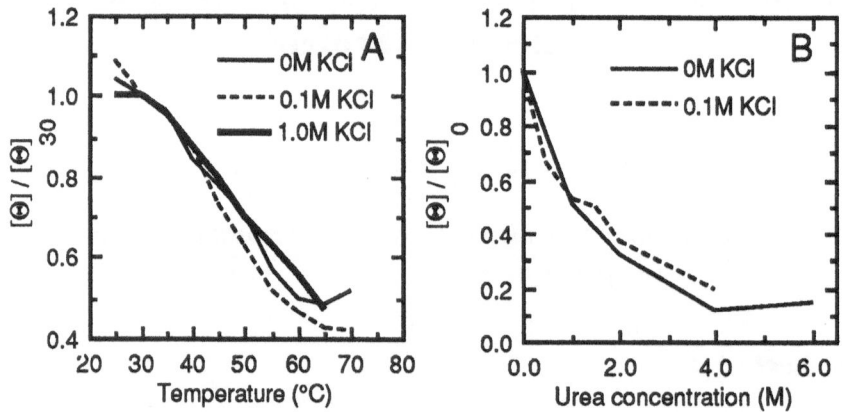

Fig. 2. Relative molar ellipticity at 222 nm of P44N with and without KCl (A) versus temperature and (B) versus urea concentration.

little α-helicity but the doubly disulfide-bridged chains of oxidized P44N were about 80% α-helix in phosphate buffer and 85% α-helix in 0.1 M KCl. Figure 1B shows that 20-50% trifluoroethanol did not significantly increase the α-helicity of P44N.

Figure 2A shows that the thermal denaturation of P44N in phosphate buffer containing 0, 0.1 or 1.0 M KCl did not show a significant increase in stability with increasing salt concentration. The melting temperature of each of these three solutions was near 45°C. A more precise value could not be determined because P44N usually precipitated at low temperatures. Figure 2B shows that denaturation of P44N with urea revealed only a small increase in stability with increasing salt concentration. P44N was half denatured by 0.5 M urea in either 0 or 0.1 M KCl.

These results with P44N show that two terminal disulfide bridges can stabilize the folded structure of a three-heptad coiled coil. Although P44N displayed protein-like thermal and chemical denaturation profiles, it was not as stable as a four-heptad coiled coil. Future design improvements may lead to three-heptad coiled coils with increased stability.

Acknowledgements

This work was supported by NIH research grant GM 42031.

References

1. Landschulz, W.H., Johnson, P.F. and McKnight, S.L., Science, 240 (1988) 1759.
2. Lau, S.Y.M., Taneja, A.K. and Hodges, R.S., J. Biol. Chem., 259 (1984) 13253.
3. Engel, M., Williams, R.W. and Erickson, B.W., Biochemistry, 30 (1991) 3161.
4. Zhou, N.E., Kay, C.M. and Hodges, R.S., Biochemistry, 32 (1993) 3178.
5. Rozzelle, J.E., Jr., Tropsha, A. and Erickson, B.W., Protein Science, in press.

Protein de novo design: Condensation of unprotected peptide blocks to topological templates via selective oxime bond formation

G. Tuchscherer[1], I. Ernest[1], K. Rose[2] and M. Mutter[1]

[1]Institute of Organic Chemistry, University of Lausanne, Rue de la Barre 2, CH - 1005 Lausanne, Switzerland
[2]Department of Medical Biochemistry, Medical Center, 1 Rue Michel Servet, CH - 1211 Geneva 4, Switzerland

Introduction

The attachment of peptide blocks to carrier molecules has become a most attractive concept for the construction of protein-like molecules. In particular, the covalent binding of secondary structure forming, amphiphilic oligopeptides to topological templates according to the TASP (Template Assembled Synthetic Proteins) approach [1, 2] represents a powerful tool for the de novo design of functional proteins. Here, we describe a novel strategy for the chemical synthesis of these branched molecules, making use of selective oxime bond formation [3, 4] of appropriately functionalized, side chain deprotected peptide blocks.

Results and Discussion

For the condensation of unprotected peptide fragments to template molecules via oxime bond formation, the two building blocks are functionalized by an aminooxyacetyl group and an aldehyde group as depicted in Figure 1. To this end, Boc-aminooxy acetic acid was coupled to the N-terminal amino acid of the growing peptide chain [H-Lys(Boc)-Arg(Pmc)-Asp(O^{t}Bu)-Ser(^{t}Bu)-resin]. After cleavage from the resin (50% TFA), the completely deprotected aminooxyacetylated peptide was purified by HPLC. In parallel, the attachment sites (ε-amino groups of lysine) on the template molecule [c(KGK-M-KGK-M); M= aminomethyl naphthoic acid as dipeptide mimetic] are transformed to aldehyde functions by reaction with glyoxylic acid 1,1 diethylacetal and subsequent hydrolysis. The condensation step of the water soluble peptide block with the template molecule is achieved under mild conditions at pH 5. Due to the kinetically stable oxime bond [3, 4], the selective attachment of the peptide via its N-terminus is achieved in good yields. The target TASP molecule (Figure 1) was purified by RP-HPLC and its chemical integrity was established by LDI-MS. Preliminary experiments using helical peptides of medium-sized chain length for the construction of 4α- helix bundle TASP molecules indicate [5] that the present methodology may overcome the problem of low solubility of fully protected peptides in convergent approaches in template-based protein de novo design.

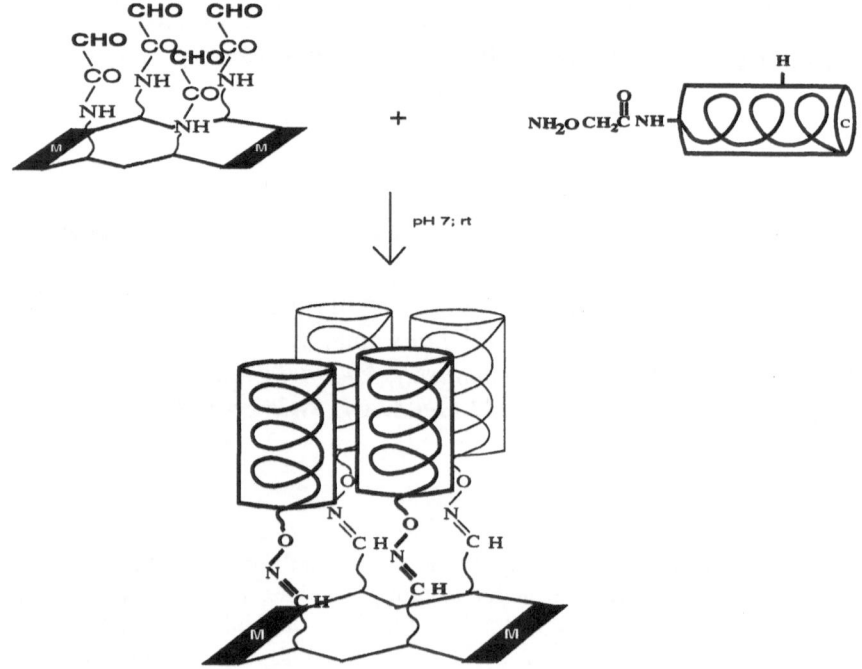

Fig. 1. *Synthesis of a TASP molecule using the oxime bond for the attachment of the peptide blocks to the topological template (see text).*

References

1. Mutter, M. and Vuilleumier, S., Angew. Chem. Int. Ed., 28 (1989) 535.
2. Mutter, M., Tuchscherer, G., Miller, C., Altmann, K.H., Carey, R., Wyss, D., Labhardt, A. and Rivier, J., J. Am. Chem. Soc., 114 (1992) 1463.
3. Rose, K., Vilaseca, L.A., Werlen, R., Meunir, A., Fisch, I., Jones, R.M.L. and Offord, R.E., Bioconj. Chem., 2 (1991) 154.
4. Fisch, I., Künzi, G., Rose, K. and Offord, R.E., Bioconj. Chem., 3 (1992) 147.
5. Tuchscherer, G., Tetrahedron Lett., 34 (1993) 8419.

The role of interchain ionic interactions in protein folding and stability of two-stranded α-helical coiled-coils

N.E. Zhou, C.M. Kay and R.S. Hodges

Department of Biochemistry and the Protein Engineering Network of Centres of Excellence, University of Alberta, Edmonton, Alberta, T6G 2H7, Canada

Introduction

Although hydrophobic interactions provide the major driving force in initiating protein folding and stabilizing protein structure, electrostatic interactions are also thought to play an essential role in molecular recognition, binding, catalysis, protein folding and the assembly of macromolecules. In this study, the role of interchain ionic interactions in protein folding and stabilization has been determined using a de novo designed two-stranded α-helical coiled-coil. This coiled-coil has been used extensively as an ideal model system for studying de novo design principles involving both intra- and inter-molecular interactions related to protein folding and stability (1-13). The two-stranded α-helical coiled-coil is characterized by a heptad repeat denoted as abcdefg, where positions a and d are normally occupied by hydrophobic residues and positions e and g by oppositely charged residues (Figure 1). The oppositely charged residues at positions g and e of adjacent heptads strongly favor the formation of interchain ion-pairs that would be expected to stabilize a parallel and in-register arrangement of the coiled-coil.

Results and Discussion

The model coiled-coil (denoted as EK) containing only interchain ionic interactions without any possible (i,i+3) and (i,i+4) intrachain ionic interactions (Figure 1) was designed. The double minima at 207 and 220 nm in the CD spectrum, the high ellipticity of -28,850 deg•cm^2•dmol^{-1} of $[\theta]_{220}$ in benign buffer (0.1 M KCl, 50 mM PO$_4$, pH 7) and no increase in helicity upon addition of the α-helix enhancing solvent TFE indicate that peptide EK is predominantly α-helical. The dimeric molecular weight of EK was determined by size-exclusion chromatography and sedimentation equilibrium experiments. These results indicate that peptide EK forms a stable coiled-coil structure in benign medium.

Two peptide analogs EE and KK containing only negatively charged Glu or positively charged Lys residues at both positions e and g were designed to exhibit interchain ionic repulsions in the homostranded coiled-coil (Figure 1). Each peptide shows little α-helical structure in benign buffer (Figure 2), whereas they are highly helical in the presence of 50% TFE, suggesting that the interchain ionic repulsions prevent the dimerization which stabilizes the helical structure in the coiled-coil by

EK: Ac-KCGALEK-KLGALEK-KAGALEK-KLGALEK-KLGALEK-amide
EE: Ac-ECGALEK-ELGALEK- EAGALEK-ELGALEK- ELGALEK-amide
KK: Ac-KCGALKE-KLGALKE-KAGALKE-KLGALKE-KLGALKE-amide

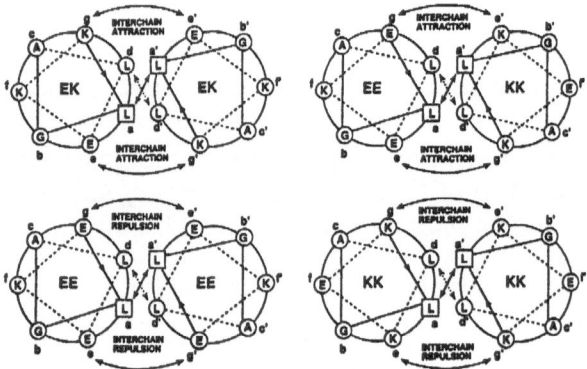

Fig. 1. Amino acid sequences of the synthetic peptides and the helical wheel representations of one heptad in the coiled-coils.

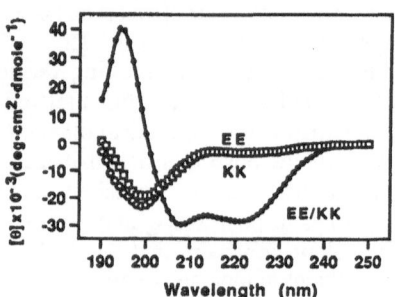

Fig. 2. CD spectra of the equimolar mixture of EE/KK and peptide EE or KK alone in 0.1 M KCl, 50 mM PO₄ buffer pH 7 at 20°C.

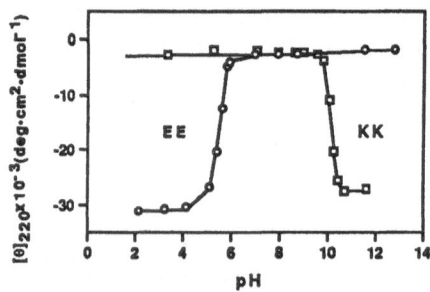

Fig. 3. pH dependence of the ellipticities of the EE and KK. All measurements were performed in 0.1 M KCl, 50 mM PO₄ buffer at 20 °C.

hydrophobic interactions. However, an equimolar mixture of these two peptides displays ~100% α-helical content under the same conditions (Figure 2). The major implication of this result is that formation of homo- and hetero-stranded coiled-coils can be controlled by the residues at positions e and g. A homodimeric coiled-coil constructed with Glu residues at e positions and Lys residues at g positions would produce interhelical ionic attractions to stabilize the homodimer. In contrast, placement of negatively charged residues in one helix and positively charged residues in the other helix at both e and g positions would result in the formation of a heterodimeric coiled-coil which is stabilized by interhelical ionic attractions while destabilizing the formation of homodimeric coiled-coils by interhelical ionic repulsions.

If a coiled-coil structure is prevented from forming by interchain ionic repulsions, protonation of the side-chains of Glu residues in peptide EE at acidic pH or deprotonation of the side-chains of Lys residues in peptide KK at basic pH would be expected to induce homostranded coiled-coil formation. The dramatic effects of pH on coiled-coil formation is demonstrated in Figure 3. At acidic and basic pH, the Glu and Lys side-chains are no longer ionized and interchain ionic repulsions would not exist which would allow the formation of the coiled-coils. The destabilizing effect of electrostatic repulsions on the formation of the coiled-coil can also be overcome by increasing salt concentration. However, the salt concentration required to induce the transition varied significantly among the different salts. In the case of peptide KK, ClO_4^- anions are more effective than Cl^- (1 M versus 3 M respectively required to induce maximum α-helix). For peptide EE, Mg^{2+} is more effective than K^+ cations. These results imply that ionic repulsions can be used advantageously in the de novo design of pH and salt sensitive proteins.

References

1. Talbot, J. and Hodges, R.S., Acc. Chem. Res., 15 (1982) 224.
2. Hodges, R.S., Current Biol., 2 (1992) 122.
3. Zhou, N.E., Zhu, B.-Y., Kay, C.M. and Hodges, R.S., Biopolymers, 32 (1992) 419.
4. Adamson, J.G., Zhou, N.E. and Hodges, R.S., Curr. Opin. Biotechnol., 4 (1993) 428.
5. Hodges, R.S., Saund, A.K., Chong, P.C.S., St-Pierre, S.A. and Reid, R.E., J. Biol. Chem., 256 (1981) 1214.
6. Hodges, R.S., Zhou, N.E., Kay, C.M. and Semchuk, P.D., Peptide Res., 3 (1990) 123.
7. O'Neil, K.T. and DeGrado,W.F., Science, 250 (1990) 646.
8. Zhou, N.E., Kay, C.M. and Hodges, R.S., J. Biol. Chem., 267 (1992) 2664.
9. Zhou, N.E., Kay, C.M. and Hodges, R.S., Biochemistry, 31 (1992) 5739.
10. Zhu, B-Y., Zhou, N.E., Kay, C.M. and Hodges, R.S., Int. J. Peptide Protein Res., 40 (1992) 171.
11. Zhu, B-Y., Zhou, N.E., Kay, C.M. and Hodges, R.S., Protein Science, 2 (1993) 383.
12. Zhou, N.E., Kay, C.M. and Hodges, R.S., Biochemistry, 32 (1993) 3178.
13. Monera, O.D., Zhou, N.E., Kay, C.M. and Hodges, R.S., J. Biol. Chem., 268 (1993) 19218.

Workshop I

Approaches and Advances in Peptide Synthesis, Purification and Analysis

Chairs: James P. Tam
Vanderbilt University
Nashville, Tennessee, U.S.A.

and

Jean Rivier
The Clayton Foundation Laboratories for Peptide Biology
The Salk Institute for Biological Studies
La Jolla, California, U.S.A.

Approaches to segment synthesis using unprotected or minimally protected peptide segments

C.F. Liu, C. Rao and J.P. Tam

Department of Microbiology and Immunology, A-5119 Medical Center North, Vanderbilt University, Nashville, TN 37232-2363, U.S.A.

Introduction

The use of the conventional segment condensation method with maximal protection strategy is limited by the low coupling efficiency and poor solubility of large, protected peptide segments. Different approaches intended to overcome these problems have been developed during the past decades [1, 2]. On the other hand, alternative approaches to the conjugation of unprotected peptides from synthetic or semisynthetic sources by surrogate peptide bonds such as hydrazone, thioether and thioester are receiving considerable attention [3, 4]. However, these approaches cannot give rise to a real peptide bond. Furthermore, the linkages in question usually are not as stable as the amide linkage in peptide chains. Enzymatic peptide synthesis by reverse proteolysis is a particular method of forming peptide bonds between unprotected peptide segments. While this method has been used mostly in the synthesis of small peptides, its application in the ligation of large peptide segments has rarely been successful. We recently developed a new, selective chemical ligation method, the domain ligation strategy, to form a peptide bond between two unprotected peptide segments. This paper outlines the steps in the design and development of this strategy.

Results and Discussion

Enzymatic coupling is the initial step in the development of our ligation strategy. For this purpose, we have developed a new class of resins (Figure 1) which allows the preparation of totally unprotected peptide esters by solid phase peptide synthesis. Since peptide esters are usually suitable substrates for kinetically controlled enzymatic peptide synthesis, they can, in principle, be used directly in enzymatic coupling to another peptide segment. We take advantage of the versatility of enzymatic coupling to effect the site-specific functionalization of an unprotected peptide segment. In this way, special functional groups can be introduced to the C-terminus of a peptide, which would specifically react with a second peptide that has a corresponding functionality on its N-terminus.

In our domain ligation strategy, an ester-aldehyde linkage should be introduced to the C-terminus of one peptide. This is made possible by enzyme-catalyzed coupling using a small amino component carrying such a linkage. The aldehyde group will react specifically with the N-terminal β-functionalized amino group of the

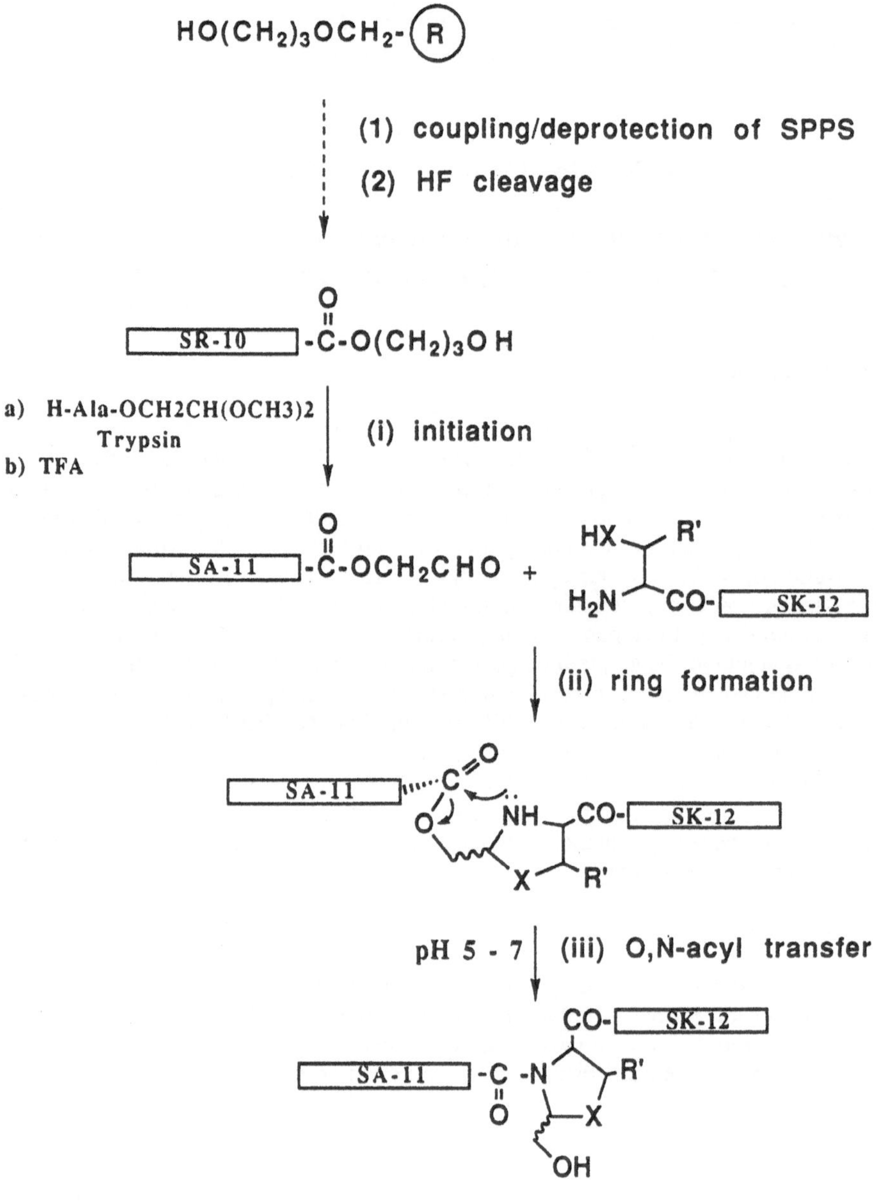

R' = H
X = S

SA-10 = SSQFQIHGPR
SK-12 = SDKHIEQYLKKK

Fig. 1. Domain Ligation Strategy - synthesis of a 24 residue peptide.

second peptide to form a ring product. The ester bond is located in a position that favors the subsequent amide bond formation through an O,N-acyl transfer reaction. As an example, a peptide of 24 residues was synthesized in high yield by the domain ligation strategy through the coupling of two segments (11+13) (Figure 1). The first 11-mer peptide was prepared by coupling a decapeptide, SR-10 hydroxypropyl ester, with alanine dimethoxyethyl ester catalyzed by trypsin in 60% DMF/H$_2$O. The obtained 11-mer with the acetal group was deprotected with TFA (5% H$_2$O) to release the aldehyde function which reacts rapidly with the N-terminal cysteine residue of the second 13-mer peptide to form a thiazolidine ring. An amide bond then forms between these two segments through the acyl rearrangement reaction under very mild conditions (pH 5-7). It is noticeable that the three steps - acetal deprotection, ring formation and rearrangement - were conducted in one reaction vessel and no intermediate purification step was needed.

In conclusion, the domain ligation strategy is a highly selective chemical ligation method to form peptide bonds between unprotected segments in aqueous solution. This strategy combines advantages of the known approaches in peptide chemistry. First, it enjoys the efficiency of the solid phase peptide synthesis approach for the preparation of large, unprotected and water-soluble peptide segments. Second, it utilizes the specificity of an enzyme for site-specific modification of the C$^\alpha$-carboxylic group. Finally, it captures the essence of the segment coupling strategy in solution synthesis by ligating purified segments to avoid chemical ambiguity. In theory, it is also applicable to the ligation of peptides or proteins produced by recombinant technology.

Acknowledgements

This work was in part supported by grants from USPHS AI 28701 and CA 36544.

References

1. Blake, J., Int. J. Pept. Protein Res., 27 (1986) 191.
2. Fotouhi, N., Galakatos, N.G. and Kemp, D.S., J. Org. Chem., 54 (1989) 2803.
3. Rose, K., Vilaseca, L.A., Werlen, R., Meunier, A., Fisch, I., Jones, R.M.L. and Offord, R.E., Bioconjugate Chem., 2 (1991) 154.
4. Schnolzer, M. and Kent, S.B.H., Science, 256 (1992) 221.

Recent progress on handles and supports for solid-phase peptide synthesis

G. Barany[1] and F. Albericio[2]

[1]Department of Chemistry, University of Minnesota, Minneapolis, MN 55455, U.S.A., and [2]Millipore Corporation, 75A Wiggins Avenue, Bedford, MA 01730, U.S.A.

One important aspect to achieving milder and/or more versatile chemical methods for solid-phase peptide synthesis is to specify the mode of attachment (*anchoring*) of the terminal residue to a polymeric *support* (resin). *Handles* are defined as bifunctional spacers, or linkers, which incorporate on one end features of a selectively removable protecting group and contain a second end which serves to achieve the required anchoring as a separate chemical step (Figure 1). This workshop presentation reviews a number of highly useful, commercially available, handles which are compatible with readily removable N^α-amino protecting groups such as the acid-labile *tert*-butyloxycarbonyl (Boc) or the base-labile 9-fluorenylmethyl-oxycarbonyl (Fmoc) functions; some very recent advances brought to the fore at this Symposium are also highlighted (see Figure 2 for structures of many but not all of the handles and/or anchoring linkages described in the following text; leading references can be tracked down by referring to names of researchers and the following general review [1]).

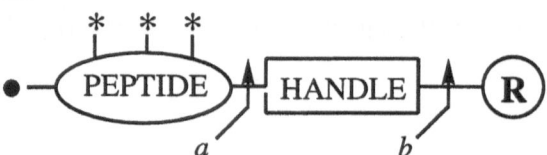

*Fig. 1. Protected (•"temporary", * "permanent" groups) peptide anchored to resin support ® via a handle. Stepwise or segment assembly to elongate the peptide chain occurs after anchoring of (usually) a protected C-terminal residue. Most commonly, linkage a is created after forming bond b which links the handle to the support. In the "preformed" handle approach, the purified C-terminal residue-handle conjugate (bond a created first) is then connected to the support (bond b created second).*

Cleavages of the appropriate anchoring linkages occur under relatively mild (in some cases *orthogonal*) conditions upon exposure to acid, base, light, fluoride ion, or palladium (0) in the presence of nucleophiles. Depending on the experimental design, the products can be peptide acids, thioacids, amides, or hydrazides; either completely free or else retaining N^α-amino and/or side-chain protection.

For the Boc/Bzl protection scheme of solid-phase synthesis, the classical *p*-alkylbenzyl esters for anchoring of peptide acids have been superseded by PAM anchors (optimally prepared in the preformed mode). For the formation of C-terminal peptide amides, MBHA resins are used. The MBHA concept involves anchoring of a C-terminal protected amino *acid* in an amide-forming reaction; the linkage produced is

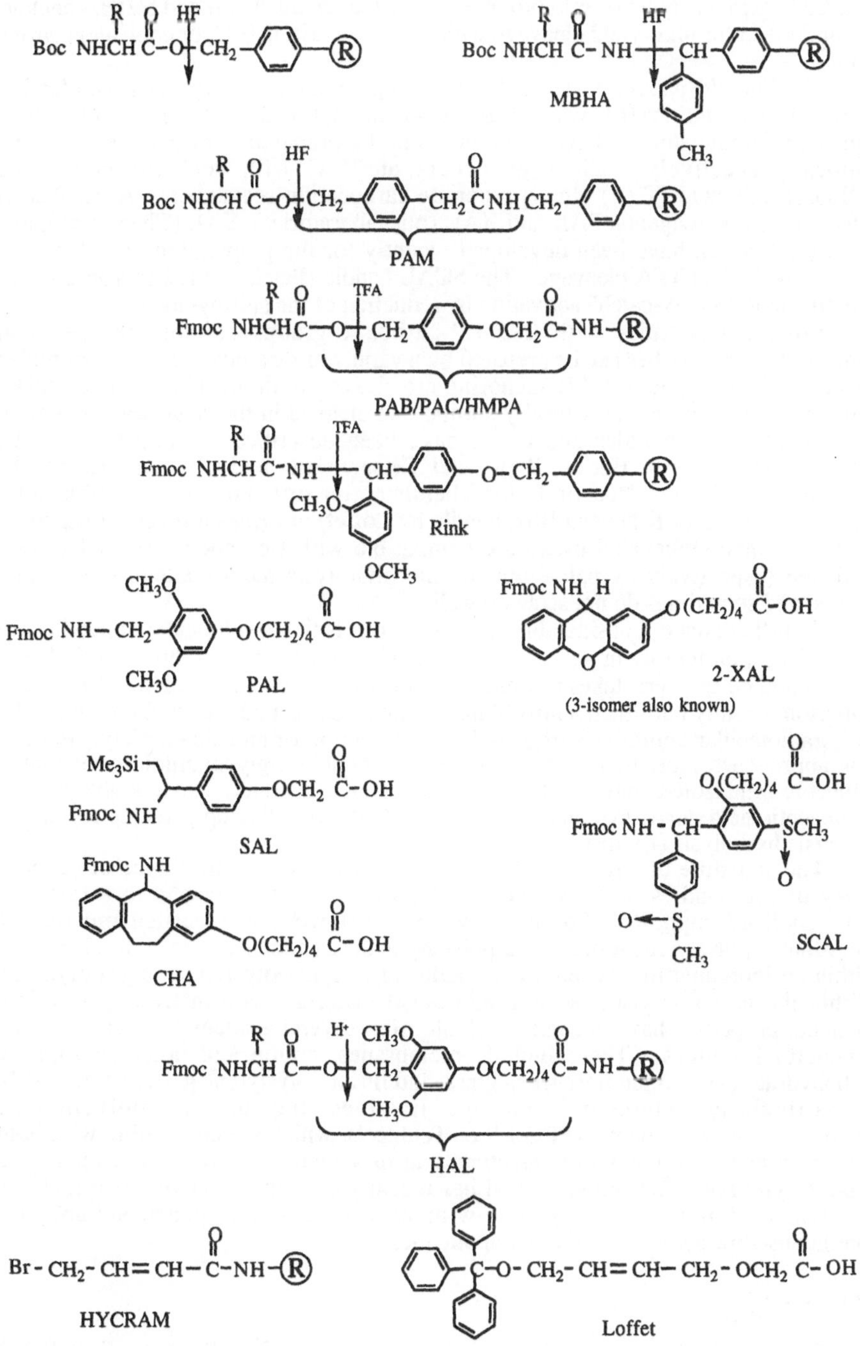

Fig. 2. Structures of some handles, anchoring linkages, and/or resin supports for solid-phase peptide synthesis.

cleaved in acid one bond over to provide the desired *amide*. PAM and MBHA anchors are optimized for high-yield cleavage with anhydrous HF at 0°C, or equivalent strong acids.

For Fmoc/*t*Bu solid-phase synthesis, Wang's *p*-alkoxybenzyl esters (preferably as handle variants PAB/PAC/HMPA shown in Figure 2) and Rink's substituted benzhydrylamine support have made possible the preparation of peptide acids and amides, respectively. Cleavage occurs at 25°C with cocktails containing trifluoroacetic acid (TFA) plus appropriate carbocation scavengers. A number of effective handle reagents: PAL and XAL (our laboratories), SAL (Chao *et al.*), and CHA (Nokihara), have been developed recently for the preparation of *C*-terminal amides with final TFA cleavage. The SCAL handle (Pátek and Lebl) is acidolyzed subsequent to "safety-catch" activation by reduction of the sulfinyl moieties.

Protected peptides with free C^α-carboxyl groups suitable for segment condensation approaches can be prepared by a variety of methods. Peptides assembled by Fmoc chemistry with HAL anchoring are cleaved in dilute TFA. The SASRIN and Riniker handles are structurally similar and cleaved in the same way, and acid-labile trityl-based handles and resins have been described recently through the independent efforts of Barlos, Rapp, and Tesser. Alternatively, peptides can be assembled by *either* Boc or Fmoc chemistry on orthogonally cleavable allyl (HYCRAM-resin of Kunz and Birr; handle by Loffet) or *ortho*-nitrobenzyl supports; Wang's α-methylphenacyl esters are compatible with Boc chemistry. Cleavages mediated respectively by palladium (0) and photolysis require attention to subtle details, and sometimes do not scale up well.

A further area of considerable current interest is the development of handles for mild release, preferably in aqueous milieus. The first success in this regard, due to Geysen and co-workers, takes advantage of diketopiperazine formation. Osborn and Robinson recently published a novel linker which cleaves under neutral conditions by an intramolecular aminolysis triggered after reduction of an azide moiety. Elegant new approaches presented at this Symposium include a glycolamidic ester that is subject to imidazole-catalyzed cleavage (Hoffmann and Frank), and a double-headed iminoacetic acid derivative that is cleaved through both diketopiperazine formation and basic hydrolysis (Krchnak *et al.*).

The signature of any solid-phase synthetic endeavor is the choice of support. Many of the handles described herein (Figure 2) can be attached to a range of functionalized supports. For many years, most work was carried out on 1% cross-linked polystyrene resins, or on polyacrylamides (these latter could be embedded within an inorganic matrix, e.g., kieselguhr, or a rigid polystyrene, e.g. Polyhipe). Within the past few years, several additional materials with interesting physico-chemical properties have become available from several academic laboratories and commercial sources. These include membranes, cotton and other appropriate carbohydrates, controlled-pore silica glass, and linear polystyrene grafted onto Kel-F. A particularly interesting concept involves the use of polyethylene glycol-polystyrene supports (PEG-PS or Tentagel), which are compatible with both batchwise and continuous-flow reactors, and may facilitate difficult chemistries in peptide synthesis. Relatedly, Meldal has recently described a novel resin that is a copolymer of polyethylene glycol with acrylamides, and hence suitable for continuous-flow synthesis and other applications.

References

1. Fields, G.B., Tian, Z. and Barany, G., In Grant, G. (Ed.), Synthetic Peptides: A User's Guide, W.H. Freeman & Co., New York, U.S.A., 1992, p.77.

Total syntheses

**J.E. Rivier, A.G. Craig, E. Mahe, A. Rabinovich,
D. Kirby, R. Kaiser, J. Porter, M.T.M. de Miranda,
A. de Miranda, G.-C. Jiang, D. Pantoja, L. Cervini,
S.L. Lahrichi, T. Goedken, W. Low, C. Miller, J. Dykert
and C. Hoeger**

*Clayton Foundation Laboratories for Peptide Biology, The Salk
Institute, 10010 N. Torrey Pines Road, La Jolla, CA 92037, U.S.A.*

Introduction

The expression "Total Synthesis" has been reserved for the description of the synthesis of native molecules in general and peptides in particular. Implicit was the fact that the synthetic replica should not only have all the physico-chemical properties of the native peptide but also all of its recognized biological activities. It was therefore expected that such synthetic peptides be (i) particularly well characterized, (ii) available in quantities that were considerably greater than those obtained by extraction and (iii) highly purified.

Most peptides are synthesized by one of four strategies: solution phase, solid phase, enzymatic or *in vitro* expression. Our laboratory has concentrated its efforts on the solid phase approach of Merrifield and on the development and enhancement of both new and well-established analytical and separation techniques which we will summarize briefly.

Results and Discussion

Synthesis. Successful SPPS has been carried out at room temperature [1]. Yet, if all steps of the synthesis could be carried out at elevated temperatures, reaction times could be shortened dramatically and it was hypothesized that poor couplings resulting from structural organization of the growing, fully protected chain would be improved. In a recent paper [2], we describe detailed protocols (Boc and Fmoc) that are compatible with high temperature with few deleterious consequences. Cycle time of 3 hr was shortened to 0.5 hr at 75°C. Although only DIC/HOBT was used as the coupling agent, it is anticipated that most coupling reagents including amino acid active esters would be compatible with high temperatures.

Purification. Purification of peptides in quantities greater than 50 mg is routinely achieved in our laboratory using RP-HPLC in two different solvent systems based on triethylammonium phosphate (TEAP) at different pH values [3, 4] and TFA [5, 6]. It was found that each system demonstrated differing selectivities on peptide mixtures [7]. Several hundred analogs of gonadotropin releasing hormone (GnRH), growth hormone releasing hormone (GHRH), corticotropin releasing factor (CRF) and neuropeptide Y (NPY) have been purified in this manner. Under careful examination (using the analytical methods described below), most of these analogs were greater than 95% pure.

For longer peptides such as rat histone H2A (1-53)NH$_2$ [8], CCK-58 analogs [9, 10] and TASP molecules [11], a strategy using ion-exchange chromatography followed by RP-HPLC was found to be extremely useful; however the degree of purity of the final preparations was difficult to fully evaluate or quantitate because of the complexity of the original mixture.

Characterization. Techniques used for the characterization of synthetic peptides have improved tremendously (gains in sensitivity and resolution allow more precise quantitation of the different components of a mixture). The last eighteen years has seen the chronological development of RP-HPLC systems compatible with quantitative elution of the peptides; this followed by the coming of age of ion exchange (IE)-HPLC, capillary zone electrophoresis (CZE), coupled with advances in both NMR and mass spectrometric (MS) techniques has permitted a more complete characterization of peptides to become a reality.

A. RP-HPLC: While most instrumentation (pumps, automatic injectors, detection systems) is now reliable, such a component as a diode-array detector with on-the-fly collection of full UV-visible spectra (190 to 600 nm) of the column effluent adds a new dimension to structure analysis. This diode-array detector is extremely useful in that it often distinguishes peptide absorbencies from reagents or contaminants, and native peptides from derivatized peptides. It also identifies peptides which contain Trp, Phe and Tyr, and allows the detection and identification of some unusual amino acids and/or peptides which contain them, such as phosphotyrosine or halogenated tryptophans. This capability saves a great deal of time by identifying absorbencies of interest and eliminating the need to further investigate HPLC fractions which are of little or no interest.

Stationary phases should be selected on the basis of their:
- ability to give quantitative recovery of the injected peptides (influential parameters include the nature of the base silica, capping and pore sizes, among others);
- selectivity (C$_{18}$ versus diphenyl for example);
- resolutive power (column length and diameter, particle size and temperature);
- resistance to aggressive buffers (polymer based versus silica based supports);
- ability to give reproducible results.

Mobile phases should be selected on the basis of their:
- compatibility with stationary phases to yield quantitative recovery;
- compatibility with the physical stability of the stationary phase;
- high transmittance at detection wavelength (preferably 210 nm and above);
- compatibility with bioassays in case the eluted fractions need to be tested;
- ability to span a wide pH range to increase selectivity, recovery and resolution. For example, very acidic peptides (such as gastrin) cannot be properly analyzed at acidic pH on a C$_{18}$ column. We have found that 0.1% TFA and triethylammonium phosphate (TEAP) buffers at pHs 2.25 to 7.5 can be extremely useful for the separation of peptides with different pI values [4].

Flow rate, gradient shapes and temperature can be optimized for particular separations.

B. Size Exclusion HPLC (SEC): SEC is used for the characterization of synthetic peptides. It allows identification and quantitation of polymeric materials (this is an important feature when analyzing sulfhydryl/disulfide containing peptides) and should be used in selected cases instead of TLC which is considerably less resolutive than RP-HPLC and does not allow easy quantitation. In other instances, it can be used to determine the apparent molecular weights of peptides and proteins [12].

C.   *Ion exchange chromatography (IEC) using FPLC:* Analytical IEC allows confirmation of the presence and quantitation of impurities often also detected by CZE and not always observed by RP-HPLC. One of the major strengths of this technique is the ease of scale up for preparative purification [8-11].

D.   *Capillary Zone Electrophoresis (CZE):* For its ability to separate very closely related compounds, CZE is unsurpassed. This is an extremely resolutive technique that allows some separations that are not otherwise seen by RP-HPLC [13]. Separations are based on charges on the peptide and influenced by buffer pH and composition. Adsorption (ionic or hydrophobic) chromatography on the wall of the capillary, will provide further avenues for optimization.

E.   *Thin Layer Chromatography (TLC):* TLC is carried out on a large number of supports. The fact that these supports are generally available in high performance columns, makes this technique somewhat obsolete. However TLC will allow detection of components that may not elute from HPLC columns (such as polymeric materials in the case of sulfhydryl/disulfide containing peptides).

F.   *Amino Acid Analysis (AAA), composition, peptide content and diastereomeric purity:* AAA of hydrolyzed samples can be reliably carried out using either post-column derivatization with ninhydrin or OPA or pre-column derivatization. Both methods have their advantages and disadvantages as the former uses IEC for the separation of the different AAs and the latter RP-HPLC for the separation of amino-protected AAs. AAA will yield, within 10% confidence limits in most cases, an estimate of the relative abundance of each amino acid in a peptide. With this information and the use of an internal standard, peptide content can be determined. Finally, the diastereomeric purity of a synthetic peptide can be checked by reacting the peptide hydrolyzate with OPA and a chiral sulfhydryl compound (3S-neomenthylthiol). The resulting diastereomers can be separated by RP-HPLC [14].

G.   *Identification and quantitation of counterions:* The determination of counterions in peptide chemistry is necessary when peptide content is to be obtained. Determination of acetate and trifluoroacetate counterions is achieved using RP-HPLC [15].

H.   *Water content and CHN analysis:* Water analysis is obtained using the Karl Fischer method and CHN analysis determined by outside analytical laboratories. Peptide, counter-ion and water contents should all match to yield the observed CHN composition [15].

I.   *Sequence analysis:* Sequence analysis is particularly important to check the primary structure of a peptide. The analyses when carried out at one or two nanomole level on a micro-sequencer will suggest failure sequences which are indicative of poor peptide assembly. The technique however is prone to the same drawbacks as those found during peptide assembly and careful analysis of the data is necessary. If the presence of any impurity is suggested by sequence analysis, it is mandatory that it be confirmed by a chromatographic technique of the whole sequence or of isolated fragments generated by partial enzymatic digestion.

J.   *Optical Rotation:* Optical rotation of a particular peptide is a physical constant that may provide a clue as to its optical integrity.

K.   *Mass spectrometry:* Liquid Secondary Ion Mass Spectrometry (LSIMS) or Laser Desorption Mass Spectrometry (LDMS) are indispensable tools for peptide/protein characterization. These techniques allow sequence analysis of peptides up to M.W. 2000 Da. The sensitivity, versatility and, in the case of LDMS, insensitivity to contaminants makes the latter extremely valuable for the screening of biological samples while the accuracy of LSIMS will enable distinction between a peptide free acid from its corresponding amide [10].

L. *Circular Dichroism and NMR:* Circular dichroism (CD) and nuclear magnetic resonance (NMR) are techniques that provide structural information; NMR, additionally can be used to check purity (>5%).

While this paper makes reference to a number of synthetic and analytical techniques used in our laboratory, it is clear that many other protocols for peptide synthesis have been successfully used and that other purification techniques (such as one that would include an affinity chromatographic step) are very powerful. We are also fully cognizant of the fact that it is important for certain investigations (structural and biological) to have peptides as pure as possible, while libraries of less defined peptides can also make valuable contributions in screening programs.

Acknowledgements

Research supported by NIH grants DK-26741, HD-13527, HL-41910, GM 22737, NIH contracts NO1-HD-3-3171 and NO1-HD-0-2906 and the Hearst Foundation.

References

1. Barany, G. and Merrifield, R.B., In Gross, E. and Meienhofer, J. (Eds.), The Peptides, Analysis, Synthesis, Biology, Academic Press, N.Y., U.S.A., 1980, p.1.
2. Rabinovich, A.K. and Rivier, J.E., In Smith, J.A. and Hodges, R.S. (Eds.), 13th American Peptide Symposium, 1993, ESCOM Science Publishers B.V., Leiden, The Netherlands, this volume.
3. Rivier, J., J. Liq. Chromatog., 1 (1978) 343.
4. Hoeger, C., Galyean, R., Boublik, J., McClintock, R. and Rivier, J., Bio-chromatography, 2 (1987) 134.
5. Rivier, J., McClintock, R., Galyean, R. and Anderson, H., J. Chromatog., 288 (1984) 303.
6. Rivier, J. and McClintock, R., J. Chromatog., 268 (1983) 112.
7. Rivier, J., Spiess, J. and Vale, W., Proc. Natl. Acad. Sci. U.S.A., 80 (1983) 4851.
8. Miller, C., Hernandez, J.-F., Craig, A.G., Dykert, J. and Rivier, J., Anal. Chim. Acta., Elsevier Science Publishers, Amsterdam, The Netherlands, 1991, p.215.
9. Miranda, M.T.M., Liddle, R.A. and Rivier, J.E., J. Med. Chem., 36 (1993) 1681.
10. Miranda, M.T.M., Craig, A.G., Miller, C., Liddle, R.A. and Rivier, J., J. Prot. Chem., 12 (1993) 533.
11. Rivier, J., Miller, C., Spicer, M., Andrews, J., Porter, J., Tuchscherer, G. and Mutter, M., In Epton, R. (Ed.), Innovation & Perspectives in Solid Phase Synthesis, Peptides, Polypeptides & Oligonucleotides, SPCC (U.K.) Ltd., Birmingham, University of Oxford, United Kingdom, 1990, p.39.
12. Grove, A., Mutter, M., Rivier, J.E. and Montal, M., J. Amer. Chem. Soc., 115 (1993) 5919.
13. Kirby, D.A., Miller, C.L. and Rivier, J.E., J. Chromatog., 648 (1993) 257.
14. Gotz, H. and Glatz, B., Hewlett-Packard Application Note; Publication number 12-5952-0029, (1989).
15. Hoeger, C., Theobald, P., Porter, J., Miller, C., Kirby, D. and Rivier, J., In Conn, P.M. (Ed.), Methods in Neurosciences, Academic Press, Orlando, Florida, 1991, p.3.

Multicyclic cystine peptides: A new method for disulfide analysis

W.R. Gray

Department of Biology, University of Utah,
Salt Lake City, UT 84112, U.S.A.

Introduction

Biologically active peptides are often cross-linked by disulfides, and their activity depends absolutely on the correct pairing. Analyzing the pattern can be extraordinarily difficult when cystines are clustered, because enzymes fail to break the peptide into definitive fragments: prolonged digestion may preferentially attack minor amounts of rearranged peptide, leading to wrong answers. An alternative approach using partial reduction to generate peptides having a subset of bonds opened [1] is complicated by thiol-disulfide exchange. Tris-[2-carboxyethyl]-phosphine (TCEP, [2]), reduces peptides at pH 3, so that exchange is minimal. Products are then separated by HPLC at pH 2, again without significant disulfide exchange. After free thiols have been alkylated, sequencer analysis is used to identify the original pairing. Some exchange may occur during alkylation, but the unrearranged product can be clearly identified. The method has worked well for many difficult test cases [3]. It is illustrated for endothelin, a peptide which is subject to rapid exchange, but for which the bridges could be unambiguously assigned.

Results and Discussion

Endothelin (C.S.C.S.S.L.M.D.K.E.C.V.Y.F.C.H.L.D.I.I.W.OH) is a 21-peptide [4] with disulfides linked [1-15; 3-11]. Figure 1a shows an HPLC elution profile of the products of partial reduction of endothelin (10 mM TCEP in 0.09 M citrate, pH 3.0, 2 min, 20°C). As was found to be characteristic of many peptides, intermediates in the reduction process elute between native (N) and fully reduced (R) molecules. Atypically, there are more intermediate peaks than the expected two, due to thiol-disulfide exchange. When reduction is carried out at pH 5, faster exchange results in an almost complete loss of peak B, with enhancement of C and D, while peak A is relatively unaffected. This suggests that A and B are the primary reduction products, while C and D arise by exchange from B.

Peptides A and B also differ greatly in their behaviour on alkylation. Even with relatively mild conditions, A gave a single major product on HPLC, and sequencer analysis showed clean labelling of Cys1 and Cys15 (Figure 2a), indicating that it was one of the expected isomers [1, 15; 3-11]. On the other hand, only the most forcing conditions were successful for peptide B. Treatment with 60 mM 4-vinylpyridine at pH 7 gave two main products, both of which were incorrectly labelled (on Cys1 and Cys11; and Cys1 and Cys15)! Best results have been obtained for B and other

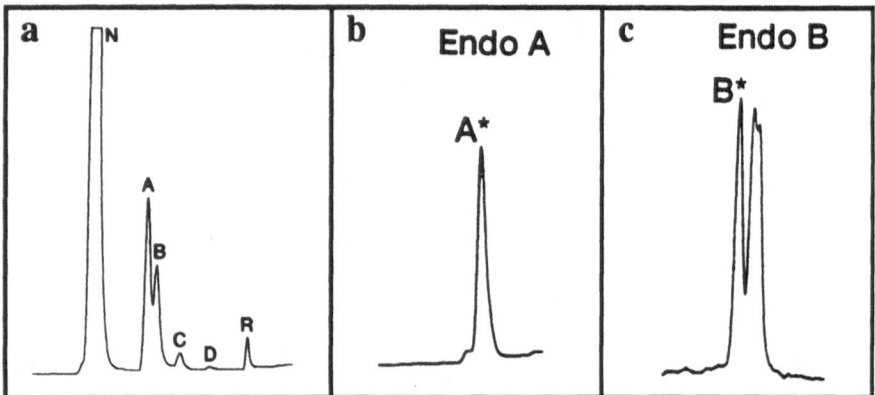

Fig. 1. Elution profiles of endothelin reduction products, and their carboxamido-methyl derivatives. Vydac C18, #218TP54 column; buffer A, 0.1% TFA in H_2O; buffer B, 0.1% TFA in 60% CH_3CN. Panel a, endothelin + 10 mM TCEP, pH 3, 2 min., 20°C (30-70% B/20 min.). Panels b & c, alkylation of peptides A and B by iodoacetamide (45-60% B/30 min.).

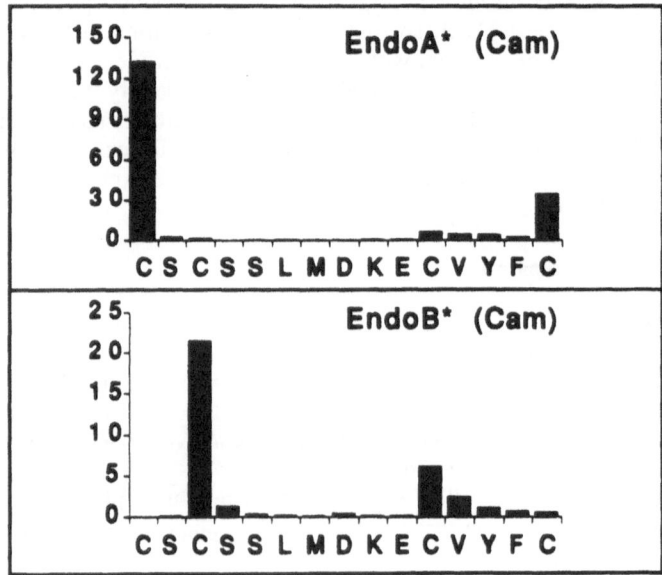

Fig. 2. Sequencer analysis of carboxamidomethylated peptides from peaks A* and B* (Fig. 1b and 1c). Ordinates represent yield (pmol) of PTH-Cys(Cam) at each cycle of the analysis.

unstable intermediates, by injecting the peptide (in 10 mM TFA from HPLC elution) into a rapidly stirred solution of iodoacetamide (2.2 M in 0.5 M tris at pH 8). The latter, prepared by heating 100 mg of iodoacetamide in 200 µl of the buffer, is supersaturated and must be used immediately after cooling. With endothelin B, this procedure resulted in appearance of a new peak in about 35% yield (Figure 1b). Sequencer analysis (Figure 2b) showed that this correctly represents the product expected for the second partial reduction isomer [1-15; 3, 11].

Partial reduction of peptides by TCEP has been applied to the problem of disulfide analysis. As illustrated for endothelin, peptides containing both thiols and disulfides can be produced and purified at low pH with minimal exchange. Rearrangement is more of a problem during alkylation, but it can be recognized and mitigated. Isomers A and B are very different in this regard, with B being among the most unstable peptides encountered. The method has been verified [3] with a variety of other difficult model peptides such as insulin, diverse conotoxins, and bacterial enterotoxin. It has now been applied to several new structures, including that of a novel conotoxin having 3 disulfides in 12 residues! Two-bridge analyses are relatively routine, usually being carried out in one day, while three-bridge analyses present a wider range of challenges. Echistatin, with 4 disulfides in 49 residues is the largest peptide analyzed to date [3]: beyond this size, direct sequencer analysis will become increasingly inefficient, and peptide fragmentation will be necessary.

Acknowledgements

Supported by BRSG Grant NSS 2 S07 RR07092. Sequencing was carried out at the Utah Cancer Center, supported by NCI f P30 CA42014. Endothelin was a gift of Dr M. Verlander, Bachem, Inc.

References

1. Gray, W.R., Luque, F.A., Galyean, R., Atherton, E., Sheppard, R.C., Stone, B.L., Reyes, A., Alford, J., McIntosh, J.M., Olivera, B.M., Cruz, L.J. and Rivier, J.E., Biochemistry, 23 (1984) 2796.
2. Burns, J.A., Butler, J.C., Moran, J. and Whitesides, G.M., J. Org. Chem., 56 (1991) 2648.
3. Gray, W.R., Protein Science, submitted for publication .
4. Kumagaye, S.-I., Kuroda, H., Nakajima, K., Watanabe, T.X., Kimura, T., Masaki, T. and Sakakibara, S., Int. J. Peptide Protein Res., 32 (1988) 519.

Multi-year, multi-center evaluation of solid-phase peptide synthesis

J.D. Young[1], R.H. Angeletti[2], S.A. Carr[3],
D.R. Marshak[4], A.J. Smith[5], J.T. Stults[6],
L.C. Williams[7], K.R. Williams[8] and G.B. Fields[9]

[1]Skyline Peptides, 1901 Shoreline Drive, #210, Alameda, CA 94501,
U.S.A., [2]Albert Einstein College of Medicine, Bronx, NY 10461,
U.S.A., [3]SmithKline Beecham Pharmaceuticals, King of Prussia,
PA 19406, U.S.A., [4]Cold Spring Harbor Laboratory, Cold Spring
Harbor, NY 11724, U.S.A., [5]Stanford University, Stanford,
CA 94305, U.S.A., [6]Genentech, Inc., South San Francisco,
CA 94080, U.S.A., [7]University of Southern California, Los Angeles,
CA 90033, U.S.A., [8]Yale University, New Haven, CT 06536,
U.S.A., and [9]University of Minnesota, Minneapolis, MN 55455,
U.S.A.

Introduction

Peptide synthesis has become one of the most important methodologies in bioorganic chemistry. The Association of Biomolecular Resource Facilities (ABRF) has ~130 member laboratories that are engaged in the synthesis and structural analysis of peptides. The ABRF Committee on Peptide Synthesis and Mass Spectrometry was formed to evaluate the quality of the synthetic methods utilized in its member laboratories for peptide synthesis. Each year for the three year time period covering 1991-1993 the ABRF member laboratories have synthesized and/or cleaved 3 different peptides [1-3]. Problems associated with peptide assembly and cleavage were evaluated by the Committee using amino acid analysis (AAA), reversed-phase high-performance liquid chromatography (RP-HPLC), capillary electrophoresis (CE), Edman degradation sequence analysis, and electrospray (ES), plasma desorption (PD), fast atom bombardment (FAB), and laser desorption (LD) mass spectrometry (MS).

Results and Discussion

The 1991 ABRF test peptide sequence was Val-Lys-Lys-Arg-Cys-Ser-Met-Trp-Ile-Ile-Pro-Thr-Asp-Asp-Glu-Ala. The sequence was designed to incorporate difficult couplings (Trp-Ile-Ile) and potential cleavage side-reactions such as alkylation of Trp and oxidation of Cys and Met. A total of 36 crude products were submitted, 18 synthesized by Boc chemistry and 18 by Fmoc chemistry. AAA indicated that all samples were correct compositionally, i.e., there were no deletion or truncation products. The average content of the desired peptide product was 44% by RP-HPLC and 53% by ESMS. As evaluated by ESMS, only 10 of the 18 Boc-synthesized crude products contained >25% of the desired product, while all 18 Fmoc-synthesized crude products fell into the same category. Examination of the nature of the non-desired products revealed 8 of the Boc-synthesized peptides contained a single dehydration and/or oxidation, while 11 of the Fmoc-synthesized peptides contained >10% of a tBu adduct. Thus, the respective cleavage conditions of Boc and Fmoc solid-phase peptide synthesis were the primary source of synthetic difficulties for the 1991 test peptide, since non-desired products were not deletions or truncations but rather the result of incomplete deprotections and/or other covalent modifications.

The 1992 ABRF test peptide sequence was Gly-Val-Arg-Gly-Asp-Lys-Gly-Asn-Pro-Gly-Trp-Pro-Gly-Ala-Pro-Tyr. The sequence incorporated numerous potential synthesis and cleavage side-reactions including dehydration of Asn and alkylation of Trp. A total of 58 crude products were supplied for the study, 16 synthesized by Boc chemistry and 42 by Fmoc chemistry. AAA showed 51 of 58 crude products to be compositionally correct (Trp was not quantitated), which was consistent with RP-HPLC analysis indicating 51 of 58 crude products having >25% of the apparent desired peptide. Deletion peptides detected by Edman degradation sequence analysis were des(Asn^8), des(Pro^{12}), des(Gly^1,Gly^7,Asn^8), and des(Gly^1,Gly^4,Trp^{11}). Assessment of product purity by ESMS, FABMS, and PDMS showed *semi-quantitative* agreement with RP-HPLC analyses. As estimated by ESMS, 30 of the 58 crude samples (52%) and 26 of the 33 purified samples (79%) contained >75% of the desired product. This represents a considerable improvement over the 1991 study, where the percentage of crude and purified products containing >75% of the desired material were 28 and 65%, respectively. Possible reasons for these improved results were any combination of (i) a synthetic peptide sequence less susceptible to side-reactions induced during peptide-resin cleavage, (ii) the greater fraction of peptides synthesized by Fmoc chemistry, where cleavage conditions are less harsh, and (iii) more rigor and care in laboratory techniques following the 1991 results.

Further examination of the 1992 samples showed 5 of the crude Fmoc syntheses and 2 of the crude Boc syntheses to contain poor yields (<25%) of the desired product. Five of the 7 poor yield crude products were the result of deletion peptides (4 from Fmoc syntheses, 1 from Boc syntheses). Of the 4 Fmoc-synthesized crude peptides that contained >10% des(Asn), 2 were assembled without Asn side-chain protection. The only Fmoc-synthesized peptide for which all three mass spectrometric techniques detected >10% dehydration was assembled without Asn side-chain protection. This dehydration was probably the result of a side reaction (nitrile formation) attributable to the use of unprotected Asn during coupling. Thus, of the 7 peptides synthesized using Fmoc-Asn, 3 had >10% of a side-reaction directly attributable to Asn incorporation. The other problem detected in Fmoc syntheses was the generation of ethanedithiol-thioanisyl or anisyl-trifluoroacetyl adducts during peptide-resin cleavage.

Impurities in Boc-synthesized peptides were mostly due to cleavage problems. Of the 12 crude samples synthesized with Trp(For), 4 had >10% of the For group still attached. Variable success was seen when deprotection of Trp(For) was attempted during HF cleavage in the presence of anisole alone or anisole plus other scavengers. Deprotection of Trp(For) was complete by treatment of the peptide-resin with ~10% piperidine-DMF for 2 h prior to HF cleavage.

Cleavage conditions were more stringently reevaluated in the 1993 study using the test sequence Lys-His-Asp-Pro-Cys-Gly-Trp-Asn-Gly-Pro-Arg-Pro-Met-Arg-Gly, which is susceptible to acidolysis of the Asp-Pro bond and incomplete Dnp removal from His in addition to the cleavage side-reactions documented in the 1991 and 1992 studies. The peptide was synthesized by the Committee in both Boc and Fmoc versions, with participating laboratories cleaving the peptide-resin and returning the crude product. A total of 46 crude products were submitted. RP-HPLC showed 1 out of the 12 Boc-based syntheses and 26 out of the 34 Fmoc-based syntheses to contain >25% of the desired product. These results suggest that peptide-resin cleavages following Boc-based syntheses may require greater attention to detail due to deleterious side-reactions.

The wide range of peptide quality allowed us to evaluate critically the strengths and weaknesses of the analytical techniques used here. RP-HPLC appeared to provide an accurate estimate of the complexity of the peptide samples and, in situations in which a sample of the desired product was available, an accurate estimate of the content of the desired product. The only exception was the failure to detect an Asn deletion peptide in the 1992 study which co-eluted with the desired product. Since a sample of the desired product is not usually available, additional analyses are required. AAA was the best technique for absolute amino acid quantitation, but was not helpful for detecting modifications of amino acid residues. Preview sequence analysis had the same strengths and weaknesses as AAA and the advantage of determining residue position. The mass spectrometric techniques examined here (ESMS, FABMS, PDMS, and LDMS) allowed for identification of peptide modifications, including residual protecting groups. Our experience in these studies suggests that peptide synthesis is susceptible to a variety of side-reactions, and that efficient characterization of synthetic peptides is best obtained by a combination of AAA, RP-HPLC, and MS, with sequencing by either Edman degradation or tandem MS being used to identify the positions of modifications and deletions.

Acknowledgements

We thank the NSF for financial support (DIR 9003100).

References

1. Smith, A.J., Young, J.D., Carr, S.A., Marshak, D.R., Williams, L.C. and Williams, K.R., In Angeletti, R.H. (Ed.), Techniques in Protein Chemistry III , Academic Press, Orlando, FL, U.S.A., 1992, p.219.
2. Fields, G.B., Carr, S.A., Marshak, D.R., Smith, A.J., Stults, J.T., Williams, L.C., Williams, K.R. and Young, J.D., In Angeletti, R.H. (Ed.), Techniques in Protein Chemistry IV Academic Press, Orlando, FL, U.S.A., 1993, p.227.
3. Fields, G.B., Angeletti, R.H., Carr, S.A., Smith, A.J., Stults, J.T., Williams, L.C. and Young, J.D., In Crabb, J. (Ed.), Techniques in Protein Chemistry V, Academic Press, Orlando, FL, U.S.A., 1993, in press.

Workshop II
An Introduction to NMR Spectroscopy of Peptides

Chairs: Brian D. Sykes
University of Alberta
Edmonton, Alberta, Canada

and

Rachel E. Klevit
University of Washington
Seattle, Washington, U.S.A.

Detecting nascent structures in solution using NMR

H.J. Dyson, J. Yao and P.E. Wright

*Department of Molecular Biology, The Scripps Research Institute,
10666 North Torrey Pines Road, La Jolla, CA 92037, U.S.A.*

Introduction

A great deal of structural information is available from the NMR spectra of macromolecules [1]. When the molecules are folded into unique or closely related families of three-dimensional structures, then the NMR information can be used in distance geometry-type calculations to give three-dimensional structures comparable to those obtainable by X-ray crystallography [2]. Certain peptide systems are sufficiently constrained to permit this type of analysis. These include folded proteins, cyclic peptides, metal substituted peptides such as zinc fingers and peptides stabilized by the presence of disulfide bonding. Linear, unconstrained peptides rarely lend themselves to the calculation of structures in the same way, since they invariably have multiple conformations in solution, giving rise to conflicting constraints which, if not treated appropriately, would lead to incorrect structures. Structure calculation *per se* is in many cases unnecessary for linear peptides, since an adequate description of the conformational preferences of the peptide can be obtained readily from the NMR spectrum in conjunction with other spectroscopic methods [3, 4].

Results and Discussion

We have studied a number of linear peptide systems in aqueous solution by NMR and CD spectroscopy (reviewed in [3, 4]). Most peptides are not folded in solution, but rather show evidence, in the form of $d_{\alpha N}(i,i+1)$ NOE connectivities, of extended conformations in solution. Even when there is evidence, for example, for helical conformations, the evidence for unfolded conformations is usually present simultaneously in the NOESY spectrum. If both sets of NOEs are included in structure calculations, together with dihedral angle constraints from coupling constants that are a population-weighted average of the components from different conformers, the result will be either structures in which one or both sets of constraints are violated, or, alternatively, in which high-energy regions of the Ramachandran plot are occupied. Since this is unlikely for an unconstrained peptide, the structures calculated by simple use of the averaged constraints are meaningless!

Nevertheless, there are some occasions when the calculation of structures for linear peptides may be instructive. In order for such a calculation to be at all meaningful, the conformational preference for the folded conformation or a closely related group of conformations must be exceptionally high (probably >70%). Even when such a high population is present, the calculation of structures may be unnecessary, for example, if the peptide apparently forms a helix in solution, since the structure of the helix is well-known and it is unlikely that local conformational details can be determined from the NMR data. The calculation of structures can, however, be of great utility when specific questions are posed by the NMR data.

An example of such a system is the Type VI turn conformation of 6-amino acid peptides of the sequence SArPArDV, where Ar is an aromatic residue. The turn conformation in the *cis* form of the peptide SYPYDV was first described in 1988 [5], and it was calculated on the basis of the increase in the percentage of the *cis* form that the minimum population of the turn conformation was ~70%. In addition, the $^3J_{HN\alpha}$ coupling constant at residue Y2 was 3.3Hz, indicative of a high turn population. Measurements using ^{15}N-labeled peptide of the $^3J_{\alpha\beta}$ coupling constants of the Y and F residues at positions 2 and 4 of the peptide SYPFDV revealed that the side chain rotameric forms were virtually fixed. The resonances of the *cis*-proline were highly upfield-shifted, and showed prominent NOEs to the aromatic rings, an indication that the aromatic rings were in close and highly specific contact with the proline ring. We therefore felt that the circumstances in this case justified the calculation of structures.

Distance geometry structures were calculated using backbone-backbone NOEs that were specific to the folded structure. All of the NOEs involving the sidechains of the proline and aromatic residues were included, together with the dihedral angle constraints obtained from the highly specific coupling constants mentioned above. Initial structures obtained showed the presence of the turn conformation as expected, but the dihedral angles were very variable, and many were present in high-energy regions of the Ramachandran plot. In addition, the well-constrained regions (Y2-F4, with the majority of the distance and dihedral angle constraints) showed as large or larger variation in the dihedral angles observed in the structures as the ends of the peptide, where there were no constraints. This was due to over-constraint, because when the distance constraints were loosened, the structures improved greatly, the dihedral angles were now present in the low-energy α and β regions of the Ramachandran plot, and the standard deviations of the dihedral angles were now much more consistent with the input data, being high at the ends and lower in the middle of the peptide. A sample of 20 of the calculated structures (Figure 1) shows that residues 2-4 have a rather similar conformation both of backbone and side chain in all structures, but the ends of the peptide are highly variable.

For the majority of linear unconstrained peptides, the calculation of structures from NMR data is probably unnecessary and may give rise to erroneous structures due to conflicting constraints associated with conformational averaging. However, when used with care in systems when the population of folded forms is sufficiently high and where specific questions can be answered, the calculation of structures can provide valuable information.

Acknowledgements

We thank Dr. Rafael Brüschweiler for help with the heteronuclear experiments. This work was supported by grant GM38794 from the National Institutes of Health.

TYR2   PRO3   PHE4

Fig. 1. Superposition of 20 structures of the peptide SYPFDV, derived from a distance geometry calculation using distance and dihedral angle constraints from NMR experiments.

References

1. Wüthrich, K., NMR of Proteins and Nucleic Acids, John Wiley & Sons, 1986, p.1.
2. Wüthrich, K., Science, 243 (1989) 45.
3. Dyson, H.J. and Wright, P.E., Curr. Opinion Struct. Biol., 3 (1993) 60.
4. Dyson, H.J. and Wright, P.E., Ann. Rev. Biophys. Biophys. Chem., 20 (1991) 519.
5. Dyson, H.J., Rance, M., Houghten, R.A., Lerner, R.A. and Wright, P.E., J. Mol. Biol., 201 (1988) 161.

Peptide structures from NMR

D.F. Mierke[1,2] and H. Kessler[1]

[1]Organisch-Chemisches Institut, Technische Universität München,
W-8046 Garching, Germany, and [2]Department of Chemistry, Clark
University, Worcester, MA 01610, U.S.A.

Introduction

Two problems with the study of conformation of peptides by NMR will be
discussed here. The first, the possibility of large, significant conformational changes
fast on the NMR timescale, is addressed by application of both NOE and coupling
constant restraints during the refinement protocol [1]. The second, the small number
of NOEs (relative to the number available in the study of proteins), has been
approached by a modified refinement procedure to be used after distance geometry
(DG) calculations.

Results and Discussion

We have previously shown that the direct application of coupling constant as a
penalty function produces a great reduction in the available conformational space [1,
2]. The penalty term, is shown below,

$$E_J = 1/2 \, K_J (\,^3J - \,^3J_{exp})^2$$

where E_J is the energy, $^3J_{exp}$ is the experimental coupling, ^3J is the coupling
calculated from the Karplus curve and K_J is the force constant which can be adjusted
according to the associated error in the A, B, and C coefficients of the Karplus curve
and the experimental coupling. The function, first proposed by Kim and Prestegard
[3], is particularly useful with multiple coupling constants about a single dihedral
angle [1, 2]. To gain information about the ψ dihedral angle, similar procedures have
been carried out for the C^{α}-H^{α} one-bond coupling [4].

The application of both NOE and J-restraints during the simulation allow for
the identification of conformational averaging. This is based on the fact that the
restraints average differently: the distances derived from the NOEs average as a
function of $<R^{-3}>$, while the coupling constants average with the Karplus curve and
the dihedral angle subtended by the coupled atoms. In the presence of conformational
averaging, both of these restraints cannot be fulfilled by the same conformation. Of
course, this requires that during the calculation the restraints are driven to zero,
something often ignored in structure refinement. If averaging is thought to take
place, each of the restraints can be applied separately producing two different averaged
conformations.

DG calculations are extremely well suited for developing structure from NMR data. In the study of peptides, DG is not often used because of the small number of NOEs. It is therefore important to incorporate as many other experimental restraints as possible. Here a modified refinement protocol including coupling constants and a hydrogen bond attraction term is proposed to partially compensate for the limited number of NOEs. The starting point is the distance driven dynamics method developed by Scheek [5]. To this are added the coupling constant penalty function, as discussed above, and a hydrogen bond function:

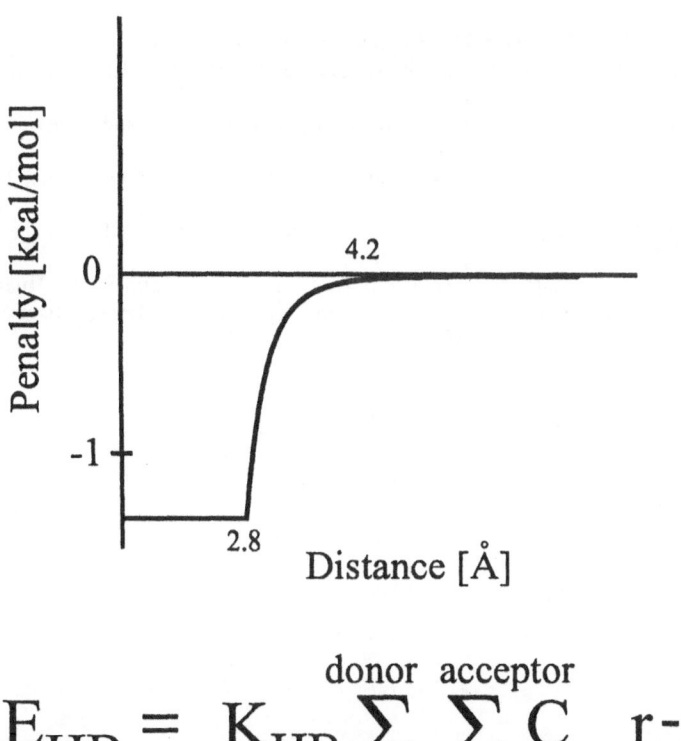

$$E_{HB} = K_{HB} \sum_{m=1}^{donor} \sum_{n=1}^{acceptor} C_{mn} r_{mn}^{-6}$$

where K_{HB} is a penalty constant and R_{ij} is the distance between the hydrogen bond acceptor and donor. The sum is run over all of the possible acceptors but only over the amide protons (donors) which have been experimentally determined (i.e., temperature coefficient, deuterium exchange) to be involved in a hydrogen bond. With this method no assumptions of the acceptor, which cannot be measured experimentally, is required. This is especially important with peptides since the use of a restraint between an amide proton and carbonyl (often used and misused [6] in protein refinement) will dominate the conformation of a peptide.

The function used here was chosen since it is short ranged; the attraction begins at ~4.2 Å. The function is turned off at 2.8 Å at which the energy and forces remain constant. The NOEs and coupling constants will bring the two atoms close together, and this function will help maintain the hydrogen bond during the refinement. The energy from the hydrogen bond term is kept small by adjustment of K_{HB}, accounting for only about 10% of the total energy of the system. But this small contribution is significant to maintain the hydrogen bonds and help produce the correct hydrogen bond geometry.

References

1. Mierke, D.F. and Kessler, H., Biopolymers, 32 (1992)1277.
2. Eberstadt, M., Mierke, D.F., Köck, M. and Kessler, H., Helv. Chim. Acta, 75 (1992) 2583.
3. Kim, Y. and Prestegard, J.H., Proteins: Structure, Function and Genetics, 8 (1990) 377.
4. Mierke, D.F., Golic-Grdadolnik, S. and Kessler, H., J. Am. Chem. Soc., 114 (1992) 8283.
5. Kaptein, R., Boelens, R., Scheek, R.M. and van Gunsteren, W.F., Biochemistry, 27 (1988) 5389.
6. Havel, T.F., In Rengugopalakrishnan, V., Carey, P.R., Smith, I.C.P., Huang, S.-G., and Storer, A.C. (Eds.), Proteins: Structure, Dynamics and Design, Escom, Leiden, The Netherlands, 1991, p.110.

Peptide structures when bound to proteins using the transferred NOE

B.D. Sykes

MRC Group in Protein Structure and Function, Protein Engineering Network of Centres of Excellence and Department of Biochemistry, University of Alberta, Edmonton, Alberta, T6G 2H7, Canada

Introduction

Central to the understanding of the interaction of peptides with proteins (or other targets) is the determination of the detailed molecular structure of the peptide-protein or peptide-target complex. For many of the goals of protein engineering, however, it is not necessary to know the complete structure of the peptide-target complex, but only the structure of the bound peptide. The structure of bound peptides can, in favorable cases, be determined by NMR transferred NOE spectroscopy even, and especially, when the size of the complex is too large to be determined by NMR methods. The method relies on the chemical exchange of nuclei on the peptide between free and bound states, and the resulting transfer of information about the structure of the bound peptide to the NMR spectrum of the free peptide which is much easier to observe both because the peptide is smaller (resulting in narrower NMR linewidths) and generally present at greater concentration. Consider the system: peptide + protein <==> peptide:protein complex, shown below, where the goal is to determine the internuclear distances between proton pairs such as H_a - H_b for the bound peptide

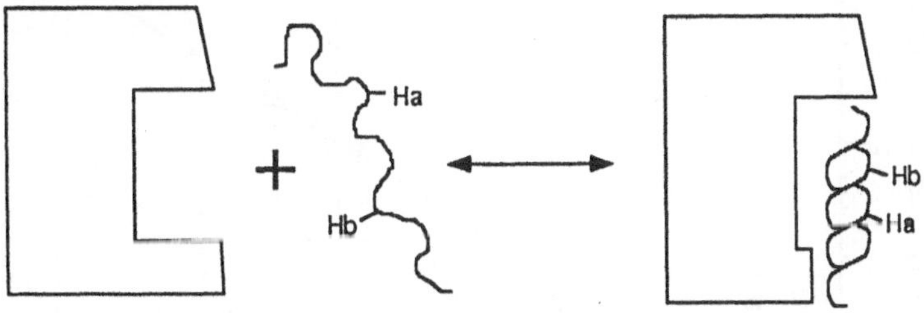

Executing a transferred NOE experiment is not more difficult than a standard NOESY experiment; the difficulty is in setting up the proper conditions under which the experiment is useful and understanding how to interpret the results.

Results and Discussion

The transferred NOE experiment is possible only under conditions of fast exchange which depends both upon the exchange rate and the NMR cross-relaxation rate between the two protons, which in turn depends upon the distance between the two protons and the size of the peptide:protein complex (Figure 1).

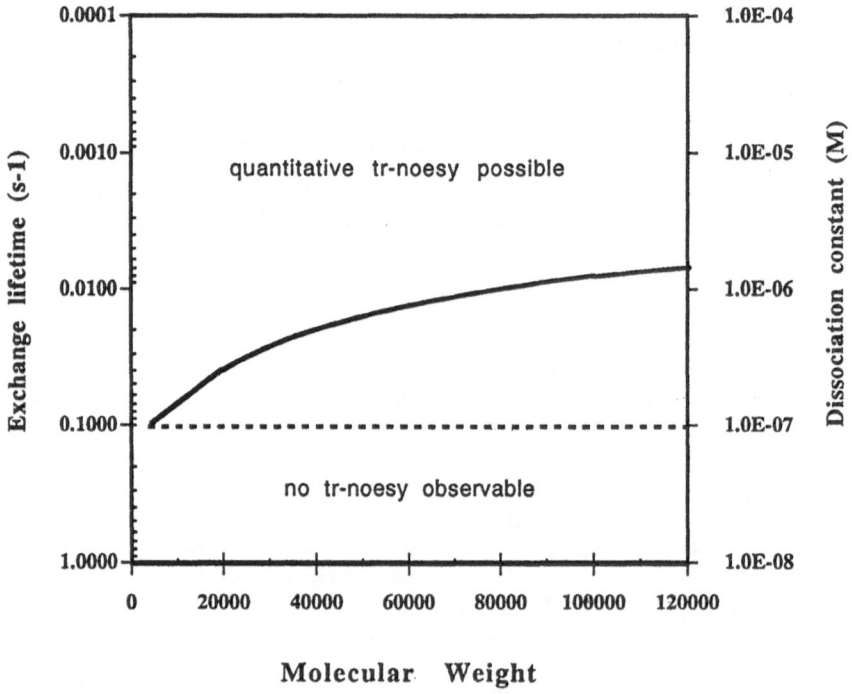

Fig. 1. *Range of applicability of the tr-NOE experiment (fr$_{ij}$ = 2.5 Å and 500 MHz).*

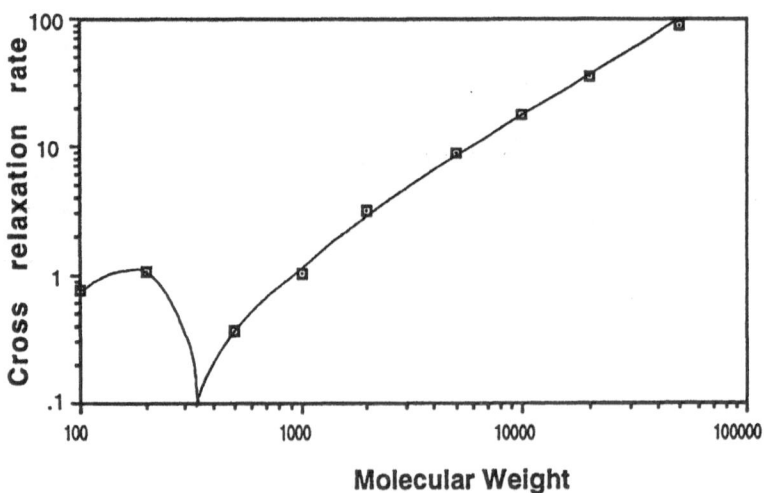

Fig. 2. *Absolute value of ^1H-^1H cross-relaxation rate vs. size of complex.*

While the tr-NOE is an average of the NOE for the free and bound peptide, it is particularly sensitive to the bound peptide because the NOE for the free peptide is very small (Figure 2). Determining accurate distances from the tr-NOE's is dependent on the mixing time of the tr-NOESY, the fraction of peptide bound, and the size of the complex [1]. Figure 3 shows that the build-up of the tr-NOE (which is proportional to r^{-6}) can be very non-linear for large MW's and long mixing times.

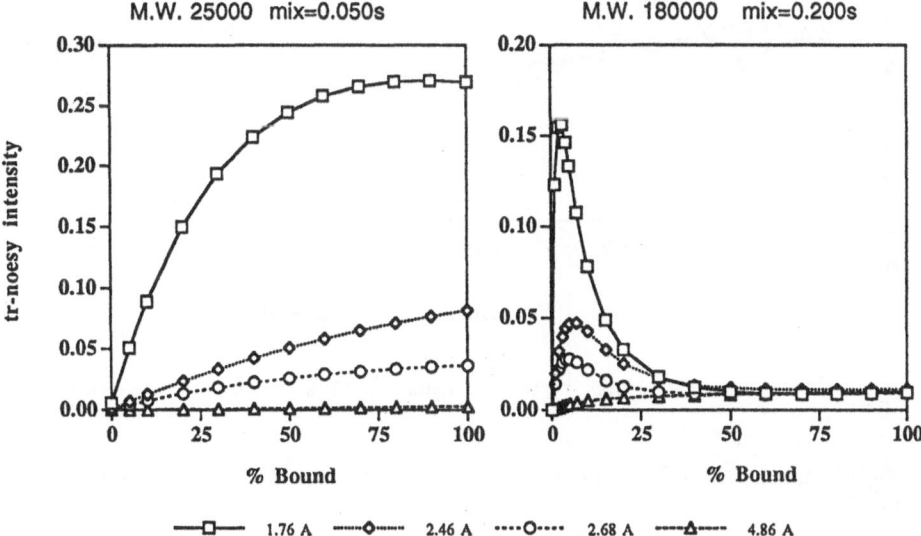

Fig. 3. Transferred NOE intensity vs fraction bound peptide for different mixing times, sizes of peptide:protein complexes, and internuclear distances.

The results in Figure 3 indicate that one should use very small amounts of the complex. However, in the limit of 0% bound, there can be no information about the bound peptide. Figure 4 shows how the tr-NOE reflects the bound distance (2.8 Å) as a function of % bound for various sizes of the free peptide (12 amino acid residues -> a correlation time of ~ 0.4 ns).

In summary, care must be used in setting up and interpreting the results of tr-NOESY experiments, not only in the manner indicated herein but also with respect to the stoichiometry and conditions of the complex formation.

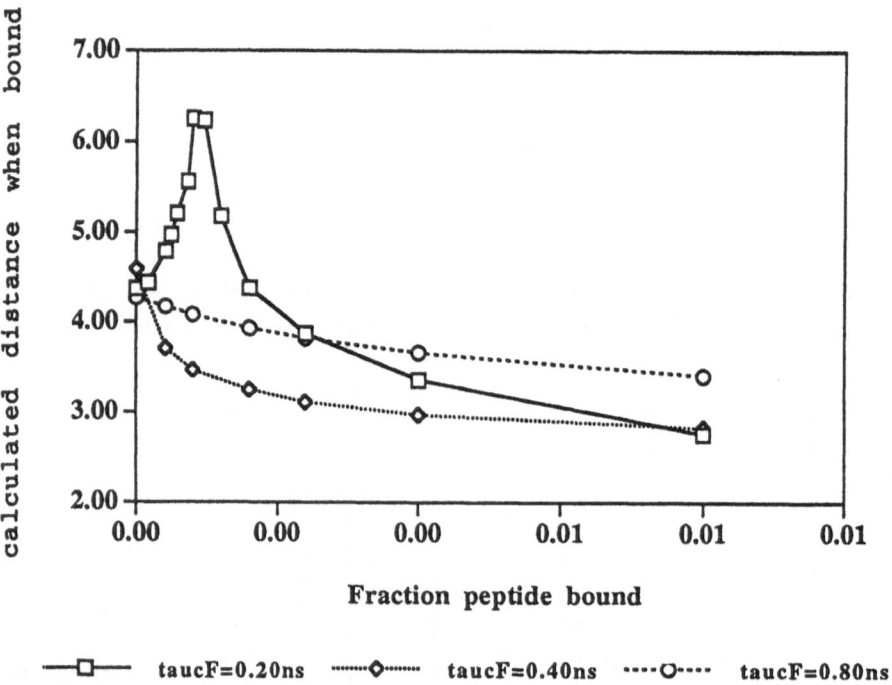

Fig. 4. *Calculated bound distance (actual is 2.8 Å) vs % bound for various sizes of the free peptide.*

References

1.   Campbell, A.P. and Sykes, B.D., J. Mag. Res., 93 (1991) 77.

Workshop III

An Introduction to Energy Minimization, Molecular Dynamics, Molecular Modeling and Conformational Analysis of Peptides

Chair: Garland R. Marshall

Washington University Center for Molecular Design
St. Louis, Missouri, U.S.A.

Peptide modeling - An overview

G.R. Marshall, D.D. Beusen and G.V. Nikiforovich

*Center for Molecular Design, Washington University,
St. Louis, MO, 63130, U.S.A.*

Introduction

Much of the study of peptides is devoted to elucidating their conformations, their dynamics and response to changes in environment, and especially their conformation when bound to receptors as a prelude to peptidomimetic design. While interaction with a receptor will certainly perturb the conformational energy surface of a flexible ligand, high affinity would suggest that the ligand binds in a conformation which is not exceptionally different from one of its low-energy minima. Mapping the energy surface of the ligand in isolation to determine the low-energy minima will, at the very least, provide a set of candidate conformations for consideration, or as starting points for further analyses. Molecular modeling has assumed an increasingly important role in this process, both as a tool in its own right and for the analysis of experimental observations.

I.   Molecular Mechanics - The Basics [1]

Molecular mechanics is based on an empirical representation of the potential energy surface and allows, in theory, the determination of the conformational ensemble which a peptide will exhibit under a given set of experimental conditions. Atoms are represented as balls connected to other atoms by springs with associated force constants. The derivation of the necessary parameters are based on both theoretical quantum calculations and calibration against experimental data. The interactions between atoms are represented by a set of equations, the force field (MM3, AMBER/OPLS, CHARMM, ECEPP, etc.), which attempts to analytically represent the potential surface of the set of atoms under consideration. Inherent in the force field representation are the vibrational spectra, relative energetics of conformational minima and rates of conformational transitions. Many other chemical properties of the system can be extracted from knowledge of the force field. Any energetic criterion for filtering candidate conformations or calculating relative populations carries with it the inherent limitations of the force field employed. The most difficult aspect of molecular mechanics is electrostatics which is intimately entangled with solvation. Various continuum models and detailed approaches with multipole representations of the electron distribution have been proposed and are being evaluated. Other difficulties arise in dealing with macromolecular systems because electrostatic interactions occur over long distances ($1/r$ dependence) and cannot be arbitrarily truncated to simplify the calculations.

A. Simulation methods [2]

The set of configurations generated by simulation methods is a relevant statistical sample of the possible arrangements of interacting molecules, yielding the macroscopic thermodynamic properties by statistical analysis of the results. In this case, the partition function is derived not by theoretical analysis of the quantum states available to the molecule, but through simulation. While it is important to minimize the number of molecules in the simulation for computational convenience, surface effects at the interface between the simulated solvent and the surrounding vacuum could seriously distort the results. In order to approximate an "infinite" liquid, a box of molecules is surrounded by mirror reflections, i.e., periodic images. Each atom in the central box has a set of related molecules in the virtual boxes surrounding the central one. The energy calculations for pairwise interactions only consider the interaction of a molecule, or its "ghost", with any other molecule, but not both.

Molecular dynamics (MD) is a deterministic process which solves Newton's equations of motion for each atom and increments the position and velocity of each atom using a small time increment. In this paradigm, atoms are essentially a collection of billiard balls with classical mechanics determining their positions and velocities at any moment in time. As the position of one atom changes with respect to the others, the forces which it experiences also change. The forces on any particular atom can be calculated using the appropriate force field. In MD, we choose a time step smaller than the period of fastest local motion to ensure that atoms move in sufficiently small increments that the position of surrounding atoms does not change significantly per incremental move. The time increment is on the order of 10^{-15} seconds which reflects the need to represent adequately atomic vibrations of similar time scale. For simulations of molecules in solvent, sufficient solvent molecules must be included to represent adequately all classes of solvent-solute interactions. This requires several hundred solvent molecules for even small solutes, and as a result simulations of more than several nanoseconds are rare. Because of the short time steps of molecular dynamics, events requiring longer times such as diffusion are difficult to simulate at the molecular level in detail. In this case, Brownian dynamics are used, and the particles (consisting of many atoms) move under the Langevin equations which govern diffusion rather than Newton's equations. Electrostatic forces are derived from the relative positions of the charged particles in the simulation by solution of the Poisson-Boltzmann equation which describes dielectric behaviour in a non-homogeneous system.

The **Monte Carlo** method is based on statistical mechanics and generates sufficient different configurations of a system by computer simulation to allow the desired structural, statistical and thermodynamic properties to be calculated as a weighted average of these properties over these configurations. Monte Carlo simulations are successfully performed by sampling only a limited set of the energetically feasible conformations, say 10^6 out of 10^{100} theoretical possibilities. One could sample all states, calculate the energy of each and then Boltzmann weight its contribution to the average. Instead, Monte Carlo operates by importance sampling and looking only at energetically feasible answers among all possibilities.

The term, Monte Carlo, comes from the random selection of the parameter (for example, coordinate or torsion angle) which determines the next configuration. The energy of the new state is compared with the old state. If it is the same, or lower, the new configuration is kept and becomes the basis for calculation of the next configuration. At this point, we can see that the procedure functions as a crude minimizer. If the successive configuration has a higher energy than the previous configuration, then it is either kept or discarded depending on the energetic difference (ΔE) and a random number, x, chosen between 0 and 1. If $\exp(-\Delta E/kT) \geq x$, then the new configuration is accepted. In other words, there is a unitary probability of accepting a move which results in an energy decrease, and an exponential probability based on the Boltzmann factor of accepting a move with a higher energy. This procedure generates trajectories which sample configurations in accord with the canonical Boltzmann distribution, and average properties can be calculated by simply averaging the properties associated with each configuration.

One aspect shared by Monte Carlo methods and molecular dynamics is the ability to cross barriers. In the case of Monte Carlo, barrier crossing occurs by random change and acceptance of higher energy states as a function of temperature. Because it is difficult to simulate systems with explicit solvent for long enough to allow conformational transitions, there is always a concern that sampling of the potential surface was insufficient. One approach to this problem is to do multiple runs from different starting configurations of the system. One can examine convergence of the ensemble averaged properties of each run to determine if adequate sampling has occurred. Obviously, the ability of the system to make transitions over activation energy barriers depends on the temperature. In order to increase the efficiency of sampling of conformational space, one can elevate the temperature for a short period, and then resume the simulation at the desired temperature, allowing time for reequilibration. This technique, called simulated annealing, is useful for overcoming activation energy barriers between local minima.

Non-Boltzmann sampling is an approach to sampling high-energy states that are infrequently reached in simulations. Often we are interested in sampling only those conformations relevant to a particular chemical question. For example, what are the energetics associated with a conformational transition from conformer A to conformer B. By definition, transition states are higher energy and would not be frequently sampled in an unconstrained simulation. Often a series of individual simulations will be conducted with constraints to focus on discrete regions of the reaction coordinate of interest. This procedure, umbrella sampling, can then be correctly reassembled by removal of the effects of the different constraining potentials for each individual simulation to generate a potential of mean force (pmf) for the reaction coordinate. Occasionally, there are sufficient crystal structure data that examination of the variation in the data will indicate the transition path (reaction coordinate) between two states of interest. A more general approach is based on the force field of the system itself and attempts to map a low energy pathway, a minimum potential energy pathway (MPEP), between the two states by a variety of mathematical techniques. The resultant MPEP on the potential energy surface is only one of many possible and since the role of entropic stabilization has not been taken into account, this is not necessarily the path with the lowest free energy of activation. It is, however, a logical starting point for simulations to determine an upper bound (as there may be another path of lower free energy) on the free energy associated with the transformation represented by the reaction coordinate.

Thermodynamic cycle perturbation [3, 4] is an approach that allows calculation of the energy difference between two states. In this method, one takes advantage of the state-function nature of a thermodynamic cycle and eliminates the parts of the simulation with long time constants, for example, diffusion. As an example, if one were interested in the difference in affinity of two ligands (L and M) for the same receptor (R) or enzyme, then the following thermodynamic cycle applies:

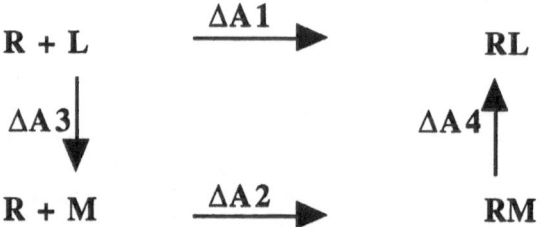

Because the thermodynamic values of the two states do not depend upon the path between the states, one can write the following equation:

$$\Delta\Delta A \quad = \quad \text{difference in affinity of L and M for R}$$
$$= \quad \Delta A2 - \Delta A1 = \Delta A4 - \Delta A3$$

By simulating the mutational paths A3 and A4, one can avoid the long simulation required for diffusion of the ligands into the receptor. One simply changes the potential functions representing ligand L to those representing ligand M during the course of the simulation making sure that the perturbations are introduced gradually and that the surrounding atoms have time to relax from the perturbation. Many interesting applications of this technique have appeared in the literature [3, 4]. Their success appears directly related to sampling problems and minimal perturbation to insure equilibration.

B. Systematic Search [5] and the Active Analog Approach [6]

Simulation methods are but a subset of the approaches used in mapping the potential surface of a ligand. It should be noted that simple minimization procedures locate the closest local minimum depending on the starting conformation. A more comprehensive, less path-dependent approach is generally required.

In contrast to most conformational searching tools which are stochastic, systematic search methods are algorithmic. All sterically allowed conformations are generated by a grid search of the torsional variables at a fixed angular increment. Systematic search methods, therefore, do not have problems in sampling and are path independent, but are combinatorial in complexity which may limit the fineness of the torsional grid and, thus compromise the results. In the Active Analog Approach, a set of distances between the postulated pharmacophoric groups, or active site points, are measured for each sterically allowed conformation. The set of distances, each of which represents a unique pattern, constitutes an OMAP. Each point of the OMAP is simply a submatrix of the distance matrix and, as such, is invariant to global translation and rotation of the molecule. If the initial assumption is valid that the same binding mode of interaction, or pharmacophoric pattern, is common to the set of molecules under consideration, then the OMAP for each active molecule must contain the pattern encrypted in the set of distances. By logically intersecting the set of OMAPs, one can determine which patterns are common to all molecules. In other words, all potential pharmacophoric, or active site, patterns consistent with the activity of the set of molecules can be found by this simple manipulation of OMAPs.

C. Correlation Approaches and 3D QSAR [6]

Because of the difficulties in direct calculation of free energies of binding and their requirements for an accurate molecular description of the structure of the binding site, alternative approaches have been developed which use a training set of observations on structure-activity to develop a correlation model which can then be used to extrapolate predicted affinities for novel compounds. This approach was originally developed for congeneric series by Hansch and others to deal with the effects of changing substituents, and is commonly referred to as quantitative structure-activity relationships, or QSAR. Generalizing this approach to three-dimensions allowed analysis of non-congeneric series assuming that appropriate orientation rules for aligning the different molecules could be developed. In the absence of three-dimensional data on receptor structure, either a pharmacophore hypothesis regarding spatial correspondence between groups in different molecules when interacting with the receptor, or an active-site hypothesis regarding essential groups within the receptor which would interact with each ligand provides a basis for deriving an alignment of ligands. Once the ligands are aligned, their potential fields can be compared and correlations drawn between differences in the fields and relative affinities. One such procedure for deriving such correlations is called Comparative Molecular Field Analysis (CoMFA) and has proven useful in predicting activities for sets of non-congeneric molecules. One can also use the structure of a known active site to help align the ligands under consideration.

II. Integrating peptide structural studies into molecular modeling

As an alternative to identifying likely bioactive conformations by computing the potential energy surface of a molecule, one can experimentally observe the conformation of the molecule, either by itself or when bound to its receptor. Experimental data usable in peptide modeling is geometric in nature: atomic coordinates, bond lengths, bond angles, internuclear distances, and torsion angles. Qualitative information such as secondary structure content and functional groups essential for ligand binding are also used. Key considerations in incorporating experimental information include: Is the observation relevant to the question at hand? Is the experimental data conformationally averaged? If so, is there useful information on relative populations of conformers? Is the absence or presence of data meaningful? What is the accuracy and/or precision of the observations?

The experimental data of most relevance to the bound conformation of a peptide is clearly derived from analyses of the ligand/receptor complex. The heterogeneity of the environment presented by the receptor is difficult to reproduce in any symmetric solution or solid-state environment, and as a result there is no guarantee that observations of peptide structure made outside the receptor environment are relevant to the ligand/receptor complex [7]. When the only experimental subject available is the isolated peptide, experimental analyses are best done in environments that closely approximate the biological environment - for example, determining the structure of a channel-forming peptide in micelles rather than DMSO. As covalent constraints are incorporated into the ligand with retention of activity, confidence in the relevance of free ligand structures may increase although even seemingly well-constrained peptides can exhibit environmentally dependent conformations [8].

X-ray crystallography is the experimental tool yielding the greatest quantity of information. Structures of ligand/receptor complexes reveal functional groups on both entities essential for binding as well as the bound conformation of the ligand which can be translated into internuclear distances and torsion angle restraints usable in subsequent modeling. X-ray analyses of free peptides can provide one or more low-energy conformers for use in modeling, although most molecular mechanics force fields are now adequate to identify a reasonable set of low-energy conformers. X-ray analysis of a free peptide is still a viable tool for determining equilibrium bond lengths and bond angles, especially for unusual amino acids. Conformational averaging is observed as reduced electron density due to partial occupancy of the unit cell, or a complete absence of density which reflects disorder. The fact that disordered regions do not produce observable data can create an artificial confidence in observed conformations. Although many molecules are available in sufficient quantity, the requirement for a crystalline sample is still a barrier to X-ray analysis.

Solution NMR is second only to X-ray in terms of information content. Several magnetic resonance experiments can provide information on the bound conformation of a ligand [9], including a straightforward NOESY evaluation of the ligand or ligand/receptor complex; transferred NOE experiments (for low affinity ligands); and isotope-directed NOE experiments. Solubility and/or aggregation problems at the sample concentrations required (0.1-2 mM) are not always appreciated, and most 3- and 4-dimensional studies require isotopically labelled material. NMR differs from X-ray in that atomic coordinates are determined using computational techniques which identify conformations consistent with the experimental restraints - interproton distances based on NOEs and torsion angles derived from coupling constants. The behavior of the NOE varies as a function of molecule size, changing from positive for small molecules to negative for proteins. At the transition point between the two extremes, NOEs are zero. This regime which often applies to peptides 3-20 residues in length; ROESY experiments are then appropriate in that cross-relaxation peaks are positive over the entire size range.

The rate at which the NOE develops, σ, is a function of the internuclear distance and the motional properties of the molecule. The r^{-6} dependence of σ means that the NOE drops off rapidly with distance, and as a result, only short distances (<5 Å) are observable. In a semi-quantitative approach, NOEs are clustered into strong, medium, and weak intensities and distance ranges assigned to each. A somewhat more quantitative approach is the isolated spin pair approximation (ISPA). This assumes that the motional properties of every proton pair in the molecule are identical, and that each proton pair is isolated from every other proton in the molecule. This approach can result in large errors (20-30%) in distances, although the problem is not as great for small molecules. An alternative means of extracting distances from NOE intensities is to solve the Bloch-Solomon equations, either by integration or by diagonalizing the matrix of peak volumes to solve for σ_{ij} [10]. Distance restraints calculated in this fashion have been shown to be more accurate (± 0.1-0.2 Å) than the ISPA treatment, although a model which describes the motion of the molecule is required. In the simplest treatments, the molecule is treated as a rigid, isotropically tumbling sphere. An analysis of the effect of this simplifying approximation on calculated distances [11] reveals that the absence of an NOE peak does not necessarily mean two protons are >5 Å from one another.

Coupling constants are converted into torsion angle restraints using the Karplus relationship and any of several calibrations that have been done for peptides and proteins. As restraints, coupling constants are most useful at their extreme values due to the degeneracy of states for mid-range values or to conformational averaging. The combined use of coupling constants and short-range NOEs can make stereospecific assignment of prochiral protons possible [12].

Extracting distances and torsion angles from NMR experiments to use as restraints in conformational searches rests on a significant assumption: the molecule is present in solution as a single conformer. The long time scale of the NMR measurement (10^{-3} -10^{-5} sec) allows contributions to the same chemical shift from conformations exchanging over a wide rate window. In addition, short-range interproton distances (for example, sequential αH-NH) can be less than the detectable limit of 5 Å over many conformational states. This differs from X-ray, where disordered regions do not generate a signal. Observable data can create an impetus to interpret it. When conformational averaging is present, this is dangerous, because the NOEs and coupling constants are population-weighted averages. Such efforts often give rise to a "virtual conformation" - a structure which satisfies the experimental restraints but is energetically unrealistic. Conformational averaging can be diagnosed by: 1) the failure of any energetically reasonable conformation to accommodate all distance and torsion angle restraints, either globally, representing complete disordering of the structure, or in local "hot spots" - regions in which restraints are poorly satisfied, due to averaging; 2) inconsistencies in coupling constant and NOE values, since these average differently; or 3) the absence of coordinated information - medium range NOEs, coupling constants, and H-bonds - over a contiguous block of amino acids necessary to define a secondary structure [14]. Most linear peptides display conformational averaging in solution [15].

Numerous strategies have evolved for dealing with conformational averaging. One could implement restraints over only that part of the molecule which appears to be well-defined and allow a conformational search to generate possible conformations of the averaged portion. Alternatively, since the r^{-6} weighting skews the observed NOE to short distances, one could argue that at any given time, a conformation is present which satisfies the observed NOE. Subsets of restraints could be randomly selected and used to generate a set of candidate conformations, each of which explains a part of the data [16]. For simulation methods, one could implement time-averaged restraints [17] over those parts which undergo averaging.

As an alternative to using the conformationally averaged NMR data as a primary filter, low-energy conformers can be used in a set of simultaneous equations to solve for the relative populations of each. Another option would be to select conformational candidates based on probabilistic information, i.e., their frequency of appearance in crystal structures. An entirely different approach would be to use the experimental data to solve for the continuous probability function about a given torsion angle [18], thereby avoiding the issue of inadequate sampling of conformational space implicit in the use of discrete candidates for averaging.

Solid-state NMR is an alternative technique for samples which are difficult to crystallize for X-ray analysis or whose motional properties make them unsuitable for solution NMR study due to increased linewidths. These studies fall into two classes [19]: those that orient the sample within the magnetic field and use the orientation dependence of chemical shift tensors and dipolar interactions to extract structural information; and those that use magic angle spinning with specialized methods to recover the rotationally-averaged, distance-dependent dipolar couplings. Orientation-dependent experiments can yield torsion angles within 10°, while distances calculated from dipolar coupling have accuracies of ±0.1-0.5 Å. In both cases, site-specific information is obtained by the incorporation of stable isotopes which are NMR-active. The large amounts of sample needed (50-200 mg) can be a barrier, but the form of the sample can range from crystalline to lyophilized powders.

Electron Spin Resonance employs stable, free radical reporter groups whose spectra are sensitive to the environment and can reflect mobility and solvent exposure of the spin-labeled site [20, 21]. The dipolar interaction between two electrons increases their relaxation rate, revealed as an increase in ESR linewidth. This perturbation has an r^{-6} distance dependence, and like the NOE experiment, quantitation of the distance requires an estimate of correlation time. The large gyromagnetic ratio of the electron means that the technique has high sensitivity, which can be translated into a decreased sample requirement and the ability to measure distances greater than the 5 Å observable by ^{1}H-^{1}H NOE. The time scale of the ESR measurement itself is fast, on the order of 10^{-8} sec, making time-resolved experiments possible. Paramagnetic species can also be used to perturb NMR linewidths, allowing distance measurements of 20 Å or more.

Circular dichroism [22] is used for identifying secondary structure content in solution environments, even for molecules which are not suitable for solution NMR. In contrast to NMR, CD yields the population of helix and sheet directly, because the signal arising from each is unique and the time scale of the measurement is rapid. The evaluation of CD spectra as summed contributions of helix and sheet was extended to turns [23], a structural class of great interest to peptidomimetic design. Those analyses revealed that although the presence of a turn could be detected by CD, exact determination of its type depends on corroborative information from another source such as NMR [24]. Quantitative determination of secondary structure content by CD continues to be subject to appropriate standards for calibration and the number and type of secondary structure descriptors [25].

Fourier transform infrared spectroscopy (FTIR) has a rapid time scale (10^{-12} sec) complementary to that of ESR and NMR. The primary lines of interest arise from the N-H stretch, which gives information on H-bonding, and the position of the amide I band, which is due mainly to backbone C=O stretching and has been correlated with secondary structure. Correlations to helix and sheet structures are the best defined; the correlation with turns is not as well characterized. In the past, the broad lines and inability to extract site-specific information have limited the impact of FTIR to qualitative structural determinations, but resolution enhancement methods have made it possible to measure line positions within 0.1 cm^{-1} [26]. Isotopic labelling allows site-specific perturbation of the FTIR spectrum [27], and as the correlation between local structure and amide I line position improves, it should be possible to extract detailed information on local conformation.

Fluorescence emission and energy transfer are additional optical techniques which provide information on the local environment and site-specific distance information through the use of appropriate reporter groups. Shifts in excitation and emission frequencies, for example, can demonstrate association of peptides with lipid in vesicles. Energy transfer is based on dipole-dipole interactions and consequently has an r^{-6} dependence; distances >20 Å are measurable. The time scale of fluorescence decay (10^{-8} sec) is such that distributions of distances can be determined with an accuracy that depends, among other things, on defining the orientation factor between the interacting dipoles [28].

Computational tools are needed to search for all possible conformations consistent with the data. Experimental restraints can be implemented in energy minimization and molecular dynamics/Monte Carlo simulations simply by adding terms to the force field which act as an attractive potential to pull the proton pairs together [29]. Entrapment in local minima is a significant problem with high constraint density, and as a result the force fields used in these calculations are modified to reduce electrostatic and van der Waals contributions. This allows atoms to pass through one another and improves the sampling properties of the algorithm. Torsional minimization using rigid bond lengths and bond angles is another method used in fitting NMR data [30]. The combinatorial difficulties with systematic search have been overcome to make this a usable tool in analyzing experimental data [31].

Distance geometry [32] is extensively used in the analysis of NMR data, largely because it is an efficient, path-independent means of generating initial structures which can be refined by other methods. There are three stages in this calculation: construction of bounds matrices and bounds smoothing; embedding; and optimization. In the first phase, matrices are created which enumerate the minimum and maximum distance between all pairs of nuclei in the molecule. Entries in these matrices originate in experimental observations or are holonomic constraints derived from the covalent structure of the molecule (bond lengths, 1-3 distances, and 1-4 distances). For most molecules, only a small fraction of the required interatomic distances are available from these sources. Additional constraints are implicit in the geometric relationships (e.g., triangle inequalities) between the distances, and in the smoothing process, entries in the bounds matrices are adjusted to satisfy these relationships. A trial distance matrix is created by randomly selecting internuclear distances within the smoothed upper and lower limits. This trial distance matrix is used in the embedding process by which Cartesian coordinates are calculated from the matrix of trial distances. The structure produced from the embedding process is only an approximate fit to the distance restraints. The optimization of x, y, and z coordinates in the final stage is with respect to the input distances and not with respect to a molecular mechanics force field.

In summary, experimental tools can provide information about low-energy conformations of peptides for use in molecular design. One should examine a ligand with a variety of experimental techniques to exploit their complementary information content and time scales. Experiments in a variety of environments which point to the same set of conformations would confirm minima. It is important to understand the assumptions in extracting information from data for use in modeling, as these restraints place bounds on the conformational search and the minima identified. Finally, it is essential to know the sampling properties of the chosen conformational searching protocol and to recognize that when the experimental restraints are not sufficient to define a structure, the results reflect the sampling biases of the technique.

III. Guidelines for Practical Applications of Molecular Modeling to Peptides

One approach to peptide design is based on the assumption that the receptor-bound (biologically active) conformation for highly potent peptide analogs resembles one of the low-energy conformers. By systematically sampling the variety of low-energy conformers for each of a set of active analogs and comparing them for three-dimensional commonality, one can deduce a common low-energy conformation accessible to all the analogs which, hopefully, resembles the bioactive conformation.

In our experience, the following steps are useful in obtaining a reasonable starting model of the biologically active conformation for a given peptide:

1. Decide on the specificity and selectivity of the peptide-receptor interaction to be modeled. In many cases, the modes of interaction of the same peptide with different specific receptors may be different, and consequently the models of the biologically active conformers should be different as well. Considerable evidence indicates that agonists and antagonists have different binding modes and, therefore, should be modeled independently.

2. Determine (or hypothesize) which groups in the peptide are important for receptor recognition and activation. This set of groups can be refined by subsequent comparison of the spatial arrangements of these functional groups in various analogs.

3. Perform energy calculations for several active analogs as well as for the peptide itself and determine sets of low-energy conformers for each. Calculations for conformationally different analogs (e.g., those with substitutions of D-, N-Me, dehydroamino acids or α-Me-amino acid residues, or those with cyclic constraints) are preferable at this step. It is essential to consider the sets of low-energy conformers for each compound, not just a single lowest-energy conformation.

4. Compare spatial arrangements of the functionally important groups for every low-energy conformer of the peptide and its active analogs. Those sets of conformers with similar spatial arrangements of the functionally important groups common to all active compounds are candidate models for the biologically active conformation. Sometimes, it is possible to include non-active analogs in the comparison as well, to eliminate common conformations from consideration. However, it should be remembered that the reasons for lack of activity may not be directly connected to conformational possibilities of the analog.

5. Use the derived model(s) to design new conformationally constrained analogs that test its predictive value. Energy calculations of the new analogs are performed independently and the model is refined against the biological results.

This approach was successfully employed for deducing a model of the receptor-bound conformation for δ-opioid peptides [33, 34]. Three δ-selective analogs of enkephalin, namely DPDPE (Tyr-*cyclo*[D-Pen-Gly-Phe-D-Pen]), JOM-13 (Tyr-*cyclo*[D-Cys-Phe-D-Pen]) and dermenkephalin (Tyr-D-Met-Phe-His-Leu-Met-Asp-NH$_2$, DRE) were selected for molecular modeling. Previous work in the field suggested a common spatial arrangement of the N-terminal α-amino group and the side chains of the Tyr[1] and Phe[4/3] residues in the receptor-bound conformation of δ-opioid peptides. Energy calculations for all three peptides were performed by the build-up procedure with the use of the ECEPP potential field. A comparison of the spatial arrangement of functionally important groups found a specific type of low-energy conformer similar to all δ-selective peptides in question, and allowed construction of a starting model for the δ-receptor-bound conformation. The most characteristic feature of the model was the placement of the Phe side chain in space corresponding to a χ_1 rotamer that is g^- for peptides containing Phe[4] and t for peptides with Phe[3] [33]. The refinement of the model was achieved by considering new "hybrid" analogs of opioid peptides with pronounced shifts in their affinities toward μ- and δ-opioid receptors [35]. Energy calculations and geometrical comparison of these analogs together with previous findings led to the suggestion that there is a unique, common spatial arrangement of the α-amino group and Tyr and Phe aromatic moieties in the DRE δ-receptor-bound conformer. It corresponds to the g^- rotamer of the Tyr and to the t rotamer of the Phe side chains [34].

The proposed model was confirmed by synthesis and biological testing of conformationally constrained analogs of deltorphin I, a δ-selective linear peptide (Tyr-D-Ala-Phe-Asp-Val-Val-Gly-NH$_2$, DT). Energy calculations performed for DT and subsequent comparison of its low-energy conformers to the proposed model showed the pronounced similarity of this model to one of the low-energy conformers of DT. The side chains of the D-Ala[2] and Val[5] residues in this particular conformer of DT are fairly close to each other, which suggested the possibility of linking positions 2 and 5 in DT by a disulfide bridge while preserving the chirality of the corresponding C$^\alpha$-atoms. Energy calculations on the [D-Cys[2], Cys[5]]-DT analog confirmed the compatibility of this cyclization with the model. Subsequently, several cyclic analogs of DT were synthesized and tested for their receptor binding and biological potency [36, 37]. The results showed that [D-Cys[2], Cys[5]]-DT was as active as linear DT at δ-receptors, but much more active at μ-receptors [36]. Thus, the δ-selectivity of the [D-Cys[2], Cys[5]]-DT analog was diminished; but it was restored in the similar [D-Cys[2], Pen[5]]-DT cyclic analog [37]. Another confirmation of the proposed model came from the synthesis of conformationally restricted analogs of the μ-selective dermorphin peptide (Tyr-D-Ala-Phe-Gly-Tyr-Pro-Ser-NH$_2$, DRM) [38, 39]. The authors [38] have fixed the Phe[3] side chain of DRM into the t rotamer, which is predicted by the model for δ-selective opioid peptides. This subtle modification alone has shifted the selectivity profile for this DRM analog from μ to δ-selectivity [38]. At the same time, [Tic[3]]-DRM, where the Phe side chain was fixed into the g^+ rotamer, was inactive on both receptor types [39].

Another example of this approach led to new conformational leads for angiotensin II (Asp[1]-Arg[2]-Val[3]-Tyr[4]-Val/Ile[5]-His[6]-Pro[7]-Phe[8], AT) [40]. Previous work demonstrated that the problem of the biologically active conformation(s) of AT and its analogs could not be resolved in terms of common backbone structures, i.e., different kinds of turns, peptide chain reversals, etc. [41]. Therefore, a common spatial arrangement of functionally important groups was sought in low-energy conformers of AT and its active analogs, with the assumption that these groups are the aromatic moieties of Tyr[4], His[6] and Phe[8] residues, as well as the C-terminal carboxyl. Geometrical comparison of the sets of low-energy backbone conformers for AT and its analogs, namely [(αMe)Phe[4]]-, [Pro[5]]-, *cyclo*[HCys[3,5]]- and *cyclo*[Cys/HCys[3], Mpt[5]]-AT, yielded a model for the receptor-bound conformation(s), which was used to design a new unusual analog, [D-Tyr[4], Pro[5]]-AT. The analog showed good affinity (IC$_{50}$ = 42.8 nM) for the AT$_1$ receptor [40] vs a pD$_2$ value of 5.77 for [D-Tyr[4]]-AT [42].

In conclusion, the examples above were obtained using the ECEPP force field, which has been calibrated specifically for peptide molecules and shown to be in good agreement with high-resolution X-ray data on proteins. The results of molecular modeling with different force fields and calculation techniques (such as build-up procedures, Monte Carlo or molecular dynamics simulations) can be different in their estimates of energy values for low-energy conformers. However, *the set of sterically accessible geometrical forms of the peptide backbone* found by different calculations are in much better agreement. A survey of publications on DPDPE(1988-92) revealed that despite the large differences in relative energies for the lowest-energy conformers, different authors found very similar allowed geometrical forms of the DPDPE backbone (in terms of a common spatial arrangement of Cα and Cβ atoms) by different protocols using a variety of force fields. Generally, the sets of low-energy geometrical forms for peptides are much more limited than the sets of low-energy conformers. At the same time, the backbone conformers for different analogs do not necessarily have to be identical, or even very similar, to ensure the similar spatial arrangement of functionally important groups which is the dominant recognition pattern.

References

1. Hagler, A.T. In Udenfriend, S. and Meienhofer, J. (Eds.), The Peptides: Analysis, Synthesis, Biology, Academic Press, New York, Vol. 7, 1985, p.213.
2. Allen, M.P. and Tildesley, D.J., Computer Simulation of Liquids, Oxford University Press, Oxford, 1989.
3. Jorgensen, W.L., Acct. Chem. Res., 22 (1989) 1184.
4. Bash, P.A., Singh, U.C., Brown, F.K., Langridge, R. and Kollman, P.K., Science, 235 (1987) 574.
5. Leach, A.R., Rev. Computat. Chem., 2 (1991) 1.
6. Marshall, G.R. and Cramer, R.D., III., Trends Pharmacol. Sci., 9 (1988) 285.
7. Wüthrich, K., von Freyberg, B., Weber, C., Wider, G., Traber, R., Widmer, H. and Braun, W., Science, 254 (1991) 953.
8. Marshall, G.R., Hodgkin, E.E., Langs, D.A., Smith, G.D., Zabrocki, J. and Leplawy, M.T., Proc. Natl. Acad. Sci. U.S.A., 87 (1990) 487.
9. Fesik, S.W., J. Med. Chem., 34 (1991) 2937.
10. Borgias, B.A., Gochin, M., Kerwood, D.J. and James, T.L., Prog. NMR Spec., 22 (1990) 83.
11. Duben, A.J. and Hutton, W.C., J. Am. Chem. Soc., 112 (1990) 5917.
12. Guntert, P., Braun, W., Billeter, M. and Wüthrich, K., J. Am. Chem. Soc., 111 (1989) 3997.
13. Jardetzky, O., Biochim. Biophys. Acta., 621 (1980) 227.
14. Wüthrich, K., NMR of Proteins and Nucleic Acids, Wiley-Interscience, New York, U.S.A., 1986.
15. Dyson, H.J. and Wright, P.E., Annu. Rev. Biophys. Biophys. Chem., 20 (1991) 519.
16. Bruschweiler, R., Blackledge, M. and Ernst, R.R., J. Biomolec. NMR., 1 (1991) 3.
17. Torda, A.E., Scheek, R.M. and van Gunsteren, W.F., J. Mol. Biol., 214 (1990) 223.
18. Dzakula, Z., Westler, W.M., Edison, A.S. and Markley, J.L., J. Am. Chem. Soc., 114 (1992) 6195.
19. Smith, S.O. and Peersen, O.B., Ann. Rev. Biophys. Biomol. Struct., 21 (1992) 25.
20. Millhauser, G.L., Trends. Biol. Sci., 17 (1992) 448.
21. Miick, S.M., Martinez, G.V., Fiori, W.R., Todd, A.P. and Millhauser, G.L., Nature, 359 (1992) 653.
22. Johnson, W.C.J., Proteins: Struct. Func. Genet., 7 (1990) 205.
23. Woody, R., In Blout, E.R., Bovey, F.A., Goodman, M. and Lotan, N. (Eds.), The Peptides, Academic Press, New York, U.S.A., 1985, p.15.
24. Smith, J.A. and Pease, L.G., CRC Crit. Rev. Biochem., 8 (1980) 314.

25. Johnson, W.C.J., Ann. Rev. Biophys. Biophys. Chem., 17 (1988) 145.
26. Arrondo, J.L.R., Muga, A., Castresana, J. and Goni, F.M., Prog. Biophys. Molec. Biol., 59 (1992) 23.
27. Tadesse, L., Nazarbaghi, R. and Walters, L., J. Am. Chem. Soc., 113 (1991) 7036.
28. Amir, D. and Haas, E., Biochem., 26 (1987) 2162.
29. Brunger, A.T. and Karplus, M., Acc. Chem. Res., 24 (1991) 54.
30. Guntert, P., Braun, W. and Wuthrich, K., J. Mol. Biol., 217 (1991) 517.
31. Beusen, D.D., Head, R.D., Clark, J.D., Hutton, W.C., Slomczynska, U., Zabrocki, J., Leplawy, M.T. and Marshall, G.R., In Schneider, C.H. (Ed.), Peptides 1992, Escom Scientific Publishers, Leiden, The Netherlands, 1993, p.79.
32. Wagner, G., Hyberts, S.G. and Havel, T.F., Annu. Rev. Biophys. Biomol. Struct., 21 (1992) 167.
33. Nikiforovich, G.V., Hruby, V.J., Prakash, O. and Gehrig, C.A., Biopolymers, 31 (1991) 941.
34. Nikiforovich, G.V. and Hruby, V.J., Biochim. Biophys. Res. Commun., 173 (1990) 521.
35. Sagan, S., Amiche, M., Delfour, A., Camus, A., Mor, A. and Nicolas, P., Biochem. Biophys. Res. Commun., 163 (1989) 726.
36. Misicka, A., Nikiforovich, G.V., Lipkowski, A.W., Horvath, R., Davis, P., Kramer, T.H., Yamamura, H.I. and Hruby, V.J., Bioorg. Med. Chem. Lett., 2 (1992) 547.
37. Misicka, A., Lipkowski, A.W., Horvath, R., Davis, P., Kramer, T.H., Yamamura, H.I., Porecca, F. and Hruby, V.J., In Schneider, C.H. (Ed.), Peptides 1992 (Proc. 22nd Eur. Pept. Symp.), Escom, Leiden, The Netherlands, 1993, p.651.
38. Tourwe, D., Verschueren, K., Van Binst, G., Davis, P., Porecca, F. and Hruby, V.J., Bioorg. Med. Chem. Lett., 2 (1992) 1305.
39. Tourwe, D., Toth, G., Lebl, M., Verschueren, K., Knapp, R.J., Davis, P., Van Binst, G., Yamamura, H.I., Burks, T.F., Kramer, T. and Hruby, V.J., In Smith, A. and Rivier, J.E. (Eds.), Peptides: Chemistry and Biology (Proceedings of the Twelfth American Peptide Symposium) Escom, Leiden, The Netherlands, 1992, p.307.
40. Nikiforovich, G.V., Plucinska, K., Zhang, W.J., Kao, J., Panek, R., Skeean, R. and Marshall, G.R., In Hodges, R.S. and Smith, J.A. (Eds.), Peptides: Chemistry, Struture and Biology (Proceedings of the Thirteenth American Peptide Symposium), Escom, Leiden, The Netherlands, 1994, this volume.
41. Plucinska, K., Kataoka, T., Yodo, M., Cody, W.L., He, J.X., Humblet, C., Lu, G.H., Lunney, E., Major, T.C., Panek, R.L., Schelkun, P., Skeean, R. and Marshall, G.R., J. Med. Chem., 36 (1993) 1902.
42. Chipens, G., Ancan, J., Afanasyeva, G., Balodis, J., Indulen, J., Klusha, V., Kudryashova, V., Liepinsh, E., Makarova, N. and Mishlyakova, N., In Loffet, A. (Ed.), Peptides 1976 (Proceedings of the 14th European Peptide Symposium), University of Brussels, Bruxelles, 1976, p.353.

Indexes

Author Index

and

Subject Index

Author index

Subject Index